JOHNSON/EVINRUDE

Outboards
1990-01 REPAIR MANUAL
ALL INLINE ENGINES, 2 AND 4-STROKE, 1-4 CYL

SELOC

Managing Partners	Dean F. Morgantini, S.A.E.
	Barry L. Beck
Executive Editor	Kevin M. G. Maher, A.S.E.
Manager-Marine/Recreation	James R. Marotta, A.S.E.
Production Managers	Melinda Possinger
	Ronald Webb

Manufactured in USA
© 2005 Seloc Publications
104 Willowbrook Lane
West Chester, PA 19382
ISBN 0-89330-052-7
2345678901 5432109876

www.selocmarine.com
1-866-SELOC55

CONTENTS

1 GENERAL INFORMATION, SAFETY AND TOOLS

HOW TO USE THIS MANUAL	1-2
BOATING SAFETY	1-4
BOATING EQUIPMENT (NOT REQUIRED BUT RECOMMENDED)	1-10
SAFETY IN SERVICE	1-12
TROUBLESHOOTING	1-13
SHOP EQUIPMENT	1-17
TOOLS	1-19
FASTENERS, MEASUREMENT, AND CONVERSIONS	1-26
SPECIFICATIONS	1-27

2 MAINTENANCE AND TUNE-UP

GENERAL INFORMATION	2-2
LUBRICATION SERVICE	2-6
ENGINE MAINTENANCE	2-15
BOAT MAINTENANCE	2-37
TUNE-UP	2-40
TIMING AND SYNCHRONIZATION	2-51
VALVE CLEARANCE (4-STROKE)	2-78
STORAGE	2-81
CLEARING A SUBMERGED MOTOR	2-83
SPECIFICATIONS	2-85

3 FUEL SYSTEM

FUEL SYSTEM BASICS	3-2
FUEL TANK AND LINES	3-6
CARBURETED FUEL SYSTEM	3-11
CARBURETOR	3-14
ELECTRONIC FUEL INJECTION (EFI)	3-41
SPECIFICATIONS	3-71

4 IGNITION AND ELECTRICAL

UNDERSTANDING AND TROUBLESHOOTING ELECTRICAL SYSTEMS	4-2
IGNITION SYSTEMS (BREAKER POINT MAGNETO)	4-8
IGNITION SYSTEMS (ELECTRONIC)	4-15
CHARGING CIRCUIT	4-69
WIRING DIAGRAMS	4-104
SPECIFICATIONS	4-141

5 LUBRICATION AND COOLING

LUBRICATION SYSTEMS (2-STROKE)	5-2
LUBRICATION SYSTEM (4-STROKE)	5-14
COOLING SYSTEM	5-21

CONTENTS

POWERHEAD	6-2	**POWERHEAD 6**
POWERHEAD BREAK-IN	6-87	
SPECIFICATIONS	6-89	

GEARCASE	7-2	**LOWER UNIT 7**
JET DRIVE	7-29	

TRIM & TILT SYSTEMS	8-2	**TRIM AND TILT 8**
TRIM & TILT WIRING	8-22	

REMOTE CONTROL	9-2	**REMOTE CONTROL 9**
CONTROL CABLES	9-14	

HAND REWIND STARTER	10-2	**HAND REWIND STARTER 10**
EMERGENCY STARTING	10-22	

MASTER INDEX	10-23	**MASTER INDEX**

MARINE TECHNICIAN TRAINING

INDUSTRY SUPPORTED PROGRAMS
OUTBOARD, STERNDRIVE & PERSONAL WATERCRAFT

- Dyno Testing • Boat & Trailer Rigging • Electrical & Fuel System Diagnostics
- Powerhead, Lower Unit & Drive Rebuilds • Powertrim & Tilt Rebuilds
- Instrument & Accessories Installation

For information regarding housing, financial aid and employment opportunities in the marine industry, contact us today:

CALL TOLL FREE
1-800-528-7995

An Accredited Institution

SM
Name _____
Address _____
City .... State ___ Zip ___
Phone _____

MARINE MECHANICS INSTITUTE
A Division of CTI
9751 Delegates Drive • Orlando, Florida 32837
2844 W. Deer Valley Rd. • Phoenix, AZ 85027

MEMBER

FINANCIAL ASSISTANCE AVAILABLE FOR THOSE WHO QUALIFY!

SAFETY NOTICE

Proper service and repair procedures are vital to the safe, reliable operation of all marine engines, as well as the personal safety of those performing repairs. This manual outlines procedures for servicing and repairing engines and drive systems using safe, effective methods. The procedures contain many NOTES, CAUTIONS and WARNINGS which should be followed, along with standard procedures, to minimize the possibility of personal injury or improper service which could damage the vehicle or compromise its safety.

It is important to note that repair procedures and techniques, tools and parts for servicing these engines, as well as the skill and experience of the individual performing the work, vary widely. It is not possible to anticipate all of the conceivable ways or conditions under which the engine may be serviced, or to provide cautions as to all possible hazards that may result. Standard and accepted safety precautions and equipment should be used during cutting, grinding, chiseling, prying, or any other process that can cause material removal or projectiles.

Some procedures require the use of tools specially designed for a specific task. Before substituting another tool or procedure, you must be completely satisfied that neither your personal safety, nor the performance of the vessel, will be endangered. All procedures covered in this manual requiring the use of special tools will be noted at the beginning of the procedure by means of an **OEM symbol**

Additionally, any procedure requiring the use of an electronic tester or scan tool will be noted at the beginning of the procedure by means of a **DVOM symbol**

Although information in this manual is based on industry sources and is complete as possible at the time of publication, the possibility exists that some manufacturers made later changes which could not be included here. While striving for total accuracy, Seloc Publishing cannot assume responsibility for any errors, changes or omissions that may occur in the compilation of this data. We must therefore warn you to follow instructions carefully, using common sense. If you are uncertain of a procedure, seek help by inquiring with someone in your area who is familiar with these motors before proceeding.

PART NUMBERS

Part numbers listed in this reference are not recommendations by Seloc Publishing for any particular product brand name, simply iterations of the manufacturer's suggestions. They are also references that can be used with interchange manuals and aftermarket supplier catalogs to locate each brand supplier's discrete part number.

SPECIAL TOOLS

Special tools are recommended by the manufacturers to perform a specific job. Use has been kept to a minimum, but, where absolutely necessary, they are referred to in the text by the part number of the manufacturer if at all possible; and also noted at the beginning of each procedure with one of the following symbols: **OEM** or **DVOM.**

The **OEM** symbol usually denotes the need for a unique tool purposely designed to accomplish a specific task, it will also be used, less frequently, to notify the reader of the need for a tool that is not commonly found in the average tool box.

The **DVOM** symbol is used to denote the need for an electronic test tool like an ohmmeter, multi-meter or, on cetain later engines, a scan tool.

These tools can be purchased, under the appropriate part number, from your local dealer or regional distributor, or an equivalent tool can be purchased locally from a tool supplier or parts outlet. Before substituting any tool for the one recommended, read the SAFETY NOTICE at the top of this page.

Providing the correct mix of service and repair procedures is an endless battle for any publisher of "How-To" information. Users range from first time do-it yourselfers to professionally trained marine technicians, and information important to one is frequently irrelevant to the other. The editors at Seloc Publishing strive to provide accurate and articulate information on all facets of marine engine repair, from the simplest procedure to the most complex. In doing this, we understand that certain procedures may be outside the capabilities of the average DIYer. Conversely we are aware that many procedures are unnecessary for a trained technician.

SKILL LEVELS

In order to provide all of our users, particularly the DIYers, with a feeling for the scope of a given procedure or task before tackling it we have included a rating system denoting the suggested skill level needed when performing a particular procedure. One of the following icons will be included at the beginning of most procedures:

EASY. These procedures are aimed primarily at the DIYer and can be classified, for the most part, as basic maintenance procedures; battery, fluids, filters, plugs, etc. Although certainly valuable to any experience level, they will generally be of little importance to a technician.

MODERATE. These procedures are suited for a DIYer with experience and a working knowledge of mechanical procedures. Even an advanced DIYer or professional technician will occasionally refer to these procedures. They will generally consist of component repair and service procedures, adjustments and minor rebuilds.

DIFFICULT. These procedures are aimed at the advanced DIYer and professional technician. They will deal with diagnostics, rebuilds and internal engine/drive components and will frequently require special tools.

SKILLED. These procedures are aimed at highly skilled technicians and should not be attempted without previous experience. They will usually consist of machine work, internal engine work and gear case rebuilds.

Please remember one thing when considering the above ratings—they are a guide for judging the complexity of a given procedure and are subjective in nature. Only you will know what your experience level is, and only you will know when a procedure may be outside the realm of your capability. First time DIYer, or life-long marine technician, we all approach repair and service differently so an easy procedure for one person may be a difficult procedure for another, regardless of experience level. All skill level ratings are meant to be used as a guide only! Use them to help make a judgement before undertaking a particular procedure, but by all means read through the procedure first and make your own decision—after all, our mission at Seloc is to make boat maintenance and repair easier for everyone whether you are changing the oil or rebuilding an engine. Enjoy boating!

ALL RIGHTS RESERVED

No part of this publication may be reproduced, transmitted or stored in any form or by any means, electronic or mechanical, including photocopy, recording, or by information storage or retrieval system, without prior written permission from the publisher.

ACKNOWLEDGMENTS

Seloc Publishing expresses appreciation to the following companies who supported the production of this book:
- Marine Mechanics Institute—Orlando, FL
- Belks Marine—Holmes, PA

Thanks to John Hartung and Judy Belk of Belk's Marine for for there assistance, guidance, patience and access to some of the motors photographed for this manual.

Seloc Publishing would like to express thanks to the fine companies who participate in the production of all our books:
- Hand tools supplied by Craftsman are used during all phases of our vehicle teardown and photography.
- Many of the fine specialty tools used in our procedures were provided courtesy of Lisle Corporation.
- Much of our shop's electronic testing equipment was supplied by Universal Enterprises Inc. (UEI).

1

GENERAL INFORMATION, SAFETY AND TOOLS

HOW TO USE THIS MANUAL	1-2
BOATING SAFETY	1-4
BOATING EQUIPMENT (NOT REQUIRED BUT RECOMMENDED)	1-10
SAFETY IN SERVICE	1-12
TROUBLESHOOTING	1-13
SHOP EQUIPMENT	1-17
TOOLS	1-19
FASTENERS, MEASUREMENTS AND CONVERSIONS	1-26
SPECIFICATIONS	1-27

Detailed Contents

BOATING EQUIPMENT (NOT REQUIRED BUT RECOMMENDED) .. 1-10
- ANCHORS 1-10
  - BAILING DEVICES 1-10
  - COMPASS 1-10
    - COMPASS PRECAUTIONS 1-11
    - INSTALLATION 1-11
    - SELECTION 1-10
  - FIRST AID KIT 1-10
  - TOOLS AND SPARE PARTS 1-12
  - VHF-FM RADIO 1-10

BOATING SAFETY 1-4
- COURTESY MARINE EXAMINATIONS 1-10
- REGULATIONS FOR YOUR BOAT 1-4
  - CAPACITY INFORMATION 1-4
  - CERTIFICATE OF COMPLIANCE ... 1-4
  - DOCUMENTING OF VESSELS 1-4
  - HULL IDENTIFICATION NUMBER .. 1-4
  - LENGTH OF BOATS 1-4
  - NUMBERING OF VESSELS 1-4
  - REGISTRATION OF BOATS 1-4
  - SALES AND TRANSFERS 1-4
  - VENTILATION 1-5
  - VENTILATION SYSTEMS 1-5
- REQUIRED SAFETY EQUIPMENT 1-5
  - FIRE EXTINGUISHERS 1-6
  - PERSONAL FLOTATION DEVICES .. 1-7
  - SOUND PRODUCING DEVICES .. 1-8
  - VISUAL DISTRESS SIGNALS 1-8
  - TYPES OF FIRES 1-5
  - VISUAL DISTRESS SIGNALS 1-8
  - WARNING SYSTEM 1-7

FASTENERS, MEASUREMENTS AND CONVERSIONS 1-26
- BOLTS, NUTS AND OTHER THREADED RETAINERS 1-26
- STANDARD AND METRIC MEASUREMENTS 1-27
- TORQUE 1-27

HOW TO USE THIS MANUAL 1-2
- AVOIDING THE MOST COMMON MISTAKES 1-3
- AVOIDING TROUBLE 1-2
- CAN YOU DO IT? 1-2
- DIRECTIONS AND LOCATIONS ... 1-2

- MAINTENANCE OR REPAIR? 1-2
- PROFESSIONAL HELP 1-3
- PURCHASING PARTS 1-3
- WHERE TO BEGIN 1-2

SAFETY IN SERVICE 1-12
- DO'S 1-12
- DON'TS 1-13

SHOP EQUIPMENT 1-17
- CHEMICALS 1-17
  - CLEANERS 1-18
  - LUBRICANTS & PENETRANTS .. 1-17
  - SEALANTS 1-18
- SAFETY TOOLS 1-17
  - EYE AND EAR PROTECTION ... 1-17
  - WORK CLOTHES 1-17
  - WORK GLOVES 1-17

SPECIFICATIONS 1-27
- CONVERSION FACTORS 1-28
- TORQUE VALUES 1-27

TOOLS 1-19
- ELECTRONIC TOOLS 1-23
  - BATTERY CHARGERS 1-24
  - BATTERY TESTERS 1-23
  - GAUGES 1-24
  - MULTI-METERS (DVOMS) 1-24
- HAND TOOLS 1-19
  - HAMMERS 1-22
  - PLIERS 1-21
  - SCREWDRIVERS 1-22
  - SOCKET SETS 1-19
  - WRENCHES 1-21
- MEASURING TOOLS 1-25
  - DEPTH GAUGES 1-26
  - DIAL INDICATORS 1-25
  - MICROMETERS & CALIPERS .. 1-25
  - TELESCOPING GAUGES 1-26
- OTHER COMMON TOOLS 1-22
- SPECIAL TOOLS 1-23

TROUBLESHOOTING 1-13
- BASIC OPERATING PRINCIPLES .. 1-13
  - 2-STROKE MOTORS 1-13
  - 4-STROKE MOTORS 1-16
  - COMBUSTION 1-16

1-2 GENERAL INFORMATION, SAFETY AND TOOLS

HOW TO USE THIS MANUAL

This manual is designed to be a handy reference guide to maintaining and repairing your Johnson or Evinrude Outboard. We strongly believe that regardless of how many or how few year's experience you may have, there is something new waiting here for you.

This manual covers the topics that a factory service manual (designed for factory trained mechanics) and a manufacturer owner's manual (designed more by lawyers than boat owners these days) covers. It will take you through the basics of maintaining and repairing your outboard, step-by-step, to help you understand what the factory trained mechanics already know by heart. By using the information in this manual, any boat owner should be able to make better informed decisions about what they need to do to maintain and enjoy their outboard.

Even if you never plan on touching a wrench (and if so, we hope that we can change your mind), this manual will still help you understand what a mechanic needs to do in order to maintain your engine.

Can You Do It?

If you are not the type who is prone to taking a wrench to something, NEVER FEAR. The procedures provided here cover topics at a level virtually anyone will be able to handle. And just the fact that you purchased this manual shows your interest in better understanding your outboard.

You may even find that maintaining your outboard yourself is preferable in most cases. From a monetary standpoint, it could also be beneficial. The money spent on hauling your boat to a marina and paying a tech to service the engine could buy you fuel for a whole weekend of boating. And, if you are really that unsure of your own mechanical abilities, at the very least you should fully understand what a marine mechanic does to your boat. You may decide that anything other than maintenance and adjustments should be performed by a mechanic (and that's your call), but if so you should know that every time you board your boat, you are placing faith in the mechanic's work and trusting him or her with your well-being, and maybe your life.

It should also be noted that in most areas a factory-trained mechanic will command a hefty hourly rate for off site service. If the tech comes to you this hourly rate is often charged from the time they leave their shop to the time that they return home. When service is performed at a boat yard, the clock usually starts when they go out to get the boat and bring it into the shop and doesn't end until it is tested and put back in the yard. The cost savings in doing the job yourself might be readily apparent at this point.

Of course, if even you're already a seasoned Do-It-Yourselfer or a Professional Technician, you'll find the procedures, specifications, special tips as well as the schematics and illustrations helpful when tackling a new job on a motor.

■ **To help you decide if a task is within your skill level, procedures will often be rated using a wrench symbol in the text. When present, the number of wrenches designates how difficult we feel the procedure to be on a 1-4 scale. For more details on the wrench icon rating system, please refer to the information under Skill Levels at the beginning of this manual.**

Where to Begin

Before spending any money on parts, and before removing any nuts or bolts, read through the entire procedure or topic. This will give you the overall view of what tools and supplies will be required to perform the procedure or what questions need to be answered before purchasing parts. So read ahead and plan ahead. Each operation should be approached logically and all procedures thoroughly understood before attempting any work.

Avoiding Trouble

Some procedures in this manual may require you to "label and disconnect . . ." a group of lines, hoses or wires. Don't be lulled into thinking you can remember where everything goes - you won't. If you reconnect or install a part incorrectly, the motor may operate poorly, if at all. If you hook up electrical wiring incorrectly, you may instantly learn a very expensive lesson.

A piece of masking tape, for example, placed on a hose and another on its fitting will allow you to assign your own label such as the letter "A", or a short name. As long as you remember your own code, you can reconnect the lines by matching letters or names. Do remember that tape will dissolve when saturated in some fluids (especially cleaning solvents). If a component is to be washed or cleaned, use another method of identification. A permanent felt-tipped marker can be very handy for marking metal parts; but remember that some solvents will remove permanent marker. A scribe can be used to carefully etch a small mark in some metal parts, but be sure NOT to do that on a gasket-making surface.

SAFETY is the most important thing to remember when performing maintenance or repairs. Be sure to read the information on safety in this manual.

Maintenance or Repair?

Proper maintenance is the key to long and trouble-free engine life, and the work can yield its own rewards. A properly maintained engine performs better than one that is neglected. As a conscientious boat owner, set aside a Saturday morning, at least once a month, to perform a thorough check of items that could cause problems. Keep your own personal log to jot down which services you performed, how much the parts cost you, the date, and the amount of hours on the engine at the time. Keep all receipts for parts purchased, so that they may be referred to in case of related problems or to determine operating expenses. As a do-it-yourselfer, these receipts are the only proof you have that the required maintenance was performed. In the event of a warranty problem (on new motors), these receipts can be invaluable.

It's necessary to mention the difference between maintenance and repair. Maintenance includes routine inspections, adjustments, and replacement of parts that show signs of normal wear. Maintenance compensates for wear or deterioration. Repair implies that something has broken or is not working. A need for repair is often caused by lack of maintenance.

For example: draining and refilling the gearcase oil is maintenance recommended by all manufacturers at specific intervals. Failure to do this can allow internal corrosion or damage and impair the operation of the motor, requiring expensive repairs. While no maintenance program can prevent items from breaking or wearing out, a general rule can be stated: MAINTENANCE IS CHEAPER THAN REPAIR.

Directions and Locations

◆ See Figure 1

Two basic rules should be mentioned here. First, whenever the Port side of the engine (or boat) is referred to, it is meant to specify the left side of the engine when you are sitting at the helm. Conversely, the Starboard means your right side. The Bow is the front of the boat and the Stern or Aft is the rear.

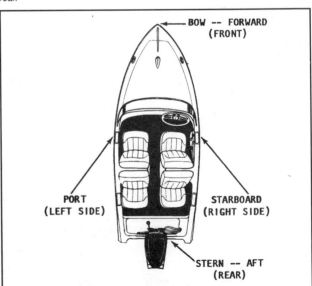

Fig. 1 Common terminology used for reference designation on boats of all size. These terms are used through out the text A

GENERAL INFORMATION, SAFETY AND TOOLS

Most screws and bolts are removed by turning counterclockwise, and tightened by turning clockwise. An easy way to remember this is: righty-tighty; lefty-loosey. Corny, but effective. And if you are really dense (and we have all been so at one time or another), buy a ratchet that is marked ON and OFF (like Snap-on® ratchets), or mark your own. This can be especially helpful when you are bent over backwards, upside down or otherwise turned around when working on a boat-mounted component.

Professional Help

Occasionally, there are some things when working on an outboard that are beyond the capabilities or tools of the average Do-It-Yourselfer (DIYer). This shouldn't include most of the topics of this manual, but you will have to be the judge. Some engines require special tools or a selection of special parts, even for some basic maintenance tasks.

Talk to other boaters who use the same model of engine and speak with a trusted marina to find if there is a particular system or component on your engine that is difficult to maintain.

You will have to decide for yourself where basic maintenance ends and where professional service should begin. Take your time and do your research first (starting with the information contained within) and then make your own decision. If you really don't feel comfortable with attempting a procedure, DON'T DO IT. If you've gotten into something that may be over your head, don't panic. Tuck your tail between your legs and call a marine mechanic. Marinas and independent shops will be able to finish a job for you. Your ego may be damaged, but your boat will be properly restored to its full running order. So, as long as you approach jobs slowly and carefully, you really have nothing to lose and everything to gain by doing it yourself.

On the other hand, even the most complicated repair is within the ability of a person who takes their time and follows the steps of a procedure. A rock climber doesn't run up the side of a cliff, he/she takes it one step at a time and in the end, what looked difficult or impossible was conquerable. Worry about one step at a time.

Purchasing Parts

◆ See Figures 2 and 3

When purchasing parts there are two things to consider. The first is quality and the second is to be sure to get the correct part for your engine. To get quality parts, always deal directly with a reputable retailer. To get the proper parts always refer to the model number from the information tag on your engine prior to calling the parts counter. An incorrect part can adversely affect your engine performance and fuel economy, and will cost you more money and aggravation in the end.

Just remember a tow back to shore will cost plenty. That charge is per hour from the time the towboat leaves their home port, to the time they return to their home port. Get the picture. . .$$$?

So whom should you call for parts? Well, there are many sources for the parts you will need. Where you shop for parts will be determined by what kind of parts you need, how much you want to pay, and the types of stores in your neighborhood.

Your marina can supply you with many of the common parts you require. Using a marina as your parts supplier may be handy because of location (just walk right down the dock) or because the marina specializes in your particular brand of engine. In addition, it is always a good idea to get to know the marina staff (especially the marine mechanic).

The marine parts jobber, who is usually listed in the yellow pages or whose name can be obtained from the marina, is another excellent source for parts. In addition to supplying local marinas, they also do a sizeable business in over-the-counter parts sales for the do-it-yourselfer.

Almost every boating community has one or more convenient marine chain stores. These stores often offer the best retail prices and the convenience of one-stop shopping for all your needs. Since they cater to the do-it-yourselfer, these stores are almost always open weeknights, Saturdays, and Sundays, when the jobbers are usually closed.

The lowest prices for parts are most often found in discount stores or the auto department of mass merchandisers. Parts sold here are name and private brand parts bought in huge quantities, so they can offer a competitive price. Private brand parts are made by major manufacturers and sold to large chains under a store label. And, of course, more and more large automotive parts retailers are stocking basic marine supplies.

Avoiding the Most Common Mistakes

There are 3 common mistakes in mechanical work:

1. Following the incorrect order of assembly, disassembly or adjustment. When taking something apart or putting it together, performing steps in the wrong order usually just costs you extra time; however, it CAN break something. Read the entire procedure before beginning disassembly. Perform everything in the order in which the instructions say you should, even if you can't immediately see a reason for it. When you're taking apart something that is very intricate, you might want to draw a picture of how it looks when assembled at one point in order to make sure you get everything back in its proper position. When making adjustments, perform them in the proper order; often, one adjustment affects another, and you cannot expect satisfactory results unless each adjustment is made only when it cannot be changed by subsequent adjustments.

■ **Digital cameras are handy. If you've got access to one, take pictures of intricate assemblies during the disassembly process and refer to them during assembly for tips on part orientation.**

2. Over-torquing (or under-torquing). While it is more common for over-torquing to cause damage, under-torquing may allow a fastener to vibrate loose causing serious damage. Especially when dealing with plastic and aluminum parts, pay attention to torque specifications and utilize a torque

Fig. 2 By far the most important asset in purchasing parts is a knowledgeable and enthusiastic parts person

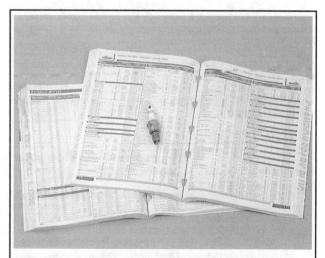

Fig. 3 Parts catalogs, giving application and part number information, are provided by manufacturers for most replacement parts

1-4 GENERAL INFORMATION, SAFETY AND TOOLS

wrench in assembly. If a torque figure is not available, remember that if you are using the right tool to perform the job, you will probably not have to strain yourself to get a fastener tight enough. The pitch of most threads is so slight that the tension you put on the wrench will be multiplied many times in actual force on what you are tightening.

3. Cross-threading. This occurs when a part such as a bolt is screwed into a nut or casting at the wrong angle and forced. Cross-threading is more likely to occur if access is difficult. It helps to clean and lubricate fasteners, then to start threading with the part to be installed positioned straight inward.

Always start a fastener, etc. with your fingers. If you encounter resistance, unscrew the part and start over again at a different angle until it can be inserted and turned several times without much effort. Keep in mind that some parts may have tapered threads, so that gentle turning will automatically bring the part you're threading to the proper angle, but only if you don't force it or resist a change in angle. Don't put a wrench on the part until it has been tightened a couple of turns by hand. If you suddenly encounter resistance, and the part has not seated fully, don't force it. Pull it back out to make sure it's clean and threading properly.

BOATING SAFETY

In 1971 Congress ordered the U.S. Coast Guard to improve recreational boating safety. In response, the Coast Guard drew up a set of regulations.

Aside from these federal regulations, there are state and local laws you must follow. These sometimes exceed the Coast Guard requirements. This section discusses only the federal laws. State and local laws are available from your local Coast Guard. As with other laws, "Ignorance of the boating laws is no excuse." The rules fall into two groups: regulations for your boat and required safety equipment on your boat.

Regulations For Your Boat

Most boats on waters within Federal jurisdiction must be registered or documented. These waters are those that provide a means of transportation between two or more states or to the sea. They also include the territorial waters of the United States.

DOCUMENTING OF VESSELS

A vessel of five or more net tons may be documented as a yacht. In this process, papers are issued by the U.S. Coast Guard as they are for large ships. Documentation is a form of national registration. The boat must be used solely for pleasure. Its owner must be a citizen of the U.S., a partnership of U.S. citizens, or a corporation controlled by U.S. citizens. The captain and other officers must also be U.S. citizens. The crew need not be.

If you document your yacht, you have the legal authority to fly the yacht ensign. You also may record bills of sale, mortgages, and other papers of title with federal authorities. Doing so gives legal notice that such instruments exist. Documentation also permits preferred status for mortgages. This gives you additional security, and it aids in financing and transfer of title. You must carry the original documentation papers aboard your vessel. Copies will not suffice.

REGISTRATION OF BOATS

If your boat is not documented, registration in the state of its principal use is probably required. If you use it mainly on an ocean, a gulf, or other similar water, register it in the state where you moor it.

If you use your boat solely for racing, it may be exempt from the requirement in your state. Some states may also exclude dinghies, while others require registration of documented vessels and non-power driven boats.

All states, except Alaska, register boats. In Alaska, the U.S. Coast Guard issues the registration numbers. If you move your vessel to a new state of principal use, a valid registration certificate is good for 60 days. You must have the registration certificate (certificate of number) aboard your vessel when it is in use. A copy will not suffice. You may be cited if you do not have the original on board.

NUMBERING OF VESSELS

A registration number is on your registration certificate. You must paint or permanently attach this number to both sides of the forward half of your boat. Do not display any other number there.

The registration number must be clearly visible. It must not be placed on the obscured underside of a flared bow. If you can't place the number on the bow, place it on the forward half of the hull. If that doesn't work, put it on the superstructure. Put the number for an inflatable boat on a bracket or fixture. Then, firmly attach it to the forward half of the boat. The letters and numbers must be plain block characters and must read from left to right. Use a space or a hyphen to separate the prefix and suffix letters from the numerals. The color of the characters must contrast with that of the background, and they must be at least three inches high.

In some states your registration is good for only one year. In others, it is good for as long as three years. Renew your registration before it expires. At that time you will receive a new decal or decals. Place them as required by state law. You should remove old decals before putting on the new ones. Some states require that you show only the current decal or decals. If your vessel is moored, it must have a current decal even if it is not in use.

If your vessel is lost, destroyed, abandoned, stolen, or transferred, you must inform the issuing authority. If you lose your certificate of number or your address changes, notify the issuing authority as soon as possible.

SALES AND TRANSFERS

Your registration number is not transferable to another boat. The number stays with the boat unless its state of principal use is changed.

HULL IDENTIFICATION NUMBER

A Hull Identification Number (HIN) is like the Vehicle Identification Number (VIN) on your car. Boats built between November 1, 1972 and July 31, 1984 have old format HINs. Since August 1, 1984 a new format has been used.

Your boat's HIN must appear in two places. If it has a transom, the primary number is on its starboard side within two inches of its top. If it does not have a transom or if it was not practical to use the transom, the number is on the starboard side. In this case, it must be within one foot of the stern and within two inches of the top of the hull side. On pontoon boats, it is on the aft crossbeam within one foot of the starboard hull attachment. Your boat also has a duplicate number in an unexposed location. This is on the boat's interior or under a fitting or item of hardware.

LENGTH OF BOATS

For some purposes, boats are classed by length. Required equipment, for example, differs with boat size. Manufacturers may measure a boat's length in several ways. Officially, though, your boat is measured along a straight line from its bow to its stern. This line is parallel to its keel.

The length does not include bowsprits, boomkins, or pulpits. Nor does it include rudders, brackets, outboard motors, outdrives, diving platforms, or other attachments.

CAPACITY INFORMATION

◆ See Figure 4

Manufacturers must put capacity plates on most recreational boats less than 20 feet long. Sailboats, canoes, kayaks, and inflatable boats are usually exempt. Outboard boats must display the maximum permitted horsepower of their engines. The plates must also show the allowable maximum weights of the people on board. And they must show the allowable maximum combined weights of people, engine(s), and gear. Inboards and stern drives need not show the weight of their engines on their capacity plates. The capacity plate must appear where it is clearly visible to the operator when underway. This information serves to remind you of the capacity of your boat under normal circumstances. You should ask yourself, "Is my boat loaded above its recommended capacity" and, "Is my boat overloaded for the present sea and wind conditions?" If you are stopped by a legal authority, you may be cited if you are overloaded.

CERTIFICATE OF COMPLIANCE

◆ See Figure 4

Manufacturers are required to put compliance plates on motorboats greater than 20 feet in length. The plates must say, "This boat," or "This

GENERAL INFORMATION, SAFETY AND TOOLS 1-5

Fig. 4 A U.S. Coast Guard certification plate indicates the amount of occupants and gear appropriate for safe operation of the vessel

equipment complies with the U. S. Coast Guard Safety Standards in effect on the date of certification." Letters and numbers can be no less than one-eighth of an inch high. At the manufacturer's option, the capacity and compliance plates may be combined.

VENTILATION

A cup of gasoline spilled in the bilge has the potential explosive power of 15 sticks of dynamite. This statement, commonly quoted over 20 years ago, may be an exaggeration; however, it illustrates a fact. Gasoline fumes in the bilge of a boat are highly explosive and a serious danger. They are heavier than air and will stay in the bilge until they are vented out.

Because of this danger, Coast Guard regulations require ventilation on many powerboats. There are several ways to supply fresh air to engine and gasoline tank compartments and to remove dangerous vapors. Whatever the choice, it must meet Coast Guard standards.

■ The following is not intended to be a complete discussion of the regulations. It is limited to the majority of recreational vessels. Contact your local Coast Guard office for further information.

General Precautions

Ventilation systems will not remove raw gasoline that leaks from tanks or fuel lines. If you smell gasoline fumes, you need immediate repairs. The best device for sensing gasoline fumes is your nose. Use it! If you smell gasoline in a bilge, engine compartment, or elsewhere, don't start your engine. The smaller the compartment, the less gasoline it takes to make an explosive mixture.

Ventilation for Open Boats

In open boats, gasoline vapors are dispersed by the air that moves through them. So they are exempt from ventilation requirements.

To be "open," a boat must meet certain conditions. Engine and fuel tank compartments and long narrow compartments that join them must be open to the atmosphere." This means they must have at least 15 square inches of open area for each cubic foot of net compartment volume. The open area must be in direct contact with the atmosphere. There must also be no long, unventilated spaces open to engine and fuel tank compartments into which flames could extend.

Ventilation for All Other Boats

Powered and natural ventilation are required in an enclosed compartment with a permanently installed gasoline engine that has a cranking motor. A compartment is exempt if its engine is open to the atmosphere. Diesel powered boats are also exempt.

VENTILATION SYSTEMS

There are two types of ventilation systems. One is "natural ventilation." In it, air circulates through closed spaces due to the boat's motion. The other type is "powered ventilation." In it, air is circulated by a motor-driven fan or fans.

Natural Ventilation System Requirements

A natural ventilation system has an air supply from outside the boat. The air supply may also be from a ventilated compartment or a compartment open to the atmosphere. Intake openings are required. In addition, intake ducts may be required to direct the air to appropriate compartments.

The system must also have an exhaust duct that starts in the lower third of the compartment. The exhaust opening must be into another ventilated compartment or into the atmosphere. Each supply opening and supply duct, if there is one, must be above the usual level of water in the bilge. Exhaust openings and ducts must also be above the bilge water. Openings and ducts must be at least three square inches in area or two inches in diameter. Openings should be placed so exhaust gasses do not enter the fresh air intake. Exhaust fumes must not enter cabins or other enclosed, non-ventilated spaces. The carbon monoxide gas in them is deadly.

Intake and exhaust openings must be covered by cowls or similar devices. These registers keep out rain water and water from breaking seas. Most often, intake registers face forward and exhaust openings aft. This aids the flow of air when the boat is moving or at anchor since most boats face into the wind when properly anchored.

Power Ventilation System Requirements

◆ See Figure 5

Powered ventilation systems must meet the standards of a natural system, but in addition, they must also have one or more exhaust blowers. The blower duct can serve as the exhaust duct for natural ventilation if fan blades do not obstruct the air flow when not powered. Openings in engine compartment, for carburetion are in addition to ventilation system requirements.

Required Safety Equipment

Coast Guard regulations require that your boat have certain equipment aboard. These requirements are minimums. Exceed them whenever you can.

TYPES OF FIRES

There are four common classes of fires:
- Class A - fires are of ordinary combustible materials such as paper or wood.
- Class B - fires involve gasoline, oil and grease.
- Class C - fires are electrical.
- Class D - fires involve ferrous metals

One of the greatest risks to boaters is fire. This is why it is so important to carry the correct number and type of extinguishers onboard.

The best fire extinguisher for most boats is a Class B extinguisher. Never use water on Class B or Class C fires, as water spreads these types of fires. Additionally, you should never use water on a Class C fire as it may cause you to be electrocuted.

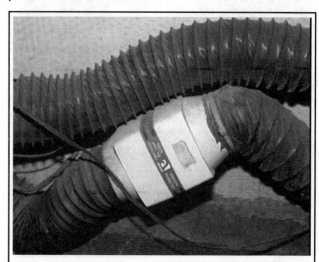

Fig. 5 Typical blower and duct system to vent fumes from the engine compartment

1-6 GENERAL INFORMATION, SAFETY AND TOOLS

FIRE EXTINGUISHERS

◆ See Figure 6

If your boat meets one or more of the following conditions, you must have at least one fire extinguisher aboard. The conditions are:
- Inboard or stern drive engines
- Closed compartments under seats where portable fuel tanks can be stored
- Double bottoms not sealed together or not completely filled with flotation materials
- Closed living spaces
- Closed stowage compartments in which combustible or flammable materials are stored
- Permanently installed fuel tanks
- Boat is 26 feet or more in length.

Contents of Extinguishers

Fire extinguishers use a variety of materials. Those used on boats usually contain dry chemicals, Halon, or Carbon Dioxide (CO2). Dry chemical extinguishers contain chemical powders such as Sodium Bicarbonate - baking soda.

Carbon dioxide is a colorless and odorless gas when released from an extinguisher. It is not poisonous but caution must be used in entering compartments filled with it. It will not support life and keeps oxygen from reaching your lungs. A fire-killing concentration of Carbon Dioxide can be lethal. If you are in a compartment with a high concentration of CO2, you will have no difficulty breathing. But the air does not contain enough oxygen to support life. Unconsciousness or death can result.

Halon Extinguishers

Some fire extinguishers and "built-in" or "fixed" automatic fire extinguishing systems contain a gas called Halon. Like carbon dioxide it is colorless and odorless and will not support life. Some Halons may be toxic if inhaled.

To be accepted by the Coast Guard, a fixed Halon system must have an indicator light at the vessel's helm. A green light shows the system is ready. Red means it is being discharged or has been discharged. Warning horns are available to let you know the system has been activated. If your fixed Halon system discharges, ventilate the space thoroughly before you enter it. There are no residues from Halon but it will not support life.

Although Halon has excellent fire fighting properties; it is thought to deplete the earth's ozone layer and has not been manufactured since January 1, 1994. Halon extinguishers can be refilled from existing stocks of the gas until they are used up, but high federal excise taxes are being charged for the service. If you discontinue using your Halon extinguisher, take it to a recovery station rather than releasing the gas into the atmosphere. Compounds such as FE 241, designed to replace Halon, are now available.

Fire Extinguisher Approval

Fire extinguishers must be Coast Guard approved. Look for the approval number on the nameplate. Approved extinguishers have the following on their labels: "Marine Type USCG Approved, Size. . ., Type. . ., 162.208/," etc. In addition, to be acceptable by the Coast Guard, an extinguisher must be in serviceable condition and mounted in its bracket. An extinguisher not properly mounted in its bracket will not be considered serviceable during a Coast Guard inspection.

Care and Treatment

Make certain your extinguishers are in their stowage brackets and are not damaged. Replace cracked or broken hoses. Nozzles should be free of obstructions. Sometimes, wasps and other insects nest inside nozzles and make them inoperable. Check your extinguishers frequently. If they have pressure gauges, is the pressure within acceptable limits? Do the locking pins and sealing wires show they have not been used since recharging?

Don't try an extinguisher to test it. Its valves will not reseat properly and the remaining gas will leak out. When this happens, the extinguisher is useless.

Weigh and tag carbon dioxide and Halon extinguishers twice a year. If their weight loss exceeds 10 percent of the weight of the charge, recharge them. Check to see that they have not been used. They should have been inspected by a qualified person within the past six months, and they should have tags showing all inspection and service dates. The problem is that they can be partially discharged while appearing to be fully charged.

Some Halon extinguishers have pressure gauges the same as dry chemical extinguishers. Don't rely too heavily on the gauge. The extinguisher can be partially discharged and still show a good gauge reading. Weighing a Halon extinguisher is the only accurate way to assess its contents.

If your dry chemical extinguisher has a pressure indicator, check it frequently. Check the nozzle to see if there is powder in it. If there is, recharge it. Occasionally invert your dry chemical extinguisher and hit the base with the palm of your hand. The chemical in these extinguishers packs and cakes due to the boat's vibration and pounding. There is a difference of opinion about whether hitting the base helps, but it can't hurt. It is known that caking of the chemical powder is a major cause of failure of dry chemical extinguishers. Carry spares in excess of the minimum requirement. If you have guests aboard, make certain they know where the extinguishers are and how to use them.

Using a Fire Extinguisher

A fire extinguisher usually has a device to keep it from being discharged accidentally. This is a metal or plastic pin or loop. If you need to use your extinguisher, take it from its bracket. Remove the pin or the loop and point the nozzle at the base of the flames. Now, squeeze the handle, and discharge the extinguisher's contents while sweeping from side to side. Recharge a used extinguisher as soon as possible.

If you are using a Halon or carbon dioxide extinguisher, keep your hands away from the discharge. The rapidly expanding gas will freeze them. If your fire extinguisher has a horn, hold it by its handle.

Legal Requirements for Extinguishers

You must carry fire extinguishers as defined by Coast Guard regulations. They must be firmly mounted in their brackets and immediately accessible.

A motorboat less than 26 feet long must have at least one approved hand-portable, Type B-1 extinguisher. If the boat has an approved fixed fire extinguishing system, you are not required to have the Type B-1 extinguisher. Also, if your boat is less than 26 feet long, is propelled by an outboard motor, or motors, and does not have any of the first six conditions described at the beginning of this section, it is not required to have an extinguisher. Even so, it's a good idea to have one, especially if a nearby boat catches fire, or if a fire occurs at a fuel dock.

A motorboat 26 feet to less than 40 feet long, must have at least two Type B-1 approved hand-portable extinguishers. It can, instead, have at least one Coast Guard approved Type B-2. If you have an approved fire extinguishing system, only one Type B-1 is required.

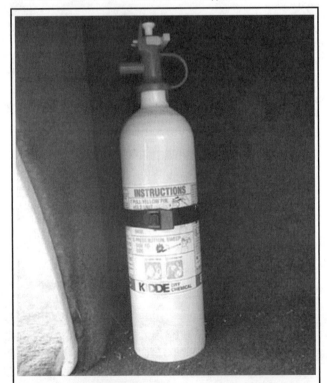

Fig. 6 An approved fire extinguisher should be mounted close to the operator for emergency use

GENERAL INFORMATION, SAFETY AND TOOLS 1-7

A motorboat 40 to 65 feet long must have at least three Type B-1 approved portable extinguishers. It may have, instead, at least one Type B-1 plus a Type B-2. If there is an approved fixed fire extinguishing system, two Type B-1 or one Type B-2 is required.

WARNING SYSTEM

Various devices are available to alert you to danger. These include fire, smoke, gasoline fumes, and carbon monoxide detectors. If your boat has a galley, it should have a smoke detector. Where possible, use wired detectors. Household batteries often corrode rapidly on a boat.

There are many ways in which carbon monoxide (a by-product of the combustion that occurs in an engine) can enter your boat. You can't see, smell, or taste carbon monoxide gas, but it is lethal. As little as one part in 10,000 parts of air can bring on a headache. The symptoms of carbon monoxide poisoning - headaches, dizziness, and nausea - are like seasickness. By the time you realize what is happening to you, it may be too late to take action. If you have enclosed living spaces on your boat, protect yourself with a detector.

PERSONAL FLOTATION DEVICES

Personal Flotation Devices (PFDs) are commonly called life preservers or life jackets. You can get them in a variety of types and sizes. They vary with their intended uses. To be acceptable, PFDs must be Coast Guard approved.

Type I PFDs

A Type I life jacket is also called an offshore life jacket. Type I life jackets will turn most unconscious people from facedown to a vertical or slightly backward position. The adult size gives a minimum of 22 pounds of buoyancy. The child size has at least 11 pounds. Type I jackets provide more protection to their wearers than any other type of life jacket. Type I life jackets are bulkier and less comfortable than other types. Furthermore, there are only two sizes, one for children and one for adults.

Type I life jackets will keep their wearers afloat for extended periods in rough water. They are recommended for offshore cruising where a delayed rescue is probable.

Type II PFDs

◆ See Figure 7

A Type II life jacket is also called a near-shore buoyant vest. It is an approved, wearable device. Type II life jackets will turn some unconscious people from facedown to vertical or slightly backward positions. The adult size gives at least 15.5 pounds of buoyancy. The medium child size has a minimum of 11 pounds. And the small child and infant sizes give seven pounds. A Type II life jacket is more comfortable than a Type I but it does not have as much buoyancy. It is not recommended for long hours in rough water. Because of this, Type IIs are recommended for inshore and inland cruising on calm water. Use them only where there is a good chance of fast rescue.

Type III PFDs

Type III life jackets or marine buoyant devices are also known as flotation aids. Like Type IIs, they are designed for calm inland or close offshore water where there is a good chance of fast rescue. Their minimum buoyancy is 15.5 pounds. They will <B>not</B> turn their wearers face up.

Type III devices are usually worn where freedom of movement is necessary. Thus, they are used for water skiing, small boat sailing, and fishing among other activities. They are available as vests and flotation coats. Flotation coats are useful in cold weather. Type IIIs come in many sizes from small child through large adult.

Life jackets come in a variety of colors and patterns - red, blue, green, camouflage, and cartoon characters. From purely a safety standpoint, the best color is bright orange. It is easier to see in the water, especially if the water is rough.

Type IV PFDs

◆ See Figures 8 and 9

Type IV ring life buoys, buoyant cushions and horseshoe buoys are Coast Guard approved devices called throwables. They are made to be thrown to people in the water, and should not be worn. Type IV cushions are often used as seat cushions. But, keep in mind that cushions are hard to hold onto in the water, thus, they do not afford as much protection as wearable life jackets.

The straps on buoyant cushions are for you to hold onto either in the water or when throwing them, they are <B>NOT</B> for your arms. A cushion should never be worn on your back, as it will turn you face down in the water.

Type IV throwables are not designed as personal flotation devices for unconscious people, non-swimmers, or children. Use them only in emergencies. They should not be used for, long periods in rough water.

Ring life buoys come in 18, 20, 24, and 30 in. diameter sizes. They usually have grab lines, but you will need to attach about 60 feet of polypropylene line to the grab rope to aid in retrieving someone in the water. If you throw a ring, be careful not to hit the person. Ring buoys can knock people unconscious

Type V PFDs

Type V PFDs are of two kinds, special use devices and hybrids. Special use devices include boardsailing vests, deck suits, work vests, and others. They are approved only for the special uses or conditions indicated on their labels. Each is designed and intended for the particular application shown on its label. They do not meet legal requirements for general use aboard recreational boats.

Hybrid life jackets are inflatable devices with some built-in buoyancy provided by plastic foam or kapok. They can be inflated orally or by cylinders of compressed gas to give additional buoyancy. In some hybrids the gas is released manually. In others it is released automatically when the life jacket is immersed in water.

The inherent buoyancy of a hybrid may be insufficient to float a person unless it is inflated. The only way to find this out is for the user to try it in the

Fig. 7 Type II PFDs are recommended for inshore/inland use on calm water (where there is a good chance of fast rescue)

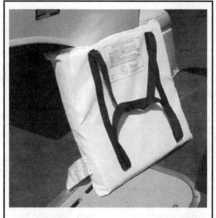

Fig. 8 Type IV buoyant cushions are thrown to people in the water. If you can squeeze air out of the cushion, it should be replaced

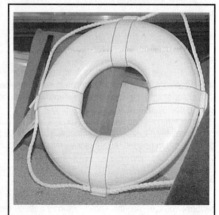

Fig. 9 Type IV throwables, such as this ring life buoy, are not designed for unconscious people, non-swimmers, or children

water. Because of its limited buoyancy when deflated, a hybrid is recommended for use by a non-swimmer only if it is worn with enough inflation to float the wearer.

If they are to count against the legal requirement for the number of life jackets you must carry, hybrids manufactured before February 8, 1995 must be worn whenever a boat is underway and the wearer must not go below decks or in an enclosed space. To find out if your Type V hybrid must be worn to satisfy the legal requirement, read its label. If its use is restricted it will say, "REQUIRED TO BE WORN" in capital letters.

Hybrids cost more than other life jackets, but this factor must be weighed against the fact that they are more comfortable than Types I, II or III life jackets. Because of their greater comfort, their owners are more likely to wear them than are the owners of Type I, II or III life jackets.

The Coast Guard has determined that improved, less costly hybrids can save lives since they will be bought and used more frequently. For these reasons, a new federal regulation was adopted effective February 8, 1995. The regulation increases both the deflated and inflated buoyancys of hybrids, makes them available in a greater variety of sizes and types, and reduces their costs by reducing production costs.

Even though it may not be required, the wearing of a hybrid or a life jacket is encouraged whenever a vessel is underway. Like life jackets, hybrids are now available in three types. To meet legal requirements, a Type I hybrid can be substituted for a Type I life jacket. Similarly Type II and III hybrids can be substituted for Type II and III life jackets. A Type I hybrid, when inflated, will turn most unconscious people from facedown to vertical or slightly backward positions just like a Type I life jacket. Type II and III hybrids function like Type II and III life jackets. If you purchase a new hybrid, it should have an owner's manual attached that describes its life jacket type and its deflated and inflated buoyancys. It warns you that it may have to be inflated to float you. The manual also tells you how to don the life jacket and how to inflate it. It also tells you how to change its inflation mechanism, recommended testing exercises, and inspection or maintenance procedures. The manual also tells you why you need a life jacket and why you should wear it. A new hybrid must be packaged with at least three gas cartridges. One of these may already be loaded into the inflation mechanism. Likewise, if it has an automatic inflation mechanism, it must be packaged with at least three of these water sensitive elements. One of these elements may be installed.

Legal Requirements

A Coast Guard approved life jacket must show the manufacturer's name and approval number. Most are marked as Type I, II, III, IV or V. All of the newer hybrids are marked for type.

You are required to carry at least one wearable life jacket or hybrid for each person on board your recreational vessel. If your vessel is 16 feet or more in length and is not a canoe or a kayak, you must also have at least one Type IV on board. These requirements apply to all recreational vessels that are propelled or controlled by machinery, sails, oars, paddles, poles, or another vessel. Sailboards are not required to carry life jackets.

You can substitute an older Type V hybrid for any required Type I, II or III life jacket provided:

   1. Its approval label shows it is approved for the activity the vessel is engaged in

   2. It's approved as a substitute for a life jacket of the type required on the vessel

   3. It's used as required on the labels

and

   4. It's used in accordance with any requirements in its owner's manual (if the approval label makes reference to such a manual.)

A water skier being towed is considered to be on board the vessel when judging compliance with legal requirements.

You are required to keep your Type I, II or III life jackets or equivalent hybrids readily accessible, which means you must be able to reach out and get them when needed. All life jackets must be in good, serviceable condition.

General Considerations

The proper use of a life jacket requires the wearer to know how it will perform. You can gain this knowledge only through experience. Each person on your boat should be assigned a life jacket. Next, it should be fitted to the person who will wear it. Only then can you be sure that it will be ready for use in an emergency. This advice is good even if the water is calm, and you intend to boat near shore.

Boats can sink fast. There may be no time to look around for a life jacket. Fitting one on you in the water is almost impossible. Most drownings occur in inland waters within a few feet of safety. Most victims had life jackets, but they weren't wearing them.

Keeping life jackets in the plastic covers they came wrapped in, and in a cabin, assure that they will stay clean and unfaded. But this is no way to keep them when you are on the water. When you need a life jacket it must be readily accessible and adjusted to fit you. You can't spend time hunting for it or learning how to fit it.

There is no substitute for the experience of entering the water while wearing a life jacket. Children, especially, need practice. If possible, give your guests this experience. Tell them they should keep their arms to their sides when jumping in to keep the life jacket from riding up. Let them jump in and see how the life jacket responds. Is it adjusted so it does not ride up? Is it the proper size? Are all straps snug? Are children's life jackets the right sizes for them? Are they adjusted properly? If a child's life jacket fits correctly, you can lift the child by the jacket's shoulder straps and the child's chin and ears will not slip through. Non-swimmers, children, handicapped persons, elderly persons and even pets should always wear life jackets when they are aboard. Many states require that everyone aboard wear them in hazardous waters.

Inspect your lifesaving equipment from time to time. Leave any questionable or unsatisfactory equipment on shore. An emergency is no time for you to conduct an inspection.

Indelibly mark your life jackets with your vessel's name, number, and calling port. This can be important in a search and rescue effort. It could help concentrate effort where it will do the most good.

Care of Life Jackets

Given reasonable care, life jackets last many years. Thoroughly dry them before putting them away. Stow them in dry, well-ventilated places. Avoid the bottoms of lockers and deck storage boxes where moisture may collect. Air and dry them frequently.

Life jackets should not be tossed about or used as fenders or cushions. Many contain kapok or fibrous glass material enclosed in plastic bags. The bags can rupture and are then unserviceable. Squeeze your life jacket gently. Does air leak out? If so, water can leak in and it will no longer be safe to use. Cut it up so no one will use it, and throw it away. The covers of some life jackets are made of nylon or polyester. These materials are plastics. Like many plastics, they break down after extended exposure to the ultraviolet light in sunlight. This process may be more rapid when the materials are dyed with bright dyes such as "neon" shades.

Ripped and badly faded fabrics are clues that the covering of your life jacket is deteriorating. A simple test is to pinch the fabric between your thumbs and forefingers. Now try to tear the fabric. If it can be torn, it should definitely be destroyed and discarded. Compare the colors in protected places to those exposed to the sun. If the colors have faded, the materials have been weakened. A life jacket covered in fabric should ordinarily last several boating seasons with normal use. A life jacket used every day in direct sunlight should probably be replaced more often.

SOUND PRODUCING DEVICES

All boats are required to carry some means of making an efficient sound signal. Devices for making the whistle or horn noises required by the Navigation Rules must be capable of a four-second blast. The blast should be audible for at least one-half mile. Athletic whistles are not acceptable on boats 12 meters or longer. Use caution with athletic whistles. When wet, some of them come apart and loose their "pea." When this happens, they are useless.

If your vessel is 12 meters long and less than 20 meters, you must have a power whistle (or power horn) and a bell on board. The bell must be in operating condition and have a minimum diameter of at least 200mm (7.9 in.) at its mouth.

VISUAL DISTRESS SIGNALS

◆ See Figure 10

Visual Distress Signals (VDS) attract attention to your vessel if you need help. They also help to guide searchers in search and rescue situations. Be sure you have the right types, and learn how to use them properly.

It is illegal to fire flares improperly. In addition, they cost the Coast Guard and its Auxiliary many wasted hours in fruitless searches. If you signal a distress with flares and then someone helps you, please let the Coast Guard or the appropriate Search And Rescue (SAR) Agency know so the distress report will be canceled.

Recreational boats less than 16 feet long must carry visual distress signals on coastal waters at night. Coastal waters are:
- The ocean (territorial sea)
- The Great Lakes
- Bays or sounds that empty into oceans
- Rivers over two miles across at their mouths upstream to where they narrow to two miles.

Recreational boats 16 feet or longer must carry VDS at all times on coastal waters. The same requirement applies to boats carrying six or fewer passengers for hire. Open sailboats less than 26 feet long without engines are exempt in the daytime as are manually propelled boats. Also exempt are boats in organized races, regattas, parades, etc. Boats owned in the United States and operating on the high seas must be equipped with VDS.

A wide variety of signaling devices meet Coast Guard regulations. For pyrotechnic devices, a minimum of three must be carried. Any combination can be carried as long as it adds up to at least three signals for day use and at least three signals for night use. Three day/night signals meet both requirements. If possible, carry more than the legal requirement.

■ The American flag flying upside down is a commonly recognized distress signal. It is not recognized in the Coast Guard regulations, though. In an emergency, your efforts would probably be better used in more effective signaling methods.

Types of VDS

VDS are divided into two groups; daytime and nighttime use. Each of these groups is subdivided into pyrotechnic and non-pyrotechnic devices.

Daytime Non-Pyrotechnic Signals

A bright orange flag with a black square over a black circle is the simplest VDS. It is usable, of course, only in daylight. It has the advantage of being a continuous signal. A mirror can be used to good advantage on sunny days. It can attract the attention of other boaters and of aircraft from great distances. Mirrors are available with holes in their centers to aid in "aiming." In the absence of a mirror, any shiny object can be used. When another boat is in sight, an effective VDS is to extend your arms from your sides and move them up and down. Do it slowly. If you do it too fast the other people may think you are just being friendly. This simple gesture is seldom misunderstood, and requires no equipment.

Daytime Pyrotechnic Devices

Orange smoke is a useful daytime signal. Hand-held or floating smoke flares are very effective in attracting attention from aircraft. Smoke flares don't last long, and are not very effective in high wind or poor visibility. As with other pyrotechnic devices, use them only when you know there is a possibility that someone will see the display.

To be usable, smoke flares must be kept dry. Keep them in airtight containers and store them in dry places. If the "striker" is damp, dry it out before trying to ignite the device. Some pyrotechnic devices require a forceful "strike" to ignite them.

All hand-held pyrotechnic devices may produce hot ashes or slag when burning. Hold them over the side of your boat in such a way that they do not burn your hand or drip into your boat.

Nighttime Non-Pyrotechnic Signals

An electric distress light is available. This light automatically flashes the international morse code SOS distress signal (••• — •••). Flashed four to six times a minute, it is an unmistakable distress signal. It must show that it is approved by the Coast Guard. Be sure the batteries are fresh. Dated batteries give assurance that they are current.

Under the Inland Navigation Rules, a high intensity white light flashing 50-70 times per minute is a distress signal. Therefore, use strobe lights on inland waters only for distress signals.

Nighttime Pyrotechnic Devices
◆ See Figure 11

Aerial and hand-held flares can be used at night or in the daytime. Obviously, they are more effective at night.

Currently, the serviceable life of a pyrotechnic device is rated at 42 months from its date of manufacture. Pyrotechnic devices are expensive. Look at their dates before you buy them. Buy them with as much time remaining as possible.

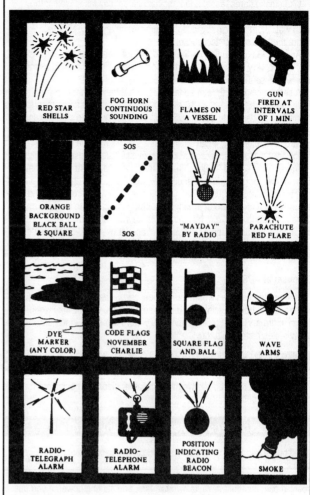

Fig. 10 Internationally accepted distress signals

Fig. 11 Moisture-protected flares should be carried onboard any vessel for use as a distress signal

1-10 GENERAL INFORMATION, SAFETY AND TOOLS

Like smoke flares, aerial and hand-held flares may fail to work if they have been damaged or abused. They will not function if they are or have been wet. Store them in dry, airtight containers in dry places. But store them where they are readily accessible.

Aerial VDSs, depending on their type and the conditions they are used in, may not go very high. Again, use them only when there is a good chance they will be seen.

A serious disadvantage of aerial flares is that they burn for only a short time; most burn for less than 10 seconds. Most parachute flares burn for less than 45 seconds. If you use a VDS in an emergency, do so carefully. Hold hand-held flares over the side of the boat when in use. Never use a road hazard flare on a boat; it can easily start a fire. Marine type flares are specifically designed to lessen risk, but they still must be used carefully.

Aerial flares should be given the same respect as firearms since they are firearms! Never point them at another person. Don't allow children to play with them or around them. When you fire one, face away from the wind. Aim it downwind and upward at an angle of about 60 degrees to the horizon. If there is a strong wind, aim it somewhat more vertically. Never fire it straight up. Before you discharge a flare pistol, check for overhead obstructions that might be damaged by the flare. An obstruction might deflect the flare to where it will cause injury or damage.

Disposal of VDS

Keep outdated flares when you get new ones. They do not meet legal requirements, but you might need them sometime, and they may work. It is illegal to fire a VDS on federal navigable waters unless an emergency exists. Many states have similar laws.

Emergency Position Indicating Radio Beacon (EPIRB)

There is no requirement for recreational boats to have EPIRBs. Some commercial and fishing vessels, though, must have them if they operate beyond the three mile limit. Vessels carrying six or fewer passengers for hire must have EPIRBs under some circumstances when operating beyond the three-mile limit. If you boat in a remote area or offshore, you should have an EPIRB. An EPIRB is a small (about 6 to 20 in. high), battery-powered, radio transmitting buoy-like device. It is a radio transmitter and requires a license or an endorsement on your radio station license by the Federal Communications Commission (FCC). EPIRBs are either automatically activated by being immersed in water or manually by a switch.

Courtesy Marine Examinations

One of the roles of the Coast Guard Auxiliary is to promote recreational boating safety. This is why they conduct thousands of Courtesy Marine Examinations each year. The auxiliarists who do these examinations are well-trained and knowledgeable in the field.

These examinations are free and done only at the consent of boat owners. To pass the examination, a vessel must satisfy federal equipment requirements and certain additional requirements of the coast guard auxiliary. If your vessel does not pass the Courtesy Marine Examination, no report of the failure is made. Instead, you will be told what you need to correct the deficiencies. The examiner will return at your convenience to redo the examination.

If your vessel qualifies, you will be awarded a safety decal. The decal does not carry any special privileges, it simply attests to your interest in safe boating.

BOATING EQUIPMENT (NOT REQUIRED BUT RECOMMENDED)

Although not required by law, there are other pieces of equipment that are good to have onboard.

Oar/Paddle (Second Means of Propulsion)

All boats less than 16 feet long should carry a second means of propulsion. A paddle or oar can come in handy at times. For most small boats, a spare trolling or outboard motor is an excellent idea. If you carry a spare motor, it should have its own fuel tank and starting power. If you use an electric trolling motor, it should have its own battery.

Bailing Devices

All boats should carry at least one effective manual bailing device in addition to any installed electric bilge pump. This can be a bucket, can, scoop, hand-operated pump, etc. If your battery "goes dead" it will not operate your electric pump.

First Aid Kit

◆ See Figure 12

All boats should carry a first aid kit. It should contain adhesive bandages, gauze, adhesive tape, antiseptic, aspirin, etc. Check your first aid kit from time to time. Replace anything that is outdated. It is to your advantage to know how to use your first aid kit. Another good idea would be to take a Red Cross first aid course.

Anchors

◆ See Figure 13

All boats should have anchors. Choose one of suitable size for your boat. Better still, have two anchors of different sizes. Use the smaller one in calm water or when anchoring for a short time to fish or eat. Use the larger one when the water is rougher or for overnight anchoring.

Carry enough anchor line, of suitable size, for your boat and the waters in which you will operate. If your engine fails you, the first thing you usually should do is lower your anchor. This is good advice in shallow water where you may be driven aground by the wind or water. It is also good advice in windy weather or rough water, as the anchor, when properly affixed, will usually hold your bow into the waves.

VHF-FM Radio

Your best means of summoning help in an emergency or in case of a breakdown is a VHF-FM radio. You can use it to get advice or assistance from the Coast Guard. In the event of a serious illness or injury aboard your boat, the Coast Guard can have emergency medical equipment meet you ashore.

■ **Although the VHF radio is the best way to get help, in this day and age, cell phones are a good backup source, especially for boaters on inland waters. You probably already know where you get a signal when boating, keep the phone charged, handy and off (so it doesn't bother you when boating right?). Keep phone numbers for a local dockmaster, coast guard, tow service or maritime police unit handy on board or stored in your phone directory.**

Compass

SELECTION

◆ See Figure 14

The safety of the boat and her crew may depend on her compass. In many areas, weather conditions can change so rapidly that, within minutes, a skipper may find himself socked in by a fog bank, rain squall or just poor visibility. Under these conditions, he may have no other means of keeping to his desired course except with the compass. When crossing an open body of water, his compass may be the only means of making an accurate landfall.

During thick weather when you can neither see nor hear the expected aids to navigation, attempting to run out the time on a given course can disrupt the pleasure of the cruise. The skipper gains little comfort in a chain of soundings that does not match those given on the chart for the expected area. Any stranding, even for a short time, can be an unnerving experience.

A pilot will not knowingly accept a cheap parachute. By the same token, a good boater should not accept a bargain in lifejackets, fire extinguishers, or compass. Take the time and spend the few extra dollars to purchase a compass to fit your expected needs. Regardless of what the salesman may tell you, postpone buying until you have had the chance to check more than one make and model.

Lift each compass, tilt and turn it, simulating expected motions of the boat. The compass card should have a smooth and stable reaction.

The card of a good quality compass will come to rest without oscillations about the lubber's line. Reasonable movement in your hand, comparable to the rolling and pitching of the boat, should not materially affect the reading.

GENERAL INFORMATION, SAFETY AND TOOLS 1-11

Fig. 12 Always carry an adequately stocked first aid kit on board for the safety of the crew and guests

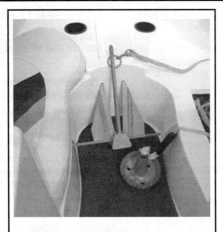

Fig. 13 Choose an anchor of sufficient weight to secure the boat without dragging

Fig. 14 Don't hesitate to spend a few extra dollars for a reliable compass

INSTALLATION

◆ See Figure 15

Proper installation of the compass does not happen by accident. Make a critical check of the proposed location to be sure compass placement will permit the helmsman to use it with comfort and accuracy. First, the compass should be placed directly in front of the helmsman, and in such a position that it can be viewed without body stress as he sits or stands in a posture of relaxed alertness. The compass should be in the helmsman's zone of comfort. If the compass is too far away, he may have to bend forward to watch it; too close and he must rear backward for relief.

Second, give some thought to comfort in heavy weather and poor visibility conditions during the day and night. In some cases, the compass position may be partially determined by the location of the wheel, shift lever and throttle handle.

Third, inspect the compass site to be sure the instrument will be at least two feet from any engine indicators, bilge vapor detectors, magnetic instruments, or any steel or iron objects. If the compass cannot be placed at least two feet (six feet would be better but on a small craft, let's get real two feet is usually pushing it) from one of these influences, then either the compass or the other object must be moved, if first order accuracy is to be expected.

Once the compass location appears to be satisfactory, give the compass a test before installation. Hidden influences may be concealed under the cabin top, forward of the cabin aft bulkhead, within the cockpit ceiling, or in a wood-covered stanchion.

Move the compass around in the area of the proposed location. Keep an eye on the card. A magnetic influence is the only thing that will make the card turn. You can quickly find any such influence with the compass. If the influence cannot be moved away or replaced by one of non-magnetic material, test to determine whether it is merely magnetic, a small piece of iron or steel, or some magnetized steel. Bring the north pole of the compass near the object, then shift and bring the south pole near it. Both the north and south poles will be attracted if the compass is demagnetized. If the object attracts one pole and repels the other, then the compass is magnetized. If your compass needs to be demagnetized, take it to a shop equipped to do the job PROPERLY.

After you have moved the compass around in the proposed mounting area, hold it down or tape it in position. Test everything you feel might affect the compass and cause a deviation from a true reading. Rotate the wheel from hard over-to-hard over. Switch on and off all the lights, radios, radio direction finder, radio telephone, depth finder and, if installed, the shipboard intercom. Sound the electric whistle, turn on the windshield wipers, start the engine (with water circulating through the engine), work the throttle, and move the gear shift lever. If the boat has an auxiliary generator, start it.

If the card moves during any one of these tests, the compass should be relocated. Naturally, if something like the windshield wipers causes a slight deviation, it may be necessary for you to make a different deviation table to use only when certain pieces of equipment are operating. Bear in mind, following a course that is off only a degree or two for several hours can make considerable difference at the end, putting you on a reef, rock or shoal.

Check to be sure the intended compass site is solid. Vibration will increase pivot wear.

Now, you are ready to mount the compass. To prevent an error on all courses, the line through the lubber line and the compass card pivot must be exactly parallel to the keel of the boat. You can establish the fore-and-aft line of the boat with a stout cord or string. Use care to transfer this line to the compass site. If necessary, shim the base of the compass until the stile-type lubber line (the one affixed to the case and not gimbaled) is vertical when the boat is on an even keel. Drill the holes and mount the compass.

COMPASS PRECAUTIONS

◆ See Figures 16, 17 and 18

Many times an owner will install an expensive stereo system in the cabin of his boat. It is not uncommon for the speakers to be mounted on the aft bulkhead up against the overhead (ceiling). In almost every case, this position places one of the speakers in very close proximity to the compass, mounted above the ceiling.

You probably already know that a magnet is used in the operation of the speaker. Therefore, it is very likely that the speaker, mounted almost under the compass in the cabin will have a very pronounced effect on the compass accuracy.

Fig. 15 The compass is a delicate instrument which should be mounted securely in a position where it can be easily observed by the helmsman

1-12 GENERAL INFORMATION, SAFETY AND TOOLS

Consider the following test and the accompanying photographs as proof:
First, the compass was read as 190 degrees while the boat was secure in her slip.
Next, a full can of soda in an aluminum can was placed on one side and the compass read as 204 degrees, a good 14 degrees off.
Next, the full can was moved to the opposite side of the compass and again a reading was observed, this time as 189 degrees, 11 degrees off from the original reading.
Finally, the contents of the can were consumed, the can placed on both sides of the compass with NO effect on the compass reading.
Two very important conclusions can be drawn from these tests.
• Something must have been in the contents of the can to affect the compass so drastically.
• Keep even innocent things clear of the compass to avoid any possible error in the boat's heading.

■ Remember, a boat moving through the water at 10 knots on a compass error of just 5 degrees will be almost 1.5 miles off course in only ONE hour. At night, or in thick weather, this could very possibly put the boat on a reef, rock or shoal with disastrous results.

Tools and Spare Parts

◆ See Figures 19 and 20

Carry a few tools and some spare parts, and learn how to make minor repairs. Many search and rescue cases are caused by minor breakdowns that boat operators could have repaired. Carry spare parts such as propellers, fuses or basic ignition components (like spark plugs, wires or even ignition coils) and the tools necessary to install them.

Fig. 16 This compass is giving an accurate reading, right?

Fig. 17 . . . well think again, as seemingly innocent objects may cause serious problems . . .

Fig. 18 . . . a compass reading off by just a few degrees could lead to disaster

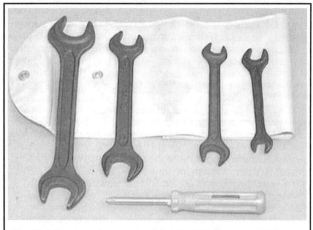

Fig. 19 A few wrenches, a screwdriver and maybe a pair of pliers can be very helpful to make emergency repairs

Fig. 20 A flashlight with a fresh set of batteries is handy when repairs are needed at night. It can also double as a signaling device

SAFETY IN SERVICE

It is virtually impossible to anticipate all of the hazards involved with maintenance and service, but care and common sense will prevent most accidents.
The rules of safety for mechanics range from "don't smoke around gasoline," to "use the proper tool(s) for the job." The trick to avoiding injuries is to develop safe work habits and to take every possible precaution. Whenever you are working on your boat, pay attention to what you are doing. The more you pay attention to details and what is going on around you, the less likely you will be to hurt yourself or damage your boat.

Do's

• Do keep a fire extinguisher and first aid kit handy.
• Do wear safety glasses or goggles when cutting, drilling, grinding or prying, even if you have 20-20 vision. If you wear glasses for the sake of vision, wear safety goggles over your regular glasses.
• Do shield your eyes whenever you work around the battery. Batteries contain sulfuric acid. In case of contact with the eyes or skin, flush the area with water or a mixture of water and baking soda; then seek immediate medical attention.
• Do use adequate ventilation when working with any chemicals or hazardous materials.
• Do disconnect the negative battery cable when working on the electrical system. The secondary ignition system contains EXTREMELY HIGH VOLTAGE. In some cases it can even exceed 50,000 volts. Furthermore, an accidental attempt to start the engine could cause the propeller or other components to rotate suddenly causing a potentially dangerous situation.
• Do follow manufacturer's directions whenever working with potentially hazardous materials. Most chemicals and fluids are poisonous if taken internally.

GENERAL INFORMATION, SAFETY AND TOOLS 1-13

- Do properly maintain your tools. Loose hammerheads, mushroomed punches and chisels, frayed or poorly grounded electrical cords, excessively worn screwdrivers, spread wrenches (open end), cracked sockets, or slipping ratchets can cause accidents.
- Likewise, keep your tools clean; a greasy wrench can slip off a bolt head, ruining the bolt and often harming your knuckles in the process.
- Do use the proper size and type of tool for the job at hand. Do select a wrench or socket that fits the nut or bolt. The wrench or socket should sit straight, not cocked.
- Do, when possible, pull on a wrench handle rather than push on it, and adjust your stance to prevent a fall.
- Do be sure that adjustable wrenches are tightly closed on the nut or bolt and pulled so that the force is on the side of the fixed jaw. Better yet, avoid the use of an adjustable if you have a fixed wrench that will fit.
- Do strike squarely with a hammer; avoid glancing blows.
- Do use common sense whenever you work on your boat or motor. If a situation arises that doesn't seem right, sit back and have a second look. It may save an embarrassing moment or potential damage to your beloved boat.

Don'ts

- Don't run the engine in an enclosed area or anywhere else without proper ventilation - EVER! Carbon monoxide is poisonous; it takes a long time to leave the human body and you can build up a deadly supply of it in your system by simply breathing in a little every day. You may not realize you are slowly poisoning yourself.
- Don't work around moving parts while wearing loose clothing. Short sleeves are much safer than long, loose sleeves. Hard-toed shoes with neoprene soles protect your toes and give a better grip on slippery surfaces. Jewelry, watches, large belt buckles, or body adornment of any kind is not safe working around any craft or vehicle. Long hair should be tied back under a hat.
- Don't use pockets for toolboxes. A fall or bump can drive a screwdriver deep into your body. Even a rag hanging from your back pocket can wrap around a spinning shaft.
- Don't smoke when working around gasoline, cleaning solvent or other flammable material.
- Don't smoke when working around the battery. When the battery is being charged, it gives off explosive hydrogen gas. Actually, you shouldn't smoke anyway, it's bad for you. Instead, save the cigarette money and put it into your boat!
- Don't use gasoline to wash your hands; there are excellent soaps available. Gasoline contains dangerous additives that can enter the body through a cut or through your pores. Gasoline also removes all the natural oils from the skin so that bone dry hands will suck up oil and grease.
- Don't use screwdrivers for anything other than driving screws! A screwdriver used as a prying tool can snap when you least expect it, causing injuries. At the very least, you'll ruin a good screwdriver.

TROUBLESHOOTING

Troubleshooting can be defined as a methodical process during which one discovers what is causing a problem with engine operation. Although it is often a feared process to the uninitiated, there is no reason to believe that you cannot figure out what is wrong with a motor, as long as you follow a few basic rules.

To begin with, troubleshooting must be systematic. Haphazardly testing one component, then another, **might** uncover the problem, but it will more likely waste a lot of time. True troubleshooting starts by defining the problem and performing systematic tests to eliminate the largest and most likely causes first.

Start all troubleshooting by eliminating the most basic possible causes. Begin with a visual inspection of the boat and motor. If the engine won't crank, make sure that the kill switch or safety lanyard is in the proper position. Make sure there is fuel in the tank and the fuel system is primed before condemning the carburetor or fuel injection system. On electric start motors, make sure there are no blown fuses, the battery is fully charged, and the cable connections (at both ends) are clean and tight before suspecting a bad starter, solenoid or switch.

The majority of problems that occur suddenly can be fixed by simply identifying the one small item that brought them on. A loose wire, a clogged passage or a broken component can cause a lot of trouble and are often the cause of a sudden performance problem.

The next most basic step in troubleshooting is to test systems before components. For example, if the engine doesn't crank on an electric start motor, determine if the battery is in good condition (fully charged and properly connected) before testing the starting system. If the engine cranks, but doesn't start, you know already know the starting system and battery (if it cranks fast enough) are in good condition, now it is time to look at the ignition or fuel systems. Once you've isolated the problem to a particular system, follow the troubleshooting/testing procedures in the section for that system to test either subsystems (if applicable, for example: the starter circuit) or components (starter solenoid).

Basic Operating Principles

◆ See Figures 21 and 22

Before attempting to troubleshoot a problem with your motor, it is important that you understand how it operates. Once normal engine or system operation is understood, it will be easier to determine what might be causing the trouble or irregular operation in the first place. System descriptions are found throughout this manual, but the basic mechanical operating principles for both 2-stroke engines (like most of the outboards covered here) and 4-stroke engines (like some outboards and like your car) are given here. A basic understanding of both types of engines is useful not only in understanding and troubleshooting your outboard, but also for dealing with other motors in your life.

All motors covered by this manual (and probably MOST of the motors you own) operate according to the Otto cycle principle of engine operation. This means that all motors follow the stages of intake, compression, power and exhaust. But, the difference between a 2- and 4-stroke motor is in how many times the piston moves up and down within the cylinder to accomplish this. On 2-stroke motors (as the name suggests) the four cycles take place in 2 movements (one up and one down) of the piston. Again, as the name suggests, the cycles take place in 4 movements of the piston for 4-stroke motors.

2-STROKE MOTORS

The 2-stroke engine differs in several ways from a conventional four-stroke (automobile or marine) engine.
1. The intake/exhaust method by which the fuel-air mixture is delivered to the combustion chamber.
2. The complete lubrication system.
3. The frequency of the power stroke.

Let's discuss these differences briefly (and compare 2-stroke engine operation with 4-stroke engine operation.)

Intake/Exhaust

◆ See Figures 23 thru 25a

Two-stroke engines utilize an arrangement of port openings to admit fuel to the combustion chamber and to purge the exhaust gases after burning has been completed. The ports are located in a precise pattern in order for them to be open and closed off at an exact moment by the piston as it moves up and down in the cylinder. The exhaust port is located slightly higher than the fuel intake port. This arrangement opens the exhaust port first as the piston starts downward and therefore, the exhaust phase begins a fraction of a second before the intake phase.

Actually, the intake and exhaust ports are spaced so closely together that both open almost simultaneously. For this reason, some 2-stroke engines utilize deflector-type pistons. This design of the piston top serves two purposes very effectively.

First, it creates turbulence when the incoming charge of fuel enters the combustion chamber. This turbulence results in a more complete burning of the fuel than if the piston top were flat. The second effect of the deflector-type piston crown is to force the exhaust gases from the cylinder more rapidly.

Loop charged motors, or as they are commonly called "loopers", differ in how the air/fuel charge is introduced to the combustion chamber. Instead of the charge flowing across the top of the piston from one side of the cylinder

1-14 GENERAL INFORMATION, SAFETY AND TOOLS

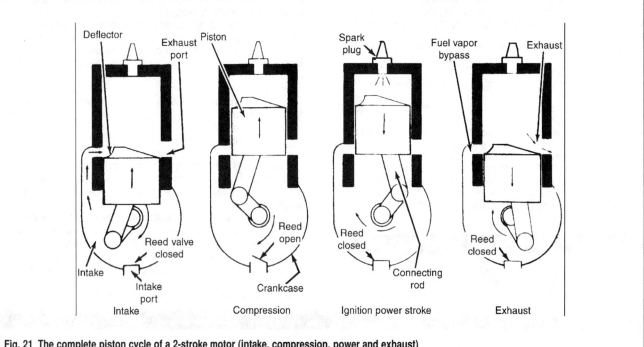

Fig. 21 The complete piston cycle of a 2-stroke motor (intake, compression, power and exhaust)

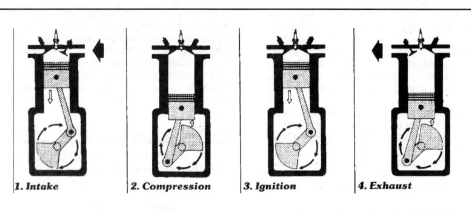

Fig. 22 The complete piston cycle of a 4-stroke motor (intake, compression, power and exhaust)

to the other (CV) the use a looping action on top of the piston as the charge is forced through irregular shaped openings cut in the piston's skirt. In an LV motor, the charge is forced out from the crankcase by the downward motion of the piston, through the irregular shaped openings and transferred upward by long, deep grooves in the cylinder wall. The charge completes its looping action by entering the combustion chamber, just above the piston, where the upward motion of the piston traps it in the chamber and compresses it for optimum ignition power.

Unlike the knife-edged deflector top pistons used in CV motors, the piston domes on Loop motors are relatively flat.

■ Over the years Johnson/Evinrude used both LV and CV configurations on many of their 2-stroke motors.

These systems of intake and exhaust are in marked contrast to individual intake and exhaust valve arrangement employed on four-stroke engines (and the mechanical methods of opening and closing these valves).

■ It should be noted here that there are some 2-stroke engines that utilize a mechanical valve train, though it is very different from the valve train employed by most 4-stroke motors. Rotary 2-stroke engines use a circular valve or rotating disc that contains a port opening around part of one edge of the disc. As the engine (and disc) turns, the opening aligns with the intake port at and for a predetermined amount of time, closing off the port again as the opening passes by and the solid portion of the disc covers the port.

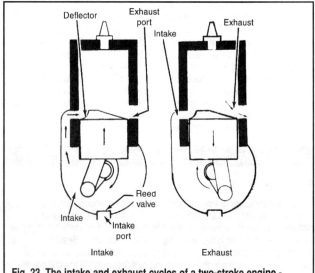

Fig. 23 The intake and exhaust cycles of a two-stroke engine - Cross flow (CV) design shown

GENERAL INFORMATION, SAFETY AND TOOLS 1-15

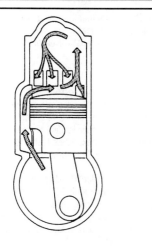

Fig. 24 Cross-sectional view of a typical loop-charged cylinder, showing charge flow while piston is moving downward

Fig. 25 Cutaway view of a typical loop-charged cylinder, depicting exhaust leaving the cylinder as the charge enters through multiple ports in the piston

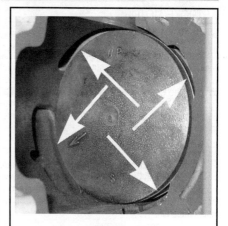

Fig. 25a The combustion chamber of a typical looper, notice the piston is far enough down the cylinder bore to reveal intake and exhaust ports

Lubrication

A 2-stroke engine is lubricated by mixing oil with the fuel. Therefore, various parts are lubricated as the fuel mixture passes through the crankcase and the cylinder. In contrast, four-stroke engines have a crankcase containing oil. This oil is pumped through a circulating system and returned to the crankcase to begin the routing again.

Power Stroke

The combustion cycle of a 2-stroke engine has four distinct phases.
1. Intake
2. Compression
3. Power
4. Exhaust

The four phases of the cycle are accomplished with each up and down stroke of the piston, and the power stroke occurs with each complete revolution of the crankshaft. Compare this system with a four-stroke engine. A separate stroke of the piston is required to accomplish each phase of the cycle and the power stroke occurs only every other revolution of the crankshaft. Stated another way, two revolutions of the four-stroke engine crankshaft are required to complete one full cycle, the four phases.

Physical Laws
◆ See Figure 26

The 2-stroke engine is able to function because of two very simple physical laws.

One: Gases will flow from an area of high pressure to an area of lower pressure. A tire blowout is an example of this principle. The high-pressure air escapes rapidly if the tube is punctured.

Two: If a gas is compressed into a smaller area, the pressure increases, and if a gas expands into a larger area, the pressure is decreased.

If these two laws are kept in mind, the operation of the 2-stroke engine will be easier understood.

Actual Operation
◆ See Figure 27

■ The engine described here is of a carbureted type. EFI or DFI (FICHT) motors operate similarly for intake of the air charge and for exhaust of the unburned gasses. Obviously though, the very nature of fuel injection changes the actual delivery of the fuel/oil charge.

Beginning with the piston approaching top dead center on the compression stroke: the intake and exhaust ports are physically closed (blocked) by the piston. During this stroke, the reed valve is open (because as the piston moves upward, the crankcase volume increases, which reduces the crankcase pressure to less than the outside atmosphere (creates a vacuum under the piston). The spark plug fires; the compressed fuel-air mixture is ignited; and the power stroke begins.

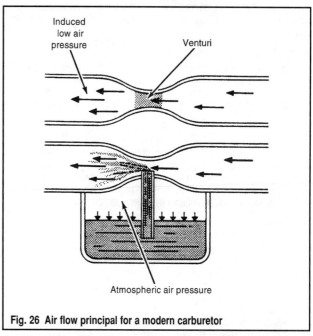

Fig. 26 Air flow principal for a modern carburetor

As the piston moves downward on the power stroke, the combustion chamber is filled with burning gases. As the exhaust port is uncovered, the gases, which are under great pressure, escape rapidly through the exhaust ports. The piston continues its downward movement. Pressure within the crankcase (again, under the piston) increases, closing the reed valves against their seats. The crankcase then becomes a sealed chamber so the air-fuel mixture becomes compressed (pressurized) and ready for delivery to the combustion chamber. As the piston continues to move downward, the intake port is uncovered. The fresh fuel mixture rushes through the intake port into the combustion chamber striking the top of the piston where it is deflected along the cylinder wall. The reed valve remains closed until the piston moves upward again.

When the piston begins to move upward on the compression stroke, the reed valve opens because the crankcase volume has been increased, reducing crankcase pressure to less than the outside atmosphere. The intake and exhaust ports are closed and the fresh fuel charge is compressed inside the combustion chamber.

Pressure in the crankcase (beneath the piston) decreases as the piston moves upward and a fresh charge of air flows through the carburetor picking up fuel. As the piston approaches top dead center, the spark plug ignites the air-fuel mixture, the power stroke begins and one complete Otto cycle has been completed.

4-STROKE MOTORS

◆ See Figure 22

The 4-stroke motor may be easier to understand for some people either because of its prevalence in automobile and street motorcycle motors today or perhaps because each of the four strokes corresponds to one distinct phase of the Otto cycle. Essentially, a 4-stroke motor completes one Otto cycle of intake, compression, ignition/power and exhaust using two full revolutions of the crankshaft and four distinct movements of the piston (down, up, down and up).

Intake

The intake stroke begins with the piston near the top of its travel. As crankshaft rotation begins to pull the piston downward, the exhaust valve closes and the intake opens. As volume of the combustion chamber increases, a vacuum is created that draws in the air/fuel mixture from the intake manifold.

Compression

Once the piston reaches the bottom of its travel, crankshaft rotation will begin to force it upward. At this point the intake valve closes. As the piston rises in the bore, the volume of the sealed combustion chamber (both intake and exhaust valves are closed) decreases and the air/fuel mixture is compressed. This raises the temperature and pressure of the mixture and increases the amount of force generated by the expanding gases during the Ignition/Power stroke.

Ignition/Power

As the piston approaches top dead center (the highest point of travel in the bore), the spark plug will fire, igniting the air/fuel mixture. The resulting combustion of the air/fuel mixture forces the piston downward, rotating the crankshaft (causing other pistons to move in other phases/strokes of the Otto cycle on multi-cylinder motors).

Exhaust

As the piston approaches the bottom of the Ignition/Power stroke, the exhaust valve opens. When the piston begins its upward path of travel once again, any remaining unburned gasses are forced out through the exhaust valve. This completes one Otto cycle, which begins again as the piston passes top dead center, the intake valve opens and the Intake stroke starts.

COMBUSTION

Whether we are talking about a 2- or 4-stroke engine, all Otto cycle, internal combustion engines require three basic conditions to operate properly,
1. Compression
2. Ignition (Spark)
3. Fuel

A lack of any one of these conditions will prevent the engine from operating. A problem with any one of these will manifest itself in hard-starting or poor performance.

Compression

An engine that has insufficient compression will not draw an adequate supply of air/fuel mixture into the combustion chamber and, subsequently, will not make sufficient power on the power stroke. A lack of compression in just one cylinder of a multi-cylinder motor will cause the motor to stumble or run irregularly.

But, keep in mind that a sudden change in compression is unlikely in 2-stroke motors (unless something major breaks inside the crankcase, but that would usually be accompanied by other symptoms such as a loud noise when it occurred or noises during operation). On 4-stroke motors, a sudden change in compression is also unlikely, but could occur if the timing belt or chain was to suddenly break. Remember that the timing belt/chain is used to synchronize the valve train with the crankshaft. If the valve train suddenly ceases to turn, some intake and some exhaust valves will remain open, relieving compression in that cylinder.

Ignition (Spark)

Traditionally, the ignition system is the weakest link in the chain of conditions necessary for engine operation. Spark plugs may become worn or fouled, wires will deteriorate allowing arcing or misfiring, and poor connections can place an undue load on coils leading to weak spark or even a failed coil. The most common question asked by a technician under a no-start condition is: "do I have spark and fuel" (as they've already determined that they have compression).

A quick visual inspection of the spark plug(s) will answer the question as to whether or not the plug(s) is/are worn or fouled. While the engine is shut **OFF** a physical check of the connections could show a loose primary or secondary ignition circuit wire. An obviously physically damaged wire may also be an indication of system problems and certainly encourages one to inspect the related system more closely.

If nothing is turned up by the visual inspection, perform the Spark Test provided in the Ignition System section to determine if the problem is a lack of or a weak spark. If the problem is not compression or spark, it's time to look at the fuel system.

Fuel

If compression and spark is present (and within spec), but the engine won't start or won't run properly, the only remaining condition to fulfill is fuel. As usual, start with the basics. Is the fuel tank full? Is the fuel stale? If the engine has not been run in some time (a matter of months, not weeks) there is a good chance that the fuel is stale and should be properly disposed of and replaced.

■ **Depending on how stale or contaminated (with moisture) the fuel is, it may be burned in an automobile or in yard equipment, though it would be wise to mix it well with a much larger supply of fresh gasoline to prevent moving your driveability problems to that motor. But it is better to get the lawn tractor stuck on stale gasoline than it would be to have your boat motor quit in the middle of the bay or lake.**

For hard starting motors, is the choke or primer system operating properly. Remember that the choke/prime should only be used for **cold** starts. A true cold start is really only the first start of the day, but it may be applicable to subsequent starts on cooler days, if the engine sat for more than a few hours and completely cooled off since the last use. Applying the primer to the motor for a hot start may flood the engine, preventing it from starting properly. One method to clear a flood is to crank the motor while the engine is at wide-open throttle (allowing the maximum amount of air into the motor to compensate for the excess fuel). But, keep in mind that the throttle should be returned to idle immediately upon engine start-up to prevent damage from over-revving.

Fuel delivery and pressure should be checked before delving into the carburetor(s) or fuel injection system. Make sure there are no clogs in the fuel line or vacuum leaks that would starve the motor of fuel.

Make sure that all other possible problems have been eliminated before touching the carburetor. It is rare that a carburetor will suddenly require an adjustment in order for the motor to run properly. It is much more likely that an improperly stored motor (one stored with untreated fuel in the carburetor) would suffer from one or more clogged carburetor passages sometime after shortly returning to service. Fuel will evaporate over time, leaving behind gummy deposits. If untreated fuel is left in the carburetor for some time (again typically months more than weeks), the varnish left behind by evaporating fuel will likely clog the small passages of the carburetor and cause problems with engine performance. If you suspect this, remove and disassemble the carburetor following procedures under Fuel System.

GENERAL INFORMATION, SAFETY AND TOOLS

SHOP EQUIPMENT

Safety Tools

WORK GLOVES

◆ See Figure 27

Unless you think scars on your hands are cool, enjoy pain and like wearing bandages, get a good pair of work gloves. Canvas or leather gloves are the best. And yes, we realize that there are some jobs involving small parts that can't be done while wearing work gloves. These jobs are not the ones usually associated with hand injuries.

A good pair of rubber gloves (such as those usually associated with dish washing) or vinyl gloves is also a great idea. There are some liquids such as solvents and penetrants that don't belong on your skin. Avoid burns and rashes. Wear these gloves.

And lastly, an option. If you're tired of being greasy and dirty all the time, go to the drug store and buy a box of disposable latex gloves like medical professionals wear. You can handle greasy parts, perform small tasks, wash parts, etc. all without getting dirty! These gloves take a surprising amount of abuse without tearing and aren't expensive. Note however, that some people are allergic to the latex or the powder used inside some gloves, so pay attention to what you buy.

EYE AND EAR PROTECTION

◆ See Figures 28 and 29

Don't begin any job without a good pair of work goggles or impact resistant glasses! When doing any kind of work, it's all too easy to avoid eye injury through this simple precaution. And don't just buy eye protection and leave it on the shelf. Wear it all the time! Things have a habit of breaking, chipping, splashing, spraying, splintering and flying around. And, for some reason, your eye is always in the way!

If you wear vision-correcting glasses as a matter of routine, get a pair made with polycarbonate lenses. These lenses are impact resistant and are available at any optometrist.

Often overlooked is hearing protection. Engines and power tools are noisy! Loud noises damage your ears. It's as simple as that! The simplest and cheapest form of ear protection is a pair of noise-reducing ear plugs. Cheap insurance for your ears! And, they may even come with their own, cute little carrying case.

More substantial, more protection and more money is a good pair of noise reducing earmuffs. They protect from all but the loudest sounds. Hopefully those are sounds that you'll never encounter since they're usually associated with disasters.

WORK CLOTHES

Everyone has "work clothes." Usually these consist of old jeans and a shirt that has seen better days. That's fine. In addition, a denim work apron is a nice accessory. It's rugged, can hold some spare bolts, and you don't feel bad wiping your hands or tools on it. That's what it's for.

When working in cold weather, a one-piece, thermal work outfit is invaluable. Most are rated to below freezing temperatures and are ruggedly constructed. Just look at what local marine mechanics are wearing and that should give you a clue as to what type of clothing is good.

Chemicals

There is a whole range of chemicals that you'll find handy for maintenance and repair work. The most common types are: lubricants, penetrants and sealers. Keep these handy. There are also many chemicals that are used for detailing or cleaning.

When a particular chemical is not being used, keep it capped, upright and in a safe place. These substances may be flammable, may be irritants or might even be caustic and should always be stored properly, used properly and handled with care. Always read and follow all label directions and be sure to wear hand and eye protection!

LUBRICANTS & PENETRANTS

◆ See Figure 30

Anti-seize is used to coat certain fasteners prior to installation. This can be especially helpful when two dissimilar metals are in contact (to help prevent corrosion that might lock the fastener in place). This is a good practice on a lot of different fasteners, BUT, NOT on any fastener that might vibrate loose causing a problem. If anti-seize is used on a fastener, it should be checked periodically for proper tightness.

Lithium grease, chassis lube, silicone grease or a synthetic brake caliper grease can all be used pretty much interchangeably. All can be used for coating rust-prone fasteners and for facilitating the assembly of parts that are a tight fit. Silicone and synthetic greases are the most versatile.

■ **Silicone dielectric grease is a non-conductor that is often used to coat the terminals of wiring connectors before fastening them. It may sound odd to coat metal portions of a terminal with something that won't conduct electricity, but here is it how it works. When the connector is fastened the metal-to-metal contact between the terminals will displace the grease (allowing the circuit to be completed). The grease that is displaced will then coat the non-contacted surface and the cavity around the terminals, SEALING them from atmospheric moisture that could cause corrosion.**

Silicone spray is a good lubricant for hard-to-reach places and parts that shouldn't be gooped up with grease.

Penetrating oil may turn out to be one of your best friends when taking something apart that has corroded fasteners. Not only can they make a job easier, they can really help to avoid broken and stripped fasteners. The most familiar penetrating oils are Liquid Wrench® and WD-40®. A newer penetrant, PB Blaster® works very well (and has become a mainstay in our shops). These products have hundreds of uses. For your purposes, they are vital!

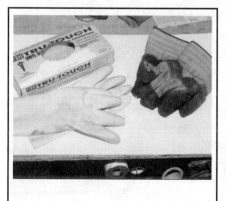

Fig. 27 Three different types of work gloves. The box contains latex gloves

Fig. 28 Don't begin major repairs without a pair of goggles for your eyes and earmuffs to protect your hearing

Fig. 29 Things have a habit of, splashing, spraying, splintering and flying around during repairs

1-18 GENERAL INFORMATION, SAFETY AND TOOLS

Before disassembling any part, check the fasteners. If any appear rusted, soak them thoroughly with the penetrant and let them stand while you do something else (for particularly rusted or frozen parts you may need to soak them a few days in advance). This simple act can save you hours of tedious work trying to extract a broken bolt or stud.

SEALANTS

◆ See Figures 31 and 32

Sealants are an indispensable part for certain tasks, especially if you are trying to avoid leaks. The purpose of sealants is to establish a leak-proof bond between or around assembled parts. Most sealers are used in conjunction with gaskets, but some are used instead of conventional gasket material.

The most common sealers are the non-hardening types such as Permatex® No.2 or its equivalents. These sealers are applied to the mating surfaces of each part to be joined, then a gasket is put in place and the parts are assembled.

■ A sometimes overlooked use for sealants like RTV is on the threads of vibration prone fasteners.

One very helpful type of non-hardening sealer is the "high tack" type. This type is a very sticky material that holds the gasket in place while the parts are being assembled. This stuff is really a good idea when you don't have enough hands or fingers to keep everything where it should be.

The stand-alone sealers are the Room Temperature Vulcanizing (RTV) silicone gasket makers. On some engines, this material is used instead of a gasket. In those instances, a gasket may not be available or, because of the shape of the mating surfaces, a gasket shouldn't be used. This stuff, when used in conjunction with a conventional gasket, produces the surest bonds.

RTV does have its limitations though. When using this material, you will have a time limit. It starts to set-up within 15 minutes or so, so you have to assemble the parts without delay. In addition, when squeezing the material out of the tube, don't drop any glops into the engine. The stuff will form and set and travel around a cooling passage, possibly blocking it. Also, most types are not fuel-proof. Check the tube for all cautions.

CLEANERS

◆ See Figures 33 and 34

There are two basic types of cleaners on the market today: parts cleaners and hand cleaners. The parts cleaners are for the parts; the hand cleaners are for you. They are **not** interchangeable.

There are many good, non-flammable, biodegradable parts cleaners on the market. These cleaning agents are safe for you, the parts and the environment. Therefore, there is no reason to use flammable, caustic or toxic substances to clean your parts or tools.

As far as hand cleaners go; the waterless types are the best. They have always been efficient at cleaning, but they used to all leave a pretty smelly odor. Recently though, most of them have eliminated the odor and added stuff that actually smells good. Make sure that you pick one that contains lanolin or some other moisture-replenishing additive. Cleaners not only remove grease and oil but also skin oil.

■ Most women already know to use a hand lotion when you're all cleaned up. It's okay. Real men DO use hand lotion too! Believe it or not, using hand lotion before your hands are dirty will actually make them easier to clean when you're finished with a dirty job. Lotion seals your hands, and keeps dirt and grease from sticking to your skin.

Fig. 30 Keep a supply of anti-seize, penetrating oil, lithium grease, electronic cleaner and silicone spray

Fig. 31 Sealants are essential for preventing leaks

Fig. 32 On some engines, RTV is used instead of gasket material to seal components

Fig. 33 Citrus hand cleaners not only work well, but they smell pretty good too. Choose one with pumice for added cleaning power

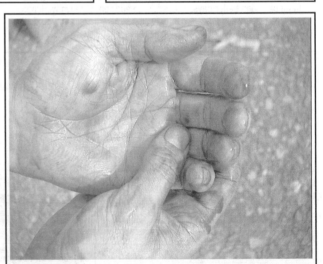

Fig. 34 The use of hand lotion seals your hands and keeps dirt and grease from sticking to your skin

GENERAL INFORMATION, SAFETY AND TOOLS

TOOLS

◆ See Figure 35

Tools; this subject could fill a completely separate manual. The first thing you will need to ask yourself, is just how involved do you plan to get. If you are serious about maintenance and repair you will want to gather a quality set of tools to make the job easier, and more enjoyable. BESIDES, TOOLS ARE FUN!!!

Almost every do-it-yourselfer loves to accumulate tools. Though most find a way to perform jobs with only a few common tools, they tend to buy more over time, as money allows. So gathering the tools necessary for maintenance or repair does not have to be an expensive, overnight proposition.

When buying tools, the saying "You get what you pay for . . ." is absolutely true! Don't go cheap! Any hand tool that you buy should be drop forged and/or chrome vanadium. These two qualities tell you that the tool is strong enough for the job. With any tool, go with a name that you've heard of before, or, that is recommended buy your local professional retailer. Let's go over a list of tools that you'll need.

Most of the world uses the metric system. However, some American-built engines and aftermarket accessories use standard fasteners. So, accumulate your tools accordingly. Any good DIYer should have a decent set of both U.S. and metric measure tools.

■ Don't be confused by terminology. Most advertising refers to "SAE and metric", or "standard and metric." Both are misnomers. The Society of Automotive Engineers (SAE) did not invent the English system of measurement; the English did. The SAE likes metrics just fine. Both English (U.S.) and metric measurements are SAE approved. Also, the current "standard" measurement IS metric. So, if it's not metric, it's U.S. measurement.

Hand Tools

SOCKET SETS

◆ See Figures 36 thru 42

Socket sets are the most basic hand tools necessary for repair and maintenance work. For our purposes, socket sets come in three drive sizes: 1/4 inch, 3/8 inch and 1/2 inch. Drive size refers to the size of the drive lug on the ratchet, breaker bar or speed handle.

A 3/8 inch set is probably the most versatile set in any mechanic's toolbox. It allows you to get into tight places that the larger drive ratchets can't and gives you a range of larger sockets that are still strong enough for heavy-duty work. The socket set that you'll need should range in sizes from 1/4 inch through 1 inch for standard fasteners, and a 6mm through 19mm for metric fasteners.

You'll need a good 1/2 inch set since this size drive lug assures that you won't break a ratchet or socket on large or heavy fasteners. Also, torque wrenches with a torque scale high enough for larger fasteners are usually 1/2 inch drive.

Plus, 1/4 inch drive sets can be very handy in tight places. Though they usually duplicate functions of the 3/8 in. set, 1/4 in. drive sets are easier to use for smaller bolts and nuts.

As for the sockets themselves, they come in shallow (standard) and deep lengths as well as 6 or 12 point. The 6 and 12 points designation refers to how many sides are in the socket itself. Each has advantages. The 6 point socket is stronger and less prone to slipping which would strip a bolt head or nut. 12 point sockets are more common, usually less expensive and can operate better in tight places where the ratchet handle can't swing far.

Standard length sockets are good for just about all jobs, however, some stud-head bolts, hard-to-reach bolts, nuts on long studs, etc., require the deep sockets.

Most marine manufacturers use recessed hex-head fasteners to retain many of the engine parts. These fasteners require a socket with a hex shaped driver or a large sturdy hex key. To help prevent torn knuckles, we would recommend that you stick to the sockets on any tight fastener and leave the hex keys for lighter applications. Hex driver sockets are available individually or in sets just like conventional sockets.

More and more, manufacturers are using Torx® head fasteners, which were once known as tamper resistant fasteners (because many people did not have tools with the necessary odd driver shape). Since Torx® fasteners have become commonplace in many DIYer tool boxes, manufacturers designed newer tamper resistant fasteners that are essentially Torx® head bolts that contain a small protrusion in the center (requiring the driver to contain a small hole to slide over the protrusion. Tamper resistant fasteners are often used where the manufacturer would prefer only knowledgeable mechanics or advanced Do-It-Yourselfers (DIYers) work.

Torque Wrenches

◆ See Figure 43

In most applications, a torque wrench can be used to ensure proper installation of a fastener. Torque wrenches come in various designs and most stores will carry a variety to suit your needs. A torque wrench should be used any time you have a specific torque value for a fastener. Keep in mind that because there is no worldwide standardization of fasteners, so charts or figure found in each repair section refer to the manufacturer's fasteners. Any general guideline charts that you might come across based on fastener size (they are sometimes included in a repair manual or with torque wrench packaging) should be used with caution. Just keep in mind that if you are using the right tool for the job, you should not have to strain to tighten a fastener.

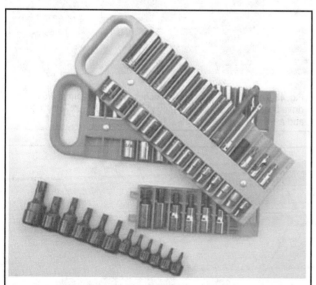

Fig. 35 Socket holders, especially the magnetic type, are handy items to keep tools in order

Fig. 36 A 3/8 in. socket set is probably the most versatile tool in any mechanic's tool box

1-20 GENERAL INFORMATION, SAFETY AND TOOLS

Fig. 37 A swivel (U-joint) adapter (left), a 1/4 in.-to-3/8 in. adapter (center) and a 3/8 in.-to-1/4 in. adapter (right)

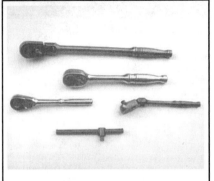

Fig. 38 Ratchets come in all sizes and configurations from rigid to swivel-headed

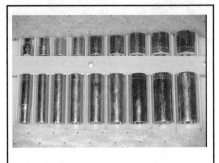

Fig. 39 Shallow sockets (top) are good for most jobs. But, some bolts require deep sockets (bottom)

Fig. 40 Hex-head fasteners require a socket with a hex shaped driver

Fig. 41 Torx® drivers . . .

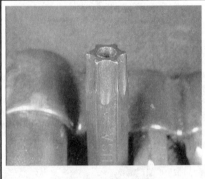

Fig. 42 . . . and tamper resistant drivers are required to remove special fasteners

BEAM TYPE
◆ See Figures 44 and 45

The beam type torque wrench is one of the most popular styles in use. If used properly, it can be the most accurate also. It consists of a pointer attached to the head that runs the length of the flexible beam (shaft) to a scale located near the handle. As the wrench is pulled, the beam bends and the pointer indicates the torque using the scale.

CLICK (BREAKAWAY) TYPE
◆ See Figures 46 and 47

Another popular torque wrench design is the click type. The clicking mechanism makes achieving the proper torque easy and most use a ratcheting head for ease of bolt installation. To use the click type wrench you pre-adjust it to a torque setting. Once the torque is reached, the wrench has a reflex signaling feature that causes a momentary breakaway of the torque wrench body, sending an impulse to the operator's hand. But be careful, as continuing the turn the wrench after the momentary release will increase torque on the fastener beyond the specified setting.

Breaker Bars
◆ See Figure 48

Breaker bars are long handles with a drive lug. Their main purpose is to provide extra turning force when breaking loose tight bolts or nuts. They come in all drive sizes and lengths. Always take extra precautions and use the proper technique when using a breaker bar (pull on the bar, don't push, to prevent skinned knuckles)

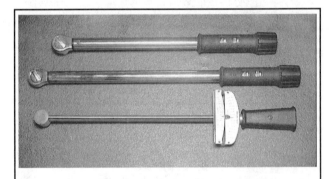

Fig. 43 Three types of torque wrenches. Top to bottom: a 3/8 in. drive beam type that reads in inch lbs., a 1/2 in. drive clicker type and a 1/2 in. drive beam type

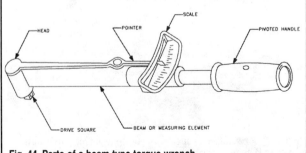

Fig. 44 Parts of a beam type torque wrench

GENERAL INFORMATION, SAFETY AND TOOLS

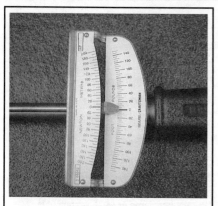

Fig. 45 A beam type torque wrench consists of a pointer attached to the head that runs the length of the flexible beam (shaft) to a scale located near the handle

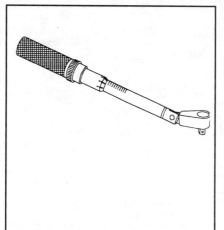

Fig. 46 A click type or breakaway torque wrench - note this one has a pivoting head

Fig. 47 Setting the torque on a click type wrench involves turning the handle until the specification appears on the dial

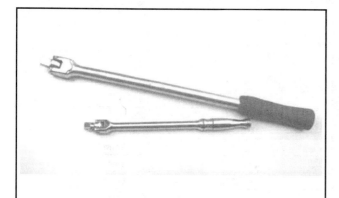

Fig. 48 Breaker bars are great for loosening large or stuck fasteners

WRENCHES

♦ See Figures 49 thru 53

Basically, there are 3 kinds of fixed wrenches: open end, box end, and combination.

Open-end wrenches have 2-jawed openings at each end of the wrench. These wrenches are able to fit onto just about any nut or bolt. They are extremely versatile but have one major drawback. They can slip on a worn or rounded bolt head or nut, causing bleeding knuckles and a useless fastener.

■ Line wrenches are a special type of open-end wrench designed to fit onto more of the fastener than standard open-end wrenches, thus reducing the chance of rounding the corners of the fastener.

Box-end wrenches have a 360° circular jaw at each end of the wrench. They come in both 6 and 12 point versions just like sockets and each type has some of the same advantages and disadvantages as sockets.

Combination wrenches have the best of both. They have a 2-jawed open end and a box end. These wrenches are probably the most versatile.

As for sizes, you'll probably need a range similar to that of the sockets, about 1/4 in. through 1 in. for standard fasteners, or 6mm through 19mm for metric fasteners. As for numbers, you'll need 2 of each size, since, in many instances, one wrench holds the nut while the other turns the bolt. On most fasteners, the nut and bolt are the same size so having two wrenches of the same size comes in handy.

■ Although you will typically just need the sizes we specified, there are some exceptions. Occasionally you will find a nut that is larger. For these, you will need to buy ONE expensive wrench or a very large adjustable. Or you can always just convince the spouse that we are talking about safety here and buy a whole (read expensive) large wrench set.

One extremely valuable type of wrench is the adjustable wrench. An adjustable wrench has a fixed upper jaw and a moveable lower jaw. The lower jaw is moved by turning a threaded drum. The advantage of an adjustable wrench is its ability to be adjusted to just about any size fastener.

The main drawback of an adjustable wrench is the lower jaw's tendency to move slightly under heavy pressure. This can cause the wrench to slip if it is not facing the right way. Pulling on an adjustable wrench in the proper direction will cause the jaws to lock in place. Adjustable wrenches come in a large range of sizes, measured by the wrench length.

PLIERS

♦ See Figure 54

Pliers are simply mechanical fingers. They are, more than anything, an extension of your hand. At least 3 pairs of pliers are an absolute necessity - standard, needle nose and slip joint.

In addition to standard pliers there are the slip-joint, multi-position pliers such as ChannelLock® pliers and locking pliers, such as Vise Grips®.

Slip joint pliers are extremely valuable in grasping oddly sized parts and fasteners. Just make sure that you don't use them instead of a wrench too often since they can easily round off a bolt head or nut.

Locking pliers are usually used for gripping bolts or studs that can't be removed conventionally. You can get locking pliers in square jawed, needle-nosed and pipe-jawed. Locking pliers can rank right up behind duct tape as the handy-man's best friend.

Fig. 49 Always use a backup wrench to prevent rounding flare nut fittings

GENERAL INFORMATION, SAFETY AND TOOLS

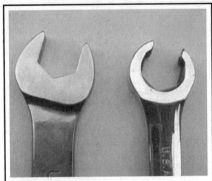

Fig. 50 Note how the flare wrench sides are extended to grip the fitting tighter and prevent rounding

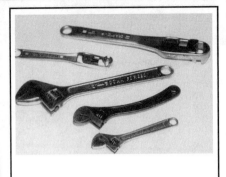

Fig. 51 Several types and sizes of adjustable wrenches

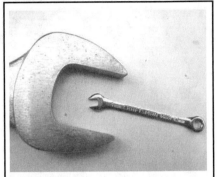

Fig. 52 You may find a nut that requires a particularly large or small wrench (it is usually available at your local tool store

INCHES	DECIMAL	DECIMAL	MILLIMETERS
1/8"	.125	.118	3mm
3/16"	.187	.157	4mm
1/4"	.250	.236	6mm
5/16"	.312	.354	9mm
3/8"	.375	.394	10mm
7/16"	.437	.472	12mm
1/2"	.500	.512	13mm
9/16"	.562	.590	15mm
5/8"	.625	.630	16mm
11/16"	.687	.709	18mm
3/4"	.750	.748	19mm
13/16"	.812	.787	20mm
7/8"	.875	.866	22mm
15/16"	.937	.945	24mm
1"	1.00	.984	25mm

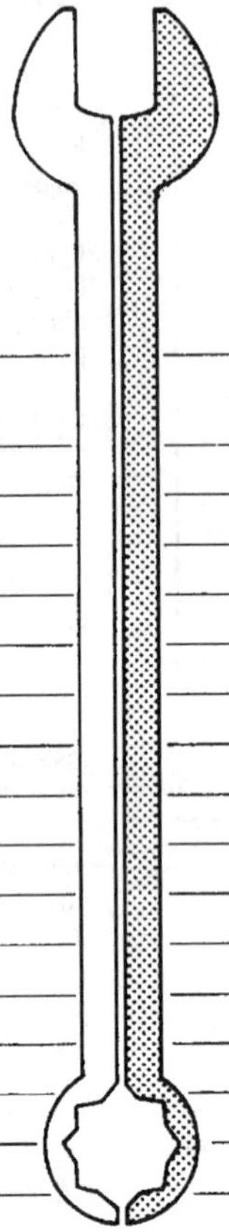

Fig. 53 Comparison of U.S. measure and metric wrench sizes

SCREWDRIVERS

You can't have too many screwdrivers. They come in 2 basic flavors, either standard or Phillips. Standard blades come in various sizes and thickness for all types of slotted fasteners. Phillips screwdrivers come in sizes with number designations from 1 on up, with the lower number designating the smaller size. Screwdrivers can be purchased separately or in sets.

HAMMERS

◆ See Figure 55

You need a hammer for just about any kind of work. You need a ball-peen hammer for most metal work when using drivers and other like tools. A plastic hammer comes in handy for hitting things safely. A soft-faced deadblow hammer is used for hitting things safely and hard. Hammers are also VERY useful with non air-powered impact drivers.

Other Common Tools

There are a lot of other tools that every DIYer will eventually need (though not all for basic maintenance). They include:
- Funnels
- Chisels
- Punches
- Files
- Hacksaw
- Portable Bench Vise
- Tap and Die Set
- Flashlight
- Magnetic Bolt Retriever
- Gasket scraper
- Putty Knife
- Screw/Bolt Extractors

GENERAL INFORMATION, SAFETY AND TOOLS 1-23

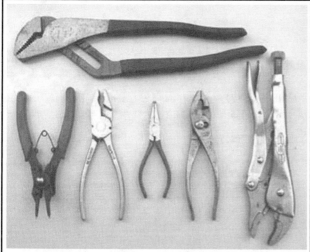

Fig. 54 Pliers and cutters come in many shapes and sizes. You should have an assortment on hand

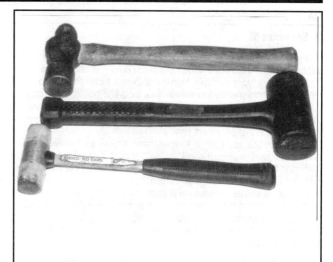

Fig. 55 Three types of hammers. Top to bottom: ball peen, rubber dead-blow, and plastic

- Prybars

Hacksaws have just one use - cutting things off. You may wonder why you'd need one for something as simple as maintenance or repair, but you never know. Among other things, guide studs to ease parts installation can be made from old bolts with their heads cut off.

A tap and die set might be something you've never needed, but you will eventually. It's a good rule, when everything is apart, to clean-up all threads, on bolts, screws or threaded holes. Also, you'll likely run across a situation in which you will encounter stripped threads. The tap and die set will handle that for you.

Gasket scrapers are just what you'd think, tools made for scraping old gasket material off of parts. You don't absolutely need one. Old gasket material can be removed with a putty knife or single edge razor blade. However, putty knives may not be sharp enough for some really stubborn gaskets and razor blades have a knack of breaking just when you don't want them to, inevitably slicing the nearest body part! As the old saying goes, "always use the proper tool for the job". If you're going to use a razor to scrape a gasket, be sure to always use a blade holder.

Putty knives really do have a use in a repair shop. Just because you remove all the bolts from a component sealed with a gasket doesn't mean it's going to come off. Most of the time, the gasket and sealer will hold it tightly. Lightly inserting a putty knife at various points between the two parts will break the seal without damage to the parts.

A small - 8-10 in. (20-25cm) long - prybar is extremely useful for removing stuck parts.

■ Never use a screwdriver as a prybar! Screwdrivers are not meant for prying. Screwdrivers, used for prying, can break, sending the broken shaft flying!

Screw/bolt extractors are used for removing broken bolts or studs that have broken off flush with the surface of the part.

Special Tools

◆ See Figure 53

Almost every marine engine around today requires at least one special tool to perform a certain task. In most cases, these tools are specially designed to overcome some unique problem or to fit on some oddly sized component.

When manufacturers go through the trouble of making a special tool, it is usually necessary to use it to ensure that the job will be done right. A special tool might be designed to make a job easier, or it might be used to keep you from damaging or breaking a part.

Don't worry, MOST maintenance procedures can either be performed without any special tools OR, because the tools must be used for such basic things, they are commonly available for a reasonable price. It is usually just the low production, highly specialized tools (like a super thin 7-point star-shaped socket capable of 150 ft. lbs. (203 Nm) of torque that is used only on

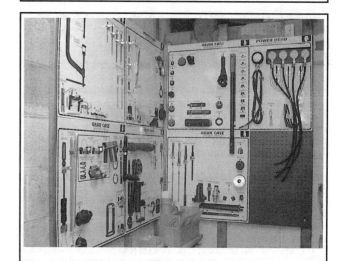

Fig. 56 Almost every marine engine around today requires at least one special tool to perform a certain task

the crankshaft nut of the limited production what-dya-callit engine) that tend to be outrageously expensive and hard to find. Hopefully, you will probably never need such a tool.

Special tools can be as inexpensive and simple as an adjustable strap wrench or as complicated as an ignition tester. A few common specialty tools are listed here, but check with your dealer or with other boaters for help in determining if there are any special tools for YOUR particular engine. There is an added advantage in seeking advice from others, chances are they may have already found the special tool you will need, and know how to get it cheaper (or even let you borrow it).

Electronic Tools

BATTERY TESTERS

The best way to test a non-sealed battery is using a hydrometer to check the specific gravity of the acid. Luckily, these are usually inexpensive and are available at most parts stores. Just be careful because the larger testers are usually designed for larger batteries and may require more acid than you will be able to draw from the battery cell. Smaller testers (usually a short, squeeze bulb type) will require less acid and should work on most batteries.

Electronic testers are available and are often necessary to tell if a sealed battery is usable. Luckily, many parts stores have them on hand and are willing to test your battery for you.

1-24 GENERAL INFORMATION, SAFETY AND TOOLS

BATTERY CHARGERS

◆ See Figure 57

If you are a weekend boater and take your boat out every week, then you will most likely want to buy a battery charger to keep your battery fresh. There are many types available, from low amperage trickle chargers to electronically controlled battery maintenance tools that monitor the battery voltage to prevent over or undercharging. This last type is especially useful if you store your boat for any length of time (such as during the severe winter months found in many Northern climates).

Even if you use your boat on a regular basis, you will eventually need a battery charger. The charger should be used anytime the boat is going to be in storage for more than a few weeks or so. Never leave the dock or loading ramp without a battery that is fully charged.

Also, some smaller batteries are shipped dry and in a partial charged state. Before placing a new battery of this type into service it must be filled and properly charged. Failure to properly charge a battery (which was shipped dry) before it is put into service will prevent it from ever reaching a fully charged state.

MULTI-METERS (DVOMS)

◆ See Figure 58

Multi-meters or Digital Volt Ohmmeter (DVOMs) are an extremely useful tool for troubleshooting electrical problems. They can be purchased in either analog or digital form and have a price range to suit any budget. A multi-meter is a voltmeter, ammeter and ohmmeter (along with other features) combined into one instrument. It is often used when testing solid state circuits because of its high input impedance (usually 10 mega-ohms or more). A brief description of the multi-meter main test functions follows:

- Voltmeter - the voltmeter is used to measure voltage at any point in a circuit or to measure the voltage drop across any part of a circuit. Voltmeters usually have various scales and a selector switch to allow the reading of different voltage ranges. The voltmeter has a positive and a negative lead. To avoid the possibility of damage to the meter, whenever possible, connect the negative lead to the negative (-) side of the circuit (to ground or nearest the ground side of the circuit) and connect the positive lead to the positive (+) side of the circuit (to the power source or the nearest power source). Luckily, most quality DVOMs can adjust their own polarity internally and will indicate (without damage) if the leads are reversed. Note that the negative voltmeter lead will always be black and that the positive voltmeter will always be some color other than black (usually red).

- Ohmmeter - the ohmmeter is designed to read resistance (measured in ohms) in a circuit or component. Most ohmmeters will have a selector switch which permits the measurement of different ranges of resistance (usually the selector switch allows the multiplication of the meter reading by 10, 100, 1,000 and 10,000). Some ohmmeters are "auto-ranging" which means the meter itself will determine which scale to use. Since the meters are powered by an internal battery, the ohmmeter can be used like a self-powered test light. When the ohmmeter is connected, current from the ohmmeter flows through the circuit or component being tested. Since the ohmmeter's internal resistance and voltage are known values, the amount of current flow through the meter depends on the resistance of the circuit or component being tested. The ohmmeter can also be used to perform a continuity test for suspected open circuits. In using the meter for making continuity checks, do not be concerned with the actual resistance readings. Zero resistance, or any ohm reading, indicates continuity in the circuit. Infinite resistance indicates an opening in the circuit. A high resistance reading where there should be little or none indicates a problem in the circuit. Checks for short circuits are made in the same manner as checks for open circuits, except that the circuit must be isolated from both power and normal ground. Infinite resistance indicates no continuity, while zero resistance indicates a dead short.

✱✱ WARNING

Never use an ohmmeter to check the resistance of a component or wire while there is voltage applied to the circuit.

- Ammeter - an ammeter measures the amount of current flowing through a circuit in units called amperes or amps. At normal operating voltage, most circuits have a characteristic amount of amperes, called "current draw" which can be measured using an ammeter. By referring to a specified current draw rating, then measuring the amperes and comparing the two values; one can determine what is happening within the circuit to aid in diagnosis. An open circuit, for example, will not allow any current to flow, so the ammeter reading will be zero. A damaged component or circuit will have an increased current draw, so the reading will be high. The ammeter is always connected in series with the circuit being tested. All of the current that normally flows through the circuit must also flow through the ammeter; if there is any other path for the current to follow, the ammeter reading will not be accurate. The ammeter itself has very little resistance to current flow and, therefore, will not affect the circuit, but, it will measure current draw only when the circuit is closed and electricity is flowing. Excessive current draw can blow fuses and drain the battery, while a reduced current draw can cause motors to run slowly, lights to dim and other components to not operate properly.

Fig. 57 The Battery Tender® is more than just a battery charger, when left connected, it keeps your battery fully charged

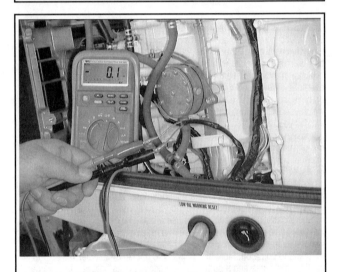

Fig. 58 Multi-meters, such as this one from UEI, are an extremely useful tool for troubleshooting electrical problems

GAUGES

Compression Gauge

◆ See Figure 59

An important element in checking the overall condition of your engine is to check compression. This becomes increasingly more important on outboards with high hours. Compression gauges are available as screw-in types and hold-in types. The screw-in type is slower to use, but eliminates the possibility of a faulty reading due to pressure escaping by the seal. A

GENERAL INFORMATION, SAFETY AND TOOLS 1-25

Fig. 59 Cylinder compression test results are extremely valuable indicators of internal engine condition

Fig. 60 Vacuum gauges are useful for troubleshooting including testing some fuel pumps

Fig. 61 You can also use the vacuum gauge on a hand-operated vacuum pump for tests

compression reading will uncover many problems that can cause rough running. Normally, these are not the sort of problems that can be cured by a tune-up.

Vacuum Gauge

◆ See Figures 60 and 61

Vacuum gauges are handy for discovering air leaks, late ignition or valve timing, and a number of other problems.

Measuring Tools

Eventually, you are going to have to measure something. To do this, you will need at least a few precision tools.

MICROMETERS & CALIPERS

Micrometers and calipers are devices used to make extremely precise measurements. The simple truth is that you really won't have the need for many of these items just for routine maintenance. But, measuring tools, such as an outside caliper can be handy during repairs. And, if you decide to tackle a major overhaul, a micrometer will absolutely be necessary.

Should you decide on becoming more involved in boat engine mechanics, such as repair or rebuilding, then these tools will become very important. The success of any rebuild is dependent, to a great extent on the ability to check the size and fit of components as specified by the manufacturer. These measurements are often made in thousandths and ten-thousandths of an inch.

Micrometers

◆ See Figure 62

A micrometer is an instrument made up of a precisely machined spindle that is rotated in a fixed nut, opening and closing the distance between the end of the spindle and a fixed anvil. When measuring using a micrometer, don't over-tighten the tool on the part as either the component or tool may be damaged, and either way, an incorrect reading will result. Most micrometers are equipped with some form of thumbwheel on the spindle that is designed to freewheel over a certain light touch (automatically adjusting the spindle and preventing it from over-tightening).

Outside micrometers can be used to check the thickness of parts such shims or the outside diameter of components like the crankshaft journals. They are also used during many rebuild and repair procedures to measure the diameter of components such as the pistons. The most common type of micrometer reads in 1/1000 of an inch. Micrometers that use a vernier scale can estimate to 1/10 of an inch.

Inside micrometers are used to measure the distance between two parallel surfaces. For example, in powerhead rebuilding work, the "inside mike" measures cylinder bore wear and taper. Inside mikes are graduated the same way as outside mikes and are read the same way as well.

Remember that an inside mike must be absolutely perpendicular to the work being measured. When you measure with an inside mike, rock the mike gently from side to side and tip it back and forth slightly so that you span the widest part of the bore. Just to be on the safe side, take several readings. It takes a certain amount of experience to work any mike with confidence.

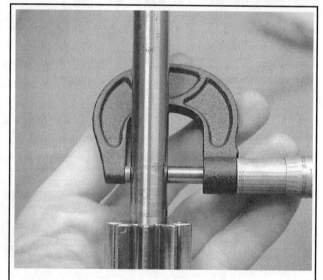

Fig. 62 Outside micrometers measure thickness, like shims or a shaft diameter

Metric micrometers are read in the same way as inch micrometers, except that the measurements are in millimeters. Each line on the main scale equals 1mm. Each fifth line is stamped 5, 10, 15 and so on. Each line on the thimble scale equals 0.01 mm. It will take a little practice, but if you can read an inch mike, you can read a metric mike.

Calipers

◆ See Figures 63, 64 and 65

Inside and outside calipers are useful devices to have if you need to measure something quickly and absolute precise measurement is not necessary. Simply take the reading and then hold the calipers on an accurate steel rule. Calipers, like micrometers, will often contain a thumbwheel to help ensure accurate measurement.

DIAL INDICATORS

◆ See Figure 66

A dial indicator is a gauge that utilizes a dial face and a needle to register measurements. There is a movable contact arm on the dial indicator. When the arm moves, the needle rotates on the dial. Dial indicators are calibrated to show readings in thousandths of an inch and typically, are used to measure end-play and runout on various shafts and other components.

Dial indicators are quite easy to use, although they are relatively expensive. A variety of mounting devices are available so that the indicator can be used in a number of situations. Make certain that the contact arm is always parallel to the movement of the work being measured.

1-26 GENERAL INFORMATION, SAFETY AND TOOLS

TELESCOPING GAUGES

♦ See Figure 67

A telescope gauge is really only used during rebuilding procedures (NOT during basic maintenance or routine repairs) to measure the inside of bores. It can take the place of an inside mike for some of these jobs. Simply insert the gauge in the hole to be measured and lock the plungers after they have contacted the walls. Remove the tool and measure across the plungers with an outside micrometer.

DEPTH GAUGES

♦ See Figure 68

A depth gauge can be inserted into a bore or other small hole to determine exactly how deep it is. One common use for a depth gauge is measuring the distance the piston sits below the deck of the block at top dead center. Some outside calipers contain a built-in depth gauge so you can save money and buy just one tool.

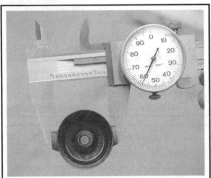

Fig. 63 Calipers are the fast and easy way to make precise measurements

Fig. 64 Calipers can also be used to measure depth . . .

Fig. 65 . . . and inside diameter measurements, to 0.001 in. accuracy

Fig. 66 This dial indicator is measuring the end-play of a crankshaft during a powerhead rebuild

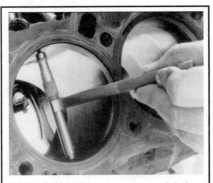

Fig. 67 Telescoping gauges are used during powerhead rebuilding procedures to measure the inside diameter of bores

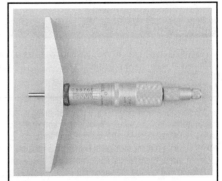

Fig. 68 Depth gauges are used to measure the depth of bore or other small holes

FASTENERS, MEASUREMENTS AND CONVERSIONS

Bolts, Nuts and Other Threaded Retainers

♦ See Figures 69 and 70

Although there are a great variety of fasteners found in the modern boat engine, the most commonly used retainer is the threaded fastener (nuts, bolts, screws, studs, etc). Most threaded retainers may be reused, provided that they are not damaged in use or during the repair.

■ Some retainers (such as stretch bolts or torque prevailing nuts) are designed to deform when tightened or in use and should not be reused.

Whenever possible, we will note any special retainers which should be replaced during a procedure. But you should always inspect the condition of a retainer when it is removed and you should replace any that show signs of damage. Check all threads for rust or corrosion that can increase the torque necessary to achieve the desired clamp load for which that fastener was originally selected. Additionally, be sure that the driver surface itself (on the fastener) is not compromised from rounding or other damage. In some cases a driver surface may become only partially rounded, allowing the driver to catch in only one direction. In many of these occurrences, a fastener may be installed and tightened, but the driver would not be able to grip and loosen the fastener again. (This could lead to frustration down the line should that component ever need to be disassembled again).

If you must replace a fastener, whether due to design or damage, you must always be sure to use the proper replacement. In all cases, a retainer of the same design, material and strength should be used. Markings on the heads of most bolts will help determine the proper strength of the fastener. The same material, thread and pitch must be selected to assure proper installation and safe operation of the motor afterwards.

Thread gauges are available to help measure a bolt or stud's thread. Most part or hardware stores keep gauges available to help you select the proper size. In a pinch, you can use another nut or bolt for a thread gauge. If the bolt you are replacing is not too badly damaged, you can select a match by finding another bolt that will thread in its place. If you find a nut that will thread properly onto the damaged bolt, then use that nut as a gauge to help select the replacement bolt. If however, the bolt you are replacing is so badly damaged (broken or drilled out) that its threads cannot be used as a gauge, you might start by looking for another bolt (from the same assembly or a similar location) which will thread into the damaged bolt's mounting. If so, the other bolt can be used to select a nut; the nut can then be used to select the replacement bolt.

In all cases, be absolutely sure you have selected the proper replacement. Don't be shy, you can always ask the store clerk for help.

✱✱ WARNING

Be aware that when you find a bolt with damaged threads, you may also find the nut or tapped bore into which it was threaded has also been damaged. If this is the case, you may have to drill and tap the hole, replace the nut or otherwise repair the threads. Never try to force a replacement bolt to fit into the damaged threads.

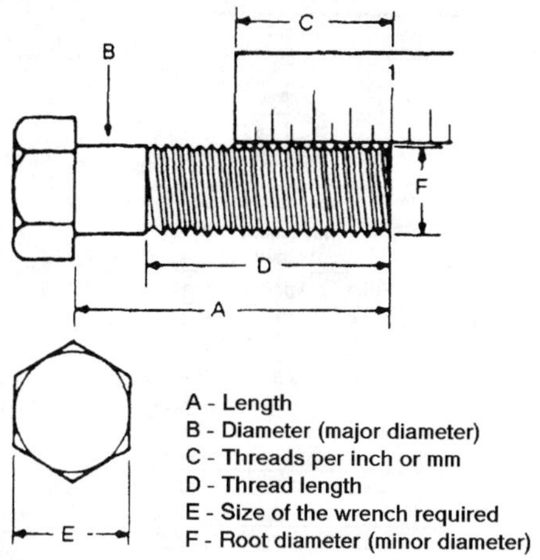

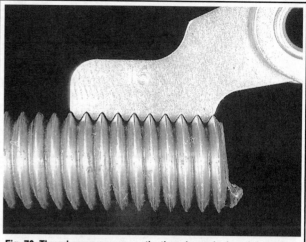

Fig. 69 Threaded retainer sizes are determined using these measurements

A - Length
B - Diameter (major diameter)
C - Threads per inch or mm
D - Thread length
E - Size of the wrench required
F - Root diameter (minor diameter)

Fig. 70 Thread gauges measure the threads-per-inch and the pitch of a bolt or stud's threads

Torque

Torque is defined as the measurement of resistance to turning or rotating. It tends to twist a body about an axis of rotation. A common example of this would be tightening a threaded retainer such as a nut, bolt or screw. Measuring torque is one of the most common ways to help assure that a threaded retainer has been properly fastened.

When tightening a threaded fastener, torque is applied in three distinct areas, the head, the bearing surface and the clamp load. About 50 percent of the measured torque is used in overcoming bearing friction. This is the friction between the bearing surface of the bolt head, screw head or nut face and the base material or washer (the surface on which the fastener is rotating). Approximately 40 percent of the applied torque is used in overcoming thread friction. This leaves only about 10 percent of the applied torque to develop a useful clamp load (the force that holds a joint together). This means that friction can account for as much as 90 percent of the applied torque on a fastener.

Standard and Metric Measurements

Specifications are often used to help you determine the condition of various components, or to assist you in their installation. Some of the most common measurements include length (in. or cm/mm), torque (ft. lbs., inch lbs. or Nm) and pressure (psi, in. Hg, kPa or mm Hg).

In some cases, that value may not be conveniently measured with what is available in your toolbox. Luckily, many of the measuring devices that are available today will have two scales so U.S. or Metric measurements may easily be taken. If any of the various measuring tools that are available to you do not contain the same scale as listed in your specifications, use the conversion factors that are provided in the Specifications section to determine the proper value.

The conversion factor chart is used by taking the given specification and multiplying it by the necessary conversion factor. For instance, looking at the first line, if you have a measurement in inches such as "free-play should be 2 in." but your ruler reads only in millimeters, multiply 2 in. by the conversion factor of 25.4 to get the metric equivalent of 50.8mm. Likewise, if a specification was given only in a Metric measurement, for example in Newton Meters (Nm), then look at the center column first. If the measurement is 100 Nm, multiply it by the conversion factor of 0.738 to get 73.8 ft. lbs.

SPECIFICATIONS

Metric Bolts						
Relative Strength Marking	4.6, 4.8			8.8		
Bolt Markings	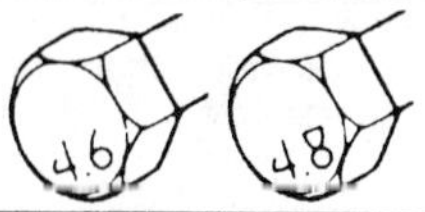			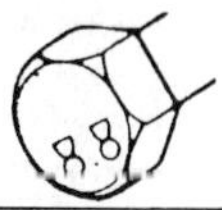		
Usage	Frequent			Infrequent		
Bolt Size	Maximum Torque			Maximum Torque		
Thread Size x Pitch (mm)	Ft-Lb	Kgm	Nm	Ft-Lb	Kgm	Nm
6 x 1.0	2–3	.2–.4	3–4	3–6	.4–.8	5–8
8 x 1.25	6–8	.8–1	8–12	9–14	1.2–1.9	13–19
10 x 1.25	12–17	1.5–2.3	16–23	20–29	2.7–4.0	27–39
12 x 1.25	21–32	2.9–4.4	29–43	35–53	4.8–7.3	47–72
14 x 1.5	35–52	4.8–7.1	48–70	57–85	7.8–11.7	77–110
16 x 1.5	51–77	7.0–10.6	67–100	90–120	12.4–16.5	130–160
18 x 1.5	74–110	10.2–15.1	100–150	130–170	17.9–23.4	180–230
20 x 1.5	110–140	15.1–19.3	150–190	190–240	26.2–46.9	160–320
22 x 1.5	150–190	22.0–26.2	200–260	250–320	34.5–44.1	340–430
24 x 1.5	190–240	26.2–46.9	260–320	310–410	42.7–56.5	420–550

1-28   GENERAL INFORMATION, SAFETY AND TOOLS

SAE Bolts

SAE Grade Number	1 or 2			5			6 or 7		
Bolt Markings Manufacturers' marks may vary—number of lines always two less than the grade number.									
Usage	Frequent			Frequent			Infrequent		
Bolt Size (inches)—(Thread)	Maximum Torque			Maximum Torque			Maximum Torque		
	Ft-Lb	kgm	Nm	Ft-Lb	kgm	Nm	Ft-Lb	kgm	Nm
1/4—20	5	0.7	6.8	8	1.1	10.8	10	1.4	13.5
—28	6	0.8	8.1	10	1.4	13.6			
5/16—18	11	1.5	14.9	17	2.3	23.0	19	2.6	25.8
—24	13	1.8	17.6	19	2.6	25.7			
3/8—16	18	2.5	24.4	31	4.3	42.0	34	4.7	46.0
—24	20	2.75	27.1	35	4.8	47.5			
7/16—14	28	3.8	37.0	49	6.8	66.4	55	7.6	74.5
—20	30	4.2	40.7	55	7.6	74.5			
1/2—13	39	5.4	52.8	75	10.4	101.7	85	11.75	115.2
—20	41	5.7	55.6	85	11.7	115.2			
9/16—12	51	7.0	69.2	110	15.2	149.1	120	16.6	162.7
—18	55	7.6	74.5	120	16.6	162.7			
5/8—11	83	11.5	112.5	150	20.7	203.3	167	23.0	226.5
—18	95	13.1	128.8	170	23.5	230.5			
3/4—10	105	14.5	142.3	270	37.3	366.0	280	38.7	379.6
—16	115	15.9	155.9	295	40.8	400.0			
7/8—9	160	22.1	216.9	395	54.6	535.5	440	60.9	596.5
—14	175	24.2	237.2	435	60.1	589.7			
1—8	236	32.5	318.6	590	81.6	799.9	660	91.3	894.8
—14	250	34.6	338.9	660	91.3	849.8			

CONVERSION FACTORS

LENGTH–DISTANCE
Inches (in.)	x 25.4	= Millimeters (mm)	x .0394	= Inches
Feet (ft.)	x .305	= Meters (m)	x 3.281	= Feet
Miles	x 1.609	= Kilometers (km)	x .0621	= Miles

VOLUME
Cubic Inches (in3)	x 16.387	= Cubic Centimeters	x .061	= in3
IMP Pints (IMP pt.)	x .568	= Liters (L)	x 1.76	= IMP pt.
IMP Quarts (IMP qt.)	x 1.137	= Liters (L)	x .88	= IMP qt.
IMP Gallons (IMP gal.)	x 4.546	= Liters (L)	x .22	= IMP gal.
IMP Quarts (IMP qt.)	x 1.201	= US Quarts (US qt.)	x .833	= IMP qt.
IMP Gallons (IMP gal.)	x 1.201	= US Gallons (US gal.)	x .833	= IMP gal.
Fl. Ounces	x 29.573	= Milliliters	x .034	= Ounces
US Pints (US pt.)	x .473	= Liters (L)	x 2.113	= Pints
US Quarts (US qt.)	x .946	= Liters (L)	x 1.057	= Quarts
US Gallons (US gal.)	x 3.785	= Liters (L)	x .264	= Gallons

MASS–WEIGHT
Ounces (oz.)	x 28.35	= Grams (g)	x .035	= Ounces
Pounds (lb.)	x .454	= Kilograms (kg)	x 2.205	= Pounds

PRESSURE
Pounds Per Sq. In. (psi)	x 6.895	= Kilopascals (kPa)	x .145	= psi
Inches of Mercury (Hg)	x .4912	= psi	x 2.036	= Hg
Inches of Mercury (Hg)	x 3.377	= Kilopascals (kPa)	x .2961	= Hg
Inches of Water (H_2O)	x .07355	= Inches of Mercury	x 13.783	= H_2O
Inches of Water (H_2O)	x .03613	= psi	x 27.684	= H_2O
Inches of Water (H_2O)	x .248	= Kilopascals (kPa)	x 4.026	= H_2O

TORQUE
Pounds–Force Inches (in-lb)	x .113	= Newton Meters (N·m)	x 8.85	= in–lb
Pounds–Force Feet (ft-lb)	x 1.356	= Newton Meters (N·m)	x .738	= ft–lb

VELOCITY
Miles Per Hour (MPH)	x 1.609	= Kilometers Per Hour (KPH)	x .621	= MPH

POWER
Horsepower (Hp)	x .745	= Kilowatts	x 1.34	= Horsepower

FUEL CONSUMPTION*
Miles Per Gallon IMP (MPG)	x .354	= Kilometers Per Liter (Km/L)		
Kilometers Per Liter (Km/L)	x 2.352	= IMP MPG		
Miles Per Gallon US (MPG)	x .425	= Kilometers Per Liter (Km/L)		
Kilometers Per Liter (Km/L)	x 2.352	= US MPG		

*It is common to covert from miles per gallon (mpg) to liters/100 kilometers (1/100 km), where mpg (IMP) x 1/100 km = 282 and mpg (US) x 1/100 km = 235.

TEMPERATURE
Degree Fahrenheit (°F) = (°C x 1.8) + 32
Degree Celsius (°C) = (°F − 32) x .56

2

MAINTENANCE & TUNE-UP

GENERAL INFORMATION	2-2
LUBRICATION SERVICE	2-6
ENGINE MAINTENANCE	2-15
BOAT MAINTENANCE	2-37
TUNE-UP	2-40
TIMING AND SYNCHRONIZATION	2-51
VALVE CLEARANCE (4-STROKE)	2-78
STORAGE	2-81
CLEARING A SUBMERGED MOTOR	2-83
SPECIFICATIONS	2-85

Contents

- BOAT MAINTENANCE ... 2-37
  - BATTERIES ... 2-37
  - FIBERGLASS HULL ... 2-39
- CLEARING A SUBMERGED MOTOR ... 2-83
- GENERAL INFORMATION ... 2-2
  - BEFORE/AFTER EACH USE ... 2-3
    - Visual Inspection ... 2-5
  - ENGINE IDENTIFICATION ... 2-2
    - Engine Serial Numbers ... 2-3
  - MAINTENANCE COVERAGE ... 2-2
  - MAINTENANCE EQUALS SAFETY ... 2-2
  - OUTBOARDS ON SAIL BOATS ... 2-2
- LUBRICATION SERVICE ... 2-6
  - ELECTRIC STARTER MOTOR PINION ... 2-6
  - ENGINE COVER LATCHES ... 2-7
  - ENGINE MOUNT CLAMP SCREWS ... 2-7
  - JET DRIVE BEARING ... 2-7
    - Recommended Lubricant ... 2-7
    - Daily Bearing Lubrication ... 2-7
    - Grease Replacement ... 2-7
  - POWER TRIM/TILT RESERVOIR ... 2-8
    - Fluid Level/Condition ... 2-8
    - Recommended Lubricant ... 2-8
  - LINKAGE, CABLES AND SHAFTS ... 2-8
  - STEERING ... 2-13
  - SWIVEL BRACKET ... 2-14
  - TILT ASSEMBLY ... 2-14
- ENGINE MAINTENANCE ... 2-15
  - ANODES (ZINCS) ... 2-35
    - Inspection ... 2-35
    - Servicing ... 2-35
  - COOLING SYSTEM ... 2-16
    - Flushing ... 2-16
  - ENGINE COVERS ... 2-15
    - Removal & Installation ... 2-15
  - ENGINE OIL (2-STROKE) ... 2-18
    - Filling ... 2-19
    - Recommendations ... 2-18
  - ENGINE OIL/FILTER (4-STROKE) ... 2-20
    - Checking Oil Level ... 2-20
    - Oil/Filter Change ... 2-21
    - Recommendations ... 2-20
  - FUEL FILTER ... 2-25
    - Carbureted Motors ... 2-26
    - EFI Motors ... 2-28
  - GEARCASE (LOWER UNIT) OIL ... 2-23
    - Checking ... 2-24
    - Draining & Filling ... 2-25
    - Recommendations ... 2-24
  - JET DRIVE IMPELLER ... 2-32
    - Checking ... 2-32
    - Inspection ... 2-32
  - PROPELLER ... 2-29
    - Inspection ... 2-29
    - Removal & Installation ... 2-30
  - RESCUEPRO ROTOR ... 2-33
  - TIMING BELT ... 2-36
    - Inspecton ... 2-36
- SPECIFICATIONS ... 2-85
  - CAPACITIES - TWO-STROKE ENGINES ... 2-90
  - CAPACITIES - FOUR-STROKE ENGINES ... 2-91
  - GENERAL ENGINE ... 2-85
  - GENERAL ENGINE SYSTEM ... 2-87
  - LUBRICATION SERVICES ... 2-89
  - MAGNETO BREAKER POINT GAP ... 2-84
  - MAINTENANCE INTERVALS ... 2-89
  - TUNE-UP SPECIFICATIONS ... 2-91
  - TWO-STROKE MOTOR FUEL:OIL RATIO ... 2-90
  - VALVE CLEARANCE ... 2-94
- STORAGE ... 2-81
  - RE-COMMISSIONING ... 2-83
  - WINTERIZATION ... 2-81
- TIMING AND SYNCHRONIZATION ... 2-51
  - HOMEMADE SYNCHRONIZATION TOOL ... 2-52
  - 2.0-3.5 HP MODELS ... 2-53
    - Idle Speed ... 2-53
  - 5 HP (109cc) MODELS ... 2-53
    - Low Speed Mixture ... 2-53
  - COLT/JUNIOR, 3/4 HP & 4 DELUXE MODELS ... 2-53
    - Camshaft Follower Pickup Point ... 2-53
    - Mixture Adjustments ... 2-54
    - Throttle Cable ... 2-54
  - 5-8 HP (164cc) ... 2-
  - STROKE MODELS
    - Low Speed & Wide Open Throttle Settings ... 2-55
    - Throttle Cable ... 2-55
    - Timing Pointer and Idle Timing (1990) ... 2-55
  - 5/6 HP (128cc) & 8/9.9 HP (211cc) 4-STROKE MODELS ... 2-56
    - Idle Speed ... 2-57
    - Low Speed Setting ... 2-56
  - 9.9/10/14/15 HP (216cc) & 9.9/10/15 HP (255cc) 2-STROKE MODELS ... 2-57
    - Cam Follower Pickup Point ... 2-58
    - Idle Speed ... 2-59
    - Low Speed Mixture ... 2-58
    - Preliminary Adjustments ... 2-58
    - Shift Lever Detent ... 2-59
    - Wide Open Throttle Stop ... 2-58
  - 9.9/15 HP (305cc) 4-STROKE MODELS ... 2-59
    - Idle Speed ... 2-60
    - Low Speed Setting ... 2-60
    - Shift Lever Detent ... 2-59
  - 20 HP (521cc) MODELS ... 2-61
    - Cam Follower Pickup Point ... 2-61
    - Idle Speed ... 2-62
    - Initial Low Speed Setting ... 2-63
    - Maximum Spark Advance ... 2-62
    - Throttle Control Rod (1990) ... 2-62
  - 18 JET & 25-35 HP (521cc) MODELS ... 2-63
    - Cam Follower Pickup Point ... 2-63
    - Idle Speed ... 2-64
    - Initial Low Speed Setting ... 2-64
    - Maximum Spark Advance ... 2-64
    - Throttle Control Rod ... 2-64
  - 25-55 HP (737cc) 2-CYL MODELS ... 2-65
    - Cam Pickup Point ... 2-66
    - Idle Speed ... 2-67
    - Initial Low Speed Setting ... 2-67
    - Maximum Spark Advance ... 2-68
    - Preliminary Adjustments ... 2-65
    - Shift Lever Detent ... 2-69
    - Throttle Cable ... 2-68
    - Throttle Control Rod ... 2-68
    - Throttle Plate Synch ... 2-66
    - Wide Open Throttle Stop ... 2-68
  - 25/35 HP (500/565cc) 3-CYL ... 2-69
    - Cam Follower Pickup ... 2-71
    - Idle Timing ... 2-70
    - Initial Low Speed ... 2-71
    - Maximum Spark Advance ... 2-71
    - Neutral Detent ... 2-71
    - Preliminary Adjustments ... 2-69
    - Timing Pointer ... 2-70
    - Throttle Plate Synch ... 2-70
    - Wide Open Throttle Positioning ... 2-71
  - 25-70 HP (913cc) 3-CYL ... 2-72
    - Cam Follower Pickup ... 2-73
    - Idle Timing ... 2-74
    - Initial Low Speed Setting ... 2-75
    - Maximum Spark Advance ... 2-75
    - Shift Lever Detent ... 2-75
    - Timing Pointer ... 2-72
    - Throttle Cam ... 2-74
    - Throttle Cable ... 2-73
    - Throttle Plate Synch ... 2-73
    - Wide Open Throttle Stop ... 2-74
  - 40/50 HP 4-STROKE MODELS ... 2-76
    - Idle Bypass Air Screw ... 2-76
    - Ignition Timing ... 2-76
    - Shift Linkage ... 2-76
  - 70 HP 4-STROKE MODELS ... 2-77
    - Idle Bypass Air Screw ... 2-77
    - Ignition Timing ... 2-77
    - Shift Linkage ... 2-77
- TUNE-UP ... 2-40
  - BREAKER POINTS IGNITION SYSTEMS ... 2-47
    - General Information ... 2-47
    - Inspection & Testing ... 2-47
    - Replacement ... 2-48
  - COMPRESSION TESTS ... 2-41
    - Leakage Check ... 2-42
    - Tune-Up Check ... 2-41
  - DE-CARBONING THE PISTONS ... 2-41
  - ELECTRONIC (CDI/UFI) IGNITION ... 2-51
    - Inspection ... 2-51
  - INTRODUCTION ... 2-40
  - SPARK PLUGS ... 2-43
    - Heat Range ... 2-43
    - Inspection & Gapping ... 2-46
    - Reading ... 2-45
    - Removal & Installation ... 2-44
  - SPARK PLUG WIRES ... 2-47
    - Removal & Installation ... 2-47
    - Testing ... 2-47
  - TUNE-UP SEQUENCE ... 2-40
- VALVE CLEARANCE (4-STROKE) ... 2-78
  - VALVE LASH ADJUSTMENT ... 2-78

MAINTENANCE & TUNE-UP

GENERAL INFORMATION (WHAT EVERYONE SHOULD KNOW ABOUT MAINTENANCE)

At Seloc, we estimate that 75% of engine repair work can be directly or indirectly attributed to lack of proper care for the engine. This is especially true of care during the off-season period. There is no way on this green earth for a mechanical engine, particularly an outboard motor, to be left sitting idle for an extended period of time, say for six months, and then be ready for instant satisfactory service.

Imagine, if you will, leaving your car or truck for six months, and then expecting to turn the key, having it roar to life, and being able to drive off in the same manner as a daily occurrence.

Therefore it is critical for an outboard engine to either be run (at least once a month), preferably, in the water and properly maintained between uses or for it to be specifically prepared for storage and serviced again immediately before the start of the season.

Only through a regular maintenance program can the owner expect to receive long life and satisfactory performance at minimum cost.

Many times, if an outboard is not performing properly, the owner will "nurse" it through the season with good intentions of working on the unit once it is no longer being used. As with many New Year's resolutions, the good intentions are not completed and the outboard may lie for many months before the work is begun or the unit is taken to the marine shop for repair.

Imagine, if you will, the cause of the problem being a blown head gasket. And let us assume water has found its way into a cylinder. This water, allowed to remain over a long period of time, will do considerably more damage than it would have if the unit had been disassembled and the repair work performed immediately. Therefore, if an outboard is not functioning properly, do not stow it away with promises to get at it when you get time, because the work and expense will only get worse, the longer corrective action is postponed. In the example of the blown head gasket, a relatively simple and inexpensive repair job could very well develop into major overhaul and rebuild work.

Maintenance Equals Safety

OK, perhaps no one thing that we do as boaters will protect us from risks involved with enjoying the wind and the water on a powerboat. But, each time we perform maintenance on our boat or motor, we increase the likelihood that we will find a potential hazard before it becomes a problem. Each time we inspect our boat and motor, we decrease the possibility that it could leave us stranded on the water.

In this way, performing boat and engine service is one of the most important ways that we, as boaters, can help protect ourselves, our boats, and the friends and family that we bring aboard.

Outboards On Sail Boats

Owners of sailboats pride themselves in their ability to use the wind to clear a harbor or for movement from Port A to Port B, or maybe just for a day sail on a lake. For some, the outboard is carried only as a last resort - in case the wind fails completely, or in an emergency situation or for ease of docking.

Therefore, in some cases, the outboard is stowed below, usually in a very poorly ventilated area, and subjected to moisture and stale air - in short, an excellent environment for "sweating" and corrosion.

If the owner could just take the time at least once every month, to pull out the outboard, clean it up, and give it a short run, not only would he/she have "peace of mind" knowing it will start in an emergency, but also maintenance costs will be drastically reduced.

Maintenance Coverage In This Manual

At Seloc, we strongly feel that every boat owner should pay close attention to this section. We also know that it is one of the most frequently used portions of our manuals. The material in this section is divided into sections to help simplify the process of maintenance. Be sure to read and thoroughly understand the various tasks that are necessary to keep your outboard in tip-top shape.

Topics covered in this section include:

1. General Information (What Everyone Should Know About Maintenance) - an introduction to the benefits and need for proper maintenance. A guide to tasks that should be performed before and after each use.

2. Lubrication Service - after the basic inspections that you should perform each time the motor is used, the most frequent form of periodic maintenance you will conduct will be the Lubrication Service. This section takes you through each of the various steps you must take to keep corrosion from slowly destroying your motor before your very eyes.

3. Engine Maintenance - the various procedures that must be performed on a regular basis in order to keep the motor and all of its various systems operating properly.

4. Boat Maintenance - the various procedures that must be performed on a regular basis in order to keep the boat hull and its accessories looking and working like new.

5. Tune-Up - also known as the pre-season tune-up, but don't let the name fool you. A complete tune-up is the best way to determine the condition of your outboard while also preparing it for hours and hours of hopefully trouble-free enjoyment.

6. Winter Storage and Spring Commissioning Checklists - use these sections to guide you through the various parts of boat and motor maintenance that protect your valued boat through periods of storage and return it to operating condition when it is time to use it again.

7. Specification Charts - located at the end of the section are quick-reference, easy to read charts that provide you with critical information such as General Engine Specifications, Maintenance Intervals, Lubrication Service (intervals and lubricant types) and Capacities.

Engine Identification

◆ See Figures 1 and 2

From 1990 to 2001 Johnson and Evinrude produced a large number of models with regards to horsepower ratings, as well a large number of trim and option variances on each of those models. In this manual, we've included all of the 1-4 cylinder inline models (of both 2 and 4-stroke designs). We chose to do this because of the many similarities these motors have to each other. But, enough differences exist that many procedures will apply only to a sub-set of these motors. When this occurs, we'll either refer to the differences within a procedure or, if the differences are significant, we'll break the motors out and give separate procedures. In order to prevent confusion, we try to sort and name the models in a way that is most easily understood.

In many cases, it is simply not enough to refer to a motor as a 9.9 hp model, since in these years Johnson/Evinrude produced four different 2-cylinder motors with that rating (the 211cc 4-stroke, the 216cc 2-stroke, the 255cc 2-stroke, and the 305cc 4-stroke). Across that same year span, Johnson/Evinrude produced and sold no fewer than 4 different 2-stroke motors rated at 25hp (the 2-cylinder, 521cc, the 2-cylinder 737cc, the 3-cylinder 913cc and the 3-cylinder 933cc). This makes proper engine identification important for everything from ordering parts to even just using the procedures in this manual.

Throughout this manual we will make reference to motors the easiest way possible. In some cases procedures will apply to all 2-strokes or all 4-strokes, in other cases, they will apply to all 1-cylinder or all 2-cylinder motor (or all 3 or 4-cylinder motors, as applicable). When it is necessary to distinguish between different types of motors with the same number of cylinders, we'll differentiate using the Hp rating or, since different motors may have the same rating, we'll use the Hp rating plus the size. In most cases, mechanical procedures will be similar or the same across different Hp ratings of the same engine family (of the same size). So it won't be uncommon to see a title or a procedure refer to 9.9/15 hp (255cc) motors or 9.9/15 hp (305cc) motors. In both cases, we would be referring to the 9.9 or 15 hp motors of a particular family, including all Rope Start, Tiller Electric or Remote Electric Models. In the case of the 9.9/15 hp (255cc) motors, we would be referring to the 2-strokes, of that size, including any Sail, Commercial or other special models.

To help with proper engine identification, all of the engines covered by this manual are listed in the General Engine and General Engine System Specifications charts at the end of this section. In these charts, the engines are listed with their respective engine families, by horsepower rating, number of cylinders, engine type (2- or 4-stroke), years of production and displacement (cubic inches and cubic centimeters or CCs).

But, whether you are trying to tell which version of a particular horsepower rated motor you have in order to follow the correct procedure or are trying to order replacement parts, the absolute best method is to start by referring to the engine serial number tag. For all models covered by this

MAINTENANCE & TUNE-UP 2-3

manual an ID tag (A, in the accompanying figure) is located on the port side of the engine clamp or swivel/tilt brackets. Most models are also equipped with an Emissions Control Information label (B, in the accompanying figure) as well.

ENGINE SERIAL NUMBERS

◆ See Figure 2

The engine serial numbers are the manufacturer's key to engine changes. These alpha-numeric codes identify the year of manufacture, the horsepower rating, gearcase shaft length and various model/option differences (such as rope start, tiller electric or remote electric models). If any correspondence or parts are required, the engine serial number must be used for proper identification.

Remember that the serial number establishes the year in which the engine was produced, which is often not the year of first installation.

The engine serial number tag contains information such as the plant in which the motor was produced, the model number or code, the serial number (a unique sequential identifier given ONLY to that one motor) as well as other useful information such as weight (mass) in Kilograms (kg).

The emissions control information label states that the motor is in compliance with EPA emissions regulations for the model year of that engine. And, more importantly, it gives tune-up specifications that are vital to proper engine performance (that minimize harmful emissions). The specifications on this label may reflect changes that are made during production runs and are often not later reflected in a company's service literature. For this reason, specifications on the label always supercede those of a print manual. Typical specifications that are found on this label will include:
- Spark plug type and gap.
- Fuel recommendations.
- Idle speed settings
- Engine timing ignition (such as wide-open throttle and/or idle timing) specification
- Engine displacement (in Cubic Inches or Cubic Centimeters, as noted on the label)

Deciphering The Model Code on 1990-98 Engines

◆ See Figure 3

Engines built for the 1990-98 model years (and all Johnson/Evinrude engines built back through 1980) will contain an 8-12 digit code for identification. If the code begins with A, B, C, H, S, T or V, it represents a model variation (a model built for use in certain countries or specifically for a boat-builder to include with their new boat). If one of these alphas is not present, the code should start with J (for Johnson) or E (for Evinrude). The next one, two or three digits will be numbers, representing the horsepower rating. The digit following the horsepower rating will be a one, two or three digit alpha code identifying the various trim/model types (such as TE for tiller electric or FRE for 4-stroke, electric start/remote). Following the model identifier may be a single alpha identifier (L, Y, X or Z) representing gearcase shaft length (a lack of this identifier would represent a 15 in. shaft length). Next, a two-digit, alpha identifier is used for the year. And lastly, the manufacturer internally uses a single check digit to designate the model run.

Refer to the accompanying illustration to interpret the various alpha identifiers found throughout the model code.

■ Starting in 1980, OMC began using the word INTRODUCES as an easy way to decipher model years. The 10 letters of that word correspond to the digits 1-9 and 0, in that order. The first letter "I" represents a 1, the second letter "N" represents a 2 and so on until "S" which represents a 0. When deciphering a model code, each of the two alpha identifiers correspond to the last two digits of the model year. A 1998 model would therefore be EC, a 1996 would be ED, and so on. For quick deciphering, right out the word INTRODUCES and then number the letters from 1-9 and then 0.

Deciphering The Model Code on 1999-01 Engines

◆ See Figure 4

Engines built for the 1999-01 model years contain a simplified version of the model code (when compared with earlier models) containing only 7-8 digits. In all cases, the identifier should start with a single alpha representing Johnson (J) or Evinrude (E). The next one, two or three digits will be numbers, representing the horsepower rating. The digit following the horsepower rating will be a single one or two digit alpha/numeric code identifying design features/model types (such as W for commercial models, T for tiller steering or 4 for 4-stroke). Following the design feature/model identifier may be a single alpha identifier (L, Y, X or Z) representing gearcase shaft length (a lack of this identifier would represent a 15 in. shaft length). Next, a two-digit, alpha identifier is used for the year and is deciphered in the same manner as all Johnson/Evinrude models numbers since 1980. Finally, in some cases, a single check digit is used by the manufacturer internally to designate the model run.

Refer to the accompanying illustration to interpret the various alpha digits found throughout the model code.

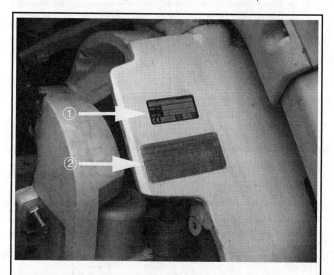

Fig. 1 A model ID tag (A) and an emission control label (B) is found on the port side of most engine clamp or swivel/tilt brackets

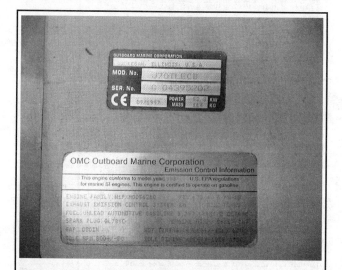

Fig. 2 The model ID tag (top) and emission label (bottom) provide critical information to identify and service the engine

Before/After Each Use

As stated earlier, the best means of extending engine life and helping to protect yourself while on the water is to pay close attention to boat/engine maintenance. This starts with an inspection of systems and components before and after each time you use your boat.

A list of checks, inspections or required maintenance can be found in the Maintenance Intervals Chart at the end of this section. Some of these inspections or tasks are performed before the boat is launched, some only after it is retrieved and the rest, both times.

2-4 MAINTENANCE & TUNE-UP

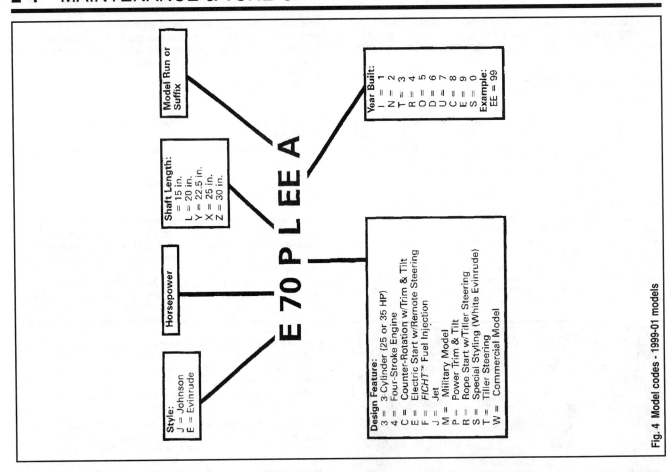

Fig. 4 Model codes - 1999-01 models

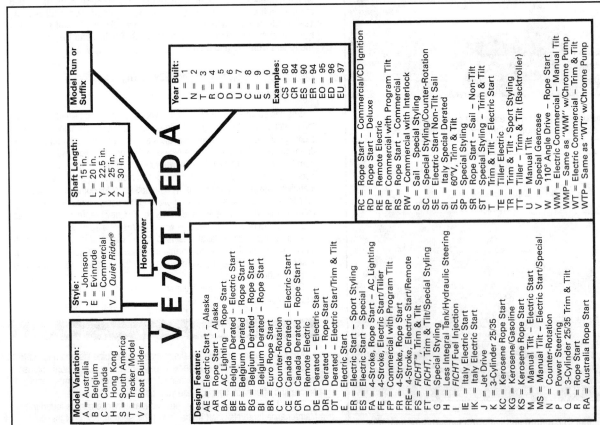

Fig. 3 Model codes - 1990-98 models

MAINTENANCE & TUNE-UP

VISUALLY INSPECTING THE BOAT AND MOTOR

◆ See Figures 5 and 6

Both before each launch and immediately after each retrieval, visually inspect the boat and motor as follows:

1. **Check the fuel and oil levels** according to the procedures in this manual. Do NOT launch a boat without properly topped off fuel and oil tanks (or without the proper crankcase oil level on 4-stroke motors). It is not worth the risk of getting stranded or of damage to the motor. Likewise, upon retrieval, check the oil and fuel levels while it is still fresh in your mind. This is a good way to track fuel consumption (one indication of engine performance). For 2-stroke motors, compare the fuel consumption to the oil consumption (a dramatic change in proportional use may be an early sign of trouble). For 4-stroke motors, oil consumption should be minimal, but all 4-stroke engines allow a small portion of oil to burn. Watch for sudden increases in the amount of oil burned and investigate further if found.

2. **Check for signs of fuel or oil leakage.** Probably as important as making sure enough fuel and oil is onboard, is the need to make sure that no dangerous conditions might arise due to leaks. Thoroughly check all hoses, fittings and tanks for signs of leakage. Oil leaks may cause the boat to become stranded, or worse, could destroy the motor if undetected for a significant amount of time. Fuel leaks can cause a fire hazard, or worse, an explosive condition. This check is not only about properly maintaining your boat and motor, but about helping to protect your life.

3. **Inspect the boat hull and engine cases** for signs of corrosion or damage. Don't launch a damaged boat or motor. And don't surprise yourself dockside or at the launch ramp by discovering damage that went unnoticed last time the boat was retrieved. Repair any hull or case damage now.

4. **Check the battery** connections to make sure they are clean and tight. A loose or corroded connection will cause charging problems (damaging the system or preventing charging). There's only one thing worse than a dead battery dockside/launch ramp and that's a dead battery in the middle of a bay, river or worse, the ocean. Whenever possible, make a quick visual check of battery electrolyte levels (keeping an eye on the level will give some warning of overcharging problems). This is especially true if the engine is operated at high speeds for extended periods of time.

5. **Check the propeller (impeller on jet drives and rotor on RescuePro® motors) and gearcase.** Make sure the propeller shows no signs of damage. A broken or bent propeller may allow the engine to over-rev and it will certainly waste fuel. The gearcase should be checked before and after each use for signs of leakage. Check the gearcase oil for signs of contamination if any leakage is noted. Also, visually check behind the propeller for signs of entangled rope or fishing lines that could cut through the lower gearcase propeller shaft seal. This is a common cause of gearcase lubricant leakage, and eventually, water contamination that can lead to gearcase failure. Even if no gearcase leakage is noted when the boat is first retrieved, check again next time before launching. A nicked seal might not seep fluid right away when still swollen from heat immediately after use, but might begin seeping over the next day, week or month as it sat, cooled and dried out.

6. **Check all accessible fasteners for tightness.** Make sure all easily accessible fasteners appear to be tight. This is especially true for the propeller nut, any anode retaining bolts, all steering or throttle linkage fasteners and the engine clamps or mounting bolts. Don't risk loosing control or becoming stranded due to loose fasteners. Perform these checks before heading out, and immediately after you return (so you'll know if anything needs to be serviced before you want to launch again.)

7. **Check operation of all controls including the throttle/shifter, steering and emergency stop/start switch and/or safety lanyard.** Before launching, make sure that all linkage and steering components operate properly and move smoothly through their range of motion. All electrical switches (such as power trim/tilt) and especially the emergency stop system(s) must be in proper working order. While underway, watch for signs that a system is not working or has become damaged. With the steering, shifter or throttle, keep a watchful eye out for a change in resistance or the start of jerky/notchy movement.

8. **Check the water pump intake grate and water indicator.** The water pump intake grate should be clean and undamaged before setting out. Remember that a damaged grate could allow debris into the system that could destroy the impeller or clog cooling passages. Once underway, make sure the cooling indicator stream is visible at all times. Make periodic checks, including one final check before the motor is shut down each time. If a cooling indicator stream is not present at any point, troubleshoot the problem before further engine operation.

9. **If equipped, check the power steering belt and fluid level.** A quick visual inspection of the power steering belt and fluid level at the end of each day will warn of problems that should be fixed before the next launch.

10. **If used in salt, brackish or polluted waters thoroughly rinse the engine (and hull), then flush the cooling system** according to the procedure in this section.

11. **Visually inspect all anodes** after each use for signs of wear, damage or to make sure they just plain didn't fall off (especially if you weren't careful about checking all the accessible fasteners the last time you launched).

12. **On EFI models, be sure to shut the battery switch off** if the engine is not going to be run for a couple of weeks or more. The Engine Control Unit (ECU) on fuel-injected motors covered by this manual will continue to draw a small amount of current from the battery, even when the motor is shut off. In order to prevent a slow drain of the entire battery, either periodically recharge the battery, or isolate it by disconnecting the cables or shutting off the battery switch when the boat is dockside or on the trailer.

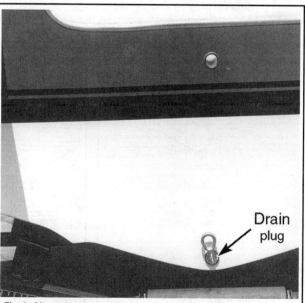

Fig. 5 Rope and fishing line entangled behind the propeller can cut through the seal, allowing water to enter and lubricant to escape

Fig. 6 Always make sure the transom plug is installed and tightened securely before a launch

2-6 MAINTENANCE & TUNE-UP

■ If the boat is not equipped with a battery switch, remove the green 30 amp fuse from the fuse holder found on the side of the engine. Of course, if this is done, tape the fuse to an obvious point so it will be installed before the next attempt to start the motor. This could save some embarrassing and frustrating troubleshooting time if the fact that it was removed becomes lost in your memory.

LUBRICATION SERVICE

An outboard motor's greatest enemy is corrosion. Face it, oil and water just don't mix and, as anyone who has visited a junkyard knows, metal and water aren't the greatest of friends either. To expose an engine to a harsh marine environment of water and wind is to expect that these elements will take their toll over time. But, there is a way to fight back and help prevent the natural process of corrosion that will destroy your beloved boat motor.

Various marine grade lubricants are available that serve two important functions in preserving your motor. Lubricants reduce friction on metal-to-metal contact surfaces and, they also displace air and moisture, therefore slowing or preventing corrosion damage. Periodic lubrication services are your best method of preserving an outboard motor.

Lubrication takes place through various forms. For all engines, internal moving parts are lubricated by engine oil, either through oil contained in the fuel/oil mixture on 2-stroke motors, or the oil contained in the engine crankcase and pumped through oil passages in 4-stroke motors. On all motors (both 2 and 4-stroke) the gearcase is filled with gear oil that lubricates the driveshaft, propshaft, gears and other internal gearcase components. The gear oil for all motors and the engine crankcase oil on 4-stroke motors should be periodically checked and replaced following the appropriate Engine Maintenance procedures. Perform these services based on time or engine use, as outlined in the Maintenance Intervals chart at the end of this section.

For motors equipped with power trim/tilt, the fluid level and condition in the reservoir should be checked periodically to ensure proper operation. Also, on these motors, correct fluid level is necessary to ensure operation of the motor impact protection system.

※※ WARNING

When equipped with power trim/tilt, proper fluid level is necessary for the built-in impact protection system. Incorrect fluid level could lead to significant gearcase damage in the event of an impact.

Most other forms of lubrication occur through the application of grease (OMC Triple-Guard, OMC EP/Wheel bearing grease, OMC Starter Pinion Lube, or their equivalents) to various points on the motor. These lubricants are either applied by hand (an old toothbrush can be helpful in preventing a mess) or using a grease gun to pump the lubricant into grease fittings (also known as zerk fittings). When using a grease gun, do not pump excessive amounts of grease into the fitting. Unless otherwise directed, pump until either the rubber seal (if used) begins to expand or until the grease just begins to seep from the joints of the component being lubricated (if no seal is used).

To ensure your motor is getting the protection it needs, perform a visual inspection of the various lubrication points at least once a week during regular seasonal operation (this assumes that the motor is being used at least once a week). Follow the recommendations given in the Lubrication Chart at the end of this section and perform the various lubricating services at least every 60 days when the boat is operated in fresh water or every 30 days when the boat is operated in salt, brackish or polluted waters. We said **at least** meaning you should perform these services more often, as discovered by your weekly inspections.

■ Jet drive models require one form of lubrication every time that they are used. The jet drive bearing should be greased, following the procedure given in this section, after every day of boating. But don't worry, it only takes a minute once you've done it before.

13. **For Pete's sake, make sure the plug is in!** We shouldn't have to say it, but unfortunately we do. If you've been boating for any length of time, you've seen or heard of someone whose backed a trailer down a launch ramp, forgetting to check the transom drain plug before submerging (literally) the boat. Always make sure the transom plug is installed and tight before a launch.

Electric Starter Motor Pinion

RECOMMENDED LUBRICANT

Use OMC Starter Pinion lubricant.

LUBRICATION

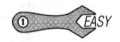

◆ See Figures 7 and 8

The starter pinion is the gear and slider assembly located on the top of the starter motor as it is mounted to the engine. When power is applied to the starter, the gear on the pinion assembly slides upward to contact and mesh with the gear teeth on the outside of the flywheel. Periodically, apply a small amount of lubricant to the sliding surface of the starter pinion in order to prevent excessive wear or possible binding on the shaft.

■ Access to the starter pinion is possible on most models by reaching under the flywheel cover using an applicator. But, in most cases, removal of the flywheel cover and/or manual starter assembly will make it much easier. If necessary refer to the Flywheel Cover or Manual Starter Assembly removal procedures for details.

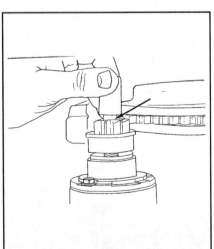

Fig. 7 Apply lubricant to the sliding surface of the electric starter pinion

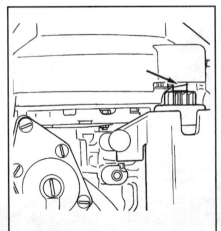

Fig. 8 In most cases, removing the flywheel cover will make access to the pinion much easier

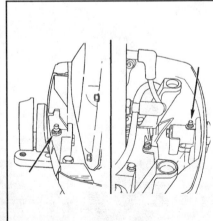

Fig. 9 If equipped, use the grease fittings to supply lubricant to the cover latches

MAINTENANCE & TUNE-UP 2-7

Engine Cover Latches

RECOMMENDED LUBRICANT

Use OMC Triple-Guard, or an equivalent water-resistant marine grease for lubrication.

LUBRICATION

♦ See Figures 9

Although the sliding surfaces of all cover latches can benefit from an application of grease, the design of the latches used on all 737cc and larger 2-stroke motors makes periodic greasing necessary to prevent the latches from binding or wearing. Depending on the latch type, either apply a small amount of grease to the metal surfaces using an applicator brush (this is typically necessary on 2-cylinder models) or use a grease gun to pump grease into the zerk fitting facing upward from the latch assembly.

Engine Mount Clamp Screws

♦ See Figure 10

RECOMMENDED LUBRICANT

Use OMC Triple-Guard, or an equivalent water-resistant marine grease for lubrication.

LUBRICATION

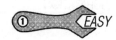

♦ See Figure 10

Many of the models covered by this manual are designed to be portable or permanently installed. Although installation and rigging will vary, if the motor is not permanently mounted in place, the threads of the engine mount clamp screws should be lubricated periodically. Apply a light coating of a suitable marine grease to the threads of both clamp screws. If necessary, apply the grease and loosen the clamp to ensure the grease is drawn through the threaded portion of the bracket, then retighten the clamp and repeat for the remaining clamp. When you are finished, be certain that the clamps are properly tightened. Also, pay extra attention to the clamps before and after the next use, to make sure they remain tightened.

Fig. 10 When equipped, be sure to apply lubricant to the threads on the engine mount clamps

Jet Drive Bearing

♦ See Figure 11

Jet drive models covered by this manual require special attention to ensure that the driveshaft bearing remains properly lubricated.

After each day of use, the jet drive bearing should be properly lubricated using a grease gun. Also, after every 30 hours of fresh water operation or every 15 hours of salt/brackish/polluted water operation, the drive bearing grease must be replaced. Follow the appropriate procedure:

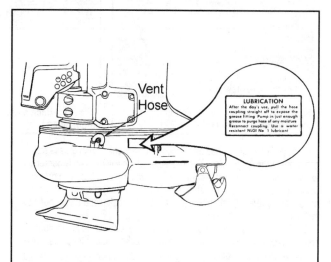

Fig. 11 Jet drive models require lubrication of the bearing after each day of use, a label on the housing usually reminds the owner

RECOMMENDED LUBRICANT

Use OMC EP/Wheel Bearing grease or an equivalent water-resistant NLGI No. 1 lubricant.

DAILY BEARING LUBRICATION

♦ See Figures 12 and 13

A grease fitting is located under a vent hose on the lower port side of the jet drive. Disconnect the hose from the fitting, then use a grease gun to apply enough grease to the fitting to **just** fill the vent hose. Basically, grease is pumped into the fitting until the old grease just starts to come out from the passages through the hose coupling, then reconnect the hose to the fitting.

■ Do not attempt to just grasp the vent hose and pull, as it is a tight fit and when it does come off, you'll probably go flying if you didn't prepare for it. The easier method of removing the vent hose from the fitting is to deflect the hose to one side and snap it free from the fitting.

GREASE REPLACEMENT

♦ See Figures 12, 13 and 14

A grease fitting is located under a vent hose on the lower port side of the jet drive. This grease fitting is utilized at the end of each day's use to add fresh grease to the jet drive bearing. But, every 30 or 15 days (depending if use is in fresh or salt/brackish/polluted waters), the grease should be completely replaced. This is very similar to the daily greasing, except that a lot more grease it used. Disconnect the hose from the fitting (by deflecting it to the side until it snaps free from the fitting), then use a grease gun to apply enough grease to the fitting until grease exiting the assembly fills the vent hose. Then, continue to pump grease into the fitting to force out all of the old

2-8 MAINTENANCE & TUNE-UP

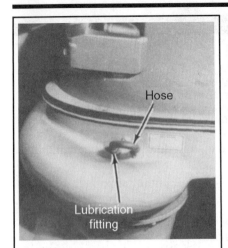

Fig. 12 The jet drive lubrication fitting is found under the vent hose

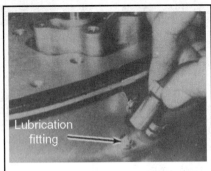

Fig. 13 Attach a grease gun to the fitting for lubrication

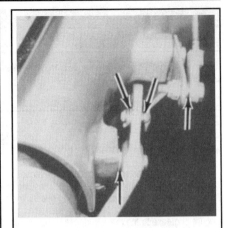

Fig. 14 Also, coat the pivot points of the jet linkage with grease periodically

grease (you can tell this has been accomplished when fresh grease starts to come out of the vent instead of old grease, which will be slightly darker due to minor contamination from normal use). When nothing but fresh grease comes out of the vent the fresh grease has completely displaced the old grease and you are finished. Be sure to securely connect the vent hose to the fitting.

Each time this is performed, inspect the grease for signs of moisture contamination or discoloration. A gradual increase in moisture content over a few services is a sign of seal wear that is beginning to allow some seepage. Very dark or dirty grease may indicate a worn seal (inspect and/or replace the seal, as necessary to prevent severe engine damage should the seal fail completely).

■ **Keep in mind that some discoloration of the grease is expected when a new seal is broken-in. The discoloration should go away gradually after one or two additional grease replacement services.**

Whenever the jet drive bearing grease is replaced, take a few minutes to apply some of that same water-resistant marine grease to the pivot points of the jet linkage.

Power Trim/Tilt Reservoir

◆ See Figure 15

✳✳ WARNING

When equipped with power trim/tilt, proper fluid level is necessary for the built-in impact protection system. Incorrect fluid level could lead to significant gearcase damage in the event of an impact.

RECOMMENDED LUBRICANT

The power trim/tilt reservoir must be kept full of OMC Power Trim/Tilt and Power Steering Fluid.

CHECKING FLUID LEVEL/CONDITION

◆ See Figure 15

The fluid in the power trim/tilt reservoir should be checked periodically to ensure it is full and is not contaminated. To check the fluid, tilt the motor upward to the full tilt position, then manually engage the tilt support for safety and to prevent damage. Remove the filler cap (they are usually threaded in position) and make a visual inspection of the fluid. It should seem clear and not milky. The level is proper if, with the motor at full tilt, the level is even with the bottom of the filler cap hole.

Linkage, Cables and Shafts (Choke, Shift, Carburetor and/or Throttle Shaft)

RECOMMENDED LUBRICANT

Use OMC Triple-Guard, or an equivalent water-resistant marine grease for lubrication.

LUBRICATION

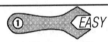

Every Johnson and Evinrude outboard uses some combination of cables and/or linkage in order to actuate the throttle plate (of the carburetor, carburetors or throttle body), the gearcase shifter and, on some smaller carbureted motors, the choke plate. Because linkage and cables contain moving parts that work in contact with other moving parts, the contact points can become worn and loose if proper lubrication is not maintained. These small parts are also susceptible to corrosion and breakage if they are not protected from moisture by light coatings of grease. Periodically apply a light coating of suitable water-resistant marine grease on each of these surfaces where either two moving parts meet or where a cable end enters a housing. For more details on grease points refer to the accompanying illustrations.

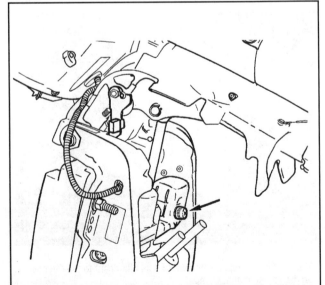

Fig. 15 Maintaining the proper power trim/tilt fluid level is critical to protecting the engine in case of an impact

MAINTENANCE & TUNE-UP

Colt/Junior and 2-6 Hp Single Cylinder Motors

◆ See Figures 16 thru 20

Apply a light coating of grease to the carburetor, choke and shift linkage at the points shown for your single cylinder motor.

3-8 Hp Two Cylinder, 2-Stroke Motors

◆ See Figures 21 thru 25

Apply a light coating of grease to the carburetor, choke and shift linkage at the points shown. On models equipped with a built-in fuel tank, check for a fuel valve and/or choke shaft assembly and grease, as necessary. Make sure all sliding, rotating or contact surfaces of the linkage are coated.

9.9/10/14/15 HP (216cc) and 9.9/10/15 HP (255cc) Two Cylinder, 2-Stroke Motors

◆ See Figures 26 and 27

Apply a light coating of grease to the carburetor, cam follower, throttle, spark advance, choke and shift linkage at the points shown. Make sure all sliding, rotating or contact surfaces of the linkage are coated.

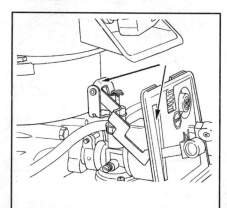

Fig. 16 Carburetor/throttle linkage and choke shaft lubrication - Colt/Junior and 2-3.5 hp motors

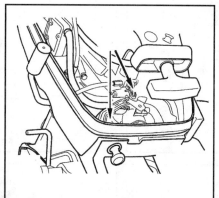

Fig. 17 Carburetor/throttle linkage and choke shaft lubrication - 5 hp 2-stroke motors

Fig. 18 If equipped, lubricate the neutral lockout cable (shown) or linkage - 5 hp 2-stroke motor shown

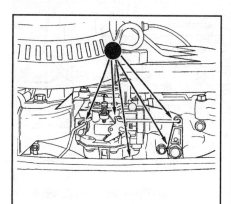

Fig. 19 Carburetor throttle and choke linkage lubrication - 5/6 hp 4-stroke motors

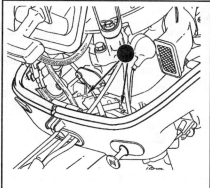

Fig. 20 Shift linkage lubrication - 5/6 hp 4-stroke motors

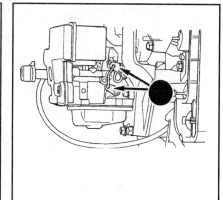

Fig. 21 Carburetor/throttle linkage lubrication - 3/4 hp motors

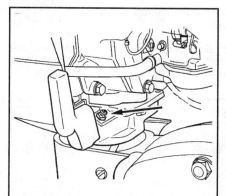

Fig. 22 The shift linkage on some models, like the 3/4 hp motors, is equipped with a grease fitting

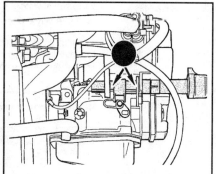

Fig. 23 Some models equipped with a built-in fuel tank utilize a fuel valve/choke shaft assembly - 3/4 hp motor shown (4 Deluxe similar)

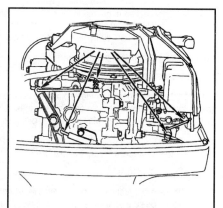

Fig. 24 Throttle and choke linkage lubrication - 5/6/8 hp (164cc) motors

2-10 MAINTENANCE & TUNE-UP

Fig. 25 Carburetor linkage lubrication - 5/6/8 hp (164cc) motors (4 Deluxe similar)

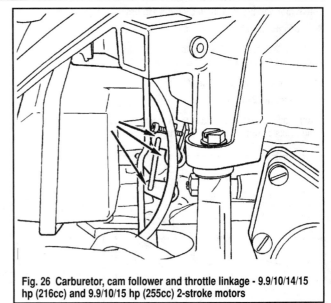

Fig. 26 Carburetor, cam follower and throttle linkage - 9.9/10/14/15 hp (216cc) and 9.9/10/15 hp (255cc) 2-stroke motors

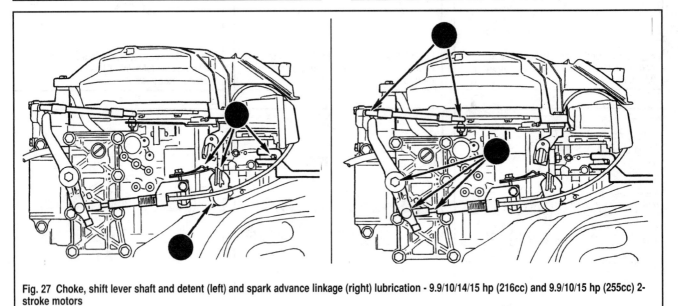

Fig. 27 Choke, shift lever shaft and detent (left) and spark advance linkage (right) lubrication - 9.9/10/14/15 hp (216cc) and 9.9/10/15 hp (255cc) 2-stroke motors

8/9.9 and 9.9/15 Hp Two Cylinder, 4-Stroke Motors

◆ See Figures 28 and 29

Apply a light coating of grease to the carburetor, throttle, choke linkages as well as the shift lever shaft and detent points shown. Make sure all sliding, rotating or contact surfaces of the linkage are coated.

18-35 Hp Two Cylinder (521cc) Motors

◆ See Figures 30 and 31

For 18-35 hp (521cc) 2-cylinder motors, be sure to apply a light coating of grease to the carburetor, throttle and shift linkages as well as the starter lockout assembly as shown. Make sure all sliding, rotating or contact surfaces of the linkage are coated.

25-55 Hp Two Cylinder (737cc) Motors

◆ See Figures 32 thru 36

Though the exact lubrication points vary slightly from model-to-model, for 737cc motors, be sure to apply a light coat of water resistant marine grease to the carburetor, throttle and shifter linkage, including the timer link and throttle shaft fittings, as applicable. Refer to the illustrations for more details.

25/35 Hp (500/565cc) Three Cylinder Motors

◆ See Figures 37 and 38

For all 25/35 hp (500/565cc) 3-cylinder motors, be sure to coat the throttle and shift linkage on the starboard side of the motor, as well as the carburetor linkage found on the port side. Refer to the illustrations for more details.

25-70 Hp (913cc) Three Cylinder, 2-Stroke Motors

◆ See Figures 39 and 40

Refer to the following illustrations to determine the applicable throttle, carburetor and shift linkage lubrication points on your 913cc 3-cylinder, 2-stroke motor.

MAINTENANCE & TUNE-UP 2-11

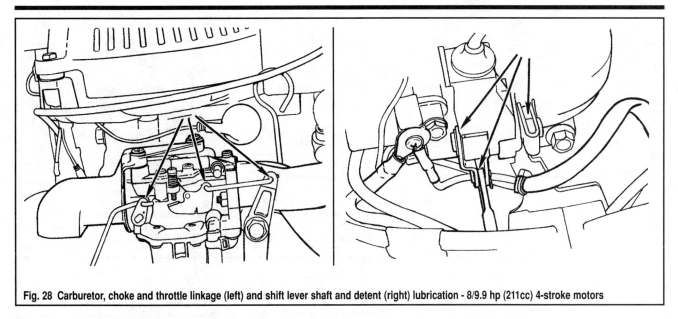

Fig. 28 Carburetor, choke and throttle linkage (left) and shift lever shaft and detent (right) lubrication - 8/9.9 hp (211cc) 4-stroke motors

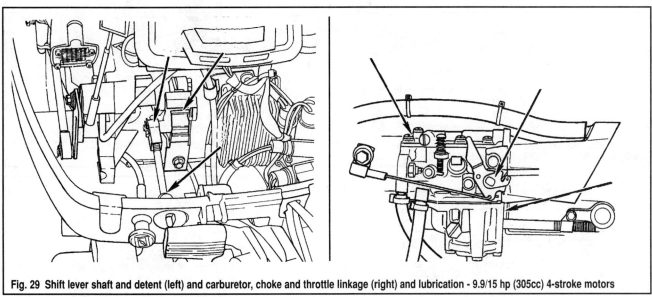

Fig. 29 Shift lever shaft and detent (left) and carburetor, choke and throttle linkage (right) and lubrication - 9.9/15 hp (305cc) 4-stroke motors

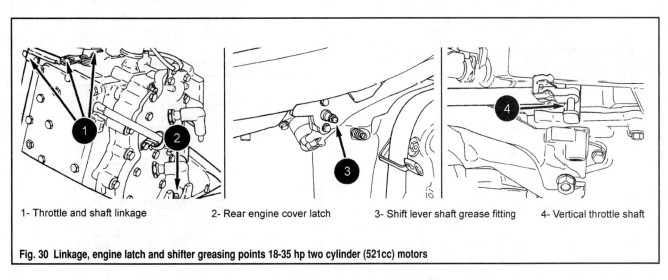

1- Throttle and shaft linkage   2- Rear engine cover latch   3- Shift lever shaft grease fitting   4- Vertical throttle shaft

Fig. 30 Linkage, engine latch and shifter greasing points 18-35 hp two cylinder (521cc) motors

2-12 MAINTENANCE & TUNE-UP

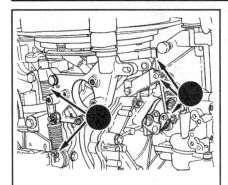

Fig. 31 Carburetor linkage, cam and shifter starter lockout greasing points 18-35 hp two cylinder (521cc) Motors

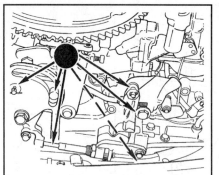

Fig. 32 Carburetor linkage and timer link greasing points on some 40-55 hp two cylinder (737cc) motors, including the 40RP, 40RW, 40WR, 45, 55WR and 55 RescuePro

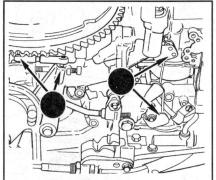

Fig. 33 Throttle and shaft linkage greasing points on some 40-55 hp two cylinder (737cc) motors, including 40RP, 40RW, 40WR, 45, 55WR and 55 RescuePro

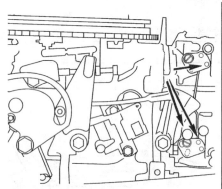
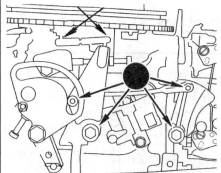

Fig. 34 Carburetor (left) and throttle (right) linkage greasing points on most 25-55 hp two cylinder (737cc) motors including the 25, 35 Jet, 40EL, 40RS, 40TL, 48, 50EL, 50SPL, 50TL and 55WML

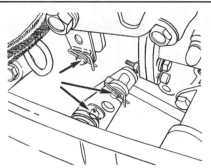

Fig. 35 Shift and throttle shaft fitting lubrication 25-50 hp two cylinder (737cc) motors including the 25, 35 Jet, 40EL, 40RS, 40TL, 48, 50EL, 50SPL and 50TL

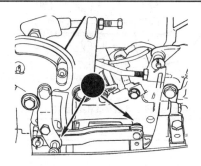

Fig. 36 Shift and throttle shaft greasing points for 55WML two cylinder (737cc) motors

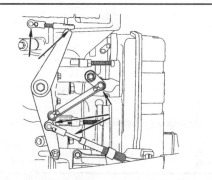

Fig. 37 Throttle and shift linkage greasing points for 25/35 hp (500/565cc) 3-cylinder motors

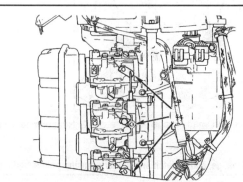

Fig. 38 Carburetor linkage lubrication points for 25/35 hp (500/565cc) 3-cylinder motors

MAINTENANCE & TUNE-UP 2-13

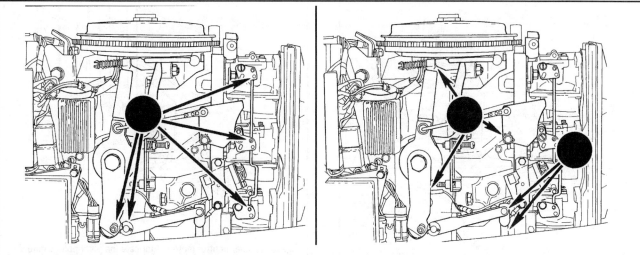

Fig. 39 Carburetor linkage, cam roller, shift shaft and control shaft/lever bushing lubrication points for 25-70 hp (913cc) 3-cylinder, 2-stroke motors (except the 65RS, 65WR and some 50-70TTL models)

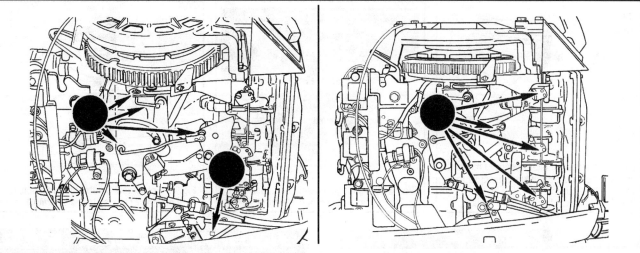

Fig. 40 Shift shaft and linkage and control lever bearing (left) along with shift and throttle cable fittings, carburetor linkage and cam follower (right) lubrication points for 65RS, 65WR and some 50-70TTL models of the (913cc) 3-cylinder, 2-stroke motor

40-70 Hp 4-Stroke Motors

◆ See Figures 41 and 42

Steering (Arm/Shaft and Friction Screw)

◆ See Figures 43 thru 46

RECOMMENDED LUBRICANT

Use OMC Triple-Guard, or an equivalent water-resistant marine grease for lubrication.

LUBRICATION

◆ See Figures 43 thru 46

All motors covered by this manual are equipped with a tiller control and/or a remote control assembly. On models equipped with a tiller, the arm's pivot point (where it attaches to the engine) should be lubricated periodically. On models with remote controls, the steering arm should be given a light coating of fresh lubricant to prevent corrosion or scoring. Many of the outboards covered by this manual (especially the portable units) are equipped with a steering friction adjustment knob/screw. Coat the exposed threads of the screw with fresh grease during lubrication services.

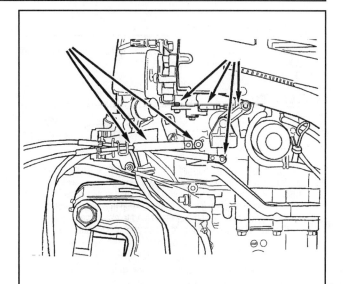

Fig. 41 Shift and throttle linkage lubrication - 40/50 hp (815cc) 3-cylinder, 4-stroke motors

2-14 MAINTENANCE & TUNE-UP

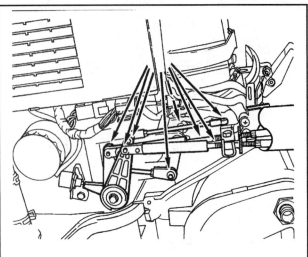

Fig. 42 Shift and throttle linkage lubrication - 70 hp (1298cc) 4-cylinder, 4-stroke motors

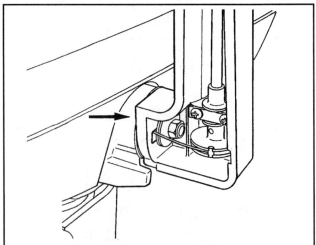

Fig. 43 On tiller control motors, lubricate the pivot point where the arm attaches to the engine - 3/4 hp 2-cylinder, 2-stroke shown (others similar)

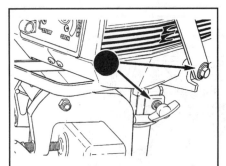

Fig. 44 Besides the tiller arm, most models utilize a steering friction knob or screw that also requires lubrication - 2-3.5 hp 1-cylinder, 2-stroke shown (others similar)

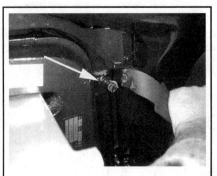

Fig. 45 Apply a light coating of grease to the threads and spring of the friction screw

Fig. 46 For remote control motors, apply a light coating of grease to the steering arm

Swivel Bracket

◆ See Figure 47

RECOMMENDED LUBRICANT

Use OMC Triple-Guard, or an equivalent water-resistant marine grease for lubrication.

LUBRICATION

◆ See Figure 47

All motors covered by this manual are equipped with at least one grease fitting on the gearcase swivel bracket. Use a grease gun to apply fresh water-resistant marine grease until a small amount of lubricant begins to seep from the swivel bracket. It is important to keep this system corrosion free in order to prevent corrosion that would lead to excessive resistance or even binding that might cause dangerous operational conditions.

Tilt Assembly (Bracket, Tube, Pin and/or Tilt Lever Shaft)

◆ See Figures 48 and 49

RECOMMENDED LUBRICANT

Use OMC Triple-Guard, or an equivalent water-resistant marine grease for lubrication.

LUBRICATION

◆ See Figures 48 and 49

Although the precise points vary from model-to-model, all motors are equipped with pivot and anchor points for the tilt system. Be sure to apply grease to all zerk fittings (check carefully, as some fittings are hidden or partially obstructed when the motor is in the full upward tilt or full downward tilt position). Apply a water-resistant marine grade grease to the fitting(s) until a small amount of grease seeps from the joints. Also, on manual tilt models, apply grease to the tilt lever and/or pin and any other metal-to-metal friction surfaces. Applying grease will prevent corrosion while also ensuring smooth operation. To make sure all surfaces are covered, apply grease with the motor in both the full tilt and full downward positions.

MAINTENANCE & TUNE-UP 2-15

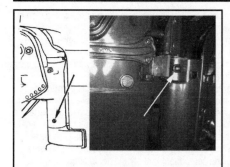

Fig. 47 Apply grease to the swivel bracket through the fitting on the port or starboard side (depending on the model)

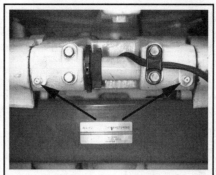

Fig. 48 Use a grease gun on all tilt assembly zerk fittings (the exact number and location vary from model-to-model) . . .

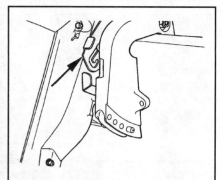

Fig. 49 . . . then apply a light coating of grease to all metal-to-metal contact areas on the tilt assembly

ENGINE MAINTENANCE

Engine Covers (Top and Lower Cases)

REMOVAL & INSTALLATION

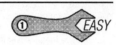

♦ See Figures 50 thru 53

Removal of the top cover is necessary for the most basic of maintenance and inspection procedures. The cover should come off before and after each use in order to perform these basic safety checks. The lower covers do not need to be removed nearly as often, but on models where they are easily removed, they should be removed at least seasonally for service and inspection procedures. Don't let a small leak or damaged cable/hose hide behind the safety of a cover.

On all models, the engine top cover is attached by some type of lever or latch. No tools are necessary to remove the cover itself. The exact shape and design of the levers vary somewhat from model-to-model, though they are usually located on the aft part of the motor, at the split line between the top cover and the lower cases.

Some of the smaller motors use a lever that is pulled outward to release the cover. A few of the very small 2-strokes and the largest 4-strokes use over-center latches that hook onto tabs on the top cover, these are normally released by pulling at the base of the latch. However, the vast majority of motors covered by this manual utilize a rotating lever that is twisted 45-90° in order to release the top cover.

No matter what design is used, be certain that the cover is fully seated and mounted tightly to the lower cases in order to prevent the possibility of it coming loose in service.

The lower covers of most motors are screwed or bolted together by fasteners found around the perimeter of one or both sides of the cover. In a few cases, such as some 5/6 hp or 8/9.9 hp 4-stroke motors, one or more of the fasteners may be hidden. These fasteners may be accessible only through access points around the cover such as through the choke handle or, in the case of many motors, through the water indicator outlet hole.

■ **Cover screws on Johnson/Evinrude outboards are usually of the Phillips or Slotted head types, but some are also of the star-headed Torx® design. For Torx® head screws, be sure to use only the proper-sized driver as an undersized driver will strip or damage to the fastener head.**

Some motors however, are equipped with 1-piece covers that are not designed for easy or convenient removal. On the 5 hp (109cc) motor, 18 Jet-35 hp (521cc) motors, some 40-55 hp (737cc) Commercial model motors, and the 25-70 hp (914cc) motors, this cover is normally a low-rise component that should not interfere with service procedures. For this reason, the cover is normally usually not removed except during a complete overhaul where the powerhead is removed from the gearcase.

In most cases, remote or tiller control cables (and choke mechanisms, if equipped) must be disconnected and/or removed from the case in order to completely remove the lower cases. But, for many procedures, the lower case can be supported out of the way (using a length of mechanic's wire or a bent wire coat hanger) with the cables still attached to the cover. You'll have to decide for yourself how much trouble it is worth to remove the covers for various maintenance procedures, but obviously they must be completely removed for major overhauls.

To separate the lower covers on 2-piece models, proceed as follows:
1. On some models the top cover seal is mounted in the groove on the top cover, for others it is placed on the top sealing surface of the lower

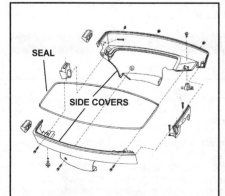

Fig. 50 Most of the engines covered by this manual utilize a 2-piece lower port and lower starboard cover assembly

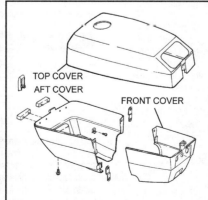

Fig. 51 The 3/4 hp and 4 Deluxe motors are unique, as they have a 2-piece front and aft cover assembly

Fig. 52 The lower covers are normally secured using screws around the perimeter. . .

2-16 MAINTENANCE & TUNE-UP

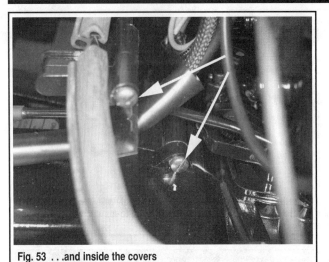

Fig. 53 ...and inside the covers

covers. On models where the seal is attached to the lower covers, carefully lift it from the covers and place it aside where it will not be damaged.

2. Locate and remove the cover retaining screws as follows:

a. On 2-3.5 hp (78cc) motors, there is a port and a starboard cover half. Remove the 4 screws from the side, then remove the 6 screws from the front and bottom.

b. On 5 hp (109cc) motors, the lower cover is a low-rise, one-piece component that should not interfere with service procedures. The cover is normally not removed except during a complete overhaul where the powerhead is removed from the gearcase.

c. On 3-4 hp and 4 hp Deluxe (87cc) motors, there are front and rear cover halves. Remove the screws securing the halves together on the inside of the covers and, check underneath as there is at least one screw mounted upward from under the rear cover half.

d. On 5/6 hp (128cc) and 8/9.9 (211cc) 4-stroke motors, there is a port and starboard cover half. Be careful on these motors, as not all screws are accessible from the outside of the case. Start by disengaging the choke knob from the carburetor link, then remove the two screws at the front of the lower cover half. Next remove the screw visible at the rear of the lower cover half. The last screw is inside the case, directly behind the water indicator tube. To access and remove that screw, disconnect the water indicator hose from the indicator, then remove the plastic indicator fitting from the case. Insert a screwdriver through the hole in the case for the indicator fitting and remove the final screw.

■ The lower of the two front cover half screws on the 5/6 hp (128cc) and 8/9.9 hp (211cc) motors can be accessed either using a stubby screwdriver, or by inserting a long screwdriver through the opening for the choke knob at the front of the case.

e. On 5-8 hp (164cc) motors, there is a port and starboard cover half. There are usually 3 screws securing the cover halves, one is inserted from the outside of the starboard cover half, while the other two are located on tabs inside the cover halves. If necessary, remove the two front plate-to-lower cover screws also.

f. On 9.9/10/14/15 hp (216cc) and 9.9/10/15 hp (255cc) motors, there is a port and starboard cover half. Be careful on these motors, as not all screws are accessible from the outside of the case. Remove the two screws and nuts from the top rear of the lower engine cover. The next screw is located inside the case, directly behind the water indicator tube. To access and remove that screw, disconnect the water indicator hose from the indicator, then remove the plastic indicator fitting from the case. Insert a screwdriver through the hole in the case for the indicator fitting and remove the screw from the lower rear of the cover. Finally, remove the 2 screws from the top front and the one screw and nut from the lower front of the lower engine cover.

g. On 9.9/15 hp (305cc) 4-stroke motors, there is a port and starboard cover half. Remove the screw securing the choke cable clamp to the powerhead, then disconnect the cable from the carburetor choke link. Remove the nut from the knob end of the choke cable, and remove the cable, this will provide access to one of the upper screws securing the lower cover. Remove the 2 upper screws from the lower cover, then remove the 5 screws from the side of the cover. Lastly, remove the water indicator hose, followed by the water indicator, then use a screwdriver inserted through the opening in the case to remove the final screw from the rear starboard side of the motor.

h. On 18 Jet-35 hp (521cc) motors, the lower cover is a low-rise, one-piece component that should not interfere with service procedures. The cover is normally not removed except during a complete overhaul where the powerhead is removed from the gearcase.

i. On 25-55 hp (737cc) non-commercial motors, there is a port and starboard cover half. Remove the 4 screws securing the halves. Two screws are mounted near the cover latch, just inside the rear of the housing. One screw is located at the top of the cover, near the front of the motor. The final screw is found outside the lower covers, just behind the steering pivot.

j. On 40-55 hp (737cc) commercial model motors, the lower cover is usually a low-rise, one-piece component that should not interfere with service procedures. The cover is normally not removed except during a complete overhaul where the powerhead is removed from the gearcase. If a 2-piece cover is encountered, refer to the previous step for 737cc non-commercial motors.

k. On 25/35 hp (500/565cc) 3-cylinder motors, there is a port and starboard cover half, but cover removal is not normally associated with maintenance or inspection for these motors, therefore the cover removal is a long and involved process. For this reason, the cover removal and installation procedure can be found as part of the powerhead removal and installation procedure for these motors.

l. On 25-70 hp (913cc) motors, the lower cover is a low-rise, one-piece component that should not interfere with service procedures. The cover is normally not removed except during a complete overhaul where the powerhead is removed from the gearcase. Refer to Powerhead Removal and Installation for more details.

m. On 40-70 hp 4-stroke motors, there is a port and starboard cover half. Remove the aft cover latch, then remove the screws from around the perimeter of the cover. There are normally 5-7 screws depending on the model.

✱✱ WARNING

Be careful to make sure that all fasteners are removed before trying to separate the covers. Absolutely never force them. If it appears that they are stuck, go back and recheck for any fasteners or screws that were missed.

3. Once the screws are removed, pull the covers back for access. Some covers will come off completely at this time, but others will still be attached to the engine due to wires, cables or hoses that are also attached to the cover. Either support the cover halves aside with these component still attached, or free any remaining components from the cover halves and remove them from the engine.

4. Installation is the reverse of the removal procedure, making sure to reattach any components that were freed from the cover or removed for access. Be careful not to pinch or damage and hoses, cables or wiring when seating the lower covers.

5. Tighten the cover screws securely, but do not over-tighten and crack the covers or strip the screw threads.

6. Make sure the top cover seal is in proper position before installing the top cover and securing the latch(es). The top cover must be a tight fit to protect the motor from excessive spray/moisture and to ensure the top cover remains properly seated in use.

Cooling System

FLUSHING THE COOLING SYSTEM

 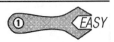

◆ See Figures 54 thru 59

The most important service that you can perform on your motor's cooling system is to flush it periodically using fresh, clean water. This should be done immediately following any use in salt, brackish or polluted waters in order to prevent mineral deposits or corrosion from clogging cooling passages. Even if you do not always boat in salt or polluted waters, get used to the flushing procedure and perform it often to ensure no silt or debris clogs your cooling system over time.

■ **Flush the cooling system after any use in which the motor was operated through suspended/churned-up silt, debris or sand.**

MAINTENANCE & TUNE-UP 2-17

Although the flushing procedure should take place right away (dockside or on the trailer), be sure to protect the motor from damage due to possible thermal shock. If the engine has just been run under high load or at continued high speeds, allow time for it to cool to the point where the powerhead can be touched. Do not pump very cold water through a very hot engine, or you are just asking for trouble. If you trailer your boat short distances, the flushing procedure can probably wait until you arrive home or wherever the boat is stored, but ideally it should occur within an hour of use in salt water. Remember that the corrosion process begins as soon as the motor is removed from the water and exposed to air.

The flushing procedure is not used only for cooling system maintenance, but it is also a tool with which a technician can provide a source of cooling water to protect the engine (and water pump impeller) from damage anytime the motor needs to be run out of the water. **Never** start or run the engine out of the water, even for a few seconds, for any reason. Water pump impeller damage can occur instantly and damage to the engine from overheating can follow shortly thereafter. If the engine must be run out of the water for tuning or testing, always connect an appropriate flushing device **before** the engine is started and leave it turned on until **after** the engine is shut off.

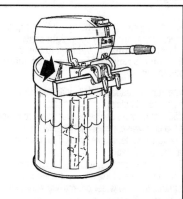
Fig. 54 All models may be flushed in a test tank. Smaller ones, in a garbage pail

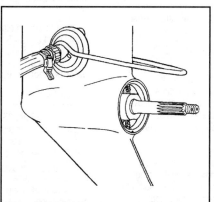
Fig. 55 The easiest way to flush most models is using a clamp-type adapter

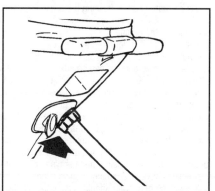
Fig. 56 Some models (like these 25/35 hp 3-cylinder motors) are equipped with an engine mounted flushing port...

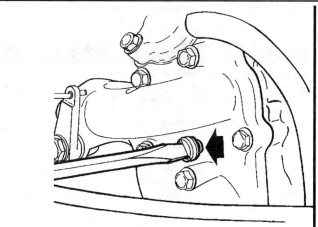
Fig. 57 ...while some others (like the 8/9.9 and 9/9/15 hp 4-strokes) have a port in the side of the powerhead that requires use of an adapter

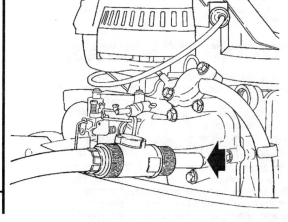

Fig. 58 A water source, such as a flushing device or test tank, must be used ANYTIME the engine is started

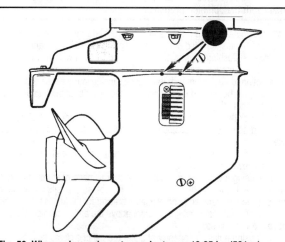
Fig. 59 When using a clamp-type adapter on 18-35 hp (521cc) motors, cover the small holes on each side of the gearcase using tape

2-18 MAINTENANCE & TUNE-UP

※※ WARNING

ANYTIME the engine is run, the first thing you should do is check the cooling stream or water indicator. All models covered by this manual are equipped with some form of a cooling stream indicator towards the aft portion of the lower engine cover. Anytime the engine is operating, a steady stream of water should come from the indicator, showing that the pump is supplying water to the engine for cooling. If the stream is ever absent, stop the motor and determine the cause before restarting.

As we stated earlier, flushing the cooling system consists of supplying fresh, clean water to the system in order to clean deposits from the internal passages. If the engine is running, the water does not normally have to be pressurized, as it is delivered through the normal water intake passages and the water pump (the system can self flush if supplied with clean water). Smaller, portable engines can be flushed by mounted them in a test tank (a sturdy, metallic 30 gallon drum or garbage pail filled with clean water). Almost all Johnson/Evinrude engines will also accept flush fittings or adapters. Most adapters are of the generic type and are designed to fit over the engine water intakes on the gearcase (and resemble a pair of strange earmuffs with a hose fitting on one side). But, other adapters (available from the manufacturer) are designed for special flushing fittings on specific motors. These special adapters attach to a cooling passage on the gearcase or powerhead. When using the later type adapter, follow the manufacturer's instructions closely regarding flushing conditions. In some cases, flushing with this type of adapter should occur only with the motor turned off, so as to prevent damage to the water pump impeller or other engine components. This varies with each motor, so be sure to check with your dealer regarding these direct to the powerhead adapters when you purchase one.

■ Most jet drive models are equipped with a flushing port mounted under a flat head screw directly above the jet drive bearing grease fitting. Use an OMC adapter (#435299) or equivalent to attach a garden hose to this port.

■ When running the engine on a flushing adapter using a garden hose, make sure the hose delivers 20-40 psi (140-300 kPa) of pressure.

Some of the smaller, portable motors covered by this manual utilize a water intake that is directly above the propeller. On these models the propeller must usually be removed before a clamp style flush adapter can be connected to the motor (unless the adapter is very thin and mounted so close to the anti-ventilation plate that it will not be hit by the propeller).

※※ CAUTION

For safety, the propeller should be removed ANYTIME the motor is run on the trailer or on an engine stand. We realize that this is not always practical when flushing the engine on the trailer, but cannot emphasize enough how much caution must be exercised to prevent injury to you or someone else. Either take the time to remove the propeller or take the time to make sure no-one or nothing comes close enough to it to become injured. Serious personal injury or death could result from contact with the spinning propeller.

When using a flushing device and a pressurized water source, most motors can be flushed tilted or in a vertical position, BUT, the manufacturer warns against flushing most motors in the tilted position with the engine running. Some models (especially most 4-strokes) can be seriously damaged by attempting to flush them with the engine running in the full tilt position. If the motor must be flushed tilted (dockside) then your best bet it to do so with the engine shut off.

1. Check the engine top case and, if necessary remove it to check the powerhead, to ensure it is cooled enough to flush without causing thermal shock.
2. Prepare the engine for flushing depending on the method you are using as follows:
   a. If using a test tank, make sure the tank is made of sturdy material, then securely mount the motor to the tank. If necessary, position a wooden plank between the tank and engine clamp bracket for thickness. Fill the tank so the water level is at least 4 in. (10cm) above the anti-ventilation plate (above the water inlet).
   b. If using a flushing adapter of either the generic clamp-type or specific port-type for your model attach the water hose to the flush test adapter and connect the adapter to the motor following the instructions that came with the adapter. If the motor is to be run (for flushing or testing),
position the outboard vertically and remove the propeller, for safety. Also, be sure to position the water hose so it will not contact with moving parts (tie the hose out of the way with mechanic's wire or wire ties, as necessary).

※※ WARNING

The fuel injected motors covered by this manual are equipped with labeled flushing ports on the gearcase. The port on the 40/50 hp (815cc) 4-stroke motors must NOT be used for flushing while the engine is running as it will restrict water supply to the powerhead and could lead to engine damage.

■ When using a clamp-type adapter, position the suction cup(s) over water intake grate(s) in such a way that they form tight seals. A little pressure seepage should not be a problem, but look to the water stream indicator once the motor is running to be sure that sufficient water is reaching the powerhead.

3. If using a clamp-type flush test adapter, follow any special instructions for your model, as noted below:
   a. On 18 Jet-35 hp (521cc) motors, use heavy duct-tape to cover the two holes on each side of the crankcase immediately below the anti-ventilation plate and just above the water intake grate. This will help ensure sufficient water pressure at the powerhead.
   b. On 40/50 hp (815cc) 4-stroke motors, use heavy duct-tape to cover the water inlet located on the underside of the anti-ventilation plate.
4. Unless using a test tank, turn the water on, making sure that pressure does not exceed 45 psi (300 kPa).
5. If using a test tank or if the motor must be run for testing/tuning procedures, start the engine and run in neutral until the motor reaches operating temperature. For most motors, the motor will continue to run at fast idle until warmed, on fuel injected motors, speed will be automatically regulated by the Engine Control Unit (ECU) at 1000 rpm for 40/50 hp models or at/below 1500 rpm for 70 hp motors.

※※ WARNING

As soon as the engine starts, check the cooling system indicator stream. It must be present and strong as long as the motor is operated. If not, stop the motor and rectify the problem before proceeding. Common problems could include insufficient water pressure or incorrect flush adapter installation.

6. Flush the motor for at least 5-10 minutes or until the water exiting the engine is clear. When flushing while running the motor, check the engine temperature (using a gauge or carefully by touch) and stop the engine immediately if steam or overheating starts to occur. Make sure that carbureted motors slow to low idle for the last few minutes of the flushing procedure.
7. Stop the engine (if running), **then** shut the water off.
8. Remove the adapter from the engine or the engine from the test tank, as applicable.
9. If flushing did not occur with the motor running (so the motor would already by vertical), be sure to place it in the full vertical position allowing the cooling system to drain. This is especially important if the engine is going to be placed into storage and could be exposed to freezing temperatures. Water left in the motor could freeze and crack the powerhead or gearcase.

Engine Oil (2-Stroke)

OIL RECOMMENDATIONS

♦ See Figure 60

Use only an NMMA (National Marine Manufacturers Association) certified TC-W3 or equivalent 2-stroke lubricant. Of course, OMC recommends using Johnson/Evinrude brand oils, since they are specially formulated to match the needs of OMC motors. In all cases, a high quality TC-W3 oils are proprietary lubricants designed to ensure optimal engine performance and to minimize combustion chamber deposits, to avoid detonation and prolong spark plug life. Use only 2-stroke type outboard oil. Never use automotive motor oil.

■ Remember, it is this oil, mixed with the gasoline that lubricates the internal parts of the 2-stroke engine. Lack of lubrication due to the wrong mix or improper type of oil can cause catastrophic powerhead failure.

MAINTENANCE & TUNE-UP 2-19

Fig. 60 This scuffed piston is an example of the damage caused by improper 2-stroke oil or mixture

FILLING

There are two methods of adding 2-stroke oil to an outboard. The first is the pre-mix method used on most low horsepower and on some commercial outboards. The second is the automatic oil injection method that automatically injects the correct quantity of oil into the engine based on throttle position and operating conditions. In both cases, the fuel ratio should be considered. This is even true on automatic oiling systems if the engine is going to be used under certain severe or high performance conditions.

Fuel:Oil Ratio

The proper fuel:oil ratio will depend upon engine operating conditions. Many of the engines covered by this manual may be equipped with an automatic oiling system (such as the VRO2 or AccuMix systems) that is designed to maintain a 50:1 ratio without adding anything to the fuel tank. But, whether or not an oiling system is used, for all Johnson/Evinrude 2-stroke engines covered by this manual, the proper fuel:oil ratio is 50:1 for **normal** operating conditions. Most manufacturers define normal as a motor operated under varying conditions from idle to wide open throttle, without excessive amounts of use at either. Unfortunately, no-one seems to put a definition to "excessive amount" either, so you'll have to use common sense. We don't think an hour of low speed trolling mixed in with some high speed operation or an hour or two of pulling a skier constitute "excessive amounts," but you'll have to make your own decision. Also necessary for defining normal operating conditions is the ambient and sea-water temperatures. The sea-water temperatures should be above 32°F (0°C) and below 68° F (20° C). Ambient air conditions should be above freezing and below the point of extreme discomfort (90-100°F).

■ The fuel:oil ratios listed here are OMC recommendations given in service literature. Because your engine may differ slightly from service manual specification, refer to your owner's manual or a reputable dealer to be certain that your mixture meets your conditions of use.

If your outboard is to be used under severe conditions including, long periods of idle, long periods of heavy load, use in severe ambient temperatures (outside the range of normal use) or under high performance (constant wide-open throttle or racing conditions) some adjustment may be necessary to the fuel:oil ratio. Proper ratios for use vary by model and oiling system:

• Most 2.0-8 hp motors (usually about 1993 and later), require a 25:1 ratio for severe and high performance conditions.
• 9.9/15 hp (255cc) and 18 jet-35 hp (521cc) motors, require a 25:1 ratio for high performance use, but OMC advises that recreational models still require only a 50:1 ratio for commercial, rental or extended severe conditions.
• 25D-55 hp (737cc) motors, require a 25:1 ratio for severe and high performance conditions. But, many of these models are equipped with VRO2 or the AccuMix oiling systems. On models equipped with either of these systems, a tank mixture of 50:1 combined with the oil system output will total the correct 25:1.

■ OMC advises that although additional oil can be mixed with gas to achieve a 25:1 ratio for 737cc engines equipped with the VR02 system, and that this should be done for high performance applications, it is not necessary for severe service. On VR02 motors used in commercial, rental or extended severe service other than high performance applications, no additional oiling is necessary.

• 25/35 hp (500/565cc) motors, require a 25:1 ratio for high performance conditions. Most of these models are equipped with an oil mixing unit. On models equipped with an oil mixing unit, a tank mixture of 50:1 combined with the oil system output will total the correct 25:1. No additional oiling is necessary for these engines when used in commercial, rental or extended severe service **other** than high performance applications.
• 25-70 hp (913cc) motors, require a 25:1 ratio for high performance conditions. Most of these models are equipped with the VRO2 oiling system. On models equipped with the automatic oiling system, a tank mixture of 50:1 combined with the oil system output will total the correct 25:1. No additional oiling is necessary for engines equipped with the VRO2 system when used in commercial, rental or extended severe service **other** than high performance applications.

■ All motors covered by this manual require a 25:1 ratio during the first 20 hours of break-in. If equipped with an oiling system, make sure the system is operating properly (by verifying that the level in the tank dropped during that 20 hours of use) before using untreated gasoline in the fuel tank.

Pre-Mix

◆ See Figure 61

Mixing the engine lubricant with gasoline before pouring it into the tank is by far the simplest method of lubrication for 2-stroke outboards. However, this method is the messiest and causes the most amount of harm to our environment.

The most important part of filling a pre-mix system is to determine the proper fuel/oil ratio. Most operating conditions require a 50:1 ratio (that is 50 parts of fuel to 1 part of oil). Consult the information in this section on Fuel:Oil Ratio and your owner's manual to determine what the appropriate ratio should be for your engine.

The procedure itself is uncomplicated, but you've got a couple options depending on how the fuel tank is set-up for your boat. To fill an empty portable tank, add the appropriate amount of oil to the tank, then add gasoline and close the cap. Rock the tank from side-to-side to gently agitate the mixture, thereby allowing for a thorough mixture of gasoline and oil. When just topping off built-in or larger portable tanks, it is best to use a separate 3 or 6 gallon (11.4 or 22.7 L) mixing tank in the same manner as the portable tank noted earlier. In this way a more exact measurement of fuel can occur in 3 or 6 gallon increments (rather than just directly adding fuel to the tank and realizing that you've just added 2.67 gallons of gas and need to

Fig. 61 Either add the oil and gasoline at the same time, or add the oil first, then add the gasoline to ensure proper mixing

2-20 MAINTENANCE & TUNE-UP

ad, uh, a little less than 8 oz of oil, but exactly how many ounces would that be?) Use of a mixture tank will prevent the need for such mathematical equations. Of course, the use of a mixing tank may be inconvenient or impossible under certain circumstances, so the next best method for topping off is to take a good guess (but be a little conservative to prevent an excessively rich oil ratio). Either add the oil and gasoline at the same time, or add the oil first, then add the gasoline to ensure proper mixing. For measurement purposes, it would obviously be more exact to add the gasoline first, then add a suitable amount of oil to match it. The problem with adding gasoline first is that unless the tank could be thoroughly agitated afterward (and that would be **really** difficult on built-in tanks), the oil might not mix properly with the gasoline. Don't take that unnecessary risk.

To determine the proper amount of oil to add to achieve the desired fuel:oil ratio, refer to the Fuel:Oil Ratio chart at the end of this section.

Oil Injection

◆ See Figure 62

Most outboard manufacturers use a mechanically driven oil pump mounted on to the powerhead that is connected to the throttle by way of a linkage arm. The system is powered by the crankshaft, which drives a gear in the pump, creating oil pressure. As the throttle lever is advanced to increase engine speed, the linkage arm also moves, opening a valve that allows more oil to flow into the oil pump.

Most mechanical-injection systems incorporate low-oil warning alarms that are also connected to an engine-overheating sensor. Also, these systems may have a built-in speed limiter. This sub-system is designed to reduce engine speed automatically when oil problems occur. This important feature goes a long way toward preventing severe engine damage in the event of an oil injection problem.

The procedure for filling these systems is simple. Most of the OMC systems use a remote oil tank and a connecting hose. The tank contains a filler cap that is removed in order to add oil to the tank. EVERY time the motor is operated, check the oil level. Whenever oil is added, place a piece of tape on the tank to mark the level and watch how fast it drops in relation to engine usage (hours and fuel consumption). Watch for changes in usage patterns that could indicate under or over oiling. Especially with a system that suddenly begins to deliver less oil, you could save yourself significant engine damage by discovering a problem that could have starved the motor for lubrication.

Should the oil hose become disconnected or suffer a break/leak, the oil prime might be lost. If so, the system should be primed **before** priming the fuel system and starting the engine. More details on servicing the oiling system are found in the Lubrication section of this manual.

■ **It is highly advisable to carry several spare bottles of 2-stroke oil with you onboard. Even in the event of an oil system failure, oil can be added to a fuel tank (in the proper ratio) in order to limp the boat and motor safely home.**

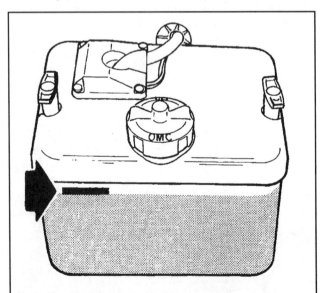

Fig. 62 Mark the oil level with a piece of tape and watch for consumption patterns

Engine Oil and Filter (4-Stroke)

OIL RECOMMENDATIONS

For all 4-stroke motors covered by this manual OMC recommends the use of Johnson or Evinrude brand Ultra 4-Stroke oil. When this oil is used, the oil can be changed after every 200 hours of operation (or at the end of each season, whichever comes first). If this oil is not available, OMC advises that a high quality oil of the correct viscosity can be substituted. For 5-15 hp motors, use an SAE 10W-30 SG or SH (or latest superceding oil type) motor oil. For the 40-70 hp motors, use an SAE 10W-40 SG or SH (or latest superceding oil type) motor oil.

■ **On the smaller 4-stroke motors covered by this manual (5-15 hp models) the manufacturer recommends switching to an alternate weight oil under certain severe operating conditions. When operating these models in conditions such as under constant heavy loads, or in sea-water temperatures above 68°F (20°C), a high-quality/high-detergent SAE 10W-40 or SAE 10W-50 should be used to provide better engine protection. Should the engine suffer from high oil consumption, even under normal operating conditions, use SAE 10W-50 to slow oil burning.**

The Society of Automotive Engineers (SAE) grade number indicates the viscosity of the engine oil; its resistance to flow at a given temperature. The lower the SAE grade number, the lighter the oil. For example, the mono-grade oils begin with SAE 5 weight, which is a thin light oil, and continue in viscosity up to SAE 80 or 90 weight, which are heavy gear lubricants. These oils are also known as "straight weight", meaning they are of a single viscosity, and do not vary with engine temperature.

Multi-viscosity oils offer the important advantage of being adaptable to temperature extremes. These oils have designations such as 10W-40, 20W-50, etc. The 10W-40 means that in winter (the "W" in the designation) the oil acts like a thin 10 weight oil, allowing the engine to spin easily when cold and offering rapid lubrication. Once the engine has warmed up, however, the oil acts like a straight 40 weight, maintaining good lubrication and protection for the engine's internal components. A 20W-50 oil would therefore be slightly heavier than and not as ideal in cold weather as the 10W-40, but would offer better protection at higher rpm and temperatures because when warm it acts like a 50 weight oil. Whichever oil viscosity you choose when changing the oil, make sure you are anticipating the temperatures your engine will be operating in until the oil is changed again.

The American Petroleum Institute (API) designation indicates the classification of engine oil used under certain given operating conditions. Only oils designated for use "Service SG, SH" or greater should be used. Oils of the SG, SH or its superseding oil type perform a variety of functions inside the engine in addition to the basic function as a lubricant. Through a balanced system of metallic detergents and polymeric dispersants, the oil prevents the formation of high and low temperature deposits and also keeps sludge and particles of dirt in suspension. Acids, particularly sulfuric acid, as well as other by-products of combustion, are neutralized. Both the SAE grade number and the APE designation can be found on top of the oil can.

CHECKING OIL LEVEL

◆ See Figures 63 thru 67

One of the most important service items for a 4-stroke engine is maintaining the proper level of fresh, clean engine oil in the crankcase. Be certain to check the oil level both before and after each time the boat is used. In order to check the oil level the motor must be placed in the full vertical position. Because it takes some time for the oil to settle (and at least partially cool), the engine must be shut off for at least 30 minutes before an accurate reading can be attained. If the boat is trailered, use the time for loading the boat onto the trailer and prepping the trailer for towing to allow the motor to cool. If the boat is kept in the water, take some time around the dock to secure lines, stow away items kept onboard and clean up the deck while waiting for the oil to settle/cool.

✱✱ WARNING

Running an engine with an improper oil level can cause significant engine damage. Although it is typically worse to run an engine with abnormally low oil, it can be just as harmful to run an engine that is overfilled. Don't take that risk, make checking the engine oil a regular part of your launch and recovery/docking routine.

MAINTENANCE & TUNE-UP 2-21

Fig. 63 On 5/6 hp (128cc) motors, the engine oil dipstick is located on the port side...

Fig. 64 ...while the oil filler cap is found on the starboard side

Fig. 65 The oil dipstick and filler cap are both found on the port side of 9.9/15 hp (305cc) motors

All motors covered by this manual are equipped with an automotive-style dipstick and oil filler cap located on the powerhead. The engine cover must be removed for access, but once removed it should be easy to locate the dipstick and filler cap if you look in the right spot (as they vary with the engine size):

- For 5/6 hp (128cc) and 8/9.9 hp (211cc) models, the oil dipstick is on the port side of the powerhead (sticking up from between the manual starter housing and carburetor for 5/6 hp motors or between the intake and carburetor on 8/9.9 hp motors.) The oil filler cap is at the top of the starboard side (on a flat boss above the ignition coil).
- For 9.9/15 hp (305cc) models, the oil dipstick is on the port side of the powerhead, slightly aft of the oil filter. The oil filler cap is slightly in front of the dipstick, on a flat boss slightly aft of and above the oil filter.
- For 40/50 hp (815cc) models, the oil dipstick is on the lower port side of the powerhead, while the oil filler cap is found on top of the rocker arm cover (at the top rear of the motor.)
- For 70 hp (1298cc) models, the oil dipstick is on the lower rear/port side of the powerhead, while the oil filler cap is found toward the bottom of the rocker arm cover (at the rear of the motor.)

1. Make sure the engine is in the full vertical position and has been shut off for at least 30 minutes. If possible, get in the habit of checking the oil with the engine cold from sitting overnight
2. Remove the engine cover.
3. Carefully pull the engine crankcase oil dipstick from the port side of the engine.
4. Wipe all traces of oil off the dipstick using a clean, lint free rag or cloth, then re-insert dipstick back into its opening until it is fully seated. Then, pull the dipstick out from the crankcase again and hold it vertically with the bottom end facing down in order to prevent a false oil reading.

■ Forget how your dad or buddy first taught you to read the level on a dipstick. It may be more convenient to hold it horizontally, but laying it down like that could allow oil to flow UPWARD giving a false high, or worse, false acceptable reading when in fact your engine needs oil. Last time we checked, oil won't flow UP a dipstick held vertically (but the high point of the oil will remain wet in contrast to the dry portion of the stick immediately above the wet line). So hold the dipstick vertically and you'll never run your engine with insufficient oil when you thought it was full.

5. If the oil level is at or slightly below the top or **FULL** mark on the dipstick, the oil level is fine. If not, add small amounts of oil through the filler cap until the level is correct. Add oil slowly, giving it time to settle into the crankcase before rechecking and again, don't overfill it either.

■ Dipstick markings will vary slightly from model-to-model. Some, like the 5/6 hp (128cc) and 8/9.9 hp (211cc) models are normally equipped with an L (low) and F (full), while the 9.9/15 hp (305cc) model dipsticks usually spell out the words low and full. The 40/50 hp motors are equipped with add and full marks that contain a crosshatched area between them. The crosshatched area is the "acceptable" operating range, but try to maintain the level towards the top of the markings. As for the 70 hp motor, the dipstick normally contains just two dots, the bottom one for add and the top for full.

6. Visually check the oil on the dipstick for water (a milky appearance will result from contamination with moisture) or a significant fuel odor. Both are signs that the powerhead likely needs overhaul to prevent damage.

Fig. 66 On 70 hp motors, the oil filler cap is towards the bottom of the rocker cove

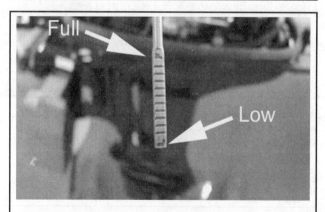

Fig. 67 Although the markings vary, dipsticks will contain a full and a low or add mark

7. Insert and properly seat the oil dipstick into the powerhead when you are finished. If removed, install the oil fill cap and rotate it until it gently locks into position.

OIL CHANGE & FILTER SERVICE

◆ See Figures 68 thru 73

Next to regular fluid level checks, the most important way to maintain a 4-stroke outboard motor is to change the engine crankcase oil (and change or clean the filter, as applicable) on a regular basis. When using Johnson or Evinrude brand Ultra 4-stroke crankcase oil this service should be performed every 200 hours. When another brand of oil is used, the manufacturer recommends that this time be cut to only 100 hours. Of course, no matter what brand is used, the oil should be changed at the end of each season,

2-22 MAINTENANCE & TUNE-UP

immediately before the motor is placed into storage. That is, as long as you use an equivalent high-quality, high-detergent oil of the proper viscosity (SAE 10W-30 for 5-15 hp motors, or SAE 10W-40 for 40-70 hp motors.) For more information regarding engine oil, refer to OIL RECOMMENDATIONS earlier in this section.

✶✶ WARNING

Research from experts who deal with these motors every day tells us that some models are especially subject to camshaft lobe wear if the engine oil is not changed regularly. During each pre-season tune-up, watch for excessive changes in valve clearance as possible signs of wear. If found, change the oil more frequently or, if oil other than the manufacturer's recommended brand is being used, try changing the type of oil too.

Whenever the engine oil is drained, the oil filter should also be serviced. The models covered by this manual utilize two types of oil filters. The smallest motors, 5/6 hp (128cc) and 8/9.9 hp (211cc) models, utilize a reusable element mounted in a housing protruding form the lower portion of the powerhead. The rest of the motors, 9.9/15 hp (305cc) and the 40-70 hp models, utilize a disposable, automotive style, spin-on filter mounted to the side of the powerhead. The best method to remove the spin-on filter (resulting in fewest skinned knuckles) is a filter wrench, and our preference is the cap style that fits over the end of the filter. When purchasing a replacement oil filter check your local marine dealer or automotive parts dealer for a cap wrench that fits the filter.

Most people who have worked on their own machines, whether that is tractors, motorcycles, cars/trucks or boat motors, will tell you that oil should be changed hot. This seems to have always been the popular method, and it works well since hot oil flows better/faster and may remove more deposits that are still held in suspension. Of course, hot oil can be messy or even a bit dangerous to work with. Coupled with the sometimes difficult method of draining oil from an outboard, this might make it better in some instances to drain the oil cold. Of course, if this is desired, you'll have to leave more time for the oil to drain completely, thereby removing as much contaminants as possible from the crankcase. The choice is really yours, but be sure to take the appropriate steps to protect yourself either way.

■ **If the engine is not being placed in storage after the oil change, it should be run to normal operating temperature (in a test tank or with a flushing device) and inspected for leaks before returning it to service. If the engine is being placed into storage it should also be run using a flush device, but be sure not to run it too long. Just start and run the engine for a few minutes to thoroughly circulate the fresh oil, then prepare it for storage by fogging the motor.**

If you decide to change oil with the engine hot, a source of cooling such as a test tank or flushing hose must be attached to the engine to prevent impeller or powerhead damage when running the engine to normal operating temperature. If you are lucky enough to store the boat (or live) close to waters in which to use the boat, you can simply enjoy a morning, evening or whole day on the water before changing the oil. The amount of time necessary to haul the boat and tow it to your work area should allow the oil to cool enough so that it won't be scalding hot, but still warm enough to flow well.

■ **Although it is not recommended for normal service, the oil CAN be drained without removing the engine cases or servicing the filter. This might be desired if too much oil was added during a routine level check or if a small sample of oil is to be removed for inspection.**

1. Prepare the engine and work area for the oil change by placing the motor in a fully vertical position over a large, flattened cardboard box (which

Fig. 68 The oil drain plug on 4-stroke motors is located near the powerhead/gearcase split line

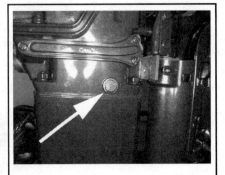

Fig. 69 This 5/6 hp crankcase drain plug is pictured with the lower covers removed

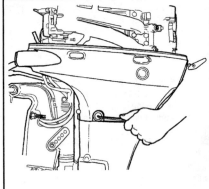

Fig. 70 Use a hex key to loosen the drain plug on fuel injected motors

Fig. 71 The oil filter element is mounted under this hex-headed cap on 5/6 and 8/9.9 hp motors

Fig. 72 Most 4-strokes utilize a spin-on oil filter - 9.9/15 hp (305cc) motor rigged for remote controls shown

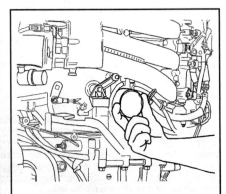

Fig. 73 The oil filter on the 40/50 hp fuel injected motors is mounted directly under the intake manifold

MAINTENANCE & TUNE-UP

can be used to catch any dripping oil missed by the drain pan). Have a drain pan, a few quarts larger than the capacity (refer to the Capacities - Four Stroke Engines Chart in this section) for the motor and a lot of clean rags or disposable shop towels handy.

■ **It may be possible to access some filters (such as on tiller control 9.9/15 hp motors) without completely removing the lower covers or by just removing one side (as on 8/9.9 hp motors). But although it may be possible, it is usually easier (and much more convenient/less messy) to just go ahead and remove both lower covers. For details, refer to the Engine Cover procedure in this section.**

2. Remove the upper engine cover and locate the oil filter. Determine if it is necessary to remove the lower side engine cover(s) based on access to the filter. If necessary or desired, refer to the Engine Cover procedure for more details.

  a. On 5/6 hp (128cc) and 8/9.9 hp (211cc) motors with reusable filter elements, the filter is found under a large hex-head cap screwed onto a bore. The bore is towards the aft, starboard side of the powerhead, directly in front of the ignition coil on 5/6 hp motors or on the aft, port side, right at the bottom of the powerhead, immediately below the rear of the intake manifold on 8/9.9 hp motors. Though the starboard lower engine cover must be removed in order to access the filter on 5/6 hp motors, the filter can be accessed by simply dislodging the port lower cover on 8/9.9 hp motors. Of course, as we've said, it might be easier to just completely remove the cover in the end.

  b. On 9.9/15 hp (305cc) and 40/50 hp motors, the disposable spin-on oil filter is found at the center, port side of the powerhead. It is just above the point where the dipstick inserts on 9.9/15 hp motors, or directly underneath the intake manifold (and behind the throttle cable/shift lever shaft) on 40/50 hp motors. On 70 hp motors, the disposable, spin-on filter is located on the lower center, starboard side of the powerhead. On these motors it is directly below the fuel injection Electronic Control Unit (ECU), but is still right behind the throttle cable/shift lever shaft.

■ **The drain plug requires a screwdriver for 5-15 hp motors or a suitably sized Allen wrench on 40-70 hp motors.**

3. Locate and remove the oil drain plug and gasket. For 5-15 hp motors the plug is on the starboard lower side of the engine just above or at the gearcase. For 40-70 hp motors the plug is on the port, lower side of the engine, towards the front of the motor. In order to improve oil flow, remove the oil fill cap.

4. Either hold the drain pan tightly against the side of the motor, or allow the oil to run down the side of the motor and drip into the pan. The later method is preferred if draining the engine cold as the oil will require more time to drain than you will want to stand there with the pan in your hand.

5. Inspect the drain plug and gasket for signs of damage. Replace the plug or gasket if any damage is found. Also, watch the draining oil for signs of contamination by moisture (a milky appearance will result), by fuel (a strong odor and thinner running oil would be present) or signs of metallic flakes/particles. A small amount of tiny metallic particles is a sign of normal wear, but large amounts or large pieces indicate internal engine damage and the need for an overhaul to determine and rectify the cause.

6. When it appears that the oil has drained, tilt the engine slightly and pivot it toward the drain plug side to ensure complete oil drainage. Clean the drain plug, the engine and the gearcase. Place a new gasket onto the drain plug then carefully thread the plug into the opening. Tighten the plug securely.

■ **Although it is not absolutely necessary to replace the gasket each time, it is a cheap way to help protect against possible leaks. We think it is a good idea.**

7. For models equipped with a reusable filter element (5/6 hp and 8/9.9 hp motors) remove and service the filter as follows:

  a. On 5/6 hp (128cc) models, unbolt the ignition coil and position it out of the way (with the wiring still connected) for access. Do not lose bolts, star washers or, most importantly, the spacer(s) located behind the coil. If necessary, refer to the Ignition Coil removal and installation procedures.

✱✱ WARNING

Failure to properly reinstall the ground strap that attaches to the ignition coil (along with the screw and star washer) can result in coil damage during engine operation.

  b. Hold a small drain pan or rag under the oil filter cap, then use a socket or wrench to loosen the cap. Unthread the cap by hand, then remove the cap and filter element assembly.

  c. Rinse the element using solvent and dry it using low pressure compressed air. If compressed air is not available, allow it to air dry for at least 15 minutes. Inspect the element for signs of clogging or damage and replace, if found.

  d. Apply a light coating of fresh 4-stroke engine oil to the filter element and O-rings, then install the element into the bore and thread the cap in place. Tighten the cap 1/4 turn after the O-ring contacts the base.

■ **Although the procedure for spin-on filters talks about placing a shop rag under the filter while it is removed, there is an alternate method to prevent a mess. If desired, loosen the filter slightly with a cap wrench, then slide a disposable Zip-Lock® or similar food storage bag completely over the filter and unthread it into the bag. Position the shop rag anyway, just to be sure to catch any stray oil that escapes. With a little practice, you'll find this method can be the best way to remove oil filters.**

8. For models equipped with a disposable, spin-on filter element (9.9/15 hp and 40-70 hp motors) remove and service the filter as follows:

  a. Position a shop rag underneath, then place the oil filter wrench onto filter element.

  b. Loosen the spin-on element by turning the filter wrench counterclockwise, then remove the wrench and finish unthreading the element by hand. Remove the filter from the powerhead and clean up any spilled oil.

  c. Make sure the rubber gasket is not stuck to the oil filter mounting surface, then use a lint free shop rag to clean all dirt and oil from mounting surface.

  d. Apply a thin coating of engine oil to the sealing ring of the new oil filter, then thread the filter onto the adapter until the sealing washer touches the mounting surface. Tighten the filter by hand an additional 1/4-2/3 turn.

9. Clean up any spilled oil, and then install the lower engine cover(s) by carefully aligning the screw holes in the two covers, while also aligning the lower cover mating surfaces. Be sure to install and tighten the screws securely. Also, don't forget to install any additional removed components such as the aft cover latch and/or the cover seal.

✱✱ WARNING

Be careful not to pinch and damage any hoses, cables or wiring when installing the engine covers.

10. Refill the engine through the oil filler cap as described under Checking Engine Oil in this section. Add the oil gradually, checking the oil level frequently. Add oil until the level reaches the upper dipstick or **Full** mark.

11. Provide a temporary cooling system to the engine as detailed under Flushing The Cooling System, then start the engine and run it to normal operating temperature while visually checking for leakage.

■ **If the engine is being placed into storage, don't run the motor too long, just long enough to use a can of fogging spray. Between the fresh oil circulated through the motor and the fogging spray coating the inside of the intake and combustion chambers you motor should sleep like a baby until next season.**

12. Stop the motor and allow it to cool, then properly re-check the oil level after it has settled again into the crankcase.

Gearcase (Lower Unit) Oil

◆ See Figures 74 and 75

Regular maintenance and inspection of the lower unit is critical for proper operation and reliability. A lower unit can quickly fail if it becomes heavily contaminated with water or excessively low on oil. The most common cause of a lower unit failure is water contamination.

Water in the lower unit is usually caused by fishing line or other foreign material, becoming entangled around the propeller shaft and damaging the seal. If the line is not removed, it will eventually cut the propeller shaft seal and allow water to enter the lower unit. Fishing line has also been known to cut a groove in the propeller shaft if left neglected over time. This area should be checked frequently.

2-24 MAINTENANCE & TUNE-UP

Fig. 74 This lower unit was destroyed because the bearing carrier froze due to lack of lubrication

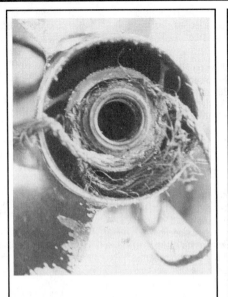

Fig. 75 Fishing line entangled behind the prop can actually cut through the seal

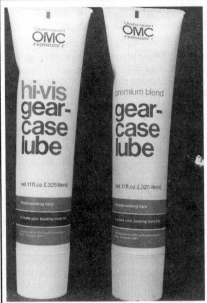

Fig. 76 Use OMC Ultra-HPF, OMC Hi-Vis or an equivalent marine gear oil

OIL RECOMMENDATIONS

◆ See Figure 76

Use only OMC Ultra-HPF or an equivalent high quality, marine gearcase lubricant that meets GL5 specifications. OMC Hi-Vis gearcase lube may be used as a substitute if Ultra-HPF is not available. In both cases, these oils are proprietary lubricants designed to ensure optimal performance and to minimize corrosion in the lower unit.

■ Remember, it is this lower unit lubricant that prevents corrosion and lubricates the internal parts of the drive gears. Lack of lubrication due to water contamination or the improper type of oil can cause catastrophic lower unit failure.

CHECKING GEARCASE OIL LEVEL & CONDITION

◆ See Figure 77    EASY

Visually inspect the gearcase before and after each use for signs of leakage. At least monthly, or as needed, remove the gearcase level plug in order to check the lubricant level and condition as follows:

1. Position the engine in the upright position with the motor shut off for at least 1 hour. Whenever possible, checking the level overnight cold will give a true indication of the level without having to account for heat expansion.
2. Disconnect the negative battery cable or remove the propeller for safety.

✳✳ CAUTION

Always observe extreme care when working anywhere near the propeller. Take steps to ensure that no accidental attempt to start the engine occurs while work is being performed or remove the propeller completely to be safe.

3. Position a small drain pan under the gearcase, then unthread the drain/filler plug at the bottom of the housing and allow a small sample (a teaspoon or less) to drain from the gearcase. Quickly install the drain/filler plug and tighten securely.
4. Examine the gear oil as follows:
    a. Visually check the oil for obvious signs of water. A small amount of moisture may be present from condensation, especially if a motor has been stored for some time, but a milky appearance indicates that either the fluid has not been changed in ages or the gearcase allowing some water to intrude. If significant water contamination is present, the first suspect is the propeller shaft seal.
    b. Dip an otherwise clean finger into the oil, then rub a small amount of the fluid between your finger and your thumb to check for the presence of debris. The lubricant should feel smooth. A **very** small amount of metallic shavings may be present, but should not really be felt. Large amounts of grit or metallic particles indicate the need to overhaul the gearcase looking for damaged/worn gears, shafts, bearings or thrust surfaces.

■ If a large amount of lubricant escapes when the level/vent plug is removed, either the gearcase was seriously overfilled on the last service, the crankcase is still too hot from the last use (and the fluid is expanded) or a large amount of water has entered the gearcase. If the later is true, some water should escape before the oil and/or the oil will be a milky white in appearance (showing the moisture contamination).

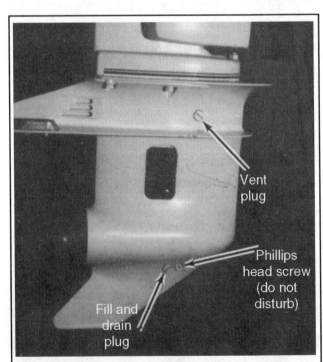

Fig. 77 Although the exact location varies slightly, the vent/level plug is always towards the top, while the fill and drain plug is towards the bottom of the gearcase

MAINTENANCE & TUNE-UP 2-25

5. Next, remove the level/vent plug from the top of the gearcase and ensure the lubricant level is up to the bottom of the level/vent plug opening. A very small amount of fluid may be added through the level plug, but larger amounts of fluid should be added through the drain/filler plug opening to make certain that the case is properly filled. If necessary, add gear oil until fluid flows from the level/vent opening. If much more than 1 oz. (29 ml) is required to fill the gearcase, check the case carefully for leaks. Install the drain/filler plugs and/or the level/vent plug, then tighten both securely.

■ **One trick that makes adding gearcase oil less messy is to install the level/vent plug BEFORE removing the pump from the drain/filler opening and threading the drain/filler plug back into position.**

6. Once fluid is pumped into the gearcase, let the unit sit in a shaded area for at least 1 hour for the fluid to settle. Recheck the fluid level and, if necessary, add more lubricant.
7. Install the propeller and/or connect the negative battery cable, as applicable.

DRAINING AND FILLING

◆ See Figures 78, 79 and 80

✳✳ CAUTION

The EPA warns that prolonged contact with used engine oil may cause a number of skin disorders, including cancer! You should make every effort to minimize your exposure to used engine oil. Protective gloves should be worn when changing the oil. Wash your hands and any other exposed skin areas as soon as possible after exposure to used engine oil. Soap and water or waterless hand cleaner should be used.

1. Place a suitable container under the lower unit.
2. Loosen the oil level/vent plug on the lower unit. This step is important! If the oil level/vent plug cannot be loosened or removed, you cannot complete lower unit lubricant service.

■ **Never remove the vent or filler plugs when the lower unit is hot. Expanded lubricant will be released through the hole.**

3. Remove the drain/filler plug from the lower end of the gear housing followed by the oil level/vent plug.
4. Allow the lubricant to completely drain from the lower unit.
5. If applicable, check the magnet end of the drain screw for metal particles. Some amount of metal is considered normal wear is to be expected but if there are signs of metal chips or excessive metal particles, the gearcase needs to be disassembled and inspected.

6. Inspect the lubricant for the presence of a milky white substance, water or metallic particles. If any of these conditions are present, the lower unit should be serviced immediately.
7. Place the outboard in the proper position for filling the lower unit. The lower unit should not list to either port or starboard and should be completely vertical.
8. Insert the lubricant tube into the oil drain hole at the bottom of the lower unit and inject lubricant until the excess begins to come out the oil level hole.

■ **The lubricant must be filled from the bottom to prevent air from being trapped in the lower unit. Air displaces lubricant and can cause a lack of lubrication or a false lubricant level in the lower unit.**

9. Oil should be squeezed in using a tube or with the larger quantities, by using a pump kit to fill the gearcase through the drain plug.

■ **One trick that makes adding gearcase oil less messy is to install the level/vent plug BEFORE removing the pump from the drain/filler opening and threading the drain/filler plug back into position.**

10. Using new gaskets (washers) install the oil level/vent plug first, then install the oil fill plug.
11. Wipe the excess oil from the lower unit and inspect the unit for leaks.
12. Place the used lubricant in a suitable container for transportation to an authorized recycling facility.

Fuel Filter

A fuel filter is designed to keep particles of dirt and debris from entering the carburetor(s) or the fuel injection system and clogging the tiny internal passages of each. A small speck of dirt or sand can drastically affect the ability of the fuel system to deliver the proper amount of air and fuel to the engine. If a filter becomes clogged, the flow of gasoline will be impeded. This could cause lean fuel mixtures, hesitation and stumbling and idle problems in carburetors. Although a clogged fuel passage in a fuel injected engine could also cause lean symptoms and idle problems, dirt can also prevent a fuel injector from closing properly. A fuel injector that is stuck partially open by debris will cause the engine to run rich due to the unregulated fuel constantly spraying from the pressurized injector.

Regular cleaning or replacement of the fuel filter (depending on the type or types used) will decrease the risk of blocking the flow of fuel to the engine, which could leave you stranded on the water. It will also decrease the risk of damage to the small passages of a carburetor or fuel injector that could require more extensive and expensive replacement. Keep in mind that fuel filters are usually inexpensive and replacement is a simple task. Service your fuel filter on a regular basis to avoid fuel delivery problems.

The type of fuel filter used on your engine will vary not only with the year

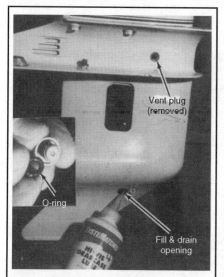

Fig. 78 Gearcase oil is pumped or squeezed into the lower unit through the filler opening, while the vent opening is removed to let air escape

Fig. 79 The exact location of the filler opening (1) and the vent opening (2) will vary slightly - small hp gearcase shown

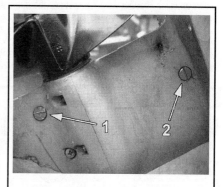

Fig. 80 Notice how the filler opening (1) is above the skeg rear on this 70 hp motor, while the vent opening (2) is still towards the front

2-26 MAINTENANCE & TUNE-UP

and model, but also with the accessories and rigging. Because of the number of possible variations it is impossible to accurately give instructions based on model. Instead, we will provide instructions for the different types of filters the manufacturer used on various families of motors or systems with which they are equipped. To determine what filter(s) are utilized by your boat and motor rigging, trace the fuel line from the tank to the fuel pump and then from the pump to the carburetors (or premix oiling system, which ever is applicable). The fuel injected motors are listed separately, as their design does not vary in the same way as the carbureted motors.

In addition to the fuel filter mounted on the engine, a filter is usually found inside or near the fuel tank. Because of the large variety of differences in both portable and fixed fuel tanks, it is impossible to give a detailed procedure for removal and installation. Most in-tank filters are simply a screen on the pick-up line inside the fuel tank. Filters of this type usually only need to be cleaned and returned to service (assuming they are not torn or otherwise damaged). Fuel filters on the outside of the tank are typically of the inline type and are replaced by simply removing the clamps, disconnecting the hoses and installing a new filter. When installing the new filter, make sure the arrow on the filter points in the direction of fuel flow.

SERVICING FUEL FILTERS ON CARBURETED MOTORS

✱✱ CAUTION

Observe all applicable safety precautions when working around fuel. Whenever servicing the fuel system, always work in a well-ventilated area. Do not allow fuel spray or vapors to come in contact with a spark or open flame. Do not smoke while working around gasoline. Keep a dry chemical fire extinguisher near the work area. Always keep fuel in a container specifically designed for fuel storage; also, always properly seal fuel containers to avoid the possibility of fire or explosion.

Integral Tank Models

◆ See Figures 81 and 82

The smallest motors covered by this manual, the Colt/Junior and 2-5 hp (78-100cc) 2-strokes are equipped with integral fuel tanks. The smallest of these motors (such as the 78cc) gravity feed the carburetor through a fuel valve in the bottom of the tank and are therefore not equipped with a fuel pump. The fuel tanks on these models are equipped with a fuel valve/filter assembly (78cc models). The filter is serviced by removing the assembly from the tank and cleaning the element or replacing it. The larger of these small, single and twin cylinder motors are equipped with a fuel pump and utilize an inline filter between the tank and the pump assembly. Inline filters like this are serviced by removing them from lines and replacing them. They are usually sealed and cannot be cleaned. The integral tank models that are equipped with fuel pumps usually are also equipped with a filter screen under the pump inlet cover. For details on how to service the filter screen, refer to Fuel Pump Filters (Remote Tank Models) in this section.

Fuel Pump Filters (Remote Tank Models)

◆ See Figures 83 thru 87

Most of the models covered by this manual are equipped with a small, flat, mechanical or vacuum (pulse) driven fuel pump mounted somewhere on the powerhead. Although the exact shape and design of this pump varies slightly from model-to-model, for this discussion, they are serviced virtually the same way and we'll refer to them as Type A fuel pumps. The various Type A Johnson/Evinrude mechanical fuel pumps normally contain a serviceable fuel filter screen mounted just underneath the fuel inlet cover. On all versions, the cover is connected to the fuel inlet hose from the fuel tank. Additionally, on all models except the smallest 4-strokes (5/6 hp 128cc and 8/9.9 hp 211cc motors) the cover is either round or a rounded square and is retained by a single bolt at the center.

The smallest 4-strokes (5/6 and 8/9.9 hp models) use four screws, but only two of the screws retain the cover, while the other two hold the fuel pump body. Of the 2 cover screws, there is one at the top and one at the bottom of the cover (looking at the pump as if the fuel inlet/outlets are on the bottom). These models may or may not be equipped with a screen under the cover, so check with your dealer before removing it. If they are not equipped with a screen, you should find an inline filter somewhere before the pump, usually in the line between the engine fuel connector and the fuel pump.

To service the fuel inlet screen on Type A fuel pumps, remove the inlet cover screw(s), then carefully separate the inlet cover, gasket or O-ring and screen from the fuel pump body. Clean the screen using a suitable solvent and blow dry with low pressure compressed air or allow it to air dry. Once the screen is dry, check it carefully for clogs or tears and replace, if necessary. Depending on the gasket material and condition it may be reused, but it is normally best (and safest) to simply replace the gasket(s).

Some of the larger 2-stroke motors covered by this manual may be equipped with a larger, more complicated fuel pump that we'll refer to as Type B. We've provided a photo for visual identification. On these pumps an inline filter is normally used somewhere between the tank and the fuel pump. Service of the inline filter is normally limited to replacement. Inline filters should be replaced annually and anytime fuel delivery/starvation problems are suspected.

■ Some Type A fuel pump motors will also be equipped with an inline filter. When present be sure to replace all inline filters at least annually.

Inline Filters

◆ See Figures 88 and 89

As noted earlier, many of the models covered by this manual are equipped with an inline filter. On some the filter is used in lieu of a fuel pump mounted or fuel tank mounted filter, but for many models the inline filter is used as an additional line of defense. We also noted earlier that most inline filters are of the sealed canister type (plastic or metallic) and cannot be cleaned, so service is usually limited to replacement. Because of the relative ease and relatively low expense of a filter (when compared with the time and hassle of a carburetor overhaul) we encourage you to replace the filter at least annually.

When replacing the filter, release the hose clamps (they are usually equipped with spring-type clamps that are released by squeezing the tabs

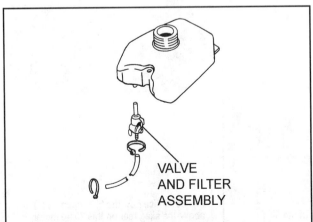

Fig. 81 Some motors with an integral fuel tank use a fuel valve/filter assembly

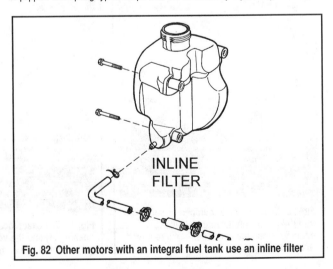

Fig. 82 Other motors with an integral fuel tank use an inline filter

MAINTENANCE & TUNE-UP 2-27

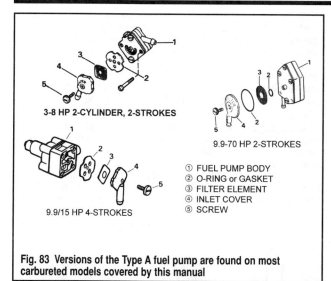

Fig. 83 Versions of the Type A fuel pump are found on most carbureted models covered by this manual

Fig. 84 Remove the bolt(s) securing the Type A pump cover...

Fig. 85 ...then remove the inlet cover to access the filter element

Fig. 86 A few of the smallest 4-strokes contain an inline filter instead of the pump element

Fig. 87 Models with Type B fuel pumps utilize inline filters

using a pair of pliers) and slide them back on the hose, past the raised portion of the filter inlet/outlet nipples. Once a clamp is released, position a small drain pan or a shop towel under the filter and carefully pull the hose from the nipple. Allow any fuel remaining in the filter and fuel line to drain into the drain pan or catch fuel with the shop towel. Repeat on the other side, noting which fuel line connects to which portion of the filter (for assembly purposes). Inline filters are usually marked with an arrow indicating fuel flow. The arrow should point towards the fuel line that runs to the motor (not the fuel tank).

Before installation of the new filter, make sure the hoses are in good condition and not brittle, cracking and otherwise in need of replacement. During installation, be sure to fully seat the hoses, then place the clamps over the raised portions of the nipples to secure them. Spring clamps will weaken over time, so replace them if they've lost their tension. If wire ties or adjustable clamps were used, be careful not to overtighten the clamp. If the clamp cuts into the hose, it's too tight; loosen the clamp or cut the wire tie (as applicable) and start again.

A few models (mostly commercial motors) are equipped with a serviceable inline filter. These can be identified by their design and shape, which varies from the typical inline filter. A typical, disposable inline filter will have a simple round canister on which the fuel lines attach to either end (or a few, with both lines on one end). The serviceable inline filters used by Johnson/Evinrude have one fuel inlet on the side and the fuel outlet on the top. Serviceable filters consist of an assembly with a knurled cap that threads onto the filter bowl or base. When servicing these filters, disconnect the hose from one end, for instance, remove the fuel pump hose from the cap nipple, then unthread the cap from the base. The filter element is removed with the cap. Clean and inspect the element in the same manner as the fuel pump filter screens described under Fuel Pump Filters (Remote Tank Models) in this section. Replace any damaged filter element. Check the filter gasket and/or O-ring for damage and replace, as necessary. When reconnecting the hose to the nipple, inspect and replace any damaged hose or clamp as you would with any other inline filter.

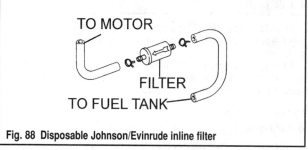

Fig. 88 Disposable Johnson/Evinrude inline filter

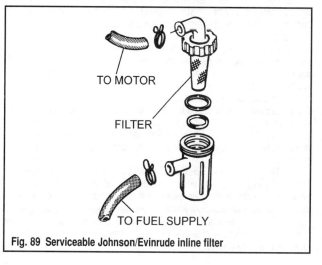

Fig. 89 Serviceable Johnson/Evinrude inline filter

MAINTENANCE & TUNE-UP

SERVICING FUEL FILTERS ON FUEL INJECTED MOTORS

✳✳ CAUTION

Observe all applicable safety precautions when working around fuel. Whenever servicing the fuel system, always work in a well-ventilated area. Do not allow fuel spray or vapors to come in contact with a spark or open flame. Do not smoke while working around gasoline. Keep a dry chemical fire extinguisher near the work area. Always keep fuel in a container specifically designed for fuel storage; also, always properly seal fuel containers to avoid the possibility of fire or explosion.

Fuel injected motors covered by this manual are equipped with two interrelated fuel circuits, the high-pressure and low-pressure systems. The low-pressure system operates essentially the same way as does a carbureted motor's fuel system. A mechanical, engine mounted fuel pump draws fuel from the tank and feeds a mechanical float controlled fuel reservoir. The difference occurs at this point as the reservoir is for the high-pressure circuit and electric high-pressure pump instead of a float bowl attached to a carburetor.

These motors utilize at least 3 fuel filters, two of these filters are inline and are replaced during normal service. One inline filter is used for each fuel circuit. An additional filter screen is mounted on the electric high-pressure fuel pump inlet. Although this screen can be replaced, it is not normally part of maintenance. The pump filter screen can only be replaced once the pump is removed from the vapor separator. Depending on the boat rigging, additional inline filters or tank filter screen may also be present.

Fuel System Pressure

On fuel injected engines, always relieve system pressure prior to disconnecting any high-pressure fuel circuit component, fitting or fuel line. For details, please refer to Fuel System Pressurization under Fuel Injection.

✳✳ CAUTION

Exercise extreme caution whenever relieving fuel system pressure to avoid fuel spray and potential serious bodily injury. Please be advised that fuel under pressure may penetrate the skin or any part of the body it contacts.

To avoid the possibility of fire and personal injury, always disconnect the negative battery cable while servicing the fuel system or fuel system components.

Always place a shop towel or cloth around the fitting or connection prior to loosening to absorb any excess fuel due to spillage. Ensure that all fuel spillage is removed from engine surfaces.

Low-Pressure Filter

◆ See Figure 90

Fuel injected motors utilize an inline, nylon canister to protect the low-pressure fuel circuit. The canister secured to a clamp bracket at the port side front of the powerhead, immediately below the intake manifold. For most applications, the filter is not serviceable and must be replaced if contaminated. But, on some motors, the canister can be opened to allow inspection, cleaning and replacement of the low-pressure filter element. In all cases, service the filter at least annually, every 100 hours of operation or if problems are suspected with the low-pressure circuit.

1. Disconnect the negative battery cable for safety.
2. Tag the hoses attached to the top and side of the fuel filter. Look for an arrow on the canister, it should indicate the fuel line that runs to the fuel vapor separator. The top line normally connects to the low-pressure fuel pump, while the line on the side of the filter usually supplies fuel to the separator.
3. The factory usually secures these fuel lines with spring-type hose clamps. Squeeze the clamp tabs and hold while sliding the clamps up the hose, past the raised nipple on the fitting. If other threaded clamps are used, loosen and slide them back. If wire ties were used, they must be cut away carefully, making sure not to damage the hose. Inspect all metallic clamps for corrosion, lack of spring tension and/or other damage. Replace any faulty or questionable hose clamps.

4. Position a container or shop rag below the fuel filter. Carefully pull the hoses from the fittings, taking care to avoid bending or breaking the fuel hose fittings on serviceable canisters. Check the hoses for cracks or brittle ends and replace any that are worn or damaged.
5. Remove the filter from the bracket on the powerhead. If the filter is serviceable, clean it using a suitable solvent, then blow it dry with low pressure compressed air (or allow it to air dry). Inspect the element of serviceable filters for clogs or tears and replace if damaged.

To install:

6. Position the new or cleaned filter on the powerhead, then attach the fuel hoses as tagged during removal. Secure the hoses using clamps. When using wire ties or threaded clamps, be sure not to over-tighten the clamps, cutting the hoses.
7. Pressurize the fuel system using the fuel primer bulb from the tank line and check for leaks. Observe the fuel hose fittings for fuel leakage and repair any fuel leaks before starting the motor. Clean up any spilled fuel.

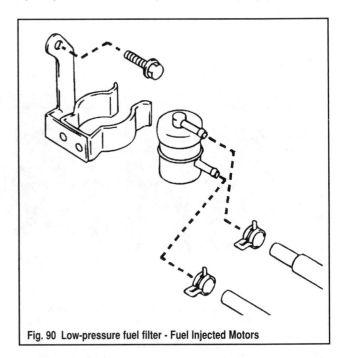

Fig. 90 Low-pressure fuel filter - Fuel Injected Motors

High-Pressure Filter

◆ See Figure 91

On fuel injected motors, the high-pressure circuit is protected by an inline filter canister found between the high-pressure pump and the fuel rail assembly. The canister is attached to the middle, port side of the powerhead, immediately above the fuel rail on 40/50 hp motors or right below the intake manifold on 70 hp motors.

1. Properly relieve the fuel system pressure as described in Fuel System Pressurization under the Fuel Injection section, then disconnect the negative battery cable for safety.

✳✳ CAUTION

Even if you leave the fuel pump fuse (70 hp motors) or fuel pump wiring harness (40/50 hp motors) disconnected it is still a good idea to disconnect the negative battery cable. Remember that sparks are a dangerous source of ignition that could ignite fuel vapors and by removing battery power from the engine components you help minimize the possibility of causing sparks while working on the motor.

2. If necessary, remove the lower engine cover for access, as described in this section under Engine Cover.

■ The filter canister is embossed with markings (IN and OUT) indicating where the fuel line comes IN from the pump or the line OUT to the fuel rail attach. With this said, we would still advice tagging the fuel lines prior to removal to help ensure ease of connection during filter installation.

MAINTENANCE & TUNE-UP 2-29

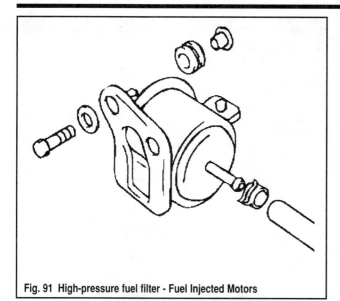

Fig. 91 High-pressure fuel filter - Fuel Injected Motors

3. Tag the fuel hoses and note the filter positioning, then remove the clamps and carefully pull the hoses from the fittings on each end of the filter. Drain any residual fuel from the hoses.

4. Remove the filter from the powerhead (on some models it may be necessary to remove the retaining bolt from the bracket first). Drain residual fuel from the filter.

To install:

5. Position the filter as noted during removal, carefully seat the hoses over the fittings and secure using the clamps. Make sure any spring-type clamps used have not lost their tension. If threaded clamps or wire ties are used, they should be snug, but not tight enough to cut the hose.

■ Upon installation, be certain to connect the hoses as tagged during removal. The hose from the fuel pump connects to the IN fitting, while the fuel rail hose connects to the fitting marked OUT. If the hoses were repositioned while servicing the filter, make sure they are routed as they were prior to removal. This will help ensure that there will be no interference with parts of the motor that could damage the hoses through heat or contact.

6. Secure the filter by tightening the mounting bolt(s).

7. Connect the negative battery cable and either reinstall the fuel pump fuse or reconnect the pump wiring harness, then pressurize the fuel system and check for leakage **before** starting the motor. For details, refer to Fuel System Pressurization, under Fuel System.

Propeller

◆ See Figures 92 thru 94

The propeller is secured to the gearcase propshaft either by a drive pin on Colt/Junior and 2-8 hp 2-stroke and 5/6 hp 4-stroke motors, or by a castellated hex nut on all other motors covered by this manual.

For models secured by a hex nut, the propeller is driven by a splined connection to the shaft and the rubber drive hub found inside the propeller. The rubber hub provides a cushioning that allows softer shifts, but more importantly, it provides some measure of protection for the gearcase components in the event of an impact. On motors where the propeller is retained by a drive pin, impact protection is provided by the drive pin itself. The pin is designed to break or shear when a specific amount of force is applied because the propeller hits something. In both cases (rubber hubs or shear pins) the amount of force necessary to break the hub or shear the pin is supposed to be just less than the amount of force necessary to cause gearcase component damage. In this way, the hope is that the propeller and hub or shear pin will be sacrificed in the event of a collision, but the more expensive gearcase components will survive unharmed. Although these systems do supply a measure of protection, this, unfortunately, is not always the case and gearcase component damage will still occur with the right impact or with a sufficient amount of force.

INSPECTION

◆ See Figures 92, 93 and 94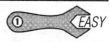

The propeller should be inspected before and after each use to be sure the blades are in good condition. If any of the blades become bent or nicked, this condition will set up vibrations in the motor. Remove and inspect the propeller. Use a file to trim nicks and burrs. Take care not to remove any more material than is absolutely necessary.

** **CAUTION**

Never run the engine with serious propeller damage, as it can allow for excessive engine speed and/or vibration that can damage the motor. Also, a damaged propeller will cause a reduction in boat performance and handling.

Also, check the rubber and splines inside the propeller hub for damage. If there is damage to either of these, take the propeller to your local marine dealer or a "prop shop". They can evaluate the damaged propeller and determine if it can be saved by rehubbing.

Additionally, the propeller should be removed every 100 hours of operation or at the end of each season, whichever comes first for cleaning, greasing and inspection. Whenever the propeller is removed, apply a fresh coating of OMC Triple-Guard or an equivalent water-resistant, marine grease to the propeller shaft and the inner diameter of the propeller hub. This is necessary to prevent possible propeller seizure onto the shaft that could lead to costly or troublesome repairs. Also, whenever the propeller is removed, any material entangled behind the propeller should be removed before any damage to the shaft and seals can occur. This may seem like a waste of time at first, but the small amount of time involved in removing the propeller is returned many times by reduced maintenance and repair, including the replacement of expensive parts.

■ **Propeller shaft greasing and debris inspection should occur more often depending upon motor usage. Frequent use in salt, brackish or polluted waters would make it advisable to perform greasing more often. Similarly, frequent use in areas with heavy marine vegetation, debris or potential fishing line would necessitate more frequent removal of the propeller to ensure the gearcase seals are not in danger of becoming cut.**

Fig. 92 This propeller is long overdue for repair or replacement

Fig. 93 Although minor damage can be dressed with a file...

Fig. 94 ...a propeller specialist should repair large nicks or damage

2-30 MAINTENANCE & TUNE-UP

Clearing the Fish Line Trap

The 8-15 hp (211-304cc) 4-stroke and 9.9-35 hp (216-521cc) 2-stroke motors covered by this manual are equipped with a special propeller thrust washer. It contains an integral fishing line trap to keep line that becomes entangled from cutting the propeller shaft seal. For models so equipped, the manufacturer recommends removing the propeller every 15-20 hours (or anytime fishing line may become entangled) in order to check and clean the trap. Whenever the propeller and thrust washer are removed, the washer must be positioned with the line trap groove facing the gearcase in order to work properly. Always note the direction of the trap groove during removal.

REMOVAL & INSTALLATION

✱✱ WARNING

Do not use excessive force when removing the propeller from the hub as excessive force can result in damage to the propeller, shaft and, even other gearcase components. If the propeller cannot be removed by normal means, consider having a reputable marine shop remove it. The use of heat or impacts to free the propeller will likely lead to damage.

■ Clean and lubricate the propeller and shaft splines using a high-quality, water-resistant, marine grease every time the propeller is removed from the shaft. This will help keep the hub from seizing to the shaft due to corrosion (which would require special tools to remove without damage to the shaft or gearcase.)

Many outboards are equipped with aftermarket propellers. Because of this, the attaching hardware may differ slightly from what is shown. Contact a reputable propeller shop or marine dealership for parts and information on other brands of propellers.

Colt/Junior and 2-8 Hp 2-Strokes and 5/6 Hp 4-Strokes

◆ See Figures 95, 96 and 97

The propeller on all Colt/Junior and 2-8 hp 2-stroke motors, as well as on 5/6 hp 4-stroke motors is secured to the propshaft using a drive pin. Be sure to always keep a spare drive pin handy when you are onboard the boat. Remember that a sheared drive pin will leave you stranded on the water. A damaged shear pin can also contribute to motor damage, exposing it to over-revving while trying to produce thrust. The pin itself is usually locked in position by a propeller cap (or the entire propeller/cap assembly on 2-3.5 hp motors and possibly some Colt/Junior models) that is in turn fastened by a cotter pin. ALWAYS replace the cotter pin once it has been removed. Remember that should the cotter pin fail, you could be diving to recover your propeller.

1. Disconnect the negative battery cable or, more likely since these motors are rarely equipped with batteries, disconnect the spark plug lead(s) from the plug(s) for safety.

✱✱ CAUTION

Don't ever take the risk of working around the propeller if the engine could accidentally be started. Always take precautions such as disconnecting the spark plug leads and, if equipped, the negative battery cable.

2. Cut the ends off the cotter pin (as that is easier than trying to straighten them in most cases). Next, free the pin by grabbing the head with a pair of needlenose pliers. Either tap on the pliers gently with a hammer to help free the pin from the propeller cap or carefully use the pliers as a lever by carefully prying back against the propeller cone. Discard the cotter pin once it is removed.

3. On all models, except the 2-3.5 hp motors:
   a. Remove the propeller cap for access to the drive pin.
   b. Grasp and remove the drive pin using the needlenose pliers.

■ If the drive pin is difficult to remove, use a small punch or a new drive pin as a driver and gently tap the pin free from the shaft.

   c. Remove the thrust washer and the propeller from the shaft by sliding them from the shaft and splines.

4. On the 2-3.5 hp motors, the end cap is normally part of the propeller:
   a. Carefully slide the propeller from the shaft and drive pin once the cotter pin is removed.
   b. Grasp and remove the drive pin using the needlenose pliers.

■ If the drive pin is difficult to remove, use a small punch or a new drive pin as a driver and gently tap the pin free from the shaft.

To install:

5. Clean the propeller hub and shaft splines, then apply a fresh coating of OMC Triple-Guard or an equivalent water-resistant, marine grease.
6. On the 2-3.5 hp motors, insert the drive pin into the propeller shaft.
7. Align the propeller, then carefully slide it over the shaft.
8. On all models, except the 2-3.5 hp motors, install the thrust washer, followed by the drive pin and the propeller end cap.
9. Install a new cotter pin, then spread the pin ends in order to form tension and secure them. Do not bend them over too far as the pin will loosen and rattle in the shaft.

All Motors Except Colt/Junior, 2-8 Hp 2-Strokes and 5/6 Hp 4-Strokes

◆ See Figure 97 thru 101

On all 8 hp and larger 4-stroke motors, as well as all 9.9 hp and larger 2-stroke motors the propeller is held in place over the shaft splines by a large castellated nut. The nut is so named because, when viewed from the side, it appears similar to the upper walls or tower of a castle.

For safety, the nut is locked in place by a cotter pin that keeps it from loosening while the motor is running. The pin passes through a hole in the propeller shaft, as well as through the notches in the sides of the castellated nut. Install a new cotter pin anytime the propeller is removed and, perhaps more importantly, make sure the cotter pin is of the correct size and is made of materials designed for marine use.

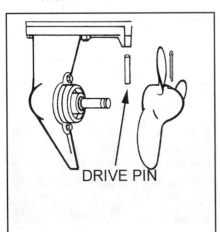

Fig. 95 Propeller mounting - 2-3.5 hp motors

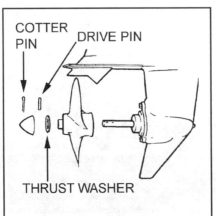

Fig. 96 If equipped with a propeller cap, remove the cotter pin to free it

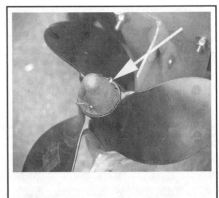

Fig. 97 If equipped with a propeller cap, remove the cotter pin to free it

MAINTENANCE & TUNE-UP 2-31

Fig. 98 Most propellers are held on the shaft by a castellated nut

Whenever working around the propeller, check for the presence of black rubber material in the drive hub and spline grease. Presence of this material normally indicates that the hub has turned inside the propeller bore (have the propeller checked by a propeller repair shop). Keep in mind that a spun hub will not allow proper torque transfer from the motor to the propeller and will allow the engine to over-rev in order to produce thrust. If the propeller has spun on the hub it has been weakened and is more likely to fail completely in use.

1. For safety, disconnect the negative cable (if so equipped) and/or disconnect the spark plug leads from the plugs (ground the leads to prevent possible ignition damage should the motor be cranked at some point before the leads are reconnected to the spark plugs).

** CAUTION

Don't ever take the risk of working around the propeller if the engine could accidentally be started. Always take precautions such as disconnecting the spark plug leads and, if equipped, the negative battery cable.

2. If the propeller is equipped with a matching propeller cap, cut the ends off the cotter pin (as that is easier than trying to straighten them in most cases). Next, free the pin by grabbing the head with a pair of needle nose pliers. Either tap on the pliers gently with a hammer to help free the pin from the propeller cap or carefully use the pliers as a lever by prying back against the propeller cap. Discard the cotter pin once it is removed.

3. Cut the ends off the cotter pin (as that is easier than trying to straighten them in most cases). Next, free the pin by grabbing the head with a pair of needlenose pliers. Either tap on the pliers gently with a hammer to help free the pin from the nut or carefully use the pliers as a lever by prying back against the castellated nut. Discard the cotter pin once it is removed.

4. Place a block of wood between the propeller and the anti-ventilation housing to lock the propeller and shaft from turning, then loosen and remove the castellated nut. Note the orientation, then remove the plain (40-70 hp 4-stroke motors only) or splined spacer (all other motors) from the propeller shaft.

5. Slide the propeller from the shaft. If the prop is stuck, use a block of wood to prevent damage and carefully drive the propeller from the shaft.

■ **If the propeller is completely seized on the shaft, have a reputable marine or propeller shop free it. Don't risk damage to the propeller or gearcase by applying excessive force.**

6. Note the direction in which the thrust washer is facing (since many models may use a thrust washer equipped with a fishing line trap that must face the proper direction if it is to protect the gearcase seal). Remove the thrust washer from the propshaft (if the washer appears stuck, tap lightly to free it from the propeller shaft).

■ **If equipped with a thrust washer mounted fishing line trap, always check the trap and clean it of any fishing line, vegetation or other debris whenever the propeller is removed. This is usually found on the 9.9/15 hp 4-stroke motors as well as 9.9-35 hp (216-521cc) motors, but can be found on other motors as well, depending on the propeller setup. When equipped, make sure the thrust washer on these motors is positioned with the fishing line trap groove facing toward the gearcase.**

7. Clean the thrust washer, propeller and shaft splines of any old grease. Small amounts of corrosion can be removed carefully using steel wool or fine grit sandpaper.

8. Inspect the shaft for signs of damage including twisted splines or excessively worn surfaces. Rotate the shaft while looking for any deflection. Replace the propeller shaft if these conditions are found. Inspect the thrust washer for signs of excessive wear or cracks and replace, if found.

To install:

9. Apply a light coat of OMC Triple-Guard or equivalent high-quality, water-resistant, marine grease to all surfaces of the propeller shaft and to the splines inside the propeller hub.

10. Position the thrust washer over the propshaft in the direction noted during removal. On all models, the shoulder should face the propeller. On washers equipped with a fishing line trap, the trap is on the opposite side of the shoulder and should face the gearcase (front of the motor).

11. Carefully slide the propeller onto the propshaft, rotating the propeller to align the splines. Push the propeller forward until it seats against the thrust washer.

12. Install the splined and/or plain spacer onto the propeller shaft, as equipped.

13. Place a block of wood between the propeller and housing to hold the prop from turning, then thread the castellated nut onto the shaft with the cotter pin grooves facing outward.

14. Tighten the castellated nut to 120 inch lbs. (14 Nm) using a suitable torque wrench, then install a new cotter pin through the grooves in the nut that align with the hole in the propshaft. If the cotter pin hole and the grooves do not align, tighten the nut additionally, just enough to align them (**do not** loosen the nut to achieve alignment.) Once the cotter pin is inserted, spread the ends sufficiently to lock the pin in place. Do not bend the ends over at

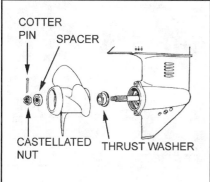

Fig. 99 Typical propeller mounting on all 8 hp and larger 4-strokes and 9.9 hp and larger 2-strokes

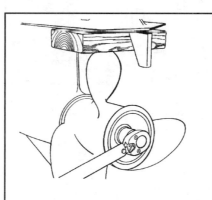

Fig. 100 Use a block of wood to keep the propeller from turning when loosening or tightening the nut

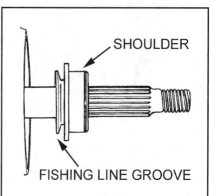

Fig. 101 On models so equipped, the fishing line groove on the thrust washer must face the gearcase

2-32 MAINTENANCE & TUNE-UP

90° or greater angles as the pin will loose tension and rattle in the slot.

15. If equipped with a propeller cap, install the cap and secure using a new cotter pin.

16. Connect the spark plug leads and/or the negative battery cable, as applicable.

Jet Drive Impeller

A jet drive motor uses an impeller enclosed in a jet drive housing instead of the propeller used by traditional gearcases. Outboard jet drives are designed to permit boating in areas prohibited to a boat equipped with a conventional propeller outboard drive system. The housing of the jet drive barely extends below the hull of the boat allowing passage in ankle deep water, white water rapids, and over sand bars or in shoal water which would foul a propeller drive.

The outboard jet drive provides reliable propulsion with a minimum of moving parts. It operates, simply stated, as water is drawn into the unit through an intake grille by an impeller. The impeller is driven by the driveshaft off the powerhead's crankshaft. Thrust is produced by the water that is expelled under pressure through an outlet nozzle that is directed away from the stern of the boat.

As the speed of the boat increases and reaches planing speed, only the very bottom of the jet drive where the intake grille is mounted facing downward remains in contact with the water.

The jet drive is provided with a reverse-gate arrangement and linkage to permit the boat to be operated in reverse. When the gate is moved downward over the exhaust nozzle, the pressure stream is deflected (reversed) by the gate and the boat moves sternward.

Conventional controls are used for powerhead speed, movement of the boat, shifting and power trim and tilt.

INSPECTION

◆ See Figure 102

The jet impeller is a precisely machined and dynamically balanced aluminum spiral. Close observation will reveal drilled recesses at exact locations used to achieve this delicate balancing. Excessive vibration of the jet drive may be attributed to an out-of-balance condition caused by the jet impeller being struck excessively by rocks, gravel or from damage caused by cavitation "burn".

The term cavitation "burn" is a common expression used throughout the world among people working with pumps, impeller blades, and forceful water movement. These "burns" occur on the jet impeller blades from cavitation air bubbles exploding with considerable force against the impeller blades. The edges of the blades may develop small dime-size areas resembling a porous sponge, as the aluminum is actually "eaten" by the condition just described.

Excessive rounding of the jet impeller edges will reduce efficiency and performance. Therefore, the impeller and intake grate (that protects it from debris) should be inspected at regular intervals.

Before and after each use, make a quick visual inspection of the intake grate and impeller, looking for obvious signs of damage. Always clear any debris such as plastic bags, vegetation or other items that sometimes become entangled in the water intake grate before starting the motor. If the intake grate is damaged, do not operate the motor, or you will risk destroying the impeller if rocks or other debris are drawn upward by the jet drive. If possible, replace a damaged grate before the next launch. This makes inspection after use all that much more important. Imagine the disappointment if you only learn of a damaged grate while inspecting the motor immediately prior to the next launch.

An obviously damaged impeller should be removed and either repaired or replaced depending on the extent of the damage. If rounding is detected, the impeller can be placed on a work bench and the edges restored to as sharp a condition as possible, using a file. Draw the file in only one direction. A back-and-forth motion will not produce a smooth edge. Take care not to nick the smooth surface of the jet impeller. Excessive nicking or pitting will create water turbulence and slow the flow of water through the pump. For more details on impeller replacement or service, please refer to the information on Jet Drives in the Gearcase section of this manual.

CHECKING IMPELLER CLEARANCE

◆ See Figures 103 and 104

Proper operation of the jet drive depends maximum thrust. In order for this to occur the clearance between the outer edge of the jet drive impeller and the water intake housing cone wall should be maintained at approximately 0.020-0.030 in. (0.5-0.8mm). This distance can be checked visually by shining a flashlight up through the intake grille and estimating the distance between the impeller and the casing cone, as indicated in the accompanying illustrations. But, it is not humanly possible to accurately measure this clearance by eye. Close observation between outings is fine to maintain a general idea of impeller condition, but, at least annually, the clearance must be measured using a set of feeler gauges. Although some gauges may be long enough to make the measurement with the intake grate installed, removal is advised for access and to allow for a more thorough inspection of the impeller itself.

✱✱ CAUTION

Whenever working around the impeller, ALWAYS disconnect the negative battery cable and/or disconnect the spark plug leads to make sure the engine cannot be accidentally started during service. Failure to heed this caution could result in serious personal injury or death in the event that the engine is started.

When checking clearance, a feeler gauge larger than the clearance

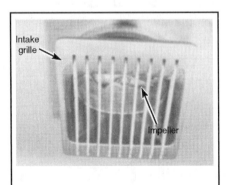

Fig. 102 Visually inspect the intake grate and impeller with each use

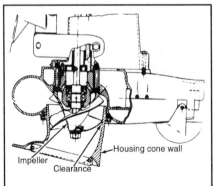

Fig. 103 Jet drive impeller clearance is the gap between the edges of the impeller and its housing

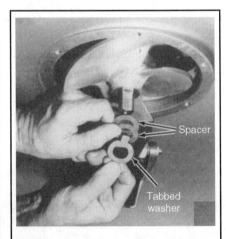

Fig. 104 Impeller clearance is adjusted by moving shims from below to above the impeller

MAINTENANCE & TUNE-UP 2-33

specification should not fit between the tips of the impeller and the housing. A gauge within specification should fit, but with a slight drag. A smaller gauge should fit without any interference whatsoever. Check using the feeler gauge at various points around the housing, while slowly rotating the impeller by hand.

After continued normal use, the clearance will eventually increase. In anticipation of this the manufacturer mounts the tapered impeller deep in a its housing, and positions spacers beneath the impeller to hold it in position. The spacers are used to position the impeller along the driveshaft with the desired clearance between the jet impeller and the housing wall. When clearance has increased, spacers are removed from underneath the impeller and repositioned behind it, dropping the impeller slightly in the housing and thereby decreasing the clearance again. Moving 1 spacer will decrease clearance approximately 0.004 in. (0.10mm).

If adjustment is necessary, refer to the Jet Drive procedures under Gearcase in this manual for impeller removal, shimming and installation procedures.

RescuePro® Rotor

◆ See Figure 105

RescuePro® models covered by this manual are equipped with a lower unit that incorporates a unique drive feature for safety purposes. Although the internals of the gearcase are very similar to the recreational models of the same engines, the gearcase differs in how it incorporates the final drive unit. Instead of the exposed propeller used by most recreational models, the gearcase of a RescuePro® motor incorporates a built-in rotor housing so that an enclosed rotor can be used in place of a propeller.

A tapered housing cover is bolted over top of the rotor to achieve jet pump like performance without sharing all of the design features of a jet drive gearcase. Shallow water jet drive housings draw water from underneath a shortened gearcase (perpendicular to movement of the boat) and achieve reverse thrust through a deflection gate. In contrast, the gearcase of a Rescue Pro® motor is virtually the same size as a typical propeller equipped recreational model. Water is drawn into the rotor housing from either side of the gearcase (parallel to the movement of the boat) and thrust is achieved by pushing the water through the tapered rotor housing cover. Reverse thrust is achieved by reversing rotor direction through the gears located inside the case, in the same fashion as a propeller-equipped motor. Exhaust is ported around the rotor housing by an exhaust pipe assembly.

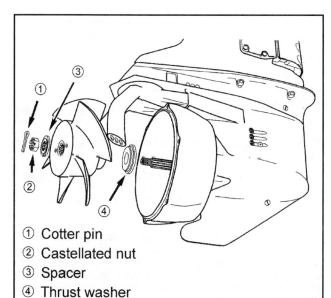

① Cotter pin
② Castellated nut
③ Spacer
④ Thrust washer

Fig. 105 RescuePro® motors are equipped with a rotor assembly instead of a propeller

INSPECTION

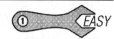

◆ See Figure 105

Rotor design and operation is similar to the impeller of a jet drive. It is a precisely weighted and balanced component that must be kept in good condition in order to achieve maximum thrust. But, because of the protective housing, the rotor is somewhat less likely to be exposed to the rocks, sand or debris that so often damage propellers and many jet drive impellers. Nonetheless, always perform a quick visual inspection of the rotor and housing before and after each use to verify its condition. Also, be sure to check the water intake of the rotor housing, making sure to keep it free of vegetation and debris.

CHECKING ROTOR CLEARANCE

◆ See Figures 105, 106 and 107

Proper operation of the rotor pump depends upon the ability to create maximum thrust. In order for this to occur the clearance between the outer edge of the rotor and the rotor housing must be less than 0.125 in. (3.2mm). This gap is checked using a feeler gauge or the special tool that comes with the OMC rotor removal tool (#345104).

Be sure to check clearance at least annually using a feeler gauge. A gauge of 0.125 in (3.2mm) should not fit between the tips of the rotor blades and the housing or should fit with a noticeable drag. A larger gauge must not fit otherwise the rotor is worn and must be replaced. Unlike the impeller used by jet drives, there is no method of shimming to decrease clearance.

In order to check the rotor clearance, the rotor housing cover must be removed for access. Also at least annually, the rotor must be removed so the shaft and hub may be greased using OMC Triple-Guard or equivalent high-quality, water-resistant, marine grease. It's best to make this one service procedure. For this detail, refer to the Removal & Installation procedure in this section.

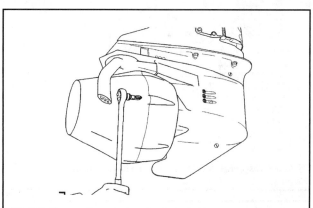

Fig. 106 Remove the rotor housing to inspect or remove the rotor

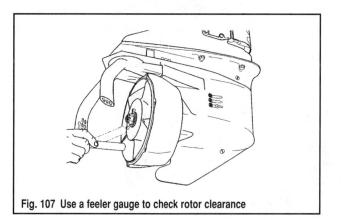

Fig. 107 Use a feeler gauge to check rotor clearance

MAINTENANCE & TUNE-UP

REMOVAL & INSTALLATION

◆ See Figures 105 thru 109

The rotor removal, greasing and installation procedure is very similar to the same actions with a standard propeller-equipped motor. The difference comes in the rotor housing which must be unbolted for access. The rotor housing must also be removed in order to check the rotor-to-housing clearance, so it is handy to conduct the removal, inspection and re-greasing procedure all at the same time.

Two special tools are necessary for this procedure. The first is a 3/16 in. ball hex driver. This is essentially a hex key with a ball on the end to allow a slight variance from a perpendicular fit in the hex-head bolt. A driver mounted in a socket, as opposed a standard ball hex key, is necessary so that a torque wrench can be used to tighten the rotor housing cover screws during installation.

The second special tool is OMC rotor removal tool (#345104). It is essentially a fixed spanner wrench that can be bolted onto the rotor in order to hold it steady during removal or installation of the castellated nut. If this tool is unavailable, one can be fabricated from a thick piece of metal stock, or aftermarket adjustable spanner wrenches of an appropriate size may be substituted as excessive torque is not required on the rotor nut.

The rotor is held in place over the shaft splines by a large castellated nut. The nut is so named because, when viewed from the side, it appears similar to the upper walls or tower of a castle.

For safety, the nut is locked in place by a cotter pin that keeps it from loosening while the motor is running. The pin passes through a hole in the rotor shaft, as well as through the notches in the sides of the castellated nut. Install a new cotter pin anytime the rotor is removed and, perhaps more importantly, make sure the cotter pin is of the correct size and is made of materials designed for marine use.

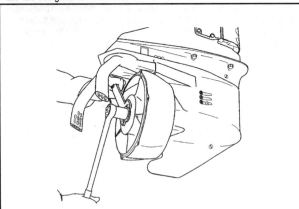

Fig. 108 A spanner or rotor tool is used to keep the rotor from turning

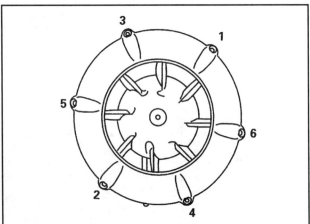

Fig. 109 Rotor housing cover screw torque sequence

1. For safety, disconnect the negative battery cable and/or disconnect the spark plug leads and ground them on the motor.

✶✶ CAUTION

Whenever working around the rotor, ALWAYS disconnect the negative battery cable and/or disconnect the spark plug leads to make sure the engine cannot be accidentally started during service. Failure to heed this caution could result in serious personal injury or death in the event that the engine is started.

2. Use a 3/16 in. ball hex driver to loosen the stator housing screws using a crisscross pattern to prevent cover warpage. Once the screws are all loosened, remove the cover and screws from the rotor housing.

■ The rotor housing cover screws are captive and cannot be removed from the cover. Loosen each screw until it is free of the rotor housing, then remove the screws along with the cover.

3. Use a feeler gauge or the feeler that accompanies the OMC rotor removal tool to check clearance between the tips of the rotor blades and the housing. The inspection tool (or 0.125 in. feeler gauge) should not move freely for the entire width of any one rotor blade, otherwise the rotor must be replaced. Turn the rotor 30° and repeat inspection at each of the blade tips.

4. Cut the ends off the cotter pin (as that is easier than trying to straighten them in most cases). Next, free the pin by grabbing the head with a pair of needle-nose pliers. Either tap on the pliers gently with a hammer to help free the pin from the nut or carefully use the pliers as a lever by carefully prying back against the castellated nut. Discard the cotter pin once it is removed.

5. Install the holding tool to the rotor, then loosen and remove the castellated nut.

6. Carefully slide the spacer and then the rotor from the shaft. If the rotor is stuck, use a block of wood to prevent damage and carefully drive the rotor from the shaft.

■ If the rotor is completely seized on the shaft, have a reputable marine or propeller shop free it. Don't risk damage to the rotor or gearcase by applying excessive force.

7. Note the direction in which the thrust washer is facing, then slide the washer from the shaft.

8. Clean the thrust washer, rotor and shaft splines of any old grease. Small amounts of corrosion can be removed carefully using steel wool or fine grit sandpaper.

9. Inspect the shaft for signs of damage including twisted splines or excessively worn surfaces. Rotate the shaft while looking for any deflection. Replace the rotor shaft if these conditions are found. Inspect the thrust washer for signs of excessive wear or cracks and replace, if found.

To install:
10. Apply a light coat of OMC Triple-Guard or equivalent high-quality, water-resistant, marine grease to all surfaces of the rotor shaft and to the splines inside the propeller hub.

11. Position the thrust washer over the shaft in the direction noted during removal.

■ During installation, the shoulder (relief) side of the thrust washer must face the rotor (aft).

12. Carefully slide the rotor onto the shaft, twisting the rotor to align the splines. Push the rotor until it seats against the thrust washer.

13. Install the spacer onto the shaft, then thread the castellated nut by hand to hold the rotor in place.

14. Position the rotor holding tool, then tighten the castellated nut to 120 inch lbs. (14 Nm) using a suitable torque wrench. Install a new cotter pin through the grooves in the nut that align with the hole in the shaft. If the cotter pin hole and the grooves do not align, tighten the nut additionally, just enough to align them (**do not** loosen the nut to achieve alignment.) Once the cotter pin is inserted, spread the ends sufficiently to lock the pin in place. Do not bend the ends over at 90° or greater angles as the pin will loose tension and rattle in the slot.

15. Install the rotor housing cover and tighten the screws using a crisscross pattern (as shown in the accompanying illustration) to 60-80 inch lbs. (7-9 Nm).

16. Connect the spark plug leads and/or the negative battery cable, as applicable.

MAINTENANCE & TUNE-UP 2-35

Anodes (Zincs)

◆ See Figure 110

The idea behind anodes (also known as sacrificial anodes) is simple: When dissimilar metals are dunked in water and a small electrical current is leaked between or amongst them, the less-noble metal (galvanically speaking) is sacrificed (corrodes).

The zinc alloy of which the anodes are made is designed to be less noble than the aluminum alloy of which your outboard is constructed. If there's any electrolysis, and there almost always is, the inexpensive zinc anodes are consumed in lieu of the expensive outboard motor.

These zincs require a little attention in order to make sure they are capable of performing their function. Anodes must be solidly attached to a clean mounting site. Also, they must not be covered with any kind of paint, wax or marine growth.

Fig. 110 Extensive corrosion of an anode suggests a problem or a complete disregard for maintenance

INSPECTION

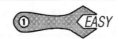

◆ See Figure 110 thru 117

Visually inspect the anodes, especially gearcase mounted ones, before and after each use. You'll want to know right away if it has become loose or fallen off in service. Periodically inspect them closely to make sure they haven't eroded too much. At a certain point in the erosion process, the mounting holes start to enlarge, which is when the zinc might fall off. Obviously, once that happens your engine no longer has any protection. Generally, a zinc anode is considered worn if it has shrunken to 2/3 or less than the original size. To help judge this, buy a spare and keep it handy (in the boat or tow vehicle for comparison).

If you use your outboard in salt water or brackish water, and your zincs never seem to wear, inspect them carefully. Paint, wax or marine growth on zincs will insulate them and prevent them from performing their function properly. They must be left bare and must be installed onto bare metal of the motor. If the zincs are installed properly and not painted or waxed, inspect around them for sings of corrosion. If corrosion is found, strip it off immediately and repaint with a rust inhibiting paint. If in doubt, replace the zincs.

On the other hand, if your zinc seems to erode in no time at all, this may be a symptom of the zincs themselves. Each manufacturer uses a specific blend of metals in their zincs. If you are using zincs with the wrong blend of metals, they may erode more quickly or leave you with diminished protection.

At least annually or whenever an anode has been removed or replaced, check the mounting for proper electrical contact using a multi-meter. Set the multi-meter to check resistance (ohms), then connect one meter lead to the anode and the other to a good, unpainted or corroded ground on the motor. Resistance should be very low or zero. If resistance is high or infinite, the anode is insulated and cannot perform its function properly.

SERVICING

◆ See Figures 111 thru 117

Depending on your boat, motor and rigging, you may have anywhere from one to four (or even more) anodes. Regardless of the number, there are some fundamental rules to follow that will give your boat and motor's sacrificial anodes the ability to do the best job protecting your boat's underwater hardware that they can.

■ **On fuel injected models, the trim tab serves as the gearcase anode. If the anode must be replaced, make an alignment mark between the anode and gearcase before removal, then transfer the mark to the replacement anode to preserve trim tab adjustment. If the boat pulls to one side after replacement and did not previously, refer to the Trim Tab Adjustment procedure in the Gearcase section to correct this condition.**

All motors covered by this manual are equipped with at least one gearcase anode, normally mounted in, on, or near the anti-ventilation plate. Some of the motors covered by this manual also have a powerhead mounted anode and/or an engine clamp bracket anode. Location of the powerhead zincs will vary slightly from motor-to-motor including mounting bosses specifically cast in the motor (9.9/15 hp 4-strokes) or on the manifold assemblies (40-70 hp 4-strokes). Most 737cc and larger motors are equipped with one or more anodes on the engine mount clamp bracket.

■ **The 25-55 hp (737cc) 2-stroke motors are equipped with unique anodes when it comes to mounting locations. Many of these motors are equipped with an anode on the propshaft, behind the propeller and thrust washer. Also, most of the motors are equipped with a clamp bracket-mounted anode, located on the inner, port side of the bracket It faces the boat's transom. Access to the clamp bracket anode on these models may be possible from underneath the bracket.**

Some people replace zincs annually. This may or may not be necessary, depending on the type of waters in which you boat and depending on whether or not the boat is hauled with each use or left in for the season. Either way, it is a good idea to remove zincs at least annually in order to make sure the mounting surfaces are still clean and free of corrosion.

The first thing to remember is that zincs are electrical components and like all electrical components, they require good clean connections. So after

Fig. 111 All motors have at least one anode mounted on the gearcase . . .

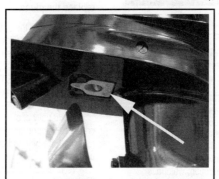

Fig. 112 . . . though some are under the anti-ventilation plate and . . .

Fig. 113 . . . others are mounted inside the gearcase (often accessed from underneath)

2-36 MAINTENANCE & TUNE-UP

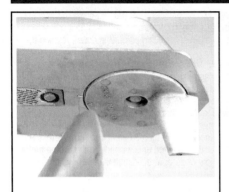
Fig. 114 Fuel injected motors utilize a trim tab anode on the gearcase

Fig. 115 Some motors, like this 9.9/15 4-stroke have bosses to mount powerhead anodes

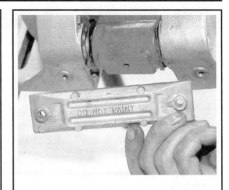
Fig. 116 Most 737cc and larger units at least one anode on the engine clamp bracket

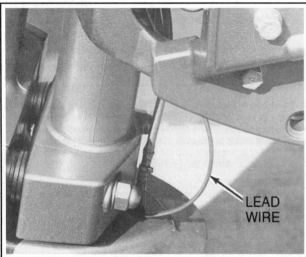
Fig. 117 Some motors use lead wires used to connect bracketed parts and assist in corrosion resistance

you've undone the mounting hardware you want to get the zinc mounting sites clean and shiny.

Get a piece of coarse emery cloth or some 80-grit sandpaper. Thoroughly rough up the areas where the zincs attach (there's often a bit of corrosion residue in these spots). Make sure to remove every trace of corrosion.

Zincs are attached with stainless steel machine screws that thread into the mounting for the zincs. Over the course of a season, this mounting hardware is inclined to loosen. Mount the zincs and tighten the mounting hardware securely. Tap the zincs with a hammer hitting the mounting screws squarely. This process tightens the zincs and allows the mounting hardware to become a bit loose in the process. Now, do the final tightening. This will insure your zincs stay put for the entire season.

Timing Belt

INSPECTON

◆ See Figure 118

All 4-stroke motors covered by this manual, except the 40/50 hp (815cc) models, use a timing belt to synchronize the camshaft and crankshaft (for correct valve timing). The timing belt is a long life component that does not require much in the way or service, but we would recommend that you inspect it at least once every year. Also, although the manufacturer provides no recommended replacement interval, experience shows that it is wise to replace it every 4-5 years or after 600-800 hours of operation, whichever comes first. Keep in mind, a timing belt that breaks or even slips a tooth will likely disable the motor, possibly stranding the boat. The 40/50 hp 4-strokes are equipped with a timing chain and do not require regular inspection.

Fig. 118 Although the timing belt is partially visible on many 4-strokes, thorough inspection requires cover removal

On some versions of 9.9/15 hp (305cc), the timing belt is visible at one point in the manual starter cover. On most 8/9.9 hp (211cc) and 5/6 hp (128cc) motors, the belt is partially visible under the manual starter cover, but a thorough inspection is much easier once the manual starter cover and/or assembly is removed. For 70 hp (1298cc) motors, the flywheel cover must be removed to inspect the belt.

1. For safety when working around the flywheel, disconnect the negative battery cable and/or disconnect the leads from the spark plugs, then ground the leads on the powerhead.

■ Although not absolutely necessary for this procedure, it is a good idea to remove the spark plugs at this time. Removing the spark plugs will relieve engine compression, making it easier to manually rotate the motor. Also, it presents a good opportunity to inspect, clean and/or replace the plugs.

2. Remove the manual starter assembly or the flywheel cover, as applicable, for better access to the timing belt.
3. Use low-pressure compressed air to blow debris out from under the camshaft pulley, flywheel and timing belt.
4. Visually check the belt for worn, cracked or oil soaked surfaces. Slowly rotate the flywheel (by hand) while inspecting all of the timing belt cogs.
5. Visually check the camshaft pulley and flywheel teeth for worn, cracked, chipped or otherwise damaged surfaces.
6. If the belt and or pulleys are damaged, replace them as described under Powerhead in this manual.
7. If removed, install the manual starter assembly or flywheel cover to the powerhead.
8. Install the spark plugs, then connect the leads followed by the negative battery cable and the engine cover.

MAINTENANCE & TUNE-UP 2-37

BOAT MAINTENANCE

Batteries

◆ See Figures 119 and 120

Batteries require periodic servicing, so a definite maintenance program will help ensure extended life.

A failure to maintain the battery in good order can prevent it from properly charging or properly performing its job even when fully charged. Low levels of electrolyte in the cells, loose or dirty cable connections at the battery terminals or possibly an excessively dirty battery top can all contribute to an improperly functioning battery. So battery maintenance, first and foremost, involves keeping the battery full of electrolyte, properly charged and keeping the casing/connections clean of corrosion or debris.

If a battery charges and tests satisfactorily but still fails to perform properly in service, one of three problems could be the cause.

1. An accessory left on overnight or for a long period of time can discharge a battery.

■ The Engine Control Unit (ECU) on fuel-injected motors covered by this manual will continue to draw a small amount of current from the battery, even when the motor is shut off. Although it will takes weeks to discharge a fully charged battery, periodically recharging the battery, or isolating it by disconnecting the cables or shutting off the battery switch when the boat is dockside or on the trailer will prevent this.

2. Using more electrical power than the stator assembly or lighting coil can replace would slowly drain the battery during motor operation, resulting in an undercharged condition.

3. A defect in the charging system. A faulty stator assembly or lighting coil, defective regulator or rectifier or high resistance somewhere in the system could cause the battery to become undercharged.

MAINTENANCE

◆ See Figures 120 thru 123

Electrolyte Level

The most common and important procedure in battery maintenance is checking the electrolyte level. On most batteries, this is accomplished by removing the cell caps and visually observing the level in the cells. The bottom of each cell has a split vent which will cause the surface of the electrolyte to appear distorted when it makes contact. When the distortion first appears at the bottom of the split vent, the electrolyte level is correct. Smaller marine batteries are sometimes equipped with translucent cases that are printed or embossed with high and low level markings on the side. On some of these, shining a flashlight through the battery case will help make it easier to determine the electrolyte level.

Fig. 119 Explosive hydrogen gas is released from the batteries in a discharged state. This one exploded when something ignited the gas. Explosions can also be caused by a spark from the battery terminals or jumper cables

Fig. 120 Ignoring a battery (and corrosion) to this extent is asking for it to fail

Fig. 121 Place a battery terminal tool over posts, then rotate back and forth . . .

Fig. 122 . . . until the internal brushes expose a fresh, clean surface on the post

Fig. 123 Clean the insides of cable ring terminals using the tool's wire brush

2-38 MAINTENANCE & TUNE-UP

During hot weather and periods of heavy use, the electrolyte level should be checked more often than during normal operation. Add distilled water to bring the level of electrolyte in each cell to the proper level. Take care not to overfill, because adding an excessive amount of water will cause loss of electrolyte and any loss will result in poor performance, short battery life and will contribute quickly to corrosion.

■ **Never add electrolyte from another battery. Use only distilled water. Even tap water may contain minerals or additives that will promote corrosion on the battery plates, so distilled water is always the best solution.**

Although less common in marine applications than other uses today, sealed maintenance-free batteries also require electrolyte level checks, through the window built into the tops of the cases. The problem for marine applications is the tendency for deep cycle use to cause electrolyte evaporation and electrolyte cannot be replenished in a sealed battery.

The second most important procedure in battery maintenance is periodically cleaning the battery terminals and case.

Cleaning

Dirt and corrosion should be cleaned from the battery as soon as it is discovered. Any accumulation of acid film or dirt will permit a small amount of current to flow between the terminals. Such a current flow will drain the battery over a period of time.

Clean the exterior of the battery with a solution of diluted ammonia or a paste made from baking soda and water. This is a base solution to neutralize any acid that may be present. Flush the cleaning solution off with plenty of clean water.

■ **Take care to prevent any of the neutralizing solution from entering the cells as it will quickly neutralize the electrolyte (ruining the battery).**

Poor contact at the terminals will add resistance to the charging circuit. This resistance will cause the voltage regulator to register a fully charged battery and thus cut down on the stator assembly or lighting coil output adding to the low battery charge problem.

At least once a season, the battery terminals and cable clamps should be cleaned. Loosen the clamps and remove the cables, negative cable first. On batteries with top mounted posts, if the terminals appear stuck, use a puller specially made for this purpose to ensure the battery casing is not damaged. NEVER pry a terminal off a battery post. These are inexpensive and available in most parts stores.

Clean the cable clamps and the battery terminal with a wire brush until all corrosion, grease, etc., is removed and the metal is shiny. It is especially important to clean the inside of the clamp thoroughly (a wire brush or brush part of a battery post cleaning tool is useful here), since a small deposit of foreign material or oxidation there will prevent a sound electrical connection and inhibit either starting or charging. It is also a good idea to apply some dielectric grease to the terminal, as this will aid in the prevention of corrosion.

After the clamps and terminals are clean, reinstall the cables, negative cable last, do not hammer the clamps onto battery posts. Tighten the clamps securely but do not distort them. To help slow or prevent corrosion, give the clamps and terminals a thin external coating of grease after installation.

Check the cables at the same time that the terminals are cleaned. If the insulation is cracked or broken or if its end is frayed, that cable should be replaced with a new one of the same length and gauge.

TESTING

◆ See Figure 124

A quick check of the battery is to place a voltmeter across the terminals. Although this is by no means a clear indication, it gives you a starting point when trying to troubleshoot an electrical problem that could be battery related. Most marine batteries will be of the 12 volt DC variety. They are constructed of 6 cells, each of which is capable of producing slightly more than two volts, wired in series so that total voltage is 12 and a fraction. A fully charged battery will normally show more than 12 and slightly less than 13 volts across its terminals. But keep in mind that just because a battery reads 12.6 or 12.7 volts does NOT mean it is fully charged. It is possible for it to have only a surface charge with very little amperage behind it to maintain that voltage rating for long under load. A discharged battery will read some value less than 12 volts, but can be brought back to 12 volts through recharging. Of course a battery with one or more shorted or un-chargeable cells will also read less than 12, but it cannot be brought back to 12+ volts after charging. For this reason, the best method to check battery condition on most marine batteries is through a specific gravity check.

A hydrometer is a device that measures the density of a liquid when compared to water (specific gravity). Hydrometers are used to test batteries by measuring the percentage of sulfuric acid in the battery electrolyte in terms of specific gravity. When the condition of the battery drops from fully charged to discharged, the acid is converted to water as electrons leave the solution and enter the plates, causing the specific gravity of the electrolyte to drop.

It may not be common knowledge but hydrometer floats are calibrated for use at 80°F (27°C). If the hydrometer is used at any other temperature, hotter or colder, a correction factor must be applied.

■ **Remember, a liquid will expand if it is heated and will contract if cooled. Such expansion and contraction will cause a definite change in the specific gravity of the liquid, in this case the electrolyte.**

A quality hydrometer will have a thermometer/temperature correction table in the lower portion, as illustrated in the accompanying illustration. By measuring the air temperature around the battery and from the table, a correction factor may be applied to the specific gravity reading of the hydrometer float. In this manner, an accurate determination may be made as to the condition of the battery.

When using a hydrometer, pay careful attention to the following points:

1. Never attempt to take a reading immediately after adding water to the battery. Allow at least 1/4 hour of charging at a high rate to thoroughly mix the electrolyte with the new water. This time will also allow for the necessary gases to be created.

2. Always be sure the hydrometer is clean inside and out as a precaution against contaminating the electrolyte.

3. If a thermometer is an integral part of the hydrometer, draw liquid into it several times to ensure the correct temperature before taking a reading.

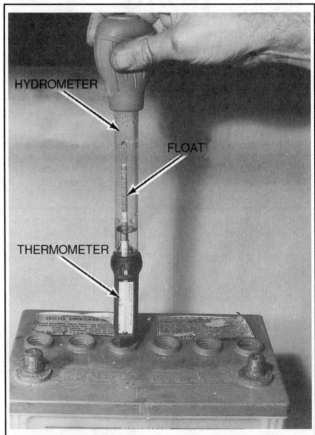

Fig. 124 A hydrometer is the best method for checking battery condition

MAINTENANCE & TUNE-UP

4. Be sure to hold the hydrometer vertically and suck up liquid only until the float is free and floating.

5. Always hold the hydrometer at eye level and take the reading at the surface of the liquid with the float free and floating.

6. Disregard the slight curvature appearing where the liquid rises against the float stem. This phenomenon is due to surface tension.

7. Do not drop any of the battery fluid on the boat or on your clothing, because it is extremely caustic. Use water and baking soda to neutralize any battery liquid that does accidentally drop.

8. After drawing electrolyte from the battery cell until the float is barely free, note the level of the liquid inside the hydrometer. If the level is within the charged (usually green) band range for all cells, the condition of the battery is satisfactory. If the level is within the discharged (usually white) band for all cells, the battery is in fair condition.

9. If the level is within the green or white band for all cells except one, which registers in the red, the cell is shorted internally. No amount of charging will bring the battery back to satisfactory condition.

10. If the level in all cells is about the same, even if it falls in the red band, the battery may be recharged and returned to service. If the level fails to rise above the red band after charging, the only solution is to replace the battery.

STORAGE

If the boat is to be laid up (placed into storage) for the winter or for more than a few weeks, special attention must be given to the battery to prevent complete discharge and/or possible damage to the terminals and wiring. Before putting the boat in storage, disconnect and remove the batteries. Clean them thoroughly of any dirt or corrosion and then charge them to full specific gravity readings. After they are fully charged, store them in a clean cool dry place where they will not be damaged or knocked over, preferably on a couple blocks of wood. Storing the battery up off the deck, will permit air to circulate freely around and under the battery and will help to prevent condensation.

Never store the battery with anything on top of it or cover the battery in such a manner as to prevent air from circulating around the filler caps. All batteries, both new and old, will discharge during periods of storage, more so if they are hot than if they remain cool. Therefore, the electrolyte level and the specific gravity should be checked at regular intervals. A drop in the specific gravity reading is cause to charge them back to a full reading.

In cold climates, care should be exercised in selecting the battery storage area. A fully-charged battery will freeze at about 60°F below zero. The electrolyte of a discharged battery, almost dead, will begin forming ice at about 19°F above zero.

■ For more information on batteries and the engine electrical systems, please refer to the Ignition and Electrical section of this manual.

Fiberglass Hull

INSPECTION AND CARE

◆ See Figures 125, 126 and 127

Fiberglass reinforced plastic hulls are tough, durable and highly resistant to impact. However, like any other material they can be damaged. One of the advantages of this type of construction is the relative ease with which it may be repaired.

A fiberglass hull has almost no internal stresses. Therefore, when the hull is broken or stove-in, it retains its true form. It will not dent to take an out-of-shape set. When the hull sustains a severe blow, the impact will be either absorbed by deflection of the laminated panel or the blow will result in a definite, localized break. In addition to hull damage, bulkheads, stringers and other stiffening structures attached to the hull may also be affected and therefore, should be checked. Repairs are usually confined to the general area of the rupture.

■ **The best way to care for a fiberglass hull is to wash it thoroughly, immediately after hauling the boat while the hull is still wet.**

A foul bottom can seriously affect boat performance. This is one reason why racers, large and small, both powerboat and sail, are constantly giving attention to the condition of the hull below the waterline.

In areas where marine growth is prevalent, a coating of vinyl, anti-fouling bottom paint should be applied if the boat is going to be left in the water for extended periods of time such as all or a large part of the season. If growth has developed on the bottom, it can be removed with a diluted solution of muriatic acid applied with a brush or swab and then rinsed with clear water. Always use rubber gloves when working with Muriatic acid and take extra care to keep it away from your face and hands. The fumes are toxic. Therefore, work in a well-ventilated area or if outside, keep your face on the windward side of the work.

■ **If marine growth is not too severe you may avoid the unpleasantness of working with muriatic acid by trying a power washer instead. Most marine vegetation can be removed by pressurized water and a little bit of scrubbing using a rough sponge (don't use anything that will scratch or damage the surface).**

Barnacles have a nasty habit of making their home on the bottom of boats that have not been treated with anti-fouling paint. Actually they will not harm the fiberglass hull but can develop into a major nuisance.

If barnacles or other crustaceans have attached themselves to the hull, extra work will be required to bring the bottom back to a satisfactory condition. First, if practical, put the boat into a body of fresh water and allow it to remain for a few days. A large percentage of the growth can be removed in this manner. If this remedy is not possible, wash the bottom thoroughly with a high-pressure fresh water source and use a scraper. Small particles of hard shell may still hold fast. These can be removed with sandpaper.

Fig. 125 The best way to care for a fiberglass hull is to wash it thoroughly

Fig. 126 If marine growth is a problem, apply a coating of anti-foul bottom paint

Fig. 127 Fiberglass, vinyl and rubber care products, like those from Meguiar's protect your boat

2-40 MAINTENANCE & TUNE-UP

TUNE-UP

Introduction to Tune-Ups

A proper tune-up is the key to long and trouble-free outboard life and the work can yield its own rewards. Studies have shown that a properly tuned and maintained outboard can achieve better fuel economy than an out-of-tune engine. As a conscientious boater, set aside a Saturday morning, say once a month, to check or replace items which could cause major problems later. Keep your own personal log to jot down which services you performed, how much the parts cost you, the date and the number of hours on the engine at the time. Keep all receipts for such items as oil and filters, so that they may be referred to in case of related problems or to determine operating expenses. These receipts are the only proof you have that the required maintenance was performed. In the event of a warranty problem on newer engines, these receipts will be invaluable.

The efficiency, reliability, fuel economy and enjoyment available from boating are all directly dependent on having your outboard tuned properly. The importance of performing service work in the proper sequence cannot be over emphasized. Before making any adjustments, check the specifications. Never rely on memory when making critical adjustments.

Before tuning any outboard, insure it has satisfactory compression. An outboard with worn or broken piston rings, burned pistons or scored cylinder walls, will not perform properly no matter how much time and expense is spent on the tune-up. Poor compression must be corrected or the tune-up will not give the desired results.

The extent of the engine tune-up is usually dependent on the time lapse since the last service. In this section, a logical sequence of tune-up steps will be presented in general terms. If additional information or detailed service work is required, refer to the section of this manual containing the appropriate instructions.

Tune-Up Sequence

A tune-up can be defined as pre-determined series of procedures (adjustments, tests and worn component replacements) that are performed to bring the engine operating parameters back to original condition. The series of steps are important, as the later procedures (especially adjustments) are dependant upon the earlier procedures. In other words, a procedure is performed only when subsequent steps would not change the result of that procedure (this is mostly for adjustments or settings that would be incorrect after changing another part or setting). For instance, fouled or excessively worn spark plugs may affect engine idle. If adjustments were made to the idle speed or mixture **before** these plugs were cleaned or replaced, the idle speed or mixture might be wrong after replacing the plugs. The possibilities of such an effect become much greater when dealing with multiple adjustments such as timing, idle speed and/or idle mixture. Therefore, be sure to follow each of the steps given here. Since many of the steps listed here are full procedures in themselves, refer to the procedures of the same name in this section for details.

A complete pre-season tune-up should be performed at the beginning of each season or when the motor is removed from storage. Operating conditions, amount of use and the frequency of maintenance required by your motor may make one or more additional tune-ups necessary during the season. Perform additional tune-ups as use dictates.

1. Before starting, inspect the motor thoroughly for signs of obvious leaks, damage and loose or missing components. Make repairs, as necessary.

2. If OMC Carbon Guard of equivalent is not used consistently with each fill-up, remove carbon from the pistons and combustion chamber after every 50 hours of operation. Refer to the De-carboning the Pistons in this section.

3. Perform a compression check to make sure the motor is mechanically ready for a tune-up. An engine with low compression on one or more cylinder should be overhauled, not tuned. A tune-up will not be successful without sufficient engine compression. Refer to the Compression Test in this section.

4. Since the spark plugs must be removed for the compression check, take the opportunity to inspect them thoroughly for signs of oil fouling, carbon fouling, damage due to detonation, etc. Clean and re-gap the plugs or, better yet, install new plugs as no amount of cleaning will precisely match the performance and life of new plugs. Refer to Spark Plugs, in this section.

5. A couple of early 1990 single cylinder models are equipped with a breaker point and condenser style ignition system. On these models it is usually a good idea to proactively replace the points at LEAST every 100 hours or at the beginning of the season, whichever comes first.

6. Visually inspect all ignition system components for signs of obvious defects. Look for signs of burnt, cracked or broken insulation. Replace wires or components with obvious defects. If spark plug condition suggests weak or no spark on one or more cylinders, perform ignition system testing to eliminate possible worn or defective components. Refer to the Ignition System Inspection procedures in this section and the Ignition and Electrical System section.

7. Remove and clean (on serviceable filters) or replace the inline filter and/or fuel pump filter, as equipped. Refer to the Fuel Filter procedures in this section. Perform a thorough inspection of the fuel system, hoses and components. Replace any cracked or deteriorating hoses.

8. Perform engine Timing and Synchronization adjustments as described in this section.

■ **Although most of the motors covered by this manual allow for certain ignition timing and carburetor adjustment procedures, none of them require the level of tuning attention that was once the norm. Many of the motors are equipped with electronic ignition systems that limit or eliminate timing adjustments. Many of the carburetors covered by this manual are U.S. EPA regulated and contain few mixture adjustments. EFI motors are all but completely controlled by the Engine Control Unit (ECU) and contain no timing or fuel adjustments.**

9. Except for jet drive models, remove the propeller (or rotor on RescuePro®) in order to thoroughly check for leaks at the shaft seal. Inspect the propeller or rotor condition, look for nicks, cracks or other signs of damage and repair or replace, as necessary. If available, install a test wheel to run the motor in a test tank after completion of the tune-up. If no test wheel is available, lubricate the shaft/splines, then install the propeller or rotor. Refer to the procedures for Propeller or RescuePro® Rotor in this section, as applicable.

10. Change the gearcase oil as directed under the Gearcase Oil procedures in this section. If you are conducting a pre-season tune-up and the oil was changed immediately prior to storage this is not necessary. But, be sure to check the oil level and condition. Drain the oil anyway if significant contamination is present.

■ **Anytime large amounts of water or debris is present in the gearcase oil, be sure to troubleshoot and repair the problem before returning the gearcase to service. The presence of water may indicate problems with the seals, while debris could a sign that overhaul is required.**

11. Check all accessible bolts and fasteners and tighten any that are loose.

12. On all carbureted models, re-torque the cylinder head bolts following the applicable steps of the cylinder the cylinder head removal and installation procedures. The cylinder head cover or rocker/valve cover must be removed for access. Each of the bolts should be loosened using the reverse of the tightening sequence, then re-torque using one or more passes of the tightening sequence, as directed. Refer to the procedures under Powerhead for details.

■ **Before installing the cylinder head (valve cover) on 4-stroke engines, check and adjust the valve lash, as necessary.**

13. On all 4-stroke models, check and adjust the valve lash clearance. Although the manufacturer does not specifically call for periodic checks of the valve clearance on 40-70 hp motors, we suggest that you make this part of your annual routine to check for premature camshaft wear that has been noticed in the field. This wear often occurs from the use of incorrect types/grade of oil or from infrequent oil changes. A sudden or drastic change in valve lash can be an early indicator, giving you time to change types of oil or the frequency of service the engine is receiving before necessitating a costly and troublesome camshaft replacement.

14. Pressurize the fuel system according to the procedures found in the Fuel System section, then check carefully for leaks.

15. Perform a test run of the engine to verify proper operation of the starting, fuel, oil and cooling systems. Although this can be performed using a flush/test adapter or even on the boat itself (if operating with a normal load/passengers), the preferred method is the use of a test tank. If possible, run the engine, in a test tank using the appropriate test wheel. Monitor the

MAINTENANCE & TUNE-UP 2-41

cooling system indicator stream to ensure the water pump is working properly. Once the engine is fully warmed, slowly advance the engine to wide-open throttle, then note and record the maximum engine speed. Refer to the Tune-Up Specifications chart to compare engine speeds with the test propeller minimum rpm specifications. If engine speeds are below specifications, yet engine compression was sufficient at the beginning of this procedure, recheck the fuel and ignition system adjustments.

De-Carboning the Pistons

A by-product of the normal combustion process, carbon will build-up on the pistons and in the combustion chambers of a motor over time. Engine tuning and condition will affect this process, as a properly tuned engine running high-quality fuels under proper conditions will reduce the amount of build-up, but not stop it completely. Generally speaking an out-of-tune motor, a motor running too rich or a motor run under extended idle conditions will increase the rate at which carbon deposits are formed. Carbon, when its presence becomes significant enough, will increase the compression ratio (by decreasing effective combustion chamber size) and will lead to detonation. Also, over time, carbon may cause piston rings to stick, which would lead to blow-by or, on 4-stroke motors, excessive oil combustion. For this reason, the original engine manufacturer recommends the use of OMC Carbon Guard fuel additive with each fill up in order to help slow this process.

The manufacturer also warns that, if this fuel additive is not used, the pistons and combustion chambers should be cleaned of deposits using OMC Engine Tuner after, at least every 50 hours of engine operation. As noted earlier, variables such as types of fuel used and patterns of usage (wide-open throttle vs. extensive idle) will also have an effect upon how often this procedure should be followed. Let your own experience (and the amount of carbon found on your spark plugs) be your guide.

1. Provide the engine with a cooling water source (either and engine flushing adapter, a test tank, or if necessary, perform this procedure with the boat and motor in the water, attached to a sturdy dock).
2. Start and run the engine at normal idle until it reaches normal operating temperature.
3. Set the engine to fast idle (except on fuel injected motors where the idle speed is controlled by the computer module). For most engines a fast idle of around 1200 rpm is sufficient.
4. For severe cases of carbon build-up, run the engine to normal operating temperature, then shut the engine off and remove the spark plugs. Lay the engine into a horizontal position and peer through the spark plug holes as you slowly turn the motor over by hand. With the pistons leveled so as to best block of the ports (on 2-strokes) cover the tops of the pistons with engine tuner and let sit for approximately one hour. After at least that amount of time, rotate the engine a couple of revolutions by hand to remove the cleaner.
5. For less severe (typical cases of carbon build-up), spray the entire contents of the OMC Engine Tuner can either through the carburetor throat(s) or through the fuel primer solenoid fogging fitting with the engine still running from Step 3 above. If working with multiple carburetors, the fogging fitting is best to use at it will ensure even distribution of the Engine Tuner. But, if the engine is not equipped with an electric primer solenoid, move the spray nozzle from carburetor-to-carburetor, back and forth in sequence until the can is emptied. Once all of the Engine Tuner has been sprayed, shut the engine **OFF** and allow the cleaner to penetrate for at least 15 minutes (but more time is permissible).
6. If removed, reconnect the flushing device or place the engine back in the water (test tank or dockside), then start the engine again and warm it to normal operating temperature. When warmed, run the engine above 1/2 throttle for at least 3-5 minutes.
7. Shut the engine off, then remove and inspect the spark plugs. Shine a small light through each spark plug bore to examine the tops of the pistons and compare the visual evidence of carbon build-up to that before the procedure. If necessary, repeat the procedure using a second can of engine tuner and following the step for severe cases.

Compression Tests

The quickest (but not necessarily most accurate) way to gauge the condition of an internal combustion engine is through a compression check. In order for an internal combustion engine to work properly, it must be able to generate sufficient compression in the combustion chamber to take advantage of the explosive force generated by the expanding gases after ignition. This is true on motors whether they are of the 2- or 4-stroke design.

If the combustion chambers, ports (2-stroke engines) or valves (4-stroke engines) are worn or damaged in some fashion as to allow pressure to escape, the engine cannot develop sufficient horsepower. Under these circumstances, combustion will not occur properly, air/fuel mixtures cannot be set to maximize power and minimize emissions. A engine with poor compression on one or more cylinders cannot given a proper tune-up, it should be overhauled.

There are two compression tests provided here, the first (TUNE-UP COMPRESSION TEST) is a quick-test used during a tune-up to determine if you should continue or stop and overhaul the motor. This test is what technicians think of when you say compression check as it measures the ability of a motor to create compression. The second (OVERHAUL LEAKAGE TEST) is a diagnostic check that is used when the engine has been partially disassembled for an overhaul (to precisely check components in the powerhead for pressure leakage) or during assembly after an overhaul (to confirm powerhead condition). The second test is also referred to as a "leak-down" test by some technicians as it measures the ability of an engine to hold pressure provided by another source and keep it from "leaking."

A compression check requires a compression gauge and a spark plug port adapter that matches the plug threads of your motor. A leakage test requires a source of pressurized air, a pressure gauge or leak-down test adapter, a spark plug port adapter and, for two-stroke motors, various plates to block of intake/exhaust ports on your engine. The low-pressures of a leakage test for a 2-stroke engine allow the use of a simple source of pressurized air such as a hand pump. The much greater pressures used on 4-stroke engines typically makes a hand pump impractical. For 4-stroke motors, use a portable air tank filled with pressurized air or best yet, an air compressor.

TUNE-UP COMPRESSION CHECK

◆ See Figure 128

When analyzing the results of a compression check, generally the actual amount of pressure measured during a compression check is not as important as the variation from cylinder-to-cylinder on the same motor. For multi-cylinder powerheads, a variation of 15 psi (100 kPa) or more is usually considered questionable. On single cylinder powerheads, a drop of 15 psi (100 kPa) from the normal compression pressure you established when it was new is cause for concern (you did do a compression test on it when it was new, didn't you?).

Ok, for the point of arguments sake let's say you bought the engine used or never checked compression the first season or so, assuming it wasn't something you needed to worry about. You're not alone. Although Johnson/Evinrude does not publish a specification for the amount of compression each of their engines should generate, a general rule of thumb

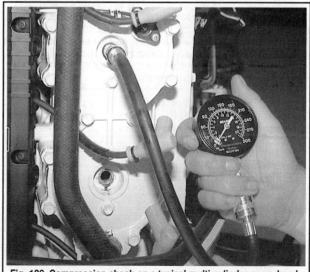

Fig. 128 Compression check on a typical multi-cylinder powerhead

that can be applied is that internal combustion engines should generate at least 100 psi (690 kPa). However, even this rule of thumb may be misleading on smaller 2-strokes, as we've seen initial specifications as low as about 70 psi (483 kPa) on units from other manufacturers (and even less if they used a form of cylinder decompression for easy starting).

Another point of comparison for your compression specifications can be similar design marine engines from other manufacturers. Other single and twin marine engines show published specifications of 115-142 psi (800-1000 kPa), while triple and 4-cylinder motors are sometimes even higher in the 185-228 psi (1300-1600 kPa) range, especially on 4-strokes.

BUT, keep in mind that these are typical specs and **not** specifications for Johnson/Evinrude motors, so don't put too much credence on your results as compared to these. Again, comparison figures with the other cylinders on the same motor (or readings when the motor was new) are most important.

When taking readings during the compression check, repeat the procedure a few times for each cylinder, recording the highest reading for that cylinder. Then, for all carbureted multi-cylinder motors covered by this manual, the compression reading on the lowest cylinder should within 15 psi (100 kPa) of the highest reading. If not, consider performing an OVERHAUL LEAKAGE CHECK to determine if the powerhead is in need of a complete or partial overhaul.

For EFI motors covered by this manual, the manufacturer states a slightly different tolerance for compression checks. On these motors, the compression reading on the lowest cylinder reading should be equal to 80% or more of the reading from the highest cylinder reading. In other words, the lowest reading should be the equal to or greater than the highest cylinder reading multiplied by 0.8. For example, if the highest reading was 150 psi (1034 kPa), then the lowest reading must be equal to or more than 150 psi x 0.8 (1035 kPa x 0.8) or 120 psi (827 kPa).

■ **If the powerhead has been in storage for an extended period, the piston rings may have relaxed. This will often lead to initially low and misleading readings. Always run an engine to normal operating temperature to ensure that the readings are accurate.**

■ **If you've never removed the spark plugs from this cylinder head before, break each one loose and retighten them, to make sure they will not seize in the head once it is warmed. Better yet, remove each one and coat the threads very lightly with some fresh anti-seize compound.**

1. Using a test tank, flush fitting adapter or other water supply, start and run the engine until it reaches normal operating temperature, then shut the engine **OFF**.

2. Disable the ignition system by removing the lanyard clip. If you do not have a lanyard, take a wire jumper lead and connect one end to a good engine ground and the other end to the metal connector inside the spark plug boot, using one jumper for each plug wire. Never simply disconnect all the plug wires.

※※ CAUTION

Removing all the spark plugs and cranking over the powerhead can lead to an explosion if raw fuel/oil sprays out of the plug holes. A plug wire could spark and ignite this mix outside of the combustion chamber if it isn't grounded to the engine. Also, on many of the ignition systems covered by this manual, cranking the engine and firing the coil without allowing the coils to discharge through the spark plug leads can lead to severe damage to the ignition system.

3. Remove all the spark plugs and be sure to keep them in order. Carefully inspect the plugs, looking for any inconsistency in coloration and for any sign of water or rust near the tip. Refer to the procedures on Spark Plugs in this section for more details.

4. Thread the compression gauge into the No. 1 spark-plug hole, taking care to not cross-thread the fitting.

5. Open the throttle to the wide open throttle position and hold it there.

■ **Some engines allow only minimal opening if the gearshift is in neutral, to guard against over-revving.**

6. Crank over the engine an equal number of times for each cylinder you test, zeroing the gauge for each cylinder.

7. If you have electric start, count the number of seconds you crank. On manual start, pull the starter rope four to five times for each cylinder you are testing.

8. Record your readings from each cylinder. When all cylinders are tested, compare the readings and determine if pressures are within the 15 psi (100 kPa) or 80% criterion, as applicable.

9. If compression readings are lower than normal for any cylinders, try a "wet" compression test, which will temporarily seal the piston rings and determine if they are the cause of the low reading. Using a can of fogging oil, fog the cylinder with a circular motion to distribute oil spray all around the perimeter of the piston. Retest the cylinder:

   a. If the compression rises noticeably in a wet test, the piston rings are sticking. You may be able to cure the problem by decarboning the powerhead.

   b. If the dry compression test was really low and no change is evident during the wet test, the cylinder is dead. The piston and/or cylinder are worn beyond specification and a powerhead overhaul or replacement is necessary.

■ **On 4-stroke engines, a problem with the valve train can also cause poor engine compression. A valve that is sticking open (due to physical damage, warpage or improper adjustment) will allow pressure to escape, lowering the compression readings.**

10. If two adjacent cylinders on a multi-cylinder engine give a similarly low reading then the problem may be a faulty head gasket. This should be suspected if there was evidence of water or rust on the spark plugs from these cylinders.

OVERHAUL LEAKAGE CHECK

Because a 2-stroke powerhead is a pump, the crankcase must be sealed against pressure created on the down stroke of the piston and vacuum created when the piston moves toward top dead center. If there are air leaks into the crankcase, insufficient fuel will be brought into the crankcase and into the cylinder for normal combustion.

■ **If it is a very small leak, the powerhead will run poorly, because the fuel mixture will be lean and cylinder temperatures will be hotter than normal.**

Air leaks are possible around any seal, O-ring, cylinder block mating surface or gasket. Always replace O-rings, gaskets and seals when service work is performed.

The 4-stroke engine also acts as a pump, drawing air/fuel mixture into the combustion chambers, but in a way slightly different from the 2-stroke engine. On 4-strokes, vacuum is created in the combustion chamber itself as the piston moves downward on the intake stroke draws the air/fuel mixture through the intake manifold and intake valve(s). Similarly, the pressure created on the exhaust stroke forces unburned gases out through the exhaust valves. Air leaks can wreak havoc for these motors as well, but again in a slightly different way. On carbureted 4-stroke motors, a lean fuel mixture can result from leaks downstream of the carburetor (leaks between the carburetor and intake manifold or between the intake manifold and engine). Air leaks can also occur on 4-stroke motors from slightly different sources, as valve trains play an important part in cylinder pressurization (a damaged or improperly adjusted intake/exhaust valve can cause problems).

If the powerhead is running, soapy water can be sprayed onto the suspected sealing areas. If bubbles develop, there is a leak at that point. Oil around sealing points and on ignition parts under the flywheel indicates a crankcase leak.

The base of the powerhead and lower crankshaft seal is impossible to check on an installed powerhead. When every test and system have been checked out and the bottom cylinder seems to be effecting performance, then the lower seal should be tested.

Adapter plates are usually available from tool manufacturers to seal the inlet, exhaust and base of the powerhead on 2-stroke motors. Adapter plates can also be manufactured by cutting metal block off plates from pieces of plate steel or aluminum. A pattern made from the gaskets can be used for an accurate shape. Seal these plates using rubber or silicone gasket making compound.

■ **Adapter plates are not necessary for leakage tests on 4-stroke motors since the camshaft can be turned to close both intake/exhaust valves, allowing for a theoretically sealed area in which to perform the leakage test (the combustion chamber).**

MAINTENANCE & TUNE-UP

1. For 2-stroke motors, prepare the powerhead for testing by removing the carburetor/manifold and exhaust manifold as necessary, then installing adapter plates over the intake ports and the exhaust ports to completely seal the powerhead. For details, refer to procedures in the Fuel System and Powerhead Overhaul sections. Into one adapter, place an air fitting which will accept a hand air pump.

■ **When installing the adapter plates on 2-stroke motors, make sure to leave the water jacket holes open.**

2. For 4-stroke motors, prepare a cylinder for testing by turning the engine (in the normal direction of rotation) until that piston is approaching top dead center of the compression stroke. This can be determined by removing the valve cover and observing the intake/exhaust valves. Both valves will close and remain so as the engine approaches the top of the compression stroke. If one valve opens as this occurs, the engine is on the exhaust stroke and crankcase must be rotated one complete revolution to bring that cylinder onto the compression stroke.

■ **Top dead center of the compression stroke is the point at which all valves should be closed on a 4-stroke motor so that power from combustion can be properly utilized. Leakage tests can only be conducted once the valves are fully seated.**

3. Using the hand pump for 2-strokes (or another regulated air source for 4-strokes), pressurize the crankcase of 2-stroke motors to about 5 psi (34.5 psi) or the combustion chamber of 4-stroke motors to 100 psi (690 kPa).

4. For 2-stroke motors, spray soapy water around the lower seal area and other sealed areas watching for bubbles that would indicate a leaking point. Turn the powerhead upside down and fill the water jacket with water. If bubbles show up in the in the water when a positive pressure is applied to the crankcase, there may be cracks or corrosion holes in the cooling system passages. These holes can cause a loss of cooling system effectiveness and lead to overheating.

5. On 4-stroke motors, listen for leakage at the carburetor/throttle body and/or exhaust (which would indicate valve sealing problems) or at the engine crankcase oil fill (which would indicate problems with the cylinder walls/compression rings). Only a very small amount of pressure should leak under normal conditions, excessive pressure leakage indicates a need for overhaul or repair. Repeat the procedure for each remaining cylinder on the 4-stroke motor.

■ **The manufacturer does not provide specifications for leak-down tests on 4-stroke motors. Industry standards vary greatly from manufacturer-to-manufacturer. Expect a small percentage of leakdown to be normal, anything from a few percent to 10 percent should not be cause for alarm. Twenty percent leakdown on a 4-stroke motor is generally considered a reason for overhaul when combined with other driveability symptoms.**

6. For 2-stroke motors, after the pressure test is completed, pull a vacuum to stress the seals in the opposite direction and watch for a pressure drop.

7. Note the leaking areas and repair or replace components, seals or gaskets, as applicable.

Spark Plugs

The spark plug performs four main functions:
- First and foremost, it provides spark for the combustion process to occur.
- It also removes heat from the combustion chamber.
- Its removal provides access to the combustion chamber (for inspection or testing) through a hole in the cylinder head.
- It acts as a dielectric insulator for the ignition system.

It is important to remember that spark plugs do not create heat, they help remove it. Anything that prevents a spark plug from removing the proper amount of heat can lead to pre-ignition, detonation, premature spark plug failure and even internal engine damage, especially in 2-stroke engines.

In the simplest of terms, the spark plug acts as the thermometer of the engine. Much like a doctor examining a patient, this "thermometer" can be used to effectively diagnose the amount of heat present in each combustion chamber.

Spark plugs are valuable tuning tools, when interpreted correctly. They will show symptoms of other problems and can reveal a great deal about the engine's overall condition. Evaluating the appearance of the spark plug's firing tip, gives visual cues to determine the engine's overall operating condition, get a feel for air/fuel ratios and even diagnose driveability problems.

As spark plugs grow older, they lose their sharp edges and material from the center and ground electrodes slowly erodes away. As the gap between these two points grows, the voltage required to bridge this gap increases proportionally. The ignition system must work harder to compensate for this higher voltage requirement and hence there are a greater rate of misfires or incomplete combustion cycles. Each misfire means lost horsepower, reduced fuel economy and higher emissions. Replacing worn out spark plugs with new ones (with sharp new edges) effectively restores the ignition system's efficiency and reduces the percentage of misfires, restoring power, economy and reducing emissions.

■ **Although spark plugs can typically be cleaned and re-gapped if they are not excessively worn, no amount of cleaning or re-gapping will return most spark plugs to original condition and it is usually best to just go ahead and replace them.**

How long spark plugs last will depend on a variety of factors, including engine compression, fuel used, gap, center/ground electrode material and the conditions in which the outboard is operated.

SPARK PLUG HEAT RANGE

◆ See Figure 129

Spark plug heat range is the ability of the plug to dissipate heat from the combustion chamber. The longer the insulator (or the farther it extends into the engine), the hotter the plug will operate; the shorter the insulator (the closer the electrode is to the engine's cooling passages) the cooler it will operate.

Selecting a spark plug with the proper heat range will ensure that the tip maintains a temperature high enough to prevent fouling, yet cool enough to prevent pre-ignition. A plug that absorbs little heat and remains too cool will quickly accumulate deposits of oil and carbon since it won't be able to burn them off. This leads to plug fouling and consequently to misfiring. A plug that absorbs too much heat will have no deposits but, due to the excessive heat, the electrodes will burn away quickly and might also lead to pre-ignition or other ignition problems.

Pre-ignition takes place when plug tips get so hot that they glow sufficiently to ignite the air/fuel mixture before the actual spark occurs. This early ignition will usually cause a pinging during heavy loads and if not corrected, will result in severe engine damage. While there are many other things that can cause pre-ignition, selecting the proper heat range spark plug will ensure that the spark plug itself is not a hot-spot source.

■ **The manufacturer recommended spark plugs are listed in the Tune-Up Specifications chart. When provided, alternate plugs for extended idle and/or extended wide-open throttle service are also listed.**

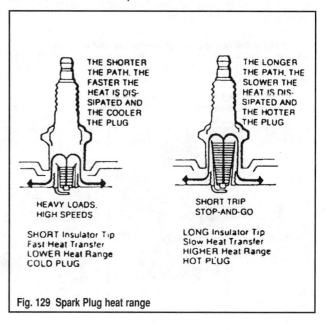

Fig. 129 Spark Plug heat range

2-44 MAINTENANCE & TUNE-UP

REMOVAL & INSTALLATION

◆ See Figures 130 thru 133

■ New technologies in spark plug and ignition system design have greatly extended spark plug life over the years. But, spark plug life will still vary greatly with engine tuning, condition and usage. In general, 2-stroke motors are a little tougher on plugs, especially if great care is not taken to maintain proper oil/fuel mixtures on pre-mix motors. On 4-stroke engines, it is not uncommon for plugs to last up to 100 hours of operation.

Typically spark plugs will require replacement once a season. The electrode on a new spark plug has a sharp edge but with use, this edge becomes rounded by wear, causing the plug gap to increase. As the gap increases, the plug's voltage requirement also increases. It requires a greater voltage to jump the wider gap and about two to three times as much voltage to fire a plug at high speeds than at idle.

■ Fouled plugs can cause hard-starting, engine mis-firing or other problems. You don't want that happening on the water. Take time, at least once a month to remove and inspect the spark plugs. Early signs of other tuning or mechanical problems may be found on the plugs that could save you from becoming stranded or even allow you to address a problem before it ruins the motor.

Tools needed for spark plug replacement include: a ratchet, short extension, spark plug socket (there are two types; either 13/16inch or 5/8inch, depending upon the type of plug), a combination spark plug gauge and gapping tool and a can of anti-seize type compound.

1. When removing spark plugs from multi-cylinder motors, work on one at a time. Don't start by removing the plug wires all at once, because unless you number them, they may become mixed up. Take a minute before you begin and number the wires with tape.
2. For safety, disconnect the negative battery cable or turn the battery switch **OFF**.
3. If the engine has been run recently, allow the engine to thoroughly cool (unless performing a compression check). Attempting to remove plugs from a hot cylinder head could cause the plugs to seize and damage the threads in the cylinder head, especially on aluminum heads!

■ To ensure an accurate reading during a compression check, the spark plugs must be removed from a hot engine. But, DO NOT force a plug if it feels like it is seized. Instead, wait until the engine has cooled, remove the plug and coat the threads lightly with anti-seize then reinstall and tighten the plug, then back off the tightened position a little less than 1/4 turn. With the plug(s) installed in this manner, re-warm the engine and conduct the compression check.

4. Carefully twist the spark plug wire boot to loosen it, then pull the boot using a twisting motion to remove it from the plug. Be sure to pull on the boot and not on the wire, otherwise the connector located inside the boot may become separated from the high-tension wire.

■ A spark plug wire removal tool is recommended as it will make removal easier and help prevent damage to the boot and wire assembly. Most tools have a wire loom that fits under the plug boot so the force of pulling upward is transmitted directly to the bottom of the boot.

5. Using compressed air (and safety glasses), blow debris from the spark plug area to assure that no harmful contaminants are allowed to enter the combustion chamber when the spark plug is removed. If compressed air is not available, use a rag or a brush to clean the area. Compressed air is available from both an air compressor or from compressed air in cans available at photography stores. In a pinch, blow up a balloon by hand and use the escaping air to blow debris from the spark plug port(s).

■ Remove the spark plugs when the engine is cold, if possible, to prevent damage to the threads. If plug removal is difficult, apply a few drops of penetrating oil to the area around the base of the plug and allow it a few minutes to work.

6. Using a spark plug socket that is equipped with a rubber insert to properly hold the plug, turn the spark plug counterclockwise to loosen and remove the spark plug from the bore.

✱✱ WARNING

Avoid the use of a flexible extension on the socket. Use of a flexible extension may allow a shear force to be applied to the plug. A shear force could break the plug off in the cylinder head, leading to costly and/or frustrating repairs. In addition, be sure to support the ratchet with your other hand - this will also help prevent the socket from damaging the plug.

7. Evaluate each cylinder's performance by comparing the spark condition. Check each spark plug to be sure they are from the same plug manufacturer and have the same heat range rating. Inspect the threads in the spark plug opening of the block and clean the threads before installing the plug.
8. When purchasing new spark plugs, always ask the dealer if there has been a spark plug change for the engine being serviced. Sometimes manufacturers will update the type of spark plug used in an engine to offer better efficiency or performance.
9. Always use a new gasket (if applicable). The gasket must be fully compressed on clean seats to complete the heat transfer process and to provide a gas tight seal in the cylinder.
10. Inspect the spark plug boot for tears or damage. If a damaged boot is found, the spark plug boot and possibly the entire wire will need replacement.
11. Check the spark plug gap prior to installing the plug. Most spark plugs do not come gapped to the proper specification.
12. Apply a thin coating of anti-seize on the thread of the plug. This is extremely important on aluminum head engines to prevent corrosion and heat from seizing the plug in the threads (which could lead to a damaged cylinder head upon removal).
13. Carefully thread the plug into the bore by hand. If resistance is felt before the plug completely bottoms, back the plug out and begin threading again.

✱✱ WARNING

Do not use the spark plug socket to thread the plugs. Always carefully thread the plug by hand or using an old plug wire/boot to prevent the possibility of crossthreading and damaging the cylinder head bore. An old plug wire/boot can be used to thread the plug if you turn the wire by hand. Should the plug begin to crossthread the wire will twist before the cylinder head would be damaged. This trick is useful when accessories or a deep cylinder head design prevents you from easily keeping fingers on the plug while it is threaded by hand.

Fig. 130 Remove the spark plug cap using a twisting motion

Fig. 131 Then loosen the plug using a spark plug socket...

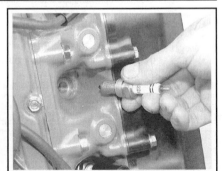

Fig. 132 ... and remove the spark plug from the cylinder head

MAINTENANCE & TUNE-UP 2-45

READING SPARK PLUGS

◆ See Figures 134 thru 139

Reading spark plugs can be a valuable tuning aid. By examining the insulator firing nose color, you can determine much about the engine's overall operating condition.

In general, a light tan/gray color tells you that the spark plug is at the optimum temperature and that the engine is in good operating condition.

Dark coloring, such as heavy black wet or dry deposits usually indicate a fouling problem. Heavy, dry deposits can indicate an overly rich condition, too cold a heat range spark plug, possible vacuum leak, low compression, overly retarded timing or too large a plug gap.

If the deposits are wet, it can be an indication of a breached head gasket, oil control from ring problems (on 4-stroke engines) or an extremely rich condition, depending on what liquid is present at the firing tip.

Also look for signs of detonation, such as silver specs, black specs or melting or breakage at the firing tip.

Compare your plugs to the illustrations shown to identify the most common plug conditions.

Fouled Spark Plugs

A spark plug is "fouled" when the insulator nose at the firing tip becomes coated with a foreign substance, such as fuel, oil or carbon. This coating makes it easier for the voltage to follow along the insulator nose and leach back down into the metal shell, grounding out, rather than bridging the gap normally.

Fuel, oil and carbon fouling can all be caused by different things but in any case, once a spark plug is fouled, it will not provide voltage to the firing tip and that cylinder will not fire properly. In many cases, the spark plug cannot be cleaned sufficiently to restore normal operation. It is therefore recommended that fouled plugs be replaced.

Signs of fouling or excessive heat must be traced quickly to prevent further deterioration of performance and to prevent possible engine damage.

Overheated Spark Plugs

When a spark plug tip shows signs of melting or is broken, it usually means that excessive heat and/or detonation was present in that particular combustion chamber or that the spark plug was suffering from thermal shock.

Fig. 133 To prevent corrosion, apply a small amount of grease to the plug and boot during installation

14. Carefully tighten the spark plug to specification using a torque wrench as follows:
- All 2-stroke motors and 70 hp 4-strokes: 17-20 ft. lbs. (23-27 Nm)
- All 5-15 hp 4-stroke motors: 14-18 ft. lbs. (19-24 Nm)
- All 40/50 hp 4-stroke motors: 11-14 ft. lbs. (15-19 Nm)

■ Whenever possible, spark plugs should be tightened to the factory torque specification. If a torque wrench is not available, and the plug you are installing is equipped with a crush washer, tighten the plug until the washer seats, then turn it 1/4 turn to crush the washer.

15. Apply a small amount of OMC Triple-Guard or a silicone dielectric grease to the ribbed, ceramic portion of the spark plug lead and inside the spark plug boot to prevent sticking, then install the boot to the spark plug and push until it clicks into place. The click may be felt or heard. Gently pull back on the boot to assure proper contact.
16. Connect the negative battery cable or turn the battery switch **ON**.
17. Test run the outboard (using a test tank or flush fitting) and insure proper operation.

Fig. 134 A normally worn spark plug should have light tan or gray deposits on the firing tip (electrode)

Fig. 135 A carbon-fouled plug, identified by soft, sooty black deposits, may indicate an improperly tuned powerhead

Fig. 136 This spark plug has been left in the powerhead too long, as evidenced by the extreme gap. Plugs with such an extreme gap can cause misfiring and stumbling accompanied by a noticeable lack of power

MAINTENANCE & TUNE-UP

Fig. 137 An oil-fouled spark plug indicates a powerhead with worn piston rings or a malfunctioning oil injection system that allows excessive oil to enter the combustion chamber

Fig. 138 A physically damaged spark plug may be evidence of severe detonation in that cylinder. Watch the cylinder carefully between services, as a continued detonation will not only damage the plug but will most likely damage the powerhead

Fig. 139 A bridged or almost bridged spark plug, identified by the build-up between the electrodes caused by excessive carbon or oil build up on the plug

Since spark plugs do not create heat by themselves, one must use this visual clue to track down the root cause of the problem. In any case, damaged firing tips most often indicate that cylinder pressures or temperatures were too high. Left unresolved, this condition usually results in more serious engine damage.

Detonation refers to a type of abnormal combustion that is usually preceded by pre-ignition. It is most often caused by a hot spot formed in the combustion chamber.

As air and fuel is drawn into the combustion chamber during the intake stroke, this hot spot will "pre-ignite" the air fuel mixture without any spark from the spark plugs.

Detonation

Detonation exerts a great deal of downward force on the pistons as they are being forced upward by the mechanical action of the connecting rods. When this occurs, the resulting concussion, shock waves and heat can be severe. Spark plug tips can be broken or melted and other internal engine components such as the pistons or connecting rods themselves can be damaged.

Left unresolved, engine damage is almost certain to occur, with the spark plug usually suffering the first signs of damage.

■ When signs of detonation or pre-ignition are observed, they are symptom of another problem. You must determine and correct the situation that caused the hot spot to form in the first place.

INSPECTION & GAPPING

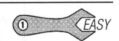

◆ See Figures 140 and 141

A particular spark plug might fit hundreds of powerheads and although the factory will typically set the gap to a pre-selected setting, this gap may not be the right one for your particular powerhead.

Insufficient spark plug gap can cause pre-ignition, detonation, even engine damage. Too much gap can result in a higher rate of misfires, noticeable loss of power, plug fouling and poor economy.

■ Refer to the Tune-Up Specifications chart for spark plug gaps.

Check spark plug gap before installation. The ground electrode (the L-shaped one connected to the body of the plug) must be parallel to the center electrode and the specified size wire gauge must pass between the electrodes with a slight drag.

Do not use a flat feeler gauge when measuring the gap on a used plug, because the reading may be inaccurate. A round-wire type gapping tool is the best way to check the gap. The correct gauge should pass through the electrode gap with a slight drag. If you're in doubt, try a wire that is one size smaller and one larger. The smaller gauge should go through easily, while the larger one shouldn't go through at all.

Wire gapping tools usually have a bending tool attached. USE IT! This tool greatly reduces the chance of breaking off the electrode and is much more accurate. Never attempt to bend or move the center electrode. Also, be

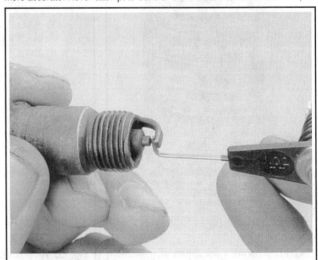

Fig. 140 Use a wire-type spark plug gapping tool to check the distance between center and ground electrodes

MAINTENANCE & TUNE-UP 2-47

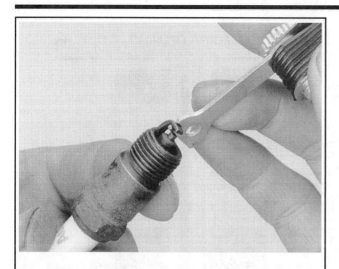

Fig. 141 Most plug gapping tools have an adjusting fitting used to bend the ground electrode

careful not to bend the side electrode too far or too often as it may weaken and break off within the engine, requiring removal of the cylinder head to retrieve it.

Spark Plug Wires

■ All Johnson/Evinrude motors, except the 40/50 hp EFI engines that utilize direct ignition coils, are equipped with secondary spark leads or spark plug wires to carry ignition voltage from the coils to the spark plugs.

TESTING

Each time you remove the engine cover, visually inspect the spark plug wires for burns, cuts or breaks in the insulation. Check the boots on the coil and at the spark plug end. Replace any wire that is damaged.

Once a year, usually when you change your spark plugs, check the resistance of the spark plug wires with an ohmmeter. Wires with excessive resistance will cause misfiring and may make the engine difficult to start. In addition worn wires will allow arcing and misfiring in humid conditions.

Remove the spark plug wire from the engine. Test the wires by connecting one lead of an ohmmeter to the coil end of the wire and the other lead to the spark plug end of the wire. Typically resistance for spark plug leads would measure approximately 7000 ohms per foot of wire. However, on carbureted Johnson/Evinrude motors, the manufacturer calls for a reading very close to or equal to zero ohms resistance. In contrast, the 70 hp EFI motors uses resistor leads that should generate 2500-4100 ohms of resistance. If a spark plug wire is found to have excessive resistance (much higher resistance than specified), or, in the case of 70 hp EFI motors, insufficient resistance, the entire set should be replaced.

■ Keep in mind that just because a spark plug wire passes a resistance test doesn't mean that it is in good shape. Cracked or deteriorated insulation will allow the circuit to misfire under load, especially when wet. Always visually check wires to cuts, cracks or breaks in the insulation. If found, run the engine in a test tank or on a flush device either at night (looking for a blueish glow from the wires that would indicate arcing) or while spraying water on them while listening for an engine stumble.

Regardless of resistance tests and visual checks, it is never a bad idea to replace spark plug leads at least every couple of years, and to keep the old ones around for spares. Think of spark plug wires as a relatively low cost item that whose replacement can also be considered maintenance.

REMOVAL & INSTALLATION

When installing a new set of spark plug wires, replace the wires one at a time so there will be no confusion. Coat the inside of the boots with OMC Triple-Guard or dielectric grease to prevent sticking. Install the boot firmly over the spark plug until it clicks into place. The click may be felt or heard. Gently pull back on the boot to assure proper contact. Repeat the process for each wire.

■ It is important to route the new spark plug wire the same as the original and install it in a similar manner on the powerhead. Improper routing of spark plug wires may cause powerhead performance problems.

Breaker Points Ignition Systems

GENERAL INFORMATION

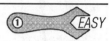

The breaker points on an outboard powerhead are an extremely important part of the ignition system. A set of points may appear to be in good condition, but they may be the source of hard starting, misfiring, or poor powerhead performance. The rules and knowledge gained from association with 4-cycle engines does not necessarily apply to a 2-cycle engine. The points should be replaced AT LEAST every 100 hours of operation (or once a year, whichever comes first).

Remember, the less an outboard engine is operated, the more care it needs. Allowing an outboard engine to remain idle will do more harm than if it is used regularly.

A breaker point set consists of two points. One is attached to a stationary bracket and does not move. The other point is attached to a moveable mount. A spring is used to keep the points in contact with each other, except when they are separated by the action of a cam built into the flywheel or machined on the crankshaft. Both points are constructed with a steel base and a tungsten cap fused to the base.

To properly diagnose magneto (spark) problems, the theory of electricity flow must be understood. The flow of electricity through a wire may be compared with the flow of water through a pipe. Consider the voltage in the wire as the water pressure in the pipe and the amperes as the volume of water. Now, if the water pipe is broken, the water does not reach the end of the pipe. In a similar manner if the wire is broken the flow of electricity is broken. If the pipe springs a leak, the amount of water reaching the end of the pipe is reduced. Same with the wire. If the installation is defective or the wire becomes grounded, the amount of electricity (amperes) reaching the end of the wire is reduced.

■ For more details on the theory of electricity flow, please refer to Understanding and Troubleshooting Electrical Systems in the Ignition and Electrical System section.

INSPECTION & TESTING

◆ See Figures 142, 143 and 144

Rough or discolored contact surfaces are sufficient reason for replacement. The cam follower will usually have worn away by the time the points have become unsatisfactory for efficient service.

Check the resistance across the contacts. If the test indicates zero resistance, the points are serviceable. A slight resistance across the points will affect idle operation. A high resistance may cause the ignition system to malfunction and loss of spark. Therefore, if any resistance across the points is indicated, the point set should be replaced.

Check the wiring carefully, inspect the points closely and adjust them accurately. The point gap setting for all breaker point ignition powerheads covered in this section (except the 2 and 2.3 hp motors) is 0.020 in. (0.51mm) for USED points, and USUALLY 0.22 in. (0.56mm) for new point sets. The point gap for the 2.0 hp powerhead is 0.008 in. (both new and used points) and the gap for the 2.3 hp powerhead is 0.014 in. (0.35mm).

2-48 MAINTENANCE & TUNE-UP

Fig. 142 Worn and corroded breaker points unfit for further service

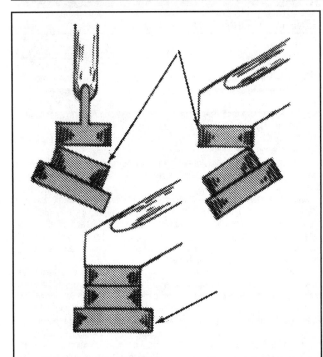

Fig. 143 Drawing to illustrate proper point alignment, bottom set, compared with exaggerated misalignment of the other two

Even though the 3.3 hp powerhead is very similar in almost every detail to the 2.3 hp, and was introduced in the same year, the point gap for the 3.3 hp engine is 0.020 in. (0.5mm) but for both new and used sets of points.

REPLACEMENT

◆ See Figures 145 thru 160

Magnetos installed on outboard engines will usually operate over extremely long periods of time without requiring adjustment or repair. However, if ignition system problems are encountered, and the usual corrective actions such as replacement of spark plugs does not correct the problem, the magneto output should be checked to determine if the unit is functioning properly.

1. Remove the hood or enough of the powerhead cover to expose the flywheel.
2. If equipped, disconnect the battery connections from the battery terminals, if a battery is used to crank the powerhead. If a hand starter is installed, remove the attaching hardware from the legs of the starter assembly and lift the starter free.
3. On hand rewind starter models, a round ratchet plate is attached to the flywheel to allow the hand starter to engage in the ratchet and thus rotate the flywheel. This plate must be removed before the flywheel nut is removed.

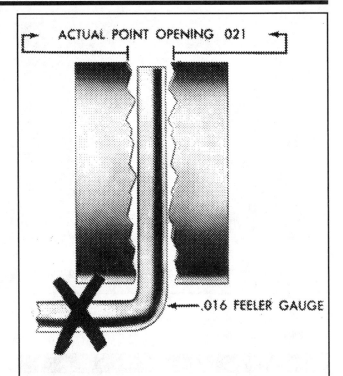

Fig. 144 Drawing to depict how a 0.016 in. feeler gauge may be inserted between a badly worn set of points and the actual opening is 0.021 in. The point set must be in good condition to obtain an accurate adjustment

4. Remove the nut securing the flywheel to the crankshaft. It may be necessary to use some type of flywheel strap to prevent the flywheel from turning as the nut is loosened.

■ After the flywheel has been removed it should be placed on the bench with the magnets facing upward. This position will help prevent small particles from becoming attached to the magnets

5. Install the proper flywheel puller using the same screw holes in the flywheel that are used to secure the ratchet plate removed earlier. Never attempt to use a puller that pulls on the outside edge of the flywheel or the flywheel may be damaged. After the puller is installed, tighten the center screw onto the end of the crankshaft. Continue tightening the screw until the flywheel is released from the crankshaft. Remove the flywheel.

✳✳ CAUTION

Do not strike the puller center bolt with a hammer in an attempt to dislodge the flywheel. Such action could seriously damage the lower seal and/or lower bearing.

Stop, and carefully observe the magneto and associated wiring layout. Study how the magneto is assembled. Take time to make notes on the wire routing. Observe how the heels of the laminated core, with the coil attached, is flush with the boss on the armature plate. These items must be replaced in their proper positions. You MIGHT want to take a digital photo of the engine with the flywheel removed: one from the top, and a couple from the sides showing the wiring and arrangement of parts.

■ The armature plate does not have to be removed to service the magneto. If it is necessary to remove the plate for other service work, such as to replace the coil or to replace the top seal, see the information in the Ignition and Electrical System section.

6. Remove the screw attaching the wires from the coil and condenser to one set of points. On engines equipped with a key switch, "kill" button, or "runaway" switch, a ground wire is also connected to this screw.
7. Using a pair of needle-nose pliers, remove the wire clip from the post protruding through the center of the points.
8. Again, with the needle-nose pliers, remove the flat retainer holding the set of points together.
9. Lift the moveable side of the points free of the other half of the set.

MAINTENANCE & TUNE-UP 2-49

Fig. 145 For access first remove the hand rewind starter...

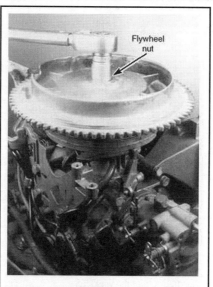

Fig. 146 ...then unbolt...

Fig. 147 ...and remove the flywheel

Fig. 148 Ignition component locations - breaker point models

Fig. 149 Remove the screw securing coil and condenser wiring

Fig. 150 Remove the wire clip from the point set post...

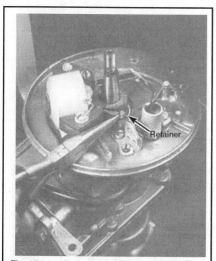

Fig. 151 ...then remove the flat retainer holding the point set together

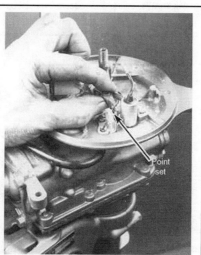

Fig. 152 Lift the moveable side of the points free...

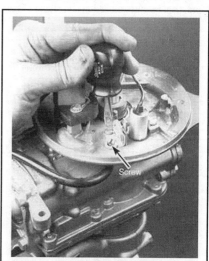

Fig. 153 ...then unscrew the fixed side of the points

MAINTENANCE & TUNE-UP

Fig. 154 Always replace the condenser...

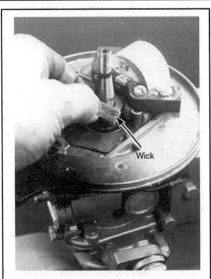
Fig. 155 ...and the wick

Fig. 156 Set the proper breaker point gap using a feeler gauge

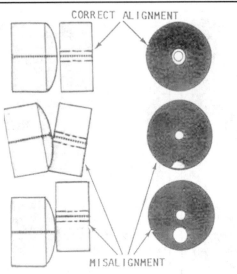
Fig. 157 Before setting the breaker point gap, the points must be properly aligned (top). Always bend the stationary point, never the breaker lever. Attempting to adjust an old worn set of points is not practical because oxidation and pitting of the points will always give a false reading

10. Remove the hold-down screw securing the non-moveable half of the point set to the armature plate.

11. Remove the hold down screw securing the condenser to the armature plate. Observe how the condenser sets into a recess in the armature plate.

■ The wick, mounted in a bracket under the coil, can be replaced without removing the armature plate. The wick should be replaced each and every time the breaker points are replaced.

12. To replace the wick, simply loosen all three coil retaining screws and remove the one screw through the wick holder. Lift the coil slightly and remove the wick and wick holder. Slide the new wick into the holder; install the holder and wick under the coil; and secure it in place with the retaining screw. Adjust the coil as described earlier in this section and tighten the three screws.

To install:

13. Install the condenser and secure it in place with the hold-down screw(s).

■ Hold the base side of the points and the flat retainer. Notice how the base has a bar at right angle to the points. Observe the hole in the bar. Observe the flat retainer. Notice that one side has a slight indentation. When the points are installed, this indentation will slip into the hole in the base bar.

14. Hold the base side of the points and slide it down over the anchor pin onto the armature plate. Install the wavy washer and hold-down screw to secure the point base to the armature plate. Tighten the hold-down screw securely.

Fig. 158 Then reinstall the flywheel...

Fig. 159 ...don't forget any washers used...

Fig. 160 ...and the hand rewind starter ratchet assembly

MAINTENANCE & TUNE-UP 2-51

15. Hold the moveable arm and slide the points down over post, and at the same post of the base points. Continue to work the points on down into the base.
16. Observe the points. The points should be together and the spring part of the moveable arm on the inside of the flat post.
17. Install the flat retainer onto the flat bar of the base points. Check to be sure the flat spring from the other side of the points is on the inside of the retainer. Push the retainer inward until the indentation slips into the hole in the base. The retainer must be horizontal with the armature plate.

Install the wire clip into the groove of the post.

✱✱ WARNING

As the coil, condenser, and "kill" switch wire are being attached to the point set, take the following precautions and adjustments: (a) The wire between the coil and the points should be tucked back under the coil and as far away from the crankshaft as possible; (b) The condenser wire leaving the top of the condenser and connected to the point set, should be bent downward to prevent the flywheel from making contact with the wire. A countless number of installations have been made only to have the flywheel rub against the condenser wire and cause failure of the ignition system; (c) Check to be sure all wires connected to the point set are bent downward toward the armature plate. The wires must not touch the plate. If any of the wires make contact with the armature plate, the ignition system will be grounded and the engine will fail to start.

18. The point spring tension is predetermined at the factory and does not require adjustment. Once the point set is properly installed, all should be well. In most cases, breaker contact and alignment will not be necessary. If a slight alignment adjustment should be required, carefully bend the insulated part of the point set.
19. Point Adjustment:

    a. Install the flywheel nut onto the end of the crankshaft. Now, rotate the crankshaft clockwise and at the same time observe the cam on the crankshaft. Continue rotating the crankshaft until the rubbing block of the point set is at the high point of the cam.

    b. At this position, use a wire gauge or feeler gauge and set the points. The point gap setting for all breaker point ignition powerheads covered in this section (except the 2 and 2.3 hp motors) is 0.020 in. (0.51mm) for USED points, and USUALLY 0.22 in. (0.56mm) for new point sets. The point gap for the 2.0 hp powerhead is 0.008 in. (both new and used points) and the gap for the 2.3 hp powerhead is 0.014 in. (0.35mm). Even though the 3.3 hp powerhead is very similar in almost every detail to the 2.3 hp, and was introduced in the same year, the point gap for the 3.3 hp engine is 0.020 in. (0.5mm) but for both new and used sets of points.

■ **A wire gauge will always give a more accurate adjustment than a feeler gauge. However, neither will guarantee an accurate adjustment on a used set of points.**

    c. Work the gauge between the points and, at the same time, turn the eccentric on the armature plate until the proper adjustment is obtained.

    d. Rotate the crankshaft a complete revolution and again check the gap adjustment. After the crankshaft has been turned and the points are on the high point of the cam, check to be sure the hold-down screw is tight against the base. There is enough clearance to allow the eccentric on the base points to turn. If the hold-down screw is tightened after the point adjustment has been made, it is very likely the adjustment will be changed.

20. Place the key in the crankshaft keyway, with the outer edge parallel to the centerline of the crankshaft, as indicated in the accompanying illustration. Apply OMC Nut Lock, or equivalent, to the cam drive pan, and then install the pin into the crankshaft. Install the cam with the side marked top facing up.
21. Slide the flywheel down over the crankshaft with the keyway in the flywheel aligned with the key on the crankshaft.
22. Rotate the flywheel clockwise and check to be sure the flywheel does not contact any part of the magneto or the wiring.
23. Place the ratchet for the starter on top of the flywheel and install the three 7/16 in. screws. On some models, a plate retainer covers these screws.
24. Coat the flywheel nut with OMC Gasket Sealing Compound, or equivalent, and then thread the flywheel nut onto the crankshaft and tighten it to the following torque value:
    - Colt & Junior: 22-25 ft. lbs. (30-40 Nm)
    - 2/2.3/3.3 hp: 29-33 ft. lbs. (40-45 Nm)
    - 3/4 hp & 4D: 30-40 ft. lbs. (40-54 Nm)
    - 5 hp thru 8 hp: 40-50 ft. lbs. (54-70 Nm)
25. After the ratchet and flywheel nut have been installed, install the hand starter over the flywheel, if one is used. Check to be sure the ratchet engages the flywheel properly.
26. If equipped, connect the battery leads to the battery terminals.

Electronic (CDI/UFI) Ignition Systems

INSPECTION

Modern electronic ignition systems have become one of the most reliable components on an outboard. There is very little maintenance involved in the operation of these ignition systems and even less to repair if they fail. Most systems are sealed and there is no option other than to replace failed components.

1. Just as a tune-up is pointless on an engine with no compression, a installing new spark plugs will not do much for an engine with a damaged ignition system. At each tune-up, visually inspect all ignition system components for signs of obvious defects. Look for signs of burnt, cracked or broken insulation. Replace wires or components with obvious defects. If spark plug condition suggests weak or no spark on one or more cylinders, perform ignition system testing to eliminate possible worn or defective components.

If trouble is suspected, it is very important to narrow down the problem to the ignition system and replace the correct components rather than just replace parts hoping to solve the problem. Electronic components can be very expensive and are usually not returnable.

Refer to the "Ignition and Electrical" section for more information on troubleshooting and repairing ignition systems.

TIMING AND SYNCHRONIZATION

◆ See Figures 161 and 162

In simple terms, synchronization is timing the fuel system to the ignition. Timing and synchronization ensures that as the throttle is advanced to increase powerhead rpm, the fuel and the ignition systems are both advanced equally and at the same rate.

Various models have unique methods of checking ignition timing. As appropriate, these differences will be explained in detail in the text.

Any time the fuel system or the ignition system on a powerhead is serviced to replace a faulty part or any adjustments are made for any reason, powerhead timing and synchronization must be carefully checked and verified.

Depending on the engine, adjustment of the timing and synchronization can be extremely important to obtain maximum efficiency. The powerhead cannot perform properly and produce its designed horsepower output if the fuel and ignition systems have not been precisely adjusted. We say, depending on the engine because some of the models covered by this manual are equipped with a single, carburetor or an EFI system both of which require few, if any adjustments on installed.

As a matter of fact, because of the EPA regulated carburetors used on most of the motors covered here, very few adjustments are possible on most carburetors. There are no periodic mixture adjustments necessary on most of the motors covered by this manual. Most high-speed jets are fixed units and the low speed mixture screws are usually sealed to prevent unnecessary tampering. However, any carburetor will require initial set-up and adjustment after disassembly or rebuilding. Also, any motor equipped with multiple carburetors will require synchronization with each other after the carburetors have been removed or separated.

Although some of the motors covered by this manual utilize fully electronically controlled ignition and timing systems, most of the 2-stroke motors allow for some form of timing adjustment. Care should be taken to ensure settings are correct during each tune-up.

Many models have timing marks on the flywheel and CDI base. A timing light is normally used to check the ignition timing with the powerhead operating (dynamically).

2-52 MAINTENANCE & TUNE-UP

Fig. 161 On some models a special timing pointer can be installed to help when checking or setting ignition timing

Fig. 162 Flywheel timing marks, aligned with a timing pointer tool

Many of the smaller models equipped with electronic ignitions do not have timing adjustments, as the system is completely controlled by the ignition module/powerpack. Most of the larger motors, excluding the EFI units, require idle and/or maximum advance timing adjustments.

■ Before making any adjustments to the ignition timing or synchronizing the ignition to the fuel system, both systems should be verified to be in good working order.

Timing and synchronizing the ignition and fuel systems on an outboard motor are critical adjustments. The following equipment is essential and is called out repeatedly in this section. This equipment must be used as described, unless otherwise instructed by the equipment manufacturer. Naturally, the equipment is removed following completion of the adjustments.

For many of the adjustments, the manufacturer recommends the use of a test wheel instead of a normal propeller in order to put a load on the engine and propeller shaft. The use of the test wheel prevents the engine from excessive rpm.

• Timing Light - During many procedures in this section, the timing mark on the flywheel must be aligned with a stationary timing mark on the engine while the powerhead is being cranked or is running. Only through use of a timing light connected to the No. 1 spark plug lead, can the timing mark on the flywheel be observed while the engine is operating

• Tachometer - A tachometer connected to the powerhead must be used to accurately determine engine speed during idle and high-speed adjustment. Engine speed readings range from 0-6,000 rpm in increments of 100 rpm. Choose a tachometers with solid state electronic circuits which eliminates the need for relays or batteries and contribute to their accuracy. For maximum performance, the idle rpm should be adjusted under actual operating conditions. Under such conditions it might be necessary to attach a tachometer closer to the powerhead than the one installed on the control panel.

• Flywheel Rotation - The instructions may call for rotating the flywheel until certain marks are aligned with the timing pointer. When the flywheel must be rotated, always move the flywheel in the indicated direction. If the flywheel should be rotated in the opposite direction, the water pump impeller vanes would be twisted. Should the powerhead be started with the pump tangs bent back in the wrong direction, the tangs may not have time to bend in the correct direction before they are damaged. The least amount of damage to the water pump will affect cooling of the powerhead

• Test Tank - Since the engine must be operated at various times and engine speeds during some procedures, a test tank or moving the boat into a body of water, is necessary. If installing the engine in a test tank, outfit the engine with an appropriate test propeller

✱✱ CAUTION

Water must circulate through the lower unit to the powerhead anytime the powerhead is operating to prevent damage to the water pump in the lower unit. Just a few seconds without water will damage the water pump impeller.

■ Remember the powerhead will not start without the emergency tether in place behind the kill switch knob.

✱✱ CAUTION

Never operate the powerhead above a fast idle with a flush attachment connected to the lower unit. Operating the powerhead at a high rpm with no load on the propeller shaft could cause the powerhead to runaway causing extensive damage to the unit.

Homemade Synchronization Tool

◆ See Figures 163 and 164

■ When making a synchronization adjustment, it is important to understand exactly what to look for and why. The critical time when the throttle shaft in the carburetor begins to move is of the utmost importance. First, realize that the instant the cam follower makes contact with the cam is not the point at which the throttle shaft starts to move. Instead, the critical instant for adjustment is when the follower hits the designated position and the throttle shaft at the carburetor begins to move.

On most motors, a considerable amount of play exists between the follower at the top of the carburetor (through the linkage) to the actual throttle shaft. Therefore, the most important consideration is to watch for movement of the throttle shaft, and not the follower. Movement of the shaft can be exaggerated by attaching a short piece of stiff wire (or a drill bit) to an alligator clip; grinding down the teeth on one side of the clip; and then attaching the clip to the throttle shaft, as shown in the illustrations. Movement of the drill bit or a jiggling of the wire will instantly indicate movement of the shaft.

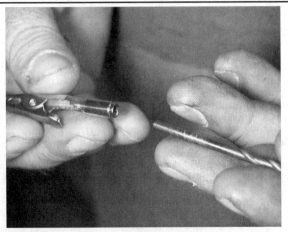

Fig. 163 Use an alligator clip and drill bit (or piece of wire) as a homemade tool

MAINTENANCE & TUNE-UP 2-53

Fig. 164 Attach the clip to the carburetor throttle shaft and watch the bit or wire for movement

2.0-3.5 Hp Models

ADJUSTMENTS

MODERATE

These are the simplest motors covered by this manual and, as such, have the fewest amounts of periodic adjustments. The Idle Speed should be checked and adjusted periodically. There is no low speed needle screw on these motors, instead, the low speed mixture is adjusted by repositioning a clip on the throttle jet needle (and doing so requires partial disassembly of the carburetor). If a change from sea level to high altitude use or vice versa requires an adjustment to be made, refer to the section on the Carbureted Fuel System for more details regarding Throttle Jet Needle Adjustment.

Idle Speed Adjustment

The 2.0-3.5 Hp models covered by this manual utilize a spring-wound idle speed screw mounted into the side of the carburetor body. Periodic checking and adjustment of this setting should be considered a part of routine maintenance.

Here's your excuse, show this page to your better half in case she (or he) doesn't believe you. It's time to take the boat out because that's how the manufacturer wants you to adjust the idle speed. Not in a test tank, not on a trailer or even attached to a dock, but with the motor mounted on a boat, underway. For safety, use an assistant to navigate while you make the adjustment.

■ **Always adjust the idle speed with the correct propeller installed.**

Refer to the Tune-Up Specifications chart for idle speed specs.

1. Set a shop tachometer to the two-cycle setting, then connect it to the primary lead for the ignition coil. Start the motor and run it at 1/2 throttle in forward gear until it warms to normal operating temperature.
2. Once the engine is fully warmed turn the idle speed adjustment screw until the engine runs in the correct rpm range. Turn the screw slowly, less than one turn at a time, allowing at least 15 seconds between adjustments for the engine to respond and the idle speed to stabilize.
3. Once adjustments are complete, confirm proper operation by running the engine at various throttle positions. If the engine is running too rich it will generally hesitate or stumble when the throttle is advanced more than 2/3rds.

5 Hp (109cc) Models

ADJUSTMENTS

These are the simplest motors covered by this manual and, as such, have the fewest amounts of periodic adjustments. The carburetor low speed setting should be checked periodically, but only adjusted if abnormal operation is noted.

Carburetor Low Speed Mixture

The adjustment is made with the boat in a body of water, with the correct propeller installed, the motor in forward gear and the boat unrestrained (not tied to the dock or trailer). For safety, you'll need an assistant to navigate while you make the adjustments. Start the engine and allow it to reach normal operating temperature.

1. Start and run the motor until normal operating temperature is reached.

✱✱ CAUTION

Water must circulate through the lower unit to the engine any time the engine is run to prevent damage to the water pump in the lower unit. Just a few seconds without water will damage the water pump.

2. With the engine running in forward gear, close the throttle and make sure the engine runs at 900-1000 rpm. If necessary, adjust the tiller control knob and/or the throttle stop screw on the top side of the carburetor to close or open the throttle slightly to achieve proper idle speed.

■ **The throttle stop screw and the low speed mixture screws are mounted on opposite sides at the top of the carburetor, don't confuse them. The mixture screw is on the same side as fuel inlet line. For more details, refer to the illustrations under Carburetor service procedures in the Fuel System section.**

3. Operate the outboard in forward at or near full throttle for a full minute. Reduce speed suddenly to a low idle and shift into neutral. The powerhead should continue to operate smoothly. If the powerhead pops or stalls, the air/fuel mixture is probably too lean. Rotate the low speed needle 1/16th turn counterclockwise, allowing about 15 seconds between adjustments, until the powerhead responds as expected. Repeat the full throttle test and sudden deceleration with the shift into neutral to check each adjustment.

■ **If the engine does not respond properly to these adjustments, check for other problems with the fuel or ignition systems. Make sure a quality fuel is being used along with the proper amount of and type of oil.**

4. Once a satisfactory adjustment has been achieved, shut **OFF** the engine.

Colt/Junior, 3/4 Hp and 4 Deluxe Models

ADJUSTMENTS

MODERATE

■ **The ignition timing is not adjustable on these models.**

The following procedures provide detailed instructions to adjust the cam follower pickup point, carburetor high speed mixture (Colt/Junior models only), carburetor low speed mixture and throttle cable. Procedures should be performed exactly as directed and in the order given to ensure proper adjustments.

Camshaft Follower Pickup Point

◆ See Figures 165 and 166

For access remove the engine covers. On some models this means the port and starboard covers. On the Colt/Junior, remove the starter housing and the fuel tank support bracket.

Movement of the throttle shaft can be exaggerated by attaching a short piece of stiff wire or drill bit to an alligator clip; grinding down the teeth on one side of the clip; and then attaching the clip to the throttle shaft, as shown under Homemade Synchronization Tool. Attach a homemade synchronization tool to the carburetor throttle shaft. Attach the clip on the port side for 3hp and 4hp models, - on the starboard side for 4D models. The wire jiggling will instantly indicate movement of the shaft.

Advance the throttle until the tip of the wire begins to move. When movement occurs, the center of the roller must align with the single mark on the cam.

■ **Some early 1990s models may have a cam which contains 2 marks. If so equipped, the roller should be centered between the two marks at the moment the synchronization tool starts to move.**

2-54 MAINTENANCE & TUNE-UP

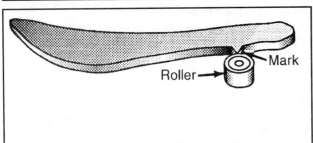

Fig. 165 Roller set at the cam mark - most models except 4D powerheads

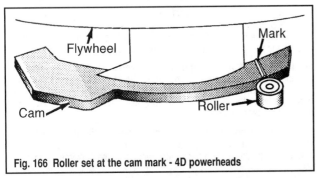

Fig. 166 Roller set at the cam mark - 4D powerheads

If not, back out the cam follower adjustment screw on the port side on the throttle shaft, with a ball hex driver (OMC #327622 or equivalent), until the throttle valve is completely closed. Now, rotate the screw clockwise until the throttle shaft just starts to rotate.

Carburetor Mixture Adjustments

Colt/Junior Models

The manufacturer recommends this adjustment will be most accurate if made on a boat that is underway. Obviously there are some safety concerns for making this adjustment in that manner. You'd need an assistant, and a relatively calm and empty body of water. If necessary, this adjustment can also be done in a test tank or on a boat that is securely affixed to a trailer or dock.

1. Temporarily install knobs onto the low speed and high speed mixture needles so they can be turned easily during this procedure.
2. With the motor on a launched craft or in a test tank, start and allow the engine to warm up normally. During warm up, run the engine at or a little more than about half throttle.

✱✱ WARNING

During warm up, take this opportunity to inspect the motor for potential fuel leaks. This is especially important if any component of the fuel system was just removed/replaced or otherwise serviced.

3. Once the engine is fully warmed, run the motor at full throttle and carefully adjust the high speed needle until the motor runs at the highest consistent rpm.
4. Now slow the engine to an idle speed of about 700-800 rpm. Next adjust the low speed needle until the motor runs at the highest consistent idle RPM. Turn the needle in small increments, allowing at least 15 seconds for the idle speed to stabilize before making further adjustments.
5. After the final low speed adjustment is made, turn the low speed needle out COUNTERCLOCKWISE 1/8th of a turn to make sure it is not TOO lean at idle speeds.
6. After the low speed needle is set, go back and READJUST the high speed needle.
7. After the final high speed adjustment is made, turn the high speed needle out COUNTERCLOCKWISE 1/8th of a turn to make sure it is not TOO lean above idle speeds. Hold the needle steady and tighten the packing nut to lock the needle in this position.
8. Install the low speed knob with the pointer up.

■ **Make sure the needle positions are not disturbed.**

9. Install the high speed knob with the pointer down.

3/4 Hp and 4 Deluxe Models
◆ See Figures 167 thru 170

To adjust the carburetor: First, adjustment is made with the boat in a body of water, with the correct propeller installed, the motor in forward gear and the boat unrestrained (not tied to the dock or trailer). For safety, you'll need an assistant to navigate while you make the adjustments. Start the engine and allow it to reach normal operating temperature.

1. To make the low speed adjustment: Loosen the two throttle cable nuts, one on each side of the throttle cable support bracket. Back out the idle speed adjustment screw, located at the rear of the powerhead just under the flywheel, - to allow the stator plate to contact the stop cast into the powerhead.

✱✱ CAUTION

Water must circulate through the lower unit to the engine any time the engine is run to prevent damage to the water pump in the lower unit. Just a few seconds without water will damage the water pump.

2. Start the powerhead and rotate the idle speed adjustment screw until the powerhead idles in the proper range according to the listing under Idle Speed (RPM) in Gear for your motor under the Tune-Up Specifications chart.
3. Adjust the low speed needle until the highest consistent powerhead speed is attained. Allow about 15 seconds between each adjustment for the engine to stabilize.

■ **Turn the screw slowly, in fractional 1/4 or 1/8 increments for each adjustment, then allow time for it to take effect.**

4. Once the engine has stabilized and runs smoothly at the highest possible idle speed, rotate the low speed needle counterclockwise 1/8th turn to prevent the powerhead from operating too lean at low speeds, particularly in neutral.
5. With the engine still shifted into forward gear, rotate the idle speed adjustment screw until the powerhead speed is reduced to the lowest setting in the idle rpm range (Tune-Up Specifications chart).
6. Operate the outboard in forward at or near full throttle for a full minute. Reduce speed suddenly to a low idle and shift into neutral. The powerhead should continue to operate smoothly. If the powerhead pops or stalls, the air/fuel mixture is probably too lean. Rotate the low speed needle 1/16th turn counterclockwise, allowing about 15 seconds between adjustments, until the powerhead responds as expected. Repeat the full throttle test and sudden deceleration with the shift into neutral to check each adjustment.

■ **If the engine does not respond properly to these adjustments, check for other problems with the fuel or ignition systems. Make sure a quality fuel is being used along with the proper amount of and type of oil.**

7. Once a satisfactory adjustment has been achieved, shut **OFF** the engine and adjust the throttle cable.

Throttle Cable
◆ See Figures 167 and 168

Adjust the throttle cable with the engine **OFF**. Rotate the throttle twist grip to the idle position. Move the stator plate against the idle speed adjustment screw. Eliminate the slack in the throttle cable by adjusting the two nuts on the cable support bracket.

5-8 Hp (164cc) 2-Stroke Models

ADJUSTMENTS

■ **The ignition timing is not adjustable on the 1991 and later versions of these models. However the 1990 versions of these motors usually used a slightly different ignition system which allowed the technician to adjust the idle timing.**

The following procedures provide detailed instructions to adjust the timing pointer/idle timing (1990 models only), throttle cable, carburetor low speed mixture and wide open throttle stop. Procedures should be performed exactly as directed and in the order given to ensure proper adjustments.

MAINTENANCE & TUNE-UP 2-55

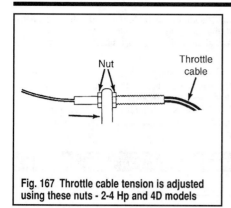

Fig. 167 Throttle cable tension is adjusted using these nuts - 2-4 Hp and 4D models

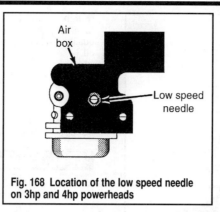

Fig. 168 Location of the low speed needle on 3hp and 4hp powerheads

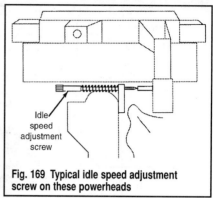

Fig. 169 Typical idle speed adjustment screw on these powerheads

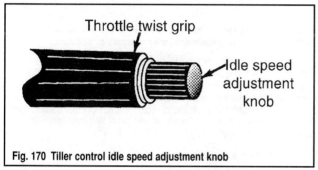

Fig. 170 Tiller control idle speed adjustment knob

Timing Pointer and Idle Timing Adjustment (1990 Models Only)

The ignition system used on most 1990 models allowed for idle timing adjustments. The first step in this procedure however is to verify that the timing pointer is properly aligned (this is necessary if the pointer itself or the manual starter has been disturbed in any way since the last time it was checked). In order to do this you'll need a dial gauge or the OMC piston stop tool (#384887).

1. Remove both of the spark plugs to relieve engine compression and to make sure the motor is not accidentally started during this procedure.
2. Center the timing pointer. Rotate the flywheel CLOCKWISE until the cast-in TDC mark is approximately 1 1/2 in. (4cm) PAST the timing pointer.
3. Install a dial gauge or the OMC piston stop tool into the No. 1 cylinder (top) spark plug hole. Adjust the center portion of the tool or the dial gauge to contact the top of the No. 1 piston. Then lock the tool in place with the lockring OR zero the gauge.
4. With the flywheel set in this position mark the flywheel (call it Mark "A") directly across from the timing pointer in this position.
5. Continue to turn the flywheel in a clockwise position until the piston either contacts the OMC lock tool OR the gauge zeros out again. Mark this spot on the flywheel (call it Mark "B").
6. Remove the lock tool or dial gauge.
7. Use a scale along the edge of the flywheel to measure the distance between the two marks you've made (A and B), then mark the centerpoint between them (call it Mark "C"). IF Mark C and the cast TDC mark on the flywheel are the same, the timing pointer is correctly aligned. HOWEVER, If not, turn the flywheel to align Mark C with the timing pointer, then hold the flywheel in this position as you loosen the screw securing the timing pointer and adjust the pointer so it now aligns with the cast-in TDC mark of the flywheel. Tighten the pointer screw.
8. Install the spark plugs and now you can set the idle timing.
9. Turn the idle speed adjustment knob (located at the end of the tiller grip) COUNTERCLOCKWISE (while facing the steering handle) to the lowest idle speed position.
10. Remove the screw retaining the throttle cable bracket to the powerhead (taking care NOT to lose the washer under the bracket).
11. IF the carburetor was rebuilt/replaced and/or the powerhead rebuilt, make the initial low speed mixture adjustment (LIGHTLY seat the idle speed screw and back it out the number of turns listed in the Carburetor Set Up Specifications chart from the Fuel System section).

12. Loosen the throttle cracking screw (the spring loaded throttle stop screw threaded vertically downward). Turn IN the screw 4 turns after it contacts the throttle valve link.
13. Connect a suitable timing light to the No. 1 cylinder spark plug wire.
14. Start and run the engine in Neutral until it reaches normal operating temperature.
15. Check the timing using a timing light. With the throttle valve at or near a closed position, adjust the idle spark stop (a slotted screw on the side of the powerhead, just below the flywheel, which secures a slotted bracket) and slide the bracket as necessary to achieve a proper idle timing specification, as listed in the Tune-Up Specifications chart in this section. Lock the bracket in place with the mounting screw.
16. Stop the engine for safety, then adjust the throttle cable bracket to the cable and linkage are preloaded TOWARD the throttle closed position by one complete turn of the throttle cable bracket.
17. Skip the Throttle Cable Adjustment and proceed with Carburetor Low Speed and Wide Open Throttle Settings, as detailed later in this section.

Throttle Cable Adjustment

◆ See Figure 170

To adjust the throttle cable:
1. Rotate the idle speed adjustment knob counterclockwise (when facing the tiller handle from the motor operator's position) to the minimum speed position. Then, rotate the twist grip throttle clockwise to the lowest throttle setting.
2. Verify the throttle cam is in contact with the idle stop.
3. Remove the bolt retaining the throttle cable bracket to the powerhead.
4. Rotate the throttle cable bracket counterclockwise through one revolution to preload the throttle in the closed position and attach the bracket to the powerhead with the single bolt. Tighten the bolt securely.

Carburetor Low Speed and Wide Open Throttle Settings

◆ See Figure 171

To adjust the carburetor the motor must be mounted on the boat and in the water or in a test tank. In both cases, the appropriate test propeller must be installed (refer to the Tune-Up specifications chart).

1. Before starting the engine - obtain a long shank, ball hex driver (like OMC #327622, or equivalent). Insert the driver through the opening in the air box (if applicable) and back out the throttle cam follower screw until the follower no longer contacts the cam. If necessary, advance the throttle to gain access to the throttle cam screw. Back off the cam screw to the lowest setting and check to be sure the follower clears the cam.
2. If the carburetor or powerhead was rebuilt/replaced and the low speed needle was not given a preliminary adjustment yet, do that now. Rotate the low speed needle, normally located on the starboard side just above the throttle shaft, clockwise until the needle is lightly seated. Back the needle out exactly the proper number of turns as designated by the Initial Low Speed Setting given in the Carburetor Set-Up Specifications Chart under the Fuel System section. Tighten the throttle "cracking" screw, identified in the accompanying illustration. Rotate the screw clockwise until the spring is fully but lightly compressed (usually about 4 complete turns after the screw tip makes contact with the throttle valve link.)

2-56 MAINTENANCE & TUNE-UP

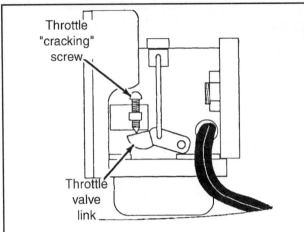

Fig. 171 Location of the throttle "cracking" screw on 5-8hp 2-stroke powerheads

3. Start the powerhead and allow it to reach normal operating temperature.

✻✻ CAUTION

Water must circulate through the lower unit to the engine any time the engine is run to prevent damage to the water pump in the lower unit. Just a few seconds without water will damage the water pump.

4. With the outboard at normal operating temperature, shift into forward gear and adjust the idle cracking screw to the center range of the Idle Speed (RPM) in Gear listed for your motor under the Tune-Up Specifications chart.

5. Adjust the low speed needle until the highest consistent powerhead speed is attained. Allow about 15 seconds between each adjustment for the engine to stabilize.

■ Turn the screw slowly, in fractional 1/4 or 1/8 increments for each adjustment, then allow time for it to take effect.

6. Once the engine has stabilized and runs smoothly at the highest possible idle speed, rotate the low speed needle counterclockwise 1/8th turn to prevent the powerhead from operating too lean at low speeds, particularly in neutral.

7. Rotate the "cracking" screw again until the powerhead idles at about the center of the specification range for Idle in Gear.

8. Stop the engine and, as necessary, remove the air intake silencer for access to the carburetor throttle shaft and throttle follower cam screw.

■ Movement of the throttle shaft can be exaggerated by attaching a short piece of stiff wire or drill bit to an alligator clip; grinding down the teeth on one side of the clip; and then attaching the clip to the throttle shaft, as shown under Homemade Synchronization Tool.

9. Attach a homemade synchronization tool to the throttle shaft, then use a ball hex (like OMC #327622 or equivalent) to rotate the throttle cam follower screw clockwise until the follower just contacts the cam and the tip of the homemade synchronization tool begins to move. Back off the throttle cam screw 1/8th turn from this point.

10. Reinstall the air silencer, then restart the motor and make sure it is fully warmed in order to check the carburetor adjustments.

✻✻ CAUTION

Water must circulate through the lower unit to the engine any time the engine is run to prevent damage to the water pump in the lower unit. Just five seconds without water will damage the water pump.

11. Operate the outboard in forward at or near full throttle for a full minute. Reduce speed suddenly to a low idle and shift into neutral. The powerhead should continue to operate smoothly. If the powerhead pops or stalls, the air/fuel mixture is probably too lean. Rotate the low speed needle 1/8th turn counterclockwise, allowing about 15 seconds between adjustments, until the powerhead responds as expected. Repeat the full throttle test and sudden deceleration with the shift into neutral to check each adjustment.

■ If the engine does not respond properly to these adjustments, check for other problems with the fuel or ignition systems. Make sure a quality fuel is being used along with the proper amount of and type of oil.

12. Once the low speed mixture setting is correct, shut the engine OFF, then check and/or adjust the wide open throttle stop as follows:
   a. Set the throttle to the wide open position.
   b. Verify the position of the throttle shaft roll pin. The roll pin should be vertical. If not, use the ball hex driver to adjust the throttle cam position until the roll pin is exactly vertical.

5/6 Hp (128cc) and 8/9.9 Hp (211cc) 4-Stroke Models

ADJUSTMENTS

■ The ignition timing is controlled by the ignition module/powerpack on these models. No adjustments are necessary or possible.

The following procedures provide detailed instructions to adjust the carburetor initial low speed setting and the idle speed. Although the idle speed should be checked periodically (and adjusted if necessary), the carburetor low speed setting (mixture adjustment) is **not** a periodic adjustment. The carburetor low speed setting should not require adjustment unless the carburetor has been removed and the setting was disturbed during cleaning or rebuilding.

Carburetor Low Speed Setting

◆ See Figures 170 and 172

The 5/6 Hp (128cc) and 8/9.9 Hp (211cc) 4-stroke models covered by this manual utilize an idle mixture screw mounted vertically in a round housing on top of the carburetor. The screw should be found under a small cap or sealed using RTV to discourage unnecessary tampering. EPA regulations require you to reseal it under RTV if adjustments are made.

■ EPA emission regulations prohibit tampering with the low speed mixture screw on these models, other than to properly follow the manufacturer's adjustment procedure.

The adjustment procedure should only be performed following a carburetor overhaul or replacement. No periodic adjustment of the low speed idle mixture screw is necessary.

The engine must be run under load for this procedure (either in a test tank or on a boat).

■ The idle speed setting procedure, which must follow this adjustment, requires that the motor operate on a boat that is unrestrained and not secured to a dock, mooring or trailer. For safety, use an assistant to navigate while you make the adjustments.

Fig. 172 Carburetor low speed mixture screw - 5/6 hp 4-stroke shown (8/9.9 hp similar)

MAINTENANCE & TUNE-UP 2-57

1. If not done during carburetor overhaul or cleaning, remove the RTV sealant, then **lightly** seat the low speed idle mixture screw. Back the screw out the initial number of turns from a lightly seated position as directed during carburetor overhaul (and listed in the Carburetor Set-Up Specifications chart under Fuel System).

※※ WARNING

Be very gentle when seating the idle mixture screw as the tapered seat is easily damaged if force is used. Turn the screw lightly and stop as soon as any resistance is felt.

2. Set a shop tachometer to the two-cycle setting, then connect it to the primary lead for the ignition coil. Start the motor and run it at 1/2 throttle in forward gear until it warms to normal operating temperature.
3. On tiller models, turn the idle speed adjustment knob counterclockwise (when facing the tiller handle from the motor operator's position) until the full slow idle position is reached.
4. With the engine running at idle in forward gear, adjust the low speed screw until the highest consistent powerhead speed is attained. Allow about 15 seconds between each adjustment for the engine to stabilize.

■ Turn the screw slowly, in fractional 1/4 or 1/8 increments for each adjustment, then allow time for it to take effect.

■ To adjust the idle mixture screw you will need OMC 910245 or an equivalent female slotted adjustment tool.

5. Once the engine has stabilized and runs smoothly at the highest possible idle speed, rotate the low speed needle counterclockwise 1/8th turn to prevent the powerhead from operating too lean at low speeds, particularly in neutral.
6. With the engine still running at normal operating temperature and in forward gear, adjust the idle speed screw to the center range of the Idle Speed (RPM) in Gear listed for your motor under the Tune-Up Specifications chart.
7. Operate the outboard in forward at or near full throttle for a full minute. Reduce speed suddenly to a low idle and shift into neutral. The powerhead should continue to operate smoothly. If the powerhead pops or stalls, the air/fuel mixture is probably too lean. Rotate the low speed needle 1/16th turn counterclockwise, allowing about 15 seconds between adjustments, until the powerhead responds as expected. Repeat the full throttle test and sudden deceleration with the shift into neutral to check each adjustment.

■ If the engine does not respond properly to these adjustments, check for other problems with the fuel or ignition systems. Make sure a quality fuel is being used.

8. Once a satisfactory adjustment has been achieved, verify engine idle speed using the Idle Speed Adjustment procedure.

■ After performing the Idle Speed Adjustment procedure, shut the engine off and reseal the low speed mixture screw using RTV silicone.

Idle Speed Adjustment

◆ See Figure 173

The 5/6 Hp (128cc) and 8/9.9 Hp (211cc) 4-stroke models covered by this manual utilize a spring-wound idle stop screw to hold the carburetor throttle plate open very slightly, even when the remote or tiller throttle is manually closed. Periodic checking and adjustment of this setting should be considered a part of routine maintenance.

Here's your excuse, show this page to your better half in case she (or he) doesn't believe you. It's time to take the boat out because that's how the manufacturer wants you to adjust the idle speed. Not in a test tank, not on a trailer or even attached to a dock, but with the motor mounted on a boat, underway. For safety, use an assistant to navigate while you make the adjustment.

■ Always adjust the idle speed with the correct propeller installed.

Refer to the Tune-Up Specifications chart for idle speed specs.
1. Set a shop tachometer to the two-cycle setting, then connect it to the primary lead for the ignition coil. Start the motor and run it at 1/2 throttle in forward gear until it warms to normal operating temperature.
2. Reduce engine speed to the slowest setting allowed by the throttle control as follows:

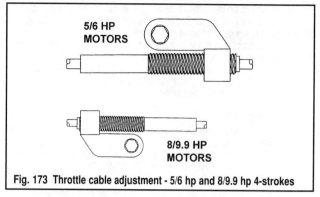

Fig. 173 Throttle cable adjustment - 5/6 hp and 8/9.9 hp 4-strokes

  a. On tiller models, turn the idle speed adjusting knob on the end of the tiller handle counterclockwise (when facing the end of the handle) to the slowest position.
  b. On remote models, manually set the throttle to the slowest position, right before the neutral shift gate.
3. Stop the engine and remove the cable retainer from the powerhead as follows:
  c. On tiller models, remove the screw and washer retaining the throttle cable bracket to the side of the powerhead.
  d. On remote models, free the cable from the anchor retainer and remove the throttle cable trunnion nut from the anchor pocket.
4. Restart the engine and observe the speed on the shop tachometer with the engine running in forward gear. Compare with the idle speeds listed in the Tune-Up Specifications chart.
5. If adjustment is necessary, either stop the engine for safety or use **extreme** caution around moving parts. Turn the spring-wound idle speed screw on the carburetor to make an adjustment. Restart the engine and recheck idle speed with the gearcase in the forward setting. Repeat this cycle until the proper setting is obtained.

※※ CAUTION

We at Seloc realize that most people probably don't shut the engine off before playing with the idle speed screw, but the manufacturer does not recommend attempting to adjust the idle speed with the engine running for safety reasons and we cannot disagree. If you ignore this caution, make sure that you take all possible precautions to prevent injury by making sure someone else is navigating. Also, keep your hands and clothing away from any hot or moving parts on the outboard.

6. Once the proper idle speed setting is confirmed, stop the engine.
7. For tiller models, twist the grip to the full slow position, then properly set the throttle cable bracket position as follows:
  a. On 5/6 hp (128cc) models, thread the throttle cable bracket on the cable until there are only two threads exposed at the rear of the bracket.
  b. On 8/9.9 hp (211cc) models, thread the throttle cable bracket on the cable until there is only one thread exposed at the front end of the bracket.
  c. On all tiller models, secure the throttle cable bracket to the powerhead using the screw and washer. Tighten the screw to 48-96 inch lbs. (5.4-10.8 Nm).
8. For remote models, pull slightly on the throttle cable casing to remove backlash, then secure the throttle cable trunnion nut into the anchor pocket, install the cover and tighten the cover screws securely.

9.9/10/14/15 Hp (216cc) and 9.9/10/15 Hp (255cc) 2-Stroke Models

ADJUSTMENTS

■ The ignition timing is not adjustable on these models.

The following procedures provide detailed instructions to adjust the cam follower pickup point, wide open throttle stop, carburetor low speed mixture (which is not a periodic adjustment but is provided here in case these adjustments are made after an overhaul), idle speed and shift lever detent.

2-58 MAINTENANCE & TUNE-UP

Procedures should be performed exactly as directed and in the order given to ensure proper adjustments. The preliminary adjustments must also be followed prior to the other specific adjustments as listed.

The idle speed portion of the timing and synchronization procedures for these units require that the engine be operated at idle rpm mounted on the boat, under load and unrestrained. For this reason, the adjustments should take place on a low-traffic body of water and only with an assistant to navigate while you make the adjustments.

Preliminary Adjustments

Perform the following procedures in the sequence given for correct timing and carburetor synchronization.

1. For remote control models, remove both remote control cables from the powerhead before commencing these adjustments.
2. For tiller models:
   a. Turn the idle adjusting knob counterclockwise (when facing the tiller handle from the motor operator's position) to the full slow position.
   b. Remove the pin securing the connector to the throttle lever on the powerhead. Turn the connector clockwise until it seats, then turn it back counterclockwise to align it with the lever (but, be sure **not** to twist is back more than 1/2 turn). Install the retaining pin.
   c. The throttle cable bracket bolted to the side of the powerhead must be flush with the inside edge of the threaded cable. If necessary, remove the bolt securing it to the powerhead, then rotate the bracket around the cable until it is even. Secure the bracket back to the powerhead using the bolt.
3. For all models, close the throttle so that the spring-wound idle speed screw on the carburetor linkage is against the stop. The center of the cam follower roller should be 1/2 the distance between the pickup mark and the end of the throttle cam. If it must be adjusted proceed as follows:
   a. Advance the throttle to the wide-open position and turn the idle speed screw (inward to move the cam closer to the roller or outward to move the cam away from the roller).
   b. Close the throttle completely and verify that the cam roller alignment is halfway between pickup mark and the end of the throttle cam. Repeat Step A, as necessary.

Cam Follower Pickup Point

◆ See Figures 174 and 175

Movement of the throttle shaft can be exaggerated by attaching a short piece of stiff wire or drill bit to an alligator clip; grinding down the teeth on one side of the clip; and then attaching the clip to the throttle shaft, as shown under Homemade Synchronization Tool. The shaft is normally accessed on the starboard side for these models. The wire jiggling will instantly indicate movement of the shaft.

■ **If the air intake silencer is installed, an access plug can be removed in order to reach the cam follower adjustment screw.**

To determine the cam follower pickup point: Advance the throttle until the tip of the wire begins to move. When movement occurs, the center of the roller must align with the single mark on the cam. If not, back out the cam follower adjustment screw, identified in the accompanying illustration, with a ball hex driver (like OMC P/N 327622 or equivalent), until the throttle valve is completely closed, and then, rotate the screw clockwise until the throttle shaft just starts to rotate.

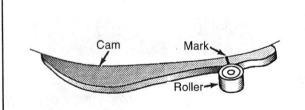

Fig. 174 Roller set at the cam mark - 9.9/10/14/15 hp (216cc) and 9.9/10/15 hp (255cc) 2-strokes

Fig. 175 Location of the cam follower adjustment screw for the cam follower pickup point - 9.9/10/14/15 hp (216cc) and 9.9/10/15 hp (255cc) 2-strokes

Wide Open Throttle Stop

To adjust the wide open throttle stop: With the powerhead not operating, turn the throttle stop screw until the tip protrudes 1/4 in. (8mm) through the throttle lever. Manually open the throttle to the wide open position. Verify the position of the throttle shaft roll pin. The roll pin should be completely vertical. If not, use a ball hex driver (like OMC #327622 or equivalent) to adjust the throttle cam position until the roll pin is exactly vertical. Turn the throttle cam screw clockwise to open the valve or counterclockwise to close the valve.

Carburetor Low Speed Mixture Adjustment

◆ See Figures 170 and 176

The carburetor low speed mixture adjustment procedure should only be performed following a carburetor overhaul or replacement. No periodic adjustment of the low speed idle mixture screw is necessary. If however adjustment is necessary, be certain to perform it at this point in the timing and synchronization procedures (after adjusting the wide open throttle stop and before adjusting the idle speed).

The engine must be run under load for this procedure (either in a test tank or on a boat).

■ **The idle speed setting procedure, which must follow this adjustment, requires that the motor operate on a boat that is unrestrained and not secured to a dock, mooring or trailer. For safety, use an assistant to navigate while you make the adjustments.**

1. If applicable, connect the remote control cables.
2. Start the engine and allow it to reach normal operating temperature.

✹✹ CAUTION

Water must circulate through the lower unit to the engine any time the engine is run to prevent damage to the water pump in the lower unit. Just a few seconds without water will damage the water pump.

3. Shift the engine into forward gear and allow it to run at slow idle (which should match the specification given in the Tune-Up Specifications chart). If not, adjust the position of the idle speed adjustment knob on the steering handle - tiller model, or adjust the idle screw - remote model, to correct the idle speed.
4. With the engine running at idle in forward gear, adjust the low speed screw until the highest consistent powerhead speed is attained. Allow about 15 seconds between each adjustment for the engine to stabilize.

■ **Turn the screw slowly, in fractional 1/4 or 1/8 increments for each adjustment, then allow time for it to take effect.**

5. Once the engine has stabilized and runs smoothly at the highest possible idle speed, rotate the low speed needle counterclockwise 1/8th turn to prevent the powerhead from operating too lean at low speeds, particularly in neutral.
6. With the engine still running at normal operating temperature and in forward gear, adjust the idle speed screw to the center range of the Idle Speed (RPM) in Gear listed for your motor under the Tune-Up Specifications chart.

MAINTENANCE & TUNE-UP 2-59

7. Operate the outboard in forward at or near full throttle for a full minute. Reduce speed suddenly to a low idle and shift into neutral. The powerhead should continue to operate smoothly. If the powerhead pops or stalls, the air/fuel mixture is probably too lean. Rotate the low speed needle 1/16th turn counterclockwise, allowing about 15 seconds between adjustments, until the powerhead responds as expected. Repeat the full throttle test and sudden deceleration with the shift into neutral to check each adjustment.

■ **If the engine does not respond properly to these adjustments, check for other problems with the fuel or ignition systems. Make sure a quality fuel is being used along with the proper mixture of oil.**

8. Once a satisfactory adjustment has been achieved, verify engine idle speed using the Idle Speed Adjustment procedure.

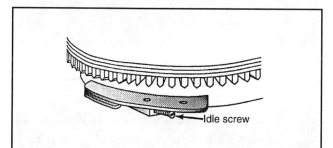

Fig. 176 Idle speed adjustment is made with the idle screw shown - remote 9.9/10/14/15 hp (216cc) and 9.9/10/15 hp (255cc) 2-strokes

Idle Speed

◆ See Figure 170

Here's your excuse, show this page to your better half in case she (or he) doesn't believe you. It's time to take the boat out because that's how the manufacturer wants you to adjust the idle speed. Not in a test tank, not on a trailer or even attached to a dock, but with the motor mounted on a boat, underway. For safety, use an assistant to navigate while you make the adjustment.

■ **Always adjust the idle speed with the correct propeller installed.**

Refer to the Tune-Up Specifications chart for idle speed specs.

■ **After a replacement or overhaul the carburetor low speed mixture screw must be set before the idle speed can be properly adjusted. If necessary, refer to that adjustment procedure first. However, the low speed setting is NOT a periodic adjustment and if no repairs were made, it should not be necessary.**

1. Start and run the engine to normal operating temperature, then with the engine running in forward gear, reduce the engine speed to slow idle.
2. On tiller models, turn the idle speed adjusting knob on the end of the tiller handle counterclockwise (when facing the end of the handle) to the slowest position.
3. Observe the speed with the engine running in forward gear. Compare with the idle speeds listed in the Tune-Up Specifications chart.
4. If adjustment is necessary, either stop the engine for safety or use **extreme** caution around moving parts. Turn the idle adjustment screw on the linkage arm (at the side of the powerhead) to make an adjustment. Restart the engine and recheck idle speed with the gearcase in the forward setting. Repeat this cycle until the proper setting is obtained. For tiller models, remove the throttle cable-to-powerhead bracket screw, then spin the bracket to adjust the cable so there is a **slight** preload against the idle speed screw when the tiller grip is held at the full slow position.

✳✳ CAUTION

We at Seloc realize that most people probably don't shut the engine off before playing with the idle speed screw, but the manufacturer does not recommend attempting to adjust the idle speed with the engine running for safety reasons and we cannot disagree. If you ignore this caution, make sure that you take all possible precautions to prevent injury by making sure someone else is navigating. Also, keep your hands and clothing away from any hot or moving parts on the outboard.

5. Operate the outboard in forward at or near full throttle for a full minute. Reduce speed suddenly to a low idle and shift into neutral. The powerhead should continue to operate smoothly. If the powerhead pops or stalls, the air/fuel mixture is probably too lean. Refer to the Carburetor Low Speed Mixture Adjustment procedure in this section.
6. Once the proper idle speed setting is confirmed and engine runs smoothly when dropped from wide open throttle to idle, stop the engine.

Shift Lever Detent

Shift lever detent adjustment is performed with the engine removed from water and with the negative battery cable and/or spark plug leads disconnected for safety.

1. Rotate the propeller or propeller shaft by hand and shift the gearcase into neutral.
2. Make sure the lower detent spring on the linkage is fully engaged with the notch on the shift lever detent.
3. If necessary, loosen the detent retaining screw, then physically move the lower detent spring to fully engage the notch, then tighten the screw securely.

9.9/15 Hp (305cc) 4-Stroke Models

ADJUSTMENTS

■ **The ignition timing is controlled by the ignition module mounted under the flywheel at the front of the motor on these models. No adjustments are necessary or possible.**

The following procedures provide detailed instructions to adjust the carburetor initial low speed setting and the idle speed. Although the idle speed should be checked periodically (and adjusted if necessary), the carburetor low speed setting is **not** a periodic adjustment. The carburetor low speed setting should not require adjustment unless the carburetor has been removed and the setting was disturbed during cleaning or rebuilding.

Shift Lever Detent Adjustment

◆ See Figure 177

Although provided as a preliminary procedure (before setting the idle mixture and idle speed), the shift lever detent should not be a periodic adjustment. Follow this procedure any time the shift components have been removed and installed or if harsh or improper shifting is noted during operation.

1. Disconnect the negative battery cable and/or disconnect the spark plugs leads for safety, then slowly rotate the propeller or propeller shaft by hand and move the shifter into neutral.
2. For remote models, remove the shift cable trunnion nut from the anchor pocket.

Fig. 177 Shift lever detent screw (1) and neutral lockout link (2) - 9.9/15 hp 4-strokes

2-60 MAINTENANCE & TUNE-UP

3. Loosen the detent spring adjustment screw (found on the lower front of the motor, directly below the manual starter handle). Disconnect the neutral lockout link from the shift lever.

4. With the gearcase in neutral, move the detent sprint until it fully engages the notch of the shift lever detent. Tighten the shift detent spring adjustment screw.

5. Use the neutral lockout link to move the starter lockout plunger (found in the manual starter housing at the top of the link) into a horizontal position. Next, adjust the neutral lockout link connector so that the lockout plunger remains horizontal once the connector is attached to the shift lever, then fasten the connector to the shift lever.

6. For remote models, remove cable backlash (if present) by pulling on the casing, then use the cable trunnion nut to locate it properly in the anchor pocket. Secure the cable to the anchor retainer and tighten the screw to 60-84 inch lbs. (7-9 Nm).

7. Connect the negative battery cable and/or reconnect the spark plugs leads.

8. Operate the engine and check for proper shift engagement. Make any corrections before returning the engine to service. Proceed with idle mixture and/or speed adjustment procedures, as necessary.

Carburetor Low Speed Setting

◆ See Figures 170, 178 and 179

The 9.9/15 Hp (305cc) 4-stroke models covered by this manual utilize an idle mixture screw mounted vertically in a round housing on top of the carburetor. The screw should be found under a small cap or sealed using RTV to discourage unnecessary tampering. EPA regulations require you to reseal it under RTV if adjustments are made.

■ EPA emission regulations prohibit tampering with the low speed mixture screw on these models, other than to properly follow the manufacturer's adjustment procedure.

The adjustment procedure should only be performed following a carburetor overhaul or replacement. No periodic adjustment of the low speed idle mixture screw is necessary.

The engine must be run under load for this procedure (either in a test tank or on a boat).

■ The idle speed setting procedure, which must follow this adjustment, requires that the motor operate on a boat that is unrestrained and not secured to a dock, mooring or trailer. For safety, use an assistant to navigate while you make the adjustments.

1. If not done during carburetor overhaul or cleaning, remove the RTV sealant, then **lightly** seat the low speed idle mixture screw. Back the screw out the initial number of turns from a lightly seated position as directed during carburetor overhaul (and listed in the Carburetor Set-Up Specifications chart under Fuel System).

✶✶ WARNING

Be very gentle when seating the idle mixture screw as the tapered seat is easily damaged if force is used. Turn the screw lightly and stop as soon as any resistance is felt.

2. Set a shop tachometer to the two-cycle setting, then connect it to the primary lead for the ignition coil. Start the motor and run it at 1/2 throttle in forward gear until it warms to normal operating temperature.

3. On tiller models, turn the idle speed adjustment knob counterclockwise (when facing the tiller handle from the motor operator's position) until the full slow idle position is reached.

4. With the engine running at idle in forward gear, adjust the low speed screw until the highest consistent powerhead speed is attained. Allow about 15 seconds between each adjustment for the engine to stabilize.

■ Turn the screw slowly, in fractional 1/4 or 1/8 increments for each adjustment, then allow time for it to take effect.

■ To adjust the idle mixture screw you will need OMC 910245 or an equivalent female slotted adjustment tool.

5. Once the engine has stabilized and runs smoothly at the highest possible idle speed, rotate the low speed needle counterclockwise 1/8th turn to prevent the powerhead from operating too lean at low speeds, particularly in neutral.

Fig. 178 The low speed mixture screw is found on top of the carburetor. . .

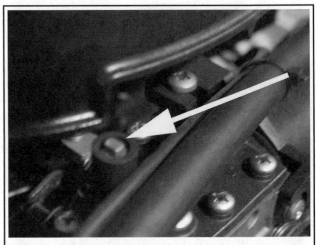

Fig. 179 . . . inside a cylindrical housing - 9.9/15 hp 4-strokes

6. With the engine still running at normal operating temperature and in forward gear, adjust the idle speed screw to the center range of the Idle Speed (RPM) in Gear listed for your motor under the Tune-Up Specifications chart.

7. Operate the outboard in forward at or near full throttle for a full minute. Reduce speed suddenly to a low idle and shift into neutral. The powerhead should continue to operate smoothly. If the powerhead pops or stalls, the air/fuel mixture is probably too lean. Rotate the low speed needle 1/16th turn counterclockwise, allowing about 15 seconds between adjustments, until the powerhead responds as expected. Repeat the full throttle test and sudden deceleration with the shift into neutral to check each adjustment.

■ If the engine does not respond properly to these adjustments, check for other problems with the fuel or ignition systems. Make sure a quality fuel is being used.

8. Once a satisfactory adjustment has been achieved, verify engine idle speed using the Idle Speed Adjustment procedure.

■ After performing the Idle Speed Adjustment procedure, shut the engine off and reseal the low speed mixture screw using RTV silicone.

Idle Speed Adjustment

◆ See Figures 180 and 181

The 9.9/15 Hp (305cc) 4-stroke models covered by this manual utilize a spring-wound idle stop screw to hold the carburetor throttle plate open very slightly, even when the remote or tiller throttle is manually closed. Periodic checking and adjustment of this setting should be considered a part of routine maintenance.

MAINTENANCE & TUNE-UP

Here's your excuse, show this page to your better half in case she (or he) doesn't believe you. It's time to take the boat out because that's how the manufacturer wants you to adjust the idle speed. Not in a test tank, not on a trailer or even attached to a dock, but with the motor mounted on a boat, underway. For safety, use an assistant to navigate while you make the adjustment.

■ **Always adjust the idle speed with the correct propeller installed.**

Refer to the Tune-Up Specifications chart for idle speed specs.

1. Set a shop tachometer to the two-cycle setting, then connect it to the primary lead for the ignition coil. Start the motor and run it at 1/2 throttle in forward gear until it warms to normal operating temperature.
2. Reduce engine speed to the slowest setting allowed by the throttle control as follows:
    a. On tiller models, turn the idle speed adjusting knob on the end of the tiller handle counterclockwise (when facing the end of the handle) to the slowest position.
    b. On remote models, manually set the throttle to the slowest position, right before the neutral shift gate.
3. Stop the engine and remove the cable retainer from the powerhead as follows:
    c. On tiller models, remove the screw and washer retaining the throttle cable bracket to the side of the powerhead.
    d. On remote models, free the cable from the anchor retainer and remove the throttle cable trunnion nut from the anchor pocket.
4. Pull on the cable casing in order to move the throttle cam away from the cam follower roller.
5. Restart the engine and observe the speed on the shop tachometer with the engine running in forward gear. Compare with the idle speeds listed in the Tune-Up Specifications chart.
6. If adjustment is necessary, either stop the engine for safety or use **extreme** caution around moving parts. Turn the spring-wound idle speed screw on the carburetor to make an adjustment. Restart the engine and recheck idle speed with the gearcase in the forward setting. Repeat this cycle until the proper setting is obtained.

Fig. 100 The spring-wound idle speed screw - 9.9/15 hp 4-strokes

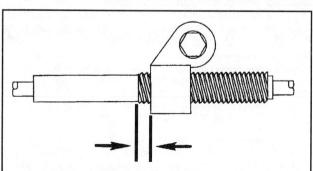

Fig. 181 On tiller models, position the throttle cable bracket 1/8 in. from end of the threads - 9.9 hp/15 4-strokes

✱✱ CAUTION

We at Seloc realize that most people probably don't shut the engine off before playing with the idle speed screw, but the manufacturer does not recommend attempting to adjust the idle speed with the engine running for safety reasons and we cannot disagree. If you ignore this caution, make sure that you take all possible precautions to prevent injury by making sure someone else is navigating. Also, keep your hands and clothing away from any hot or moving parts on the outboard.

7. Once the proper idle speed setting is confirmed, stop the engine.
8. On tiller models, twist the grip to the full slow position, then properly set the throttle cable bracket position by threading the throttle cable bracket on the cable until it is 1/8 in. (3mm) from the end of the housing threads. Then, pull slightly on cable casing to remove backlash and secure the throttle cable bracket using the screw and washer. Tighten the screw to 48-96 inch lbs. (5.4-10.8 Nm).
9. For remote models, pull slightly on the throttle cable casing to remove backlash, then secure the throttle cable trunnion nut into the anchor pocket, install the cover and tighten the cover screws securely. Verify that there is a 1/32 in. (0.8mm) gap between the throttle cam and cam follower roller on the engine. If adjustment is necessary on tiller models, remove the throttle bracket screw and turn the bracket on the housing threads until the proper gap is achieved.

20 Hp (521cc) Models

Some of the timing and synchronization procedures for these units require operating the motor at idle rpm under load and at wide-open throttle. Therefore, the outboard **must** be placed in a test tank or a body of water with the boat well secured to the dock or in a slip (except for the idle speed setting which must be conducted on an unrestrained boat). Never attempt to make the load adjustments or run the engine at wide open throttle with a flush attachment connected to the lower unit. The powerhead operating at high rpm with such a device, would likely cause a runaway condition from a lack of load on the propeller, causing extensive damage.

Mount the outboard unit in a test tank or on a boat well secured. Connect a fuel line from a fuel source to the powerhead. Remove the powerhead top cover to gain access to adjustment locations.

■ **If a test tank must be used for most of the settings, the final idle speed adjustment will have to wait until the motor is installed on a boat for a proper loaded adjustment.**

On remote control models, the control cables must be removed from the engine before performing these adjustments. Obviously, the cables should be reconnected before the idle speed setting.

The following procedures provide detailed instructions to adjust the cam follower pickup point, throttle control rod (1990 only), maximum spark advance and idle speed. The carburetor low speed mixture adjustment (which is not a periodic adjustment) is also provided here in case this adjustment is necessary due to carburetor replacement, overhaul or otherwise un-resolvable idle problems after the other adjustments are completed. Procedures should be performed exactly as directed and in the order given to ensure proper adjustments.

The idle speed portion of the timing and synchronization procedures for these units require that the engine be operated at idle rpm mounted on the boat, under load and unrestrained. For this reason, the adjustments should take place on a low-traffic body of water and only with an assistant to navigate while you make the adjustments.

ADJUSTMENTS

Perform the following procedures in the sequence given for correct timing and carburetor synchronization.

Cam Follower Pickup Point

◆ See Figures 182 and 183

Movement of the throttle shaft can be exaggerated by attaching a short piece of stiff wire or drill bit to an alligator clip; grinding down the teeth on one side of the clip; and then attaching the clip to the throttle shaft, as shown

2-62 MAINTENANCE & TUNE-UP

under Homemade Synchronization Tool. The shaft is normally accessed on the starboard side for these models. The wire jiggling will instantly indicate movement of the shaft.

1. For tiller models, turn the idle speed adjusting knob on the end of the tiller handle counterclockwise (when facing the end of the handle) to the slowest position.
2. On remote models, turn the arm stop screw until the arm touches the screw's mounting bracket.
3. To determine the cam follower pickup point: Advance the throttle by moving the throttle control lever, until the tip of the wire (or drill bit) begins to move. When movement occurs, the center of the roller must align between the two marks on the cam. If the roller does not align, loosen the screw securing the adjustment lever to the throttle shaft. Move the adjustment lever up or down to make the adjustment.

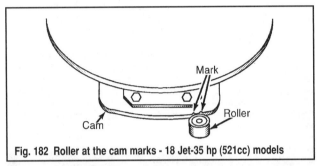

Fig. 182 Roller at the cam marks - 18 Jet-35 hp (521cc) models

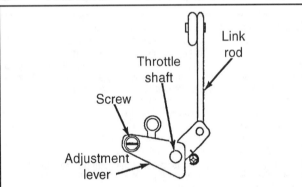

Fig. 183 If the roller does not align between the two cam marks, reposition the adjustment lever - 18 Jet-35 hp (521cc) models

Throttle Control Rod (1990 Models Only)

1. While rotating the propeller shaft slowly by hand, move the shift lever to the FORWARD gear.
2. Advance the throttle lever to contact the stop on the cylinder block. At this point the carburetor throttle plate should be horizontal and the throttle shaft must be free-moving. Also the off-set on the pivot block must face toward the control rod collar.
3. If necessary, move the control rod collar so to ensure the carburetor plate is exactly horizontal with the throttle lever is fully advanced. Tighten the collar screw securely after it is repositioned.
4. Retard the linkage and recheck the adjustment.

Maximum Spark Advance

◆ See Figure 184

To check the maximum spark advance: the outboard must be operated with the proper test wheel. Do not operate the powerhead with a propeller and especially don't use a flush adapter while performing this adjustment.

1. Connect a timing light to the top cylinder.
2. Start the powerhead and allow it to reach normal operating temperature. Boat movement must be restrained - secured to the dock.

※※ CAUTION

Water must circulate through the lower unit to the engine any time the engine is run to prevent damage to the water pump in the lower unit. Just a few seconds without water will damage the water pump.

3. With the engine at normal operating temperature, run the motor in forward gear at Wide Open Throttle (WOT). Aim the timing light at the timing marks on the flywheel perimeter. On electric start models, the mark on the timing pointer (located next to the starter motor) should align between the appropriate marks mark on the flywheel. On manual start models, the triangular pointer embossed on the manual starter housing should align between the appropriate marks on the flywheel.

■ **The maximum spark advance specifications are found in the Tune-Up Specifications chart.**

4. If the marks do not align, shut down the engine. Loosen the nut on the timing adjustment screw, identified in the accompanying illustrations. Turn the adjustment screw to correct the timing. Rotating the screw through one revolution clockwise retards the timing one degree. Rotating the screw through one revolution counterclockwise advances the timing one degree. Tighten the nut on the timing adjustment screw to hold this new adjustment.
5. After adjustment, recheck the timing to ensure it did not change while tightening the hold-down nut.

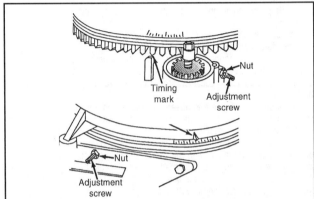

Fig. 184 Timing marks and adjustment points for electric start (top) and a manual start (bottom) 18 Jet-35 hp (521cc) motors

Idle Speed

◆ See Figures 170, 185 and 186

The 20 Hp (521cc) models covered by this manual utilize either a threaded throttle cable bracket (tiller models through 1996) or a spring-wound throttle arm stop screw to hold and adjust idle speed (remote models and 1997 and later tiller models). Periodic checking and, when necessary, adjustment of this setting should be considered a part of routine maintenance.

Here's your excuse, show this page to your better half in case she (or he) doesn't believe you. It's time to take the boat out because that's how the manufacturer wants you to adjust the idle speed. Not in a test tank, not on a trailer or even attached to a dock, but with the motor mounted on a boat, underway. For safety, use an assistant to navigate while you make the adjustment.

■ **Always adjust the idle speed with the correct propeller installed.**

Refer to the Tune-Up Specifications chart for idle speed specs.

1. On remote models, if the control cables were disconnected for other adjustments, reconnect them now.
2. Start the engine and allow it to reach normal operating temperature.

※※ CAUTION

Water must circulate through the lower unit to the engine any time the engine is run to prevent damage to the water pump in the lower unit. Just a few seconds without water will damage the water pump.

3. If servicing a tiller model, rotate the idle adjustment knob on the tiller grip counterclockwise (when facing the steering handle) to the full slow position.
4. The engine should idle between 650 and 700 rpm in forward gear. Ideally, always try to set the idle in the middle of the range - 675 rpm.
5. If idle speed is incorrect, adjust by moving the throttle cable bracket on tiller models through 1996 or using the throttle arm stop screw on remote models and 1997 or later tiller models.

MAINTENANCE & TUNE-UP 2-63

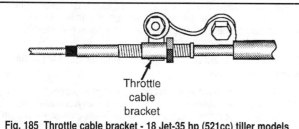

Fig. 185 Throttle cable bracket - 18 Jet-35 hp (521cc) tiller models (thru 1996)

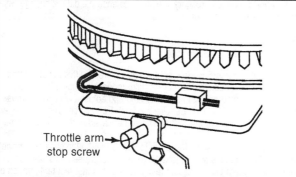

Fig. 186 Location of the throttle arm screw for idle adjustment - 18 Jet-35 hp (521cc) remote and 1997 or later tiller models

■ On tiller models through 1996, if the throttle cable bracket is repositioned for idle speed adjustment, check the Cam Follower Pickup Point to verify it has not changed and adjust, as necessary. For details refer to the procedure in this section.

6. To check the carburetor mixture adjustment, operate the powerhead with the outboard in forward gear at or near wide open throttle for a full minute. Quickly reduce powerhead speed to an idle of 700-800 rpm and shift into neutral. The powerhead should continue to operate smoothly. If the powerhead pops or stalls, the air/fuel mixture is probably too lean. Rotate the low speed needle 1/16th turn counterclockwise, allowing about 15 seconds between adjustments, until the powerhead responds as expected. Repeat the wide open throttle and quick deceleration test to verify smooth operation. If this does not help, verify the condition of the fuel and oil (as well as type used and mixture), make sure there are no other problems (such as fuel or ignition system trouble) and if no problems are found, reset the Carburetor Initial Low Speed Setting.

Carburetor Initial Low Speed Setting

Although not a periodic adjustment, periodic inspections could reveal the need for this adjustment to correct for otherwise unresolved performance issues. Generally speaking however, this adjustment should not be necessary unless the carburetor has been replaced or overhauled.

The engine must be run under load for this procedure (either in a test tank or on a boat).

1. Start the engine and allow it to reach normal operating temperature.

✱✱ CAUTION

Water must circulate through the lower unit to the engine any time the engine is run to prevent damage to the water pump in the lower unit. Just a few seconds without water will damage the water pump.

2. If servicing a tiller model, rotate the idle adjustment knob on the tiller grip counterclockwise (when facing the steering handle) to the full slow position.
3. Shift the engine into forward gear and allow it to run at slow idle (which should match the specification given in the Tune-Up Specifications chart). If not, adjust the idle speed according to the procedure in this section.
4. With the engine running at idle in forward gear, adjust the low speed screw until the highest consistent powerhead speed is attained. Allow about 15 seconds between each adjustment for the engine to stabilize.

■ Turn the screw slowly, in fractional 1/4 or 1/8 increments for each adjustment, then allow time for it to take effect.

5. Once the engine has stabilized and runs smoothly at the highest possible idle speed, rotate the low speed needle counterclockwise 1/8th turn to prevent the powerhead from operating too lean at low speeds, particularly in neutral.
6. With the engine still running at normal operating temperature and in forward gear, adjust the idle speed screw to the center range of the Idle Speed (RPM) in Gear listed for your motor under the Tune-Up Specifications chart.
7. Operate the outboard in forward at or near full throttle for a full minute. Reduce speed suddenly to a low idle and shift into neutral. The powerhead should continue to operate smoothly. If the powerhead pops or stalls, the air/fuel mixture is probably too lean. Rotate the low speed needle 1/16th turn counterclockwise, allowing about 15 seconds between adjustments, until the powerhead responds as expected. Repeat the full throttle test and sudden deceleration with the shift into neutral to check each adjustment.

■ If the engine does not respond properly to these adjustments, check for other problems with the fuel or ignition systems. Make sure a quality fuel is being used along with the proper mixture of oil.

8. Once a satisfactory adjustment has been achieved, verify engine idle speed using the Idle Speed Adjustment procedure.

18 Jet and 25-35 Hp (521cc) Models

Some of the timing and synchronization procedures for these units require operating the motor at idle rpm under load and at wide-open throttle. Therefore, the outboard **must** be placed in a test tank or a body of water with the boat well secured to the dock or in a slip (except for the idle speed setting which must be conducted on an unrestrained boat). Never attempt to make the load adjustments or run the engine at wide open throttle with a flush attachment connected to the lower unit. The powerhead operating at high rpm with such a device, would likely cause a runaway condition from a lack of load on the propeller, causing extensive damage.

Mount the outboard unit in a test tank or on a boat well secured. Connect a fuel line from a fuel source to the powerhead. Remove the powerhead top cover to gain access to adjustment locations.

■ If a test tank must be used for most of the settings, the final idle speed adjustment will have to wait until the motor is installed on a boat for a proper loaded adjustment.

On remote control models, the control cables must be removed from the engine before performing these adjustments. Obviously, the cables should be reconnected before the idle speed setting.

The following procedures provide detailed instructions to adjust the cam follower pickup point, throttle control rod, maximum spark advance and idle speed. The carburetor low speed mixture adjustment (which is not a periodic adjustment) is also provided here in case this adjustment is necessary due to carburetor replacement, overhaul or otherwise un-resolvable idle problems after the other adjustments are completed. Procedures should be performed exactly as directed and in the order given to ensure proper adjustments.

The idle speed portion of the timing and synchronization procedures for these units require that the engine be operated at idle rpm mounted on the boat, under load and unrestrained. For this reason, the adjustments should take place on a low-traffic body of water and only with an assistant to navigate while you make the adjustments.

ADJUSTMENTS

Perform the following procedures in the sequence given for correct timing and carburetor synchronization.

Cam Follower Pickup Point

◆ See Figures 182 and 183

Movement of the throttle shaft can be exaggerated by attaching a short piece of stiff wire or drill bit to an alligator clip; grinding down the teeth on one side of the clip; and then attaching the clip to the throttle shaft, as shown under Homemade Synchronization Tool. The shaft is normally accessed on the starboard side for these models. The wire jiggling will instantly indicate movement of the shaft.

2-64 MAINTENANCE & TUNE-UP

1. For tiller models, turn the idle speed adjusting knob on the end of the tiller handle counterclockwise (when facing the end of the handle) to the slowest position.
2. On remote models through 1996, turn the arm stop screw until the arm touches the screw's mounting bracket.
3. On all 1997-2001 models, loosen the locknut, then turn arm stop screw counterclockwise until the arm touches the screw's mounting bracket.
4. To determine the cam follower pickup point: Advance the throttle by moving the throttle control lever, until the tip of the wire (or drill bit) begins to move. When movement occurs, the center of the roller must align between the two marks on the cam. If the roller does not align, loosen the screw securing the adjustment lever to the throttle shaft. Move the adjustment lever up or down to make the adjustment.

Throttle Control Rod

♦ See Figure 187

Before making any adjustment to the throttle control rod, be sure the offset on the pivot block faces toward the control rod collar.

✶✶ CAUTION

Before touching the propeller, always disconnect the negative battery cable and/or remove the spark plugs leads for safety.

1. Rotate the propeller shaft and at the same time, move the shift lever into the forward position.
2. Loosen the control rod collar screw.
3. Advance the throttle lever until the lever makes contact with the stop cast into the cylinder block. Now, move the throttle control rod forward until the carburetor throttle plate is horizontal.
4. Slide the control rod collar backwards until it makes contact with the pivot block, then tighten the screw on the collar to hold this new adjustment.
5. Verify the adjustment by observing the carburetor throttle plate, it must be precisely horizontal when the throttle lever is fully advanced. Repeat the adjustment, as necessary.

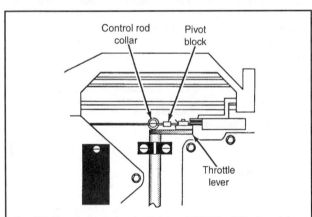

Fig. 187 Throttle control rod - 18 Jet and 25-35 hp (521cc) models

Maximum Spark Advance

♦ See Figure 184

To check the maximum spark advance: the outboard must be operated with the proper test wheel. Do not operate the powerhead with a propeller and especially don't use a flush adapter while performing this adjustment.
1. Connect a timing light to the top cylinder.
2. Start the powerhead and allow it to reach normal operating temperature. Boat movement must be restrained - secured to the dock.

✶✶ CAUTION

Water must circulate through the lower unit to the engine any time the engine is run to prevent damage to the water pump in the lower unit. Just a few seconds without water will damage the water pump.

3. With the engine at normal operating temperature, run the motor in forward gear at Wide Open Throttle (WOT). Aim the timing light at the timing marks on the flywheel perimeter. On electric start models, the mark on the timing pointer (located next to the starter motor) should align between the appropriate marks mark on the flywheel. On manual start models, the triangular pointer embossed on the manual starter housing should align between the appropriate marks on the flywheel.

■ **The maximum spark advance specifications are found in the Tune-Up Specifications chart.**

4. If the marks do not align, shut down the engine. Loosen the nut on the timing adjustment screw, identified in the accompanying illustrations. Turn the adjustment screw to correct the timing. Rotating the screw through one revolution clockwise retards the timing one degree. Rotating the screw through one revolution counterclockwise advances the timing one degree. Tighten the nut on the timing adjustment screw to hold this new adjustment.
5. After adjustment, recheck the timing to ensure it did not change while tightening the hold-down nut.

Idle Speed

♦ See Figures 170, 185 and 186

The 18 Jet and 25-35 Hp (521cc) models covered by this manual utilize either a threaded throttle cable bracket (tiller models through 1996) or a spring-wound throttle arm stop screw to hold and adjust idle speed (remote models as well as 1997 and later tiller models). Periodic checking and, when necessary, adjustment of this setting should be considered a part of routine maintenance.

Here's your excuse, show this page to your better half in case she (or he) doesn't believe you. It's time to take the boat out because that's how the manufacturer wants you to adjust the idle speed. Not in a test tank, not on a trailer or even attached to a dock, but with the motor mounted on a boat, underway. For safety, use an assistant to navigate while you make the adjustment.

■ **Always adjust the idle speed with the correct propeller installed.**

Refer to the Tune-Up Specifications chart for idle speed specs.
1. On remote models, if the control cables were disconnected for other adjustments, reconnect them now.
2. Start the engine and allow it to reach normal operating temperature.

✶✶ CAUTION

Water must circulate through the lower unit to the engine any time the engine is run to prevent damage to the water pump in the lower unit. Just a few seconds without water will damage the water pump.

3. If servicing a tiller model, rotate the idle adjustment knob on the tiller grip counterclockwise (when facing the steering handle) to the full slow position.
4. The engine should idle between 650 and 700 rpm in forward gear. Ideally, always try to set the idle in the middle of the range - 675 rpm.
5. If idle speed is incorrect, adjust by moving the throttle cable bracket on tiller models through 1996 or using the throttle arm stop screw on all other models (remote models and tiller models 1997 or later).

■ **On tiller models through 1996, if the throttle cable bracket is repositioned for idle speed adjustment, check the Cam Follower Pickup Point to verify it has not changed and adjust, as necessary. For details refer to the procedure in this section.**

6. To check the carburetor mixture adjustment, operate the powerhead with the outboard in forward gear at or near wide open throttle for a full minute. Quickly reduce powerhead speed to an idle of 700-800 rpm and shift into neutral. The powerhead should continue to operate smoothly. If the powerhead pops or stalls, the air/fuel mixture is probably too lean. Rotate the low speed needle 1/8th turn counterclockwise, allowing about 15 seconds between adjustments, until the powerhead responds as expected. Repeat the wide open throttle and quick deceleration test to verify smooth operation. If this does not help, verify the condition of the fuel and oil (as well as type used and mixture), make sure there are no other problems (such as fuel or ignition system trouble) and if no problems are found, reset the Carburetor Initial Low Speed Setting.

Carburetor Initial Low Speed Setting

Although not a periodic adjustment, periodic inspections could reveal the need for this adjustment to correct for otherwise unresolved performance issues. Generally speaking however, this adjustment should not be necessary unless the carburetor has been replaced or overhauled.

■ Before disturbing the factory setting, make an alignment mark on the carburetor body. This mark is used as a point of reference during the adjustment procedure.

The engine must be run under load for this procedure (either in a test tank or on a boat).

1. Start the engine and allow it to reach normal operating temperature.

✴✴ CAUTION

Water must circulate through the lower unit to the engine any time the engine is run to prevent damage to the water pump in the lower unit. Just a few seconds without water will damage the water pump.

2. If servicing a tiller model, rotate the idle adjustment knob on the tiller grip counterclockwise (when facing the steering handle) to the full slow position.
3. Shift the engine into forward gear and allow it to run at fast idle of 700-800 rpm for at least 3 minutes.
4. With the engine running at idle in forward gear, observe the running conditions as follows:
   a. If the engine is running rich, it will show a rough or unsteady idle.
   b. If the engine is running lean, it will sneeze or backfire.
5. If necessary, adjust the low speed mixture screw as follows to obtain a smooth idle:
   a. For rich mixtures, noting the reference mark made earlier, turn the needle 1/8th turn clockwise, allowing about 15 seconds between adjustments, until the highest consistent rpm is reached.
   b. For lean mixtures, noting the reference mark made earlier, turn the needle 1/8th turn counterclockwise, allowing about 15 seconds between adjustments, until the highest consistent rpm is reached.
6. With the engine still running at normal operating temperature and in forward gear, adjust the idle speed screw to the center range of the Idle Speed (RPM) in Gear listed for your motor under the Tune-Up Specifications chart.
7. Operate the outboard in forward at or near full throttle for a full minute. Reduce speed suddenly to a low idle and shift into neutral. The powerhead should continue to operate smoothly. If the powerhead pops or stalls, the air/fuel mixture is probably too lean. Rotate the low speed needle 1/8th turn counterclockwise, allowing about 15 seconds between adjustments, until the powerhead responds as expected. Repeat the full throttle test and sudden deceleration with the shift into neutral to check each adjustment.

■ If the engine does not respond properly to these adjustments, check for other problems including:

- Improper engine temperatures
- Incorrect linkage adjustments
- Insufficient exhaust backpressure
- Other problems with the fuel or ignition systems
- Quality and mixture of fuel and oil

8. Once a satisfactory adjustment has been achieved, verify engine idle speed using the Idle Speed Adjustment procedure.

25-55 Hp (737cc) 2-Cylinder Models

Some of the timing and synchronization procedures for these units require operating the motor at idle rpm under load and at wide-open throttle. Therefore, the outboard **must** be placed in a test tank or a body of water with the boat well secured to the dock or in a slip (except for the idle speed setting which must be conducted on an unrestrained boat). Never attempt to make the load adjustments or run the engine at wide open throttle with a flush attachment connected to the lower unit. The powerhead operating at high rpm with such a device, would likely cause a runaway condition from a lack of load on the propeller, causing extensive damage.

Mount the outboard unit in a test tank or on a boat well secured. Connect a fuel line from a fuel source to the powerhead. Remove the powerhead top cover to gain access to adjustment locations (on some models this means only the top cover, but on models equipped with split lower covers, they must usually be removed as well).

■ If a test tank must be used for most of the settings, the final idle speed adjustment will have to wait until the motor is installed on a boat for a proper loaded adjustment.

The following procedures provide detailed instructions to adjust the throttle plate synchronization, cam pickup point, idle speed, throttle control rod, wide open throttle stop, throttle cable installation, maximum spark advance and, for some models, shift lever adjustment. The carburetor low speed mixture adjustment (which is not a periodic adjustment) is also provided here in case this adjustment is necessary due to carburetor replacement, overhaul or otherwise un-resolvable idle problems after the other adjustments are completed. Procedures should be performed exactly as directed and in the order given to ensure proper adjustments.

■ The preliminary adjustments must be performed before starting these adjustment procedures.

The idle speed portion of the timing and synchronization procedures for these units require that the engine be operated at idle rpm mounted on the boat, under load and unrestrained. For this reason, the adjustments should take place on a low-traffic body of water and only with an assistant to navigate while you make the adjustments.

ADJUSTMENTS

◆ See Figures 188 and 189

Perform the following procedures in the sequence given in order to ensure correct timing and carburetor synchronization. Start the procedures by following these preliminary adjustments:

Preliminary Adjustments

◆ See Figures 188 and 189

1. Either disconnect the throttle cable from the lever (except 40RW, 40RP, 40WR, 45 and 55 Commercial models) or remove the bolt and washer securing the throttle cable bracket to the powerhead (40RW, 40RP, 40WR, 45 and 55 Commercial models).
2. For 40/48/50 hp models through 1993, check that the throttle control rod is approximately 7 13/16 in. (19.8cm) in length when measured to the center of the socket. If necessary, disconnect the socket and turn it inward or outward to achieve the proper preliminary length. For later models, OMC changed the procedures slightly and adjust the throttle control rod length based on a gap measured later in the procedure.
3. Measure the length of the short spark control rod from the ball joint center at one end to the ball joint center at the other end of the rod, as shown in one of the accompanying illustrations. This distance varies slightly by year and model. If the distance between centers is not as specified, carefully pry the forward ball joint free of the spark lever cam and rotate the rod end to correct the length of the rod. Spark control rod specifications are as follows:
   - For motors through 1992, the spark control rod on 40/48/50 hp models should be about 2 1/16 in. (5.3cm) long. For 45/55 commercial models through 1992 there is NO preliminary spark control rod spec or adjustment.
   - For 1993 motors, the specifications are as follows: 25D/40R/40BA/48 hp models (these have a 4-amp charging system) the spark control rod should be about 2 1/16 in. (5.3cm) long. For 40E/40TE/40TL/40TTL/50BE/50TE/50TL/50JE models (these have a 12-amp charging system) the spark control rod should be about 2 5/8 in. (6.7cm) long. For 1993 45/55 commercial models there is NO preliminary spark control rod spec or adjustment.
   - For 1994 and later motors spark control rod adjustment varies only with the charging system. It should be either 2-1/2 in. (6.4cm) on models equipped with a 12-amp charging system or 2 in. (5.8cm) on models equipped with a 4-amp charging system (most commercial models in this year range have a 4-amp system).
4. On all except the 40RW, 40RP, 40WR, 45 and 55 Commercial models, disconnect the throttle control rod socket from the throttle cam (unless already done on models through 1993 in a previous step).
5. On 40RW, 40RP, 40WR, 45 and 55 Commercial models proceed as follows, depending upon the year:
   - For models through 1993 back out the idle speed and WOT stop screws, then turn the twist grip to the full open and full closed positions. Make sure there is 1/4 in. (6mm) gap between the roller and the end of the slot in the throttle lever in both positions. If necessary, adjust this by loosening the jam nut and rotating the thumb wheel at the throttle cable bracket. Re-tighten the jam nut once the throttle lever is positioned correctly.

2-66 MAINTENANCE & TUNE-UP

- For 1994 and later models, loosen the locknut, then rotate the nut and the thumb wheel clockwise until seated on the rod.

6. For all models EXCEPT 45/55 Commercial motors through 1993, check the initial idle screw adjustment dimension (distance from the underside of the screw head to the spark lever cam). The initial adjustment must be 1/2 in. (12mm), if necessary loosen the idle speed screw locknut, and rotate the screw to set the dimension. Tighten the locknut to keep the screw from moving.

7. Proceed with the adjustments, starting with throttle plate synchronization.

Throttle Plate (Cam) Synchronization

◆ See Figures 188, 189 and 190

To synchronize the throttle plate (cam):
1. Loosen the adjustment screw for the throttle cam follower roller, then move the follower away from the throttle cam.
2. Loosen the carburetor lever adjustment screw for the upper carburetor.
3. Rotate both throttle shafts slightly and permit them to snap shut. Apply a gentle pressure on the adjustment tab, as shown, and tighten the adjustment screw.
4. Check to be sure that both throttle shafts begin to rotate at exactly the same time.

Cam Pickup Point

◆ See Figures 188 and 189

To adjust the cam pickup point: Verify that both carburetor throttle plates are closed, then check to be sure the throttle cam follower roller makes contact with the throttle cam at the embossed mark on the cam. The mark must align with the center of the roller. If not, loosen the adjustment screw and reposition the roller until it makes contact at the mark on the cam. Tighten the screw to hold the new adjustment.

The throttle cam on 40RW, 40RP, 40WR, 45 and 55 Commercial models have 2 embossed marks facing the follower roller. Make sure that the LOWER of the 2 marks aligns with the center of the roller otherwise use the adjustment screw to center the follower. On these models through 1993

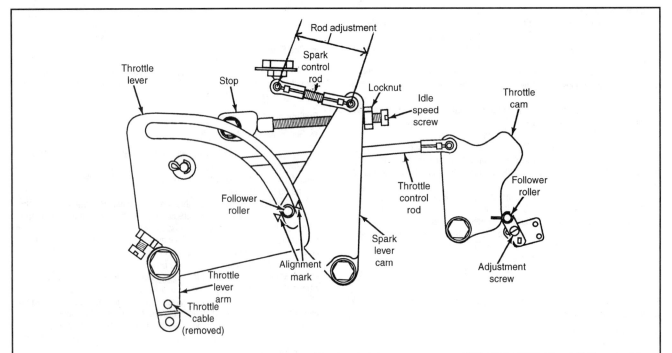

Fig. 188 Throttle, idle and spark linkage - 737cc 2-cylinder 25, 40, 40RS, 48 and 50 Hp motors, except 40RW, 40RP, 40WR, 45 and 55 Commercial models

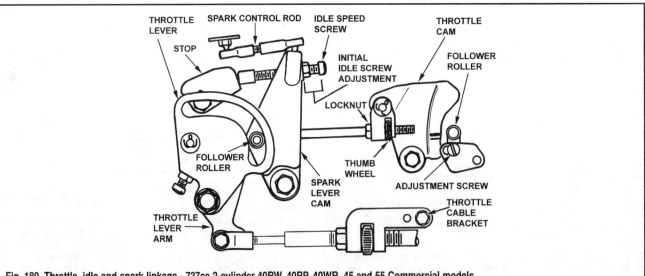

Fig. 189 Throttle, idle and spark linkage - 737cc 2-cylinder 40RW, 40RP, 40WR, 45 and 55 Commercial models

MAINTENANCE & TUNE-UP 2-67

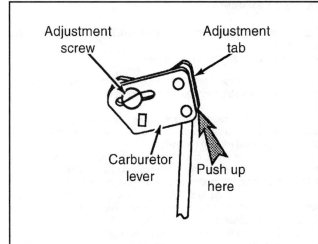

Fig. 190 Press up gently on the tab while tightening the adjusting screw - carb synchronization on 737cc models

OMC recommended that you also check the pickup timing (Idle Timing) using a timing light and, if necessary, adjusting it at this point by loosening the jam nut on the spark lever cam-to-throttle cam link rod and using the thumb wheel to change the length of the link. Rotate the TOP of the thumb wheel toward the crankcase to INCREASE pickup timing and away from the crankcase to DECREASE pickup timing. Tighten the jam nut after any adjustment.

■ We're not sure why, but OMC changed their recommended order of procedures here. On models through 1993 they wanted you to adjust the Maximum Spark Advance next (followed by the Wide Open Throttle Stop on Commercial Models only) and THEN return to adjust Idle Speed. However starting in 1994 they recommended going to Idle Speed next and later adjusting the Maximum Spark Advance. We've tried it both ways and honestly think it doesn't matter, but we mention it here so that you can follow the original recommendations on earlier models IF you prefer.

Idle Speed

◆ See Figures 188 and 189

Here's your excuse, show this page to your better half in case she (or he) doesn't believe you. It's time to take the boat out because that's how the manufacturer wants you to adjust the idle speed. Not in a test tank, not on a trailer or even attached to a dock, but with the motor mounted on a boat, underway. For safety, use an assistant to navigate while you make the adjustment.

■ Always adjust the idle speed with the correct propeller installed.

Refer to the Tune-Up Specifications chart for idle speed specs.
1. Connect an accurate shop tachometer, then start the engine and allow it to reach normal operating temperature.

✴✴ CAUTION

Water must circulate through the lower unit to the engine any time the engine is run to prevent damage to the water pump in the lower unit. Just a few seconds without water will damage the water pump.

2. Once the engine has fully warmed, operate the engine at idle with the gearcase in the forward position. Note the engine speed reading on the tachometer as compared to the specs listed in the Tune-Up Specifications chart. Anything in this range is acceptable, but ideally idle should be set toward the center of this range.
3. If adjustment is necessary, either stop the engine for safety or use **extreme** caution around moving parts. Loosen the idle speed screw locknut, then turn the screw to make an adjustment. Restart the engine and recheck idle speed with the gearcase in the forward setting. Repeat this cycle until the proper setting is obtained.

✴✴ CAUTION

We at Seloc realize that most people probably don't shut the engine off before playing with the idle speed screw, but the manufacturer does not recommend attempting to adjust the idle speed with the engine running for safety reasons and we cannot disagree. If you choose to ignore this caution, make sure that you take all possible precautions to prevent injury by making sure someone else is navigating. Also, keep your hands and clothing away from any hot or moving parts on the outboard.

4. Once the proper idle speed setting is confirmed, use a timing light to check and confirm that idle timing is 0-2° or 1-5° After Top Dead Center (ATDC), depending upon the year and model. If idle timing is out of specification, perform the carburetor low speed mixture adjustment to optimize carburetor settings. If even after adjustment, proper idle timing cannot be set, check the engine temperature, powerhead condition, for sufficient exhaust backpressure and, if used, vacuum balance hose condition.

Carburetor Initial Low Speed Setting

On these models the low speed mixture for each carburetor is controlled by an adjustable needle.

Although not a periodic adjustment, periodic inspections could reveal the need for this adjustment to correct for otherwise unresolved performance issues. Generally speaking however, this adjustment should not be necessary unless the carburetor has been replaced or overhauled.

■ Before disturbing the factory setting, make an alignment mark on the carburetor body. This mark is used as a point of reference during the adjustment procedure.

The engine must be run under load for this procedure (either in a test tank or on a boat).
1. Start the engine and allow it to reach normal operating temperature.

✴✴ CAUTION

Water must circulate through the lower unit to the engine any time the engine is run to prevent damage to the water pump in the lower unit. Just a few seconds without water will damage the water pump.

2. If servicing a tiller model, rotate the idle adjustment knob on the tiller grip counterclockwise (when facing the steering handle) to the full slow position.
3. Shift the engine into forward gear and allow it to run at idle for at least 3 minutes.
4. With the engine running at idle in forward gear, observe the running conditions as follows:
   a. If the engine is running rich, it will show a rough or unsteady idle.
   b. If the engine is running lean, it will sneeze or backfire.
5. If necessary, adjust the low speed mixture screw as follows to obtain a smooth idle:
   a. For rich mixtures, noting the reference mark made earlier, turn the needle 1/8th turn clockwise, allowing about 15 seconds between adjustments, until the highest consistent rpm is reached.
   b. For lean mixtures, noting the reference mark made earlier, turn the needle 1/8th turn counterclockwise, allowing about 15 seconds between adjustments, until the highest consistent rpm is reached.
6. Repeat this procedure for the remaining carburetor.
7. With the engine still running at normal operating temperature and in forward gear, adjust the idle speed screw to the center range of the Idle Speed (RPM) in Gear listed for your motor under the Tune-Up Specifications chart.
8. Operate the outboard in forward at or near full throttle for a full minute. Reduce speed suddenly to a low idle and shift into neutral. The powerhead should continue to operate smoothly. If the powerhead pops or stalls, the air/fuel mixture is probably too lean. Rotate the low speed needle 1/8th turn counterclockwise, allowing about 15 seconds between adjustments, until the powerhead responds as expected. Repeat the full throttle test and sudden deceleration with the shift into neutral to check each adjustment.

■ If the engine does not respond properly to these adjustments, check for other problems including:

MAINTENANCE & TUNE-UP

- Improper engine temperatures
- Incorrect linkage adjustments
- Insufficient exhaust backpressure
- A problem with the pulse equalization hose
- Other problems with the fuel or ignition systems
- Quality and mixture of fuel and oil

Throttle Control Rod (1994 and Later Models)

◆ See Figures 188 and 189

On 1993 and earlier models the throttle control rod was already adjusted either during preliminary procedures or during other adjustments. However, starting in 1994 OMC changed the recommendation for when and how to adjust this rod, so take a moment now on later-model motors to adjust the rod as follows:

■ The throttle control rod adjustment is performed statically (with the engine NOT running).

1. Hold the idle speed screw against its stop on the engine (making sure that both throttle plates are closed), and check the gap between the throttle cam and follower roller using a feeler gauge, it should be 0.020 in. (0.5mm).

■ When checking gaps, a feeler gauge of the proper size should pass through the gap with a slight drag while the next size up should not fit and the next size down should not drag.

2. If the throttle control rod gap is incorrect, adjust the length of the rod as follows:
   a. On all except 40RW, 40RP, 40WR, 45 and 55 Commercial models, pry the rod free of the throttle cam and rotate the rod end until the correct length is achieved in order to produce the specified gap. Then, reconnect the control rod socket to the throttle cam and verify the adjustment.
   b. On 40RW, 40RP, 40WR, 45 and 55 Commercial models, loosen the throttle rod locknut, then turn the thumbwheel (counterclockwise when dealing with a fully seated throttle cam) to obtain the proper gap. Tighten the locknut and verify the adjustment.

Wide Open Throttle Stop

◆ See Figures 188, 189 and 191

■ The Wide Open Throttle (WOT) stop adjustment is performed statically (with the engine NOT running).

To check the WOT stop position: With the powerhead not operating, advance the throttle lever to the full throttle (or wide open) position. At this point the carburetor throttle shaft roll pins (found on the opposite side of the carburetors from the linkage) should be completely vertical indicating the throttle valves are all the way open. If not, loosen the locknut on the stop screw at the lower corner of the throttle lever. Rotate the screw as necessary to bring the roll pins into vertical alignment, then tighten the locknut on the stop screw to hold this new adjustment. Verify the positioning of the screw did not change while tightening the locknut.

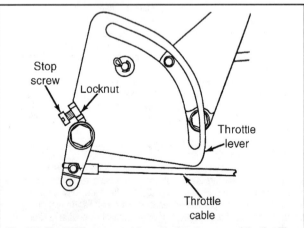

Fig. 191 The WOT stop screw is on the lower corner of the throttle lever - 737cc motors except 40RW, 40RP, 40WR, 45 and 55 Commercial models shown, others similar

Throttle Cable Installation

Tiller Control Models, Except 40RW, 40RP, 40WR, 45 and 55 Commercial Motors

◆ See Figures13122g188"> re-reference

1. Turn the idle speed adjustment knob counterclockwise (when facing the tiller handle from the motor operator's position) until the full slow idle position is reached.
2. Connect the throttle cable to the upper hole in the throttle lever arm and secure using the pin and cotter clip. Make sure the clip is installed parallel to the raised ridge on the lever.
3. Hold the tiller throttle grip in the full slow position, and with the idle speed screw on its stop, pull firmly on the cable to remove any backlash. Rotate the cable bracket to align with the mounting screw hole.
4. Preload the linkage by threading the cable 2 turns toward the cable end, then install the bracket using the flat washer between the bracket and intake manifold. Tighten the retaining bolt to 36 inch lbs. (4 Nm).
5. Install the cable retainer plate and tighten the bolts securely.
6. Check the throttle cable bracket adjustment by rotating the throttle twist several times and verifying that the idle speed screw touches the stop with the throttle grip in the slow position.

Remote Control Models, Except 40RW, 40RP, 40WR, 45 and 55 Commercial Motors

◆ See Figure 188

1. If the gearcase was serviced, check the neutral detent adjustment as directed under Gearcase.
2. Attach the throttle cable to the arm and tighten the locknut securely.
3. Verify that the fast idle lever is down in the run position.
4. Have an assistant slowly rotate the propeller shaft by hand while you shift the remote handle from neutral to forward and then 1/2 the distance back to neutral. This positions the control as necessary to obtain proper adjustment.
5. Move the engine throttle lever so that the spark lever stop screw is against the stop. Then, attach the throttle cable casing guide to the throttle lever lower hole using the pin and cotter clip. The clip must be installed parallel to the raised rib.
6. Pull firmly on the throttle cable to remove backlash, then install the trunnion nut in the anchor pocket.
7. Install the cable retainer and screw to hold everything in position, then check cable adjustment to ensure proper idle speed (that it is not too high) and shifting (to prevent excessive effort). Move the remote control all the way forward and then back into neutral. Verify that the idle stop screw is against the stop. If not, remove backlash by adjusting the trunnion nut.
8. Once adjustment is correct, securely tighten the cable retainer screw.

40RW, 40RP, 40WR, 45 and 55 Commercial Motors

◆ See Figure 189

1. Turn the idle speed adjustment knob counterclockwise (when facing the tiller handle from the motor operator's position) until the full slow idle position is reached.
2. Loosen the locknut on the throttle cable bracket.
3. Hold the tiller throttle grip in the full slow position and turn the bracket thumbwheel until the forward bracket hole aligns with the mounting hole. Install the screw through the forward mounting hole, placing the washer between the bracket and the powerhead, then tighten the screw to 36 inch lbs. (4 Nm).
4. Rotate the thumbwheel 2 revolutions clockwise to preload the linkage, then securely tighten the locknut.
5. Verify the throttle cable adjustment by rotating the throttle twist several times and verifying that the idle speed screw touches the stop with the throttle grip in the slow position.

Maximum Spark Advance

◆ See Figures 188 and 189

To check the maximum spark advance, the outboard must be operated with the proper test wheel. Do not operate the powerhead with a propeller, or flush adapter while performing this adjustment.

MAINTENANCE & TUNE-UP

** CAUTION

Water must circulate through the lower unit to the engine any time the engine is run to prevent damage to the water pump in the lower unit. Just a few seconds without water will damage the water pump.

1. Connect a timing light to the top cylinder.

■ For a clear reference, mark the flywheel timing grid at the 19° mark.

2. Start the engine and run it at idle until it reaches operating temperature.
3. Once the motor is fully warmed, run it at a minimum of 5000 rpm with the outboard in forward gear. Aim the timing light at the timing marks on the flywheel perimeter. The mark on the timing pointer should align between the 18° and 20° marks on the flywheel.
4. If the timing requires adjustment, shut down the powerhead for safety and alter the length of the spark control lever to correct the timing, as follows:
   a. Pry off the forward ball joint on the spark control rod from the ball joint on the spark lever cam.
   b. Rotate the rod end clockwise to advance the timing, or counterclockwise to retard the timing.

■ Two complete revolutions of the rod end equals approximately one degree change in the timing. Also, keep in mind that if you make an adjustment of more than 2 revolutions, the idle speed adjustment and throttle control length must both be readjusted to ensure proper motor operation.

   c. Snap the rod end onto the spark lever cam ball joint when the adjustment is complete.
5. Restart the engine and verify proper maximum spark advance timing.
6. Double-check the idle speed. IF the forward ball joint was turned more than 2 full revolutions, it is quite likely that the Idle Speed and Throttle Control Rod adjustments must be performed again. THIS might be one of the reasons that OMC originally had the Maximum Spark Advance adjustment BEFORE the Idle Speed adjustment (at least as their recommendations went for models through 1993). It is likely that other benefits from placing the Maximum Spark Advance towards the ends of the adjustments outweighed the minor inconvenience of resetting these two adjustments.

Shift Lever Detent (40RW, 40RP, 40WR, 45 and 55 Commercial Models)

1. Loosen the detent spring screw.
2. Move the shift lever until the gearcase neutral shift position is felt (and verified by turning the propeller shaft slowly by hand).
3. Center the detent spring in the detent plate notch, then tighten the screw to 60 inch lbs. (7 Nm).

25/35 Hp (500/565cc) 3-Cylinder Models

Some of the timing and synchronization procedures for these units require operating the motor at idle rpm under load and at wide-open throttle. Therefore, the outboard **must** be placed in a test tank or a body of water with the boat well secured to the dock or in a slip (except for the idle timing setting which must be conducted on an unrestrained boat). Never attempt to make the load adjustments or run the engine at wide open throttle with a flush attachment connected to the lower unit. The powerhead operating at high rpm with such a device, would likely cause a runaway condition from a lack of load on the propeller, causing extensive damage.

The necessity to conduct the idle timing procedure on an unrestrained boat so early in the timing and synchronizing sequence makes it impractical to truly tune these motors without having them mounted on a boat. If at all possible, avoid the use of a test tank for these adjustments. Remove the engine top cover to gain access to adjustment locations.

■ If a test tank must be used for most of the settings, all settings from the idle timing on should be checked and adjusted again once the motor is finally mounted on a boat.

The following procedures provide detailed instructions to set the timing pointer (if disturbed), then adjust the throttle plate synchronization, idle timing, cam follower pickup, maximum spark advance, wide open throttle positioning and neutral detent adjustment. The carburetor low speed mixture adjustment (which is not a periodic adjustment) is also provided here in case this adjustment is necessary due to carburetor replacement, overhaul or otherwise un-resolvable idle problems after certain of the other adjustments

are completed. Procedures should be performed exactly as directed and in the order given to ensure proper adjustments.

■ The preliminary adjustments must be performed before starting these adjustment procedures.

The idle timing portion of the synchronization procedures for these units require that the engine be operated at idle rpm mounted on the boat, under load and unrestrained. For this reason, the adjustments should take place on a low-traffic body of water and only with an assistant to navigate while you make the adjustments.

ADJUSTMENTS

♦ See Figure 192

Perform the following procedures in the sequence given in order to ensure correct timing and carburetor synchronization. Start the procedures by following these preliminary adjustments:

Preliminary Adjustments

♦ See Figure 192

These preliminary adjustments will ensure that the linkage us in the correct position before starting the timing and synchronization procedures.

** CAUTION

For safety, the manufacturer recommends disconnecting the spark plug leads whenever the procedure does not call for running the motor. This will prevent the possibility that the engine could be accidentally started while someone's hands are near a moving part.

1. Measure the length of the idle stop screw from the contact point at the throttle arm to the beginning of the screw boss (the throttle arm side of the screw boss). This length should be 3/4 in. (19mm), if necessary loosen the locknut on the face of the screw boss, then adjust the screw and tighten the nut to hold it in place.
2. For tiller models:
   a. Remove the pivot screw that holds the throttle cable bracket to the support arm.

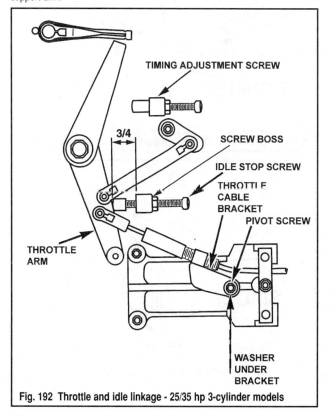

Fig. 192 Throttle and idle linkage - 25/35 hp 3-cylinder models

2-70 MAINTENANCE & TUNE-UP

    b. Turn the idle speed adjustment knob on the tiller handle counterclockwise to the full slow position. Make sure that the knob is turned fully to the slow setting and that the throttle arm is contacting the idle stop screw.

    c. Twist the throttle cable bracket in order to align it with the mounting screw hole, then twist the bracket 1 complete revolution forward toward the cable end in order to preload the cable.

    d. Install the bracket and pivot screw placing the flat washer underneath the bracket, between the bracket and the cable support arm.

  3. For remote models, disconnect the throttle cable from the engine.

Setting the Timing Pointer

◆ See Figures 193 and 194

A timing pointer is mounted to the flywheel cover, at the mouth of an access opening directly above the carburetors. If the timing pointer bracket has been disturbed use this procedure to check and set the pointer positioning (thereby ensuring the accuracy of the idle timing and maximum spark advance procedures to follow).

  1. Remove the 3 spark plugs from the cylinder head using the procedures found under Spark Plugs in this section.

  2. Loosen the screw fastening the timing pointer, then center the pointer and retighten the screw to hold it in position.

  3. Slowly rotate the flywheel clockwise until the cast Top Dead Center (TDC) mark is about 1 in. (25.4mm) **past** the timing pointer. One way to rotate the engine is to SLOWLY pull on the hand starter, but be careful not to pull too hard or fast, otherwise you'll kick the motor around, right past the timing marks.

※※ WARNING
Under NO circumstances should you EVER rotate the flywheel counterclockwise. If you do there is a good chance that the water pump impeller vanes will become damaged.

  4. Install the OMC Piston Stop Tool (#384887) or equivalent, into the spark plug bore for the top (No.1) cylinder. Adjust the tool to a point where it makes contact with the piston.

  5. Hold the piston firmly against the tool piston stop tool and make a mark on the flywheel inline with the timing pointer. Label this mark 1 or A to distinguish it from the next mark.

  6. Rotate the flywheel in a clockwise direction until the piston contacts the tool again, then make a second mark (label it 2 or B) on the flywheel inline with the timing pointer, then remove the piston stop tool.

  7. Use the molded marks to locate the exact midway point between the first and second marks and place a mark (label it 3 or C) at this location, since it represents actual TDC for where the timing pointer is currently set.

■ **The piston stop tool is basically just an adjustable rod with a locknut, mounted through spark plug threads. A substitute could be fabricated with a little creativity (using the casing of a spark plug or even a plug thread chaser). The key to the tool is that it can be locked in place (using the locknut) at some early point in the piston's downstroke. The first mark you make on the flywheel translates into this random point. Then, the piston is brought the rest of the way down in its travel and back up again (by rotating the flywheel) until it reaches the very same height in the bore (and contacts the tool again, but this time on the way up). A second mark, represents the exact same physical point in the bore, on the exact opposite side of Top Dead Center (TDC) or the very top of the piston travel. By locating the exact point midway between these to marks, you've found the spot on the flywheel that corresponds perfectly to Top Dead Center with regards to the pointer's current position.**

  8. Rotate the flywheel clockwise to align your TDC mark (labeled 3 or C) with the timing pointer, since you know that will place the flywheel and crankshaft at TRUE Top Dead Center, then, holding the flywheel in this position, loosen the timing pointer screw again. Slide the pointer away from the center mark and align it with the cast TDC mark on the flywheel itself. Voila, your pointer is now aligned with the flywheel TDC mark WHILE the crankshaft and flywheel is TRULY at TDC. Tighten the timing pointer retaining screw securely as it is now set for accurate readings on the flywheel again.

  9. Install the spark plugs.

Throttle Plate (Cam) Synchronization

  1. Remove the air intake silencer for access.

  2. Use a 5/16 ball-hex driver to turn the cam follower adjustment screw counterclockwise until the cam and follower are not in contact with each other.

  3. Loosen the two carburetor link screws and verify that the throttle plates are completely closed (by peering through the throttle bores).

  4. Very, very lightly push upward on the carburetor link tabs and tighten the screws.

Idle Timing

◆ See Figure 192

Here's your excuse, show this page to your better half in case she (or he) doesn't believe you. It's time to take the boat out because that's how the manufacturer wants you to adjust the idle speed. Not in a test tank, not on a trailer or even attached to a dock, but with the motor mounted on a boat, underway. For safety, use an assistant to navigate while you make the adjustment.

■ **Always adjust the idle speed with the correct propeller installed.**

Refer to the Tune-Up Specifications chart for idle speed specs.

  1. Disable the motor's QuikStart cold starting feature in order to get an accurate timing reading. To do this, disconnect the white/black lead connector from the temperature switch on the powerhead, then use an alligator clip jumper wire to connect the ignition module side of the circuit to a clean engine ground.

  2. Connect an accurate shop tachometer, then connect an inductive timing light to the No. 1 (top) cylinder spark plug lead.

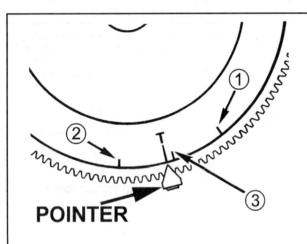

Fig. 193 Centering the timing pointer (marks 1 and 2 are used to find 3 which is TRUE TDC where the pointer is currently aligned)

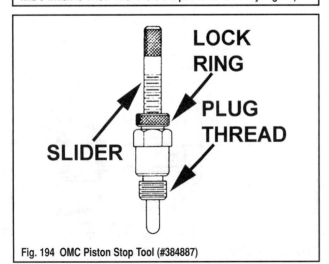

Fig. 194 OMC Piston Stop Tool (#384887)

MAINTENANCE & TUNE-UP

> ✳✳ **CAUTION**
>
> Water must circulate through the lower unit to the engine any time the engine is run to prevent damage to the water pump in the lower unit. Just a few seconds without water will damage the water pump.

3. Start the engine and wait at least 10 seconds for the QuikStart system to return the timing to normal, then shift the engine into forward gear. Visually check the throttle arm to make sure it is against the stop, then check the idle timing by aiming the timing light at the timing pointer and the flywheel. On these motors, idle timing should be 4° After Top Dead Center (ATDC).

4. If adjustment is necessary, either stop the engine for safety or use **extreme** caution around moving parts. Loosen the idle speed screw locknut (located just in front of the screw boss), then turn the screw to make an adjustment. Turning the screw clockwise will advance the idle timing, while turning the screw counterclockwise will retard the idle timing. Restart the engine and recheck idle speed with the gearcase in the forward setting. Repeat this cycle until the proper setting is obtained.

> ✳✳ **CAUTION**
>
> We at Seloc realize that most people probably don't shut the engine off before playing with the idle speed screw, but the manufacturer does not recommend attempting to adjust the idle speed with the engine running for safety reasons and we cannot disagree. If you choose to ignore this caution, make sure that you take all possible precautions to prevent injury by making sure someone else is navigating. Also, keep your hands and clothing away from any hot or moving parts on the outboard.

5. Once the proper idle timing setting is confirmed, use the tachometer to check idle rpm. This procedure should result in an idle speed of 700-800 rpm (as specified for these motors), depending on propeller selection and condition. If the speed is incorrect, check the following:

  a. If the speed it too high, check the induction system for air leaks.

  b. If the speed is too low, and engine components and systems are operating properly, decrease the idle timing by one or two degrees (say from 4 ATDC to 3 or 2 ATDC) to achieve the desired rpm.

  c. If idle speed is inconsistent or the engine runs rough or spits, and no problems can be found, suspect an incorrect carburetor mixture problem, refer to Carburetor Initial Low Speed Setting adjustment in this section.

Carburetor Initial Low Speed Setting

On these models the low speed mixture for each carburetor is controlled by an adjustable needle.

Although not a periodic adjustment, periodic inspections could reveal the need for this adjustment to correct for otherwise unresolved performance issues. Generally speaking however, this adjustment should not be necessary unless the carburetor has been replaced or overhauled.

The engine must be run under load for this procedure (either in a test tank or on a boat).

1. Verify that the low speed screw was properly set to factory specifications before starting this procedure. To start this procedure, the underside of the screw head should be 0.390 in. (9.9mm) above the carburetor cover.

2. Start the engine and allow it to reach normal operating temperature.

> ✳✳ **CAUTION**
>
> Water must circulate through the lower unit to the engine any time the engine is run to prevent damage to the water pump in the lower unit. Just a few seconds without water will damage the water pump.

3. If servicing a tiller model, rotate the idle adjustment knob on the tiller grip counterclockwise (when facing the steering handle) to the full slow position.

4. Shift the engine into forward gear and allow it to run at slow idle of about 700-800 rpm.

5. With the engine running at idle in forward gear, adjust the low speed screw until the highest consistent powerhead speed is attained. Allow about 15 seconds between each adjustment for the engine to stabilize.

■ Turn the screw slowly, in fractional 1/4 or 1/8 increments for each adjustment, then allow time for it to take effect.

6. Once the engine has stabilized and runs smoothly at the highest possible idle speed, rotate the low speed needle counterclockwise 1/8th turn to prevent the powerhead from operating too lean at low speeds, particularly in neutral.

7. With the engine still running at normal operating temperature and in forward gear, adjust the idle speed screw to run the engine at a minimum of 650 rpm in forward gear.

8. Operate the outboard in forward at or near full throttle for a full minute. Reduce speed suddenly to a low idle (700-800 rpm) and shift into neutral. The powerhead should continue to operate smoothly. If the powerhead pops or stalls, the air/fuel mixture is probably too lean. Rotate the low speed needle 1/16th turn counterclockwise, allowing about 15 seconds between adjustments, until the powerhead responds as expected. Repeat the full throttle test and sudden deceleration with the shift into neutral to check each adjustment.

■ If the engine does not respond properly to these adjustments, check for other problems with the fuel or ignition systems. Make sure a quality fuel is being used along with the proper mixture of oil.

9. Once a satisfactory adjustment has been achieved, verify engine idle speed again using the idle timing procedure.

Cam Follower Pickup

◆ See Figure 192

1. Check the throttle arm to make sure it is against the stop screw.

2. Use a grease pencil to make a mark on the crankcase, directly below and inline with the 4° ATDC mark on the timer base grid. The grid is located just under the flywheel cover at the very front of the motor (right above the regulator/rectifier and round wiring connector).

3. Advance the throttle to align the mark on the crankcase with the 0° mark on the timing base grid, then use a 5/64 in. ball hex driver to adjust the cam follower screw so it **just** touches the cam.

4. Close the throttle, then open it again until the mark on the crankcase aligns with the 0° mark on the timing grid and verify that the cam follower is again **just** touching the cam.

Maximum Spark Advance

■ For the ignition timing maximum spark advance specifications, please refer to the Tune-Up Specifications chart.

1. Advance the throttle fully to the wide open position, then check the timing pointer, it must align with the correct timing mark on the timer base grid. The grid is located just under the flywheel cover at the very front of the motor (right above the regulator/rectifier and round wiring connector).

2. If adjustment is necessary, loosen the locknut on the timing adjustment screw, then rotate the screw to align the mark on the crankcase with the correct timing specification on the base grid. Each complete turn clockwise will advance timing approximately 2°, while each complete turn counterclockwise will retard timing about 2°.

3. Tighten the timing adjustment screw locknut after adjustment is complete.

4. Verify the adjustment by opening and closing the throttle repeatedly and checking to make sure mark on the crankcase consistently aligns with the correct specification on the timer base grid each time the throttle is advanced to the WOT position.

Wide Open Throttle Positioning

1. Disconnect the drain hose and remove the air intake silencer for access.

2. Advance the throttle to the Wide Open Throttle (WOT) position and hold while visually checking the throttle plates to make sure they are completely horizontal (peer in through the carburetor throttle bores and make sure the plates are completely open).

3. If the plates are not in the WOT position, check the linkage or incorrect assembly, binding or interference.

4. Attach the drain hose and install the air intake silencer.

Neutral Detent Adjustment

1. Rotate the propeller shaft slowly by hand while moving the shift lever into the neutral position.

2. Move the shift lever back and forth slightly to verify the neutral detent position, then loosen both neutral detent screws (found side-by-side at the base of the lever).

2-72 MAINTENANCE & TUNE-UP

3. Locate the middle of the neutral position by first moving the lever forward to the limit of free-play and mark this position with a line on the neutral detent cam. Then move the shift lever backward to the other limit of free-play and mark this position with a line as well. Find the physical middle point between the two lines and place a mark at this position.

4. Move the shift lever to the middle mark and tighten the two detent screws to 60-84 inch lbs. (7-9 Nm).

25-70 HP (913cc) 3-Cylinder, 2-Stroke Models

Some of the timing and synchronization procedures for these units require operating the motor at idle rpm under load and at wide-open throttle. Therefore, the outboard **must** be placed in a test tank or a body of water with the boat well secured to the dock or in a slip (except for the idle timing setting which must be conducted on an unrestrained boat). Never attempt to make the load adjustments or run the engine at wide open throttle with a flush attachment connected to the lower unit. The powerhead operating at high rpm with such a device, would likely cause a runaway condition from a lack of load on the propeller, causing extensive damage.

The necessity to conduct the idle timing procedure on an unrestrained boat so early in the timing and synchronizing sequence on RS, TTL, WR models and 25SIK makes it impractical to truly tune these motors without having them mounted on a boat. If at all possible, avoid the use of a test tank for these adjustments. Remove the powerhead cowling to gain access to adjustment locations.

■ **If a test tank must be used for most of the settings, all settings from the idle timing on should be checked and adjusted again once the motor is finally mounted on a boat.**

The following procedures provide detailed instructions to set the timing pointer (if disturbed). Then, adjust the throttle cable (on RS, TTL, WR models and 25SIK), throttle plate synchronization, cam follower pickup, the throttle cam, wide open throttle positioning, idle timing (performed before the throttle cam and wide open throttle adjustments on RS, TTL, WR models and 25SIK) and maximum spark advance. The RS, TTL, WR models and 25SIK also have a neutral detent adjustment. The carburetor low speed mixture adjustment (which is not a periodic adjustment) is also provided here in case this adjustment is necessary due to carburetor replacement, overhaul or otherwise un-resolvable idle problems after certain of the other adjustments are completed. Procedures should be performed exactly as directed and in the order given (or noted) to ensure proper adjustments.

The idle timing portion of the synchronization procedures for these units require that the engine be operated at idle rpm mounted on the boat, under load and unrestrained. For this reason, the adjustments should take place on a low-traffic body of water and only with an assistant to navigate while you make the adjustments.

ADJUSTMENTS

■ **For remote control models, remove the throttle cable from the control arm and the anchor pocket prior to beginning these procedures.**

Setting the Timing Pointer

◆ See Figures 193, 194 and 195

A timing pointer is mounted to the top of the engine, under the flywheel cover. If the timing pointer or intake manifold has been disturbed use this procedure to check and set the pointer positioning (thereby ensuring the accuracy of the idle timing and maximum spark advance procedures to follow).

■ **Most models are equipped with two timing mark grids, be certain to use the correct grid for the motor on which you are working. Remote control models and TTL models all use the second grid (flywheel rotates in a clockwise direction). The 65RS models use the grid marked "rope."**

1. Remove the 3 spark plugs from the cylinder head using the procedures found under Spark Plugs in this section.

2. Loosen the screw fastening the timing pointer, then center the pointer and retighten the screw to hold it in position.

Fig. 195 Most 913cc 3-cylinder motors use the second timing grid

3. Slowly rotate the flywheel clockwise until the cast Top Dead Center (TDC) mark is about 1 1/2 in. (40mm) **past** the timing pointer.

✱✱ WARNING

Under NO circumstances should you rotate the flywheel counterclockwise. If you do there is a good chance that the water pump impeller vanes will become damaged or will when the motor is started next time. IF for some reason the flywheel is ever rotated counterclockwise, slowly and carefully rotate it back clockwise, by hand, for a couple of turns. This SHOULD reset the impeller vanes protecting them from the possibility of further damage during the next engine start-up.

4. Install the OMC Piston Stop Tool (#384887) or a dial gauge, into the spark plug bore for the top (No.1) cylinder. Adjust the tool using the slider to a point where it makes contact with the piston, then lock the tool in place using the lock ring (or zero the gauge).

5. Hold the piston firmly against the piston stop tool and make a mark on the flywheel inline with the timing pointer. Label this mark "1" to distinguish it from the next mark.

6. Rotate the flywheel in a clockwise direction until the piston contacts the tool again, then make a second mark (label it "2") on the flywheel inline with the timing pointer, then remove the piston stop tool.

7. Using a flexible scale, measure along the flywheel to locate the exact midway point between the first and second marks and place a mark at this location labeled "3". This mark represents TDC to the timing pointer's current setting. If this mark and the cast TDC mark on the flywheel align, then the pointer position is correct.

■ **The piston stop tool is basically just an adjustable rod with a locknut, mounted through spark plug threads. A substitute could be fabricated with a little creativity (using the casing of a spark plug or even a plug thread chaser). The key to the tool is that it can be locked in place (using the locknut) at some early point in the piston's downstroke. The first mark you make on the flywheel translates into this random point. Then, the piston is brought the rest of the way down in its travel and back up again (by rotating the flywheel) until it reaches the very same height in the bore (and contacts the tool again, but this time on the way up). A second mark, represents the exact same physical point in the bore, on the exact opposite side of Top Dead Center (TDC) or the very top of the piston travel. By locating the exact point midway between these to marks, you've found the spot on the flywheel that corresponds perfectly to Top Dead Center with regards to the pointer's current position.**

8. If adjustment is necessary (the "3" mark and the cast TDC do not align), rotate the flywheel clockwise to align this TDC mark with the timing pointer. Then, holding the flywheel in this position, loosen the timing pointer screw again. Slide the pointer away from the center mark and align it with the cast TDC mark on the flywheel itself. Tighten the timing pointer retaining screw securely as it is now set for accurate readings on the flywheel again.

9. Install the spark plugs.

MAINTENANCE & TUNE-UP 2-73

Throttle Cable Adjustment (RS, TTL, WR Models and 25SIK only)

1. Turn the idle speed adjustment knob on the tiller handle counterclockwise to the full slow position.
2. Check to verify that the throttle cable bracket is secured to the powerhead using the forward mounting hole.
3. Loosen the idle speed and wide open throttle screws (threaded upward against the throttle cam, the idle speed screw is the top of the two screws mounted side-by-side). Back the screws out until they are flush with the anchor block.
4. Turn the twist grip to the full open and then full closed positions. Check the clearance between the throttle cam roller and the end of the throttle cam slot in each position. The roller should have about a 1/4 in. (6mm) gap between itself and the respective end in each position.
5. If adjustment is necessary, loosen the locknut, then rotate the thumbwheel until the correct gaps are achieved. Tighten the locknut after adjustment.

Throttle Plate Synchronization

◆ See Figures 196 and 197

1. Loosen the cam follower screw, then push the follower away from the throttle cam. The follower cannot touch the cam during this procedure or the adjustment will be incorrect.
2. Loosen the upper and lower carburetor tab linkage screws, then verify that all throttle plates are closed.

■ If necessary, remove the air intake for a view down the throttle bore of each carburetor.

3. With all throttle plates closed, apply a light upward pressure to the upper and lower carb linkage tabs while tightening each tab's screw.

■ For proper results, tighten the lower carburetor's tab screw first, then the upper carburetor's tab screw.

Fig. 196 The cam follower screw is used to adjust the follower in relation to the throttle cam

Fig. 197 Apply light pressure upward (arrow) to each carb linkage tab while tightening its screw

Cam Follower Pickup

◆ See Figures 196, 198 and 199

1. Carefully snap the throttle link coupler free of the ball on the throttle cam.
2. While holding the cam follower against the throttle cam, move the throttle cam until the embossed mark aligns with the center of the cam follower. Tighten the cam follower screw to hold it in this position.

■ Some 1990 models may be equipped with a throttle cam which has 2 alignment marks. IF SO, use the shorter embossed mark which is normally located lower on the cam than the longer mark. Align the shorter mark with the center of the cam follower.

3. Connect an amplifier tool (as described under Homemade Synchronization Tool in this section) to the linkage tab on the top carburetor.
4. Verify the adjustment by advancing the throttle cam and watching the amplifier tool for movement. The throttle cam mark should align with the roller **just** as the end of the amplifier tool begins to move.

The throttle cam on some 1990 and 1991 models (like models equipped with a dual marked throttle cam or the 65RS) contain a throttle cam whose link rod is adjusted through a jam nut and thumbwheel. On those models OMC recommended that you also check the pickup timing (Idle Timing) using a timing light at this point in the procedure and adjust it now, as necessary (instead of later during the Idle Timing procedure later in this section). To adjust the Idle or Pickup timing on these models, loosen the jam nut on the spark lever cam-to-throttle cam link rod and use the thumb wheel to change the length of the link. Rotate the TOP of the thumb wheel toward the crankcase to INCREASE pickup timing and away from the crankcase to DECREASE pickup timing. Tighten the jam nut after any adjustment.

Fig. 198 View of the throttle link coupler and throttle cam ball

Fig. 199 Attach a homemade amplifier tool to the top carb linkage tab

2-74 MAINTENANCE & TUNE-UP

Throttle Cam

◆ See Figure 198

The manufacturer recommends performing the Idle Timing procedure BEFORE the Throttle Cam and Wide Open Throttle Stop adjustments on all RS, TTL, WR and 25SIK motors. It is unclear whether that has something to do with the necessity of reconnecting the throttle cable on remote models that forces the Throttle Cam procedure to come first or if there is another benefit to performing the procedures in that order. Suffice it to say that you will not go wrong on the RS, TTL, WR and 25SIK if you skip ahead to the Idle Timing procedure and come back to Throttle Cam and Wide Open Throttle Stop afterwards.

■ This procedure does not apply to models equipped with a dual marked throttle cam or the 65RS through 1991.

1. For remote control models, reconnect the throttle cable as follows:
   a. Verify that the fast idle lever is down in the run position.
   b. Have an assistant slowly rotate the propshaft by hand while you shift the remote handle from neutral to forward and then 1/2 the distance back to neutral. This positions the control as necessary to obtain proper adjustment.
   c. Move the engine throttle lever tightly against the throttle stop screw. Then, attach the throttle cable casing guide to the throttle lever pin using the washer and locknut, tighten the locknut securely.
   d. Pull firmly on the throttle cable to remove backlash, then install the trunnion nut in the anchor pocket.
   e. Install the cable retainer and screw to hold everything in position, then check cable adjustment to ensure proper idle speed (that it is not too high) and shifting (to prevent excessive effort). Move the remote control all the way forward and then back into neutral. Verify that the idle stop screw is against the stop. If not, remove backlash by adjusting the trunnion nut.
   f. Once adjustment is correct, securely tighten the cable retainer screw.
2. For all models, verify that the throttle arm is firmly against the idle stop.
3. Lightly hold the throttle link coupler against the throttle cam ball using light pressure. Rotate the cam follower and check clearance between the follower and the throttle cam. The adjustment is correct if the follower **does not** touch the cam and the clearance between the two components is 0.010 in. (0.25mm).
4. If adjustment is necessary, turn the shell of the throttle link coupler further onto the link in order to increase clearance, or further off the link to decrease clearance. Once the proper gap is achieved, carefully snap the shell back onto the ball.
5. Test the adjustment by slowly raising the remote or tiller control to fast idle or run, then returning the control to low idle and check that the throttle arm is against the idle stop while the same 0.010 in. (0.25mm) of clearance exists between the roller and throttle cam.

Wide Open Throttle Stop

◆ See Figure 200

1. With the engine **not running** manually advance the throttle lever to the Wide Open Throttle (WOT) position.
2. With the throttle lever at WOT, check each of the roll pins in the carburetor shafts (located under a spring, on the shafts behind each of the linkage tabs. The roll pins should be fully vertical, but NOT over-rotated past vertical.
3. If adjustment is necessary, turn the WOT screw to achieve the proper setting.

Idle Timing

◆ See Figure 201

■ This procedure applies to all models OTHER than those equipped with a dual marked throttle cam or the 65RS through 1991, on which the Idle or Pickup timing should have been adjusted earlier, during the Cam Follower Pickup adjustment.

Here's your excuse, show this page to your better half in case she (or he) doesn't believe you. It's time to take the boat out because that's how the manufacturer wants you to adjust the idle speed. Not in a test tank, not on a trailer or even attached to a dock, but with the motor mounted on a boat, underway. For safety, use an assistant to navigate while you make the adjustment.

■ Always adjust the idle speed with the correct propeller installed.

Refer to the Tune-Up Specifications chart for idle speed specs. If not done already, reconnect the spark plug leads.

1. Connect an accurate shop tachometer, then connect a timing light to the No. 1 (top) cylinder spark plug lead.
2. Start the engine and allow it to reach normal operating temperature.

※※ CAUTION

Water must circulate through the lower unit to the engine any time the engine is run to prevent damage to the water pump in the lower unit. Just a few seconds without water will damage the water pump.

3. Once the engine has fully warmed, operate the engine at idle with the gearcase in the forward position.
4. Make sure the throttle arm is against the idle stop, then aim the timing light at the pointer and check the result against the idle timing spec in the Tune-Up Specifications chart in this section.
5. If adjustment is necessary, either stop the engine for safety or use **extreme** caution around moving parts. Adjust the idle timing by turning the idle timing screw found under the flywheel (as shown in the accompanying photo). Turning the screw inward (clockwise) advances timing, while backing the screw out (counterclockwise) retards idle timing. Restart the engine and recheck idle timing with the gearcase in the forward setting. Repeat this cycle until the proper setting is obtained.

※※ CAUTION

We at Seloc realize that most people probably don't shut the engine off before playing with the idle speed screw, but the manufacturer does not recommend attempting to adjust the idle speed with the engine running for safety reasons and we cannot disagree. If you choose to ignore this caution, make sure that you take all possible precautions to prevent injury by making sure someone else is navigating. Also, keep your hands and clothing away from any hot or moving parts on the outboard.

Fig. 200 When properly adjusted, the roll pins in the carburetor shafts will be vertical at WOT

Fig. 201 Idle timing is adjusted using this screw found under the flywheel

MAINTENANCE & TUNE-UP

6. Once the proper idle timing setting is confirmed, use the tachometer to check idle rpm. This procedure should result in an idle speed within spec (as listed in the Tune-Up Specifications chart in this section), depending on propeller selection and condition. If the speed is incorrect, check the following:

   a. If the speed it too high, check the induction system for air leaks.

   b. If the speed is too low, and engine components and systems are operating properly, decrease the idle timing by one or two degrees (say from 4 ATDC to 3 or 2 ATDC) to achieve the desired rpm.

   c. If idle speed is inconsistent or the engine runs rough or spits, and no problems can be found, suspect an incorrect carburetor mixture problem, refer to Carburetor Initial Low Speed Setting adjustment in this section.

■ **If this procedure was followed before Throttle Cam and Wide Open Throttle Stop adjustments on RS, TTL, WR and 25SIK models, perform those adjustments now.**

Carburetor Initial Low Speed Setting

On these models the low speed mixture for each carburetor is controlled by an adjustable needle.

Although not a periodic adjustment, periodic inspections could reveal the need for this adjustment to correct for otherwise unresolved performance issues. Generally speaking however, this adjustment should not be necessary unless the carburetor has been replaced or overhauled.

■ **Before disturbing the factory setting, make an alignment mark on the carburetor body. This mark is used as a point of reference during the adjustment procedure.**

The engine must be run under load for this procedure (either in a test tank or on a boat).

1. Start the engine and allow it to reach normal operating temperature.

✲✲ CAUTION

Water must circulate through the lower unit to the engine any time the engine is run to prevent damage to the water pump in the lower unit. Just a few seconds without water will damage the water pump.

2. If servicing a tiller model, rotate the idle adjustment knob on the tiller grip counterclockwise (when facing the steering handle) to the full slow position.

3. Shift the engine into forward gear and allow it to run at idle for at least 3 minutes.

4. With the engine running at idle in forward gear, observe the running conditions as follows:

   a. If the engine is running rich, it will show a rough or unsteady idle.

   b. If the engine is running lean, it will sneeze or backfire.

5. If necessary, adjust the low speed mixture screw as follows to obtain a smooth idle:

   a. For rich mixtures, noting the reference mark made earlier, turn the needle 1/8th turn clockwise, allowing about 15 seconds between adjustments, until the highest consistent rpm is reached.

   b. For lean mixtures, noting the reference mark made earlier, turn the needle 1/8th turn counterclockwise, allowing about 15 seconds between adjustments, until the highest consistent rpm is reached.

6. Repeat this procedure for each of the remaining carburetors.

7. With the engine still running at normal operating temperature and in forward gear, adjust the Idle speed screw to the center range of the Idle Speed (RPM) in Gear listed for your motor under the Tune-Up Specifications chart.

8. Operate the outboard in forward at or near full throttle for a full minute. Reduce speed suddenly to a low idle and shift into neutral. The powerhead should continue to operate smoothly. If the powerhead pops or stalls, the air/fuel mixture is probably too lean. Rotate the low speed needle 1/8th turn counterclockwise, allowing about 15 seconds between adjustments, until the powerhead responds as expected. Repeat the full throttle test and sudden deceleration with the shift into neutral to check each adjustment.

■ **If the engine does not respond properly to these adjustments, check for other problems including:**

- Improper engine temperatures
- Incorrect linkage adjustments
- Insufficient exhaust backpressure
- A problem with the pulse equalization hose
- Other problems with the fuel or ignition systems
- Quality and mixture of fuel and oil

■ **If this procedure was followed before Throttle Cam and Wide Open Throttle Stop adjustments on RS, TTL, WR and 25SIK models, perform those adjustments now.**

Maximum Spark Advance

◆ See Figure 202

■ **For the ignition timing maximum spark advance specifications, please refer to the Tune-Up Specifications chart.**

To check the maximum spark advance, the outboard must be operated with the proper test wheel. Do not operate the powerhead with a propeller, or flush adapter while performing this adjustment.

✲✲ CAUTION

Water must circulate through the lower unit to the engine any time the engine is run to prevent damage to the water pump in the lower unit. Just a few seconds without water will damage the water pump.

1. Connect a timing light to the top cylinder.

2. Start the engine and run it at idle until it reaches operating temperature.

3. Once the motor is fully warmed, run it at a minimum of 5000 rpm with the outboard in forward gear. Aim the timing light at the timing marks on the flywheel perimeter. The mark on the timing pointer should align at the correct degree mark on the flywheel (refer to the Tune-Up Specifications chart in this section).

4. If the timing requires adjustment, shut down the powerhead for safety and adjust as follows:

   a. For all models, except the RS, TTL, WR and 25SIK, loosen the locknut on the max timing adjustment screw (found under the flywheel, just behind the starter), then turn the screw to adjust the timing. Turning the screw 1 turn counterclockwise will advance the timing approximately 1°. Likewise, turning the screw clockwise 1 turn will retard timing approximately 1°. Tighten the nut and recheck the max timing setting.

   b. For RS, TTL, WR and 25SIK models, timing is adjusted by repositioning the nut in the slotted cam roller. Loosen the roller screw and rotate the nut until the specified timing is reached. Secure the cam, then recheck the max timing setting.

Shift Lever Detent Adjustment (RS, TTL, WR Models and 25SIK only)

1. Rotate the propeller shaft slowly by hand while moving the shift lever into the neutral position.

2. The lower detent spring should be fully engaged in the notch on the shift lever detent. If not, loosen the detent spring screw and move the lower spring to fully engage the notch, then tighten the screw securely.

Fig. 202 The maximum spark advance is adjusted using a slotted head screw and locknut

2-76 MAINTENANCE & TUNE-UP

40/50 Hp 4-Stroke Models

ADJUSTMENTS

One of the great benefits of a fuel injected motor is that most of the functions that are mechanical on a carbureted motor (and therefore subject to wear and adjustment) are electronically monitored and adjusted to maximize engine performance. The fuel and ignition systems are all but completely controlled by the Engine Control Unit (ECU) on these models. The ECU is a computer control module that accepts input from various sensors mounted around the engine and makes both ignition timing and fuel mapping decisions based on those inputs.

Ignition timing can be checked using a timing light, but there are no adjustments. Should it be found out of specification, the electronic engine control system should be checked for problems. Of course, don't get into the trap of assuming every problem that arises is electronic. Although the ECU does an incredible job of regulating engine operation on these motors, it is subject to the same mechanical limitations of any motor. Mechanical problems will often manifest themselves in symptoms of the electronic engine control system and can lead frustration during troubleshooting if you concentrate only on the electronics.

Shift linkage on these motors are still mechanical (no drive by wire systems here, yet) and therefore some adjustments are possible. But, do not expect to make periodic adjustments on the shift linkage. Attention should only be necessary after service has been performed and these components have been at least disconnected and reinstalled. Again, a procedure is provided here to check and adjust the linkage, but a good adage applies here, "if it ain't broke, don't fix it." Don't go looking for problems in the linkage just to have something to adjust, enjoy the fact that your buddies are still working on their tenth adjustment procedure after you install new spark plugs and launch for the first time that season.

There is one mechanical setting that should be checked and adjusted annually (or after every 200 hours of operation). The idle air bypass air screw is a mechanically adjustable passage that provides air to the motor beyond what is controlled by the ECU. This system is used to set a base idle speed by allowing a certain amount of air to bypass the ECU controls. Although the manufacturer recommends annual checking of this setting, don't get too excited yet, these motors have proven to be very reliable when maintained properly and this setting does not require adjustment often.

Shift Linkage

The shift cable/linkage is adjusted during initial engine rigging, whenever the linkage is disturbed, when the gearcase has been serviced or if incorrect adjustment is suspected to cause an engine malfunction. Periodic adjustments should not be necessary as a part of normal engine maintenance, though verifying proper operation can be a part of a through tune-up.

1. For safety, disconnect the negative battery cable and remove the propeller from the gearcase. For details, refer to the Propeller procedures in this section.
2. Remove the locking pin, then pull the shift linkage from the post on the lever.
3. Move the shift linkage lever back and forth to find the neutral detent.

■ **The propeller shaft turns freely in both directions in neutral.**

4. Use a piece of masking tape and felt marker to mark neutral lever position on the powerhead.
5. Move the shift linkage lever forward until the detent is felt when the forward gear engages. Rotate the propeller shaft to engage the clutch to the forward gear.
6. Place another piece of tape on the powerhead or a new mark on the original piece that represents the forward shift lever position.
7. Move the lever rearward until the detent is felt when the reverse gear engages. Rotate the propeller shaft to fully engage the clutch to the reverse gear.
8. Place another piece of tape or another mark on the original to mark the reverse lever position.
9. Using the marks as reference, place the lever in neutral and measure the distance from neutral to each of the other two marks. The amount of movement, forward or reverse, necessary to engage a gear from neutral should be the same in both directions.

10. If adjustment is necessary (more movement is necessary in one direction than the other) proceed as follows:
   a. Locate the lower to upper shift connector (it can be found on the front edge of the driveshaft housing).
   b. Loosen the locknut, then rotate the shift connector until there is equal shift selector movement in both directions.
   c. Thread the connector onto the lower shaft for at least 6 threads, then tighten the locknut.
   d. Check the lever for equal movement in both directions. Repeat the adjustment, if necessary.
11. Place the shift linkage lever in neutral, then rotate the propeller shaft to verify.
12. Reconnect the linkage to the post of the linkage lever, then slip the locking pin into the hole in the post.
13. Remove the locking pin and pull the cable connector from the arm post. Inspect the alignment of the tip on shift arm to the plunger on the neutral switch. The shift lever tip must align with the switch plunger. If necessary, adjust it as follows:
   a. Loosen the locknut on the lever end of the linkage, then rotate the linkage connector until the lever and plunger are aligned.
   b. Verify that the linkage connector threads onto the linkage at least 0.314 in. (8mm), then tighten the locknut.
   c. Install the linkage to the post and secure using the locking pin.
   d. Check the adjustment to make sure it is correct.
14. Place the remote control shift selector in neutral.
15. Remove excessive slack by pushing in lightly on the shift cable during adjustment. Loosen the locknut and rotate the cable connector until it aligns with the shift arm post. The cable end must thread onto the shift cable at least 0.314 in. (8mm). Tighten the locknut, then install the cable to the arm and secure using the locking pin. Check alignment of the lever and neutral switch. Readjust the shift cable as necessary, then tighten all fasteners.
16. Install the propeller and connect the negative battery cable.
17. Operate the engine and check for proper shift engagement.
18. Verify proper operation of the neutral switch. The engine must not start if in gear. Do not return the motor to service if unless all systems test properly.

Checking Ignition Timing

The ECU controls both the fuel and ignition systems. The ECU adjusts ignition timing to optimize engine operation based on input from sensors such as the Manifold Absolute Pressure (MAP) sensor, Closed Throttle Position (CTP) switch, Cylinder Temperature (CT) sensor and the Crankshaft Position (CKP) sensors.

At initial start-up (while cranking), each all ignition coils fire simultaneously each time a piston reaches 7 degrees Before Top Dead Center (BTDC). Once engine speed rises above 440 rpm, the ECU will begin ignition timing based on programmed mapping.

After the engine starts and runs at fast idle, ignition timing will remain fixed at 9 degrees BTDC with the motor running in neutral above 1200 rpm.

During idling/trolling, the ECU will vary ignition timing to help stabilize idle speed. The ECU will control ignition timing at 5-13 degrees BTDC with engine speed between 800-900 rpm.

For normal operation including acceleration, deceleration and engine speeds in gear, above idle, the ECU will follow various ignition timing mapping programs. The ECU will maintain timing between 0-32 degrees BTDC on 40 hp motors or 0-25 BTDC for 50 hp motors.

To check ignition timing:
1. Connect the timing light according to the tool manufacturer's instructions.
2. Run the engine either at idle in neutral using a cooling water supply or mounted on a boat/in a test tank and under the various conditions noted above.

■ **Timing marks can be found along with a pointer on the flywheel cover at the top of the motor.**

3. If proper fixed timing is noted during fast idle operation, the ECU is controlling engine timing.

Idle Bypass Air Screw Adjustment

Idle speed on the 40/50 hp EFI motors is controlled electronically through the Idle Air Control (IAC) valve. The valve is a stepper motor that can be used by the ECU to allow greater amounts of air into the engine in order to produce fast idle (for quick engine warm-up). Once the engine reaches normal operating temperature, the IAC valve usually closes and all idle air is supplied through the IAC bypass. Therefore, during warm engine operation, the idle bypass air screw adjustment determines the amount of air circumventing the otherwise closed IAC valve.

Perform the warm engine idle speed and bypass air screw check and adjustment annually or during every other tune-up (every 200 hours of operation).

■ **Incorrect idle air bypass screw adjustment may affect the IAC dashpot during deceleration, as well as fast-idle operation during engine warm-up.**

1. Connect a shop tachometer following the manufacturer's instructions and provide a cooling water source.
2. Place the gear selector in neutral, then start the engine and allow it to run until it fully reaches normal operating temperature.
3. Place the throttle control in the idle position and keep the gear selector in neutral.
4. Disconnect the IAC valve hose from the air silencer, then block air flow by using a golf tee, plastic plug or piece of tape to seal the hose.
5. If used, remove the rubber plug from the idle speed screw opening on top of the boss, just below the overhang in the air intake silencer.
6. If idle speed requires adjustment, slowly turn the idle speed screw until the engine reaches 800 rpm. Turning the screw clockwise reduces air flow (decreases rpm), while turning the counterclockwise increases air flow (increases rpm).
7. Advance the throttle to 2000 rpm, then slowly return it to idle.
8. Allow a few minutes for the idle to stabilize, then note the idle speed. Readjust the idle speed as necessary.
9. Unblock the IAC valve hose and shift the engine into forward gear, then note the idle speed. Compare the idle speed with the specification given in the Tune-Up Specifications chart in this section. If idle speed is incorrect, check the IAC valve for proper operation.
10. Shift the engine into neutral and shut the motor down.
11. Remove the tachometer and install the rubber idle air screw plug.

70 Hp 4-Stroke Models

ADJUSTMENTS

One of the great benefits of a fuel injected motor is that most of the functions that are mechanical on a carbureted motor (and therefore subject to wear and adjustment) are electronically monitored and adjusted to maximize engine performance. The fuel and ignition systems are all but completely controlled by the Engine Control Unit (ECU) on these models. The ECU is a computer control module that accepts input from various sensors mounted around the engine and makes both ignition timing and fuel mapping decisions based on those inputs.

Ignition timing can be checked using a timing light, but there are no adjustments. Should it be found out of specification, the electronic engine control system should be checked for problems. Of course, don't get into the trap of assuming every problem that arises is electronic. Although the ECU does an incredible job of regulating engine operation on these motors, it is subject to the same mechanical limitations of any motor. Mechanical problems will often manifest themselves in symptoms of the electronic engine control system and can lead frustration during troubleshooting if you concentrate only on the electronics.

Shift linkage on these motors are still mechanical (no drive by wire systems here, yet) and therefore some adjustments are possible. But, do not expect to make periodic adjustments on the shift linkage. Attention should only be necessary after service has been performed and these components have been at least disconnected and reinstalled. Again, a procedure is provided here to check and adjust the linkage, but a good adage applies here, "if it ain't broke, don't fix it." Don't go looking for problems in the linkage just to have something to adjust, enjoy the fact that your buddies are still working on their tenth adjustment procedure after you install new spark plugs and launch for the first time that season.

There is one mechanical setting that should be checked and adjusted annually (or after every 200 hours of operation). The idle air bypass air screw is a mechanically adjustable passage that provides air to the motor beyond what is controlled by the ECU. This system is used to set a base idle speed by allowing a certain amount of air to bypass the ECU controls. Although the manufacturer recommends annual checking of this setting,

don't get too excited yet, these motors have proven to be very reliable when maintained properly and this setting does not require adjustment often.

Shift Linkage

The shift cable/linkage is adjusted during initial engine rigging, whenever the linkage is disturbed, when the gearcase has been serviced or if incorrect adjustment is suspected to cause an engine malfunction. Periodic adjustments should not be necessary as a part of normal engine maintenance, though verifying proper operation can be a part of a through tune-up.

Shift linkage is adjusted in the same manner as the 40/50 hp motors covered earlier in this section. Please refer to that procedure for details on checking and adjusting the shift linkage.

Checking Ignition Timing

The ECU controls both the fuel and ignition systems. The ECU adjusts ignition timing to optimize engine operation based on input from sensors such as the Manifold Absolute Pressure (MAP) sensor, Closed Throttle Position (CTP) switch, Cylinder Temperature (CT) sensor and the Crankshaft Position (CKP) sensors.

At initial start-up (while cranking), each all ignition coils fire simultaneously each time a piston reaches 5 degrees Before Top Dead Center (BTDC). Once engine speed rises above 440 rpm, the ECU will begin ignition timing based on programmed mapping.

After the engine starts and runs at fast idle, ignition timing will remain fixed. The ECU will fix timing 5 degrees BTDC with the motor running in neutral above 1000 rpm.

During idling/trolling, the ECU will vary ignition timing to help stabilize idle speed. The ECU will maintain ignition timing at 6-14 degrees BTDC with a stable engine speed of 700 rpm.

For normal operation including acceleration, deceleration and engine speeds in gear, above idle, the ECU will follow various ignition timing mapping programs. The ECU will maintain timing between 10-36 degrees BTDC.

To check ignition timing:
1. Connect the timing light according to the tool manufacturer's instructions.
2. Run the engine either at idle in neutral using a cooling water supply or mounted on a boat/in a test tank and under the various conditions noted above.

■ **Timing marks can be found along with a pointer on the flywheel cover at the top of the motor.**

3. If proper fixed timing is noted during fast idle operation, the ECU is controlling engine timing.

Idle Bypass Air Screw Adjustment

The idle speed on the 70 hp motor is controlled electronically through the Idle Air Control (IAC) valve that protrudes from the side of the intake at the front, center of the motor. The valve is a stepper motor that is be used by the ECU to allow greater amounts of air into the engine in order to produce fast idle (for quick engine warm-up). Once the engine reaches normal operating temperature, the IAC valve closes and idle air is supplied through the IAC bypass (a small brass inlet on the intake manifold that looks like a hose connector).

During warm operation, the idle bypass air screw adjustment determines the amount of air circumventing the closed IAC valve. The screw, located in the intake manifold downstream of the throttle body and adjacent to the brass air inlet, is set at the factory and sealed to prevent unnecessary tampering or adjustment.

Perform the warm engine idle speed and bypass air screw check annually or at least after every 200 hours of operation.

1. Connect a shop tachometer following the manufacturer's instructions and provide a cooling water source.
2. Shift the gearcase into neutral, then start the engine and allow it to run until it reaches normal operating temperature.
3. With the engine still running in neutral, place the throttle control in the idle position.
4. Disconnect the IAC valve hose from the intake air silencer. Block air flow using a plastic plug, golf tee or tape to seal the hose.
5. If adjustment is necessary, remove the rubber plug from the idle speed screw opening, then turn the idle speed screw slowly until the idle speed reaches 600 rpm. Turn the screw clockwise to reduce air flow (decrease engine rpm), or counterclockwise to increase air flow (increase engine rpm).

2-78 MAINTENANCE & TUNE-UP

6. Advance the throttle to 2000 rpm, then slowly return it to idle.
7. Allow a few minutes for the idle to stabilize then note the idle speed. Readjust the as necessary.
8. Unblock the IAC valve hose and shift the engine into forward gear. Note the idle speed. Compare the idle speed with the spec found in the Tune-Up Specifications chart in this section. If idle speed is incorrect, check the IAC valve for proper operation.
9. Shift the engine to neutral, then shut down the powerhead.
10. Remove the tachometer and reinstall the rubber idle air screw plug.

VALVE CLEARANCE (4-STROKE MODELS)

Valve Lash

◆ See Figures 203 thru 206

Valve lash is the clearance between the rocker arm and the valve tappet or valve stem, as equipped. This distance is critical, as too little clearance can hold the valve open or keep it from fully contacting the valve seat, preventing it from cooling by transmitting heat to the cylinder head through contact. This would lead to a burnt valve, requiring overhaul and replacement. Also, too much valve clearance might keep the valve from opening fully, preventing the engine from making maximum horsepower. Of course poor driveability is arguably better than burning a valve. There's an older mechanic's saying that applies here, "a tappy valve, is a happy valve." That's not to say you necessarily WANT your valves a little too loose and tapping, but whenever your in doubt, leave a valve a **little** too loose rather than a little too tight.

Adjusting valve lash is a part of normal maintenance to keep the valve clearance within a specified range accounting for the normal wear and tear of valve train components. Valve clearance is always measured when the rocker or tappet for the valve being measured is fully released meaning that it is in a position where it contacts the base of the camshaft and not any part of the raised camshaft lobe. For this reason it is necessary to find when the piston for the valve being measured is a TDC of the compression stroke. Remember that TDC of the compression stroke is the point during the 4-stroke engine cycle when the piston travels upward and both valves close in order to seal the combustion chamber and compress the air/fuel mixture.

Although timing marks can be used to determine TDC, always double-check the timing marks by watching the valves for the cylinder on which you are working as the timing mark on the flywheel or camshaft approaches the mark on the engine. During a normal cycle, the exhaust valve and then the intake valve will close as the piston begins its travel upward. If instead the exhaust valve opens, the engine is one full turn of the crankshaft away from the start of the compression stroke. Both valves must be closed, and remain closed as the piston comes to the top of the cylinder and the timing marks align. If so, that piston is on TDC and the valves for that cylinder may be adjusted.

ADJUSTMENT

5-15 Hp Motors

◆ See Figure 203

A set of flat feeler gauges is the only tool that is absolutely necessary to check valve clearance on these motors, but, adjustment is much easier when using OMC adjustment tool (#341444) or an equivalent tappet adjustment tool to rotate the valve adjuster.

Valve specifications are for an overnight cold engine. It is best to check and/or adjust the valves with the powerhead at approximately 20° C (68° F).

1. For safety, disconnect the negative battery cable and/or remove the spark plug(s) and ground the spark plug lead(s). For twin cylinder motors, be sure to tag the spark plug leads before disconnection.
2. If necessary for additional clearance, remove the lower engine covers as described in this section.
3. Remove the manual starter assembly as described in the section on Manual Starters.
4. For 5/6 hp (128cc) and 8/9.9 (211cc) motors, tag and disconnect the 3 hoses from the rocker cover.

✳✳ CAUTION

When disconnecting the fuel hoses, use a rag to catch all spilled fuel. Use extreme care when working around fuel and fumes as both are highly flammable. Keeping all potential sources of spark or ignition (no smoking and avoid sparks) out of the work area. Refer to the Fuel System section for more details.

5. For 9.9/15 hp (305cc) motors, disconnect the hose from the side of the rocker cover and, if necessary, remove the spark plug wires from the retaining clips at the top of the cover. On some models, other hoses may be run near the cover and could interfere with removal. Either pull these hoses

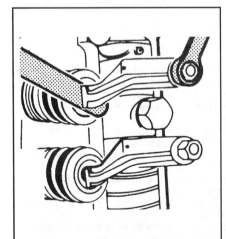

Fig. 203 Adjusting valve clearance between a rocker and valve stem - all models except 40/50 hp motors

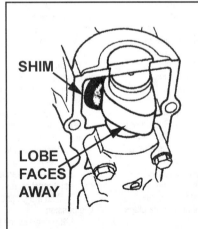

Fig. 204 To check valve lash on the 40/50 hp motors, face the camshaft lobe away from the shim. . .

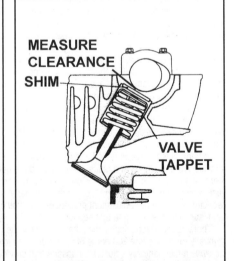

Fig. 205 . . . then measure between the shim and camshaft using a feeler gauge

MAINTENANCE & TUNE-UP 2-79

back out of the way while they are still attached or tag and disconnect the hoses before proceeding.

6. Support the rocker arm cover while removing the four (5/6 and 8/9.9 hp) or eight (9.9/15 hp) cover bolts using the reverse of the rocker arm cover torque sequence. For details, please refer to the Powerhead overhaul section.

7. Carefully pull the rocker arm cover from the cylinder head. Inspect the gasket for damage.

8. Rotate the flywheel clockwise to align the timing mark(s) for valves being checked, as follows:

   a. For 5/6 hp (128cc) and 8/9.9 hp (211cc) models, turn the flywheel slowly until the triangular mark on the flywheel aligns with the raised mark on the manual starter boss. At the same time, the mark on the camshaft pulley should align with the protrusion on the cylinder head. Both valves should be closed on the 5/6 hp motor, confirming that the single cylinder is a TDC. For 8/9.9 hp motors, both valves for cylinder No. 1 should be closed, confirming that No. 1 cylinder is at TDC. The exhaust valve should be open on cylinder No. 2 for these motors).

   b. For 9.9/15 hp motors, turn the flywheel slowly until the pointer on the cam pulley aligns with the protrusion on the manual starter boss. At the same time, the white mark on the lower belt will align with the crankcase to cylinder block split line on the port side of the powerhead. In this position the valves for the No. 2 piston will both be closed, confirming that the No. 2 piston is a TDC of the compression stroke.

9. Measure the clearance of the cylinder intake and exhaust valves. Insert feeler gauges of various sizes between the rocker arm and the valve stem for both valves being checked at this point. The size gauge that passes between the arm and stem with a slight drag indicates the valve clearance. Compare the clearance measured with the Valve Clearance Specifications Chart. To determine which valve is an intake and which is an exhaust, observe the position of the valves in relation to the other components on the powerhead. The intake valves are adjacent to the ports for the intake manifold attached to the powerhead and the exhaust valves are adjacent to the exhaust ports. On these motors, the intake and exhaust valves are situated as follows:

   • On 5/6 hp (128cc) and 8/9.9 hp (211cc) motors, the intake valve is the upper rocker while the exhaust valve is the lower rocker.

   • On 9.9/15 hp (305cc) motors the intake valve is on bottom of each cylinder while the exhaust valve rocker is on top.

10. If adjustment is necessary, proceed as follows:

    a. Loosen the locknut, then turn the adjusting screw until the clearance is correct.

    b. Hold the screw and tighten the locknut to 10-20 inch lbs. (1.1-2.3 Nm) for all models, except the 1999-01 5/6 hp and 8/9.9 hp motors on which the locknuts should be tightened to 100-115 inch lbs. (11-13 Nm).

11. For 8-15 hp motors, rotate the flywheel clockwise one full revolution or 360 degrees in order to turn the camshaft pulley one half of a revolution or 180 degrees. The pointer should now face opposite the timing mark on the starter boss. In this position the opposite cylinder will be at TDC:

    c. For 8/9.9 hp (211cc) motors, the valves for the No. 2 cylinder should now be closed. Adjust those valves in the same manner as the previous cylinder.

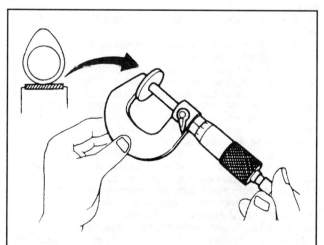

Fig. 206 Use a micrometer to measure old shims when determining the proper replacements

    d. For 9.9/15 hp (305cc) motors, the valves for the No. 1 cylinder should now be closed. Adjust those valves in the same manner as the previous cylinder.

12. Install a new rocker arm cover gasket as follows:

    a. For 5/6 hp (128cc) motors coat the gasket with OMC Triple-Guard or an equivalent marine grease.

    b. For 8/9.9 hp (211cc) motors, coat the gasket with OMC Gasket Sealing Compound.

    c. For 9.9/15 hp (305cc) motors, leave the gasket dry.

    d. For all models, coat the cover screw threads with a light coating of OMC Gasket Sealing Compound.

13. Install the rocker arm cover and tighten the bolts using the proper crossing pattern as directed under the Powerhead section to 84-106 inch lbs. (10-12 Nm).

14. Reconnect any hoses that were removed.

15. Install the lower engine covers.

16. Install the manual starter.

17. Install the spark plug(s) and lead(s) and, if applicable, connect the negative battery cable.

18. Provide a water source, then start the engine and check for oil leaks at the rocker arm cover mating surfaces.

19. Install the upper engine cover.

40/50 Hp Motors

♦ See Figures 204, 205 and 206

The 40/50 hp motors utilize replaceable shims to adjust valve clearance. A set of flat feeler gauges is all that is required to **check** valve clearance. But, if adjustments are necessary, you will need a micrometer for shim measurement, an assortment of shims (or a trip during the procedure to purchase the proper size shims) and OMC #345832 or an equivalent tappet holder tool.

■ **For these motors, the No. 1 cylinder is at TDC of the compression stroke when the mark on the crankshaft aligns with the mark on the crankcase and the marks on the camshaft sprockets align with the protrusions on the cylinder head. At TDC of a compression stroke, the base of the camshaft lobe will be directly above the valve tappet shim (the raised portion of the lobe will face away from the valve tappet/shim).**

Valve specifications are for an overnight cold engine. It is best to check and/or adjust the valves with the powerhead at approximately 20° C (68° F).

1. For safety, disconnect the negative battery cable.
2. Remove the spark plugs as described in this section.
3. Remove the lower engine covers as described in this section.
4. Remove the flywheel cover as described under Powerhead.
5. Rotate the engine clockwise to bring the No. 1 cylinder to TDC of the compression stroke and thereby relieving mechanical pressure from the fuel pump arm. Remove the low pressure fuel pump from the cylinder head cover as described in the Fuel System section.
6. Remove the breather hose and ignition coils from the cylinder head cover.
7. Disconnect the harness from the camshaft position sensor. For details, please refer to Electronic Engine Controls.
8. Loosen and remove the cylinder head cover bolts in the opposite of the tightening sequence as detailed under Powerhead.
9. Remove the cover along with the gasket and O-rings from the powerhead. Be careful not to damage the cover or any O-rings that might be reused. Inspect the gasket for damaged surfaces and replace, if necessary.
10. Check that the flywheel is still in the No. 1 TDC position by making sure the camshaft lobes still face out and directly opposite the valve tappets for that cylinder. Insert feeler gauges of various sizes between the tappet shim and the camshaft lobe. The size gauge that passes between the shim and lobe with a slight drag indicates the valve clearance. Record the clearance for each valve.
11. The firing order on this motor is 1-3-2, so the No. 3 cylinder should come up to TDC next and the No. 2 cylinder last, then the cycle will return back to the beginning of the firing order. Rotate the flywheel clockwise until the No. 3 camshaft lobe tips are facing out and directly opposite the valve tappets. Insert feeler gauges of various sizes between the tappet shim and the camshaft lobe. The size gauge that passes between the shim and lobe with a slight drag indicates the valve clearance. Record the valve clearance for each valve.

MAINTENANCE & TUNE-UP

12. Rotate the flywheel clockwise until the No. 2 camshaft lobe tips are facing out and directly opposite the valve tappets. Insert feeler gauges of various sizes between the tappet shim and the camshaft lobe. The size that passes between the shim and lobe with a slight drag indicates the valve clearance. Record the valve clearance for each valve.

13. Compare the clearances with the Valve Clearance Specifications chart in this section.

14. If incorrect valve clearance is noted, substitute the proper sized shim, determined as follows:

   a. If valve clearance is greater than spec, subtract the valve clearance specification (using the dead middle of the range) listed in the specification chart from the measured clearance. Obtain a replacement shim, that much thicker than the current shim.

   b. If valve clearance is less than specification, subtract the measured clearance from the valve clearance specification (again, using the dead middle of the range) listed in the chart. Obtain a replacement shim that much smaller than the original.

15. If shim replacement is necessary to obtain the proper valve lash:

   a. Rotate the flywheel clockwise until the camshaft lobe tip for the selected valve is opposite the tappet.

   b. Carefully rotate the tappet within its bore until the notch is facing toward the opposite camshaft. The notch must be accessible for tappet shim removal.

   c. Rotate the flywheel clockwise one complete revolution or until the camshaft lobe tip contacts the tappet shim (opening the valve). Remove the bolts from the camshaft cap next to the selected valve.

   d. Place OMC #345832 or an equivalent tappet retainer over the camshaft cap. Each end of the retainer is marked either **IN** (for intake) or **EX** (for exhaust) and the appropriate end must face inward toward the center of the cylinder head (toward the notch on the tappet from which the shim will be removed). Thread the camshaft cap bolts through the retainer and the camshaft cap. Tighten the bolts securely, ensuring the fingers of the tool contact the barrel portion of the tappet and not the shim itself.

   e. Rotate the flywheel clockwise one half revolution or until the camshaft lobe tip rotates 90° away from the tappet. Insert a screwdriver (with tape covering the blade to prevent scoring or damage) into the tappet notch and carefully pry the shim from the tappet. Use a magnet to pull the shim from the tappet. Do not use your fingers.

   f. Measure the shim thickness using a micrometer. Shims are available in sizes from 0.086-0.118 in. (2.18-3.0mm) in 0.001 in. (0.02mm) increments. Select the correct shim thickness as described earlier.

■ On new shims, the thickness can be identified by the number present on the shim's face, divided by 100. Move the decimal point 2 places to the left and you've got your size. For example a new shim labeled 258 on the face is 2.58mm thick. The label can only be trusted on new shims, that have not been in service, as normal wear during service might change shim thickness overtime.

   g. Place the selected shim into the tappet with the numbered side facing down.

   h. Make sure the shim seats fully against the step within the tappet, then rotate the flywheel counterclockwise one half revolution or until the camshaft lobe tip contacts the shim. Remove the bolts from the camshaft cap and tappet retainer.

   i. Carefully pull the retainer from the cap, then install camshaft cap bolts and tighten evenly to 84-90 inch lbs. (9.5-10.2 Nm).

16. Repeat for each of the valves whose lash was noted out of specification.

17. Apply a light coating of GM Silicone Rubber Sealer (or equivalent RTV gasket sealant) to the cylinder head cover mating surfaces of the cylinder head (the mounting areas for the O-rings and gasket).

18. Install the cover to the cylinder head, then tighten the cover bolts using the sequence provided under Powerhead to 40-54 inch lbs. (4.5-6.0 Nm).

19. Connect the harness to the camshaft position sensor.

20. Connect the breather hose.

21. Install the spark plugs and the ignition coils.

22. Install the fuel pump, flywheel cover and lower engine covers as detailed in the appropriate procedures.

23. Connect the negative battery cable, then connect a flushing device and start the engine to check for oil leaks at the rocker arm cover mating surfaces.

24. Install the engine top case.

70 Hp Motors

◆ See Figure 203

A set of flat feeler gauges is necessary to check valve clearance on these motors.

Valve specifications are for an overnight cold engine. It is best to check and/or adjust the valves with the powerhead at approximately 20° C (68° F).

■ The No. 1 cylinder is at TDC of the compression stroke on the 70 hp motor, when the holes in the crankshaft pulley belt guides align with the protrusion on the cylinder block and the No. 1 mark on the camshaft pulley aligns with the raised boss on the cylinder head. At TDC of a compression stroke, the base of the camshaft lobe will be touching the rocker arm (the raised portion of the lobe will face away from the rocker).

1. Disconnect the negative battery cable for safety.

2. Tag and disconnect the spark plug wires, then remove the spark plugs. For details, please refer to the Spark Plug procedure in this section.

✳✳ WARNING

Ground the spark plug leads to prevent damage if the engine is cranked while they are disconnected.

3. Remove the lower engine covers as described in this section.

4. Remove the flywheel cover as described under Powerhead.

5. Rotate the engine clockwise to bring the No. 1 cylinder to TDC of the compression stroke and thereby relieving mechanical pressure from the fuel pump arm. Remove the low pressure fuel pump from the cylinder head cover as described in the Fuel System section.

6. Remove the breather hose from the cylinder head cover.

7. Remove the ignition coils from the cylinder head. For details, refer to the Ignition System section.

8. Support the rocker arm cover and remove the six cover bolts using a crossing pattern.

9. Pull the rocker arm cover from the cylinder head, then carefully remove the cover gasket.

10. Make sure the flywheel is still in the No. 1 TDC position. The raised portion of the camshaft lobes should face away from and not be in contact with the rockers. If the position is correct, both the valves for the No. 1 cylinder will be closed.

11. Measure the clearance intake and exhaust valves for the No. 1 cylinder, also measure the No. 2 cylinder intake valve and the No. 3 cylinder exhaust valve. Insert feeler gauges of various sizes between the rocker arm and the valve stem for each valve. The size that passes between the arm and stem with a slight drag indicates the valve clearance. Compare the clearance measured with the Valve Clearance Specification chart in this section.

12. If lash is out of specification on one or more valve, adjust it as follows:

   a. Loosen the locknut, then turn the adjusting screw until the clearance is correct.

   b. Hold the screw while tightening the locknut to 12-13 ft. lbs. (16-17 Nm).

   c. Recheck the valve clearance to make sure the adjuster wasn't turned while tightening the locknut.

13. Rotate the flywheel clockwise one full revolution (360 degrees) so the No. 4 TDC mark aligns with the raised boss on the cylinder head (and the No. 1 cylinder is now on its exhaust stroke). The camshaft rotates at 1/2 the rate of the crankshaft/flywheel, so rotating the flywheel as directed will turn the camshaft sprocket only 180 degrees. This places the No. 1 TDC mark exactly 1/2 a turn away from the previous location. At this point, both of the valves for the No. 4 cylinder should be closed.

14. Measure the clearance of the No. 4 cylinder intake and exhaust valves, the No 2 exhaust valve and the No. 3 intake valve. Insert feeler gauges of various sizes between the rocker arm and the valve stem for each valve. The size that passes between the arm and stem with a slight drag indicates the valve clearance. Compare the clearance measured with the specs listed in the Valve Clearance Specifications chart in this section. If necessary, adjust the valve clearance as described earlier.

15. Rotate the flywheel clockwise one full revolution (360 degrees), until the No. 1 TDC mark again aligns with the raised boss on the cylinder head.

16. Install the rocker arm cover using a new gasket, then tighten the bolts using a crisscross pattern to 84-96 inch lbs. (10-11 Nm).

MAINTENANCE & TUNE-UP

17. Install the fuel pump, ignition coils, flywheel cover and lower engine covers as described in the appropriate procedures.
18. Reconnect the breather hose to the rocker cover.
19. Install the spark plugs and connect the leads.
20. Connect the negative battery cable, then connect a flushing device and start the engine to check for oil leaks at the rocker arm cover mating surfaces.
21. Install the engine top case.

STORAGE (WHAT TO DO BEFORE AND AFTER)

Winterization

◆ See Figure 207

Taking extra time to store the boat and motor properly at the end of each season or before any extended period of storage will greatly increase the chances of satisfactory service at the next season. Remember, that next to hard use on the water, the time spent in storage can be the greatest enemy of an outboard motor. Ideally, outboards should be used regularly. If weather in your area allows it, don't store the motor, enjoy it. Use it, at least on a monthly basis. It's best to enjoy and service the boat's steering and shifting mechanism several times each month. If a small amount of time is spent in such maintenance, the reward will be satisfactory performance, increased longevity and greatly reduced maintenance expenses.

But, in many cases, weather or other factors will interfere with time for enjoying a boat and motor. If you must place them in storage, take time to properly winterize the boat and outboard. This will be your best shot at making time stand still for them.

For many years there was a widespread belief simply shutting off the fuel at the tank and then running the powerhead until it stops constituted prepping the motor for storage. Right? Well, WRONG!

First, it is not possible to remove all fuel in the carburetor or fuel injection system by operating the powerhead until it stops. Considerable fuel will remain trapped in the float chamber and other passages, especially in the lines leading to carburetors. The only guaranteed method of removing all fuel is to take the physically drain the carburetors from the float bowls. On EFI systems, disassembling the fuel injection components to drain the fuel is impractical so properly mixing fuel stabilizer becomes that much more important. Actually, the manufacturer recommends prepping all of the motors using fuel stabilizer as opposed to draining the fuel system, but on carbureted motors, you always have the option.

Proper storage involves adequate protection of the unit from physical damage, rust, corrosion and dirt. The following steps provide an adequate maintenance program for storing the unit at the end of a season.

PREPPING FOR STORAGE

Where to Store Your Boat and Motor

Ok, a well lit, locked, heated garage and work area is the best place to store you precious boat and motor, right? Well, we're probably not the only ones who wish we had access to a place like that, but if you're like most of us, we place our boat and motor wherever we can.

Of course, no matter what storage limitations are placed by where you live or how much space you have available, there are ways to maximize the storage site.

If possible, select an area that is dry. Covered is great, even if it is under a carport or sturdy portable structure designed for off-season storage. Many people utilize canvas and metal frame structures for such purposes. If you've got room in a garage or shed, that's even better. If you've got a heated garage, God bless you, when can we come over? If you do have a garage or shed that's not heated, an insulated area will help minimize the more extreme temperature variations and an attached garage is usually better than a detached for this reason. Just take extra care to make sure you've properly inspected the fuel system before leaving your boat in an attached garage for any amount of time.

If a storage area contains large windows, mask them to keep sunlight off the boat and motor otherwise, use a high-quality, canvas cover over the boat, motor and if possible, the trailer too. A breathable co●er is best to avoid the possible build-up of mold or mildew, but a heavy duty, non-breathable cover will work too. If using a non-breathable cover, place wooden blocks or length's of 2 x 4 under various reinforced spots in the cover to hold it up off the boat's surface. This should provide enough room for air to circulate under the cover, allowing for moisture to evaporate and escape.

Whenever possible, avoid storing your boat in industrial buildings or parks areas where corrosive emissions may be present. The same goes for storing your boat too close to large bodies of saltwater. Hey, on the other hand, if you live in the Florida Keys, we're jealous again, just enjoy it and service the boat often to prevent corrosion from causing damage.

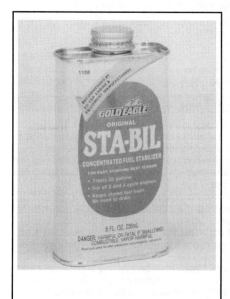

Fig. 207 Add fuel stabilizer to the system anytime it will be stored without complete draining

Fig. 208 Fogging oil can be added through a service fitting when equipped with an electric primer

Fig. 209 The 25/35 hp (500/565cc) 2-stroke motors can be fogged through the maintenance valve on the intake

2-82 MAINTENANCE & TUNE-UP

Finally, when picking a place to store your motor, consider the risk or damage from fire, vandalism or even theft. Check with your insurance agent regarding coverage while the boat and motor is stored.

Storage Checklist (Preparing the Boat and Motor)

◆ See Figure 207, 208 and 209

The amount of time spent and number of steps followed in the storage procedure will vary with factors such as the length of planed storage time, the conditions under which boat and motor are to be stored and your personal decisions regarding storage.

But, even considering the variables, plans can change, so be careful if you decide to perform only the minimal amount of preparation. A boat and motor that has been thoroughly prepared for storage can remain so with minimum adverse affects for as short or long a time as is reasonably necessary. The same cannot be said for a boat or motor on which important winterization steps were skipped.

■ Always store a Johnson/Evinrude motor, especially a 4-stroke model, vertically on the boat or on a suitable engine stand. Do not lay a 4-stroke motor down for any length of time, as engine oil will seep past the rings causing extreme smoking upon startup. At best, burning that oil will promote spark plug and combustion chamber fouling, at worst it could cause a partial hydro-lock condition that could even mechanically damage the powerhead.

1. Thoroughly wash the boat motor and hull. Be sure to remove all traces of dirt, debris or marine life. Check the water stream fitting, water inlet(s) and, on jet models, the impeller grate for debris. If equipped, inspect the speedometer opening at the leading edge of the gearcase or any other gearcase drains for debris (clean debris with a compressed air or a piece of thin wire).

■ The manufacturer recommends the use of OMC 2+4 Fuel Conditioner when treating the fuel systems on Johnson/Evinrude motors. OMC 2+4 is normally used in a ratio of 1.0 oz. (30 ml) for every gallon (3.8 L), but follow the directions on the bottle if they differ. On carbureted motors, you have the option of draining the fuel system instead, either using the float bowl drains on the carburetor(s) or by removing the carburetor(s) completely from the motor. For more details on carburetor service, please refer to the Fuel System section.

2. Stabilize the engine's fuel supply using a high quality fuel stabilizer and take this opportunity to thoroughly flush the engine cooling system at the same time as follows:

   a. Add an appropriate amount of fuel stabilizer to the fuel tank and top off to minimize the formation of moisture through condensation in the fuel tank. For EFI motors prepare a fuel storage mixture as directed. Use of 6.0 gal. (23 L) gas tank to mix 5.0 gal. (19 L) of gas, 2.0 qt. (1.9 L) of OMC Storage Fogging Oil and 2.5 oz. (74 ml) of OMC 2+4 Fuel Conditioner or equivalent storage fluids. Connect this tank to the engine in order to provide a treated fuel mixture to the engine for storage.

   b. Attach a flushing attachment as a cooling water/flushing source. For details, please refer to the information on Flushing the Cooling System, in this section.

   c. Start and run the engine at about 1500 rpm for approximately 5 minutes on carbureted models or 10 minutes on EFI models. This will ensure the entire fuel supply system contains the appropriate storage mixtures.

   d. Stop the engine and remove the flushing source, keeping the outboard perfectly vertical. Allow the cooling system to drain completely, especially if the outboard might be exposed to freezing temperatures during storage.

※※ WARNING

NEVER keep the outboard tilted when storing in below-freezing temperatures as water could remain trapped in the cooling system. Any water left in cooling passages might freeze and could cause severe engine damage by cracking the powerhead or gearcase.

3. Drain and refill the engine gearcase while the oil is still warm (for details, refer to the Gearcase Oil procedures in this section). Take the opportunity to inspect for problems now, as storage time should allow you the opportunity to replace damaged or defective seals. More importantly, remove the old, contaminated gear oil now and place the motor into storage with fresh oil to help prevent internal corrosion.

4. On 4-stroke models, drain the engine crankcase oil while it is still warm. Replace the oil filter. Refill the crankcase and gearcase with fresh oil (for details, refer to the Oil and Filter Change procedures in this section). On EFI motors (or carbureted motors than will not be run for fogging), reconnect the flush fitting, then start and run the motor again, but only for a few minutes to evenly distribute the fresh oil across internal bearing surfaces.

※※ WARNING

Besides treating the fuel system to prevent evaporation or clogging from deposits left behind, coating all bearing surfaces in the motor with FRESH, clean oil is the most important step you can take to protect the engine from damage during storage. NEVER leave the engine filled with used oil, that likely contains moisture and, in the case of 4-stroke crankcase oil, acids and other damaging byproducts of combustion that will damage engine bearings over time.

5. Fog the motors using one of the following methods (as appropriate for your engine, or as desired):

   a. For EFI motors (or other motors that will not be run again until re-commissioning), Tag and disconnect the spark plug leads, then remove the spark plugs as described under Spark Plugs. Spray a generous amount of fogging oil into the spark plug ports. Turn the flywheel slowly by hand (clockwise, in the normal direction of rotation) to distribute the fogging oil evenly across the cylinder walls. On electric start models, the starter can be used to crank the motor over in a few short bursts, but make sure the spark plugs leads remain disconnected and grounded to the powerhead (away from the spark plug ports) to prevent accidental combustion. If necessary, re-spray into each cylinder when that cylinder's piston reaches the bottom of its travel. Reinstall and tighten the spark plugs, but leave the leads disconnected to prevent further attempts at starting until the motor is ready for re-commissioning.

■ On motors equipped with a rope start handle, the rope can be used to turn the motor slowly and carefully using the rope starter. For other models, turn the flywheel by hand or using a suitable tool, but be sure to ALWAYS turn the engine in the normal direction of rotation (normally clockwise on these motors).

   b. For carbureted motors, the engine can also be fogged by spraying the can of fogging oil either down the carburetor throat(s) or through a fogging fitting while the engine is running. Reconnect the cooling water source and follow the instructions on the can of fogging oil. Johnson/Evinrude electric fuel primers are normally equipped with a fogging fitting. Some models are equipped with a separate fitting, such as the 25/35 hp (500/565cc) motors that use an intake manifold mounted maintenance valve to conveniently attach a spray can of fogging oil.

■ Even if a carbureted motor is fogged while running, it is a good idea to follow the steps of the fogging procedure for EFI motors. Remove the spark plugs, spray additional fogging oil directly into the combustion chambers and manually turn the engine over by hand to distribute that oil. Better safe than seized, we always say!

6. For models equipped with portable fuel tanks, disconnect and relocate them to a safe, well-ventilated, storage area, away from the motor. Drain any fuel lines that remain attached to the tank.

7. Remove the battery or batteries from the boat and store in a cool dry place. If possible, place the battery on a smart charger or Battery Tender®, otherwise, trickle charge the battery once a month to maintain proper charge.

※※ WARNING

Remember that the electrolyte in a discharged battery has a much lower freezing point and is more likely to freeze (cracking/destroying the battery case) when stored for long periods in areas exposed to freezing temperatures. Although keeping the battery charged offers one level or protection against freezing; the other is to store the battery in a heated or protected storage area.

8. For models equipped with a boat mounted fuel filter or filter/water canister, clean or replace the boat mounted fuel filter at this time. If the fuel system was treated, the engine mounted fuel filters should be left intact, so the sealed system remains filled with treated fuel during the storage period.

9. For 2-stroke motors with external oil tanks, if possible, leave the oil supply line connected to the motor. This is the best way to seal moisture out of the system. If the line must be disconnected for any reason (such as to remove the motor or oil tank from the boat), seal the line by sliding a snug fitting cap over the end. Most motors equipped with remote oil tanks are equipped with a cap, mounted somewhere on the engine, such as on the fuel line, near the fuel pump. Top off the oil tank to displace moisture-laden air and help prevent contamination of the oil in storage.

MAINTENANCE & TUNE-UP

10. For EFI motors or motors equipped with a gearcase speedometer pickup, disconnect the speedometer hose from the upper most connector and blow all water from the gearcase speedometer pickup. If compressed air is available, use less than 25 psi (167 kPa) of air pressure in order to prevent damage to the system.

11. Perform a complete lubrication service following the procedures in this section.

12. Remove the propeller (or rotor on RescuePro® models) and check thoroughly for damage. Clean the propeller shaft and apply a protective coating of grease. On Jet models, thoroughly inspect the impeller and check the impeller clearance. Refer to the procedures in this section.

13. Check the motor for loose, broken or missing fasteners. Tighten fasteners and, again, use the storage time to make any necessary repairs.

14. Inspect and repair all electrical wiring and connections at this time. Make sure nothing was damaged during the season's use. Repair any loose connectors or any wires with broken, cracked or otherwise damaged insulation.

15. Clean all components under the engine cover and apply a corrosion preventative spray.

16. Too many people forget the boat and trailer, don't be one of them.
    a. Coat the boat and outside painted surfaces of the motor with a fresh coating of wax, then cover it with a breathable cover
    b. If possible place the trailer on stands or blocks so the wheels are supported off the ground.
    c. Check the air pressure in the trailer tires. If it hasn't been done in a while, remove the wheels to clean and repack the wheel bearings.

17. Sleep well, since you know that your baby will be ready for you come next season.

Re-commissioning

REMOVAL FROM STORAGE

The amount of service required when re-commissioning the boat and motor after storage depends on the length of non-use, the thoroughness of the storage procedures and the storage conditions.

At minimum, a thorough spring or pre-season tune-up and a full lubrication service is essential to getting the most out of your engine. If the engine has been properly winterized, it is usually no problem to get it in top running condition again in the springtime. If the engine has just been put in the garage and forgotten for the winter, then it is doubly important to perform a complete tune-up before putting the engine back into service. If you have ever been stranded on the water because your engine has died and you had to suffer the embarrassment of having to be towed back to the marina you know how it can be a miserable experience. Now is the time to prevent that from occurring.

Take the opportunity to perform any annual maintenance procedures that were not conducted immediately prior to placing the motor into storage. If the motor was stored for more than one off-season, pay special attention to inspection procedures, especially those regarding hoses and fittings. Check the engine gear oil for excessive moisture contamination. The same goes for engine crankcase oil on 4-strokes or oil tanks on 2-strokes, so equipped. If necessary, change the gearcase or engine oil to be certain no bad or contaminated fluids are used.

■ **Although not absolutely necessary, it is a good idea to ensure optimum cooling system operation by replacing the water pump impeller at this time.**

Other items that require attention include:
1. Install the battery (or batteries) if so equipped.
2. Inspect all wiring and electrical connections. Rodents have a knack for feasting on wiring harness insulation over the winter. If any signs of rodent life are found, check the wiring carefully for damage, do not start the motor until damaged wiring has been fixed or replaced.
3. On 2-strokes with a remote oil tank, if the line was disconnected, remove the cover and reconnect the line, then prime the system to ensure proper operation once the motor is started.
4. If not done when placing the motor into storage clean and/or replace the fuel filters at this time. This is usually the case on EFI motors, as the filters are often not replaced before filling the system with the storage fuel mixture.
5. If the fuel tank was emptied, or if it must be emptied because the fuel is stale fill the tank with fresh fuel. Keep in mind that even fuel that was treated with stabilizer will eventually become stale, especially if the tank is stored for more than one off-season. Pump the primer bulb and check for fuel leakage or flooding at the carburetor or vapor separator tank. For EFI motors, pressurize the high pressure fuel circuit turning the ignition on (and listening to verify that the fuel pump runs for a few seconds). Inspect the fuel rail and fittings under the engine top case for leaks.
6. Attach a flush device or place the outboard in a test tank and start the engine. Run the engine at idle speed and warm it to normal operating temperature. Check for proper operation of the cooling, electrical and warning systems.

✱✱ CAUTION

Before putting the boat in the water, take time to verify the drain plug is installed. Countless number of spring boating excursions have had a very sad beginning because the boat was eased into the water only to have the boat begin to fill with it.

CLEARING A SUBMERGED MOTOR

Unfortunately, because an outboard is mounted on the exposed transom of a boat, and many of the outboards covered here are portable units that are mounted and removed on a regular basis, an outboard can fall overboard. Ok, it's relatively rare, but it happens often enough to warrant some coverage here. The best way to deal with such a situation is to prevent it, by keeping a watchful eye on the engine mounting hardware (bolts and/or clamps). But, should it occur, here's how to salvage, service and enjoy the motor again.

In order to prevent severe damage, be sure to recover an engine that is dropped overboard or otherwise completely submerged as soon as possible. It is really best to recover it immediately. But, keep in mind that once a submerged motor is recovered exposure to the atmosphere will allow corrosion to begin etching highly polished bearing surfaces of the crankshaft, connecting rods and bearings. For this reason, not only do you have to recover it right away, but you should service it right away too. Make sure the motor is serviced within 3 hours of initial submersion.

OK, maybe now you're saying "3 hours, it will take me that long to get it to a shop or to my own garage." Well, if the engine cannot be serviced immediately (or sufficiently serviced so it can be started), re-submerge it in a tank of fresh water to minimize exposure to the atmosphere and slow the corrosion process. Even if you do this, do not delay any more than absolutely necessary, service the engine as soon as possible. This is especially important if the engine was submerged in salt, brackish or polluted water as even submersion in fresh water will not preserve the engine indefinitely. Service the engine, at the **MOST** within a few days of protective submersion.

After the engine is recovered, vigorously wash all debris from the engine using pressurized freshwater.

■ **If the engine was submerged while still running, there is a good chance of internal damage (such as a bent connecting rod). Under these circumstances, don't start the motor, follow the beginning of this procedure to try turning it over slowly by hand, feeling for mechanical problems. If necessary, refer to Powerhead Overhaul for complete disassembly and repair instructions.**

✱✱ WARNING

NEVER try to start a recovered motor until at least the first few steps (the ones dealing with draining the motor and checking to see it if is hydro-locked or damaged) are performed. Keep in mind that attempting to start a hydro-locked motor could cause major damage to the powerhead, including bending or breaking a connecting rod.

If the motor was submerged for any length of time it should be thoroughly disassembled and cleaned. Of course, this depends on whether water

MAINTENANCE & TUNE-UP

intruded into the motor or not. To determine this check the crankcase oil (on 4-strokes), and check the gearcase oil (on all motors) for signs of contamination.

The extent of cleaning and disassembly that must take place depends also on the type of water in which the engine was submerged. Engines totally submerged, for even a short length of time, in salt, brackish or polluted water will require more thorough servicing than ones submerged in fresh water for the same length of time. But, as the total length of submerged time or time before service increases, even engines submerged in fresh water will require more attention. Complete powerhead disassembly and inspection is required when sand, silt or other gritty material is found inside the engine cover.

Many engine components suffer the corrosive effects of submersion in salt, brackish or polluted water. The symptoms may not occur for some time after the event. Salt crystals will form in areas of the engine and promote significant corrosion.

Electrical components should be dried and cleaned or replaced, as necessary. If the motor was submerged in salt water, the wire harness and connections are usually affected in a shorter amount of time. Since it is difficult (or nearly impossible) to remove the salt crystals from the wiring connectors, it is best to replace the wire harness and clean all electrical component connections. The starter motor, relays and switches on the engine usually fail if not thoroughly cleaned or replaced.

To ensure a through cleaning and inspection:

1. Remove the engine cover and wash all material from the engine using pressurized freshwater. If sand, silt or gritty material is present inside the engine cover, completely disassemble and inspect the powerhead.

2. Tag (except on single cylinder motors) and disconnect the spark plugs leads. Be sure to grasp the spark plug cap and not the wire, then twist the cap while pulling upward to free it from the plug. Remove the spark plugs. For more details, refer to the Spark Plug procedure in this section.

3. Disconnect the fuel supply line from the engine, then drain and clean all fuel lines. Depending on the circumstances surrounding the submersion, inspect the fuel tank for contamination and drain, if necessary.

4. Support the engine horizontally with the spark plug port(s) facing downward, allowing water, if present, to drain. Force any remaining the water out by slowly rotating the flywheel by hand about 20 times or until there are no signs of water. If there signs of water are present, spray some fogging oil into the spark plug ports before turning the flywheel. This will help dislodge moisture and lubricate the cylinder walls.

✳✳ WARNING

When attempting to turn the flywheel for the first time after the submersion, be sure to turn it SLOWLY, feeling for sticking or binding that could indicate internal damage from hydro-lock. This is a concern, especially if the engine was cranked before the spark plug(s) were removed to drain water or if the engine was submerged while still running.

5. On carbureted models, drain the carburetor(s). The best method to thoroughly drain/clean the carburetor is to remove and disassemble it. For details refer to the Carburetor procedures under Fuel System.

6. Support the engine in the normal upright position. Check the engine gearcase oil for contamination. Refer to the procedures for Gearcase Oil in this section. The gearcase is sealed and, if the seals are in good condition, should have survived the submersion without contamination. But, if contamination is found, look for possible leaks in the seals, then drain the gearcase and make the necessary repairs before refilling it. For more details, refer to the section on Gearcases.

7. On 4-stroke motors, drain the crankcase engine oil and change the filter. Refer to the procedures in this section. If contaminated oil drains from the crankcase, flush the crankcase using a quart or two of fresh four-stroke engine oil (by pouring it into the motor as normal, but allowing it to drain as well) before refilling the crankcase.

8. Remove all external electrical components for disassembly and cleaning. Spray all connectors with electrical contact cleaner, then apply a small amount of dielectric grease prior to reconnection to help prevent corrosion. For electric start models, remove, disassemble and clean the starter components. For details on the electrical system components, refer to the Ignition and Electrical section.

9. Reassemble the motor and mount the engine or place it in a test tank. Start and run the engine for 1/2 hour. If the engine won't start, remove the spark plugs again and check for signs of moisture on the tips. If necessary, use compressed air to clean moisture from the electrodes or replace the plugs.

10. Stop the engine, then recheck the gearcase oil (and engine crankcase oil on 4-strokes).

11. Perform all other lubrication services.

12. Try not to let it get away from you (or anyone else) again!

SPECIFICATIONS

Magneto Breaker Point Gap Specifications

Model (Hp)	No. of Cyl	Year	Displace cu. in. (cc)	Breaker Point Gap Used Points in. (mm)	Breaker Point Gap New Points in. (mm)	Condenser Type/Capacity
Colt/Junior	1	1990	7.8 (128)	0.020 (0.51)	0.022 (0.056)	0.18-0.22 Mfd.
2	1	1993-95	4.75 (77.8)	0.008 (0.20)	0.008 (0.20)	0.22-0.26 Mfd.
2.3	1	1991-95	4.75 (77.8)	0.014 (0.35)	0.014 (0.35)	0.22-0.26 Mfd.
3.3	1	1991-95	4.75 (77.8)	0.020 (0.51)	0.020 (0.50)	0.22-0.26 Mfd.
3	1	1990	5.28 (87)	0.020 (0.51)	0.022 (0.056)	0.18-0.22 Mfd.
4	1	1990	5.28 (87)	0.020 (0.51)	0.022 (0.056)	0.18-0.22 Mfd.

MAINTENANCE & TUNE-UP

General Engine Specifications

Model (Hp)	No. of Cyl	Engine Type	Year	Displacement cu. in. (cc)	Bore and Stroke in. (mm)	Gear Ratio	Appx Weight lb. (kg)
Colt/Junior	1	2-stroke	1990	2.64 (43)	1.57 x 1.38 (40 x 35)	12:25 (0.48)	24 (11)
2	1	2-stroke	1993-01	4.75 (77.8)	1.89 x 1.69 (48 x 43)	13:24 (0.54)	30 (13)
2.3	1	2-stroke	1991-01	4.75 (77.8)	1.89 x 1.69 (48 x 43)	13:24 (0.54)	27-30 (12-13)
3.3	1	2-stroke	1991-01	4.75 (77.8)	1.89 x 1.69 (48 x 43)	13:24 (0.54)	27-32 (12-14)
3.5	1	2-stroke	2001	4.75 (77.8)	1.89 x 1.69 (48 x 43)	13:24 (0.54)	30-32 (13-14)
5	1	2-stroke	1999-01	6.7 (109)	2.16 x 1.80 (55 x 48)	12:23 (0.52)	46-47 (21-22)
5	1	4-stroke	1997-01	7.8 (128)	2.22 x 2.01 (57 x 51)	13:29 (0.45)	68-72 (31-33)
6	1	4-stroke	1997-01	7.8 (128)	2.22 x 2.01 (57 x 51)	13:29 (0.45)	68-72 (31-33)
3	2	2-stroke	1990-97	5.28 (87)	1.57 x 1.37 (40 x 35)	12:25 (0.48)	33-35 (15-16)
4	2	2-stroke	1990-01	5.28 (87)	1.57 x 1.37 (40 x 35)	12:25 (0.48)	33-37 (15-17)
4 Deluxe	2	2-stroke	1990-96	5.28 (87)	1.57 x 1.38 (40 x 35)	13:29 (0.45)	50-53 (23-24)
5	2	2-stroke	1990-98	10 (164)	1.94 x 1.70 (49 x 43)	13:29 (0.45)	56-61 (25-27)
6	2	2-stroke	1990-01	10 (164)	1.94 x 1.70 (49 x 43)	13:29 (0.45)	56-61 (25-27)
6.5	2	2-stroke	1990-93	10 (164)	1.94 x 1.70 (49 x 43)	13:29 (0.45)	56-61 (25-27)
8	2	2-stroke	1990-01	10 (164)	1.94 x 1.70 (49 x 43)	13:29 (0.45)	56-64 (25-29)
8	2	4-stroke	1996-01	12.87 (211)	2.22 x 1.65 (57 x 42)	12:29 (0.41)	82-91 (37-41)
9.9	2	4-stroke	1997-98	12.87 (211)	2.22 x 1.65 (57 x 42)	12:29 (0.41)	82-101 (37-46)
9.9	2	2-stroke	1990-92	13.2 (216)	2.19 x 1.76 (56 x 45)	12:29 (0.41)	72-88 (33-40)
10	2	2-stroke	1990-92	13.2 (216)	2.19 x 1.76 (56 x 45)	12:29 (0.41)	72-80 (33-36)
10 comm	2	2-stroke	1990-91	13.2 (216)	2.19 x 1.76 (56 x 45)	14:29 (0.48)	72-80 (33-36)
14	2	2-stroke	1990-91	13.2 (216)	2.19 x 1.76 (56 x 45)	14:29 (0.48)	72-80 (33-36)
15	2	2-stroke	1990-92	13.2 (216)	2.19 x 1.76 (56 x 45)	12:29 (0.41)	72-80 (33-36)
9.9	2	2-stroke	1993-01	15.6 (255)	2.38 x 1.76 (60 x 45)	12:29 (0.41)	72-88 (33-40)
10	2	2-stroke	1993-98	15.6 (255)	2.38 x 1.76 (60 x 45)	12:29 (0.41)	72-80 (33-36)
15	2	2-stroke	1993-01	15.6 (255)	2.38 x 1.76 (60 x 45)	12:29 (0.41)	72-80 (33-36)
9.9	2	4-stroke	1995-01	18.61 (305)	2.56 x 1.81 (65 x 46)	12:29 (0.41)	99-119 (45-54)
15	2	4-stroke	1995-01	18.61 (305)	2.56 x 1.81 (65 x 46)	12:29 (0.41)	99-114 (45-52)
18 Jet	2	2-stroke	1995-97	31.1 (521)	3.00 x 2.25 (76 x 57)	13:28 (0.46)	114-120 (52-55)
20	2	2-stroke	1990-01	31.8 (521)	3.00 x 2.25 (76 x 57)	13:28 (0.46)	114-120 (52-55)
25	2	2-stroke	1990-01	31.8 (521)	3.00 x 2.25 (76 x 57)	①	114-120 (52-55)
28	2	2-stroke	1990-97	31.8 (521)	3.00 x 2.25 (76 x 57)	①	114-120 (52-55)
30	2	2-stroke	1990-01	31.8 (521)	3.00 x 2.25 (76 x 57)	①	114-120 (52-55)
35	2	2-stroke	1990-97	31.8 (521)	3.00 x 2.25 (76 x 57)	13:28 (0.46)	114-120 (52-55)
25D	2	2-stroke	1990-95	45 (737)	3.19 x 2.82 (81 x 72)	12:29 (0.41)	180-185 (81-84)
30	2	2-stroke	1998-01	45 (737)	3.19 x 2.82 (81 x 72)	12:29 (0.41)	180-190 (81-86)
40	2	2-stroke	1990-01	45 (737)	3.19 x 2.82 (81 x 72)	②	180-202 (81-92)
45	2	2-stroke	1990-98	45 (737)	3.19 x 2.82 (81 x 72)	12:32 (0.38)	184-202 (83-92)
48	2	2-stroke	1990-96	45 (737)	3.19 x 2.82 (81 x 72)	12:29 (0.41)	180-185 (81-84)
50	2	2-stroke	1990-01	45 (737)	3.19 x 2.82 (81 x 72)	12:29 (0.41)	180-190 (81-86)
55	2	2-stroke	1990-01	45 (737)	3.19 x 2.82 (81 x 72)	12:32 (0.38)	184-202 (83-92)

General Engine Specifications

Model (Hp)	No. of Cyl	Engine Type	Year	Displacement cu. in. (cc)	Bore and Stroke in. (mm)	Gear Ratio	Appx Weight lb. (kg)
25	3	2-stroke	1995-01	30.5 (500)	2.35 x 2.34 (61 x 60)	13:28 (0.46)	151-170 (69-77)
35	3	2-stroke	1995-01	34.5 (565)	2.50 x 2.34 (64 x 60)	13:28 (0.46)	151-170 (69-77)
40	3	4-stroke	1999-01	49.7 (815)	2.80 x 2.70 (71 x 69)	11:25 (0.44)	238 (108)
50	3	4-stroke	1999-01	49.7 (815)	2.80 x 2.70 (71 x 69)	11:25 (0.44)	238 (108)
25	3	2-stroke	1995	56.1 (913)	3.19 x 2.34 (81 x 60)	12:29 (0.41)	230 (104)
40	3	2-stroke	1999-01	56.1 (913)	3.19 x 2.34 (81 x 60)	12:29 (0.41)	230 (104)
50	3	2-stroke	1995-98	56.1 (913)	3.19 x 2.34 (81 x 60)	12:29 (0.41)	230-253 (104-115)
60	3	2-stroke	1990-01	56.1 (913)	3.19 x 2.34 (81 x 60)	12:29 (0.41)	230-253 (104-115)
65	3	2-stroke	1990-01	56.1 (913)	3.19 x 2.34 (81 x 60)	12:29 (0.41)	230-253 (104-115)
70	3	2-stroke	1990-01	56.1 (913)	3.19 x 2.34 (81 x 60)	12:29 (0.41)	230-253 (104-115)
70	4	4-stroke	1998-01	79.2 (1298)	2.91 x 2.97 (74 x 76)	12:29 (0.41)	343 (156)

① Gear ratio of 13:28 (0.46) except on 25 commercial, 28 and 30 SPL models through 1997 which are 12:21 (0.57)

② Gear ratio of 12:29 (0.41) on all models, including the 40RS commercial, except on 40RW, 40RP and all other commercial models which are 12:32 (0.38)

MAINTENANCE & TUNE-UP

General Engine System Specifications

Model (Hp)	No. of Cyl	Engine Type	Year	Displace cu. in. (cc)	Oil Injection System ①	Ignition System	Starting System	Cooling System	Fuel System	Charging System ②	Battery ③ cca (mca)
Colt/Junior	1	2-stroke	1990	2.64 (43)	pre-mix	Magneto	R/MC	IMP / UG	1 - 1bc	NA	NA
2	1	2-stroke	1993-95	4.75 (77.8)	pre-mix	Magneto ⑧	R/MC	IMP / LG	1 - 1bc	NA	NA
	1	2-stroke	1996-01	4.75 (77.8)	pre-mix	Mag CD	R/MC	IMP / LG	1 - 1bc	NA	NA
2.3	1	2-stroke	1991-95	4.75 (77.8)	pre-mix	Magneto ⑧	R/MC	IMP / LG	1 - 1bc	NA	NA
	1	2-stroke	1996-01	4.75 (77.8)	pre-mix	Mag CD	R/MC	IMP / LG	1 - 1bc	NA	NA
3.3	1	2-stroke	1991-95	4.75 (77.8)	pre-mix	Magneto ⑧	R/MC	IMP / LG	1 - 1bc	NA	NA
	1	2-stroke	1996-01	4.75 (77.8)	pre-mix	Mag CD	R/MC	IMP / LG	1 - 1bc	NA	NA
3.5	1	2-stroke	2001	4.75 (77.8)	pre-mix	Mag CD	R/MC	IMP / LG	1 - 1bc	NA	NA
5	1	2-stroke	1999-01	6.7 (109)	pre-mix	Mag CD	R/MC	IMP / UG	1 - 1bc	NA	NA
5	1	4-stroke	1997-01	7.8 (128)	not applicable	Mag CD	R/MC	IMP / UG /TC	1 - 1bc	NA	NA
6	1	4-stroke	1997-01	7.8 (128)	not applicable	Mag CD	R/MC	IMP / UG /TC	1 - 1bc	NA	NA
3	2	2-stroke	1990-97	5.28 (87)	pre-mix	Mag CD ⑨	R/MC	IMP / UG	1 - 1bc	NA	NA
4	2	2-stroke	1990-01	5.28 (87)	pre-mix	Mag CD ⑨	R/MC	IMP / UG	1 - 1bc	NA	NA
4 Deluxe	2	2-stroke	1990-96	5.28 (87)	pre-mix	Mag CD	R/MC	IMP / UG	1 - 1bc	NA	NA
5	2	2-stroke	1990-98	10 (164)	pre-mix	Mag CD	R/MC	IMP / UG /TC	1 - 1bc	NA	NA
6	2	2-stroke	1990-01	10 (164)	pre-mix	Mag CD	R/MC	IMP / UG /TC	1 - 1bc	4 amp NR	360 (465) ⑥
6.5	2	2-stroke	1990-93	10 (164)	pre-mix	Mag CD	R/MC	IMP / UG /TC	1 - 1bc	4 amp NR	360 (465) ⑥
8	2	2-stroke	1990-01	10 (164)	pre-mix	Mag CD	R/MC	IMP / UG /TC	1 - 1bc	4 amp NR	360 (465) ⑥
8	2	4-stroke	1996-01	12.87 (211)	not applicable	Mag CD	R/MC	IMP / UG /TC	1 - 1bc	NA	NA
9.9	2	4-stroke	1997-98	12.87 (211)	not applicable	Mag CD	R/MC, TE/MC or RE/EP	IMP / UG /TC	1 - 1bc	4 amp NR	360 (465) ⑥
9.9	2	2-stroke	1990-92	13.2 (216)	pre-mix	Mag CD	R/MC, TE/MC or RE/EP	IMP / UG /TC	1 - 1bc	4 amp NR	360 (465) ⑥
10	2	2-stroke	1990-92	13.2 (216)	pre-mix	Mag CD	R/MC or R/MP	IMP / UG /TC	1 - 1bc	4 amp NR	360 (465) ⑥
14	2	2-stroke	1990-91	13.2 (216)	pre-mix	Mag CD	R/MC or R/MP	IMP / UG /TC	1 - 1bc	4 amp NR	360 (465) ⑥
15	2	2-stroke	1990-92	13.2 (216)	pre-mix	Mag CD	R/MC, TE/MC or RE/EP	IMP / UG /TC	1 - 1bc	4 amp NR	360 (465) ⑥
9.9	2	2-stroke	1993-01	15.6 (255)	pre-mix	Mag CD	R/MC, TE/MC or RE/EP	IMP / UG /TC	1 - 1bc	4 amp NR	360 (465) ⑥
10	2	2-stroke	1993-98	15.6 (255)	pre-mix	Mag CD	R/MC, TE/MC or RE/EP	IMP / UG /TC	1 - 1bc	4 amp NR	360 (465) ⑥
15	2	2-stroke	1993-01	15.6 (255)	pre-mix	Mag CD	R/MC, TE/MC or RE/EP	IMP / UG /TC	1 - 1bc	4 amp NR	360 (465) ⑥
9.9	2	4-stroke	1995-01	18.61 (305)	not applicable	Mag CD	R/MC, TE/MC or RE/EP	IMP / UG /TC	1 - 1bc	12 amp FR	360 (465) ⑥
15	2	4-stroke	1995-01	18.61 (305)	not applicable	Mag CD	R/MC, TE/MC or RE/EP	IMP / UG /TC	1 - 1bc	6 amp NR	360 (465) ⑥
18 Jet	2	2-stroke	1995-97	31.1 (521)	pre-mix	Mag CD	R/MP, TE/MP or RE/EP	IMP / UG /TC	1 - 1bc	4 amp NR	360 (465) ⑥
20	2	2-stroke	1990-01	31.8 (521)	pre-mix	Mag CD	R/MP, TE/MP or RE/EP	IMP / UG /TC	1 - 1bc	4 amp NR	360 (465) ⑥
25	2	2-stroke	1990-01	31.8 (521)	pre-mix	Mag CD	R/MP, TE/MP or RE/EP	IMP / UG /TC	1 - 1bc	4 amp NR	360 (465) ⑥
28	2	2-stroke	1990-97	31.8 (521)	pre-mix	Mag CD	R/MP, TE/MP or RE/EP	IMP / UG /TC	1 - 1bc	4 amp NR	360 (465) ⑥
30	2	2-stroke	1990-01	31.8 (521)	pre-mix	Mag CD	R/MP, TE/MP or RE/EP	IMP / UG /TC	1 - 1bc	4 amp NR	360 (465) ⑥
35	2	2-stroke	1990-97	31.8 (521)	pre-mix	Mag CD	R/MP, TE/MP or RE/EP	IMP / UG /TC	1 - 1bc	4 amp NR	360 (465) ⑥

MAINTENANCE & TUNE-UP

General Engine System Specifications

Model (Hp)	No. of Cyl	Engine Type	Year	Displace cu. in. (cc)	Oil Injection System ①	Ignition System	Starting System	Cooling System	Fuel System	Charging System ②	Battery ③ cca (mca)
25D	2	2-stroke	1990-95	45 (737)	VRO or AM	Mag CD	R/MP, TE/EP or RE/EP	IMP / UG /TC	2 - 1bc	4-amp NR ④	360 (465) ⑥
30	2	2-stroke	1998-01	45 (737)	VRO or AM	Mag CD	RE/EP	IMP / UG /TC	2 - 1bc	12-amp FR ④	360 (465) ⑥
40	2	2-stroke	1990-01	45 (737)	VRO or AM	Mag CD	R/MP, TE/EP or RE/EP	IMP / UG /TC	2 - 1bc	12-amp FR ④	360 (465) ⑥
45	2	2-stroke	1990-98	45 (737)	VRO or AM	Mag CD	R/MP	IMP / UG /TC	2 - 1bc	60 watts ④	NA
48	2	2-stroke	1990-96	45 (737)	VRO or AM	Mag CD	TE/EP or RE/EP	IMP / UG /TC	2 - 1bc	4-amp NR ④	360 (465) ⑥
50	2	2-stroke	1990-01	45 (737)	VRO or AM	Mag CD	R/MP, TE/EP or RE/EP	IMP / UG /TC	2 - 1bc	12-amp FR ④	360 (465) ⑥
55	2	2-stroke	1990-01	45 (737)	VRO or AM	Mag CD	R/MP, TE/EP or RE/EP	IMP / UG /TC	2 - 1bc	60 watts ④	NA
25	3	2-stroke	1995-01	30.5 (500)	Mixing Unit	Mag CD	R/MC, TE/MC or RE/EP	IMP / UG /TC	3 - 1bc	12-amp FR	360 (465) ⑥
35	3	2-stroke	1995-01	34.5 (565)	Mixing Unit	Mag CD	R/MC, TE/MC or RE/EP	IMP / UG /TC	3 - 1bc	12-amp FR	360 (465) ⑥
40	3	4-stroke	1999-01	49.7 (815)	not applicable	ECU Cont	RE	IMP / UG /TC	EFI	17-amp FR	550 (620) ⑦
50	3	4-stroke	1999-01	49.7 (815)	not applicable	ECU Cont	RE	IMP / UG /TC	EFI	17-amp FR	550 (620) ⑦
25	3	2-stroke	1995	56.1 (913)	VRO	Mag CD	RE/EP	IMP / UG /TC	3 - 1bc	12-amp FR	360 (465) ⑥
40	3	2-stroke	1999-01	56.1 (913)	VRO	Mag CD	RE/EP	IMP / UG /TC	3 - 1bc	12-amp FR	360 (465) ⑥
50	3	2-stroke	1995-98	56.1 (913)	VRO	Mag CD	TE/MP or RE/EP	IMP / UG /TC	3 - 1bc	12-amp FR	360 (465) ⑥
60	3	2-stroke	1990-01	56.1 (913)	VRO	Mag CD	TE/MP or RE/EP	IMP / UG /TC	3 - 1bc	12-amp FR	360 (465) ⑥
65	3	2-stroke	1990-01	56.1 (913)	VRO	Mag CD	R/MP, TE/MP or RE/EP	IMP / UG /TC	3 - 1bc	12-amp FR ⑤	360 (465) ⑥
70	3	2-stroke	1990-01	56.1 (913)	VRO	Mag CD	TE/MP or RE/EP	IMP / UG /TC	3 - 1bc	12-amp FR	360 (465) ⑥
70	4	4-stroke	1998-01	79.2 (1298)	not applicable	ECU Cont	RE	IMP / UG /TC	EFI	17-amp FR	550 (620) ⑦

1bc: 1 barrel carburetor
AM: AccuMix system (2 fuel pumps and an engine mounted oiling chamber)
ECU Cont: Fully transistorized, ECU controlled, battery powered ignition
EFI: Electronic Fuel Injection
EP: Electric Primer
FR: Fully-Regulated (equipped with regulator/rectifier)
IMP: Impeller pump
LG: Lower gearcase mounted
Mag CD: Magneto Powered Capacitor Discharge
MC: Manual Choke
MP: Manual Primer
NA: Not applicable
NR: Non-Regulated (equipped with rectifier)
R: Rope
RE: Remote Electric Start
TC: Thermostatically controlled
TE: Tiller Electric Start
UG: Upper gearcase mounted
VRO: OMC's VRO automatic oiling system (VRO pump and external oil tank)

① Most common oiling system noted but will vary with installation, most 2-stroke models can be set-up for pre-mix or an automatic oiling system
② The most common, referenced charging system (may be optional on some models, ratings shown is minimum)
③ When equipped with electric start or optional battery charging system
④ If not equipped with regulator or rectifier, but equipped with AC wiring, system output is 60 watts
⑤ 65 hp WM, WMLW, WMYW and WE models are equipped with a 6-amp NR charging system, all other models are 12-amp
⑥ Minimum recommended battery for electric start models: 360 cca (465 mca) w/ 90 min reserve (50 AH)
⑦ Minimum recommended battery for electric start models: 550 cca (620 mca) w/ 90 min reserve (60 AH)

MAINTENANCE & TUNE-UP

Maintenance Intervals Chart

Component	Each Use	Monthly or As Needed	First 20-Hour Check	Every 12mths/100hrs	Off Season
Anode(s)	I			I	I
Battery condition and connections (if equipped)*	I (condition/connections)	I (charge / fluid level)		I	I
Breaker points and condenser (Magneto only)			I	R	R
Boat hull*	I			I	I
Bolts and nuts (all accessible fasteners)*			I	I	I
Case finish (wash and wax)	C (salt / brackish / polluted water)	C		C	C
Cylinder head bolts (except 40-70 hp 4-strokes)			T	T	
Electricical wiring and connectors*			I	I	I
Emergency stop switch, clip &/or lanyard*	I	I	I	I	I
Engine crankcase oil (4-stroke motors)	I		R	R ①	R
Engine crankcase oil filter (4-stroke motors)			R / C (as applicable)	R / C (as applicable)	R / C (as applicable)
Engine mounting clamps/bolts		I		I	I
Flush cooling system	if in salt / brackish / polluted water	P		P	P
Fuel filter (clean or replace, as applicable)			P	P	P
Fuel hose and system components*		I	I	I	I
Gear oil	I (for signs of leakage)	I (level and condition)	R	R	R
Impeller clearance/intake grate (jet models)	I (intake grate for debris or damage)	I (visually inspect impeller)	I	I	I
Jet drive bearing lubrication	L (fill vent hose after each day)	I (L every 10 hours)	L	L	L
Lubrication points		②	②	②	②
Oil system hose and components (2-stroke)*	I	I	I	I	I
Pistons (Decarbon)				③	
Power steering belt, fluid, filter (if equipped)	I (quick-check of belt/fluid)		I	R (500 hours)	I
Power trim and tilt (if equipped)		I (check fluid level monthly)	I	I	I
Propeller (or rotor on RescuePro motors)	I	I	I	I	I
Propeller (or rotor) shaft and nut	I		I	I, L / T	I, L / T
Remote control*	I	I	I	I	I
Spark plugs		R (as needed)	I	I	I
Steering cable*	I	L (as needed)	L	L	L
Steering friction	I	A (as needed)		I	I
Timing belt (4-stroke, except 40-50 hp motors)			I	I ④	I
Tune-up		A (as needed)		I (annually)	Perform pre-season tune-up
Valve clearance			I	I (annually)	Perform with pre-season tune-up
Water pump intake grate and indicator	I				

A-Adjust
C-Clean
I-Inspect and Clean, Adjust, Lubricate or Replace, as necessary
L-Lubricate
R-Replace
T-Tighten

* Denotes possible safety item (although, all maintenance inspections/service can be considered safety related when it means not being stranded on the water should a component fail.)
① Interval extended to every 200 hours if using Johnson/Evinrude crankcase oil
② Varies with use, generally every 30 days when used in salt, brackish or polluted water and every 60 days when used in fresh water (refer to Lubrication Chart for more details)
③ Every 50 hours is OMC Carbon Guard additive is NOT used consistently with fuel.
④ Replace every 800 hours or 4 years, whichever comes first. Note that 70 hp 4-stroke is an interference motor and a broken belt could allow severe engine damage.

Lubrication Chart

Component	Applicable Models	Recommended Lubricant	Minimum Frequency	
			Fresh Water	Salt, Polluted or Brackish Water
Choke linkage	2-15 Hp motors	OMC Triple-Guard or equivalent marine grease	every 60 days ①	every 30 days ①
Electric starter motor pinion	Electric start models only	OMC Start pinion lube	every 60 days ①	every 30 days ①
Engine cover latches	737cc and larger 2-strokes	OMC Triple-Guard or equivalent marine grease	every 60 days ①	every 30 days ①
Engine mount clamp screws	1 and 2-Cylinder motors	OMC Triple-Guard or equivalent marine grease	every 60 days ①	every 30 days ①
Jet drive bearing, lubrication	Jet drive models	OMC EP/Wheel Bearing or equivalent water-resistant NLGI No. 1 grease	after every use	after every use
Jet drive bearing, grease replacement	Jet drive models	OMC EP/Wheel Bearing or equivalent water-resistant NLGI No. 1 grease	every 30 hours	every 15 hours
Power trim/tilt reservoir	3 and 4-Cylinder motors	OMC Power trim/tilt and power steering fluid	every 30 days	every 30 days
throttle shafts, cables and/or linkage	All	OMC Triple-Guard or equivalent marine grease	every 60 days ①	every 30 days ①
Steering (remote arm/ tiller shaft pivot and friction screw)	All	OMC Triple-Guard or equivalent marine grease	every 60 days ①	every 30 days ①
Swivel bracket	All	OMC Triple-Guard or equivalent marine grease	every 60 days ①	every 30 days ①
Tilt assembly (bracket, tube, pin and/or tilt lever shaft)	All	OMC Triple-Guard or equivalent marine grease	every 60 days ①	every 30 days ①

① Lubrication points should be checked weekly or with each use, whichever is LESS frequent. Based upon individual motor/use needs frequency of actual lubrication should occur at recommended intervals or more often during season. Perform all lubrication procedures immediately prior to extended motor storage.

MAINTENANCE & TUNE-UP

TWO-STROKE MOTOR FUEL:OIL RATIO CHART

Desired Fuel:Oil Ratio	Amount of oil needed when mixed with:				
	3 G (11.4 L) of Gas	6 G (22.7 L) of Gas	18 G (68.1 L) of Gas	30 G (114 L) of Gas	45 G (171 L) of Gas
100:1 (1% oil)	4 fl. oz. (118 mL)	8 fl. oz. (236 mL)	24 fl. oz. (708 mL)	40 fl. oz. (1180 mL)	60 fl. oz. (1770 mL)
50:1 (2% oil)	8 fl. oz. (236 mL)	16 fl. oz. (473 mL)	48 fl. oz. (1419 mL)	80 fl. oz. (2360 mL)	120 fl. oz. (3.54 L)
25:1 (4% oil)	16 fl. oz. (473 mL)	32 fl. oz. (946 mL)	96 fl. oz. (2838 mL)	160 fl. oz. (4.73 L)	240 fl. oz. (7.1 L)

NOTE: Fuel:Oil ratios listed here are for calculation purposes. Refer to the fuel:oil recommendations for your engine before mixing. Remember that a pre-mix system designed to produce a 50:1 ratio will produce a 25:1 ratio if a 50:1 ratio is already in the fuel tank feeding the motor.

Capacities - Two-Stroke Engines

Model (Hp)	No. of Cyl	Year	Displace cu. in. (cc)	Gear Oil Oz (mL)	Injection Oil Ratio ④
Colt/Junior	1	1990	2.64 (43)	1.2 (35)	50:1
2	1	1993-01	4.75 (77.8)	3 (90)	50:1
2.3	1	1991-01	4.75 (77.8)	3 (90)	50:1
3.3	1	1991-01	4.75 (77.8)	3 (90)	50:1
3.5	1	2001	4.75 (77.8)	3 (90)	50:1
5	1	1999-01	6.7 (109)	6.4 (190)	50:1
3	2	1990-97	5.28 (87)	2.7 (80)	50:1
4	2	1990-01	5.28 (87)	2.7 (80)	50:1
4 Deluxe	2	1990-96	5.28 (87)	11 (325)	50:1
5	2	1990-98	10 (164)	11 (325)	50:1
6	2	1990-01	10 (164)	11 (325)	50:1
6.5	2	1990-93	10 (164)	11 (325)	50:1
8	2	1990-01	10 (164)	11 (325)	50:1
9.9	2	1990-92	13.2 (216)	9 (260)	50:1
10	2	1990-92	13.2 (216)	9 (260)	50:1
10 Comm	2	1990-91	13.2 (216)	8 (245)	50:1
	2	1992	13.2 (216)	9 (260)	50:1
14	2	1990-91	13.2 (216)	8 (245)	50:1
15	2	1990-92	13.2 (216)	9 (260)	50:1
9.9	2	1993-01	15.6 (255)	9 (260)	50:1
10 Comm / 15 Comm	2	1993-98	15.6 (255)	9 (260)	50:1
15	2	1993-01	15.6 (255)	9 (260)	50:1
18 Jet	2	1995-97	31.1 (521)	11 (330)	50:1
20	2	1990-01	31.8 (521)	11 (330)	50:1
25	2	1990-01	31.8 (521)	①	50:1
28	2	1990-97	31.8 (521)	①	50:1
30	2	1990-01	31.8 (521)	①	50:1
35	2	1990-97	31.8 (521)	11 (330)	50:1
25D	2	1990-95	45 (737)	16.4 (485)	50:1
30	2	1998-01	45 (737)	②	50:1
40	2	1990-01	45 (737)	③	50:1
45	2	1990-98	45 (737)	22 (650)	50:1
48	2	1990-96	45 (737)	16.4 (485)	50:1
50	2	1990-01	45 (737)	16.4 (485)	50:1
55	2	1990-01	45 (737)	22 (650)	50:1
25	3	1995-01	30.5 (500)	11 (330)	50:1
35	3	1995-01	34.5 (565)	11 (330)	50:1
25	3	1995	56.1 (913)	22 (650)	50:1
40	3	1999-01	56.1 (913)	22 (650)	50:1
50	3	1995-98	56.1 (913)	22 (650)	50:1
60	3	1990-01	56.1 (913)	22 (650)	50:1
65	3	1990-01	56.1 (913)	22 (650)	50:1
70	3	1990-01	56.1 (913)	22 (650)	50:1

① 11 fl. oz. (330 mL) for all models except 1990-1997 25 commercial, 28 and 30 SPL which are 8 fl. oz (245 mL).
② Varries with gearcase, either 16.4 fl. oz. (485 mL) or 22 fl. oz (650 mL), fill slowly as you approach the 16 oz mark and continue until full.
③ Varries with model, 16.4 fl. oz. (485 mL) on most models, including the 40RS commercial, but 22 fl. oz (650 mL) on the 40RW, 40RP and all other commercial models.
④ Injection oil ratio based on normal operating conditions, some severe or high performance applications may need higher ratio, refer to information on Fuel Recommendations under Maintenance

MAINTENANCE & TUNE-UP

Capacities - Four-Stroke Engines

Model (Hp)	No. of Cyl	Year	Displace cu. in. (cc)	Engine Oil U.S. (Metric)	Gear Oil fl. oz. (mL)
5	1	1997-01	7.8 (128)	27 fl. oz. (800 mL)	11 (325)
6	1	1997-01	7.8 (128)	27 fl. oz. (800 mL)	11 (325)
8	2	1996-01	12.87 (211)	33.8 fl. oz. (1000 mL)	9.0 (260)
9.9	2	1997-98	12.87 (211)	33.8 fl. oz. (1000 mL)	9.0 (260)
9.9	2	1995-01	18.61 (305)	33.8 fl. oz. (1272 mL)	9.0 (260)
15	2	1995-01	18.61 (305)	33.8 fl. oz. (1272 mL)	9.0 (260)
40	3	1999-01	49.7 (815)	2.5 qt. (2.4 L)	21 (610)
50	3	1999-01	49.7 (815)	2.5 qt. (2.4 L)	21 (610)
70	4	1998-01	79.2 (1298)	4.8 qt. (4.5 L)	19 (560)

Tune-Up Specifications

Model (Hp)	No. of Cyl	Engine Type	Year	Displace cu. in. (cc)	Spark Plug Make	Spark Plug Type	Gap In. (mm)	Ignition Timing Degrees Idle	Ignition Timing Degrees Max	Idle Speed RPM (In Gear)	WOT MAX RPM (In Gear)	OMC Test Prop*	Min. Test RPM*
Colt/Junior	1	2-stroke	1990	2.64 (43)	Champion	RJ6C or J6C	0.030 (0.8)	not adj.	not adj.	625-675	4500-5500	329568	3900
2	1	2-stroke	1993-95	4.75 (77.8)	Champion	L87YC or L77JC4	0.030 (0.8)	not adj.	not adj.	1100-1300	4000-5000	115208	4500
	1	2-stroke	1996-98	4.75 (77.8)	Champion	QL87YC or L87YC	0.030 (0.8)	not adj.	not adj.	1100-1300	4000-5000	115208	4500
	1	2-stroke	1999-01	4.75 (77.8)	Champion	QL87YC or L87YC	0.030 (0.8)	not adj.	not adj.	1000-1200	4000-5000	115297	4500
2.3	1	2-stroke	1991-92	4.75 (77.8)	Champion	QL77JC4 or L77JC4	0.030 (0.8)	not adj.	not adj.	1100-1300	4200-5200	115297	4800
	1	2-stroke	1993-95	4.75 (77.8)	Champion	L87YC or L77JC4	0.030 (0.8)	not adj.	not adj.	1100-1300	4200-5200	115208	4800
	1	2-stroke	1996-98	4.75 (77.8)	Champion	QL87YC or L87YC	0.030 (0.8)	not adj.	not adj.	1100-1300	4200-5200	115208	4800
	1	2-stroke	1999-01	4.75 (77.8)	Champion	QL87YC or L87YC	0.030 (0.8)	not adj.	not adj.	1000-1200	4200-5200	115297	4800
3.3	1	2-stroke	1991-92	4.75 (77.8)	Champion	QL77JC4 or L77JC4	0.030 (0.8)	not adj.	not adj.	1000-1200	4300-5000	115306	4900
	1	2-stroke	1993-95	4.75 (77.8)	Champion	L87YC or L77JC4	0.030 (0.8)	not adj.	not adj.	1000-1200	4500-5500	115306	5000
	1	2-stroke	1996-01	4.75 (77.8)	Champion	QL87YC or L87YC	0.030 (0.8)	not adj.	not adj.	1000-1200	4500-5500	115306	5000
3.5	1	2-stroke	2001	4.75 (77.8)	Champion	QL87YC or L87YC	0.030 (0.8)	not adj.	not adj.	1000-1200	4500-5500	115306	5000
5	1	2-stroke	1999-01	6.7 (109)	NGK	BPR74S-10	0.035-0.039 (0.9-1.0)	not adj.	not adj.	900-1000	4600-5400	㉙	4400
5	1	4-stroke	1997-98	7.8 (128)	Champion	P10Y	0.027 (0.7)	not adj.	not adj.	850-950	4500-5500	390239	4700
	1	4-stroke	1999-01	7.8 (128)	Champion	P10Y	0.027 (0.7)	not adj.	not adj.	1150-1250	5500-6500	444508	5500
6	1	4-stroke	1997-98	7.8 (128)	Champion	P10Y	0.027 (0.7)	not adj.	not adj.	850-950	5000-6000	390239	4700
	1	4-stroke	1999-01	7.8 (128)	Champion	P10Y	0.027 (0.7)	not adj.	not adj.	1150-1250	5500-6500	444508	5500
3	2	Magneto	1990	5.28 (87)	Champion	RL82C or L82C	0.030 (0.8)	not adj.	not adj.	600-650	4500-5500	317738	4400
	2	CDI	1990-95	5.28 (87)	Champion	QL77JC4 or L77JC4 ①	0.030 (0.8)	not adj.	not adj.	700-800	4500-5500	317738	4400
	2	2-stroke	1996	5.28 (87)	Champion	QL86C ②	0.030 (0.8)	not adj.	not adj.	700-800	4500-5500	317738	4400
	2	2-stroke	1997	5.28 (87)	Champion	QL86C or L86C	0.030 (0.8)	not adj.	not adj.	700-800	4500-5500	317738	4400
4	2	Magneto	1990	5.28 (87)	Champion	RL82C or L82C	0.030 (0.8)	not adj.	not adj.	600-650	4500-5500	317738	4400
	2	CDI	1990-95	5.28 (87)	Champion	QL77JC4 or L77JC4 ①	0.030 (0.8)	not adj.	not adj.	700-800	4500-5500	317738	4400
	2	2-stroke	1996	5.28 (87)	Champion	QL86C ②	0.030 (0.8)	not adj.	not adj.	700-800	4500-5500	317738	4400
	2	2-stroke	1997-98	5.28 (87)	Champion	QL86C or L86C	0.030 (0.8)	not adj.	not adj.	700-800	4500-5500	317738	4400
	2	2-stroke	1999-01	5.28 (87)	Champion	QL86C or L86C	0.030 (0.8)	not adj.	not adj.	800-1000	4500-5500	317738	4400
4 Deluxe	2	2-stroke	1990-95	5.28 (87)	Champion	QL77JC4 or L77JC4 ①	0.030 (0.8)	not adj.	not adj.	600-650	4500-5500	390123	5100
	2	2-stroke	1996	5.28 (87)	Champion	QL86C ②	0.030 (0.8)	not adj.	not adj.	600-650	4500-5500	390123	5100
5	2	2-stroke	1990-93	10 (164)	Champion	QL77JC4 or L77JC4 ①	0.030 (0.8)	not adj.	not adj.	650-700	4500-5500	390239	4500
	2	2-stroke	1994-95	10 (164)	Champion	QL77JC4 or L77JC4 ①	0.030 (0.8)	not adj.	not adj.	675-725	4500-5500	390239	4500
	2	2-stroke	1996	10 (164)	Champion	QL86C ③	0.030 (0.8)	not adj.	not adj.	675-725	4500-5500	390239	4500
	2	2-stroke	1997-98	10 (164)	Champion	QL86C or L86C ④	0.030 (0.8)	not adj.	not adj.	675-725	4500-5500	390239	4500
6	2	2-stroke	1990-95	10 (164)	Champion	QL77JC4 or L77JC4 ①	0.030 (0.8)	not adj.	not adj.	⑤	4500-5500	390239	4800
	2	2-stroke	1996	10 (164)	Champion	QL86C ③	0.030 (0.8)	not adj.	not adj.	⑤	4500-5500	390239	4800
	2	2-stroke	1997-01	10 (164)	Champion	QL86C or L86C ④	0.030 (0.8)	not adj.	not adj.	⑤	4500-5500	390239	4800
6.5	2	2-stroke	1990-93	10 (164)	Champion	QL77JC4 or L77JC4 ①	0.030 (0.8)	not adj.	not adj.	⑤	4500-5500	390239	4800
8	2	2-stroke	1990-95	10 (164)	Champion	QL77JC4 or L77JC4 ①	0.030 (0.8)	not adj.	not adj.	⑤	5000-6000	390239	⑥
	2	2-stroke	1996	10 (164)	Champion	QL86C ③	0.030 (0.8)	not adj.	not adj.	⑤	5000-6000	390239	⑥
	2	2-stroke	1997-01	10 (164)	Champion	QL86C or L86C ④	0.030 (0.8)	not adj.	not adj.	⑤	5000-6000	390239	⑥

MAINTENANCE & TUNE-UP

Tune-Up Specifications

Model (Hp)	No. of Cyl	Engine Type	Year	Displace cu. in. (cc)	Spark Plug Make	Spark Plug Type	Gap In. (mm)	Ignition Timing Degrees Idle	Ignition Timing Degrees Max	Idle Speed RPM (In Gear)	WOT MAX RPM (In Gear)	OMC Test Prop*	Min. Test RPM*
8	2	4-stroke	1996	12.87 (211)	Champion	P10Y	0.027 (0.7)	not adj.	not adj.	850-950	4200-5200	386537	4700
	2	4-stroke	1997	12.87 (211)	Champion	P10Y	0.027 (0.7)	not adj.	not adj.	850-950	4500-5500	386537	4700
	2	4-stroke	1998	12.87 (211)	Champion	P10Y	0.027 (0.7)	not adj.	not adj.	850-950	5000-6000	386537	4700
	2	4-stroke	1999-01	12.87 (211)	Champion	P10Y	0.027 (0.7)	not adj.	not adj.	950-1050	5000-6000	386537	4800
9.9	2	4-stroke	1997-98	12.87 (211)	Champion	P10Y	0.027 (0.7)	not adj.	not adj.	850-950	5000-6000	386537	4700
9.9	2	2-stroke	1990-92	13.2 (216)	Champion	QL77JC4 or L77JC4 ⑦	0.030 (0.8)	not adj.	not adj.	650-700	5000-6000	386537	4800
10	2	2-stroke	1990-92	13.2 (216)	Champion	QL77JC4 or L77JC4 ⑦	0.030 (0.8)	not adj.	not adj.	650-700	4500-5500	386537	4800
14	2	2-stroke	1990-91	13.2 (216)	Champion	QL77JC4 or L77JC4 ⑦	0.030 (0.8)	not adj.	not adj.	650-900	5000-6000	395096	5250
15	2	2-stroke	1990-92	13.2 (216)	Champion	QL77JC4 or L77JC4 ⑦	0.030 (0.8)	not adj.	not adj.	650-700	5500-7000	386537	6100
9.9	2	2-stroke	1993-95	15.6 (255)	Champion	QL77JC4 or L77JC4 ⑦	0.030 (0.8)	not adj.	not adj.	675-725	5000-6000	435750	5200
	2	2-stroke	1996-01	15.6 (255)	Champion	QL82C ⑩	0.030 (0.8)	not adj.	not adj.	675-725	5000-6000	⑧	⑨
10	2	2-stroke	1993-95	15.6 (255)	Champion	QL77JC4 or L77JC4 ⑦	0.030 (0.8)	not adj.	not adj.	⑪	5000-6000	⑧	5200
	2	2-stroke	1996-98	15.6 (255)	Champion	QL82C ⑩	0.030 (0.8)	not adj.	not adj.	⑪	5000-6000	⑧	4900
15	2	2-stroke	1993-95	15.6 (255)	Champion	QL77JC4 or L77JC4 ⑦	0.030 (0.8)	not adj.	not adj.	⑪	5000-6000	⑧	5700
	2	2-stroke	1996-01	15.6 (255)	Champion	QL82C ⑩	0.030 (0.8)	not adj.	not adj.	⑪	5000-6000	⑧	5700
9.9	2	4-stroke	1995-97	18.61 (305)	Champion	RA8HC	0.030-0.040 (0.8-1.0)	not adj.	not adj.	800-900	㉙	340177	⑫
	2	4-stroke	1998	18.61 (305)	Champion	RA8HC	0.030-0.040 (0.8-1.0)	not adj.	not adj.	950-1050	㉙	340177	⑫
	2	4-stroke	1999	18.61 (305)	Champion	RA8HC	0.030-0.040 (0.8-1.0)	not adj.	not adj.	750-850	㉙	340177	⑫
	2	4-stroke	2000-01	18.61 (305)	Champion	RA8HC	0.030-0.040 (0.8-1.0)	not adj.	not adj.	850-950	㉙	340177	⑫
15	2	4-stroke	1995-01	18.61 (305)	Champion	RA8HC	0.030-0.040 (0.8-1.0)	not adj.	not adj.	800-900	5000-6000	340177	5000
	2	4-stroke	1998	18.61 (305)	Champion	RA8HC	0.030-0.040 (0.8-1.0)	not adj.	not adj.	950-1050	5000-6000	340177	5000
	2	4-stroke	1999	18.61 (305)	Champion	RA8HC	0.030-0.040 (0.8-1.0)	not adj.	not adj.	750-850	5000-6000	340177	5000
	2	4-stroke	2000-01	18.61 (305)	Champion	RA8HC	0.030-0.040 (0.8-1.0)	not adj.	not adj.	850-950	5000-6000	340177	5000
18 Jet	2	2-stroke	1995	31.8 (521)	Champion	QL77JC4 or L77JC4 ①	0.030 (0.8)	not app.	33-35 BTDC	650-700	4500-5500	not app.	not app.
	2	2-stroke	1996-97	31.8 (521)	Champion	QL82C	0.030 (0.8)	not app.	33-35 BTDC	650-700	4500-5500	not app.	not app.
20	2	2-stroke	1990-95	31.8 (521)	Champion	QL77JC4 or L77JC4 ①	0.030 (0.8)	not app.	33-35 BTDC	650-700	4500-5500	386891	4550
	2	2-stroke	1996-97	31.8 (521)	Champion	QL82C	0.030 (0.8)	not app.	33-35 BTDC	650-700	4500-5500	386891	4550
	2	2-stroke	1998-01	31.8 (521)	Champion	QL77JC4 or QL82C	0.030 (0.8)	not app.	33 BTDC	650-700	4500-5500	386891	4550
25	2	2-stroke	1990-95	31.8 (521)	Champion	QL77JC4 or L77JC4 ③	0.030 (0.8)	not app.	28-31 BTDC	650-700	4500-5500	⑤	4800
	2	2-stroke	1996	31.8 (521)	Champion	QL82C	0.030 (0.8)	not app.	28-31 BTDC	650-700	4500-5500	⑤	4800
	2	2-stroke	1997	31.8 (521)	Champion	QL82C ⑭	0.030 (0.8)	not app.	28-31 BTDC	650-700	4500-5500	⑤	4800
	2	2-stroke	1998-99	31.8 (521)	Champion	QL77JC4 or QL82C ⑭	0.030 (0.8)	not app.	30 BTDC	650-700	4500-5500	434505	4800
	2	2-stroke	2000-01	31.8 (521)	Champion	QL77JC4 or QL82C	0.030 (0.8)	not app.	30 BTDC	650-700	4500-5500	434505	4800
28	2	2-stroke	1990-95	31.8 (521)	Champion	QL77JC4 or L77JC4 ①	0.030 (0.8)	not app.	28-31 BTDC	650-700	4500-5500	398948	4800
	2	2-stroke	1996-97	31.8 (521)	Champion	QL82C	0.030 (0.8)	not app.	28-31 BTDC	650-700	4500-5500	398948	4800
30	2	2-stroke	1990-95	31.8 (521)	Champion	QL77JC4 or L77JC4 ①	0.030 (0.8)	not app.	28-31 BTDC	650-700	5200-5800	434505	5400
	2	2-stroke	1996-97	31.8 (521)	Champion	QL82C	0.030 (0.8)	not app.	28-31 BTDC	650-700	5200-5800	⑯	⑯
	2	2-stroke	1998-01	31.8 (521)	Champion	QL77JC4 or QL82C	0.030 (0.8)	not app.	30 BTDC	650-700	5200-5800	434505	5400

Tune-Up Specifications

Model (Hp)	No. of Cyl	Engine Type	Year	Displace cu. in. (cc)	Spark Plug Make	Spark Plug Type	Gap In. (mm)	Ignition Timing Degrees Idle	Ignition Timing Degrees Max	Idle Speed RPM (In Gear)	WOT MAX RPM (In Gear)	OMC Test Prop*	Min. Test RPM*
35	2	2-stroke	1990-95	31.8 (521)	Champion	QL77JC4 or L77JC4 ①	0.030 (0.8)	not app.	28-31 BTDC	650-700	5200-5800	434505	5400
	2	2-stroke	1996-97	31.8 (521)	Champion	QL82C	0.030 (0.8)	not app.	28-31 BTDC	650-700	5200-5800	⑯	⑯
25D	2	2-stroke	1990-93	45 (737)	Champion	QL78C ①	0.030 (0.8)	0-2 ATDC	18-20 BTDC	725-775	4500-5500	433638	5000
	2	2-stroke	1994-95	45 (737)	Champion	QL78C ①	0.030 (0.8)	0-2 ATDC	18-20 BTDC	775-825	4500-5500	433638	5000
30	2	2-stroke	1998-01	45 (737)	Champion	QL78C ①	0.030 (0.8)	0-2 ATDC	18-20 BTDC	775-825	4500-5500	n/a	n/a
40	2	2-stroke	1990-92	45 (737)	Champion	QL78C ⑰	0.030 (0.8)	1-5 ATDC	18-20 BTDC	725-775	4500-5500	⑲	⑲
	2	2-stroke	1993-95	45 (737)	Champion	QL78C ⑰	0.030 (0.8)	0-2 ATDC	18-20 BTDC	775-825	4500-5500	⑲	⑲
	2	2-stroke	1996-01	45 (737)	Champion	QL78YC	0.030 (0.8)	0-2 ATDC	18-20 BTDC	775-825	4500-5500	⑲	⑲
45	2	2-stroke	1990-95	45 (737)	Champion	QL16V or L16V ⑳	0.030 (0.8)	0-2 ATDC	18-20 BTDC	775-825	4500-5500	382861	5200
	2	2-stroke	1996-98	45 (737)	Champion	QL78YC	0.030 (0.8)	0-2 ATDC	18-20 BTDC	775-825	4500-5500	382861	5200
48	2	2-stroke	1990-92	45 (737)	Champion	QL78C ①	0.030 (0.8)	1-5 ATDC	18-20 BTDC	725-775	4500-5500	432968	5200
	2	2-stroke	1993-95	45 (737)	Champion	QL78C ①	0.030 (0.8)	0-2 ATDC	18-20 BTDC	775-825	4500-5500	432968	5200
	2	2-stroke	1996	45 (737)	Champion	QL78YC	0.030 (0.8)	0-2 ATDC	18-20 BTDC	775-825	4500-5500	432968	5200
50	2	2-stroke	1990-92	45 (737)	Champion	QL78C ⑰	0.030 (0.8)	1-5 ATDC	18-20 BTDC	725-775	4500-5500	432968	5200
	2	2-stroke	1993-95	45 (737)	Champion	QL78C ⑰	0.030 (0.8)	0-2 ATDC	18-20 BTDC	775-825	4500-5500	432968	5200
	2	2-stroke	1996-01	45 (737)	Champion	QL78YC ㉑	0.030 (0.8)	0-2 ATDC	18-20 BTDC	775-825	4500-5500	432968	5200
55	2	2-stroke	1990-95	45 (737)	Champion	QL16V or L16V ⑳	0.030 (0.8)	0-2 ATDC	18-20 BTDC	775-825	4500-5500	382861	5200
	2	2-stroke	1996-01	45 (737)	Champion	QL78YC	0.030 (0.8)	0-2 ATDC	18-20 BTDC	775-825	4500-5500	382861	5200
25	3	2-stroke	1995-99	30.5 (500)	Champion	QL86C or L86C ㉒	0.030 (0.8)	4 ATDC	19-21 BTDC	700-800	5200-5800	434505	4200
	3	2-stroke	2000-01	30.5 (500)	Champion	QL82C	0.030 (0.8)	4 ATDC	20 BTDC	700-800	5200-5800	434505	4200
35	3	2-stroke	1995-99	34.5 (565)	Champion	QL86C or L86C ㉒	0.030 (0.8)	4 ATDC	21-23 BTDC	700-800	5200-5800	434505	5000
	3	2-stroke	2000-01	34.5 (565)	Champion	QL82C	0.030 (0.8)	4 ATDC	22 BTDC	700-800	5200-5800	434505	5000
25	3	2-stroke	1995	56.1 (913)	Champion	QL77JC4 or L77JC4 ①	0.030 (0.8)	8-10 ATDC	16-18 BTDC	㉓	5000-6000	437770	4800
40	3	2-stroke	1999-01	56.1 (913)	Champion	QL78YC	0.030 (0.8)	4 ATDC	17 BTDC	750-850	4500-5500	n/a	n/a
50	3	2-stroke	1995	56.1 (913)	Champion	QL77JC4 or L77JC4 ①	0.030 (0.8)	6-8 ATDC	16-18 BTDC	㉓	5000-6000	386665	4600
	3	2-stroke	1996-97	56.1 (913)	Champion	QL78YC ㉑	0.030 (0.8)	7-9 ATDC	16-18 BTDC	㉓	5000-6000	386665	4600
	3	2-stroke	1998	56.1 (913)	Champion	QL78YC ㉑	0.030 (0.8)	8 ATDC	17 BTDC	750-850	5000-6000	386665	4600
60	3	2-stroke	1990-94	56.1 (913)	Champion	QL77JC4 or L77JC4 ①	0.030 (0.8)	1-3 ATDC	18-20 BTDC	㉓	5000-6000	386665	5000
	3	2-stroke	1994-95	56.1 (913)	Champion	QL77JC4 or L77JC4 ①	0.030 (0.8)	3-5 ATDC	16-18 BTDC	㉓	5000-6000	386665	5000
	3	2-stroke	1996-97	56.1 (913)	Champion	QL78YC	0.030 (0.8)	3-5 ATDC	16-18 BTDC	㉓	5000-6000	386665	5000
	3	2-stroke	1998-01	56.1 (913)	Champion	QL78YC	0.030 (0.8)	4 ATDC	17 BTDC	750-850	5000-6000	386665	5000
65	3	2-stroke	1990-94	56.1 (913)	Champion	QL16V or L16V ⑳	0.030 (0.8)	3-5 ATDC	18-20 BTDC	㉓	4500-5500	386665	5550
	3	2-stroke	1994-95	56.1 (913)	Champion	QL16V or L16V ⑳	0.030 (0.8)	6-8 ATDC	16-18 BTDC	㉓	4500-5500	386665	5550
	3	2-stroke	1996-01	56.1 (913)	Champion	QL78YC ㉑	0.030 (0.8)	(24)	17 BTDC	㉓	4500-5500	386665	5550
70	3	2-stroke	1990-94	56.1 (913)	Champion	QL77JC4 or L77JC4 ①	0.030 (0.8)	3-5 ATDC	18-20 BTDC	㉓	5000-6000	386665	5700
	3	2-stroke	1994-95	56.1 (913)	Champion	QL77JC4 or L77JC4 ①	0.030 (0.8)	3-5 ATDC	16-18 BTDC	㉓	5000-6000	386665	5700
	3	2-stroke	1996-01	56.1 (913)	Champion	QL78YC	0.030 (0.8)	4 ATDC	17 BTDC	㉓	5000-6000	386665	5700

MAINTENANCE & TUNE-UP

Tune-Up Specifications

Model (Hp)	No. of Cyl	Engine Type	Year	Displace cu. in. (cc)	Spark Plug Make	Spark Plug Type	Gap In. (mm)	Ignition Timing Degrees Idle	Ignition Timing Degrees Max	Idle Speed RPM (In Gear)	WOT MAX RPM (In Gear)	OMC Test Prop*	Min. Test RPM*
40	3	4-stroke	1999-01	49.7 (815)	NGK	DCPR6E	0.035-0.039 (0.9-1.0)	㉕	㉕	800-900	5200-5800	n/a	n/a
50	3	4-stroke	1999-01	49.7 (815)	NGK	DCPR6E	0.035-0.039 (0.9-1.0)	㉕	㉕	800-900	5900-6500	n/a	n/a
70	4	4-stroke	1998-01	79.2 (1298)	NGK	BPR6ES	0.030 (0.8)	㉕	㉕	700	5200-5800	386665	5000

* Note: Test propeller and rpm not applicable to Jet models   n/a: not available   not adj: not adjustable   not app: not a not app: not applicable

① For sustained high speed operation on models other than the 8RC or 8RCL, use Champion QL16V or L16V fixed gap spark plug
② For extended idle use Champion L90C
③ For extended idle on models other than the 8RC use Champion L92YC
④ For extended idle on models other than the 8RC or 8WR, use Champion L82YC
⑤ Idle speed is 650-700 rpm on all models through 1993 or 675-725 rpm on all 1994 and later models except 6SL, 8SRL and 8RX which are 850-900 rpm through 1993 or 875-925 rpm for 1994 and later models
⑥ Test prop rpm is 5300 on all 8 hp models except the 8SRL and 8RX which are 4850 rpm
⑦ For sustained high speed operation use Champion fixed gap spark plugs QL16V or L16V fixed except on commercial models which use QL78V or L78V
⑧ Use test prop part no. 340177 for all models except 15 KC commmercial models and all 1999 and later commercial models which use 386537
⑨ Test prop rpm is 4900 on most 9.9 2-stroke models, except 9.9 SEL models which are 3500 rpm
⑩ For extended idle use Champion QL86C or L86C
⑪ Idle speed in gear is 675-725 rpm for all 2-stroke 15 hp models, except the KC which is 1000-1200 rpm for 1995 models or 900-1100 rpm for later models
⑫ Test prop rpm is 3500 on all 4-stroke 9.9 rope start models or 4000 on 9.9 electric start models
⑬ Specification is for non-commercial under normal conditions, for other conditions and applications note the following:
   a. non-commercial motors under sustained high speed use fixed gap plugs QL16V or L16V
   b. 25 hp Commercial motors under normal conditions use fixed gap plugs QL16V or L16V
   c. 25 hp Commercial motors under sustained high speed use fixed gap plugs QL78V or L78V
⑭ Specification is for non-commercial motors, for 25 hp commercial motors use the fixed gap plugs QL16V or L16V
⑮ 25 hp motors use test prop no. 434505, except 1995-1997 25 hp commercial models which use 396561
⑯ Standard models use test prop no. 434505 for a test rpm of 5400, SPL models use test prop no. 398948 for a test rpm of 4800
⑰ Specification is for standard 40/50 hp models, on these model use fixed gap plugs QL16V or L16V for sustained high speed operation, for other 40 hp models:
   a. For 40RS models use QL16V or L16V for normal operation or QL78C for sustained high speed operation
   b. For 40RW or 40RP models use QL16V or L16V for normal operation or QL78V or L78V for sustained high speed operation
⑱ Use test prop no. 432968 for standard 40 hp and 40RS commercial models, or prop no. 382861 for 40RW, 40RP, WR or other commercial models
⑲ Test prop speed is 4900 rpm for standard 40 hp models, while speed on all commercial models such as the 40RS, 40RW, 40RP and 40WR is 5200 rpm
⑳ For sustained high speed operation use Champion fixed gap spark plugs QL78V or L78V
㉑ For extended idle use Champion QL82C
㉒ For extended idle use Champion QL86C or L86C
㉓ Idle speed achieved in gear at proper timing setting as follows:
   a. For tiller control models 750-850 rpm
   b. For remote control models 600-700 rpm for 1990-97 models, 750-850 for 1998-2001 models
㉔ Idle timing is 4 degrees ATDC on all models, except the 65RS and 65WR on which idle timing is 7 degrees ATDC
㉕ Ignition timing is ECU controlled and varies upon engine operating condition between 0-32 degrees for 40 hp motors, 0-25 for 50 hp motors and 5-36 for 70 hp motors
㉖ Use standard propeller.
㉗ For rope start models max rpm is 5000-6000 rpm, for electric start models max rpm is 4000-5000 rpm

Valve Clearance Specifications - Four-Stroke Engines

Model (Hp)	No. of Cyl	Year	Displace cu. in. (cc)	Intake in. (mm)	Exhaust in. (mm)
5	1	1997-01	7.8 (128)	0.003 (0.08)	0.005 (0.12)
6	1	1997-01	7.8 (128)	0.003 (0.08)	0.005 (0.12)
8	2	1996-01	12.87 (211)	0.003 (0.08)	0.005 (0.12)
9.9	2	1997-98	12.87 (211)	0.003 (0.08)	0.005 (0.12)
9.9	2	1995-01	18.61 (305)	0.004 (0.10)	0.006 (0.15)
15	2	1995-01	18.61 (305)	0.004 (0.10)	0.006 (0.15)
40	3	1999-01	49.7 (815)	0.007-0.009 (0.18-0.24)	0.007-0.009 (0.18-0.24)
50	3	1999-01	49.7 (815)	0.007-0.009 (0.18-0.24)	0.007-0.009 (0.18-0.24)
70	4	1998-01	79.2 (1298)	0.005 (0.13)	0.006 (0.15)

3

FUEL SYSTEM

FUEL SYSTEM BASICS	3-2
FUEL TANK AND LINES	3-6
CARBURETED FUEL SYSTEM	3-11
CARBURETOR	3-14
ELECTRONIC FUEL INJECTION (EFI)	3-41
SPECIFICATIONS	3-71

CARBURETED FUEL SYSTEM ... 3-11
DESCRIPTION AND OPERATION ... 3-11
- Basic Functions ... 3-11
TROUBLESHOOTING ... 3-12
FUEL PUMP ... 3-34
- Overhaul ... 3-37
- Removal & Installation ... 3-26
- Testing ... 3-34
MANUAL FUEL PRIMER ... 3-38
- Assembly/Installation ... 3-39
- Cleaning and Inspection ... 3-39
- Function Test ... 3-38
- Primer Check ... 3-38
- Removal/Disassembly ... 3-38
ELECTRIC FUEL PRIMER ... 3-39
- Function Test ... 3-39
- Removal And Installation ... 3-40
- Solenoid Check ... 3-40
CARBURETOR
2.0-3.5 HP MOTORS ... 3-14
- Overhaul ... 3-15
- Removal & Installation ... 3-14
- Throttle Jet Needle Adjustment ... 3-16
5 HP (109cc) MOTORS ... 3-16
- Overhaul ... 3-17
- Removal & Installation ... 3-16
COLT/JUNIOR, HP & 4 DELUXE MOTORS ... 3-4 ... 3-18
- Overhaul ... 3-18
- Removal & Installation ... 3-18
5/6 HP 4-STROKE MOTORS ... 3-20
- Overhaul ... 3-21
- Removal & Installation ... 3-20
8/9.9 & 9.9/15 HP 4-STROKE MOTORS ... 3-22
- Overhaul ... 3-23
- Removal & Installation ... 3-22
9.9/10/14/15 HP (216/255cc) MOTORS THRU 1993 & ALL 18 JET-35 HP (521cc) MOTORS ... 3-24
- Overhaul ... 3-27
- Removal & Installation ... 3-27
25/35 HP (500/565cc) MOTORS ... 3-28
- Overhaul ... 3-29
- Removal & Installation ... 3-29
ALL EXCEPT ABOVE, INCLUDING 5/6/6.5/8 (164cc), 9.9/10/15 (255cc, 1994 AND LATER), 25-55 (737cc) & 25-70 (913cc) MOTORS ... 3-30
- Overhaul ... 3-32
- Removal & Installation ... 3-30
CLEANING & INSPECTION ... 3-33
- Inspecting the Emulsion Tube ... 3-34
- Removing Core Plugs ... 3-34
ELECTRONIC FUEL INJECTION (EFI) ... 3-41
AIR INTAKE SILENCER AND FLAME ARRESTER ... 3-51
- Removal & Installation ... 3-51
CAMSHAFT POSITION (CMP) SENSOR ... 3-60
- Removal & Installation ... 3-61
- Testing ... 3-60
CLOSED THROTTLE POSITION (CTP) SWITCH ... 3-67
- Removal & Installation ... 3-67
- Testing ... 3-67
CRANKSHAFT POSITION (CKP) SENSOR ... 3-61
- Removal & Installation ... 3-62
- Testing ... 3-62
DESCRIPTION & OPERATION ... 3-41
EFI SYSTEM RELAY ... 3-67
- Testing & Service ... 3-67
ENGINE CONTROL UNIT (ECU) ... 3-62
- Removal & Installation ... 3-63
ENGINE SYMPTOM DIAGNOSTIC CHARTS ... 3-47
- Engine Cranks But Won't Run ... 3-47
- Engine Idles Improperly ... 3-48
- Engine Overheats ... 3-50
- Engine Runs Rough/Lacks Power ... 3-49
- Engine Won't Crank ... 3-47
FUEL RAIL AND INJECTORS ... 3-58
- Injector Operational Test ... 3-59
- Injector Resistance Test ... 3-59
- Injector Signal Test ... 3-59
- Removal & Installation ... 3-59
IDLE AIR CONTROL (IAC) VALVE ... 3-68
- Removal & Installation ... 3-68
- Testing ... 3-68
INTAKE MANIFOLD ... 3-52
- Removal & Installation ... 3-52
LOW PRESSURE FUEL PUMP ... 3-53
- Overhaul ... 3-55
- Pump Pressure Test ... 3-54
- Removal & Installation ... 3-54
MANIFOLD ABSOLUTE PRESSURE (MAP) SENSOR ... 3-66
- Removal & Installation ... 3-66
- Testing ... 3-66
NEUTRAL (SAFETY) SWITCH ... 3-69
- Removal & Installation ... 3-70
- Testing ... 3-69
SELF DIAGNOSTIC SYSTEM ... 3-43
- Reading & Clearing Codes ... 3-43
SENSOR AND CIRCUIT RESISTANCE/OUTPUT TESTS ... 3-45
- Testing EFI Components ... 3-45
TEMPERATURE SENSORS ... 3-63
- Removal & Installation ... 3-65
- Testing ... 3-64
THROTTLE BODY ... 3-51
- Removal & Installation ... 3-51
TROUBLESHOOTING ... 3-41
VAPOR SEPARATOR TANK & FUEL PUMP ... 3-55
- Removal & Installation ... 3-57
- Testing ... 3-56
FUEL SYSTEM BASICS ... 3-2
FUEL ... 3-2
- Alcohol-Blended Fuels ... 3-3
- Checking ... 3-3
- High Altitude Operation ... 3-3
- Octane Rating ... 3-3
- Recommendations ... 3-3
- Vapor Pressure ... 3-3
FUEL SYSTEM PRESSURIZATION ... 3-4
- Pressurizing (Checking For Leaks) ... 3-5
- Relieving Pressure (EFI Only) ... 3-5
FUEL SYSTEM SERVICE CAUTIONS ... 3-2
FUEL TANK AND LINES ... 3-6
FUEL LINES AND FITTINGS ... 3-8
- Service ... 3-9
- Testing ... 3-8
FUEL TANK ... 3-6
- Service ... 3-6
SPECIFICATIONS ... 3-71
CARBURETOR SET-UP ... 3-71

3-2 FUEL SYSTEM

FUEL SYSTEM BASICS

** CAUTION

If equipped, disconnect the negative battery cable ANYTIME work is performed on the engine, especially when working on the fuel system. This will help prevent the possibility of sparks during service (from accidentally grounding a hot lead or powered component). Sparks could ignite vapors or exposed fuel. Disconnecting the cable on electric start motors will also help prevent the possibility fuel spillage if an attempt is made to crank the engine while the fuel system is open.

** CAUTION

Fuel leaking from a loose, damaged or incorrectly installed hose or fitting may cause a fire or an explosion. ALWAYS pressurize the fuel system and run the motor while inspecting for leaks after servicing any component of the fuel system.

The carburetion or fuel injection, and the ignition principles of engine operation must be understood in order to perform troubleshoot and repair an outboard motors fuel system or to perform a proper tune-up on carbureted motors.

If you have any doubts concerning your understanding of engine operation, it would be best to study

The Basic Operating Principles of an engine as detailed under Troubleshooting in Section 1, before tackling any work on the fuel system.

The fuel systems used on engines covered by this manual range from single carburetors to multiple carburetors or electronic fuel injection. The carbureted motors utilize various means of enriching fuel mixture for cold starts, including a manual choke, manual primer or electric primer solenoid. Similarly, the 2-stroke motors covered by this manual may require pre-mixing of the fuel and oil or may be equipped with the Variable Rate Oiling (VRO2) or AccuMix automatic oiling systems. Refer to the General Engine System Specifications chart in Section 2 for more details as to what systems were commonly used on what motors, but keep in mind that additional systems, such as the AccuMix were available as accessories for most motors. For details on the VRO2 and AccuMix oiling systems, please refer to the section on Lubrication and Cooling.

Fuel System Service Cautions

There is no way around it. Working with gasoline can provide for many different safety hazards and requires that extra caution is used during all steps of service. To protect yourself and others, you must take all necessary precautions against igniting the fuel or vapors (which will cause a fire at best or an explosion at worst).

** CAUTION

Take extreme care when working with the fuel system. NEVER smoke (it's bad for you anyhow, but smoking during fuel system service could kill you much faster!) or allow flames or sparks in the work area. Flames or sparks can ignite fuel, especially vapors, resulting in a fire at best or an explosion at worst.

For starters, disconnect the negative battery cable EVERY time a fuel system hose or fitting is going to be disconnected. It takes only one moment of forgetfulness for someone to crank the motor, possibly causing a dangerous spray of fuel from the opening. This is especially true on the high-pressure fuel circuit of EFI motors, where just turning the key to on will energize the fuel pump.

Gasoline contains harmful additives and is quickly absorbed by exposed skin. As an additional precaution, always wear gloves and some form of eye protection (regular glasses help, but only safety glasses can really protect your eyes).

■ Throughout service, pay attention to ensure that all components, hoses and fittings are installed them in the correct location and orientation to prevent the possibility of leakage. Matchmark components before they are removed as necessary.

Because of the dangerous conditions that result from working with gasoline and fuel vapors always take extra care and be sure to follow these guidelines for safety:
- Keep a Coast Guard-approved fire extinguisher handy when working.
- Allow the engine to cool completely before opening a fuel fitting. Don't all gasoline to drip on a hot engine.
- The first thing you must do after removing the engine cover is to check for the presence of gasoline fumes. If strong fumes are present, look for leaking or damage hoses, fittings or other fuel system components and repair.
- Do not repair the motor or any fuel system component near any sources of ignition, including sparks, open flames, or anyone smoking.
- Clean up spilled gasoline right away using clean rags. Keep all fuel soaked rags in a metal container until they can be properly disposed of or cleaned. NEVER leave solvent, gasoline or oil soaked rags in the hull.
- Don't use electric powered tools in the hull or near the boat during fuel system service or after service, until the system is pressurized and checked for leaks.
- Fuel leaking from a loose, damaged or incorrectly installed hose or fitting may cause a fire or an explosion. ALWAYS pressurize the fuel system and run the motor while inspecting for leaks after servicing any component of the fuel system.

Fuel

◆ See Figure 1

Fuel recommendations have become more complex as the chemistry of modern gasoline changes. The major driving force behind the many of the changes in gasoline chemistry was the search for additives to replace lead as an octane booster and lubricant. These additives are governed by the types of emissions they produce in the combustion process. Also, the replacement additives do not always provide the same level of combustion stability, making a fuel's octane rating less meaningful.

In the 1960's and 1970's, leaded fuel was common. The lead served two functions. First, it served as an octane booster (combustion stabilizer) and second, in 4-stroke engines, it served as a valve seat lubricant. For 2-stroke engines, the primary benefit of lead was to serve as a combustion stabilizer. Lead served very well for this purpose, even in high heat applications.

For decades now, all lead has been removed from the refining process. This means that the benefit of lead as an octane booster has been eliminated. Several substitute octane boosters have been introduced in the place of lead. While many are adequate in an automobile engine, most do not perform nearly as well as lead did, even though the octane rating of the fuel is the same.

Fig. 1 Damaged piston, possibly caused by; using too-low an octane fuel; using fuel that had "soured" or by insufficient oil (in fuel on 2-strokes or in a crankcase on 4-strokes)

FUEL SYSTEM 3-3

OCTANE RATING

◆ See Figure 1

A fuel's octane rating is a measurement of how stable the fuel is when heat is introduced. Octane rating is a major consideration when deciding whether a fuel is suitable for a particular application. For example, in an engine, we want the fuel to ignite when the spark plug fires and not before, even under high pressure and temperatures. Once the fuel is ignited, it must burn slowly and smoothly, even though heat and pressure are building up while the burn occurs. The unburned fuel should be ignited by the traveling flame front, not by some other source of ignition, such as carbon deposits or the heat from the expanding gasses. A fuel's octane rating is known as a measurement of the fuel's anti-knock properties (ability to burn without exploding). Essentially, the octane rating is a measure of a fuel's stability.

Usually a fuel with a higher octane rating can be subjected to a more severe combustion environment before spontaneous or abnormal combustion occurs. To understand how two gasoline samples can be different, even though they have the same octane rating, we need to know how octane rating is determined.

The American Society of Testing and Materials (ASTM) has developed a universal method of determining the octane rating of a fuel sample. The octane rating you see on the pump at a gasoline station is known as the pump octane number. Look at the small print on the pump. The rating has a formula. The rating is determined by the R+M/2 method. This number is the average of the research octane reading and the motor octane rating.
• The Research Octane Rating is a measure of a fuel's anti-knock properties under a light load or part throttle conditions. During this test, combustion heat is easily dissipated.
• The Motor Octane Rating is a measure of a fuel's anti-knock properties under a heavy load or full throttle conditions, when heat buildup is at maximum.

In general, 2-stroke engines tend to respond more to the motor octane rating than the research octane rating, because a 2-stroke engine has a power stroke (with heat buildup) every revolution. Therefore, in a 2-stroke outboard motor, the motor octane rating of the fuel is one of the best indications of how it will perform.

VAPOR PRESSURE

Fuel vapor pressure is a measure of how easily a fuel sample evaporates. Many additives used in gasoline contain aromatics. Aromatics are light hydrocarbons distilled off the top of a crude oil sample. They are effective at increasing the research octane of a fuel sample but can cause vapor lock (bubbles in the fuel line) on a very hot day. If you have an inconsistent running engine and you suspect vapor lock, use a piece of clear fuel line to look for bubbles, indicating that the fuel is vaporizing.

One negative side effect of aromatics is that they create additional combustion products such as carbon and varnish. If your engine requires high octane fuel to prevent detonation, de-carbon the engine more frequently with an internal engine cleaner to prevent ring sticking due to excessive varnish buildup.

ALCOHOL-BLENDED FUELS

When the Environmental Protection Agency mandated a phase-out of the leaded fuels in January of 1986, fuel suppliers needed an additive to improve the octane rating of their fuels. Although there are multiple methods currently employed, the addition of alcohol to gasoline seems to be favored because of its favorable results and low cost. Two types of alcohol are used in fuel today as octane boosters, methanol (wood alcohol) or ethanol (grain alcohol).

When used as a fuel additive, alcohol tends to raise the research octane of the fuel, so these additives will have limited benefit in an outboard motor. There are, however, some special considerations due to the effects of alcohol in fuel.
• Since alcohol contains oxygen, it replaces gasoline without oxygen content and tends to cause the air/fuel mixture to become leaner.
• On older outboards, the leaching affect of alcohol will, in time, cause fuel lines and plastic components to become brittle to the point of cracking. Unless replaced, these cracked lines could leak fuel, increasing the potential for hazardous situations.
• When alcohol blended fuels become contaminated with water, the water combines with the alcohol then settles to the bottom of the tank. This leaves the gasoline (and the oil for 2-stroke models using premix) on a top layer.

■ Modern outboard fuel lines and plastic fuel system components have been specially formulated to resist alcohol leaching effects.

HIGH ALTITUDE OPERATION

At elevated altitudes there is less oxygen in the atmosphere than at sea level. Less oxygen means lower combustion efficiency and less power output. As a general rule, power output is reduced three percent for every thousand feet above sea level.

On carbureted engines, re-jetting for high altitude does not restore lost power, it simply corrects the air-fuel ratio for the reduced air density and makes the most of the remaining available power. The most important thing to remember when re-jetting for high altitude is to reverse the jetting when return to sea level. If the jetting is left lean when you return to sea level conditions, the correct air/fuel ratio will not be achieved (the motor will run very lean) and possible powerhead damage may occur.

RECOMMENDATIONS

According to the fuel recommendations that come with your outboard, there is no engine in the product line that requires more than 87 octane when rated by the R+M/2 or. Most Johnson/Evinrude engines need only 87 octane or less. An 89 or higher octane rating generally means middle to premium grade unleaded. Premium unleaded is more stable under severe conditions but also produces more combustion products. Therefore, when using premium unleaded, more frequent de-carboning is necessary.

■ Check the emissions label found on your motor as it will normally list the minimum required fuel octane rating for your specific model.

CHECKING FOR STALE/CONTAMINATED FUEL

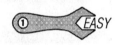

◆ See Figures 2, 3, 4 and 5

Outboard motors often sit weeks at a time making them the perfect candidate for fuel problems. Gasoline has a short life, as combustibles begin evaporating almost immediately. Even when stored properly, fuel starts to deteriorate within a few months, leaving behind a stale fuel mixture that can cause hard-starting, poor engine performance and even lead to possible engine damage.

Further more, as gasoline evaporates it leaves behind gum deposits that can clog filters, lines and small passages. Although the sealed high-pressure fuel system of an EFI motor is less susceptible to fuel evaporation, the low-pressure fuel systems of all engines can suffer the affects. Carburetors, dues to their tiny passages and naturally vented designs are the most susceptible components on non-EFI motors.

As mentioned under Alcohol-Blended fuels, modern fuels contain alcohol, which is hydroscopic (meaning it absorbs water). And, over time, fuel stored in a partially filled tank or a tank that is vented to the atmosphere will absorb water. The water/alcohol settles to the bottom of the tank, promoting rust (in metal tanks) and leaving a non-combustible mixture at the bottom of a tank that could leave a boater stranded.

One of the first steps to fuel system troubleshooting is to make sure the fuel source is not at fault for engine performance problems. Check the fuel if the engine will not start and there is no ignition problem.

Stale or contaminated fuels will often exhibit an unusual or even unpleasant unusual odor.

■ The best method of disposing stale fuel is through a local waste pickup service, automotive repair facility or marine dealership. But, this can be a hassle. If fuel is not too stale or too badly contaminated, it may be mixed with greater amounts of fresh fuel and used to power lawn/yard equipment or even an automobile (if greatly diluted so as to prevent misfiring, unstable idle or damage to the automotive engine). But we feel that it is much less of a risk to have a lawn mower stop running because of the fuel problem than it is to have your boat motor quit or refuse to start.

Most carburetors are equipped with a float bowl drain screw that can be used to drain fuel from the carburetor for storage or for inspection. Some models are equipped with an orifice plug behind the drain screw and this must usually be removed as well. For EFI models, a fuel system drain is found on the vapor separator tank, but access to the drain may require removal of the intake manifold and other interfering components.

For some motors, it may be easier to drain a fuel sample from the hoses leading to or from the low pressure fuel filter or fuel pump. Removal and

3-4 FUEL SYSTEM

Fig. 2 Carburetor float bowls are normally equipped with a drain screw

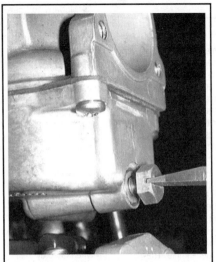

Fig. 3 To drain the carburetor, remove the drain screw...

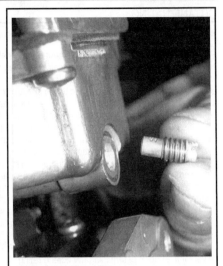

Fig. 4 ...and, if equipped, the orifice plug from the float bowl

installation instructions for the fuel filters are provided in the Maintenance Section, while fuel pump procedures are found in this section. To check for stale or contaminated fuel:

1. Disconnect the negative battery cable for safety. Secure it or place tape over the end so that it cannot accidentally contact the terminal and complete the circuit.

✱✱ CAUTION

Throughout this procedure, clean up any spilled fuel to prevent a fire hazard.

2. For carbureted motors, remove the float bowl drain screw (and orifice plug, if equipped), then allow a small amount of fuel to drain into a glass container.

■ **If there is no fuel present in the carburetor, disconnect the inlet line from the fuel pump and use the fuel primer bulb to obtain a sample as on EFI motors.**

3. For EFI motors, disconnect the fuel supply hose from the pump or low pressure fuel filter (as desired), then squeeze the fuel primer bulb to obtain a small sample of fuel. Place the sample in a clear glass container and reconnect the hose.

■ **If a sample cannot be obtained from the fuel filter or pump supply hose, there is a problem with the fuel tank-to-motor fuel circuit. Check the tank, primer bulb, fuel hose, fuel pump, fitting or inlet needle on carbureted models.**

4. Check the appearance and odor of the fuel. An unusual smell, signs of visible debris or a cloudy appearance (or even the obvious presence of water) points to a fuel that should be replaced.

5. If contaminated fuel is found, drain the fuel system and dispose of the fuel in a responsible manner, then clean the entire fuel system. On EFI models, this includes draining the vapor separator tank, then properly draining the high-pressure fuel system by relieving system pressure according to the instructions in this section.

■ **If debris is found in the fuel system, clean and/or replace all fuel filters.**

6. When finished, reconnect the negative battery cable, then properly pressurize the fuel system and check for leaks.

Fuel System Pressurization

When it comes to safety and outboards, the condition of the fuel system is of the utmost importance. The system must be checked for signs of damage or leakage with every use and checked, especially carefully when portions of the system have been opened for service.

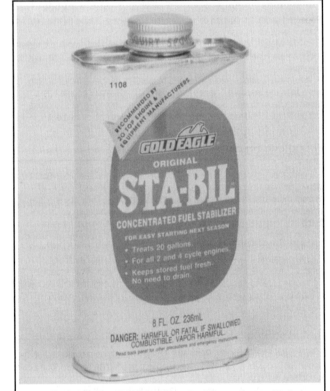

Fig. 5 Commercial additives, such as Sta-bil, may be used to help prevent "souring"

The best method to check the fuel system is to visually inspect the lines, hoses and fittings once the system has been properly pressurized.

Furthermore, EFI motors are equipped with two inter-related fuel circuits, a low pressure circuit that is similar to the circuit that feeds carburetors on other motors and a high pressure circuit that feeds the fuel injection system. As its name implies, the high pressure circuit contains fuel under pressure that, if given the chance, will spray from a damaged/loose hose or fitting. When servicing components of the high pressure system, the fuel pressure must first be relieved in a safe and controlled manner to help avoid the potential explosive and dangerous conditions that would result from simply opening a fitting and allowing fuel to spray uncontrolled into the work area.

FUEL SYSTEM

RELIEVING FUEL SYSTEM PRESSURE (EFI MOTORS ONLY)

Before servicing the high pressure fuel circuit or related components, including the vapor separator tank, high pressure filter, fuel rail, injector and related lines, the pressure must be released. Failure to do so in a proper manner could lead to high pressure fuel spray, excessive concentrations of vapors and an extremely dangerous, potentially explosive condition.

40/50 Hp EFI Models

1. Turn the key switch to **OFF**.

2. Tag, then disconnect the wiring (primary lead wire) from each ignition coil.

3. Disconnect the high pressure fuel pump wiring from the vapor separator by pushing down on the connector's lock tab, then pulling the connector free.

4. Use the key switch to crank the engine in 3 second bursts for 10-20 times. This will dissipate the fuel pressure in the lines. After the first couple of bursts, start squeezing the high pressure line to determine when the pressure is released. Once the hose is soft to the touch, crank the engine a few more times to ensure pressure is gone.

■ **Even after most or all of the pressure has been dissipated, there may still be some liquid fuel left in the lines. Always wrap a shop rag around fittings before they are disconnected to catch any escaping fuel.**

5. Unless necessary for service procedures or for safety, reconnect the ignition coil primary leads.

6. Disconnect the negative battery cable for safety during service, or leave the fuel pump wiring disconnected until the maintenance or repairs have been completed.

■ **We still recommend disconnecting the negative battery cable, especially if any work will be one or around electrical components. Any work on or near the gearcase, propeller or other potentially hazardous moving parts is also good reason to keep the battery disconnected.**

7. After maintenance or repairs are finished, fully pressurize the high and low pressure fuel circuits and thoroughly check the system for leakage.

70 Hp EFI Models

1. Locate and remove the 15 amp fuse for the high pressure fuel pump circuit from the fuse holder.

■ **The 15 amp fuel pump fuse is found on the port side of the powerhead, mounted between the No. 1 and No. 2 intake manifold runners (if necessary, follow the pink wire back from the fuel pump connector at the vapor separator tank to the fuse holder).**

2. With power removed from the fuel pump, use the engine to dissipate pressure from the high pressure circuit using one of 3 possible ways, as follows:

a. If the engine operates (is not being repaired for a no start or no run condition), start the engine and allow it to run until it stalls. Then restart or crank the engine 3 more times, to make sure fuel line pressure is dissipated.

b. If the engine cranks, but does not start or remain run properly, crank the engine 5-10 times, in 3 second long bursts in order to dissipate fuel pressure.

✱✱ CAUTION

When releasing fuel pressure using the screw on the top of the fuel rail, use extreme caution to prevent fuel from spraying uncontrolled into the work area. There must be NO open flames, sparks or other sources of ignition. It is imperative that there is proper ventilation in order to dissipate vapors. Wear safety glasses to protect your eyes, gloves to protect your skin and, finally, keep extra rags handy, as one might not do the trick.

c. If the engine does not crank at all, place a shop rag over the screw on top of the fuel rail and slowly loosen it. Once all pressure is relieved, tighten the screw to 28-30 ft. lbs. (38-41 Nm).

■ **Even after most or all of the pressure has been dissipated, there may still be some liquid fuel left in the lines. Always wrap a shop rag around fittings before they are disconnected to catch any escaping fuel.**

3. Disconnect the negative battery cable for safety during service, or leave the fuel pump fuse disconnected until the maintenance or repairs have been completed.

■ **We still recommend disconnecting the negative battery cable, especially if any work will be one or around electrical components. Any work on or near the gearcase, propeller or other potentially hazardous moving parts is also good reason to keep the battery disconnected.**

4. After maintenance or repairs are finished, fully pressurize the high and low pressure fuel circuits and thoroughly check the system for leakage.

PRESSURIZING THE FUEL SYSTEM (CHECKING FOR LEAKS)

✱✱ CAUTION

Fuel leaking from a loose, damaged or incorrectly installed hose or fitting may cause a fire or an explosion. ALWAYS pressurize the fuel system and run the motor while inspecting for leaks after servicing any component of the fuel system.

Carbureted Models

Carbureted engines covered by this manual are only equipped with a low pressure fuel system, making pressure release before service an non-issue. But, even a low pressure fuel system should be checked following repairs to make sure that no leaks are present. Only by checking a fuel system under normal operating pressures can you be sure of the system's integrity.

Most carbureted engines (except some integral tank models with gravity feed) utilize a fuel primer bulb mounted inline between the fuel tank and engine. On models so equipped, the bulb can be used to pressurize that portion of the fuel system. Squeeze the bulb until it and the fuel lines feel firm with gasoline. At this point check all fittings between the tank and motor for signs of leakage and correct, as necessary.

Once fuel reaches the engine it is the job of the fuel pump(s) to distribute it to the carburetors. On 4-stroke motors and pre-mix 2-stroke motors the fuel is pumped directly from the pump to the carburetor. On 2-strokes equipped with the AccuMix oiling system, the one fuel pump draws fuel from the tank, while the other pumps it through the mixing unit to the carburetor(s). When equipped with the VRO2 system, a traditional fuel pump and the VRO pump (consisting of a fuel and oil pump, as well as a fuel/oil mixing unit) is responsible for feeding an fuel/oil mixture to the carburetors.

No matter what system you are inspecting, start and run the motor with the engine top case removed, then check each of the system hoses, fittings and gasket-sealed components to be sure there is no leakage after service.

EFI Models

EFI models covered by this manual utilize 2 fuel circuits. A low pressure circuit consisting of a fuel tank, primer bulb, low pressure fuel pump and low pressure filter and low pressure fuel line to the vapor separator tank all operate in the same manner as the low pressure fuel system of a carbureted motor. The high pressure circuit consists of the electric fuel pump (integral with the vapor separator tank), the high pressure filter, the fuel rail/injectors and the high pressure lines.

Although it is necessary to pressurize and inspect both systems after repairs have been performed on the motor, it is especially important to properly check the high pressure circuit. Leaks from the high pressure circuit will (as you might expect) be under much greater pressures leading to even more potentially hazardous conditions than a low pressure leak. That's not to say the a low pressure leak isn't dangerous, but a high pressure leak can be even more so.

1. Pressurize and check the low pressure circuit as follows: Make sure the fuel tank is sufficiently full to provide an uninterrupted fuel source, then squeeze the bulb until it begins to feel firm. Check the low pressure lines, fittings and components for signs of leakage before continuing.

3-6 FUEL SYSTEM

2. Pressurize the high pressure fuel circuit as follows: Make sure the negative battery cable is connected (if removed for service), then turn the key switch to **ON** for 3 seconds and then **OFF** again for 3 seconds. Repeat the key switch cycle 3-4 times, while listening at the vapor separator to hear the high pressure pump run each time the key is turned to the **ON** position. If the pump does not run, check the fuel pump and circuit as described in this section under Vapor Separator Tank and High Pressure Pump in this section.

Once pressurized, check the high pressure lines, fittings and components for signs of leakage.

3. Start the engine, then allow it to idle it for a few seconds, while continuing to scan all fuel system components for signs of leakage.
4. Stop the motor and recheck the fittings.
5. Repair any leakage, then recheck the fuel system integrity.

FUEL TANK AND LINES

※※ CAUTION

If equipped, disconnect the negative battery cable ANYTIME work is performed on the engine, especially when working on the fuel system. This will help prevent the possibility of sparks during service (from accidentally grounding a hot lead or powered component). Sparks could ignite vapors or exposed fuel. Disconnecting the cable on electric start motors will also help prevent the possibility fuel spillage if an attempt is made to crank the engine while the fuel system is open.

※※ CAUTION

Fuel leaking from a loose, damaged or incorrectly installed hose or fitting may cause a fire or an explosion. ALWAYS pressurize the fuel system and run the motor while inspecting for leaks after servicing any component of the fuel system.

If a problem is suspected in the fuel supply, tank and/or lines, by far the easiest test to eliminate these components as possible culprits is to substitute a known good fuel supply. This is known as running a motor on a test tank (as opposed to running a motor IN a test tank, which is an entirely different concept). If possible, borrow a portable tank, fill it with fresh gasoline (or gas and oil for pre-mix 2-strokes) and connect it to the motor.

■ When using a test fuel tank, make sure the inside diameter of the fuel hose and fuel fittings is at least 5/16 in. (8mm) or larger.

Fuel Tank

◆ See Figures 6 and 7

There are 3 different types of fuel tanks that might be used along with these Johnson/Evinrude motors. The very smallest motors covered by this manual may be equipped with an integral fuel tank mounted to the powerhead. But, most motors, even some of the smaller ones, may be rigged using either a portable fuel tank or a boat mounted tank. In both cases, a tank that is not mounted to the engine itself is commonly called a remote tank.

■ Although many Johnson/Evinrude dealers rig boats using OMC fuel tanks, there are many other tank manufacturers and tank designs may vary greatly. Your outboard might be equipped with a tank from the engine manufacturer, the boat manufacturer or even another tank manufacturer. Although components used, as well as the techniques for cleaning and repairing tanks are similar for almost all fuel tanks be sure to use caution and common sense. If the design varies from the instructions included here, stop and assess the situation instead of following the instructions blindly. If we reference 2 or 4 screws for something and the component is still tight after removing that many, look for another or for another means of securing the component, don't force it. Refer to a reputable marine repair shop or marine dealership when parts are needed for aftermarket fuel tanks.

Whether or not your boat is equipped with a boat mounted, built-in tank depends mostly on the boat builder and partially on the initial engine installer. Boat mounted tanks can be hard to access (sometimes even a little hard to find if parts of the deck must be removed). When dealing with boat mounted tanks, look for access panels (as most manufacturers are smart or kind enough to install them for tough to reach tanks). At the very least, all manufacturers must provide access to fuel line fittings and, usually, the fuel level sender assembly.

No matter what type of tank is used, all must be equipped with a vent (either a manual vent or an automatic one-way check valve) which allows air in (but should prevent vapors from escaping). An inoperable vent (one that is blocked in some fashion) would allow the formation of a vacuum that could prevent the fuel pump from drawing fuel from the tank. A blocked vent could cause fuel starvation problems. Whenever filling the tank, check to make sure air does not rush into the tank each time the cap is loosened (which could be an early warning sign of a blocked vent).

If fuel delivery problems are encountered, first try running the motor with the fuel tank cap removed to ensure that no vacuum lock will occur in the tank or lines due to vent problems. If the motor runs properly with the cap removed but stall, hesitates or misses with the cap installed, you know the problem is with the tank vent system.

SERVICE

Integral Fuel Tanks
◆ See Figure 8

On small powerheads equipped with an integral fuel tank above the flywheel, or elsewhere on the powerhead, the filter in the tank may become plugged preventing proper fuel flow. To check fuel flow, disconnect the fuel line at the carburetor and fuel should flow from the line if the shut-off valve is open. If fuel is not present, the usual cause is a plugged filter in the fuel tank. It is very difficult to determine if a porcelain-type filter is plugged. The general appearance that the filter is satisfactory may be a false indication. Therefore, if the filter is suspected, the best remedy is replacement. The cost is very modest and this one area is thus eliminated as a problem source.

Would you believe, many times a lack of fuel at the carburetor is caused because the vent on the fuel tank was not opened? As mentioned under Fuel Tank in this section, it is important to make sure the vent is open at all times during engine operation to prevent formation of a vacuum in the tank that could prevent fuel from reaching the fuel pump or carburetor.

To remove the tank, proceed as follows:
1. Remove the upper and lower engine covers (as applicable) and as detailed under Engine Covers (Top and Lower Cases) in the Engine Maintenance section.
2. If equipped, make sure the fuel tank valve is shut off.

■ For models not equipped with a fuel tank mounted shut off valve, have a shop rag, large funnel and gas can handy to drain the fuel into when the hose is removed from the carburetor.

3. For all models except the 1999 and later 5 hp 2-stroke, trace the fuel line from the tank to the carburetor, then disconnect it from the carburetor. For models without a shut off valve, immediately direct the hose into the funnel and gas can to prevent or minimize fuel spillage.
4. For 1999 and later 5 hp 2-strokes (and on other models, AS necessary), remove the 3 bolts securing the manual starter, then move it out of the way.
5. Remove the screws fastening the tank to the powerhead, additional details are available on some models as follows:
   a. For 2.0-3.5 hp models, remove the 2 screws securing the tank to the powerhead.
   b. For 3/4 hp models, remove the 3 screws securing the tank to the mounting brackets.
   c. For 5 hp models, remove the rear fuel tank retaining screw.
6. Carefully lift the fuel tank from the powerhead. On 5 hp motors, you will have to compress the spring clamp holding the fuel line to the shut off valve, then slide the clamp back and pull the line from the valve. Immediately direct the fuel line into the funnel/gas can and allow the tank to drain.
7. Check the tank, filler cap and gasket for wear or damage. Replace any components as necessary to correct any leakage (if found.)
8. On models with a tank-mounted filter element, remove the filter and clean or replace as necessary. For more information refer to Fuel Filters in the Engine Maintenance section.

FUEL SYSTEM 3-7

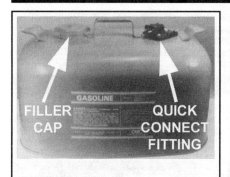

Fig. 6 These engines are most commonly rigged with portable fuel tanks such as this example

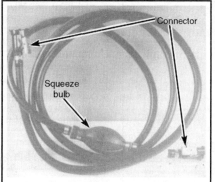

Fig. 7 Remote tanks are connected to the motor using a fuel line with a primer bulb

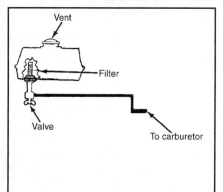

Fig. 8 Typical integral fuel tank and fuel supply circuit

9. If cleaning is necessary, flush the tank with a small amount of solvent or gasoline, then drain and dispose of the flammable liquid properly.

10. If equipped, check the tank cushions for excessive deterioration, wear or damage and replace, as necessary.

11. When installing the tank, be sure all fuel lines are connected properly and that the retaining screws are securely bolted in place.

12. Refill the tank and pressure test the system by opening the fuel valve, then starting and running the engine.

Portable Fuel Tanks

◆ See Figures 9 thru 13

Modern fuel tanks are vented to prevent vapor-lock of the fuel supply system, but are normally vented by a one-way valve to prevent pollution through the evaporation of vapors. A squeeze bulb is used to prime the system until the powerhead is operating. Once the engine starts, the fuel pump, mounted on the powerhead pull fuel from the tank and feeds the carburetor(s) or EFI high pressure fuel circuit, as applicable. The pickup unit in the tank is usually sold as a complete unit, but without the gauge and float.

To disassemble and inspect or replace tank components, proceed as follows:

1. For safety, remove the filler cap and drain the tank into a suitable container.
2. Disconnect the fuel supply line from the tank fitting.
3. To replace the pickup unit, first remove the screws (normally 4) securing the unit in the tank. Next, lift the pickup unit up out of the tank.
4. Remove the Phillips screws (usually 2) securing the fuel gauge to the bottom of the pickup unit and set the gauge aside for installation onto the new pickup unit.

■ If the pickup unit is not being replaced, clean and check the screen for damage. It is possible to bend a new piece of screen material around the pickup and solder it in place without purchasing a complete new unit.

5. If equipped with a level gauge assembly, check for smooth, non-binding movement of the float arm and replace if binding is found. Check the float itself for physical damage or saturation and replace, if found.

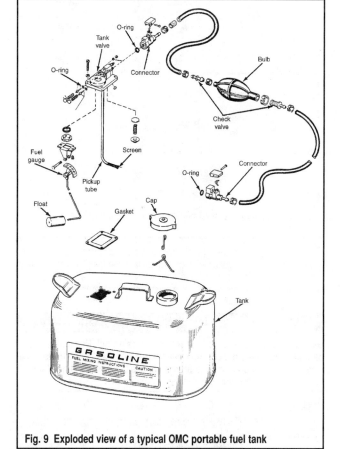

Fig. 9 Exploded view of a typical OMC portable fuel tank

Fig. 10 To service the tank, disconnect the fuel line from the quick-connect fitting...

Fig. 11 ... then remove the screws holding the pickup and float assembly to the tank

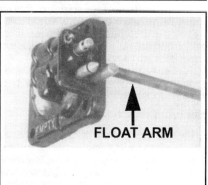

Fig. 12 When removing the pickup and float, be careful not to damage the float arm

3-8 FUEL SYSTEM

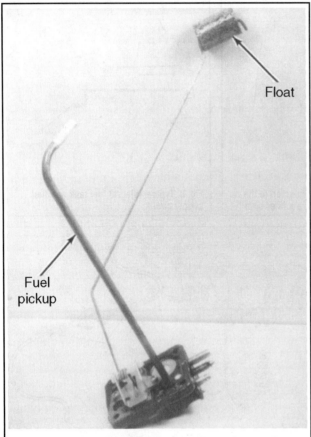

Fig. 13 The assembly must be tilted to remove the float arm from the tank

6. Check the fuel tank for dirt or moisture contamination. If any is found use a small amount of gasoline or solvent to clean the tank. Pour the solvent in and slosh it around to loosen and wash away deposits, then pour out the solvent and recheck. Allow the tank to air dry, or help it along with the use of an air hose from a compressor.

✷✷ CAUTION

Use extreme care when working with solvents or fuel. Remember that both are even more dangerous when their vapors are concentrated in a small area. No source of ignition from flames to sparks can be allowed in the workplace for even an instant.

To install:

7. Attach the fuel gauge to the new pickup unit and secure it in place with the Phillips screws.
8. Clean the old gasket material from fuel tank and, if being used, the old pickup unit. Position a new gasket/seal, then work the float arm down through the fuel tank opening, and at the same time the fuel pickup tube into the tank. It will probably be necessary to exert a little force on the float arm in order to feed it all into the hole. The fuel pickup arm should spring into place once it is through the hole.
9. Secure the pickup and float unit in place with the attaching screws.
10. If removed, connect the fuel tank, then pressurize the fuel system and check for leaks.

Boat Mounted Fuel Tanks

The other type of remote fuel tank sometimes used on these models (usually only on the larger models covered by this manual) is a boat mounted built-in tank. Depending on the boat manufacturer, built-in tanks may vary greatly in actual shape/design and access. All should be of a one-way vented to prevent a vacuum lock, but capped to prevent evaporation design.

Most boat manufacturers are kind enough to incorporate some means of access to the tank should fuel lines, fuel pickup or floats require servicing.

But, the means of access will vary greatly from boat-to-boat. Some might contain simple access panels, while others might require the removal of one or more minor or even major components for access. If you encounter difficulty, seek the advice of a local dealer for that boat builder. The dealer or his/her techs should be able to set you in the right direction.

✷✷ CAUTION

Observe all fuel system cautions, especially when working in recessed portions of a hull. Fuel vapors tend to gather in enclosed areas causing an even more dangerous possibility of explosion.

Fuel Lines and Fittings

◆ See Figure 7

In order for an engine to run properly it must receive an uninterrupted and unrestricted flow of fuel. This cannot occur if improper fuel lines are used or if any of the lines/fittings are damaged. Too small a fuel line could cause hesitation or missing at higher engine rpm. Worn or damaged lines or fittings could cause similar problems (also including stalling, poor/rough idle) as air might be drawn into the system instead of fuel. Similarly, a clogged fuel line, fuel filter or dirty fuel pickup or vacuum lock (from a clogged tank vent as mentioned under Fuel Tank) could cause these symptoms by starving the motor for fuel.

If fuel delivery problems are suspected, check the tank first to make sure it is properly vented, then turn your attention to the fuel lines. First check the lines and valves for obvious signs of leakage, then check for collapsed hoses that could cause restrictions.

■ If there is a restriction between the primer bulb and the fuel tank, vacuum from the fuel pump may cause the primer bulb to collapse. Watch for this sign when troubleshooting fuel delivery problems.

✷✷ CAUTION

Only use the proper fuel lines containing suitable Coast Guard ratings on a boat. Failure to do so may cause an extremely dangerous condition should fuel lines fail during adverse operating conditions.

TESTING

Fuel Line Quick Check

◆ See Figure 7

Stalling, hesitation, rough idle, misses at high rpm are all possible results of problems with the fuel lines. A quick visual check of the lines for leaks, kinked or collapsed lengths or other obvious damage may uncover the problem. If no obvious cause is found, the problem may be due to a restriction in the line or a problem with the fuel pump.

If a fuel delivery problem due to a restriction or lack of proper fuel flow is suspected, operate the engine while attempting to duplicate the miss or hesitation. While the condition is present, squeeze the primer bulb rapidly to manually pump fuel from the tank to (and through) the fuel pump to the carburetors (or EFI vapor separator tank). If the engine then runs properly while under these conditions, suspect a problem with a clogged restricted fuel line, a clogged fuel filter or a problem with the fuel pump.

Checking Fuel Flow at Motor

◆ See Figures 7, 14 and 15

To perform a more thorough check of the fuel lines and isolate or eliminate the possibility of a restriction, proceed as follows:
1. For safety, disconnect the spark plug leads, then ground each of them to the powerhead to prevent sparks and to protect the ignition system.
2. Disconnect the fuel line from the engine quick-connector. Place a suitable container over the end of the fuel line to catch the fuel discharged. Insert a small screwdriver into the end of the line to hold the valve open.
3. Squeeze the primer bulb and observe if there is satisfactory fuel flow from the line. If there is no fuel discharged from the line, the check valve in the squeeze bulb may be defective, or there may be a break or obstruction in the fuel line.

FUEL SYSTEM 3-9

4. If there is a good fuel flow, reconnect the tank-to-motor fuel supply line and disconnect the fuel line from the carburetor(s) or EFI vapor separator, directing that line into a suitable container. Crank the powerhead. If the fuel pump is operating properly, a healthy stream of fuel should pulse out of the line. If sufficient fuel does not pulse from the line, compare flow at either side of the inline fuel filter (if equipped) or check the fuel pump.

5. Continue cranking the powerhead and catching the fuel for about 15 pulses to determine if the amount of fuel decreases with each pulse or maintains a constant amount. A decrease in the discharge indicates a restriction in the line. If the fuel line is plugged, the fuel stream may stop. If there is fuel in the fuel tank but no fuel flows out the fuel line while the powerhead is being cranked, the problem may be in one of several areas:

6. Plugged fuel line from the fuel pump to the carburetor(s).
7. Defective O-ring in fuel line connector into the fuel tank.
8. Defective O-ring in fuel line connector into the engine.
9. Defective fuel pump.
10. The line from the fuel tank to the fuel pump may be plugged; the line may be leaking air; or the squeeze bulb may be defective.
11. Defective fuel tank.
12. If the engine does not start even though there is adequate fuel flow from the fuel line, the fuel inlet needle valve and the seat may be gummed together and prevent adequate fuel flow into the float bowl or EFI vapor separator tank.

Checking the Primer Bulb
◆ See Figures 7, 14 and 16

The way most outboards are rigged, fuel will evaporate from the system during periods of non-use. Also, anytime quick-connect fittings on portable tanks are removed, there is a chance that small amounts of fuel will escape and some air will make it into the fuel lines. For this reason, outboards are normally rigged with some method of priming the fuel system through a hand-operated pump (primer bulb).

When squeezed, the bulb forces fuel from inside the bulb, through the one-way check valve toward the motor filling the carburetor float bowl(s) or EFI vapor separator tank with the fuel necessary to start the motor. When the bulb is released, the one-way check valve on the opposite end (tank side of the bulb) opens under vacuum to draw fuel from the tank and refill the bulb.

When using the bulb, squeeze it gently as repetitive or forceful pumping may flood the carburetor (or overfill the EFI vapor separator tank. The bulb is operating normally if a few squeezes will cause it to become firm, meaning the float bowl/tank is full, and the float valve is closed. If the bulb collapses and does not regain its shape, the bulb must be replaced.

For the bulb to operate properly, both check valves must operate properly and the fuel lines from the check valves back to the tank or forward to the motor must be in good condition (properly sealed). To check the bulb and check valves use hand operated vacuum/pressure pump (available from most marine or automotive parts stores):

1. Remove the fuel hose from the tank and the motor, then remove the clamps for the quick-connect valves at the ends of the hose.

■ Most quick-connect valves are secured to the fuel supply hose using disposable plastic ties that must be cut and discarded for removal. If equipped, spring-type or threaded metal clamps may be reused, but be sure they are in good condition first. Do not overtighten threaded clamps and crack the valve or cut the hose.

2. Carefully remove the quick-connect valve from the motor side of the fuel line, then place the end of the line into the filler opening of the fuel tank. Gently pump the primer bulb to empty the hose into the fuel tank.

■ Be careful when removing the quick-connect valve from the fuel line as fuel will likely still be present in the hose and will escape (drain or splash) if the valve is jerked from the line. Also, make sure the primer bulb is empty of fuel before proceeding.

3. Next, remove the quick-connect valve from the tank side of the fuel line, draining any residual fuel into the tank.

■ For proper orientation during testing or installation, the primer bulb is marked with an arrow that faces the engine side check valve.

4. Securely connect the pressure pump to the hose on the tank side of the primer bulb. Using the pump, slowly apply pressure while listening for air escaping from the end of the hose that connects to the motor. If air escapes, both one-way check valves on the tank side and motor side of the prime bulb are opening.

5. If air escapes prior to the motor end of the hose, hold the bulb, check valve and hose connections under water (in a small bucket or tank). Apply additional air pressure using the pump and watch for escaping bubbles to determine what component or fitting is at fault. Repair the fitting or replace the defective hose/bulb component.

6. If no air escapes, attempt to draw a vacuum form the tank side of the primer bulb. The pump should draw and hold a vacuum without collapsing the primer bulb, indicating that the tank side check valve remained closed.

7. Securely connect the pressure pump to the hose on the motor side of the primer bulb. Using the pump, slowly apply pressure while listening for air escaping from the end of the hose that connects to the motor. This time, the check valve on the tank side of the primer bulb should remain closed, preventing air from escaping or from pressurizing the bulb. If the bulb pressurizes, the motor side check valve is allowing pressure back into the bulb, but the tank side valve is operating properly.

8. Replace the bulb and/or check valves if they operate improperly.

SERVICE

◆ See Figures 15 thru 18

Whenever work is performed on the fuel system, check all hoses for wear or damage. Replace hoses that are soft and spongy or ones that are hard and brittle. Fuel hoses should be smooth and free of surface cracks, and they should definitely not have split ends (there's a bad hair joke in there, but we won't sink that low). Do not cut the split ends of a hose and attempt to

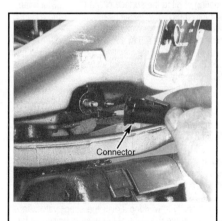

Fig. 14 Remove the fuel supply line quick-disconnect fitting from the engine to check fuel flow

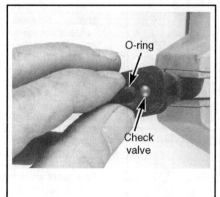

Fig. 15 Fuel quick-connector with O-ring and check valve visible

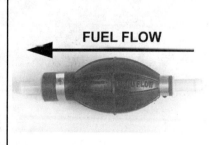

Fig. 16 The primer bulb contains an arrow that indicates the direction of fuel flow (points toward the motor)

3-10 FUEL SYSTEM

reuse it, whatever caused the split (most likely time and deterioration) will cause the new end to follow soon. Fuel hoses are safety items, don't scrimp on them, instead, replace them when necessary. If one hose is too old, check the rest, as they are likely also in need of replacement.

■ **When replacing fuel lines, make sure the inside diameter of the fuel hose and fitting is at least 5/16 in. (8mm) or larger. Also, be certain to use only marine fuel line the meets or exceeds United States Coast Guard (USCG) A1 or B1 guidelines.**

When replacing fuel lines only use Johnson/Evinrude replacement hoses or other marine fuel supply lines that meet United States Coast Guard (USCG) requirements A1 or B1 for marine applications. All lines must be of the same inner diameter as the original to prevent leakage and maintain the proper seal that is necessary for fuel system operation.

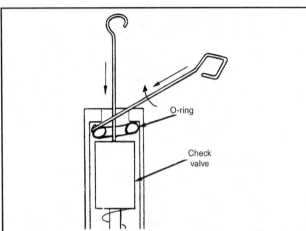

Fig. 17 Use two picks, punches or other small tool to replace quick-connect O-rings. One to push the valve and the other work the O-ring free

■ **Using a smaller fuel hose than specified could cause fuel starvation problems leading to misfiring, hesitation, rough idling and possibly even engine damage.**

The USCG ratings for fuel supply lines have to do with whether or not the lines have been testing regarding length of time it might take for them to succumb to flame (burn through) in an emergency situation. A line is "A" rated if it passes specific requirements regarding burn-through times, while "B" rated lines are not tested in this fashion. The A1 and B1 lines (normally recommended on Johnson/Evinrude applications) are capable of containing liquid fuel at all times. The A2 and B2 rated lines are designed to contain fuel vapor, but not liquid.

✱✱ CAUTION

To help prevent the possibility of significant personal injury or death, do not substitute "B" rated lines when "A" rated lines are required. Similarly, DO NOT use "A2" or "B2" lines when "A1" or "A2" lines are specified.

Various styles of fuel line clamps may be found on these motors. Many applications will simply secure lines with plastic wire ties or special plastic locking clamps. Although some of the plastic locking clamps may be released and reconnected, it is usually a good idea to replace them. Obviously wire ties are cut for removal, which requires that they be replaced.

Some applications use metal spring-type clamps, that contain tabs which are squeezed allowing the clamp to slid up the hose and over the end of the fitting so the hose can be pulled from the fitting. Threaded metal clamps are nice since they are very secure and can be reused, but do not overtighten threaded clamps as they will start to cut into the hose and they can even damage some fittings underneath the hose. Metal clamps should be replaced anytime they've lost tension (spring type clamps), are corroded, bent or otherwise damaged.

■ **The best way to ensure proper fuel fitting connection is to use the same size and style clamp that was originally installed (unless of course the "original" clamp never worked correctly, but in those cases, someone probably replaced it with the wrong type before you ever saw it).**

To avoid leaks, replace all displaced or disturbed gaskets, O-rings or seals whenever a fuel system component is removed.

On most installations, the fuel line is provided with quick-disconnect fittings at the tank and at the powerhead. If there is reason to believe the problem is at the quick-disconnects, the hose ends can be replaced as an assembly, or new O-rings may be installed. A supply of new O-rings should be carried on board for use in isolated areas where a marine store is not available (like dockside, or worse, should you need one while on the water). For a small additional expense, the entire fuel line can be replaced and eliminate this entire area as a problem source for many future seasons. (If the fuel line is replaced, keep the old one around as a spare, just in case).

If an O-ring must be replaced, use two small punches, picks or similar tools, one to push down the check valve of the connector and the other to work the O-ring out of the hole. Apply just a drop of oil into the hole of the connector. Apply a thin coating of oil to the surface of the O-ring. Pinch the O-ring together and work it into the hole while simultaneously using a punch to depress the check valve inside the connector.

The primer squeeze bulb can be replaced in a short time. A squeeze bulb assembly kit, complete with the check valves installed, may be obtained from the local Johnson/Evinrude dealer. The replacement kit will also include two tie straps to secure the bulb properly in the line.

An arrow is clearly visible on the squeeze bulb to indicate the direction of fuel flow. The squeeze bulb must be installed correctly in the line because the check valves in each end of the bulb will allow fuel to flow in only one direction. Therefore, if the squeeze bulb should be installed backwards, in a moment of haste to get the job done, fuel will not reach the carburetor or EFI vapor separator tank.

To replace the bulb, first unsnap the clamps on the hose at each end of the bulb. Next, pull the hose out of the check valves at each end of the bulb. New clamps are included with a new squeeze bulb.

If the fuel line has been exposed to considerable sunlight, it may have become hardened, causing difficulty in working it over the check valve. To remedy this situation, simply immerse the ends of the hose in boiling water for a few minutes to soften the rubber. The hose will then slip onto the check valve without further problems. After the lines on both sides have been installed, snap the clamps in place to secure the line. Check a second time to be sure the arrow is pointing in the fuel flow direction, towards the powerhead.

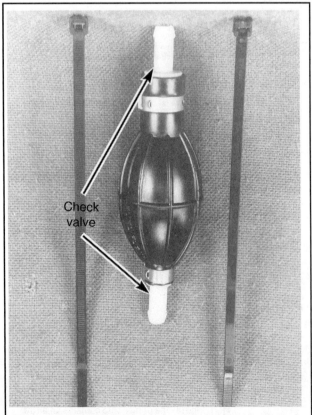

Fig. 18 A squeeze bulb kit usually includes the bulb, 2 check valves, and two tie straps

FUEL SYSTEM 3-11

CARBURETED FUEL SYSTEM

✳✳ CAUTION

If equipped, disconnect the negative battery cable ANYTIME work is performed on the engine, especially when working on the fuel system. This will help prevent the possibility of sparks during service (from accidentally grounding a hot lead or powered component). Sparks could ignite vapors or exposed fuel. Disconnecting the cable on electric start motors will also help prevent the possibility fuel spillage if an attempt is made to crank the engine while the fuel system is open.

✳✳ CAUTION

Fuel leaking from a loose, damaged or incorrectly installed hose or fitting may cause a fire or an explosion. ALWAYS pressurize the fuel system and run the motor while inspecting for leaks after servicing any component of the fuel system.

Carbureted motors covered by this manual are equipped with either one, two or three single barrel carburetors that are used to feed an air/fuel (4-stroke) or air/fuel/oil (2-stroke) mixture to the combustion chambers. Although on initial inspection some of the larger carbureted motors may look somewhat complicated, they're actually very basic, especially when compared with other modern fuel systems (such as an automotive or marine fuel injection system).

The entire system essentially consists of a fuel tank, a fuel supply line, and a mechanical fuel pump assembly mounted to the powerhead (and on some 2-stroke motors a oil/fuel mixing system) all designed to feed the carburetor with the fuel necessary to power the motor.

Cold starting is enhanced by the use of a choke plate on the smallest motors (generally speaking, tiller control models up to and including 15 horsepower), a manual primer (usually found on tiller models 18 hp and larger) or an electric fuel primer solenoid (generally all remote models).

For information on fuels, tanks and lines please refer to the sections on Fuel System Basics and Fuel Tanks and Lines.

The most important fuel system maintenance that a boat owner can perform is to provide and to stabilize fuel supplies before allowing the system to sit idle for any length of time more than a few weeks. The next most important item is to provide the system with fresh gasoline if the system has stood idle for any length of time, especially if it was without fuel system stabilizer during that time.

If a sudden increase in gas consumption is noticed, or if the engine does not perform properly, a carburetor overhaul, including cleaning or replacement of the fuel pump may be required.

Description and Operation

◆ See Figures 19 and 20

BASIC FUNCTIONS

The Role of a Carburetor

◆ See Figures 19 and 20

The carburetor is merely a metering device for mixing fuel and air in the proper proportions for efficient engine operation. At idle speed, an outboard engine requires a mixture of about 8 parts air to 1 part fuel. At high speed or under heavy duty service, the mixture may change to as much as 12 parts air to 1 part fuel.

Carburetors are wonderful devices that succeed in precise air/fuel mixture ratios based on tiny passages, needle jets or orifices and the variable vacuum that occurs as engine rpm and operating conditions vary.

Because of the tiny passages and small moving parts in a carburetor (and the need for them to work precisely to achieve exact air/fuel mixture ratios) it is important that the fuel system integrity is maintained. Introduction of water (that might lead to corrosion), debris (that could clog passages) or even the presence of unstabilized fuel that could evaporate over time can cause big problems for a carburetor. Keep in mind that when fuel evaporates it leaves behind a gummy deposit that can clog those tiny passages, preventing the carburetor (and therefore preventing the engine) from operating properly.

Float Systems

◆ See Figures 19 and 20

Ever lift the tank lid off the back of your toilet. Pretty simple stuff once you realize what's going on in there. A supply line keeps the tank full until a valve opens allowing all or some of the liquid in the tank to be drawn out through a passage. The dropping level in the tank causes a float to change position, and, as it lowers in the tank it opens a valve allowing more pressurized liquid back into the tank to raise levels again. OK, we were talking about a toilet right, well yes and no, we're also talking about the float bowl on a carburetor. The carburetor uses a more precise level, uses vacuum to draw out fuel from the bowl through a metered passage and, most importantly, store gasoline instead of water, but otherwise, they basically work in the same way.

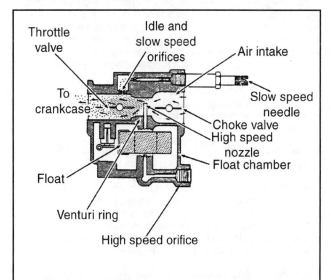

Fig. 19 Fuel flow through the venturi, showing principle and related parts controlling intake and outflow (carburetor with manual choke circuit shown)

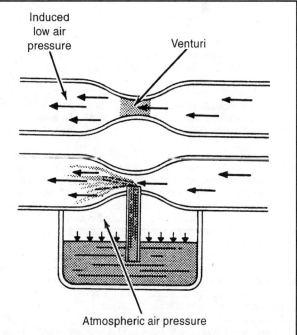

Fig. 20 Air flow principle of a modern carburetor, demonstrates how the low pressure induced behind the venturi draws fuel through the high speed nozzle

FUEL SYSTEM

A small chamber in the bottom of the carburetor serves as a fuel reservoir. A float valve admits fuel into the reservoir to replace the fuel consumed by the engine.

Fuel level in each chamber is extremely critical and must be maintained accurately. Accuracy is obtained through proper adjustment of the float. This adjustment will provide a balanced metering of fuel to each cylinder at all speeds. Improper levels will lead to engine operating problems. Too high a level can promote rich running and spark plug fouling, while excessively low float bowl fuel levels can cause lean conditions, possibly leading to engine damage.

Following the fuel through its course from carburetor float bowl to the combustion chamber of the cylinder, will provide an appreciation of exactly what is taking place. At the carburetor, fuel from the pump or fuel/oil mixing unit fuel passes through the inlet passage to the needle and seat, and then into the float chamber (reservoir). A float in the chamber rides up and down on the surface of the fuel. After fuel enters the chamber and the level rises to a predetermined point, a tang on the float closes the inlet needle and fuel entering the chamber is cutoff. When fuel leaves the chamber as the powerhead operates, the fuel level drops and the float tang allows the inlet needle to move off its seat and fuel once again enters the chamber. In this manner a constant reservoir of fuel is maintained in the chamber to satisfy the demands of the engine at all speeds.

A fuel chamber vent hole is located near the top of the carburetor body to permit atmospheric pressure to act against the fuel in each chamber. This pressure assures an adequate fuel supply to the various operating systems of the engine.

Air/Fuel Mixture

◆ See Figures 19 and 20

A suction effect is created each time the piston moves upward in the cylinder of a 2-stroke motor, or conversely, every OTHER time the piston moves downward in a 4-stroke motor (each time the piston moves downward on the intake stroke, with the intake valve open). This suction draws air through the throat of the carburetor. A restriction in the throat, called a venturi, controls air velocity and has the effect of reducing air pressure at this point.

The difference in air pressures at the throat and in the fuel chamber, causes the fuel to be pushed out metering jets extending down into the fuel chamber. When the fuel leaves the jets, it mixes with the air passing through the venturi. This air/fuel mixture should then be in the proper proportion for burning in the cylinder/s for maximum engine performance.

In order to obtain the proper air/fuel mixture for all engine speeds, high- and low-speed orifices or needle valves are installed. On most modern powerheads the high-speed needle valve has been replaced with a fixed high-speed orifice (to more discourage tampering and to help maintain proper emissions under load). There is no adjustment with the orifice type. The needle valves are used to compensate for changing atmospheric conditions. The low-speed needles, on the other hand, are still provided so that air/fuel mixture can be precisely adjusted for idle conditions other than what occurs at atmospheric sea-level. Although the low speed needle should not normally require periodic adjustment, it can be adjusted to compensate for high-altitude (river/lake) operation or to adjust for component wear within the fuel system.

Powerhead operation at sea level compared with performance at high altitudes is quite noticeable. A throttle valve controls the volume of air/fuel mixture drawn into the powerhead. A cold engine requires a richer fuel mixture to start and during the brief period it is warming to normal operating temperature. Either a choke valve is placed ahead of the metering jets and venturi to provide the extra amount of air required for start and while the engine is cold, or an enrichment system is used to provide extra fuel through additional passages.

When the choke valve is closed or the enrichment system is actuated, a very rich fuel mixture is drawn into the engine. This mixture will help wake-up a cold motor, but will quickly foul the plugs on a warm engine so it should only be used for cold starts.

The throat of the carburetor is usually referred to as the "barrel." Carburetors installed on engines covered in this manual all have a single metering jet with a single throttle and, if used, a single choke plate. Single barrel carburetors are fed by one float and chamber.

So, as far as carburetors go, these are relatively easy carburetors to understand, rebuild or adjust.

Troubleshooting the Carbureted Fuel System

COMMON PROBLEMS

The last step fuel system troubleshooting is to adjust or rebuild and then adjust the carburetor. We say it is the last step, because it is the most involved repair procedures on the fuel system and should only be performed after all other possible causes of fuel system trouble have been eliminated.

Fuel Delivery

◆ See Figures 21 and 22

Many times fuel system troubles are caused by a plugged fuel filter, a defective fuel pump, or by a leak in the line from the fuel tank to the fuel pump. Aged fuel left in the carburetor and the formation of varnish could cause the needle to stick in its seat and prevent fuel flow into the bowl. A defective choke may also cause problems. Would you believe, a majority of starting troubles, which are traced to the fuel system, are the result of an empty fuel tank or aged fuel.

If fuel delivery problems are suspected, refer to the testing procedures in Fuel Tank and lines to make sure the tank vent is working properly and that there are not leaks or restrictions that would prevent fuel from getting to the pump and/or carburetor(s).

A blocked low-pressure fuel filter causes hard starting, stalling, misfire or poor performance. Typically the engine malfunction worsens with increased engine speed. This filter prevents contaminants from reaching the low-pressure fuel pump. Models covered by this manual are usually equipped with a fuel filter screen under the pump inlet cover and/or (especially for integral tank models) on the fuel tank outlet (outlet valve or pickup tube, as applicable). Refer to the Fuel Filter in the section on Maintenance and Tune-Up for more details on checking, cleaning or replacing fuel filters.

Sour Fuel

Fuel will begin to sour in a matter of weeks, and within a couple of months, will cause engine starting problems. Therefore, leaving the motor setting idle with fuel in the carburetor, lines, or tank during the off-season, usually results in very serious problems. A fuel additive such as Sta-Bil® may be used to prevent gum from forming during storage or prolonged idle periods.

Refer to the information on Fuel System Basics in this section, specifically the procedure under Fuel entitled Checking For Stale/Contaminated Fuel will provide information on how to determine if stale fuel is present in the system. If draining the system of contaminated fuel and refilling it with fresh fuel does not make a difference in the problem, look for restrictions or other problems with the fuel delivery system. If stale fuel was left in the tank/system for a long period of time and evaporation occurred, there is a good chance that the carburetor is gummed (tiny passages are clogged by deposits left behind when the fuel evaporated). If no fuel delivery problems are found, the carburetor(s) should be removed for disassembly and cleaning.

■ **Although there are some commercially available fuel system cleaning products that are either added to the fuel mixture or sprayed into the carburetor throttle bores, the truth is that although they can provide some measure of improvement, there is not substitute for a thorough disassembly and cleaning. The more fuel which was allowed to evaporate, the more gum or varnish that may have been left behind and the more likely that only a disassembly will be able to restore proper performance.**

Choke/Enrichment Problems

◆ See Figure 23

When the engine is hot, the fuel system can cause starting problems. After a hot engine is shut down, the temperature inside the fuel bowl may rise to 200°F (94°C) and cause the fuel to actually boil. All carburetors are vented to allow this pressure to escape to the atmosphere. However, some of the fuel may percolate over the main nozzle.

If the choke should stick in the open position or the enrichment circuit (manual or electric) fail to operate, while the engine is cold, it will be hard to start. Likewise, if the choke should stick in the closed position or the enrichment circuit remains activated during normal engine operating temperatures, the engine will flood making it very difficult to start or, once started, making it buck or hesitate, especially at lower speeds.

FUEL SYSTEM 3-13

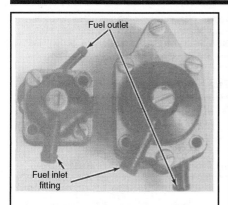

Fig. 21 Typical Johnson/Evinrude fuel pumps...

Fig. 22 ...most are equipped with filter elements under the pump inlet covers

Fig. 23 Fouled spark plug, possibly caused by over-choking or a malfunctioning enrichment circuit

In order for this raw fuel to vaporize enough to burn, considerable air must be added to lean out the mixture. Therefore, one remedy is to make sure the choke is open or the enrichment circuit is off and open the throttle to the fully open position (to allow in maximum air) and hold it there while the engine is cranked. If this doesn't work, the only remedy remaining is to remove the spark plugs and ground the leads, then crank the powerhead through about ten revolutions to blow out raw fumes. Then, clean the plugs; install the plugs again; and start the engine.

If the needle valve and seat assembly is leaking, an excessive amount of fuel may enter the intake manifold in the following manner: After the powerhead is shut down, the pressure left in the fuel line will force fuel past the leaking needle valve. This extra fuel will raise the level in the fuel bowl and cause fuel to overflow into the intake manifold.

A continuous overflow of fuel into the intake manifold may be due a defective float or overpriming the system using the primer bulb which would cause an extra high level of fuel in the bowl and overflow into the intake manifold.

Rough Engine Idle

If a powerhead does not idle smoothly, the most reasonable approach to the problem is to perform a tune-up to eliminate such areas as faulty spark plugs and timing or synchronization out of adjustment.

Other problems that can prevent an engine from running smoothly include an air leak in the intake manifold; uneven compression between the cylinders; and sticky or broken reeds on 2-stroke motors.

Of course any problem in the carburetor affecting the air/fuel mixture will also prevent the engine from operating smoothly at idle speed. These problems usually include too high a fuel level in the bowl; a heavy float; leaking needle valve and seat; defective choke or enrichment circuit; and improper idle (low-speed) needle valve adjustments.

"Sour" fuel (fuel left in a tank without a preservative additive) will cause an engine to run rough and idle with great difficulty.

As with all troubleshooting procedures, start with the easiest items to check/fix and work towards the more complicated ones.

Excessive Fuel Consumption

◆ See Figures 24, 25 and 26

Excessive fuel consumption can result from one of three conditions, or a combination of all three.
1. Inefficient engine operation.
2. Damaged condition of the hull, outdrive or propeller, including excessive marine growth.
3. Poor boating habits of the operator.

If the fuel consumption suddenly increases over what could be considered normal, then the cause can probably be attributed to the engine or boat and not the operator (unless he/she just drastically changed the manner in which the boat is operated).

Marine growth on the hull can have a very marked effect on boat performance. This is why sail boats always try to have a haul-out as close to race time as possible. While you are checking the bottom take note of the propeller condition. A bent blade or other damage will definitely cause poor boat performance.

If the hull and propeller are in good shape, then check the fuel system for possible leaks. Check the line between the fuel pump and the carburetor while the engine is running and the line between the fuel tank and the pump when the engine is not running. A leak between the tank and the pump many times will not appear when the engine is operating, because the suction created by the pump drawing fuel will not allow the fuel to leak. Once the engine is turned off and the suction no longer exists, fuel may begin to leak.

If a minor tune-up has been performed and the spark plugs and engine timing/synchronization are properly adjusted, then the problem most likely is in the carburetor, indicating an overhaul is in order. Check for leaks at the needle valve and seat. Use extra care when making any adjustments affecting the fuel consumption, such as the float level.

Engine Surge

If the engine operates as if the load on the boat is being constantly increased and decreased, even though an attempt is being made to hold a

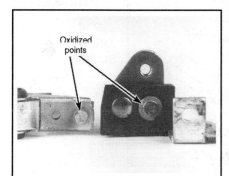

Fig. 24 If spark plugs are not to blame for poor performance, the point set used on the smallest motors may require attention

Fig. 25 Marine growth on the lower unit will create "drag" and seriously hamper boat performance

Fig. 26 A corroded hub on a small engine propeller. Hub and propeller damage will also cause poor performance

3-14 FUEL SYSTEM

constant engine speed, the problem can most likely be attributed to the fuel pump. Refer to Fuel Tank and Lines in this section for information on checking the lines for restrictions and checking fuel flow. Also, refer to Fuel Pump under Carbureted Fuel System for more information on fuel pump operation and service.

Carburetor

This section provides complete detailed procedures for removal and installation (including initial bench adjustments), overhaul (disassembly/assembly) and cleaning and inspecting, for the various carburetors installed on powerheads covered in this manual. Although there are similarities between the carburetors used on each motor, small differences from model-to-model make it best to cover them in multiple procedures, sorted by the models on which that carburetor is found.

SERVICE (REMOVAL, OVERHAUL & INSTALLATION)

■ **Good shop practice dictates a carburetor repair kit be purchased and new parts be installed any time the carburetor is disassembled.**

Make an attempt to keep the work area clean and organized. Be sure to cover parts after they have been cleaned. This practice will prevent foreign matter from entering passageways or adhering to critical parts.

Be sure to have a rag handy to catch spilled fuel, as some fuel is bound to still be present in the lines and the float bowl. Take this opportunity to closely inspect the fuel lines and replace any that are damaged or deteriorated.

During removal or overhaul procedures, always matchmark hoses or connections prior to removal to ensure proper assembly and installation. Following a rebuild a complete and the initial bench settings, perform the complete Timing and Synchronization procedure as detailed in the Maintenance and Tune-Up section.

■ **To avoid leaks, replace all displaced or disturbed gaskets, O-rings or seals whenever a fuel system component is removed. This is especially true when rebuilding a carburetor.**

2.0-3.5 Hp Motors

◆ See Figure 27

This carburetor is a single-barrel, float feed type with a manual choke. Fuel to the carburetor is normally gravity fed from a fuel tank mounted on top of the powerhead.

REMOVAL & INSTALLATION

◆ See Figures 27

1. Remove the spark plug lead to prevent accidental starting of the engine. Shut off the fuel supply at the fuel valve on the side of the engine.
2. Remove the engine covers as described under Engine Maintenance in the Maintenance and Tune-Up section.
3. Remove the knobs from the throttle and choke levers.
4. If necessary, disconnect the stop switch wiring.
5. Remove the 2 screws holding the air silencer to the carburetor, then remove the silencer from the powerhead.

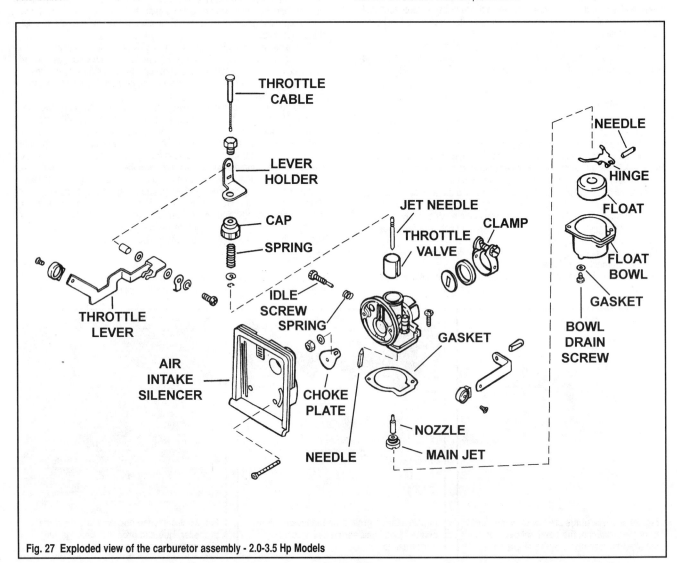

Fig. 27 Exploded view of the carburetor assembly - 2.0-3.5 Hp Models

FUEL SYSTEM 3-15

6. Place a small shop rag under the fuel line connection at the carburetor (to catch any escaping fuel still in the line), then disconnect the hose from the carburetor.

7. Loosen the carburetor clamp screw, then carefully remove the carburetor from the engine.

To install:

8. Position the new O-ring the carburetor throat, then slide the carburetor onto the mount and secure by tightening the clamp.

9. Connect the fuel tank supply hose to the carburetor and secure to the fitting using a plastic wire tie.

10. Install the air intake silencer and secure using the retaining screws.
11. Install the knobs to the throttle and choke levers.
12. If removed, connect the stop switch wiring.
13. Install the engine covers as detailed under Engine Maintenance.
14. Check engine idle speed and adjust, as necessary. For details, refer to the procedures found under Timing and Synchronization in the Maintenance and Tune-Up section.

OVERHAUL

 MODERATE

◆ See Figures 27 thru 35

1. Position the carburetor over a small drain basin, then remove the float bowl drain screw from the bottom of the carburetor and drain any fuel remaining in the bowl. Discard the drain screw gasket.

2. Remove the 2 float chamber screws, then turn the carburetor upside down and carefully lift the float chamber from the carburetor body.

3. Lift the float from the carburetor body, then remove and discard the float chamber gasket.

4. Remove the main jet and nozzle (needle jet) from the center underside of the carburetor body.

5. Count and record the number of turns required to lightly seat the idle adjustment screw. The number of turns will give a rough adjustment during installation. Back out the idle speed screw and spring from the side of the carburetor body. Normally you discard the screw, but save the spring (a new screw is usually provided in the carburetor rebuild kit to ensure a damaged screw is not used again.)

6. Remove the float pin and hinge, then remove the float valve.
7. Remove the valve seat/fuel inlet nozzle from the carburetor body.
8. Remove the throttle lever, then remove the cap (mixing chamber) from the top of the carburetor body.
9. Compress the spring, then unhook the throttle cable from the throttle valve. Withraw the jet needle and retainer from the throttle valve.

■ It is not necessary to remove the E-clip from the jet needle, unless replacement is required or if the engine is to be operated at a significantly different elevation.

10. Thoroughly clean and visually check all of the carburetor components as detailed under Cleaning & Inspection in this section.

To assemble:

11. Before starting, make sure all components are completely clean and serviceable. Compare parts from the replacement kit to the parts removed from the carburetor. With the exception of wear or damage that might occur on the old parts (requiring their replacement in the first place) the new components should be identical. If you have any questions, check with a local dealer to check parts against a current part catalog before proceeding.

■ To ensure proper operation and durability, replace all displaced or disturbed gaskets, O-rings or seals when rebuilding a carburetor regardless of their appearance.

12. If the adjustment clip was removed form the jet needle, refer to Throttle Jet Needle Adjustment in this section before proceeding.

13. Prepare the jet needle and carburetor cap assembly as follows:

    a. Reassemble the jet needle to the throttle valve, then place the jet retainer over the E-clip. Align the retainer slot with the slot in the throttle valve.

    b. Position the spring over the throttle cable, then compress the spring and slide the cable anchor through the slot and into the pocket at the bottom of the valve.

Fig. 28 Parts usually included in a carburetor repair kit.

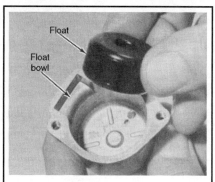

Fig. 29 View of the float bowl and float (removed from the carburetor body)

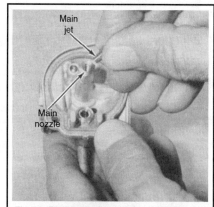

Fig. 30 Remove the main jet and nozzle from the carburetor body...

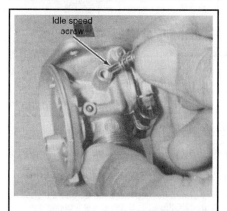

Fig. 31 ...then remove the idle speed screw and spring

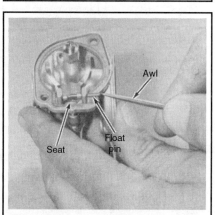

Fig. 32 Use an awl to gently remove or install the float pin

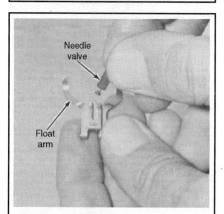

Fig. 33 The needle valve fits into a groove on the float arm

3-16 FUEL SYSTEM

c. Position the throttle valve so the slot slides over the alignment pin, then slide the assembly into the carburetor body and secure the cover.

14. Install the throttle lever, placing it in front of the holder, then position the support tab on the holder (at the hole) and tighten the retaining screw.

15. Install the valve seat/fuel inlet nozzle to the carburetor body. Do NOT over-tighten needles or jets as this will likely cause distortion and problems with air/fuel metering.

16. Carefully insert the float valve into the valve seat, then attach the float hinge and pin.

17. Install the main nozzle and jet.

18. Install the low speed adjustment screw and spring onto the side of the carburetor body. For initial adjustment, tighten the screw until just lightly seated, then back it out the same number of turns that were necessary to seat it during removal. If the number of turns was not recorded, back the screw out 1- 3/4 turns as a rough adjustment. A finer adjustment will be made with the engine in the water at the end of the installation procedures.

19. Position the new float chamber gasket on the carburetor body, then check float height adjustment by measuring the distance from the top of the gasket to the top of the float hinge. Height should be 0.090 in. (2.3mm). Carefully, bend the hinge, if necessary, to achieve the required measurement.

20. Install the float, followed by the float chamber. Tighten the float chamber screws securely.

21. Install the float bowl drain screw using a new gasket and tighten securely.

22. Adjust the idle speed as detailed under the Engine Maintenance section.

23. If spark plug fouling or excessively rich operation is noted (the motor cannot reach maximum rpm or hesitates/stumbles whenever the throttle is opened past 2/3rds, check and adjust the Throttle Jet Needle as specified in this section.

THROTTLE JET NEEDLE ADJUSTMENT

◆ See Figures 27 and 36

The carburetor on these motors uses a throttle valve that is equipped with a tapered needle to vary fuel flow with throttle opening. Adjustments are made by raising or lowering the needle in relation to the throttle valve (which is done by repositioning the clip on the needle).

If the E-clip on the jet needle is lowered, the carburetor will cause the engine to operate "rich". Raising the E-clip will cause the engine to operate "lean".

If the mixture is too rich, the engine will experience a decrease in rpm as the throttle is advanced beyond 2/3 of full throttle. If the mixture is too lean, the engine temperature will rapidly increase. Since there are no temperature monitoring devices on these small hp engines, the operator must be aware of operating temperature at all times. A lean motor will also likely spit or pop when throttle is rapidly reduced from a wide-open position to or near idle.

If the outboard is to be operated at higher altitude - raise E-clip to compensate for the lower oxygen contents found in the air at elevations.

■ The carburetor must be partially disassembled in order to access the needle and clip. Refer to the Overhaul procedure in this section for more details.

5 Hp (109cc) Motors

◆ See Figure 37

This carburetor is a single-barrel, float feed type with a manual choke. Fuel to the carburetor is normally supplied via pump from either an integral tank mounted on top of the powerhead or from a remote tank.

REMOVAL & INSTALLATION

◆ See Figure 37

1. Remove the spark plug lead to prevent accidental starting of the engine.

2. Loosen or release the clamp on the fuel line attached to the side of the carburetor.

■ How to release the fuel line clamp will vary depending on the style clamp used to retain the fuel line. Plastic ties must be cut and removed from the line while metallic clamps are loosened and slid back up the line away from the carburetor and off the fitting.

3. Place a small shop rag under the fuel line connection at the carburetor (to catch any escaping fuel still in the line), then disconnect the hose from the carburetor.

4. Loosen the set screw securing the throttle cable end to the top of the carburetor, then pull the cable free of the carburetor throttle linkage.

5. Remove the 2 carburetor retaining bolts, then move the choke lever to the full choke position and remove the carburetor. Tilt the carburetor as necessary to disconnect the choke linkage, then remove it from the motor.

6. Remove the carburetor insulator and keep for reuse, but remove and discard the gaskets.

7. Make sure the gasket mating surfaces are clean and free of all gasket traces. Be careful not to damage the surfaces when cleaning them.

To install:

8. Inspect the insulator for wear, deterioration or cracks and replace, if necessary.

■ Do not use any sealant on the carburetor insulator gaskets.

9. Position new gaskets on each side of the carburetor insulator, then install the carburetor (with the choke lever in the full choke position to

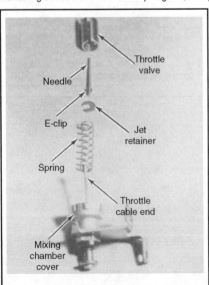

Fig. 34 Exploded view of the carburetor cap (mixing chamber) and jet needle assembly

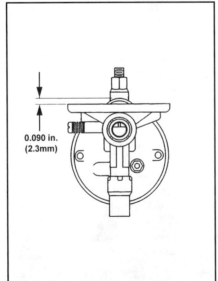

Fig. 35 The top of the float hinge must be 0.090 in. (2.3mm) above the float bowl gasket

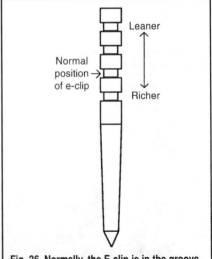

Fig. 36 Normally, the E-clip is in the groove shown. If the clip is moved up into a higher groove, the air/fuel mixture will become leaner. If the clip is moved down to a lower groove, the mixture will richen.

FUEL SYSTEM 3-17

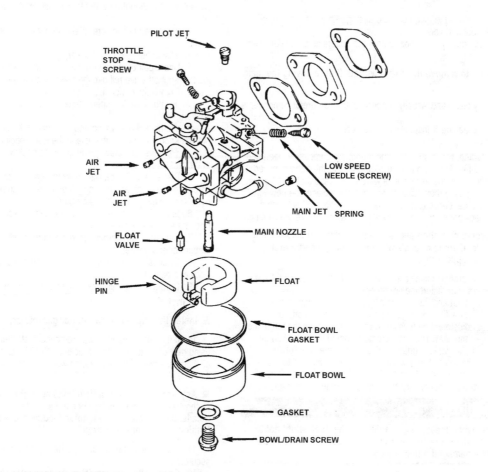

Fig. 37 Exploded view of the carburetor assembly - 5 Hp (109cc) Models

facilitate connection of the choke linkage). Tilt the carburetor as necessary to connect the choke linkage, then position the carburetor, insulator and gaskets to the motor.

10. Secure the carburetor assembly using the two retaining bolts. Tighten them securely.
11. Connect the throttle cable to the carburetor, then tighten the set screw.
12. Connect the fuel supply hose to the carburetor and secure to the fitting using the metallic clamp or a new plastic tie, as applicable.
13. Check the Carburetor Low Speed Adjustment. For details, refer to the procedures found under Timing and Synchronization in the Maintenance and Tune-Up section.

OVERHAUL

◆ See Figures 37 and 38

1. Remove the low speed needle and spring protruding at an angle from the top, side of the carburetor.
2. Remove the pilot jet mounted vertically into the top of the carburetor.
3. Use a 1/4 in. bladed flat-head screwdriver to carefully remove the air jets from the cast pockets in the front opening of the carburetor throat (venturi). Mark each of the air jets as they are removed for identification during installation.

■ Keep the two air jets separated and identified to ensure they are installed in the correct sides of the venturi. The two air jets are normally of different sizes and mixing them up can lead to significant engine performance problems.

4. Position the carburetor over a small drain basin, then remove the float bowl/drain screw from the bottom of the carburetor and drain any fuel remaining in the bowl. Discard the screw gasket.

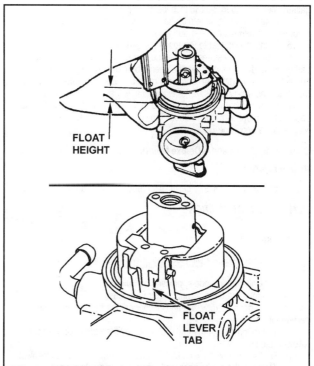

Fig. 38 Float height measurement and adjustment (adjust by bending the float lever tab)

FUEL SYSTEM

5. Remove the float bowl and gasket from the underside of the carburetor body.
6. Carefully remove the hinge pin by pressing with a small awl or pick, then remove the float and valve.
7. Remove the main jet and nozzle from the underside of the carburetor body.

■ **Do not attempt to remove the nozzle without first removing the main jet.**

8. Thoroughly clean and visually check all of the carburetor components as detailed under Cleaning & Inspection in this section.

To assemble:

9. Before starting, make sure all components are completely clean and serviceable. Compare parts from the replacement kit to the parts removed from the carburetor. With the exception of wear or damage that might occur on the old parts (requiring their replacement in the first place) the new components should be identical. If you have any questions, check with a local dealer to check parts against a current part catalog before proceeding.

■ **To ensure proper operation and durability, replace all displaced or disturbed gaskets, O-rings or seals when rebuilding a carburetor regardless of their appearance.**

10. Gently and carefully install the main nozzle and the main jet. Do not over-tighten and crack or distort the nozzle or jet.
11. Position the float and valve, then secure using the hinge pin.
12. Position the new float bowl gasket on the carburetor body, then check float height adjustment by measuring the distance from the top of the gasket to the top of the float. This measurement should be taken with the carburetor inverted, the gasket surface horizontal and the float lightly contacting the needle valve. A vernier caliper is the best method to make this measurement. Float height should be 0.047-0.055 in. (12-14mm). Carefully, bend the float lever tab, if necessary, to achieve the required measurement.

✽✽ WARNING

Be VERY careful when measuring or adjusting the float height not to force the float needle valve downward into the seat. The valve or the seat will likely be damaged if this occurs.

13. Make sure the gasket is in the carburetor body groove, then install the float bowl. Secure the bowl using the bowl/drain screw and a new gasket, then tighten to 84-138 inch lbs. (10-16 Nm).
14. Install the each air jets to the appropriate cast pocket of the carburetor venturi (as marked during removal).
15. Install the pilot jet to the top of the carburetor.
16. Install the low speed needle and spring to the top, side of the carburetor body. Gently tighten the needle by hand until it just **lightly** contacts the valve (DO NOT force the needle or it will become distorted), then back the screw out 2 1/2 turns for initial low speed adjustment.
17. Install the carburetor as directed in this section, then check the low speed adjustment as detailed under Timing and Synchronization in the Maintenance and Tune-Up section.

Colt/Junior, 3-4 Hp and 4 Deluxe Motors

◆ See Figure 39

This carburetor is a single-barrel, float feed type with a manual choke.

REMOVAL & INSTALLATION

◆ See Figure 39

1. Remove the spark plug leads to prevent accidental starting of the engine.
2. For Colt/Junior models, proceed as follows:
   a. Remove the hand rewind starter for access.
   b. If not done already, tag and disconnect the fuel line from the carburetor.
   c. At the top rear of the carburetor, remove the cam follower screw, follower and link.
3. For 3-4 hp models (except 4 Deluxe models), proceed as follows:
   a. Remove the choke knob and the stop button retainer nut.
   b. Remove the engine covers as detailed in the Engine Maintenance section.
   c. Cut the wire tie securing the fuel line to the lower cover. Be very careful when cutting the tie, not to cut, nick or otherwise damage the fuel line.
   d. Remove the two screws securing the air silencer, then remove the silencer.
   e. Disconnect the choke linkage.
   f. Use a pair of pliers or a pair of cutters to carefully grasp and remove the fuel shut-off valve roll pin, then withdraw the valve.
4. On 4 Deluxe models:
   a. Remove the low speed adjustment knob by pulling it straight outward.
   b. Remove the snapring from the rear of the choke knob shaft, then pull the knob out through the hole in the lower engine cover.
   c. Remove the screw securing the cam follower and link to the top side of the carburetor.
5. Place a small shop rag under the fuel line connection(s) at the carburetor (to catch any escaping fuel still in the line), then disconnect the hose(s) from the bottom of the carburetor.
6. Except on 4 Deluxe models, remove the screw, cam follower and link from the side of the carburetor.
7. Remove the 2 carburetor retaining nuts, then carefully remove the carburetor and gasket.
8. If necessary, remove the intake manifold bolts, then remove the leaf plate and gaskets.
9. Carefully clean all gasket mating surfaces of any remaining material.

To install:

■ **Install the intake and carburetor gaskets dry. Do not use sealer.**

10. If the intake manifold was removed, install the manifold and leaf plate using new gaskets. Tighten the bolts to 96-120 inch lbs. (10.8-13.6 Nm) on all models, except the 4 Deluxe, on which the bolts are tightened to 60-84 in. lbs. (7-9 Nm).

■ **Anytime the intake manifold bolts are tightened the ignition module-to-flywheel air gap must be checked and, if necessary, adjusted. For details, please refer to the Ignition Module procedures found in the Ignition and Electrical section.**

11. Install the carburetor using a new gasket, then tighten the nuts securely.
12. Connect the fuel hose(s) to the carburetor. Use a wire tie to secure the fuel hose(s) over the fitting(s).
13. Install the cam follower and link to the side of the carburetor using the retaining screw.
14. For 4 Deluxe models:
   a. Install the choke knob and secure using the snapring.
   b. Install the low speed adjustment knob.
15. On 3-4 hp motors (Except for 4 Deluxe models), proceed as follows:
   a. Make sure both O-rings are in place and in good condition (better yet, replace them) in the grooves on the fuel shut-off valve, then install the valve. Install the roll pin to secure the valve.
   b. Connect the choke assembly and choke shaft.
   c. Install the lower covers and the control panel, then install the choke knob and stop switch nut.
16. For Colt/Junior models, proceed as follows:
   a. Reconnect the cam follower screw, follower and link.
   b. After the necessary preliminary synchronization adjustments (which might be easier with the hand rewind starter still out of the way) reconnect the fuel line and install the hand rewind starter assembly.
17. Check the carburetor synchronization and adjustments. For details, refer to the procedures found under Timing and Synchronization in the Maintenance and Tune-Up section.

OVERHAUL

◆ See Figures 39 thru 43

1. Position the carburetor over a small drain basin, then remove the 4 float bowl screws from the underside of the carburetor body flange. Carefully separate the float bowl from the carburetor body and drain any fuel remaining in the bowl.
2. Remove the low speed packing nut, then back out and remove the needle. Remove the valve packing washers from the bore, but be carefully not to damage the threads.

FUEL SYSTEM 3-19

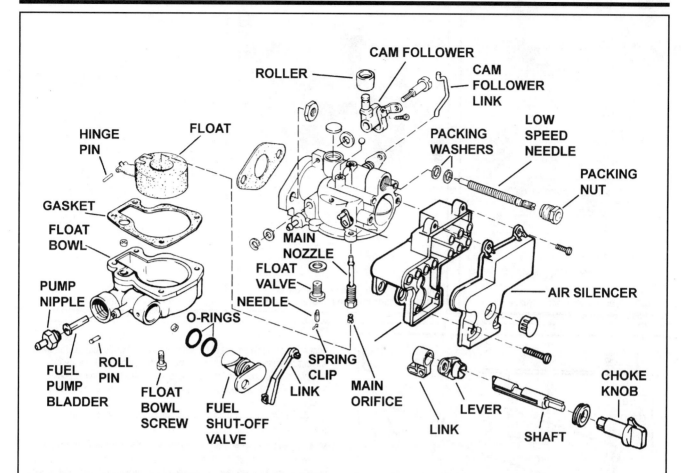

Fig. 39 Exploded view of the carburetor assembly - 3-4 Hp Models (4 Deluxe and Colt/Junior similar. Note the version found on the Colt/Junior normally has a high speed needle in place of the shut off valve in the float bowl and no fuel pump nipple/bladder)

3. Use a pair of pliers or small pair of cutters to carefully grasp and pull the roll pin from the side of the carburetor body (freeing the fuel shut-off valve). If equipped, grasp and carefully pull the shut-off valve from the carburetor, then remove and discard the O-rings.

4. If equipped, unthread the fuel pump nipple, then remove the fuel pump bladder from the bore in the side of the float bowl.

5. Remove the hinge pin using a small awl or pick, then remove the float, float valve and spring clip. Unscrew and remove the valve seat and gasket.

6. If equipped, remove the high speed orifice plug, then carefully remove the high speed nozzle. On Colt/Junior models this function is normally performed by a high speed needle installed in the float bowl bore in place of the shut off valve on other models. Like the low speed needle it is installed under a packing nut and threaded through packing and a washer.

7. Thoroughly clean and visually check all of the carburetor components as detailed under Cleaning & Inspection in this section.

To assemble:

8. Before starting, make sure all components are completely clean and serviceable. Compare parts from the replacement kit to the parts removed from the carburetor. With the exception of wear or damage that might occur on the old parts (requiring their replacement in the first place) the new components should be identical. If you have any questions, check with a local dealer to check parts against a current part catalog before proceeding.

■ **To ensure proper operation and durability, replace all displaced or disturbed gaskets, O-rings or seals when rebuilding a carburetor regardless of their appearance.**

9. If applicable, slowly install the high speed nozzle into the carburetor by hand and tighten lightly. Do not over-tighten and crack or distort the nozzle. Install the orifice in the nozzle.

10. Install the float valve seat using a new gasket, then position the float valve, float, spring clip and hinge pin.

11. Check float height adjustment using OMC float gauge (#324891). This measurement should be taken with the carburetor inverted, the gasket surface horizontal and the float valve lightly closed (by the float weight). Use the cutout marked "2 thru 6 HP" on the float gauge, the top of the float must rest between the notches on the gauge. Carefully bend the metal float arm, if necessary, to achieve the required measurement.

✱✱ WARNING

Be VERY careful when measuring or adjusting the float height not to force the float needle valve downward into the seat. The valve or the seat will likely be damaged if this occurs.

12. Invert the carburetor so the top is now facing upward and check the float drop setting (to ensure the float valve can open completely for high rpm/full throttle operation). Measure the distance from the bottom lip of the carburetor body to the lowest corner of the float. The float drop must be 1 1/8 -1 1/2 in. (28-38mm). If not, carefully bend the tab on the float arm, where it contacts the float seat.

13. Install the fuel pump bladder and nipple, as equipped.

14. If equipped, install 2 new O-rings in the grooves on the fuel shut-off valve, then coat the O-rings lightly with fresh, clean, 2-stroke engine oil. Insert the valve into the bore on the float bowl twisting slowly from side-to-side as it is inserted in order to protect the O-rings. Secure the valve in position using the roll pin.

15. Install the float bowl with a new gasket and tighten the retaining screws to 15-22 inch lbs. (1.6-2.4 Nm) using a crisscross pattern.

16. Install the cam follower linkage.

17. Position new valve packing washers into the low speed needle bore, then install packing nut, but do not tighten it against the washers at this time. Carefully thread the low speed needle into the bore and tighten slowly using light pressure until it is just lightly seated, then back it out 1 turn to the initial low speed setting (1 1/4 turns on Colt/Junior models).

3-20 FUEL SYSTEM

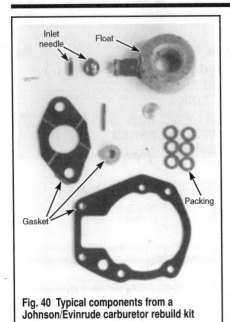

Fig. 40 Typical components from a Johnson/Evinrude carburetor rebuild kit

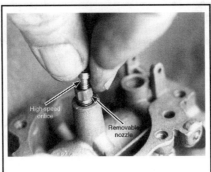

Fig. 41 The main (high speed) orifice usually threads into the removable main nozzle

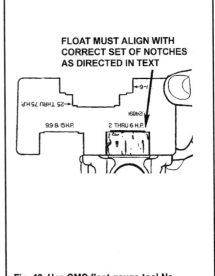

Fig. 42 Use OMC float gauge tool No. 324891 to check float height

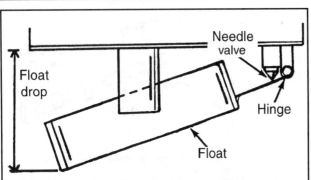

Fig. 43 Check the float drop dimension to prevent fuel starvation at high rpm

✱✱ WARNING

Use extreme care when threaded needles into the carburetor passages. Tightening them against their valve seats with anything more than extremely light force will likely distort or damage the valve and/or the valve seat.

18. Gently tighten the packing nut against the packing washers until the needle can just be turned by finger pressure.
19. On Colt/Junior models, position new valve packing and washers into the high speed needle bore, then install packing nut, but do not tighten it against the washers at this time. Carefully thread the high speed needle into the bore and tighten slowly using light pressure until it is just lightly seated, then back it out 1 turn to the initial high speed setting.
20. Install the carburetor, then check and adjust the Timing and Synchronization settings.

5/6 Hp 4-Stroke Motors

◆ See Figure 44

This carburetor is a single-barrel, float feed type with an accelerator pump and a manual choke.

REMOVAL & INSTALLATION

◆ See Figures 44 and 45

1. Remove the spark plug leads to prevent accidental starting of the engine.

2. Remove the manual starter and/or the lower engine covers, if necessary for additional access. Although the carburetor can be removed without removing either of these components completely from the motor, removal makes access a lot easier and it is probably worth the additional time. Refer to the procedures under Manual Starter and Engine Covers, as necessary.
3. Remove the air horn by disconnecting the hose, then removing the retaining screws.

■ Most hoses are secured by plastic wire ties on these models, for removal carefully cut the ties using a pair of cutters or dikes. Be careful when cutting the ties not to cut, nick or otherwise damage the hoses.

4. Disconnect the hoses from the carburetor water and/or fuel fittings, as applicable. Have a rag or small drain basin handy to drain residual fuel from the carburetor fuel hose.
5. If necessary for clearance, remove the oil pressure sensor wire from the sending unit.

■ Access to the carburetor flange nuts is tight on some models. Either use a short, thin wrench or the OMC carburetor flange nut removal tool (#342211).

6. Remove the carburetor flange nuts, then remove the carburetor, gaskets and insulator. As the carburetor assembly is separated from the motor, carefully disconnect the carburetor and throttle linkage.
7. Separate the insulator from the carburetor, then carefully clean all gasket mating surfaces of any remaining material.

To install:

■ Install the intake and carburetor gaskets dry. Do not use sealer.

8. Position a new gasket on either side of the insulator, then position the insulator to the intake manifold.
9. Connect the throttle and choke linkage as the carburetor is positioned onto the insulator and carburetor retaining studs, then install the carburetor to the intake manifold and tighten the nuts securely.
10. If removed, reconnect the oil pressure sensor wiring.
11. Install the air horn and tighten the screws securely.
12. Connect the air horn, carburetor fuel and carburetor water hoses, as applicable, then secure using new wire ties.
13. If removed, install the manual starter assembly and/or the lower engine covers.
14. Gently squeeze the primer bulb while checking for fuel leakage. Correct any fuel leaks before returning the engine to service.
15. Perform the necessary timing and synchronization procedures from the Maintenance and Tune-Up Section. If the carburetor was repaired or rebuilt, be sure to perform the initial low speed adjustment procedure

FUEL SYSTEM 3-21

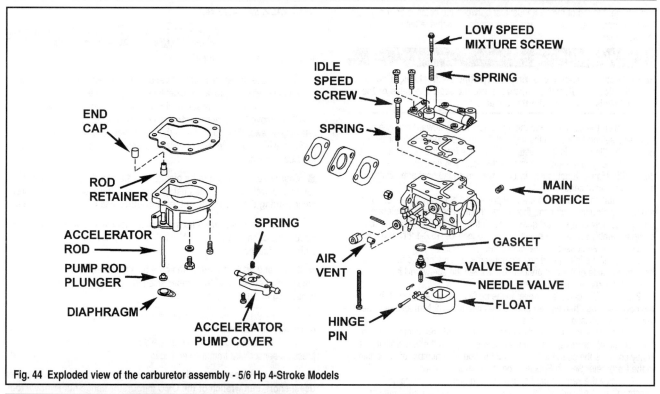

Fig. 44 Exploded view of the carburetor assembly - 5/6 Hp 4-Stroke Models

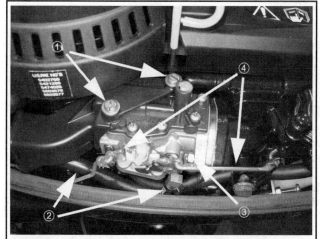

Fig. 45 Disconnect carburetor components in the following order (1) air horn, (2) hoses, (3) retaining nuts and (4) linkage - 5/6 hp 4-stroke motors (note most but not all connections visible)

OVERHAUL

◆ See Figures 42, 43 and 44

1. Position the carburetor over a small drain basin, then remove the float bowl drain screw from the bottom of the carburetor and drain any fuel remaining in the bowl. Discard the drain screw gasket.
2. Remove the low speed (idle mixture) screw and spring from the small tower in the top of the carburetor cover.
3. Remove the 7 carburetor cover screws, then lift the cover and gasket from the carburetor.
4. Invert the carburetor, then remove the 8 float bowl and accelerator pump housing screws (6 short and 2 long, keep track of the short screw locations for installation purposes).
5. Lift the float bowl and gasket from the underside of the carburetor body. Remove the accelerator rod, rod retainer and the end cap from the accelerator pump body on the float bowl.

6. Using a small pick or awl, carefully remove the float hinge pin, then remove the float and float valve from the carburetor body. Remove the inlet valve seat and the gasket.

■ Use OMC #317002 or an equivalent orifice plug screwdriver to remove the main orifice (high speed jet) from these carburetors.

7. Remove the main orifice (high speed jet) from the side of the carburetor body.
8. Remove the idle speed screw and spring mounted vertically in the boss on the side of the carburetor body.
9. Remove the air vent from the side of the carburetor body.
10. Remove the accelerator pump cover from the bottom of the float bowl, then remove the spring, diaphragm and pump rod plunger.
11. Thoroughly clean and visually check all of the carburetor components as detailed under Cleaning & Inspection in this section.

To assemble:

12. Before starting, make sure all components are completely clean and serviceable. Compare parts from the replacement kit (especially the gaskets) to the parts removed from the carburetor. With the exception of wear or damage that might occur on the old parts (requiring their replacement in the first place) the new components should be identical. If you have any questions, check with a local dealer to check parts against a current part catalog before proceeding.

■ To ensure proper operation and durability, replace all displaced or disturbed gaskets, O-rings or seals when rebuilding a carburetor regardless of their appearance.

13. Install the accelerator rod plunger, diaphragm and spring to the float bowl with the accelerator pump cover. Secure using the screw(s).
14. Press the air vent carefully into the side of the carburetor body, then carefully thread the main orifice (high speed jet) into place on the opposite side of the carburetor body.
15. Install the float valve seat using a new gasket. Carefully insert the float valve into the valve seat, while installing the float using the hinge pin.

■ When using the OMC float gauge on these models, DO NOT use the cutout marked "2 THRU 6 HP" as it is intended for 2-stroke models only.

16. Check float height adjustment using OMC float gauge (No. 324891). This measurement should be taken with the carburetor inverted, the gasket surface horizontal and the float valve lightly closed (by the float weight). Contrary to common sense, use the cutout marked "9.9 & 15 HP" on the

3-22 FUEL SYSTEM

float gauge, the top of the float must rest between the notches on the gauge. Carefully bend the metal float arm, if necessary, to achieve the required measurement.

⚠ WARNING

Be VERY careful when measuring or adjusting the float height not to force the float needle valve downward into the seat. The valve or the seat will likely be damaged if this occurs.

17. Invert the carburetor so the top is now facing upward and check the float drop setting (to ensure the float valve can open completely for high rpm/full throttle operation. Measure the distance from the bottom lip of the carburetor body to the lowest corner of the float. The float drop must be 1-1 3/8 in. (25-35mm). If not, carefully bend the tab on the float arm, where it contacts the float seat.

18. Install the end cap so it will be positioned between the accelerator rod and the throttle shaft cam, then install the float bowl and accelerator pump assembly using a new gasket. Install the float bowl and accelerator pump assembly screws, then tighten to 8-10 inch lbs. (0.8-1.2 Nm) following the sequence embossed on the float bowl.

19. Install the idle speed adjustment screw and spring into the vertical boss on the side of the carburetor body. Thread the screw until it **just** contacts the idle adjustment lever.

20. Install the carburetor cover to the body using a new gasket. Install the cover screws, then tighten to 8-10 inch lbs. (0.8-1.2 Nm) following the sequence embossed on the cover.

21. Install the low speed (idle) mixture screw and spring into the round boss on top of the cover. Thread the screw slowly into the bore until it just lightly contacts the seat, then back it off the specified number of turns listed in the Carburetor Set-Up Specifications table in this section.

22. Install the carburetor, then perform the necessary timing and synchronization procedures from the Maintenance and Tune-Up Section (including the idle speed and mixture adjustments).

8/9.9 and 9.9/15 Hp 4-Stroke Motors

◆ See Figures 46 and 47

The carburetors found on these models are single-barrel, float feed types. The two versions found on 15 hp models, are equipped with accelerator pumps.

REMOVAL & INSTALLATION

◆ See Figures 46, 47 and 48

1. Remove the spark plug leads and/or disconnect the negative battery cable (if equipped) to prevent accidental starting of the engine.
2. Remove the manual starter, if necessary for additional access.
3. Remove the air horn by disconnecting the hose, then removing the retaining screws. (For 8/9.9 motors there is at least one screw on top of the housing, while 9.9/15 hp motors utilize one screw on top and one screw on the side of the carburetor.)

■ Most hoses are secured by plastic wire ties on these models, for removal carefully cut the ties using a pair of cutters or dikes. Be careful when cutting the ties not to cut, nick or otherwise damage the hoses.

4. Disconnect the hoses from the carburetor fuel and water fittings. Have a rag or small drain basin handy to drain residual fuel from the carburetor fuel hose.

■ On most 9.9/15 hp models the choke cable clamp is retained by a Torx® head screw. Use the proper sized Torx® head driver when loosening or tightening the screw.

5. On 9.9/15 hp motors, remove the choke cable clamp screw, then disengage the cable from the link.
6. If necessary for clearance on 8/9.9 hp motors, remove the oil pressure sensor wire from the sending unit.

■ Access to the carburetor flange nuts is tight on some models. Either use a short, thin wrench or the OMC carburetor flange nut removal tool (#342211).

7. Remove the carburetor flange nuts, then remove the carburetor, gaskets and insulator. On 8/9.9 hp motors, carefully disconnect the carburetor and throttle linkage as the carburetor assembly is separated from the motor.
8. Separate the insulator from the carburetor, then carefully clean all gasket mating surfaces of any remaining material.

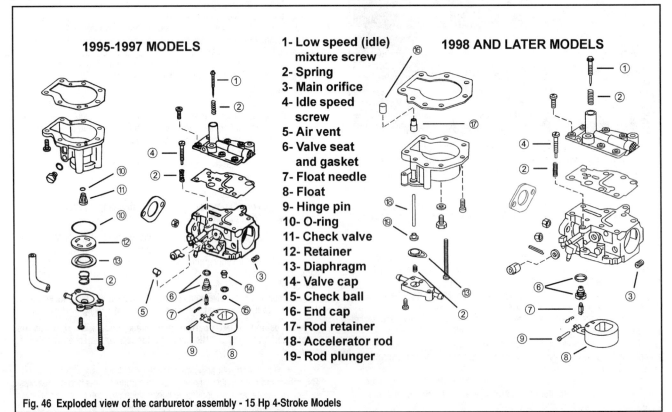

Fig. 46 Exploded view of the carburetor assembly - 15 Hp 4-Stroke Models

FUEL SYSTEM 3-23

Fig. 48 Remove the choke clamp screw (1), then disconnect the choke link (2) from the carburetor - 9.9/15 Hp 4-Stroke Models

To install:

■ Install the intake and carburetor gaskets dry. Do not use sealer.

9. Position a new gasket on either side of the insulator, then position the insulator to the intake manifold.

■ On 8/9.9 hp motors, connect the throttle and choke linkage as the carburetor is positioned over the studs.

10. Position the carburetor over the retaining studs and against the insulator (connecting the throttle and choke linkage in the process on 8/9.9 hp motors).
11. Install and securely tighten the carburetor retaining nuts.
12. For 9.9/15 hp motors, install the choke cable in the link, then secure the cable to the powerhead using the clamp and tighten the bolt to 60-84 inch lbs. (7-9 Nm).
13. Install the air horn and tighten the retaining screw(s).

14. Connect the hoses and secure using new wire ties
15. If removed, reconnect the oil pressure sensor wiring.
16. If removed, install the manual starter assembly.
17. Gently squeeze the primer bulb while checking for fuel leakage. Correct any fuel leaks before returning the engine to service.
18. Perform the necessary timing and synchronization procedures from the Maintenance and Tune-Up Section. If the carburetor was repaired or rebuilt, be sure to perform the initial low speed adjustment procedure.

OVERHAUL

◆ See Figures 42, 43 and 49

1. Position the carburetor over a small drain basin, then remove the float bowl drain screw from the bottom or side of the carburetor float bowl (as applicable) and drain any fuel remaining in the bowl. Discard the drain screw gasket.
2. Remove the low speed (idle) mixture screw and spring from the round boss on top of the carburetor.
3. Remove the 7 screws securing the carburetor cover to the body, then lift the cover and gasket from the carburetor.

■ Most models covered here utilize 7 float bowl screws, except 1998 and later 15 hp models which use 6 screws. Also, the 1998 and later 15 hp models utilize both short and long screws, so pay attention to screw placement in order to ease assembly.

4. Invert the carburetor, then remove the float bowl (and, if applicable, accelerator pump housing) screws. Separate the float bowl from the carburetor body, then remove the gasket.
5. On 15 hp models:
   a. For motors through 1997, remove the check ball, gasket and check valve cap from the carburetor body.
   b. For 1998 and later motors, remove the accelerator rod, rod retainer and the end cap from the accelerator pump body on the float bowl.
6. Use a small pick or awl to remove the hinge pin, then carefully remove the float and float valve from the bottom of the carburetor. Remove the float valve and gasket.

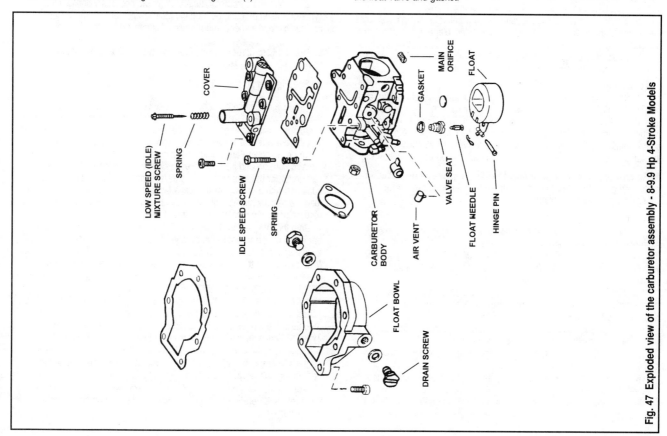

Fig. 47 Exploded view of the carburetor assembly - 8-9.9 Hp 4-Stroke Models

3-24 FUEL SYSTEM

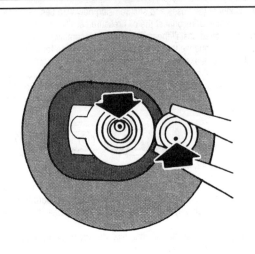

Fig. 49 During assembly, position the valve cap with the orifice hole 180° away from the idle fuel pickup tube - 15 Hp motors through 1997

■ Use OMC #317002 or an equivalent orifice plug screwdriver to remove the main orifice (high speed jet) from these carburetors.

7. Remove the main orifice (high speed jet) from the side of the carburetor body.
8. Remove the idle speed screw and spring mounted vertically in the boss on the side of the carburetor body.
9. On all motors, except for 1998 and later 15 hp motors, remove the air vent from the side of the carburetor body.
10. On 15 hp models:
   a. For motors through 1997, loosen the bolts, then remove the cover from the underside of the float bowl. Remove the spring, diaphragm, retainer, large O-ring, check valve assembly and small O-ring.
   b. For 1998 and later motors, remove the accelerator pump cover from the bottom of the float bowl, then remove the spring, diaphragm and pump rod plunger.
11. Thoroughly clean and visually check all of the carburetor components as detailed under Cleaning & Inspection in this section.

To assemble:

12. Before starting, make sure all components are completely clean and serviceable. Compare parts from the replacement kit to the parts removed from the carburetor. With the exception of wear or damage that might occur on the old parts (requiring their replacement in the first place) the new components should be identical. If you have any questions, check with a local dealer to check parts against a current part catalog before proceeding.

■ To ensure proper operation and durability, replace all displaced or disturbed gaskets, O-rings or seals when rebuilding a carburetor regardless of their appearance.

13. On 15 hp motors:
   a. For motors through 1997, install the small O-ring, check valve assembly, large O-ring, retainer, diaphragm and spring into the bottom of the float bowl, then secure using the cover. Make sure the retainer is positioned with the flat side against the check valve assembly. The diaphragm is then positioned so the outer rib fits in the cover groove.
   b. For 1998 and later motors, install the accelerator rod plunger, diaphragm and spring to the float bowl with the accelerator pump cover. Secure using the screw(s).
14. On all models, except for 1998 and later 15 hp motors, carefully press the air vent into the side of the carburetor body.
15. Carefully thread the main orifice (high speed jet) into the carburetor body.
16. Install the float valve seat using a new gasket. Carefully insert the float valve into the valve seat, while installing the float using the hinge pin.
17. Check float height adjustment using OMC float gauge (No. 324891). This measurement should be taken with the carburetor inverted, the gasket surface horizontal and the float valve lightly closed (by the float weight). On all 8-15 hp 4-strokes, use the cutout marked "9.9 & 15 HP" on the float gauge, the top of the float must rest between the notches on the gauge. Carefully bend the metal float arm, if necessary, to achieve the required measurement.

✱✱ WARNING

Be VERY careful when measuring or adjusting the float height not to force the float needle valve downward into the seat. The valve or the seat will likely be damaged if this occurs.

18. Invert the carburetor so the top is now facing upward and check the float drop setting (to ensure the float valve can open completely for high rpm/full throttle operation). Measure the distance from the bottom lip of the carburetor body to the lowest corner of the float. The float drop must be 1-1 3/8 in. (25-35mm). If not, carefully bend the tab on the float arm, where it contacts the float seat.
19. On 15 hp motors:
   a. For motors through 1997, install the check ball, gasket and valve cap. The cap must be positioned with the orifice hole 180° away from the idle fuel pickup tube.
   b. For 1998 and later motors, install the end cap so it will be positioned between the accelerator rod and the throttle shaft cam when the float bowl and accelerator pump cover assembly is installed to the carburetor body.
20. Position the float bowl to the carburetor body using a new gasket. Install and tighten the screws to 8-10 inch lbs. (0.8-1.2 Nm) following the sequence embossed on the float bowl.
21. Install the idle speed adjustment screw and spring into the vertical boss on the side of the carburetor body. Thread the screw until it **just** contacts the idle adjustment lever.
22. Install the cover to the carburetor body using a new gasket. Install the cover screws, then tighten to 8-10 inch lbs. (0.8-1.2 Nm) following the sequence embossed on the cover.
23. Install the low speed (idle) mixture screw and spring into the round boss on top of the cover. Thread the screw slowly into the bore until it just lightly contacts the seat, then back it off the specified number of turns listed in the Carburetor Set-Up Specifications table in this section.
24. Install the carburetor, then perform the necessary timing and synchronization procedures from the Maintenance and Tune-Up Section (including the idle speed and mixture adjustments).

9.9/10/14/15 Hp (216cc/255cc) Motors Thru 1993 and All 18 Jet Thru 35 Hp (521cc) Motors

◆ See Figures 50 thru 53

A few slightly different carburetors may be found on these models over the years covered here, but they generally fall into 2 categories (with slight variances available on each of the categories). Both carburetors were used on different versions of these models across the covered year span. If the motor on which you are working contains a one-piece carburetor body with a metallic housing and integral cover, we'll refer to it here as a Type "A" carburetor (or the less common Type "A" Variance). Some models are equipped with a carburetor that uses a separate (removable) cover that is made of a black plastic material, we'll refer to that model as a Type "B" carburetor (or the less common Type "B" Variance).

Both the Type A and B carburetors used on these models are single-barrel, float feed types. Neither model is equipped with an idle speed screw, as idle speed is controlled through positioning of the throttle cable bracket (tiller models) or the linkage throttle arm stop screw (remote models).

Main fuel metering is accomplished by a removable main orifice (main jet) USUALLY mounted in the float chamber, directly behind the main fuel inlet nipple (though the Type "B" Variance mounts it in a nozzle well instead). Low speed (idle) mixture is adjusted using a low speed screw (needle) that mounts in the top, side of the carburetor body on both types (except the Type "A" Variance, where it mounts in the front of the carburetor body).

Since all of these motors should be equipped with electric or manual primer's to assist in cold starts, a fuel primer nipple is also mounted to the carburetor, on the opposite side as the idle mixture screw.

Most service and adjustment procedures are identical on Type A and B carburetors. Differences for the carb types or their variances are noted in text where necessary.

FUEL SYSTEM 3-25

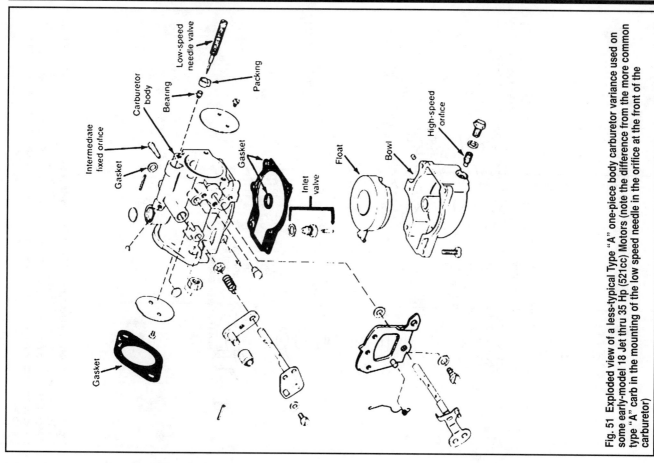

Fig. 51 Exploded view of a less-typical Type "A" one-piece body carburetor variance used on some early-model 18 Jet thru 35 Hp (521cc) Motors (note the difference from the more common type "A" carb in the mounting of the low speed needle in the orifice at the front of the carburetor)

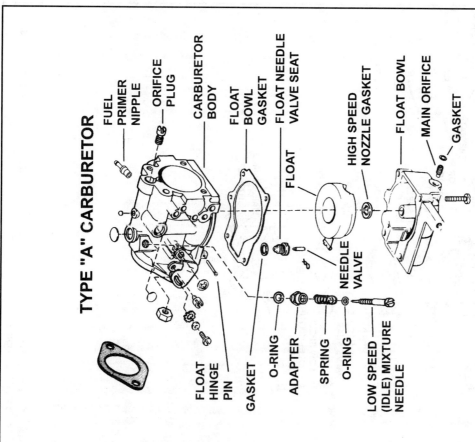

Fig. 50 Exploded view of a typical Type "A" one-piece body carburetor used on many 18 Jet thru 35 Hp (521cc) Motors

3-26 FUEL SYSTEM

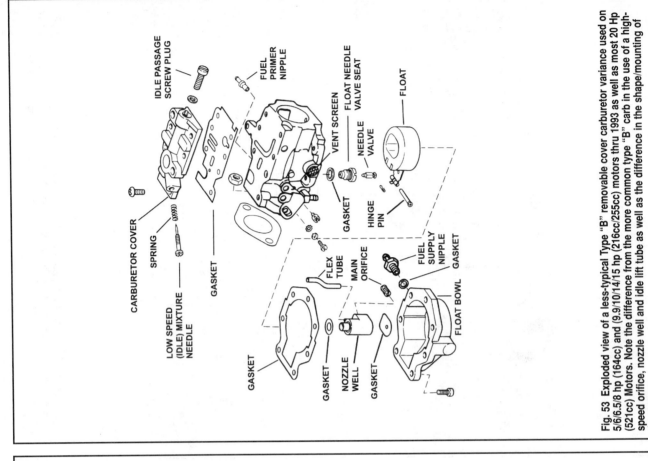

Fig. 53 Exploded view of a less-typical Type "B" removable cover carburetor variance used on 5/6/6.5/8 hp (164cc) and (9.9/10/14/15 hp (216cc/255cc) motors thru 1993 as well as most 20 Hp (521cc) Motors. Note the difference from the more common type "B" carb in the use of a high-speed orifice, nozzle well and idle lift tube as well as the difference in the shape/mounting of the cover

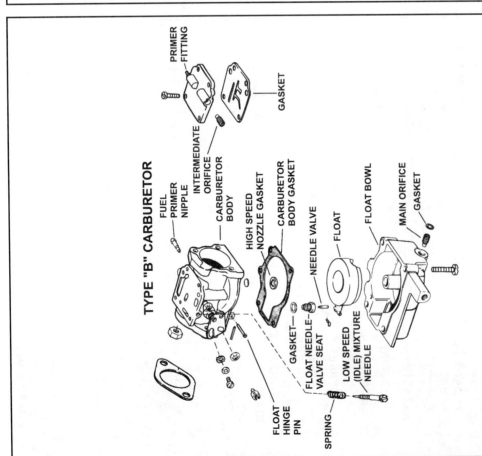

Fig. 52 Exploded view of a typical Type "B" removable cover carburetor used on some 18 Jet thru 35 Hp (521cc) Motors

FUEL SYSTEM

REMOVAL & INSTALLATION

◆ See Figures 50 thru 53

1. Remove the spark plug leads and/or disconnect the negative battery cable (if equipped) to prevent accidental starting of the engine.
2. If equipped, remove the electric starter bracket for access.
3. For 9.9/10/14/15 hp models, proceed as follows:
   a. Remove the manual starter for access. On 216cc models, hold the starter pinion securely against the starter base to keep the recoil spring retained, then loosen the mounting screw and remove the assembly with the screw. Position the assembly on aide on 216cc models and install a 3/8-16 nut on the starter mounting screw.
   b. For 216cc models, remove the low speed adjustment knob.
   c. Loosen the 4 (216cc) or 2 (255cc) retaining screws and remove the air intake silencer cover. On 216cc models you'll have to tilt the base in order to get it past the low speed needle.
   d. Remove the choke knob detent and chock knob (216cc models) or the choke lever (255cc models).
   e. Remove the cam follower screw and washer (216cc models) or cam follower O-ring (255cc models).
4. For 18 Jet and larger models, proceed as follows:
   a. Carefully push the throttle link out of the cam follower.

■ Most hoses are secured by plastic wire ties on these models, for removal carefully cut the ties using a pair of cutters or dikes. Be careful when cutting the ties not to cut, nick or otherwise damage the hoses.

   b. Disconnect the fuel hose(s) from the carburetor. On 18 Jet and larger models this includes the primer fuel hose from the nipple on the side of the carburetor body.
5. Remove the carburetor flange nuts, then remove the carburetor and gasket from the intake manifold.
6. Turn the carburetor slightly for access, then use a pair of cutters to carefully remove the wire tie from the carburetor fuel supply line. Position a rag or small drain basin to catch any fuel left in the line, then disconnect the line from the carburetor and remove the carburetor completely from the motor.
7. Carefully clean the gasket mating surfaces of any remaining material.

To install:

■ Install the carburetor gasket dry. Do not use sealer.

8. Position a new gasket over the intake manifold studs.
9. Connect the fuel inlet line to the carburetor, then secure using a new wire tie.
10. Position the carburetor over the intake manifold studs and tighten the flange nuts securely.
11. For 18 Jet and larger models, proceed as follows:
    a. Connect the throttle link to the cam follower.
    b. Install the primer hose to the nipple on the upper side of the carburetor, then secure using a new wire tie.
12. For 9.9/10/14/15 hp models, proceed as follows:
    a. Install the cam follower screw and washer (216cc models) or cam follower O-ring (255cc models).
    b. Install the choke knob detent and chock knob (216cc models) or the choke lever (255cc models).
    c. Install the air intake silencer cover.
    d. For 216cc models, install the low speed adjustment knob.
    e. Install the manual starter assembly.
13. Gently squeeze the primer bulb while checking for fuel leakage. Correct any fuel leaks before returning the engine to service.
14. Perform the necessary timing and synchronization procedures from the Maintenance and Tune-Up Section. If the carburetor was repaired or rebuilt, be sure to perform the carburetor initial low speed setting procedure.

OVERHAUL

◆ See Figures 42, 43 and 50 thru 62

■ Because of design similarities, this overhaul procedure is also applicable to the carburetors used on 5/6/6.5/8 hp (164cc) models through 1993.

1. Position the carburetor over a small drain basin, then remove the main fuel inlet nipple, followed for most models (except the variance of Type B carburetors) by the main orifice (high speed jet) from the float bowl. Drain any fuel remaining from the bowl through the opening.
2. If equipped, remove the fuel primer nipple from the upper side of the carburetor body.
3. Remove either the intermediate air bleed orifice from the cover of Type B carburetors, the orifice plug from the top front of Type A carburetors, or the low speed needle and packing from the top front of the Type A variance carburetors.
4. Invert the carburetor, then remove the float bowl retaining screws and carefully separate the bowl from the bottom of the carburetor body. Remove the gasket and all traces of gasket material from the mating surfaces.
5. Using an awl or small pick, carefully remove the hinge pin, then remove the float, float needle from the bottom of the carburetor body. Unthread the needle valve seat and remove the gasket.
6. For Type B variance carburetors CAREFULLY cut the idle lift tube away from the 2 nipples and discard (a new tube will be used during installation), then remove the high speed orifice and nozzle well. Remove the lower nozzle well gasket (which sits between the well and float bowl).
7. Remove the high speed nozzle well gasket from the center underside of the carburetor body.
8. If not done already (such as on Type A variance carburetors), remove the low speed needle and spring. On Type A carburetors, remove the O-rings and adapter as well.
9. On Type B carburetors, remove the 4 screws (5 or 7 screws on the variance carb) retaining the cover to the carburetor body, then remove the cover and gasket. Carefully remove all traces of gasket material from the mating surfaces.
10. Thoroughly clean and visually check all of the carburetor components as detailed under Cleaning & Inspection in this section.

To assemble:

11. Before starting, make sure all components are completely clean and serviceable. Compare parts from the replacement kit to the parts removed from the carburetor. With the exception of wear or damage that might occur on the old parts (requiring their replacement in the first place) the new components should be identical. If you have any questions, check with a local dealer to check parts against a current part catalog before proceeding.

■ To ensure proper operation and durability, replace all displaced or disturbed gaskets, O-rings or seals when rebuilding a carburetor regardless of their appearance.

Fig. 54 On models so equipped, unthread the intermediate orifice. .

Fig. 55 . . .and remove it from the carburetor cover

Fig. 56 Remove the carburetor float bowl . . .

3-28 FUEL SYSTEM

Fig. 57 . . . and gasket for access to the float components

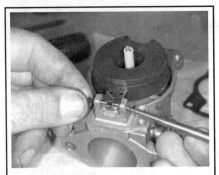

Fig. 58 Use a small pick or awl to push gently on the hinge pin. . .

Fig. 59 . . . then remove the pin, freeing the float and needle valve

Fig. 60 With the float and needle removed, unthread the valve seat

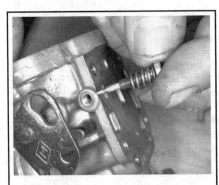

Fig. 61 Unthread the low speed needle and spring

Fig. 62 If equipped, remove the carburetor cover and gasket

12. Starting with the carbureted inverted on the work surfaces, install a new high speed nozzle gasket.

13. For Type B variance carburetors, install the high speed orifice into the nozzle well, then position the nozzle well and well-to-carburetor body gasket onto the carburetor. Connect the nozzle well to the carburetor using a NEW idle lift tube.

14. Install the inlet valve seat using a new gasket. Tighten the valve seat lightly, do not over-tighten and crack or distort it.

15. Carefully install the float and needle valve assembly, using the hinge pin to secure them.

16. Check float height adjustment using OMC float gauge (#324891) as follows:

   a. For 18J and 25 hp or larger motors this measurement should be taken with the carburetor inverted, the gasket surface horizontal and the float valve lightly closed (by the float weight). Use the side of the gauge marked "V-4 & V-6" on one side of the cutout and marked "25 THRU 75 HP" on the other side of the cutout.

   b. For 20 hp and smaller models only this measurement should be taken with the part of the gauge cutout marked "9.9 & 15 HP" or the portion of the cutout marked "2 THRU 6 HP" on the same side.

17. The top of the float must rest between the notches on the gauge. Carefully bend the metal float arm, if necessary, to achieve the required measurement.

✳✳ WARNING

Be VERY careful when measuring or adjusting the float height not to force the float needle valve downward into the seat. The valve or the seat will likely be damaged if this occurs.

18. Invert the carburetor so the top is now facing upward and check the float drop setting (to ensure the float valve can open completely for high rpm/full throttle operation). Measure the distance from the bottom lip of the carburetor body to the lowest corner of the float. The float drop must be 1 1/8 -1 5/8 in. (28-41mm) for 18J and 25 hp or larger models or 1 - 1 1/3/8 in. (25-35mm) for all 20 hp and smaller models. If not, carefully bend the tab on the float arm, where it contacts the float seat.

19. On all except Type B variance carburetors (for which you would have already installed the main orifice), install the main orifice (high speed jet), gasket and fuel nipple to the float bowl. Do not overtighten and distort the main orifice.

20. Apply a coating of OMC Locquic Primer and OMC screw lock or equivalent threadlocking compound to the 4 (or 5, as applicable) float bowl retaining screws. Position a new gasket on the underside of the carburetor body, then install the float bowl and tighten the screws to 25-35 inch lbs. (2.8-4.0 Nm) for all 25 hp and larger motors or to 8-10 inch lbs. (0.8-1.2 Nm) for 20 hp and smaller motors using a crisscross pattern.

21. Install the intermediate air bleed orifice (Type B), orifice plug (Type A, non variance), as applicable.

22. On Type A carburetors as applicable, install the low speed needle adapter using a new O-ring.

23. Install the low speed needle and spring or packing, as applicable. On Type A carburetors (non-variance), make sure a new O-ring is installed on the needle. Thread the screw into the bore slowly until it just lightly contacts the seat, then back it the correct number of turns as listed under Initial Low Speed Setting in the Carburetor Specifications chart in this section as a starting point for the initial low speed setting.

24. On Type B carburetors, install the carburetor cover to the body using a new gasket. Tighten the retaining screws to 15-22 inch lbs. (1.6-2.4 Nm) for all 25 hp and larger motors or to 8-10 inch lbs. (0.8-1.2 Nm) for 20 hp and smaller motors using a crisscross pattern.

25. Install the carburetor, then perform the necessary timing and synchronization procedures from the Maintenance and Tune-Up Section (including the idle speed and mixture adjustments).

25/35 Hp (500/565cc) Motors

◆ See Figure 63

The carburetors used on these models are single-barrel, float feed types. They are unitized, with a removable float bowl and carburetor cover. Main fuel metering is accomplished by a removable main orifice (main jet) mounted in the float chamber, directly behind the main fuel inlet nipple or drain plug. Low speed (idle) mixture is adjusted using a low speed screw (needle) that mounts in the side of the carburetor cover.

FUEL SYSTEM 3-29

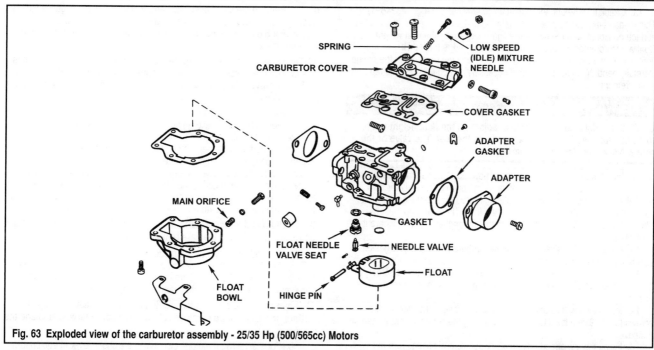

Fig. 63 Exploded view of the carburetor assembly - 25/35 Hp (500/565cc) Motors

REMOVAL & INSTALLATION

◆ See Figure 63

1. Remove the spark plug leads and/or disconnect the negative battery cable (if equipped) to prevent accidental starting of the engine.
2. Release the air intake silencer from the motor by turning the retainer knob located at the center of the silencer, then disconnecting the drain hose from the nipple at the bottom.
3. If equipped, disconnect the choke link from the carburetor choke lever.
4. Disconnect the throttle linkage from the throttle arm.

■ **On these models the carburetor mounting screws require a driver, usually of a Torx® head design.**

5. Remove the carburetor mounting screws, then remove the carburetor and gasket from the intake manifold.

■ **When removing the plastic wire ties on these models use a pair of cutters or dikes, but be careful not to cut, nick or otherwise damage the hoses.**

6. Turn the carburetor slightly for access, then use a pair of cutters to carefully remove the wire tie from the carburetor fuel supply line. Position a rag or small drain basin to catch any fuel left in the line, then disconnect the line from the carburetor and remove the carburetor completely from the motor.
7. Carefully clean the gasket mating surfaces of any remaining material.

To install:

8. Apply a coating of OMC Locquic Primer and OMC nut lock or equivalent threadlocking compound to the 2 carburetor retaining screws.
9. Holding the carburetor near the mounting in one hand, attach the fuel supply hose and secure using a new wire tie.

■ **Install the carburetor gasket dry. Do not use sealer.**

10. Insert the 2 carburetor bolts through the mounting flange, then place a new gasket over the bolts. Position the carburetor and gasket to the intake manifold, then tighten the bolts to 60-84 inch lbs. (7-9 Nm)
11. Snap the throttle linkage into the fitting on the throttle arm, then pull back gently to check for a secure connection.
12. If equipped, install the choke linkage to the choke lever.
13. Gently squeeze the primer bulb while checking for fuel leakage. Correct any fuel leaks before returning the engine to service.
14. Attach the drain hose to the nipple on the bottom of the housing, then install the air intake silencer to the carburetors and secure by turning the lock at the center of the silencer.

15. Perform the necessary timing and synchronization procedures from the Maintenance and Tune-Up Section. If the carburetor was repaired or rebuilt, be sure to perform the carburetor initial low speed setting procedure.

OVERHAUL

◆ See Figures 42, 43 and 56 thru 63

1. Remove the 2 screws holding the air intake silencer adapter to the carburetor, then remove the adapter and gasket from the carburetor assembly.
2. Remove the low speed (idle) mixture needle and spring from the carburetor body cover.
3. Remove the 8 screws from the carburetor body cover, then remove the cover and gasket from the carburetor assembly.
4. Remove the 7 float bowl screws, then remove the float bowl and gasket from the carburetor assembly. Carefully drain any remaining fuel from the float bowl as it is removed.
5. Using a small pick or awl, carefully remove the hinge pin, then remove the float and float needle valve from the underside of the carburetor body.
6. Remove the float needle valve seat and discard the gasket.
7. Remove the main (high speed) orifice from the float bowl.
8. Thoroughly clean and visually check all of the carburetor components as detailed under Cleaning & Inspection in this section. Carefully remove all traces of gasket from the mating surfaces on the carburetor body, adapter, cover and float bowl.

To assemble:

9. Before starting, make sure all components are completely clean and serviceable. Compare parts from the replacement kit to the parts removed from the carburetor. With the exception of wear or damage that might occur on the old parts (requiring their replacement in the first place) the new components should be identical. If you have any questions, check with a local dealer to check parts against a current part catalog before proceeding.

■ **To ensure proper operation and durability, replace all displaced or disturbed gaskets, O-rings or seals when rebuilding a carburetor regardless of their appearance.**

10. Install the main (high speed) orifice.
11. Install the needle valve seat using a new gasket.
12. Install the float and float needle valve, then carefully secure using the hinge pin.

■ **When using the OMC float gauge on these models, DO NOT use the cutout marked "25 THRU 75 HP" as it is for the other 2-stroke models of this horsepower range only.**

FUEL SYSTEM

13. Check float height adjustment using OMC float gauge (#324891). This measurement should be taken with the carburetor inverted, the gasket surface horizontal and the float valve lightly closed (by the float weight). Contrary to common sense, use the cutout marked "**9.9 & 15 HP**" on the float gauge, the top of the float must rest between the notches on the gauge. Carefully bend the metal float arm, if necessary, to achieve the required measurement.

✴✴ WARNING

Be VERY careful when measuring or adjusting the float height not to force the float needle valve downward into the seat. The valve or the seat will likely be damaged if this occurs.

14. Invert the carburetor so the top is now facing upward and check the float drop setting (to ensure the float valve can open completely for high rpm/full throttle operation). Measure the distance from the bottom lip of the carburetor body to the lowest corner of the float. The float drop must be 1-1 3/8 in. (25-35mm). If not, carefully bend the tab on the float arm, where it contacts the float seat.
15. Install the air intake silencer to the carburetor using a new gasket, secure using the retaining screws.
16. Install the float bowl using a new gasket, then tighten the screws to 17-19 inch lbs. (1.9-2.1 Nm) following the sequence embossed on the float bowl.
17. Install the carburetor cover using a new gasket, then tighten the screws to 17-19 inch lbs. (1.9-2.1 Nm) following the sequence embossed on the cover.
18. Install the low speed (idle) mixture needle and spring. Turn the needle inward slowly by hand until the underside of the screw head sits 0.390 in. (9.9mm) above the boss of the cover. Use a straightedge placed against the cover boss where the spring contacts it, and measure out to the underside of the screw head.
19. Install the carburetor, then perform the necessary timing and synchronization procedures from the Maintenance and Tune-Up Section (including the idle speed and mixture adjustments).

All Except Above, Including 5/6/6.5/8 (164cc), 9.9/10/15 (255cc, 1994 and later), 25-55 (737cc) and 25-70 (913cc) Motors

◆ See Figures 64 and 65

The carburetors used on these models are single-barrel, float feed types. They are unitized, with a removable float bowl and carburetor cover. The unitized design allows the same basic carburetor body to be adapted for a wide variety of models. Although some design features (especially linkage) will vary from model-to-model, the basic operating principles remain the same.

Main fuel metering is accomplished by a removable main orifice (main jet) mounted in the float chamber, directly behind the main fuel inlet nipple or drain plug. Low speed (idle) mixture is adjusted using a low speed screw (needle) that mounts in the side of the carburetor cover.

Because of slight differences in the float bowls, covers and related components, we'll divide these carburetors further by model as:
- 15 Hp motors and below
- 25 Hp motors or larger

Although the service procedures are virtually the same on these models, we've included illustrations and specific steps where there are differences using these identifications.

■ Because of a few unique components found on the 5/6/6.5/8 hp motors through 1993, please refer to the procedures for 9.9/10/14/15 Hp (216cc/255cc) Motors Thru 1993 and All 18 Jet Thru 35 Hp (521cc) Motors with regards to OVERHAUL only.

REMOVAL & INSTALLATION

◆ See Figures 64 thru 67

■ The fuel hose fittings are delicate on these models, especially on the 25 hp and larger motors). To protect the fittings, gently push the hoses from them instead of grasping and pulling on the hose itself. If pushing won't free the hose, use a utility knife to carefully slit the hose from the end to a point at or near the fitting flange, then peel the hose from the fitting and replace it upon reinstallation.

1. Remove the spark plug leads and/or disconnect the negative battery cable (if equipped) to prevent accidental starting of the engine.
2. For 5/6/6.5/8 hp motors, remove the manual starter from the top of the powerhead.

■ On some models, access to the air intake silencer or carburetors is easier with the lower engine covers removed. For details, please refer to Engine Covers (Top and Lower Cases) under Engine Maintenance for more details.

3. Remove the air intake silencer:
   a. For 15 hp and smaller motors, remove the two retaining screws, then separate the silencer from the engine.
   b. For 25 hp and larger motors, first remove the bolts (usually 7 or 4 depending on them model) from the perimeter of the silencer cover, then remove the cover and discard the gasket. Next remove the silencer base screws (at least 6) mounted to the silencer base at the carburetor throats. Remove the silencer base, disconnecting the drain hose as it is pulled away from the motor, then discard both the base screws and gasket.

■ On 25 hp and larger motors the air intake silencer base screws must be replaced each time they are removed.

4. Proceed as follows as per model:
   a. For 5/6/6.5/8 hp motors, remove the bellcrank retaining screw, then cut the wire tie or release the clamp and remove the fuel hose from the carburetor. Have a rag handy to catch any escaping fuel.

■ When removing the plastic wire ties on these models use a pair of cutters or dikes, but be careful not to cut, nick or otherwise damage the hoses.

   b. For 9.9/10/15 hp motors, disconnect the choke lever (if equipped), then remove the cam follower O-ring.
   c. For 25 hp and larger motors, cut the fuel hose wire ties at the carburetors, then disconnect the fuel hoses. Have a rag handy to catch any escaping fuel.
5. Remove the carburetor mounting nuts, lock washers and the carburetor(s) from the powerhead.
6. For 15 hp and smaller motors, reposition the carburetor for better access, then cut the wire ties and disconnect the fuel supply hose from the carburetor.
7. For models equipped with an electric primer (such as some remote 9.9/10/15 hp motors and all 25 hp and larger motors) disconnect the fuel primer hose(s) from the carburetor(s).
8. If necessary on 5/6/6.5/8 hp motors, remove the cam follower and linkage.
9. Remove the carburetor flange gasket(s).
10. Turn the carburetor slightly for access, then use a pair of cutters to carefully remove the wire tie from the carburetor fuel supply line. Position a rag or small drain basin to catch any fuel left in the line, then disconnect the line from the carburetor and remove the carburetor completely from the motor.
11. Carefully clean the gasket mating surfaces of any remaining material.

To install:
12. Position a new gasket (or new gaskets on multi-carburetor motors) over the carburetor mounting studs.

■ Install the carburetor gasket dry. Do not use sealer.

13. If it was easier to access fuel supply or primer lines with the carburetor(s) off the mounting studs, attach those lines now and secure using new wire ties.
14. Install the carburetor(s) over the mounting studs using the nuts and lock washers, then tighten the nuts securely.
15. If any fuel hoses were not installed prior to positioning the carburetor(s), connect them now and secure using a new wire tie.
16. Proceed as follows, depending on the model:
   a. For 5/6/6.5/8 hp motors, if removed, install the cam follower linkage, then install the bellcrank retaining screw.
   b. For 9.9/10/15 hp motors, install the cam follower on the post, positioning the O-ring on the post to secure the follower. If equipped, connect the choke lever.
   c. For 25 hp and larger motors, snap the throttle linkage into place.
17. Gently squeeze the primer bulb while checking for fuel leakage. Correct any fuel leaks before returning the engine to service.
18. Except for 25 hp and larger motors, install the air intake silencer assembly. For 25 hp and larger motors, perform the next step first, as access to some components will be restricted if the silencer is installed before making adjustments.

FUEL SYSTEM 3-31

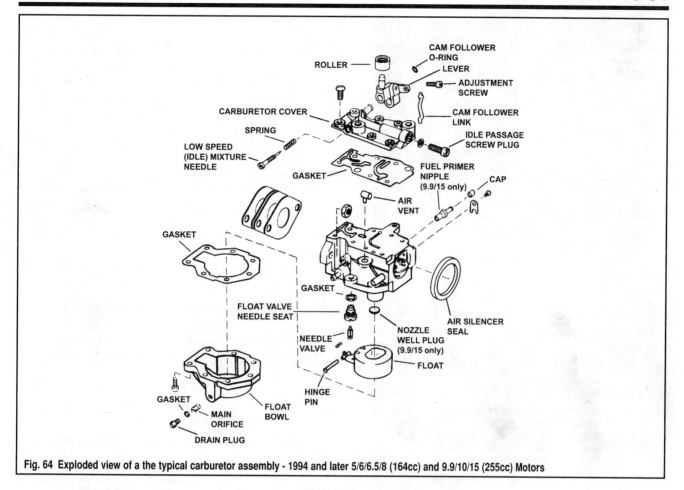

Fig. 64 Exploded view of a the typical carburetor assembly - 1994 and later 5/6/6.5/8 (164cc) and 9.9/10/15 (255cc) Motors

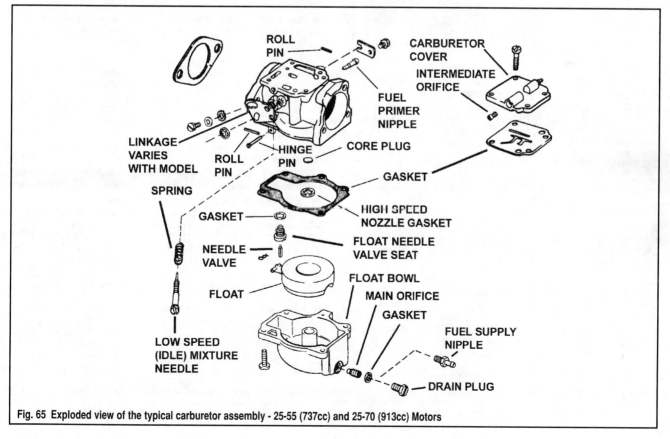

Fig. 65 Exploded view of the typical carburetor assembly - 25-55 (737cc) and 25-70 (913cc) Motors

3-32 FUEL SYSTEM

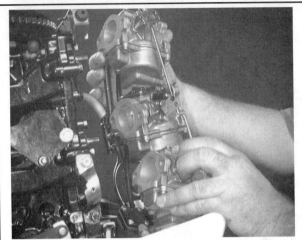

Fig. 66 Photo showing all multiple carburetors being removed as an assembly

Fig. 67 In this case, all 3 carburetors and the electric primer were removed together from a 70 hp motor

19. Perform the necessary timing and synchronization procedures from the Maintenance and Tune-Up Section. If the carburetor was repaired or rebuilt, be sure to perform the carburetor initial low speed setting procedure.

20. For 25 hp and larger motors, connect the drain hose to the nipple on the bottom of the silencer base, then install the base using new screws and a new gasket. Install the air silencer cover using a new gasket.

OVERHAUL

◆ See Figures 42, 43 and 56 thru 65

■ Because of a few unique components found on the 5/6/6.5/8 hp motors through 1993, please refer to the procedures for 9.9/10/14/15 Hp (216cc/255cc) Motors Thru 1993 and All 18 Jet Thru 35 Hp (521cc) Motors with regards to OVERHAUL only.

1. Position the carburetor over a small drain basin, then remove the float bowl drain screw or fuel inlet nipple from the bottom side of the float bowl and drain any fuel remaining in the bowl. Discard the drain screw or nipple gasket.

2. On most models the main orifice (high speed jet) is mounted in the float bowl directly behind the fuel inlet nipple or drain screw. If accessible, remove the orifice at this time. If not accessible, wait until the float bowl is removed.

3. For 25 hp and larger motors, remove the intermediate air bleed orifice from the side of the carburetor cover.

4. Remove the screws (usually 7 for 15 hp and smaller motors or 5 for 25 hp and larger motors) securing the float bowl the carburetor body, then remove the float bowl and gasket.

5. Using a pick or awl carefully push out the float hinge pin, then remove the float and float needle valve from the underside of the carburetor body. Remove the float valve needle seat and gasket.

6. On 25 hp and larger motors, remove the high speed nozzle gasket from the center underside of the carburetor body (some smaller motors are equipped with a nozzle plug at this point, do NOT remove the plug unless otherwise directed).

7. Unthread the low speed (idle) mixture screw from the top of the carburetor body or carburetor cover, as applicable. Remove the needle and spring.

8. Remove the screws (usually 6 for 15 hp and smaller motors or 4 for 25 hp and larger motors) securing the carburetor cover to the body, then remove the cover and gasket.

9. If not done earlier, remove the main orifice.

10. If necessary on 15 hp and smaller motors, remove the air vent from the vertical mount on the side of the carburetor body.

11. Thoroughly clean and visually check all of the carburetor components as detailed under Cleaning & Inspection in this section. Carefully remove all traces of gasket material from the various mating surfaces.

To assemble:

12. Before starting, make sure all components are completely clean and serviceable. Compare parts from the replacement kit to the parts removed from the carburetor. With the exception of wear or damage that might occur on the old parts (requiring their replacement in the first place) the new components should be identical. If you have any questions, check with a local dealer to check parts against a current part catalog before proceeding.

■ To ensure proper operation and durability, replace all displaced or disturbed gaskets, O-rings or seals when rebuilding a carburetor regardless of their appearance.

13. For 25 hp and larger motors, install a new high speed nozzle gasket. Also, install the intermediate air bleed orifice to the carburetor cover.

14. For 15 hp and smaller motors, If removed, carefully press the air vent into the vertical boss on the side of the carburetor body.

15. Install the main orifice (high speed jet).

16. Install the drain plug or the fuel supply nipple to the float bowl, as applicable.

17. Install the needle valve seat using a new gasket.

18. Install the float and float needle valve, then carefully secure using the hinge pin.

■ When using the OMC float gauge on these models, use the cutout marked "25 THRU 75 HP" for 25 hp and larger motors or the cutout marked "9.9 & 15 HP" for 15 hp and smaller motors.

19. Check float height adjustment using OMC float gauge (No. 324891). This measurement should be taken with the carburetor inverted, the gasket surface horizontal and the float valve lightly closed (by the float weight). Use the cutouts marked "9.9 & 15" for 15 hp and smaller motors. For all 25 hp and larger motors, use the cutout marked "V-4 & V-6" on one side of the cutout and marked "25 THRU 75 HP" on the other side of the cutout. In either case, the top of the float must rest between the notches on the gauge. Carefully bend the metal float arm, if necessary, to achieve the required measurement.

✱✱ WARNING

Be VERY careful when measuring or adjusting the float height not to force the float needle valve downward into the seat. The valve or the seat will likely be damaged if this occurs.

20. Invert the carburetor so the top is now facing upward and check the float drop setting (to ensure the float valve can open completely for high rpm/full throttle operation). Measure the distance from the bottom lip of the carburetor body to the lowest corner of the float. The float drop must be 1-1 3/8 in. (25-35mm) for 15 hp and smaller motors or 1 1/8 -1 5/8 in. (28-41mm) for 25 hp and larger motors. If not, carefully bend the tab on the float arm, where it contacts the float seat.

21. For 25 hp and larger motors, apply a coating of OMC Locquic primer and OMC screw lock to the float bowl retaining screw threads.

22. Install the float bowl using a new gasket, then tighten the bolts as follows:

 a. For 15 hp and smaller motors, tighten the bolts to 8-10 inch lbs. (0.8-1.2 Nm) following the pattern embossed on the float bowl.

 b. For 25 hp and larger motors, tighten the bolts to 25-35 inch lbs. (2.8-4.0 Nm) using a crisscross pattern.

FUEL SYSTEM 3-33

23. Install the carburetor cover using a new gasket, then tighten the bolts as follows:

    c. For 15 hp and smaller motors, tighten the bolts to 8-10 inch lbs. (0.8-1.2 Nm) following the pattern embossed on the float bowl

    d. For 25 hp and larger motors, tighten the bolts to 15-22 inch lbs. (1.6-2.4 Nm) using a crisscross pattern.

24. Install the low speed (idle) mixture needle and spring to the carburetor body or cover (as applicable). Thread the needle in slowly until it just lightly contacts the valve seat, then back the screw out the number of turns listed for initial low speed setting in the Carburetor Set-Up Specifications chart in this section.

25. Install the carburetor, then perform the necessary timing and synchronization procedures from the Maintenance and Tune-Up Section (including the idle speed and mixture adjustments).

CLEANING & INSPECTION

◆ See Figures 68 thru 71

Start by making sure that all components (even those being discarded) are spread out on a clean work surface for inspection and comparison to replacement parts. Make sure that gaskets are of the same patterns. If a gasket differs, determine if the gasket will function by holding it to each side of the gasket mating surface it will seal. Make sure that the replacement doesn't block or cover any passage that the original did not. If it differs, seek the advice of your parts supplier, they should be able to hunt down the correct gasket or the reason it is now acceptable (possibly its a superceding part).

Never dip rubber parts, plastic parts, diaphragms, or pump plungers in carburetor cleaner. These parts should be cleaned only in solvent safe for plastic or rubber components and then immediately blown dry with low pressure (less than 25 psi or 172 kPa) compressed air.

Fig. 68 Lay out all original parts for comparison to the rebuild kit

✷✷ CAUTION

Always take precautions when working with solvent and compressed air. Protect your eyes and your skin from chemical burns.

Place all metal parts in a screen-type tray and dip them in carburetor cleaner until they appear completely clean, then blow them dry with compressed air.

■ **If compressed air and/or a small solvent tank is not available, use a commercially available carburetor and choke cleaner (such as that sold by OMC) to clean components and blow out carburetor passages.**

Blow out all passages in the castings with low pressure (less than 25 psi or 172 kPa) compressed air. Whenever possible, apply air in the same direction as normal air or fuel flow. Check all parts and passages to be sure they are not clogged or contain any deposits. Never use a piece of wire or any type of pointed instrument to clean drilled passages or calibrated holes in a carburetor (wire could remove metal from the inner surface of a calibrated passage, changing calibration from spec and causing performance problems). If necessary, use something that is small, but softer than the metal of the passages to clean them, like a piece of straw from a broom.

■ **If debris or contamination is found in the carburetor, inspect and clean the entire system upstream of the carburetor. Chances are water or debris contamination will be present in other components as well. Failure to clean the entire fuel system could result in clogging a newly rebuilt carburetor shortly after it is reinstalled.**

Move the throttle shaft back and forth to check for wear. If the shaft appears to be too loose, replace the complete throttle body because individual replacement parts are normally not available.

Inspect the main body, airhorn, and venturi cluster gasket surfaces for cracks and burrs which might cause a leak.

Check the float for deterioration. Check to be sure the float spring has not been stretched. If any part of the float is damaged, the unit must be replaced. Check the float arm needle contacting surface and replace the float if this surface has a groove worn in it.

Inspect the tapered section of the idle adjusting needles and replace any that have developed a groove.

As previously mentioned, most of the parts which should be replaced during a carburetor overhaul are included in overhaul kits available from your local marine dealer. One of these kits will contain a matched fuel inlet needle and seat. This combination should be replaced each time the carburetor is disassembled as a precaution against leakage (which could lead to flooding during normal operation).

Before assembly, use a syringe filled with isopropyl alcohol to check all drillings and passages.

Many of the carburetors used on Johnson/Evinrude motors are equipped with an emulsion pickup tube in the center of the venturi. If equipped, check it for leaks according to the procedure in this section.

Many of the carburetors used on Johnson/Evinrude motors are equipped with one or more plugs on the carburetor body (such as the vent well core plug found on the underside of most unitized carburetors for the 25-55 (737cc) and 25-70 (913cc) motors). If leaks or problems are suspected in passages or vent wells behind core plugs, they can be removed for inspection. Refer to the core plug procedure in this section.

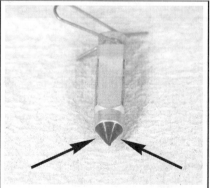

Fig. 69 Pay close attention to the needle tips during inspection

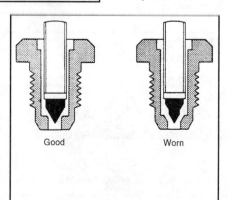

Fig. 70 The float needle valve and seat must not be worn or deformed

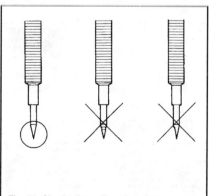

Fig. 71 Check all needle valves for grooves, pitting or damage

3-34 FUEL SYSTEM

Inspecting the Emulsion Pickup Tube

◆ See Figure 72

Many of the carburetors used on Johnson/Evinrude motors are equipped with an emulsion pickup tube in the center of the venturi. If equipped, check it for leaks using a syringe filled with isopropyl alcohol as follows:

1. Invert the carburetor body (so it is upside down), then fill the idle circuit with isopropyl alcohol.
2. Check the top of the emulsion tube (the bottom of the tube now that the carburetor is inverted) for leaks of the isopropyl alcohol.
3. If leakage is found at the tube, drain the alcohol and thoroughly blow the area dry using low pressure compressed air. Seal the tube by applying a drop of OMC Ultra Lock or equivalent sealant.

Fig. 72 Check the emulsion tube for leaks while the carburetor is inverted, if necessary seal leaks with a drop of OMC Ultra Lock applied here

Removing Core Plugs for Inspection

◆ See Figures 73, 74 and 75

Many of the carburetors used on Johnson/Evinrude motors are equipped with one or more plugs on the carburetor body. For example, the unitized carburetor found on 25-55 (737cc) and 25-70 (913cc) motors usually contains a vent well core plug on the underside of the carburetor body.

If leaks or problems are suspected in passages or vent wells behind core plugs or lead shot, they can be removed for inspection.

■ Most, but not all, plugs are illustrated in the exploded views included with each set of carburetor procedures. Refer to the exploded views first when deciding whether or not to remove any plug. Also, it is advisable to obtain the replacement before removal (to be sure one is available).

To remove a core plug, pierce it with a 1/8 in. (3mm) or slightly smaller punch, then carefully pry the plug out of the bore. Inspect the area as needed.

Before assembling the carburetor, position the plug to the bore with the convex (tapered side) facing upward. Drive the plug into the bore using a flat end punch (one that is close to the size of the plug, but slightly smaller) and a plastic mallet. Once it is installed, apply a coating of Gasoila sealer (or equivalent) to the rim of the plug.

If the carburetor is equipped with an emulsion pickup tube in the center of the venturi, check it according to the procedure in this section. Also, for the carburetors found on 25-55 (737cc) and 25-70 (913cc) motors, remove the vent well core plug for inspection, then install a new plug before assembly, again, refer to the procedure in this section.

Fuel Pump

◆ See Figure 76

Carbureted models are equipped with a diaphragm-displacement type fuel pump. The pump is mounted somewhere on the side or end of the powerhead (intake manifold for most 2-strokes and rocker cover for most 4-strokes). On 2-stroke motors, the diaphragm is actuated by the cycle of crankcase pressure alternately receiving pressure and vacuum. On 4-stroke motors, the pump is mounted in such a manner as to mechanically actuate the diaphragm through a plunger that contacts a camshaft lobe.

Engines equipped with the VRO2 oiling system are equipped with a separate VRO2 fuel/oil pump assembly that is covered in the Lubrication and Cooling Section. For fuel system testing and VRO2 pump troubleshooting, refer to the information under the VR02 system in the Lubrication and Cooling Section.

The 25/35 hp 3-cylinder motors covered here utilize a pre-oiling system that depends upon the combined operation of 2 diaphragm-displacement pumps (a suction and a pressure pump), mounted side-by-side on the powerhead.

Although repair and overhaul procedures are virtually the same on all diaphragm-displacement fuel pumps, the unique interrelation of the 2 pumps used by the 25/25 hp motors requires different testing procedures.

■ Have a shop towel and a suitable container handy when testing or servicing a fuel pump as fuel will likely spill from hoses disconnected during these procedures. To ensure correct assembly and hose routing, mark the orientation of the fuel pump and hoses before removal.

TESTING

◆ See Figure 77

The problem most often seen with fuel pumps is fuel starvation, hesitation or missing due to inadequate fuel pressure/delivery. In extreme cases, this might lead to a no start condition as all but total failure of the pump prevents fuel from reaching the carburetor(s). More likely, pump failures are not total, and the motor will start and run fine at idle, only to miss, hesitate or stall at speed when pump performance falls short of the greater demand for fuel at high rpm.

Before replacing a suspect fuel pump, be absolutely certain the problem is the pump and NOT with fuel tank, lines or filter. A plugged tank vent could create vacuum in the tank that will overpower the pump's ability to create vacuum and draw fuel through the lines. An obstructed line or fuel filter could

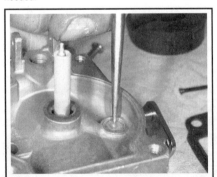

Fig. 73 To remove a core plug, pierce it with a punch...

Fig. 74 ...then insert the punch for leverage and pry

Fig. 75 A new plug is installed using a larger, flat driver

FUEL SYSTEM 3-35

also keep fuel from reaching the pump. Any of these conditions could partially restrict fuel flow, allowing the pump to deliver fuel, but at a lower pressure/rate. A pump delivery or pressure test under these circumstances would give a low reading that might be mistaken for a faulty pump. Before testing the fuel pump, refer to the testing procedures found under Fuel Lines and Fitting to ensure there are no problems with the tank, lines or filter.

If inadequate fuel delivery is suspected and no problems are found with the tank, lines or filters, a conduct a quick-check to see how the pump affects performance. Use the primer bulb to supplement fuel pump. This is done by operating the motor under load and otherwise under normal operating conditions to recreate the problem. Once the motor begins to hesitate, stumble or stall, pump the primer bulb quickly and repeatedly while listening for motor response. Pumping the bulb by hand like this will force fuel through the lines to the carburetor, regardless of the fuel pump's ability to deliver fuel. If the engine performance problem goes away while pumping the bulb, and returns when you stop, there is a good chance you've isolated the fuel pump as the culprit. Perform a pressure test to be certain, then repair or replace the pump assembly.

■ On 2-stroke models, low vacuum supply from the crankcase or insufficient vacuum at the pump itself due to bad seals can also be the culprit for poor fuel delivery.

✳✳ WARNING

Never run a motor without cooling water. Use a test tank, a flush/test device or launch the craft. Also, never run a motor at speed without load, so for tests running over idle speed, make sure the motor is either in a test tank with a test wheel or on a launched craft with the normal propeller installed.

All Models Except the 25/35 Hp 3-Cylinder Motors

Pump Pressure Test

◆ See Figure 77

By far the most accurate way to test the fuel pump is using a low pressure fuel gauge while running the engine at various speeds, under load. To prevent the possibility of severe engine damage from over-speed, the test must be conducted under load, either in a test tank (with a proper test propeller) or mounted on the boat with a suitable propeller.

1. Test the Fuel Lines and Fittings as detailed in this section to be sure there are no vacuum/fuel leaks and no restrictions that could give a false low reading.
2. Make sure the fuel filter(s) is(are) clean and serviceable.
3. Start and run the engine in forward gear, at idle, until normal operating temperature is reached. Then shut the motor down to prepare for the test.
4. Remove the fuel tank cap to make sure there is no pressure in the tank (the fuel tank vent must also be clear to ensure there is no vacuum). Check the tank location, for best results, make sure the tank is not mounted any more than 30 in. (76mm) below the fuel pump mounting point. On portable tanks, reposition them, as necessary to ensure accurate readings.

■ The fuel outlet line from the fuel pump may be disconnected at either the pump or the carburetor whichever provides easier access. If you disconnect it from the pump itself you might have to provide a length of fuel line (depending on whether or not the gauge contains a length of line to connect to the pump fitting).

5. Disconnect the fuel output hose from the carburetor or fuel pump, as desired.
6. Connect a fuel pressure gauge inline between the pump and the carburetor(s).
7. Run the engine at or around each of the following speeds and observe the pressure on the gauge:
   - At 600 rpm, the gauge should read about 1 psi (7 kPa).
   - At 2500-3000 rpm, the gauge should read about 1.5 psi (10 kPa).
   - At 4500 rpm, the gauge should read about 2.5 psi (17 kPa).
8. If readings are below specification and other causes such as fuel line or filter restrictions have been eliminated, repair or replace the pump.

Pump Leak Test

Pressurize the fuel system using the primer bulb. Squeeze repeatedly, but slowly, until the bulb is firm, then check the pump body and connections for leaks. On 2-stroke motors, remove the pump from the powerhead, leaving the fuel lines connected. Observe the vacuum port at the rear of the pump (where is connects to the port on the powerhead). The leakage of any fuel at this point indicates a damaged diaphragm.

Repair or replace any pump that shows signs of leakage.

25/35 Hp 3-Cylinder Motors

◆ See Figure 77

The 25/35 hp 3-cylinder motors covered here are equipped with an integral pre-oiling system. Proper operation of the pre-oiling system depends upon the combined operation of 2 diaphragm-displacement pumps. These pumps are mounted side-by-side in the powerhead. The first pump, mounted to the right while facing them is known as the suction pump since it draws fuel from the tank. The other pump, mounted to its left, is known as the pressure pump since it delivers fuel under pressure to the mixing unit and finally to the carburetors.

To ensure proper functionality, 2 tests must be performed on this system. The first (Suction Pump Test) will determine the ability of the pump on the right to draw fuel from the tank. The second (Pressure Pump Test) will determine the ability of the pump on the left to supply pressurized fuel to the mixing unit and carburetors.

Because these tests require operation at wide-open throttle, they must be conducted with the engine under load, either in a test tank (with a test propeller) or mounted on a launched craft (with a suitable propeller installed).

✳✳ WARNING

To prevent the possibility of severe engine damage due to over-speed running, NEVER conduct these tests on a engine flush-device.

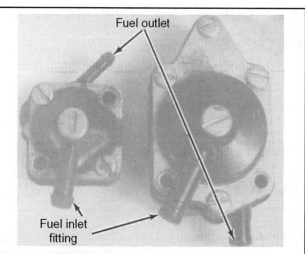

Fig. 76 Typical diaphragm-displacement fuel pumps used on Johnson/Evinrude outboards

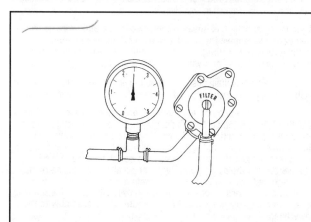

Fig. 77 Use a fuel pressure gauge connected in the pump outlet line to check operating pressure - typical Johnson/Evinrude pump shown

3-36 FUEL SYSTEM

Suction Pump Test

This test checks the pump mounted to the right (while facing the two pumps) and its ability to draw fuel from the tank.

1. Start and then run the engine at idle, in forward gear, until is reaches normal operating temperature, then shut the engine down to install a vacuum gauge.
2. Disconnect the fuel tank supply line from the suction pump inlet fitting, then connect a vacuum gauge.
3. Start and run the engine at wide-open throttle while observing the gauge. Vacuum should read no more than 4 in. Hg (102mm Hg).
4. If the reading is out of specification, check the fuel delivery system for restrictions. For more details, refer to Fuel Lines and Fittings in this section.
5. If no restrictions are found, repair or replace the pump, as necessary.
6. If pump operation is within spec, but a problem is still suspected in the system, perform the Pressure Pump Test.

Pressure Pump Test
◆ See Figure 77

This test checks the pump mounted to the left (while facing the two pumps) and its ability to supply pressurized fuel to the mixing unit and carburetors.

1. Start and run the engine at idle, in forward gear, until is reaches normal operating temperature, then shut the engine down to install a vacuum gauge.
2. Disconnect the fuel pump-to-inline filter hose from the inlet fitting, then connect fuel pressure gauge inline between the pump outlet fitting and the filter.
3. Start and then run the engine at wide-open throttle while observing the gauge. Fuel pressure on the gauge should no less than 3 psi (21 kPa).
4. If the reading is below specification, repair or replace the pump, as necessary.

REMOVAL & INSTALLATION

◆ See Figures 78 and 79

 EASY

✱✱ WARNING

The 2 fuel pumps used by 25/35 hp (500/565cc) motors are NOT interchangeable. Severe damage to the powerhead could occur from improper oiling if these pumps are mixed-up and installed in the wrong positions.

1. For safety, either disconnect the negative battery cable (if so equipped) and/or disconnect the spark plug lead(s) and ground them to the powerhead.
2. Locate the fuel pump on the powerhead and determine if it will be easier to remove the lower engine covers. On some models equipped with split (2-piece) lower covers, it is easier to access the pump if the lower engine covers are removed. For details, refer to the Engine Cover procedure under the Engine Maintenance section.

■ On most models, fuel hoses are retained by a plastic wire tie (which must be cut to remove the hose). Use a pair or dikes or cutters to carefully remove the wire tie. Be sure not to cut, nick or otherwise damage the fuel hose or it will have to be replaced.

3. Place a small drain basin or a shop rag under the fuel line fittings (to catch escaping fuel), then tag and disconnect the fuel hoses from the pump.

■ The fuel pumps used on these motors are equipped with 2 or 3 sets of bolts visible on the surface of the pump. On most models, a round inlet cover is mounted to the center of the pump with a single bolt. Then, of the remaining bolts, 2 are usually used to secure the pump to the powerhead and the balance is used to bolt the halves of the body together around the diaphragm. But some smaller models use pumps with a squared housing that do not contain a separate inlet housing cover screw. On these models there are usually 2 bolts for the housing cover and body assembly, and 2 bolts mounting the assembly to the powerhead.

4. Loosen the pump mounting bolts (the bolts that thread not through the inlet cover, but the body of the pump and into the powerhead). If in doubt as to which bolts secure the pump, look at the back of the pump (as can be seen at the pump-to-powerhead seam line) to see which bolts continue through the pump assembly and into the powerhead. These are the only bolts that should be loosened for pump removal.

■ On the 5 sided pumps used by most 2-stroke models covered by this manual, there are 2 mounting bolts at the bottom of the pump assembly. On squared pumps, 2 of the four visible bolts are usually used for mounting, they are usually slightly larger than the cover/body assembly bolts.

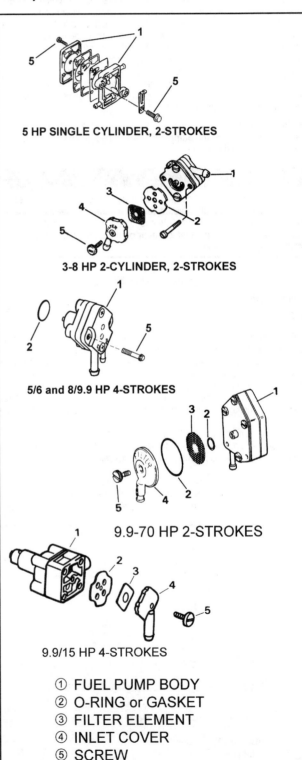

Fig. 79 Various fuel pumps used on these Johnson/Evinrude outboards

FUEL SYSTEM 3-37

5. If necessary, remove the cover screw or screws and disassemble the fuel pump for inspection or overhaul, as applicable. For details, refer to the Fuel Pump, Overhaul procedure in this section.

6. Clean the mating surface of the pump and powerhead of any remaining gasket material (2-strokes), dirt, or debris. Be careful not to damage the surface as that could lead to vacuum leaks on 2-stroke motors or oil leaks on 4-strokes.

To install:

7. Apply a coating of OMC nut lock to the fuel pump retaining screws. On 4-stroke models, apply a light coating of OMC Triple Guard or an equivalent marine grade grease to the new pump O-ring.

8. Position a new gasket (2-strokes) or a new O-ring (4-strokes) and install the pump to the powerhead using the retaining screws. Tighten the screws to specification as follows:
- For 2-stroke motors: 24-36 inch lbs. (2.8-4.0 Nm)
- For 5/6 and 8/9.9 hp 4-stroke motors: 35-56 inch lbs. (4-6 Nm)
- For 9.9/15 hp 4-stroke motors: 30-35 inch lbs. (3.4-4.0 Nm)

9. Connect the fuel lines as noted during removal and secure using the clamp or new wire ties, as applicable.

10. Gently squeeze the primer bulb while checking for fuel leakage. Correct any fuel leaks before returning the engine to service.

11. Connect the negative battery cable and/or spark plug lead(s).

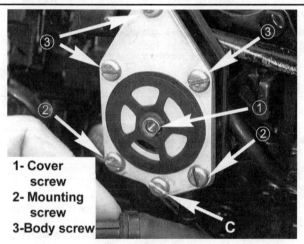

1- Cover screw
2- Mounting screw
3- Body screw

Fig. 78 Example of pump screw identification - 9.9-70 hp 2-strokes

OVERHAUL

◆ See Figures 79 and 80

MODERATE

Most of the fuel pumps used on these motors may be disassembled for overhaul. Of course, the pumps are of a fairly simple design with relatively few moving parts. Check with your local parts supplier to make sure that an overhaul kit containing the necessary parts are available for your model. In most cases, the parts are limited to the diaphragm(s), gasket(s) and a fuel inlet screen (if equipped).

If overhaul is required due to damage from contamination or debris (as opposed to simple deterioration) disassemble and clean the rest of the fuel supply system prior to the fuel pump. Failure to replace filters and clean or replace the lines and fuel tank, could result in damage to the overhauled pump after it is placed back into service.

All diaphragms and seals should be replaced during assembly, regardless of their condition. Check for fuel leakage after completing the repair and verify proper operating pressures before returning the motor to service.

✱✱ WARNING

No sealant should be used on fuel pump components unless otherwise specifically directed. If small amounts of a dried sealant were to break free and travel through the fuel supply system it could easily clog passages (especially the small, metered orifices and needle valves of the carburetor).

1. Remove the fuel pump from the powerhead as detailed in this section.

2. Matchmark the fuel pump cover, housing and base to ensure proper assembly.

■ To ease inspection and assembly, lay out each piece of the fuel pump as it is removed. In this way, keep track of each component's orientation in relation to the entire assembly.

3. On all 2-stroke motors, proceed as follows:
   a. For all except the 5 hp single cylinder motor, remove the center cover screw, then remove the fuel inlet cover, filter element (screen) and gasket or O-rings (as applicable).

■ The 5 hp, single cylinder, 2-stroke motor uses a 3-piece diaphragm assembly.

   b. Remove the pump housing-to-base screws (usually 2 flat-head screws) and carefully separate the housing from the base, removing the diaphragm(s) from the center.

4. For 5/6 hp and 8/9.9 4-stroke motors, proceed as follows:
   a. Remove the pump cover screws, then separate the pump cover, reed valve, housing and base.
   b. Push inward on the plunger and diaphragm assembly, then hold it down while rotating 90° (releasing the diaphragm from the base). Lift up slightly on the plunger/diaphragm, allowing spring pressure to gently separate the components. Remove the plunger, spring and diaphragm from the base.

5. For 9.9/15 hp 4-stroke motors, disassemble the pump as follows:
   a. Remove the center cover screw, then remove the fuel inlet cover, filter element (screen) and gasket.
   b. Remove the 2 pump assembly screws, then carefully separate the inner cover, diaphragm, housing and base components.

6. Clean the metallic components thoroughly using solvent and carefully remove all traces of gasket material.

7. Inspect the diaphragm closely for cracks or tears.

■ It is advisable to replace the diaphragm ANYTIME the fuel pump is disassembled to ensure reliability and proper performance.

8. Inspect the fuel pump body for cracks. Check gasket surfaces for nicks, scratches, or irregularities. Inspect the mating surfaces of the fuel cover, body and base using a straight edge to ensure that they are not warped from heat or other damage Replace warped or damaged components.

9. On 4-stroke motors, check the spring for damage or lack of tension.

To assemble:

10. Assemble the components of the fuel pump housing, base and diaphragm noting the following:
   a. Use new gaskets. Make sure each gasket and the components it seals are aligned properly.
   b. Align the matchmarks made during disassembly to ensure proper component mounting.
   c. On 5/6 and 8/9.9 hp 4-stroke motors, rotate the diaphragm 90 degrees after insertion to lock it in place. Be certain that the reed valve is in position between the housing and cover.
   d. On 9.9/15 hp 4-stroke motors, be sure that the round supports (found on either side of the diaphragm) are positioned with the convex side (the side that bends outward) facing towards the diaphragm.

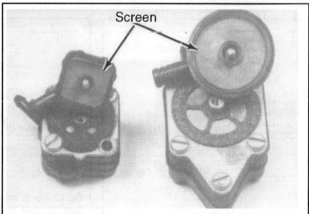

Fig. 80 Many of the pumps used on these motors are equipped with a filter screen under the cover

3-38 FUEL SYSTEM

11. Apply a coating of OMC Nut Lock, or an equivalent threadlocking compound to the pump housing and base screws, then install and tighten them to specification:
   - For 2-stroke motors: 24-36 inch lbs. (2.8-4.0 Nm)
   - For 5/6 and 8/9.9 hp 4-stroke motors: 35-56 inch lbs. (4-6 Nm)
   - For 9.9/15 hp 4-stroke motors: 30-35 inch lbs. (3.4-4.0 Nm)

12. If equipped, install the fuel pump cover and filter screen using a new gasket or O-ring(s), then tighten the center screw securely.

13. Install the fuel pump, then check for leaks and for proper operation.

Manual Fuel Primer

◆ See Figure 81

Some of these 18 Jet-70 hp 2-stroke Johnson/Evinrude outboards are equipped with a manual fuel primer system to aid with cold starts. The basic design of the manual primer is that of a small, hand-operated plunger-type pump. The primer works by drawing fuel into the pump housing through a fuel line with a one-way check valve when the shaft is withdrawn. The fuel is then forced out, toward the motor, through a second one-way check valve when the shaft is pushed back inward.

The primer performs the same function of a choke (aiding cold starting by making sure the engine receives a richer fuel mixture), but by opposite means. Whereas a choke reduces the amount of air provided to the combustion chamber (thus increasing the fuel portion of the air/fuel ratio), a primer works on the fuel side of the ratio by manually increasing the amount of fuel. The extra fuel provided by the manual primer enriches the air/fuel mixture for cold start purposes only. Use of the primer on an engine that is at or near operating temperature can flood the motor preventing starting.

TESTING

◆ See Figure 81

An inoperable manual primer will cause hard start or possibly even a no start condition during attempts to start a cold motor. The colder the ambient temperature, the more trouble an inoperable primer will cause. A primer with internal leakage (allowing fuel to bypass the air/fuel metering system) will cause rich running conditions that could include hesitation, stumbling, rough running, especially at idle and lead spark plug fouling.

Function Test

If the motor is operable, but trouble is suspected with the primer system, perform a function test with the engine running. Although this test can be conducted on a flush-fitting, engine speed will reach 2000 rpm and it is much safer to conduct the test in a test tank or with the boat launched.

1. Start and run the engine until it reaches normal operating temperature.
2. Once the engine warms, set the throttle so it runs at 2000 rpm.
3. Pump the manual primer knob and observe engine operation. If the primer is operating correctly, the engine should run rich and speed should drop to about 1000 rpm.

4. If the primer seems ineffective, stop the engine, then remove the primer hose from its fitting(s). Check each fitting for clogs using a syringe filled with isopropyl and a clear vinyl 1/8 in. inner diameter hose. Attach the hose to the fitting being checked and press lightly on the syringe. Fluid will move through the fitting unless it is clogged.

5. If any clogs are found, use a thin pick to carefully clean the fitting. OMC makes a cleaning tool for this purpose (#326623).

6. Be sure to check the primer hose T fitting for clogs as well.

7. If no clogs are found, perform the Primer Check procedure to see if the problem lies within the primer assembly itself.

Primer Check

If you suspect the manual primer system is not functioning correctly (and no clogs were found in the lines or fittings), check the primer as follows:

1. Remove the fuel line from the primer fitting at the carburetor.
2. Place the end of the fuel line just removed into a suitable container. Squeeze the fuel tank primer bulb to make sure the carburetor bowls are full of fuel.
3. Operate the primer choke lever twice. If fuel squirts from the disconnected fuel line into the container, the manual primer system is functioning correctly. If not, a kinked or restricted fuel line may be the problem, or if no kinks/clogs are found, the primer is at fault. Check the primer nipple to ensure the nipple is free of obstructions.

■ The most probable cause of a malfunctioning primer system is internal leakage past the O-rings. Therefore if the primer itself is still suspected, proceed to Servicing the Manual Primer.

SERVICING THE MANUAL PRIMER

◆ See Figures 81, 82 and 83

Removal/Disassembly

◆ See Figures 81, 82 and 83

1. Disconnect and plug the inlet and outlet fuel lines to prevent loss of fuel and contamination.
2. Remove the primer lever from the plunger. Back off the large nut securing the assembly to the lower cowling and lift the assembly free of the powerhead.
3. Pry the retaining clip from the primer body housing. Pull out the end cap, plunger, and spool valve assembly. Slide the end cap from the plunger. Remove and discard the O-ring around the end cap.

■ Observe the 3 small O-rings, 2 on the spool valve and 1 around the plunger shaft. These 3 O-rings are made from a special material and must be replaced with a genuine OMC replacement part. Just mating O-ring sizes will not work!

4. Remove and discard the 3 O-rings.
5. Remove the large washer and spring from the plunger shaft.

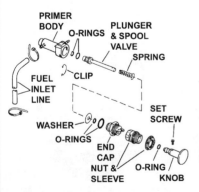

Fig. 81 Exploded view of a manual primer assembly

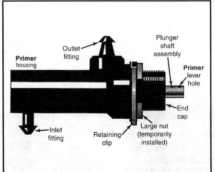

Fig. 82 The primer choke valve removed from the powerhead. The large nut is temporarily installed onto the threads of the end cap for safe keeping

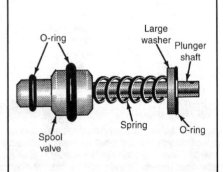

Fig. 83 Plunger shaft assembly removed from the choke housing with major parts identified

FUEL SYSTEM 3-39

Cleaning and Inspection

◆ See Figures 81, 82 and 83

1. Inspect the grooves of the spool valve and the shaft of the plunger for any scratches or burrs. Polish away any imperfections using crocus cloth. If a smooth finish cannot be obtained without removing excessive material, replace the spool valve and plunger assembly.
2. Check the plunger where the cross hole meets the inside hole. The slightest burrs around the cross hole will cause rapid O-ring wear. Remove any burrs and polish using crocus cloth.
3. Inspect the condition of the plunger spring, replace as required.
4. Test each of the one-way valves (there is 1 at each fuel fitting) by blowing through them in turn. Each valve is functioning correctly if it allows air to pass one direction, but not in the other direction. If a valve allows air to be drawn both in and out, the valve is defective. Individual valves are not serviceable. The primer body must be replaced.

■ The valves can also be checked using a syringe filled with isopropyl alcohol and a length of tube. Squeeze the syringe lightly to force alcohol through the hose, although it is permissible for a drop or two to pass through the wrong direction of a check valve, a steady stream indicates the valve has failed and must be replaced.

Assembly/Installation

◆ See Figures 81, 82 and 83

1. Install the two new O-rings around the spool valve. Slide the spring, followed by the large washer and the third O-ring, over the plunger.
2. Install a new O-ring over the end cap and place the end cap over the plunger end.
3. If desired, bench test the assembly before installation, as follows:
   a. Connect a 5 in. (127mm) long piece of hose to the large nipple on the primer assembly, then place the other end of the hose in a container of alcohol.
   b. Connect a length of hose to the small primer nipple and place the other end in a small container (preferable a graduated cylinder or measuring cup).
   c. Hold the primer horizontally (the same way it would be installed on the motor) and pump the plunger 10 times. The primer should deliver approximately 10cc of alcohol to the graduated cylinder total as a result of the 10 strokes.
   d. If the pump does not deliver sufficient volume, disassemble it again and check for torn, missing or dislodged O-rings.
4. Insert the assembly into the primer housing and install the retaining clip to secure everything together.
5. Slide the assembled primer into the opening in the lower cowling and thread the large nut over the protruding threads. Tighten the nut securely.
6. Install the fuel lines to the appropriate fittings and snap the choke lever into the vertical hole in the plunger.
7. Pressurize the fuel system using the fuel tank supply line primer bulb and thoroughly inspect for leaks.

Electric Fuel Primer

◆ See Figure 84

During engine operation, fuel from the fuel tank, passes through the fuel line, to the fuel pump, and into the carburetor. From the carburetor the fuel and air mixture passes through the crankcase and into the cylinder. Some (mostly remote electric start) 9.9-70 hp motors are equipped with an electric fuel primer system to aid with cold starts.

The primer performs the same function of a choke (aiding cold starting by making sure the engine receives a richer fuel mixture), but by opposite means. Whereas a choke reduces the amount of air provided to the combustion chamber (thus increasing the fuel portion of the air/fuel ratio), a primer works on the fuel side of the ratio by manually increasing the amount of fuel. The extra fuel provided by the manual primer enriches the air/fuel mixture for cold start purposes only. Manual or electronic activation of the primer on an engine that is at or near operating temperature can flood the motor preventing starting.

The basic design of the electric primer is that of a solenoid valve and, as such, does not pump fuel, but instead opens or closes a passage that allows fuel from a pressurized supply to flow. The circuit is designed to receive pressurized fuel from the fuel pump or the primer bulb. When power is

Fig. 84 Typical Johnson/Evinrude electric fuel primer assembly

applied to the circuit, the solenoid energizes, opening the valve. Once the circuit is deactivated, an internal spring closes the valve once again.

The primer system injects fuel directly into the cylinder. The system is controlled by the "push-in" type key switch. As the key is pushed in, the solenoid is activated. The electric primer system consists of a solenoid valve, distribution lines, and injection nozzles.

The electric primer is equipped with a manual lever that can be used in the event of battery or solenoid failure. When the lever is rotated by hand it will physically move the solenoid valve to the open position, allowing fuel from the pump or primer bulb to flow through the priming system. When used, the valve must be manually closed immediately following engine start-up to prevent spark plug fouling.

✳✳ CAUTION

If the fuel tank has been exposed to direct sunlight, pressure may have developed inside the tank. Therefore, when the solenoid lever is moved to the manual position, an excessive amount of fuel may be forced into the cylinders. As a safety precaution, under possible fuel tank pressure conditions, the fuel tank cap should be opened slightly to allow the pressure to escape before attempting to start the engine.

The electric primer assembly is also equipped with a maintenance or fogging fitting. Under a removable plastic cap on one end or side of the valve is a Schraeder valve. The valve is provided as an easy way to add fogging oil or OMC Engine Tuner to the combustion chambers.

TESTING

◆ See Figure 84

An inoperable electric primer will cause hard start or possibly even a no start condition during attempts to start a cold motor. The colder the ambient temperature, the more trouble an inoperable primer will cause. A primer with internal leakage (weak spring and/or stuck open valve allowing a constant supply of additional fuel) will cause rich running conditions that could include hesitation, stumbling, rough running, especially at idle and lead spark plug fouling.

Function Test

◆ See Figures 84 and 85

If the motor is operable, but trouble is suspected with the primer system, perform a function test with the engine running. Although this test can be conducted on a flush-fitting, engine speed will reach 2000 rpm and it is much safer to conduct the test in a test tank or with the boat launched.

1. Start and run the engine until it reaches normal operating temperature.

3-40 FUEL SYSTEM

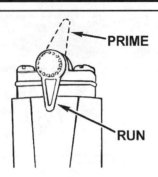

Fig. 85 The electric primer is also equipped with a manual override valve

2. Once the engine warms, set the throttle so it runs at 2000 rpm.
3. Push inward on the keyswitch. If the primer is operating correctly, the engine should run rich and speed should drop to about 1000 rpm.
4. If the keyswitch makes no difference in operation, move the manual lever on the solenoid housing to the open or prime position. The engine should now run rich and speed should drop to about 1000 rpm. If it does not respond, suspect a clog in the system. If it responds this time, but not when the switch was depressed, suspect a problem with the activation circuit. Refer to the Solenoid Test to decide if the problem is the solenoid motor or the activation circuit. Refer to the wiring diagrams to help troubleshoot the circuit, as necessary.
5. If a clog is suspected, stop the engine, then remove the primer hose from its fitting(s). Check each fitting for clogs using a syringe filled with isopropyl and a clear vinyl 1/8 in. inner diameter hose. Attach the hose to the fitting being checked and press lightly on the syringe. Fluid will move through the fitting unless it is clogged.
6. If any clogs are found, use a thin pick to carefully clean the fitting. OMC makes a cleaning tool for this purpose (#326623).
7. If no clogs are found, perform the Solenoid Check procedure to see if the problem lies within the solenoid motor itself or the circuit.

✱✱ CAUTION

When checking the fuel lines of the primer system, pay particular attention to any evidence of a crack in a fuel line that may permit fuel to escape and cause a very hazardous condition.

Solenoid Check

◆ See Figures 84, 86 and 87

Use an ohmmeter to check condition of the solenoid to determine if a system fault is localized to the solenoid motor or if you must instead troubleshoot the balance of the activation circuit. Refer to the wiring diagrams to assist with circuit troubleshooting, as necessary.
1. Disconnect the primer solenoid wiring, then connect an ohmmeter to the solenoid between the purple/white stripe wire and the black ground wire. The ohmmeter should indicate 5.5 plus/or minus 1.5 ohms (4.0-7.0 ohms). If the reading is not within the prescribed range, the solenoid is defective and must be replaced.

■ Remember that ohmmeter readings will vary with temperature and test specifications are designed for ambient/component temperatures around 68°F (20°F).

2. Use a syringe filled with isopropyl alcohol to lightly pressurize the fuel inlet fitting. With the no power applied to the circuit and the lever in the run position (facing the primer body) no fluid should pass through the inlet valve.
3. Move the lever to the prime position (facing away from the primer body) or apply power to the circuit, fluid should come out of the outlet fitting(s).
4. If necessary disassemble the primer for component inspection or replacement.

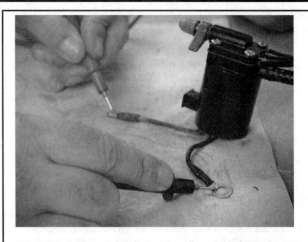

Fig. 86 Use a DVOM or ohmmeter to probe across the primer solenoid terminals

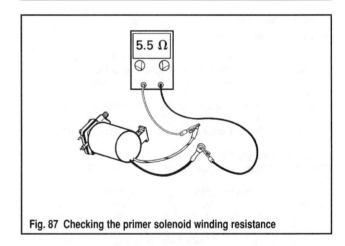

Fig. 87 Checking the primer solenoid winding resistance

REMOVAL AND INSTALLATION

◆ See Figures 84, 88 and 89

1. Disconnect the negative battery cable for safety.

■ Wrap the end of each fuel line using a shop rag to catch any remaining spray when the line is removed from the fitting.

2. Note the fuel line routing, then tag and disconnect the lines from the primer fitting.
3. Disconnect the solenoid wiring (usually a bullet connector and a wire terminal).
4. Loosen the retainer(s) securing the primer and bracket assembly to the powerhead, then remove the primer from the motor.
5. If necessary, remove the 4 solenoid cover screws, then disassemble the primer for component inspection or replacement.

To install:
6. If necessary, reassemble the primer and tighten the 4 cover screws securely.
7. Position the primer solenoid and secure using the bracket retainer(s).
8. Connect the wiring, followed by the fuel lines. Secure the fuel lines in the same positions noted during removal. Be sure to install new wire ties if any were cut and removed for line repositioning.
9. Pump the fuel tank supply line primer bulb in order to pressurize the system, then check for leaks. Make sure pressure is felt in the bulb.
10. Connect the negative battery cable.

FUEL SYSTEM 3-41

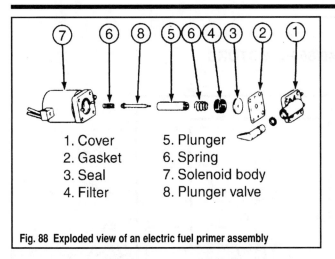

Fig. 88 Exploded view of an electric fuel primer assembly

1. Cover
2. Gasket
3. Seal
4. Filter
5. Plunger
6. Spring
7. Solenoid body
8. Plunger valve

Fig. 89 If overhaul is necessary, remove the 4 solenoid cover screws in order to disassemble it

ELECTRONIC FUEL INJECTION (EFI)

✴✴ CAUTION

Disconnect the negative battery cable ANYTIME work is performed on the engine, especially when working on the fuel system. This will help prevent the possibility of sparks during service (from accidentally grounding a hot lead or powered component). Sparks could ignite vapors or exposed fuel. Disconnecting the cable on EFI motors will also help prevent the possibility fuel spillage if the key is turned to start while the high-pressure fuel system is open.

✴✴ CAUTION

Fuel leaking from a loose, damaged or incorrectly installed hose or fitting may cause a fire or an explosion. ALWAYS pressurize the fuel system and run the motor while inspecting for leaks after servicing any component of the fuel system.

Description and Operation

◆ See Figures 90 and 91

The fuel system used on 40-70 hp 4-stroke models is an electronically control multi-port fuel injection system, not unlike that used on modern automobiles. If fact, all of the EFI motors covered here are built by Suzuki for Johnson/Evinrude and, the 70 hp motor is essentially a marinized version of the motor found in the Suzuki Sidekick and Geo Tracker for many years.

The EFI system itself can be segmented into 3 inter-related sub-systems:
- Low pressure fuel circuit
- High pressure fuel circuit
- Electronic engine controls

The low pressure circuit delivers fuel from the tank to the vapor separator via a mechanical diaphragm-displacement fuel pump, very similar in form and function to those used on Johnson/Evinrude carburetor 4-stroke motors. The low pressure circuit may be rigged with a portable or built-in fuel tank and normally contains both an inline filter and a fuel primer bulb. The role of the low pressure circuit is to keep the high pressure circuit electric pump supplied with sufficient fuel to meet engine operating demands.

The high pressure circuit is necessary to ensure proper operation of the fuel injectors (whose operation depends upon a constant supply of highly pressurized fuel). The circuit uses a high pressure pump mounted within the vapor separator tank to build system pressure and maintain it through a fuel pressure regulator that vents excess fuel/pressure back into the vapor separator tank. Fuel injectors, mounted between a fuel rail assembly and the intake manifold or cylinder head, deliver the high pressure fuel directly before the cylinder head mounted intake valves for each cylinder. The high pressure circuit is activated by the engines Electronic Control Unit (ECU) for a few seconds every time the ignition is turned ON and constantly during engine cranking or operation. Fuel injectors, also controlled by the ECU, are electronic solenoid valves that open against internal spring pressure when activated (allowing fuel to spray from the nozzle tips). Injector activation occurs sequentially, immediately before each intake valve is ready to open.

The electronic engine control system monitors and controls engine operation in order to properly meter fuel delivery to match operating conditions. The role of the ECU and the fuel injectors is to do electronically what the carburetor does mechanically on other motors. The precise control made possible by the ECU's microprocessors allows an EFI engine to increase both reliability and performance while simultaneously decreasing harmful emissions. And, those are all good things, right?

The electronic engine control system monitors engine operation through a number of electronic switches and sensors. The role of the sensors and switches is to translate mechanical information such as engine temperature, speed, throttle position or even the exact position of the pistons (on what portion of each stroke, each piston is) into electronic data for use by the ECU. The system is equipped with the following sensors:
- Camshaft position (CMP) sensor
- Crankshaft position (CKP) sensor
- Closed throttle position (CTP) switch
- Cylinder temperature (CT) sensor
- Intake air temperature (IAT) sensor
- Manifold air pressure (MAP) sensor

In fuel injection terms this type of system is known as a speed-density injection system. This is because the basic engine fuel mapping decisions are made by the ECU based on a comparison between the engine rpm (speed) and the manifold pressure (air density). The basic, pre-programmed, fuel mapping is then modified based upon input from the remaining sensors. Specifically, cylinder and air temperatures are taken into account. When a CTP switch signal is received (the throttle is closed), the ECU uses a fuel delivery strategy specifically intended for idling/trolling.

Besides the amount of fuel delivered (controlled through the fuel injectors) the ECU can also control the amount of air delivered during idle conditions. This is accomplished through the Idle Air Control (IAC) valve.

Finally, the ECU uses inputs from the CMP sensor(s) to control the fully electronic ignition system and make all ignition timing adjustments.

Troubleshooting Electronic Fuel Injection

On carbureted outboards fuel is metered through needles and valves that react to changes in engine vacuum as the amount of air drawn into the motor increases or decreases. The amount of air drawn into carbureted motors is controlled through throttle plates that effectively increase or decrease the size of the carburetor throat (as they are rotated open or closed).

In contrast, fuel injected engines use a computer control module to regulate the amount of fuel introduced to the motor. The module or Electronic Control Unit (ECU) monitors input from various engine sensors in order to receive precise data on items like engine position (where each piston is on its 4-stroke cycle), engine speed, engine and air temperatures, manifold pressure and throttle position. Analyzing the data from these sensors tells the engine exactly how much air is drawn into the motor at any given moment and allows the ECU to determine how much fuel is required.

3-42 FUEL SYSTEM

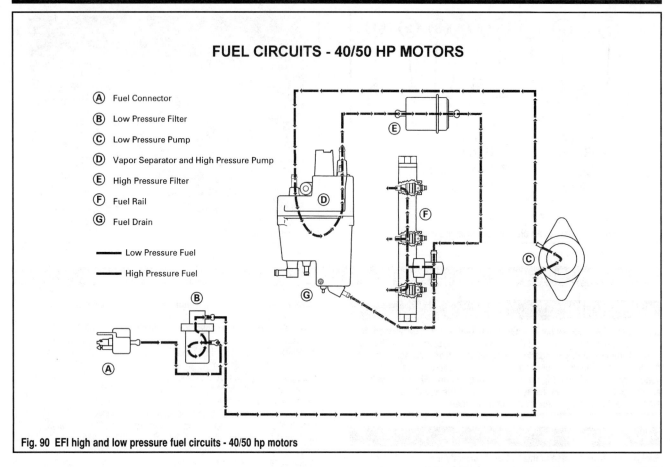

Fig. 90 EFI high and low pressure fuel circuits - 40/50 hp motors

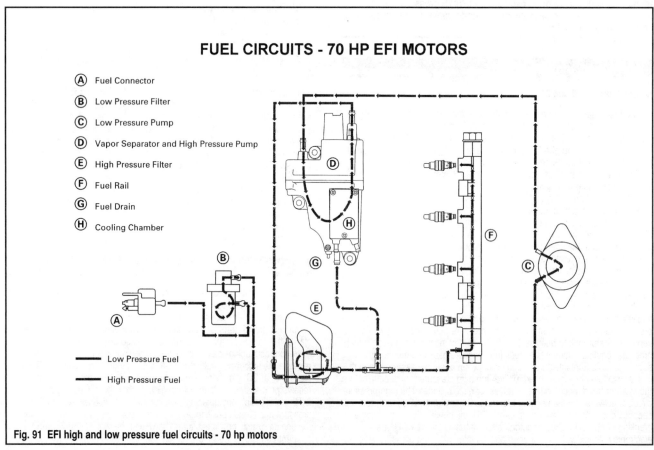

Fig. 91 EFI high and low pressure fuel circuits - 70 hp motors

FUEL SYSTEM 3-43

The ECU will energize (open) the fuel injectors for the precise length of time required to spray the amount of fuel needed for that intake stroke. In actuality, the injectors are not just activated and held-open as much as they are pulsed, opened and closed rapidly for the correct total amount of time necessary to spray the desired amount of fuel. This electronically controlled, precisely metered fuel spray or "fuel injection" is the heart of a modern fuel injection system and the main difference between a fuel injected and carburetor motor.

Troubleshooting a fuel injected motor contains similarities to carbureted motors. Mechanically, the powerhead of a 4-stroke fuel injected motor operates in the same way as a carbureted motor. There still must be good engine compression and mechanical timing for either engine to operate properly. Wear or physical damage will have virtually the same affect upon either motor. Furthermore, the low pressure fuel system that supplies fuel to the reservoir in the vapor separator tank operates in the same manner as the fuel circuit that supplies gasoline to the carburetor float bowl.

The major difference in troubleshooting engine performance on EFI motors is the presence of the ECU and electronic engine controls. The complex interrelation of the sensors used to monitor engine operation and the ECU used to control both the fuel injection and ignition systems makes logical troubleshooting all that much more important.

Before beginning troubleshooting on an EFI motor, make sure the basics are all true. Make sure the engine mechanically has good compression (refer to the Compression Check procedure that is a part of a regular Tune-Up). Make sure the fuel is not stale. Check for leaks or restrictions in the Lines and Fittings of the low pressure fuel circuit, as directed in this section under Fuel Tank and Lines. EFI systems cannot operate properly unless the circuits are complete and a sufficient voltage is available from the battery and charging systems. A quick-check of the battery state or charge and alternator output with the engine running will help determine if these conditions are adversely affecting EFI operation.

■ **Loose/corroded connections or problems with the wiring harness cause a large percentage of the problems with EFI systems. Before getting too far into engine diagnostics, check each connector to make sure they are clean and tight. Visually inspect the wiring harness for visible breaks in the insulation, burn spots or other obvious damage.**

In order to help find electronic problems with the EFI system, the ECU contains a self-diagnostic system that constantly monitors and compares each of the signals from the various sensors. Should a value received by the ECU from one or more sensors fall outside certain pre-determined ranges the ECU will determine there is a problem with that sensor's circuit. Basically the ECU compares signals received from different sensors to each other and to real world possible values and makes a decision if it thinks one must be lying. For instance, if the ECU receives camshaft position sensor signals that show the engine is rotating, but also receives a signal from the crankshaft position sensor that says the engine is stopped, it knows one is wrong. Similarly, if it receives a ridiculous signal, say the intake air temperature suddenly provides a signal above 338°F/170°C, the ECU will know there is something wrong with that signal. Depending on the severity of the fault or faults, the engine will continue to run, substituting fixed values for the sensors that are considered out of range. Under these circumstances, engine performance and economy may become drastically reduced.

■ **Check for diagnostic codes as described in this section BEFORE disconnecting the battery. If codes are not present, yet problems persist, use the symptom charts to help determine what further components or systems to check.**

When a fault is present the ECU will store a diagnostic code in memory. The ECU will illuminate the Check Engine light in the gauge package and sound the warning horn to alert the operator. Codes can be retrieved by following specific procedures and then used to help determine what components and circuits should be checked for trouble. Remember that a fault code doesn't automatically mean that a component (such as a sensor) is bad, it means that the signal received from the sensor circuit is missing or out of range. This can be caused by loose or corroded connections, problems with the wiring harness, problems with mechanical components (that are actually causing this condition to be true), or a faulty sensor.

In addition, troubleshooting charts are provided based on symptoms that should be used to help narrow down problems in engine performance. If no diagnostic codes are present, refer to these charts (after performing the basic checks mentioned earlier) to help determine what further components or systems to check.

Once components or circuits that require testing have been identified, use the testing procedures found in this section (or other sections, as applicable) and the wiring diagrams to test components and circuits until the fault has been determined.

Keep in mind that although a haphazard approach might find the cause of problems, only a systematic approach will prevent wasted time and the possibility of unnecessary component replacement. In some cases, installing an electronic component into a faulty circuit that damaged or destroyed the previous component, will instantly destroy the replacement. For various reasons, including this possibility, most parts suppliers do not accept returns on electrical components.

Self Diagnostic System

◆ See Figures 92 thru 95

READING & CLEARING CODES

◆ See Figures 92 thru 95

■ **Certain electrical equipment such as stereos and communication radios can interfere with the electronic fuel injection system. To be certain there is no interference, shut these devices off when troubleshooting. If a check engine light illuminates immediately after installing or re-rigging an existing accessory, reroute the accessory wiring to prevent interference.**

When the electronic engine control system detects a problem with one of its circuits, the ECU will activate the check engine light found in the gauge pack and sound the warning horn. As a result of most faults, the ECU will ignore the circuit signal and enter a fail-safe mode designed to keeps the boat and motor from becoming stranded. During fail-safe operation the ECU will provide a fixed substitute value for the faulty circuit. During fail-safe operation the engine will run, but usually with reduced performance (power and economy). Substitute values are as follows:
- MAP sensor: 319-475mmHg @ 750-4000 rpm.
- CMP sensor: 1 simultaneous injection for all cylinders for every 2 crankshaft rotations (determined by CKP sensor signals) and the firing of each spark plug 1 time every crankshaft rotation.
- CT or EMT sensors: 140°F (60°C)
- IAT sensor: 113°F (45°C)

Once a malfunction ceases (has been fixed or the circuit signal returns to something within the anticipated normal range), the warning system and engine operation will return to normal.

The gauge check engine light is not only used as a warning, but it is also used to read the diagnostic trouble codes. A code is normally displayed by using short flashes of the check engine light. The manufacturer has diagnostic software for the 4-stroke EFI engines that can also be used to

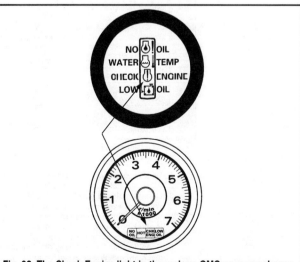

Fig. 92 The Check Engine light in the various OMC gauge packages is used to alert the operator if the ECU detects an electronic fault

3-44 FUEL SYSTEM

Priority	Failed Circuit	Code No.	Gauge LED	S.L.O.W.
1	MAP Sensor 1 (electrical)	34	Check Engine	No
2	CKP Sensor(s)	42	Check Engine	No
3	IAC Valve/Idle Air Screw	31	Check Engine	No
4	CMP Sensor	24	Check Engine	No
5	CTP Switch	22	Check Engine	No
6	Cylinder Temperature Sensor	14*	Check Engine	No
7	IAT Sensor	23	Check Engine	No
8	MAP Sensor 2 (vacuum loss)	32	Check Engine	No
9	Rectifier/Regulator (overcharging)	11	Check Engine	No
10	Manifold Temperature Sensor	15*	Check Engine	No

* Denotes that code is set by a faulty circuit/sensor NOT an overheat condition.

Fig. 93 Diagnostic trouble codes - 40/50 hp EFI motors

Priority	Failed Circuit	Code No.	Gauge LED	S.L.O.W.
1	MAP Sensor 1 (electrical)	34	Check Engine	No
2	CKP Sensor(s)	42	Check Engine	No
3	IAC Valve/Idle Air Screw	31	Check Engine	No
4	CMP Sensor	24	Check Engine	No
5	CTP Switch	22	Check Engine	No
6	Cylinder Temperature Sensor	14	Check Engine	No
	Cylinder Overheat	14	Temp	Yes
7	IAT Sensor	23	Check Engine	No
8	MAP Sensor 2 (vacuum loss)	32	Check Engine	No
9	Rectifier/Regulator (overcharging)	11	Check Engine	No
10	Manifold Overheat	15	Temp	Yes
none	Low Battery Voltage	none	Check Engine	No
none	Over-Rev Sensor	none	none	Yes
none	Oil Pressure Switch	none	No Oil	Yes
none	Neutral Start Switch - Engine	none	none	No

Fig. 94 Diagnostic trouble codes - 70 hp EFI motors

Code	Failed Circuit	Condition	Tool	Normal ①
34	MAP Sensor 1 ECU Terminal - D7	• No signal (with engine running) • Receiving an out of range 1.45-33.86 in. Hg (4.90-114.35 kPa) (0.05-4.84V) signal (with engine running)	DVM	②
42	CKP Sensor ECU Terminals - E1, E3, E4	• No signal from any CKP sensor while receiving 8 signals from other CKP sensors • No signal from one CMP sensor while receiving 6 signals from CMP sensor (one CKP sensor fail)	PRV POS50	4.5V or more at cranking
31	IAC Valve / Bypass Air Screw Adjustment ③	• IAC valve operates at 90% duty or higher when CTP switch is "ON" • IAC valve operates at 10% duty or lower when CTP switch is "ON"	Diagnostic Software	20% ± 5%
24	CMP Sensor ECU Terminal - E2	• No signal while receiving 12 signals from CKP sensors	PRV SEN5	3.5V ④
22	CTP Switch ECU Terminal - D8	• Receiving "ON" signal when engine speed is 2500 RPM or higher and intake manifold pressure is 14.96 in. Hg (50.52 kPa) or higher	DVM	0V @ idle 5V @ Off-idle
14	Cylinder Temperature Sensor ECU Terminal - D3	• No signal • Receiving an out of range -50 to 338° F (-46 to +170° C) (0.10 - 4.84V) signal	DVM	1.2V @ 122° F (50° C)
23	IAT Sensor ECU Terminal - D6	• No signal • Receiving an out of range -50 to 336° F (-46 to +169° C) (0.10 - 4.88V) signal	DVM	2.4V @ 72° F (22° C)
32	MAP Sensor 2 (Sensor Hose) ECU Terminal - D7	• Receiving unchanging signal regardless engine speed change ⑤	DVM	②
11	Rectifier/Regulator (Overcharging)	• Receiving 16 volts or higher signal	DVM	14.3V @ 1500 RPM
15	Exhaust Manifold Temperature Sensor ECU Terminal - C9	• No signal • Receiving an out of range -50 to 338° F (-46 to +170° C) (0.10-4.84V) signal	DVM	1.2V @ 122° F (50° C)

DVM = Digital Volt Ohmmeter (DVOM)

1 - Normal values will vary slightly due to ambient temperatures and brand/type of DVOM used.
2 - Approximately 1.4v at idle, 2.2v at 2500 rpm or 4v at full throttle (WOT)
3 - Code/Conditions may be caused by IAC valve failure OR by incorrect bypass screw adjustment. When IAC valve is always closed or there is insufficient bypass air, the ECU will increase IAC valve duty cycle in an attempt to maintain idle/trolling speeds. The opposite is also true if the IAC valve is stuck open or there is too much bypass air (ECU will decrease duty cycle).
4 - Value while engine is running.
5 - Condition can be caused by disconnected, kinked or clogged MAP hose or a clogged filter for the inet manifold. Will cause rich engine operation.

Fig. 95 Trouble Codes and conditions that set them vs. normal operating parameters - 40/50 hp Motors

retrieve codes and other stored ECU operational data. If available, a diagnostic connector is located on the front starboard side of the engine. Follow the instructions that come with the software pack to use the diagnostic connector.

Proceed as follows to display stored codes:

1. If the engine is running, shut the engine **OFF** using the ignition key switch.
2. Turn the ignition switch to **ON** without starting the engine. All 4 LEDs in the gauge will illuminate and then go out, one at a time.
3. If present, a trouble code will begin flashing on the Check Engine LED. Count the flashing to determine the code(s). If more than one is present, codes will flash in the order of priority listed in the accompanying code charts.

■ **Interpret codes by counting the flashes. There will be a short pause between digits of the 2-digit code. A longer pause indicates the start of a different number. A code will flash 3 times, then, if there is more than one code, the next code will flash 3 times and the sequence will repeat.**

4. Once all the codes are counted/recorded, watch the sequence again to verify then shut the ignition switch **OFF**
5. Compare the codes with the accompanying code tables to determine the defective circuit. Remember that a codes does NOT necessarily mean a given component is at fault, it means that the ECU sees signal that is out of the normal, predetermined operating range. The problem may also lie with the wiring, another system or component that would make a circuit read out of specification.

■ **Make sure all connections for that circuit and related components of the electrical system are clean and tight before troubleshooting the circuit. Bad wiring or connections can cause out of range signals and set trouble codes.**

6. Test the sensor or system components as described in this section. When troubleshooting, always start with the easiest checks/fixes and work toward the more complicated.
7. Once the problem has been corrected, connect a source of cooling water to the flushing system, then start and run the engine to clear the codes and verify proper operation. If the problems are gone, all LEDs should extinguish within 3 minutes of starting the motor.

Sensor and Circuit Resistance/Output Tests

◆ See Figures 96 thru 99

TESTING EFI COMPONENTS

◆ See Figures 96 thru 99

Before conducting resistance or voltage checks on a component/circuit, make sure all electrical connections in the system are clean and tight. Visually check the wiring for obvious breaks, defects or other problems.

Keep in mind that although a haphazard approach might find the cause of problems, only a systematic approach will prevent wasted time and the possibility of unnecessary component replacement. In some cases, installing an electronic component into a faulty circuit that damaged or destroyed the previous component, will instantly destroy the replacement. For various reasons, including this possibility, most parts suppliers do not accept returns on electrical components.

Make sure you perform each test procedure and especially, make all test connections as described. If a component tests out of range (faulty), but the reading is close to the service limit, bring the component to your marine dealer so they can verify your result before purchasing the replacement. In some cases, your dealer may be willing to check a sensor against the reading on a new one to verify whether or not your sensor is actually faulty.

Although resistance checks are relatively easy to conduct, their results can be difficult to interpret. Remember that resistance in any electrical component or circuit will vary with temperature. Also, ohmmeters will vary with quality, meaning that readings can vary on the same component between different meters. The specifications provided in this manual are based upon the use of a high-quality digital multi-meter applied to a component that is currently at about 68°F (20°).

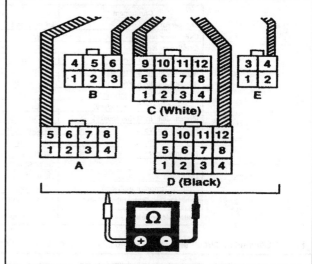

Fig. 96 ECU harness connector terminal identification (pin out) - view facing connectors after they are removed from the back of the ECU

■ **More information on electrical test equipment is can be found in the General Information section, while additional information on basic electrical theory and electrical troubleshooting can be found in the Electrical section.**

As a general rule, the resistance of a circuit or component will increase (rise) as temperature increases. Although this is true for most resistor and windings, some manufacturers use negative temperature coefficient sensors for certain functions. A negative temperature coefficient sensor reacts in an opposite manner, meaning that its resistance decreases (lowers) as temperature increases (rises) and vice-versa.

Most components of the EFI system are checked using resistance or voltages checks. This means that a voltmeter or ohmmeter is connected to the component or circuit under the proper circumstances.

For resistance checks, no voltage must be applied to the circuit (except that as provided by the meter to perform the check). Typically, the circuit or component wiring is isolated, by disconnecting the wiring from the component itself or somewhere else in the circuit, then applying the meter probes across 2 terminals that connect through the component. When conducting resistance checks, you **must** isolate the ECU from the meter in order to prevent the possibility of damage to the control module. By identifying the proper circuit you can conduct tests of the entire component circuit right from the harness connectors that attach to the ECU. Otherwise, use the wiring diagrams and the test procedures to determine other points to disconnect the wiring and test the circuit/component.

We recommend you work in one of two directions. Either test the component first, then work your way back through the component wiring toward the ECU, or disconnect the harness connectors from the ECU, testing the entire circuit first, then working your way toward the component. A bad test reading at the ECU harness connector must be verified by making sure it is not the result of a bad wiring harness or connection along the way to the component. Likewise, a good reading at a component (when there is a bad reading on the circuit as noted by the presence of a trouble code or by a reading at the ECU harness) must be verified by checking the circuit wiring for trouble.

When checking the circuit wiring, isolate the harness by disconnecting it from both the component and the ECU, then using an ohmmeter to check resistance from one connector to the next on each wire. Wiggle the wire and connectors while conducting the test to see if reading fluctuate. Although no specifications are provided for wiring resistance, all sensor/switch wiring for these motors should have very little resistance. Also, be sure to check wiring for shorts to ground (by checking resistance between one terminal and a good engine ground), and for shorts to power (by checking a terminal using a voltmeter and connecting the other probe to a good engine ground or the negative battery cable).

When conducting voltage tests, the circuit must be complete in order to receive a reading. In most cases, this means the wiring must be left connected and the meter probes should be carefully inserted through the back of the connector (called back-probing) in order to get a reading without

3-46 FUEL SYSTEM

Component	Terminal Probe	Ω @ 68° F (20° C)
CKP Sensor No. 1	E4 (R/B) to D1 (B/W)	
CKP Sensor No. 2	E3 (W/B) to D1 (B/W)	168 - 252 Ω
CKP Sensor No. 3	E1 (R/W) to D1 (B/W)	
Ignition Coil No. 1 (Primary)	A5 (O) to B5 (Gr)	
Ignition Coil No. 2 (Primary)	A1 (Bl) to B5 (Gr)	1.9 - 2.5 Ω
Ignition Coil No. 3 (Primary)	A3 (G) to B5 (Gr)	
Ignition Coil No. 1 (Secondary)	B5 (Gr) to No. 1 spark plug cap	
Ignition Coil No. 2 (Secondary)	B5 (Gr) to No. 2 spark plug cap	8.1 - 11.1 kΩ
Ignition Coil No. 3 (Secondary)	B5 (Gr) to No. 3 spark plug cap	
Fuel Injector No. 1	A4 (O/B) to B5 (Gr)	
Fuel Injector No. 2	A7 (B/Y) to B5 (Gr)	11.0 - 16.5 Ω
Fuel Injector No. 3	A8 (R/W) to B5 (Gr)	
IAC Valve	B4 (B/R) to B5 (Gr)	21.5 - 32.3 Ω
IAT Sensor	D6 (Lg/B) to D1 (B/W)	32° F (0° C) : 5.3 - 6.6 kΩ
Cylinder Temperature Sensor	D3 (Lg/W) to D1 (B/W)	77° F (25° C) : 1.8 - 2.3 kΩ
Exh. Manifold Temperature Sensor	C9 (V/W) to D1 (B/W)	122° F (50° C) : 0.73 - 0.96 kΩ
		135° F (75° C) : 0.33 - 0.45 kΩ
		(Thermistor characteristic)
Starter Motor Relay	D11 (Y/G) to Ground	3.5 - 5.1 Ω

Fig. 97 Component/circuit test values for 40/50 hp motors when testing at the ECU harness connectors

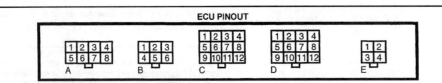

Fig. 98 ECU terminal identification (pin out) - Rear of control module shown

Terminal	Color	Circuit	Standard Voltage	Condition / Remarks
A1	Bl	No. 2 Ignition (−)	Approx. 12V	Key switch ON
A2	B	Ground for power source	−	−
A3	G	No. 3 Ignition (−)	Approx. 12V	Key switch ON
A4	B/Y	No. 2 Fuel injector (−)	Approx. 12V	Key switch ON
A5	O	No. 1 Ignition (−)	Approx. 12V	Key switch ON
A6	B	Ground for power source	−	−
A7	R/W	No. 3 Fuel injector (−)	Approx. 12V	Key switch ON
A8	O/B	No. 1 Fuel injector (−)	Approx. 12V	Key switch ON
B1	B/W	Fuel Pump (−)	Approx. 0V	For 3 seconds after Key switch ON or while cranking engine
			Approx. 12V	Other than above (Key switch ON)
B2	P/B	Ground for ECU main relay	−	−
B3	B	Ground for ECU	−	−
B4	B/R	IAC valve solenoid (−)	Approx. 12V	Key switch ON
B5	Gr	ECU power source	Approx. 12V	Key switch ON
B6	B	Ground for ECU	−	−
C1	Bl/R	Emergency stop switch	Approx. 5V	Key switch ON, clip ON
			Approx. 0V	Key switch ON, clip OFF
C2	Y/B	Tachometer	−	−
C3	G/W	CHECK ENGINE lamp	−	−
C4	G/Y	TEMP lamp	−	−
C5	−	Not used	−	−
C6	Bl/B	No OIL lamp	−	−
C7	Bl/W	Not used	−	−
C8	Br	Neutral switch	Approx. 0V	Key switch ON, in NEUTRAL
			Approx. 5V	Key switch ON, in gear
C9	V/W	Exh. manifold temperature sensor	0.14 - 4.75V	Key switch ON
C10	B	PC communication	−	−
C11	B	PC communication	−	−
C12	O	Not used	−	−
D1	B/W	Ground for sensors	−	−
D2	R	Power source for MAP sensor	Approx. 5V	Key switch ON
D3	Lg/W	Cylinder temperature sensor	0.14 - 4.75V	Key switch ON
D4	Bl	Oil pressure switch	Approx. 5V	While engine running
			Approx. 0V	Other than above (Key switch ON)
D5	B/G	O₂ feedback/PC communication	−	−
D6	Lg/B	IAT sensor	0.14 - 4.75V	Key switch ON
D7	W	MAP sensor	0.20 - 4.50V	Key switch ON
D8	Lg/R	CTP switch	Approx. 5V	Key switch ON, throttle Off-idle
			Approx. 0V	Key switch ON, throttle fully closed
D9	B	PC communication	−	−
D10	−	Not used	−	−
D11	Y/G	Engine start switch (signal)	6 - 12V	While engine cranking
			Approx. 0V	Other than above (Key switch ON)
D12	O/Y	PC communication	−	−
E1	R/W	CKP sensor No. 3	−	−
E2	O/G	CMP sensor	Approx. 1V or 5V	Key switch ON
E3	W/B	CKP sensor No. 2	−	−
E4	R/B	CKP sensor No. 1	−	−

Fig. 99 ECU pin-out circuit identification and normal voltage readings - values for completed circuit only (back-probed at the harness connectors, with all connectors attached to the ECU)

FUEL SYSTEM 3-47

opening and disrupting circuit operation. This is true for the ECU pin-out voltages (circuit actuation and reference voltages) provided for 40/50 hp motors. Take great care when testing circuits, especially those on an ECU. Connecting the wrong circuits through the meter can cause significant damage, even destroy an ECU. Also, take care not to damage the wire or connector when backprobing. Never force the meter probe into the connector, use a little finesse.

On 40/50 hp motors, the manufacturer has provided ECU harness connector, circuit identifier and test values. In a few cases, these values for tests of the entire circuit vary slightly than tests conducted at the component itself, be sure to use the values given in the accompanying chart when testing at the ECU harness for these engines.

On 70 hp motors, the manufacturer does NOT provide information specifically for the ECU harness testing. That does not mean you cannot conduct tests in the same manner, but you may need to make a slight allowance in the specifications for readings taken on 70 hp motors. When conducting testing on the at the ECU harness connectors, use the values given under the individual test procedures for those components. Also, use the wiring diagrams and wire colors provided to help verify ECU harness connections.

Engine Symptom Diagnostic Charts

ENGINE WON'T CRANK

System/Component	Causes/Effects	Actions	Code
Neutral Switch in Engine	Faulty or disconnected	Repair as necessary	none
Neutral Start Switch in Control Box	Faulty or disconnected	Repair as necessary	none
Starter Relay	Faulty or disconnected	Repair as necessary	none
Broken Wire or Connection	No power to starter relay primary or connection to ground	Repair as necessary - remember to look for bad ground connections	none
Starter Motor	Failed	Test and repair as necessary	none
Main Battery Switch	Not turned on	•	none
30A Fuse in Engine	No power to key switch	Replace and look for overloaded circuit	none
Key Switch	Failed	Replace	none

ENGINE CRANKS BUT WON'T RUN - 40/60 HP

System/Component	Causes/Effects	Actions	Code
No Fuel	Tank empty	•	none
	Restricted or leaking fuel delivery system	Vacuum test boat system(4 in./Hg max) - check for damaged primer bulb, antisiphon valve or poor connections that leak air	none
	Failed low pressure pump	Engine will run if primer bulb is constantly squeezed - replace as necessary	none
	Hose not connected	•	none
	One or both fuel filters completely plugged	Replace	none
	Vapor separator needle stuck closed	No or low pressure at fuel rail. Repair vapor separator	none
	Blown high pressure pump fuse (15A)	Replace	none
	Failed high pressure pump (pump should run for 3 seconds at key on) or plugged intake screen on pump	(1) Verify fuel pump fuse (2) Repair separator/fuel pump assembly	none
	Failed pressure regulator	Low pressure at fuel rail - replace fuel regulator in vapor separator	none
Crankshaft Sensor (CKP)	Loss of signal. No spark and no injector signal to that cylinder	Engine may start and run by using the warm-up lever - test and replace as necessary. See component test	42
Camshaft Sensor (CMP)	Loss of signal. See default values - this section. Shop tach will read double.	Engine may start and run by using the warm-up lever. Check CMP signal or Resistance	24
MAP Sensor (MAP2)	Hose off or plugged - No signal change. See default Values - this section	Engine may start and run by using the warm-up lever, but will run very rich and rough at lower speeds - test and replace as necessary	32
Failed ECU	Would probably affect other engine functions	Replace	none
IAC Valve	Disconnected or plugged	Hard to start - Low idle speed	
Main Relay	Cranks but won't start	Check also; connections, key switch, ECU signal	
Emergency Stop Switch	In stop position	•	
No B+ to Injectors	Failed main relay, fuse, wire, or connection	Check each component	
No B+ to Coils	Same as above	Check each component	
Mechanical Failure	Valve timing, valve clearance, powerhead damage, etc.	Repair as necessary	

3-48 FUEL SYSTEM

ENGINE IDLES INCORRECTLY - 40/50 HP

System/Component	Causes/Effects	Actions	Code
MAP Sensor MAP 1 - Open Circuit	Idle may be slightly high - Goes lean above 3000 RPM	Check for disconnected wires or failed sensor.	34
MAP Sensor MAP 2 - Hose Off	Hard to start - extremely rough at idle or may die - very rich idle	Check for disconnected or damaged hose - if OK test sensor	32
MAP Sensor MAP 2 - Hose Plugged	Lower than normal idle speed - @ 600 RPM. Won't accelerate	Check for pinched hose or plugged filter	32
Idle Air Control (IAC) - Open Circuit	Lower than normal idle speed - tends to die when throttle is quickly returned to idle	Check for disconnected or failed IAC valve. Use Diagnostic Software to verify IAC duty cycle (15-25%) after repair.	31
Idle Air Control (IAC) Plugged Hose	(1) Requires warm-up lever to start (2) Slow idle (3) Will set code in gear	Check for pinched hose, plugged silencer, or incorrectly connected hose. Use Diagnostic Software to verify IAC duty cycle (15-25%) after repair	none
Closed Throttle Switch (CTP) - Open Circuit	(1) Will return to idle speed immediately - not gradually decrease from 1200 RPM (2) Will idle faster than normal	Check for disconnected or damaged switch. Check for voltage change at ECU D8 terminal.	none
Idle Air Screw Misadjusted	Probably no bad running characteristics but IAC duty cycle will be out of normal range	Adjust idle air screw to attain 15-25% IAC duty cycle - abnormal IAC duty cycle can also be caused by other problems such as a vacuum leak or mechanical damage	maybe 31
Vacuum Leak at Intake Manifold	Faster than normal idle - IAC duty cycle will be extremely low	Spray around intake gasket areas with starter fluid while idling - engine will react when starter fluid hits leak	maybe 31
Overcooling of Engine	Faster than normal idle similar to a cold started engine	Check for a failed thermostat or debris in thermostat. Verify using Diagnostic Software or pyrometer	none
Incorrect Valve Adjustment	Excessive tappet noise or uneven compression	Adjust valve lash	none
Failed Injector	Low, rough idle - dies going into gear	Test injector signal and resistance	maybe 24 or 42
Failed Coil	Same as failed injector	Test coil signal and resistance	maybe 24 or 42
CKP Sensor	Low, rough idle	Loss of both injector and ignition signals on an affected cylinder. Test signal and resistance	42

ENGINE CRANKS BUT WON'T RUN - 70 HP

System/Component	Causes/Effects	Actions	Code
No Fuel	Tank empty		none
	Restricted or leaking fuel delivery system	Vacuum test boat system - check for damaged primer bulb, anti-siphon valve or poor connections that leak air	none
	Failed low pressure pump	Engine will run if primer bulb is constantly squeezed - replace as necessary	none
	Hose not connected		none
	One or both fuel filters completely plugged	Replace	none
	Vapor separator needle stuck closed	No or low pressure at fuel rail - repair	none
	Blown high pressure pump fuse (15A)	Replace and look for overloaded circuit (pump draws excessive current - check using ammeter)	none
	Failed high pressure pump (pump should run for 3 sec. at key on) or plugged intake screen on pump	Replace separator/fuel pump assembly	none
	Failed pressure regulator	Low pressure at fuel rail	none
Crankshaft Sensor (CKP)	Failed component or broken wire	Engine may start and run by using the warm-up lever - test and replace as necessary. Dashboard tach will read ½ actual RPM but Diagnostic Software will show correct RPM. Test using ohmmeter. Verify using Diagnostic Software	42
Camshaft Sensor (CMP)	Failed component or broken wire	Engine may start and run by using the warm-up lever - test using ohmmeter and replace as necessary. Verify using Diagnostic Software	24
MAP Sensor (MAP2)	Hose off	Engine may start and run by using the warm-up lever, but will run very rough at lower speeds - test and replace as necessary	32
Failed ECU	Would probably affect other engine functions	Replace	none
Mechanical Failure	Valve timing, valve clearance, powerhead damage, etc	Repair as necessary	

FUEL SYSTEM 3-49

ENGINE IDLES INCORRECTLY - 70 HP

System/Component	Causes/Effects	Actions	Code
MAP Sensor MAP 1	May idle slightly high - 800 to 850 RPM	Check for disconnected wires or failed sensor. Verify using Diagnostic Software	34
MAP Sensor MAP 2	Hard to start - extremely rough at idle or may die	Check for disconnected or damaged hose - if OK replace sensor	32
MAP Sensor MAP 2	Lower than normal idle speed - @ 600 RPM	Check for pinched hose or plugged pulse limiter	32
Idle Air Control (IAC) Disconnected Wire	Lower than normal idle speed - @ 600 RPM	Check for disconnected IAC valve. Use Diagnostic Software to verify IAC duty cycle after repair	31
Idle Air Control (IAC) Plugged Hose	(1) Requires warm-up lever to start (2) Slow idle – 550-600 RPM (3) Will set code in gear	Check for pinched hose, plugged silencer, or incorrectly connected hose. Use Diagnostic Software to verify IAC duty cycle after repair	31
Closed Throttle Switch (CTP)	(1) Will return to idle speed immediately - not gradually decrease from 1200 RPM (2) Will idle faster than normal – @ 800 RPM	Check for disconnected or damaged switch. Check continuity and use Diagnostic Software to verify	maybe 22
Idle Air Screw Misadjusted	Probably no bad running characteristics but IAC duty cycle will be out of normal range	Adjust idle air screw - abnormal IAC duty cycle can also be caused by other problems such as a vacuum leak or mechanical damage	none
Vacuum Leak at Intake Manifold	Faster than normal idle - IAC duty cycle will be extremely low	Spray around intake gasket areas with starter fluid while idling - engine will react when starter fluid hits leak	maybe 31
Overcooling of Engine	Faster than normal idle similar to a cold started engine	Check for a failed thermostat or debris in thermostat. Verify using diagnostic Software or pyrometer	none
Incorrect Valve Adjustment	Excessive tappet noise or uneven compression	Adjust valve lash	none

ENGINE RUNS ROUGH/NO POWER - 40/50 HP

System/Component	Causes/Effects	Actions	Code
Low Fuel Pressure at Injectors	(1) Worn electric fuel pump or restricted fuel system (2) Failed pressure regulator in vapor separator	Test fuel pressure at fuel rail - should be 34 PSI (234 kPa) or more and steady	none
Failed Injector	Improper fuel delivery to one cylinder	Verify injector signal and resistance	none
Failed Low Pressure Pump	Engine should run if primer bulb is constantly squeezed	Check for fuel leakage around diaphragm drive piston	none
Fouled Spark Plug or Failed Coil	No spark to one cylinder	Inspect spark plugs. Use timing light to verify coil output. (Requires Stevens coil extension lead, P/N CL-30). Check coil signal and resistance	none
MAP Sensor 1 - Open Circuit	Engine goes lean above 3000 RPM. Elevated idle speed	Check for Voltage change at ECU D2 terminal as vacuum changes	34
MAP Sensor 2 - No signal change	Hose off or pulse limiter is plugged	Engine may start with warm-up lever raised but will run very rich and rough at low speed	32
CTP Switch Stuck On	Timing will remain at 5° BTDC when engine speed exceeds 1000 RPM	Verify using timing light or Diagnostic Software - check for pinched wire or faulty switch	22
Rig is Overpropped	WOT RPM is less than recommended when water-testing	Install correct pitch propeller	none
Engine is in S.L.O.W. (caused by:)	(1) Overheat (2) No oil (3) Extended over-revving (4) Neutral switch stuck on	(1) Overheat - verify water flow (2) No oil - verify oil level (3) Over-revving - install correct propeller (4) Repair as necessary	none none 22 none
Internal Mechanical Problem		Check compression, gear lube, etc.	
ECU Calibration	Incorrect fuel calibration	Recalibrate using O₂ sensor and Diagnostic Software. See Software User's Guide	

FUEL SYSTEM

ENGINE OVERHEATS - 40/50 HP

System/Component	Causes/Effects	Actions	Code
Faulty Thermostat	Overheat light, horn, and S.L.O.W.	Inspect thermostat for blockage or damage. Test in hot water - should open at @ 122° F (50° C)	none
Pressure Relief Valve	Overheat light, horn, and S.L.O.W.	Check for damaged or plugged relief valve or relief valve assembled backwards. Normal system pressure is about 5.5 PSI (40 kPa)	none
Exhaust Temp Sensor	Overheat light, horn, and S.L.O.W.	Check for both items listed above - on late models only	none
Water Discharge Hose	Blocked or pinched	Remove hose from powerhead adapter and run engine briefly while watching for considerable water flow	none
Damaged Water Pump	Worn or plugged	Run engine for short time with flushing plug removed from midsection - considerable water discharge should be observed	none
Missing Flushing Plug	Considerable water discharge from fresh water flushing port	Replace plug	none
Air in Cooling Water	Engine mounted too high allowing air into sub-water intake	Lower engine on transom or block auxiliary subwater intake using Plug, P/N 5031618	none
False Warning	Failed sensor or lead shorted to ground	Verify overheat using pyrometer - repair as necessary	none
Blocked Water Intakes	Plugged	Clean as necessary	none

ENGINE RUNS ROUGH/NO POWER - 70 HP

System/Component	Causes/Effects	Actions	Code
Low Fuel Pressure at Injectors	(1) Worn electric fuel pump or restricted fuel system (2) Failed pressure regulator in vapor separator	Use adapter OMC P/N 5000000. Test fuel pressure at fuel rail - should be 34 PSI (234 kPa) or more and steady	none
Failed Injector	Improper fuel delivery to one cylinder	Verify injector signal - replace as necessary	none
Failed Low Pressure Pump	Engine will run if primer bulb is constantly squeezed	Verify cam lobe is not damaged - replace as necessary	none
Fouled Spark Plug or Failed Coil	No spark to one or two cylinders	Verify coil output	none
MAP Sensor	MAP2 - hose off or pulse limiter is plugged	Engine may start with warm-up lever raised but will run very rough at low speed	32
CTP Switch (on engine) Stuck On	Timing will remain at 5° BTDC when engine speed exceeds 1000 RPM	Verify using Diagnostic Software - check for pinched wire or faulty switch	none
Rig is Overpropped	WOT RPM is less than recommended when water-testing	Install correct pitch propeller	none
Engine is in S.L.O.W. (caused by:)	(1) Overheat (2) No oil (3) Extended over-revving (4) Neutral switch stuck on	(1) Overheat - verify water flow (2) No oil - verify oil level (3) Over revving - install correct propeller (4) Repair as necessary	14 none none none
Internal Mechanical Problem		Check compression, gear lube, etc.	none

ENGINE OVERHEATS - 70 HP

System/Component	Causes/Effects	Actions	Code
Faulty Thermostat	Overheat light, horn, and S.L.O.W.	Inspect thermostat for blockage or damage. Test in hot water - should open at @ 122°F (50°C)	14
Pressure Relief Valve	Overheat light, horn, and S.L.O.W.	Check for damaged or plugged relief valve or relief valve assembled backwards. Normal system pressure is about 5.5 PSI (40 kPa)	14 or 15
Exhaust Temp Sensor	Overheat light, horn, and S.L.O.W.	Check for both items listed above - on late models only	15
Water Discharge Hose	Blocked or pinched	Remove fitting from powerhead adapter and run engine briefly while watching for considerable water flow	14 or 15
Damaged Water Pump	Worn or plugged	Run engine for short time with flushing plug removed from midsection - considerable water discharge should be observed	14 or 15
Missing Flushing Plug	Considerable water discharge from fresh water flushing port	Replace plug	14 or 15
Air in Cooling Water	Engine mounted too high allowing air into sub-water intake	Lower engine on transom or block sub-water intake	14 or 15
False Warning	Failed sensor or lead shorted to ground	Verify overheat using pyrometer - repair as necessary	14 or 15
Blocked Water Intakes	Plugged	Clean as necessary	

FUEL SYSTEM 3-51

Air Intake Silencer and Flame Arrester

◆ See Figures 109 and 110

The air intake silencer and flame arrester assembly attaches to the throttle body at the front of the powerhead. As their names imply, the air intake silencer is designed to reduce mechanical noise emitted from the engine while the flame arrester is designed to protect the engine cases and external components from the possibility of a backfire through the manifold and throttle body.

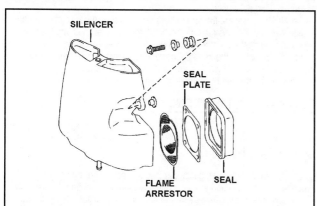

Fig. 109 Exploded view of the air intake silencer and flame arrester assembly - 40/50 Hp Motors

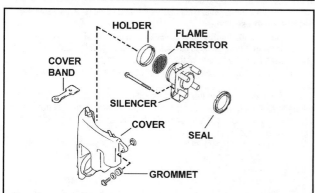

Fig. 110 Exploded view of the air intake silencer and flame arrester assembly - 70 Hp Motors

REMOVAL & INSTALLATION

40/50 Hp Models

◆ See Figure 109

For the 40 and 50 hp motors, a plastic silencer housing is mounted over the flame arrester, seal plate and seal. A small drain valve located in the bottom of the silencer housing prevents condensation build-up.

The silencer housing serves as a mounting point for the Intake Air Temperature (IAT) sensor and the Idle Air Control (IAC) valve air hose.

1. Remove the flywheel cover as described in the Powerhead section if needed for additional access.
2. Disconnect the IAC valve hose from the bottom center of the silencer housing.
3. Disengage the wiring connector from the IAT sensor.
4. Remove the intake silencer mounting bolt(s).
5. Carefully pull the silencer housing away from the throttle body. Wipe the inside of the silencer housing using a rag and a small amount of suitable solvent.
6. If necessary, remove the flame arrester, seal plate and seal. Whether removed or not, inspect the flame arrester and seal for signs of damage, and replace, as necessary.

To install:
7. If removed, install the seal, seal plate and flame arrester.
8. Align the cover and throttle body, then carefully push the cover onto the throttle body and install the retaining bolt(s).
9. Inspect the wiring connector terminals and clean, if necessary. Connect the wiring harness to the IAT sensor.
10. Connect the IAC valve hose to the silencer fitting.
11. If removed, install the flywheel cover as described in the Powerhead section.

70 Hp Models

◆ See Figure 110

For 70 hp motors, a separate plastic silencer cover is mounted over top a flame arrester, holder and the air intake silencer housing. The silencer housing itself serves as a mounting point for the vent hoses from the vapor separator and the crankcase.

1. Remove the flywheel cover as described in the Powerhead section if needed for additional access.
2. Disconnect the crankcase and vapor separator breather hoses from the silencer cover.
3. Remove the bolt holding the silencer cover, then remove the cover from the silencer. Wipe the inside of the silencer cover clean with a shop towel and suitable solvent.
4. If necessary, remove the flame arrester and holder. Whether removed or not, inspect the flame arrester for signs of damage, and replace, as necessary.
5. Remove the bolts securing the intake silencer housing itself and remove it from the engine.

To install:
6. Align the bolt holes, then install intake silencer housing and secure using the retaining bolts.
7. If removed install the flame arrester and holder.
8. Align the cover and intake silencer opening, then install the cover and secure using the mounting bolt.
9. Connect the breather hoses to the intake silencer cover.
10. If removed, install the flywheel cover as described in the Powerhead section.

Throttle Body

◆ See Figures 111 and 112

Like the throttle bore of a carburetor, the throttle body assembly controls engine speed by mechanically controlling the amount of air allowed to enter the engine. For all EFI models, it is found at the front of the intake manifold, behind the air intake silencer.

The Closed Throttle Position (CTP) switch that is mounted to the throttle body was adjusted at the factory to signal the Engine Control Unit (ECU) when the throttle is closed. When the engine is operating at idle, the throttle plate closes most of the way, but is held open very slightly by the CTP switch. The air that passes by the throttle plate, along with the air metered through the Idle Air Control (IAC) valve and idle air bypass screw passage determines engine speed at idle.

For 70 hp motors, the Intake Air Temperature (IAT) sensor is also mounted to the throttle body assembly. For 40/50 hp motors, the IAT sensor is found in the intake silencer housing.

REMOVAL & INSTALLATION

◆ See Figures 111 and 112

1. Remove the air intake silencer and flame arrester assembly as described in this section.
2. Disengage the wiring connector from the CTP switch.

■ **When disconnecting wiring, be sure to note routing for installation purposes.**

3. On 70 hp motors, disengage the wiring connector from the IAT sensor.
4. Separate the throttle rod from the throttle lever by carefully prying it from the pivot.

3-52 FUEL SYSTEM

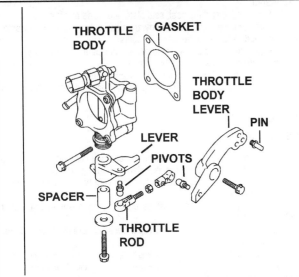

Fig. 111 Exploded view of the throttle body assembly - 40/50 Hp Motors

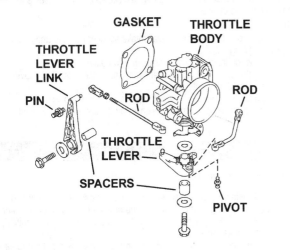

Fig. 112 Exploded view of the throttle body assembly - 70 Hp Motors

5. Remove the throttle body retaining bolts, then remove the assembly from the powerhead.
6. Remove and discard the gasket, then carefully clean the mating surfaces of all debris.
7. If necessary, remove the throttle lever retaining bolt, washer, spacer and throttle lever from the throttle body assembly.

To install:
8. If removed, install the throttle lever to the throttle body using the spacer, washer and bolt. Tighten the bolt securely.
9. Install the throttle body using a new gasket, then tighten the bolts securely.
10. Apply a light coating of grease to all pivot points. Reconnect the throttle rod to the throttle lever.

■ Grease all points, even any pivot point not disconnected.

11. Engage the harness connectors to the IAT sensor and/or CTP switch, as applicable.

■ Be sure to route all wiring as noted during removal to avoid interference.

12. Install the air intake silencer and flame arrester assembly as described in this section.

INTAKE MANIFOLD

◆ See Figures 113 and 114

On EFI motors the intake manifold consists of a common plenum connected to separate intake runners (one for each cylinder). The intake manifold is used a mounting or anchoring point for multiple other components on these models as well.

For all EFI motors, the vapor separator tank and the throttle body, as well as the Manifold Absolute Pressure (MAP) sensor vacuum hose are attached to the manifold assembly.

On 40/50 hp motors, the fuel rail is integral with (is a cast portion of) the intake manifold. For this reason, the high pressure fuel lines and the fuel injectors are attached to the intake manifold as well.

For 70 hp motors, the Idle Air Control (IAC) valve and IAC air silencer, along with the idle air bypass screw and high pressure fuel filter are all mounted to the manifold assembly as well.

REMOVAL & INSTALLATION

◆ See Figures 113 thru 116

Use the exploded views accompanying this procedure and other procedures in this section to assist with manifold removal and disassembly.
1. For safety, properly relieve the fuel system pressure as described under Fuel System Pressurization in this section.
2. Remove the lower engine covers as described under Engine Coves in the Engine Maintenance section.
3. Remove the air intake silencer cover and flame arrester assembly, as described in this section.
4. Remove the throttle body assembly, as described in this section.
5. Loosen the water hose clamps, then disconnect the water inlet and outlet hoses from the bottom of the vapor separator tank.
6. For 70 hp motors disconnect the hoses from the low pressure fuel pump.
7. Drain the fuel from the vapor separator tank. Loosen the screw on the bottom of the tank, then use the drain hose to empty the fuel from the vapor tank into an appropriate container (the fuel can be poured back into the main boats fuel tank).
8. Tag and disconnect all fuel hoses from the vapor separator tank. Position a shop rag to catch any spillage, then drain each hose into a suitable container as it is removed from the fitting.
9. For 70 hp motors, remove the high pressure fuel filter bracket from the manifold.
10. If necessary, remove the fuel rail and/or fuel injectors as described in this section.

■ For 40/50 hp models, the fuel rail is integral with the manifold. For these models, tag and disconnect the lines, then disengage the injector wiring from the manifold/rail assembly.

11. Tag or note any remaining hose or wire connections, then remove them from the intake manifold.
12. For 70 hp motors either remove the oil dipstick and guide tube or remove the dipstick and take great care when removing the manifold. Leaving the guide tube in place will force you to pull the manifold carefully over the guide without damaging the tube.

※※ WARNING

To prevent the possibility of warping and damaging the intake manifold, work slowly when loosing the fasteners. Although not absolutely necessary, it is a good idea to a few passes in the reverse of the torque sequence to properly loosen the fasteners.

13. Support the intake manifold, then loosen the nuts and bolts gradually in the reverse of the torque sequence (working from the front of the powerhead and moving toward the rear; starting from the outside and moving toward the inside). Once you are certain all fasteners are removed, carefully pry the intake from the cylinder head.
14. If necessary, remove the vapor separator tank from the manifold as detailed in this section.
15. Remove the screws, then lift the water jacket cover from starboard side of the manifold.

FUEL SYSTEM 3-53

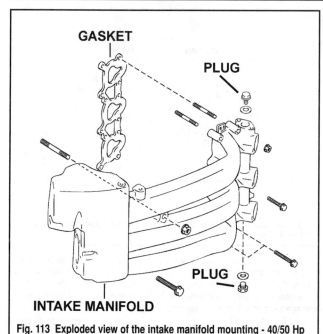

Fig. 113 Exploded view of the intake manifold mounting - 40/50 Hp Motors

16. Remove all gasket material from the mating surfaces, being careful

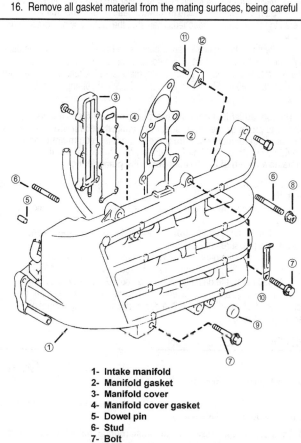

1- Intake manifold
2- Manifold gasket
3- Manifold cover
4- Manifold cover gasket
5- Dowel pin
6- Stud
7- Bolt
8- Nut
9- Cushion
10- Clamp
11- Anode screw
12- Anode

Fig. 114 Exploded view of the intake manifold mounting - 70 Hp Motors

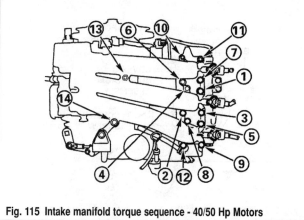

Fig. 115 Intake manifold torque sequence - 40/50 Hp Motors

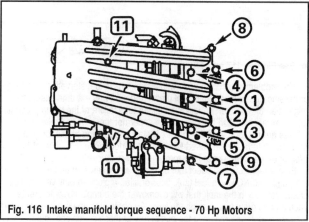

Fig. 116 Intake manifold torque sequence - 70 Hp Motors

not to score or otherwise damage them, then use a quick drying solvent to clean all removed components.

To install:

17. Install the water jacket cover to the starboard side of the manifold and secure using the screws.
18. If removed, install the vapor separator tank to the manifold assembly.
19. Position a new intake manifold gasket, then install the manifold using the retainers. Carefully thread and hand-tighten the retainers, then tighten them to the following specifications, using multiple passes of the torque sequence:
 • 40/50 hp motors, tighten all large screws/nuts to 16-18 ft. lbs. (22-24 Nm) and the small screws/nuts 96-108 inch lbs. (11-12 Nm).
 • 70 hp motors: all screws/nuts to 16-18 ft. lbs. (22-24 Nm)
20. Route all hoses and wiring as noted during removal. Reconnect and secure the fuel, water and/or vent hoses, as applicable. Engage all wiring connectors.
21. Install the throttle body assembly.
22. Install the air intake silencer and flame arrester assembly.
23. Install the lower engine covers.
24. Properly pressurize the fuel system, then check for fuel leakage **before** starting and running the motor. Repair all fuel leaks before proceeding any further.

Low Pressure Fuel Pump

◆ See Figures 117 and 118

The low pressure fuel pump operates in an identical fashion to the pump found on 4-stroke carbureted engines. It draws a steady fuel supply from the tank and feeds a float bowl. The difference comes in what happens in the float bowl, since it is mounted in a vapor separator tank with a high pressure pump instead of under the throttle bore of a carburetor.

In any case, the low pressure, mechanical fuel pump is mounted on the rocker cover at the rear of the powerhead. The pump is actuated by a lobe on the camshaft.

3-54 FUEL SYSTEM

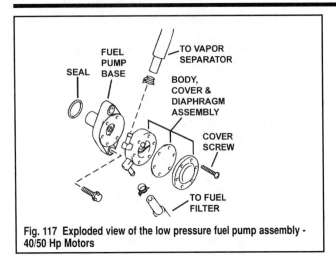

Fig. 117 Exploded view of the low pressure fuel pump assembly - 40/50 Hp Motors

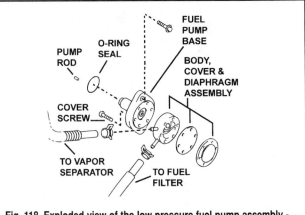

Fig. 118 Exploded view of the low pressure fuel pump assembly - 70 Hp Motors

TESTING

◆ See Figures 117 and 118

The problem most often seen with fuel pumps is fuel starvation, hesitation or missing due to inadequate fuel pressure/delivery. In extreme cases, this might lead to a no start condition as all but total failure of the pump prevents fuel from reaching and filling the vapor separator tank). More likely, pump failures are not total, and the motor will start and run fine at idle, only to miss, hesitate or stall at speed when pump performance falls short of the greater demand for fuel at high rpm.

Before replacing a suspect fuel pump, be absolutely certain the problem is the pump and NOT with fuel tank, lines or filter. A plugged tank vent could create vacuum in the tank that will overpower the pump's ability to create vacuum and draw fuel through the lines. An obstructed line or fuel filter could also keep fuel from reaching the pump. Any of these conditions could partially restrict fuel flow, allowing the pump to deliver fuel, but at a lower pressure/rate. A pump delivery or pressure test under these circumstances would give a low reading that might be mistaken for a faulty pump. Before testing the fuel pump, refer to the testing procedures found under Fuel Lines and Fitting to ensure there are no problems with the tank, lines or filter.

If inadequate fuel delivery is suspected and no problems are found with the tank, lines or filters, a conduct a quick-check to see how the pump affects performance. Use the primer bulb to supplement fuel pump. This is done by operating the motor under load and otherwise under normal operating conditions to recreate the problem. Once the motor begins to hesitate, stumble or stall, pump the primer bulb quickly and repeatedly while listening for motor response. Pumping the bulb by hand like this will force fuel through the lines to the vapor separator tank, regardless of the fuel pump's ability to draw and deliver fuel. If the engine performance problem goes away while pumping the bulb, and returns when you stop, there is a good chance you've isolated the low pressure fuel pump as the culprit. Perform a pressure test to be certain, then repair or replace the pump assembly.

✱✱ WARNING

Never run a motor without cooling water. Use a test tank, a flush/test device or launch the craft. Also, never run a motor at speed without load, so for tests running over idle speed, make sure the motor is either in a test tank with a test wheel or on a launched craft with the normal propeller installed.

Pump Pressure Test

◆ See Figure 117

By far the most accurate way to test the fuel pump is using a low pressure fuel gauge while running the engine at various speeds, under load. To prevent the possibility of severe engine damage from over-speed, the test must be conducted under load, either in a test tank (with a proper test propeller) or mounted on the boat with a suitable propeller. Unfortunately the manufacturer does not provide operating specifications for the low pressure fuel pump, so the specifications given here are taken from similar design fuel circuits and cannot be considered absolute. If possible, ask a dealer about your findings before condemning the low pressure pump. Also, perform the quick-check (pumping the primer bulb by hand) to override the fuel pump and see what affects that has on the condition.

1. Test the Fuel Lines and Fittings as detailed in this section to be sure there are no vacuum/fuel leaks and no restrictions that could give a false low reading.
2. Make sure the tank and inline fuel filter are clean and serviceable.
3. Start and run the engine in forward gear, at idle, until normal operating temperature is reached. Then shut the motor down to prepare for the test.
4. Remove the fuel tank cap to make sure there is no pressure in the tank (the fuel tank vent must also be clear to ensure there is no vacuum). Check the tank location, for best results, make sure the tank is not mounted any more than 30 in. (76mm) below the fuel pump mounting point. On portable tanks, reposition them, as necessary to ensure accurate readings.

■ The fuel outlet line from the fuel pump may be disconnected at either the pump or the vapor separator tank whichever provides easier access. If you disconnect it from the pump itself you might have to provide a length of fuel line (depending on whether or not the gauge contains a length of line to connect to the pump fitting).

5. Disconnect the fuel output hose from the vapor tank or fuel pump, as desired.
6. Connect a fuel pressure gauge inline between the pump and the tank.
7. Run the engine at or around each of the following speeds and observe the pressure on the gauge:
   - At 600 rpm, the gauge should read about 1 psi (7 kPa).
   - At 2500-3000 rpm, the gauge should read about 1.5 psi (10 kPa).
   - At 4500 rpm, the gauge should read about 2.5 psi (17 kPa).
8. If readings are much different than specification and all other tests point to the fuel pump, speak with a dealer to confirm your findings then replace the pump.

REMOVAL & INSTALLATION

◆ See Figures 117 and 118

The pump is located on the rocker arm cover on the cylinder head (at the rear of the powerhead assembly).

■ To ensure proper assembly and hose routing, mark the fuel pump relative to the powerhead before removal.

1. Disconnect the negative battery cable for safety.
2. Set the engine at Top Dead Center (TDC) to ensure the fuel pump rod or arm is not pre-loaded.

■ For details on finding TDC of the No. 1 cylinder refer to the Valve Adjustment procedure found in the Maintenance and Tune-Up section. Of course, only follow the portions of the procedure that you need to, and there should be no need to remove the valve cover.

FUEL SYSTEM 3-55

3. Tag and disconnect the hoses.

■ If hoses are removed completely or replaced, be sure to make a note of hose routing for installation purposes. Hoses must be carefully positioned to prevent interference with other components, as interference could wear away at hoses over time, eventually causing a hazardous fuel leak.

4. Loosen the spring clamps (using pliers) and reposition them back on the fuel hoses. With a small drain basin to catch any escaping fuel, carefully disconnect the hoses from the fittings. If a hose is stuck on the fitting, use a small blade to carefully cut and peel it free of the fitting. Be careful not to damage the fittings with the blade.

✱✱ WARNING

Use extreme care when disconnecting the hoses to prevent damaging or breaking the fittings on the fuel pump assembly. Replace hoses that are worn (spongy, hard or brittle).

5. Loosen the pump mounting bolts, then slowly pull the pump from the rocker cover.

■ It is not necessary to remove the pump pushrod from the powerhead except for replacement.

6. To replace the pump pushrod, use a pair of pliers or a magnet to pull it from the opening in the rocker cover.
7. Remove and discard the O-ring seal from the fuel pump or rocker arm surface.
8. Carefully clean and thoroughly inspect the fuel pump mounting surfaces. Keep in mind that dirty or damaged surfaces can cause oil leaks.
9. Inspect spring clamps for corrosion or a lack of spring tension. Replace damaged, worn or questionable clamps.
10. Replace hoses that are worn (spongy, hard or brittle).
11. If necessary, disassemble the pump for overhaul as detailed in this section.

To install:
12. If removed, install the new pushrod by applying oil to the surfaces and then sliding it into the bore in the pump mounting boss.
13. Apply a coating of fresh engine oil to the new pump O-ring seal, then install the seal to the pump assembly.
14. Position the pump carefully onto the rocker cover, then install and tighten the retaining bolts to 84 inch lbs. (10 Nm).
15. Connect the fuel lines as noted during removal and secure the hoses using clamps.
16. Connect the negative battery cable, then properly pressurize the fuel system and check for leakage. Pump the primer bulb until it becomes firm, then check of the fuel fittings and lines that were disconnected for any signs of weepage.
17. Correct any fuel leaks before starting or running the engine, then run the motor and recheck.

OVERHAUL

◆ See Figures 117 and 118   **MODERATE**

If overhaul is required due to damage from contamination or debris (as opposed to simple deterioration) disassemble and clean the rest of the fuel supply system prior to the fuel pump. Failure to replace filters and clean or replace the lines and fuel tank, could result in damage to the overhauled pump after it is placed back into service.

All diaphragms and seals should be replaced during assembly, regardless of their condition. Check for fuel leakage after completing the repair and verify proper operating pressures before returning the motor to service.

✱✱ WARNING

No sealant should be used on fuel pump components unless otherwise specifically directed. If small amounts of a dried sealant were to break free and travel through the fuel supply system it could easily clog passages (especially the small, metered orifices and needle valves of the fuel injectors).

1. Remove the fuel pump from the powerhead as detailed in this section.

2. Matchmark the fuel pump cover, housing and base to ensure proper assembly.

■ To ease inspection and assembly, lay out each piece of the fuel pump as it is removed. In this way, keep track of each component's orientation in relation to the entire assembly.

3. Remove the 6 cover screws from the fuel pump, then carefully lift the outer cover from the pump body. If necessary, pry gently using a small prytool covered with tape to avoid damaging the gasket surface.

✱✱ WARNING

Do not disturb the diaphragm that is attached to the pump cover unless the diaphragm is going to be replaced. During installation the diaphragm surface molded into the shape of the cover, and, if it is removed for any reason it must be installed EXACTLY in the same position (which is pretty darn tough and not worth the effort).

4. If you are replacing the diaphragms, remove them from the outer cover and pump body. To do so you'll have to remove the pump body from the mounting base, but, matchmark them to ensure proper installation, then carefully pry them apart.
5. Note the position of the diaphragm tab (again this is for assembly purposes), then slowly push the plunger in until the spring fully compresses. Hold the plunger compressed while rotating the upper portion of the plunger assembly on the pump mounting base. Turn it about 90°, until the pin in the plunger aligns with the mounting base slot. Continue to hold pressure on the plunger.
6. Remove the pun from the mounting base using a small magnet, then release the pressure on the plunger, slowly allowing the plunger to push outward. Lift the plunger and spring from the mounting base. Pull the piston and spring from the opposite side of the base.
7. Clean the fuel pump using a suitable solvent, then dry all components with compressed air.
8. Inspect the fuel pump cover, body and base using a straight edge. Inspect their gasket surfaces for scratches, voids or any irregularities. Replace warped or damaged components.
9. Inspect the fuel pump body for cracks. Replace damaged components.
10. Inspect the pump check valves for bent, cracked or corroded surfaces. The check valves are not normally replaceable, if damage or defects are found, replace the pump body.

To assemble:
11. Insert the large spring into the top of the body and the small spring into the bottom, then align the slotted portion of the plunger with the hole in the piston and the slot in the mounting base. Push on the plunger and piston, compressing the spring. Then hold the spring tension.
12. With the spring still compressed, install the pin through the openings. Rotate the plunger 90° to face the pin opposite from the mounting base slot, then slowly release spring tension, making sure the plunger and piston do not bind in the body.

■ When assembling, use the screw holes in the gaskets and diaphragms to ensure proper orientation.

13. Install new gaskets and, if disturbed, worn or in anyway damaged, install new diaphragms.
14. Align the matchmarks made before disassembly, then install the outer cover and secure using the covers screws. Tighten the screws to 84 inch lbs. (10 Nm).
15. Install the low pressure fuel pump, as described in this section.

Vapor Separator Tank and High Pressure Fuel Pump

The first major difference between an EFI and carbureted system (as far as fuel delivery is concerned) comes at the fuel vapor separator tank. The vapor separator is mounted on the powerhead (port side) of the motor, directly behind and attached to the intake manifold.

The separator tank functions as a bizarre cross between a very large float bowl and a very tiny gas tank. It receives fuel from the low pressure pump via a float and needle valve assembly (in the same manner as a carburetor's float bowl). The level is maintained within the vapor separator tank so that is serves as a reservoir for the high pressure electric fuel pump mounted in the separator cover. In addition, the separator tank also provides an outlet for excessive fuel pressure generated by the electric fuel pump.

FUEL SYSTEM

■ The float and needle valve are serviced in the same manner as a carburetor's float bowl. They can be accessed once the vapor separator cover/fuel pump assembly is removed.

To help prevent fuel problems like vapor lock or hot soak, the separator is water cooled by the engine cooling circuit. On 40/50 hp motors, fuel vapors are vented to the flywheel cover, but on 70 hp models, the vapors are vented to the air intake silencer so they can be drawn into the throttle body and burned when the engine is running.

The tank cover contains the high pressure fuel pump assembly. Although the cover/pump assembly utilizes a replaceable pickup filter screen, they are otherwise unserviceable and must be replaced as an assembly should problems arise.

The fuel pressure regulator mounted in the bottom of the separator tank. On most applications and the fuel pressure regulator may be removed from within the tank should the regulator require replacement.

A fuel reservoir drain screw can be found on the bottom of the tank and can be used to drain the tank of fuel, but the cooling water circuit will self-drain whenever the engine is shut off and left in a perfectly vertical position.

TESTING

◆ See Figures 119 and 120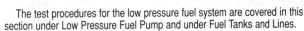

The test procedures for the low pressure fuel system are covered in this section under Low Pressure Fuel Pump and under Fuel Tanks and Lines.

Perform the Electric Pump Circuit Tests if the engine refuses to start and/or you suspect the high pressure circuit is not building any fuel pressure. The pressure tests are used to locate the culprit when power is applied to the pump, and the pump seems to be running, but is still not supplying sufficient pressure.

✱✱ CAUTION

The high pressure fuel system, as it name might imply, is capable of spraying fuel under extreme pressure. This means that the fuel will spray free under high pressure if a fitting is opened without first relieving pressure (which makes for good fuel atomization and a highly combustible condition). It will also spray fuel if the pump is actuated for any reason while a fitting is disconnected. These could lead to extremely dangerous work conditions. Do not allow ANY source of ignition (sparks, flames, etc) anywhere near the work area when servicing the fuel system.

Electrical Pump Circuit Tests

Use this test to check for circuit operation (including proper pump voltage supply and proper pump motor operation).

QUICK TEST

1. Place a stethoscope (or if one is not available, use a short length of hose or a small wooden dowel as a substitute tool) on the vapor separator tank behind the intake manifold.
2. Listen on the stethoscope or substitute while an assistant turns the ignition keyswitch to **ON**, without starting the motor. You should hear the pump run for about 2-3 seconds.
3. Turn the switch **OFF** for at least 30 seconds, then repeat.
4. Again, you should hear the pump run for about 2-3 seconds. This should occur anytime the keyswitch is turned **ON** in such as manner as described with 30 second delays between each cycle.

VOLTAGE TEST

If the pump fails to operate in the Quick Test, check the voltage supply as follows:

1. Disconnect the wire harness from the top side of the vapor separator tank.
2. Set the DVOM to the 20 VDC scale, then connect the positive meter test lead to the pink wire terminal in the engine harness connector. Connect the negative meter test lead to the black/white wire terminal of the connector.
3. Observe the meter while an assistant cycles the ignition key switch to the ON position. If the connections are correct and the circuit is working properly, the meter should indicate 6-12 volts for 2-3 seconds, then it should indicate about 0 volts. If so, the circuit is operating properly.

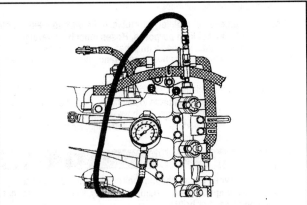

Fig. 119 To test the fuel pump and high pressure system, connect a fuel pressure gauge to the top of the fuel rail using an adapter

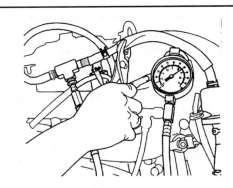

Fig. 120 To isolate and only check the electric pump, connect a gauge inline between the pump and filter, then pinch off the line after the gauge

4. If the specified voltage could not been found across the pump harness connectors, move the positive meter test lead to the battery terminal of the starter relay and repeat the previous step. If the meter still fails to show voltage during the cycle, check the 15 amp fuse (if blown, check all wiring and connections before installing the replacement).

5. If still no voltage is generated, and the 15 amp fuse is intact, check the 30 amp fuse, system relay, neutral switch, ignition key switch and all wiring in the circuit.

6. **ONLY** after all components and wiring in the system test correctly should you suspect the ECU. On 40/50 hp motors, refer to the information on Sensor and Circuit Resistance/Output Tests to see if there is any information that can be used in the ECU pinout and normal reference values.

7. When finished testing, be sure to reconnect harness to the vapor separator tank.

PUMP PRESSURE TEST

◆ See Figures 119 and 120

■ Perform the Quick Test found under Electric Pump Circuit Tests to ensure that the fuel pump is operating before attempting to check its pressure.

The Pump Pressure Test is used to determine if the electric fuel pump is delivering fuel at the fuel pressures necessary for proper engine operation. But, use this test with caution as it only determines if the pump is capable of building sufficient pressure. It does not test whether or not the pump and regulator continue to deliver sufficient fuel pressure under all possible engine operating conditions.

■ Remember that the low and high pressure fuel systems may test within specification at or near idle speeds only to fall below spec once higher rpm and load demands are placed on them. Leaks or restrictions (including partially blocked fuel filters) may cause fuel starvation problems only at higher engine rpm.

FUEL SYSTEM

If pressure tests fine, but engine performance suggests inadequate fuel supply under certain conditions, operate the engine under those conditions with a fuel pressure gauge still attached to determine if the fuel delivery system is responsible.

A special hose/adapter (#5000900) or equivalent, and a fuel pressure gauge of at least 46 psi (317 kPa) capacity are necessary for pressure testing. Make sure the gauge and test hose/adapter are sealed in order to prevent the possibility of a hazardous fuel leak.

1. Relieve the fuel system pressure, as detailed in this section, leaving the fuel pump harness disengaged (40/50 hp motors) or the 15 amp fuse removed (70 hp motors) for safety while installing the test gauge.
2. Either disengage the harness connector from each of the three ignition coils (40/50 hp motors) or disconnect the secondary spark plug leads and ground them on the powerhead (70 hp motors), as applicable.
3. Remove the bolt from the top of the fuel rail at the intake manifold, then install an adapter, hose and pressure gauge to the fuel rail.
4. Reconnect the fuel pump wiring harness or install the 15 amp fuse, as applicable.
5. Squeeze the primer bulb to fill the vapor separator tank with fuel.
6. Turn the ignition switch **ON** and listen for the fuel pump. It should run for about 3 seconds. Turn the ignition switch back **OFF** and wait a few seconds, then turn it on again. Repeat the cycle 3 or 4 times to ensure the pump runs sufficiently to build maximum pressure in the system. Watch the gauge as you cycle the switch. The gauge should indicate a minimum of 34 psi (234 kPa) of fuel pressure while the pump is running:

■ **Fuel pressure may drop slightly when the pump stops running, but should stabilize in less than 20 seconds and should hold steady for at least 5 minutes.**

   a. If the pump does not run, perform the pump Voltage Test
   b. If the pressure was insufficient, install the gauge using a "T" adapter between the electric fuel pump outlet and the high pressure filter, then pinch the hose between the "T" and the filter while cycling the switch. If pressure is now sufficient, the problem is no the pump and is probably a leaking injector. If pressure still does not build to specification, the pump is faulty.
   c. If the pressure was sufficient, but drops too much or too quickly, suspect either a leaking injector or a bad fuel pressure regulator. Test the pressure regulator to see if that is the problem (inspect spark plugs to help determine if there is a leaky injector. The excessive fuel allowed into the combustion chamber by a leaky injector should lead to some extra carbon fouling).

7. Connect the harness connectors to the coils (40/50 hp motors) or the spark plug leads (70 hp motors), as applicable.
8. Connect a cooling system flushing adapter (or alternate source of cooling water) to the motor.
9. Start and run the engine at idle while watching the fuel pressure gauge. The pressure must be at least 34 psi (234 kPa) and should remain steady.
10. Vary the engine speed, again the gauge should remain steady. If system pressures are sufficient and steady, the pump and fuel system is operating properly. If the pressure drops, check for a restricted line or filter. If no restrictions are found and the low pressure system is operating properly, check the pressure regulator. If the regulator is operating correctly, suspect the high pressure pump.
11. Relieve the fuel system pressure once again, leaving the fuel pump fuse or connector removed for safety, then remove the gauge and adapter.
12. Pressurize the fuel system and check for leaks.

PRESSURE REGULATOR TEST

A regulated air supply (from a compressor or a hand-pump), along with a pressure gauge and a length of fuel hose are necessary to test the fuel pressure regulator.
1. Relieve the fuel system pressure, as detailed in this section.
2. Position a small basin to catch any fuel that might escape, then disconnect the fuel supply hose from the regulator pressure nipple (located at the bottom of the vapor separator tank assembly).
3. Attach the regulated air source (compressor or hand-pump) to the regulator pressure nipple using a length of fuel supply hose.
4. Apply 45 psi (310 kPa) of air pressure to the regulator nipple and listen for air discharging from the regulator. If using an air compressor, shut the compressor off once 45 psi (310 kPa) is applied (if using a hand-pump, stop pumping). The pressure should stabilize at 34 psi (240 kPa) or more.
5. The fuel pressure regulator should be replaced if insufficient pressure is obtained or if it does not stabilize at a sufficient pressure once the pump is turned off.

REMOVAL & INSTALLATION

◆ See Figures 121 and 122

1. Relieve the fuel system pressure, as detailed in this section.
2. Drain the fuel from the vapor separator tank into an approved container using the drain screw located on the bottom of the tank and a drain hose.
3. On 70 hp motors, disengage the harness connector from the high pressure fuel pump by pressing downward on the lock tab and pulling the connector free.
4. Carefully free the fuel pump fuse holder (15 amp) from the retaining clip.
5. Remove the intake manifold, as detailed under the Intake Manifold procedure in this section.
6. Loosen and remove the mounting screws (normally 3), then remove the vapor separator as an assembly from the back of the intake manifold.
7. Loosen and remove the cover screws (usually 6), then carefully separate the cover and pump assembly from the reservoir tank. Remove and discard the O-ring from the cover.
8. Invert the cover to visually check the small, round, fuel pump pickup screen. Replace the screen if it is damaged or contaminated:
   a. Remove the screw securing bracket to the bottom of the pump assembly (the bracket is shaped something like a 1/2 moon), then lift out the support bracket.
   b. Carefully grasp the filter screen and pull it from the pump. If it is stuck, carefully pry the screen free.
   c. Taking care not to damage the replacement screen or the pump inlet, push the new screen into the pump.
   d. Install the 1/2 moon support bracket and securely tighten the retaining screw.

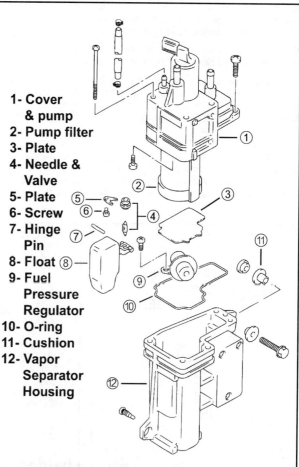

1- Cover & pump
2- Pump filter
3- Plate
4- Needle & Valve
5- Plate
6- Screw
7- Hinge Pin
8- Float
9- Fuel Pressure Regulator
10- O-ring
11- Cushion
12- Vapor Separator Housing

Fig. 121 Exploded view of the vapor separator tank and high pressure fuel pump assembly

3-58 FUEL SYSTEM

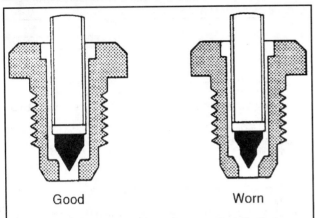

Fig. 122 Just like when servicing carburetors, check the float needle and valve seat for wear or damage

9. If necessary, remove the pressure regulator from the bottom of the vapor separator tank. Remove the plastic deflector plate from the bottom of the separator tank for access, then remove retaining screw and separate the regulator from the tank. Remove and discard the old O-ring from the regulator fitting.

10. If necessary, remove the float, valve and needle assembly for inspection or replacement:

   a. Using a small pick or awl, carefully remove the float hinge pin.
   b. Carefully lift the float and inlet valve needle from the underside of the tank cover.
   c. Remove the screw, plate and valve seat from the underside of the tank cover.

11. Clean vapor separator tank, cover and any parts which were removed that are not being replaced (such as components of the float/valve assembly) using a suitable solvent. Use low pressure compressed air to carefully blow out each of the passages.

■ Make sure all passages, especially the cooling passage on the side of the separator are clean and free of all debris or obstructions.

12. If disassembled for inspection, check the inlet valve seat and needle for damage. Make sure the surfaces are not grooved, pitted, worn, cracked or otherwise unsuitable for further service. Also, check the float and pin for damage or excessive wear. Replace any worn or defective components in the float valve assembly.

To install:

13. If removed, carefully install the valve seat, then install the float and needle valve securing them with the hinge pin.

14. If removed, apply a small amount of fuel to the new O-ring and install the fuel pressure regulator. Tighten the retaining bolt securely and install the deflector plate.

15. Apply a light coating of OMC Adhesive M (or an equivalent sealant) to the very outer edge of the cover lip.

■ Do not apply any sealant to the O-ring, groove or inner edge of the mating surfaces.

16. Install the cover and fuel pump assembly to the tank, then tighten securely tighten the retaining screws.

17. Secure to the vapor separator tank to the intake manifold using the retaining screws.

18. Install the intake manifold as detailed in this section.

19. Secure the fuel pump fuse holder to the retaining clip and connect the fuel pump wiring harness.

20. Properly pressurize the fuel system, as detailed in this section and check for leaks. Correct any leaks before returning the engine to service.

Fuel Rail and Injectors

◆ See Figures 123 and 124

A fuel injector is a small, solenoid valve that is designed to open against spring pressure when power is applied to the circuit. An internal spring snaps the valve closed the instant that power is removed from the circuit.

Fuel injectors are supplied with a constant supply of high pressure fuel. Because the pressure is held constant, the amount of fuel that sprays through the injector is a function of time (the less each injector is actuated, the less fuel is delivered, the more each injector is actuated, the more fuel is delivered).

Each cylinder is equipped with an individual electronic fuel injector to deliver metered amounts of fuel, matching engine operating conditions. The exact fuel metering made possible by the fuel injection system is responsible for the EFI engine's ability to maximize both engine performance and fuel economy.

The most important fuel injection system maintenance is a combination of periodic filter changes and the use of fuel stabilizer if the motor is stored for any amount of time (more than a few weeks). This is true because the passages inside a fuel injector are very small, and are easily clogged by dirt or debris in the fuel system.

The fuel injectors are installed in a fuel rail assembly so they spray fuel directly behind the intake valves. For 40/50 hp motors, the fuel rail is an integral portion (cast into) the intake manifold. The injectors are fastened to the fuel rail by clips on these models. For 70 hp motors, fuel rail is bolted to the powerhead, securing the injectors between them.

TESTING

◆ See Figures 123, 124 and 125

The injectors are mounted on the rear and port side of the powerhead to the intake manifold and fuel rail.

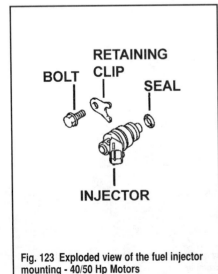

Fig. 123 Exploded view of the fuel injector mounting - 40/50 Hp Motors

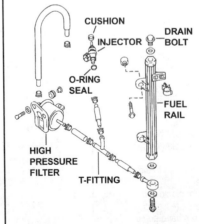

Fig. 124 Exploded view of the fuel injector and fuel rail assembly mounting - 70 Hp Motors

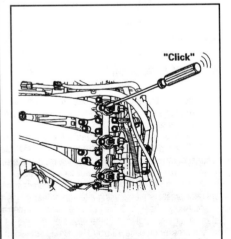

Fig. 125 One quick check for an injector is to listen or feel for operating with the engine cranking or running

FUEL SYSTEM 3-59

Injector Operational Test

◆ See Figure 125

The fastest way to check for inoperable injectors is to listen or feel for solenoid operation. If available, use a mechanic's stethoscope, but because solid matter transmits sound and vibration, a long screwdriver can also be used to amplify and/or feel each injector.

  1. Provide the motor with a source of cooling water.
  2. Start and run the engine at idle.
  3. Position the stethoscope or screwdriver against the body of each fuel injector.

■ **If using a screwdriver to amplify the sounds of the injector, place your ear near the handle or hold the driver lightly while feeling for the light tapping of the solenoid valve.**

  4. If an injector is operating properly you will hear or feel a slight clicking from it. This tells you that the valve is opening and closing.
  5. If there is a noticeably different noise or no clicking is felt at all from an individual injector, perform the Injector Resistance Test and check the wiring between the injector and ECU. If resistance is within specification, check the harness for opens or shorts. Before replacing an ECU, substitute a known good injector.

■ **For test purposes, fuel injectors can be switched from cylinder-to-cylinder to see if the problem follows the injector or remains behind. If the problem remains behind, look to the harness and signal for trouble. If the problem follows the injector, the problem IS the injector.**

Injector Resistance Test

Another quick-check of a fuel injector is made using an ohmmeter to measure the resistance of the winding inside the injector itself. It is important to remember that a correct reading does not mean the injector is operating. Mechanical damage or clogs within the injector could prevent it from opening or closing properly which would lead to engine performance problems.

■ **The injector does not need to be removed from the engine to check its resistance.**

The combined fuel injector and circuit resistance can be checked at the ECU wiring harness connectors. Information on connector views, pinouts and additional specifications are found under Sensor and Circuit Resistance Tests/Output in this section.

  1. In order to protect the test equipment, disconnect the negative battery cable.
  2. Push inward on the wire clip, then disengage the wire harness connector from the injector.
  3. Set the DVOM to the resistance scale, then apply the meter probes across the 2 terminals on the top of the injector. The meter should read between 11.0-16.6 ohms. Replace the injector if resistance exceeds specification.
  4. If the injector is inoperable, but resistance is within specification, check the circuit and signal to help determine if it is a circuit fault or a mechanical fault within the injector itself.
  5. When finished, align the tabs on the connector with the slot and connect the wiring harness to the injector.

Injector Signal Test

If an injector appears to be inoperable and resistance is within or close to specification, check the signal circuit using a DVOM, a Peak-Reading Voltmeter (PRV) or a fuel injector test light (noid). Test lights are available from aftermarket sources such as Snap-On's ® Blue Point® division, part No. FID-8339. Inexpensive noid lights should be available from other sources as well, just check carefully to make sure the connector on the light will mate properly with the injector wiring harness.

Refer to the wiring diagrams to identify the wires leading to the injectors.

  1. Decide how you will proceed based on the equipment that is available.
  2. If one or more injector test lights are available:
   a. Push inward on the wire clip, then disengage the wire harness connector from the injector(s) to be tested.
   b. Connect a noid light to the wiring harness being tested.
   c. Crank the engine, watching the light. If the circuit is operating properly the light will blink as the engine is cranked. Check the main power relay or the 30 amp fuse if no power is applied to the injector light while cranking. On 40/50 hp motors, a faulty CKP for that cylinder can also cause this condition.

■ **Information on connector views, pinouts and additional specifications are found under Sensor and Circuit Resistance Tests/Output in this section.**

  3. If a PRV is available:
   a. Reconnect the injector wiring (if removed) and backprobe the circuit. Connecting the black meter lead to the positive battery terminal (either at the battery or at the starter solenoid), then connect the red lead either to the colored (not gray) wire at the injector or the appropriate terminal at the ECU harness connector. On 40/50 hp motors, ECU testing occurs at connector/terminals A4, A7 and A8.

■ **A small amount of dielectric grease can be applied to the probe to ease insertion into the back of the connector. Better yet, OMC test probe No. 342677 can be used to ease circuit backprobing and help protect the harness from physical damage.**

   b. Crank the motor while reading the meter. The meter should display 6-10 volts at peak if the ECU is operating properly and there are no problems with the harness.
   c. If the signal is not received at the injector end of the harness, check continuity of the harness between the ECU and the injector (be sure to disconnect the harness from the ECU first for safety). Check the main power relay or the 30 amp fuse if no power is applied to the injector light while cranking. On 40/50 hp motors, a faulty CKP for that cylinder can also cause this condition.
  4. If only a standard DVOM is available:
   a. Disconnect the wiring harness from both the injector and the ECU in order to protect the ECU itself, then check circuit continuity between the harness and the injector.
   b. Check the main power relay and the 30 amp fuse to make sure power is applied to the circuit.

■ **Information on ECU pinout testing and additional specifications are found under Sensor and Circuit Resistance/Output Tests in this section.**

REMOVAL & INSTALLATION

◆ See Figures 123 and 124

■ **New O-rings and grommets should be installed anytime the injectors are removed. Pay close attention to all hose routing and connections prior to removal.**

40/50 Hp Models

◆ See Figure 123

  1. Relieve the fuel system pressure as detailed in this section and disconnect the negative battery cable for safety.
  2. Remove the lower engine covers for access. For details, refer to Engine Covers under the Engine Maintenance section.
  3. Use compressed air (from a compressor or hand pump) to blow debris from the injectors and intake manifold.

■ **If no compressed air is available, use a soft brush or some spray engine degreaser and rag to clean debris from the area around the injector mounting.**

  4. Push inward on the wire clip, then disengage the wire harness connectors from the injectors.
  5. Remove the screws and clips securing the injectors to the intake manifold.
  6. Carefully withdraw each injector from the intake manifold (use a rag to catch any escaping fuel). If necessary, gently rock the injector from side-to-side to help free it from the manifold.
  7. Remove and discard the injector O-ring seals.

To install:

  8. Injectors must be completely clean before installation, if necessary clean all dirt or debris from the injectors using solvent and lint free towels.

3-60 FUEL SYSTEM

9. Lubricate the new O-rings using clean engine oil, then slide them over the injector bodies.

■ **Always use new O-rings for safety (to make sure no leaks develop from the use of old O-rings).**

10. Insert each injector into the manifold openings, taking care not to nick, pinch or otherwise damage the O-rings. Once installed, rotate the injector body to position the connector so it is 180° away from the retainer clip bolt.

11. Install the injector retaining clips and tighten the bolts to 98-108 inch lbs. (11-12 Nm).

12. Pressurize the fuel system, as detailed in this section and then check for leaks. Repair all fuel leaks before returning the motor to service.

70 Hp Models

◆ See Figure 124

1. Relieve the fuel system pressure as detailed in this section and disconnect the negative battery cable for safety.
2. Remove the lower engine covers for access. For details, refer to Engine Covers under the Engine Maintenance section.
3. Use compressed air (from a compressor or hand pump) to blow debris from the injectors and cylinder head.

■ **If no compressed air is available, use a soft brush or some spray engine degreaser and rag to clean debris from the area around the injector mounting.**

4. Push inward on the wire clip, then disengage the wire harness connectors from the injectors.

■ **The fuel rail may be completely removed from the engine or it may be unbolted and pulled back for access to the injectors with the fuel supply line still attached.**

5. If the fuel rail is being completely removed from the motor:
   a. Position a rag to catch any remaining fuel in the lines.
   b. Remove the bolt at the bottom of the rail
   c. Disconnect the fuel supply hose connector from the rail.
6. Remove the 2 retaining bolts and spacers from the fuel rail assembly, then pull the assembly and the injectors from the engine. Carefully free each injector from the fuel rail, using a gently rocking motion, as necessary.
7. Remove and discard the O-ring seals from the injectors and/or the injector openings in the fuel rail and manifold.

To install:

8. Injectors must be completely clean before installation, if necessary clean all dirt or debris from the injectors using solvent and lint free towels.
9. Lubricate the new O-rings using clean engine oil, then slide them over the injector bodies.

■ **Always use new O-rings for safety (to make sure no leaks develop from the use of old O-rings).**

10. Insert each injector into the fuel rail or the manifold openings (whichever you feel will be easier), taking care not to nick, pinch or otherwise damage the O-rings. Once installed, rotate each injector body to position the harness connector facing the cylinder head.

11. Install the fuel rail, carefully seating the injectors into the cylinder head or the rail (depending on your choice in the previous step).

12. Install the fuel rail bolts and spacers, then tighten to 16.5 ft. lbs. (23 Nm).

13. If the fuel rail was completely removed from the motor, carefully position the fuel line so it will not be kinked, pinched or damaged by other components, then secure it to the fuel rail. Tighten the union bolt to 29 ft. lbs. (40 Nm).

14. Pressurize the fuel system, as detailed in this section and then check for leaks. Repair all fuel leaks before returning the motor to service.

Camshaft Position (CMP) Sensor

◆ See Figures 126 and 127

The Camshaft Position (CMP) sensor is used to determine engine positioning, relative to the four-stroke engine cycle moment. The ECU uses the CMP signal, along with the Crankshaft Position Sensor (CKP) signals to set fuel injection and ignition timing, as well as injector duration, fuel pump operation, tachometer operation and IAC valve operation.

The sensor is a Hall-effect type, designed to generate an electrical signal based upon a magnetic force passed through its field. On 70 hp motors, the signal is generated by a reluctor bar is fitted to the top edge of the camshaft pulley. On 40/50 hp motors, the reluctor bar is pressed into the top end of the intake port side (camshaft).

The sensor is mounted at the rear, top, port edge of the rocker cover on 40/50 hp motors (near the No. 1 cylinder ignition coil). For 70 hp motors the sensor is wired as part of the CMP/CKP sensor assembly, but is mounted separately from the CKP sensors, on the top, starboard side, rear of the cylinder head (near the camshaft pulley).

TESTING

◆ See Figures 126 and 127

If the ECU detects a CMP circuit fault it will make fuel and ignition timing decisions based upon signals from the crankshaft position sensors. Engine performance is usually not impacted, but the ECU will set fault code No. 24

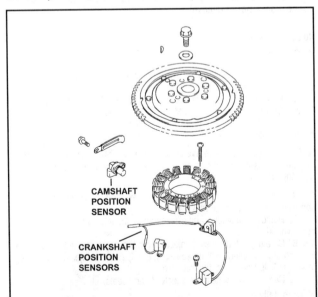

Fig. 126 Exploded view of the CMP and CKP sensor mounting - 40/50 Hp Motors

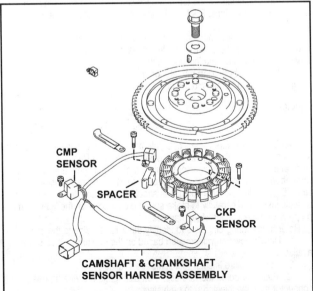

Fig. 127 Exploded view of the CMP and CKP sensor mounting - 70 Hp Motors

FUEL SYSTEM 3-61

and display the Check Engine light. In some instances, failure of a CMP sensor will cause the tachometer to display 1/2 engine speed.

Sensor removal is not required for testing, but keep in mind that resistance readings for this component will vary somewhat with temperature. Specifications are based upon testing the sensor when it is at approximately 68°F (20°C).

1. Disconnect the negative battery cable for safety and in order to protect the test meter.
2. Follow the camshaft position sensor wires from the sensor either to the sensor connector on 40/50 hp motors or to the 4-pin connector at the ECU. Refer to the wiring diagrams for more information on these sensor harness colors.
3. Disconnect the sensor harness from the sensor or the ECU, as desired.

■ On 70 hp motors, the sensor is wired to the 4-pin connector located at one end of the ECU (connector "E"). Check terminals No. 1 (black wire) and No. 2 (orange/green wire). More information on connector views and pinouts are found under Sensor and Circuit Resistance/Output Tests in this section.

4. Set the DVOM to the resistance scale, then connect the probes to the sensor wire terminals or the ECU wiring harness terminals for the sensor and record the meter reading. CMP resistance should be about 168-252 ohms. If resistance is outside specifications, check the wiring harness for damage before replacing the sensor.

✱✱ WARNING

DO NOT attach a DVOM to the ECU itself. Some circuits of the ECU can be damaged or destroyed by applying voltage from an ohmmeter or a DVOM set to the resistance scale.

5. Connect the sensor wire harness to the sensor or ECU, as applicable. Be sure to route the wires carefully, avoiding interference with other components.
6. Connect the negative battery cable.

REMOVAL & INSTALLATION

♦ See Figures 126 and 127

Prior to removal, note the wire harness routing (a digital camera can be very handy for these matters) to ensure proper positioning during installation. On 70 hp motors, a feeler gauge will be necessary to properly adjust the air gap while installing the sensor.

40/50 Hp Models

♦ See Figure 126

The sensor is mounted at the rear, top, port edge of the rocker cover, near the No. 1 cylinder ignition coil.
1. Disconnect the negative battery cable for safety.
2. Carefully bend the clamp back to free the sensor wire, then unplug the harness connector from the sensor.
3. Remove the retaining bolt, then lift the sensor and clamp from the rocker arm cover.
4. Use a suitable solvent to clean dirt or debris from the sensor and mounting surface. Clean the harness connector terminals of debris or corrosion.

To install:
5. Insert the tip of the sensor into the opening, then seat the sensor on the rocker cover and secure using the bolt and clamp.
6. Engage the harness connector to the CMP sensor.
7. With the wire harness properly routed, bend the clamp to secure the harness.
8. Connect the negative battery cable.
9. Check for proper ignition system operation.

70 Hp Models

♦ See Figure 127

The sensor is wired as part of the CMP/CKP sensor assembly, but is mounted separately from the CKP sensors, on the top, starboard side, rear of the cylinder head (near the camshaft pulley). Each sensor of the CKP assembly contains an individual signal lead, but use a common ground wire. The CMP may be replaced separately, but you will have to splice it into the CMP/CKP harness.
1. Disconnect the negative battery cable for safety.
2. Remove the flywheel as detailed under Flywheel in the Powerhead section.
3. Remove the CKP/CMP sensor assembly as detailed under Crankshaft Position (CKP) Sensor, Removal & Installation, in this section. Remove both CMP sensor mounting screws, then lift the sensor and mounting block from the cylinder head.
4. If only the CMP sensor is being replaced, compare the replacement sensor to the current assembly in order to determine the proper point for splicing in the ground wire, then cut the old sensor ground wire free of the harness.
5. Clean the sensor mounting location and the threads of the mounting screws.

To install:
6. If the only the CMP sensor (and not the entire harness) was replaced, carefully splice the new sensor ground wire to the CMP/CKP harness. The best method for splicing is to twist and solder the wires, then seal the connection using heat shrink material. Solderless connectors of the proper size can be used, but are more likely to fail if not sized and connected properly.
7. Place the mounting block and sensor onto the cylinder head. Apply a coating of OMC Nut Lock or and equivalent threadlock to the threads and install the sensor mounting screws. Do not tighten the screws at this time.
8. Slowly rotate the flywheel clockwise by hand until the reactor plate (raised section) on the camshaft pulley aligns with the sensor, then slide a 0.030 in. (0.75mm) feeler gauge between the sensor and the reactor plate.
9. To adjust the sensor air gap, push the sensor toward the camshaft pulley using light finger pressure, then hold it in that position while tightening the screws. Check the air gap to make sure it has not changed. An 0.030 in. (0.75mm) feeler gauge should pass between the pulley and sensor with a slight drag (the next larger size should not fit and the next smaller size should pass with no drag).
10. If necessary, loosen the sensor retaining screws and readjust the gap, then retighten the screws.
11. Install the balance of the CKP sensor assembly.
12. Connect the negative battery cable.
13. Check for proper ignition system operation.

Crankshaft Position (CKP) Sensor

♦ See Figures 126 and 127

The CKP sensors are similar in both design and function to the CMP sensor. The CKP sensors convert changes in an electromagnetic field to a voltage signal used by the ECU to help control engine operation. Each CKP sensor is positioned to give the ECU information regarding where a piston (40/50 hp motors) or pair of pistons (70 hp motors) are in the cylinder.

The major difference between the CMP and CKP is the fact that the crankshaft completes two full turns for each single turn of the camshaft. This is necessary since the 4 phases of a 4-stroke engine require the piston to travel up and down in the cylinder 2 times. This limits the amount of information that the ECU can gain from the CKP sensors, since they will generate the same signal for 2 different strokes of the 4-stroke cycle. Without a CMP sensor signal as well, the ECU cannot determine if a given signal means that cylinder is at TDC of the compression stroke or the exhaust stroke.

■ For more details on the basic operation of a 4-stroke engine, refer to the information on Engine Operation under General Information.

As stated earlier, the 40/50 hp motors manual a CKP sensor for each cylinder (3). The sensors are mounted directly under the flywheel, evenly spaced apart around the top of the top of the engine. They are set at points around the flywheel so as to determine piston positioning for one individual cylinder. If one of the CKP sensors fail, the ECU will maintain engine operation based on the signals from the CMP and remaining CKP sensors, but both fuel and ignition will not be provided to the cylinder for the failed sensor.

The 70 hp motors utilize 2 CKP sensors that provide information on paired cylinders. One sensor provides a signal the ECU interprets as piston positioning for the No. 1 and No. 4 cylinders, while the other sensor provides a signal for the No. 2 and No. 3 cylinders. On 70 hp motors the two signals

FUEL SYSTEM

from the CKP sensors are redundant (meaning that they provide the ECU with the same basic information, since the ECU can calculate positioning of one pair of cylinders if it knows the position of the other pair).

For this reason, if once sensor fails the ECU can maintain engine operation based on the other CKP sensor (and the CKP sensor) signal. The manufacturer designed the system with 2 sensors for fast starts and as a fail-safe should one sensor cease functioning while the boat is underway (to get you home).

TESTING

◆ See Figures 126 and 127

The CKP sensors are mounted to the top of the powerhead underneath and around the perimeter of the flywheel. Two or three sensors are used depending on the model (3 sensors on 40/50 hp motors and 2 sensors on 70 hp motors).

For 40/50 hp motors, a faulty sensor will cause the Engine Control Unit (ECU) to stop spark and fuel delivery to the corresponding cylinder. A fault with more than one sensor will cause the engine to cease operation (ECU will prevent spark and fuel to all 3 cylinders).

For 70 hp motors, a fault with one sensor causes the ECU to base engine timing on the remaining CKP sensor signal. Normal operation will continue with 1 faulty CKP sensor on these motors, but, like the 40/50 hp models, a fault on 2 sensor circuits will cause the engine to stop running.

■ **It is not necessary to remove the CKP sensor(s) for testing. As with all resistance readings, keep in mind that test results will vary with temperature. The specifications for this test were determined at an ambient temperature of about 68°F (20°C).**

1. Disconnect the negative battery cable for safety and in order to protect the test meter.

■ **Depending on the model, the CKP harness traced back from the assembly and disconnected inline before the main wiring harness for testing and/or at the ECU wiring harness connectors, as noted. Wire colors should be the same at either point.**

2. Locate and unplug the appropriate ECU connector(s) and corresponding terminals in order to connect an ohmmeter to conduct a resistance check. For more details, refer to the wiring diagram and/or the connector views and pinouts found under Sensor and Circuit Resistance/Output Testing, in this section.

   a. For 40/50 hp motors, use the ECU harness connectors "E" (small 4-wire connector) and "D" (large, black, 12-wire connector) to test the following connections using a DVOM set to the resistance scale:
   - No. 1 cylinder sensor: D1 (black/white wire) and E4 (red/black wire)
   - No. 2 cylinder sensor: D1 (black/white wire) and E3 (white/black wire)
   - No. 3 cylinder sensor: D1 (black/white wire) and E1 (red/white wire)

■ **Information on connector views, pinouts and additional specifications are found under Sensor and Circuit Resistance/Output Tests in this section.**

   b. For 70 hp motors, use the ECU harness connector "E" from the end of the ECU to test the following connections using a DVOM set to the resistance scale:
   - Port side CKP sensor: E1 (black wire) and E4 (red/blue wire)
   - Starboard side sensor: E1 (black wire) and E3 (white/black wire)

3. The DVOM should read approximately 168-252 ohms. If resistance is outside specifications, check the wiring harness for damage before replacing the sensor.

※※ **WARNING**

DO NOT attach a DVOM to the ECU itself. Some circuits of the ECU can be damaged or destroyed by applying voltage from an ohmmeter or a DVOM set to the resistance scale.

4. Connect the sensor wire harness to the sensor or ECU, as applicable. Be sure to route the wires carefully, avoiding interference with other components.
5. Connect the negative battery cable.

REMOVAL & INSTALLATION

◆ See Figures 126 and 127

There are 3 sensors wired together in the CKP (40/50 hp motors) or CKP/CMP (70 hp motors) assembly. If one sensor has and the engine has been in service for some time, it is probably advisable (and easier) to simply replace the whole sensor assembly (all 3 sensors).

But, individual replacement sensors are normally available. Although each of the sensors utilize a unique signal wire, they all share the same ground wire. If a single sensor is to be replaced, trace the wire back to the common ground and cut it out at a position where the replacement can be spliced (be sure to have the replacement on hand before performing this to ensure it is a logical point). The best method for splicing is to twist and solder the wires, then seal them using heat shrink, but if desired solderless crimp connectors may be used. The major concern with solderless crimp connectors are that they must be of the proper size and crimped properly to ensure connection. Most crimp connectors are also designed to seal with heat.

Prior to removal note the wire routing (or take a picture with a digital camera) for reference during installation. Make sure that wires are routed properly to prevent interference, especially with moving components such as the flywheel.

Although no adjustment is necessary on 40/50 hp motors, a feeler gauge must be used to properly adjust the air gap on the sensors for 70 hp motors to ensure proper operation.

1. Disconnect the negative battery cable for safety.
2. If necessary, remove the flywheel as detailed under Flywheel in the Powerhead section for better access.
3. Remove the 6 retaining bolts for the CKP or CKP/CMP sensors (2 bolts per sensor) and remove the assembly from the powerhead. If only one sensor is being replaced, it may be easier to remove them all and perform the splicing on a workbench, but that's your call.
4. Clean the sensor mounts, as well as the mounting screw threads.

To install:

5. Carefully route the sensor wires into position as noted proper to removal, then position the sensors onto the powerhead with the protrusions facing the flywheel.
6. Apply a coating of OMC Nut Lock or equivalent threadlock to the threads of the sensor mounting screws.
7. Position the sensors to the mounting points and install the retaining screws. For 40/50 hp motors, tighten the screws securely. For 70 hp motors, position the flywheel and adjust the sensors before tightening the screws as follows:

   a. If removed, install the flywheel and hand-tighten the retaining bolt.
   b. Slide a 0.030 in. (0.75mm) feeler gauge between the sensor and the raised section of the flywheel (reactor plate). Push on the sensor lightly by finger against the gauge and flywheel to adjust the air gap. Holding the sensor lightly in this position and tighten the mounting screws, then double-check for correct gap. Loosen the screws and readjust, as necessary.

■ **When using a feeler gauge to check the air gap, make sure that a 0.030 in. (0.75mm) gauge passes between the sensor and flywheel with only a slight drag. The next larger gauge should not pass and the next smaller gauge should pass without any interference.**

8. Make sure all sensor screws are tightened securely and the harnesses are properly routed. If any clamps were used to retain the harness, bend them over in position.
9. Connect the harness wiring.
10. If removed, finish installing the flywheel, as detailed in the Powerhead section. Once the flywheel is bolted in position, rotate it slowly by hand to make sure that no interference occurs between it and any sensors or wiring.
11. Connect the negative battery cable, then check for proper ignition system operation.

Engine Control Unit (ECU)

◆ See Figure 128

The Engine Control Unit (ECU) controls all functions of the EFI and ignition systems. Problems with the ECU are rare, but when they occur can cause a no-start, stumbling, misfire, hesitation, rough idle or incorrect speed limiting through improper control of the ignition and/or fuel injection systems.

FUEL SYSTEM 3-63

Unfortunately, solid state components like the ECU cannot be directly tested in many ways. One exception comes with checking reference voltages, which can be performed using the information found under Sensor and Circuit Resistance/Output Tests on 40/50 hp motors. In most cases, ECU testing involves a process of elimination, testing all other possible causes of a symptom. Condemn the ECU only if all other components that could cause a problem have been eliminated. Remember that many of the circuits used by the ECU for information or for direct control of the motor are sensitive to changes in resistance. Simple problems such as loose, dirty or corroded connectors, even pinched wires or interference caused by marine radios or other electronic accessories can cause symptoms making an otherwise good ECU seem bad.

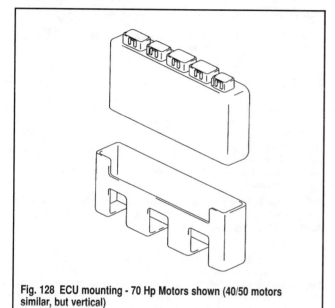

Fig. 128 ECU mounting - 70 Hp Motors shown (40/50 motors similar, but vertical)

REMOVAL & INSTALLATION

◆ See Figure 128

The ECU is mounted on the starboard side of the powerhead in a rubber mounted electrical parts holder. On 40/50 hp motors, the ECU is mounted vertically, on 70 hp motors it is mounted horizontally with the connectors facing upward.

Make sure all other possible causes of improper operation have been eliminated before replacing this component. The ECU tends to be expensive and is normally non-returnable. Also, make sure that no other problems exist in the EFI system and wiring harnesses, because circuit problems could exist that destroy or damage an ECU could also instantly destroy or damage the replacement.

1. Disconnect the negative battery cable to prevent the possibility of damage to the ECU.

✴✴ WARNING

Static electricity can instantly damage or destroy the solid state control modules. In order to prevent this possibility, ground yourself by touching a metallic component on the motor immediately prior to touching the ECU. If you feel a small static shock, you may have prevented damage to the control unit.

2. If equipped, remove the plastic cover from the electrical component holder for access.
3. Tag and disconnect the 5 wiring harness connectors fastened to the ECU. Push the retaining tabs on each connector DOWN while pulling back on the connector housing to free them from the ECU.

■ **Three of the five connectors can only be installed on one of the ECU terminals, but there are two 12-pin connectors (one black and one white). Make sure the 12-pin terminals are tagged to ensure proper installation.**

4. If equipped, remove the screws retaining the ECU.

■ **Note any ground wires connected at the retaining bolts. To prevent problems with the unit (including failures due to poor grounding) make sure ground wires are connected during installation.**

5. Remove the ECU from the powerhead.

To install:

6. Clean dirt and corrosion from the mounting location.

■ **Make sure any ground wires that are used are installed properly. Take care when routing all ground or harness wiring to prevent interference with other components.**

7. Position the ECU to the powerhead, then install any fasteners and tighten securely.
8. Engage the 5 wiring harness connectors to the back of the ECU.
9. If equipped, install the plastic cover to the electrical component holder.
10. Connect the negative battery cable and check for proper operation.

Temperature Sensors (CT, EM and IAT)

◆ See Figures 129 thru 132

Signals from the Intake Air Temperature (IAT) and Cylinder Temperature (CT) sensor are used by the EFI system to manage engine operation. The IAT sensor is used by the ECU to help determine air/fuel ratios. The CT sensor is used to inform the ECU once the engine has warmed to normal operating temperature. During operation signals from both the CT sensor and the Exhaust Manifold (EM) sensor are also used by the engine overheat or Speed Limited Overheat Warning (S.L.O.W.) system.

Temperature sensors for modern fuel injection systems are normally thermistors, meaning that they are variable resistors or electrical components that change their resistance value with changes in temperature. The Johnson/Evinrude temperature sensors are Negative Temperature Coefficient (NTC) thermistors. Whereas the resistance of most thermistors (and most electrical circuits) increases with temperature increases (or lowers as the temperature goes down), an NTC sensor operates in an opposite manner. The resistance of an NTC thermistor goes down as temperature rises (or goes up when temperature goes down).

On 40/50 hp motors the sensors are located:
- CT: the center, top of the powerhead, immediately behind the flywheel For 70 hp motors, the CT is found just below the thermostat housing, behind the electric parts holder.
- EM: toward the top of the (drum roll please) exhaust manifold (surprised?), on the starboard side of the engine.

■ **The EM is on the opposite end of the manifold from the anode on these models.**

- IAT: the bottom of the air intake silencer for 40 and 50 hp motors, or just behind the flame arrester screen in the throttle body on 70 hp motors.

On 70 hp motors the sensors are located:
- CT: immediately below the thermostat housing, behind the electric component holder.
- EM: toward the top of the (drum roll please) exhaust manifold (surprised?), on the starboard side of the engine.

■ **The EM is mounted only inches above the anode on these motors, of course there is little chance of confusing the two since no wiring harness is attached to the anode.**

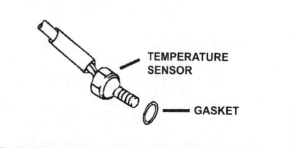

Fig. 129 Typical Cylinder Temperature (CT) an Exhaust Manifold (EM) - 40/50 Hp Motors

FUEL SYSTEM

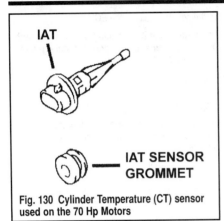

Fig. 130 Cylinder Temperature (CT) sensor used on the 70 Hp Motors

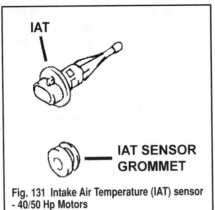

Fig. 131 Intake Air Temperature (IAT) sensor - 40/50 Hp Motors

Fig. 132 Intake Air Temperature (IAT) sensor - 70 Hp Motors

- IAT: in the throttle body, behind the flame arrester screen.

■ Because fuel mapping decisions are made using input from the IAT and CT sensors, incorrect operation can result in excessive exhaust smoke, spark plug fouling or other poor engine performance can result from an incorrect air/fuel mixture caused by a defective IAT signal. Problems with the CT circuit can lead to hard starting or problems during warm-up operation.

TESTING

◆ See Figures 133 and 134

Temperature sensors are among the easiest components of the EFI system to check for proper operation. That is because the operation of an NTC thermistor is basically straightforward. In general terms, raise the temperature of the sensor and resistance should go down. Lower the temperature of the sensor and resistance should go up. The only real concern during testing is to make clean test connections with the probe and to use accurate (high quality) testing devices including a DVOM and a relatively accurate thermometer or thermosensor.

A quick check of the circuit and/or sensor can be made by disconnecting the sensor wiring and checking resistance (comparing specifications to the ambient temperature of the motor and sensor at the time of the test). Keep in mind that this test can be misleading as it could mask a sensor that reads incorrectly at other temperatures. Of course, a cold engine can be warmed and checked again in this manner.

More detailed testing involves removing the sensor and suspending it in a container of liquid (either water or oil, as desired), then slowly heating the liquid (without using a flame if the liquid of choice is oil) while watching sensor resistance changes on a DVOM. This method allows you to check for problems in the sensor as it heats across it's entire operating range.

■ Information on ECU connector views, pinouts and additional specifications for testing at the ECU harness are found under Sensor and Circuit Resistance/Output Tests in this section.

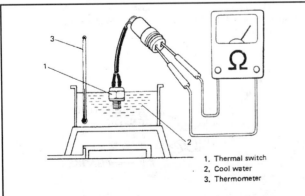

Fig. 133 Resistance of these temperature sensors should be high when the temperature is low.

Quick Test

A quick check of a temperature sensor can be made using a DVOM set to the resistance scale and applied across the sensor terminals. The DVOM can be connected to directly to the sensor (IAT), the sensor pigtail (CT or EM) or at the appropriate ECU harness connector. Use a thermometer or a thermosensor to determine ambient engine/sensor temperature before checking resistance.

■ Isolate the ECU from the sensor harness by disconnecting sensor or ECU harness connector before attempting to check resistance. Even the small voltage provided by a DVOM in order to check a circuit's resistance can damage an ECU.

Even if the sensor tests ok cold, the sensor might read incorrectly hot (or anywhere in between). If trouble is suspected, reconnect the circuit, then start and run the engine to normal operating temperature. After the engine is fully warmed, shut the engine **OFF** and recheck the sensor hot. If the sensor checks within specification hot, it is still possible that another point during warm-up could be causing a problem, but not likely. The sensor can be removed and checked using the Comprehensive Test in this section or other causes for the symptoms can be checked. If the sensor was checked directly, be sure to check the wiring harness between the ECU and the sensor for continuity. Excessive resistance due to loose connections or damage in the wiring harness can cause the sensor to read out of range.

When checking the sensor be sure to make the correct connections as follows:

- CT: Connect the meter to the violet (40/50 hp motors) or light green (70 hp motors) and black wires of the sensor pigtail or to the light green/white and black/white ECU harness wires, as desired.
- EM: Connect the meter to the violet and black wires of the sensor pigtail or to the violet/white and black/white ECU harness wires, as desired.
- IAT: Connect the meter directly the sensor terminals (there is no pigtail on these sensors) or to the light green/black and black/white ECU harness wires, as desired.

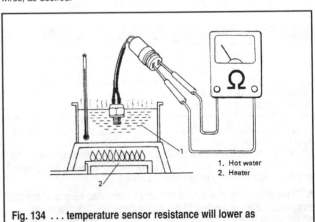

Fig. 134 ... temperature sensor resistance will lower as temperature rises

FUEL SYSTEM

■ When testing 40/50 hp motors through the ECU harness, the manufacturer provides resistance specifications that are slightly different than those listed here. Unfortunately they do not give an explanation for these difference (although it is likely that they are allowing slightly for additional resistance in the wiring harness and the terminal connectors). For specific connector and terminal identification, as well as resistance specifications for testing through the ECU harness refer to Sensor and Circuit Resistance/Output Tests in this section. To complicate matters further, the manufacturer does NOT provide alternate resistance specifications for these tests on 70 hp motors, so use the specs listed here, even if testing through the circuit (but allow for a small, about 5%, variance in readings through the harness before replacing a sensor).

Resistance specifications for these temperature sensors are as follows:
- 5,100-6,000 ohms @ 32°F (0°C)
- 1,900-2,100 ohms @ 77°F (25°C)
- 760-900 ohms @ 122°F (50°C)
- 340-420 ohms @ 135°F (75°C)

Comprehensive Test

◆ See Figures 133 and 134

It is important to EFI operation that the temperature sensor (especially the IAT and CT) provides accurate signals across the entire operating range and not just when fully hot or fully cold. For this reason, it is best to test the sensor by watching output or resistance constantly as the sensor is heated from a cold temperature to the upper end of the engine's operating range. The easiest way to do this is to use the OMC Diagnostic Software installed in a suitable IBM compatible laptop. If the software is not available, the next best (and most often used solution) is to suspend the sensor in a container of water or oil, connect a DVOM and slowly heat the liquid while watching resistance on the meter.

To perform this check, you will need a high quality (accurate) DVOM, a thermometer (or thermosensor, some multi-meters are available with thermosensor adapters), a length of wire, a metal or laboratory grade glass container and a heat source (such as a hot plate or camp stove). A DVOM with alligator clip style probes will make this test a lot easier, otherwise alligator clip adapters can be used, but check before testing to make sure they do not add significant additional resistance to the circuit. This check is performed by connecting the two alligator clips together and checking for a very low or 0 resistance reading. If readings are higher than 0, record the value to subtract from the sensor resistance readings compensating for the use of the alligator clips.

※※ CAUTION

When using a camp stove or source of open flame to heat the container oil should not be used as the suspension liquid to protect against the possibility of an oil fire. Use water when flame is involved.

1. Remove the temperature switch as detailed in this section.
2. Suspend the sensor and the thermometer or thermosensor probe in a container of cool water or four-stroke engine oil.

■ To ensure accurate readings make sure the temperature sensor and the thermometer are suspended in the liquid and are not touching the bottom or sides of the container.

3. Set the DVOM to the resistance scale, then attach the probes to the sensor or sensor pigtail terminals, as applicable.

■ To ensure test accuracy, do not allow the sensor/switch or thermometer to touch the bottoms or sides of the container during the test as the temperature of the container may vary somewhat from the liquid contained within and sensor held in suspension.

4. Allow the temperature of the sensor and thermometer to stabilize, then note the temperature and the resistance reading.
5. Use the hot plate or camp stove to slowly raise the temperature. Continue to note resistance readings as the temperature rises to 77°F (25°C), 122°F (50°C) and finally to 135°F (75°C). The meter should show a steady decrease in resistance that is proportional to the rate at which the liquid is heated. Extreme peaks or valleys in the sensor signal should be rechecked to see if they are results of sudden temperature increases or a possible problem with the sensor.

6. Compare the readings to the following specifications:
- 5,100-6,000 ohms @ 32°F (0°C)
- 1,900-2,100 ohms @ 77°F (25°C)
- 760-900 ohms @ 122°F (50°C)
- 340-420 ohms @ 135°F (75°C)

7. Sensors that read well outside specifications (more than about a 5% variance in reading) should be replaced to ensure proper engine operation.

REMOVAL & INSTALLATION

Cylinder Temperature (CT) Sensor

◆ See Figures 129 and 130

1. Disconnect the negative battery cable for safety.
2. Locate the CT sensor (if necessary refer to the wiring diagrams to help trace the wires from the ECU harness) as follows:
- 40/50 hp motors: the sensor is mounted on the center, top of the powerhead (behind the flywheel/in front of the thermostat).
- 70 hp motors: the sensor is mounted on the starboard side of the powerhead (behind the electrical component holder/below the thermostat housing). If necessary, remove or reposition the electrical component holder for additional access.

3. Pull upward on the retaining tab, then disengage the wiring harness connector from the sensor. Note the wire routing for installation purposes. Make sure the terminals on the sensor and in the wiring harness are clean and free of corrosion.
4. Either unthread and remove the sensor from the powerhead (40/50 hp motors) or remove the retaining bolt and plate, then remove the sensor from the powerhead (70 hp motors). Be sure to clean the sensor threads and the threads in the powerhead before installation.
5. If removed, remove and discard the gasket or seal.
6. Installation is the reverse of removal, but take care not to overtighten the sensor or mounting bolt (as applicable) and damage the sensor or the mounting threads in the powerhead. Push the wiring connector until the retaining tab snaps in place, then pull back gently to ensure proper connection.

Exhaust Manifold (EM) Temperature Sensor

◆ See Figure 129

1. Disconnect the negative battery cable for safety.

※※ CAUTION

The exhaust manifold becomes extremely hot during operation, to prevent burns, make sure the engine has cooled sufficiently before removing the sensor.

2. Locate the EM sensor on the top of the exhaust manifold (if necessary refer to the wiring diagrams to help trace the wires from the ECU harness).
3. Pull upward on the retaining tab, then disengage the wiring harness connector from the sensor. Note the wire routing for installation purposes. Make sure the terminals on the sensor and in the wiring harness are clean and free of corrosion.
4. Unthread and remove the sensor from the manifold. Be sure to clean the sensor threads and the threads in the manifold before installation.

※※ WARNING

Repeated heating and cooling cycles of the motor during normal service will often seize components in the exhaust manifold, including this sensor. If it is difficult to remove, DO NOT force it or it could break off in the manifold. Instead, apply a few drops of a penetrating oil such as PB Blaster or WD-40 and give it a few minutes to work before attempting to loosen the sensor again. If the sensor still won't budge, warm the engine to normal operating temperature and loosen the sensor as it cools (using GREAT care to prevent burning yourself).

5. If removed, remove and discard the gasket or seal.
6. Installation is the reverse of removal, but take care not to overtighten and damage the sensor or the mounting threads in the manifold. Push the wiring connector until the retaining tab snaps in place, then pull back gently to ensure proper connection.

3-66 FUEL SYSTEM

Intake Air Temperature (IAT) Sensor

◆ See Figures 131 and 132

1. Disconnect the negative battery cable for safety.
2. Locate the IAT sensor (if necessary refer to the wiring diagrams to help trace the wires from the ECU harness) as follows:
   - 40/50 hp motors: the sensor is mounted in the bottom of the silencer cover, on the starboard side.
   - 70 hp motors: the sensor is mounted to the top, side of the throttle body (behind the flame arrester screen).
3. Push the retaining tab downward and disengage the wiring harness connector from the sensor. Note the wire routing for installation purposes. Make sure the terminals on the sensor and in the wiring harness are clean and free of corrosion.
4. Either unthread and remove the sensor from the throttle body (70 hp motors) or grasp the sensor and gently pull it free of the grommet in the silencer cover (40/50 hp motors).
5. For 70 hp motors, remove and discard the sensor gasket. Clean the sensor threads and the threads in the throttle body before installation.
6. For 40/50 hp motors, inspect the sensor grommet (mounted in the silencer cover) for damage or deterioration and replace, if necessary.
7. Installation is the reverse of removal. Do not overtighten the sensor on 70 hp motors and damage the sensor or the throttle body. On 40/50 hp motors, make sure the sensor is fully seated and secure in the grommet. Push the wiring connector until the retaining tab snaps in place, then pull back gently to ensure proper connection.

Manifold Absolute Pressure (MAP) Sensor

◆ See Figure 135

During normal EFI operation, the majority of fuel mapping decisions (injector on-time strategy) is made based upon signals from the MAP sensor. The sensor detects atmospheric pressure (air density) when the key is turned on (such as during engine starting) or cranking. Once the engine starts, the sensor provides intake manifold air pressure.

The sensor is mounted to the top of the intake manifold runners (port side) on 40/50 hp motors or attached to the starter motor bracket (front, port side above the throttle body) for 70 hp motors. On all models, a short rubber hose connects the sensor to a pulse limiter (mounted behind the throttle body) that is used to dampen vacuum signal changes.

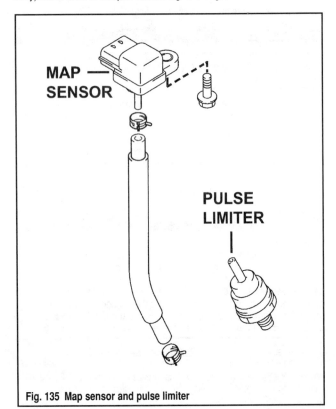

Fig. 135 Map sensor and pulse limiter

TESTING

◆ See Figures 96 and 135

The single largest input on which the ECU relies for fuel mapping is the MAP sensor signal regarding intake manifold pressure (during engine operation). The sensor cannot function correctly unless the vacuum hose is attached without leaks.

If problems occur with the sensor circuit (out of range operation), there is excessive smoke form the exhaust, spark plug fouling or other driveability problems, check the condition and connection of the vacuum hose before performing electronic tests on the sensor and circuit.

If trouble codes are present, keep in mind that a code 32 refers to a suspected possible problem with the hose (as determined by the ECU if it sees a steady MAP sensor signal regardless of varying rpm signals from the CKP sensors). When the ECU sets a code 34, it either see no signal from the sensor at all or a signal that is out of range. The most likely causes of a code 34 are circuit, not vacuum hose, problems.

The voltage output check is the best way to test the sensor circuit. Because the wiring must be connected in order to conduct a voltage check, testing is performed by carefully probing through the back of the appropriate connector (back-probe). Testing can take place at the back of the component wiring connector or at the back of the appropriate ECU harness connector. When installed, the harness connector view would be a mirror image of the connector views shown under the Sensor and Circuit Resistance/Output Test. If necessary, refer to the wiring diagrams to check circuit wire colors and confirm terminal positioning.

■ **Make sure all sensor wiring/connections are in clean and tight before replacing the sensor due to voltage output check test results.**

A DVOM equipped with a small test probe or a test probe from the OMC test probe kit (#342677) and a hand vacuum pump are necessary to perform this test. Hand vacuum pumps should be available at most marine or automotive supply stores (they are used for testing and for certain maintenance tasks such as drawing hydraulic fluid through a system).

1. Locate ECU connector D, the black 12-pin connector attached to the back of the ECU, just to the side of the white 12-pin connector located at the center of the 5 connectors. On 70 hp motors, remove the cover from the electrical component holder for access to the ECU wiring.
2. Using a DVOM set to the low DC voltage scale, insert the back of a small probe (or a test probe from OMC test probe kit {#342677}) through the back of the connector D terminal 7. D7 should have a white wire. Connect the other voltmeter probe to a good engine ground.
3. Disconnect the vacuum hose from the MAP sensor, then attach a hand vacuum pump so that specific pressures can be applied to the sensor while testing for specific voltage outputs.
4. Turn the ignition key **ON** and apply the following vacuum levels, looking for the appropriate voltage readings on the DVOM:
   - 0 in. Hg (0 kPa): 4.0 volts
   - 11.8 in. Hg (40 kPa): 2.42 volts
   - 24 in. Hg (80 kPa): 0.84 volts
5. Replace the MAP sensor if the voltage does read as specified under each of the test vacuum conditions.

REMOVAL & INSTALLATION

◆ See Figures 135

The sensor is mounted to the top of the intake manifold runners (port side) on 40/50 hp motors or attached to the starter motor bracket (front, port side above the throttle body) for 70 hp motors. On all models, a short rubber hose connects the sensor to a pulse limiter (mounted behind the throttle body) that is used to dampen vacuum signal changes.

1. Disconnect the negative battery cable for safety.
2. Locate the map sensor on the intake manifold (40/50 hp motors) or starter motor bracket (70 hp motors), as applicable.

■ **Note the wire and hose routing before disconnecting anything from the sensor to ensure proper installation.**

3. Push down on the locktab while pulling back on the harness connector to disconnect it from the sensor.

FUEL SYSTEM 3-67

4. Disconnect the vacuum hose from the sensor.
5. Carefully remove sensor the mounting screws, then remove the sensor from the intake manifold or powerhead bracket, as applicable.
6. Clean the sensor and manifold/bracket mating surfaces.
7. Installation is essentially the reverse of removal, but take care to make sure the sensor bolt, vacuum hose and wiring connector are all secure. Do not over-tighten the mounting bolt that could damage the sensor housing or strip the threads in the manifold or starter bracket. Pull back on the wiring connector after the tab engages to make sure it is locked in place. Most of all, make sure the vacuum hose is properly connected with no leaks.

Closed Throttle Position (CTP) Switch

◆ See Figure 136

The Closed Throttle Position (CTP) switch signals the ECU when the throttle is in the closed position, indicating that the ECU should control idle speed. When the throttle is closed the ECU will actuate the Idle Air Control (IAC) solenoid valve and alter ignition timing in order to control idle speed.

Once the throttle is opened, the signal from the switch alerts the ECU to increase fuel delivery and advance spark timing to improve throttle response.

The switch is located on the top of the throttle body for 40/50 hp motors or on the bottom of the throttle body on 70 hp motors.

TESTING

Problems with the CTP circuit will usually cause hesitation when the throttle is advanced quickly off idle, but trouble with the circuit can also cause a high idle speed or a rough idle.

1. Disconnect the negative battery cable.
2. Locate the throttle body on the front, port side of the powerhead, then locate the CTP switch and wiring on the throttle body. If necessary, trace the light green/red wire to the switch.
3. Push the retaining tab downward and disengage the wiring harness connector from the sensor. Note the wire routing for installation purposes. Make sure the terminals on the sensor and in the wiring harness are clean and free of corrosion.
4. Set a DVOM to the resistance scale, then connector the red (positive) meter lead to a good engine ground and the black (negative) meter lead to the switch wire terminal.
5. Slowly open and close the throttle while watching output on the meter, is should read as follows:
   - With the throttle closed and the switch button fully depressed, the meter should show continuity.
   - As the throttle is just cracked open, and the switch button is released, the meter should show no continuity.
6. Replace the throttle body if the switch does not operate properly.

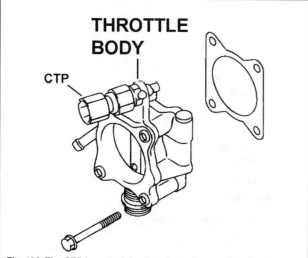

Fig. 136 The CTP is part of the throttle body assembly, installed and adjusted at the factory - 40/50 hp motors shown

REMOVAL & INSTALLATION

The CTP switch is mounted to the throttle body and calibrated at the factory. Do NOT disturb the switch or attempt to replace it, no parts are available for service. If the switch fails, the throttle body assembly must be replaced.

Fuel Pressure Regulator

Like the electric fuel pump, the fuel pressure regulator is part of the vapor separator tank assembly. The regulator itself is mounted inside the tank and can be replaced when servicing the tank if the cover/pump assembly is removed. If problems are suspected with the high pressure fuel system, refer to the regulator and pump pressure testing procedures found under the Vapor Separator Tank and High Pressure Fuel Pump procedures in this section.0

EFI System Relay

◆ See Figure 137

A main system relay is mounted in the upper left of the electrical parts holder (near the ECU) on the starboard side of the powerhead. On 70 hp motors, the electrical component holder cover must be removed for access. The main relay provides power to the ECU, fuel injectors, ignition coils, IAC valve, diagnostic connector, and for 40/50 hp motors, the CMP sensor. Power is applied only when the ignition keyswitch is turned to the **ON** position.

TESTING & SERVICE

◆ See Figure 137

Problems with the system relay will normally keep the engine from starting, but intermittent problems (that come and go) could cause the engine to stall periodically. Intermittent faults are often caused by heat related conditions, so if the engine stalls with no other symptoms, but restarts only after it has cooled, check the system relay.

Although the manufacturer provides various tests for the relay, the easiest and safest test is to remove the relay (which helps protect test equipment during testing). In order to properly test the relay in this fashion you will need a DVOM, jumper wires with alligator clips and fresh (charged) 12-volt battery.

The solenoid terminals are identified as follows:
- B+ Battery (12-volt) positive solenoid, white wire.
- B- Battery (12-volt) ECU negative (ground), pink/black wire.
- L Load (12-volt) power feed to components, large gray wire.
- S Keyswitch signal, small gray wire.

1. Disconnect the negative battery cable for safety, then remove the main relay from the powerhead and disconnect it from the wiring harness.
2. Position the relay on the workbench with the connector retainer tab facing up as shown in the accompanying figure. Use the art to help identify the relay terminals as follows:

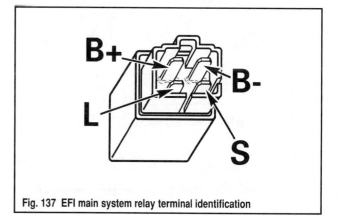

Fig. 137 EFI main system relay terminal identification

3-68 FUEL SYSTEM

- B+ Upper left terminal (when relay tab is on top).
- B- Upper right terminal (when relay tab is on top).
- L Lower left terminal (when relay tab is on top).
- S Lower right terminal (when relay tab is on top).

3. With the DVOM set to the resistance scale, connect the positive meter test lead to the B+ terminal, then connect the negative meter test lead to the L terminal. No continuity should be noted on the DVOM.

4. Move the meter leads to the B- and S terminals. The meter must now read 80-120 ohms resistance.

5. Remove the meter leads, now connect relay terminal S to the positive terminal of a fully charged 12-volt battery. Connect another jumper lead to the negative terminal of the battery.

6. Connect the DVOM (still set to check resistance) to the B+ and L terminals again. With the meter connected and the 12-volt battery positive terminal jumped to the S terminal, jump the negative terminal of the battery to the remaining relay terminal B-. When power is applied to S and B- the internal switch in the relay should close causing B+ and L to show continuity.

7. Replace the system relay if B+ and L fail to show continuity when power is applied to the other 2 terminals or if they show continuity when power is removed.

8. Clean the relay and harness connector terminals, then plug the relay back into the harness connector. Secure the relay back into it's mounting, and if applicable, install the electrical component cover.

9. Connect the negative battery cable and verify proper relay operation.

Idle Air Control (IAC) Valve and Silencer

◆ See Figures 138 and 139

Although some air for idle operation is provided by the idle air bypass screw and the CTP switch which holds the throttle plate open slightly, additional air is provided by the Idle Air Control (IAC) valve. The ECU uses the IAC valve control idle speed through the additional air metered to the engine through the valve passages. Air for the IAC valve is drawn through a dedicated IAC valve air silencer mounted in a clamp to the powerhead in the same vicinity as the valve.

The IAC valve is located at the front of the engine. It is just starboard of the throttle body on 40/50 hp motors. For 70 hp motors, the valve is just above and slightly port of the throttle body, mounted directly to the intake manifold (visible from the front of the engine, just below the top, port side of the air silencer).

TESTING

◆ See Figures 138 and 139

A faulty IAC valve can cause varied idle problems (including rough, slow or high idles), but it can also cause the engine to stall when shifted into gear. Because the IAC valve operates by allowing small amounts of air through dedicated passages, the affects of a faulty valve are usually inconsequential at higher speeds.

■ **Information on ECU connector views, pinouts and additional specifications for testing the IAC valve at the ECU harness are found under Sensor and Circuit Resistance/Output Tests in this section.**

1. Disconnect the negative battery cable for safety.
2. Locate the valve mounted near the throttle body and air intake silencer (40/50 hp motors) or intake manifold (70 hp motors). If necessary, trace the red/black and gray wires from the appropriate ECU connectors (refer to the wiring diagrams for more details) to the harness connector.
3. Release the lock tab (by pushing downward on 40/50 hp motors or pulling upward on 70 hp motors), and carefully pull the harness wiring connector from the valve (40/50 hp motors) or the IAC valve pigtail (70 hp motors).
4. Using a DVOM set to the resistance scale, connect the leads across the IAC valve (or pigtail, as applicable) terminals. Resistance should be 21.5-32.3 ohms on 40/50 hp models or 4.8-7.2 ohms on 70 hp models. Replace the IAC valve if resistance is not within specification.
5. When finished, reconnect the wire harness to the valve (routing the wires to avoid interference if they were moved significantly), then reconnect negative battery cable.

REMOVAL & INSTALLATION

◆ See Figures 138 and 139

The IAC valve is located at the front of the engine. It is just starboard of the throttle body on 40/50 hp motors. For 70 hp motors, the valve is just above and slightly port of the throttle body, mounted directly to the intake manifold (visible from the front of the engine, just below the top, port side of the air silencer).

1. Disconnect the negative battery cable for safety.
2. If necessary for additional access, remove the air intake silencer (40/50 hp motors) or silencer cover (70 hp motors) as detailed in this section under Air Intake Silencer and Flame Arrester.
3. Locate the IAC valve, then note the routing and connection points of the hoses and wires for installation purposes.
4. On 40/50 hp motors, disconnect the hoses from the valve.
5. Release the lock tab (by pushing downward on 40/50 hp motors or pulling upward on 70 hp motors), and carefully pull the harness wiring connector from the valve.
6. Remove the valve as follows:
   a. For 40/50 hp motors, remove the screws, then pull the valve and bracket from the powerhead.
   b. For 70 hp motors, remove the screws and retainer, then pull the valve and flange. Remove the rubber gasket (valve seat) from the valve flange or intake manifold.
7. Carefully clean the mating surfaces using solvent.

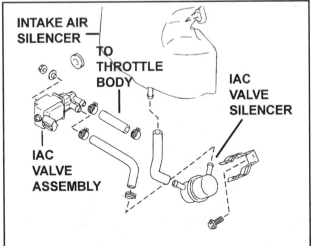

Fig. 138 Exploded view of the IAC valve and silencer mounting - 40/50 Hp Motors

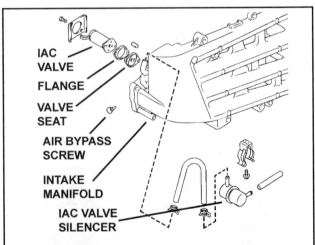

Fig. 139 Exploded view of the IAC valve and silencer mounting - 70 Hp Motors

FUEL SYSTEM 3-69

8. If desired, remove the IAC valve silencer (a round plastic housing) as follows:

   a. For 40/50 hp motors, the silencer is located directly under the throttle body at the front of the engine. The silencer hoses attach to the IAC valve (already disconnected at this point) and the air intake silencer (which also may be disconnected already if the main air intake silencer was removed in Step 2 for additional access. To remove the IAC valve silencer, carefully pull it from the retaining clamp.

   b. For 70 hp motors, the silencer is located under the intake runners at the port front of the engine. One of the silencer hoses attaches to the intake manifold (disconnect it from the manifold or the silencer, whichever is easier), while the other is simply routed to provide intake air. The second hose (open to the atmosphere) may be secured in a wire tie or clamp (if so, disconnect it from the anchor point or from the silencer, as desired). Carefully remove the silencer from the retaining clamp.

To install:

9. If removed, install the IAC valve silencer to the retaining clamp. Reposition hoses for connection to the valve and main silencer (40/50 hp motors) or the intake manifold and anchor point (70 hp motors), as applicable.

10. Install the IAC valve to the mounting bracket. On 70 hp motors, be sure to install the flange and valve seat (rubber gasket).

11. Tighten the mounting screws securely.

12. Connect the hoses and wiring as noted during removal to avoid interference with other components. When installing the harness connector, push gently until the lock tab engages, then give a slight tug back on the harness to ensure it is locked in place. On 40/50 hp motors, one hoses attaches the air intake silencer, if it was removed, connect the hose as it is reinstalled.

13. If removed, install the air intake silencer or silencer cover, as applicable.

14. Connect the negative battery cable.

Neutral (Safety) Switch

◆ See Figures 140, 141 and 142

The neutral switch signals the ECU when the engine is shifted into or out of neutral gear. The ECU uses the signal to prevent starting (but cutting operation of the fuel injectors and ignition system) if an attempt is made to start the engine in gear.

The switch is mounted so plunger contacts the shifter linkage. The plunger should be actuated (depressed) only when the shifter is in Neutral (as can be verified by manually rotating the prop shaft). The harness connector is found behind the upper side of the vapor separator (40/50 hp motors) or directly behind the ECU (70 hp motors). The ECU must be removed from the holder and repositioned to access the harness connector on 70 hp motors (this usually can be accomplished with the wiring connected, as long as care is taken not to pull on the wires and terminals).

TESTING

◆ See Figures 140 and 141

Problems with the neutral (safety) switch could allow the engine to start in gear (a potentially dangerous condition) or could cause a no-start condition in neutral.

The quickest way to test the neutral switch is to attempt to start the motor in gear, if it starts, there's a problem with the circuit. Alternately, the switch should be checked (along with the starter solenoid) if the battery, fuses and connections are good, but a no-start condition exists in neutral.

To isolate and test the switch itself:

1. Disconnect the negative battery cable for safety.
2. Remove the lower engine covers as described under Engine Covers in the Engine Maintenance section.
3. For 70 hp motors, remove the holder from the electrical component cover, then remove the ECU from the holder for access to the switch harness.
4. Locate the neutral switch wiring harness connector (behind the upper side of the vapor separator tank on 40/50 hp motors, or behind the ECU on 70 hp motors). If the connector is hard to find, first locate the switch itself (on the lower starboard side of the powerhead, next to the shift linkage), then trace the switch wires (yellow/green and brown) back to the harness connector.
5. Pull upward on the retaining tab, then disengage the wiring harness connector from the switch pigtail connector. Note the wire routing for installation purposes. Make sure the terminals on the switch pigtail connector and in the wiring harness are clean and free of corrosion.
6. Using a DVOM set to read resistance, connect the positive meter lead to the pigtail terminal for the yellow/green wire and the negative test lead to the pigtail terminal for the brown wire.

✳✳ WARNING

To prevent damage to the gear shifter assembly, have an assistant slowly rotate the propeller shaft by hand when attempting to move the shifter into or out of gear.

7. Place the remote control handle in neutral. The switch plunger should be depressed and the meter should show continuity.
8. Place the remote control the forward and then the reverse gear positions. The switch plunger should release and the meter should shown no continuity with the engine in either gear.
9. Adjust the switch if the plunger is not where it should be in the noted positions. Replace the switch if the meter shows incorrect readings when the plunger is actuated as noted.
10. When finished with tests and, if necessary, repairs, reconnect the engine wire harness to the neutral switch pigtail connector. Route the wires to avoid interference.

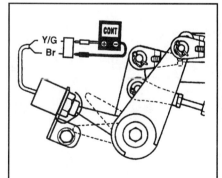

Fig. 140 The neutral safety switch should only show continuity when the shift linkage is in neutral - 70 Hp Motors shown

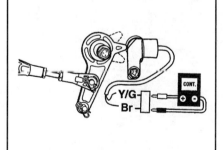

Fig. 141 The switch plunger should be fully depressed (for switch continuity) with the shift linkage in neutral - 40/50 Hp Motors shown

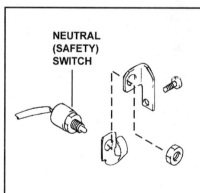

Fig. 142 Exploded view of the switch mounting - 40/50 Hp Motors shown (70 hp motors similar, but the switch is mounted vertically instead of horizontally)

3-70 FUEL SYSTEM

11. For 70 hp motors, reposition the ECU into the holder, then install the cover.
12. Install the lower engine covers and reconnect the negative battery cable.

REMOVAL & INSTALLATION

◆ See Figure 142

The switch is mounted so plunger contacts the shifter linkage. The plunger should be actuated (depressed) only when the shifter is in Neutral (as can be verified by manually rotating the prop shaft). Adjustment is accomplished by repositioning the switch in a position where the shifter linkage properly actuates the plunger ONLY in neutral. The switch bracket should be positioned in such a way as to eliminate the need for adjustment unless the switch, bracket or linkage have been physically damaged in some way.

1. Disconnect the negative battery cable for safety.
2. Remove the lower engine covers as described under Engine Covers in the Engine Maintenance section.
3. Locate the neutral switch on the lower starboard side of the powerhead, next to the shift linkage. If the switch is hard to find, locate the wiring harness connector (behind the upper side of the vapor separator tank on 40/50 hp motors, or behind the ECU on 70 hp motors), then trace the switch wires (yellow/green and brown) back through the pigtail to the switch.

※※ WARNING

To prevent damage to the gear shifter assembly, have an assistant slowly rotate the propeller shaft by hand when attempting to move the shifter into or out of gear.

4. To remove the switch, position the shifter in forward or reverse (so the switch plunger is released), then loosen the switch locknut and remove the switch from the bracket.

To install:

5. Installation is essentially the reverse of removal. Before installing the lower covers or other components removed in order to access the switch, observe switch plunger operation (the plunger should be depressed only with the shifter in neutral). Temporarily connect the negative battery cable and verify switch operation. If necessary, reposition the switch and bracket slightly to ensure proper operation, the remove the negative battery cable again for safety.
6. Complete the installation and connect the negative battery cable.

FUEL SYSTEM 3-71

SPECIFICATIONS

Carburetor Set-Up Specifications

Model (Hp)	No. of Cyl	Engine Type	Year	Displace cu. in. (cc)	Initial Low Speed Setting	Float Setting In. (mm)	OMC Float Guage
Colt/Junior	1	2-stroke	1990	2.64 (43)	1 1/4 Low / 1 High	1 1/8-1 1/2 (28-38)	324891
2	1	2-stroke	1993-01	4.75 (77.8)	1 3/4 turns / 3rd groove	Hinge top 0.090 (2.3)	n/a
2.3	1	2-stroke	1991-01	4.75 (77.8)	1 3/4 turns / 3rd groove	Hinge top 0.090 (2.3)	n/a
3.3	1	2-stroke	1991-01	4.75 (77.8)	1 3/4 turns / 3rd groove	Hinge top 0.090 (2.3)	n/a
3.5	1	2-stroke	2001	4.75 (77.8)	1 3/4 turns / 3rd groove	Hinge top 0.090 (2.3)	n/a
5	1	2-stroke	1999-01	6.7 (109)	2 1/2 turns	15/32 (12)	n/a
5	1	4-stroke	1997	7.8 (128)	6 turns	1-1 3/8 (25-35)	324891
	1	4-stroke	1998-2001	7.8 (128)	1 5/8 turns	1-1 3/8 (25-35)	324891
6	1	4-stroke	1997-01	7.8 (128)	6 turns	1-1 3/8 (25-35)	324891
	1	4-stroke	1998-2001	7.8 (128)	4 turns	1-1 3/8 (25-35)	324891
3	2	2-stroke	1990-97	5.28 (87)	1 turn	1 1/8-1 1/2 (28-38)	324891
4	2	2-stroke	1990-01	5.28 (87)	1 turn	1 1/8-1 1/2 (28-38)	324891
4 Deluxe	2	2-stroke	1990-96	5.28 (87)	1 turn	1 1/8-1 1/2 (28-38)	324891
5	2	2-stroke	1990-93	10 (164)	2 1/2 turns	1-1 3/8 (25-35)	324891
	2	2-stroke	1994-98	10 (164)	3 turns	1-1 3/8 (25-35)	324891
6	2	2-stroke	1990-93	10 (164)	2 1/2 turns	1-1 3/8 (25-35)	324891
	2	2-stroke	1994-01	10 (164)	3 turns	1-1 3/8 (25-35)	324891
6.5	2	2-stroke	1990-93	10 (164)	2 1/2 turns	1-1 3/8 (25-35)	324891
8	2	2-stroke	1990-93	10 (164)	2 1/2 turns	1-1 3/8 (25-35)	324891
	2	2-stroke	1994-01	10 (164)	3 turns	1-1 3/8 (25-35)	324891
8	2	4-stroke	1996-01	12.87 (211)	5 turns	1-1 3/8 (25-35)	324891
9.9	2	4-stroke	1997-98	12.87 (211)	5 turns	1-1 3/8 (25-35)	324891
9.9	2	2-stroke	1990-92	13.2 (216)	2 1/2 turns	1-1 3/8 (25-35)	324891
10	2	2-stroke	1990-92	13.2 (216)	2 1/2 turns	1-1 3/8 (25-35)	324891
14	2	2-stroke	1990-91	13.2 (216)	2 1/2 turns	1-1 3/8 (25-35)	324891
15	2	2-stroke	1990-92	13.2 (216)	2 1/2 turns	1-1 3/8 (25-35)	324891
9.9	2	2-stroke	1993-01	15.6 (255)	3 turns	1-1 3/8 (25-35)	324891
10	2	2-stroke	1993-98	15.6 (255)	3 turns	1-1 3/8 (25-35)	324891
15	2	2-stroke	1993-01	15.6 (255)	3 turns	1-1 3/8 (25-35)	324891
9.9	2	4-stroke	1995-01	18.61 (305)	4 turns	1-1 3/8 (25-35)	324891
15	2	4-stroke	1995-01	18.61 (305)	4 turns	1-1 3/8 (25-35)	324891
18 Jet	2	2-stroke	1995-97	31.1 (521)	2 1/2 turns	1 1/8-1 5/8 (28-41)	324891
20	2	2-stroke	1990-93	31.8 (521)	2 1/2 turns	1-1 3/8 (25-35)	324891
	2	2-stroke	1994-01	31.8 (521)	3 turns	1-1 3/8 (25-35)	324891
25	2	2-stroke	1990-01	31.8 (521)	2 1/2 turns	1 1/8-1 5/8 (28-41)	324891
28	2	2-stroke	1990-97	31.8 (521)	2 1/2 turns	1 1/8-1 5/8 (28-41)	324891
30	2	2-stroke	1990-01	31.8 (521)	2 1/2 turns	1 1/8-1 5/8 (28-41)	324891
35	2	2-stroke	1990-97	31.8 (521)	2 1/2 turns	1 1/8-1 5/8 (28-41)	324891

FUEL SYSTEM

Carburetor Set-Up Specifications

Model (Hp)	No. of Cyl	Engine Type	Year	Displace cu. in. (cc)	Initial Low Speed Setting	Float Setting In. (mm)	OMC Float Guage
25D	2	2-stroke	1990-95	45 (737)	2 to 2 1/2 turns	1 1/8-1 5/8 (28-41)	324891
30	2	2-stroke	1998-01	45 (737)	2 1/2 turns	1 1/8-1 5/8 (28-41)	324891
40	2	2-stroke	1990-01	45 (737)	①	1 1/8-1 5/8 (28-41)	324891
45	2	2-stroke	1990-98	45 (737)	2 1/2 turns	1 1/8-1 5/8 (28-41)	324891
48	2	2-stroke	1990-94	45 (737)	2 1/2 turns	1 1/8-1 5/8 (28-41)	324891
	2	2-stroke	1995-96	45 (737)	3 1/2 turns	1 1/8-1 5/8 (28-41)	324891
50	2	2-stroke	1990-94	45 (737)	2 1/2 turns	1 1/8-1 5/8 (28-41)	324891
	2	2-stroke	1995-01	45 (737)	3 1/2 turns	1 1/8-1 5/8 (28-41)	324891
55	2	2-stroke	1990-01	45 (737)	3 1/2 turns	1 1/8-1 5/8 (28-41)	324891
25	3	2-stroke	1995-01	30.5 (500)	②	1-1 3/8 (25-35)	324891
35	3	2-stroke	1995-01	34.5 (565)	②	1-1 3/8 (25-35)	324891
25	3	2-stroke	1995	56.1 (913)	2 3/4 turns	1 1/8-1 5/8 (28-41)	324891
40	3	2-stroke	1999-01	56.1 (913)	1 3/4 turns	1 1/8-1 5/8 (28-41)	324891
50	3	2-stroke	1995	56.1 (913)	2 3/4 turns	1 1/8-1 5/8 (28-41)	324891
	3	2-stroke	1995-98	56.1 (913)	1 3/4 turns	1 1/8-1 5/8 (28-41)	324891
60	3	2-stroke	1990-01	56.1 (913)	2 to 2 3/4 turns	1 1/8-1 5/8 (28-41)	324891
65	3	2-stroke	1990-01	56.1 (913)	2 to 2 3/4 turns	1 1/8-1 5/8 (28-41)	324891
70	3	2-stroke	1990-01	56.1 (913)	2 to 2 3/4 turns	1 1/8-1 5/8 (28-41)	324891

n/a: not applicable

Initial low speed setting turn(s): back (counterclockwise) from a *lightly* seated position

① All 40 hp models are 2 to 2 1/2 turns, except the 40RW which is 3 1/2 turns

② Low speed screw is initially installed so the underside of the screw head is 0.390 in. (9.9mm) above the carburetor cover

4

IGNITION & ELECTRICAL

CHARGING CIRCUIT	**4-57**
AC LIGHTING COIL	4-66
Removal & Installation	4-67
Testing	4-66
BATTERY	4-67
CHARGING SYSTEM IDENTIFICATION	4-58
GENERAL INFORMATION	4-57
RECTIFIER	4-62
Removal & Installation	4-63
Testing	4-63
REGULATOR/RECTIFIER	4-64
Removal & Installation	4-66
Testing	4-64
SERVICE PRECAUTIONS	4-57
STATOR/BATTERY CHARGE COIL	4-61
Removal & Installation	4-62
Testing	4-61
TROUBLESHOOTING	4-58
Testing	4-59
IGNITION SYSTEMS (BREAKER POINT MAGNETO)	**4-8**
CLEANING & INSPECTION	4-13
FLYWHEEL & BREAKER POINT	4-11
Installation	4-15
Removal	4-11
GENERAL INFORMATION	4-8
IGNITION COIL & ARMATURE	4-11
Installation	4-14
Removal	4-11
IGNITION SERVICE	4-10
General Information	4-10
TOP SEAL	4-13
Replacement	4-13
TROUBLESHOOTING	4-8
IGNITION SYSTEMS (ELECTRONIC)	**4-15**
CHARGE COIL	4-24
Removal & Installation	4-29
Testing	4-24
DESCRIPTION AND OPERATION	4-15
IGNITION COILS	4-51
Description & Operation	4-51
Removal & Installation	4-54
Testing	4-52
POWER COIL	4-39
Removal & Installation	4-41
Testing	4-39
POWER PACK (IGNITION MODULE) - CARB ONLY	4-45
Removal & Installation	4-51
Testing	4-46
SENSOR/TRIGGER COIL	4-41
Removal & Installation	4-45
Testing	4-41
TROUBLESHOOTING	4-16
SPECIFICATIONS	**4-141**
IGNITION TESTING SPECIFICATIONS - CARB MOTORS	4-141
STARTING CIRCUIT	**4-69**
DESCRIPTION AND OPERATION	4-69
STARTER MOTOR	4-86
Cleaning And Inspection	4-95
Disassembly/Assembly	4-89
Removal & Installation	4-87
Testing	4-87
STARTER MOTOR SOLENOID/RELAY SWITCH	4-97
Removal & Installation	4-98
Testing The Solenoid	4-97
TROUBLESHOOTING	4-69
Testing (Carb)	4-70
Testing (EFI)	4-85
UNDERSTANDING AND TROUBLE-SHOOTING ELECTRICAL SYSTEMS	**4-2**
BASIC ELECTRICAL THEORY	4-2
ELECTRICAL COMPONENTS	4-2
ELECTRICAL TESTING	4-6
PRECAUTIONS	4-7
TEST EQUIPMENT	4-4
TROUBLESHOOTING	4-6
WIRE AND CONNECTOR REPAIR	4-7
WARNING SYSTEM	4-98
DESCRIPTION AND OPERATION	4-98
Carbureted Models	4-99
EFI Models	4-99
ENGINE TEMPERATURE SENSORS (EFI)	4-101
ENGINE TEMPERATURE SWITCHES (CARB)	4-101
Removal & Installation	4-102
Testing	4-101
OIL LEVEL SWITCH (OIL INJECTED 2-STROKES)	4-103
OIL PRESSURE SWITCH (4-STROKE ONLY)	4-102
Removal & Installation	4-103
Testing	4-103
WARNING HORN OR BUZZER	4-103
Testing	4-103
WARNING SYSTEM TROUBLE-SHOOTING	4-99
WIRING DIAGRAMS	**4-104**
INDEX	4-104

UNDERSTANDING AND TROUBLE-SHOOTING ELECTRICAL SYSTEMS	4-2
IGNITION SYSTEMS (BREAKER POINT MAGNETO)	4-8
IGNITION SYSTEMS (ELECTRONIC)	4-15
CHARGING CIRCUIT	4-69
WIRING DIAGRAMS	4-104
SPECIFICATIONS	4-141

4-2 IGNITION AND ELECTRICAL SYSTEMS

UNDERSTANDING AND TROUBLESHOOTING ELECTRICAL SYSTEMS

Basic Electrical Theory

◆ See Figure 1

For any 12-volt, negative ground, electrical system to operate, the electricity must travel in a complete circuit. This simply means that current (power) from the positive terminal (+) of the battery must eventually return to the negative terminal (-) of the battery. Along the way, this current will travel through wires, fuses, switches and components. If, for any reason, the flow of current through the circuit is interrupted, the component fed by that circuit would cease to function properly.

Perhaps the easiest way to visualize a circuit is to think of connecting a light bulb (with two wires attached to it) to the battery - one wire attached to the negative (-) terminal of the battery and the other wire to the positive (+) terminal. With the two wires touching the battery terminals, the circuit would be complete and the light bulb would illuminate. Electricity would follow a path from the battery to the bulb and back to the battery. It's easy to see that with wires of sufficient length, our light bulb could be mounted nearly anywhere on the boat. Further, one wire could be fitted with a switch inline so that the light could be turned on and off without having to physically remove the wire(s) from the battery.

The normal marine circuit differs from this simple example in two ways. First, instead of having a return wire from each bulb to the battery, the current travels through a single ground wire that handles all the grounds for a specific circuit. Secondly, most marine circuits contain multiple components that receive power from a single circuit. This lessens the overall amount of wire needed to power components.

HOW DOES ELECTRICITY WORK: THE WATER ANALOGY

Electricity is the flow of electrons - the sub-atomic particles that constitute the outer shell of an atom. Electrons spin in an orbit around the center core of an atom. The center core is comprised of protons (positive charge) and neutrons (neutral charge). Electrons have a negative charge and balance out the positive charge of the protons. When an outside force causes the number of electrons to unbalance the charge of the protons, the electrons will split off the atom and look for another atom to balance out. If this imbalance is kept up, electrons will continue to move and an electrical flow will exist.

Many people find electrical theory easier to understand when using an analogy with water. In a comparison with water flowing through a pipe, the electrons would be the water and the wire is the pipe.

The flow of electricity can be measured much like the flow of water through a pipe. The unit of measurement used is amperes, frequently abbreviated as amps (a). You can compare amperage to the volume of water flowing through a pipe (for water that would mean a measurement of mass usually measured in units delivered over a set amount of time such as gallons or liters per minute). When connected to a circuit, an ammeter will measure the actual amount of current flowing through the circuit. When relatively few electrons flow through a circuit, the amperage is low. When many electrons flow, the amperage is high.

Water pressure is measured in units such as pounds per square inch (psi). The electrical pressure is measured in units called volts (v). When a voltmeter is connected to a circuit, it is measuring the electrical pressure.

The actual flow of electricity depends not only on voltage and amperage, but also on the resistance of the circuit. The higher the resistance, the higher the force necessary to push the current through the circuit. The standard unit for measuring resistance is an ohm (Ω). Resistance in a circuit varies depending on the amount and type of components used in the circuit. The main factors that determine resistance are:

• Material - some materials have more resistance than others. Those with high resistance are said to be insulators. Rubber materials (or rubber-like plastics) are some of the most common insulators used, as they have a very high resistance to electricity. Very low resistance materials are said to be conductors. Copper wire is among the best conductors. Silver is actually a superior conductor to copper and is used in some relay contacts, but its high cost prohibits its use as common wiring. Most marine wiring is made of copper.

• Size - the larger the wire size being used, the less resistance the wire will have (just as a large diameter pipe will allow small amounts of water to just trickle through). This is why components that use large amounts of electricity usually have large wires supplying current to them.

• Length - for a given thickness of wire, the longer the wire, the greater the resistance. The shorter the wire, the less the resistance. When determining the proper wire for a circuit, both size and length must be considered to design a circuit that can handle the current needs of the component.

• Temperature - with many materials, the higher the temperature, the greater the resistance (positive temperature coefficient). Some materials exhibit the opposite trait of lower resistance with higher temperatures (these are said to have a negative temperature coefficient). These principles are used in many engine control sensors (especially those found on EFI systems).

OHM'S LAW

There is a direct relationship between current, voltage and resistance. The relationship between current, voltage and resistance can be summed up by a statement known as Ohm's law.

Voltage (E) is equal to amperage (I) times resistance (R): $E = I \times R$
Other forms of the formula are $R = E/I$ and $I = E/R$

In each of these formulas, E is the voltage in volts, I is the current in amps and R is the resistance in ohms. The basic point to remember is that if the voltage of a circuit remains the same, as the resistance of that circuit goes up, the amount of current that flows in the circuit will go down.

The amount of work that electricity can perform is expressed as power. The unit of power is the watt (w). The relationship between power, voltage and current is expressed as:

Power (W) is equal to amperage (I) times voltage (E): $W = I \times E$

This is only true for direct current (DC) circuits; the alternating current formula is a tad different, but since the electrical circuits in most vessels are DC type, we need not get into AC circuit theory.

Electrical Components

POWER SOURCE

◆ See Figure 2

Typically, power is supplied to a vessel by two devices: The battery and the stator (or battery charge coil). The stator supplies electrical current anytime the engine is running in order to recharge the battery and in order to operate electrical devices of the vessel. The battery supplies electrical power during starting or during periods when the current demand of the vessel's electrical system exceeds stator output capacity (which includes times when the motor is shut off and stator output is zero).

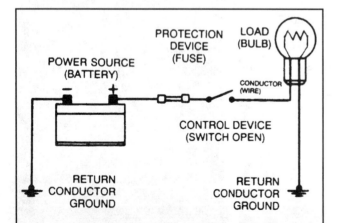

Fig. 1 This example illustrates a simple circuit. When the switch is closed, power from the positive (+) battery terminal flows through the fuse and the switch, and then to the light bulb. The electricity illuminates the bulb and the circuit is completed through the ground wire back to the negative (-) battery terminal.

IGNITION AND ELECTRICAL SYSTEMS

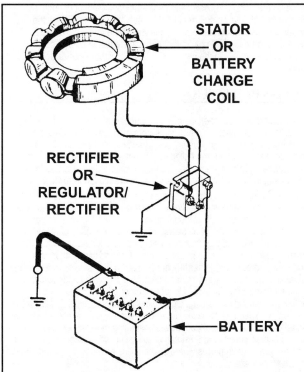

Fig. 2 Functional diagram of a typical charging circuit showing the relationship between the stator (battery charge coil), rectifier (or regulator/rectifier) and battery

The Battery

In most modern vessels, the battery is a lead/acid electrochemical device consisting of six 2-volt subsections (cells) connected in series, so that the unit is capable of producing approximately 12 volts of electrical pressure. Each subsection consists of a series of positive and negative plates held a short distance apart in a solution of sulfuric acid and water.

The two types of plates in each battery cell are of dissimilar metals. This sets up a chemical reaction, and it is this reaction which produces current flow from the battery when its positive and negative terminals are connected to an electrical load. Power removed from the battery in use is replaced by current from the stator and restores the battery to its original chemical state.

The Stator

Alternators and generators are devices that consist of coils of wires wound together making big electromagnets. The coil is normally referred to as a stator or battery charge coil. Either, one group of coils spins within another set (or a set of permanently charged magnets, usually attached to the flywheel, are spun around a set of coils) and the interaction of the magnetic fields generates an electrical current. This current is then drawn off the coils and fed into the vessel's electrical system.

■ Some vessels utilize a generator instead of an alternator. Although the terms are often misused and interchanged, the main difference is that an alternator supplies alternating current that is changed to direct current for use on the vessel, while a generator produces direct current. Alternators tend to be more efficient and that is why they are used on almost all modern engines.

GROUND

Two types of grounds are used in marine electric circuits. Direct ground components are grounded to the electrically conductive metal through their mounting points. All other components use some sort of ground wire that leads back to the battery. The electrical current runs through the ground wire and returns to the battery through the ground or negative (-) cable; if you look, you'll see that the battery ground cable connects between the battery and a heavy gauge ground wire.

■ A large percentage of electrical problems can be traced to bad grounds.

If you refer back to the basic explanation of a circuit, you'll see that the ground portion of the circuit is just as important as the power feed. The wires delivering power to a component can have perfectly good, clean connections, but the circuit would fail to operate if there was a damaged ground connection. Since many components ground through their mounting or through wires that are connected to an engine surface, contamination from dirt or corrosion can raise resistance in a circuit to a point where it cannot operate.

PROTECTIVE DEVICES

◆ See Figure 3

Problems can occur in the electrical system that will cause large surges of current to pass through the electrical system of your vessel. These problems can be the fault of the charging circuit, but more likely would be a problem with the operating electrical components that causes an excessively high load. An unusually high load can occur in a circuit from problems such as a seized electric motor (like a damaged starter) or the excessive resistance caused by a bad ground (from loose or damaged wires or connections). A short to ground that bypasses the load and allows the battery to quickly discharge through a wire can also cause current surges.

If this surge of current were to reach the load in the circuit, the surge could burn it out or severely damage it. It can also overload the wiring, causing the harness to get hot and melt the insulation. To prevent this, fuses, circuit breakers and/or fusible links are connected into the supply wires of the electrical system. These items are nothing more than a built-in weak spot in the system. When an abnormal amount of current flows through the system, these protective devices work as follows to protect the circuit:

• Fuse - when an excessive electrical current passes through a fuse, the fuse blows (the conductor melts) and opens the circuit, preventing current flow.

• Circuit Breaker - a circuit breaker is basically a self-repairing fuse. It will open the circuit in the same fashion as a fuse, but when the surge subsides, the circuit breaker can be reset and does not need replacement. Most circuit breakers on marine engine applications are self-resetting, but some that operate accessories (such as on larger vessels with a circuit breaker panel) must be reset manually (just like the circuit breaker panels in most homes).

• Fusible Link - a fusible link (fuse link or main link) is a short length of special, high temperature insulated wire that acts as a fuse. When an excessive electrical current passes through a fusible link, the thin gauge wire inside the link melts, creating an intentional open to protect the circuit. To repair the circuit, the link must be replaced. Some newer type fusible links are housed in plug-in modules, which are simply replaced like a fuse, while older type fusible links must be cut and spliced if they melt. Since this link is very early in the electrical path, it's the first place to look if nothing on the vessel works, yet the battery seems to be charged and is otherwise properly connected.

Fig. 3 Fuses and circuit breakers may be found in a central location, or mounted to individual holders in the wiring harness

4-4 IGNITION AND ELECTRICAL SYSTEMS

✱✱ CAUTION

Always replace fuses, circuit breakers and fusible links with identically rated components. Under no circumstances should a component of higher or lower amperage rating be substituted. A lower rated component will disable the circuit sooner than necessary (possibly during normal operation), while a higher rated component can allow dangerous amounts of current that could damage the circuit or component (or even melt insulation causing sparks or a fire).

SWITCHES & RELAYS

◆ See Figure 4

Switches are used in electrical circuits to control the passage of current. The most common use is to open and close circuits between the battery and the various electric devices in the system. Switches are rated according to the amount of amperage they can handle. If a sufficient amperage rated switch is not used in a circuit, the switch could overload and cause damage.

Some electrical components that require a large amount of current to operate use a special switch called a relay. Since these circuits carry a large amount of current, the thickness of the wire in the circuit is also greater. If this large wire were connected from the load to the control switch, the switch would have to carry the high amperage load and the space needed for wiring in the vessel would be twice as big to accommodate the increased size of the wiring harness. A relay is used to prevent these problems.

Think of relays as essentially "remote controlled switches." They allow a smaller current to throw the switch that operates higher amperages devices. Relays are composed of a coil and a set of contacts. When current is passed through the coil, a magnetic field is formed that causes the contacts to move together, closing the circuit. Most relays are normally open, preventing current from passing through the main circuit until power is applied to the coil. But, relays can take various electrical forms depending on the job for which they are intended. Some common circuits that may use relays are horns, lights, starters, electric fuel pumps and other potentially high draw circuits.

LOAD

Every electrical circuit must include a load (something to use the electricity coming from the source). Without this load, the battery would attempt to deliver its entire power supply from one pole to another. This is called a short circuit. All this electricity would take a short cut to ground and cause a great amount of damage to other components in the circuit (including the battery) by developing a tremendous amount of heat. This condition could develop sufficient heat to melt the insulation on all the surrounding wires and reduce a multiple wire cable to a lump of plastic and copper. A short can allow sparks that could ignite fuel vapors or other combustible materials in the vessel, causing an extremely hazardous condition.

WIRING & HARNESSES

The average vessel contains miles of wiring, with hundreds of individual connections. To protect the many wires from damage and to keep them from becoming a confusing tangle, they are organized into bundles, enclosed in plastic or taped together and called wiring harnesses. Different harnesses serve different parts of the vessel. Individual wires are color coded to help trace them through a harness where sections are hidden from view.

Marine wiring or circuit conductors can be either single strand wire, multi-strand wire or printed circuitry. Single strand wire has a solid metal core and is usually used inside such components as stator coil windings, motors, relays and other devices. Multi-strand wire has a core made of many small strands of wire twisted together into a single conductor. Most of the wiring in a marine electrical system is made up of multi-strand wire, either as a single conductor or grouped together in a harness. All wiring is color coded on the insulator, either as a solid color or as a colored wire with an identification stripe. A printed circuit is a thin film of copper or other conductor that is printed on an insulator backing. Occasionally, a printed circuit is sandwiched between two sheets of plastic for more protection and flexibility. A complete printed circuit, consisting of conductors, insulating material and connectors is called a printed circuit board. Printed circuitry is used in place of individual wires or harnesses in places where space is limited, such as behind 1-piece instrument clusters.

Since marine electrical systems are very sensitive to changes in resistance, the selection of properly sized wires is critical when systems are repaired. A loose or corroded connection or a replacement wire that is too small for the circuit will add extra resistance and an additional voltage drop to the circuit.

The wire gauge number is an expression of the cross-section area of the conductor. Vessels from countries that use the metric system will typically describe the wire size as its cross-sectional area in square millimeters. In this method, the larger the wire, the greater the number. Another common system for expressing wire size is the American Wire Gauge (AWG) system. As gauge number increases, area decreases and the wire becomes smaller. Using the AWG system, an 18 gauge wire is smaller than a 4 gauge wire. A wire with a higher gauge number will carry less current than a wire with a lower gauge number. Gauge wire size refers to the size of the strands of the conductor, not the size of the complete wire with insulator. It is possible, therefore, to have two wires of the same gauge with different diameters because one may have thicker insulation than the other.

It is essential to understand how a circuit works before trying to figure out why it doesn't. An electrical schematic shows the electrical current paths when a circuit is operating properly. Schematics break the entire electrical system down into individual circuits. In most schematics no attempt is made to represent wiring and components as they physically appear on the vessel; switches and other components are shown as simply as possible. But, this is usually **not** the case on Johnson/Evinrude schematics and some of the wiring diagrams provided here. So, when using a Johnson/Evinrude schematic if the component in question is represented by something more than a small square or rectangle with a label, it is likely a true representation of the component. On most schematics, the face views of harness connectors show the cavity or terminal locations in all multi-pin connectors to help locate test points.

Test Equipment

Pinpointing the exact cause of trouble in an electrical circuit is usually accomplished by the use of special test equipment, but the equipment does not always have to be expensive. The following sections describe different types of commonly used test equipment and briefly explains how to use them in diagnosis. In addition to the information covered below, be sure to read and understand the tool manufacturer's instruction manual (provided with most tools) before attempting any test procedures.

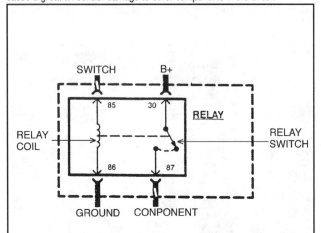

Fig. 4 Relays are composed of a coil and a switch. These two components are linked together so that when one is operated it actuates the other. The large wires in the circuit are connected from the battery to one side of the relay switch (B+) and from the opposite side of the relay switch to the load (component). Smaller wires are connected from the relay coil to the control switch for the circuit and from the opposite side of the relay coil to ground

IGNITION AND ELECTRICAL SYSTEMS 4-5

JUMPER WIRES

♦ See Figure 5

✹✹ CAUTION

Never use jumper wires made from a thinner gauge wire than the circuit being tested. If the jumper wire is of too small a gauge, it may overheat and possibly melt. Never use jumpers to bypass high resistance loads in a circuit. Bypassing resistances, in effect, creates a short circuit. This may, in turn, cause damage and fire. Jumper wires should only be used to bypass lengths of wire or to simulate switches.

Jumper wires are simple, yet extremely valuable, pieces of test equipment. They are basically test wires that are used to bypass sections of a circuit. Although jumper wires can be purchased, they are usually fabricated from lengths of standard marine wire and whatever type of connector (alligator clip, spade connector or pin connector) that is required for the particular application being tested. In cramped, hard-to-reach areas, it is advisable to have insulated boots over the jumper wire terminals in order to prevent accidental grounding. It is also advisable to include a standard marine fuse in any jumper wire. This is commonly referred to as a fused jumper. By inserting an in-line fuse holder between a set of test leads, a fused jumper wire is created for bypassing open circuits. Use a 5-amp fuse to provide protection against voltage spikes.

Jumper wires are used primarily to locate open electrical circuits, on either the ground (-) side of the circuit or on the power (+) side. If an electrical component fails to operate, connect the jumper wire between the component and a good ground. If the component operates only with the jumper installed, the ground circuit is open. If the ground circuit is good, but the component does not operate, the circuit between the power feed and component may be open. By moving the jumper wire successively back from the component toward the power source, you can isolate the area of the circuit where the open is located. When the component stops functioning, or the power is cut off, the open is in the segment of wire between the jumper and the point previously tested.

You can sometimes connect the jumper wire directly from the battery to the hot terminal of the component, but first make sure the component uses a full 12 volts in operation. Some electrical components, such as fuel injectors or sensors, are designed to operate on smaller voltages like 4 or 5 volts, and running 12 volts directly to these components can damage or destroy them.

TEST LIGHTS

♦ See Figure 6

The test light is used to check circuits and components while electrical current is flowing through them. It is used for voltage and ground tests. To use a 12-volt test light, connect the ground clip to a good ground and probe connectors the pick where you are wondering if voltage is present. The test light will illuminate when voltage is detected. This does not necessarily mean that 12 volts (or any particular amount of voltage) is present; it only means that some voltage is present. It is advisable before using the test light to touch its ground clip and probe across the battery posts or terminals to make sure the light is operating properly and to note how brightly the light glows when 12 volts is present.

✹✹ WARNING

Do not use a test light to probe electronic ignition, spark plug or coil wires, as the circuit is much, much higher than 12 volts. Also, never use a pick-type test light to probe wiring on electronically controlled systems unless specifically instructed to do so. Whenever possible, avoid piercing insulation with the test light pick, as you are inviting shorts or corrosion and excessive resistance. But, any wire insulation that is pierced by necessity, must be sealed with silicone and taped after testing.

Like the jumper wire, the 12-volt test light is used to isolate opens in circuits. But, whereas the jumper wire is used to bypass the open to operate the load, the 12-volt test light is used to locate the presence or lack of voltage in a circuit. If the test light illuminates, there is power up to that point in the circuit; if the test light does not illuminate, there is an open circuit (no power). Move the test light in successive steps back toward the power source until the light in the handle illuminates. The open is between the probe and the point that was previously probed.

The self-powered test light is similar in design to the 12-volt test light, but contains a 1.5 volt penlight battery in the handle. It is most often used in place of a multimeter to check for open or short circuits when power is isolated from the circuit (thereby performing a continuity test).

The battery in a self-powered test light does not provide much current. A weak battery may not provide enough power to illuminate the test light even when a complete circuit is made (especially if there is high resistance in the circuit). Always make sure that the test battery is strong. To check the battery, briefly touch the ground clip to the probe; if the light glows brightly, the battery is strong enough for testing.

■ **A self-powered test light should not be used on any electronically controlled system or component. Even the small amount of electricity transmitted by the test light is enough to damage many electronic components.**

MULTI-METERS

♦ See Figure 7

Multi-meters are extremely useful for troubleshooting electrical problems. They can be purchased in either analog or digital form and have a price range to suit nearly any budget. A multi-meter is a voltmeter, ammeter and ohmmeter (along with other features) combined into one instrument. It is often used when testing solid state circuits because of its high input impedance (usually 10 megaohms or more). A high-quality digital multi-meter or Digital Volt Ohm Meter (DVOM) helps to ensure the most accurate test results and, although not absolutely necessary for electronic components such as EFI systems and charging systems, is highly recommended. A brief description of the main test functions of a multi-meter follows:

• Voltmeter - the voltmeter is used to measure voltage at any point in a circuit, or to measure the voltage drop across any part of a circuit. Voltmeters

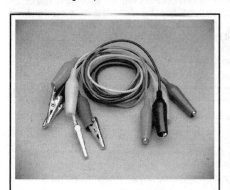

Fig. 5 Jumper wires are simple, but valuable pieces of test equipment

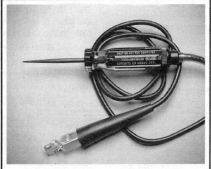

Fig. 6 A 12-volt test light is used to detect the presence of voltage in a circuit

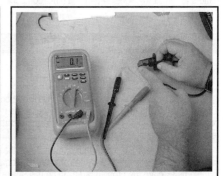

Fig. 7 Multi-meters are probably the most versatile and handy tools for diagnosing faulty electrical components or circuits

usually have various scales and a selector switch to allow metering and display of different voltage ranges. The voltmeter has a positive and a negative lead. To avoid damage to the meter, connect the negative lead to the negative (-) side of the circuit (to ground or nearest the ground side of the circuit) and connect the positive lead to the positive (+) side of the circuit (to the power source or the nearest power source). This is mostly a concern on analog meters, as DVOMs are not normally adversely affected (as they are usually designed to take readings even with reverse polarity and display accordingly). Note that the negative voltmeter lead will always be black and that the positive voltmeter will always be some color other than black (usually red).

- Ohmmeter - the ohmmeter is designed to read resistance (measured in ohms) in a circuit or component. Most ohmmeters will have a selector switch which permits the measurement of different ranges of resistance (usually the selector switch allows the multiplication of the meter reading by 10, 100, 1,000 and 10,000). Most modern ohmmeters (especially DVOMs) are auto-ranging which means the meter itself will determine which scale to use. Since ohmmeters are powered by an internal battery, the ohmmeter can be used like a self-powered test light. When the ohmmeter is connected, current from the ohmmeter flows through the circuit or component being tested. Since the ohmmeter's internal resistance and voltage are known values, the amount of current flow through the meter depends on the resistance of the circuit or component being tested. The ohmmeter can also be used to perform a continuity test for suspected open circuits. When using the meter for continuity checks, do not be concerned with the actual resistance readings. Zero resistance, or any ohm reading, indicates continuity in the circuit. Infinite resistance indicates an opening in the circuit. A high resistance reading where there should be none indicates a problem in the circuit. Checks for short circuits are made in the same manner as checks for open circuits, except that the circuit must be isolated from both power and normal ground. Infinite resistance indicates no continuity, while zero resistance indicates a dead short.

✱✱ WARNING

Never use an ohmmeter to check the resistance of a component or wire while there is voltage applied to the circuit. Voltage in the circuit can damage or destroy the meter.

- Ammeter - an ammeter measures the amount of current flowing through a circuit in units called amperes or amps. At normal operating voltage, most circuits have a characteristic amount of amperes, called current draw that can be measured using an ammeter. By referring to a specified current draw rating, then measuring the amperes and comparing the two values, you can determine what is happening within the circuit to aid in diagnosis. An open circuit, for example, will not allow any current to flow, so the ammeter reading will be zero. A damaged component or circuit will have an increased current draw, so the reading will be high. The ammeter is always connected in series with the tested circuit. All of the current that normally flows through the circuit must also flow through the ammeter; if there is any other path for the current to follow, the ammeter reading will not be accurate. The ammeter itself has very little resistance to current flow and, therefore, will not affect the circuit, but it will measure current draw only when the circuit is closed and electricity is flowing. Excessive current draw can blow fuses and drain the battery, while a reduced current draw can cause motors to run slowly, lights to dim and other components to not operate properly.

Troubleshooting Electrical Systems

When diagnosing a specific problem, organized troubleshooting is a must. The complexity of a modern marine vessel demands that you approach any problem in a logical, organized manner. There are certain troubleshooting techniques, however, which are standard:

- **Establish when the problem occurs.** Does the problem appear only under certain conditions? Were there any noises, odors or other unusual symptoms? Isolate the problem area. To do this, make some simple tests and observations, then eliminate the systems that are working properly. Check for obvious problems, such as broken wires and loose or dirty connections. Always check the obvious before assuming something complicated is the cause.

- **Test for problems systematically to determine the cause once the problem area is isolated.** Are all the components functioning properly? Is there power going to electrical switches and motors? Performing careful, systematic checks will often turn up most causes on the first inspection, without wasting time checking components that have little or no relationship to the problem.

- **Test all repairs after the work is done to make sure that the problem is fixed.** Some causes can be traced to more than one component, so a careful verification of repair work is important in order to pick up additional malfunctions that may cause a problem to reappear or a different problem to arise. A blown fuse, for example, is a simple problem that may require more than another fuse to repair. If you don't look for a problem that caused a fuse to blow, a shorted wire (for example) may go undetected and cause the new fuse to blow right away (if the short is still present) or during subsequent operation (as soon as the short returns if it is intermittent).

Experience shows that most problems tend to be the result of a fairly simple and obvious cause, such as loose or corroded connectors, bad grounds or damaged wire insulation that causes a short. This makes careful visual inspection of components during testing essential to quick and accurate troubleshooting.

Electrical Testing

VOLTAGE

◆ See Figure 8

This test determines the voltage available from the battery and should be the first step in any electrical troubleshooting procedure after visual inspection. Many electrical problems, especially on electronically controlled systems, can be caused by a low state of charge in the battery. Many circuits cannot function correctly if the battery voltage drops below normal operating levels.

Loose or corroded battery cable terminals can cause poor contact that will prevent proper charging and full battery current flow.

1. Set the voltmeter selector switch to the 20V position.
2. Connect the meter negative lead to the battery's negative (-) post or terminal and the positive lead to the battery's positive (+) post or terminal.
3. Turn the ignition switch **ON** to provide a small load.
4. A well charged battery should register over 12 volts. If the meter reads below 11.5 volts, the battery power may be insufficient to operate the electrical system properly. Check and charge or replace the battery as detailed under Engine Maintenance before further tests are conducted on the electrical system.

VOLTAGE DROP

◆ See Figure 9

When current flows through a load, the voltage beyond the load drops. This voltage drop is due to the resistance created by the load and also by small resistances created by corrosion at the connectors (or by damaged insulation on the wires). Since all voltage drops are cumulative, the maximum allowable voltage drop under load is critical, especially if there is more than one load in the circuit.

1. Set the voltmeter selector switch to the 20 volts position.
2. Connect the multi-meter negative lead to a good ground.
3. Operate the circuit and check the voltage prior to the first component (load).
4. There should be little or no voltage drop in the circuit prior to the first component. If a voltage drop exists, the wire or connectors in the circuit are suspect.
5. While operating the first component in the circuit, probe the ground side of the component with the positive meter lead and observe the voltage readings. A small voltage drop should be noticed. This voltage drop is caused by the resistance of the component.
6. Repeat the test for each component (load) down the circuit.
7. If an excessively large voltage drop is noticed, the preceding component, wire or connector is suspect.

IGNITION AND ELECTRICAL SYSTEMS

RESISTANCE

◆ See Figure 10

※※ WARNING

Never use an ohmmeter with power applied to the circuit. The ohmmeter is designed to operate on its own power supply. The normal 12-volt electrical system voltage will damage or destroy many meters!

1. Isolate the circuit from the vessel's power source.
2. Ensure that the ignition key is **OFF** when disconnecting any components or the battery.
3. Where necessary, also isolate at least one side of the circuit to be checked, in order to avoid reading parallel resistances. Parallel circuit resistances will always give a lower reading than the actual resistance of either of the branches.
4. Connect the meter leads to both sides of the circuit (wire or component) and read the actual measured ohms on the meter scale. Make sure the selector switch is set to the proper ohm scale for the circuit being tested, to avoid misreading the ohmmeter test value.

■ The resistance reading of most electrical components will vary with temperature. Unless otherwise noted, specifications given are for testing under ambient conditions of 68°F (20°C). If the component is tested at higher or lower temperatures, expect the readings to vary slightly. When testing engine control sensors or coil windings with smaller resistance specifications (less than 1000 ohms) it is best to use a high quality DVOM and be especially careful of your test results. Whenever possible, double-check your results against a known good part before purchasing the replacement. If necessary, bring the old part to the marine parts dealer and have them compare the readings to prevent possibly replacing a good component.

OPEN CIRCUITS

◆ See Figure 11

This test already assumes the existence of an open in the circuit and it is used to help locate position of the open.
1. Isolate the circuit from power and ground.
2. Connect the self-powered test light or ohmmeter ground clip to the ground side of the circuit and probe sections of the circuit sequentially.
3. If the light is out or there is infinite resistance, the open is between the probe and the circuit ground.
4. If the light is on or the meter shows continuity, the open is between the probe and the end of the circuit toward the power source.

SHORT CIRCUITS

◆ See Figure 12

■ Never use a self-powered test light to perform checks for opens or shorts when power is applied to the circuit under test. The test light can be damaged by outside power.

1. Isolate the circuit from power and ground.
2. Connect the self-powered test light or ohmmeter ground clip to a good ground and probe any easy-to-reach point in the circuit.
3. If the light comes on or there is continuity, there is a short somewhere in the circuit.
4. To isolate the short, probe a test point at either end of the isolated circuit (the light should be on or the meter should indicate continuity).
5. Leave the test light probe engaged and sequentially open connectors or switches, remove parts, etc. until the light goes out or continuity is broken.
6. When the light goes out, the short is between the last two circuit components that were opened.

Wire And Connector Repair

Almost anyone can replace damaged wires, as long as the proper tools and parts are available. Wire and terminals are available to fit almost any need. Even the specialized weatherproof, molded and hard shell connectors used on Johnson/Evinrude engines are usually available for purchase individually.

Be sure the ends of all the wires are fitted with the proper terminal hardware and connectors. Wrapping a wire around a stud is not a permanent solution and will only cause trouble later. Replace wires one at a time to avoid confusion. Always route wires in the same manner of the manufacturer.

When replacing connections, make absolutely certain that the connectors are certified for marine use. Automotive wire connectors may not meet United States Coast Guard (USCG) specifications.

■ If connector repair is necessary, only attempt it if you have the proper tools. Weatherproof and hard shell connectors require special tools to release the pins inside the connector. Attempting to repair these connectors with conventional hand tools will damage them. See a Johnson/Evinrude dealer about the proper connector terminal tools available from the manufacturer for these engines.

Electrical System Precautions

- Wear safety glasses when working on or near the battery.
- Don't wear a watch with a metal band when servicing the battery or

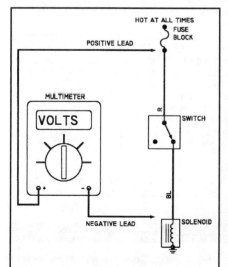

Fig. 8 A voltage check determines the amount of battery voltage available and, as such, should be the first step in any troubleshooting procedure

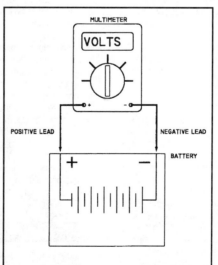

Fig. 9 Voltage drops are due to resistance in the circuit, from the load or from problems with the wiring

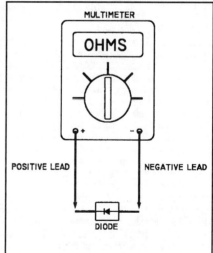

Fig. 10 Resistance tests must be conducted on portions of the circuit, isolated from battery power

4-8 IGNITION AND ELECTRICAL SYSTEMS

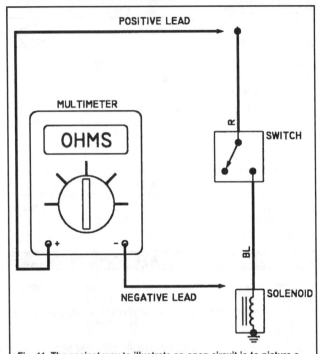

Fig. 11 The easiest way to illustrate an open circuit is to picture a circuit in which the switch is turned OFF (creating an opening in the circuit) that prevents power from reaching the load

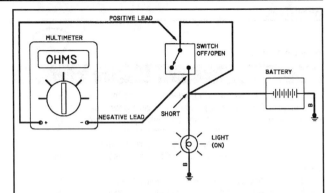

Fig. 12 In this illustration a load (the light) is powered when it should not be (since the switch should be creating an open condition), but a short to power (battery) is powering the circuit. Shorts like this can be caused by chaffed wires with worn or broken insulation

starter. Serious burns can result if the band completes the circuit between the positive battery terminal (or a hot wire) and ground.

• Be absolutely sure of the polarity of a booster battery before making connections. Remember that even momentary connection of a booster battery with the polarity reversed will damage charging system diodes. Connect the cables positive-to-positive, and negative (of the good battery)- to-a good ground on the engine (away from the battery to prevent the possibility of an explosion if hydrogen vapors are present from the electrolyte in the discharged battery). Connect positive cables first (starting with the discharged battery), and then make the last connection to ground on the body of the booster vessel so that arcing cannot ignite hydrogen gas that may have accumulated near the battery. • Disconnect both vessel battery cables before attempting to charge a battery.

• Never ground the alternator or generator output or battery terminal. Be cautious when using metal tools around a battery to avoid creating a short circuit between the terminals.

• When installing a battery, make sure that the positive and negative cables are not reversed.

• Always disconnect the battery (negative cable first) when charging.

• Never smoke or expose an open flame around the battery. Hydrogen gas is released from battery electrolyte during use and accumulates near the battery. Hydrogen gas is **highly** explosive.

IGNITIONS SYSTEMS (BREAKER POINT MAGNETO)

General Information

◆ See Figure 13 and 14

As of 1996 all Johnson/Evinrude outboard engines use some form of a pointless electronic ignition system, however prior to that various models of the 1 hp motors use a point type magneto. Specifically, all Colt/Junior models, certain versions of 1990 3/4 hp (87cc) and certain versions of 1991-95 2/2.3/3.3 hp (77.8cc) models.

Read and Believe. A battery installed to crank the powerhead does not mean the engine is equipped with a battery-type ignition system. A magneto system uses the battery only to crank the powerhead. Once the powerhead is running, the battery has absolutely no effect on engine operation. Therefore, if the battery is low and fails to crank the powerhead properly for starting, the powerhead may be cranked manually, started, and operated. Under these conditions, the key switch must be turned to the on position or the powerhead will not start by hand cranking.

A magneto system is a self-contained unit. The unit does not require assistance from an outside source for starting or continued operation. Therefore, as previously mentioned, if the battery is dead, the engine may be cranked manually and the powerhead started.

This ignition system uses a mechanically switched, collapsing field to induce spark at the plug. A magnet moving by a coil produces current in the primary coil winding. The current in the primary winding creates a magnetic field. When the points are closed the current goes to ground. As the breaker points open the primary magnetic field collapses across the secondary field. This induces (transforms) a high voltage potential in the secondary coil winding. This high voltage current travels to the spark plug and jumps the gap.

The point type ignition system contains a condenser that works like a sponge in the circuit. Current that is flowing through the primary circuit tries to keep going. When the breaker point switch opens the current will arc over the widening gap. The condenser is wired in parallel with the points. The condenser absorbs some of the current flow as the points open. This reduces arc over and extends the life of the points.

The flywheel-type magneto unit consists of an armature plate and a permanent magnet built into the flywheel. The ignition coil, condenser and breaker points are mounted on the armature plate.

As the pole pieces of the magnet pass over the heels of the coil, a magnetic field is built up about the coil, causing a current to flow through the primary winding.

At the proper time, the breaker points are separated by action of a cam designed into the collar of the flywheel and the primary circuit is broken. When the circuit is broken, the flow of primary current stops and causes the magnetic field about the coil to break down instantly. At this precise moment, an electrical current of extremely high voltage is induced in the fine secondary windings of the coil. This high voltage is conducted to the spark plug where it jumps the gap between the points of the plug to ignite the compressed charge of air-fuel mixture in the cylinder.

The carburetion and ignition principles of two-cycle engine operation must be understood in order to perform a proper tune-up on or troubleshoot an outboard motor.

If you have any doubts concerning your understanding of two-cycle engine operation, it would be best to study the operation theory section in the General Information, Safety and Tools section, before tackling any work on the ignition system.

Troubleshooting

Always attempt to proceed with the troubleshooting in an orderly manner. The "shotgun" approach will only result in wasted time, incorrect diagnosis, replacement of unnecessary parts, and frustration.

Begin the ignition system troubleshooting with the wiring harness and the spark plugs and then continue through the system until the source of trouble is located.

IGNITION AND ELECTRICAL SYSTEMS

Fig. 13 Typical magneto ignition system installed on a single cylinder powerhead

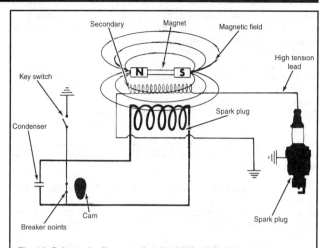

Fig. 14 Schematic diagram of a simple single cylinder magneto ignition system with principle parts identified

Remember, a magneto system is a self-contained unit. Therefore, if the engine has a key switch and wire harness, remove them from the powerhead and then make a test for spark. If a good spark is obtained with these two items disconnected, but no spark is available at the plug when they are connected, then the trouble is in the harness or the key switch. If a test is made for spark at the plug with the harness and switch connected, check to be sure the key switch is turned to the on position.

CHECKING THE WIRING HARNESS & KEYSWITCH

◆ See Figure 15 and 16

※※ WARNING

These next two paragraphs may well be the most important words in this section. Misuse of the wiring harness is the most single cause of electrical problems with outboard power plants.

A wiring harness is used between the key switch and the powerhead. This harness seldom contains wire of sufficient size to allow connecting accessories. Therefore, anytime a new accessory is installed, new wiring should be used between the battery and the accessory.

A separate fuse panel must be installed on the control panel. To connect the fuse panel, use one red and one black No. 10 gauge wire from the battery. If a small amount of 12 volt current should be accidentally attached

Fig. 15 A coil destroyed when 12 volts was connected into the magneto wiring system. Mechanics report in 85% of the cases, the damage occurs when an accessory is connected through the key switch

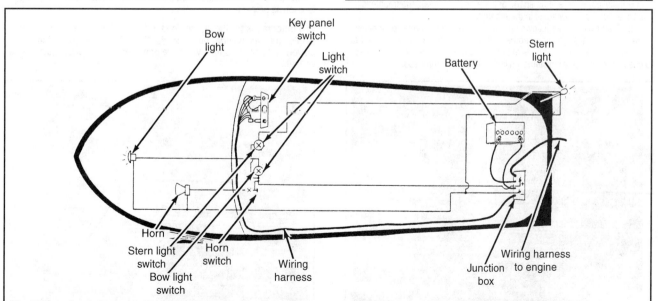

Fig. 16 Functional diagram to illustrate proper hookup of accessories through a junction box. If a junction is not installed on the boat, connect accessories directly to the battery. Never connect accessories through the key switch

4-10 IGNITION AND ELECTRICAL SYSTEMS

to the magneto system, the coil may be damaged or destroyed. Such a mistake in wiring can easily happen if the source for the 12 volt accessory is taken from the key switch. Therefore, again let it be said, never connect accessories through the key switch.

A magneto key switch operates in reverse of any other type key switch. When the key is moved to the off position, the circuit is closed between the magneto and ground. In some cases, when the key is turned to the off position the points are grounded. For this reason, an automotive type switch must never be used, because the circuit would be opened and closed in reverse, and if 12-volts should reach the coil, the coil will be destroyed.

CHECKING SPARK PLUGS & CHECKING FOR SPARK

◆ See Figures 17 and 18

1. Check the plug wires to be sure they are properly connected. Check the entire length of the wire, from the plug to the magneto under the armature plate. If the wire is to be removed from the spark plug, always use a pulling and twisting motion as a precaution against damaging the connection.

2. Attempt to remove the spark plugs by hand. This is a rough test to determine if the plug is tightened properly. You should not be able to remove the plug without using the proper socket size tool. Remove the spark plugs and identify from which cylinder they were removed.

If the spark plugs have been removed and the problem cannot be determined, but the plug appears to be in satisfactory condition, electrodes, etc., then replace the plugs in the spark plug openings.

A conclusive spark plug test should always be performed with the spark plugs installed. A plug may indicate satisfactory spark when it is removed and tested but under a compression condition, may fail. An example would be the possibility of a person being able to jump a given distance on the ground, but if a strong wind is blowing, his distance might be reduced by half. The same is true with the spark plug. Under good compression in the cylinder, the spark may be too weak to ignite the fuel properly.

Therefore, to test the spark plug under compression, replace it in the engine and tighten it to the proper torque value. Another reason for testing for spark with the plugs installed is to duplicate actual operating conditions regarding flywheel speed. If the flywheel is rotated with the pull cord with the plugs removed, the flywheel will rotate much faster because of the no-compression condition in the cylinder, giving the false indication of satisfactory spark.

3. Use a spark tester and check for spark at each cylinder. If a spark tester is not available, hold the plug wire about 1/4 in. from the powerhead. Turn the flywheel with a pull starter or electrical starter and check for spark. A strong spark over a wide gap must be observed when testing in this manner, because under compression a strong spark is necessary in order to ignite the air-fuel mixture in the cylinder. This means it is possible to think you have a strong spark, when in reality the spark will be too weak when the plug is installed. If there is no spark, or if the spark is weak, the trouble is most likely under the flywheel in the magneto.

4. Each cylinder has its own ignition system in a flywheel-type ignition system. This means if a strong spark is observed on one cylinder and not at another, only the weak system is at fault. However, it is always a good idea to check and service all systems while the flywheel is removed.

CHECKING THE CONDENSER

◆ See Figures 19 and 20

In simple terms, a condenser is composed of two sheets of tin or aluminum foil laid one on top of the other, but separated by a sheet of insulating material such as waxed paper, etc. The sheets are rolled into a cylinder to conserve space and then inserted into a metal case for protection and to permit easy assembly.

The purpose of the condenser is to absorb or store the secondary current built up in the primary winding at the instant the breaker points are separated. By absorbing or storing this current, the condenser prevents excessive arcing and the useful life of the breaker points is extended. The condenser also gives added force to the charge produced in the secondary winding as the condenser discharges.

Modern condensers seldom cause problems, therefore, it is not necessary to install a new one each time the points are replaced. However, if the points show evidence of arcing, the condenser may be at fault and should be replaced. A faulty condenser may not be detected without the use of special test equipment. The modest cost of a new condenser justifies its purchase and installation to eliminate this item as a source of trouble.

CHECKING THE BREAKER POINTS

■ Complete details for breaker point service including inspection, removal, installation and gapping can be found under Tune-Up in the Maintenance and Tune-Up section.

CHECKING THE IGNITION COIL

■ The ignition coil functions in the same basic manner on the Magneto Breaker Point ignition as it does for Electronic Ignitions. For details on Testing the Ignition Coil, please refer to Ignition Coils, in the Ignition System (Electronic) section.

Flywheel & Breaker Point Ignition Component Service

GENERAL INFORMATION

Magnetos installed on outboard engines will usually operate over extremely long periods of time without requiring adjustment or repair. However, if ignition system problems are encountered and the usual corrective actions such as replacement of spark plugs does not correct the problem, the magneto output should be checked to determine if the unit is functioning properly.

■ Photographs for the illustrations in this section were taken of both single and twin cylinder powerheads. Naturally, the single cylinder powerheads covered here will have only one set of breaker points, one coil, etc.

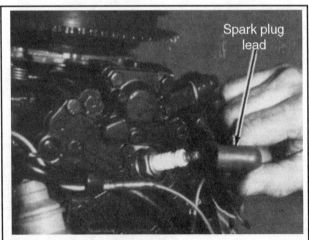

Fig. 17 Disconnect the spark plug lead...

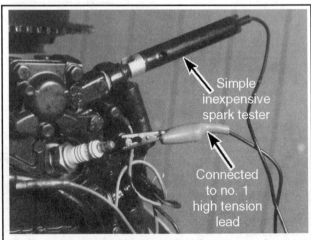

Fig. 18 ...then use a spark tester to check for proper spark

IGNITION AND ELECTRICAL SYSTEMS 4-11

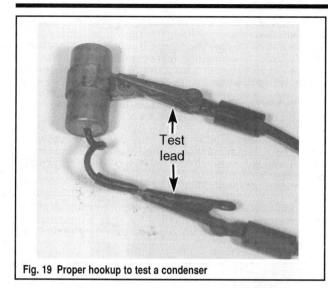

Fig. 19 Proper hookup to test a condenser

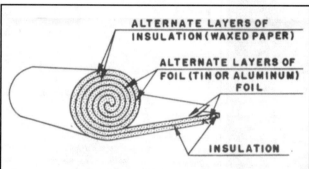

Fig. 20 Rough sketch to illustrate how the waxed paper, aluminum foil, and insulation are rolled in a typical condenser

FLYWHEEL & BREAKER POINT REMOVAL

◆ See Figure 21

■ Because removal of the flywheel along with breaker point and condenser replacement is necessary for periodic maintenance on these motors we've included this procedure under Tune-Up in the Maintenance and Tune-Up section of this manual. Please refer there for details.

1. Remove the hood or enough of the powerhead cover to expose the flywheel.
2. If equipped, disconnect the battery connections from the battery terminals, if a battery is used to crank the powerhead. If a hand starter is installed, remove the attaching hardware from the legs of the starter assembly and lift the starter free.
3. On hand rewind starter models, a round ratchet plate is attached to the flywheel to allow the hand starter to engage in the ratchet and thus rotate the flywheel. This plate must be removed before the flywheel nut is removed.
4. Remove the nut securing the flywheel to the crankshaft. It may be necessary to use some type of flywheel strap to prevent the flywheel from turning as the nut is loosened.
5. Install the proper flywheel puller using the same screw holes in the flywheel that are used to secure the ratchet plate removed earlier. Never attempt to use a puller that pulls on the outside edge of the flywheel or the flywheel may be damaged. After the puller is installed, tighten the center screw onto the end of the crankshaft. Continue tightening the screw until the flywheel is released from the crankshaft. Remove the flywheel.

✱✱ CAUTION

Do not strike the puller center bolt with a hammer in an attempt to dislodge the flywheel. Such action could seriously damage the lower seal and/or lower bearing.

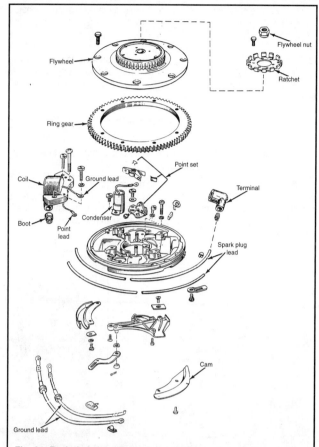

Fig. 21 Exploded view of a typical magneto system. Only one coil and set of points are shown

Stop, and carefully observe the magneto and associated wiring layout. Study how the magneto is assembled. Take time to make notes on the wire routing. Observe how the heels of the laminated core, with the coil attached, is flush with the boss on the armature plate. These items must be replaced in their proper positions. You MIGHT want to take a digital photo of the engine with the flywheel removed: one from the top, and a couple from the sides showing the wiring and arrangement of parts.

■ The armature plate does not have to be removed to service the magneto. If it is necessary to remove the plate for other service work, such as to replace the coil or to replace the top seal, see the information in the Ignition and Electrical System section.

6. Remove the screw attaching the wires from the coil and condenser to one set of points. On engines equipped with a key switch, "kill" button, or "runaway" switch, a ground wire is also connected to this screw.
7. Using a pair of needle-nose pliers, remove the wire clip from the post protruding through the center of the points.
8. Again, with the needle-nose pliers, remove the flat retainer holding the set of points together.
9. Lift the moveable side of the points free of the other half of the set.
10. Remove the hold-down screw securing the non-moveable half of the point set to the armature plate.
11. Remove the hold down screw securing the condenser to the armature plate. Observe how the condenser sets into a recess in the armature plate.

IGNITION COIL & ARMATURE REMOVAL

◆ See Figures 21 and 22 thru 26

■ It is not necessary to remove the armature plate unless the top seal or the coil is to be replaced.

4-12 IGNITION AND ELECTRICAL SYSTEMS

The armature plate does not have to be removed to service the magneto breaker points or condenser. However, removal IS necessary to replace the ignition coil or top seal.

■ Photographs for the illustrations in this section were taken of both single and twin cylinder powerheads. Naturally, the single cylinder powerheads covered here will have only one set of breaker points, one coil, etc.

1. Remove the Flywheel as detailed under Flywheel & Breaker Point Removal.

2. Disconnect the advance arm connecting the armature plate with the power shaft on the side of the powerhead.

3. Next, remove the wires connecting the underside of the armature plate with the "kill" switch. If a "kill" switch is not installed, these wires are connected to the wiring harness plug. The wires of most units have a quick-disconnect fitting. Remove the wires from the vacuum (runaway) switch, if one is installed.

4. Observe the four screws in a square pattern through the armature plate. Two of these screws pass through the laminated core and the armature plate into the powerhead retainer. The other two pass just through the plate. Loosen these four screws. After the screws are loose, lift the armature plate up the crankshaft and clear of the engine. If any oil is present on top of the armature plate, or on the points, the top seal must be replaced.

■ Notice how the coil has a laminated core. The coil cannot be separated, that is, the laminations from the core.

5. Turn the armature plate over and notice how the high-tension leads are installed on the plate in a recess. The routing of the wires is misleading. The wire to the No. 1 spark plug is not connected to the No. 1 coil as might be expected.

6. Remove the three screws attaching the coils to the armature plate.

7. Hold the armature plate and separate the coil from the plate. As the coil is separated from the plate, observe the high-tension lead to the spark plug inside the coil. Work the small boot, if used, and the high-tension lead from the coil.

Fig. 22 The armature plate assembly must be removed for ignition coil or top seal service

Fig. 23 Remove the retaining screws and lift the plate from the top of the motor

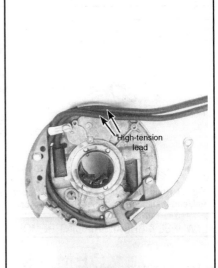

Fig. 24 Before removing the coil, note the wire routing underneath the plate

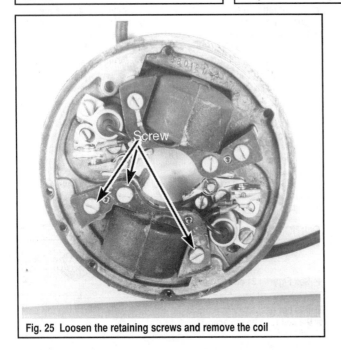

Fig. 25 Loosen the retaining screws and remove the coil

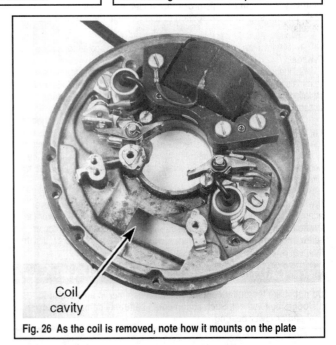

Fig. 26 As the coil is removed, note how it mounts on the plate

IGNITION AND ELECTRICAL SYSTEMS

TOP SEAL REPLACEMENT

◆ See Figures 27, 28 and 29

Replacement of the top seal on a Johnson/Evinrude engine is not a difficult task, with the proper tools: a seal remover and seal installer. Never attempt to remove the seal with screwdrivers, punch, pick, or other similar tool. Such action will most likely damage the collars in the powerhead.

Obtain an OMC/BOMBARDIER Seal Remover (#387780). A 1 1/8 in. open end wrench is needed to hold the remover portion of the tool, while a 3/8 in. open end wrench is used on the top bolt.

1. To remove the seal, first, work the point cam up and free of the driveshaft.
2. Next, remove the Woodruff key from the crankshaft. A pair of side-cutters is a handy tool for this job. Grasp the Woodruff key with the side-cutters and use the leverage of the pliers against the crankshaft to remove the key.
3. Work the special tool into the seal. Observe how the special tool is tapered and has threads. Continue working and turning the tool until it has a firm grip on the inside of the seal.
4. Now, tighten the center screw of the puller against the end of the crankshaft and the seal will begin to lift from the collars. Continue turning this center screw until the seal can be raised manually from the crankshaft.
5. To install the new seal: Coat the inside diameter of the seal with a thin layer of oil. Apply OMC/BOMBARDIER sealer to the outside diameter of the seal.
6. Slide the seal down the crankshaft and start it into the recess of the powerhead. Use the special tool and drive the seal completely into place in the recess.
7. Install the Woodruff key into the crankshaft. On some models, a pin was used to locate the cam for the points. If the pin was used, install it at this time. Observe the difference to the sides of the cam. On almost all cams, the word top is stamped on one side. Also, on some cams, the groove does not go all the way through. Therefore, it is very difficult to install the cam incorrectly, with the wrong side up.
8. Slide the cam down the crankshaft with the word top facing upward. Continue working the cam down the crankshaft until it is in place over the Woodruff key or pin.

CLEANING & INSPECTION

◆ See Figures 21 and 30 thru 32

Inspect the flywheel for cracks or other damage, especially around the inside of the center hub. Check to be sure metal parts have not become attached to the magnets. Verify each magnet has good magnetism by using a screwdriver or other tool.

Thoroughly clean the inside taper of the flywheel and the taper on the crankshaft to prevent the flywheel from "walking" on the crankshaft while the engine is running.

Check the top seal around the crankshaft to be sure no oil has been leaking onto the armature plate. If there is any evidence the seal has been

Fig. 27 For access to the top seal, first lift off the point cam from the crankshaft...

Fig. 28 ...then work a tapered/threaded seal remover into the seal

Fig. 29 Oil and install a new seal using a suitable driver...

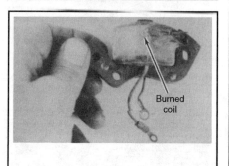

Fig. 30 A coil burned where the high-tension lead enters the coil on the bottom side. Arcing caused the damage

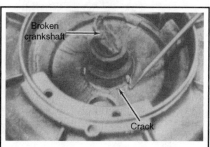

Fig. 31 A broken crankshaft and cracked flywheel damaged when the engine was operated at a high rpm with a flush attachment and garden hose connected to the lower unit

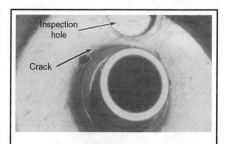

Fig. 32 Cracks in the flywheel hub caused by metal fatigue due to flywheel construction and the inspection hole. This hole is no longer incorporated in late-model flywheels

4-14 IGNITION AND ELECTRICAL SYSTEMS

leaking, it must be replaced, as outlined earlier in this section.

Test the armature plate to verify it is not loose. Attempt to lift each side of the plate. There should be little or no evidence of movement.

Clean the surface of the armature plate where the points and condenser attach. Install a new condenser into the recess and secure it with the hold-down screw.

■ Because Inspection & Testing of the breaker point is part for periodic maintenance on these motors we've included this procedure under Tune-Up in the Maintenance and Tune-Up section of this manual. Please refer there for details.

IGNITION COIL & ARMATURE INSTALLATION

◆ See Figures 21 and 33 thru 39

■ Photographs for the illustrations in this section were taken of both single and twin cylinder powerheads. Naturally, the single cylinder powerheads covered here will have only one set of breaker points, one coil, etc.

1. To install a new coil, first turn the armature plate over, and loosen the spark plug lead wires, and push them through the armature plate. Now, work the leads into the coil.

2. After the leads are into the coil, work the small boot up onto the coil. Apply a coating of rubber seal material underneath the boot, if a boot is used.

3. Start the three screws through the laminated core into the armature plate, but do not tighten them. If the powerhead being serviced has a second coil, install the other coil in the same manner.

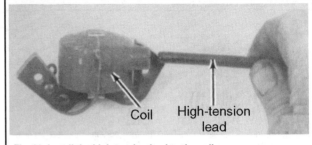

Fig. 33 Install the high tension lead to the coil...

4. Check to be sure the spark plug (high-tension) leads are properly positioned in the coil and are securely attached to the bottom side of the armature plate.

5. To adjust the coil: A special ring tool is required that fits down over the armature plate. This tool will properly locate the coil in relation to the flywheel. Install this special tool over the armature plate. Push outward on the coil and secure the two outer screws.

6. If a special ring tool is not available, and in an emergency, hold a straight edge against the boss on the armature plate and bring the heel of the laminated core out square against the edge of the boss on the armature plate. The ground wire for the coil should be attached under the head of the top screw passing through the laminated core.

7. Slide the armature plate down over the crankshaft and onto the powerhead.

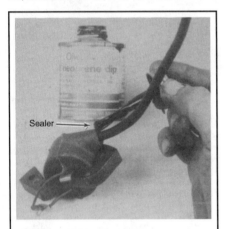

Fig. 34 ...then apply a coating of rubber seal material

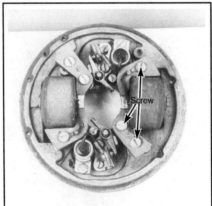

Fig. 35 ...then apply a coating of rubber seal material

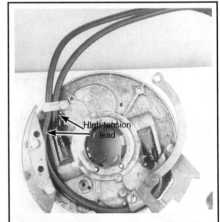

Fig. 36 ...then route the wires and noted during removal

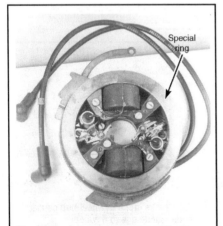

Fig. 37 Use a locating ring to position the coil on the plate before tighten the mounting screws

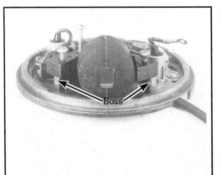

Fig. 38 If a locating ring is not available, use a straight edge to ensure the heel of the coil laminated core sits square against the plate bosses

Fig. 39 Position the assembled plate to the motor

IGNITION AND ELECTRICAL SYSTEMS 4-15

8. Align the screw holes in the armature plate with the holes in the powerhead retainer. After the armature plate is in place, install and tighten the two screws securing the armature plate to the retainer. Now, take up on the three screws through the laminated core closest to the crankshaft. Tighten the screws securely. Attach the advance arm from the magneto to the tower shaft arm.

■ The points must be assembled as they are installed. One side of each point set has the base and is non-moveable. The other side of the set has a moveable arm. A small wire clip and a flat retainer are included in each point set package.

IGNITION SYSTEMS (ELECTRONIC)

■ The VAST majority of the motors covered here (except a handful of 1990-95 single-cylinder motors) utilize an electronic ignition system. When an electronic system is used, for all carbureted motors, the system is either a version of the magneto powered Capacitor Discharge Ignition (CDI) system commonly referred to as CD II (as in CD TWO) by the OEM or a version of the similar Johnson/Evinrude Under Flywheel Ignition (UFI) system. On all EFI motors, the ignition is a fully transistorized, battery powered, electronic ignition system.

For testing and service purposes the CDI, UFI and EFI Ignition systems are virtually the same, with slight differences in components or specifications that vary by year and model. Individual components will vary slightly from model-to-model (especially between rope start and electric start versions of some models or models with 6-amp charging systems when compared to the same models equipped with 12-amp systems). But, even with slight variances in the system's source of power and trigger signals, the basic theories of system operation and service are virtually the same for all models. Differences are noted under suitable headings or sub-steps.

The ignition system's main purpose is to provide the spark necessary for engine combustion, and to do so at the proper time. It does so by converting a low voltage power source (such as the low voltage alternating current produced by the stator or charge coil on carbureted models or the 12-volt DC battery power from EFI motors) into a high voltage DC current. This is accomplished in the primary circuit of the ignition coil. Power is then conducted from the primary circuit, through the ignition coil's secondary circuit to the spark plugs.

On all electronic ignition models except the 40/50 hp EFI motors, this conduction is through secondary spark plug leads (spark plug wires). However, on 40/50 hp EFI motors individual ignition coils are mounted directly to the top of each spark plug, eliminating the need for secondary leads.

This section provides information for troubleshooting and repairing the ignition system.

BREAKER POINT & FLYWHEEL INSTALLATION

◆ See Figure 21

■ Because installation of the flywheel along with breaker point and condenser replacement is necessary for periodic maintenance on these motors we've included this procedure under Tune-Up in the Maintenance and Tune-Up section of this manual. Please refer there for details.

Description and Operation

CAPACITOR DISCHARGE IGNITION (CDI)/UNDER FLYWHEEL IGNITION (UFI) - CARBURETED MOTORS

◆ See Figures 40 thru 45

In its simplest form, a CDI/UFI ignition system is composed of the following elements:
- Power Source (Magneto stator or charge coil)
- Signal Source (Signal coil either separate from the Ignition Module for CDI or integrated for UFI)
- Ignition coil (to transfer low voltage signals to the high voltage current necessary for proper spark)
- Spark plugs

Other components such as main switches, stop switches, or computer systems may be included, though, these items are not necessary for basic CDI operation.

To understand basic CDI operation, it is important to understand the basic theory of induction. Induction theory states that if we move a magnet (magnetic field) past a coil of wire (or the coil by the magnet), it will generate AC current in the coil. This current is used to feed the ignition system.

The amount of current produced depends on several factors:
- How fast the magnet moves past the coil
- The size of the magnet (strength)
- How close the magnet is to the coil
- Number of turns of wire and the size of the windings

On most models, current for the ignition system is provided to the power pack (ignition module) and stored in a capacitor located inside the box. As a charge coil produces current, a capacitor stores it.

■ The 3/4 Hp (except 4 Deluxe) 2-cylinder, 2-stroke motors utilize a charge coil that is built into the UFI power pack/ignition module.

Fig. 40 Power for CDI systems comes from the magneto comprised of the flywheel-mounted magnets and . . .

Fig. 41 . . . the charge coil windings mounted under the flywheel, like in this stator

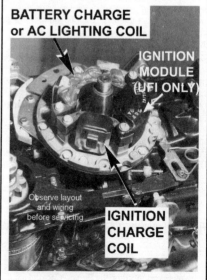

Fig. 42 Some models utilize a separate charge coil (note the location of the Ignition Module means this is a UFI motor)

4-16 IGNITION AND ELECTRICAL SYSTEMS

Fig. 43 The ignition coil uses the low voltage from the charge coil to induct high voltage to fire the spark plugs

Fig. 44 On most models, the ignition coil is controlled by either a remote mounted (CDI)...

Fig. 45 ... or an under the flywheel (UFI) mounted ignition module

For almost all models, at a specific time in the magneto's revolution, the magnets go past the sensor or trigger coil. A sensor coil is essentially a smaller version of a charge coil that produces a smaller current. The current from the sensor coil is used by the power pack to fire the capacitor (that explains why the sensor coil is often referred to as a trigger coil). The current from the capacitor flows out to the ignition coil and spark plug. The sensor acts much like the points found on ignitions systems in days gone by.

■ **The 5 hp single-cylinder motor is the only model covered here that does not use any sensor coil signal (even internal). This is only possible because the single-cylinder motor is a 2-stroke that requires the spark plug to fire each time the charge coil produces voltage, so no separate controlling circuit is necessary. The charge coil on this model feeds the primary side of the combined power pack/ignition coil directly with no other components contained in the system.**

The ignition coil is a step-up transformer. It turns the relatively low voltage entering the primary windings into high voltage at the secondary windings. This occurs due to a phenomenon known as induction.

The high voltage generated in the secondary windings leaves the ignition coil and goes to the spark plug. The spark in turn ignites the air-fuel charge in the combustion chamber.

Once the complete cycle has occurred, the spinning magneto repeats the process again and again.

All but the most simple Johnson/Evinrude engines contain additional components to enhance functionality of the ignition system. Although the primary power for the ignition system comes from the charge coil, most motors also contain another set of windings called the power coil. When equipped with power coil is usually used to supply current for additional features such as the Speed Limiting Operator Warning (S.L.O.W.) circuits integrated into the power pack. On some models (mostly on smaller rope start motors) the charge and power coils are separate units mounted next to each other on an ignition plate directly under the flywheel. On most motors (pretty much all of the larger motors and many of the smaller units, including most electric start or models equipped with lighting coils) the power and/or charge coils are often combined with the charging system's battery charge coil in a one-piece stator assembly.

The spark advance is handled by a variety of means, depending largely on whether or not it is adjustable (refer to Timing and Synchronization for more details on your model). Most later-models utilize an electronic advance for the spark system that is controlled via the power pack. All 4-stroke models covered here are completely electronically controlled.

Main switches, engine stop switches, and the like are usually connected on the wire in between the power pack box and the ignition coil. When the main switch or stop switch is turned to the OFF position, the switch is closed. This closed switch short-circuits the charge coil current to ground rather than sending it through the power pack. With no charge coil current through the power pack, there is no spark and the engine stops or, if the engine is not running, no spark is produced preventing it from starting.

FULLY TRANSISTORIZED, BATTERY POWERED IGNITION - EFI MOTORS

The fully transistorized direct ignition system used on 40/50 hp motors is completely controlled by the Engine Control Unit (ECU). The computer control module or ECU sets and adjusts ignition timing based on engine speed for efficiency and reliability. No end user adjustments are necessary or possible on this system, which only requires periodic cleaning, inspection and replacement of the spark plugs for service. Problems that can occur with the system are usually traced to poor wire connections or a defective stop circuit.

As the name Battery Powered might imply, current to power the system comes from the battery, through the ECU, system relay and ignition switch. On 40/50 hp motors, one crankshaft position sensor for each cylinder provides the ECU with information on engine rpm and piston positioning. For 70 hp motors, 2 crankshaft position sensors and a camshaft position sensor provide signals to the ECU. The ECU will then control the ignition coil primary circuit based upon pre-programmed ignition timing to match the fuel-mapping and operating conditions. When the ECU interrupts the current to the primary coil winding, high voltage is induced in the secondary coil circuit in order to fire the spark plugs.

On 40/50 hp motors, an individual ignition coil is mounted directly over top of each spark plug, eliminating the need for secondary spark plug leads. The system found on 70 hp motors, utilizes two ignition coils, one for cylinders 1 and 4, and one for cylinders 2 and 3. When an individual coil fires, it fires both spark plugs simultaneously, even though only one cylinder is at TDC. In automotive terms, the type of ignition used on 70 hp motors is called a waste-spark system since each time a coil fires it simultaneously fires both a plug at TDC of compression and a plug at TDC of exhaust. But, this is really a misnomer, since no real voltage is actually wasted, the majority of the available voltage is used by the plug at TDC of compression.

Anytime the ignition switch is turned to the **OFF** position the system relay will de-energize, removing power from the ECU and coil.

Troubleshooting the Ignition System

Don't waste your time with haphazard testing. The only way to ensure success (and the only way to avoid the possibility of accidentally replacing a good part) is to perform ignition system testing in a logical and systematic manner.

An engine must have 3 things to run properly, Compression, Fuel and Spark. If compression is not an issue, then fuel system and ignition system checks are normally next.

Begin all electronic troubleshooting procedures by ensuring that wiring and connections are in good condition. For EFI systems, check battery condition to make sure it is at a sufficient voltage to operate the system at start-up (and that the charging system is producing sufficient voltage to charge and maintain battery voltage levels during operation).

Before conducting any tests, double-check all wire colors and locations with the Wiring Diagrams provided in this section.

The quickest and most important check after the basic condition of the wiring harness has been verified is to perform a Spark Check, as detailed in this section to determine if further ignition system troubleshooting is warranted, or if problems are instead fuel or compression related.

Weak or no spark conditions found during the Spark Check will lead you to additional circuit or component testing procedures, depending on the results and the specific systems utilized on the motor being tested.

IGNITION AND ELECTRICAL SYSTEMS 4-17

✳✳ CAUTION

During ignition system testing, ALWAYS follow the steps of our procedures and any tool manufacturer's instructions closely to avoid injury or possible damage to the engine's ignition system. Make sure that spark tests are only conducted with spark plugs installed in the engine to prevent the possible combustion of fuel vapors leaking out of the cylinders.

✳✳ WARNING

If, during testing, the engine is to be run at speeds over 2000 rpm, it must be mounted in a test tank using a suitable test wheel or placed in the water. Running the engine on a flushing device at speeds above 2000 rpm could allow the motor to run overspeed and possibly suffer severe mechanical damage.

IGNITION SYSTEM TESTING

◆ See Figures 46 thru 49

Spark Test (All Models)

◆ See Figures 46 thru 49

The first question normally asked when an engine does not run properly, or at all "Is there spark and fuel?" should be answered before in depth troubleshooting is conducted on any system. Since engine mechanical problems that might cause a lack of compression do not usually happen without warning or without an accompanying noise or other symptom, compression is usually taken for granted. If you are unsure of engine compression, refer to the tests provided under Engine Maintenance to ensure there is no mechanical problem before proceeding.

If there is no problem with engine compression, then this Spark Test is the quickest way to determine whether or not the ignition system requires further attention.

Because modern ignition systems are capable of extremely high voltages (30,000-50,000 and sometimes even higher voltage outputs) we must insist that you abandon the spark test you might have witnessed as a child. It is no longer advisable to simply remove a spark plug and hold it against the powerhead (even using an insulated tool) while the motor is cranked to see if a spark jumps the gap. Besides being dangerous to the person holding the spark plug, this form of testing also jeopardizes the solid state components of the ignition system that can be destroyed by excessively high or infinite resistance in output circuits. Never operated the ignition system on these motors unless the spark plugs are properly attached or their leads are properly grounded to the powerhead.

With all that said, a spark gap tester is required for this test. There are many types of spark testers available from both marine and automotive parts outlets. The most basic (and least expensive) consists of what looks somewhat like a spark plug and a large alligator clip that is used to fasten the device to an engine ground or a jumper with an alligator clip for the same reason. Most spark testers will be equipped with some form of an adjustable gap (often a threaded, screw-type adjuster is built into the tester) so either the spark plug gap or a test gap for your motor can be used to simulate the proper electrical load on the system. The most convenient type of a tester, for multi-cylinder motors is a unit in a self-contained housing that allows the connection of all spark plug leaks at the same time. Of course, you also have the option of buying one small, inexpensive tester per cylinder too. If you only have one tester connection for a multi-cylinder motor, remove and ground the spark plug leads for the other cylinders, then move the tester from lead to lead for each test.

✳✳ CAUTION

The extremely high voltages generated by the secondary ignition circuit (for spark) can cause serious injury or even death from electric shock. Never touch bare wires or wire connections during testing or while the engine is running. Also, for additional safety, never perform tests while you or the powerhead is wet (sure a water supply or test tank must be used when cranking or running the motor, just keep it off you and the outside of the powerhead, and DO NOT stand in puddles). Finally, we know you're not reading the fuel system section right now, but spark and fuel don't mix, unless it's inside the combustion chamber, so NEVER perform tests on the ignition system if you can smell fuel vapors. Ventilate the area thoroughly before proceeding.

■ The 40/50 hp EFI motors are equipped with an individual ignition coil mounted directly above each spark plug, without secondary spark plug leads. The spark tester(s) should be connected directly to the ignition coil spark plug connector on these models.

1. Except for jet drive or RescuePro® models, remove the propeller for safety. Refer to the procedures found for Propeller under Engine Maintenance.

2. For 2-4-cylinder motors, tag and disconnect the spark plug wires to ensure proper connection after the test is finished. Be sure to note the wiring routing as well as the final connection point.

✳✳ CAUTION

Leave all spark plugs installed and tightened in the powerhead to ensure that no raw fuel or fuel vapors are expelled from the spark plug bores and ignited by the testing procedure. Spark plug leads not connected to a tester (if you are using a single connection tester on a multi-cylinder motor) must be grounded to prevent damage to the ignition system (for safety, ground them as far away from the spark plug bores as possible).

Fig. 46 Twist and disconnect the spark plug lead(s) from the spark plug(s) . . .

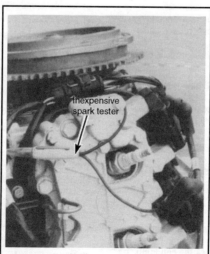

Fig. 47 . . . in order to connect a spark tester to the leads . . .

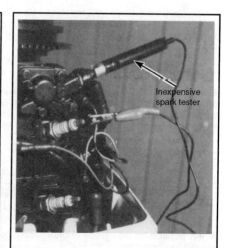

Fig. 48 . . . or to the coil, as directed by the tool manufacturer

4-18 IGNITION AND ELECTRICAL SYSTEMS

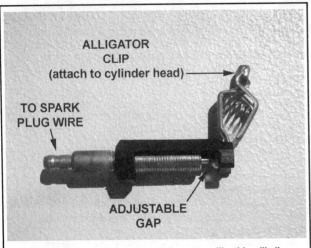

Fig. 49 The use of multiple inexpensive testers like this will allow you to check all cylinders simultaneously

3. Connect the spark tester alligator clip or ground lead to a good grounding point on the powerhead, such as a cylinder head bolt.

4. Connect the spark plug connector (lead on most models, the ignition coil spark plug connector itself on 40/50 hp EFI motors) to the spark tester.

5. Set the spark tester gap as follows:
- 2-stroke motors up to 4 hp (except 4 Deluxe) and 5 hp single cylinder: 3/8 in. (9mm)
- 2-stroke motors, 4 Deluxe and larger (except 5 hp single cylinder): 1/2 in. (12mm)
- 4-stroke carbureted motors: 3/8 in. (9mm)
- 4-stroke EFI motors: 7/16 in. (11mm)

■ To prevent the possibility of high voltage arcing, try to route tester wiring at least 2 in. (51mm) away from any metallic surface.

6. Make sure the shifter is in **Neutral** then operate the starter and observe the spark tester. Display on testers will vary slightly, so refer to the instructions that came with the tester. On simple mechanical gap testers, watch make sure the ignition produces a strong blue spark from each of the spark leads (coils).

7. If using a single test connection on multi-cylinder motors, repeat the procedure for each of the remaining cylinders. Results must be good on all cylinders or further testing is required.

8. Refer to the testing procedures according to the motor being tested:

2.0-3.5 (78cc) Hp Motors

If the spark tester shows good test results, but problems are still suspected, perform the running output test for the Power Pack found under Power Pack (Ignition Module) in this section to ensure performance under load.

If the spark tester shows no spark or poor spark on these models, perform the Stop Circuit Test found in this section.

■ It is possible for the ignition to show good output on the spark test and still have problems. If the engine spits, pops or backfires during startup, the ignition system may be out of time. Check the condition and mounting of the flywheel as directed under Powerhead.

5 Hp (109cc) Motors

If the spark tester shows no spark when cranking, perform the Stop Circuit Test found in this section.

■ It is possible for the ignition to show good output on the spark test and still have problems. If the engine spits, pops or backfires during startup, the ignition system may be out of time. Check the condition and mounting of the flywheel as directed under Powerhead.

3-4 Hp (87cc) Motors (Except 4 Deluxe)

If the spark tester shows good test results on both cylinders, but problems are still suspected, perform the running output test for the Power Pack found under Power Pack (Ignition Module) in this section to ensure performance under load.

If the spark tester shows no spark or poor spark on these models, perform the Stop Circuit Test found in this section.

■ It is possible for the ignition to show good output on the spark test and still have problems. If the engine spits, pops or backfires during startup, the ignition system may be out of time. Check the following:

- Routing of the coil primary wires
- Routing of the spark plug leads
- Condition and mounting of the flywheel as directed under Powerhead
- Timing and synchronization adjustments

4 Deluxe (87cc) And 5/6/8 Hp (164cc) Motors

If the spark tester shows good test results, but problems are still suspected, perform the running output test for the Power Pack found under Power Pack (Ignition Module) in this section to ensure performance under load.

If the spark tester shows good test results on ONE cylinder, perform the cranking output test for the Power Pack, found under Power Pack (Ignition Module) in this section.

If the spark tester shows no spark poor spark on these models, perform the Stop Circuit Test found in this section.

■ It is possible for the ignition to show good output on the spark test and still have problems. If the engine spits, pops or backfires during startup, the ignition system may be out of time. Check the following:

- Routing of the coil primary wires
- Routing of the spark plug leads
- Condition and mounting of the flywheel as directed under Powerhead
- Timing and synchronization adjustments

5/6 Hp (128cc) And 8/9.9 Hp (211cc) 4-Stroke Motors

If the spark tester shows good test results, but problems are still suspected, perform the running output test for the Power Pack found under Power Pack (Ignition Module) in this section to ensure performance under load.

If the spark tester shows good test results on ONE cylinder of 8/9.9 hp motors, perform the Ignition Coil resistance tests (or test the ignition coil using a dynamic coil tester, if available).

If the spark tester shows no spark on these models, perform the Stop Circuit Test found in this section.

■ It is possible for the ignition to show good output on the spark test and still have problems. If the engine spits, pops or backfires during startup, the ignition system may be out of time. Check the following:

- Power Pack cranking and running outputs
- Timing belt condition and timing mark alignment
- Timing wheel and timing wheel key

9.9/15 Hp (305cc) 4-Stroke Motors

If the spark tester shows good test results, but problems are still suspected, perform the running output test under Power Pack (Ignition Module) in this section to ensure performance under load.

If the spark tester shows good test results, but for only ONE cylinder, perform the Ignition Coil resistance tests (or test the ignition coil using a dynamic coil tester, if available).

If the spark tester shows no spark on these models, perform the Stop Circuit Test for the stop button and keyswitch, found in this section.

■ It is possible for the ignition to show good output on the spark test and still have problems. If the engine spits, pops or backfires during startup, the ignition system may be out of time. Check the following:

IGNITION AND ELECTRICAL SYSTEMS 4-19

- Power Pack cranking and running outputs
- Timing belt condition and timing mark alignment
- Timing wheel and timing wheel key

9.9/10/14/15 Hp (216cc), 9.9/10/15 Hp (255cc) And 18 Jet-35 Hp (521cc) Motors

If the spark tester shows good test results, but problems are still suspected, perform the running output test under Power Pack (Ignition Module) in this section to ensure performance under load.

If the spark tester shows good test results, but for only ONE cylinder, perform the Power Pack cranking output portion of the test.

If the spark tester shows no spark on these models, perform the Stop Circuit Test for the stop button and keyswitch, found in this section.

■ It is possible for the ignition to show good output on the spark test and still have problems. If the engine spits, pops or backfires during startup, the ignition system may be out of time. Check the following:

- Coil primary wire routing
- Spark plug secondary wire routing
- Flywheel condition and location
- Timing and synchronization adjustments

25/35 Hp (500/565cc) Motors

◆ See Figure 50

Various components of the ignition system should be checked, when interpreting the results of the Spark Test. Refer to the accompanying chart to help determine what additional components or systems should be tested depending on the observed results of the Spark Test. If the tester shows good results, but problems are still suspected, other systems should be checked to help decide whether or not the ignition system is actually at fault for a performance problem.

■ It is possible for the ignition to show good output on the spark test and still have problems. If the engine spits, pops or backfires during startup, the ignition system may be out of time. Check the following:

- Coil primary wire routing
- Spark plug secondary wire routing
- Spark Sensor and Encoder

25-55 Hp (737cc) Motors

If the spark tester shows good test results, but problems are still suspected, perform the running output test under Power Pack (Ignition Module) in this section to ensure performance under load.

If the spark tester shows good test results, but for only ONE cylinder, proceed as follows:

- For Rope start models or models with a 4-amp charging system, perform the cranking output portion of the Power Pack test, as detailed in this section.
- For models equipped with a 12-amp charging system, test the Sensor/Trigger Coil, as detailed in this section.

If the spark tester shows no spark on these models, perform the Stop Circuit Test for the stop button and keyswitch, found in this section.

■ It is possible for the ignition to show good output on the spark test and still have problems. If the engine spits, pops or backfires during startup, the ignition system may be out of time. Check the following:

- Coil primary wire routing
- Spark plug secondary wire routing
- Flywheel condition and location
- Timing and synchronization adjustments

25-70 Hp (913cc) Motors

◆ See Figure 51

Various components of the ignition system should be checked, when interpreting the results of the Spark Test. Refer to the accompanying chart to help determine what additional components or systems should be tested depending on the observed results of the Spark Test. If the tester shows good results, but problems are still suspected, other systems should be checked to help decide whether or not the ignition system is actually at fault for a performance problem.

■ It is possible for the ignition to show good output on the spark test and still have problems. If the engine spits, pops or backfires during startup, the ignition system may be out of time. Check the following:

- Coil primary wire routing
- Spark plug secondary wire routing
- Flywheel condition and location
- Timing and synchronization adjustments

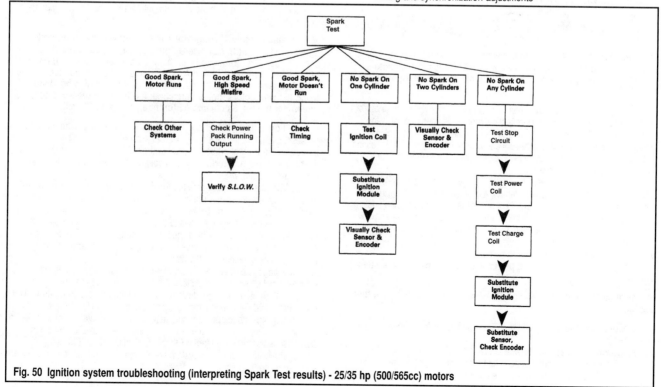

Fig. 50 Ignition system troubleshooting (interpreting Spark Test results) - 25/35 hp (500/565cc) motors

4-20 IGNITION AND ELECTRICAL SYSTEMS

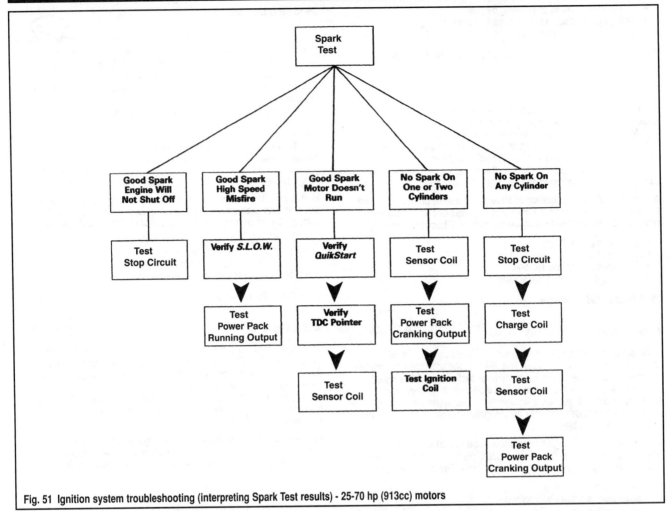

Fig. 51 Ignition system troubleshooting (interpreting Spark Test results) - 25-70 hp (913cc) motors

40/50 Hp (815cc) And 70 Hp (1298cc) EFI Motors

■ A peak-reading DVOM will be necessary for this test. Also, the OMC test probe (#342677) is useful to help prevent damage to connectors when back-probing them to check their signal without interrupting the circuit.

Unlike the ignition system used on carbureted Johnson/Evinrude motors (whose diagnostic process involves testing various sources of power) the system used on EFI motors is comparatively simple. Power for the EFI ignition system comes from the battery and is controlled by the ECU. Testing is therefore a simple matter of verifying battery power and input/output signals for ignition components.

The EFI system relay is used to supply voltage to the ignition coils (as well as the fuel injection system). Usually, a problem with the relay prevents the engine from starting, but intermittent problems can cause engine stalling (especially right after the engine warms up). When this occurs, the engine will usually not restart unless the keyswitch is switched off long enough so the relay cools. Refer to the Electronic Fuel Injection (EFI) section for details on testing the System Relay.

If the spark test shows a lack of proper spark, check the ignition system signals as follows:

■ Leave the spark tester installed as detailed in the Spark Test in this section.

1. Set a peak-reading DVOM to indicate positive voltage on the 50 volts scale.
2. Connect the meter as follows:
  a. For 40/50 hp motors, connect the black meter probe to the positive terminal of the battery. Then, back-probe each of the following ECU connector terminals: **A1**, **A3** and **A5** by carefully inserting the red meter lead (or a test probe attached to the red lead) into the back of the appropriate ECU connector.

■ Information on ECU connector and pin-outs can be found under Sensor and Circuit Resistance/Output Tests in the Electronic Fuel Injection section. Connector A is the 8-pin connector found at one end of the ECU. Terminals A1 and A5 are on one end of the connector, while A3 is two terminals over from A1. The terminals can also be identified using the wire colors that are attached: A1 (blue, for the No. 2 coil), A3 (green, for the No. 3 coil) and A5 (orange, for the No. 1 coil).

  b. For 70 hp motors, it is easiest to back-probe the ignition coil wiring harness. Connect the black meter lead to a good engine ground, then use the red lead (or a test probe attached to the red lead) to carefully back-probe the blue wire (No. 1 and No. 2 ignition coil) or the orange wire (No. 3 and No. 4 ignition coil).
3. With the meter connected to the ECU or ignition coil wiring harness as directed, note the meter reading. The meter should show 0 volts with the ignition keyswitch **ON** and battery voltage with the keyswitch **OFF**.
4. Move the black meter lead to the positive battery connector of the starter solenoid (the point where the positive battery cable connects to the starter solenoid). Then crank the engine and watch the reading on the DVOM, it should indicate 6-10 volts:
  • If voltage is present as indicated, the ECU signal is good. If the cylinder's still show no spark, check the secondary coil resistance.
5. If there is no voltage, disconnect the wiring harness from the ECU (to protect the module) and ignition coil, then check continuity on the primary wire circuit. Next check primary and secondary coil resistance, as detailed in this section. Finally, on 40/50 hp motors, check the Crankshaft Position (CKP) Sensor, as detailed in the Electronic Fuel Injection (EFI) section.
6. If all wiring and other system components test ok, but there is still no spark, replace the ECU.

IGNITION AND ELECTRICAL SYSTEMS 4-21

Stop Circuit Test

◆ See Figure 52

■ Refer to the schematics found in the Wiring Diagram section for more details on the stop circuit.

The Stop Circuit Test is used to determine if an ignition fault is caused by the ignition or stop switch side of the circuit, or if the problem comes from the charge coil, power pack and ignition coil side of the circuit. Problems with the stop circuit usually cause a no start fault, but intermittent faults may occur that cause the engine to cut out.

The Spark Test is normally performed before checking the stop circuit in order to determine if spark is occurring at all. If there is spark at one or more cylinders of a multi-cylinder motor it is not likely that the stop circuit is the culprit. Of course, checking it is a relatively fast and easy way to eliminate concern over it.

2.0-3.5 (78cc) Hp Motors

■ Leave the spark tester installed as detailed in the Spark Test in this section.

1. Disengage the connector for the black lead between the stop button and the power pack, then crank the motor and watch the tester. If spark now appears or improves with the black lead disconnected, the problem is with the stop circuit; proceed with the next step. If there is still no spark, perform the Sensor Coil test as detailed in this section.
2. With the engine off, connect the ohmmeter probes between the stop button black lead terminal and a good engine ground. The meter must show high or infinite resistance.
3. Press the stop button and observe the meter reading, it should show low or zero resistance with the button held in the down position.
4. Replace the stop button if the circuit does not perform as noted.
5. Reconnect all wiring as removed for testing.

5 Hp (109cc) Motors

■ Leave the spark tester installed as detailed in the Spark Test in this section.

1. Disengage the bullet connectors for the blue/red and black leads between the stop button and the ignition coil, then crank the motor and watch the tester.
2. If spark now appears only with these leads disconnected, the problem is with the stop circuit. Replace the switch assembly.
3. If there is still no spark, perform the Charge Coil test as detailed in this section.
4. Reconnect all wiring as removed for testing.

3-4 Hp (87cc) Motors (Except 4 Deluxe)

■ Leave the spark tester installed as detailed in the Spark Test in this section.

1. Disengage the 6-pin connector (it's a Packard style connector) from the power pack (ignition module) at the flywheel assembly.

■ 1990 models are normally not equipped with the referenced 6-pin connector. On an early-model motor, simply disconnect the black lead directly from the ignition module instead of from the connector used on later model motors.

2. Use a thin wire, like a paper clip, about 1/4 in. (6mm) into the slot at the top of the terminal for the black/yellow lead (connector terminal E) to release the terminal. Gently pull the lead out of the back of the connector body.
3. Reconnect the 6-pin Packard connector to the power pack (ignition module) without the black/yellow lead. Put a piece of tape over the black/yellow terminal to make sure it does not ground on the powerhead.
4. Crank the motor and watch the spark tester. If spark appears only with the black/yellow lead disconnected, it appears the problem is with the stop circuit. Proceed with the next step. If there is still no spark, perform the Sensor Coil test as detailed in this section.
5. Disengage the 6-pin connector from the power pack (ignition module) again. Install the black/yellow lead back into the harness connector. Gently push the terminal in through the rear of the connector until the tab locks in place, then pull back on the wire with light pressure to be sure it is secured.
6. With the engine off, connect the ohmmeter probes between terminal E (connector black/yellow lead) and a good engine ground. The meter must show high or infinite resistance.
7. Press the stop button and observe the meter reading, it should show low or zero resistance with the button held in the down position.
8. Replace the stop button if the circuit does not perform as noted.
9. Reconnect all wiring as removed for testing.

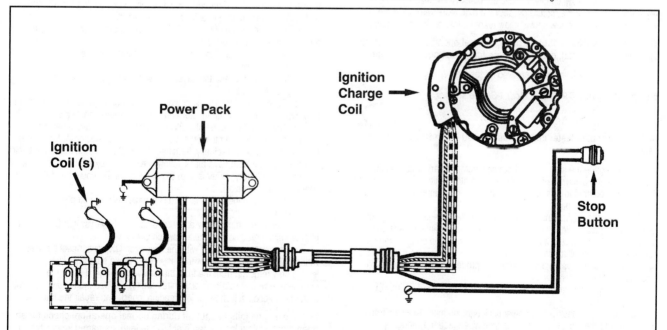

Fig. 52 Typical stop circuit - the stop button connects the power pack to ground, interrupting the charge coil voltage supplied to the ignition coil primary circuits, through the power pack

4-22 IGNITION AND ELECTRICAL SYSTEMS

4 Deluxe (87cc) And 5/6/8 Hp (164cc) Motors

■ Leave the spark tester installed as detailed in the Spark Test in this section.

1. Disengage the 5-pin connector (it's an Amphenol style connector) between the power pack (ignition module) and the ignition charge coil mounting plate located under the flywheel.

■ The 5-8 hp motors through 1992 and the 4D through 1993 were equipped with the UFI version of this ignition system and therefore did not use the 5-pin amphenol connector. So on those models not equipped with the connector, trace the stop circuit wire to the retainer and the SINGLE PIN amphenol connector. Disengage the 1-pin connector and use that lead for this test.

2. Use 4 jumper wires to reconnect terminals **A**, **B**, **C** and **D** with the halves of the wiring harness still disconnected. In effect, this removes the black wire, terminal **E** for the stop switch from the circuit.
3. Crank the motor and watch the spark tester. If spark appears on both cylinders only with the black lead disconnected, it appears the problem is with the stop circuit. Proceed with the next step. If there is no spark at ONE cylinder, perform the output test on the Power Pack (Ignition Module), as detailed in this section. If there is still no spark at both cylinders, perform the Charge Coil test, as detailed in this section.
4. Remove the jumper wires from the 5-pin connector.
5. With the engine not running or cranking, check the stop button/safety lanyard switch as follows:
   a. Connect the ohmmeter probes between terminal **E** (connector for the black lead) and a good engine ground.
   b. With the safety lanyard clip installed and the switch in the run position, the meter must show high or infinite resistance.
   c. Press the stop button and/or remove the safety lanyard while observing the meter reading, it should show low or zero resistance with the button held in the down position or the safety lanyard removed.
   d. Replace the stop button if the circuit does not perform as noted.
6. If equipped, check the keyswitch with the engine not running or cranking, as follows:
   a. Connect the ohmmeter probes between terminal **E** (connector for the black lead) and a good engine ground.
   b. Make sure the safety lanyard is installed and the stop switch is in the run position.
   c. The meter must show a low reading with the keyswitch **OFF** and a high or infinite reading with the keyswitch **ON**.
   d. With the keyswitch **ON**, remove the black/yellow lead from the keyswitch harness. If the meter now shows a high reading, replace the keyswitch. If the meter now shows a low reading, test the wiring harness.

■ If the engine fails to shut off during normal operation, check for an open black/yellow lead, damaged keyswitch or damaged power pack.

7. Reconnect all wiring as removed for testing.

5/6 Hp (128cc) And 8/9.9 Hp (211cc) 4-Stroke Motors

■ Leave the spark tester installed as detailed in the Spark Test in this section.

1. Disengage the stop button 1-pin connector (it's an Amphenol style connector for the black wire found between the stop switch and power pack). Although the other stop switch wire is also black, it is connected to a ground terminal at the power pack lower mounting bolt.
2. Crank the engine and observe the spark tester again. If the spark jumps (at all gaps on 8/9.9 motors), but did not in a preliminary test, the problem is the stop circuit. If there is no spark at one gap on 8/9.9 motors, check the Ignition Coil as detailed in this section. If there is no spark on 5/6 hp motors or at all gaps on 8/9.9 hp motors, proceed with the Power Coil testing as detailed in this section.
3. If equipped, check the keyswitch with the engine not running or cranking, as follows:
   a. Follow the black/yellow power pack (ignition module) wiring from the power pack to the 1-pin connector and then disengage the wiring.
   b. Connect the probes of a DVOM set to read resistance between a good engine ground and the keyswitch side of the connector for the black/yellow wire.
   c. Make sure the safety lanyard is installed and the stop switch is in the run position.
   d. The meter must show a low reading with the keyswitch **OFF** and a high or infinite reading with the keyswitch **ON**.
   e. If resistance was low with the keyswitch **ON**, disconnect harness wiring harness from the keyswitch (especially, make sure the black/yellow lead is now removed from the from the keyswitch terminal M). If the ohmmeter now shows a high or infinite reading, replace the keyswitch. If the meter connection still shows a low reading, check and repair the harness.

■ If the engine fails to shut off during normal operation, check for an open black/yellow lead, damaged keyswitch or damaged power pack.

4. Reconnect all wiring as removed for testing.

9.9/15 Hp (305cc) 4-Stroke Motors

■ Leave the spark tester installed as detailed in the Spark Test in this section.

1. Disengage the stop button 3-pin connector (it's an Amphenol style connector found between the start button or remote harness and power pack).

■ For rope start models, terminals B and C of the 3-pin stop button connector are bridged on one side of the connector. The wiring on the engine side of the connector runs to both the power pack (ignition module) and the stop switch.

2. Crank the engine and observe the spark tester again. If the spark now jumps both gaps, but did not in a preliminary test, the problem is the stop circuit; proceed with the ohmmeter tests given in this procedure to isolate a problem with the stop button or keyswitch. If there is no spark at one gap, check the ignition coil as detailed in this section. If there is no spark at all gaps, then proceed with the Power Coil testing as detailed in this section.
3. Check the stop button with the engine not running or cranking, as follows:
   a. Install the safety clip/lanyard and make sure the switch is in the **RUN** position.
   b. Using a DVOM set to measure resistance between the terminal **C** (black wire) on the button side of the 3-pin stop button connector and a good engine ground. There should be high or infinite resistance with the lanyard and the button in the **RUN** position.
   c. Press the stop button inward and observe the meter. Resistance between terminal **C** and ground should become very low to zero with the button depressed, then go back up when released.
   d. Remove the safety lanyard, the meter should show continuity to ground (resistance should drop again).
   e. If the stop button does not respond as indicated, replace the stop button assembly for safety.
4. If equipped, check the keyswitch with the engine not running or cranking, as follows:
   a. Follow the black/yellow power pack (ignition module) wiring from the power pack to the 3-pin connector and then disengage the wiring.
   b. Connect the probes of a DVOM set to read resistance between a good engine ground and the keyswitch side of the connector for terminal **B** the black/yellow wire. (Actually, terminals **B** and **C** that come out of the back of that connector both have black/yellow wires, but they are bridged in the harness, so in theory, either is fine, though terminal **B** should be at the center of the connector).
   c. Make sure the safety lanyard is installed and the stop switch is in the run position.
   d. The meter must show a low reading with the keyswitch **OFF** and a high or infinite reading with the keyswitch **ON**.
   e. If resistance was low with the keyswitch **ON**, disconnect harness wiring harness from the keyswitch (especially, make sure the black/yellow lead is now removed from the from the keyswitch terminal M). If the ohmmeter now shows a high or infinite reading, replace the keyswitch. If the meter connection still shows a low reading, check and repair the harness.

■ If the engine fails to shut off during normal operation, check for an open black/yellow lead, open black/white lead, damaged keyswitch or damaged power pack.

5. Reconnect all wiring as removed for testing.

IGNITION AND ELECTRICAL SYSTEMS

9.9/10/14/15 Hp (216cc), 9.9/10/15 Hp (255cc) And 18 Jet-35 Hp (521cc) Motors

■ Leave the spark tester installed as detailed in the Spark Test in this section.

1. Disengage the 5-pin connector (it's an Amphenol style connector) between the power pack (ignition module) and the ignition charge coil mounting plate located under the flywheel.

■ Motors through 1991 and some 1992 models were equipped with the UFI version of this ignition system and therefore did not use the 5-pin amphenol connector. So on those models not equipped with the connector, trace the stop circuit wire to the retainer and the SINGLE PIN amphenol connector. Disengage the 1-pin connector and use that lead for this test.

2. Use 4 jumper wires to reconnect terminals **A**, **B**, **C** and **D** with the halves of the wiring harness still disconnected. In effect, this removes the black or black/yellow wire, terminal **E** for the stop switch or remote keyswitch from the circuit, as applicable.

3. Crank the motor and watch the spark tester. If spark appears on both cylinders only with the black (or black/yellow) lead disconnected, it appears the problem is with the stop circuit. Proceed with the next step. If there is no spark at ONE cylinder, perform the output test on the Power Pack (Ignition Module), as detailed in this section. If there is still no spark at both cylinders, perform the Charge Coil test, as detailed in this section.

4. Remove the jumper wires from the 5-pin connector.

5. With the engine not running or cranking, check the stop button/safety lanyard switch as follows:

   a. Except for remote models, connect the ohmmeter probes between terminal **E** (connector for the black lead) and a good engine ground. For remote motors, refer to the wiring diagram, then disengage the bullet connector for the tan or tan/red wire running to the stop button, connect the ohmmeter probes between the switch side of the button and a good engine ground.

   b. With the safety lanyard clip installed and the switch in the run position, the meter must show high or infinite resistance.

   c. Press the stop button and/or remove the safety lanyard while observing the meter reading, it should show low or zero resistance with the button held in the down position or the safety lanyard removed.

   d. Replace the stop button if the circuit does not perform as noted.

6. If equipped, check the keyswitch with the engine not running or cranking, as follows:

   a. Connect the ohmmeter probes between terminal **E** (connector for the black/yellow lead) and a good engine ground.

   b. Make sure the safety lanyard is installed and the stop switch is in the run position.

   c. The meter must show a low reading with the keyswitch **OFF** and a high or infinite reading with the keyswitch **ON**.

   d. With the keyswitch **ON**, remove the black/yellow lead from the keyswitch harness. If the meter now shows a high reading, replace the keyswitch. If the meter now shows a low reading, test the wiring harness.

■ If the engine fails to shut off during normal operation, check for an open black/yellow lead, damaged keyswitch or damaged power pack.

7. Reconnect all wiring as removed for testing.

25/35 Hp (500/565cc) Motors

■ Leave the spark tester installed as detailed in the Spark Test in this section.

1. Disengage the 4-pin connector (it's an Amphenol style connector) located between the power pack (ignition module) and the start button.

2. Use 4 jumper wires to reconnect terminals **A**, **C** and **D** with the halves of the wiring harness still disconnected. In effect, this removes the black/yellow (black on one side of the connector for some models) wire, terminal **B** for the stop switch or remote keyswitch from the circuit, as applicable.

3. Crank the motor and watch the spark tester. If spark appears on both cylinders only with the black/yellow lead disconnected, it appears the problem is with the stop circuit. Test the stop button and/or the keyswitch, as equipped. If there is still no spark, perform the Power Coil test, as detailed in this section.

4. Remove the jumper wires from the 4-pin connector.

5. Except for remote models, check the stop button/safety lanyard switch with the engine not running or cranking as follows:

   a. Connect the ohmmeter probes between terminal **B** (connector for the black/yellow or black lead) and a good engine ground.

   b. With the safety lanyard clip installed and the switch in the run position, the meter must show high or infinite resistance.

   c. Press the stop button and/or remove the safety lanyard while observing the meter reading, it should show low or zero resistance with the button held in the down position or the safety lanyard removed.

   d. Replace the stop button if the circuit does not perform as noted.

6. For remote models, check the keyswitch with the engine not running or cranking, as follows:

   a. Connect the ohmmeter probes between terminal **B** (connector for the black/yellow lead) and a good engine ground.

   b. If equipped, make sure the safety lanyard is installed.

   c. The meter must show a low reading with the keyswitch **OFF** and a high or infinite reading with the keyswitch **ON**.

   d. With the keyswitch **ON**, remove the black/yellow lead from the keyswitch harness. If the meter now shows a high reading, replace the keyswitch. If the meter now shows a low reading, test the wiring harness.

■ If the engine fails to shut off during normal operation, check for an open black/yellow lead, damaged keyswitch or damaged power pack.

7. Reconnect all wiring as removed for testing.

25-55 Hp (737cc) Motors

■ Leave the spark tester installed as detailed in the Spark Test in this section.

1. Disengage the 5-pin connector (it's an Amphenol style connector) between the power pack (ignition module) and the ignition charge coil mounting plate located under the flywheel.

■ Motors through 1991 and some 1992-93 models were equipped with the UFI version of this ignition system and therefore did not use the 5-pin amphenol connector. So on those models not equipped with the connector, trace the stop circuit wire to the retainer and the SINGLE PIN amphenol connector. Disengage the 1-pin connector and use that lead for this test.

2. Use 4 jumper wires to reconnect terminals **A**, **B**, **C** and **D** with the halves of the wiring harness still disconnected. In effect, this removes the black/yellow wire, terminal **E** for the stop switch or remote keyswitch from the circuit, as applicable.

3. Crank the motor and watch the spark tester. If spark appears on both cylinders only with the black/yellow) lead disconnected, it appears the problem is with the stop circuit. Proceed with testing of the stop button or keyswitch, as applicable. If there is no spark at ONE cylinder, perform the output test on the Power Pack (Ignition Module), as detailed in this section. If there is still no spark at both cylinders, perform the Charge Coil test, as detailed in this section.

4. Remove the jumper wires from the 5-pin connector.

5. With the engine not running or cranking, check the stop button/safety lanyard switch as follows:

   a. Connect the ohmmeter probes between terminal **E** (connector for the black/yellow lead) and a good engine ground.

   b. With the safety lanyard clip installed and the switch in the run position, the meter must show high or infinite resistance.

   c. Press the stop button and/or remove the safety lanyard while observing the meter reading, it should show low or zero resistance with the button held in the down position or the safety lanyard removed.

   d. Replace the stop button if the circuit does not perform as noted.

6. If equipped, check the keyswitch with the engine not running or cranking, as follows:

   a. Connect the ohmmeter probes between terminal **E** (connector for the black/yellow lead) and a good engine ground.

   b. Make sure the safety lanyard is installed and the stop switch is in the run position.

   c. The meter must show a low reading with the keyswitch **OFF** and a high or infinite reading with the keyswitch **ON**.

   d. With the keyswitch **ON**, remove the black/yellow lead from the keyswitch harness. If the meter now shows a high reading, replace the keyswitch. If the meter now shows a low reading, test the wiring harness.

4-24 IGNITION AND ELECTRICAL SYSTEMS

■ If the engine fails to shut off during normal operation, check for an open black/yellow lead, damaged keyswitch or damaged power pack.

7. Reconnect all wiring as removed for testing.

25-70 Hp (913cc) Motors

■ Leave the spark tester installed as detailed in the Spark Test in this section.

1. Disengage the 5-pin connector (it's an Amphenol style connector) located between the power pack (ignition module) and the ignition charge coil mounting plate located under the flywheel.

■ Power packs on these models are normally equipped with more than one connector, in all cases, disengage the 5-pin connector that contains the black/yellow wire coming from the power pack.

2. Use 4 jumper wires to reconnect terminals **A**, **B**, **C** and **D** with the halves of the wiring harness still disconnected. In effect, this removes the black/yellow wire, terminal **E** for the stop switch or remote keyswitch from the circuit, as applicable.
3. Crank the motor and watch the spark tester. If spark appears on both cylinders only with the black/yellow) lead disconnected, it appears the problem is with the stop circuit. Proceed with testing of the stop button or keyswitch, as applicable. If there is SOME spark present at one or more cylinder, but the system is not operating properly, check the Sensor Coil, as detailed in this section. If there is still no spark from at cylinder, perform the Charge Coil test, as detailed in this section.
4. Remove the jumper wires from the 5-pin connector.
5. With the engine not running or cranking, check the stop button/safety lanyard switch as follows:
    a. Connect the ohmmeter probes between terminal **E** (connector for the black/yellow or black lead) and a good engine ground.
    b. With the safety lanyard clip installed and the switch in the run position, the meter must show high or infinite resistance.
    c. Press the stop button and/or remove the safety lanyard while observing the meter reading, it should show low or zero resistance with the button held in the down position or the safety lanyard removed.
    d. Replace the stop button if the circuit does not perform as noted.
6. If equipped, check the keyswitch with the engine not running or cranking, as follows:
    a. Connect the ohmmeter probes between terminal **E** (connector for the black/yellow lead) and a good engine ground.
    b. If equipped, make sure the safety lanyard is installed and the stop switch is in the run position.
    c. The meter must show a low reading with the keyswitch **OFF** and a high or infinite reading with the keyswitch **ON**.
    d. With the keyswitch **ON**, remove the black/yellow lead from the keyswitch harness. If the meter now shows a high reading, replace the keyswitch. If the meter now shows a low reading, test the wiring harness.

■ If the engine fails to shut off during normal operation, check for an open black/yellow lead, open black/white lead, a damaged keyswitch or damaged power pack.

7. Reconnect all wiring as removed for testing.

Charge Coil

◆ See Figures 40, 41 and 42

All carbureted Johnson/Evinrude motors are equipped with an ignition system charge coil that is used to generate voltage to power the ignition system. The coil is mounted under the flywheel (centered directly underneath on most models) to interact with one or more permanent magnets attached to the underside of the flywheel assembly. As the flywheel rotates, the magnetic force cuts through the coil assembly, generating an electric current that is utilized by the ignition system.

■ The 3/4 Hp (except 4 Deluxe) 2-cylinder, 2-stroke motors utilize a charge coil that is built into the power pack/ignition module. For this reason, the charge coil cannot be tested or replaced separately. For more details, please refer to the heading for this engine under Power Pack.

On some motors, the charge coil is an individual component, mounted side-by-side with a battery charge coil or accessory lighting coil that can be removed or replaced separately. In many instances, the semi-circular bracket that contains an individual charge coil also contains a power coil. On even more Johnson/Evinrude outboards, the charge coil is part of a round one-piece stator that also supplies current to the charging system and must be serviced as an assembly.

Refer to the wiring diagrams provided in this section to help determine the system on your engine. Each diagram shows whether a one-piece stator or an individual charge coil is used.

■ EFI motors are equipped with a one-piece stator. The battery powered transistorized ignition system used on EFI motors does not utilize an ignition charge coil circuit.

TESTING

◆ See Figures 53, 54 and 55

Charge coil testing typically encompasses making sure the wiring is in good condition, then making verifying coil output (using a peak-reading DVOM) while cranking the motor and/or checking resistance (using an ohmmeter) across the coil windings when the motor is not turning. The charge coil and wiring can also be checked for shorts to ground either using the peak-reading DVOM while the motor is cranking or the ohmmeter when the motor is at rest.

Problems with the ignition charge coil usually cause a no start condition. But, a partial short of the coil winding can cause hard starting and/or an ignition misfire. The most reliable tests for the charge coil are to dynamically check the output with the engine cranking (or in some cases, running). Of course, dynamic tests require a digital multimeter capable of reading and displaying peak voltage values (also known as a peak-reading voltmeter). If one is not available, specifications are usually available to statically check the coil winding using an ohmmeter. But, before replacing the coil based only on static test that shows borderline readings, remember that resistance readings will vary with temperature and the specifications provided here are based on a component temperature of about 68°F (20°C). Also, remember that a coil may test within specification statically, but show intermittent faults under engine operating conditions, especially at normal operating temperature.

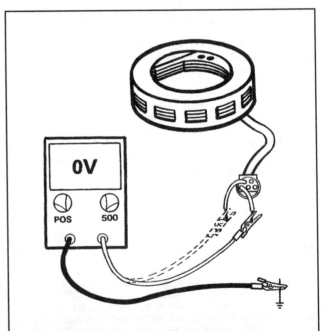

Fig. 53 Charge coil testing on CDI models involves checking between the appropriate harness terminals and a good engine ground to make sure there are no shorts to ground. This test can usually be conducted with a voltmeter (engine cranking) and/or an ohmmeter (engine static)

IGNITION AND ELECTRICAL SYSTEMS   4-25

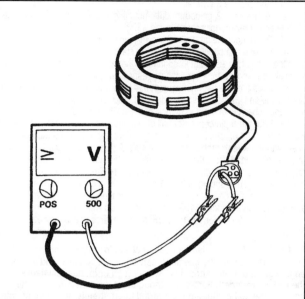

Fig. 54 Charge coil testing on CDI models also involves either checking voltage output while the motor is cranking and/or checking resistance across coil windings while the motor is at rest - cranking test on a typical stator equipped motor shown

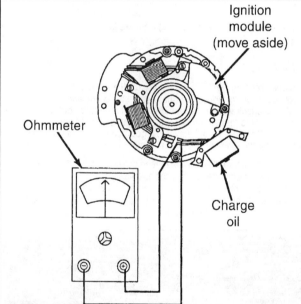

Fig. 55 All checks can be made at the wiring harness, but this shows a resistance check on a UFI model being made to a partially removed charge coil

2.0-3.5 (78cc) Hp Motors

■ A peak-reading DVOM is necessary for the ground and output portions of this test. If one is not available, the resistance check will give some idea of coil condition, but be careful as a bad coil could test within specification but short out and fail intermittently under load. Likewise, a good coil may test out of spec due to variances in meters and test temperatures.

1. Check the charge coil ground first as follows:
    a. Set the peak reading voltmeter to read positive on the 500 volts scale.
    b. Disengage the brown and brown/yellow leads between the charge coil and power pack.
    c. Connect the meter between the brown lead from the charge coil and a good engine ground, then crank the engine and observe the readings.
    d. Connect the meter between the brown/yellow lead from the charge coil and a good engine ground, then crank the engine and observe the readings.
    e. Any voltage from this portion of the test indicates that the charge coil is shorted to ground. Repair the wire which is grounded or replace the charge coil, as necessary. If there is no reading, perform the next step to check coil output.
2. Check the charge coil output next as follows:
    a. Set the peak reading voltmeter to read positive on the 500 volts scale.
    b. Now connect the meter probes across the 2 charge coil leads. Connect one probe to the brown lead and the other to the brown/yellow lead.
    c. Crank the motor and observe charge coil output on the meter. If the reading is 180 volts or higher, perform the cranking power pack test as detailed in this section under Power Pack (Ignition Module), Testing. If the reading is less than 180 volts, check the condition of the wiring and connectors (if no problems are found with the wiring, perform the resistance check in the next step.)
3. With the engine not running or cranking, connect the probes from a DVOM set to read resistance across the two terminals of the charge coil (brown and brown/yellow). The meter must show 550-670 ohms resistance at a temperature of about 68°F (20°C) or the charge coil is out of specification.

■ An ohmmeter may also be used to check for a grounded coil using a static resistance test by checking between each coil lead and a good engine ground. Meter should show very high or infinite resistance, or the coil is grounded. Remember though that this test can be misleading because the coil might short to ground intermittently only under engine operation.

4. Verify test results and connections, then replace the charge coil, if the reading is well out of specification.
5. Reconnect all wiring as removed for testing.

3-4 Hp (87cc) Motors

On these motors the functions of the charge coil are integrated into the power pack/ignition module that is mounted along the edge of the flywheel. Because of the internal circuitry of the coil windings, no separate provisions are provided to test or replace the charge coil. Refer to Power Pack in this section for information on testing or servicing the primary circuit of the ignition system.

5 Hp (109cc) Motors

The charge coil on this model feeds the primary side of the combined power pack/ignition coil directly with no other components contained in the system.

■ A peak-reading DVOM is necessary for the dynamic (cranking) portions of this test. If one is not available, the resistance check will give some idea of coil condition, but be careful as a bad coil could test within specification but short out and fail intermittently under load. Likewise, a good coil may test out of spec due to variances in meters and test temperatures.

1. Check the charge coil ground first as follows:
    a. Set the peak reading voltmeter to read positive on the 500 volts scale.
    b. Disengage the blue/red and black leads between the charge coil and ignition coil
    c. Connect the meter red lead to the black/orange (possibly black/red on some models) lead from the charge coil and the black meter lead to a good engine ground, then crank the engine and observe the readings.
    d. The meter should show 130 volts or more. If there reading is low or there is no reading at all, replace the charge coil.
    e. Now connect the meter probes across the 2 charge coil leads. Connect the red probe to the black/orange (possibly black/red on some models) lead and the black probe to the black lead.
    f. Crank the motor and observe charge coil output on the meter. If the reading is 130 volts or higher, the ignition system is performing normally. If the reading is less than 130 volts, check for continuity between the black lead and a good engine ground. If the lead is open (no continuity with ground) the engine will not shut down when the stop button is actuated. If

4-26 IGNITION AND ELECTRICAL SYSTEMS

there are no problems with the wiring and output is still below 130 volts, replace the charge coil.

2. A resistance check can be used to statically check the charge coil, but keep in mind that a coil could test within spec statically, but show a fault dynamically (especially once operating temperatures are reached). The advantage of the resistance check is that not everyone has access to a peak-reading DVOM.

    a. Disengage the bullet connectors for the black and the black/orange (possibly black/red on some models) leads coming directly from the charge coil.

    b. Connect the leads of an ohmmeter across the two charge coil leads. Resistance should be 100-150 ohms for a component at about 68°F (20°C) or the charge coil is out of specification.

■ An ohmmeter may also be used to check for a grounded coil using a static resistance test by checking between each coil lead and a good engine ground. Meter should show very high or infinite resistance, or the coil is grounded. Remember though that this test can be misleading because the coil might short to ground intermittently only under engine operation.

3. Verify test results and connections, then replace the charge coil, if the reading is well out of specification.

4. Reconnect all wiring as removed for testing.

4 Deluxe (87cc) and 5/6/8 Hp (164cc) Motors

■ The 5-8 hp motors through 1992 and the 4D through 1993 were equipped with the UFI version of this ignition system. Since the charge coil is connected to the Ignition Module and cranking tests would involve checking the voltage on those lines (which cannot be accessed with the flywheel installed) there is no way to perform cranking tests on those years. Instead you'll only be able to remove the flywheel (for access to the wiring) and conduct static ohmmeter tests.

■ A peak-reading DVOM is necessary for the basic ground and the dynamic (cranking) portion of this test. If one is not available, the resistance check will give some idea of coil condition, but be careful as a bad coil could test within specification but short out and fail intermittently under load. Likewise, a good coil may test out of spec due to variances in meters and test temperatures.

1. For most models (not equipped with UFI system), check the charge coil ground first as follows:

    a. Set the peak reading voltmeter to read negative on the 500 volts scale.

    b. Disengage the 5-pin connector (it's an Amphenol style connector) between the power pack (ignition module) and the ignition charge coil mounting plate located under the flywheel.

    c. Connect the meter red lead to terminal **A** (brown lead from the charge coil) and the black meter lead to a good engine ground, then crank the engine and observe the readings.

    d. Connect the meter red lead to terminal **D** (brown/yellow lead from the charge coil) and the black meter lead to a good engine ground, then crank the engine and observe the readings.

    e. Voltage present at either test indicates that the charge coil or the wiring harness is shorted to ground. Locate and repair the short or replace the charge coil assembly.

2. For most models (not equipped with UFI system), check the charge coil output as follows:

    a. Now connect the red meter probe to terminal **D** (brown/yellow lead from the charge coil) and the black meter probe to terminal **A** (brown lead from the charge coil).

    b. Crank the motor and observe charge coil output on the meter. If the reading is 230 volts or higher, the charge coil is performing normally, to continue troubleshooting an ignition system problem, perform the Sensor Coil testing as detailed in this section. If the reading is less than 230 volts, check the condition of the wiring and connectors. If there are no problems with the wiring and output is still below 230 volts, proceed with the resistance test in the next step (but charge coil replacement is probably necessary).

3. For models equipped with UFI system (specifically this should include 5-8 hp motors through 1992 and the 4D through 1993), access the charge coil, as follows:

    a. Remove the manual starter assembly, as detailed in the Hand Rewind Starter section.

    b. Remove the Flywheel.

    c. Remove the 2 ignition module mounting screws, then reposition the module as necessary for access to the wiring.

    d. Disconnect the brown and brown/yellow lead bullet connectors.

4. A resistance check can be used to statically check the charge coil (and the wiring harness), but keep in mind that a coil could test within spec statically, but show a fault dynamically (especially once operating temperatures are reached). Of course, this is your only option for testing the charge coil on UFI motors, so do your best. And, one small advantage of the resistance check is that not everyone has access to a peak-reading DVOM.

    a. With the 5-pin connector still disengaged from the previous steps (or the flywheel and other components removed for access on UFI models directly to the wires), probe across terminals **A** (brown lead) and **D** (brown/yellow lead) from the charge coil. On UFI models the terminal references are not applicable, but the wire colors should be the same.

    b. Resistance should be 800-1000 ohms for a component at about 68°F (20°C) or the charge coil is out of specification. Please note that the specification given is for all 1991 and later models with either CDI or UFI, however the specification for 1990 UFI models differs and is only 535-585 ohms.

■ An ohmmeter may also be used to check for a grounded coil using a static resistance test by checking between each coil lead and a good engine ground. Meter should show very high or infinite resistance, or the coil is grounded. Remember though that this test can be misleading because the coil might short to ground intermittently only under engine operation.

5. Verify test results and connections, then replace the charge coil, if the reading is well out of specification.

6. Reconnect all wiring as removed for testing. On UFI models, remember the orange/blue primary lead must be connected to the No. 1 ignition coil.

5/6 Hp (128cc) and 8/9.9 Hp (211cc) 4-Stroke Motors

◆ See Figure 56

■ A peak-reading DVOM is necessary for the basic ground and the dynamic (cranking) portion of this test. If one is not available, the resistance check will give some idea of coil condition, but be careful as a bad coil could test within specification but short out and fail intermittently under load. Likewise, a good coil may test out of spec due to variances in meters and test temperatures.

1. Set a peak-reading DVOM to read positive in the 500 volts scale, then check for a grounded charge coil as follows:

    a. Disconnect the 2 bullet connectors for the brown wires (usually brown/white and brown/black) coming from the charge coil underneath the flywheel and running to the power pack.

Fig. 56 Most of the wiring for the charge coil, power coil, battery charge coil and sensor coil on these models is attached through bullet connectors found on the back of the power pack - 5/6 hp and 8/9.9 hp 4-strokes

IGNITION AND ELECTRICAL SYSTEMS 4-27

b. Connect the black meter probe to a good engine ground and the red probe to one of the brown charge coil leads. Crank the engine and observe, then note the reading.

c. Move the red meter probe to the other brown charge coil lead, then crank the engine and observe the meter again.

d. Any voltage reading from either test connection indicates a short in the charge coil or harness. Either, locate and repair the problem with the harness, or replace the charge coil.

2. Check the charge coil cranking output using the peak-reading voltmeter, set to the same scale. Connect the meter probes to the two brown leads from the charge coil and crank the engine. Note the meter reading, it should be at least 200 volts for 5/6 hp motors or 300 volts for 8/9.9 hp motors:

• If the cranking voltage is within spec, but ignition problems still exist, check the Power Pack output, as detailed in this section.

• If the cranking voltage is below the specification, check the wiring/connectors, then, if the wiring appears to be in good shape, check coil resistance.

3. With the engine not running or cranking, connect the probes from a DVOM set to read resistance across the two brown leads from the charge coil. The meter must show 800-1000 ohms resistance at a temperature of about 68°F (20°C) or the charge coil is out of specification.

■ An ohmmeter may also be used to check for a grounded coil using a static resistance test by checking between each coil lead and a good engine ground. Meter should show very high or infinite resistance, or the coil is grounded. Remember though that this test can be misleading because the coil might short to ground intermittently only under engine operation.

4. Verify test results and connections, then replace the power coil, if the reading is well out of specification.

5. Reconnect all wiring as removed for testing.

9.9/15 Hp (305cc) 4-Stroke Motors

The charge coil is incorporated into stator assembly on these models. Although the charge and power coil winding can be tested separately, the entire stator assembly must be replaced if either is faulty.

■ **A peak-reading DVOM is necessary for the basic ground and the dynamic (cranking) portion of this test. If one is not available, the resistance check will give some idea of coil condition, but be careful as a bad coil could test within specification but short out and fail intermittently under load. Likewise, a good coil may test out of spec due to variances in meters and test temperatures.**

1. Set a peak-reading DVOM to read positive in the 500 volts scale, then check for a grounded charge coil as follows:

a. Disengage the 4-pin connector (it's a Packard style connector) located between the stator and the power pack (it contains two brown or one brown and one brown/yellow charge coil wires).

b. Connect the black meter probe to a good engine ground and the red probe to one of the brown charge coil leads. Crank the engine and observe, then note the reading.

c. Move the red meter probe to the other brown (or brown/yellow) charge coil lead, then crank the engine and observe the meter again.

d. Any voltage reading from either test connection indicates a short in the charge coil or harness. Either, locate and repair the problem with the harness, or replace the charge coil.

2. Check the charge coil cranking output using the peak-reading voltmeter, set to the same scale. Connect the meter probes to the two brown leads from the charge coil and crank the engine. Note the meter reading, it should be at least 300 volts for 1995-96 motors or 220 volts for 1997 and later motors:

• If the cranking voltage is within spec, but ignition problems still exist, check the Power Pack output, as detailed in this section.

• If the cranking voltage is below the specification, check the wiring/connectors, then, if the wiring appears to be in good shape, check coil resistance.

3. With the engine not running or cranking, connect the probes from a DVOM set to read resistance across the two brown (or brown and brown/yellow) leads from the charge coil. The meter must show 1010-1230 ohms resistance for 1995-97 rope start models, or 720-880 ohms resistance for all other models. Remember that these specifications are for a temperature of about 68°F (20°C).

■ An ohmmeter may also be used to check for a grounded coil using a static resistance test by checking between each coil lead and a good engine ground. Meter should show very high or infinite resistance, or the coil is grounded. Remember though that this test can be misleading because the coil might short to ground intermittently only under engine operation.

4. Verify test results and connections, then replace the power coil, if the reading is well out of specification.

5. Reconnect all wiring as removed for testing.

9.9/10/14/15 Hp (216cc), 9.9/10/15 Hp (255cc) and 18 JET-35 Hp (521cc) Motors

■ Motors through 1991 and some 1992 models were equipped with the UFI version of this ignition system. Since the charge coil is connected to the Ignition Module and cranking tests would involve checking the voltage on those lines (which cannot be accessed with the flywheel installed) there is no way to perform cranking tests on those years/models. Instead you'll only be able to remove the flywheel (for access to the wiring) and conduct static ohmmeter tests.

■ **A peak-reading DVOM is necessary for the basic ground and the dynamic (cranking) portion of this test. If one is not available, the resistance check will give some idea of coil condition, but be careful as a bad coil could test within specification but short out and fail intermittently under load. Likewise, a good coil may test out of spec due to variances in meters and test temperatures.**

1. For most models (not equipped with UFI system), check the charge coil ground first as follows:

a. Set the peak reading voltmeter to read negative on the 500 volts scale for 9.9/15 hp motors or to the positive 500 volts scale for 18 jet-35 hp motors.

b. Disengage the 5-pin connector (it's an Amphenol style connector) between the power pack (ignition module) and the ignition charge coil mounting plate located under the flywheel.

c. Connect the meter red lead to terminal **A** (brown lead from the charge coil) and the black meter lead to a good engine ground, then crank the engine and observe the readings.

d. Connect the meter red lead to terminal **D** (brown/yellow lead from the charge coil) and the black meter lead to a good engine ground, then crank the engine and observe the readings.

e. Voltage present at either test indicates that the charge coil or the wiring harness is shorted to ground. Locate and repair the short or replace the charge coil assembly.

2. For most models (not equipped with UFI system), check the charge coil output as follows:

a. Now connect the red meter probe to terminal **D** (brown/yellow lead from the charge coil) and the black meter probe to terminal **A** (brown lead from the charge coil).

b. Crank the motor and observe charge coil output on the meter. If the reading is 230 volts or higher, the charge coil is performing normally, to continue troubleshooting an ignition system problem, perform the Sensor Coil testing as detailed in this section. If the reading is less than 230 volts, check the condition of the wiring and connectors. If there are no problems with the wiring and output is still below 230 volts, proceed with the resistance test in the next step (but charge coil replacement is probably necessary).

3. For models equipped with UFI system (specifically this should be all 1990-91 models and some 1992 models), access the charge coil, as follows:

a. Remove the manual starter assembly, as detailed in the Hand Rewind Starter section.

b. Remove the Flywheel.

c. Remove the 2 ignition module mounting screws, then reposition the module as necessary for access to the wiring.

d. Disconnect the brown and brown/yellow lead bullet connectors.

4. A resistance check can be used to statically check the charge coil (and the wiring harness), but keep in mind that a coil could test within spec statically, but show a fault dynamically (especially once operating temperatures are reached). Of course, this is your only option for testing the charge coil on UFI motors, so do your best. And, one small advantage of the resistance check is that not everyone has access to a peak-reading DVOM.

a. With the 5-pin connector still disengaged from the previous steps (or using the wires disconnected from the UFI module as applicable), probe across terminals **A** (brown lead) and **D** (brown/yellow lead) from the charge coil.

4-28 IGNITION AND ELECTRICAL SYSTEMS

b. Resistance will vary by year and model (and ignition type). Please refer to the Ignition Testing Specifications chart for more details. Keep in mind that specifications are for a component at about 68°F (20°C).

■ An ohmmeter may also be used to check for a grounded coil using a static resistance test by checking between each coil lead and a good engine ground. Meter should show very high or infinite resistance, or the coil is grounded. Remember though that this test can be misleading because the coil might short to ground intermittently only under engine operation.

5. Verify test results and connections, then replace the charge coil, if the reading is well out of specification.

6. Reconnect all wiring as removed for testing. On UFI models, remember the orange/blue primary lead must be connected to the No. 1 ignition coil.

25/35 Hp (500/565cc) Motors

The charge coil is incorporated into stator assembly on these models. Although the charge and power coil winding can be tested separately, the entire stator assembly must be replaced if either is faulty.

■ A peak-reading DVOM is necessary for the basic ground and the dynamic (cranking) portion of this test. If one is not available, the resistance check will give some idea of coil condition, but be careful as a bad coil could test within specification but short out and fail intermittently under load. Likewise, a good coil may test out of spec due to variances in meters and test temperatures.

1. Set a peak-reading DVOM to read positive in the 500 volts scale, then check for a grounded charge coil as follows:
   a. Disengage the 4-pin connector (it's a Packard style connector) located between the stator and the power pack (it contains two brown or one brown and one brown/yellow charge coil wires).
   b. Connect the black meter probe to a good engine ground and the red probe to one of the brown charge coil leads. Crank the engine and observe, then note the reading.
   c. Move the red meter probe to the other brown (or brown/yellow) charge coil lead, then crank the engine and observe the meter again.
   d. Any voltage reading from either test connection indicates a short in the charge coil or harness. Either, locate and repair the problem with the harness, or replace the charge coil.

2. Check the charge coil cranking output using the peak-reading voltmeter, set to the same scale. Connect the meter probes to the two brown leads from the charge coil and crank the engine. Note the meter reading, it should be at least 300 volts for motors through 1998 motors or 220 volts for 1999 and later motors:
   • If the cranking voltage is within spec, but ignition problems still exist, check the Power Pack output, as detailed in this section.
   • If the cranking voltage is below the specification, check the wiring/connectors, then, if the wiring appears to be in good shape, check coil resistance.

3. With the engine not running or cranking, connect the probes from a DVOM set to read resistance across the two brown (or brown and brown/yellow) leads from the charge coil. The meter must show 1010-1230 ohms resistance for 1995-97 rope start models, or 720-880 ohms resistance for all other models. Remember that these specifications are for a temperature of about 68°F (20°C).

■ An ohmmeter may also be used to check for a grounded coil using a static resistance test by checking between each coil lead and a good engine ground. Meter should show very high or infinite resistance, or the coil is grounded. Remember though that this test can be misleading because the coil might short to ground intermittently only under engine operation.

4. Verify test results and connections, then replace the power coil, if the reading is well out of specification.

5. Reconnect all wiring as removed for testing.

25-55 Hp (737cc) Motors

Rope start models or models with a 4-amp charging system are equipped with an ignition charge coil as well as separate battery charge and sensor/trigger coils. Each of these components can be tested and replaced separately. On models with 12-amp charging systems (including all electric start and remote models) the ignition charge coil is incorporated into stator assembly. Although the charge coil and battery charge coil windings can be tested separately, the entire stator assembly must be replaced if either is faulty.

■ The quickest way to check whether one of these models is equipped with a 4-amp or 12-amp charging system is to trace the yellow wires (for the battery charge coil or stator) from under the flywheel. If they connect (through a rectangular junction block) to a rectangular, finned electronic component (a regulator/rectifier), the motor has a 12-amp charging system. If, however they run unregulated to AC lighting, or to a small cylindrical component mounted on a triangular bracket (a rectifier) the motor contains the 4-amp charging system.

■ Motors through 1991 and some 1992-93 models were equipped with the UFI version of this ignition system. Since the charge coil is connected to the Ignition Module and cranking tests would involve checking the voltage on those lines (which cannot be accessed with the flywheel installed) there is no way to perform cranking tests on those years/models. Instead you'll only be able to remove the flywheel (for access to the wiring) and conduct static ohmmeter tests.

■ A peak-reading DVOM is necessary for the basic ground and the dynamic (cranking) portion of this test. If one is not available, the resistance check will give some idea of coil condition, but be careful as a bad coil could test within specification but short out and fail intermittently under load. Likewise, a good coil may test out of spec due to variances in meters and test temperatures.

1. For most models (not equipped with UFI system), check the charge coil ground first as follows:
   a. Set the peak reading voltmeter to read positive on the 500 volts scale.
   b. Disengage the 5-pin connector (it's an Amphenol style connector) between the power pack (ignition module) and the ignition charge coil mounting plate located under the flywheel.
   c. Connect the meter red lead to terminal **A** (brown lead from the charge coil) and the black meter lead to a good engine ground, then crank the engine and observe the readings.
   d. Connect the meter red lead to terminal **D** for rope/4-amp models or terminal **B** for 12-amp models (brown/yellow lead from the charge coil) and the black meter lead to a good engine ground, then crank the engine and observe the readings.
   e. Voltage present at either test indicates that the charge coil or the wiring harness is shorted to ground. Locate and repair the short or replace the charge coil assembly.

2. For most models (not equipped with UFI system), check the charge coil output as follows:
   a. Now connect the red meter probe to terminal **D** for rope/4-amp models or terminal **B** for 12-amp models (brown/yellow lead from the charge coil) and the black meter probe to terminal **A** (brown lead from the charge coil).
   b. Crank the motor and observe charge coil output on the meter. If the reading is 230 volts or higher, the charge coil is performing normally, to continue troubleshooting an ignition system problem, perform the Sensor Coil testing as detailed in this section. If the reading is less than 230 volts, check the condition of the wiring and connectors. If there are no problems with the wiring and output is still below 230 volts, proceed with the resistance test in the next step (but charge coil replacement is probably necessary).

3. For models equipped with UFI system (specifically this should be all models through 1991 and some 1992-93 models), access the charge coil, as follows:
   a. Remove the manual starter assembly, as detailed in the Hand Rewind Starter section.
   b. Remove the Flywheel.
   c. Remove the 2 ignition module mounting screws, then reposition the module as necessary for access to the wiring.
   d. Disconnect the brown and brown/yellow lead bullet connectors.

4. A resistance check can be used to statically check the charge coil (and the wiring harness), but keep in mind that a coil could test within spec statically, but show a fault dynamically (especially once operating temperatures are reached). Of course, this is your only option for testing the charge coil on UFI motors, so do your best. And, one small advantage of the resistance check is that not everyone has access to a peak-reading DVOM.

IGNITION AND ELECTRICAL SYSTEMS 4-29

a. With the 5-pin connector still disengaged from the previous steps (if applicable), probe across terminals **A** (brown lead) and **D** for rope/4-amp models or terminal **B** for 12-amp models (brown/yellow lead) from the charge coil. On UFI models the terminal references are not applicable, but the wire colors should be the same.

b. Resistance should be 800-1000 ohms for rope start/4-amp models or 750-950 ohms for 12-amp models. Keep in mind that specifications are for a component at about 68°F (20°C).

■ **An ohmmeter may also be used to check for a grounded coil using a static resistance test by checking between each coil lead and a good engine ground. Meter should show very high or infinite resistance, or the coil is grounded. Remember though that this test can be misleading because the coil might short to ground intermittently only under engine operation.**

5. Verify test results and connections, then replace the charge coil, if the reading is well out of specification.

6. Reconnect all wiring as removed for testing. On UFI models, remember the orange/blue primary lead must be connected to the No. 1 ignition coil.

25-70 Hp (913cc) Motors

On these models the ignition charge and power coils are incorporated into 1-piece stator assembly. Although the charge coil and battery charge coil windings can be tested separately, the entire stator assembly must be replaced if either is faulty.

When testing the charge coil, some specifications vary between models equipped with 6-amp charging systems and those equipped with 12-amp systems. Most models are equipped with a 12-amp system, except certain commercial models, usually including the 65RS and 65WR, but this might also include some 65WE, 65WM, 65WMLW and the 65WMYW. The easiest way to tell the difference between a 12-amp and 6-amp system is that the 6-amp system is **not** regulated. If a finned rectangular box is attached to the powerhead and connected to the yellow battery charge coil wiring coming out from the stator assembly (after connecting to a terminal strip), the motor is equipped with a 12-amp, regulated charging system.

■ **A peak-reading DVOM is necessary for the basic ground and the dynamic (cranking) portion of this test. If one is not available, the resistance check will give some idea of coil condition, but be careful as a bad coil could test within specification but short out and fail intermittently under load. Likewise, a good coil may test out of spec due to variances in meters and test temperatures.**

1. Check the charge coil ground first as follows:
   a. Set the peak reading voltmeter to read positive on the 500 volts scale.
   b. Disengage the 5-pin connector (it's an Amphenol style connector) between the power pack (ignition module) and the stator (ignition charge coil) located under the flywheel.
   c. Connect the meter red lead to terminal **A** (brown lead from the charge coil) and the black meter lead to a good engine ground, then crank the engine and observe the readings.
   d. Connect the meter red lead to terminal **B** (brown/yellow lead from the charge coil) and the black meter lead to a good engine ground, then crank the engine and observe the readings.
   e. Voltage present at either test indicates that the charge coil or the wiring harness is shorted to ground. Locate and repair the short or replace the charge coil assembly.

2. Check the charge coil output as follows:
   a. Now connect the red meter probe to terminal **B** (brown/yellow lead from the charge coil) and the black meter probe to terminal **A** (brown lead from the charge coil).
   b. Crank the motor and observe charge coil output on the meter. If the reading is 250 volts or higher, the charge coil is performing normally, to continue troubleshooting an ignition system problem, perform the Sensor Coil testing as detailed in this section. If the reading is less than 250 volts, check the condition of the wiring and connectors. If there are no problems with the wiring and output is still below 250 volts, proceed with the resistance test in the next step (but charge coil replacement is probably necessary).

3. A resistance check can be used to statically check the charge coil (and the wiring harness), but keep in mind that a coil could test within spec statically, but show a fault dynamically (especially once operating temperatures are reached). The advantage of the resistance check is that not everyone has access to a peak-reading DVOM.

a. With the 5-pin connector still disengaged from the previous steps, probe across terminals **A** (brown lead) and **B** (brown/yellow lead) from the charge coil.

b. Resistance should be 455-505 ohms for 1990-91 models, or for later models it will vary by system output and is either 360-440 ohms for 6-amp models or 750-950 ohms for 12-amp models. Keep in mind that specifications are for a component at about 68°F (20°C).

■ **An ohmmeter may also be used to check for a grounded coil using a static resistance test by checking between each coil lead and a good engine ground. Meter should show very high or infinite resistance, or the coil is grounded. Remember though that this test can be misleading because the coil might short to ground intermittently only under engine operation.**

4. Verify test results and connections, then replace the charge coil, if the reading is well out of specification.

5. Reconnect all wiring as removed for testing.

REMOVAL & INSTALLATION

◆ See Figures 57 thru 60

The charge coil may be an individual component mounted to an ignition base plate or it may be incorporated into the windings on the 1-piece stator assembly. In all cases, the manual starter cover or flywheel cover, and the flywheel must be removed for access. Refer to the procedures found in the Manual Starter section and Powerhead section (on Flywheel) for more details.

■ **Prior to removal, make a quick sketch or take a photograph (aren't digital cameras great?) of the wire harness routing. During installation, be sure to position all wires as noted before removal so as to prevent possible interference with moving parts. Remember that contact with a component such as the flywheel could break wear through the insulation or the wires themselves. An exposed portion of wire could short damaging the coil windings or other components in the system.**

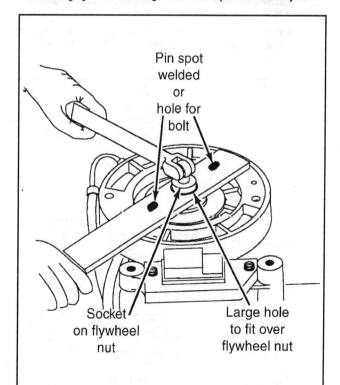

Fig. 57 The flywheel retaining bolt or nut is usually removed while holding the flywheel steady using a homemade removal tool or a strap wrench

4-30 IGNITION AND ELECTRICAL SYSTEMS

Fig. 58 If necessary, use a puller to help free the flywheel

Fig. 59 Remove the stator or individual coil retaining bolts. . .

Fig. 60 . . . then lift the coil from the mounting plate

2.0-3.5 Hp (78cc) Motors

◆ See Figure 61

1. Remove the Manual Starter and then remove the Flywheel, as detailed under the appropriate section.
2. Disengage the brown and blue wires connectors for the charge coil and the sensor/trigger coil.
3. For access to the wiring clip, remove the 2 Philip's head screws securing the ignition plate, then remove the plate from the motor.
4. Invert the ignition plate, then loosen the wiring clip screw and free the wiring from the clip.
5. Turn the ignition plate over again for access to the charge coil and/or sensor/trigger coil retaining screws.

■ The manufacturer makes a locating tool to ensure proper positioning of the charge and/or sensor/trigger coils once they have been removed from the ignition plate. If the tool is not available, make alignment marks on the plate to ensure installation of the coils in the proper positions. The best method is to trace part of the component's outline to ensure exact positioning.

6. Remove the 2 screws securing the charge coil and/or the 2 screw securing the sensor/trigger coil to the ignition plate, as desired, then remove the coil from the plate.

To install:
7. Clean the ignition plate and mounting area of dirt, corrosion or debris.
8. Position the charge coil and/or sensor coil on the ignition plate, positioning the wiring under the clip and through the braided sleeve.
9. Apply a coating of OMC Screw Lock or equivalent threadlock to the threads of the wire clip, coil and ignition plate retaining screws.
10. Make sure the wiring is properly positioned under the clip, then install and tighten the clip retaining screw to 15-19 inch lbs. (1.7-2.2 Nm).
11. If available, position an OMC Locating Tool (# 342674) using the studs and nuts that come with the tool. The **S** mark on the tool should align with the sensor and the **C** mark should align with the charge coil.

■ When using the locator tool, the mark second from the left on the sensor should align with the mark second from the left on the ignition plate.

12. Install the charge and/or sensor coils to the ignition plate, aligning with the marks made earlier, and/or the marks on the locator tool, as applicable. Install and tighten the retaining screws to 19-25 inch lbs. (2.2-2.8 Nm) for the charge coil and/or to 15-19 inch lbs. (1.7-2.2 Nm) for the sensor coil.
13. If used, remove the locator tool.
14. Install the ignition plate to the engine while routing the wiring harness, then install and tighten the retaining screws to 27-44 inch lbs. (3-5 Nm).
15. Install the flywheel, followed by the manual starter cover.

3-4 Hp (87cc) Motors

On these motors the functions of the charge coil are integrated into the power pack/ignition module that is mounted along the edge of the flywheel. Because of the internal circuitry of the coil windings, no separate provisions are provided to test or replace the charge coil. Refer to Power Pack in this section for information on testing or servicing the primary circuit of the ignition system.

5 Hp (109cc) Motors

◆ See Figure 62

1. Remove the Manual Starter and then remove the Flywheel, as detailed under the appropriate section.
2. Disengage the bullet connectors for the charge coil wiring.

■ Be sure to note routing of the charge coil wiring to ensure proper installation. Remember that wires improperly routed could contact moving parts (like the flywheel), and rub through the installation causing shorts that could destroy the coil (or just cutting the wires).

■ Make a small alignment mark before removing the coil to ensure installation with the exact proper orientation.

3. Remove the screw, washer and clamp securing the charge coil to the engine.
4. Remove the charge coil assembly.
5. Installation is essentially the reverse of the removal. Be sure to align the marks made during removal when securing the coil. Also, make sure the wiring is routed exactly as noted to prevent interference with moving components.

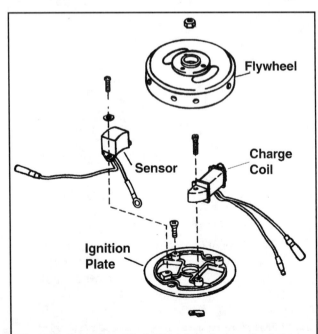

Fig. 61 Exploded view of the ignition plate, charge coil and sensor coil mounting - 2.0-3.5 Hp (78cc) Motors

IGNITION AND ELECTRICAL SYSTEMS 4-31

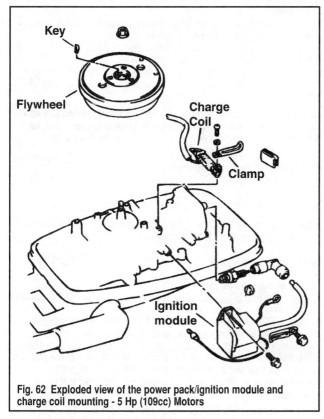

Fig. 62 Exploded view of the power pack/ignition module and charge coil mounting - 5 Hp (109cc) Motors

6. Install the flywheel, followed by the manual starter cover.

5/6 HP (128CC) AND 8/9.9 HP (211CC) 4-STROKE MOTORS

◆ See Figures 56 and 63

These motors are equipped with a combination charge/power coil and a sensor coil. Some versions are also equipped with a battery charge coil. All of the coil windings may be tested separately and, with the exception of the charge/power coil assembly, maybe replaced separately.

■ **If only the sensor/trigger coil is being serviced only the manual starter must be removed, the flywheel does not have to be removed for access.**

1. On electric start models, disconnect the negative battery cable for safety.
2. Remove the Manual Starter and then remove the Flywheel, as detailed under the appropriate section.
3. Disengage the wiring for the component being removed (orange and brown wires for the charge and power coil assembly). If the battery charge coil (yellow wires) or sensor/trigger coil (white/black) wires are being removed also, disconnect their wiring too at this time.

■ **It may be necessary to cut one or more plastic wire ties. If so, take note of their location to ensure proper harness restraint by securing the wires with a new tie during installation.**

4. Mark the positioning of the charge/power coil relative to the mounting base.
5. Remove the charge/power coil mounting screws, then lift the coil from the top of the motor, while carefully guiding the coil wires.
6. If the sensor/trigger coil and/or, the battery charge coil (if equipped) must be removed, mark their alignment, then loosen and remove the screws in the same manner as directed for the charge/power coil. Remove the coil from the motor.

To install:

7. Clean the coil mounting location(s), as well as the threads of the mounting screws of dirt corrosion or debris.
8. Position the coil on the powerhead, carefully aligning the marks made earlier.

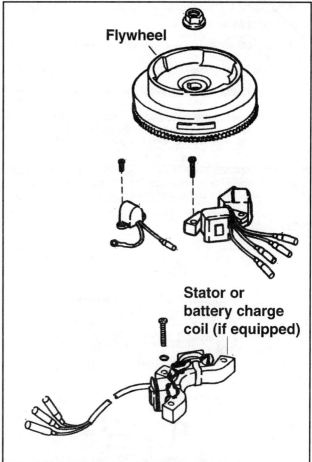

Fig. 63 Exploded view of the ignition charge/power coil, sensor coil and, battery charge coil (if equipped) mounting - 5-9.9 Hp (128-211cc) 4-stroke motors

9. Apply a coat of OMC Nut Lock or equivalent threadlock to the coil retaining screws, then install and finger-tighten the screws.
10. If the OMC Stator Alignment Guide Kit (# 342670) and the Locating Ring (# 334994) are available, use them to check coil alignment. Place the bridge on the locating ring and secure using the set-screws on either end of the bridge. Install the correct end of the guide bushing (from the locating kit) over the tapered end of the crankshaft (aligning the keyway), then place the bridge and ring over the bushing. Push on the power/charge or battery charge coil until it lightly contacts the edge of the locating ring.
11. Once the power/charge or battery charge coil is aligned perfectly with the alignment marks made before removal and/or the alignment ring from the OMC kit, tighten the retaining screws to 36-60 inch lbs. (4-6 Nm).
12. If the sensor/trigger coil was removed, install the flywheel, then install the sensor coil and adjust the air gap as follows:
    a. Apply a coating of OMC Nut Lock or an equivalent threadlock to the sensor/trigger coil mounting screw threads.
    b. Position the sensor to the mounting boss and finger-tighten the screws.
    c. Slowly turn the flywheel clockwise by hand until the raised sensor plate (that protrudes from the side of the flywheel) is facing the sensor.

■ **The proper air gap for the sensor/trigger coil is 0.009-0.011 in. (0.23-0.28mm) for 5/6 hp motors or to 0.03-0.05 in. (0.8-1.2mm) for 8/9.9 hp motors.**

    d. Hold a feeler gauge (of about 0.008 in. in size for 5/6 hp motors or 0.04 in. in size for 8/9.9 motors) between the sensor and flywheel sensor plate. Push the sensor lightly against the gauge and tighten the mounting screws to 36-60 inch lbs. (4-6 Nm), then recheck the gap to ensure it did not change while tightening the fasteners.
13. Install the manual starter cover, and if equipped, connect the negative battery cable.

4-32 IGNITION AND ELECTRICAL SYSTEMS

9.9/15 Hp (305cc) 4-Stroke Motors

◆ See Figures 64 and 65

The charge coil windings in these models are incorporated into the stator coil assembly.

1. On electric start models, disconnect the negative battery cable for safety.
2. Remove the Manual Starter, as detailed in the Manual Starter section.
3. Remove the Flywheel, as detailed in the Powerhead section.
4. Disengage the wiring connector from the ignition module.
5. If equipped, disengage the wiring connector for the rectifier or the regulator/rectifier.

■ It may be necessary to cut one or more plastic wire ties. If so, take note of their location to ensure proper harness restraint by securing the wires with a new tie during installation.

6. Remove the stator retaining screws, then remove the stator coil from the powerhead.

To install:

7. Clean the coil mounting location(s), as well as the threads of the mounting screws of dirt corrosion or debris.
8. Position the stator coil on the powerhead, and align the coil with the 3 holes in the powerhead for the mounting screws.
9. Apply a coat of OMC Nut Lock or equivalent threadlock to the coil retaining screws, then install and tighten the screws to 60-84 inch lbs. (7-9 Nm).

■ Make sure the wiring is routed properly to avoid interference with moving components. If any wire ties were present and removed, be sure to replace them with new wire ties in the same position(s).

10. Connect the stator wiring to the ignition module, and, if equipped, the rectifier or regulator/rectifier assembly.
11. Install the flywheel.
12. Install the manual starter cover.
13. If equipped, connect the negative battery cable.

4 Deluxe (87cc), 5/6/8 Hp (164cc), 9.9/10/14/15 Hp (216cc), 9.9/10/15 Hp (255cc) and 18 Jet-35 Hp (521cc) Motors

◆ See Figures 66 thru 75   **OEM**

Most of these models (both CDI and UFI versions) are equipped with ignition and charging system components that are mounted independently to the ignition plate. On these models, the charge coil, sensor coil (or Ignition Module on UFI) and, if equipped, battery charge coil may be removed and replaced individually. But, on some 9.9-15 hp motors (usually electric start models only) the charge coil windings (as well as the windings for the sensor and battery charge coils) are incorporated into the stator coil assembly. Although the functions of the individual coil windings may be tested separately on these models, a fault with one will require replacement of the entire stator.

■ Some early-models through about 1992 or 1993 MAY be equipped with the Under Flywheel Ignition (UFI) where the ignition module is mounted on the ignition plate next to the charge coil and/or stator coil. For these models, Power Pack (Ignition Module) replacement is the same basic procedure as Coil replacement, and as such is also covered in this procedure.

In most cases, a wire cover is mounted to the underside of the ignition plate that will prevent the safe removal of the wiring harness for the component being serviced, unless the ignition plate itself is unbolted and removed from the top of the powerhead. In some cases, you will be able to sneak the wiring through the cover, but it is usually lot easier to simply remove the screws and lift the ignition plate from the powerhead to ease removal and installation.

The OMC Locating Ring (# 334994) should be used to ensure proper component alignment whenever installing a charge, sensor or battery charge coil assembly to the ignition plate. If this tool is not available, it is possible to make matchmarks on the plate to help align the new component. Be sure to trace as much of the old component's outline as possible, to ensure the best possible placement. If the new component does not function properly after installation, check alignment using the proper locating ring.

■ There is another old trick to get around the use of the OMC Locating Ring. Place the coil or module on the ignition plate and just barely tighten the mounting screws with the outside edge of the module/coil flush with the machined surface of the stator plate. Apply about three layers of masking tape to achieve about a total thickness of 0.025 in. (0.64mm) of tape across the flywheel magnets. Slip the flywheel down over the crankshaft and rotate the flywheel slowly CLOCKWISE, then remove the flywheel and inspect the tape. If the magnets have dug into the masking tape, the module must be moved inward slightly. If there is absolutely no sign of the magnets making even the slightest contact with the tape, the module must be moved outward just a few thousands of an inch (mm). Once some evidence of the tape making the smallest amount of contact with the protrusion on the module is achieved, consider the module adjusted and secure the mounting screws securely to prevent shifting during powerhead operation. Make a final check to ensure the module did not shift during final tightening of the screws.

1. On electric start models, disconnect the negative battery cable for safety.
2. Remove the Manual Starter, as detailed in the Manual Starter section.

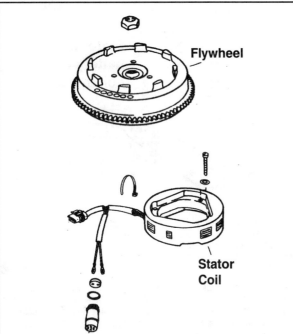

Fig. 64 The stator coil includes the charge coil, power coil and, on electric start models, the battery charge coil windings - 9.9/15 Hp (305cc) motors

Fig. 65 Disengage the stator coil wiring from the ignition module (1) and, if equipped, the regulator/rectifier (2)

IGNITION AND ELECTRICAL S...

3. Remove the Flywheel, as detailed in the Powerhead section.

4. Identify the coil (or UFI ignition module) that is being removed and, especially if a locating ring tool is not available, trace the outline of the coil on the ignition plate to help ensure proper alignment during installation.

5. Loosen the screws retaining the coil (charge, sensor, battery charge or stator) that is being removed. On some UFI assemblies it will be easier to loosen and swing another component out of the way to remove the desired component (for example when removing the stator or charge coils it is normally easier to remove one ignition module retainer and loosen the other, then swing the module out of the way).

6. Loosen the retaining screws, then remove the wire clamp/retainer plate from the top of the ignition plate.

■ It is not necessary to remove the ignition plate in all instances. However for certain oil replacements on CDI units or for stator coil replacement on UFI units you normally need to remove the plate for access to the wiring.

7. Loosen the ignition plate retaining screws (usually 4 bolts on UFI or 5 or 6 on CDI), then carefully lift the ignition plate from the powerhead and invert it for access to the wires or, if equipped, the wire cover.

8. If equipped, loosen the screw, then lift the wire cover from the underside of the ignition plate.

■ It may be necessary to cut one or more plastic wire ties when removing components like a stator or battery charge coil. If so, take note of their location to ensure proper harness restraint by securing the wires with a new tie during installation.

9. Disengage the wiring connector(s) for ... removed. For most charge and sensor coils, i... terminals are connected from the coil assembl... Disconnect these leads and remove them from ... charge coils or stators, the wires may be route... individual bullet connectors or attached to ring-t... component and model. Although the wire colors ... usually as follows:
- Charge coil: brown and brown/yellow
- Sensor coil: white or white/black, and black/white
- Battery charge coil: yellow, yellow blue and yellow gray
- Stator: all of the above

■ If necessary, refer to the schematics found in the Wiring Diagram section to help identify the wiring for the component being removed. Also, refer to the testing information in this section to help identify wire colors.

10. Remove the coil from the ignition plate and the engine.

To install:

11. Clean the coil mounting location(s), as well as the threads of the mounting screws of dirt corrosion or debris. Coat the threads of the coil retaining screws with OMC Nut Lock, or equivalent threadlocking compound.

12. Loosely install the new coil on the ignition plate to hold it in position when installing the leads.

13. Route the leads as noted during removal and reconnect them to the blade, wiring connectors or ring-terminals, as applicable. If the leads were removed from the spiral wrap, reposition them as noted during removal.

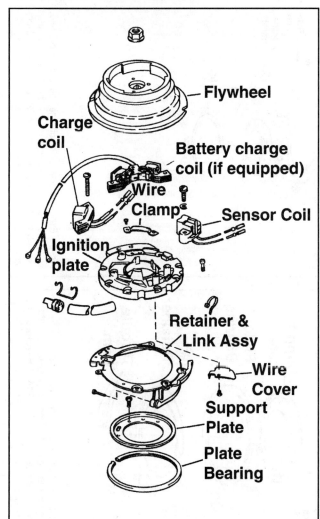

Fig. 66 Exploded view of the charge coil, sensor coil and, if equipped, battery charge coil mounted to the ignition plate assembly - CDI equipped 4 Deluxe (87cc) and 5/6/8 Hp (164cc) motors

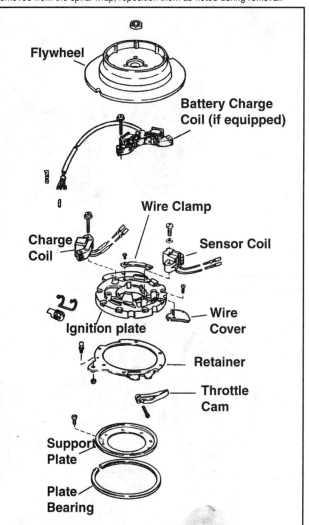

Fig. 67 Exploded view of the charge coil, sensor coil and, if equipped, battery charge coil mounted to the ignition plate assembly - CDI equipped rope start 9.9/10/14/15 Hp (216cc) and 9.9/10/15 Hp (255cc) motors

IGNITION AND ELECTRICAL SYSTEMS

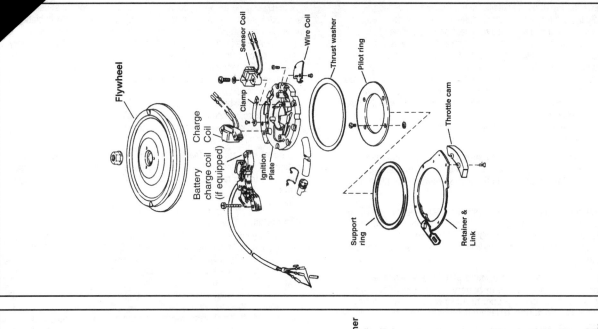

Fig. 70 Exploded view of the charge coil, sensor coil and, if equipped, battery charge coil mounted to the ignition plate assembly - CDI equipped late model 20-30 Hp (521cc) motors

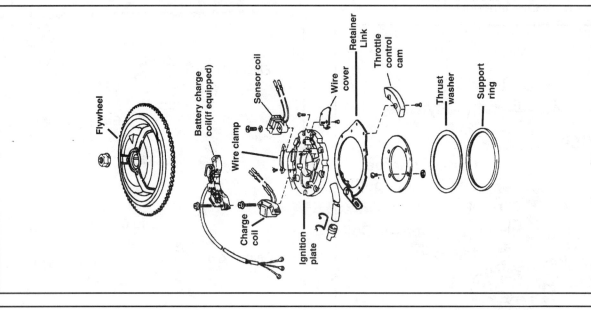

Fig. 69 Exploded view of the charge coil, sensor coil and, if equipped, battery charge coil mounted to the ignition plate assembly - CDI equipped early model 18 Jet-35 Hp (521cc) motors

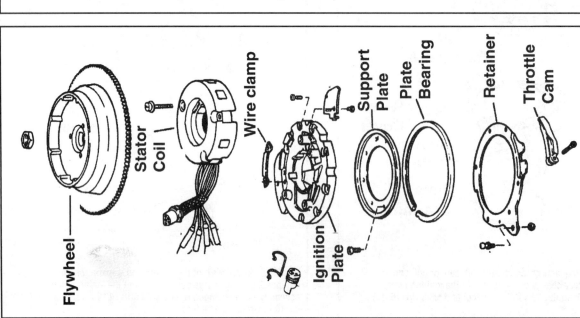

Fig. 68 Exploded view of the stator coil mounting (incorporating the functions of the charge, sensor and battery coils) - CDI equipped electric start 9.9/10/14/15 Hp (216cc) and 9.9/10/15 Hp (255cc) motors

IGNITION AND ELECTRICAL SYSTEMS 4-35

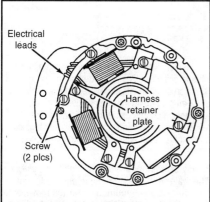

Fig. 71 Most ignition plates use a wiring clamp/retainer plate - UFI shown, CDI similar

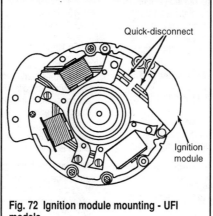

Fig. 72 Ignition module mounting - UFI models

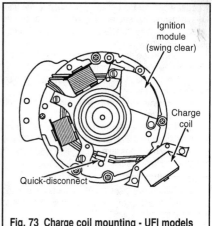

Fig. 73 Charge coil mounting - UFI models

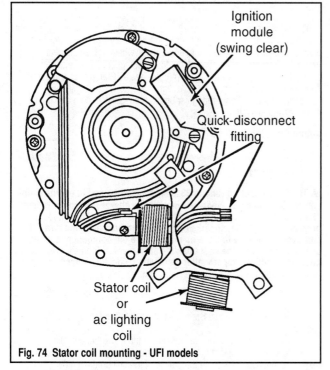

Fig. 74 Stator coil mounting - UFI models

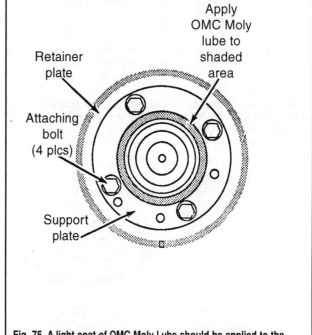

Fig. 75 A light coat of OMC Moly Lube should be applied to the friction surface of the ignition plate - UFI shown, CDI similar

14. Make sure that all wiring is properly routed. If equipped make sure the wires are run under the cover, then secure the cover to the underside of the ignition plate, using the retaining screw.

15. Make sure the wiring is properly routed under the clamp/retainer plate, then secure the clamp to the top of the ignition plate.

■ On most models there is a friction surface at the crankcase boss area toward the center of the ignition plate which should be lightly coated with OMC Moly Lube before reinstallation. Also some models contain a groove for the bearing assembly which should be lightly coated with the same lube.

16. Position the ignition plate to the powerhead while aligning the 4-6 mounting bolt holes (as applicable). Make sure there is sufficient slack on the ignition/charging system component leads to prevent damage, but not such excessive movement as to cause damage.

17. Coat the ignition plate retaining screw threads using OMC Nut Lock or an equivalent threadlocking compound, then install and tighten the screws to 25-35 inch lbs. (2.8-4.0 Nm).

18. Make sure the new coil is aligned to the marks traced prior to removal. On all components except a 1-piece stator coil assembly, if an OMC Locating Ring (# 34994) is available, position the ring over the ignition plate and push the coil against it to ensure proper positioning. When the coil is properly aligned with the locator tool or pre-removal alignment marks, tighten the mounting screws to 15-22 inch lbs. (1.6-2.4 Nm).

■ Double check the positioning of all wires that were disturbed during service. If a battery charge coil or stator coil was replaced, make sure the wires are routed properly and, any wire ties that were removed, are replaced with new ties.

19. Install the flywheel.
20. Install the manual starter cover.
21. If equipped, connect the negative battery cable.

25/35 Hp (500/565cc) Motors

◆ See Figure 76

These models are equipped with a 1-piece stator assembly that incorporates the functions of the charge coil, power coil and, if equipped, the battery charge coil windings. Although the functions of the individual coil windings may be tested separately on these models, a fault with one will require replacement of the entire stator.

A separate optical sensor assembly is mounted underneath the stator, while an encoder ring for the sensor is attached to the flywheel.

1. On electric start models, disconnect the negative battery cable for safety.
2. Remove the Manual Starter, as detailed in the Manual Starter section.
3. Remove the Flywheel, as detailed in the Powerhead section. The

4-36 IGNITION AND ELECTRICAL SYSTEMS

encoder wheel for the sensor is attached to the underside of the flywheel. Always inspect the encoder for dirt, debris or damage. If replacement is necessary, be sure to note encoder orientation before removal and match it upon installation.

4. Disengage the stator wiring connector and, if necessary, the sensor wiring connector from the ignition module.

5. On all models except rope start motors, disengage the stator wiring connector from the regulator/rectifier wiring.

6. Loosen the 3 screws retaining the stator coil to the powerhead, then remove the stator assembly.

7. If the optical sensor must be replaced, disconnect the ball socket (if available, the OMC Ball Socket Remover (#342226) will accomplish this easily and without possible damage). Loosen the 3 sensor retaining screws and then remove the sensor from the powerhead.

To install:

8. Clean the sensor and/or stator coil mounting location(s), as applicable, of dirt, corrosion or debris. Remember to clean the threads of the mounting screws as well.

9. If the sensor assembly was removed, position it to the powerhead. Coat the sensor retaining screw threads using OMC Licquic Primer, followed by OMC nut lock (or equivalent threadlocking compound), then install the screws and tighten to 24-36 inch lbs. (2.7-4.1 Nm). Connect the sensor ball link (OMC Ball Socket Installer {#342225} will make this job easier).

10. Place the stator coil in position on the powerhead, aligning the mounting holes, then install the retaining screws and tighten securely.

11. Route the wiring as noted prior to removal, then engage the connectors for the sensor and stator to the ignition module, and if equipped, connect the stator-to-regulator/rectifier connector.

■ Double check the positioning of all wires that were disturbed during service. Make sure the wires are routed properly and, any wire ties that were removed, are replaced with new ties.

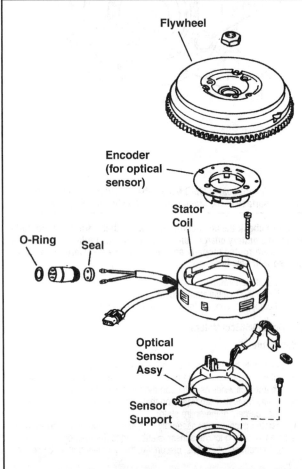

Fig. 76 Exploded view of the ignition/charging system stator and optical sensor/encoder assembly mounting - 25/35 Hp (500/565cc) motors

12. Install the flywheel.
13. Install the manual starter cover.
14. If equipped, connect the negative battery cable.

25-55 Hp (737cc) Motors

CDI Rope start models, or models with a 4-amp charging system, are equipped with an ignition charge coil as well as separate battery charge and sensor/trigger coils. Each of these components can be tested and replaced separately. On models with 12-amp charging systems (including all electric start and remote models) the ignition charge coil is incorporated into stator assembly. Although the charge, power and battery charge coil windings can each be tested separately, the entire stator assembly must be replaced if either is faulty.

■ Some 1990-93 models may be equipped with a UFI ignition system as opposed to the CDI. The major difference on these models is that the Power Pack (Ignition Module) is also mounted under the flywheel in proximity to react to spinning flywheel magnets. To service the coils or ignition module on these models please refer to the procedures in this section for the Charge Coil on the 4 Deluxe (87cc), 5/6/8 Hp (164cc), 9.9/10/14/15 Hp (216cc), 9.9/10/15 Hp (255cc) and 18 Jet-35 Hp (521cc) Motors.

The quickest way to check whether one of these models is equipped with a 4-amp or 12-amp charging system is to trace the yellow wires (for the battery charge coil or stator) from under the flywheel. If they connect (through a rectangular junction block) to a rectangular, finned electronic component (a regulator/rectifier), the motor has a 12-amp charging system. If however, they run unregulated to AC lighting, or to a small cylindrical component mounted on a triangular bracket (a rectifier), the motor contains a 4-amp charging system.

4-Amp Charging Systems

◆ See Figure 77

■ The OMC Locating Ring (# 334994) should be used to ensure proper component alignment whenever installing a charge, sensor or battery charge coil to the ignition plate. If this tool is not available, it is possible to make matchmarks on the plate to help align the new component. Be sure to trace as much of the old component's outline as possible, to ensure the best possible placement. If the new component does not function properly after installation, check alignment using the proper locating ring.

1. On electric start models, disconnect the negative battery cable for safety.
2. Remove the Manual Starter, as detailed in the Manual Starter section.
3. Remove the Flywheel, as detailed in the Powerhead section.
4. Identify the coil that is being removed and, especially if a locating ring tool is not available, trace the outline of the coil on the ignition plate to help ensure proper alignment during installation.
5. Loosen the screws retaining the coil (charge, sensor or battery charge) that is being removed.
6. Loosen the retaining screws, then remove the wire clamp from the top of the ignition plate.
7. Loosen the ignition plate retaining screws (usually 5), then carefully lift the ignition plate from the powerhead and invert it for access to the wire cover.
8. Loosen the screw, then lift the wire cover from the underside of the ignition plate.

■ It may be necessary to cut one or more plastic wire ties when removing components a battery charge coil. If so, take note of their location to ensure proper harness restraint by securing the wires with a new tie during installation.

9. Disengage the wiring connector(s) for the component(s) being removed. For most charge and sensor coils, individual leads with blade terminals are connected from the coil assembly to the wiring harness. Disconnect these leads and remove them from the spiral wrap. On battery charge coils, the wires may be routed into individual bullet connectors or attached to ring-terminals depending on the model. Although the wire colors may vary slightly, they are usually as follows:
- Charge coil: brown and brown/yellow
- Sensor coil: white/black, and black/white

IGNITION AND ELECTRICAL SYSTEMS 4-37

- Battery charge coil: yellow, yellow blue and yellow gray

■ **If necessary, refer to the schematics found in the Wiring Diagram section to help identify the wiring for the component being removed. Also, refer to the testing information in this section to help identify wire colors.**

10. Remove the coil from the ignition plate and the engine.

To install:

11. Clean the coil mounting location(s), as well as the threads of the mounting screws of dirt corrosion or debris. Coat the threads of the coil retaining screws with OMC Nut Lock, or equivalent threadlocking compound.

12. Loosely install the new coil on the ignition plate to hold it in position when installing the leads.

13. Route the leads as noted during removal and reconnect them to the blade, wiring connectors or ring-terminals, as applicable. If the leads were removed from the spiral wrap, reposition them as noted during removal.

14. Make sure that all wiring is routed under the cover and then secure the cover to the underside of the ignition plate, using the retaining screw.

15. Make sure the wiring is properly routed under the clamp and then secure the clamp to the top of the ignition plate.

16. Position the ignition plate to the powerhead while aligning the 5 mounting bolt holes. Make sure there is sufficient slack on the ignition/charging system component leads to prevent damage, but not such excessive movement as to cause damage.

17. Coat the ignition plate retaining screw threads using OMC Nut Lock or an equivalent threadlocking compound, then install and tighten the screws to 25-35 inch lbs. (2.8-4.0 Nm).

18. Make sure the new coil is aligned to the marks traced prior to removal. If an OMC Locating Ring # 34994 is available, position the ring over the ignition plate and push the coil against it to ensure proper positioning. When the coil is properly aligned, tighten the mounting screws to 15-22 inch lbs. (1.6-2.4 Nm).

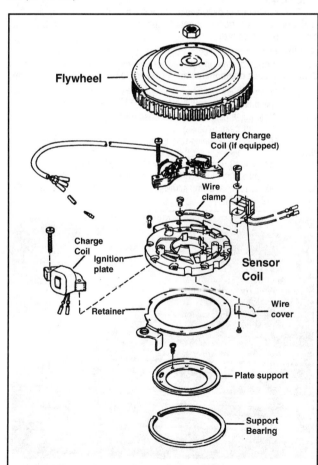

Fig. 77 Models equipped with the 4-amp charging system have a separately serviceable charge, sensor, and if equipped, battery charge/AC lighting coil - 25-55 Hp (737cc) motors

■ **Double check the positioning of all wires that were disturbed during service. If a battery charge coil or stator coil was replaced, make sure the wires are routed properly and, any wire ties that were removed, are replaced with new ties.**

19. Install the flywheel.
20. Install the manual starter cover.
21. If equipped, connect the negative battery cable.

12-Amp Charging Systems

◆ See Figures 78, 79 and 80

These models are equipped with a 1-piece stator assembly that incorporates the functions of the charge, power and battery charge coil windings. Although the functions of the individual coil windings may be tested separately, a fault with one will require replacement of the entire stator.

The timer base mounted inside and slightly underneath the stator assembly, performs the functions of the sensor/trigger coil.

1. On electric start models, disconnect the negative battery cable for safety.
2. Remove the Manual Starter, as detailed in the Manual Starter section.
3. Remove the Flywheel, as detailed in the Powerhead section.
4. Disconnect the yellow and yellow/gray ring stator ring terminals from the terminal board on the powerhead.
5. Disengage the 5-pin Amphenol style connector in the wiring that runs between the stator and power pack. Use a small pick or wire terminal tool to release the terminal locktab and then carefully pull the black/yellow wire (for the stop circuit) free from terminal **E** of the connector (on the stator side).
6. Loosen the 3 screws retaining the stator coil to the stator mounting plate on the powerhead, then remove the stator assembly.
7. If the timer base must be replaced, disengage the 3-pin Amphenol style connector for the sensor wiring, then remove the 3 timer base retaining clamp screws and separate the base from the adapter. If necessary, remove the timer base retainer (noting the orientation of the tabs in relation to the timer base lever for installation purposes), loosen the 3 stator mounting plate retaining screws and remove the mounting plate from the powerhead.

To install:

8. Clean the stator mounting plate, timer base and/or stator coil mounting location(s), as applicable, of dirt, corrosion or debris. Remember to clean the threads of the mounting screws as well.

■ **If the timer base was removed, check the bushing in the base, as well as the boss for dirt, metal chips or scratches. Clean or replace components as necessary to ensure proper operation.**

9. If the stator mounting plate was removed, align it with the threaded holes in the powerhead, making sure the slotted portion of the plate faces the starboard side of the motor. Install the plate and tighten the mounting screws to 48-60 inch lbs. (5.5-7 Nm).

10. If the timer base was removed install it as follows:

    a. Apply a light coating of OMC Moly Lube or equivalent assembly lubricant on the powerhead boss, timer base bushing, the groove in the timer base and the timer base retaining ring.

    b. Install the retainer into the timer base groove so that the tabs on the retainer are to the right of the lever (as noted during removal).

    c. Install the timer base, positioning the 3 retainer clamps over the mounting holes.

    d. Apply a coating of OMC Nut Lock or equivalent threadlocking compound to the retainer clamp screws, then install and tighten them to 25-35 inch lbs. (2.8-4 Nm).

    e. Reconnect the 3-pin Amphenol wiring connector for the timer base.

11. Place the stator coil in position on the powerhead mounting plate and around the timer base, then route the wiring as noted during removal.

12. Apply a coating of OMC Nut Lock or equivalent threadlocking compound to the threads of the stator mounting screws, then install the screws and tighten to 120-144 inch lbs. (14-16 Nm).

13. Secure the yellow and yellow/gray ring terminal connectors to the proper spots on the terminal bar.

14. Install the black/yellow wire into the 5-pin Amphenol connector for the stator wiring, then engage the connector half with the half from the power pack.

■ **Double check the positioning of all wires that were disturbed during service. Make sure the wires are routed properly and, any wire ties that were removed, are replaced with new ties.**

4-38 IGNITION AND ELECTRICAL SYSTEMS

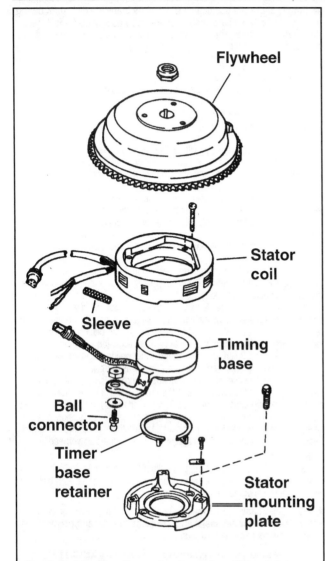

Fig. 78 Models with a 12-amp charging system utilize a 1-piece stator that contains windings for the charge, power and battery charge coils - 25-55 Hp (737cc) motors

15. Install the flywheel.
16. Install the manual starter cover.
17. If equipped, connect the negative battery cable.

25-70 Hp (913cc) Motors

◆ See Figures 79 and 81

These models are equipped with a 1-piece stator assembly that incorporates the functions of the charge, power and battery charge coil windings. Although the functions of the individual coil windings may be tested separately, a fault with one will require replacement of the entire stator.

The timer base mounted inside and slightly underneath the stator assembly, performs the functions of the sensor/trigger coil.

1. On electric start models, disconnect the negative battery cable for safety.
2. Remove the Manual Starter, as detailed in the Manual Starter section.
3. Remove the Flywheel, as detailed in the Powerhead section.
4. Disconnect the yellow and yellow/gray ring stator ring terminals from the terminal board on the powerhead.
5. Disengage the 5-pin Amphenol style connector in the wiring that runs between the stator and power pack (it contains brown and orange wires). Use a small pick or wire terminal tool to release the terminal locktab and then carefully pull the black/yellow wire (for the stop circuit) free from terminal **E** of the connector (on the stator side).
6. Loosen the 3 screws retaining the stator coil to the stator mounting plate on the powerhead, then remove the stator assembly.
7. If the timer base must be replaced, disengage the 4- or 5-pin Amphenol style connector for the sensor wiring (it contains blue, green, purple and white wires), then remove the 3 timer base retaining clamp screws and separate the base from the adapter. If necessary, remove the timer base retainer.

To install:

8. Clean the timer base and/or stator coil mounting location(s), as applicable, of dirt, corrosion or debris. Remember to clean the threads of the mounting screws as well.

■ **If the timer base was removed, check the bushing in the base, as well as the boss on the powerhead for dirt, metal chips or scratches. Clean or replace components as necessary to ensure proper operation.**

9. If the timer base was removed install it as follows:
   a. Apply a light coating of OMC Moly Lube or equivalent assembly lubricant on the powerhead boss, timer base bushing, the groove in the timer base and the timer base retaining ring.
   b. Compress the retaining ring lightly and install the timer base assembly.
   c. Position the 3 retainer clamps over the mounting holes.
   d. Apply a coating of OMC Nut Lock or equivalent threadlocking compound to the retainer clamp screws, then install and tighten them to 25-35 inch lbs. (2.8-4 Nm).

Fig. 79 If necessary, remove the timer base (sensor/trigger coil) retaining clamp screws...

Fig. 80 ...then lift the timer base from the powerhead

IGNITION AND ELECTRICAL SYSTEMS

e. Reconnect the 4 or 5-pin Amphenol wiring connector for the timer base.
10. Place the stator coil in position on the powerhead and around the timer base, then route the wiring as noted during removal.
11. Apply a coating of OMC Nut Lock or equivalent threadlocking compound to the threads of the stator mounting screws, then install the screws and tighten to 120-144 inch lbs. (14-16 Nm).
12. Secure the yellow and yellow/gray ring terminal connectors to the proper spots on the terminal bar.
13. Install the black/yellow wire into the 5-pin Amphenol connector for the stator wiring, then engage the connector half with the half from the power pack.

■ Double check the positioning of all wires that were disturbed during service. Make sure the wires are routed properly and, any wire ties that were removed, are replaced with new ties.

14. Install the flywheel.
15. Install the manual starter cover.
16. If equipped, connect the negative battery cable.

Power Coil

Many carbureted Johnson/Evinrude motors are equipped with an ignition system power coil that is used to generate voltage for various ignition system functions. On most motors equipped with a power coil voltage generated from the coil is provided to the power pack (ignition module) in order to operate the Speed Limiting Operator Warning (S.L.O.W.) system. But a lack of a power coil does not mean the motor is not equipped with the S.L.O.W. system, as power for the system can be provided from other sources. For instance, the 9.9/15 hp (255cc) and 18 Jet-35 hp (521cc) 2-strokes are not equipped with a power coil, but are often equipped with the S.L.O.W. system. Power for system operation comes from other sources on these models.

Like the ignition charge coil, the power coil is mounted directly under the flywheel to interact with one or more permanent magnets attached to the underside of the flywheel assembly. As the flywheel rotates, the magnetic force cuts through the coil assembly, generating an electric current that is utilized by the ignition system.

When equipped, the power coil is mounted either to the semi-circular bracket containing the charge coil or it is integrated into the round, one-piece stator assembly that powers the charging system as well.

Refer to the wiring diagrams provided in this section to help determine the system on your engine. Each diagram shows whether a round one-piece stator or a semi-circular charge/power coil assembly is used. Also, refer to the Ignition Testing Specifications chart in this section. A lack of testing specs for a power coil on your motor indicates that it is not used.

■ EFI motors are always equipped with a one-piece stator assembly.

TESTING

◆ See Figures 82, 83 and 84

Power coil testing typically encompasses making sure the wiring is in good condition, then verifying coil output and/or checking resistance across the coil windings.

Problems with the ignition power coil can result in a no-spark condition, but can also account for hard starting and/or an ignition misfire. The most reliable tests for the power coil are to dynamically check the output with the engine cranking (or in some cases, running). Of course, dynamic tests require a digital multi-meter capable of reading and displaying peak voltage values (also known as a peak-reading voltmeter). If one is not available, specifications are usually available to statically check the coil winding using

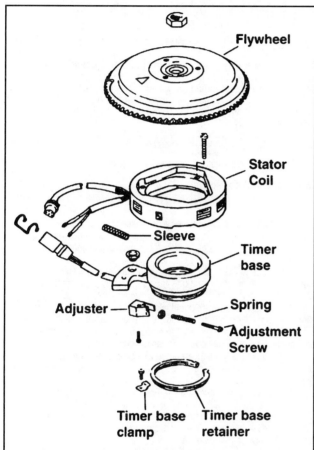

Fig. 81 Exploded view of a typical stator coil and timer base mounting - 25-70 Hp (913cc) motor

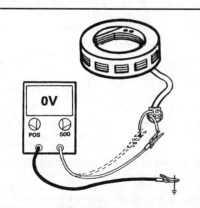

Fig. 82 Power coil windings are tested in the same manner as other coil windings, using a voltmeter and/or ohmmeter to check for shorted windings...

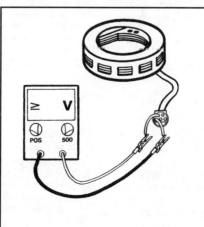

Fig. 83 ...using a voltmeter to check cranking power output (stator shown)

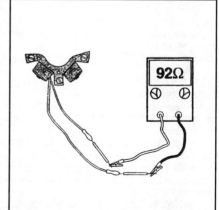

Fig. 84 ...and using an ohmmeter to check resistance (combination charge/power coil shown)

4-40 IGNITION AND ELECTRICAL SYSTEMS

an ohmmeter. But, before replacing the coil based only on static test that shows borderline readings, remember that resistance readings will vary with temperature and the specifications provided here are based on a component temperature of about 68°F (20°C). Also, remember that a coil may test within specification statically, but show intermittent faults under engine operating conditions, especially at normal operating temperature.

5/6 Hp (128cc) and 8/9.9 Hp (211cc) 4-Stroke Motors

◆ See Figure 56

These motors are equipped with a combination charge/power coil and a sensor coil. Some versions are also equipped with a battery charge coil. All of the coil windings may be tested separately and, with the exception of the charge/power coil assembly, maybe replaced separately.

1. Disengage the 2 bullet connectors for the orange wires (one may be orange/black) coming from the power coil (underneath the flywheel) and running to the power pack/ignition module.

2. Set a peak-reading DVOM to read positive in the 500 volts scale, then check for a grounded power coil as follows:
   a. Connect the black meter probe to a good engine ground and the red probe to one of the orange power coil leads. Crank the engine and observe, then note the reading.
   b. Move the red meter probe to the other orange power coil lead, then crank the engine and observe the meter again.
   c. Any voltage reading from either test connection indicates a short in the power coil or harness. Either, find and repair the problem with the harness, or replace the power coil.

3. Check the power coil cranking output using the peak-reading voltmeter, set to the same scale. Connect the meter probes to the two orange leads from the power coil and crank the engine. Note the meter reading, it should be at least 60 volts for 5/6 hp motors or 100 volts for 8/9.9 hp motors:
   - If the cranking voltage is within spec, but ignition problems still exist, perform the Charge Coil test, as detailed in this section.
   - If the cranking voltage is below the specification, check the wiring/connectors, then, if the wiring appears to be in good shape, check coil resistance.

4. With the engine not running or cranking, connect the probes from a DVOM set to read resistance across the two orange leads from the power coil. The meter must show 82-102 ohms resistance at a temperature of about 68°F (20°C) or the charge coil is out of specification.

■ An ohmmeter may also be used to check for a grounded coil using a static resistance test by checking between each coil lead and a good engine ground. Meter should show very high or infinite resistance, or the coil is grounded. Remember though that this test can be misleading because the coil might short to ground intermittently only under engine operation.

5. Verify test results and connections, then replace the power coil, if the reading is well out of specification.
6. Reconnect all wiring as removed for testing.

9.9/15 Hp (305cc) 4-Stroke Motors

These models are equipped with a one-piece stator assembly that incorporates the functions of both the charge and power coils. Although the coils can be tested separately, replacement involves the entire stator assembly.

1. Disengage the 4-pin connector (it's a Packard style connector) located between the stator and power pack. The connector contains two orange stator power coil leads on models through 1997 or one orange and one orange/black lead on 1998 and later models, coming from the power coil (underneath the flywheel) and running to the power pack/ignition module.

2. Set a peak-reading DVOM to read positive in the 500 volts scale, then check for a grounded power coil as follows:
   a. Connect the black meter probe to a good engine ground and the red probe to one of the orange power coil leads. Crank the engine and observe, then note the reading.
   b. Move the red meter probe to the other orange (or orange/black) power coil lead, then crank the engine and observe the meter again.
   c. Any voltage reading from either test connection indicates a short in the power coil or harness. Either, find and repair the problem with the harness, or replace the power coil.

3. Check the power coil cranking output using the peak-reading voltmeter, set to the same scale. Connect the meter probes to the two orange (or orange and orange/black) leads from the power coil and crank the engine. Note the meter reading. Cranking output should be at least 100 volts for 1995-96 models, 70 volts for 1997-98 models, or 30 volts for 1999 and later models.
   - If the cranking voltage is within spec, but ignition problems still exist, perform the Charge Coil test, as detailed in this section.
   - If the cranking voltage is below the specification, check the wiring/connectors, then, if the wiring appears to be in good shape, check coil resistance.

4. With the engine not running or cranking, connect the probes from a DVOM set to read resistance across the two orange (or orange and orange/black) leads from the power coil. The meter must show 76-92 ohms resistance for 1995-97 rope start models or 52-62 ohms resistance for all other models. Remember that these specifications are for a temperature of about 68°F (20°C).

■ An ohmmeter may also be used to check for a grounded coil using a static resistance test by checking between each coil lead and a good engine ground. Meter should show very high or infinite resistance, or the coil is grounded. Remember though that this test can be misleading because the coil might short to ground intermittently only under engine operation.

5. Verify test results and connections, then replace the power coil, if the reading is well out of specification.
6. Reconnect all wiring as removed for testing.

25/35 Hp (500/565cc) Motors

These models are equipped with a one-piece stator assembly that incorporates the functions of both the charge and power coils. Although the coils can be tested separately, replacement involves the entire stator assembly.

1. Disengage the 4-pin connector (it's a Packard style connector) located between the stator and power pack. The connector contains two orange stator power coil leads on models through 1997 or one orange and one orange/black lead on 1998 and later models, coming from the power coil (underneath the flywheel) and running to the power pack/ignition module.

2. Set a peak-reading DVOM to read positive in the 500 volts scale, then check for a grounded power coil as follows:
   a. Connect the black meter probe to a good engine ground and the red probe to one of the orange power coil leads. Crank the engine and observe, then note the reading.
   b. Move the red meter probe to the other orange (or orange/black) power coil lead, then crank the engine and observe the meter again.
   c. Any voltage reading from either test connection indicates a short in the power coil or harness. Either, find and repair the problem with the harness, or replace the power coil.

3. Check the power coil cranking output using the peak-reading voltmeter, set to the same scale. Connect the meter probes to the two orange (or orange and orange/black) leads from the power coil and crank the engine. Note the meter reading. Cranking output should be at least 100 volts for 1995-98 models or 30 volts for 1999 and later models.
   - If the cranking voltage is within spec, but ignition problems still exist, perform the Charge Coil test, as detailed in this section.
   - If the cranking voltage is below the specification, check the wiring/connectors, then, if the wiring appears to be in good shape, check coil resistance.

4. With the engine not running or cranking, connect the probes from a DVOM set to read resistance across the two orange (or orange and orange/black) leads from the power coil. The meter must show 76-92 ohms resistance for 1995-97 rope start models or 52-62 ohms resistance for all other models. Remember that these specifications are for a temperature of about 68°F (20°C).

■ An ohmmeter may also be used to check for a grounded coil using a static resistance test by checking between each coil lead and a good engine ground. Meter should show very high or infinite resistance, or the coil is grounded. Remember though that this test can be misleading because the coil might short to ground intermittently only under engine operation.

5. Verify test results and connections, then replace the power coil, if the reading is well out of specification.
6. Reconnect all wiring as removed for testing.

IGNITION AND ELECTRICAL SYSTEMS 4-41

REMOVAL & INSTALLATION

◆ See Figures 57 thru 60

When equipped the power coil windings are either part of the charge coil or the 1-piece stator coil assembly. For removal and installation procedures, please refer to Charge Coil, in this section.

Sensor/Trigger Coil

◆ See Figures 85 and 86

■ The ignition timing functions of the sensor/trigger coil are performed by the crankshaft and, if applicable, the camshaft position sensors on EFI motors. Please refer to the section on Electronic Fuel Injection for information on testing or servicing the sensors on EFI motors.

Most Johnson/Evinrude engines are equipped with some form of a sensor or trigger coil. Various forms of the sensor exist, and not all can be accurately described as a coil (such as the optical sensors used on the 25/35 hp 3-cylinder 2-strokes). But on most carbureted motors some form of sensor is used to generate a voltage based on the rotating magnets in the flywheel. This signal is used by the power pack to fire the ignition coil(s) at the appropriate times. The nickname Trigger Coil comes from this function.

■ On EFI motors, the Crankshaft Position (CKP) and Camshaft Position (CMP) sensors perform the functions of the sensor coil. For more information on testing and service of these sensors, please refer to the Electronic Fuel Injection (EFI) section.

Fig. 85 Sensor coils take many forms, like this individual sensor coil (5/6 hp 4-stroke shown)...

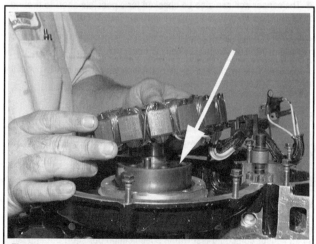

Fig. 86 ...or this timer base assembly found on some 25-55 hp (737cc) motors and all 25-70 hp (913cc) motors

The sensor coil is therefore normally mounted somewhere under the flywheel. On a few motors, such as the 9.9/15 hp 4-stroke, as well as the early 2-strokes equipped with a UFI system, its function is incorporated into the power pack assembly that is mounted under the flywheel. On other motors, the coil is a separate component that can be removed, tested or replaced individually. On a few motors, the sensor coil is referred to as a timer base, because it is mounted under other ignition system components, like the stator on 25-55 hp (737cc) motors with a 12-amp charging system or all 25-70 hp (913cc) motors. The 3-4 hp (87cc) motors, except the 4 Deluxe, utilize a sensor coil/timer base assembly mounted directly underneath the flywheel. The power pack on these models is mounted to the edge of the sensor coil/timer base.

■ The manufacturer provides no information on testing the optical sensors used on 25/35 hp 3-cylinder 2-strokes, other than to say to make sure the sensors are kept clean and free of dirt or debris to ensure proper function.

Signals from the sensor coil are generated at certain points in flywheel revolution, thus alerting the power pack to exact engine positioning. In turn, the power pack uses the signal to determine proper spark timing. In this manner, a spark at the plug may be accurately timed by the timing marks on the flywheel relative to the magnets in the flywheel and to provide as many as 100 sparks per second for a powerhead operating at 6000 rpm.

■ The simple ignition system of the 5-cylinder, single cylinder, 2-stroke engine does not use a sensor/trigger coil or a power pack, firing the ignition coil instead each time the flywheel magnet passes by the charge coil.

TESTING

◆ See Figures 87, 88 and 89

Most motors (except the 25/35 hp motors equipped with optical sensors) are equipped with a sensor coil that operates in the same manner as a charge, power or battery charge coil (generating voltage using the electromagnetic force created by magnets attached to the flywheel). Therefore, on most motors, testing is conducted in the same manner as the charge, power or battery charge coils. A voltmeter is used to perform dynamic checks (watching for voltage while cranking the motor checking for shorts to ground and/or for proper output), while an ohmmeter is used for static checks (testing the windings for proper resistance and checking for shorts to ground).

2.0-3.5 (78cc) Hp Motors

■ A peak-reading DVOM like the Stevens CD-77 is necessary for this test.

1. Set the peak reading voltmeter to read positive sensor readings on the 5 volts scale (on the Stevens meter that is accomplished by setting POS and 5, or SEN and 5).
2. Disengage the blue lead between the sensor coil and the power pack.
3. Connect the black probe from the DVOM to a good engine ground, then attach the red meter lead to the sensor coil blue lead.
4. Crank the engine and observe the meter and proceed as follows:
• If the meter reads 1 volt or higher, sensor output it probably good, test the Charge Coil as detailed in this section.
• If the meter shows less than 1 volt, first check the wiring and connectors. If the wiring and connectors appear in good condition, proceed with the ohmmeter test in the next step.
5. With the engine not running or cranking, connect the ohmmeter leads between the sensor coil blue lead and a good engine ground. The meter must show 40-60 ohms resistance at a temperature of about 68°F (20°C) or the sensor coil is out of specification.
6. Verify test results and connections, then replace the sensor coil, if the reading is well out of specification.
7. Reconnect all wiring as removed for testing.

4-42 IGNITION AND ELECTRICAL SYSTEMS

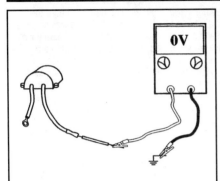

Fig. 87 Sensor coils are tested using a voltmeter (or an ohmmeter) to check for shorts to ground

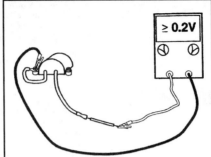

Fig. 88 A peak-reading voltmeter is necessary to sensor coil output while the motor is cranking (individual coil assembly shown)

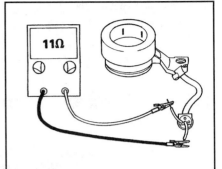

Fig. 89 An ohmmeter is necessary for checking resistance across the sensor coil windings (timer base shown)

3-4 Hp (87cc) Motors (Except 4 Deluxe)

■ **A peak-reading DVOM like the Stevens CD-77 is necessary for this test.**

1. Set the peak reading voltmeter to read positive sensor readings on the 5 volts scale (on the Stevens meter that is accomplished by setting POS and 5, or SEN and 5).
2. Disengage the 6-pin harness connector from the power pack (ignition module) at the flywheel.

■ **1990 models did not use the 6-pin harness connector, so the test must be conducted directly using the 2 sensor coil leads directly at the ignition module. Identify and disconnect the 2 sensor coil leads from the module, then continue with the testing procedure.**

3. Connect the black probe from the DVOM to a good engine ground, then attach the red meter lead to the sensor coil blue/white lead (terminal **C** of the 6-pin connector). Then move the red meter lead to the sensor coil white lead (terminal **C** of the 6-pin connector). Any reading indicates that the sensor coil or leads are grounded, repair the wiring or replace the coil. If there is no reading, proceed to the next step for a dynamic check.
4. Attach the meter probes to the 2 sensor leads (terminals **C** and **D**, the blue/white and white leads of the 6-pin connector).
5. Crank the engine and observe the meter and proceed as follows:
• If the meter reads 4 volts or higher, sensor output it probably good, test the Power Pack output while cranking, as detailed in this section.
• If the meter shows less than 4 volts, first check the wiring and connectors. If the wiring and connectors appear in good condition, proceed with the ohmmeter test in the next step.
6. With the engine not running or cranking, leave the DVOM probes connected between the sensor coil leads (blue/white and white leads off the 6-pin connector). The meter must show 85-115 ohms resistance at a temperature of about 68°F (20°C) or the sensor coil is out of specification.

■ **The sensor coil can also be checked for a possible short to ground by probing between a good engine ground and each of the two sensor coil leads. If the meter shows continuity between either lead and ground the wiring or the sensor coil itself is grounded. Repair the harness or replace the sensor coil, as necessary.**

7. Verify test results and connections, then replace the sensor coil/timer base, if the reading is well out of specification.
8. Reconnect all wiring as removed for testing.

4 Deluxe (87cc) and 5/6/8 Hp (164cc) Motors

■ **The 5-8 hp motors through 1992 and the 4D through 1993 were equipped with the UFI version of this ignition system. Since the ignition module is mounted directly to the ignition plate Under the Flywheel (as the name UFI suggests) so that internal coil windings can react directly to flywheel movement, it does NOT use a separate signal coil. Therefore, on these models, if the sensor coil circuit seems to be the issue, and all other components test normally, you should suspect the Ignition Module itself.**

This test is for the signal coil on CDI equipped models.

■ **A peak-reading DVOM like the Stevens CD-77 is necessary for this test.**

1. Set the peak reading voltmeter to provide negative sensor readings on the 5 volts scale (on the Stevens meter that is accomplished by setting NEG and 5, or SEN and 5).
2. Disengage the 5-pin connector (it's an Amphenol style connector) between the power pack (ignition module) and the ignition charge coil mounting plate located under the flywheel.
3. Check for a grounded sensor coil as follows:
  a. Connect the meter red lead to terminal **B** (white/black lead from the sensor coil) and the black meter lead to a good engine ground, then crank the engine and observe the meter reading.
  b. Connect the meter red lead to terminal **C** (black/white lead from the sensor coil) and the black meter lead to a good engine ground, then crank the engine and observe the meter reading.
  c. Any voltage reading indicates that the sensor coil or wiring harness is shorted to ground. Either repair the harness or replace the sensor coil assembly.
4. Check the sensor coil output as follows:
  a. Attach the meter probes to the 2 sensor leads (terminals **B** and **C**, the white/black and black/white leads of the 5-pin connector).
  b. Crank the engine and observe the meter and proceed as follows:
• If the meter reads 1.5 volts or higher, sensor output it probably good, test the Power Pack output while cranking, as detailed in this section.
• If the meter shows less than 1.5 volts, first check the wiring and connectors. If the wiring and connectors appear in good condition, proceed with the ohmmeter test in the next step.
5. With the engine not running or cranking, leave the DVOM probes connected between the sensor coil leads (white/black and black/white leads off the 5-pin connector). The meter must show 30-50 ohms resistance at a temperature of about 68°F (20°C) or the sensor coil is out of specification.

■ **The sensor coil can also be checked for a possible short to ground by probing between a good engine ground and each of the two sensor coil leads. If the meter shows continuity between either lead and ground the wiring or the sensor coil itself is grounded. Repair the harness or replace the sensor coil, as necessary.**

6. Verify test results and connections, then replace the sensor coil, if the reading is well out of specification.
7. Reconnect all wiring as removed for testing.

5/6 Hp (128cc) and 8/9.9 Hp (211cc) 4-Stroke Motors

◆ See Figure 56

On these motors, the sensor/trigger coil is mounted on top of the powerhead, just to the outside and along the edge of the flywheel.

■ **A peak-reading DVOM is necessary for the basic ground and the dynamic (cranking) portion of this test. If one is not available, the resistance check will give some idea of coil condition, but be careful as a bad sensor coil could test within specification but short out and fail intermittently under load. Likewise, a good coil may test out of spec due to variances in meters and test temperatures.**

IGNITION AND ELECTRICAL SYSTEMS 4-43

1. Set the peak reading voltmeter to provide negative sensor readings on the 5 volts scale (on the Stevens meter that is accomplished by setting NEG and 5, or SEN and 5).

2. Disconnect the black sensor ground wire between the starter mount plate and the sensor/trigger coil, then disengage the bullet connector for the white/black sensor wire that runs to the power pack.

3. Check for a grounded sensor. Connect the black meter probe to a good engine ground and the red probe to the sensor end of the white/black bullet connector. Crank the engine and observe the reading. A reading (of any value) indicates that the charge coil is grounded. Either, locate and repair the ground or replace the sensor assembly, as necessary.

4. Check the sensor output during engine cranking. Reconnect the black sensor ground strap between the sensor and starter mount plate, then connect the black meter probe to the ground wire. Connect the red meter probe to the sensor end of the white/black bullet. Crank the engine and observe the peak voltage:
 • If the meter reads 0.2 volts or higher, sensor output it probably good, test the Power Pack output while cranking, as detailed in this section.
 • If the meter shows less than 0.2 volts, first check the wiring and connectors. If the wiring and connectors appear in good condition, proceed with the ohmmeter test in the next step.

5. With the engine not running or cranking, leave the DVOM probes connected between the sensor coil lead (white/black and black ground wire). The meter must show 132-162 ohms resistance at a temperature of about 68°F (20°C) or the sensor coil is out of specification.

■ The sensor coil can also be checked for a possible short to ground by first disconnecting the black wire from the starter mount plate, then probing between a good engine ground and the white/black sensor coil lead. Be sure to place some tape over the end of the black wire to ensure it does not contact and ground anywhere else, which would give a false reading. If the meter shows continuity between the white/black sensor lead and ground (with the black lead isolated) either the wiring or the sensor coil itself is grounded. Repair the harness or replace the sensor coil, as necessary.

6. Verify test results and connections, then replace the sensor coil, if the reading is well out of specification.
7. Reconnect all wiring as removed for testing.

9.9/10/14/15 Hp (216cc), 9.9/10/15 Hp (255cc) and 18 JET-35 Hp (521cc) Motors

■ Motors through 1991 and some 1992 models were equipped with the UFI version of this ignition system. Since the ignition module is mounted directly to the ignition plate Under the Flywheel (as the name UFI suggests) so that internal coil windings can react directly to flywheel movement, it does NOT use a separate signal coil. Therefore, on these models, if the sensor coil circuit seems to be the issue, and all other components test normally, you should suspect the Ignition Module itself.

This test is for the signal coil on CDI equipped models.

■ A peak-reading DVOM like the Stevens CD-77 is necessary for this test.

1. Set the peak reading voltmeter to provide negative sensor readings on the 5 volts scale (on the Stevens meter that is accomplished by setting NEG and 5, or SEN and 5) for 9.9/15 hp motors. For 18 Jet-35 hp motors, set the meter to read positive sensor readings on the 5 volts scale (on the Stevens meter that is accomplished by setting POS and 5, or SEN and 5).

2. Disengage the 5-pin connector (it's an Amphenol style connector) between the power pack (ignition module) and the ignition charge coil mounting plate located under the flywheel.

3. Check for a grounded sensor coil as follows:
 a. Connect the meter red lead to terminal **B** (white/black lead from the sensor coil) and the black meter lead to a good engine ground, then crank the engine and observe the meter reading.
 b. Connect the meter red lead to terminal **C** (black/white lead from the sensor coil) and the black meter lead to a good engine ground, then crank the engine and observe the meter reading.
 c. Any voltage reading indicates that the sensor coil or wiring harness is shorted to ground. Either repair the harness or replace the sensor coil assembly.

4. Check the sensor coil output as follows:
 a. Attach the meter probes to the 2 sensor leads (terminals **B** and **C**, the white/black and black/white leads of the 5-pin connector).
 b. Crank the engine and observe the meter and proceed as follows:
 • If the meter reads 1.5 volts or higher, sensor output it probably good, test the Power Pack output while cranking, as detailed in this section.
 • If the meter shows less than 1.5 volts, first check the wiring and connectors. If the wiring and connectors appear in good condition, proceed with the ohmmeter test in the next step.

5. With the engine not running or cranking, leave the DVOM probes connected between the sensor coil leads (white/black and black/white leads off the 5-pin connector). The meter must show 30-50 ohms resistance at a temperature of about 68°F (20°C) or the sensor coil is out of specification.

■ The sensor coil can also be checked for a possible short to ground by probing between a good engine ground and each of the two sensor coil leads. If the meter shows continuity between either lead and ground the wiring or the sensor coil itself is grounded. Repair the harness or replace the sensor coil, as necessary.

6. Verify test results and connections, then replace the sensor coil, if the reading is well out of specification.
7. Reconnect all wiring as removed for testing.

25/35 Hp (500/565cc) Motors

The manufacturer does not provide any information for testing the optical sensor used on the 3-cylinder 25 hp (500cc) and 35 hp (565cc) 2-stroke models. If trouble is suspected with the assembly, the sensor lens should be inspected for signs of dirt, contamination or physical damage.

25-55 Hp (737cc) Motors

■ Motors through 1991 and some 1992-93 models were equipped with the UFI version of this ignition system. Since the ignition module is mounted directly to the ignition plate Under the Flywheel (as the name UFI suggests) so that internal coil windings can react directly to flywheel movement, it does NOT use a separate signal coil. Therefore, on these models, if the sensor coil circuit seems to be the issue, and all other components test normally, you should suspect the Ignition Module itself.

This test is for the signal coil on CDI equipped models.

■ A peak-reading DVOM like the Stevens CD-77 is necessary for this test.

1. Set the peak reading voltmeter to provide positive sensor readings on the 5 volts scale (on the Stevens meter that is accomplished by setting POS and 5, or SEN and 5).

■ The 12-amp models are equipped with 2 separate connectors between the power pack and the stator/sensor mounting plate (located under the flywheel). Although one is a 5-pin connector similar to the type used by rope/4-amp models, it does not contain the wiring for the sensor on 12-amp models.

2. Disengage the 5-pin connector for rope/4-amp models or the 3-pin connector for 12-amp models located between the power pack (ignition module) and the ignition sensor coil mounting plate located under the flywheel. In both cases, it is an Amphenol style connector.

3. Check for a grounded sensor coil as follows:
 a. For rope/4-amp models, connect the meter red lead to terminal **B** (white/black lead from the sensor coil) and the black meter lead to a good engine ground, then crank the engine and observe the meter reading. Then connect the meter red lead to terminal **C** (black/white lead from the sensor coil) and the black meter lead to a good engine ground, then crank the engine and observe the meter reading.
 b. For 12-amp models, connect the meter red lead to terminal **A** (blue lead from sensor coil) and the black meter lead to a good engine ground, then crank the engine and observe the meter reading. Repeat this step for terminals **B** (white lead from sensor coil) and **C** (green lead from sensor coil), cranking the engine and observing the meter each time for any reading.
 c. Any voltage reading indicates that the sensor coil or wiring harness is shorted to ground. Either repair the harness or replace the sensor coil assembly.

4. Check the sensor coil output as follows:

4-44 IGNITION AND ELECTRICAL SYSTEMS

a. Attach the red meter probe to the sensor lead for terminal **B** and the black meter probe to terminal **C** (as identified earlier), then crank the engine and note the meter reading. For 12-amp models, move the black meter probe from terminal **C** to terminal **A**, then crank the engine and note the meter reading:
- If the meter reads 1.5 volts or higher for rope/12-amp models or 0.5 volts or higher for 12-amp models, sensor output it probably good. If further ignition troubleshooting is necessary, test the Power Pack output while cranking, as detailed in this section.
- If the meter shows less than 1.5 volts or 0.5 volts respectively, first check the wiring and connectors. If the wiring and connectors appear in good condition, proceed with the ohmmeter test in the next step.

5. With the engine not running or cranking, connect the DVOM probes between the sensor coil leads for terminal **B** and terminal **C**, then note the resistance reading on the meter. For 12-amp models, move one probe from terminal **C** to terminal **A** and note the reading. The meter must show 30-50 ohms resistance for rope/4-amp models or 22-32 ohms resistance across both noted testing combinations for 12-amp models. Remember the specifications are for a component at a temperature of about 68°F (20°C).

■ The sensor coil can also be checked for a possible short to ground by probing between a good engine ground and each of the sensor coil leads. If the meter shows continuity between any lead and ground the wiring or the sensor coil itself is grounded. Repair the harness or replace the sensor coil, as necessary.

6. Verify test results and connections, then replace the sensor coil, if the reading is well out of specification.
7. Reconnect all wiring as removed for testing.

25-70 Hp (913cc) Motors

When testing the sensor/trigger coil, specifications vary between models that are equipped with the Quickstart system, as opposed to models that are not equipped with this system. Most models are in fact equipped with this system, except certain 65 hp commercial models, usually including the 65RS and 65WR, as well as most versions of the 65WE, 65WMLW and the 65WMLC. The easiest way to tell the difference between models equipped with Quickstart and those that are not, is to examine the wiring coming from the sensor/trigger coil assembly located under the flywheel. Models equipped with Quickstart have two 5-pin connectors coming from the stator and sensor assemblies, while models that are not equipped with Quickstart have a 5-pin connector for the stator and a 4-pin connector for the sensor.

Models With Quickstart

■ A peak-reading DVOM like the Stevens CD-77 is necessary for this test.

1. Set the peak reading voltmeter to provide positive sensor readings on the 5 volts scale (on the Stevens meter that is accomplished by setting POS and 5, or SEN and 5).

■ These models equipped with 2 separate connectors between the power pack and the stator/sensor mounting plate (located under the flywheel). Both are 5-pin connectors. The connector with orange and brown wires is for the power/charge coil windings in the stator, while the connector with the blue, purple and green wires is for the sensor/trigger coil windings.

2. Disengage the 5-pin connector (it's an Amphenol style connector) for the sensor/trigger coil assembly.
3. Check for a grounded sensor coil or wire at each terminal of the sensor connector by connecting the meter red probe to the terminal and black probe to ground, then cranking the motor while watching the meter. Repeat this for each of the terminals. Any voltage reading from one of the connections indicates that the sensor coil or wiring harness is shorted to ground. Either repair the harness or replace the sensor coil assembly.
4. Check the sensor coil output as follows:
   a. Attach the black meter probe to the sensor lead for terminal **E** (white wire) and the red meter probe to terminal **A** (blue wire), **B** (purple wire), or **C** (green wire), then crank the engine and note the meter reading. Repeat for each of the 3-terminals noted.
   - If the meter reads 1.5 volts or higher, continue with the next sub-step.
   - If the meter shows less than 1.5 volts, first check the wiring and connectors. If the wiring and connectors appear in good condition, proceed with the ohmmeter test in the next numeric step.
   b. Use 2 jumpers wires to reconnect the wiring for terminals **D** (black/white wire) and **E** (white wire) completing those two circuits across the disconnected halves of the 5-pin connector.
   c. Attach the black meter probe to a good engine ground and the red meter probe to terminal **A** (blue wire), **B** (purple wire), or **C** (green wire), then crank the engine and note the meter reading. Repeat for each of the 3-terminals noted.
   - If the meter reads 1.5 volts or higher, but ignition system trouble is still suspected, perform the cranking output test for the Power Pack, as detailed in this section.
   - If the meter shows less than 1.5 volts, first check the wiring and connectors. If the wiring and connectors appear in good condition, replace the timer base.

5. With the engine not running or cranking, connect the DVOM red lead to terminal **E** (white wire). Then connect the black meter probe alternately to terminal **A** (blue wire), **B** (purple wire), or **C** (green wire), noting the resistance readings across terminal **E** and each of the other terminals. Resistance readings will vary slightly depending on the meter used, but in all cases a high-quality DVOM is necessary. Some examples of readings on different meters include:
- Stevens AT-101: 270-330 ohms for models through 1998 or 250-350 ohms for 1999 and later models.
- Merc-O-Tronic M-700: 630-770 ohms for models through 1998 or 750-850 ohms for 1999 and later models.
- Fluke 29 Series II: no specification provided for models through 1998 or 850-950 ohms for 1999 and later models.

■ Remember that resistance specifications are for a component at a temperature of about 68°F (20°C).

6. The sensor coil can also be checked for a possible short to ground by probing between a good engine ground and each of the sensor coil leads. If the meter shows continuity between any lead and ground the wiring or the sensor coil itself is grounded. Repair the harness or replace the sensor coil, as necessary.
7. Verify test results and connections, then replace the sensor coil, if the reading is well out of specification.
8. Reconnect all wiring as removed for testing.

Models Without Quickstart

■ A peak-reading DVOM like the Stevens CD-77 is necessary for this test.

1. Set the peak reading voltmeter to provide positive sensor readings on the 5 volts scale (on the Stevens meter that is accomplished by setting POS and 5, or SEN and 5).

■ These models equipped with 2 separate connectors between the power pack and the stator/sensor mounting plate (located under the flywheel). One is a 5-pin connector (with orange and brown wires) for the power/charge coil windings in the stator, while the other is a 4-pin connector (with the blue, purple and green wires) for the sensor/trigger coil windings.

2. Disengage the 4-pin connector (it's an Amphenol style connector) for the sensor/trigger coil assembly.
3. Check for a grounded sensor coil or wire at each terminal of the sensor connector by connecting the meter red probe to the terminal and black probe to ground, then cranking the motor while watching the meter. Repeat this for each of the terminals. Any voltage reading from one of the connections indicates that the sensor coil or wiring harness is shorted to ground. Either repair the harness or replace the sensor coil assembly.
4. Check the sensor coil output as follows:
   a. Attach the black meter probe to the sensor lead for terminal **D** (white wire) and the red meter probe to terminal **A** (blue wire), **B** (green wire), or **C** (purple wire), then crank the engine and note the meter reading. Repeat for each of the 3-terminals noted.
   - If the meter reads 0.3 volts or higher, but ignition system trouble is still suspected, perform the cranking output test for the Power Pack, as detailed in this section.
   - If the meter shows less than 0.3 volts, first check the wiring and connectors. If the wiring and connectors appear in good condition, perform the resistance test to verify the sensor coil must be replaced.

IGNITION AND ELECTRICAL SYSTEMS   4-45

5. With the engine not running or cranking, connect the DVOM black lead to terminal **D** (white wire). Then connect the red meter probe alternately to terminal **A** (blue wire), **B** (green wire), or **C** (purple wire), noting the resistance readings across terminal **D** and each of the other terminals. Resistance readings should be 8-14 ohms for a component at a temperature of about 68°F (20°C).

■ The sensor coil can also be checked for a possible short to ground by probing between a good engine ground and each of the sensor coil leads. If the meter shows continuity between any lead and ground the wiring or the sensor coil itself is grounded. Repair the harness or replace the sensor coil, as necessary.

6. Verify test results and connections, then replace the sensor coil, if the reading is well out of specification.
7. Reconnect all wiring as removed for testing.

REMOVAL & INSTALLATION

◆ See Figures 85 and 86

The sensor coil takes various forms on the carbureted versions of these engines. On some motors an individual sensor is mounted on the perimeter of the flywheel or on the ignition plate with other ignition/charging system coils (like the charge, power and/or battery charge coils) directly underneath the flywheel. On some models (3-4 hp, except 4 Deluxe models), a sensor coil/timer base assembly is mounted directly under the flywheel. On still other models, like the 25-55 hp (737cc) motors with a 12-amp charging system or all 25-70 hp (913cc) motors, the sensor coil/timer base assembly is located under the stator.

■ A few models do not use individually serviceable sensor coils. These include the 5 hp (109cc) single cylinder, 2-stroke motor and the 9.9/15 hp (305cc) 4-stroke motor.

All Except 3-4 Hp (87cc) Motors (Including the 4-Deluxe)

For all models equipped with a serviceable sensor/trigger coil or sensor coil/timing base assembly, the coil is accessible once the flywheel and, in the case of most timer bases, when the 1-piece stator assembly is removed.

For these models, please refer to the Removal & Installation procedures for the Charge Coil in this section for more details.

3-4 Hp (87cc) Motors (Not Including the 4-Deluxe)

◆ See Figure 90

The 3-4 Hp (87cc) Johnson/Evinrude motors (not including the 4-Deluxe which is equipped with multiple coils mounted to an ignition plate and a remote power pack), do not use a separately serviceable charge coil. Instead, the functions of the charge coil are incorporated into the power pack that is mounted under the flywheel, directly adjacent to the timer base assembly. If the sensor is not functioning properly, or is otherwise damaged, the timer base must be replaced.

1. Remove the Manual Starter, as detailed in the Manual Starter section.
2. Remove the Flywheel, as detailed in the Powerhead section.
3. Disengage the sensor coil/timer base-to-power pack/ignition module wiring (white and white/blue wires).
4. Remove the 3 screws securing the sensor coil/timer base to the support plate and plate retainer.
5. Lift the sensor coil/timer base from the powerhead.

To install:

6. Clean the sensor coil/timer base mounting location, as well as the threads of the mounting screws of dirt corrosion or debris.
7. Position the sensor coil/timer base to the support plate, and align the base with the 3 holes in the plate retainer for the mounting screws.
8. Apply a coat of OMC Nut Lock or equivalent threadlock to the timer base retaining screws, then install and tighten the screws to 25-35 inch lbs. (2.8-4.0 Nm).

■ Make sure the wiring is routed properly to avoid interference with moving components. If any wire ties were present and removed, be sure to replace them with new wire ties in the same position(s).

9. Connect sensor coil/timer base-to-power pack/ignition module wiring.

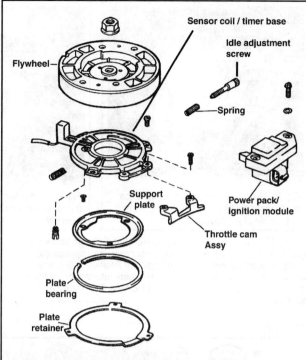

Fig. 90 Exploded view of the sensor coil/timer base and power pack mounting - 3-4 hp (87cc) models (except 4 Deluxe)

10. Install the flywheel.
11. Install the manual starter cover.

Power Pack (Ignition Module) - Carbureted Engines Only

◆ See Figures 91, 92 and 93

On carbureted motors, current from the charge coil used to run the ignition, is controlled by a solid-state Power Pack or Ignition Module. When equipped with the Speed Limiting Operator Warning (S.L.O.W.) system, additional circuits are provided for warning system control and speed limiting operation.

When ignition timing is controlled electronically functions such as timing advance are also performed by the solid-state circuitry in the power pack.

■ Ignition functions on EFI motors, are managed by the Electronic Control Unit (ECU). For more details on those models, refer to the Electronic Fuel Injection section.

Most motors use a remote mounted power pack/ignition module assembly. Wiring runs from the charge (and power coil or stator) and the sensor assembly to the remote mounted power pack. Wires from the power pack are then connected to the primary side of the ignition coil to power the coils and control coil firing.

■ The power pack on 5 hp single-cylinder motor is combined with the ignition coil and cannot be serviced separately. On these models, please refer to the procedures under Ignition Coil for testing and service of the ignition coil/power pack assembly.

Some models are equipped with a power pack/ignition module that is mounted under the flywheel because the module also incorporates the function of another component. On 3-4 hp (87cc) motors (except 4 Deluxe), the power pack incorporates the windings and functions of the charge coil, generating power for the system. For this reason, the module positioning is very sensitive on these models and the air gap between the module and the flywheel must be adjusted anytime the module is removed.

The power pack used on 9.9/15 hp (305cc) four-stroke motors is mounted under the flywheel because it incorporates the functions of the timing sensor, however module positioning is fixed on these models and not adjustable.

4-46 IGNITION AND ELECTRICAL SYSTEMS

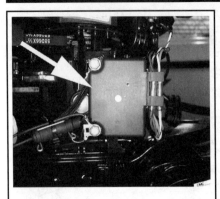

Fig. 91 Most power pack/ignition modules on Johnson/Evinrude engines are mounted to the side of the powerhead...

Fig. 92 ...but some modules, like on this 9.9/15 hp 4-stroke, are mounted under the flywheel

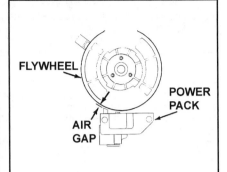

Fig. 93 The ignition module on 3-4 hp (87cc) motors (except 4 Deluxe) must be adjusted to set the air gap with the flywheel

■ On EFI motors, the Electronic Control Unit (ECU) performs the functions of the power pack, completely controlling ignition coil function based on input from the Camshaft Position (CMP) and/or Crankshaft Position (CKP) sensors. For more information on the ECU for these models, please refer to the Electronic Fuel Injection (EFI) section.

TESTING

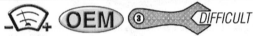

◆ See Figures 94 and 95

A peak-reading voltmeter must be used to check the power pack output with the engine cranking or running. The power pack (ignition module) used on carbureted Johnson/Evinrude motors is a solid state component. For this reason, there area no static methods to check it (such as a resistance check using an ohmmeter).

The cranking test for most motors requires the use of a special test load adapter. The manufacturer recommends an adapter produced by the Stevens Test Equipment Company, No. PL-88. If this special adapter is not available, you can make one using a 10-ohm, 10-watt resistor from an electronics store (at the time of publication this was available from Radio Shack® as part #271-132.)

The engine running test requires a terminal adapter on the primary side of the ignition coil so you can connect the meter without disrupting the circuit. If a terminal extender is not available, you can make one out of an old primary side ignition wire (the wire that is normally connected between the power pack and coil). Cut a small section of the insulation away from an old wire, taking great care not to damage or break any strands of the wire under the insulation (as this would raise resistance in the circuit, possibly rendering false results). Install the wire in place of the one that is currently installed between the power pack and coil. During testing, connect the voltmeter probe to the exposed portion of the wire, using great care to make sure neither shorts to ground on the powerhead.

2.0-3.5 (78cc) Hp Motors

◆ See Figures 94 and 95

■ A peak-reading DVOM is necessary for this test, as well as certain test adapters. You can refer to the information provided under the Testing head for fabricating test adapters from other sources, but there's no way around the peak-reading voltmeter to ensure accurate testing.

1. Check the power pack cranking output first as follows:
   a. Set the peak reading voltmeter to read positive on the 500 volts scale.
   b. Twist and pull the primary lead from the power pack off the ignition coil, then install the terminal extender and reconnect the primary lead (or install the homemade primary test lead, as applicable).

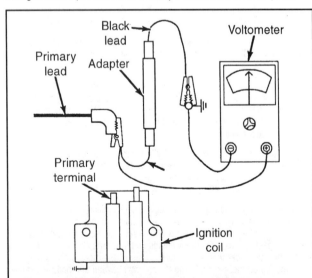

Fig. 94 Checking the power pack output while cranking usually involves connecting an adapter (Stevens PL-88 or equivalent) and a voltmeter to the primary lead from the power pack

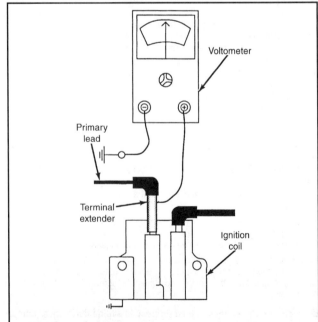

Fig. 95 The running output test requires that the circuit is completed so a terminal extender (or modified primary wire that allows signal probing) is necessary to get a voltage reading

IGNITION AND ELECTRICAL SYSTEMS 4-47

    c. Connect the black DVOM lead to a good engine ground, then connect the red meter lead to the bare portion of the terminal extender (or test lead).
    d. Crank the engine and observe meter reading, it should be 180 volts or higher.
    e. If the reading was less than 180 volts, disconnect the wiring from the terminal extender and install the Stevens PL-88 load-adapter to the wire from the power pack. Connect the other end of the adapter to a good engine ground (when using the Stevens adapter, connect the red end to the power pack wiring and the black end to ground). Crank the engine and observe the meter readings as follows:
- If the meter now reads 180 volts or higher, test the Ignition Coil as detailed in this section.
- If the meter still reads below 180 volts, check the primary wire and spring clip for wear or damage and replace if any is found. If the wiring is not faulty, replace the power pack.

  2. Check the power pack under load, at engine speeds where intermittent problems are noted as follows:
    a. Connect a source of cooling water to the engine, as detailed under Flushing the Cooling System in the Engine Maintenance section.

✹✹ WARNING

If the engine is to be run at speeds over 2000 rpm, it must be mounted in a test tank using a suitable test wheel or placed in the water to prevent possible overspeed and damage which could occur if the engine was run on a flushing device.

    b. Install a terminal extender to the primary terminal of the ignition coil.
    c. Set the peak-reading DVOM to read positive 500 volts scale, then connect the black meter lead to a good engine ground and the red meter lead to the metal portion of the terminal extender (or bare wire of the homemade test lead).
    d. Double-check to make sure the test leads are routed so they will not be damaged by rotating parts, then start and run the engine. If the engine is to be run above high-idle, allow it to warm before advancing the throttle.
    e. With the engine running, power pack output should be at least 220 volts and should remain steady and a constant engine speed. If a problem has been noted at a particular engine speed, attempt to duplicate the symptom while observing the meter. Output must not dip below 220 volts.
    f. If the meter reads less than 220 volts, the power pack is faulty and should be replaced.

  3. Verify test results and connections, then replace the power pack, if necessary.
  4. Reconnect all wiring as removed for testing.

3-4 Hp (87cc) Motors (Except 4 Deluxe)

◆ See Figure 95

■ **A peak-reading DVOM is necessary for this test, as well as certain test adapters. You can refer to the information provided under the Testing head for fabricating 2 terminal extending test leads from other sources, but there's no way around the peak-reading voltmeter to ensure accurate testing.**

  1. Check the power pack cranking output first as follows:
    a. Set the peak reading voltmeter to read negative on the 500 volts scale.
    b. Twist and pull the primary leads from the power pack off the ignition coil, then install 2 terminal extenders and reconnect the primary leads (or install 2 homemade primary test leads, as applicable).
    c. Connect the black DVOM lead to a good engine ground, then connect the red meter lead to the bare portion of a terminal extender (or test lead) for one of the cylinders.
    d. Crank the engine and observe meter reading, it should be 125 volts or higher.
    e. If the reading was less than 125 volts, disconnect the wiring from the terminal extender and connect the red meter lead directly to the spring clip inside the primary lead boot. Crank the engine again and observe the meter readings as follows:
- If the meter now reads 125 volts or higher, test the Ignition Coil as detailed in this section.
- If the meter still reads below 125 volts, check the primary wire and spring clip for wear or damage and replace if any is found. If the wiring is not faulty, replace the power pack.

■ **If you are now testing the other coil for the other cylinder, reconnect the primary lead to the terminal extender that was just removed.**

  2. Repeat the previous step to check cranking output on the other cylinder.
  3. Check the power pack under load, at engine speeds where intermittent problems are noted as follows:
    a. Connect a source of cooling water to the engine, as detailed under Flushing the Cooling System in the Engine Maintenance section.

✹✹ WARNING

If the engine is to be run at speeds over 2000 rpm, it must be mounted in a test tank using a suitable test wheel or placed in the water to prevent possible overspeed and damage which could occur if the engine was run on a flushing device.

    b. Install terminal extenders to the primary terminals of the ignition coil.
    c. Set the peak-reading DVOM to read negative 500 volts scale, then connect the black meter lead to a good engine ground and the red meter lead to the metal portion of the terminal extender for the No. 1 cylinder (or bare wire of the homemade test lead).

■ **The primary lead for the No. 1 cylinder is colored orange/blue, while the primary lead for the No. 2 cylinder is colored orange/green.**

    d. Double-check to make sure the test leads are routed so they will not be damaged by rotating parts, then start and run the engine. If the engine is to be run above high-idle, allow it to warm before advancing the throttle.
    e. With the engine running, power pack output should be at least 150 volts and should remain steady at a constant engine speed. If a problem has been noted at a particular engine speed, attempt to duplicate the symptom while observing the meter. Output must not dip below 150 volts.
    f. If the meter reads 125 volts or more, test the Ignition Coil as detailed in this section. If the meter reads less than 125 volts, the power pack is faulty and should be replaced.

  4. Verify test results and connections, then replace the power pack, if necessary.
  5. Repeat the engine running portion of the test for the No. 2 ignition coil.
  6. Reconnect all wiring as removed for testing.

4 Deluxe (87cc) and 5/6/8 Hp (164cc) Motors

◆ See Figures 94 and 95

■ **A peak-reading DVOM is necessary for this test, as well as certain test adapters. You can refer to the information provided under the Testing head for fabricating a test load adapter (for the cranking test) and 2 terminal extending test leads (for the running test) from other sources, but there's no way around the peak-reading voltmeter to ensure accurate testing.**

  1. Check the power pack cranking output first as follows:
    a. Set the peak reading voltmeter to read negative on the 500 volts scale.
    b. Twist and pull the primary lead from the power pack off the ignition coil whose input signal from the power pack is being tested. Install the Stevens PL-88 load adapter connecting the red end of the adapter to the primary wire lead and the black end of the adapter to a good engine ground.
    c. Connect the black DVOM lead to a good engine ground also, then connect the red meter lead to the red end of the test adapter (where it connects to the primary wire).
    d. Crank the engine and observe meter reading, it should be 175 volts or higher. Record the results, remove the test adapter and reconnect the wiring. Then repeat the test for the other primary lead.
    e. Interpret the meter readings as follows:
- If the meter reads 175 volts or higher, but the ignition is not performing properly for the spark test, check each Ignition Coil as detailed in this section.
- If one primary lead has no output, replace the Power Pack, as detailed in this section.
- If both primary leads have no output, test the Charge Coil, as detailed in this section. If the charge coil output is within spec, the Power Pack must be replaced, as detailed in this section.

  2. To check the power pack under load, at engine speeds where intermittent problems are noted, proceed as follows:

4-48 IGNITION AND ELECTRICAL SYSTEMS

a. Connect a source of cooling water to the engine, as detailed under Flushing the Cooling System in the Engine Maintenance section.

※※ WARNING

If the engine is to be run at speeds over 2000 rpm, it must be mounted in a test tank using a suitable test wheel or placed in the water to prevent possible overspeed and damage which could occur if the engine was run on a flushing device.

b. Install terminal extenders to the primary terminals of the ignition coil. When the terminal extenders (or test leads) are installed, make sure the ignition coils are connected to the proper terminals. The orange/blue primary lead from the power pack should attach to the No. 1 cylinder ignition coil.

c. Set the peak-reading DVOM to read negative 500 volts scale, then connect the black meter lead to a good engine ground and the red meter lead to the metal portion of the terminal extender for the No. 1 cylinder (or bare wire of the homemade test lead).

■ The primary lead for the No. 1 cylinder is colored orange/blue, while the primary lead for the No. 2 cylinder is colored orange.

d. Double-check to make sure the test leads are routed so they will not be damaged by rotating parts, then start and run the engine. If the engine is to be run above high-idle, allow it to warm before advancing the throttle.

e. With the engine running, power pack output should be at least 200 volts and should remain steady at a constant engine speed. If a problem has been noted at a particular engine speed, attempt to duplicate the symptom while observing the meter. Output must not dip below 200 volts.

f. If the meter reads less than 200 volts, test the Charge Coil as detailed in this section (if the charge coil is good, replace the power pack).

g. Repeat the engine running test for the other ignition coil primary lead.

3. Verify test results and connections, then replace the power pack, if necessary.

4. Reconnect all wiring as removed for testing.

5/6 Hp (128cc) and 8/9.9 Hp (211cc) 4-Stroke Motors

◆ See Figures 94 and 95

■ A peak-reading DVOM is necessary for this test, as well as certain test adapters. You can refer to the information provided under the Testing head for fabricating a test load adapter (for the cranking test) and a terminal extending test lead (for the running test) from other sources, but there's no way around the peak-reading voltmeter to ensure accurate testing.

1. Check the power pack cranking output first as follows:
   a. Set the peak reading voltmeter to read positive on the 500 volts scale.
   b. Twist and pull the primary lead from the power pack off the ignition coil. Install the Stevens PL-88 load adapter connecting the red end of the adapter to the primary wire lead and the black end of the adapter to a good engine ground.
   c. Connect the black DVOM lead to a good engine ground also, then connect the red meter lead to the red end of the test adapter (where it connects to the primary wire).
   d. Crank the engine and observe meter reading, it should be 100 volts or higher for 1996 motors or 50 volts or higher for 1997 and later motors. Record the results, remove the test adapter and reconnect the wiring.
   e. Interpret the meter readings as follows:
   • If the meter reads at or higher than specification, but the ignition is not performing properly for the spark test, check the Ignition Coil as detailed in this section.
   • If the primary lead has no output, check the Charge Coil as detailed in this section. If the charge coil is good, then replace the Power Pack, as detailed in this section.

2. To check the power pack under load, at engine speeds where intermittent problems are noted, proceed as follows:
   a. Connect a source of cooling water to the engine, as detailed under Flushing the Cooling System in the Engine Maintenance section.

※※ WARNING

If the engine is to be run at speeds over 2000 rpm, it must be mounted in a test tank using a suitable test wheel or placed in the water to prevent possible overspeed and damage which could occur if the engine was run on a flushing device.

b. Install a terminal extender to the primary terminal of the ignition coil.

c. Set the peak-reading DVOM to read positive 500 volts scale, then connect the black meter lead to a good engine ground and the red meter lead to the metal portion of the terminal extender.

d. Double-check to make sure the test leads are routed so they will not be damaged by rotating parts, then start and run the engine. If the engine is to be run above high-idle, allow it to warm before advancing the throttle.

e. With the engine running, power pack output should be at least 220 volts for 5/6 hp motors or 240 volts for 8/9.9 hp motors, and should remain steady at a constant engine speed. If a problem has been noted at a particular engine speed, attempt to duplicate the symptom while observing the meter. Output must not dip below the specified voltage output for engine running conditions.

f. If the meter reads less than specified, test the Charge Coil as detailed in this section (if the charge coil is good, replace the power pack).

3. Verify test results and connections, then replace the power pack, if necessary.

4. Reconnect all wiring as removed for testing.

9.9/15 Hp (305cc) 4-Stroke Motors

◆ See Figures 94 and 95

■ A peak-reading DVOM is necessary for this test, as well as certain test adapters. You can refer to the information provided under the Testing head for fabricating a test load adapter (for the cranking test) and a terminal extending test lead (for the running test) from other sources, but there's no way around the peak-reading voltmeter to ensure accurate testing.

1. Check the power pack cranking output first as follows:
   a. Set the peak reading voltmeter to read positive on the 500 volts scale.
   b. Twist and pull the primary lead from the power pack off the ignition coil. Install the Stevens PL-88 load adapter connecting the red end of the adapter to the primary wire lead and the black end of the adapter to a good engine ground.
   c. Connect the black DVOM lead to a good engine ground also, then connect the red meter lead to the red end of the test adapter (where it connects to the primary wire).
   d. Crank the engine and observe meter reading, it should be 100 volts or higher. Record the results, remove the test adapter and reconnect the wiring.
   e. Interpret the meter readings as follows:
   • If the meter reads at or higher than specification, but the ignition is not performing properly for the spark test, check the Ignition Coil as detailed in this section.
   • If the primary lead has no output, check the Charge Coil as detailed in this section (if not already tested earlier). If the charge coil is good, then replace the Power Pack, as detailed in this section.

2. To check the power pack under load, at engine speeds where intermittent problems are noted, proceed as follows:
   a. Connect a source of cooling water to the engine, as detailed under Flushing the Cooling System in the Engine Maintenance section.

※※ WARNING

If the engine is to be run at speeds over 2000 rpm, it must be mounted in a test tank using a suitable test wheel or placed in the water to prevent possible overspeed and damage which could occur if the engine was run on a flushing device.

b. Install a terminal extender to the primary terminal of the ignition coil.

c. Set the peak-reading DVOM to read positive 500 volts scale, then connect the black meter lead to a good engine ground and the red meter lead to the metal portion of the terminal extender.

d. Double-check to make sure the test leads are routed so they will not be damaged by rotating parts, then start and run the engine. If the engine is to be run above high-idle, allow it to warm before advancing the throttle.

IGNITION AND ELECTRICAL SYSTEMS 4-49

e. With the engine running, power pack output should be at least 240 volts and should remain steady at a constant engine speed. If a problem has been noted at a particular engine speed, attempt to duplicate the symptom while observing the meter. Output must not dip below the specified voltage output for engine running conditions.

f. If the meter reads less than specified, test the Charge Coil (if not done already), as detailed in this section (if the charge coil is good, replace the power pack).

3. Verify test results and connections, then replace the power pack, if necessary.

4. Reconnect all wiring as removed for testing.

9.9/10/14/15 Hp (216cc), 9.9/10/15 Hp (255cc) and 18 JET-35 Hp (521cc) Motors

◆ See Figures 94 and 95

■ A peak-reading DVOM is necessary for this test, as well as certain test adapters. You can refer to the information provided under the Testing head for fabricating a test load adapter (for the cranking test) and 2 terminal extending test leads (for the running test) from other sources, but there's no way around the peak-reading voltmeter to ensure accurate testing.

1. Check the power pack cranking output first as follows:
  a. Set the peak reading voltmeter to read as follows, depending upon the year/model/ignition system:
  - 9.9-15 hp motors with UFI to positive on the 500 volts scale
  - 9.9-15 hp motors with CDI to negative on the 500 volts scale
  - 18 Jet-35 hp motors with UFI or CDI to positive on the 500 volts scale

  b. Twist and pull the primary lead from the power pack off the ignition coil whose input signal from the power pack is being tested. Install the Stevens PL-88 load adapter connecting the red end of the adapter to the primary wire lead and the black end of the adapter to a good engine ground.

  c. Connect the black DVOM lead to a good engine ground also, then connect the red meter lead to the red end of the test adapter (where it connects to the primary wire).

  d. Crank the engine and observe meter reading, it should be 175 volts or higher. Record the results, remove the test adapter and reconnect the wiring. Then repeat the test for the other primary lead.

  e. Interpret the meter readings as follows:
  - If the meter reads 175 volts or higher, but the ignition is not performing properly for the spark test, check each Ignition Coil as detailed in this section.
  - If one primary lead has no output, replace the Power Pack, as detailed in this section.
  - If both primary leads have no output, test the Charge Coil, as detailed in this section. If the charge coil output is within spec, the Power Pack must be replaced, as detailed in this section.

2. To check the power pack under load, at engine speeds where intermittent problems are noted, proceed as follows:
  a. Connect a source of cooling water to the engine, as detailed under Flushing the Cooling System in the Engine Maintenance section.

※※ WARNING

If the engine is to be run at speeds over 2000 rpm, it must be mounted in a test tank using a suitable test wheel or placed in the water to prevent possible overspeed and damage which could occur if the engine was run on a flushing device.

  b. Install terminal extenders to the primary terminals of the ignition coil. When the terminal extenders (or test leads) are installed, make sure the ignition coils are connected to the proper terminals. The orange/blue primary lead from the power pack should attach to the No. 1 cylinder ignition coil.

  c. Set the peak reading voltmeter to read as follows, depending upon the year/model/ignition system:
  - 9.9-15 hp motors with UFI to positive on the 500 volts scale
  - 9.9-15 hp motors with CDI to negative on the 500 volts scale
  - 18 Jet-35 hp motors with UFI or CDI to positive on the 500 volts scale

  d. Then connect the black meter lead to a good engine ground and the red meter lead to the metal portion of the terminal extender for the No. 1 cylinder (or bare wire of the homemade test lead).

■ The primary lead for the No. 1 cylinder is colored orange/blue, while the primary lead for the No. 2 cylinder is colored orange.

  e. Double-check to make sure the test leads are routed so they will not be damaged by rotating parts, then start and run the engine. If the engine is to be run above high-idle, allow it to warm before advancing the throttle.

  f. With the engine running, power pack output should be at least 200 volts and should remain steady at a constant engine speed. If a problem has been noted at a particular engine speed, attempt to duplicate the symptom while observing the meter. Output must not dip below 200 volts.

  g. If the meter reads less than 200 volts, test the Charge Coil as detailed in this section (if the charge coil is good, replace the power pack).

  h. Repeat the engine running test for the other ignition coil primary lead.

3. Verify test results and connections, then replace the power pack, if necessary.

4. Reconnect all wiring as removed for testing.

25/35 Hp (500/565cc) Motors

◆ See Figures 94 and 95

■ A peak-reading DVOM is necessary for this test, as well as certain test adapters. You can refer to the information provided under the Testing head for fabricating a test load adapter (for the cranking test) and a terminal extending test lead (for the running test) from other sources, but there's no way around the peak-reading voltmeter to ensure accurate testing.

1. Check the power pack cranking output first as follows:
  a. Set the peak reading voltmeter to read positive on the 500 volts scale.

  b. Twist and pull the primary lead from the power pack off the ignition coil. Install the Stevens PL-88 load adapter connecting the red end of the adapter to the primary wire lead and the black end of the adapter to a good engine ground.

  c. Connect the black DVOM lead to a good engine ground also, then connect the red meter lead to the red end of the test adapter (where it connects to the primary wire).

  d. Crank the engine and observe meter reading, it should be 100 volts or higher. Record the results, remove the test adapter and reconnect the wiring. Repeat the cranking output test for each of the remaining ignition coil outputs from the power pack.

  e. Interpret the meter readings as follows:
  - If the meter reads at or higher than specification, but the ignition is not performing properly for the spark test, check the Ignition Coil as detailed in this section.
  - If the primary lead has no output, check the Charge Coil as detailed in this section (if not already tested earlier). If the charge coil is good, then visually inspect the optical sensor for dirt, debris or damage. If no visible sensor damage is found, the only method for testing whether the problem is the sensor, encoder or power pack is to substitute a known good component (one at a time) until the system operates properly.

2. To check the power pack under load, at engine speeds where intermittent problems are noted, proceed as follows:
  a. Connect a source of cooling water to the engine, as detailed under Flushing the Cooling System in the Engine Maintenance section.

※※ WARNING

If the engine is to be run at speeds over 2000 rpm, it must be mounted in a test tank using a suitable test wheel or placed in the water to prevent possible overspeed and damage which could occur if the engine was run on a flushing device.

  b. Install a terminal extender to the primary terminal of the ignition coil whose power pack signal is being tested.

  c. Set the peak-reading DVOM to read positive 500 volts scale, then connect the black meter lead to a good engine ground and the red meter lead to the metal portion of the terminal extender.

  d. Double-check to make sure the test leads are routed so they will not be damaged by rotating parts, then start and run the engine. If the engine is to be run above high-idle, allow it to warm before advancing the throttle.

  e. With the engine running, power pack output should be at least 240 volts and should remain steady at a constant engine speed. If a problem has been noted at a particular engine speed, attempt to duplicate the symptom while observing the meter. Output must not dip below the specified voltage output for engine running conditions. Repeat the running output test on the power pack signals for the remaining engine coils.

f. If the meter reads less than specified, test the Charge Coil (if not done already), as detailed in this section. If the charge coil is good, then visually inspect the optical sensor for dirt, debris or damage. If no visible sensor damage is found, the only method for testing whether the problem is the sensor, encoder or power pack is to substitute a known good component (one at a time) until the system operates properly.

3. Reconnect all wiring as removed for testing.

25-55 Hp (737cc) Motors

◆ See Figures 94 and 95

■ A peak-reading DVOM is necessary for this test, as well as certain test adapters. You can refer to the information provided under the Testing head for fabricating a test load adapter (for the cranking test) and 2 terminal extending test leads (for the running test) from other sources, but there's no way around the peak-reading voltmeter to ensure accurate testing.

1. Set the peak reading voltmeter to read on the appropriate scale, depending on the model as follows:
   • For all UFI models, set the meter to read positive on the 500 volts scale.
   • For all CDI models through 1998 except the 40RE, 40BA or any 12-amp model, set the meter to read negative on the 500 volts scale.
   • For all 1999 and later models except the 40R, or any 12-amp model, set the meter to read negative on the 500 volts scale.
   • For all CDI 12-amp models, as well as any 40RE or 40BA through 1998, any 1999 and later 40R model, set the meter to read positive on the 500 volts scale.

2. Check the power pack cranking output first as follows:
   a. Twist and pull the primary lead from the power pack off the ignition coil whose input signal from the power pack is being tested. Install the Stevens PL-88 load adapter connecting the red end of the adapter to the primary wire lead and the black end of the adapter to a good engine ground.
   b. Connect the black DVOM lead to a good engine ground also, then connect the red meter lead to the red end of the test adapter (where it connects to the primary wire).
   c. Crank the engine and observe meter reading, it should be 150 volts or higher. Record the results, remove the test adapter and reconnect the wiring. Then repeat the test for the other primary lead.
   d. Interpret the meter readings as follows:
   • If the meter reads 150 volts or higher, but the ignition is not performing properly for the spark test, check each Ignition Coil as detailed in this section.
   • If one primary lead has no output, replace the Power Pack, as detailed in this section.
   • If both primary leads have no output, test the Charge Coil, as detailed in this section. If the charge coil output is within spec, the Power Pack must be replaced, as detailed in this section.

3. To check the power pack under load, at engine speeds where intermittent problems are noted, proceed as follows:
   a. Connect a source of cooling water to the engine, as detailed under Flushing the Cooling System in the Engine Maintenance section.

✱✱ WARNING

If the engine is to be run at speeds over 2000 rpm, it must be mounted in a test tank using a suitable test wheel or placed in the water to prevent possible overspeed and damage which could occur if the engine was run on a flushing device.

b. Install terminal extenders to the primary terminals of the ignition coil. When the terminal extenders (or test leads) are installed, make sure the ignition coils are connected to the proper terminals. The orange/blue primary lead from the power pack should attach to the No. 1 cylinder ignition coil.
c. With the peak-reading DVOM set as directed at the beginning of this test, connect the black meter lead to a good engine ground and the red meter lead to the metal portion of the terminal extender for the No. 1 cylinder (or bare wire of the homemade test lead).
d. Double-check to make sure the test leads are routed so they will not be damaged by rotating parts, then start and run the engine. If the engine is to be run above high-idle, allow it to warm before advancing the throttle.
e. With the engine running, power pack output should be at least 175 volts and should remain steady at a constant engine speed. If a problem has been noted at a particular engine speed, attempt to duplicate the symptom while observing the meter. Output must not dip below 175 volts.
f. If the meter reads less than 175 volts, test the Charge Coil as detailed in this section (if the charge coil is good, replace the power pack).
g. Repeat the engine running test for the other ignition coil primary lead.

4. Verify test results and connections, then replace the power pack, if necessary.

5. Reconnect all wiring as removed for testing.

25-70 Hp (913cc) Motors

◆ See Figures 94 and 95

■ A peak-reading DVOM is necessary for this test, as well as certain test adapters. You can refer to the information provided under the Testing head for fabricating a test load adapter (for the cranking test) and 3 terminal extending test leads (for the running test) from other sources, but there's no way around the peak-reading voltmeter to ensure accurate testing.

1. Check the power pack cranking output first as follows:
   a. Set the peak reading voltmeter to read positive on the 500 volts scale.
   b. Twist and pull the primary lead from the power pack off the ignition coil whose input signal from the power pack is being tested. Install the Stevens PL-88 load adapter connecting the red end of the adapter to the primary wire lead and the black end of the adapter to a good engine ground.
   c. Connect the black DVOM lead to a good engine ground also, then connect the red meter lead to the red end of the test adapter (where it connects to the primary wire).
   d. Crank the engine and observe meter reading, it should be 190 volts or higher. Record the results, remove the test adapter and reconnect the wiring. Then repeat the test for the other primary leads.
   e. Interpret the meter readings as follows:
   • If the meter reads 190 volts or higher for most models however it is 230 volts or higher for 1990-91 models, but the ignition is not performing properly for the spark test, check each Ignition Coil as detailed in this section.
   • If one primary lead has no output, replace the Power Pack, as detailed in this section.
   • If all primary leads have low output, test the Charge Coil, followed by, as detailed in this section. If all the primary leads have no output, test the sensor coil. If the charge coil and sensor coil outputs are within spec, the Power Pack must be replaced, as detailed in this section.

2. To check the power pack under load, at engine speeds where intermittent problems are noted, proceed as follows:
   a. Connect a source of cooling water to the engine, as detailed under Flushing the Cooling System in the Engine Maintenance section.

✱✱ WARNING

If the engine is to be run at speeds over 2000 rpm, it must be mounted in a test tank using a suitable test wheel or placed in the water to prevent possible overspeed and damage which could occur if the engine was run on a flushing device.

b. Install terminal extenders to the primary terminals of the ignition coils. When the terminal extenders (or test leads) are installed, make sure the ignition coils are connected to the proper terminals. The orange/blue primary lead from the power pack should attach to the No. 1 cylinder ignition coil, the orange/greed primary lead is for the No. 2 cylinder coil, while the orange/green lead is for the No 3 cylinder.
c. Make sure the peak-reading DVOM is still set to read positive on the 500 volts scale. Then connect the black meter lead to a good engine ground and the red meter lead to the metal portion of the terminal extender for the No. 1 cylinder (or bare wire of the homemade test lead).
d. Double-check to make sure the test leads are routed so they will not be damaged by rotating parts, then start and run the engine. If the engine is to be run above high-idle, allow it to warm before advancing the throttle.
e. With the engine running, power pack output should be at least 220 volts for most models or 250 volts for 1990-91 models, and should remain steady at a constant engine speed. If a problem has been noted at a particular engine speed, attempt to duplicate the symptom while observing the meter. Output must not dip below 220 or 250 volts (respectively). Note the reading.

IGNITION AND ELECTRICAL SYSTEMS 4-51

  f. Repeat the engine running test for the other ignition coil primary leads.

  g. If the meter reads less than 220 or 250 volts (respectively) for one or more cylinders, test the Charge Coil as detailed in this section. If the meter shows no output for one or more cylinders, test the Sensor Coil, as detailed in this section. If all other ignition components are good, replace the power pack.

3. Verify test results and connections, then replace the power pack, if necessary.

4. Reconnect all wiring as removed for testing.

REMOVAL & INSTALLATION

On most CDI models, power pack removal and installation is pretty straight-forward. You simply disconnect the wiring and unbolt the module. But, pay close attention to wiring harness position, especially any ground straps that may be present. Failure to reattach a ground strap properly could cause ignition system problems at best or module damage at worst.

■ Some 1990-93 models may be equipped with a UFI ignition system as opposed to the CDI. The major difference on these models is that the Power Pack (Ignition Module) is also mounted under the flywheel in proximity to react to spinning flywheel magnets. To service the coils or ignition module on ALL UFI models please refer to the procedures in this section for the Charge Coil on the 4 Deluxe (87cc), 5/6/8 Hp (164cc), 9.9/10/14/15 Hp (216cc), 9.9/10/15 Hp (255cc) and 18 Jet-35 Hp (521cc) Motors.

On 3-4 Hp (87cc) Johnson/Evinrude motors (except the 4-Deluxe), module air gap must be adjusted anytime the unit is removed.

All CDI Except 3-4 Hp (87cc) Motors (Including the 4-Deluxe)

◆ See Figures 91 and 92

■ The power pack on 5 hp single-cylinder motor is combined with the ignition coil and cannot be serviced separately. On these models, please refer to the procedures under Ignition Coil for testing and service of the ignition coil/power pack assembly.

1. If equipped, disconnect the negative battery cable for safety and to protect components during service.

2. For 9.9/15 hp motors, remove the manual starter cover for access and so the lower timing belt guide can be properly aligned. Rotate the crankshaft slowly by hand (in the normal clockwise direction of rotation) until the lower belt guide opening is aligned over the power pack timing sensors. This will allow the sensors to clear the wheel during removal.

■ The power packs on some motors are equipped with a ground strap secured by a star washer and retaining bolt. Pay close attention whenever this strap is removed to make sure the ground strap and star washer is retained and positioned properly during installation.

3. Tag and disconnect the power pack/ignition module wiring. This is especially important on models that contain multiple primary leads, as they must be reconnected to the proper terminals on the power pack assembly during installation.

■ Note the routing of all wiring harnesses, especially if wire ties must be cut so that wires can be repositioned. During installation, the wiring must be routed in an identical manner to make sure no contact occurs with moving components (which could rub through the insulation, shorting or cutting out ignition circuits).

4. Loosen the power pack mounting bolt/screws, then remove the power pack/ignition module from the motor.

5. Clean the mounting area of the power pack and powerhead of any dirt, debris, oil or corrosion. Be especially sure that the mounting point(s) and ring terminal(s) for any ground strap is clean and free of dirt, debris or corrosion to ensure to electrical contact.

To install:

6. Position the power pack to the powerhead. Make sure any ground straps attached to the module itself are in position with their appropriate washers (usually star washers).

7. Install and tighten the retaining bolts/screws securely.

8. Apply a light coating of dielectric grease to the terminal(s) for the primary ignition circuit (the terminal for the wire running to the ignition coil).

9. Connect the power pack wiring as tagged during removal. Be certain that wires, if disturbed, are routed properly to prevent interference with and damage from other components.

10. For 9.9/15 hp (305cc) 4-stroke motors, install the manual starter cover assembly.

11. If equipped, connect the negative battery cable.

CDI 3-4 Hp (87cc) Motors (Not Including the 4-Deluxe)

◆ See Figure 93

The 3-4 Hp (87cc) Johnson/Evinrude motors (not including the 4-Deluxe which is equipped with a remote power pack), do not use a separately serviceable charge coil. Instead, the power pack, mounted under the flywheel, performs the functions of the charge coil and the ignition module. For this reason, the module positioning is very sensitive on these models and the air gap between the module and the flywheel must be adjusted anytime the module is removed.

1. Unplug the wiring harness from the ignition module located on the perimeter of the flywheel.

2. Remove the mounting screws and then remove the ignition module from the powerhead.

3. Clean the mounting boss and bolt threads of dirt, corrosion or debris.

To install:

4. Apply a light coating of OMC Gasket sealing compound to the threads of the ignition module mounting screws.

5. Install the ignition module, threading the screws by hand until just finger-tight.

6. Adjust the ignition module air gap (clearance between the module pickup and the flywheel) using a **non-metallic** feeler gauge that is 0.013-0.017 inch (0.33-0.43mm). Place the gauge between the flywheel and module, then lightly press the module against the gauge while tightening the mounting screws to 60-84 inch lbs. (7-9 Nm).

7. Remove the gauge and then double-check the gap. A slightly thicker gauge should not pass between the module and the flywheel, while a slightly smaller gauge should pass with no drag. The proper sized gauge should pass, with a slight drag. If necessary, loosen the screws and reposition the module until the proper adjustment is achieved, then retighten the screws.

8. Plug the wiring harness back into the module.

Ignition Coils

DESCRIPTION & OPERATION

◆ See Figure 96

Besides the spark plugs and wires, the ignition coil is the last major link in the chain that produces spark for ignition. Coils of various size, shape and design are used on Johnson/Evinrude engines. Some are equipped with a 1 primary input and a 1 secondary output terminal. Others are equipped with 1 primary and 2 secondary terminals, and even more may be equipped with 2 primary and 2 secondary terminals. It all depends on how many spark plugs are attached to the coil and whether the motor is designed to fire them individually or simultaneously.

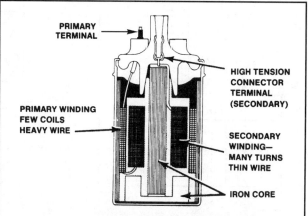

Fig. 96 Conventional ignition coil. All coils work on the same principal: power is transformed from low voltage on the primary side to high voltage on the secondary side

4-52 IGNITION AND ELECTRICAL SYSTEMS

The primary circuit of an ignition coil is connected to the Power Pack (Ignition Module) on carbureted models or the Electronic Control Unit (ECU) on EFI models. Low voltage power is fed and cut from the primary circuit in a manner that induces a high voltage discharge in the secondary winding. When power is cut to the primary circuit, the secondary winding discharges a burst of high voltage through the secondary lead. The voltage then travels through the spark plug and jumps the gap at the plug's tip. The actual voltage jump is the spark referred to when discussing ignition system operation. This spark is what ignites the air/fuel mixture in the combustion chamber and causes the engine to produce power.

■ **The 40/50 hp EFI motors do not use secondary leads (spark plug wires) since each ignition coil is mounted over and connected directly to its spark plug.**

TESTING

◆ See Figures 97, 98 and 99

Although the best test for an ignition coil is performed using a dynamic ignition coil tester (which will show problems that might occur under load and might not be revealed by static test), the simple fact that not everyone has access to a coil tester makes resistance checks useful.

If you do have access to an ignition coil tester, follow the manufacturer's instructions closely in order to prevent damage to the test equipment or the coil assembly. For all Johnson/Evinrude carbureted engines, be sure not to exceed the maximum test amperage ratings. These ratings will vary slightly with test equipment, but two representative values would be 1.1 amps for a Stevens ignition coil tester or 1.5 amps for the Merc-O-Tronic ignition coil tester.

When checking ignition coils, keep in mind that there are two circuits, the primary winding circuit and the secondary winding circuit. Unless coil design prevents it, both need to be checked.

■ **The combined ignition coil/power pack module used on Johnson/Evinrude 5 hp single-cylinder 2-stroke motors prevents you from testing of the primary circuit, but the secondary winding should be checked in the normal manner.**

The tester connection procedure for a continuity check will depend on how the coil is constructed. Generally, the primary circuit is the small gauge wire, while the secondary circuit contains the high tension or plug lead. The primary circuit is connected to the Power Pack (carbureted models) or ECU (EFI models), while the secondary circuit is connected to the spark plug.

Problems with an ignition coil usually cause a no spark condition for the connected spark plug(s). But, a partial internal short or a cracked/damaged coil case that allows voltage leakage could cause an ignition misfire that appears only under certain conditions. If this is the case, the best test for the coil is to use a dynamic ignition tester and try to recreate those conditions.

■ **An ignition coil with an internal or external short will prevent the creation of a strong, blue spark at the plug. If this is suspected, listen closely to the coil during operation for audible clicking noises that may be an indication of a short. External shorts are often visible at night as they cause a blue arc from the coil body to short.**

To ensure the best possible results with resistance testing, remember that specifications will change with temperature and with meters. Whenever possible, use a high quality DVOM for coil resistance testing. The specifications provided are for ignition coils tested when they are 68°F (20°C). That means you can't simply stop a motor that has been operating for 1/2 hour and take a resistance test just because it is 68°F (20°C) outside today. You'd have to wait for the coil to cool off first.

In most cases the ignition coil does not have to be removed from the powerhead for testing, as long as the wiring can be disconnected for access to the terminals.

■ **On method for checking ignition coils on EFI motor is through the wiring harness (once the appropriate harness connectors are removed from the back of the ECU). For details on testing components through the wiring harness on these EFI motors, including wire colors, connector and terminal identifications and specifications, please refer to Sensor and Circuit Resistance/Output Tests in the Electronic Fuel Injection section.**

■ **If there is any question as to the location of a coil, find a spark plug in the powerhead and trace the secondary ignition wire (spark plug lead) back to the coil.**

Primary Coil Winding

◆ See Figure 97

1. If equipped, disconnect the negative battery cable to protect the test equipment and for safety.
2. Note the routing and connection points then disconnect the primary wire or wires from the ignition coil. This is easy enough on electronic ignition systems, but requires some work to access the coil on breaker point systems.

■ **When testing multiple ignition coils on the same motor, it is a good idea to only disconnect and test one coil at a time, this will help prevent confusion when reconnecting the wiring.**

3. Select the resistance scale on the DVOM.

■ **The combined ignition coil/power pack module used on Johnson/Evinrude 5 hp single-cylinder 2-stroke motors prevents you from testing of the primary circuit, but the secondary winding should be checked in the normal manner.**

4. On all carbureted motors (except the 5 hp single-cylinder 2-stroke), test the primary side of the coil as follows:

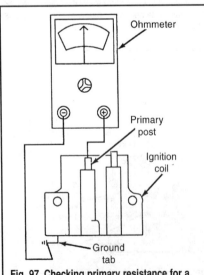

Fig. 97 Checking primary resistance for a coil on a typical carbureted motor with one primary and one secondary terminal

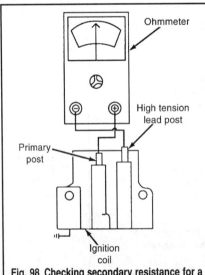

Fig. 98 Checking secondary resistance for a coil on a typical carbureted motor with one primary and one secondary terminal

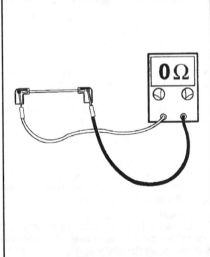

Fig. 99 On carbureted motors, spark plug lead resistance should be very low, close to zero

IGNITION AND ELECTRICAL SYSTEMS 4-53

  a. Connect the black meter probe to a good ground (on the engine if the coil is installed or to a tab on the coil itself if it is removed from the motor).

  b. Connect the red meter probe to the primary terminal (the small terminal on which the wire coming from the power pack normally connects).

  c. Resistance should be 0.05-0.15 ohms for all CDI/UFI 2-stroke engines, except the 25/35 hp (500/565cc) 3-cylinder motors through 1999. On the aforementioned 25/35 hp 3-cylinder 2-strokes through 1999 and on all carbureted 4-stroke motors, the coil primary resistance should be 0.23-0.15. On breaker point ignition 2-stroke engines the primary coil resistance should be 0.7-1.1 ohms.

  5. On 40/50 hp EFI motors, either take the reading across the 2 terminals of the wiring connector (primary harness) or use the appropriate terminals of the ECU harness connector (only after they've been disconnected from the ECU). When testing through the harness, touch the positive meter probe to the ignition coil terminals of the ECU connector **A** that connect to the orange, blue or green engine harness wires. Touch the negative meter probe to the coil terminal of ECU connector **B** that connects to the gray engine harness wire. Either way, primary ignition coil resistance should be 1.9-2.5 ohms.

■ **For more information on terminal pin numbers, connector views and testing through the ECU wire harness, please refer to Sensor and Circuit Resistance/Output Tests in the Electronic Fuel Injection section.**

  6. On 70 hp motors, conduct the test through the ECU wiring harness or at the connector terminals on the coil itself. To test directly at the coil, touch the meter probes to the two coil terminals that normally connect to the blue and gray or the orange and gray wires of the engine harness connector (depending on which of the 2 coils you are testing). To test through the ECU harness, refer to the wiring diagrams in this section and the ECU harness connector/pinout information found in the Electronic Fuel Injection section under Sensor and Circuit Resistance/Output Tests. Touch the positive meter probe to ECU harness terminal that connects to the orange or blue engine harness wires. Touch the negative meter probe to the harness terminal that connects to the gray engine harness wire. Either way, primary ignition coil resistance should be 1.9-2.5 ohms.

  7. If tests are out of specification, make sure that only the wires/connectors were probed. When checking with a ground on the motor, make sure the ground was good (painted and insulated surfaces will prevent you from getting a good ground). When testing through any portion of the harness, carefully trace the wires in the harness to make sure there are no loose, damaged, corroded or broken wires or terminals in the harness that could give a false negative test result.

  8. Replace the coil(s) if readings are out of specification and no other causes can be located.

  9. Reconnect the wiring when testing and/or repairs are complete.

  10. If equipped, connect the negative battery cable and verify proper system operation.

Secondary Coil Winding

◆ See Figures 98 thru 103

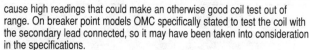

On most carbureted models (except those with breaker point ignitions) it is best to remove the spark plug leads before checking the ignition coils. Spark plug leads on these models should have little or no resistance when in proper condition, but age, deterioration or problems with the connectors may cause high readings that could make an otherwise good coil test out of range. On breaker point models OMC specifically stated to test the coil with the secondary lead connected, so it may have been taken into consideration in the specifications.

In contrast, EFI motors are tested with the secondary leads (70 hp motors) or connector (40/50 hp motors) installed. In both cases, any resistance of the secondary lead or connector has been taken into consideration with the resistance specification. The secondary leads for 70 hp motors are resistor leads and should produce 2500-4100 ohms resistance when in proper condition. The leads may be checked individually as well to ensure they are within specification.

  1. If equipped, disconnect the negative battery cable to protect the test equipment and for safety.

  2. Select the resistance scale on the DVOM.

■ **When testing multiple ignition coils on the same motor, it is a good idea to only disconnect and test one coil at a time, this will help prevent confusion when reconnecting the wiring.**

  3. Disconnect the appropriate wiring and connect the DVOM probes to the coil terminals as follows:

  a. On breaker point ignition 2-stroke motors, connect one meter probe to the coil primary connector and the other to the spark plug lead. The resistance should be 4500-5100 ohms.

  b. On 8-15 hp carbureted 4-stroke motors, connect the meter probes to the two spark plug towers on the coil itself. Resistance should be 2000-2600 ohms.

  c. On all other carbureted motors (2- and 4-strokes), connect one meter probe to the coil primary connector tower and the other to the adjacent secondary tower. If there is a second spark plug tower (for coils that fire 2 spark plugs), note the reading, then move both meter leads to the primary and secondary tower pair for the other spark plug. Resistance should be 225-325 ohms except for the 1995-99 25/35 hp (500/565cc) 2-strokes on which the resistance should be 2000-2600 ohms.

■ **On 40/50 hp EFI motors the ignition coil must be removed from the engine (from its mounting directly above the spark plug) in order to provide access to the secondary connector. However, the primary connector does not need to be removed if testing is desired through the ECU harness. For more information on terminal pin numbers, connector views and testing through the ECU wire harness, please refer to Sensor and Circuit Resistance/Output Tests in the Electronic Fuel Injection section.**

  d. For 40/50 hp EFI motors, connect the red meter probe to the ignition coil terminal that normally connects to the gray engine harness wire. Touch the black meter test probe to the coil terminal that normally connects directly to the spark plug. Resistance should be 8,100-11,100 ohms.

■ **On 70 hp EFI motors, the ignition coil secondary circuit specification is for the coil winding and the spark plug resistor leads, so for these motors, disconnect both secondary wires from the spark plugs instead of the ignition coil.**

  e. For 70 hp EFI motors, connect the meter probes across the two spark plug secondary lead caps. Resistance should be 15,000-28,000 ohms. If resistance is too high, remove the spark plug leads and check to make sure they are within specification. The spark plug leads should generate 2500-4100 ohms resistance. If the leads are out of specification, replace them and retest the coil to determine if the coil itself should be replaced as well.

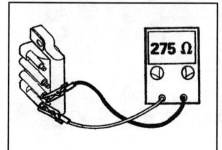

Fig. 100 Checking secondary resistance on a 2-stroke motor that uses 2 primary and 2 secondary leads for a single coil

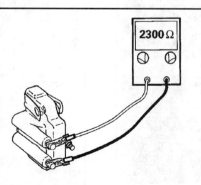

Fig. 101 Checking secondary coil resistance for 8-15 hp 4-stroke motors

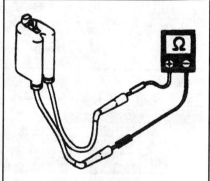

Fig. 102 Checking secondary coil resistance on 70 hp EFI motors

4-54 IGNITION AND ELECTRICAL SYSTEMS

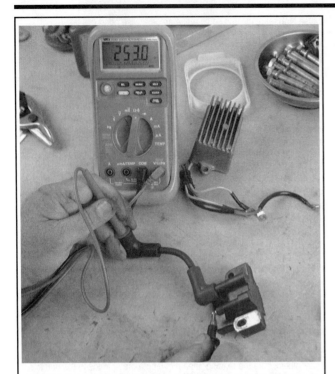

Fig. 103 Since the secondary lead resistance on carbureted motors should be at or near zero, the spark plug end of the lead may be probed instead of the tower itself-note this coil tests in spec for most 2-stroke motors

4. Confirm all connections and readings, especially if the coils test out of specification. Replace the ignition coil(s) if the Primary or Secondary readings are well out of specification and all test connections/conditions were conducted properly.

5. Reconnect the wiring after testing and/or repairs. Be sure to route the wires carefully so as to prevent contact with moving components.

6. If equipped, connect the negative battery cable.

REMOVAL & INSTALLATION

◆ See Figures 104 and 105

On most Johnson/Evinrude motors the ignition coil(s) is(are) found on the rear side of the motor (starboard or port depending on the model). If coil location is not readily apparent, locate a spark plug wire and trace the wire back to the coil. If necessary, refer to the schematics under Wiring Diagrams to help identify the primary wiring for motors equipped with multiple coils.

Fig. 104 The ignition coil is normally found on the rear side of the motor (port or starboard, depending on the model)

■ If the engine has more than one spark plug and/or more than one primary lead, be sure to tag the wiring before disconnecting it from the ignition coil(s) in order to ensure proper installation and operation.

Carbureted Motors

◆ See Figures 106 thru 114

■ The ignition coil contains an integral ignition module on 5 hp single-cylinder motors, therefore there is no traditional primary side wiring from a separate power pack. But, in addition to the secondary spark plug lead, wiring from the charge coil is attached to the ignition coil in order to power the primary circuit.

1. If equipped, disconnect the negative battery cable for safety.
2. On multi-cylinder motors, tag and disconnect the wiring (spark plug leads and primary leads) from the ignition coil(s).

■ If more than one coil is being removed, either replace them one at a time to avoid confusion or tag all wiring before removal. Failure to reconnect the primary and secondary wiring in the same order as originally installed will usually cause incorrect timing and minor-to-severe performance problems including backfiring, stumbling or even a no-start.

3. Check the coil mounting bolts for spacers, ground wires and flat, fiber or star washers. Note the location of each on a piece of paper (or if the mounting matches one of the provided illustrations). The washers, spacers and, most importantly ground straps, must be installed in the proper order/orientation during installation.

4. Carefully remove the ignition coil mounting bolts (usually 2 on most coils), keeping each of the washers and ground straps in order as they are removed. If necessary, reposition the bolt back into the powerhead through the various washers and straps (marking at what point the coil is installed) or spread them out on a secure work surface to remind you of their proper order.

5. Remove the coil from the powerhead.

To install:

6. Carefully clean any traces of dirt, debris or corrosion from the mating surfaces of the coil and powerhead. This is especially important to ensure proper electrical contact for the ground strap.

7. On 5/6 hp (128cc) and 8/9.9 hp (211cc) 4-stroke motors, apply a light coating of OMC nut lock to the threads of the ignition coil mounting screws.

Fig. 105 When the motor is equipped with more than one coil, be careful not to mix up the primary or secondary wires

IGNITION AND ELECTRICAL SYSTEMS 4-55

8. Position the ignition coil to the powerhead with the washers and ground strap(s) in the order noted during removal. Make sure the ground strap has good electrical contact with the powerhead. (This can be quickly checked using an ohmmeter if there is any concern).

■ Although there is no diagram for the 5/6 hp (128cc) motors, make sure to install the lower of the 2 coil retaining bolts through the star washer, followed by the ground strap, coil body and finally the washer/spacer.

9. Tighten the coil retaining screws to 84-106 inch lbs. (9-12 Nm) for 5/6 hp (128cc) and 8/9.9 hp (211cc) 4-stroke motors, or to 48-96 inch lbs. (5.4-10.8 Nm) for all other carbureted 2- and 4-stroke motors.

10. Apply a light coating OMC electrical grease or other suitable dielectric grease to the primary and secondary terminals, then reconnect the wiring as noted during removal.

■ On multi-cylinder models with more than one primary lead, the No. 1 cylinder primary lead is normally orange/blue. When equipped with a separate No. 2 cylinder primary lead, 2-cylinder motors may utilize an orange or orange/green lead. On 3-cylinder motors, the No. 1 primary lead is orange/blue, while the No. 2 is normally orange/purple and the No. 3 is normally orange/green.

Fuel Injected Motors

40/50 Hp EFI Models

The 40/50 hp EFI motors are equipped with a direct ignition system that mounts the ignition coils to the rocker cover, directly over and connected to the spark plugs. This design uses an individual coil to fire each cylinder and eliminates the need for spark plug wires.

1. Disconnect the negative battery cable for safety.

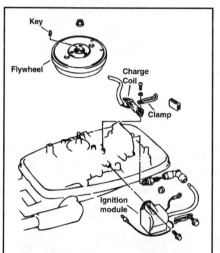

Fig. 106 Exploded view of the combination power pack/ignition coil mounting found on 5 Hp (109cc) Motors

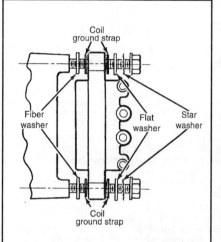

Fig. 107 Ignition coil mounting (washer and ground strap orientation) - 3/4 hp (87cc) motors (except 4 Deluxe)

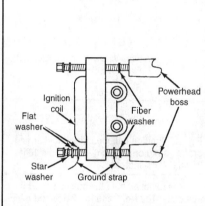

Fig. 108 Ignition coil mounting (washer and ground strap orientation) - 4 Deluxe motorsIgnition coil mounting (washer and ground strap orientation) - 4 Deluxe motors

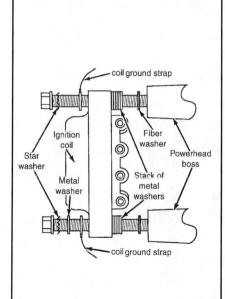

Fig. 109 Ignition coil mounting (washer and ground strap orientation) - 5/6/8 hp (164cc) motors

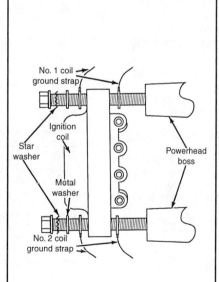

Fig. 110 Ignition coil mounting (washer and ground strap orientation) - 9.9-15 2-stroke (216cc) motors and some early-model 18 jet-35 hp (521cc) motors

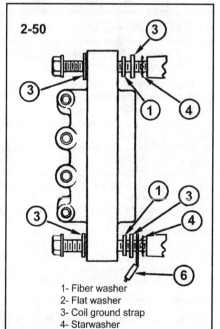

1- Fiber washer
2- Flat washer
3- Coil ground strap
4- Starwasher
5- J-Clamp
6- Stop switch ground lead

Fig. 111 Ignition coil mounting (washer and ground strap orientation) - 9.9-15 2-stroke (255cc) motors

4-56 IGNITION AND ELECTRICAL SYSTEMS

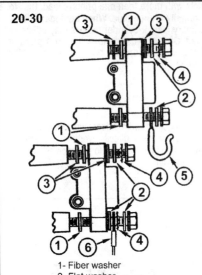
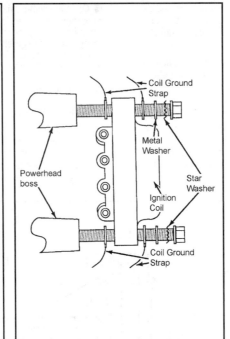
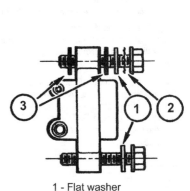

20-30

1- Fiber washer
2- Flat washer
3- Coil ground strap
4- Starwasher
5- J-Clamp
6- Stop switch ground lead

Fig. 112 Ignition coil mounting (washer and ground strap orientation) - late-model 18 jet-35 hp (521cc) motors

Fig. 113 Ignition coil mounting (washer and ground strap orientation) - 25-55 hp (737cc) motors

1 - Flat washer
2 - Starwasher
3 - Ignition coil ground strap

Fig. 114 Ignition coil mounting (washer and ground strap orientation) - 25-70 hp (913cc) motors

■ Before removal, note the positioning of the coils and wiring harness connectors. The No. 1 and No. 2 (top and middle) cylinder's wiring connectors point toward the starboard side of the motor at a 45° angle up from vertical. Similarly, the wiring connector for the No. 3 (bottom) cylinder points toward the port side at a 45° angle up from vertical.

2. Free the wires for the No. 2 and No. 3 ignition coil wires from the clamp(s), as necessary.

3. Push downward on the locktab to release the connector and then carefully unplug the wiring connector from the ignition coil. If one or both of the remaining coils must be removed at the same time, tag the wiring and repeat for the remaining coil(s).

■ Replace coils one at a time to prevent the wiring from becoming mixed. But, if more than one coil must be removed at a time, the wiring should be tagged before removal.

4. Loosen the coil-to-rocker cover bolt, then carefully twist and pull the ignition coil from the spark plug.

To install:

5. Make sure the spark plug connector in the coil housing is clean and free of debris or corrosion.

6. Clean the mounting surface and the coil retaining bolt threads of dirt, debris or corrosion.

7. Place a small amount of dielectric grease over the end of the spark plug and then gently snap the coil into position over the selected spark plug. Twist the coil until the connector aligns as noted during removal (45° up from vertical to the starboard side of the motor for No. 1 and No. 2 cylinder coils or port for the No. 3 cylinder coil).

8. Secure the ignition coil using the retaining bolt and tighten securely.

9. Connect the wiring harness to the ignition coil by aligning the terminals and pushing gently until the locktab snaps into position.

10. Repeat, as necessary, for the remaining coils.

11. Connect the negative battery cable.

70 Hp EFI Models

On the 70 hp EFI motor, 2 ignition coils are mounted on the aft, starboard side of the cylinder head.

1. Disconnect the negative battery cable for safety.

2. Push downward on the locktab to release the connector and then carefully unplug the primary wiring connector from the ignition coil. If both coils must be removed at the same time, tag the wiring to make sure each harness connector is plugged into the proper coil upon installation.

3. Tag and disconnect the spark plug leads from the plugs. Grasp the lead boot and pull it from the spark plug while twisting gently from side to side.

4. Use a 10mm socket to loosen the coil retaining bolts, then carefully remove the coil with the attached secondary leads from the cylinder head brackets.

■ The top mounting bolt for the upper coil secures a metal wire harness guide/protector. Note the positioning of the guide bracket for installation purposes.

5. If on or more of the secondary spark plug leads must be removed from the coil (for replacement or isolated testing), use a small tool to carefully release the lead retainer prongs, then pull the lead from the coil. If the wire separator is still installed, free the wire from the separator.

To install:

6. Make sure the spark plug connector in the coil housing or the secondary spark plug lead (whichever is exposed) is clean and free of debris or corrosion.

7. Clean the mounting surface and the coil retaining bolt threads of dirt, debris or corrosion.

8. If one or more of the secondary spark plug leads were removed from the coil, install the lead(s) as follows:
   a. Make sure the small O-ring is in position between the plastic connector that secures the lead to the ignition coil, and the ignition coil itself.
   b. Use a small amount of silicone spray to lubricate the lead, connector and retainer.
   c. Slide the retainer up the lead at least 2 in. (51mm) for clearance, then slide the O-ring up against the retainer.
   d. Push the lead into the coil until it seats (the lead should push inward about 2 in. or 51mm, positioning the O-ring and retainer close to the coil.
   e. Slide the retainer and O-ring toward the coil until the retainer locks the lead into position. If removed, repeat for the other lead.

9. Position the coil onto the cylinder head bracket (with the metal wiring harness guide/protector positioned properly on the top bolt or the top coil), then tighten the mounting bolts securely.

10. Connect the wiring harness to the ignition coil by aligning the terminals and pushing gently until the locktab snaps into position.

11. Place a small amount of dielectric grease over the end of the spark plugs or lubricate the inside of the boots with a small amount of silicone spray to ease installation, then connect the secondary leads to the plugs as tagged during removal.

12. Connect the negative battery cable.

IGNITION AND ELECTRICAL SYSTEMS 4-57

CHARGING CIRCUIT

General Information

◆ See Figures 115, 116 and 117

A charging system is standard on all tiller or remote electric start models. The charging system provides current to maintain the battery and to operate other engine/boat mounted electric components.

The charging systems used electric start models have various outputs (4-amp, 5-amp, 6-amp, 12-amp) matched to appropriate load capacities for the motor and loads from accessories due to anticipated boat size. The systems can be broken down into two basic types, unregulated and regulated (the difference being that regulated systems protect the battery from overcharging and boiling away electrolyte or possibly even overheating and cracking). An AC lighting coil is standard on some rope start models, but may be added as an option to most. However, because the voltage produced by this system is Alternating Current (AC), and not Direct Current (DC, which is necessary for 12-volt DC marine batteries), models equipped only with AC lighting coils cannot charge marine batteries.

Regardless of the system used, all types start by generating an AC current in the same fashion. A stator coil, battery charge coil or AC lighting coil is mounted directly under the flywheel in order to utilize its mechanical spinning motion when the engine is operating. Permanent magnets attached to the flywheel generate a spinning magnetic field that cuts through the coil windings producing an alternating current. The coil wiring (yellow and/or yellow with various tracer colors on all models), delivers current (through a ring-terminal board on most larger 2-stroke motors) to charging system components or to the AC lighting system.

On charging systems (as opposed to AC lighting systems), current is then passed through a series of diodes (electrical components that only pass current in one direction) contained in the rectifier or rectifier/regulator. The diodes convert the current to DC in order to charge the battery and operate DC voltage accessories. On non-regulated systems, charging rates are controlled only by engine speed, as a rectifier only contains circuitry to convert the current, not control it. For this reason, the total output on a non-regulated system is normally lower than a regulated system in order to help prevent the possibility of battery overcharging. However, extended high-speed operation will risk overcharging on some models, so be diligent with battery maintenance (checking and topping-off cell fluid levels) on these models. A non-regulated system can be identified easily by the cylindrical rectifier mounted on a small, triangular bracket.

Regulated charging systems utilize a rectifier/regulator that not only converts current to DC, but limits charging system output to about 14.6 volts maximum output to prevent battery overcharging. Visually, the rectifier is identified easily by tracing the yellow (or yellow with tracers) wires from the stator coil (through the terminal bar on some models) to the rectangular, finned regulator assembly.

Servicing charging systems is not difficult if you follow a few basic rules. Always start by verifying the problem. If the complaint is that the battery will not stay charged do not automatically assume that the charging system is at fault. Something as simple as an accessory that draws current with the key off will convince anyone they have a bad charging system. Another culprit is the battery. Remember to clean and service your battery regularly. Battery abuse is the number one charging system problem. The second most common cause of charging system complaints comes from loose or corroded wiring connections, mostly at the battery, but problems can be caused at any charging system harness connection.

On regulated systems, the regulator/rectifier is the brains of the charging system. This assembly controls current flow in the charging system. If battery voltage is low the regulator sends the available current from the rectifier to the battery. If the battery is fully charged the regulator diverts most of the current from the rectifier back to the lighting coil through ground.

Do not expect the regulator/rectifier to send current to a fully charged battery. You may find that you must pull down the battery voltage below 12.5 volts to test charging system output. Running the power trim and tilt will reduce the battery voltage, or running other accessories, especially a spot light or other high-power accessory load will reduce the battery voltage quickly.

When equipped, the regulator/rectifier is the most complex item to troubleshoot. You can avoid troubleshooting the regulator/rectifier by checking around it. Check the battery and charge or replace it as needed. Check the amp output of the coil. If amperage is low check the coil for proper resistance and insulation to ground. If all tests and wiring are good, the regulator is likely the culprit, but this can be verified using a variable load tester, if available.

SERVICE PRECAUTIONS

For safety and to prevent damage to the ECU on EFI motors or Power Pack and, when equipped, regulator/rectifier on carbureted motors the following precautionary measures must be taken when working with the electrical system:

- Wear safety glasses when working on or near the battery.
- Don't wear a watch with a metal band when servicing the battery. Serious burns can result if the band completes the circuit between the positive battery terminal and ground.
- When installing a battery, make sure that the positive and negative cables are not reversed.
- Be absolutely sure of the polarity of a booster battery before making connections. Connect the cables positive-to-positive, and negative-to-negative. Connect positive cables first, and then make the last connection to ground on the body of the booster vehicle so that arcing cannot ignite hydrogen gas that may have accumulated near the battery. Even momentary

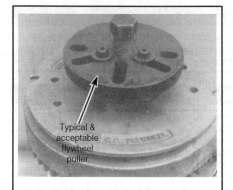

Fig. 115 A universal puller tool can be used to access the charging coil (stator, battery charge or AC lighting)

Fig. 116 When equipped an AC lighting or battery charge coil is normally mounted on the ignition plate, adjacent or across from the charge coil (UFI ignition model shown)

Fig. 117 Most electric start models are equipped with a 1-piece stator whose windings incorporate the battery charge and ignition charge coils (or only the battery charge coils on EFI motors)

4-58 IGNITION AND ELECTRICAL SYSTEMS

connection of a booster battery with the polarity reversed will damage alternator diodes. This also applies to using a battery charger. Reversed polarity will burn out the alternator and regulator in a matter of seconds.
- Always disconnect the battery ground cable before disconnecting the alternator lead.
- Always disconnect the battery (negative cable first) when charging it.
- Never ground the alternator or generator output or battery terminal. Be cautious when using metal tools around a battery to avoid creating a short circuit between the terminals. Be just as careful when working around a hot wire (a circuit to which current is currently applied). Because terminal grounding can occur with any hot wire, not just the battery cable, the negative battery cable should be disconnected before servicing any part of the electrical system.

Charging System Identification

◆ See Figure 118

One of three basic systems can be found on these motors.
1. AC lighting system
2. Unregulated charging system
3. Regulated charging system

Most motors may be equipped with 2 or more of these systems depending on the model. For the most part, rope start models will be equipped with no charging system, but may be equipped with an AC lighting coil. Tiller or remote electric models are equipped with a charging system of some sort, but amperage output and whether or not it is regulated various with model or options.

The most common motor and system combinations are listed below, but keep in mind that systems can vary. If in doubt, trace the yellow (or yellow with tracer) wires coming from the coil (stator, battery charge or AC lighting) under the flywheel. Follow the wires back to their first connection (unless it is ring terminal board, in which case, you should continue to follow them to the first component).

If the wires run only to a boat connection or electrical accessory on rope start models, you've got an AC lighting coil. If the wires run to a smooth cylindrical shaped component mounted to a small triangular bracket on the powerhead (it's a rectifier), you've got an un-regulated charging system (look at bulleted model list to determine output). If the wires run to a finned, rectangular box (a regulator/rectifier) mounted to the powerhead, you've got a regulated charging system (again, check the bulleted model list to help determine output).

- 6/8 hp (164cc) 2-stroke sail models: 4-amp, non-regulated (battery charge coil and rectifier equipped)
- 9.9 hp (211cc) 4-stroke tiller or remote electric models: 4-amp, non-regulated (battery charge coil and rectifier equipped)
- 9.9-15 hp (216cc) and 1993 9.9-15 hp (255cc) 2-stroke sail and tiller or remote electric models; 4-amp, non-regulated (battery charge coil or stator, and rectifier equipped)
- 1994 or later 9.9-15 hp (255cc) 2-stroke sail and tiller or remote electric models: 5-amp, non-regulated (battery charge coil or stator, and rectifier equipped)
- 9.9 hp (305) 4-stroke tiller or remote electric high-thrust models: 12-amp, fully-regulated (stator and regulator/rectifier equipped)
- 15 hp (305) 4-stroke tiller or remote electric models: 6-amp, non-regulated (stator and rectifier equipped)
- 18 Jet-35 hp (521cc) rope with AC: 60 watts AC, non-rectified (AC lighting coil)
- 18 Jet-35 hp (521cc) tiller or remote electric models: 4-amp, non-regulated (battery charge coil and rectifier equipped)
- 25-55 hp (737cc) rope with AC: 60 watts AC, non-rectified (AC lighting coil)
- 25-55 hp (737cc) tiller or remote electric models without VRO2 or power trim/tilt: 4-amp, non-regulated (battery charge coil and rectifier equipped)
- 25-55 hp (737cc) tiller or remote electric models with power trim/tilt and/or VRO2: 12-amp, fully-regulated (stator and regulator/rectifier equipped)
- 25/35 hp (500/565cc) rope with AC: 71 watts AC, non-rectified (AC lighting coil integrated into stator)
- 25/35 hp (500/565cc) tiller or remote electric models: 12-amp, fully-regulated (stator and regulator/rectifier equipped)
- 25-70 hp (913cc) rope with AC: 100 watts AC, non-rectified (AC lighting coil integrated into stator)
- 65 hp (913cc) commercial electric model without VRO2 and all 1990-91 25-70 hp (913cc) models: 6-amp, fully-regulated (stator and regulator/rectifier equipped)
- 1992 or later 25-70 hp (913cc) tiller or remote electric models with power trim/tilt and/or VR02: 12-amp, fully-regulated (stator and regulator/rectifier equipped)
- 40-70 hp EFI 4-stroke motors: 17 amp, fully-regulated (stator and regulator/rectifier equipped)

■ For more details on charging system components and model identification information, refer to the schematic located in the Wiring Diagram section.

Troubleshooting

◆ See Figure 119

Don't waste your time with haphazard testing. The only way to ensure success (and the only way to avoid the possibility of accidentally replacing a good part) is to perform charging system testing in a logical and systematic manner.

Begin all electronic troubleshooting procedures by ensuring that wiring and connections are in good condition. Check battery condition and connections to make sure there is sufficient voltage to operate the system at start-up.

Before conducting any tests, double-check all wire colors and locations with the Wiring Diagrams provided in this section.

✳✳ CAUTION

During system testing, ALWAYS follow the steps of our procedures and any tool manufacturer's instructions closely to avoid injury or possible damage to the engine's charging system.

✳✳ WARNING

If, during testing, the engine is to be run at speeds over 2000 rpm, it must be mounted in a test tank using a suitable test wheel or placed in the water. Running the engine on a flushing device at speeds above 2000 rpm could allow the motor to run overspeed and possibly suffer severe mechanical damage.

Servicing charging systems is not difficult if you follow a few basic rules. Always start by verifying the problem.
- Do not automatically assume that the charging system is at fault.
- A small draw with the key off, a battery with a low electrolyte level, or an overdrawn system can cause the same symptoms.
- It has become common practice on outboard engines to overload the electrical system with accessories. This places an excessive demand on the charging system. If the system is "overdrawn at the amp bank" then no amount of parts changing will fix it.

The charging system should be inspected if:• The battery is undercharged (there is insufficient power to crank the starter)

Fig. 118 The most obvious sign of a motor's charging system is the presence of a regulator/rectifier (which indicates the motor has a fully-regulated system)

IGNITION AND ELECTRICAL SYSTEMS   4-59

Where to Look	Cause	Procedure
Battery	1. Battery defective or worn out 2. Low electrolyte level 3. Terminal connections loose or corroded 4. Excessive electrical load	1. Check condition and charge 2. Add water and recharge 3. Clean and tighten 4. Evaluate accessory loads
Wiring	1. Connections loose or corroded 2. Stator leads shorted or grounded 3. Circuit wiring grounded	1. Clean and tighten 2. Perform ohmmeter tests 3. Perform ohmmeter tests
Coil (Stator, Battery charge, or AC Lighting)	1. Damaged stator windings 2. Weak flywheel magnets 3. Damaged stator leads	1. Perform ohmmeter tests 2. Perform running output tests 3. Perform ohmmeter tests
Rectifier	1. Damaged wiring or diodes	1. Perform rectifier ohmmeter tests
Rectifier/regulator	1. Inoperative rectifier/regulator	1. Perform variable load tests

Fig. 119 When inspecting and troubleshooting the charging system, check each of the components, as applicable. Charging systems are equipped with either a rectifier or a regulator/rectifier. AC lighting coil models are not equipped with either, nor are they equipped with a battery

- The battery is overcharged (electrolyte level is low and/or boiling out)
- The voltmeter on the instrument panel (if equipped) indicates improper charging (either high or low) voltage

A thorough, systematic approach to troubleshooting will pay big rewards. Build your troubleshooting check list with the most likely offenders at the top. Do not be tempted to throw parts at a problem without systematically troubleshooting the system first.

The starting point for all charging system problems begins with the inspection of the battery and related wiring. The battery must be in good condition and fully charged before system testing. Perform a visual check of the battery, wiring and fuses. Are there any new additions to the wiring? An excellent clue might be, "Everything was working OK until I added that live well pump." With a comment like this you would know where to check first (is the pump operating properly or did it increase total draw on the system past capacity).

1. Test the battery thoroughly. Check the electrolyte level, the wiring connections and perform a load test to verify condition.

2. Perform a fuse and Red wire (positive battery cable) check with the voltmeter. Verify the ground at the rectifier or regulator/rectifier. Do you have 12 volts and a good fuse? While you are at the Red wire, check alternator output with an ammeter. Be sure the battery is down around 12 volts.

3. Do a draw test if it fits the symptoms. Many times a battery that will not charge overnight or week-to-week has a constant electrical draw applied that is always sapping a small amount of power. Put a test lamp or ammeter in the line with everything off and look for a draw.

4. A similar problem can be a system that is simply overdrawn. The electrical system cannot keep up with the demand. Do a consumption survey. More amps out than the alternator can return will require a different strategy.

5. Next, go to the source. Check the stator (or battery charge) coil for correct resistance and shorts to ground.

6. If all these tests fail to pinpoint the problem and you have verified low or no output to the battery then replace the rectifier or regulator/rectifier.

On fully-regulated systems there is really only one cause for undercharging (unless wiring is damaged/corroded or the battery is cannot accept a charge). If other components are in good shape, the only cause for undercharging on a regulated system is that the regulator/rectifier is not working. Refer to the information under regulator/rectifier for more testing hints.

✴✴ WARNING

Never operate a marine motor without providing a source of cooling water such as a test tank or flush fitting. It takes less than a minute to damage the water pump impeller that could lead to much more expensive powerhead failures later during normal motor usage. Use a suitable test wheel whenever possible, but remember that for safety the propeller should not be installed on motors when running on a flush fitting.

TESTING

✴✴ WARNING

To protect the motor and test equipment, perform voltage tests with the leads connected and the terminals exposed to accommodate test lead connection (unless specifically directed otherwise by our instructions). All electrical components must be securely grounded to the powerhead any time the engine is cranked or started or certain components could be damaged.

Charging System Quick Test

You can use a DVOM and a tachometer to get a quick idea whether or not the charging system is doing its job. For part of this test, the engine is run while a suitable source of water is supplied to the cooling system (such as a test tank or a flush fitting).

Fully regulated charging systems are designed to keep system voltage output around 14.5 volts in order to charge the battery to a point near 13 volts. Any regulated charging system that is putting out less than 12 volts or more than 15 volts indicates a problem. Undercharging could be a problem with a coil winding/wiring or the regulator/rectifier, while overcharging is usually the fault of the regulator/rectifier, but perform additional component tests as directed in this section to be sure.

4-60 IGNITION AND ELECTRICAL SYSTEMS

■ Remember that charging system problems are often caused by problems with the wiring harness. Loose and corroded connections can cause excessive resistance in the charging circuit. Similarly damaged wiring such as insulation that exposes the wire core because of burning, melting or wear from interference with moving components like the flywheel can cause shorts to ground preventing voltage from reaching the rectifier or regulator/rectifier.

✳✳ WARNING

Never operate a marine motor without providing a source of cooling water such as a test tank or flush fitting. It takes less than a minute to damage the water pump impeller that could lead to much more expensive powerhead failures later during normal motor usage. Use a suitable test wheel whenever possible, but remember that for safety the propeller should not be installed on motors when running on a flush fitting.

1. Make sure the ignition switch and all electrical boat/motor accessories are turned **OFF**.
2. Connect a shop tachometer to the engine, following the instructions provided from the tool manufacturer.
3. With the engine not running, connect the red DVOM lead to the positive battery terminal and the black lead to a good engine ground. Read and record the voltage. If the battery is fully charged (reads more than 12.6 volts) apply a load to draw the system down slightly for testing purposes. Use boat/motor mounted accessories for a few minutes such as search lights, trolling motors, power trim/tilt to drain the battery slightly, but not below 11.0 volts, then note the reading.
4. Start and run the engine until it warms, then allow it to run at slow idle. Again, check the DVOM and record the voltage produced by the charging system at idle.
5. Advance the throttle until the engine is running at or above 2000 rpm. Check the DVOM and record the voltage produced by the charging system above idle.
6. If the charging system is working at all, the voltage readings taken with the motor running at idle and above 2000 rpm should show an increase of **at least** 0.3 volt over the static reading of the battery prior to running the motor. If no voltage increase is noted, test stator/battery charge coil and the rectifier or regulator/rectifier as detailed in this section.
7. If the charging system performed properly, but battery the battery is discharging (and the battery tests ok), take additional voltage readings, with the engine running above 2000 rpm, running at idle and engine stopped, but this time, turn on **ALL** boat and motor accessories. Again, the voltage must be at least 0.3 volt higher with the engine running than stopped or the stator/charge coil and the rectifier or regulator/rectifier should be tested. If component tests show that there is no problem, the system is being overtaxed by accessories (more amps are being withdrawn from the system than are available at the proverbial amp bank, don't use all those accessories at once).

Charging System Running Output Test

◆ See Figure 120

Using an ammeter and a tachometer you can determine if the charging system is operating to specification (generating sufficient amperage). This test requires running the engine across a wide portion of its powerband and is really best conducted with the motor in a test tank or attached to a boat that is firmly secured to a dock or navigated by an assistant. We don't recommend running an engine close to wide open throttle on a flush fitting as damage could occur if the engine is run overspeed.

1. Determine proper amperage output for the motor being tested. Refer to the information on Charging System Identification in this section.
2. Make sure the ignition switch and all electrical boat/motor accessories are turned **OFF**.
3. Connect a shop tachometer to the engine, following the instructions provided from the tool manufacturer.
4. With the engine not running, connect the red DVOM lead to the positive battery terminal and the black lead to a good engine ground. Read and record the voltage. If the battery is fully charged (reads more than 12.6 volts) apply a load to draw the system down slightly for testing purposes. Use boat/motor mounted accessories for a few minutes such as search lights, trolling motors, power trim/tilt to drain the battery slightly, but not below 11.0 volts, then note the reading.
5. Disconnect the red rectifier lead or the red regulator/rectifier lead from the terminal board.
6. Connect an ammeter with at least 0-40 amp capacity in series with the rectifier red lead and the wiring harness red lead. (connect the ammeter inline into the harness).

✳✳ WARNING

Make sure no portion of the wiring (red lead, harness or ammeter leads) contacts engine ground while the engine is running or arcing will likely occur.

7. Start and run the engine until it warms, then allow it to run at slow idle. While the engine is running at idle compare the reading on the ammeter with the specified amperage output for that engine rpm on the accompanying charts.
8. Once the engine has warmed fully, slowly advance the throttle noting the ammeter readings at each 500 or 1000 rpm increase in speed. Output should increase linearly according to the output chart for the system being tested. If there is no output or output is incorrect, perform the stator/battery charge coil and rectifier or regulator/rectifier tests in this section, as applicable.

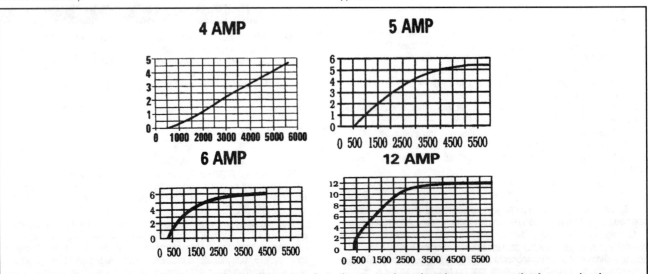

Fig. 120 Charging system amperage output at engine rpm-be sure to refer to the correct charge based on amperage rating for your charging system

IGNITION AND ELECTRICAL SYSTEMS 4-61

Electric System Current Draw Test

Using an ammeter you can determine if the total load of all electrical accessories on the boat/motor exceed the total capability of the charging system to produce power. The ammeter should be rated for at least 3-5 amps more than the charging system rating, meaning a minimum of 7 amps on the smallest, unregulated charging systems or as much as 20 amps for the most powerful systems (which are found on EFI motors).

■ To help determine the rated output for the charging system of the motor being tested, please refer to the information on Charging System Identification in this section.

1. Install an ammeter as follows:
    a. Unless an inductive ammeter is being used, disconnect the negative cable, followed by the positive cable from the battery. Install the ammeter between the positive battery post and cable, then reconnect the negative battery cable.
    b. If using an inductive ammeter, attach the inductive pickup clamp to the positive battery cable following any instructions provided by the tool manufacturer.
2. Turn the ignition keyswitch **ON** to the **RUN** position, but do not start the engine. Turn on all boat/motor accessories (radios, spot lights, trolling motors etc) and note the reading on the ammeter.
3. If the reading exceeds charging system output, turn off accessories one at a time (noting the difference in the reading each time) to determine the individual loads provided by each accessory. This will tell you what accessories can be used together in order to prevent discharging the battery while the motor is running.

Stator/Battery Charge Coil

◆ See Figure 117

All models equipped with a marine battery and charging system are equipped with either a 1-piece stator (that incorporates the windings of the battery charge and ignition charge coils) or a separate battery charge coil. Whether or not the coil windings are integrated into a stator, they can be tested separately using the proper winding leads.

The stator or battery charge coil consists of a series of metal windings that are used generate Alternating Current (AC) voltage current that is then converted to Direct Current (DC) by a rectifier or regulator/rectifier. Current is generated with the help a magnetic field generated by permanent magnets that are attached to the flywheel and which rotate past the coil winding repeatedly during engine operation. The rectified DC current is used to recharge the marine battery and supply voltage to the boat and motor electrical accessories whenever the engine is running.

In all cases the stator or battery charge coil is mounted directly underneath the flywheel. Testing occurs through the wiring harness, but removal and replacement of the coil requires removal of the flywheel for access.

TESTING

◆ See Figure 121

■ Before testing, make sure all connections and mounting bolts are clean and tight. Many charging system problems are related to loose and corroded terminals or bad grounds. Don't overlook the engine ground connection to the body.

Since most stator or battery charge coil test values are near or below 1 ohm, testing requires the use of a high quality Digital Volt Ohmmeter (DVOM). As with all resistance testing, care must be taken as specifications are determined by testing components that are precisely 68°F (20°C) and readings taken on components at other temperatures will vary. Also, keep in mind that certain intermittent problems (such as certain internal shorts) might only appear when the coil is put under a load during normal operation or at certain operating temperatures. Therefore, dynamic checks such as those provided under the Charging System Running Output Test and Charging System Quick Test should be used to verify problems with the system before suspecting a coil based only on this resistance check.

During testing the coil winding resistance is checked across various combinations of the stator coil wiring (yellow or yellow with trace color wires) in order to see if the circuit contains the proper amount of resistance and is therefore complete, yet containing no internal shorts between coils. Each stator lead is then checked between the coil and a good engine ground to make sure no coil winding is shorted to ground. If opens or shorts are found, the wiring harness should be checked for damage (when present, visible signs of broken or worn insulation are usually the cause of the problem) before deciding to replace the coil itself.

1. Disconnect the negative battery cable for safety and to protect the test equipment.
2. On EFI motors, when applicable, carefully free the stator/battery charge coil wires from the spiral wrap.
3. Trace the stator yellow (and yellow with a trace color) wiring from under the edge of the flywheel to the harness or terminal board connectors, as applicable. Disconnect the stator wiring for access while testing with the DVOM. Inspect the terminals for corrosion or damage and clean or replace them, as necessary.

■ As usual on outboards, take note of wire routing before disconnecting them. Wires must always be returned to the original position/routing in order to prevent interference with and damage from moving components.

4. Using the DVOM set to its lowest resistance scale, probe across the appropriate lead pairs for the coil windings and compare the readings to specification, as follows:
   - **6/8 hp** (164cc) 2-stroke sail models (with 4-amp, non-regulated systems), connect the black meter probe to the yellow/blue lead and the red probe to the yellow lead and note the reading, then move the red lead to the yellow/gray lead. Both readings should be 0.50-0.60 ohms.
   - **9.9 hp** (211cc) 4-stroke tiller or remote electric models (with 4-amp, non-regulated systems), connect one probe to the yellow/blue lead and the other probe to the yellow lead. The meter should show 0.40-0.80 ohms.
   - **9.9-15 hp** (216cc) and 1993 **9.9-15 hp** (255cc) 2-stroke sail and tiller or remote electric models (with 4-amp, non-regulated systems), connect the black meter probe to the yellow/blue lead and the red probe to the yellow lead and note the reading, then move the red lead to the yellow/gray lead. In either combination, the meter should show 0.22-0.32 ohms for 1990 models or 0.50-0.60 ohms for 1991-93 models.
   - 1994 or later **9.9-15 hp** (255cc) 2-stroke sail and tiller or remote electric models (with 5-amp, non-regulated systems), first connect the probes to the 2 yellow/gray leads. The ohmmeter should show little or no resistance, otherwise repair the wiring between the 2 yellow/gray connector or replace the stator. Next, connect the probes to the yellow and yellow/blue leads and note the reading. Move one probe from the yellow/blue lead to the yellow/gray lead. Both readings between the yellow and either of the other 2 stator wires should be 0.80-0.90 ohms. Finally, take a resistance reading across the yellow/blue and yellow/gray lead, resistance should be 1.5-1.7 ohms.

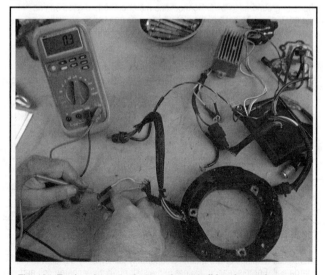

Fig. 121 Testing the stator/battery charge coil involves using an ohmmeter to check resistance across various terminals - note removal is NOT necessary (or even preferred) for testing

4-62 IGNITION AND ELECTRICAL SYSTEMS

- **9.9 hp** (305) 4-stroke tiller or remote electric high-thrust models (with 12-amp, fully-regulated systems), connect one probe to the yellow/gray lead and the other probe to the yellow lead. The meter should show 0.45-0.54 ohms.
- **15 hp** (305) 4-stroke tiller or remote electric models (with 6-amp, non-regulated systems), connect one probe to the yellow/gray lead and the other probe to the yellow lead. The meter should show 1.38-1.68 ohms.
- **18 Jet-35 hp** (521cc) tiller or remote electric models and 25-55 hp (737cc) tiller or remote electric models without VRO2 or power trim/tilt (all with 4-amp, non-regulated systems). Connect the black meter probe to the yellow/blue lead and the red probe to the yellow lead and note the reading, then move the red lead to the yellow/gray lead. Both readings should be 0.50-0.60 ohms.
- **25-55 hp** (737cc) tiller or remote electric models without VRO2 or power trim/tilt (with 4-amp, non-regulated systems), connect the black meter probe to the yellow/blue lead and the red probe to the yellow lead and note the reading, then move the red lead to the yellow/gray lead. In either combination, the meter should show 0.22-0.32 ohms for 1990 models or 0.50-0.60 ohms for 1991 and later models.
- **25-55 hp** (737cc) tiller or remote electric models with power trim/tilt and/or VRO2 (with 12-amp, fully-regulated systems), connect one probe to the yellow/gray lead and the other probe to the yellow lead. The meter should show 0.40-0.80 ohms.
- **25/35 hp** (500/565cc) tiller or remote electric models (with 12-amp, fully-regulated systems), connect one probe to the yellow/gray lead and the other probe to the yellow lead. The meter should show 0.45-0.54 ohms.
- **65 hp** (913cc) commercial electric model without VRO2 and all 1990-91 **25-70 hp** (913cc) models (with 6-amp, fully-regulated systems), connect one probe to the yellow/gray lead and the other probe to the yellow lead. The meter should show 1.3-1.5 ohms.
- 1992 or later **25-70 hp** (913cc) tiller or remote electric models with power trim/tilt and/or VR02 (with 12-amp, fully-regulated systems), connect one probe to the yellow/gray lead and the other probe to the yellow lead. The meter should show 0.40-0.60 ohms.
- **40-70 hp** EFI 4-stroke motors: (with 17-amp, fully-regulated systems), connect one probe to any of the stator harness terminals for one of the 3 yellow wires and the other to one of the 2 remaining terminals. Note the reading, then move the probe to the 3rd terminal and note the reading. Finally move the probe that was not moved in the second step to the other terminal from the first check and note the reading (in this way you will have checked across all possible combinations of the 3 terminals). In all cases, the meter should show 0.56-0.84 ohms for 40/50 hp EFI motors or 0.30-0.50 ohms for 70 hp EFI motors

5. If resistance readings across any pair of the appropriate terminals are out of specification, recheck connections to ensure proper testing readings (and make sure the components are being tested at the proper temperature). Double-check the wiring harness for any problems and, if none are found, replace the stator or battery charge coil, as applicable.

6. Check the stator or battery charge coil winding and harness for shorts to ground as follows. Connect one meter probe to a good engine ground, then probe each of the stator/battery charge coil wires (yellow or yellow with tracer) in turn using the other probe. Resistance should be high or infinite in all cases, or there is a short to ground. If a short is present either repair the wiring harness or replace the coil, as applicable.

7. Once the tests and/or repairs are completed, be sure to route the wires as noted before removal to ensure there is no interference with moving components (such as the flywheel).

8. If applicable on EFI motors, protect the wires by enclosing them back into the spiral wrap.

9. Connect the negative battery cable.

REMOVAL & INSTALLATION

◆ See Figures 122 and 123

Carbureted Motors

On all carbureted Johnson/Evinrude outboards the battery charge coil windings are either integrated into the 1-piece stator that also contains the charge coil windings or are mounted separately immediately adjacent to the charge coil. For this reason, access to the stator or battery charge coil and removal/replacement procedures are covered in the Charge Coil, Removal & Installation procedures under the Ignition System portion of this section. Please refer there for details.

Fuel Injected Motors

◆ See Figures 122 and 123

On EFI motors, the stator coil is not directly part of the ignition system (as the ignition draws power from the battery instead of a charge coil winding in the stator dedicated to the ignition system as occurs in most carbureted models). The 1-piece stator used on all EFI motors however is removed and installed in essentially the same manner. In all cases, the flywheel must be removed for access.

1. Disconnect the negative battery cable for safety.
2. Remove the Manual Starter, as detailed in the Manual Starter section.
3. Remove the Flywheel, as detailed in the Powerhead section.
4. Disconnect the stator wiring from the wiring harness connector for the 3 yellow wires running to the regulator/rectifier.

■ **Note the wire routing to ensure proper installation without interference with any moving components.**

5. Loosen the screws retaining the stator coil to the stator mounting plate on the powerhead, then remove the stator assembly.
6. Clean the coil mounting location, the threads or the mounting screws.

To install:

7. Clean the stator mounting plate of dirt, corrosion or debris. Remember to clean the threads of and powerhead bores for the stator mounting screws as well.
8. Place the stator coil in position on the powerhead mounting plate.
9. Apply a coating of OMC Nut Lock or equivalent threadlocking compound to the threads of the stator retaining screws, then install and tighten the screws to 24-36 inch lbs. (3-4 Nm).
10. Route the stator wiring as noted during removal to prevent damage, then reconnect the wiring harness.

■ **Double check the positioning of all wires that were disturbed during service. Make sure the wires are routed properly and, any wire ties that were removed, are replaced with new ties.**

11. Install the flywheel.
12. Install the manual starter cover.
13. Connect the negative battery cable.

Rectifier

◆ See Figure 124

Carbureted models that are equipped with a battery charging system are either non-regulated or fully-regulated depending on model design parameters. About half of the charging systems on 18 Jet-55 hp (737cc) motors, as well as most of the charging systems on smaller motors are of the non-regulated type. Though, even with this said, you should check the information under Charging System Identification to be sure of the specified charging system for your model, since there are exceptions.

Non-regulated systems tend to be equipped with a separate battery charge coil, as opposed to a 1-piece stator assembly, though there are some exceptions in the 9.9-15 hp 2-stroke engine family. These systems are equipped with a rectifier whose purpose is to convert the Alternating Current (AC) voltage produced by the battery charge coil windings into Direct Current (DC) voltage for use in charging the battery.

These systems are known as non-regulated because there is no electronic control for the amount of current put out. Whenever the engine is running, voltage is produced to charge the battery. These systems tend to have lower amperage output (4, 5 or 6-amps) in order to help reduce the possibility of overcharging the battery during engine operation. Of course, extended full-throttle operation on a boat with an otherwise fully-charged battery and no other electrical accessories could lead to overcharging and boiling away of electrolyte. This makes periodic battery maintenance even more important on these models to prevent the possibility of premature battery failure.

■ **The rectifier can be visually identified by following the yellow (and yellow with tracer) wiring for the battery charge coil from underneath the flywheel (through the terminal board when used) to a smooth, round component mounted to a small triangular base (the rectifier). If these wires instead are connected to a finned, rectangular box, the motor is equipped with a regulated system and uses a regulator/rectifier instead of the rectifier we discuss here.**

IGNITION AND ELECTRICAL SYSTEMS 4-63

Fig. 122 A universal puller usually makes flywheel removal much easier

Fig. 123 The stator (or battery charge coil) is mounted directly underneath the flywheel

Fig. 124 A wire terminal strip is attached to the back of some rectifiers

TESTING

◆ See Figures 124 thru 126

Rectifiers use diodes to act as one-way electrical check valves, thereby converting Alternating Current (AC) to Direct Current (DC). Testing is performed by using an ohmmeter to verify that the rectifier will only pass current in one direction. Essentially the test is conducted by connecting the ohmmeter probes, one to a common ground and one to a lead (terminal) coming from the rectifier. Using the high ohm scale, look for a reading, then reverse the leads. This time there should be no reading (infinity). This means the diode is good. Repeat the test on the other leads. Also test between the red lead and the other two leads (terminals). A normal diode will show a reading. For step-by-step instructions, proceed as follows:

1. Disconnect the negative battery cable for safety and to protect the test equipment.

■ **The 9.9 hp (211cc) 4-stroke only uses 3 leads, red, yellow and yellow/blue. Also, unlike most other rectifier equipped models, the wiring on this engine is attached with 2 bullet connectors and one ring terminal (red lead).**

2. To ensure proper installation, note the wire harness routing and all of the connection points, then disconnect the rectifier lead ring terminals from the terminal board (most models) or from the bullet connector(s), ring terminals or other wiring harness connector (as applicable on certain 9.9-15 hp models).

■ **If necessary, refer to the schematics found under Wiring Diagrams in this section for more details on wire colors and connections.**

3. Connect the black probe of a DVOM set to read resistance to a good engine ground, then connect the red probe to the yellow/blue lead. Note the reading on the meter, then reverse the leads (or, if the DVOM is equipped, press the reverse polarity button on the meter) and note the reading. The readings must be high in one direction and low in the other, indicating the diode inside the rectifier is good. If readings are the same (high or low) in both directions, the diode is damaged and the rectifier must be replaced.

4. Repeat the previous step for the yellow lead, then for all models except the 9.9 hp (211cc) 4-stroke which only uses two leads, repeat the step again for the yellow/gray lead.

5. Connect the black DVOM probe to the red rectifier lead, then connect the red probe to the rectifier yellow/blue lead. Reverse the leads or press the polarity button and compare the results. The readings must be high in one direction and low in the other, indicating the diode inside the rectifier is good. If readings are the same (high or low) in both directions, the diode is damaged and the rectifier must be replaced.

6. Repeat the previous step connecting the red probe to the yellow lead, then for all models except the 9.9 hp (211cc) 4-stroke which only uses two leads, repeat the step again for the yellow/gray lead.

7. If test results vary, verify all test connections and results, then replace the rectifier.

8. After testing and/or repairs, route the wiring and secure it to the terminal board or connectors as noted during removal.

9. Reconnect the negative battery cable.

REMOVAL & INSTALLATION

◆ See Figure 124

To help locate the smooth round metal cased rectifier (mounted to a triangular bracket), trace the yellow (and yellow with tracer) wires from under the flywheel either through the terminal board (most motors) or through the bullet connectors (9.9 hp 4-strokes) to the rectifier.

1. Disconnect the negative battery cable for safety.

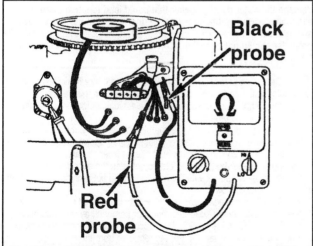

Fig. 125 Rectifier testing involves checking between various yellow leads and ground for continuity in only one direction...

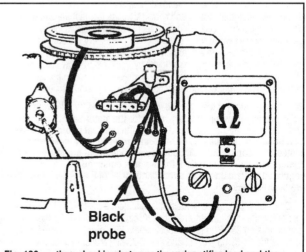

Fig. 126 ...then checking between the red rectifier lead and the various yellow leads for continuity in only one direction

4-64 IGNITION AND ELECTRICAL SYSTEMS

■ **The 9.9 hp (211cc) 4-stroke only uses 3 leads, red, yellow and yellow/blue. Also, unlike most other rectifier equipped models, the wiring on this engine is attached with 2 bullet connectors and one ring terminal (red lead).**

2. To ensure proper installation, note the wire harness routing and all of the connection points, then disconnect the rectifier lead ring terminals from the terminal board (most models) or from the bullet connector(s), ring terminals or other wiring harness connector (as applicable on certain 9.9-15 hp models).

■ **If necessary, refer to the schematics found under Wiring Diagrams in this section for more details on wire colors and connections.**

3. Remove the mounting screws from the triangular bracket, then remove the rectifier from the powerhead.

■ **The rectifier mounting screws on some models require the use of an appropriate sized Torx® bit.**

To install:
4. Clean the rectifier mounting area, along with the powerhead and screw threads of any dirt, corrosion or debris.
5. Inspect the rectifier terminal wiring and connectors for damage or corrosion and repair, clean or replace, as necessary.
6. Position the rectifier and tighten the mounting bolts securely.
7. Properly route and reconnect the rectifier wiring as noted during removal.
8. Reconnect the negative battery cable.

Regulator/Rectifier

◆ See Figure 118

Carbureted models that are equipped with a battery charging system are either non-regulated or fully-regulated depending on model design parameters. About half of the charging systems on 18 Jet-55 hp (737cc) motors, as well as most of the charging systems on larger motors (including all EFI motors) are of the fully-regulated type. Though, even with this said, you should check the information under Charging System Identification to be sure of the specified charging system for your model, since there are exceptions.

Fully-regulated systems tend to be equipped with a 1-piece stator assembly, though there are some exceptions in the 9.9-15 hp 2-stroke engine family. These systems are equipped with a regulator/rectifier that serves a dual purpose. Just like a rectifier, the regulator/rectifier first and foremost converts the Alternating Current (AC) voltage produced by the battery charge coil windings of the stator into Direct Current (DC) voltage for use in charging the battery. Then, the regulator portion of the circuitry is used to control the amount of voltage/current that is supplied to the battery, in order to prevent overcharging. This type of a charging system works on the same principles and in the same basic manner as the charging system in your car or truck.

These systems tend to have higher amperage output (6-, 12- or 17-amps) in order to meet the larger demands of boat and motor electrical accessories that come with larger motors and boats. Also, the fact that the charging circuit is regulated or controlled to protect the battery allows the system to generate more power without the worry that the battery will be damaged.

Although the location and appearance of the regulator/rectifier varies slightly by model, it can be easily found by tracing the yellow stator battery charge coil wires from underneath the flywheel (and through the terminal board on most larger carbureted models) to the regulator/rectifier assembly.

■ **The regulator/rectifier can be visually identified by following the yellow (and yellow with tracer) wiring for the stator battery charge coil from underneath the flywheel (through the terminal board when used) to a finned, rectangular box (the regulator/rectifier). If these wires instead are connected a smooth, round component mounted to a small triangular base (the rectifier), the motor is equipped with a non-regulated system and uses a rectifier instead of the regulator/rectifier we discuss here.**

On smaller carbureted models, it is mounted to the front of the engine, just below the rope starter on 9.9 hp (305cc) motors, or next to the carburetors on 25/35 hp (500/565cc) motors. For larger carbureted motors such as the 25-55 hp (737cc) and the 25-70 hp (913cc) motors, the regulator/rectifier is normally mounted to the aft of the starboard side of the motor (on the opposite side from the electric starter motor).

Like the larger carbureted model, on all EFI motors the rectifier/regulator is located on the starboard side of the engine. For 40/50 hp models, it is mounted immediately above the low pressure fuel filter on the aft, starboard side of the powerhead. For 70 hp models it is mounted inside the electrical component cover (which must be removed for access) on the starboard side of the powerhead.

■ **If necessary, refer to the schematics provided under Wiring Diagrams in this section to help locate the regulator/rectifier and wiring necessary for this test.**

TESTING

◆ See Figures 127, 128 and 129

The only reliable way to test the regulator/rectifier is observe its affect on charging system output when a battery is fully charged as opposed to when a battery is discharged. The safest and easiest method to perform this test is using an ammeter to monitor charging system output in amps while a carbon pile load tester (such as the Stevens LB-85 or the Snap-On MT540D) is used to draw down power from the battery.

■ **It may be possible to simulate the affect of the carbon pile load tester by performing the test using different batteries, one fully or near fully charged and the other near fully discharged. The only problem is how to start the motor with the near fully discharged. The best method would be to use the auxiliary rope start on motors so equipped, though it might be possible to jump-start the motor. Connecting a booster battery just long enough to get the motor running should not appreciably charge the "discharged" battery so test readings would still likely be valid.**

✱✱ WARNING

We do not recommend using a dual-battery switch to change between the charged and the discharged battery while the motor is running, especially on EFI units as voltage spikes could damage solid state electronic modules like the ECU on EFI motors or the power pack on carbureted engines.

Perform this test with the motor in a test tank or with the motor properly mounted on a boat that is either firmly secured to a dock or underway and navigated by an assistant. The test will involve running the engine under load at speeds of 3000-5000 rpm depending on the model. Running the engine at these speeds using a flush fitting could allow the motor to become damaged from overspeed operation.

✱✱ WARNING

Never operate a marine motor without providing a source of cooling water such as a test tank or flush fitting. It takes less than a minute to damage the water pump impeller that could lead to much more expensive powerhead failures later during normal motor usage. Use a suitable test wheel whenever possible, but remember that for safety the propeller should not be installed on motors when running on a flush fitting.

1. Determine proper amperage output for the motor being tested. Refer to the information on Charging System Identification in this section.
2. On 70 hp EFI motors, remove the electrical component cover for access to the regulator/rectifier.
3. The regulation circuit of the charging system cannot work properly unless it receives a signal of battery voltage during operation. Before proceeding, use a DVOM to check the battery power signal to the regulator is working. The test should be performed between the appropriate by checking for battery power with the key on, engine not running at the proper wire connector as follows:
  • 9.9 hp (305cc) and 25/35 hp (500/565cc) carbureted motors, check the purple lead bullet connector for the regulator from the keyswitch harness.
  • 25-55 hp (737cc) and 25-70 hp (913cc) carbureted motors, check the purple lead at the terminal strip (for the regulator from the keyswitch harness.)
  • 40-70 hp EFI engines, check the regulator white wire.

IGNITION AND ELECTRICAL SYSTEMS 4-65

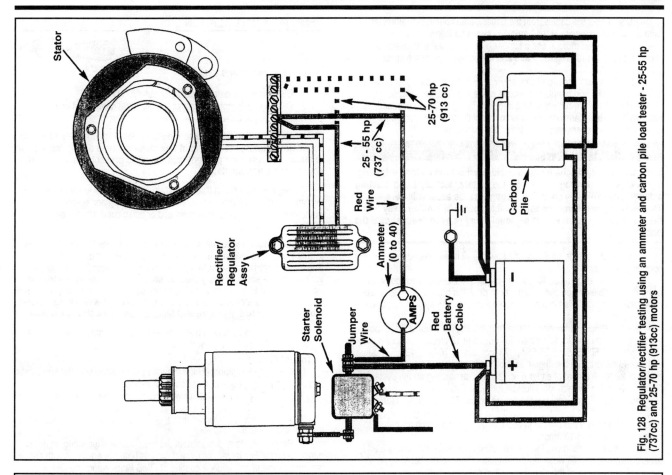

Fig. 128 Regulator/rectifier testing using an ammeter and carbon pile load tester - 25-55 hp (737cc) and 25-70 hp (913cc) motors

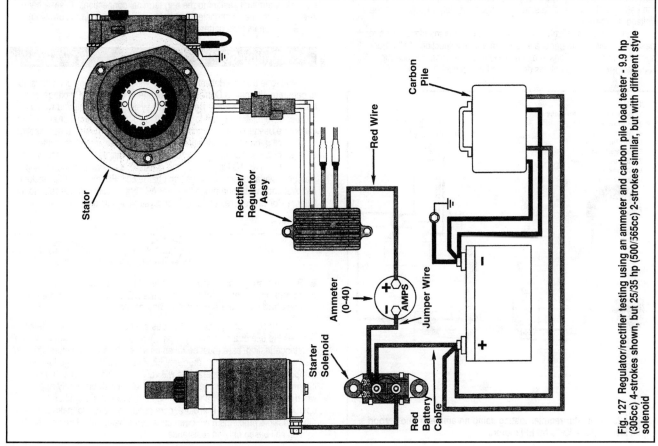

Fig. 127 Regulator/rectifier testing using an ammeter and carbon pile load tester - 9.9 hp (305cc) 4-strokes shown, but 25/35 hp (500/565cc) 2-strokes similar, but with different style solenoid

4-66 IGNITION AND ELECTRICAL SYSTEMS

4. Disconnect the cables from the battery, then connect an ammeter (with a minimum capacity of 0-40 amps) in series as follows:
• For carbureted motors, use a jumper wire to install the ammeter in series between the red regulator wire and the battery side of the starter solenoid/relay.
• For EFI motors use a pair jumper wires to install the ammeter in series between the male and female connectors of the regulator white wire.

5. Reconnect the battery cables and then install a carbon pile load tester to the battery terminals. Be sure to follow the tool manufacturer's instructions when connecting the carbon pile load tester.

** CAUTION

Excessive battery discharge rates could overheat the battery, releasing highly explosive hydrogen gas from the battery electrolyte and creating a dangerous explosive condition. Regardless, be especially sure to keep sparks, open flame, or any possible source of ignition away from the test area to prevent severe personal injury or even death should the battery explode.

6. Start and run the engine at idle speed until it reaches normal operating temperature
7. Once the engine has fully warmed, run it at the specified rpm while using the variable carbon pile load tester to draw the battery down at a rate equivalent to the stator's full output, then watch charging system output on the ammeter:
• 9.9 hp (305cc) carbureted motors, run the engine at about 5000 rpm, watching the ammeter for full or near full output of 12 amps.
• 25/35 hp (500/565cc), 25-55 hp (737cc) and 25-70 hp (913cc) carbureted motors, run the engine at about 4500 rpm, watching the ammeter for full or near full output of either 6 or 12 amps, depending on the system. Refer to Charging System Identification for more details.
• 40-70 hp EFI engines, run the engine first at 700 rpm looking for about 10 amps output, then advance the throttle to 1200 rpm looking for about 16 amps. Next advance the throttle to about 5000 rpm and watch for output to drop slightly, to about 13 amps.

8. Using the carbon pile load tester, slowly decrease the battery load toward zero amps while watching the ammeter. The meter should show a decrease in output from the charging system as the draw decreases. Check the battery voltage, it should stabilize at approximately 14.5 volts as the voltage decreases.
9. If output is too high or if output is too low and the stator has already passed a static ohmmeter check, perform a final inspection of all other system wiring/connections and, if no problems are found, replace the regulator/rectifier assembly as detailed in this section.

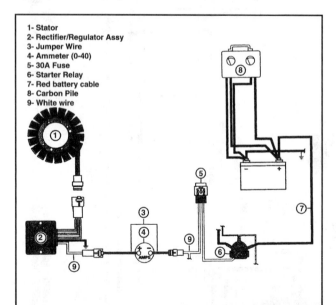

1- Stator
2- Rectifier/Regulator Assy
3- Jumper Wire
4- Ammeter (0-40)
5- 30A Fuse
6- Starter Relay
7- Red battery cable
8- Carbon Pile
9- White wire

Fig. 129 Regulator/rectifier testing using an ammeter and carbon pile load tester - 40-70 hp EFI motors

REMOVAL & INSTALLATION

◆ See Figures 130 and 131

1. Disconnect the battery cables for safety.
2. On 70 hp EFI motors, remove the electrical component cover for access to the regulator/rectifier.

■ Note wire routing before disconnecting any ring terminals or wiring connectors. Noting the routing at this stage will help during installation to make sure wires are properly positioned to prevent interference with any moving components.

3. Tag and disconnect the regulator/rectifier terminal leads at the terminal board (most larger carbureted motors) or tag and disengage the wiring connectors (usually Amphenol and/or bullet connectors) from the wiring harnesses.

■ If necessary, refer to the schematics found under Wiring Diagrams in this section to assist with wire terminal/connector identification. Regulator/Rectifier assemblies are normally connected to two or more yellow (and in some cases yellow w/ some tracer color) wires from the stator/terminal strip. They are also connected to either a red power and a purple battery signal wire, and possibly a ground wire, on carbureted models or a white signal wire and black ground wire on EFI models.

4. Remove the mounting bolts (usually 2), then remove the regulator/rectifier from the powerhead.

To install:

5. Clean the regulator/rectifier mounting area, along with the powerhead and screw threads of any dirt, corrosion or debris.
6. Inspect the ring terminal or wiring connectors for damage or corrosion and repair, clean or replace, as necessary.
7. Position the regulator/rectifier, then install and tighten the mounting bolts to 60-84 inch lbs. (7-9 Nm).
8. Properly route and reconnect the regulator/rectifier wiring as noted during removal. For larger carbureted models (that are equipped with ring terminal connections), apply a coating of OMC Black Neoprene Dip or equivalent terminal sealant to the ring terminal connections (this will help keep them securely connected and help to protect them from corrosion).
9. Reconnect the battery cables.

AC Lighting Coil

Some rope start models are equipped with an AC lighting coil that is used to generate Alternating Current (AC) voltage to operated AC voltage powered boat mounted accessories. Because this system is not equipped with a rectifier or regulator/rectifier to convert this voltage to Direct Current (DC) it cannot be used to charge marine batteries or run DC powered accessories.

The coil generates electricity in the same manner as a stator/battery charge coil or ignition charge coil as detailed in this section. In fact, when equipped with an AC lighting coil, the coil itself is mounted to the ignition plate, directly adjacent to the ignition charge coil. For more details, including removal and installation procedures for AC lighting coils, please refer to the Charge Coil procedures under Ignition System, in this section.

TESTING

■ Before testing, make sure all connections and mounting bolts are clean and tight. Many problems with the AC lighting system are related to loose and corroded terminals or bad grounds.

Since most AC charge coil test values are near or below 1 ohm, testing requires the use of a high quality Digital Volt Ohmmeter (DVOM). As with all resistance testing, care must be taken as specifications are determined by testing components that are precisely 68°F (20°C) and readings taken on components at other temperatures will vary. Also, keep in mind that certain intermittent problems (such as certain internal shorts) might only appear when the coil is put under a load during normal operation or at certain operating temperatures. Therefore, care should be taken to make sure all other possible problems have been considered before condemning the AC lighting coil based only on resistance test results.

IGNITION AND ELECTRICAL SYSTEMS

During testing the coil winding resistance is checked across various combinations of the AC lighting coil wiring (yellow or yellow with trace color wires) in order to see if the circuit contains the proper amount of resistance and is therefore complete, yet containing no internal shorts between coils. Each lead is then checked between the coil and a good engine ground to make sure no coil winding is shorted to ground. If opens or shorts are found, the wiring harness should be checked for damage (when present, visible signs of broken or worn insulation are usually the cause of the problem) before deciding to replace the AC lighting coil itself.

1. Disconnect the negative battery cable for safety and to protect the test equipment.

■ As usual on outboards, take note of wire routing before disconnecting them. Wires must always be returned to the original position/routing in order to prevent interference with and damage from moving components.

2. Trace the AC lighting coil yellow (and yellow with a trace color) wiring from under the edge of the flywheel to the harness or terminal board connectors, as applicable. Disconnect the coil wiring for access while testing with the DVOM. Inspect the terminals for corrosion or damage and clean or replace them, as necessary.

■ Most AC lighting coils are equipped with 3 wires (yellow, yellow/gray and yellow/blue), but some motors contain coils whose winding are only connected to 2 wires (yellow and yellow/gray). Don't be alarmed if your motor is equipped with either.

3. Using the DVOM set to its lowest resistance scale, probe across the appropriate lead pairs for the coil windings and compare the readings to specification, as follows:

 a. Connect one meter probe to the AC lighting coil yellow/gray wire and the other to the yellow wire. Resistance should be 0.81-0.91 ohm.

 b. On most models (equipped with 3 coil leads), move one meter probe from the yellow wire to the yellow blue wire (leaving the other probe still connected to the yellow/gray wire). Resistance should be 1.19-1.23 ohms.

4. If resistance readings across either pair of the noted terminals are out of specification, recheck connections to ensure proper testing readings (and make sure the components are being tested at the proper temperature). Double-check the wiring harness for any problems and, if none are found, replace the AC lighting coil.

5. Check the coil windings and harnesses for shorts to ground as follows. Connect one meter probe to a good engine ground, then probe each of the AC lighting coil wires (yellow, yellow/gray and, if applicable, yellow/blue) using the other probe. Resistance should be high or infinite in all cases, or there is a short to ground. If a short is present either repair the wiring harness or replace the coil, as applicable.

6. Once the tests and/or repairs are completed, be sure to route the wires as noted before removal to ensure there is no interference with moving components (such as the flywheel).

7. Connect the negative battery cable.

REMOVAL & INSTALLATION

 MODERATE

When equipped, the AC lighting coil is mounted immediately adjacent to the charge coil on the ignition plate (found under the flywheel). Because removal and installation procedures are the same, please refer to Charge Coil, Removal & Installation procedures that are found under the Ignition System portion of this section.

Battery

The battery is one of the most important parts of the electrical system. In addition to providing electrical power to start the engine (and for the ignition system on EFI motors), it also provides power for operation of the running lights, radio, and electrical accessories.

Because of its job and the consequences (failure to perform in an emergency), the best advice is to purchase a well-known brand, with an extended warranty period, from a reputable dealer.

The usual warranty covers a pro-rated replacement policy, which means the purchaser would be entitled to a consideration for the time left on the warranty period if the battery should prove defective before its time.

Do not consider a battery of less Cold Cranking Amperage (CCA) or Amp Hour (AH) rating than the battery that was originally installed for your motor. In fact, due to the increased resistance that will occur in circuits over time (from things like corrosion or internal wire strands that wear and break inside the insulation), it is advisable to buy a replacement battery with higher capacity than the original (but do not go overboard, pun intended).

■ Original minimum battery CCA ratings are provided in the General Engine System Specifications chart found in the General Information and Maintenance section.

MARINE BATTERIES

◆ See Figure 132

Because marine batteries are required to perform under much more rigorous conditions than automotive batteries, they are constructed differently than those used in automobiles or trucks. Therefore, a marine battery should always be the No. 1 unit for the boat and other types of batteries used only in an emergency (or possibly as a second battery).

Marine batteries have a much heavier exterior case to withstand the violent pounding and shocks imposed on it as the boat moves through rough water and in extremely tight turns. The plates are thicker and each plate is securely anchored within the battery case to ensure extended life. The caps are spill proof to prevent acid from spilling into the bilges when the boat heels to one side in a tight turn, or is moving through rough water. Because of these features, the marine battery will recover from a low charge condition and give satisfactory service over a much longer period of time than any type intended for automotive use.

Fig. 130 To remove the regulator on most models, first disconnect the wiring (1) from the terminal strip (2) or harness connections (as applicable)...

Fig. 131 ... then unbolt and remove the regulator/rectifier from the powerhead

Fig. 132 A fully charged battery, filled to the proper level with electrolyte, is the heart of the ignition and electrical systems. Engine cranking and efficient performance of electrical items depend on a full rated battery

4-68 IGNITION AND ELECTRICAL SYSTEMS

✱✱ WARNING

Never use a maintenance-free battery with an outboard engine that is not voltage regulated. The charging system will continue to charge as long as the engine is running and it is possible that the electrolyte could boil out rendering the battery useless.

BATTERY CONSTRUCTION

◆ See Figure 133

A battery consists of a number of positive and negative plates immersed in a solution of diluted sulfuric acid. The plates contain dissimilar active materials and are kept apart by separators. The plates are grouped into elements. Plate straps on top of each element connect all of the positive plates and all of the negative plates into groups.

The battery is divided into cells holding a number of the elements apart from the others. The entire arrangement is contained within a hard plastic case. The top is a one-piece cover and contains the filler caps for each cell. The terminal posts protrude through the top where the battery connections for the boat are made. Each of the cells is connected to its neighbor in a positive-to-negative manner with a heavy strap called the cell connector.

BATTERY RATINGS

◆ See Figure 134

Three different methods are used to measure and indicate battery electrical capacity:
- Amp/hour (AH) rating
- Cold Cranking Amp (CCA) performance
- Reserve capacity

The AH rating of a battery refers to the battery's ability to provide a set amount of amps for a given amount of time under test conditions at a constant temperature. Therefore, if the battery is capable of supplying 4 amps of current for 20 consecutive hours, the battery is rated as an 80 amp/hour battery. The amp/hour rating is useful for some service operations, such as slow charging or battery testing.

CCA performance is measured by cooling a fully charged battery to 0°F (-17°C) and then testing it for 30 seconds to determine the maximum current flow. In this manner the cold cranking amp rating is the number of amps available to be drawn from the battery before the voltage drops below 7.2 volts.

The illustration depicts the amount of power in watts available from a battery at different temperatures and the amount of power in watts required of the engine at the same temperature. It becomes quite obvious - the colder the climate, the more necessary for the battery to be fully charged.

Reserve capacity of a battery is considered the length of time, in minutes, at 80°F (27°C), a 25 amp current can be maintained before the voltage drops below 10.5 volts. This test is intended to provide an approximation of how long the engine, including electrical accessories, could operate satisfactorily if the stator assembly or lighting coil did not produce sufficient current. A typical rating is 100 minutes.

If possible, the new battery should have a power rating equal to or higher than the unit it is replacing.

BATTERY LOCATION

Every battery installed in a boat must be secured in a well protected, ventilated area. If the battery area lacks adequate ventilation, hydrogen gas, which is given off during charging may gather in a concentrated quantity, and could cause a hazardous condition as it is very explosive.

BATTERY SERVICE

Details regarding cleaning the battery, checking fluid level and testing it and maintaining a proper charge while the battery is in storage can be found under Batteries in the Boat Maintenance section.

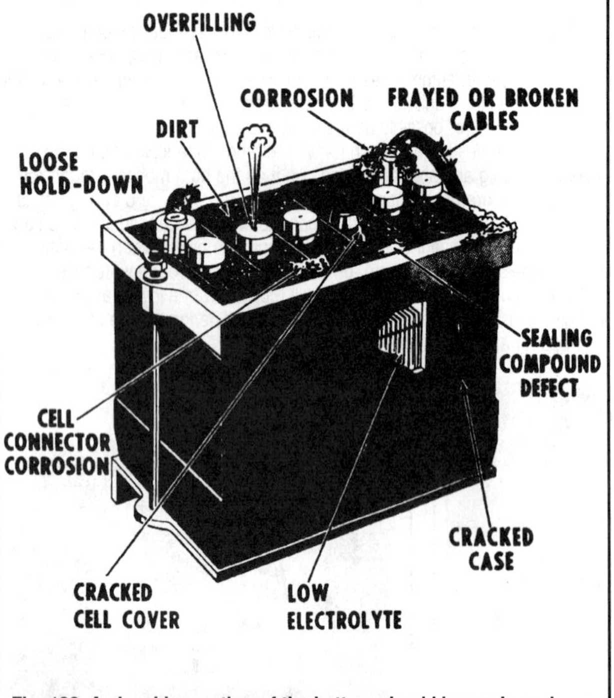

Fig. 133 A visual inspection of the battery should be made each time the boat is used. Such a quick check may reveal a potential problem in its early stages. A dead battery in a busy waterway or far from assistance could have serious consequences

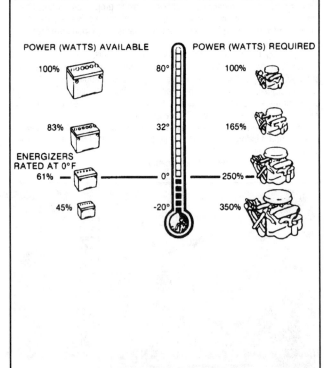

Fig. 134 Comparison of battery efficiency and engine demands at various temperatures

IGNITION AND ELECTRICAL SYSTEMS

STARTING CIRCUIT

Description and Operation

◆ See Figure 135

In the early days, all outboards were started by pulling on a rope wrapped around the flywheel. As time passed and owners were reluctant to use muscle power (or came up short especially with larger and larger motors), it was necessary to replace the rope starter with some form of power cranking system. Today, most small engines may be started by pulling on a rope, but many smaller most larger engines are also equipped with an electric starter system.

The system used to replace the rope starter is an electric starter motor coupled with a mechanical gear mesh between the starter motor and the powerhead flywheel, similar to the method used to crank an automobile engine.

As the name implies, the sole purpose of the starter circuit is to control operation of the starter motor causing it to crank the powerhead until the engine catches and runs. The circuit usually includes a solenoid or magnetic switch that connects the motor to the battery when the circuit is actuated and disconnects the motor from the battery when the circuit is deactivated. The operator controls the solenoid switch with a key switch or starter button, depending on the model.

■ **Some smaller models, like the 9.9-15 hp 2-strokes, do not use a solenoid. The starter button is located inline between power and the starter itself. In this unique case, the starter button switch physically closes (directly performing the job of the solenoid) to provide power to the starter.**

A neutral safety or, on some smaller models, a mechanical starter lockout is installed to permit operation of the starter motor only if the shift control lever is in neutral. When used, the switch is a safety device that prevents accidental engine starts when the motor is in gear.

The starter motor itself is an electric component consisting of a wound coil that draws a heavy current from the battery. It is designed for short periods of usage when cranking the engine for startup. To prevent overheating the motor, cranking should not be continued for more than 30-seconds without allowing the motor to cool for at least three minutes. Actually, this time can be spent in making preliminary checks to determine why the engine fails to start.

Power is transmitted from the starter motor to the powerhead flywheel through a Bendix drive. This drive has a pinion gear mounted on screw threads. When the motor is operated, the pinion gear moves upward and meshes with the teeth on the flywheel ring gear.

When the powerhead starts, the pinion gear is driven faster than the shaft, and as a result, it screws out of mesh with the flywheel. A rubber cushion is built into the Bendix drive to absorb the shock when the pinion meshes with the flywheel ring gear. The parts of the drive must be properly assembled for efficient operation. If the drive is removed for cleaning, take care to assemble the parts as noted during removal (and shown in accompanying illustrations). If the screw shaft assembly is reversed, it will strike the splines and the rubber cushion will not absorb the shock.

The sound of the motor during cranking is a good indication of whether the starter motor is operating properly or not. Naturally, temperature conditions will affect the speed at which the starter motor is able to crank the engine. The speed of cranking a cold engine will be much slower than when cranking a warm engine. An experienced operator will learn to recognize the favorable sounds of the powerhead cranking under various conditions.

Troubleshooting

If the starter motor spins, but fails to crank the engine, the cause is usually a corroded or gummy Bendix drive. The drive should be at least lubricated or it should be removed, cleaned, and given an inspection.

If the starter motor cranks the engine too slowly, the following are possible causes and the corrective actions that may be taken:
- Battery charge is low. Charge the battery to full capacity.
- High resistance connections at the battery, solenoid, or motor. Clean and tighten all connections.
- Undersize battery cables. Replace cables with sufficient size.
- Battery cables too long (which creates too high a resistance in the circuit). Relocate the battery to shorten the run to the solenoid.
- Binding mechanical problem with the powerhead or gearcase.

If the starter does not crank the motor at all, before wasting too much time troubleshooting the starter motor circuit, the following checks should be made. Many times, the problem will be corrected. Make sure the:
- Battery is fully charged.
- Shift control lever is in neutral.
- Main fuse (if used in the starter circuit, refer to Starter Circuit Testing or Wiring Diagrams, in this section for more information) is good (not blown).
- All electrical connections are clean and tight.
- Wiring in good condition, insulation not worn or frayed.

Also, keep in mind that mechanical problems could exist that would cause the powerhead to freeze or crank slowly even though the starter motor circuit is in excellent condition
- A tight or frozen powerhead
- Hydro-locked motor (see Clearing a Submerged Motor, in the Maintenance and General Information section).

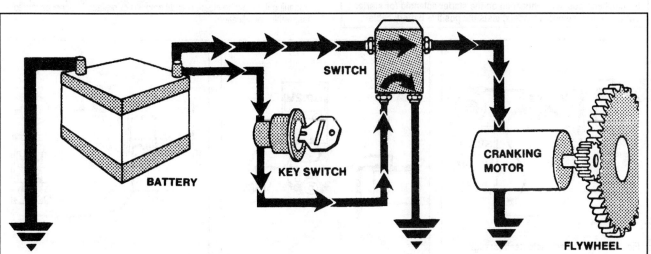

Fig. 135 A typical starting system converts electrical energy into mechanical energy to turn the engine. The components are: Battery, to provide electricity to operate the starter; Ignition switch, to control the energizing of the starter relay or solenoid; Starter relay or solenoid switch, to make and break the circuit between the battery and starter; Starter, to convert electrical energy into mechanical energy to rotate the engine; Starter drive gear, to transmit the starter rotation to the engine flywheel

4-70 IGNITION AND ELECTRICAL SYSTEMS

- Water in the lower unit.

If no obvious cause has been found and the starter motor does not operate, test the circuit and/or components as detailed in this section.

STARTING SYSTEM TESTING (CARBURETED MODELS)

Voltage Drop Test (Starter Turns Slowly)

◆ See Figures 136 thru 139

If the preliminary checks covered under Troubleshooting the Starting System do not reveal the problem, either perform the voltage drop test covered here or the No Load Current Draw Test covered under Starter Motor.

■ For more details on the theory behind Voltage Drop Testing, please refer to the information found in Understanding and Troubleshooting Electrical Systems, found in this section.

Excessively high resistance within the starter circuit (from problems with battery cables, the starter solenoid or any of the wiring/grounds) can impede the amount of voltage available to the starter. This may cause the starter to turn slowly or even not at all. In addition, a slow cranking speed will keep the ignition charge coil from reaching full output, thereby weakening spark during attempts to start the motor. This might cause the ignition system to appear faulty when diagnosing a hard or no-start condition.

A voltage drop test is used to check the amount of voltage consumed by the circuit. Remember also that intermittent problems from internally broken or frayed cables to loose or corroded connections can cause slow cranking. If readings vary when testing a portion of the circuit, grasp and flex cables attached to the point of testing to make sure no intermittent connections are adversely affecting the test or circuit operation.

1. Disconnect and ground the spark plug leads to the powerhead for safety (to keep the engine from starting during testing) and to protect the ignition system.

2. If not done already, perform all preliminary checks as listed under Troubleshooting the Starting System. The battery must be fully charged, all connections must be clean and tight. All wiring and electrical components must at least appear to be in good condition.

3. Using a DVOM set to read voltage, connect the red probe to the positive battery post and the black probe to positive post of the starter solenoid terminal (the point at which the red cable from the battery connects to the starter solenoid). Then crank the starter while watching the DVOM. A reading of more than 0.3 volt indicates high resistance in the positive battery cable (meaning it should be cleaned and tightened or replaced, as necessary to test within specification).

4. Connect the DVOM black probe to the starter side of the solenoid terminal (the terminal whose wiring runs to the starter motor itself). **With the engine cranking**, touch the red DVOM probe to the positive post of the starter solenoid terminal (again, the point at which the red cable from the battery connects to the solenoid). In order to prevent damage to the DVOM, **do not** allow the red probe to make contact with the post **unless** the motor is cranking (otherwise, voltage may try to short around the solenoid, through the meter). A reading of more than 0.2 volt indicates high resistance in the starter solenoid itself. If such a reading is noted on the meter, replace the starter solenoid.

5. Connect the DVOM red probe to the starter side of the solenoid terminal (again, the terminal whose wiring runs to the starter motor itself). Connect the black probe directly to the terminal on the starter motor (the other end of the wire where the red probe is connected). Actuate the starter while watching the DVOM. A reading of more than 0.2 volt indicates high resistance in the solenoid-to-motor cable (meaning it should be cleaned and tightened or replaced, as necessary to test within specification).

6. Connect the DVOM red probe to the negative battery cable, where it connects to the powerhead, then connect the black probe to the negative battery post. Actuate the starter while watching the DVOM. A reading of more than 0.3 volt, indicates high resistance in the negative battery cable (meaning it should be cleaned and tightened or replaced, as necessary to test within specification).

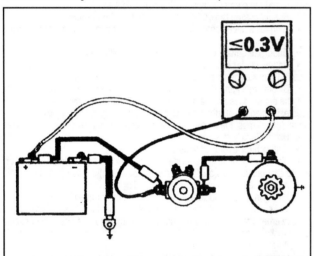

Fig. 136 Checking voltage drop between the positive battery terminal and positive connection on the starter solenoid (or starter button on non-solenoid models) tests the positive battery cable

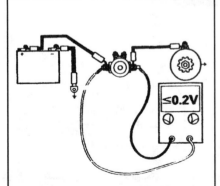

Fig. 137 Test the solenoid itself for high resistance by checking drop on either side of it (solenoid positive terminal and solenoid starter terminal) - NEVER make this connection unless the motor is cranking (the same test could be made at the starter button on non-solenoid models)

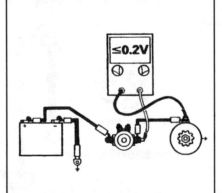

Fig. 138 Probing at both ends of the solenoid-to-starter cable and cranking the motor will check for excessive resistance in this cable/connection

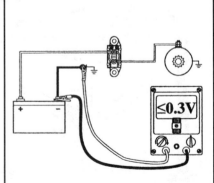

Fig. 139 Checking voltage drop between the negative battery terminal and the powerhead ground connection tests the negative battery cable-typical 4-stroke solenoid pictured

IGNITION AND ELECTRICAL SYSTEMS 4-71

Starter Circuit Testing (Starter Does Not Turn)

◆ See Figures 140 thru 172

If the starter does not operate at all, check the starter circuit using a DVOM (or a test light if one if not available) to check for voltage at points throughout the circuit. The accompanying charts will take you step-by-step through each starter circuit in the most logical sequence for that circuit. On some, that means checking for battery voltage first at the starter and then working your way back until power is found (meaning the problem in the circuit is between that point and the previously tested point). While on other models, sub-circuits are checked in a specific order, eliminating them one by one, but not necessarily beginning at the starter and working your way back to the battery. Illustrations accompany each chart that can be used to help determine proper testing points.

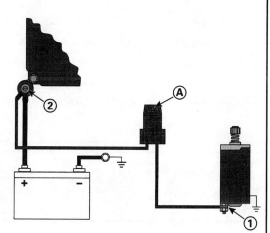

⚠ Disconnect starter cable at ① to avoid accidental starting while testing.

[Note] Engine must be in NEUTRAL, and battery must be fully charged.

Step	Procedure	Results
A. Check voltage on cable at ①.	Connect voltmeter or test light to cable at ① and to ground. Press start button Ⓐ.	a. If meter shows 12 volts, or test light goes ON, test starter. b. If meter shows 0 volts or test light remains OFF, go to Step A
B. Check voltage at terminal ②.	Connect voltmeter or test light to terminal ② and to ground.	a. If meter shows 12 volts or test light goes ON, test start switch and wires. b. If meter shows 0 volts or test light remains OFF, test battery cables and connections.

Fig. 140-141 Starter circuit testing - 9.9 hp (211cc) 4-stroke tiller control models

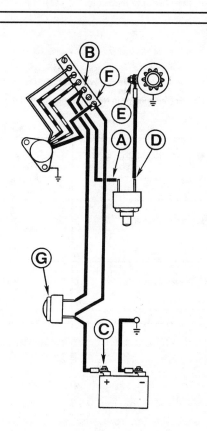

Note: Battery must be fully charged

Step	Procedure	Results
1. Check voltage between A and a good ground.	Connect meter red lead to A (neutral start switch) Connect meter black lead to an engine ground Press start button, voltage should be present at A	If no voltage is present, go to step 2
2. Check voltage between B and a good ground.	Connect meter red lead to B (wire terminal) Connect meter black lead to an engine ground Again, press the start button, voltage should be present at B	If voltage is now present, the lead between A and B is open and must be fixed If still no voltage, go to step 3
3. Check the start button G and leads for continuity	Disconnect the start switch leads at C, F and B. Connect an ohmmeter lead to A and the other to B. Watch the meter	Meter should show very low reading or 0 ohms otherwise the switch or leads are open and must be replaced If switch/leads are good at this point, go to step 4
4. Continue to check the start button G and leads for continuity	Connect an ohmmeter lead to A and the other to C. Press the start button and watch the meter.	Meter should show very low reading or 0 ohms otherwise the switch or leads are open and must be replaced
5. Check voltage at D (neutral start switch)	Set the meter to read voltage. Connect meter red lead to D (starter side of neutral switch). Then connect black lead to an engine ground. Press the start button G and watch for battery voltage on the meter	If voltage is not present, check for loose connections and/or replace/adjust the neutral switch (as necessary).
6. Check voltage at the starter E	Connect meter red lead to E Connect meter black lead to an engine ground Press the start button G and watch for battery voltage on the meter	If voltage is no present, check for open between E and D. If voltage IS present but starter does not operate, check/replace the starter motor itself.

Fig. 142-143 Starter circuit testing - 9.9-15 hp (216cc) 2-stroke models

4-72 IGNITION AND ELECTRICAL SYSTEMS

⚠ Avoid accidental starting while testing; disconnect the starter cable at point ①.

Note Engine must be in NEUTRAL throughout test procedure and battery must be fully charged.

Step	Procedure	Results
A. Check voltage between Ⓐ and Ⓑ.	Connect voltmeter red lead to Ⓐ. Connect meter black lead to Ⓑ.	a. If meter shows 12 volts, go to Step **B**. b. If meter shows 0 volts, test battery.
B. Check voltage between Ⓐ and Ⓒ.	Connect meter black lead to Ⓒ.	a. If meter shows 12 volts, go to Step **C**. b. If meter shows 0 volts, test negative battery cable.
C. Check voltage between Ⓓ and Ⓒ.	Connect meter red lead to Ⓓ.	a. If meter shows 12 volts, go to Step **D**. b. If meter shows 0 volts, test positive battery cable.
D. Check voltage between Ⓓ and Ⓔ.	Connect meter black lead to Ⓔ (powerhead).	a. If meter shows 12 volts, go to Step **E**. b. If meter shows 0 volts, clean and retighten rectifier screw Ⓒ.
E. Check voltage between Ⓕ and Ⓔ.	Connect meter red lead to Ⓕ. Push start button Ⓖ.	a. If meter shows 12 volts, go to Step **F**. b. If meter shows 0 volts, check start switch, wires, and connector Ⓓ.
F. Check operation of starter.	Connect start switch output lead Ⓕ to starter terminal Ⓗ. Push start button Ⓖ.	a. If starter runs, check condition of pinion Ⓘ. b. If starter does not run, service starter.

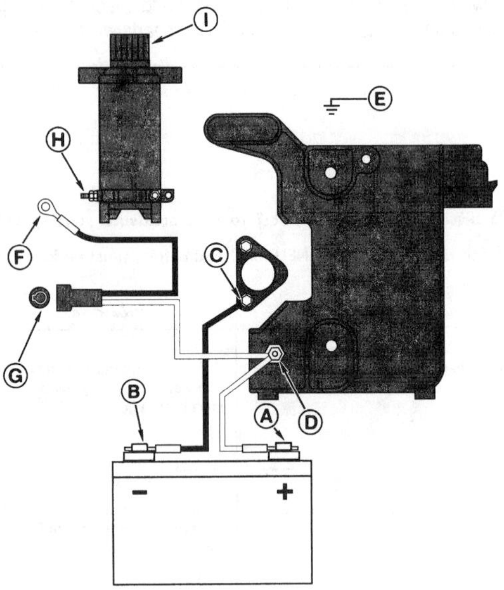

Fig. 144-145 Starter circuit testing - 9.9-15 hp (255cc) 2-stroke models

⚠ Avoid accidental starting while testing; disconnect the starter cable at point ①.

Note Engine must be in NEUTRAL throughout test procedure and battery must be fully charged.

Step	Procedure	Results
A. Check voltage to cable at ①.	Connect voltmeter between cable at ① and ground. Press start button Ⓑ.	a. If meter shows 12 volts, test starter. b. If meter shows 0 volts, go to Step **B**.
B. Check voltage at ②.	Connect voltmeter between ② and ground.	a. If meter shows 12 volts, go to Step **C**. b. If meter shows 0 volts, test battery and cables.
C. Check voltage at ③.	Connect voltmeter between ③ and ground. Press start button Ⓑ.	a. If meter shows 12 volts, go to Step **D**. b. If meter shows 0 volts, test fuse, start button, and wiring.
D. Check voltage at ④.	Connect voltmeter between ④ and ground. Press start button Ⓑ.	a. If meter shows 12 volts, replace starter cable. b. If meter shows 0 volts, check solenoid and ground wire.

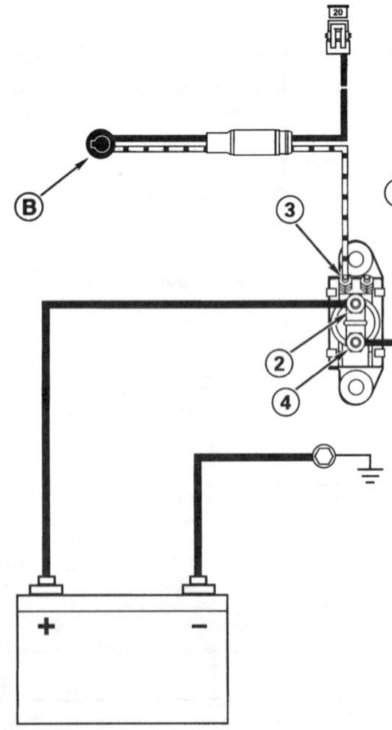

Fig. 146-147 Starter circuit testing - 9.9/15 hp (305cc) 4-stroke tiller control models

IGNITION AND ELECTRICAL SYSTEMS   4-73

Note: Battery must be fully charged.

Step	Procedure	Result
A. Check voltage between ground and ①. ⚠ Steps **A** thru **F** – Remove starter to solenoid cable from ⑦ to prevent starter engagement while making checks.	Remove black lead from ground at ①. Connect voltmeter between ① and common engine ground. Turn key to START position. Voltmeter should show battery voltage.	a. If no reading, go to Step **B** b. If meter reads battery voltage, reconnect black lead to ground and go to Step **F**
B. Check voltage between ground and ②. **Note:** Steps **B** thru **F** – Turn key OFF before connecting and disconnecting meter. Turn key to START position *after* connecting.	Connect meter at ②. Turn key to START position.	a. If meter reads battery voltage, lead is open between ① and ②. b. If no reading, go to Step **C**
C. Check voltage between ground and ③.	Connect meter at ③. Turn key to START position.	a. If meter reads battery voltage, solenoid is faulty. b. If no reading, go to Step **D**
D. Check voltage between ground and ④.	Connect meter at ④. Turn key to START position.	a. If meter reads battery voltage, lead is open between ③ and ④ or neutral start switch Ⓐ is open or improperly adjusted. b. If no reading, go to Step **E**
E. Check voltage between ground and ⑤.	Connect meter at ⑤. Turn key to OFF position.	a. If meter reads battery voltage, check key switch. b. If no reading at ⑤, check for open lead or open fuse between ⑤ and ⑥. c. Connect meter at ⑥. If no reading, check for open lead between ⑥ and battery "+" terminal. If meter reads battery voltage, go to Step **G**
F. Check voltage between ground and ⑦.	Connect voltmeter at ⑦. Turn key to START position.	a. If no reading, solenoid is faulty. b. If solenoid clicks and meter reads battery voltage, go to Step **G**
G. Check voltage between ground and ⑧.	Reconnect starter to solenoid cable at ⑦. Connect meter at ⑧. Turn key to START position.	a. If meter reads battery voltage and starter motor does not turn, check starter motor. b. If no reading, check for broken cable or poor connection.

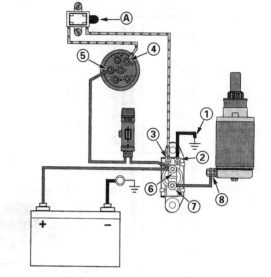

Fig. 148-149 Starter circuit testing - 9.9/15 hp (305cc) 4-stroke remote control models

---

⚠ Avoid accidental starting while testing; disconnect the starter cable at point ①.

Note: Engine must be in NEUTRAL throughout test procedure and battery must be fully charged.

Step	Procedure	Results
A. Check voltage to cable at ①.	Connect voltmeter Ⓓ between cable at ① and ground. Press start button Ⓑ.	a. If meter shows 12 volts, test starter. b. If meter shows 0 volts, go to Step **B**
B. Check voltage at ②.	Connect voltmeter Ⓓ between ② and ground.	a. If meter shows 12 volts, go to Step **C** b. If meter shows 0 volts, test battery and cables.
C. Check voltage at ③.	Connect voltmeter Ⓓ between ③ and ground. Press start button Ⓑ.	a. If meter shows 12 volts, go to Step **D** b. If meter shows 0 volts, test start button and wiring.
D. Check voltage at ④.	Disconnect yellow/red lead ④ from neutral start switch Ⓐ. Connect voltmeter Ⓓ between lead ④ and ground. Press start button Ⓑ.	a. If meter shows 12 volts, test neutral start switch Ⓐ and proceed to Step **E** b. If meter shows 0 volts, replace solenoid or lead.
E. Check voltage at ⑤.	Reconnect yellow/red lead ④. Connect voltmeter Ⓓ between ⑤ and ground. Press start button Ⓑ.	a. If meter shows 0 volts, replace solenoid. b. If meter shows 12 volts, replace starter cable.

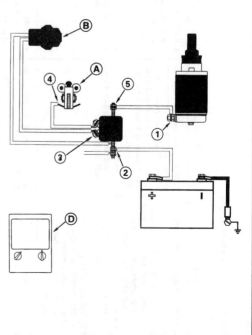

Fig. 150-151 Starter circuit testing - 18 Jet-35 hp (521cc) tiller control models (note late-model solenoid shown, tests are the same on earlier model with round-body solenoid)

4-74 IGNITION AND ELECTRICAL SYSTEMS

Note: Battery must be fully charged.

Step	Procedure	Result
A. Check voltage between ground and ① ⚠ **Steps A thru F** - Remove starter to solenoid cable from ⑦ to prevent starter engagement while making checks.	Remove black lead from ground at ①. Connect voltmeter between ① and common engine ground. Turn key to START position. Voltmeter should show battery voltage.	a. If no reading, go to Step **B**. b. If meter reads battery voltage, reconnect black lead to ground and go to Step **F**.
B. Check voltage between ground and ② **Note** Steps **B** thru **F** - Turn key OFF before connecting and disconnecting meter. Turn key to START position **after** connecting.	Connect meter at ② Turn key to START position.	a. If meter reads battery voltage, lead is open between ① and ②. b. If no reading, go to Step **C**.
C. Check voltage between ground and ③	Connect meter at ③ Turn key to START position.	a. If meter reads battery voltage, solenoid is faulty. b. If no reading, go to Step **D**.
D. Check voltage between ground and ④	Connect meter at ④ Turn key to START position.	a. If meter reads battery voltage, lead is open between ③ and ④ or neutral start switch Ⓐ is open or improperly adjusted. b. If no reading, go to Step **E**.
E. Check voltage between ground and ⑤	Connect meter at ⑤ Turn key to OFF position.	a. If meter reads battery voltage, check key switch. b. If no reading at ⑤ check for open lead or open fuse between ⑤ and ⑥ c. Connect meter at ⑥ If no reading, check for open lead between ⑥ and battery "+" terminal. If meter reads battery voltage, go to Step **G**.
F. Check voltage between ground and ⑦	Connect voltmeter at ⑦ Turn key to START position.	a. If no reading, solenoid is faulty. b. If solenoid clicks and meter reads battery voltage, go to Step **G**
G. Check voltage between ground and ⑧	Reconnect starter to solenoid cable at ⑦ Connect meter at ⑧ Turn key to START position.	a. If meter reads battery voltage and starter motor does not turn, check starter motor. b. If no reading, check for broken cable or poor connection.

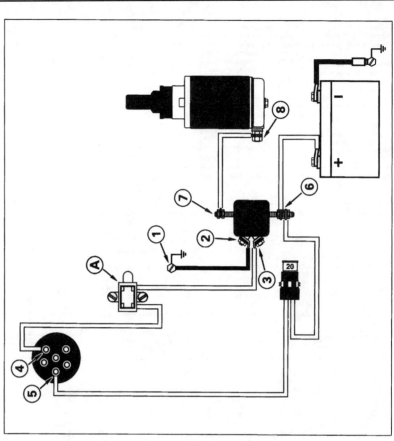

Fig. 152-153 Starter circuit testing - 18 Jet-35 hp (521cc) remote control models (note late-model solenoid shown, tests are the same on earlier model with round-body solenoid)

IGNITION AND ELECTRICAL SYSTEMS 4-75

Note: Battery must be fully charged.

Step	Procedure	Result
A. Check voltage between ground and ①. ⚠ **Steps A thru F** – **Remove starter to solenoid cable from ⑦ to prevent starter engagement while making checks.**	Remove black lead from ground at ①. Connect voltmeter between ① and common engine ground. Turn key to START position. Voltmeter should show battery voltage.	a. If no reading, go to Step **B**. b. If meter reads battery voltage, reconnect black lead to ground and go to Step **F**.
B. Check voltage between ground and ②. **Note:** Steps **B** thru **F** – Turn key OFF before connecting and disconnecting meter. Turn key to START position **after** connecting.	Connect meter at ②. Turn key to START position.	a. If meter reads battery voltage, lead is open between ① and ②. b. If no reading, go to Step **C**.
C. Check voltage between ground and ③.	Connect meter at ③. Turn key to START position.	a. If meter reads battery voltage, solenoid is faulty. b. If no reading, go to Step **D**.
D. Check voltage between ground and ④.	Connect meter at ④. Turn key to START position.	a. If meter reads battery voltage, lead is open between ③ and ④ or neutral start switch ④ is open or improperly adjusted. b. If no reading, go to Step **E**.
E. Check voltage between ground and ⑤.	Connect meter at ⑤. Turn key to OFF position.	a. If meter reads battery voltage, check key switch. b. If no reading at ⑤, check for open lead or open fuse between ⑤ and ⑥. c. Connect meter at ⑥. If no reading, check for open lead between ⑥ and battery "+" terminal. If meter reads battery voltage, go to Step **G**.
F. Check voltage between ground and ⑦.	Connect voltmeter at ⑦. Turn key to START position.	a. If no reading, solenoid is faulty. b. If solenoid clicks and meter reads battery voltage, go to Step **G**
G. Check voltage between ground and ⑧.	Reconnect starter to solenoid cable at ⑦. Connect meter at ⑧. Turn key to START position.	a. If meter reads battery voltage and starter motor does not turn, check starter motor. b. If no reading, check for broken cable or poor connection.

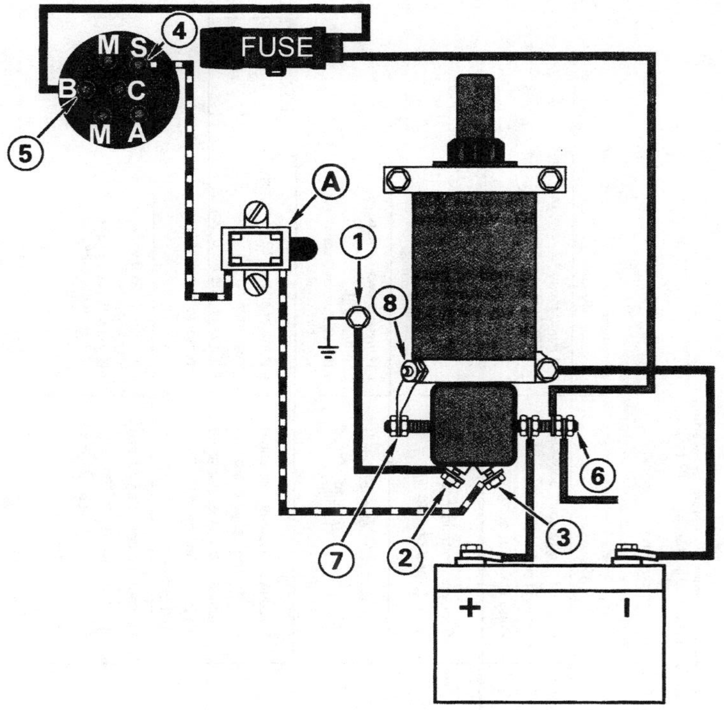

Fig. 154-155 Starter circuit testing - 25/35 hp (500/565cc) models

4-76 IGNITION AND ELECTRICAL SYSTEMS

⚠ Disconnect starter cable at point ① before proceeding with this check to avoid accidental starting while testing.

Note Engine must be in "Neutral" throughout test procedure.

Step	Procedure	Results
1 Check voltage to cable at point ①.	Connect voltmeter between cable at point ① and ground. Press start button Ⓐ.	• If meter shows 12 volts, test starter. • If meter shows 0 volts, go to step **2**.
2 Check voltage at point ②.	Connect voltmeter between point ② and ground.	• If meter shows 12 volts, go to step **3**. • If meter shows 0 volts, test battery and cables.
3 Check voltage at point ③.	Connect voltmeter between point ③ and ground. Press start button.	• If meter shows 12 volts, go to step **4**. • If meter shows 0 volts, test start button and wiring.
4 Check voltage at point ④.	Disconnect small solenoid lead ④ from engine ground. Connect voltmeter between lead ④ and ground. Press start button Ⓐ.	• If meter shows 12 volts, proceed to step **5**. • If meter shows 0 volts, replace solenoid or lead.
5 Check voltage at point ⑤.	Reconnect small solenoid lead ④. Connect voltmeter between point ⑤ and ground. Press start button Ⓐ.	• If meter shows 0 volts, replace solenoid. • If meter shows 12 volts, replace starter cable.

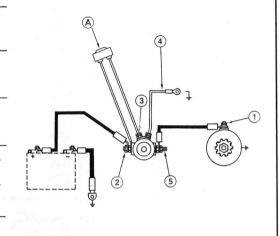

Fig. 156-157 Starter circuit testing - 25-55 hp (737cc) 1990 tiller models

⚠ Avoid accidental starting while testing; disconnect the starter cable at point ① before proceeding with this check.

Note Engine must be in NEUTRAL throughout test procedure. Battery must be fully charged.

Step	Procedure	Results
1 Check voltage to cable at point ①.	Connect voltmeter between cable at point ① and ground. Press start button Ⓐ.	• If meter shows 12 volts, test starter. • If meter shows 0 volts, go to step **2**.
2 Check voltage at point ②.	Connect voltmeter between point ② and ground.	• If meter shows 12 volts, go to step **3**. • If meter shows 0 volts, test battery and cables.
3 Check voltage at point ③.	Connect voltmeter between point ③ and ground. Press start button.	• If meter shows 12 volts, go to step **4**. • If meter shows 0 volts, test start button and wiring.
4 Check voltage at point ④.	Disconnect solenoid ground lead ④ from engine ground. Connect voltmeter between lead ④ and point ③. Press start button Ⓐ.	• If meter shows 12 volts, proceed to step **5**. • If meter shows 0 volts, replace solenoid or lead.
5 Check voltage at point ⑤.	Reconnect solenoid ground lead ④. Connect voltmeter between point ⑤ and ground. Press start button Ⓐ.	• If meter shows 0 volts, replace solenoid. • If meter shows 12 volts, replace starter cable.

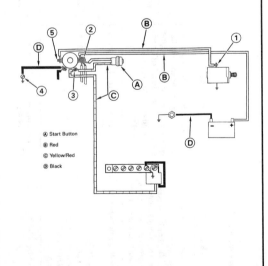

Ⓐ Start Button
Ⓑ Red
Ⓒ Yellow/Red
Ⓓ Black

Fig. 158-159 Starter circuit testing - 25-55 hp (737cc) 1991-92 tiller models

IGNITION AND ELECTRICAL SYSTEMS 4-77

⚠ Avoid accidental starting while testing; disconnect the starter cable at point ① before proceeding with this check.

Note: Engine must be in NEUTRAL throughout test procedure. Battery must be fully charged.

Step	Procedure	Result
1 Check voltage between ① and ground.	Connect voltmeter between ① and ground. Press start button Ⓐ.	• If meter shows 12V, test starter. • If meter shows 0V, go to Step **2**.
2 Check voltage between ② and ground.	Connect voltmeter between ② and ground. Press start button Ⓐ.	• If meter shows 12V, test cable to starter. • If meter shows 0V, go to Step **3**.
3 Check voltage between ③ and ground.	Connect voltmeter between ③ and ground.	• If meter shows 0V, check battery cable and battery. • If meter shows 12V, go to Step **4**.
4 Check voltage between ④ and ground.	Connect voltmeter between ④ and ground. Turn key switch Ⓑ ON. Press start button Ⓐ.	• If meter shows 12V, go to Step **8**. • If meter reads 0V, go to Step **5**.
5 Check voltage between ⑤ and ground.	Connect voltmeter between ⑤ and ground. Turn key switch Ⓑ ON.	• If meter shows 12V, replace start button Ⓐ and/or wire. • If meter shows 0V, go to Step **6**.
6 Check voltage between ⑥ and ground.	Connect voltmeter between ⑥ and ground.	• If meter shows 12V, go to Key Switch Check, this section. Check wire to terminal strip. • If meter shows 0V, go to Step **7**.
7 Check voltage between ⑦ and ground.	Connect voltmeter between ⑦ and ground.	• If meter shows 12V, replace fuse, holder, or wire to key switch. • If meter shows 0V, replace wire from solenoid terminal ③ to fuse ⑦.
8 Check voltage between ⑧ and ground.	Disconnect black ground wires from ⑧. Connect voltmeter between solenoid terminal ⑧ and ground. Turn key switch Ⓑ ON. Push start switch Ⓐ.	• If meter shows 12V, repair/replace black wire between ⑧ and ground. • If meter shows 0V, disconnect all leads from solenoid, and test solenoid. Go to **Solenoid Check**, this section.

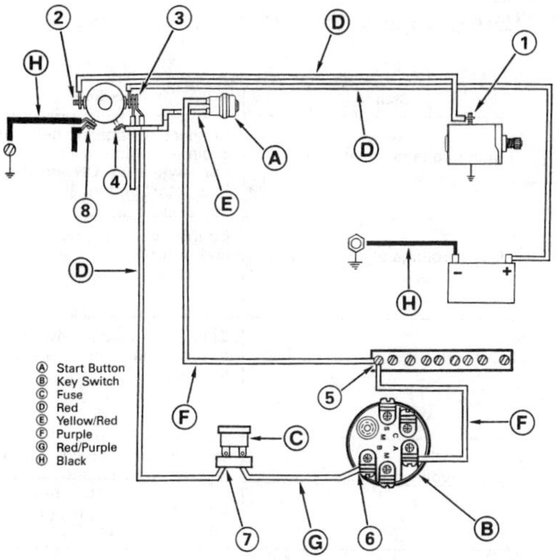

Fig. 160-161 Starter circuit testing - 25-55 hp (737cc) 1993-95 tiller models

---

⚠ To avoid accidental starting of engine, disconnect starter cable at ①.

Note: Engine must be in NEUTRAL throughout test procedure. Battery must be fully charged.

Step	Procedure	Results
1 Check voltage to cable at ①.	Connect voltmeter between cable at ① and ground. Press start button Ⓐ.	• If meter shows 12 volts, test starter. • If meter shows 0 volts, go to Step **2**.
2 Check voltage at ②.	Connect voltmeter between ② and ground.	• If meter shows 12 volts, go to Step **3**. • If meter shows 0 volts, test battery and cables.
3 Check voltage at ③.	Connect voltmeter between ③ and ground. Press start button.	• If meter shows 12 volts, go to Step **4**. • If meter shows 0 volts, test start button and wiring.
4 Check voltage at ④.	Disconnect solenoid ground lead ④ from engine ground. Connect voltmeter between lead ④ and ③. Press start button Ⓐ.	• If meter shows 12 volts, proceed to Step **5**. • If meter shows 0 volts, replace solenoid or lead.
5 Check voltage at ⑤.	Reconnect solenoid ground lead ④. Connect voltmeter between ⑤ and ground. Press start button Ⓐ.	• If meter shows 0 volts, replace solenoid. • If meter shows 12 volts, replace starter cable.

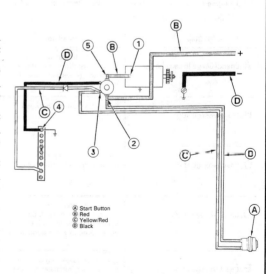

Fig. 162-163 Starter circuit testing - 40-70 hp (913cc) tiller models through 1993 (note, models after 1991 did not normally use the yellow/red wire from the solenoid to the terminal strip)

4-78 IGNITION AND ELECTRICAL SYSTEMS

⚠ To avoid accidental starting of engine, disconnect starter cable at ①.

Note Engine must be in NEUTRAL throughout test procedure. Battery must be fully charged.

Step	Procedure	Results
1. Check voltage to cable at ①.	Connect voltmeter between cable at ① and ground. Turn key switch to ON position. Press start button Ⓐ.	• If meter shows 12 volts, test starter. • If meter shows 0 volts, go to Step **2**.
2. Check voltage at ②.	Connect voltmeter between ② and ground.	• If meter shows 12 volts, go to Step **3**. • If meter shows 0 volts, test battery and cables.
3. Check voltage at ③.	Connect voltmeter between ③ and ground. Press start button.	• If meter shows 12 volts, go to Step **4**. • If meter shows 0 volts, test start button, key switch, and wiring.
4. Check voltage at ④.	Disconnect solenoid ground lead ④ from engine ground. Connect voltmeter between lead ④ and engine ground. Press start button Ⓐ.	• If meter shows 12 volts, proceed to Step **5**. • If meter shows 0 volts, replace solenoid or lead.
5. Check voltage at ⑤.	Reconnect solenoid ground lead ④. Connect voltmeter between ⑤ and ground. Press start button Ⓐ.	• If meter shows 0 volts, replace solenoid. • If meter shows 12 volts, replace starter cable.

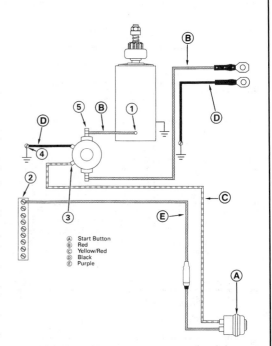

Ⓐ Start Button
Ⓑ Red
Ⓒ Yellow/Red
Ⓓ Black
Ⓔ Purple

Fig. 164-165 Starter circuit testing - 40-70 hp (913cc) 1994-95 tiller models

⚠ To avoid accidental starting of engine, disconnect starter cable at ①.

Note Engine must be in NEUTRAL throughout test procedure. Battery must be fully charged.

Step	Procedure	Results
A. Check voltage to cable at ①.	Connect voltmeter between cable at ① and ground. Turn key switch to ON position. Press start button Ⓐ.	• If meter shows 12 volts, test starter. • If meter shows 0 volts, go to Step **B**
B. Check voltage at ②.	Connect voltmeter between ② and ground.	• If meter shows 12 volts, go to Step **C** • If meter shows 0 volts, test battery and cables.
C. Check voltage at ③.	Connect voltmeter between ③ and ground. Press start button.	• If meter shows 12 volts, go to Step **D** • If meter shows 0 volts, test start button, key switch, and wiring.
D. Check voltage at ④.	Disconnect solenoid ground lead ④ from engine ground. Connect voltmeter between lead ④ and engine ground. Press start button Ⓐ.	• If meter shows 12 volts, proceed to Step **E** • If meter shows 0 volts, replace solenoid or lead.
E. Check voltage at ⑤.	Reconnect solenoid ground lead ④. Connect voltmeter between ⑤ and ground. Press start button Ⓐ.	• If meter shows 0 volts, replace solenoid. • If meter shows 12 volts, replace starter cable.

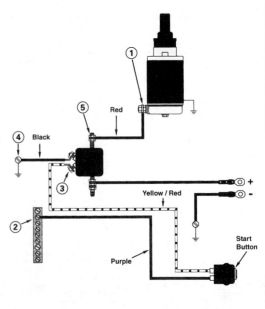

Fig. 166-167 Starter circuit testing - 25-70 hp (913cc) 1996-98 tiller control models

IGNITION AND ELECTRICAL SYSTEMS 4-79

⚠ **Avoid accidental starting while testing; disconnect the starter cable at point ① before proceeding with this check.**

Note Engine must be in NEUTRAL throughout test procedure. Battery must be fully charged.

Step	Procedure	Result
A. Check voltage between ① and ground.	Connect voltmeter between ① and ground. Press start button Ⓐ.	• If meter shows 12V, test starter. • If meter shows 0V, go to Step **B**
B. Check voltage between ② and ground.	Connect voltmeter between ② and ground. Press start button Ⓐ.	• If meter shows 12V, test cable to starter. • If meter shows 0V, go to Step **C**
C. Check voltage between ③ and ground.	Connect voltmeter between ③ and ground.	• If meter shows 0V, check battery cable and battery. • If meter shows 12V, go to Step **D**
D. Check voltage between ④ and ground.	Connect voltmeter between ④ and ground. Turn key switch Ⓑ ON. Press start button Ⓐ.	• If meter shows 12V, go to Step **H** • If meter reads 0V, go to Step **E**
E. Check voltage between ⑤ and ground.	Connect voltmeter between ⑤ and ground. Turn key switch Ⓑ ON.	• If meter shows 12V, replace start button Ⓐ and/or wire. • If meter shows 0V, go to Step **F**
F. Check voltage between ⑥ and ground.	Connect voltmeter between ⑥ and ground.	• If meter shows 12V, go to **Check Key Switch**, this section. Check wire to terminal strip. • If meter shows 0V, go to Step **G**
G. Check voltage between ⑦ and ground.	Connect voltmeter between ⑦ and ground.	• If meter shows 12V, replace fuse, holder, or wire to key switch. • If meter shows 0V, replace wire from solenoid terminal ③ to fuse ⑦.
H. Check voltage between ⑧ and ground.	Disconnect black ground wires from ⑧ Connect voltmeter between solenoid terminal ⑧ and ground. Turn key switch Ⓑ ON. Push start switch Ⓐ.	• If meter shows 12V, repair/replace black wire between ⑧ and ground. • If meter shows 0V, disconnect all leads from solenoid, and test solenoid. Go to **Check solenoid**, this section.

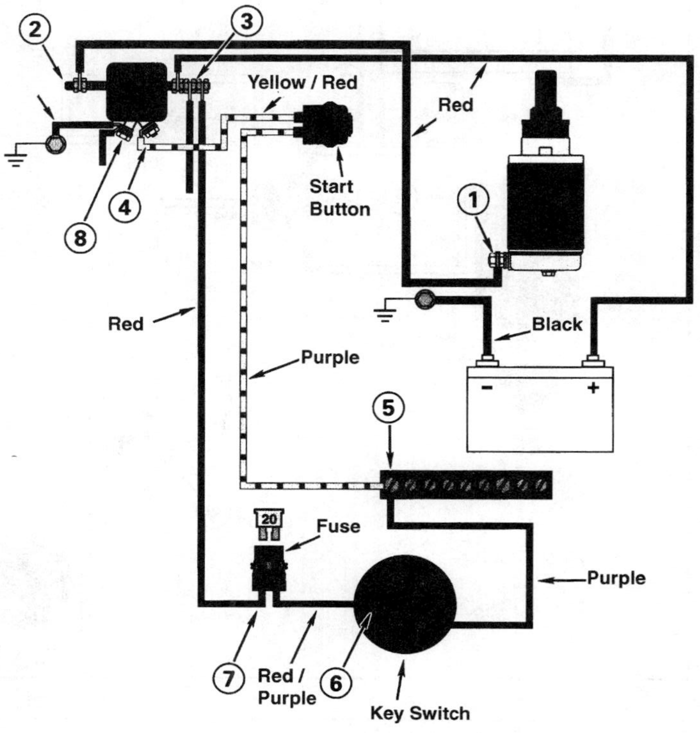

Fig. 168-169 Starter circuit testing - 25-55 hp (737cc) 1996 and later tiller models and 40-70 hp (913cc) 1999 and later tiller models

IGNITION AND ELECTRICAL SYSTEMS

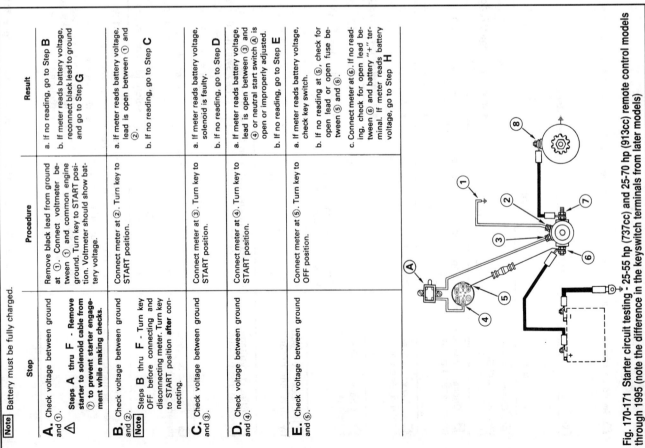

Step	Procedure	Result
Note Battery must be fully charged.		
A. Check voltage between ground and ①. ⚠ **Steps A thru F - Remove starter to solenoid cable from** ⑦ **to prevent starter engagement while making checks.**	Remove black lead from ground at ①. Connect voltmeter between ① and common engine ground. Turn key to START position. Voltmeter should show battery voltage.	a. If no reading, go to Step **B**. b. If meter reads battery voltage, reconnect black lead to ground and go to Step **G**.
B. Check voltage between ground and ②. **Note** Steps **B** thru **F** - Turn key OFF before connecting and disconnecting meter. Turn key to START position **after** connecting.	Connect meter at ②. Turn key to START position.	a. If meter reads battery voltage, lead is open between ① and ②. b. If no reading, go to Step **C**.
C. Check voltage between ground and ③.	Connect meter at ③. Turn key to START position.	a. If meter reads battery voltage, solenoid is faulty. b. If no reading, go to Step **D**.
D. Check voltage between ground and ④.	Connect meter at ④. Turn key to START position.	a. If meter reads battery voltage, lead is open between ④ and ④ or neutral start switch ④ is open or improperly adjusted. b. If no reading, go to Step **E**.
E. Check voltage between ground and ⑤.	Connect meter at ⑤. Turn key to OFF position.	a. If meter reads battery voltage, check key switch. b. If no reading at ⑤, check for open lead or open fuse between ⑤ and ⑥. c. Connect meter at ⑥. If no reading, check for open lead between ⑥ and battery "+" terminal. If meter reads battery voltage, go to Step **H**.
F. Check voltage between ground and ⑦.	Connect voltmeter at ⑦. Turn key to START position.	a. If no reading, solenoid is faulty. b. If solenoid clicks and meter reads battery voltage, go to Step **G**.
G. Check voltage between ground and ⑧.	Reconnect starter to solenoid cable at ⑦. Connect meter at ⑧. Turn key to START position.	a. If meter reads battery voltage and starter motor does not turn, check starter motor. b. If no reading, check for broken cable or poor connection.

Fig. 170-171 Starter circuit testing - 25-55 hp (737cc) and 25-70 hp (913cc) remote control models through 1995 (note the difference in the keyswitch terminals from later models)

Fig. 172 Starter circuit testing - 25-55 hp (737cc) and 25-70 hp (913cc) remote control models 1996 and later (note the difference in the keyswitch terminals from earlier models)

IGNITION AND ELECTRICAL SYSTEMS 4-81

Testing the Starter Button and Neutral Safety Switch (Tiller Models)

Although the wire colors and mounting positions vary slightly, all Johnson/Evinrude starter buttons operate in the same basic fashion. The button is a spring-loaded, continuity switch that completes the circuit (has continuity) when held down (depressed) and breaks the circuit (has no continuity) when released.

Refer to the schematics in the Wiring Diagrams section for details.

1. Disconnect the negative battery cable for safety (and to prevent accidental starting when testing or if switch leads are accidentally connected when they are removed from the switch or harness).
2. Locate the starter button and trace the wiring looking for the easiest disconnection/testing point. Disconnect the harness for access to the switch wires.
3. Connect a DVOM set to read resistance across the two switch terminals or wires leading to the switch. Switch harness wires are normally yellow/red and red/purple for 4-stroke or yellow/red and red or yellow/red and purple for most 2-stroke motors, except 9.9/15 hp 2-strokes on which both leads are red.
4. Observe the meter with the button depressed and the button released. The DVOM should show no continuity (infinite resistance) when the button is released, then must show continuity (low resistance) when the button is held down.
5. Reconnect the wiring once the testing and/or repairs are completed.
6. For tiller models equipped with a neutral safety switch the switch operates in the same manner as the starter switch (completing or breaking the circuit depending upon the position of the switch plunger). Test is the same way using a DVOM or Ohmmeter.
7. Connect the negative battery cable.

Testing Starter Switches (Isolating Wiring Problems) on 1995 and earlier non-MWS Equipped Models

◆ See Figure 173

Starting in 1996 Johnson/Evinrude introduced the new Modular Wiring System (MWS) for all Remote models. Prior to the introduction of the MWS system remote models were wired with another common proprietary system, the tests for which are included here. HOWEVER, an adaptor was available to rig older engines to the MWS system if necessary, so use caution when testing. Basically if the connectors match the descriptions in this section, you should be good to go.

The following tests can all be performed using an ohmmeter or a continuity test light to test for continuity. All tests must be started with these preliminary steps

1. Disconnect the cables from both terminals on the battery.
2. Disconnect the throttle and shift cables at the powerhead.
3. Remove the retainer securing the harness connector together at the powerhead.

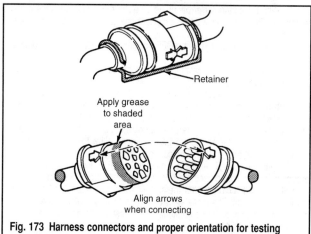

Fig. 173 Harness connectors and proper orientation for testing

4. Disengage the harness connector. Note the arrow on both halves. During the tests this arrow must always be on top - at the 12 o'clock position - for the test probes to be inserted in the correct openings, per the illustrations.

Emergency Stop Switch Circuit

◆ See Figure 174

1. Install the safety clip and lanyard behind the stop switch button.
2. Obtain an ohmmeter or a continuity light. Set the meter scale for continuity reading. A continuity test light may also be used.
3. Turn the key to the on position.
4. With the connector arrow at the 12 o'clock position, insert the ohmmeter probes into the 1 o'clock and 6 o'clock female openings. The meter must show no continuity.
5. Keep the test probes in the same openings and remove the emergency stop switch clip from behind the knob. The ohmmeter must indicate continuity.
6. If the tests are not satisfactory, check the wiring carefully. Final solution - replace the emergency stop switch.

Key Switch Circuit

◆ See Figures 174, 175 and 176

An ohmmeter or a continuity light may be used to make the following tests.

1. Install the safety switch clip behind the knob.

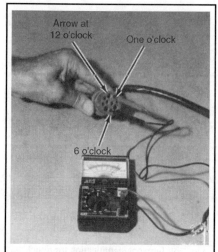

Fig. 174 Checking the emergency stop switch and keyswitch circuits

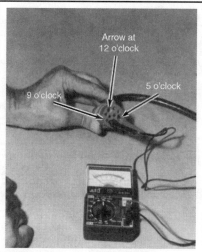

Fig. 175 Checking the cold start and ignition on functions of the keyswitch

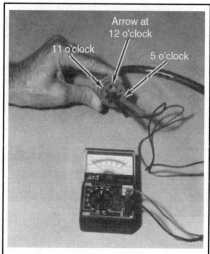

Fig. 176 Checking the neutral start switch circuit

4-82 IGNITION AND ELECTRICAL SYSTEMS

2. With the connector arrow at the 12 o'clock position, leave the meter test probes in the 1 o'clock and 6 o'clock openings, as in the previous tests, illustration **B**.

3. Turn the key switch to the OFF position. The meter should indicate continuity.

4. Now, turn the key switch to the on position. The meter should indicate no continuity.

5. Keep the connector arrow at the 12 o'clock position, and move the meter probes to the 5 o'clock and 9 o'clock positions in order to check the cold start and ignition on functions of the switch .With probes in place, move the key switch to the on and start positions and at the same time push the key inward. The meter should indicate continuity. Allow the key to move outward and the meter should indicate no continuity.

6. With the control handle in the neutral position, insert the meter probes into the 5 o'clock and 11 o'clock positions in order to check the neutral start switch circuit. After the probes are in place, turn the key switch to the start position. The meter should show continuity. Release the key and the meter should indicate no continuity. If the test results are not satisfactory, check the wiring and test the switch.

Neutral Start Switch Circuit

◆ See Figure 175

An ohmmeter or a continuity light may be used to make the following tests.

1. Check to be sure the control handle is in the neutral position. Insert the meter probes in the 5 o'clock and 11 o'clock positions. Turn the key to the start position and the meter should indicate continuity.

2. Move the control handle to the forward position and again turn the key to the start position. The meter should indicate no continuity.

3. Move the control handle to the reverse position and turn the key to the start position. The meter should indicate no continuity.

4. If the test results are not satisfactory, check the wiring and test the neutral start switch.

Trim/Tilt Switch Circuit

◆ See Figures 177 and 178

An ohmmeter or a continuity light may be used to make the following tests.

1. At the outboard, disconnect the control handle trim/tilt harness from the trim/tilt motor harness at the connector.

2. Make contact with the meter test probes to the upper center pin and the lower left pin, as indicated in the illustration for the first connection. Depress the trim/tilt switch for the up direction. The meter should indicate continuity. Release the switch and the meter should indicate no continuity.

3. Make contact with the meter probes to the upper center pin and the lower right pin, as show in the illustration for the second connection.

4. Depress the trim/tilt switch for the down direction. The meter should indicate continuity. Release the switch and the meter should indicate no continuity.

5. If the test results are not satisfactory check the wire location or replace the switch.

Keyswitch Test

◆ See Figure 179

This test directly checks function of factory Johnson/Evinrude rigged keyswitch assemblies that is used with the 1995 and earlier models (PRE-MWS). For details on terminal locations refer to the accompanying illustration.

A DVOM set to read resistance is used by this test to check continuity, though a self-powered test light could also be used.

1. Disconnect the negative battery cable from the battery for safety.
2. Disassemble the remote control as necessary for access to the keyswitch terminals. Tag (for proper installation) and disconnect the wiring from the terminals.
3. Install the stop switch clip and safety lanyard.
4. Make sure the keyswitch is turned to **OFF**.
5. Probe across terminals B (for Battery) and A, there should be no continuity.
6. Turn the keyswitch to **ON**, there should now be continuity across terminals A and B.
7. Turn the keyswitch to **START**, there should still be continuity across terminals A and B.
8. With the keyswitch held in **START**, move one probe from terminal A to terminal S, there should still be continuity.
9. Move both probes to the M terminals. There should be NO continuity with the keyswitch in the **START** or **ON** positions.
10. Turn the keyswitch **OFF**, there should now be continuity across the M terminals.
11. Move the probes to terminals B and C, then turn the keyswitch to **ON** and push inward. There should be continuity once the switch is pushed inward in this position.
12. With the probes still connected across terminals B and C, turn the keyswitch to **START** and push inward. There should be continuity once the switch is pushed inward in this position as well.
13. Recheck connections and test conditions if the switch fails any part of the test, then replace the switch if found defective. If the switch is not faulty, recheck the harness or look for other problems.
14. When tests or repairs are completed, assembly the remote control assembly.
15. Connect the negative battery cable.

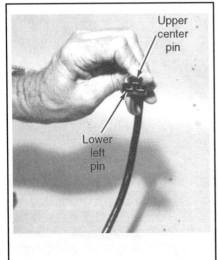

Fig. 177 First connection for trim/tilt switch testing

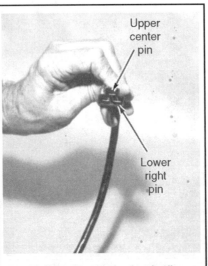

Fig. 178 Second connection for trim/tilt switch testing

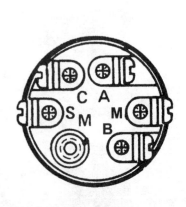

Fig. 179 Terminal identification for direct keyswitch testing on 1995 and earlier models

IGNITION AND ELECTRICAL SYSTEMS

Testing Starter Switches (Isolating Wiring Problems) on 1996 and Later MWS Equipped Models

Keyswitch Circuit Test

◆ See Figures 180 thru 183

This test checks function of factory Johnson/Evinrude rigged keyswitch assemblies through the Modular Wiring System (MWS). For more details on terminal locations and wiring colors refer to the accompanying illustrations or the schematics for the MWS remote control and harness found in the Wiring Diagram section.

A DVOM set to read resistance is used by this test to check continuity, though a self-powered test light could also be used.

1. Disconnect the negative battery cable from the battery for safety.
2. Install the stop switch clip and safety lanyard.
3. Locate and disengage the MWS 6-pin remote harness connector inline between the remote and the engine (the harness contains yellow/red, black/white, black, red, purple and purple/white wires).
4. Make sure the keyswitch is turned to **OFF**.
5. Probe the remote control side of the harness terminals 4 and 6 (black/yellow and black wires) as identified in the accompanying illustration, there should be continuity.
6. Turn the keyswitch to **ON**, while continuing to probe terminals 4 and 6. There should now be no continuity with the keyswitch **ON**.
7. Connect the probes across terminals 1 and 2 (purple/yellow and red/purple wires), there should be no continuity until the keyswitch is pushed inward while in the **ON** and **START** positions.
8. Move the control handle into **Neutral**.
9. Connect the probes across terminals 2 and 5 (red/purple and yellow/red wires), there should be no continuity if the key is released, but there should be continuity when the key is turned to the **START** position.
10. If the circuit does not test properly, check for problems with the harness wires or connectors. If no problems are found with the harness, test the keyswitch directly, as detailed in this section, before replacing the switch. If the switch is not faulty, recheck the harness and test connections.

Keyswitch Test

◆ See Figures 184 thru 187

This test directly checks function of factory Johnson/Evinrude rigged keyswitch assemblies that is used with the Modular Wiring System (MWS). For more details on terminal locations and associated wiring colors refer to the accompanying illustrations or the schematics for the MWS remote control and harness found in the Wiring Diagram section.

A DVOM set to read resistance is used by this test to check continuity, though a self-powered test light could also be used.

1. Disconnect the negative battery cable from the battery for safety.
2. Disassemble the remote control as necessary for access to the keyswitch terminals. Tag (for proper installation) and disconnect the wiring from the terminals.
3. Install the stop switch clip and safety lanyard.
4. Make sure the keyswitch is turned to **OFF**.
5. Probe across terminals A and B (purple and red/purple wires), there should be no continuity.
6. Turn the keyswitch to **ON**, there should now be continuity across terminals A and B.
7. Turn the keyswitch to **START**, there should still be continuity across terminals A and B.
8. With the keyswitch held in **START**, move one probe from terminal A to terminal S (yellow/red wire), there should still be continuity.
9. Move both probes to the M terminals (black and black/yellow). There should be NO continuity with the keyswitch in the **START** or **ON** positions.
10. Turn the keyswitch **OFF**, there should now be continuity across the M terminals.
11. Move the probes to terminals B and C (red/purple and purple/white), then turn the keyswitch to **ON** and push inward. There should be continuity once the switch is pushed inward in this position.
12. With the probes still connected across terminals B and C, turn the keyswitch to **START** and push inward. There should be continuity once the switch is pushed inward in this position as well.

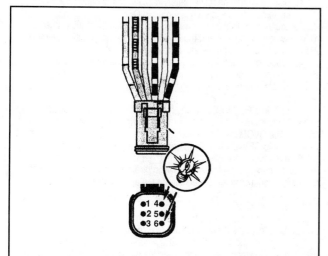

Fig. 180 There should be continuity across terminals 4 and 6 with the keyswitch OFF

Fig. 181 There should be no continuity across terminals 4 and 6 with the keyswitch ON

Fig. 182 There should be continuity across terminals 1 and 2 with the keyswitch pushed inward while it is in the ON or START position

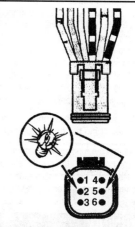

Fig. 183 There should be continuity across terminals 2 and 5 with the keyswitch turned to START and the remote lever in the Neutral position

4-84 IGNITION AND ELECTRICAL SYSTEMS

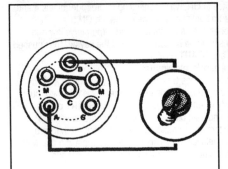

Fig. 184 Ignition keyswitch testing and terminal identification - with the keyswitch OFF the M terminals are connected, but A and B should show no continuity

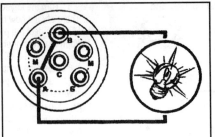

Fig. 185 With the keyswitch ON the M terminals are disconnected, while A and B should be continuous

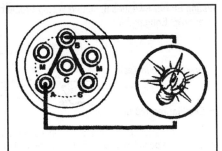

Fig. 186 With the keyswitch held in the START position, terminals A and B, as well as B and S should show continuity

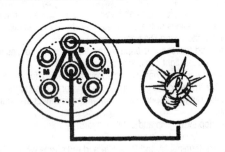

Fig. 187 Pressing inward on the keyswitch when it is in ON or, as shown in START, while allow for continuity between B and C (applying power to the electric primer, when equipped)

13. Recheck connections and test conditions if the switch fails any part of the test, then replace the switch if found defective. If the switch is not faulty, recheck the harness or look for other problems.
14. When tests or repairs are completed, assembly the remote control assembly.
15. Connect the negative battery cable.

Neutral Safety Switch And Circuit Test

  MODERATE

◆ See Figure 188

This test checks function of remote control mounted neutral safety switch assembly through the Modular Wiring System (MWS) harness. For details on terminal locations and wiring colors refer to the accompanying illustrations or the schematics for the MWS remote control and harness found in the Wiring Diagram section.

A DVOM set to read resistance is used by this test to check continuity, though a self-powered test light could also be used.

The switch itself can also be tested directly, by checking continuity across the switch contacts while the plunger is released and while the plunger is depressed. Continuity should only exist when the plunger is held downward.

■ For models equipped with an engine mounted switch, test the switch directly following the final step of this test procedure.

Keep in mind that the function of the neutral safety switch is to prevent the motor from being started (to keep the starter circuit open) unless the gearcase is in **Neutral**. This is a safety feature that is meant to prevent injury on land during testing/tune-ups or on the water (to keep the boat from lurching suddenly when the motor is started).

✲✲ WARNING

The gearbox should not be forced into gear without propshaft rotation. If necessary, have an assistant slowly turn the shaft by hand when shifting into FORWARD or REVERSE gear.

✲✲ CAUTION

Be sure to disconnect the negative battery cable and to remove/ground the spark plug leads to prevent accidental starting which could lead to serious injury or death, especially for your assistant.

1. Disconnect the negative battery cable from the battery for safety.
2. Make sure the keyswitch is turned to **OFF**.
3. Locate and disengage the MWS 6-pin remote harness connector inline between the remote and the engine (the harness contains yellow/red, black/white, black, red, purple and purple/white wires).
4. Probe the remote control side of the harness terminals 2 and 5 (red/purple and yellow/red wires) as identified in the accompanying illustration, there should be continuity only with the shifter in **Neutral** and the keyswitch held in the **START** position.
5. Shift into **FORWARD** while an assistant rotates the propshaft slowly by hand to prevent damage, then turn and hold the keyswitch in the **START** position. The DVOM must now show no continuity. Repeat this step, moving the shifter into **REVERSE**, the meter must still show no continuity.
6. If readings vary, check the wiring harness and switch itself.
7. To check the neutral safety switch directly (mounted in the remote assembly or on the motor, depending on the application):
   a. Locate the switch (by removing the engine cover or disassembling the remote control unit for access).
   b. Disconnect the wiring from the switch (usually 2 yellow/red leads), then probe across the 2 switch terminals using a DVOM.
   c. Depress and release the switch while watching the meter. The switch must shown continuity only when the plunger is depressed and should show no continuity as soon as the plunger is released.

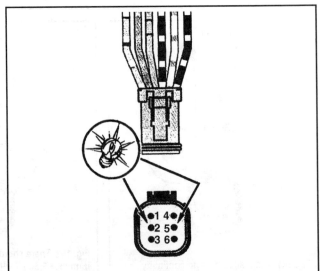

Fig. 188 Testing the remote mounted neutral safety switch through the MWS harness

d. The switch should be replaced if it does not function properly.

■ In some applications the switch mounting is adjustable. If so, when installed, make sure the switch plunger is only held down when in **NEUTRAL**. The plunger must be extended when in gear (either **FORWARD** or **REVERSE**).

8. Once testing and/or repairs are finished, reconnect the wiring.
9. Connect the negative battery cable.

STARTING SYSTEM TESTING (EFI MODELS)

Perform all preliminary checks as noted under Troubleshooting the Starting System. Also, refer to the engine troubleshooting charts found in the Electronic Fuel Injection (EFI) System section, specifically ENGINE WON'T CRANK for troubleshooting hinds. If no obvious causes of a no or slow cranking condition are found, perform the voltage and switch component testing in the order provided here.

• Test the voltage at the starter motor, then the starter solenoid relay and finally the keyswitch.
• If voltage readings are good, check the function of the neutral safety switch, then the starter solenoid relay and finally the keyswitch.

Voltage Testing

Checking at the Starter Motor

1. For safety and to prevent accidental engine starting, disconnect the leads from the spark plugs (70 hp motors) and ground them or disconnect and ground the primary leads (40/50 hp motors) from the coils.
2. If not already, place the gear shifter into **Neutral**. If shifting is difficult, have an assistant manually rotate the propshaft slowly by hand until the shift lever moves freely.
3. Locate the starter motor, then find where the red power cable attaches to it. Pull the rubber insulator cover back slightly for access and connect a DVOM red probe to the red power cable at the point where it attaches to the starter motor terminal. Connect the black probe to a good engine ground.
4. Operate the keyswitch while observing the meter for a reading of at least 9.5 volts and proceed as follows:
   a. If power is present, but the starter is not cranking, move the black probe to the top of the starter casing and recheck. If there is no reading or voltage drops, check the starter for a bad ground that is preventing the circuit from properly completing, clean the starter mounting surface to achieve a better ground. If the voltage reading does not change, repair or replace the starter, as necessary.
   b. If no power is present, check for voltage at the starter solenoid relay as the next step in determining the problem with the circuit. At this point, it could be the fuse, solenoid, neutral safety switch, or keyswitch.
   c. If insufficient power is present, check the battery, cables and connections.

Checking at the Starter Solenoid Relay

The solenoid relay is the remote controlled electric switch that is used by the starting circuit to apply power to the starter motor. It can be located by following the red starter cable back from the starter motor to the solenoid itself. For 40/50 hp motors the solenoid relay is found on the starboard side of the powerhead, adjacent to the electric starter. For 70 hp motors it is found on the starboard side of the powerhead, just aft of the air intake silencer cover.

1. If not done already, disconnect the leads from the spark plugs (70 hp motors) and ground them or disconnect and ground the primary leads (40/50 hp motors) from the coils for safety and to prevent accidental engine starting.
2. If not already, place the gear shifter into **Neutral**. If shifting is difficult, have an assistant manually rotate the propshaft slowly by hand until the shift lever moves freely.
3. Connect a DVOM red probe to the battery cable side of the relay (terminal to which the red battery cable is attached), then connect the black meter probe to a good engine ground. There must be a reading of at least 12 volts at this point, or the cable, battery or connections are faulty. Clean, repair or replace, as necessary before proceeding.

4. Move the red probe to the starter cable side of the relay (terminal to which the red cable that runs to the starter motor is attached). Operate the keyswitch while watching the meter, it must read at least 9.5 volts. Otherwise the solenoid is faulty or no signal is being received to trigger the solenoid.
5. Disconnect the yellow/green wire from the solenoid, then connect the red DVOM probe to the wire itself. Move the black probe to a good engine ground. Operate the keyswitch while watching the meter, if a reading of 12 volts or higher is not noted, the signal is not reaching the solenoid. The problem could be in the neutral safety switch or keyswitch (and any of the related wiring). In this case, proceed with checking Voltage at the Ignition Keyswitch, as detailed in this section.

■ If the signal voltage is present when checking between the yellow/green wire and a good engine ground, move the black meter probe to the solenoid wire terminal for the small black wire that attaches to the point on the solenoid next to the yellow/green wire. Repeat the check, if voltage is not present, the solenoid has a bad ground. Either clean and tighten the terminals or repair or replace the wire to provide a suitable ground (or replace the solenoid relay if that does not fix the problem).

6. If signal voltage is present and no bad ground is located, yet the solenoid relay is not closing the internal switch to apply power to the starter, you've found the problem (it's the solenoid).

Checking at the Ignition Keyswitch

The EFI motors sold by Johnson/Evinrude are rigged from the factory for remote control installation. When installed in this fashion, a remote or dash mounted ignition keyswitch is used. Long jumpers may be necessary to test remote models. Some EFI motors however, may be equipped with accessory tiller controls and, when so equipped would use an ignition keyswitch mounted to the motor. The switch however is tested in the same fashion as the remote switch.

Before switch testing can occur, the remote or tiller control housing may require partial disassembly for access to the switch itself. Before going through this trouble, make sure all other possible causes of a no or slow crank condition have been thoroughly checks. Wires, connections and fuses should be checked for damage or corrosion.

For remote models turn the ignition keyswitch to **ON** while watching the instruments for backlights or other indications that power has been applied. If there is not power when the keyswitch is turned, check for a blown fuse or faulty wiring problem.

1. For safety and to prevent accidental engine starting, disconnect the leads from the spark plugs (70 hp motors) and ground them or disconnect and ground the primary leads (40/50 hp motors) from the coils.
2. Either remove the lower tiller cover for access to switch wiring (tiller models) or disassemble the remote for access to the switch (remote mounted keyswitch models) or access the switch from under the dash (dash mounted remote keyswitch models), as applicable.
3. Disconnect the white wire from the harness.
4. Using a DVOM set to read DC voltage, connect the red probe to the engine side of the white wire and a the black probe to a good engine ground. Reconnect the negative battery cable and check for at least 12 volts on the meter. A lack of voltage (or insufficient voltage) indicates a problem with the battery, fuse, wiring or connections. Check for damaged components and loose/corroded connections. If the correct voltage is present, proceed with component testing to find the culprit, start with the neutral safety switch, then the solenoid relay and finally the ignition keyswitch.
5. Once tests or repairs are completed, reconnect the white switch to the harness, routing the wiring as noted during removal.

Switch Component Testing

Checking the Neutral Safety Switch

◆ See Figure 189

Johnson/Evinrude EFI motors are equipped with an engine mounted neutral safety switch. The switch is designed to open the signal circuit to the starter relay whenever the gearcase is shifted into **FORWARD** or **REVERSE**. The switch is mounted in such as manner as to allow direct contact with the shift linkage. When the linkage is in **Neutral**. A tab on the linkage depresses and holds the switch plunger in a downward position, providing for switch continuity. In any position other than **Neutral** the linkage tab is no longer in

4-86 IGNITION AND ELECTRICAL SYSTEMS

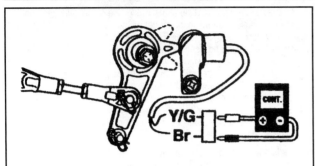

Fig. 189 The neutral safety switch must show continuity only when the plunger is held downward (in NEUTRAL)

contact with the switch plunger which moves back outwards, opening the circuit.

Testing is a simple matter of connecting a DVOM across the two switch contacts and checking response to gear linkage and plunger movement.

1. Disconnect the negative battery cable for safety.
2. Locate the neutral safety switch by following the yellow/green wire back from the starter solenoid to the neutral switch.
3. Disengage the harness connector, removing both wires (yellow/green and brown) from the switch pigtail.
4. Connect the DVOM across the two switch pigtail terminals.
5. Using an assistant to slowly rotate the propshaft (thereby preventing binding and damage to the gearcase/shifter linkage), place the shit linkage in **Neutral**. Visibly check to ensure that the switch plunger is held downward and observe the meter, there should be continuity.
6. Shift the gearcase into **FORWARD** and **REVERSE** respectively, observing the meter each time. The plunger must be released as the linkage is moved into either gear and the meter should show no continuity.
7. Replace the switch if it does not close the circuit ONLY when the plunger is held downward.
8. Once tests and/or repairs are completed, reconnect the switch harness and the negative battery cable.

Checking the Solenoid Relay

◆ See Figure 190

A DVOM can be used to check the coil winding of the starter solenoid. In addition a fully-charged 12-volt battery and a set of jumper leads can be used to perform a functional check of the solenoid by applying power to the signal circuit and watching the DVOM to see if the main switch circuit (starter power feed circuit) closes.

To protect the test equipment, all engine wiring must be disconnected from the solenoid relay before performing these tests.

■ **Be sure to perform both parts of the test and replace the solenoid relay if it fails either one.**

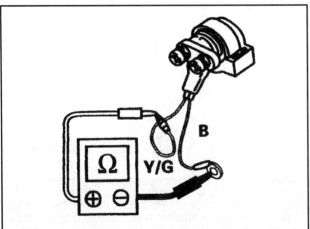

Fig. 190 Testing solenoid relay coil winding resistance across the signal circuit terminals

1. Disconnect the negative battery cable for safety.
2. Tag and disconnect the wiring from the solenoid. If desired, remove the solenoid completely from the powerhead.
3. Set the DVOM to read resistance, then connect the probes across the yellow/green and black wire connections for the solenoid relay signal circuit. Resistance should be about 3.5-5.1 ohms.

■ **Remember like with all resistance testing, resistance specifications and readings vary with temperature. The specifications provided are for a component being tested at 68°F (20°C).**

4. Move the DVOM probes to the battery input and starter motor output terminals, then use the jumpers to apply battery power to the yellow/green and black wire (signal) connections for the solenoid. If it is functioning properly, an audible click should be heard as the solenoid closes the switch contacts for the starter power circuit and the DVOM should now show continuity between the battery input and starter output terminals.
5. If the solenoid relay fails either test, recheck to make sure connections were proper, then replace the solenoid.
6. After testing or repairs, reconnect the wiring as tagged and reconnect the negative battery cable.

Checking the Ignition Keyswitch

The ignition keyswitch provides battery voltage to operate dash gauges, the starting circuit and engine control circuits on EFI models. Access to the switch leads is required for testing, which will require partial disassembly of the remote control or accessory tiller control bracket on most motors.

These motors are normally rigged using the standard Johnson/Evinrude keyswitch assembly and, on remote models, the Modular Wiring System (MWS). Testing is conducted in the same manner as keyswitch and circuit testing for carbureted models. For this reason, please refer to the Keyswitch Circuit Test and the Keyswitch test under Testing Starter Switches (Isolating Wiring Problems) in the Starting System Troubleshooting (Carbureted Models) section for details.

Starter Motor

◆ See Figures 191 and 192.

As the name implies, the sole purpose of the cranking motor circuit is to control operation of the electric starter motor to crank (turn) the powerhead flywheel and crankshaft until the engine is operating. The circuit usually includes a solenoid relay (magnetic switch) to connect or disconnect the motor from the battery. When equipped, the operator controls the solenoid relay using either a key switch or a starter button.

A neutral safety switch (or sometimes a mechanical lockout) is usually installed into the circuit to permit operation of the cranking motor only if the shift control lever is in neutral. This switch is a safety device to prevent accidental engine startup when the motor is in gear.

The starter is an electric motor consisting of a series of windings that draw a heavy current from the battery. It is designed to be used only for short periods of time. To prevent overheating the motor, cranking should not be continued for more than 30-seconds without allowing the motor to cool for at least three minutes. Actually, this time can be spent in making preliminary checks to determine why the engine fails to start.

Power is transmitted from the cranking motor to the powerhead flywheel through a Bendix drive that utilizes a pinion gear mounted on screw threads. When the motor is operated, the pinion gear moves upward and meshes with the teeth on the flywheel ring gear. When the powerhead starts, the pinion gear is driven faster than the shaft, and as a result, it screws out of mesh with the flywheel. A rubber cushion is built into the Bendix drive to absorb the shock when the pinion meshes with the flywheel ring gear. The parts of the drive must be properly assembled for efficient operation. If the drive is removed for cleaning, take care to assemble the parts as shown in the accompanying illustrations in this section. If the screw shaft assembly is reversed, it will strike the splines and the rubber cushion will not absorb the shock.

The sound of the motor during cranking is a good indication of whether the starter is operating properly or not. Naturally, temperature conditions will affect the speed at which the cranking motor is able to crank the engine. The speed of cranking a cold engine will be much slower than when cranking a warm engine. An experienced operator will learn to recognize the favorable sounds of the powerhead cranking under various conditions.

IGNITION AND ELECTRICAL SYSTEMS

TESTING

Starter No-Load Current Draw Test

◆ See Figure 193

■ Johnson/Evinrude does not provide specifications for a no-load current draw test on EFI motors. If the starter motor is suspect on these models (fails to operate properly even though sufficient voltage is available at the motor wiring) the starter should be removed, disassembled and inspected, as detailed in this section.

On carbureted Johnson/Evinrude outboards, if starter motor is suspect (failed to operate properly even though sufficient voltage is available at the motor wiring and no ground problem was located), the motor can be removed and checked using a no-load current draw test. An ammeter of 0-50 amp capacity, a vibration tachometer (such as a Frahm Reed tachometer) and a fully-charged battery of suitable capacity will be necessary for this test. On all models except 25/35 hp (500/565cc) motors, a battery of 350 CCA should be sufficient, but for 25/35 hp (500/565cc) motors the manufacturer recommends using a 500 CCA battery. For safety, you should also use a bench vise mounted to a suitable working surface to hold the motor securely in place during testing. Though people have been known to simply place the starter on the floor and hold it securely underfoot, but this can get pretty awkward when trying to read the various meters used during testing.

■ An inductive (clamp-on type) ammeter is easiest to use for this test.

To use a Frahm Reed or other suitable vibration tachometer, simply hold it against the starter housing while the motor is running. If one if not available, a stroboscopic tachometer may be used, provided a reference mark is placed on the drive gear (pinion).

■ Most hobby shops will sell tachometers designed for use on model plane engines that would be held against the drive unit to measure rotational speed.

1. Remove the starter from the engine as detailed in this chapter.
2. Mount the starter motor in a suitable bench vise (or a mounting fixture to which the motor can be bolted to hold it securely during testing).

✱✱ WARNING

If possible, use a soft-jawed vise or some form of rubber padding to protect the starter housing. Be careful not to overtighten and damage the housing by crushing it in the vise.

3. Connect a voltmeter across the battery terminals to monitor voltage.

✱✱ WARNING

Make sure the cables used to connect the starter motor are of sufficient gauge (at least as thick as the battery cables normally connected to the starter motor, in fact those cables can be used). Also, make sure the cables are not too long. Use of cables with insufficient gauge or of too great a length will increase resistance in the circuit leading to false readings and possibly overheating the components used in the test.

■ If jumper cables of some sort are used to connect the battery to the starter motor, make sure that sufficient electrical connections are provided. Cables held loosely against the starter will increase resistance in the test circuit, and could lead to false results.

4. Connect the positive terminal of the battery to the positive wiring terminal on the starter motor itself (the point where the red cable normally attaches to the motor when installed). If using a conventional ammeter you'll have to place the meter inline between the 2 points (connect the positive battery terminal to the ammeter positive probe, then wire the negative meter probe to the starter motor positive terminal). If using an inductive meter, connect the positive terminal of the battery directly to the positive terminal on the starter, then clamp the meter pickup somewhere on the cable that was used.

✱✱ CAUTION

Use great care when connecting the cable from the negative terminal of the battery to the starter motor. Make sure there are NO sparks created anywhere near the battery (that could ignite the explosive vapors emitted by battery electrolyte).

5. Connect a suitable jumper cable to the negative terminal of the battery, then make sure the voltmeter, ammeter and tachometer are all ready to conduct the test.
6. Connect the jumper from the negative battery terminal to the starter motor casing, then check each of the meters.

■ Voltage must not drop below 12 volts DC or the battery is either insufficiently charged for this test or defective.

7. With voltage maintained between 12.0-12.4 VDC make sure the starter motor reaches the proper cranking RPM without exceeding amperage rating for the engine model being tested, as follows:
   • For 9.9 hp (211cc) 4-stroke and 9.9-15 hp (216/255cc) 2-stroke motors, the starter must rotate 7000-9200 rpm while drawing no more than 7 amps.
   • For 9.9/15 hp (305cc) 4-stroke or 18-35 hp (521cc) and 25/35 hp (500/565cc) 2-stroke motors, the starter must rotate 6500-7500 rpm while drawing no more than 30 amps.
   • For 25-55 hp (737cc) and 25-70 hp (913cc) motors, the starter must rotate 5700-8000 rpm while drawing no more than 32 amps.

8. If the motor exceeds current draw and/or fails to achieve proper speed, the motor must be overhauled or replaced. Refer to the procedures found in this section.

REMOVAL & INSTALLATION

◆ See Figure 191 and 194 thru 197

✱✱ WARNING

Be careful when servicing the starter motor assembly not to drop or strike the housing as this could crack or damage the permanent magnets contained with the motor.

Fig. 191 When equipped, the electric starter motor is mounted vertically to the powerhead, next to the flywheel

Fig. 192 When actuated, the starter pinion gear meshes with the flywheel ring gear teeth in order to turn the flywheel

Fig. 193 Starters can be tested for operation using a fully-charged 12-volt battery and a set of jumper cables

IGNITION AND ELECTRICAL SYSTEMS

Fig. 194 On some riggings, cables or hoses may be attached to the starter - 4-stroke carbureted motor shown

Fig. 195 All carbureted 4-strokes have at least one bolt mounted vertically underneath the flywheel cover/manual starter

Fig. 196 Once accessed, disconnect the wiring and remove the mounting bolts...

Fig. 197 ...then remove the starter assembly from the powerhead

Except 25/35 Hp (500/565cc) Motors

◆ See Figures 191 and 194 thru 197

1. Disconnect the negative battery cable, followed by the positive cable at the battery for safety.
2. If necessary for access on carbureted models, remove the flywheel cover or manual starter assembly

■ On 4-stroke carbureted models the flywheel cover/manual starter assembly must usually be removed in order to access and remove the starter bolt(s) mounted vertically through the top of the starter bracket and into the powerhead.

3. For EFI motors, remove the flywheel cover, followed by the flywheel. For details, refer to the Flywheel procedure in the Powerhead section.
4. If necessary for access to the starter wiring or, if used, lower mounting bolt, remove one or both of the lower engine covers.

■ On 4-stroke carbureted models, the lower port engine over must be removed in order to access both the lower mounting bolt and the starter positive terminal wiring.

5. For 25-70 hp (913cc) 2-strokes and all EFI 4-stroke motors, remove the air intake silencer for improved access.

6. For 18-35 hp (521cc) motors, tag and disconnect all leads from the starter solenoid. Then, right above the solenoid, remove the vertical throttle shaft clamp screws and remove the clamp.
7. To free the starter from the motor, remove the starter retaining bolts and disconnect the red positive power cable from the starter terminal. If access to the wiring terminal is clear, disconnect it first, but on some models it is easier to unbolt and pivot the starter for access to the wiring. For more details on your model proceed as follows:

• On 9.9 hp (211cc) 4-stroke motors, disconnect the wiring lead from the bottom of the starter motor and the 2 retaining bolts mounted vertically through the top of the motor.

• On 9.9-15 hp (216/255cc) 2-stroke motors, if equipped, unbolt the rectifier from the starter motor bracket and position it aside with the wiring attached. Then, disconnect the wiring lead from the bottom of the starter motor and the 2 retaining bolts mounted horizontally through top of the motor housing and into the powerhead.

• On 9.9/15 hp (305cc) 4-stroke motors, loosen the terminal nut and disconnect the red positive cable wiring from the bottom side of the starter motor. Remove the screw holding the bottom starter mounting bracket to the powerhead, then remove the 2 screws (on mounted horizontally and the other mounted vertically) from the top of the starter assembly.

• On 18-35 hp (521cc) 2-stroke motors, remove the screws and washers securing the starter mounting bracket to the side of the powerhead. Take note of the fact that the lower screw also secures a ground strap and a star washer. Next, remove the locknut and washer securing the mounting bracket to the front of the powerhead. Remove the motor with the red lead still attached, from the powerhead. If necessary, the red lead can be removed from the motor at this time.

• On 25-55 hp (737cc) and 25-70 hp (913cc) motors, disconnect the red positive cable from the starter, then remove the 2 side bracket and 1 front bracket retaining screws.

• On EFI motors, tag and disconnect the positive and negative cables from the starter motor/bracket. Remove the starter band clamp, then remove the 2 mounting bolts from the top of the starter.

8. Using a suitable solvent, clean all dirt, debris or corrosion from the starter mounting surfaces of the starter housing and powerhead. Clean all bolt/stud threads of any dirt, debris or corrosion.
9. If used, inspect rubber insulators for wear, damage or decay and replace as necessary.

To install:

10. Apply a light coating of OMC Nut Lock or an equivalent threadlock to the threads of the retaining bolts and, if applicable, nut(s).
11. If it is easier to access the wiring with the starter unbolted from the powerhead, connect the wiring at this point.
12. Position the starter to the powerhead and secure using the retainers. Tighten the starter mounting bolts and, if applicable, nuts to:

IGNITION AND ELECTRICAL SYSTEMS

- On 9.9-15 hp (216/255cc) 2-stroke motors, tighten the retainers to 10-12 ft. lbs. (14-16 Nm).
- On 18-35 hp (521cc) 2-stroke motors, tighten the retainers to 60-84 inch lbs. (7-9 Nm).
- On 9.9 hp (211cc) and 9.9/15 hp (305cc) 4-stroke motors, as well as 25-55 hp (737cc) and 25-70 hp (913cc) 2-stroke motors, tighten the retainers to 14-16 ft. lbs. (19-22 Nm).
- On EFI motors, tighten the retainers to 15-18 ft. lbs. (20-24 Nm).

13. If not done before positioning the starter motor, connect the red positive cable to the starter motor terminal (and on EFI motors, the black ground wire to the ground terminal/bracket). Tighten the retaining nut securely. Johnson/Evinrude does not provide torque specs for the terminal on all motors, but when applicable, tighten the terminal nut to:
- On 9/15 hp (305cc) 4-stroke motors, tighten the nut to 14-16 ft. lbs. (19-22 Nm).
- On 25-55 hp (737cc) and 25-70 hp (913cc) motors, tighten the nut to 10-12 ft. lbs. (14-16 Nm).
- On EFI motors, tighten the nut to 8-10 ft. lbs. (11-14 Nm).

14. For all carbureted motors, apply a light coating of OMC Black Neoprene Dip or a weather-strip sealant over all wiring terminals to protect them from moisture and corrosion, except the red, positive lead on 18-35 hp (521cc) 2-strokes.

■ Johnson/Evinrude specifically excludes the red positive lead terminal on 18-35 hp (521cc) 2-strokes from being coated with sealant. They do not, however, provide any explanation why.

15. For 25-70 hp (913cc) 2-strokes, use new air silencer base screws when installing the silencer and base onto the carburetors. Tighten the screws to 60-84 inch lbs. (7-9 Nm).
16. The balance of the installation procedure is the reverse of the removal.
17. Connect the positive battery cable, followed by the negative.

25/35 Hp (500/565cc) Motors

On these engines the electric starter motor is mounted to the starboard side of the powerhead, but its location might not be obvious at first glance. That's because it is mounted behind an electrical bracket box, sandwiched between the oil tank reservoir and the shift linkage. The quickest way to spot it is to look for the small toothed pinion gear attached to the top of the starter motor assembly, it just peeks out from underneath the flywheel/manual starter cover.

1. Disconnect the negative battery cable, followed by the positive cable at the battery for safety.
2. Disconnect all ground cables and wires from the ground stud at the front, lower corner of the electrical bracket that obscures the starter motor.
3. Tag all connections on the starter solenoid (located at the base of the starter motor, directly below the ground stud from the previous step). Disconnect the positive battery cable from the large post on the solenoid, then, disconnect the yellow/red neutral safety switch wire and the ground wire from the starter solenoid posts. Finally, disconnect the red lead from the remaining large solenoid terminal.
4. Remove the ground stud (from the second step) at the forward, bottom of the electrical bracket, then remove the retaining bolt from the top, aft corner (opposite corner from the ground stud). Reposition the bracket for access to the starter motor assembly.
5. Remove the top oil reservoir mounting screw and pull the reservoir out slightly from the powerhead for access to the aft starter motor mounting bolt.
6. Remove the 2 bolts securing the top of the starter motor horizontally to the side of the powerhead.
7. Using a suitable solvent, clean all dirt, debris or corrosion from the starter mounting surfaces of the starter housing and powerhead. Clean all bolt/stud threads of any dirt, debris or corrosion.

To install:

8. Apply a light coating of OMC Locquic Primer, followed by OMC Nut Lock or an equivalent threadlock to the threads of the 2 starter retaining bolts.
9. Install the starter to the powerhead and tighten the retaining bolts to 15-17 ft. lbs. (20-23 Nm).
10. Install the top oil reservoir mounting screw and tighten to 6-9 ft. lbs. (8-12 Nm).
11. Reposition and secure electrical bracket using the retaining screw and ground stud. Tighten the screw and stud to 60-84 inch lbs. (7-9 Nm).
12. Connect the wiring to the starter solenoid as tagged during removal and tighten the nuts securely (but do not over-tighten and damage the terminals). Position the protective cover over the positive battery cable connection.
13. Connect the ground straps to the ground stud and tighten the nut securely.
14. Apply a light coating of OMC Black Neoprene Dip or a weather-strip sealant over all wiring terminals (except the positive battery terminal connection with the protective cover) to protect them from moisture and corrosion.
15. Connect the positive battery cable, followed by the negative.

OVERHAUL

 DIFFICULT

◆ See Figures 198 thru 213

Although the working internal components of all Johnson/Evinrude starter motors are virtually the same, slight differences in housings, end-caps and mounting brackets used on a few motors makes for subtle differences in the most logical order of disassembly/assembly. For this reason, we've provided a few different disassembly and assembly procedures, based on engine models. When it comes to cleaning and inspection however, a starter motor is a starter motor, except that specifications will vary with engine model. A single cleaning and inspection procedure is provided.

Fig. 198 To disassemble a typical Johnson/Evinrude starter, begin by removing the protective cap. . .

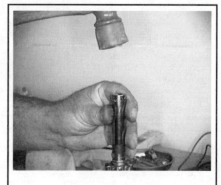

Fig. 199 . . . then use driver to push the cup and spring down revealing the snapring

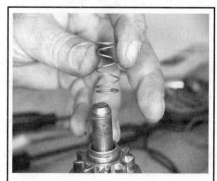

Fig. 200 Once the snapring and cup are removed, lift the spring and lower cup/spacer from the armature shaft

4-90 IGNITION AND ELECTRICAL SYSTEMS

Fig. 201 Remove the pinion gear and...

Fig. 202 ...if equipped, remove pinion gear base from the helical coils on the armature shaft

Fig. 203 Matchmark the end caps (commutator and drive) to the starter housing for assembly purposes

Fig. 204 Remove the starter housing through-bolts (usually 2)...

Fig. 205 ...then remove the end caps (commutator and drive, the drive end cap is shown)

Fig. 206 The brush assemblies are located in the commutator end cap, if removed, note the wire routing

Fig. 207 Exploded view of a typical starter. Lay out each of your starter's components in such a manner to ease assembly

Fig. 208 Take care when inserting the armature in the starter housing, as permanent magnets will pull it inward strongly

Fig. 209 During assembly, position the brushes as noted during removal, and place the brush springs under each contact...

Fig. 210 ...then use a modified putty knife to hold the brushes in position as the armature and starter housing are lowered onto the commutator cap

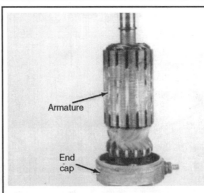
Fig. 211 An alternative procedure for holding the brushes that will work on most starters is to position the armature over the commutator...

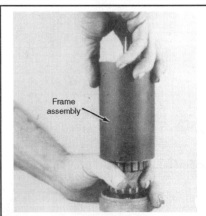
Fig. 212 ...then carefully slide the starter housing over the assembly (remember the magnets will pull strongly)

IGNITION AND ELECTRICAL SYSTEMS

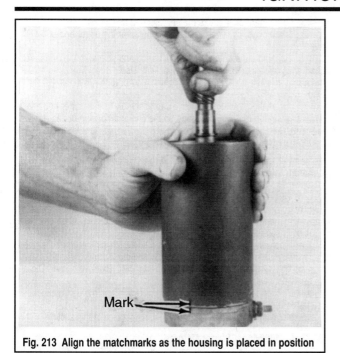

Fig. 213 Align the matchmarks as the housing is placed in position

Disassembly/Assembly

9.9 Hp (211cc) 4-Stroke And 9.9-15 Hp (211/255cc) 2-Stroke Starter Motors

DIFFICULT

◆ See Figures 211 thru 216

■ Many of the early model 2-strokes, including 18-35 hp (521cc), 25-55 hp (737cc) and 25-70 hp (913cc) models are equipped with a nut retained drive assembly, almost identical to the drive unit found on these models. If so, refer to this procedure for order of disassembly and for methods to remove the drive assembly.

1. Remove the starter from the powerhead as detailed in this section.
2. Mount the starter with the pinion gear sideways in a soft-jawed vise to ease disassembly. Do not overtighten the vise and damage the starter motor housing.
3. Matchmark the 2 end caps (drive cap on top and the commutator cap assembly on the bottom) to the starter housing.
4. Remove the 2 starter housing through-bolts from the bottom (commutator cap) end of the starter.

■ Take care not to disturb or loose the springs under the 2 ground brushes and insulated brushes when removing the commutator cap.

5. Remove the starter from the vise and hold it vertically (with the pinion gear towards the top as with normal installation), then carefully remove the commutator cap from the bottom of the starter.

■ If the commutator or drive cap is difficult to remove, gently tap on the end of the cap using a plastic or soft-faced mallet to help free it. Keep in mind that the permanent magnets contained in the housing will resist removal of the drive cap and armature from the housing.

6. Remove the drive cap and armature from the starter housing.
7. If necessary, hold the pinion gear from turning using a pair of sliding jaw pliers (or hold the armature directly using a strap wrench), then use a deep socket or wrench to loosen the nut securing the spring and spacer over the pinion gear. Discard the old pinion nut. Unthread the pinion gear from the armature shaft, then remove the armature from the drive cap.
8. Clean and inspect all components as detailed in this section under Cleaning and Inspection.
9. Replace any worn or unserviceable components.

To assemble:

10. Apply a single drop of SAE No. 10 oil on the armature shaft bearing surface, then apply a light coating of OMC Starter Pinion Lube to the helical threads and the bearing surface above the threads on the armature shaft.
11. Position new gaskets on the commutator and/or drive caps.

■ Johnson/Evinrude allows for 2 slightly different methods of assembling the starter from this point forward. Although either series of steps should work for either of the engines on which these starters are used, the specifications differ slightly between the actual starter assemblies used on 2-strokes and those used on 4-strokes. We'll provide both procedures and recommend that you follow the one for your motor, but if you have trouble, try the other (just remember to use the specs that apply to your motor).

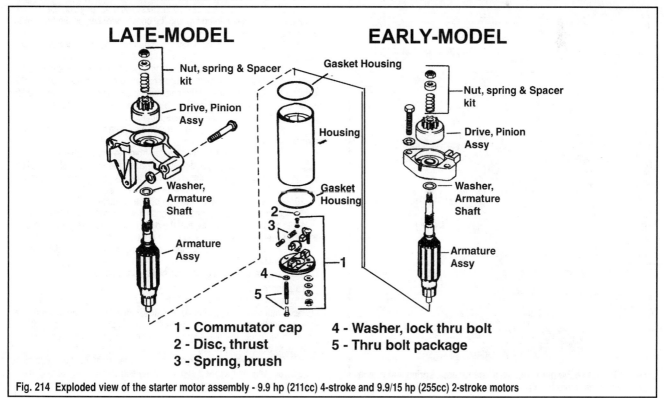

Fig. 214 Exploded view of the starter motor assembly - 9.9 hp (211cc) 4-stroke and 9.9/15 hp (255cc) 2-stroke motors

4-92 IGNITION AND ELECTRICAL SYSTEMS

12. For starters on 2-stroke motors, proceed as follows:
  a. Assemble the commutator cap, complete with brushes and springs onto the lower portion of the armature shaft.

✲✲ CAUTION

Protect your fingers when installing the armature shaft into the starter housing. Remember that the permanent magnets contained within the housing will excerpt extreme force on the components, pulling them together and crushing any fingers left in between!

  b. Carefully slide the starter housing over the armature, while aligning the matchmarks made on the commutator cap and the housing.
  c. If removed, position the thrust washer over the armature shaft, then install the drive cap (with a new gasket) over the armature shaft and onto the starter housing (while aligning the matchmarks made during disassembly).
  d. Apply a single drop of oil to each of the 2 starter through-bolts, then gently insert them through the end cap. Tighten the 2 through-bolts to 30-40 inch lbs. (3.4-4.6 Nm), then seal the bolts using OMC Black Neoprene Dip or an equivalent weather strip sealant.
  e. Install the pinion gear, spring and spacer to the top of the armature shaft, then secure using a new pinion nut. Hold the pinion gear and armature shaft from turning using a pair of sliding jaw pliers and tighten the pinion nut to 150-170 inch lbs. (17-19 Nm)

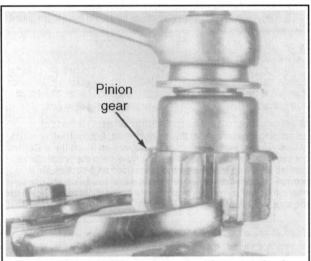

Fig. 215 On 211cc and 255cc 2-strokes, the pinion gear and spring assembly is secured using a nut. . .

Fig. 216 . . .once the nut is removed, the retainer, spring and pinion gear can be removed from the armature shaft

13. For starters on 4-stroke motors, proceed as follows:
  a. If removed, position the thrust washer over the upper end of the armature shaft.
  b. Assemble the drive cap, pinion, spring and spacer to the armature shaft, then secure using a new pinion locknut. Use a strap wrench to hold the armature shaft from turning and tighten the nut to 20-25 ft. lbs. (27-34 Nm).
  c. Install the armature and drive cap assembly to the starter housing while aligning the matchmarks made on the cap and housing during removal.

✲✲ WARNING

Remember the permanent magnets of the starter housing will pull the armature in forcibly, so watch your fingers during assembly.

  d. Use a pair of external snapring pliers to hold the brushes apart while you carefully position the armature and starter housing over the commutator cap. Align the matchmarks on the commutator cap and housing before the housing is seated.
  e. Apply a single drop of oil to each of the 2 starter through-bolts, then gently insert them through the end cap. Tighten the 2 thru-bolts to 95-110 inch lbs. (11-12 Nm), then seal the bolts using OMC Black Neoprene Dip or an equivalent weather strip sealant.
14. Perform the Starter No-Load Current Draw Test in order to check starter operation.
15. Install the starter to the powerhead, as detailed in this section.

25/35 Hp (500/565cc) Starter Motors

◆ See Figures 209, 210 and 217

■ If a brush holder is not available, install the armature to the commutator cap BEFORE assembling the drive cap components. In this alternative method of assembly, the armature is held in position on top of the brushes while the starter housing is lowered over the assembly, THEN the drive cap and components are installed.

1. Remove the starter from the powerhead as detailed in this section.
2. If not done already, remove the nut fastening the solenoid strap to the starter post.
3. If not done already, remove the 2 screws securing the solenoid and bracket to the bottom of the starter motor.
4. Check the end caps for matchmarks, and if not present, Matchmark the 2 end caps (drive cap on top and the commutator cap on the bottom) to the starter housing.
5. Remove the 2 starter housing through-bolts from the bottom of the starter.
6. Carefully remove the commutator cap from the bottom of the starter, trying not to disturb or loose the brush springs. The brush kits (brush and terminal sets) are mounted to the commutator cap. They can be removed for inspection, cleaning or replacement, but first take note of the correction orientation of each brush set and wire terminal.
7. Remove the drive cap and armature as an assembly from the top of the starter housing.
8. Remove the plastic protective cap from the Bendix shaft.
9. Gently push the cupped spacer down on the armature shaft until the snapring is exposed, then carefully remove the retaining ring with a pair of snapring pliers.
10. Pull the cup, spring and spacer off the armature shaft.
11. Unthread the pinion gear and base from the armature shaft.
12. Separate the armature and the drive cap. If necessary, remove the thrust washer from the upper end of the armature shaft.
13. Clean and inspect all components as detailed in this section under Cleaning and Inspection.
14. Replace any worn or unserviceable components.

To assemble:

15. Apply a single drop of SAE No. 10 oil on the armature shaft bearing surface, then apply a light coating of OMC Starter Pinion Lube to the helical threads and the sliding surface above the threads of the armature shaft.
16. If removed, install the thrust washer over the top of the armature shaft, then place the drive cap over top of the armature shaft.
17. Thread the base and pinion gear over the armature shaft, then slide the spacer (with the recessed end facing the end of the shaft), spring and cup (with the deeper recess facing the end of the shaft) over the shaft.

IGNITION AND ELECTRICAL SYSTEMS 4-93

■ The spacer and the cup form a housing or retainer for the spring assembly, the openings in each must face the spring itself so the spring can seat in them.

18. Push downward on the cup in order to expose the snapring groove in the armature shaft and then install the snapring. Pull the cup upward over the snapring to verify that the ring seats in both the shaft and cup grooves to allow proper Bendix travel. If there is interference, use a small pry tool to gently compress and seat the ring.

19. Gently snap the protective plastic cover into place.

20. Insert the armature into the starter housing while aligning the matchmarks made on the drive cap and housing during removal. Put the assembly down and prepare the commutator cap.

21. Apply a light coating of OMC wheel bearing grease to the armature bearing surfaces in the commutator cap.

■ If removed, apply OMC Locquic Primer, followed by OMC Screw Lock or equivalent threadlock to the brush card screws.

22. If removed, install the brush plate assembly to the commutator cap with the leads properly routed. Be sure to position the brush sets as noted during removal (as reversing the positive and negative brushes will cause the motor to run backwards). Place each brush spring into the bore in the brush plate, then position each brush just above its spring.

■ A brush holder can be fabricated by cutting a slot in putty knife. Make the slot just thick enough so that the knife can be slid around the bottom of the armature shaft, or more correctly, the armature shaft can be positioned over the knife and commutator cap, then the knife can be withdrawn from between the 2 components. This will hold the brushes down against spring pressure until armature is in place.

23. Position a brush holder tool (modified putty knife) over the brushes to hold them down into the commutator cap, then carefully lower the armature (with the drive cap and starter housing assembly) onto the commutator cap while aligning the matchmarks. As the armature is seated, slowly withdraw the putty knife from between the cap and armature, making sure the brushes remain positioned properly as the assembly is seated.

24. Apply OMC Locquic Primer, followed by OMC Screw Lock or equivalent threadlock to the threaded portion of the 2 starter housing thru-bolts and then apply a single drop of oil to the shafts of each bolt.

25. Gently insert the starter housing through-bolts into the housing from the commutator cap and tighten to 50-60 inch lbs. (5.6-9 Nm)

26. Install the starter solenoid and bracket to the starter. Attach the solenoid strap to the starter post. Tighten the screws and nut securely, then coat the strap, post and nut using OMC Black Neoprene Dip or an equivalent weather strip sealant.

27. Perform the Starter No-Load Current Draw Test in order to check starter operation.

28. Install the starter motor, as detailed in this section.

9.9/15 Hp (305cc), 18-35 Hp (521cc), 25-55 Hp (737cc) And 25-70 Hp (913cc) Starter Motors

◆ See Figures 198 thru 213, 218 and 219

■ Many of the early model 2-strokes are equipped with a nut retained drive assembly, almost identical to the drive unit found on 9.9 hp (211cc) 4-stroke and 9.9-15 hp (211/255cc) 2-stroke motors. To find out if the motor on which you are working if of this type, simply check the top of the starter motor pinion shaft. If there is a nut on the end, instead of a protective cap and lockring, refer to the 9.9-15 hp 2-stroke procedures found earlier in this section for order of disassembly and for methods to remove the drive assembly.

1. Remove the starter from the powerhead as detailed in this section.
2. Mount the starter with the pinion gear facing upward (in the same direction the starter motor is normally installed on the powerhead) in a soft-jawed vise to ease disassembly. Do not over-tighten the vise and damage the starter motor housing.
3. Carefully pry the protective cap from the groove in the spacer.
4. Gently push the cupped spacer down on the armature shaft until the snapring is exposed, then carefully remove the retaining ring with a pair of snapring pliers.

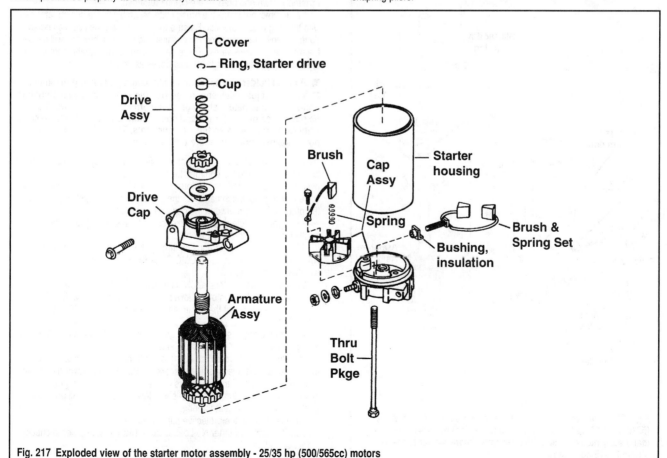

Fig. 217 Exploded view of the starter motor assembly - 25/35 hp (500/565cc) motors

4-94 IGNITION AND ELECTRICAL SYSTEMS

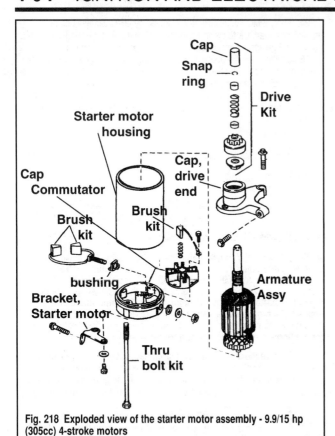

Fig. 218 Exploded view of the starter motor assembly - 9.9/15 hp (305cc) 4-stroke motors

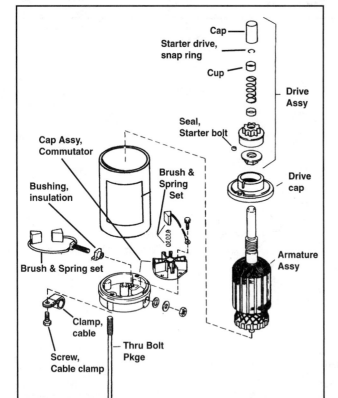

Fig. 219 Exploded view of the starter motor assembly - 18-35 hp (521cc), 25-55 hp (737cc) and 25-70 hp (913cc) motors (Note that many early models use a drive assembly more similar to the 2-stroke 9.9/15 hp motors)

5. Pull the cup, spring and spacer off the armature shaft.
6. Unthread the pinion gear and base from the armature shaft.
7. If necessary on 4-stroke motors, remove the screws holding the mounting bracket to the bottom of starter motor (to the commutator cap).
8. Matchmark the 2 end caps (drive cap on top and the commutator cap on the bottom) to the starter housing.
9. Remove the 2 starter housing through-bolts from the top or bottom of the starter (depending on the model).

■ On models where the through-bolts thread from the bottom (commutator cap) it may be necessary to remove the starter housing from the vise and reposition it for access.

10. Separate the end caps from the housing and carefully withdraw the armature assembly.
11. The brush kits (brush and terminal sets) are mounted to the commutator cap. They can be removed for inspection, cleaning or replacement, but first take note of the correction orientation of each brush set and wire terminal.
12. Clean and inspect all components as detailed in this section under Cleaning and Inspection.
13. Replace any worn or unserviceable components.

To assemble:
14. Apply a single drop of SAE No. 10 oil on the armature shaft bearing surface, then apply a light coating of OMC Starter Pinion Lube to the helical threads of the armature shaft.

■ If a brush holder is not available, install the armature to the commutator cap BEFORE assembling the drive cap components. In this alternative method of assembly, the armature is held in position on top of the brushes while the starter housing is lowered over the assembly, THEN the drive cap and components are installed.

15. If removed, install the thrust washer over the top of the armature shaft, then place the drive cap over top of the armature shaft.
16. Insert the armature into the starter housing while aligning the matchmarks made on the drive cap and housing during removal. Put the assembly down and prepare the commutator cap.
17. If removed, install the brush plate assembly to the commutator cap with the long lead in the slot. Be sure to position the brush sets as noted during removal (as reversing the positive and negative brushes will cause the motor to run backwards). Place each brush spring into the bore in the brush plate, position each brush just above its spring.

■ A brush holder can be fabricated by cutting a slot in putty knife. Make the slot just thick enough so that the knife can be slid around the bottom of the armature shaft, or more correctly, the armature shaft can be positioned over the knife and commutator cap, then the knife can be withdrawn from between the 2 components. This will hold the brushes down against spring pressure until armature is in place.

18. Position a brush holder tool (modified putty knife) over the brushes to hold them down into the commutator cap, then carefully lower the armature (with the drive cap and starter housing assembly) onto the commutator cap while aligning the matchmarks. As the armature is seated, slowly withdraw the putty knife from between the cap and armature.
19. Apply a single drop of oil to each of the 2 starter through-bolts, then gently insert them through the end cap.
20. Tighten the starter housing through-bolts to 95-100 inch lbs. (11-12 Nm), then seal the bolts using OMC Black Neoprene Dip or an equivalent weatherstrip sealant.
21. On 4-stroke motors, apply a coating of OMC nut lock or equivalent threadlock to the threads of the lower bracket retaining screw. Install the bracket to the bottom of the commutator cap and tighten to 95-100 inch lbs. (11-12 Nm).
22. Thread the base and pinion gear over the armature shaft, then slide the spacer, spring and cup over the shaft.
23. Push downward on the cup in order to expose the snapring groove in the armature shaft, then install the snapring. Pull the cup upward over the snapring to verify that the ring seats in both the shaft and cup grooves to allow proper Bendix travel. If there is interference, use a small prytool to gently compress and seat the ring.
24. Gently snap the protective cap into place.
25. Perform the Starter No-Load Current Draw Test in order to check starter operation.
26. Install the starter motor, as detailed in this section.

IGNITION AND ELECTRICAL SYSTEMS 4-95

EFI Engine Starter Motors

♦ See Figures 211 thru 213 and 220

1. Remove the starter from the powerhead as detailed in this section.
2. Mount the starter with the pinion gear facing upward (in the same direction the starter motor is normally installed on the powerhead) in a soft-jawed vise to ease disassembly. Do not overtighten the vise and damage the starter motor housing.

✳✳ CAUTION

Wear safety glasses when removing the pinion stop set, as the components are held against spring pressure and could fly free once the snapring is removed.

3. Grasp the stopper using a pair of pliers and push it downward one the armature shaft toward the pinion drive until the snapring is exposed. Carefully pry the snapring from the slot in the shaft, then slowly release the spring pressure while pulling the stopper and spring upward and off the shaft.
4. Unthread the pinion gear counterclockwise from the armature helical splines.
5. Matchmark the end covers (drive on top and commutator on bottom) to the starter housing with a permanent marker for reference during assembly.
6. Reposition the starter housing in the vise to access the through-bolts on the bottom, then remove the bolts.
7. While holding the starter upright (facing the normal direction of installation with the drive end on top), pull the commutator (lower) cap from the starter housing, being careful not to disturb the brushes until the you can note the orientation of the wiring.

■ It may be necessary to tap LIGHTLY on the commutator cover with a rubber mallet in order to free it from the housing.

8. Remove the thrust washers from the armature shaft or lower cover.
9. Turn the starter housing sideways, then tap lightly (using the rubber mallet) on the exposed end of the armature shaft to free the armature and upper cover from the starter housing.

■ Before proceeding, note the arrangement and orientation of the thrust washers on the drive end of the armature shaft. They must be reinstalled in the same order, facing the same direction.

10. Remove the thrust washers from the drive end of the armature shaft. Mark them and wire them together to ensure installation in the original order and directions.
11. If disassembly of the commutator cap and brushes is necessary:
   a. Remove the nut from the large terminal and then slide the washers and insulators from the terminal.
   b. Matchmark the brush plate and commutator cover to ensure correct orientation during assembly. c. Remove the 2 screws securing the brush plate to the commutator cover (from the underside of the cover), then carefully lift out the brush plate, brush set and spring.
12. Clean and inspect all components as detailed in this section under Cleaning and Inspection.
13. Replace any worn or unserviceable components.

To assemble:
14. Apply a light coating of OMC Moly Lube or equivalent marine grease to the armature shaft bushing surfaces in the commutator and drive covers.

✳✳ WARNING

Be sure not to apply excessive amounts of grease to the bushing surfaces. No grease should be allowed near the brushes or brush contact surfaces.

15. Apply a light coat of OMC Starter Pinion Lube to the helical threads on the base of the armature shaft (where the shaft meets the armature body).
16. Align the matchmarks made during disassembly, then install the brush plate into the commutator cover. Tighten all fasteners securely. Be certain that the large terminal and insulator fit securely into the commutator cover opening.
17. If used, place the thrust washer(s) onto the lower end of the armature shaft, then install the armature shaft into the commutator plate, while holding the brushes in position. Manually hold the brushes away while attaching the armature to the cover. Release the brushes and inspect them for damage before proceeding. If the brushes become dislodged, pull the assembly apart again and fabricate a tool to hold the brushes in place during installation. Bend a stiff piece of thin rod into a shape like the letter **U**, then use the ends of the tool to hold them in position.
18. Hold the armature tightly into the commutator cover, then align the matchmarks made earlier and carefully slide the starter housing into position.
19. Install the thrust washers over the top of the armature shaft using the same order and orientation as noted during removal.
20. Slide the drive cover over the armature and into contact with the starter housing while aligning the matchmarks made earlier.
21. Install the 2 through-bolts and tighten them to 36-60 inch lbs. (4-7 Nm).
22. Thread the pinion gear onto the armature shaft, then position the spring and stopper.
23. While wearing safety glasses, push the pinion stopper downward toward the starter housing in order to expose the slot in the armature shaft, then install the snapring.

■ Make sure the snapring fully engages the armature shaft and stopper.

24. Install the starter motor, as detailed in this section, then check operation.

Cleaning and Inspection

♦ See Figures 221 thru 229

■ NEVER attempt to clean the starter drive assembly with solvent while the components are installed. Solvent could wash dirt into the bearings and commutator that would eventually lead to starter failure.

1. Remove the starter motor from the powerhead, as detailed in this section.
2. Disassemble the starter motor, laying out each of the components in a logical order to ease inspection and assembly, as detailed in this section.

■ Work on only one component at a time, then place it back into your logical layout on the wor surface.

3. Use compressed air to remove brush material or debris from the armature and commutator cover.

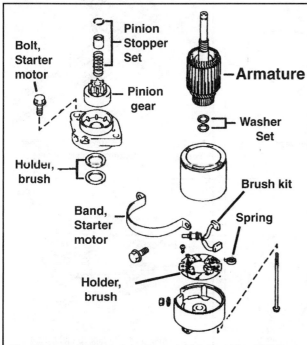

Fig. 220 Exploded view of the starter motor assembly - 40-70 hp EFI 4-stroke motors

4-96 IGNITION AND ELECTRICAL SYSTEMS

4. Use a mild solvent to clean all starter motor components, except the brush plate and brushes.
5. Clean the brush plate and brushes using an electrical contact cleaner.
6. Visually check the pinion drive for damaged (chipped, cracked or worn) teeth and replace it if needed.
7. Inspect the helical splines at the base of the armature shaft (where the armature shaft meets the body) for damage or corrosion. Thread the pinion drive on and off of the shaft splines. Replace the pinion drive and/or armature if the pinion drive does move smoothly on the threads.
8. Inspect the entire armature assembly for damage, wear or corrosion and, replace, if necessary.
9. Carefully secure the armature in a vise with soft jaws (otherwise, use wooden blocks or rubber pads to protect the armature in the vise). Tighten the vise just sufficiently to secure the armature, but not so tight as to damage it.
10. With the armature in the vise, carefully polish the commutator using 300-grit emery cloth. Rotate the armature often in the vise to polish it evenly around the circumference.

✱✱ WARNING

Only polish the commutator by hand as power tools would likely remove too much material making it unserviceable. Proceed slowly and evenly removing only a minimal amount of material.

11. Check the commutator for shorts or open windings using a DVOM or ohmmeter:
   a. Probe between the commutator segments and the core (laminated section) of the armature. There should be no continuity between any commutator segment and any laminated section (if continuity is present, there is a shorted winding and the commutator must be replaced).
12. Probe between the armature shaft and each commutator segment. There should be no continuity should between any commutator segment and the armature shaft (if continuity is present, there is a shorted winding and the commutator must be replaced).
13. Probe between the between each commutator segment (place one meter probe on a segment and the other probe against an adjacent segment, then move the probes sequentially around the segments until all are

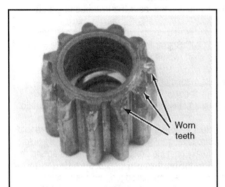
Fig. 221 Check the pinion gear for damaged or worn teeth and replace, if necessary

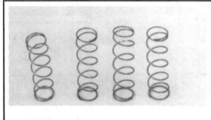
Fig. 222 Compare the color and length of the brush springs. Replace the set if any are stretched or bluish in color

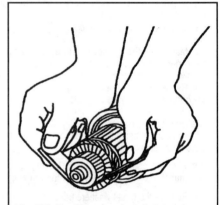
Fig. 223 Use 300 grit emery cloth to carefully polish the commutator surface

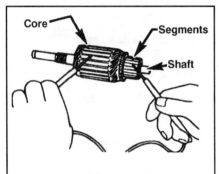
Fig. 224 Use an ohmmeter to check the commutator for shorts between the commutator and core or shaft. . .

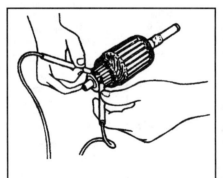
Fig. 225 . . .and for opens between segments

Fig. 226 For starter motors on EFI engines, measure the commutator outside diameter. . .

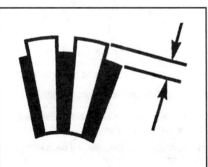
Fig. 227 . . .followed by the mica undercut. . .

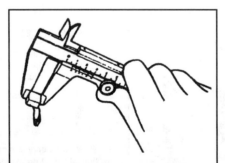
Fig. 228 . . . and finally, the brush length. Replace components if under spec for EFI motors

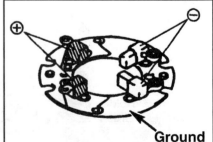
Fig. 229 Use an ohmmeter to make sure there is no continuity between positive (+) and, negative (-) or the brush plate ground

IGNITION AND ELECTRICAL SYSTEMS 4-97

checked). There must be continuity between segment pairs (if continuity is **not** present, there is an open winding and the armature must be replaced).

14. Use a small file to carefully undercut the mica (lengthwise cuts that are located between the commutator segments). Remove all metal particles using compressed air.

15. Visually inspect the brushes, springs and brush plate for damage (chipped or broken surfaces), dirt or corrosion. If any brush springs are weak, lack tension, or are discolored, replace all the springs as a set. Brushes or brush plates that show damage must be replaced.

16. Additional specifications are available from the manufacturer for the starters used on

EFI motors, measure the following dimensions to help determine if the starter components are serviceable:

   a. Using a sliding or dial caliper or outside diameter micrometer, measure the outer diameter of the armature commutator (after polishing) at several locations. The commutator is manufactured with an outer diameter of 1.30 in (33mm), but must be replaced if it has become less than 1.26 in. (32mm) due to wear or polishing.

   b. Using a depth micrometer, measure the depth of each mica undercut. The commutator is manufactured with 0.020-0.030 in. (0.5-0.8mm) deep undercuts, but must be replaced if wear causes undercut depth to become less than 0.008 in. (0.2mm).

   c. Using the same sliding caliper or outside micrometer as in Step A, earlier, measure the length of each brush. New brushes should be about 0.49 in. (12.5mm), and old brushes must be replaced once they are less than 0.35 in. (9.0mm).

   d. Using the ohmmeter, check the brush holder for proper continuity. Probe between the positive terminals of the brush holder and the negative terminals, then between positive and the base plate ground. The brush plate must be replaced if continuity exists between the positive terminals and either negative or ground.

17. Check the permanent magnets in the starter housing for dirt, debris or corrosion and clean, as necessary. Make sure none of the magnets are loose or damaged (cracked or visibly deformed). The starter housing must be replaced if it or the magnets have been damaged.

18. Check the bearing surfaces on the armature and inside the bushing (commutator and drive caps) for discoloration and/or excessive or uneven wear. Replace any questionable bearings/bushings using a suitable puller and driver. Replace the armature its bearing surfaces are rough or uneven.

19. Assemble the starter and install it to the powerhead as detailed in the procedures in this section.

Starter Motor Solenoid/Relay Switch

◆ See Figure 230

When the starter button or ignition keyswitch is actuated, current flows to and energizes the starters solenoid/relay coil. The energized coil closes a set of contacts that allows high current from the battery terminal of the starter to connect with the power output terminal of the relay and reach the starter motor.

Fig. 230 Typical Johnson/Evinrude starter solenoid with signal terminal identification - used on most 2-strokes, 4-stroke models similar but vary in terminal

The solenoid/relay is normally mounted to the powerhead in the immediate vicinity of the starter motor. If its location is not immediately apparent, remember that the solenoid is essentially a remote controlled switch located inline between the battery and starter. Follow either the positive battery cable from the battery to the solenoid or the red, positive cable, from the starter back to the solenoid.

■ On 25/35 hp (500/565cc) motors the solenoid is actually mounted to a bracket on the base of the starter itself, but because the starter is hidden by an electrical bracket, this might not be readily apparent.

TESTING THE SOLENOID

◆ See Figure 230 and 231

■ Solenoid testing on EFI models is part of the step-by-step troubleshooting process for the entire starting system. Also, it varies, very slightly, from the carbureted models covered here, therefore on EFI motors, please refer to Checking the Solenoid Relay found under Starting System Testing (EFI Models), earlier in this section.

For all carbureted models, a DVOM can be used to check the coil winding of the starter solenoid for continuity only under specific conditions. To perform this functional check of the solenoid, a fully-charged 12-volt battery and a set of jumper leads will also be necessary. The test essentially involves checking to make sure the main switch (battery power-to-starter power terminal) is closed when the solenoid signal circuit is not activated, then manually activating the circuit using the battery and jumpers to make sure the main switch closes.

For safety and to ensure proper results, be sure to tag and disconnect all wiring from the starter solenoid/relay before performing this test.

1. Disconnect the negative battery cable for safety.
2. Tag and disconnect the wiring from the solenoid. If desired, remove the solenoid completely from the powerhead.
3. Set the DVOM to read resistance, then connect the probes across the 2 large terminals on either end or either side of the solenoid. These are the main switch terminals to which the positive battery cable and the red starter cable normally attach. The meter should show no continuity. If the

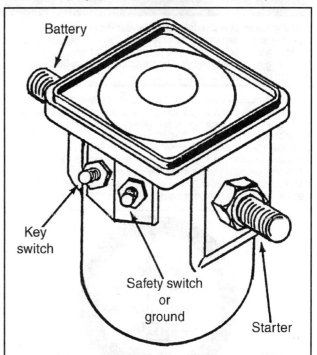

Fig. 231 On all Johnson/Evinrude solenoids, the larger terminals are for the battery and starter cables, while the smaller terminals connect to the signal wires - typical late-model 2-stroke solenoid shown (4-stroke and early-models vary in shape, but should share the same BASIC terminal layout, though sides may switch)

4-98 IGNITION AND ELECTRICAL SYSTEMS

switch is closed with no power applied to the signal terminals, the solenoid is stuck closed and must be replaced.

4. Using the set of jumper wires, apply battery voltage to the solenoid signal terminals. Use one jumper wire to connect from the positive battery cable to one signal terminal (the small terminals to which the ignition switch and ground wiring normally connect) and the other jumper wire to connect the negative battery cable to the other terminal.

■ Listen as the battery connections are made to the signal terminals of the solenoid. An audible click should be heard as the main power switch closes.

5. With the battery power applied to the signal switch of the solenoid, recheck the meter for continuity between the 2 large terminals. The meter must now show very little or no resistance at all, as the main switch must be closed (which would normally allow battery power to reach the starter motor). If the solenoid/relay switch does not close, it must be replaced.

6. After testing or repairs, reconnect the wiring as tagged and reconnect the negative battery cable.

REMOVAL & INSTALLATION

◆ See Figure 230

Removal and installation of the solenoid itself is a relatively simple matter of disconnecting the wiring and unbolting the solenoid. Sounds easy right? Well, it is, as long as you remember to tag the wiring before disconnecting it. Look, we all do this, we go, I'll remember it, then the phone rings or someone walks in or you bang you big toe on the trailer, whatever. Point is, take a few pieces of tape and a few seconds to tag the wiring before you start and you'll have no problem installing the solenoid, whenever you finish.

WARNING SYSTEM

Description and Operation

◆ See Figures 232 and 233

※※ WARNING

Read and follow all warnings/cautions in your owner's manual regarding engine warning system operation. The warning system is meant to protect both your engine (by protecting it from potentially severe engine damage) and you (by protecting you from becoming stranded on the water).

Most Johnson/Evinrude models, except some 2-strokes, 8 hp and smaller, are equipped with one or more warning systems designed to alert the boater should a malfunction occur in various engine operating systems (such as engine cooling or oil delivery). Most Johnson/Evinrude engines (especially late-models) are equipped with some version of the Speed Limiting Operational Warning (S.L.O.W) system. This system monitors one or more parameters, depending on the model. On most models it monitors engine temperature (on all 2-strokes and most remote control 4-strokes) and/or oil pressure (on all 4-stroke models), and/or oil level in the oil tank (oil injected

■ On 25/35 hp (500/565cc) motors the solenoid is actually mounted to a bracket on the base of the starter itself, but because the starter is hidden by an electrical bracket, this might not be readily apparent. Although it may be possible to access the relay with the electrical bracket installed, it may be easier to remove the ground wires from the ground stud, then remove the ground stud and the bolt at the upper corner of the bracket, then reposition the bracket for clearance. For more information, please refer to the Starter removal and installation procedure for these engines.

1. Disconnect the negative battery cable, followed by the positive battery cable for safety.

2. Locate the starter solenoid by following the red positive cable back from the starter or forward from the battery.

3. Tag the wiring to ensure proper and easy installation. If necessary, sketch a quick diagram of each wire color and terminal location. Disconnect the wiring and move it gently aside for clearance.

■ The retaining screws used on Johnson/Evinrude starter solenoids are of varying designs. Some motors, mostly 4-strokes, may be equipped with Torx® head screws. Be sure to use the proper sized driver for the screws on your solenoid or solenoid bracket.

4. Remove the solenoid or solenoid bracket retaining screws, then remove the solenoid from the powerhead or mounting bracket.

5. Installation is the reverse of the removal procedure. Tighten the screws and wiring terminal nuts securely. If the wiring is not equipped with a plastic/rubber protective cover, apply a light coating of OMC Black Neoprene or equivalent weather stripping adhesive over the wires to protect them from moisture and corrosion.

2-strokes). The system is used to warn the boater of trouble before damage to the engine can occur.

The system will normally trigger if high temperature or low oil pressure (4-stroke)/low oil level (2-stroke) is detected, depending on the model. A much more comprehensive version of the system is incorporated into the Electronic Control Unit (ECU) on EFI engines. For fuel injected engines, the system monitors engine speed as well as oil pressure and both coolant and exhaust component temperatures.

All EFI and most remote control carbureted engines are equipped with either a stand-alone Johnson/Evinrude System Check engine monitor gauge or a tachometer housing that contains the same 4 trouble-lights at the base. The System Check monitor acts as a dash mounted gauge/horn to alert the operator in the event of certain potentially damaging operating conditions.

Gauge circuits monitor conditions such as a lack of oil, excessive water temperature, electronic fault detection (check engine light condition) or low oil (meaning low oil in a reservoir that feeds an automatic oiling system on a 2-stroke engine). On EFI motors, in addition to the information on oil pressure and water temperature provided by powerhead mounted sensors, the Check Engine light is connected to a warning circuit for the ECU designed to alert the boat operator should a malfunction occur in the electronic engine control system.

Fig. 232 The System Check monitor in normally installed in the boat dash as a stand-alone gauge...

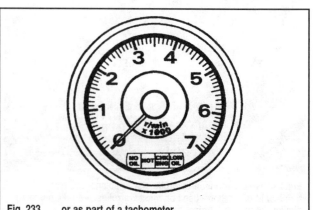

Fig. 233 ...or as part of a tachometer

IGNITION AND ELECTRICAL SYSTEMS

CARBURETED MODELS

◆ See Figures 232 and 233

All carbureted outboards are equipped with a water pump indicator stream. Always be sure that a strong stream of water exits the fitting at the lower rear area of the motor cover while the engine is operating. If the stream of water is ever absent or weak, shut the motor down immediately and inspect the cooling system. Check the stream more frequently anytime engine is operated in water clouded with sand or debris.

On carbureted engines, if the S.L.O.W. system is activated, the ignition module will gradually reduce and limit the engine speed to 2000-2500 rpm, depending on the model. This will occur if 4-stroke oil pressure drops to a dangerously low level. This will also occur if the oil level in the tank drops to a pre-determined level on certain 2-strokes. On most remote control 4-strokes, as well as all 2-strokes equipped with the system it will also occur if the engine temperature exceeds specification as follows:

- 4-stroke motors (carbureted): approximately 240°F (116°C)
- 18-35 hp (521cc) 2-stroke motors: approximately 180°F (82°C)
- 25/35 hp (500/565cc) 2-stroke motors: approximately 250°F (116°C)
- 25-55 hp (737cc) and 23-70 hp (913cc) 2-stroke motors: approximately 207°F (97°C)

In order for normal operation to resume, engine speed must drop below 1500 rpm and the condition that caused system activation must cease. On some models, the motor must actually be shut down and restarted to resume normal operation (this is includes all 25-55 hp [737cc] and 23-70 hp [913cc] 2-stroke motors with 12 amp charging systems).

EFI MODELS

◆ See Figures 232 and 233

On EFI models, the ECU monitors input from multiple sensors to determine if the S.L.O.W. system mode should be actuated. It will automatically trigger the operation in this mode if oil pressure drops below or temperature (from coolant or exhaust sensors) rises above predetermined levels. It will also actuate the system for excessive engine speed.

Excessive engine speed is defined as more than 3000 rpm in Neutral or, in gear, more than 6500 rpm (on 40 and 70 hp motors) or 7000 rpm (on 50 hp motors) for more than 10 seconds. In S.L.O.W. mode, the ECU will supply only an intermittent signal to the fuel injectors to gradually reduce engine speed to 3000 rpm. Once engine speed drops below 3000 rpm, the ECU will resume normal operation for speeds up to that rpm. If the condition that caused activation ceases (temperature goes down or oil pressure resumes) and engine speed has been reduced to idle, normal operation throughout the engine rpm range will resume.

A warning buzzer is normally located within the remote control unit (but also may be located with the tiller control housing or behind the dash). The buzzer is actuated by the ECU, which provides a ground circuit. When the ignition switch is first turned on, battery voltage is applied to the buzzer and if the circuit is completed through the ECU at any time, the buzzer sounds.

The oil pressure warning circuit is comprised of the oil pressure switch, the No Oil warning light in the System Check gauge, the buzzer and ECU. Anytime oil pressure falls below approximately 10-19 psi (70-130 kPa) the oil pressure switch will close sending a signal to the ECU. When low oil pressure is detected, the ECU will illuminate the oil light and sound the warning buzzer in a cyclical on - off tone.

The overheat warning circuit is comprised of an engine temperature sensor, as well as an exhaust manifold temperature sensor. The overheat warning will be activated anytime the occurs receives a signal of excessive temperature from either sensor. For 40/50 hp models this occurs at temperatures of 250°F (121°C) or more. For 70 hp models the ECU is looking for temperatures of 234°F (111°C) or more. When an overheat signal is detected, the ECU will illuminate the Water Temp light and sound the buzzer.

■ On EFI motors, the ECU unit can also activate an overheat warning if it detects a rapid temperature increase immediately following a cold-start or if it detects excessive engine speed before the engine reaches normal operating temperature. Both conditions are designed to force the operator to allow the engine to reach normal operating temperature before attempting to operate the motor at high speeds.

Besides actuating the No Oil and Water Temp lights, the ECU will limit engine speed during certain conditions (to protect the powerhead) by intermittently switching off the fuel injectors. Anytime the warning system is activated by low oil pressure, normal operation will resume after the engine is shut down and restarted (provided that oil pressure returns). When a warning circuit is activated due to temperature, normal operation will resume once temperature drops below system activation levels.

If the Check Engine warning light illuminates, first check the battery condition as detailed under Batteries in the General Information and Maintenance section. If the battery is good, the ECU may have detected a fault within the electronic engine control system (and stored a diagnostic trouble code). If the ECU detects a fuel injection or electronic engine control fault, it will store a Service Code, sound the horn for 10 seconds and illuminate the gauge LED for a minimum of 30 seconds. The light will go out if the fault does not remain present, otherwise the LED will remain illuminated until the fault goes away or the key is turned off (whichever comes first). If the fault is still present the next time the engine is started, the LED will illuminate once again to alert the operator that problem is still present. The Check Engine light can also be used to output the Service Code. For more information, please refer to Reading and Clearing Codes in the section for Electronic Fuel Injection (EFI).

Warning System Troubleshooting

TESTING WARNING CIRCUITS

System Check Gauge

◆ See Figures 232 and 233

Self-Test Mode

◆ See Figures 232 and 233

The System Check gauge enters a self-test mode each time the key is turned to the **ON** or **RUN** position. When power is first applied, the internal gauge electronics gauge (or the circuits in the ECU) will sound the warning horn for a 1/2 second and illuminate all four gauge LEDs. Then, the electronics will turn off each LED in sequence. Each self-test helps the operator be sure that the warning horn and all LEDs are functioning, as well as reassuring that the electronic control circuits for the system are operating correctly.

Should battery voltage drop below 7 VDC anytime during engine operation or if the ignition switch is simply left in the **ON** or **RUN** position, the gauge may automatically re-enter the self-test mode.

If the self test mode does not occur normally, check the following:
- If 1-3 of the LEDS do not illuminate, the control circuits may be at fault and, if so the gauge or ECU, must be replaced. Make sure the control circuits are at fault by checking the related circuits/components.
- If all 4 LEDs do not illuminate, test the power to the gauge as follows:

1. Locate and disconnect the system gauge 8-pin connector, then use a voltmeter to check for power on the purple lead (using the black lead to provide ground) with the ignition keyswitch turned to **ON** or **RUN**. If battery voltage is not supplied under these conditions, recheck connecting the black lead directly to ground (if power is now present, repair/replace the open in the black lead). If voltage is not available under either condition, check the purple lead, 20-amp fuse and/or ignition keyswitch and repair or replace components, as necessary.

2. If the purple lead indicates battery voltage and the black lead is continuous with ground, but not LEDs illuminate when the harness is reconnected, replace the system check gauge or the ECU.

■ The high cost of an ECU would make us seek a second opinion if we were not completely sure of our test procedures and that nothing was overlooked in making harness connections.

- If the warning horn does not sound for 1/2 second when the keyswitch is first turned to **ON** or **RUN**

3. If available, substitute a known good horn. If the replacement horn beeps, the problem is found. More likely, if a known good horn is not available, disconnect the 8-pin connector at the system check gauge. Turn the ignition switch to **ON** or **RUN** and use a suitable jumper lead to ground the connector's tan/blue lead to the black lead. If the warning horn now sounds, replace the system check gauge.

4. If the warning still does not sound, leave the gauge disconnected and now disengage the warning horn from its 2-pin connector. Test the tan/blue lead for continuity between the warning horn (2-pin) and gauge (8-pin) connectors. If no continuity is found, repair or replace the tan/blue lead (and/or connectors) whichever is necessary.

4-100 IGNITION AND ELECTRICAL SYSTEMS

5. Next, test the purple lead in the 2-pin connector (on the wiring harness side) for battery voltage whenever the ignition switch is turned to **ON** or **RUN**. Repair or replace the purple lead, 20-amp fuse and/or ignition switch as necessary. Reconnect the 2-pin connector when finished.

6. If the circuits check out, but the horn will still not sound you can check the horn using jumpers to connect it directly to a 12-volt battery, then replace the horn if it will not sound.

Operational Mode

◆ See Figures 232 and 233

The System Check gauge (or ECU) circuitry enters operational mode each time the self-test mode is complete and the engine is started. When in the operational mode, anytime a sensor activates (the switch closes providing a ground signal) or anytime the ECU detects a fault in a monitored signal for EFI motors, the corresponding LED will illuminate and the warning horn will sound for ten seconds. The LED will remain illuminated for a minimum of 30 seconds, even if the problem only occurred momentarily and then disappeared. This gives the operator time to react to the horn and check what LED circuit was illuminated. If the problem remains, the LED will remain illuminated until the condition goes away or the keyswitch is turned **OFF**

When the sensor deactivates (opens from ground on the switch type sensors used by carbureted motors), the LED will remain illuminated for an additional 30 seconds, then go out.

■ **Should additional sensors activate while an LED is already illuminated (say an overheating engine, suddenly reaches the point of no oil (low pressure or low level in the remote tank), the warning horn will sound again for ten seconds and the second LED will illuminate.**

If the warning system activates immediately take steps to identify and rectify the problem and/or protect the motor:
• If the NO OIL LED illuminates:
 1. On 4-stroke motors, this indicates that oil pressure has dropped below safe levels. Stop the engine immediately and check the oil level. If the oil level appears good, perform a oil pressure check using a gauge before restarting the motor. If you are stranded on the water by this, you can remove the oil pressure sensor from the motor and start the engine momentarily to check for a stream of pressurized oil from the open bore (catch it, if present using a small bucket to prevent un-necessary pollution.). If some oil stream is present, the motor may be operated at idle to limp the boat home. If oil is present, it is possible that oil flow is just below spec (and dangerous for high-speed operation) or that there is a problem with the circuit (meaning that oil pressure might be within spec and high-speed operation would be ok, but who wants to take that risk). Limp in and check out the oil delivery system.
 2. On 2-stroke motors, this indicates that an oil delivery problem has occurred with the VRO2 system. Stop the engine and check oil level in the tank, check for kinked oil delivery hoses or problems with the oil line (leaks or blockages). Also, check the VRO2 pump for proper operation. If necessary, determine how much fuel is left in your tank and how much oil is on hand, it may be possible to mix the oil directly into the fuel tank in order to limp the engine home without damaging the powerhead. Refer to the information on 2-Stroke Engine Oil and Pre-Mixing under General Information and Maintenance in order to help determine proper ratios.

✲✲ WARNING

Operating an engine with insufficient oiling (too little oil for 2-strokes or insufficient oil pressure on 4-strokes) can result in SEVERE powerhead damage requiring a total overhaul or replacement.

• If the WATER TEMP LED illuminates, reduce the engine speed to idle and immediately check the cooling water indicator stream. If present, carefully check to see if the water from the stream is hotter than usual (you've of course checked it before during normal operation, ok if you haven't take a guess, if it's steaming, it's too hot). Stop the motor and allow it to cool, then limp home at idle. If the stream is absent, check for clogging at the water intake, and if not found, start signaling for a tow. Remember that even if you let the motor cool thoroughly for an hour or so, overheating will occur again quickly (possibly warping or damaging the powerhead beyond repair) in no time at all without cooling water. When back at port, inspect the water pump impeller.

• If the CHECK ENGINE LED illuminates on EFI motors, the ECU has detected an out-of-range sensor signal in the engine control system and stored a diagnostic trouble code. For most faults, the engine will continue to run, but with diminished performance until the circuit is repaired and the proper signal restored. For more details on troubleshooting please refer to the information in the Electronic Fuel Injection (EFI) section.

• If the LOW OIL LED illuminates on 2-stroke motors with oiling systems, immediately check the level in the oil tank. This LED is designed to illuminate when the tank reaches 1/4 capacity. But keep in mind that a stuck float may have kept the signal from illuminating until it was jarred loose by action of the boat and the level could be lower. Of course, as a responsible boater you **ALWAYS** check the level in the oil tank visually before setting out, so it is not likely that you've used THAT much oil today right?

Diagnostic Mode (Carbureted Models)

◆ See Figures 232, 233 and 234

The System Check gauge electronics on carbureted models feature a gauge diagnostic mode which is entered automatically by turning the ignition switch to the **ON** or **RUN** position without starting the motor. When this is done, the gauge will go through the usual self-test mode, then enter the diagnostic mode automatically. In this mode the warning horn circuits are disabled so a technician can manually activate a circuits to manually check it's LED.

If a switch is activated in this mode (by simulating sensor fault conditions or by manually connecting the sensor lead to ground), the appropriate LED will illuminate as long as the sensor is activated.

Keep in mind that the diagnostic mode does not operate unless the self-test mode has satisfactorily completed.

To test various LEDs of the System Check gauge, turn the ignition keyswitch to **ON** or **RUN** and wait for it to complete the self-check mode, then proceed as follows:

• To check the NO OIL LED proceed according to instructions for your motor:
 1. For carbureted 4-stroke motors, disconnect the tan/yellow wire from the oil pressure switch, then use a jumper wire to ground the lead (connect it to a good engine ground). The NO OIL led must illuminate and remain so as long as the tan lead is jumpered to ground. Otherwise, trace and repair the tan/yellow lead.
 2. For 2-strokes other than 25/35 hp (500/565cc) motors, remove the 4 screws securing the pickup assembly to the oil tank, then carefully lift the assembly out of the tank and hold it upright. As the float drops the LOW OIL led must illuminate and remain so as long as float is toward the bottom of its travel. If the LED does not illuminate, disengage the 2-pin harness connector at the pickup and connect the 2 wires of the harness together using a jumper. If the LED now illuminates, replace the pickup. If the LED still does not illuminate, trace the wiring and repair the problem with the tan/black and black wires.
 3. For 25/35 hp (500/565cc) 2-stroke motors, remove the pickup assembly from the integral oil tank, and hold it horizontally. Slowly move the

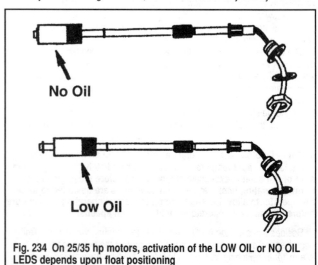

Fig. 234 On 25/35 hp motors, activation of the LOW OIL or NO OIL LEDS depends upon float positioning

IGNITION AND ELECTRICAL SYSTEMS 4-101

float down the pickup tube until it comes within an inch or two (about 25-50mm) from the lower end of its travel (see the accompanying illustration). The LOW OIL led must illuminate and remain so as long as float is toward the bottom (but not actually AT the bottom) of its travel. If the LED does not illuminate, disengage the 3-pin harness connector at the pickup and connect the tan/black and black leads of the harness together using a jumper. If the LED now illuminates, replace the pickup. If the LED still does not illuminate, trace the wiring and repair the problem with the tan/black and black wires.

• To check the WATER TEMP LED, disconnect the tan lead from the temperature switch harness connector, then use a jumper wire to ground the lead (connect it to a good engine ground). The WATER TEMP led must illuminate and remain so as long as the tan lead is jumpered to ground. Otherwise, trace and repair the tan lead.

• To check the LOW OIL LED proceed according to instructions for your motor:

4. For 2-strokes other than 25/35 hp (500/565cc) motors, disengage the 4-pin connector for the VRO2 pump. Connect the tan/yellow and black wires of the harness together using a jumper. If the LED now illuminates, check the VRO2 pump warning circuit, as detailed under the VRO2 Oil System section. If the LED still does not illuminate, trace the wiring and repair the problem with the tan/yellow and black wires.

5. For 25/35 hp (500/565cc) 2-stroke motors, remove the pickup assembly from the integral oil tank, and hold it horizontally. Slowly move the float down the pickup tube until it comes to the VERY end of its travel its travel (see the accompanying illustration). The NO OIL led must illuminate and remain so as long as float at the very bottom of its travel. If the LED does not illuminate, disengage the 3-pin harness connector at the pickup and connect the tan/yellow and black leads of the harness together using a jumper. If the LED now illuminates, replace the pickup. If the LED still does not illuminate, trace the wiring and repair the problem with the tan/yellow and black wires.

Diagnostic Mode (EFI Models)

If the CHECK ENGINE LED illuminates on EFI motors, the ECU has detected an out-of-range sensor signal in the engine control system and stored a diagnostic trouble code. For most faults, the engine will continue to run, but with diminished performance until the circuit is repaired and the proper signal restored. For more details on troubleshooting, including information on how to read and interpret stored diagnostic trouble codes please refer to the information in the Electronic Fuel Injection (EFI) section.

Engine Temperature Sensors (EFI Engines)

The engine temperature sensors used on EFI motors are an integral part of the Electronic Fuel Injection (EFI) system. As such, details on testing, as well as removal and installation instructions can be found in that section.

Engine Temperature Switches (Carbureted Engines)

A temperature switch is used on most remote control carbureted 4-strokes as well as all 18 hp and larger 2-strokes (and possibly some smaller models). The switch is used to activate the warning horn or buzzer to warn of potential overheating conditions when a specific temperature is reached.

In addition, some 25/35 hp (500/565cc) and 25-70 hp (913cc) models are equipped with a second warm-up temperature switch. The function of the warm up switch is to monitor at what point the motor has reached normal operating temperature. This is used by the power pack on these models to control spark advance functions.

TESTING

◆ See Figure 235

■ A DVOM (or ohmmeter or even a self-powered test light), a thermometer (or thermo-sensor adapter for the DVOM), a metal or laboratory grade glass container and a length of mechanic's wire are necessary for this test. Also, you will need a heat source, but for safety (since oil is used), you should not use an open flame, instead make sure you have access to an electric burner (your stove?) or a hot plate. To prevent the possibility of burns (and to ease testing), use probes that can be attached (clipped) to the test points.

The temperature switch used on carbureted motors differs slightly from the sensors found on EFI motors in the way that it operates. Whereas the sensor (on EFI motors) is a thermistor or variable resistor that changes its resistance value based on temperature, and thereby is capable of sending different electrical signals across a range of temperatures, the switch (on carbureted motors) is simply an on/off type. The temperature switch is used only by the warning system and only to turn on the LED and/or warning horn.

All temperature switches on these motors are designed to remain open until a specific temperature (that varies with model) is reached. When the switch is open, the circuit that is fed power anytime the engine is running cannot complete. Once a specific temperature is reached, the switch closes, grounding and completing the circuit, and activating the LED and/or warning horn (or in the case of warm-up switches, signaling the power pack that the engine has reached operating temperature).

Temperature switches are tested in the same basic fashion as the sensors on EFI motors, only the specifications observed during the test differ. The best method to test the switch is to remove it from the powerhead, suspend it in a warm liquid (oil) and to use an ohmmeter and a temperature probe to determine at what temperature the circuit closes.

✱✱ WARNING

When testing the switch avoid the use of an open flame to heat the container as it will be filled with oil. If possible, use an electric stove or a hot plate.

If the switch may have been removed/replaced previously, make sure the proper switch for your motor. Johnson/Evinrude OE switches will have different lead wire colors based on switch calibration:

• 4-stroke motors (carbureted) use a switch with a tan/blue lead
• 18-35 hp (521cc) 2-stroke motors: use either a switch with a tan lead or one with a tan/red lead, depending on the model/calibration
• 25/35 hp (500/565cc) 2-stroke motors: use a warning switch with a tan/blue lead and, if equipped, a these motors use a warm-up switch with a white/black lead
• 25-55 hp (737cc) motors: use a warning switch with either a tan lead or one with a tan/blue lead, depending on the model/calibration
• 23-70 hp (913cc) 2-stroke motors use a warning switch with either a tan lead or one with a tan/blue lead, depending on the model/calibration. Also these motors may be equipped with a warm-up switch that uses a white/black lead as well

1. Remove the temperature switch from the powerhead, as detailed in this section.
2. Connect a DVOM set to read resistance or check continuity with one lead on the switch terminal and the other to the metallic switch body. There should be no continuity with a switch that is cold or at ambient temperatures.

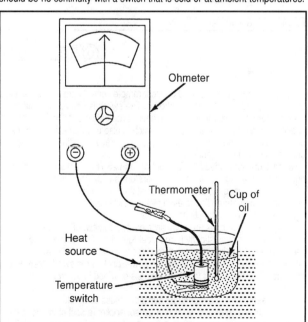

Fig. 235 Temperature switches are tested by suspending them in an oil bath, heating the oil and watching when the switch closes

4-102 IGNITION AND ELECTRICAL SYSTEMS

■ If there is continuity at ambient temperatures, the switch is shorted or stuck closed and must be replaced, in these cases, the horn/buzzer would normally not shut off or otherwise there is a problem with the circuit or your test method.

3. Using a small wooden dowel across the container, suspend the switch in a container of cool automotive oil. To ensure the switch remains at the same temperature as the liquid (and not the temperature of the container) it must only touch the oil (and not touch the bottom or sides of the container).

4. Suspend the thermometer or the thermosensor in the same fashion as the switch. Again, to ensure accurate reading it cannot touch the bottoms or sides of the container. When using a thermometer give it a few minutes to stabilize at the oil's temperature.

5. Using the heat source, slowly raise the temperature of the liquid.

6. Watch the DVOM (or listen to the continuity checker) for the point at which the circuit closes, then check the thermometer/thermo-sensor reading and compare with the specification for your switch (as listed below by the switch wiring lead colors). The switch must close at the specified temperature for its motor to ensure proper operation of the warning system (or warm-up system if testing warm-up switches):
- Tan lead switch: 197-209°F (92-98°C)
- Tan/blue lead switch: 234-246°F (112-120°C)
- Tan/red lead switch: 174-186°F (79-85°C)
- White/black lead (warm-up) switch through 1993: 93-99°F (34-38°C)
- White/black lead (warm-up) switch 1994 or later: 102-108°F (39-43°C)

7. Remove the heat source, and allow the switch to cool, watch the DVOM (or listen to the continuity checker) for the point at which the circuit opens, then check the thermometer/thermo-sensor reading and compare with the specification for your motor. The switch must open at approximately the specified temperature for its motor to ensure proper operation of the warning system (or warm-up system if testing warm-up switches):
- Tan lead switch: 155-185°F (68-86°C)
- Tan/blue lead switch: 192-222°F (88-106°C)
- Tan/red lead switch: 140-170°F (62-74°C)
- White/black lead (warm-up) switch (all years): 87-93°F (30-34°C)

8. A switch that does not close until well after the specified temperature MUST be replaced to prevent the potential of powerhead damage should an overheating condition occur (or to ensure proper operation of the warm-up circuit when testing warm-up switches). A switch that closes a little early can still be used, unless it is so early that the horn or buzzer never shuts off (that would get old real fast, wouldn't it?). Similarly, a switch that remains closed too long after the motor has cooled can cause false overheating signals or, worse, on warm-up switches might prevent the motor from operating properly.

9. After testing or repairs, install the switch and reconnect the wiring.

REMOVAL & INSTALLATION

◆ See Figure 236

A temperature switch is used on most remote control carbureted 4-strokes as well as all 18 hp and larger 2-strokes. The switch is used to activate the warning horn or buzzer to warn of potential overheating conditions when a specific temperature is reached. The switch is also used as a signal to the power pack to activate the S.L.O.W. system operation (limiting engine speed until the problem is corrected).

In addition, some 25/35 hp (500/565cc) and 25-70 hp (913cc) models are equipped with a second warm-up temperature switch. The function of the warm-up switch is to monitor at what point the motor has reached normal operating temperature. This is used by the power pack on these models to control spark advance functions.

Temperature switches are normally mounted somewhere on the powerhead, but varies slightly from model-to-model. The best method is to trace the lead (usually tan or tan with a tracer, except warm-up switches which are always white/black) from the power pack to the switch itself. For more details, please refer to the schematics provided in the Wiring Diagrams section.

1. Disengage the temperature sensor harness connector.
2. Using an open-end wrench or sensor socket (a socket that is made or modified to prevent damage to the sensor terminals or wiring), carefully unthread the sensor from the powerhead.
3. Clean the sensor and powerhead threads.

To install:

4. Carefully thread the sensor into the opening by hand making sure not to cross-thread it.
5. Tighten the sensor securely, but take care not to overtighten and damage the sensor or the mounting threads.
6. Connect the sensor wiring harness, making sure all wires are routed to prevent interference with moving components.
7. Connect the motor to a source of cooling water, then start and run the engine. If the sensor was threaded into a water jacket, make sure there is no leakage.

Oil Pressure Switch (4-Stroke Models Only)

◆ See Figure 237

Johnson/Evinrude 4-stroke motors utilize an oil pressure warning circuit to protect the powerhead in case of a drop in oil pressure (due to excessively low oil level or pump failure). The pressure warning circuit consists of a normally closed oil pressure switch, a warning light (on remote models) and a warning horn or buzzer. When the ignition switch is turned on, battery voltage is applied to the horn or buzzer and the circuit is completed through the closed oil pressure switch. The horn/buzzer will normally sound to verify that the system is operational. When the engine is started, creating sufficient oil pressure, the switch is opened, shutting off the horn/buzzer. But, if at anytime during engine operation, the oil pressure drops below a predetermined level, the switch will close again, activating the warning buzzer and light by providing a ground circuit.

Fig. 236 The temperature sensor is mounted on the powerhead, to locate it, trace the tan wire from the power pack - 9.9 hp 4-stroke shown

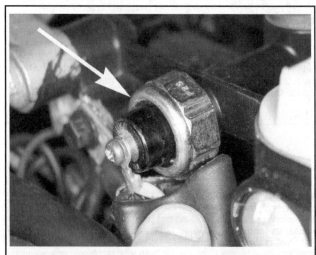

Fig. 237 Although oil pressure switch location varies slightly by model, it is always threaded into an oil passage on the powerhead

IGNITION AND ELECTRICAL SYSTEMS

Switch mounting varies slightly by model, locate your switch as follows:
- For 5/6 hp (128cc) models, the switch is located under a rubber boot at the upper port side of the powerhead, immediately in front of the air intake for the carburetor.
- For 8/9.9 hp (211cc) models, the switch is located under a rubber boot, at the upper port side of the powerhead, immediately in front of the oil dipstick and behind the carburetor.
- For 9.9/15 hp (305cc) models, the switch is located under a rubber boot at the upper port side of the powerhead, immediately in front of the oil fill and right above the oil filter.
- For 40/50 hp (815cc) EFI models, the switch is located at the upper port side of the powerhead, right below the flywheel.
- For 70 hp (1298cc) EFI models, the switch is located at the starboard side of the powerhead (behind the electric component holder and ECU).

■ **For all models, verify switch location using the wire harness colors. For details, refer to the schematics provided in the Wiring Diagrams section.**

TESTING

◆ See Figure 237

Testing requires a DVOM and a hand held vacuum/pressure pump.
1. Remove the oil pressure switch from the powerhead, for details refer to the procedure in this section.
2. Connect the pressure fitting of the pump to a length of hose at the other end of the hose to the pressure fitting on the switch. Use a clamp to secure the hose to the fitting.
3. Connect the red probe of a DVOM (set to read resistance) to the switch wire terminal and the black probe to the switch body. The switch is normally closed, meaning that with no pressure applied and proper test connections made, the meter must show continuity (little or no resistance).
4. Apply pressure to the switch using the gauge while watching the meter. The switch must open (showing very high/infinite resistance) at the proper pressure.
- For 5/6 hp (128cc) motors, the switch must open at approximately 15 psi (140 kPa).
- For 8/9.9 hp (211cc) motors, the switch must open at approximately 10 psi (70 kPa).
- For 9.9/15 hp (305cc) motors, the switch must open at approximately 20 psi (140 kPa).
- For 40/50 hp (815cc) and 70 hp (1298cc) EFI motors, the switch must open at approximately 10-19 psi (70-130 kPa)

5. Although the switch can be used if it does not open until a slightly higher pressure, it MUST be replaced if it opens sooner to protect the powerhead in case of diminished oiling capacity. A switch that opens too soon could allow the engine to run with insufficient oil pressure.
6. When testing or repairs are completed, install the oil pressure switch to the powerhead.

REMOVAL & INSTALLATION

◆ See Figure 237

1. Disengage the wiring harness from the oil pressure switch.
2. Using an open-end wrench or a suitable socket (a sensor socket or socket modified to prevent damage to the switch terminals or wiring, as equipped), carefully unthread the switch from the powerhead.
3. Clean all sealant residue from the switch threads, making sure not dirt, debris or residue enters the oil passage in the powerhead.

To install:
4. Wrap a layer of Teflon® tape around the switch threads.
5. Carefully thread the switch into the opening by hand, then tighten the switch until snug (but do not over-tighten and damage the switch or powerhead threads). If a torque wrench and sensor socket is available, tighten the switch to 120-168 inch lbs. (13.5-19 Nm) for carbureted motors or to 9.5 ft. lbs. (13 Nm) for EFI motors.
6. Reconnect the switch wiring. Route the harness to prevent interference with moving components.
7. Provide a source of cooling water, then start the engine check for oil leakage and proper switch operation.

Oil Level Switch (Oil Injected 2-Strokes)

Johnson/Evinrude 2-stroke motors that are equipped with oil reservoir tanks and oil injection systems use one or more sensors to ensure that the motor is always receiving a supply of 2-stroke oil. Two sensors are normally used for this circuit, one in the reservoir tank that is used to signal when the reservoir reaches approximately 1/4 or less capacity and one in the VRO2 pump that signals if the oil feed line should loose pressure. For models equipped with the VRO2 system, please refer to the information on Oil Injection Systems for details on the VRO2 warning circuit.

The 25/35 hp (500/565cc) motors utilize a powerhead mounted oil reservoir tank. The oil level switch on these models is a multi-position switch incorporated into the tank pickup assembly. One switch position will actuate the LOW OIL warning circuit and LED, while a second position (lower on the pickup) will actuate the NO OIL warning circuit and LED.

To test the oil level switch warning circuits, refer to the Diagnostic Mode (Carbureted Models) section found under Testing Warning Circuits.

Warning Horn or Buzzer

TESTING

Testing the warning horn or buzzer is a relatively simple matter of using two sets of jumper leads to temporarily connect a fully-charged 12-volt battery across the horn/buzzer terminals. For safety, always connect jumpers to the battery first and THEN to the horn or buzzer. This prevents the possibility of sparks/arcing near the battery (which could ignite explosive gases released by batteries).

Location of the warning horn or buzzer may vary with the boat and motor rigging, but it is normally found within the control box or behind the dash on remote control models. For tiller models, the horn or buzzer is normally found within the tiller control housing or under the engine cover.

Since access to the wire terminals is required for testing, you may need to partially disassembly the remote or tiller control housings.

To test the horn or buzzer, use the jumper wires to apply 12-volts across the horn/buzzer terminals. Replace the unit if it fails to emit a suitable warning tone.

4-104 IGNITION AND ELECTRICAL SYSTEMS

WIRING DIAGRAM INDEX

1-Cylinder Motors

1990-95 Breaker Point Models 4-105
1996-01 Rope Start/CDI 2.0-3.5 Hp (78cc) 1-Cyl 4-105
1999-01 Rope Start 5 Hp (109cc) 1-Cyl/2-Stroke 4-106
1997-01 Rope Start 5/6 hp (128cc) 1-Cyl/4-Stroke 4-106

2-Cylinder Motors

1990 4 Hp (87cc) Excel 4/Ultra 4 2-Cyl 4-107
1990-01 Rope Start 3/4 Hp (87cc) 2-Cyl 4-107
1990-01 Rope Start 6/8 Hp (164cc) Sail 2-Cyl 4-108
1990-93 Rope Start 4Deluxe (87cc), 5/6/8 (164cc) & 9.9-15 Hp (216/255cc) 2-Cyl/2-Stroke w/UFI & AC power 4-109
1991-01 Rope Start 4Deluxe (87cc), 5/6/6.5/8 Hp (164cc) and 9.9-15 Hp (255cc) 2-Cyl/2-Stroke w/CDI 4-109
1996-01 Rope Start 8/9.9 Hp (211cc) 2-Cyl/4-Stroke 4-110
1997-98 Tiller Electric 9.9 Hp (211cc) 2-Cyl/4-Stroke 4-110
1997-98 Remote Electric 9.9 Hp (211cc) 2-Cyl/4-Stroke 4-111
1990-92 Tiller Electric 9.9-15 Hp (211cc) and Electric Sail 2-Cyl/2-Stroke w/UFI 4-111
1990-92 Tiller Electric 9.9-15 Hp (211cc) 2-Cyl/2-Stroke w/UFI 4-112
1993-01 Tiller Electric 9.9-15 Hp (255cc) and Electric Sail 2-Cyl/2-Stroke 4-112
1993-01 Remote Electric 9.9-15 Hp (255cc) 2-Cyl/2-Stroke 4-113
1995-01 Rope Start 9.9/15 Hp (305cc) 2-Cyl/4-Stroke 4-114
1995-01 Tiller Electric 9.9 Hp (305cc) High Thrust 2-Cyl/4-Stroke 4-114
1995 Remote Electric 9.9 Hp (305cc) High Thrust 2-Cyl/4-Stroke 4-115
1996-01 Remote Electric 9.9 Hp (305cc) High Thrust 2-Cyl/4-Stroke .4-115
1995-01 Remote Electric 15 Hp (305cc) 2-Cyl/4-Stroke 4-116
1990-92 Rope Start 20-30 Hp (521cc) 2-Cyl w/UFI & AC power 4-117
1990-92 Tiller Electric 20-30 Hp (521cc) 2-Cyl w/UFI 4-118
1990-92 Remote Electric 20-30 Hp (521cc) 2-Cyl w/UFI 4-118
1993-01 Rope Start 18 Jet and 20-35 Hp (521cc) 2-Cyl w/CDI & AC .4-119
1993-01 Tiller Electric 18 Jet and 20-35 Hp (521cc) 2-Cyl w/CDI ...4-119
1993-01 Remote Electric 18 Jet and 20-35 Hp (521cc) 2-Cyl 4-120
1990-93 Rope Start 25D-55 Hp (737cc) 2-Cyl w/UFI & AC 4-121
1990-93 Tiller Electric 25D-55 Hp (737cc) 2-Cyl w/UFI 4-121
1990-93 Remote Electric 25D-55 Hp (737cc) 2-Cyl 4-122
1992-01 Rope Start 25D-55 Hp (737cc) 2-Cyl w/CDI & AC 4-123
1992-95 Remote Electric 25DE and 48E (737cc) 2-Cyl w/CDI & VRO, Manual Trim/Tilt 4-124
1992-95 Remote Electric 25DTL (737cc) 2-Cyl w/CDI, VRO & PTT ..4-124
1996-98 Remote Electric 48 Hp (737cc) E Model and 55 Hp (737cc) Com 2-Cyl 4-125
1992-98 Tiller Electric 40-55 Hp (737cc) 2-Cyl 4-125
1992-95 Electric Start 40/50 Hp (737cc) TTL 2-Cyl w/CDI, VRO & PTT 4-126
1996-01 Electric Start 40 Hp (737cc) TTL 2-Cyl w/PTT 4-127
1992-01 Remote Electric 40-55 Hp (737cc) 2-Cyl 4-128

3-Cylinder Motors

1996-01 Rope Start 25/35 Hp (500/565cc) 3-Cyl 4-130
1996-01 Tiller Electric 25/35 Hp (500/565cc) 3-Cyl 4-130
1996-01 Remote Electric 25/35 Hp (500/565cc) 3-Cyl 4-131
1999-01 Remote Electric 40/50 Hp (815cc) 3-Cyl/4-Stroke w/PTT ..4-132
1990-01 Rope 65 Hp (913cc) Commercial 3-Cyl w/AC 4-132
1990-92 Electric 60-70 Hp (913cc) TTL 3-Cyl w/PTT 4-133
1993-98 Electric 50-60 Hp (913cc) TTL 3-Cyl/2-Stroke w/PTT 4-133
1990-92 Remote Electric 60-70 Hp (913cc) EL 3-Cyl w/pre-mix & manual trim/tilt 4-134
1993-98 Remote Electric 50-70 Hp (913cc) EL and 65 Hp (913cc) WMLE Com 3-Cyl/2-Stroke 4-135
1993-01 Remote Electric 65 Hp (913cc) WML, WMLW & WE Com 3-Cyl 4-136
1990-92 Remote Electric 60-70 Hp (913cc) TL & TX 3-Cyl w/PTT 4-137
1993-95 Remote Electric 50-70 Hp (913cc) D, DT, PL, TL, TX & TY 3-Cyl w/PTT 4-137
1996-01 Remote Electric 40-70 Hp (913cc) D, DT, PL, TL, TX & TY Model 3-Cyl w/PTT 4-138

4-Cylinder Motors

1999-01 Remote Electric 70 Hp (1298cc) 4-Cyl w/Electric Trim/Tilt ...4-139

Johnson/Evinrude Remotes

1996-01 Remote Control/Keyswitch for MWS Wiring Harness 4-140
1996-01 MWS Instrument Wiring Harness for Remotes 4-140
1996-01 Remote Control/Keyswitch for MWS Wiring Harness 4-140

IGNITION AND ELECTRICAL SYSTEMS

1-Cylinder Motors

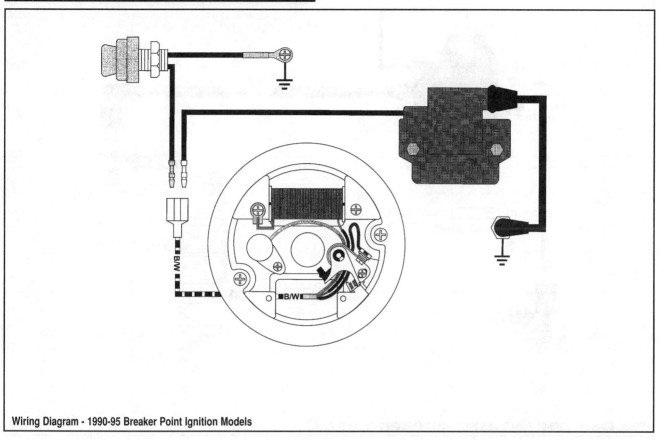

Wiring Diagram - 1990-95 Breaker Point Ignition Models

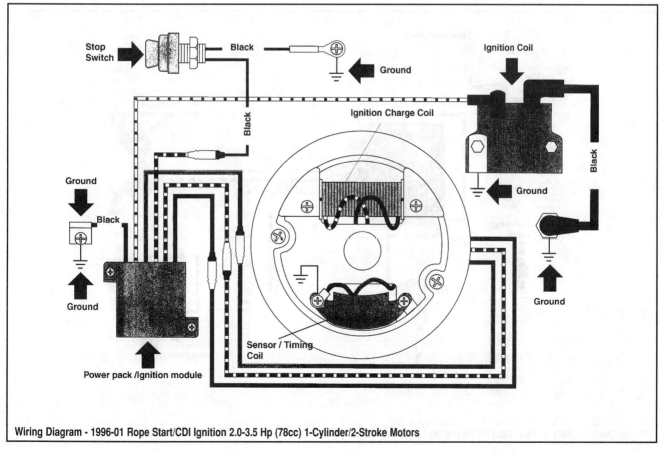

Wiring Diagram - 1996-01 Rope Start/CDI Ignition 2.0-3.5 Hp (78cc) 1-Cylinder/2-Stroke Motors

4-106 IGNITION AND ELECTRICAL SYSTEMS

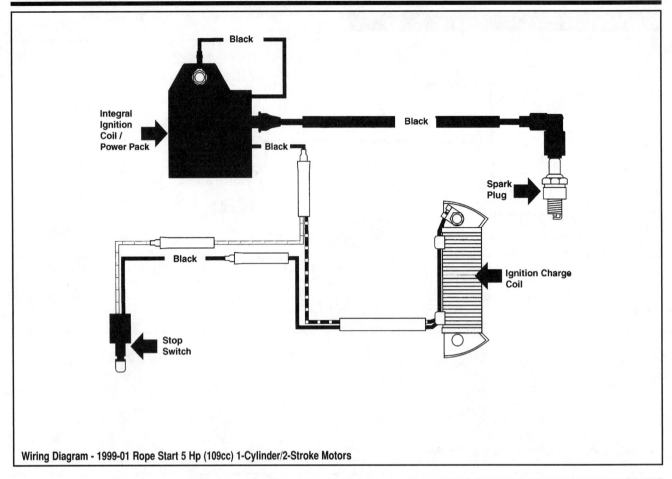

Wiring Diagram - 1999-01 Rope Start 5 Hp (109cc) 1-Cylinder/2-Stroke Motors

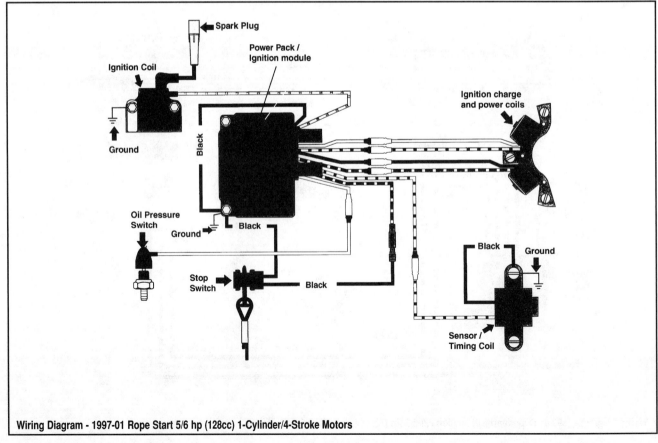

Wiring Diagram - 1997-01 Rope Start 5/6 hp (128cc) 1-Cylinder/4-Stroke Motors

IGNITION AND ELECTRICAL SYSTEMS

2-Cylinder Motors

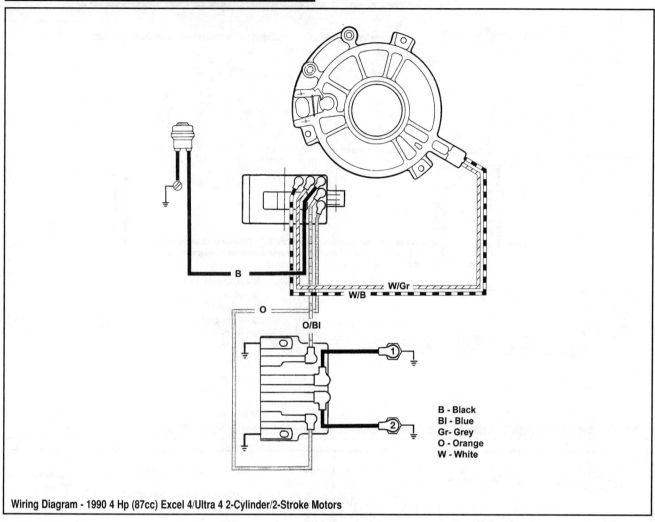

Wiring Diagram - 1990 4 Hp (87cc) Excel 4/Ultra 4 2-Cylinder/2-Stroke Motors

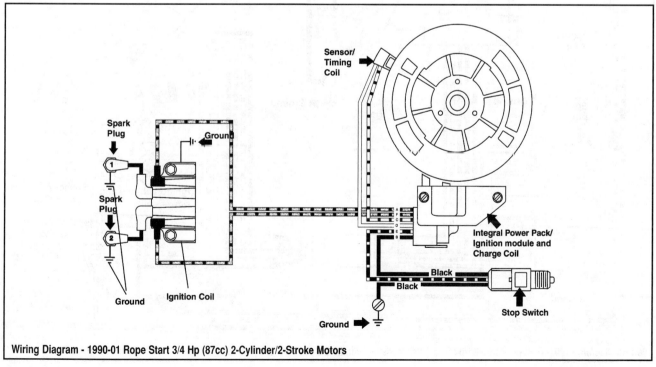

Wiring Diagram - 1990-01 Rope Start 3/4 Hp (87cc) 2-Cylinder/2-Stroke Motors

4-108 IGNITION AND ELECTRICAL SYSTEMS

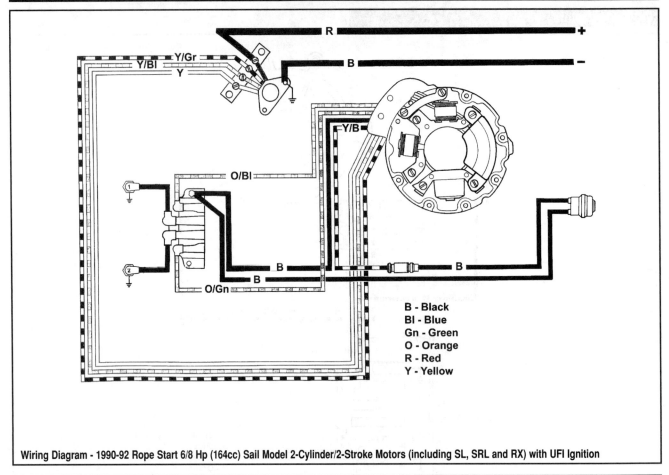

Wiring Diagram - 1990-92 Rope Start 6/8 Hp (164cc) Sail Model 2-Cylinder/2-Stroke Motors (including SL, SRL and RX) with UFI Ignition

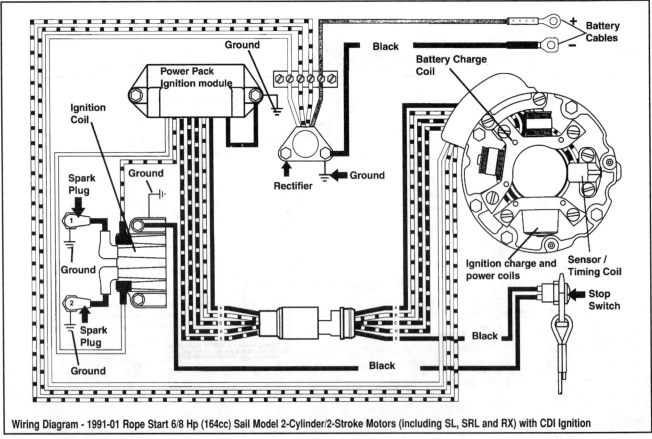

Wiring Diagram - 1991-01 Rope Start 6/8 Hp (164cc) Sail Model 2-Cylinder/2-Stroke Motors (including SL, SRL and RX) with CDI Ignition

IGNITION AND ELECTRICAL SYSTEMS

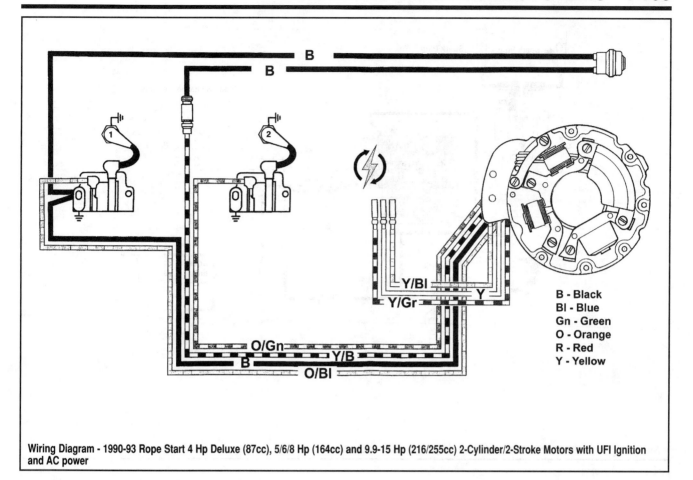

Wiring Diagram - 1990-93 Rope Start 4 Hp Deluxe (87cc), 5/6/8 Hp (164cc) and 9.9-15 Hp (216/255cc) 2-Cylinder/2-Stroke Motors with UFI Ignition and AC power

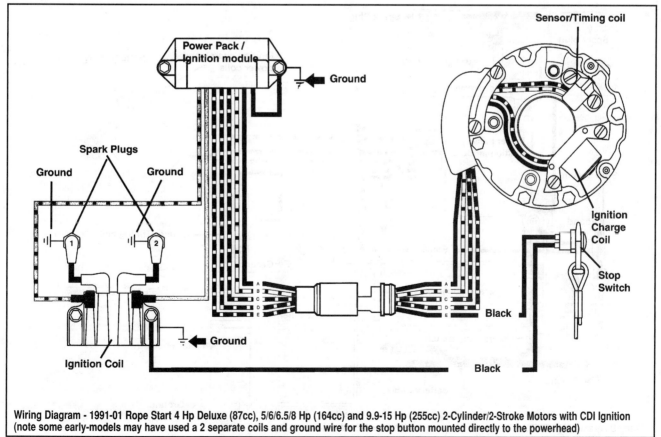

Wiring Diagram - 1991-01 Rope Start 4 Hp Deluxe (87cc), 5/6/6.5/8 Hp (164cc) and 9.9-15 Hp (255cc) 2-Cylinder/2-Stroke Motors with CDI Ignition (note some early-models may have used a 2 separate coils and ground wire for the stop button mounted directly to the powerhead)

4-110 IGNITION AND ELECTRICAL SYSTEMS

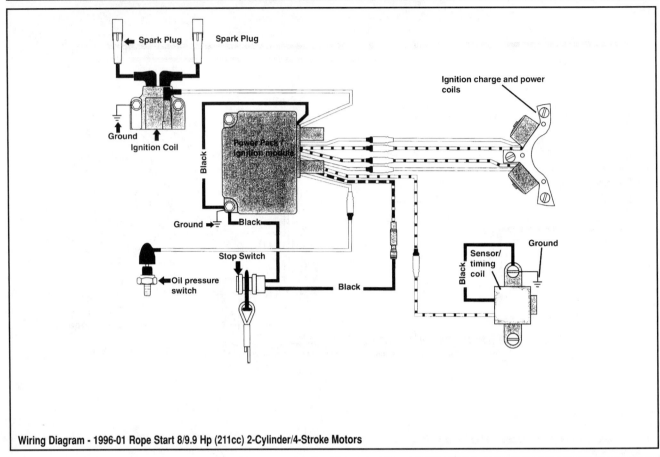

Wiring Diagram - 1996-01 Rope Start 8/9.9 Hp (211cc) 2-Cylinder/4-Stroke Motors

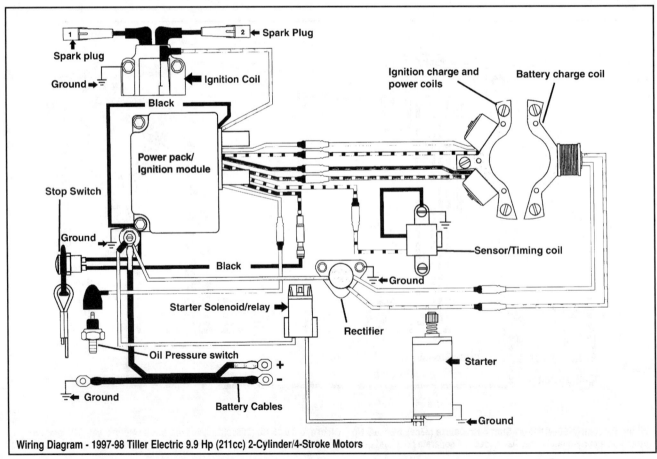

Wiring Diagram - 1997-98 Tiller Electric 9.9 Hp (211cc) 2-Cylinder/4-Stroke Motors

IGNITION AND ELECTRICAL SYSTEMS 4-111

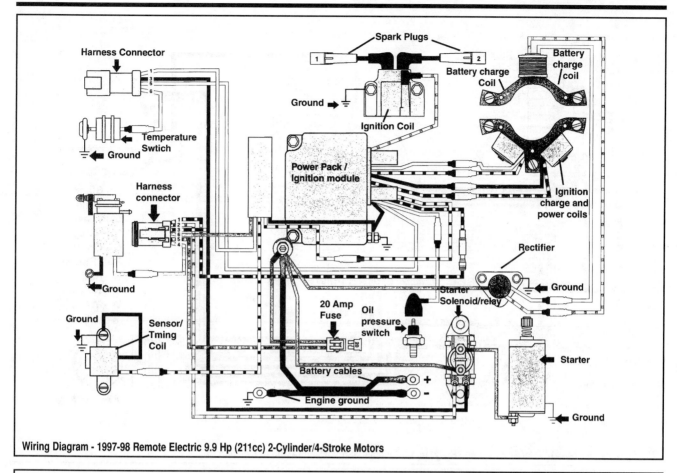

Wiring Diagram - 1997-98 Remote Electric 9.9 Hp (211cc) 2-Cylinder/4-Stroke Motors

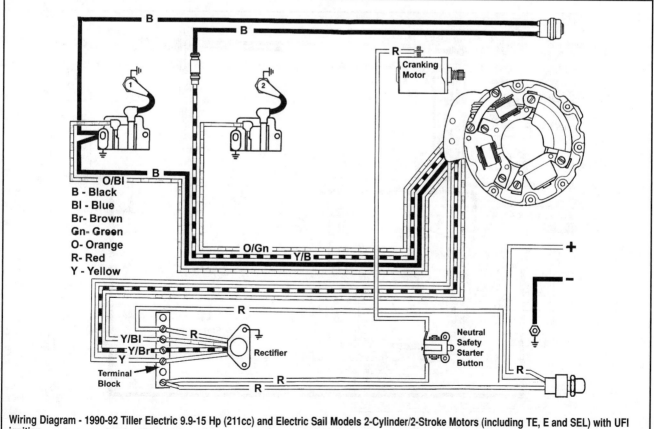

Wiring Diagram - 1990-92 Tiller Electric 9.9-15 Hp (211cc) and Electric Sail Models 2-Cylinder/2-Stroke Motors (including TE, E and SEL) with UFI ignition

4-112 IGNITION AND ELECTRICAL SYSTEMS

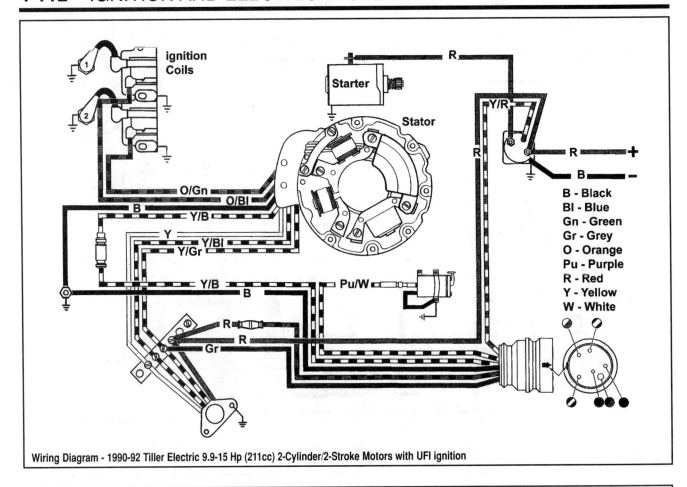

Wiring Diagram - 1990-92 Tiller Electric 9.9-15 Hp (211cc) 2-Cylinder/2-Stroke Motors with UFI ignition

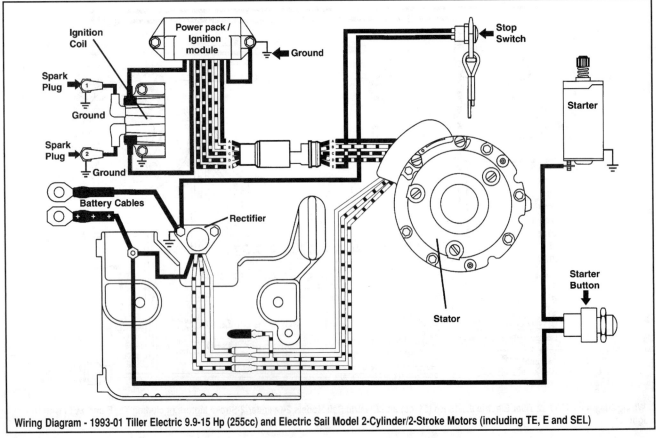

Wiring Diagram - 1993-01 Tiller Electric 9.9-15 Hp (255cc) and Electric Sail Model 2-Cylinder/2-Stroke Motors (including TE, E and SEL)

IGNITION AND ELECTRICAL SYSTEMS 4-113

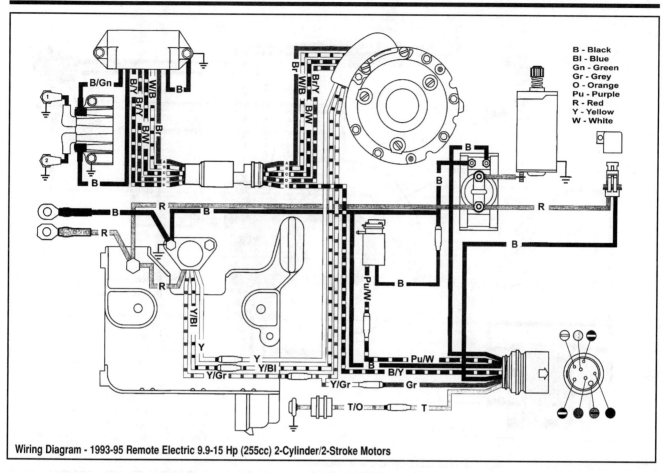

Wiring Diagram - 1993-95 Remote Electric 9.9-15 Hp (255cc) 2-Cylinder/2-Stroke Motors

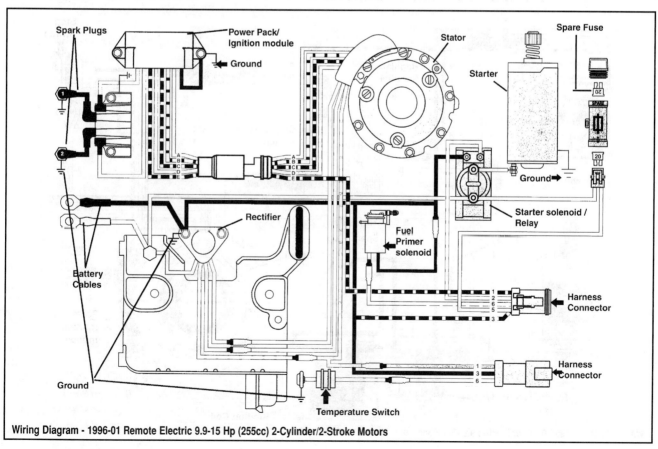

Wiring Diagram - 1996-01 Remote Electric 9.9-15 Hp (255cc) 2-Cylinder/2-Stroke Motors

4-114 IGNITION AND ELECTRICAL SYSTEMS

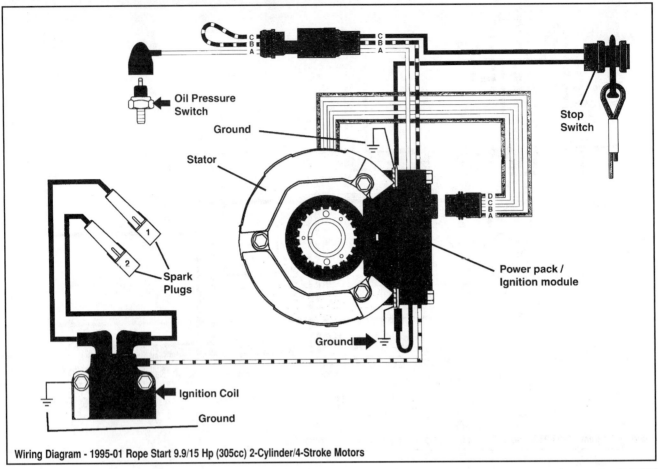

Wiring Diagram - 1995-01 Rope Start 9.9/15 Hp (305cc) 2-Cylinder/4-Stroke Motors

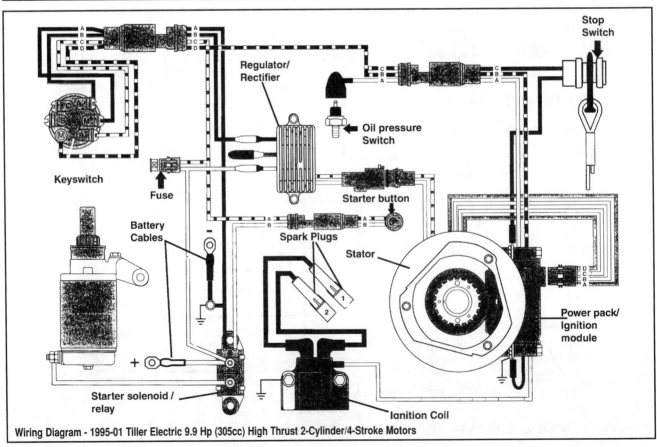

Wiring Diagram - 1995-01 Tiller Electric 9.9 Hp (305cc) High Thrust 2-Cylinder/4-Stroke Motors

IGNITION AND ELECTRICAL SYSTEMS 4-115

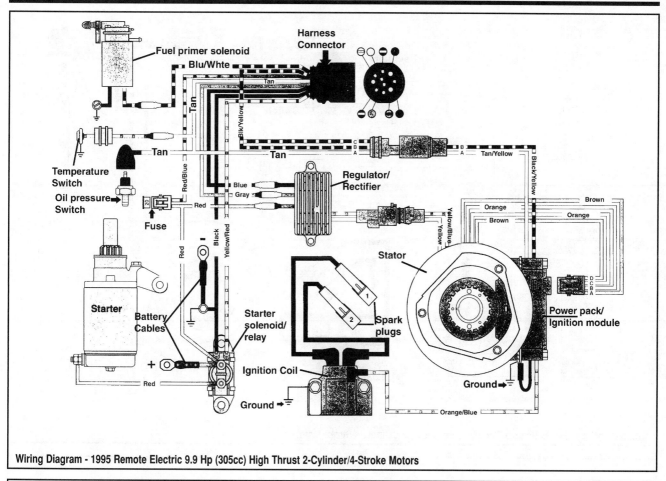

Wiring Diagram - 1995 Remote Electric 9.9 Hp (305cc) High Thrust 2-Cylinder/4-Stroke Motors

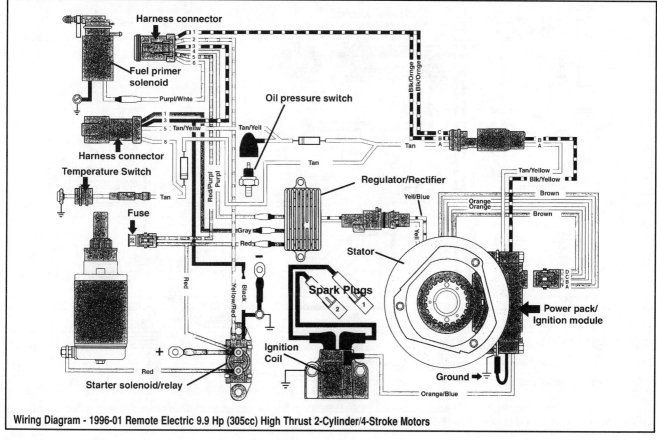

Wiring Diagram - 1996-01 Remote Electric 9.9 Hp (305cc) High Thrust 2-Cylinder/4-Stroke Motors

4-116 IGNITION AND ELECTRICAL SYSTEMS

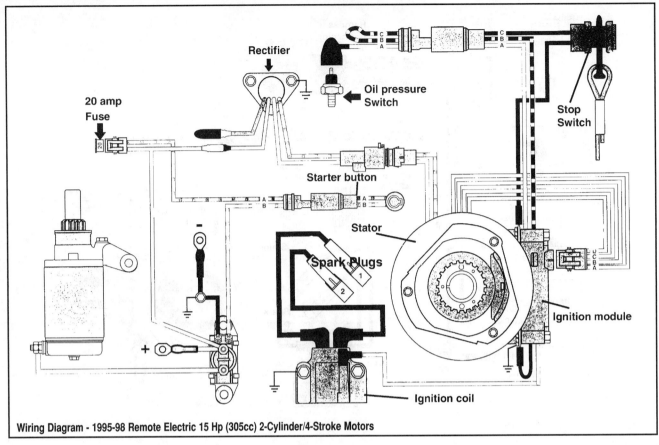

Wiring Diagram - 1995-98 Remote Electric 15 Hp (305cc) 2-Cylinder/4-Stroke Motors

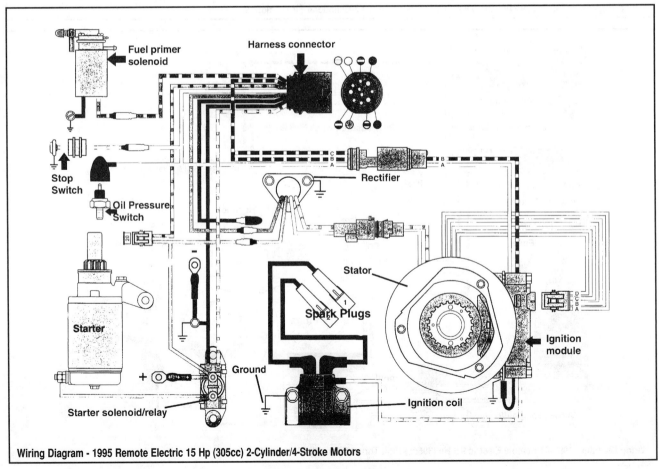

Wiring Diagram - 1995 Remote Electric 15 Hp (305cc) 2-Cylinder/4-Stroke Motors

IGNITION AND ELECTRICAL SYSTEMS 4-117

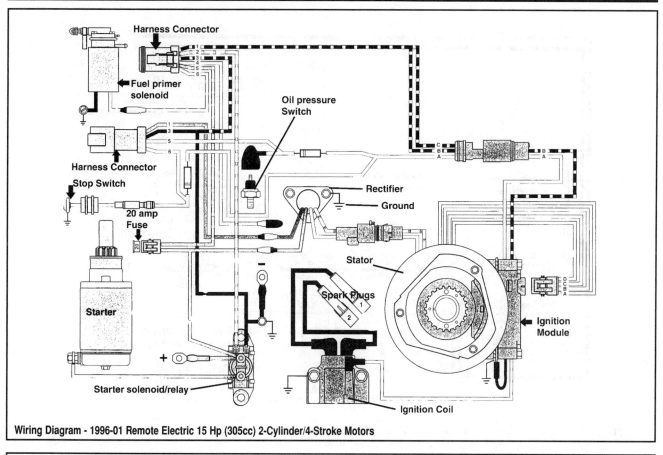

Wiring Diagram - 1996-01 Remote Electric 15 Hp (305cc) 2-Cylinder/4-Stroke Motors

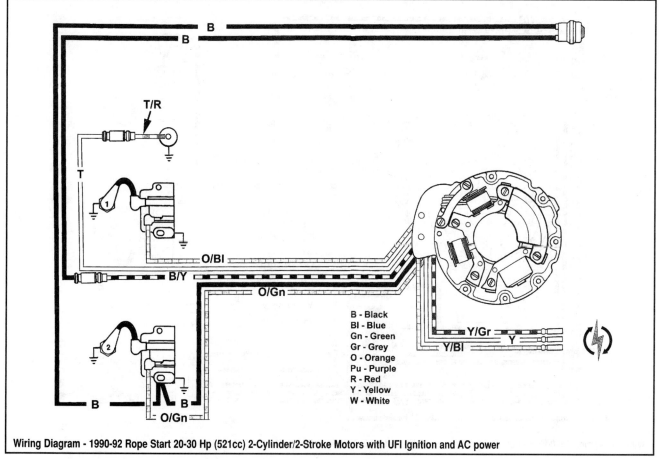

Wiring Diagram - 1990-92 Rope Start 20-30 Hp (521cc) 2-Cylinder/2-Stroke Motors with UFI Ignition and AC power

4-118 IGNITION AND ELECTRICAL SYSTEMS

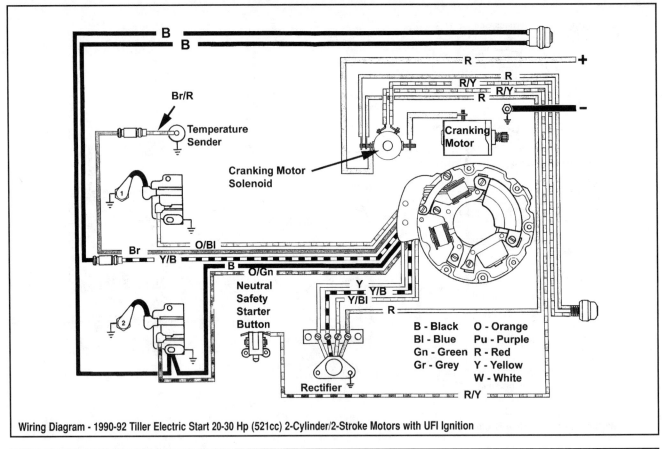

Wiring Diagram - 1990-92 Tiller Electric Start 20-30 Hp (521cc) 2-Cylinder/2-Stroke Motors with UFI Ignition

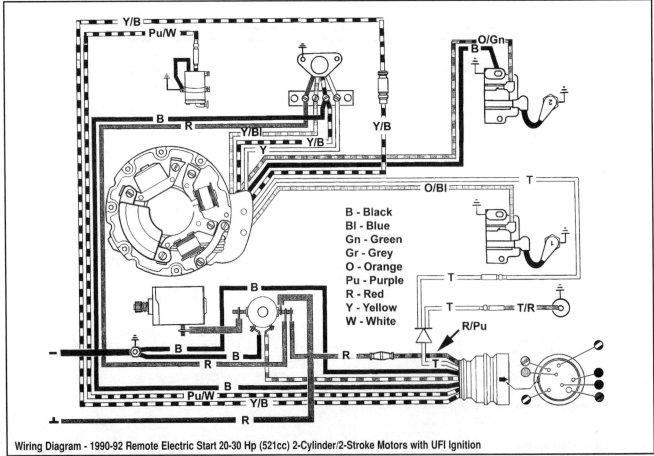

Wiring Diagram - 1990-92 Remote Electric Start 20-30 Hp (521cc) 2-Cylinder/2-Stroke Motors with UFI Ignition

IGNITION AND ELECTRICAL SYSTEMS 4-119

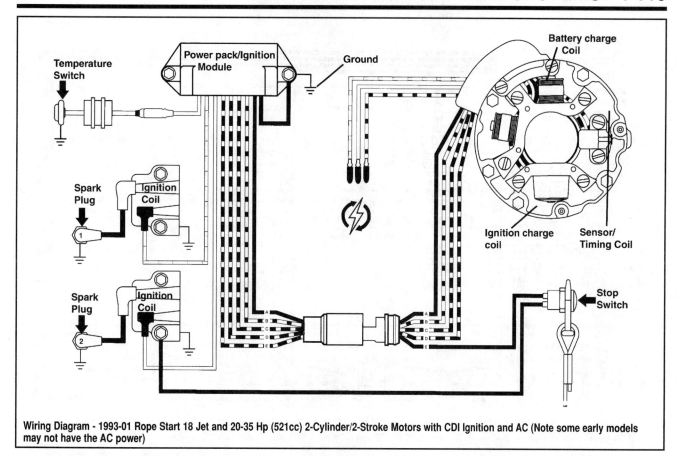

Wiring Diagram - 1993-01 Rope Start 18 Jet and 20-35 Hp (521cc) 2-Cylinder/2-Stroke Motors with CDI Ignition and AC (Note some early models may not have the AC power)

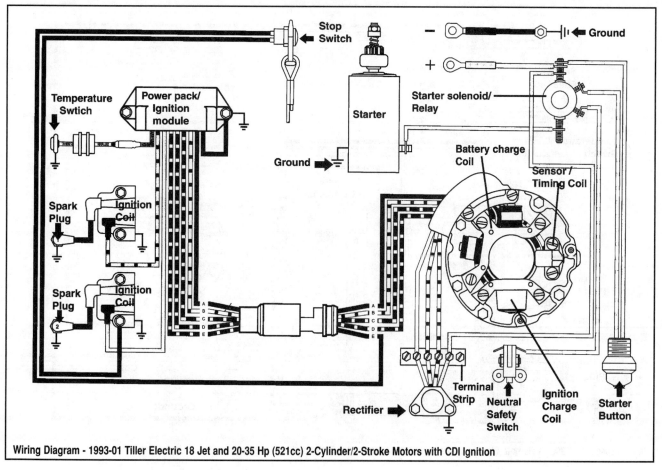

Wiring Diagram - 1993-01 Tiller Electric 18 Jet and 20-35 Hp (521cc) 2-Cylinder/2-Stroke Motors with CDI Ignition

4-120 IGNITION AND ELECTRICAL SYSTEMS

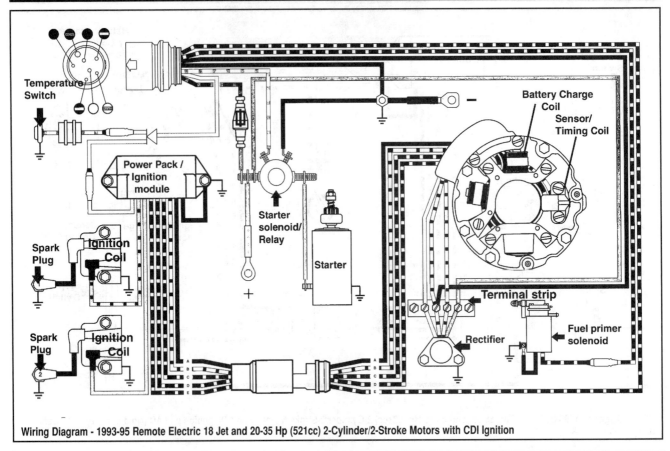

Wiring Diagram - 1993-95 Remote Electric 18 Jet and 20-35 Hp (521cc) 2-Cylinder/2-Stroke Motors with CDI Ignition

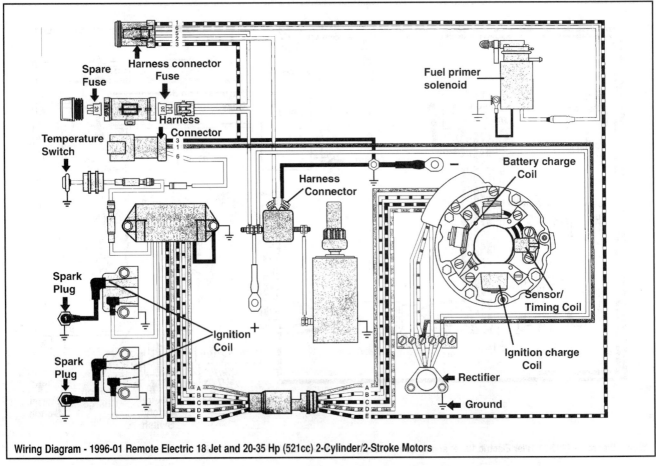

Wiring Diagram - 1996-01 Remote Electric 18 Jet and 20-35 Hp (521cc) 2-Cylinder/2-Stroke Motors

IGNITION AND ELECTRICAL SYSTEMS 4-121

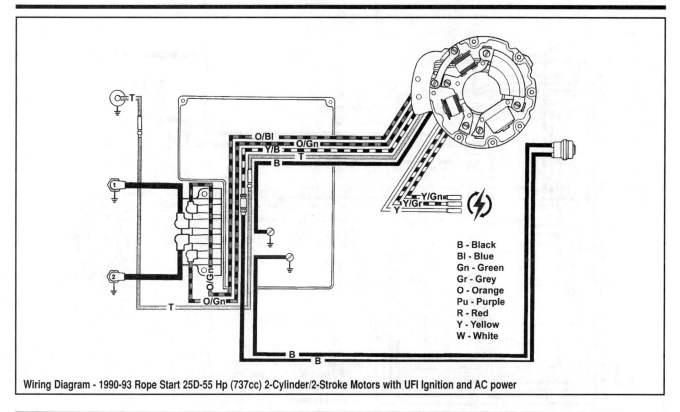

Wiring Diagram - 1990-93 Rope Start 25D-55 Hp (737cc) 2-Cylinder/2-Stroke Motors with UFI Ignition and AC power

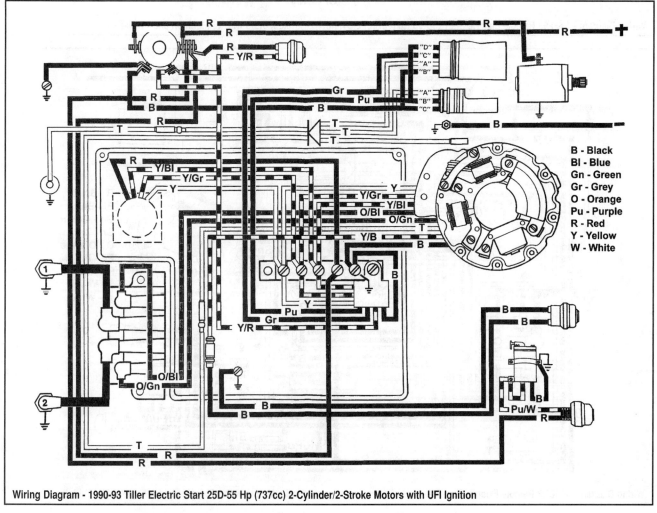

Wiring Diagram - 1990-93 Tiller Electric Start 25D-55 Hp (737cc) 2-Cylinder/2-Stroke Motors with UFI Ignition

4-122 IGNITION AND ELECTRICAL SYSTEMS

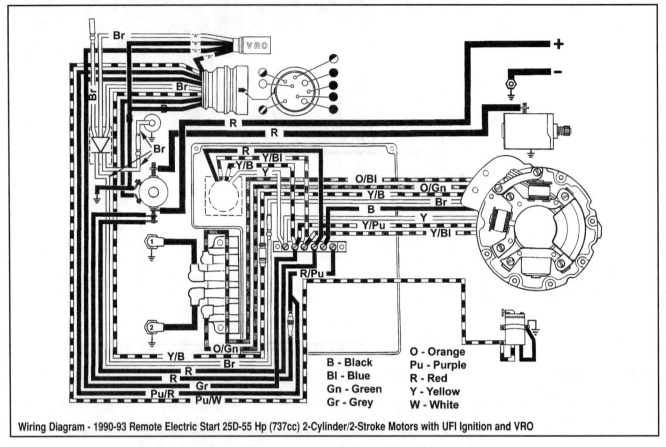

Wiring Diagram - 1990-93 Remote Electric Start 25D-55 Hp (737cc) 2-Cylinder/2-Stroke Motors with UFI Ignition and VRO

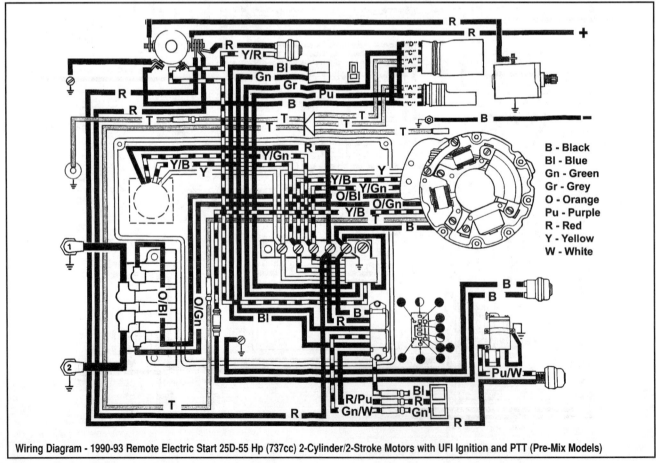

Wiring Diagram - 1990-93 Remote Electric Start 25D-55 Hp (737cc) 2-Cylinder/2-Stroke Motors with UFI Ignition and PTT (Pre-Mix Models)

IGNITION AND ELECTRICAL SYSTEMS 4-123

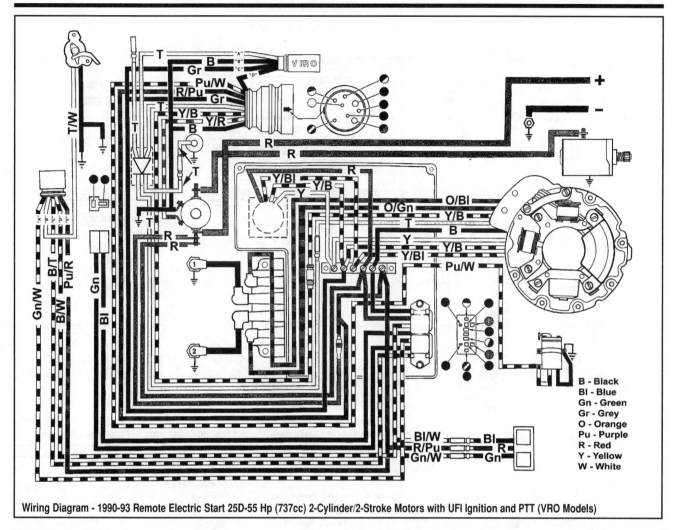

Wiring Diagram - 1990-93 Remote Electric Start 25D-55 Hp (737cc) 2-Cylinder/2-Stroke Motors with UFI Ignition and PTT (VRO Models)

B - Black
Bl - Blue
Gn - Green
Gr - Grey
O - Orange
Pu - Purple
R - Red
Y - Yellow
W - White

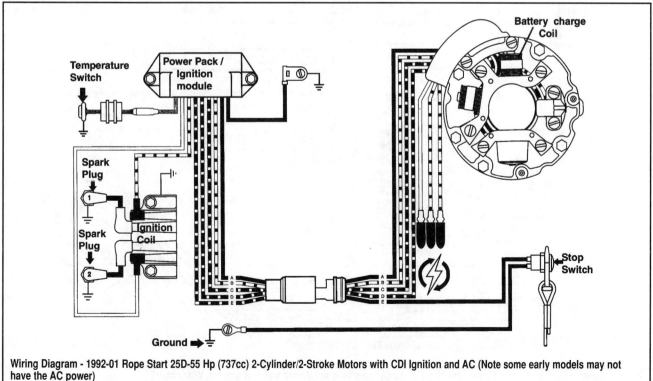

Wiring Diagram - 1992-01 Rope Start 25D-55 Hp (737cc) 2-Cylinder/2-Stroke Motors with CDI Ignition and AC (Note some early models may not have the AC power)

4-124 IGNITION AND ELECTRICAL SYSTEMS

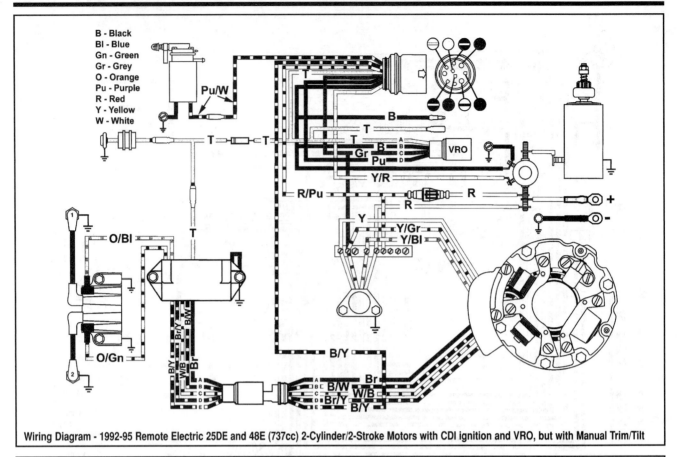

Wiring Diagram - 1992-95 Remote Electric 25DE and 48E (737cc) 2-Cylinder/2-Stroke Motors with CDI ignition and VRO, but with Manual Trim/Tilt

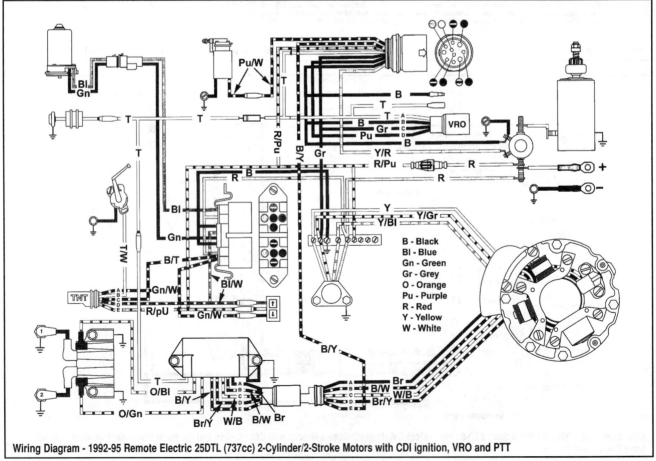

Wiring Diagram - 1992-95 Remote Electric 25DTL (737cc) 2-Cylinder/2-Stroke Motors with CDI ignition, VRO and PTT

IGNITION AND ELECTRICAL SYSTEMS 4-125

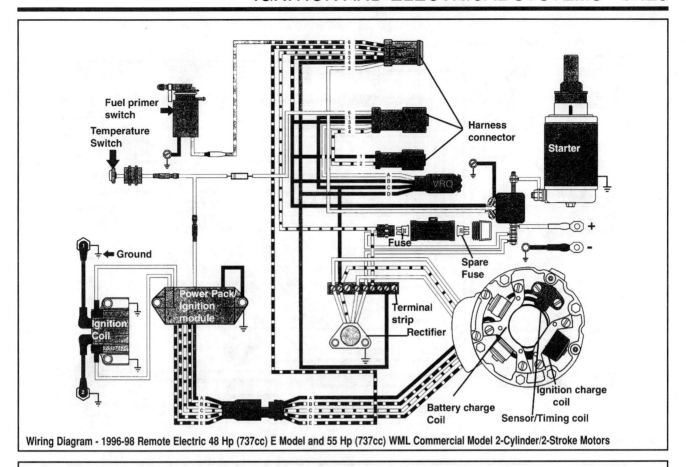

Wiring Diagram - 1996-98 Remote Electric 48 Hp (737cc) E Model and 55 Hp (737cc) WML Commercial Model 2-Cylinder/2-Stroke Motors

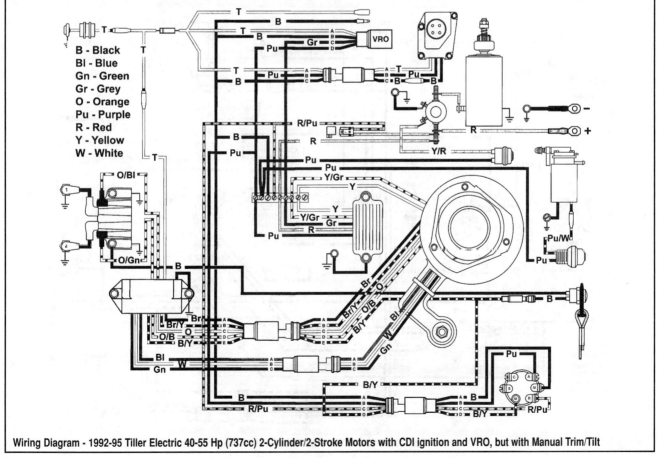

Wiring Diagram - 1992-95 Tiller Electric 40-55 Hp (737cc) 2-Cylinder/2-Stroke Motors with CDI ignition and VRO, but with Manual Trim/Tilt

4-126 IGNITION AND ELECTRICAL SYSTEMS

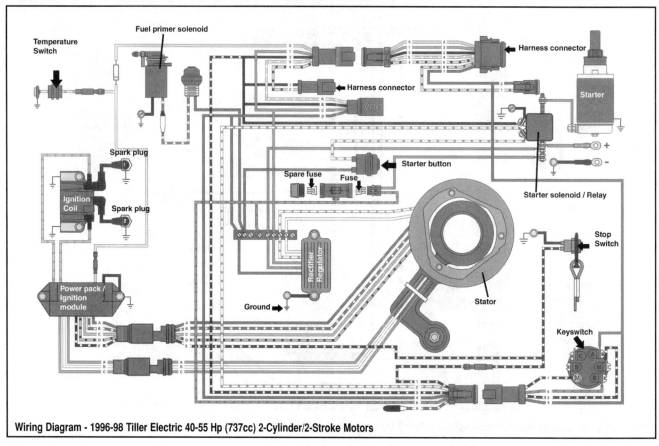

Wiring Diagram - 1996-98 Tiller Electric 40-55 Hp (737cc) 2-Cylinder/2-Stroke Motors

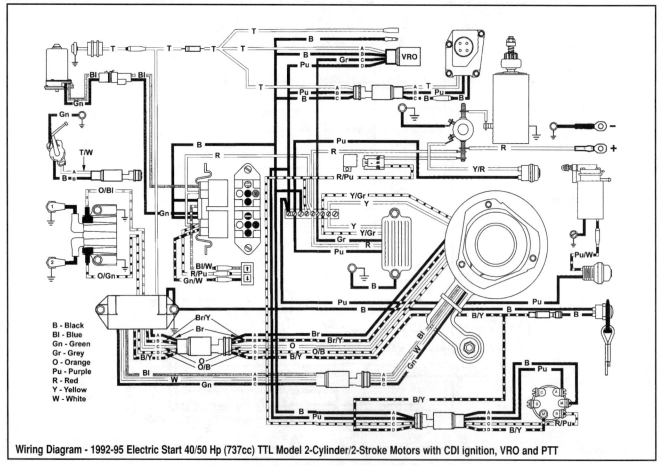

Wiring Diagram - 1992-95 Electric Start 40/50 Hp (737cc) TTL Model 2-Cylinder/2-Stroke Motors with CDI ignition, VRO and PTT

IGNITION AND ELECTRICAL SYSTEMS 4-127

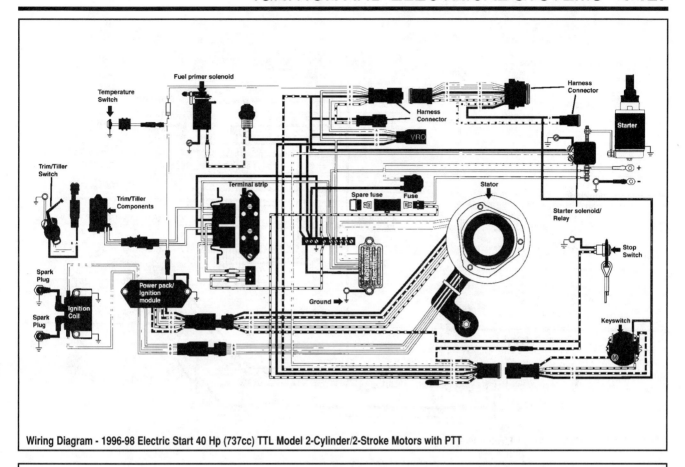

Wiring Diagram - 1996-98 Electric Start 40 Hp (737cc) TTL Model 2-Cylinder/2-Stroke Motors with PTT

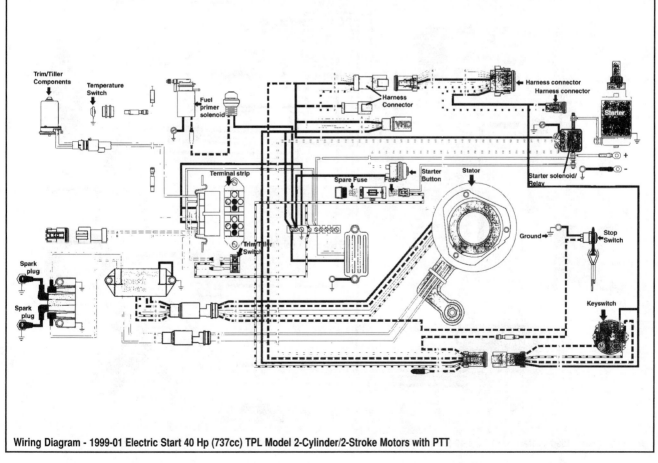

Wiring Diagram - 1999-01 Electric Start 40 Hp (737cc) TPL Model 2-Cylinder/2-Stroke Motors with PTT

4-128 IGNITION AND ELECTRICAL SYSTEMS

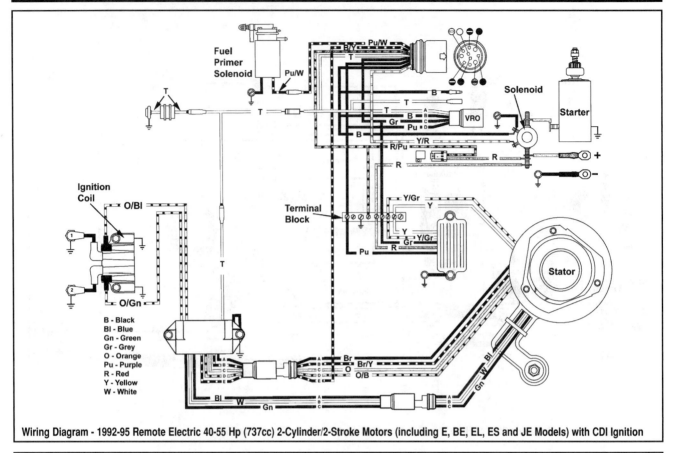

Wiring Diagram - 1992-95 Remote Electric 40-55 Hp (737cc) 2-Cylinder/2-Stroke Motors (including E, BE, EL, ES and JE Models) with CDI Ignition

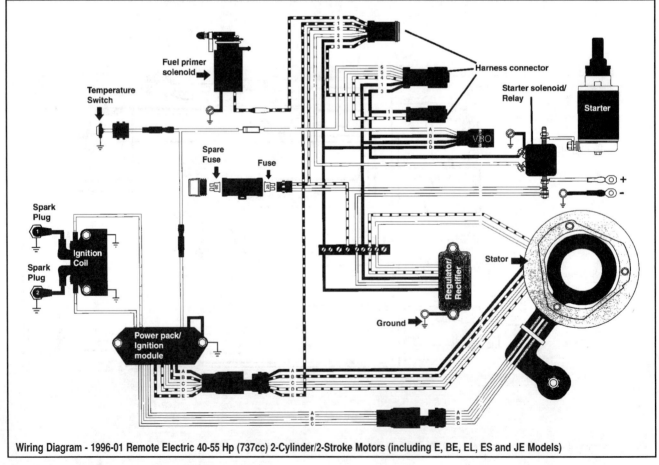

Wiring Diagram - 1996-01 Remote Electric 40-55 Hp (737cc) 2-Cylinder/2-Stroke Motors (including E, BE, EL, ES and JE Models)

IGNITION AND ELECTRICAL SYSTEMS 4-129

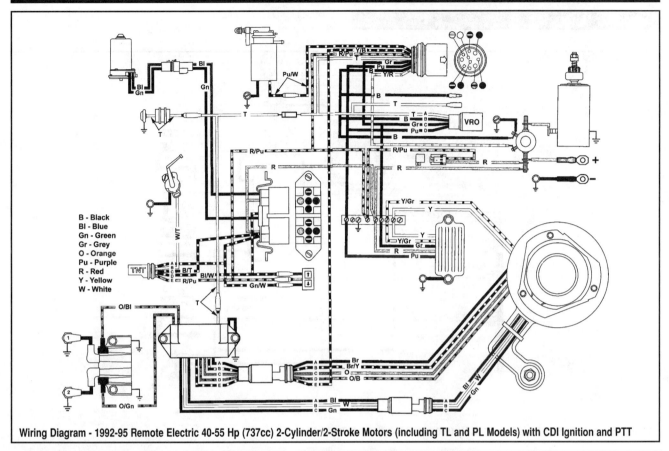

Wiring Diagram - 1992-95 Remote Electric 40-55 Hp (737cc) 2-Cylinder/2-Stroke Motors (including TL and PL Models) with CDI Ignition and PTT

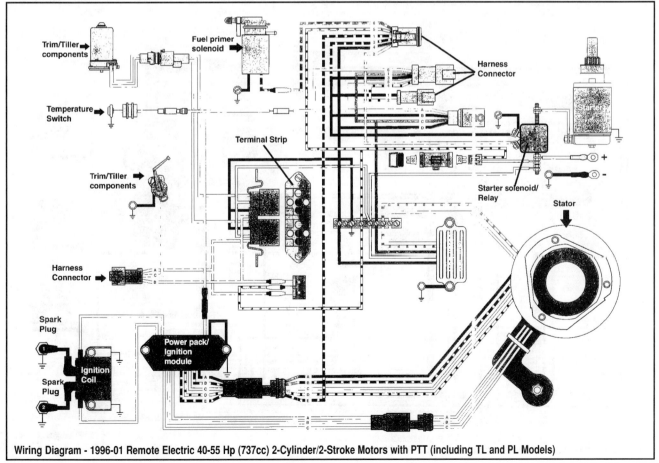

Wiring Diagram - 1996-01 Remote Electric 40-55 Hp (737cc) 2-Cylinder/2-Stroke Motors with PTT (including TL and PL Models)

4-130 IGNITION AND ELECTRICAL SYSTEMS

3-Cylinder Motors

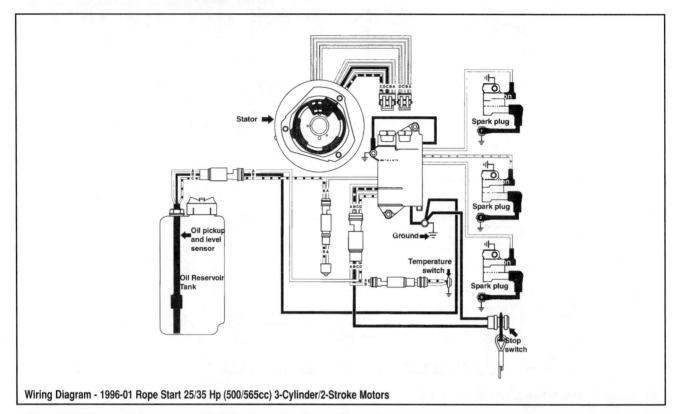

Wiring Diagram - 1996-01 Rope Start 25/35 Hp (500/565cc) 3-Cylinder/2-Stroke Motors

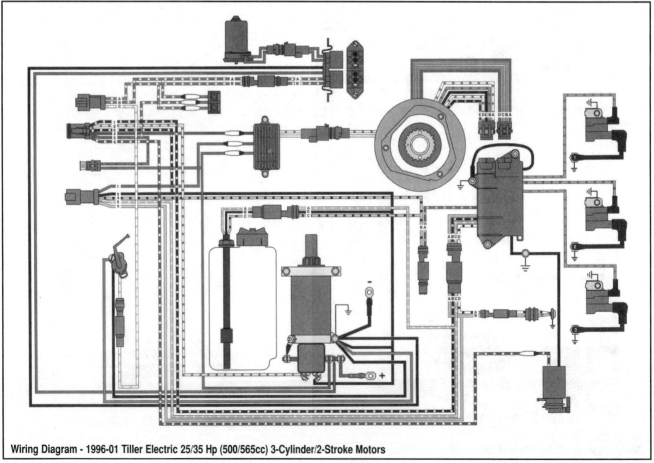

Wiring Diagram - 1996-01 Tiller Electric 25/35 Hp (500/565cc) 3-Cylinder/2-Stroke Motors

IGNITION AND ELECTRICAL SYSTEMS 4-131

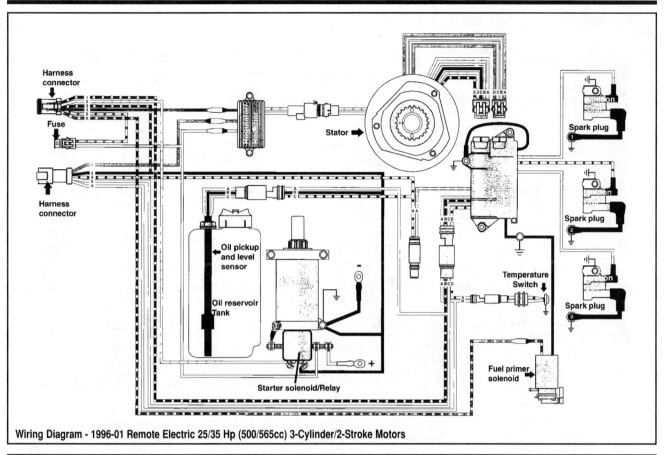

Wiring Diagram - 1996-01 Remote Electric 25/35 Hp (500/565cc) 3-Cylinder/2-Stroke Motors

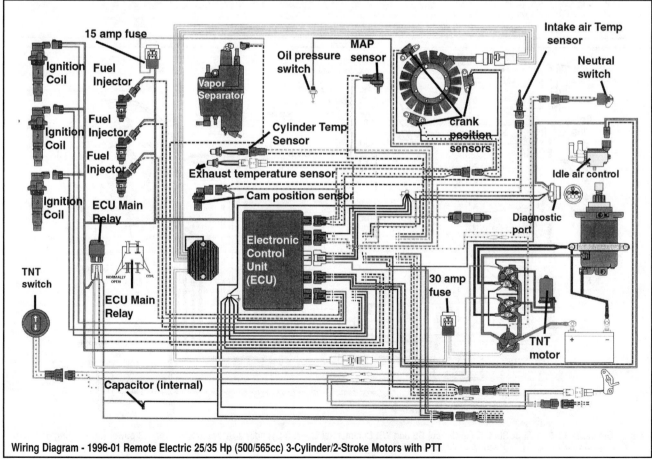

Wiring Diagram - 1996-01 Remote Electric 25/35 Hp (500/565cc) 3-Cylinder/2-Stroke Motors with PTT

4-132　IGNITION AND ELECTRICAL SYSTEMS

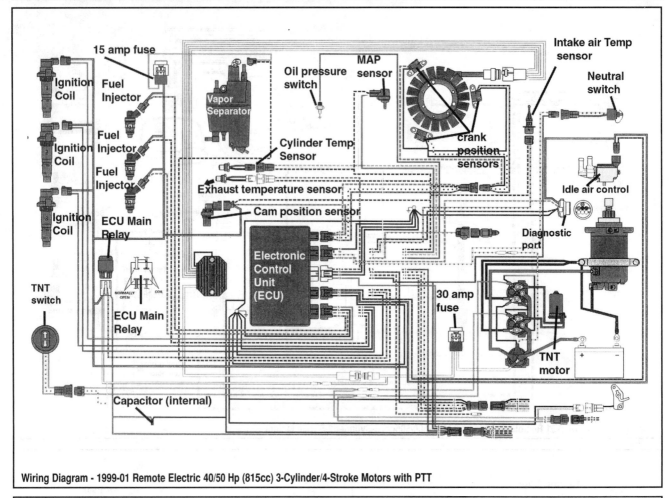

Wiring Diagram - 1999-01 Remote Electric 40/50 Hp (815cc) 3-Cylinder/4-Stroke Motors with PTT

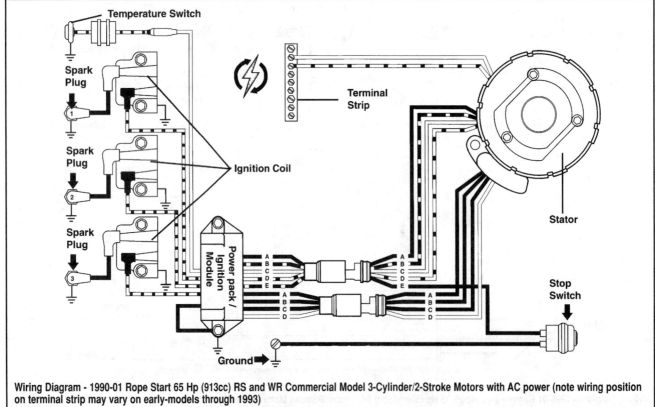

Wiring Diagram - 1990-01 Rope Start 65 Hp (913cc) RS and WR Commercial Model 3-Cylinder/2-Stroke Motors with AC power (note wiring position on terminal strip may vary on early-models through 1993)

IGNITION AND ELECTRICAL SYSTEMS

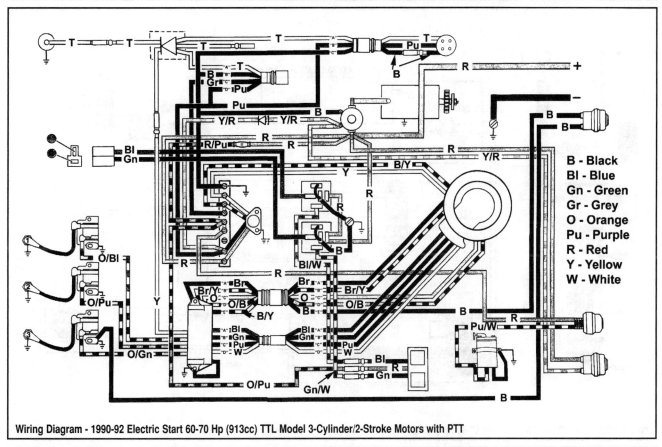

Wiring Diagram - 1990-92 Electric Start 60-70 Hp (913cc) TTL Model 3-Cylinder/2-Stroke Motors with PTT

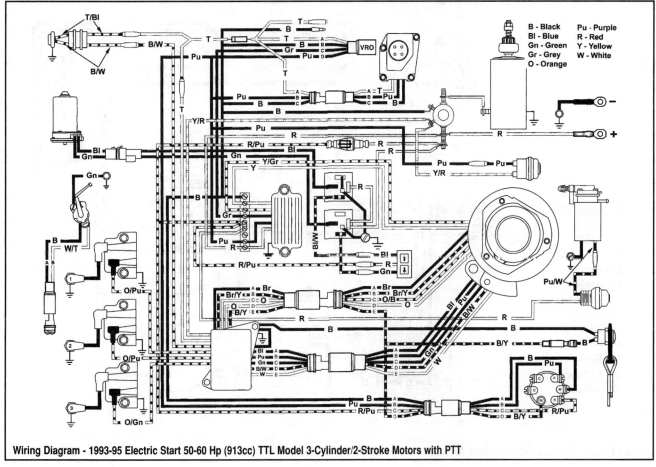

Wiring Diagram - 1993-95 Electric Start 50-60 Hp (913cc) TTL Model 3-Cylinder/2-Stroke Motors with PTT

4-134 IGNITION AND ELECTRICAL SYSTEMS

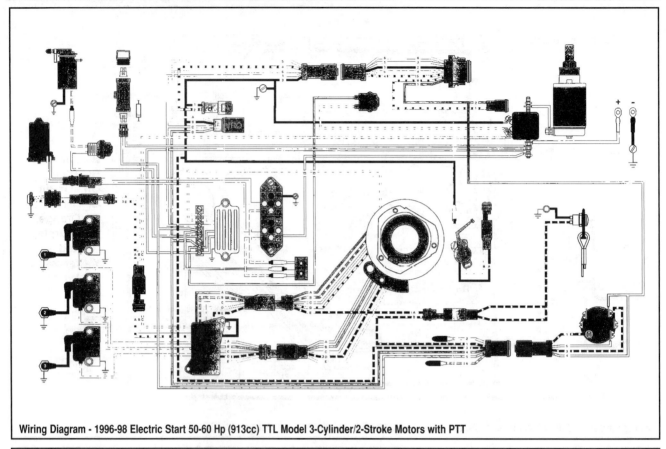

Wiring Diagram - 1996-98 Electric Start 50-60 Hp (913cc) TTL Model 3-Cylinder/2-Stroke Motors with PTT

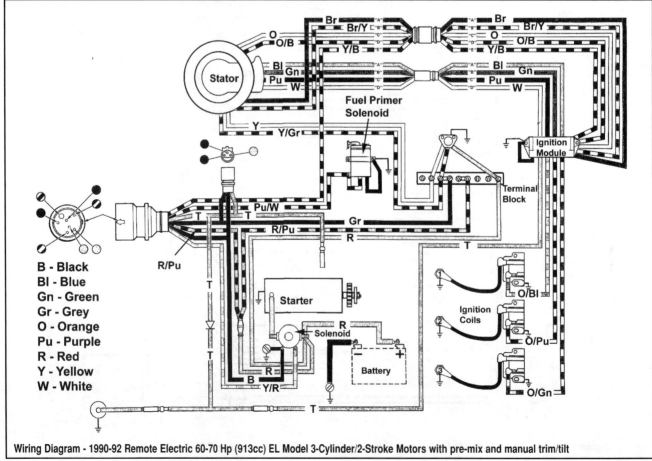

Wiring Diagram - 1990-92 Remote Electric 60-70 Hp (913cc) EL Model 3-Cylinder/2-Stroke Motors with pre-mix and manual trim/tilt

IGNITION AND ELECTRICAL SYSTEMS 4-135

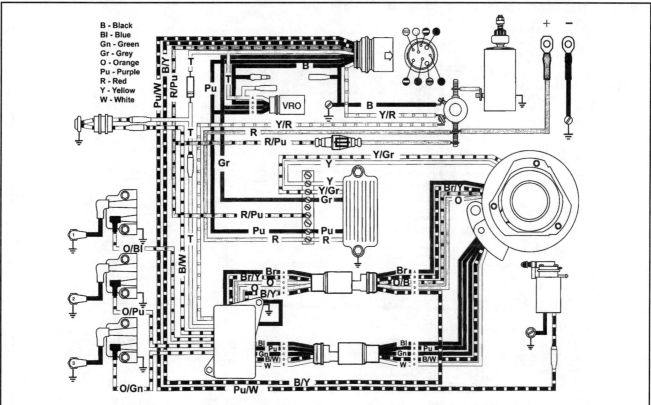

Wiring Diagram - 1993-95 Remote Electric 50-70 Hp (913cc) EL Model and 65 Hp (913cc) WMLE Commercial Model 3-Cylinder/2-Stroke Motors with manual trim/tilt

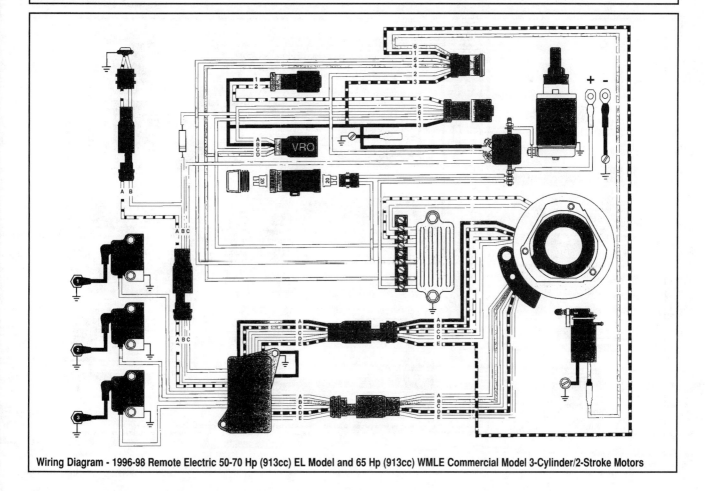

Wiring Diagram - 1996-98 Remote Electric 50-70 Hp (913cc) EL Model and 65 Hp (913cc) WMLE Commercial Model 3-Cylinder/2-Stroke Motors

4-136 IGNITION AND ELECTRICAL SYSTEMS

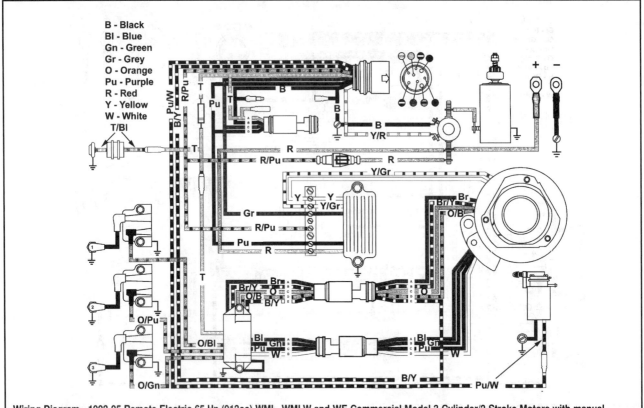

Wiring Diagram - 1993-95 Remote Electric 65 Hp (913cc) WML, WMLW and WE Commercial Model 3-Cylinder/2-Stroke Motors with manual trim/tilt

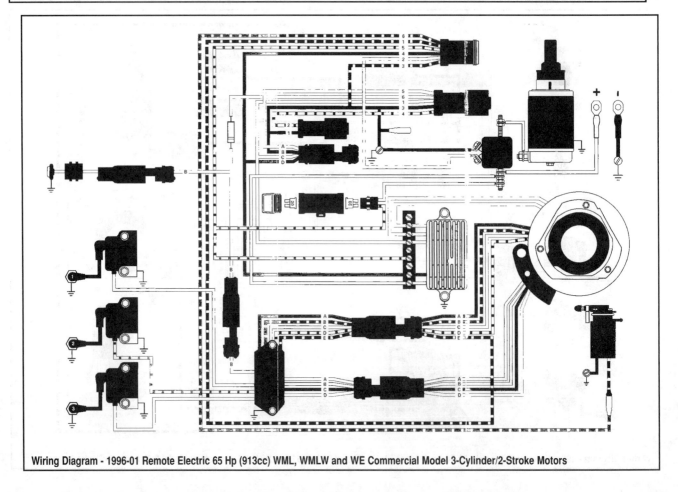

Wiring Diagram - 1996-01 Remote Electric 65 Hp (913cc) WML, WMLW and WE Commercial Model 3-Cylinder/2-Stroke Motors

IGNITION AND ELECTRICAL SYSTEMS

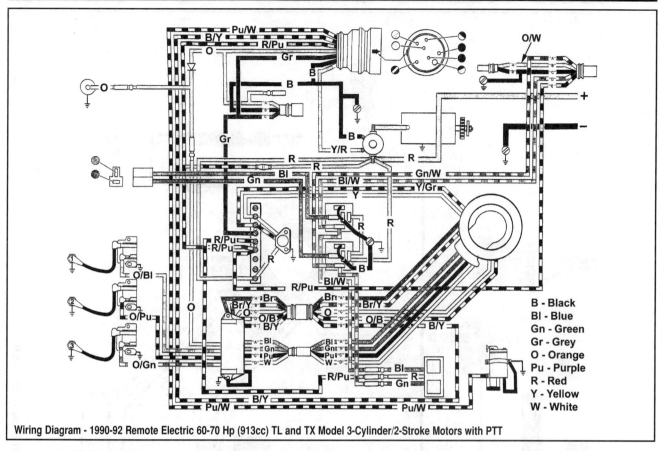

Wiring Diagram - 1990-92 Remote Electric 60-70 Hp (913cc) TL and TX Model 3-Cylinder/2-Stroke Motors with PTT

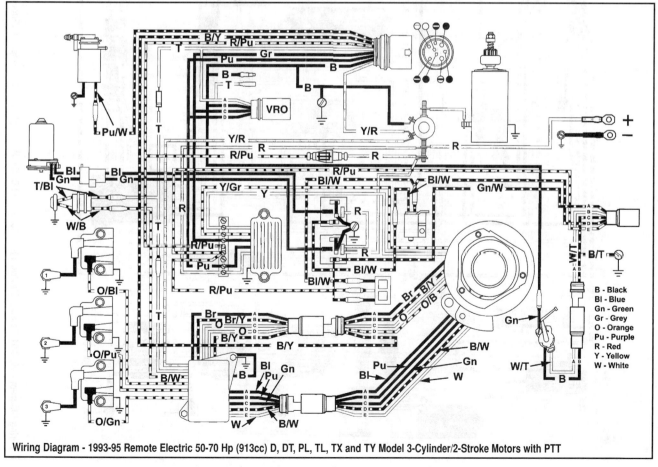

Wiring Diagram - 1993-95 Remote Electric 50-70 Hp (913cc) D, DT, PL, TL, TX and TY Model 3-Cylinder/2-Stroke Motors with PTT

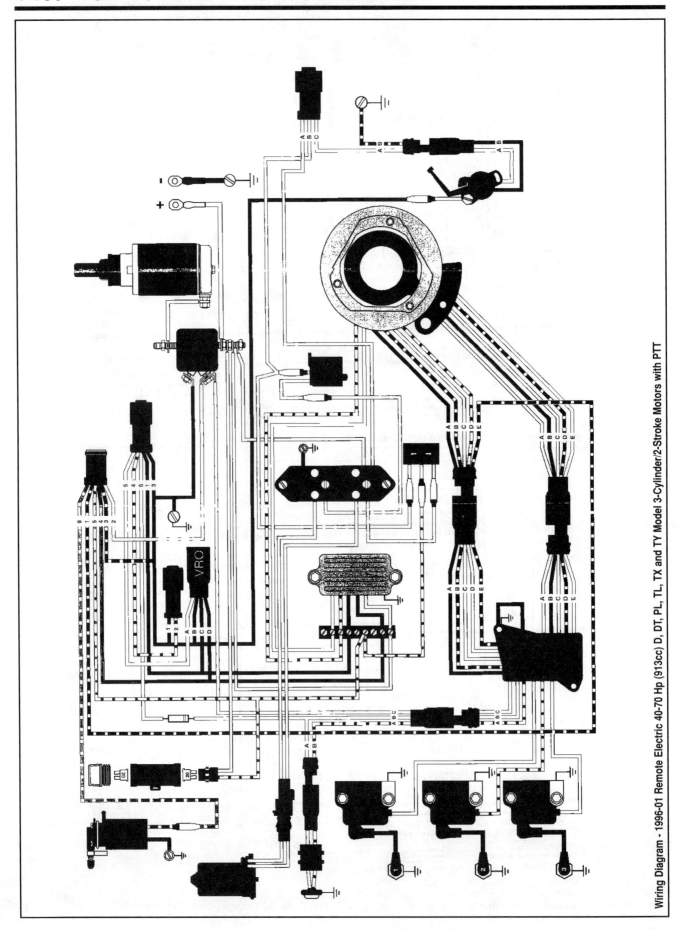

Wiring Diagram - 1996-01 Remote Electric 40-70 Hp (913cc) D, DT, PL, TL, TX and TY Model 3-Cylinder/2-Stroke Motors with PTT

IGNITION AND ELECTRICAL SYSTEMS

4-Cylinder Motors

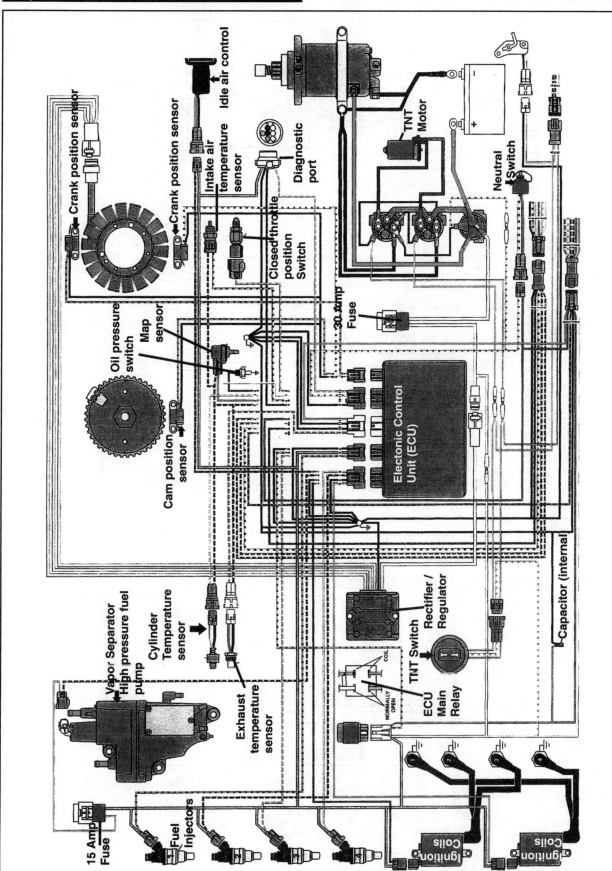

Wiring Diagram - 1999-01 Remote Electric 70 Hp (1298cc) 4-Cylinder/4-Stroke Motors with Electric Trim and Tilt

4-140 IGNITION AND ELECTRICAL SYSTEMS

Johnson/Evinrude Remotes

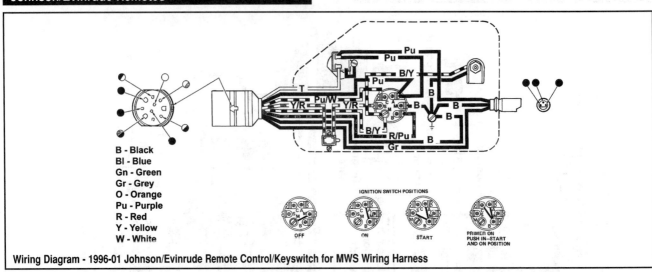

Wiring Diagram - 1996-01 Johnson/Evinrude Remote Control/Keyswitch for MWS Wiring Harness

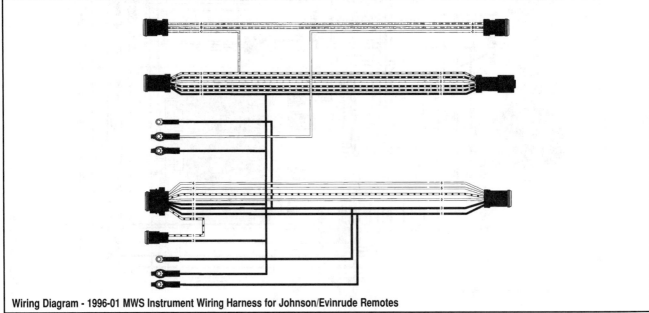

Wiring Diagram - 1996-01 MWS Instrument Wiring Harness for Johnson/Evinrude Remotes

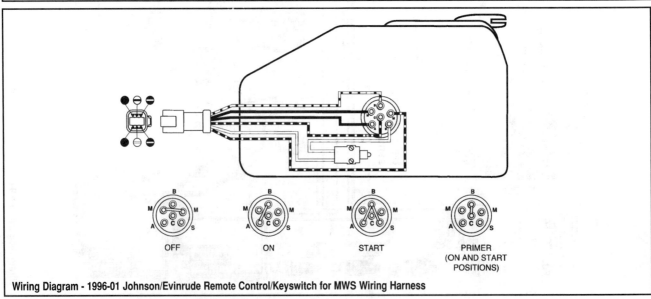

Wiring Diagram - 1996-01 Johnson/Evinrude Remote Control/Keyswitch for MWS Wiring Harness

IGNITION AND ELECTRICAL SYSTEMS 4-141

SPECIFICATIONS

Ignition Testing Specifications - Carbureted Motors

Model (Hp)	No. of Cyl	Engine Type	Year	Displace cu. in. (cc)	Ignition Coil Primary Resistance	Ignition Coil Secondary Resistance	Charge Coil Resistance	Charge Coil Min. Volts Cranking	Power Coil Resistance	Power Coil Min. Volts Cranking	Sensor/Trigger Coil Resistance	Sensor/Trigger Coil Min. Volts Cranking	Power Pack Min. Volts Cranking	Power Pack Min. Volts Running
Colt/Junior	1	2-stroke	1990	2.64 (43)	0.7-1.1	4500-5100	N/A	N/A	N/A	N/A	N/A	N/A	N/A	N/A
2.0-3.5	1	Magneto	1991-95	2.64 (43)	0.7-1.1	4500-5100	N/A	N/A	N/A	N/A	N/A	N/A	N/A	N/A
		CDI	1996-01	4.75 (77.8)	0.05-0.15	550-670	N/A	180	N/A	N/A	40-60	1	180	220
5	1	2-stroke	1999-01	6.7 (109)	N/A	225-325	100-150	130	N/A	N/A	N/A	N/A	N/A	N/A
5/6	1	4-stroke	1997-01	7.8 (128)	0.23-0.33	800-1000	82-102	200	82-102	60	132-162	0.2	50	220
3-4	2	Magneto	1990	5.28 (87)	0.7-1.1	4500-5100	N/A	N/A	N/A	N/A	N/A	N/A	N/A	N/A
	2	CDI	1990-01	5.28 (87)	0.05-0.15	225-325	N/A	N/A	N/A	N/A	85-115	4	125	150
4 Deluxe	2	UFI	1990	5.28 (87)	0.05-0.15	535-585	N/A	N/A	N/A	N/A	N/A	N/A	175	200
	2	UFI	1991-93	5.28 (87)	0.05-0.15	800-1000	N/A	N/A	N/A	N/A	N/A	N/A	175	200
	2	CDI	1994-96	5.28 (87)	0.05-0.15	800-1000	N/A	230	N/A	N/A	30-50	1.5	175	200
5/6/6.5/8	2	UFI	1990	10 (164)	0.05-0.15	535-585	N/A	N/A	N/A	N/A	N/A	N/A	175	200
	2	UFI	1991-92	10 (164)	0.05-0.15	800-1000	N/A	N/A	N/A	N/A	N/A	N/A	175	200
	2	CDI	1991-01	10 (164)	0.05-0.15	800-1000	N/A	N/A	N/A	N/A	30-50	1.5	175	200
8/9.9	2	4-stroke	1996	12.87 (211)	0.23-0.33	2000-2600	82-102	300	82-102	100	132-162	0.2	100	240
	2	4-stroke	1997-01	12.87 (211)	0.23-0.33	2000-2600	82-102	300	82-102	100	132-162	0.2	50	240
9.9-15	2	UFI	1990	13.2 (216)	0.05-0.15	225-325	N/A	N/A	N/A	N/A	N/A	N/A	175	200
	2	UFI	1991-92	13.2 (216)	0.05-0.15	800-1000	N/A	N/A	N/A	N/A	N/A	N/A	175	200
	2	CDI	1992	13.2 (216)	0.05-0.15	515-635	N/A	230	N/A	N/A	30-50	1.5	200	230
9.9-15	2	2-stroke	1993-01	15.6 (255)	0.05-0.15	225-325	①	230	N/A	N/A	30-50	1.5	175	200
9.9/15	2	4-stroke	1995-96	18.6 (305)	0.23-0.33	2000-2600	②	300	③	100	N/A	N/A	100	240
	2	4-stroke	1997	18.6 (305)	0.23-0.33	2000-2600	②	220	③	70	N/A	N/A	100	240
	2	4-stroke	1998	18.6 (305)	0.23-0.33	2000-2600	720-880	220	52-62	70	N/A	N/A	100	240
	2	4-stroke	1999-01	18.6 (305)	0.23-0.33	2000-2600	720-880	220	52-62	30	N/A	N/A	100	240
18 Jet-35	2	UFI	1990	13.2 (216)	0.05-0.15	225-325	535-585	N/A	N/A	N/A	N/A	N/A	175	200
	2	UFI	1991-92	13.2 (216)	0.05-0.15	800-1000	N/A	N/A	N/A	N/A	N/A	N/A	175	200
	2	CDI	1992	13.2 (216)	0.05-0.15	515-635	N/A	230	N/A	N/A	30-50	1.5	200	230
	2	2-stroke	1993-01	31.8 (521)	0.05-0.15	800-1000	N/A	230	N/A	N/A	30-50	1.5	175	200
25D-55	2	UFI	1990-91	45 (737)	0.05-0.15	225-325	535-585	N/A	N/A	N/A	N/A	N/A	175	200
	2	UFI	1992-93	45 (737)	0.05-0.15	800-1000	N/A	N/A	N/A	N/A	N/A	N/A	175	200
	2	CD/4amp	1992-93	45 (737)	0.05-0.15	800-1000	N/A	230	N/A	N/A	30-50	1.5	200	230
	2	CD/12amp	1992-93	45 (737)	0.05-0.15	750-950	360-440	230	N/A	N/A	13-17	0.5	230	250
	2	2-stroke	1994-01	45 (737)	0.05-0.15	225-325	④	230	⑤	N/A	⑥	⑦	150	175

Ignition Testing Specifications - Carbureted Motors

Model (Hp)	No. of Cyl	Engine Type	Year	Displace cu. in. (cc)	Ignition Coil Primary Resistance	Ignition Coil Secondary Resistance	Charge Coil Resistance	Charge Coil Min. Volts Cranking	Power Coil Resistance	Power Coil Min. Volts Cranking	Sensor/Trigger Coil Resistance	Sensor/Trigger Coil Min. Volts Cranking	Power Pack Min. Volts Cranking	Power Pack Min. Volts Running
25/35	3	2-stroke	1995-97	31/35 (500/565)	0.23-0.33	2000-2600	⑧	300	⑨	100	OS / TNA	OS / TNA	100	240
	3	2-stroke	1998	31/35 (500/565)	0.23-0.33	2000-2600	720-880	300	52-62	100	OS / TNA	OS / TNA	100	240
	3	2-stroke	1999	31/35 (500/565)	0.23-0.33	2000-2600	720-880	220	52-62	30	OS / TNA	OS / TNA	100	240
	3	2-stroke	2000-01	31/35 (500/565)	0.05-0.15	225-325	720-880	220	52-62	30	OS / TNA	OS / TNA	100	240
25-70	3	2-stroke	1990-91	56.1 (913)	0.05-0.15	225-325	455-505	250	455-505	N/A	8-14	0.3	230	250
	3	6-amp	1992-96	56.1 (913)	0.05-0.15	225-325	360-440	250	360-440	N/A	8-14	0.3	190	220
	3	12-amp	1992-96	56.1 (913)	0.05-0.15	225-325	750-950	250	360-440	N/A	⑫	⑫	190	200
	3	2-stroke	1997-98	56.1 (913)	0.05-0.15	225-325	⑩	⑪	360-440	N/A	⑬	⑬	190	220
	3	2-stroke	1999-01	56.1 (913)	0.05-0.15	225-325	⑩	⑪	360-440	N/A	⑭	⑭	190	220

NOTE: All resistance tests are based on a high-quality DVOM testing components at an ambient temperature of 68 degrees F (20 degrees C). Testing under other conditions, such as using lower quality meters or at testing different temperatures could lead to false results.

OS - optical sensor   TNA - test not available

NA - not applicable or not available on this model

① Specification is 800-1000 ohms for rope start models, or 680-840 ohms for electric start models
② Specification is 1010-1230 ohms for rope start models, or 720-880 ohms for electric start models
③ Specification is 76-92 ohms for rope start models, or 52-62 ohms for electric start models
④ Specification is 800-1000 ohms for rope start and 4-amp models, or 750-950 ohms for 12-amp models
⑤ The power coil is used on 12-amp models only, specification is 360-440 ohms
⑥ Specification is 30-50 ohms for rope start and 4-amp models, or 22-32 ohms for 12-amp models
⑦ Specification is 1.5 ohms for rope start and 4-amp models, or 0.5 ohms for 12-amp models
⑧ Specification is 1010-1230 ohms for rope start models, or 720-880 ohms for electric start models
⑨ Specification is 76-92 ohms for rope start models, or 52-62 ohms for electric start models
⑩ Specification is 360-440 ohms for 6-amp models, or 750-950 ohms for 12-amp models
⑪ Specification is 250 volts output cranking/275 volts output running on 6-amp models, or 250 volts cranking for both normal and advance sensor coils on 6-amp models, or 250 volts cranking for both normal and advance coils on Quickstart, or for both normal and advance coils on Quickstart models
⑫ Specification is 8-14 ohms on models without Quickstart, or for both normal and advance coils on Quickstart models, it varries by DVOM: 270-330 ohms on a Stevens AT-101 meter or 630-770 ohms on a Merc-O-Tronic M-700 meter
⑬ Specification is 0.3 volts output cranking/1.0 volts output running on models without Quickstart, or 1.5 volts cranking for both the normal and advance sensor coils on Quickstart models
⑭ Specification is 8-14 ohms on models without Quickstart, or for both normal and advance coils on Quickstart models, it varries by DVOM: 250-350 ohms on a Stevens AT-101 meter, 750-850 ohms on a Merc-O-Tronic M-700 meter, or 850-950 ohms on a Fluke 29 Series II meter

5 LUBRICATION & COOLING

LUBRICATION SYSTEMS (2-STROKE) 5-2
LUBRICATION SYSTEM (4-STROKE) . 5-14
COOLING SYSTEM 5-21

LUBRICATION SYSTEMS (2-STROKE) 5-2
ACCUMIX OIL INJECTION SYSTEM 5-2
- OIL RESERVOIR CANISTER 5-2
  - Assembly 5-3
  - Cleaning And Inspection............ 5-3
  - Removal 5-2
INTEGRAL OIL MIXING UNIT SYSTEM (25/35 HP 3-CYL) 5-11
- OIL TANK AND MIXING UNIT 5-12
  - Overhaul 5-13
  - Removal/Installation................ 5-12
- SENDING UNIT 5-12
  - Removal/Installation................ 5-12
- TROUBLESHOOTING.................. 5-11
  - Sending Unit Tests................. 5-11
VARIABLE RATIO OIL (VRO2) OIL INJECTION SYSTEM 5-3
- PULSE LIMITER 5-7
- SYSTEM VERIFICATION & TROUBLESHOOTING................. 5-3
  - LOW OIL Warning Circuit 5-3
  - NO OIL Warning Circuit 5-4
  - VRO2 Fuel/Oil Pump Circuits 5-4
  - VRO2 Fuel Pump Delivery 5-5
  - VRO2 Oil Pump Delivery 5-6
  - VRO2 Pump Self-Priming............. 5-4
- VRO2 PICKUP AND OIL SUPPLY HOSE .. 5-7
- VRO2 PUMP....................... 5-8
  - Overhaul 5-8
  - Removal and Installation 5-8

LUBRICATION SYSTEM (4-STROKE) 5-14
- DESCRIPTION AND OPERATION........ 5-14
- OIL PRESSURE 5-14
  - Testing....................... 5-14
- OIL PRESSURE SWITCH AND WARNING SYSTEM...................... 5-21
- OIL PUMP........................ 5-17
  - Removal, Overhaul & Installation 5-17
- OIL PUMP........................ 5-17
- REMOVAL, OVERHAUL & INSTALLATION . 5-17
  - 5/6 Hp (128cc)..................... 5-17
  - 8/9.9 Hp (211cc)................... 5-18
  - 9.9/15 Hp (305cc).................. 5-18
  - 40/50 Hp (815oo) EFI................ 5-19
  - 70 Hp (1298cc) EFI 5-20

COOLING SYSTEM 5-21
- COOLING SYSTEM SCHEMATICS........ 5-40
  - Colt/Junior motors 5-40
  - 2.0-3.5 hp (78cc) 5-41
  - 3-4 hp (87cc) 5-41
  - 4 deluxe (87cc) 5-42
  - 5 hp (109cc)..................... 5-42
  - 5/6 hp (128cc) 4-stroke 5-43
  - 5/6/8 hp (164cc)................... 5-43
  - 8/9.9 hp (211cc) 4-stroke 5-44
  - 9.9-15 hp (216/255cc) 5-44
  - 9.9/15 hp (305cc) 4-stroke 5-45
  - 18 jet-35 hp (521cc)................ 5-45
  - 25/35 hp (500/565cc) 5-46
  - 45/55 hp (737cc) Commercial.......... 5-46
  - 25-55 hp (737cc) 5-47
  - 25-70 hp (913cc) 5-47
  - 40/50 hp (815cc) EFI 5-48
  - 70 hp (1298cc) EFI................. 5-48
- DESCRIPTION & OPERATION 5-22
- THERMOSTAT 5-37
  - Removal & Installation 5-37
- TROUBLESHOOTING.................. 5-23
  - Cooling System 5-24
  - Thermostat...................... 5-25
- WATER PRESSURE RELIEF VALVE...... 5-39
  - Removal & Installation 5-39
- WATER PUMP 5-25
  - Inspection & Overhaul 5-36
  - Removal & Installation 5-25
- Water Pump 5-25
- REMOVAL & INSTALLATION 5-25
  - Colt/Junior 5-25
  - 2.0-3.5 Hp (78cc) 5-26
  - 3-4 Hp (87cc)..................... 5-26
  - 4 Deluxe (87cc) and 5/6/8 Hp (164cc) ... 5-28
  - 5 Hp (109cc) 5-29
  - 5/6 Hp (128cc), 8/9.9 Hp (211cc), 9.9-15 Hp (216/255cc) and 9.9/15 Hp (305cc)... 5-29
  - 10/14 Hp (216/255cc), 25 Commercial And 28 Hp (521cc) 2-Cyl.............. 5-31
  - 18 Jet-35 Hp (521cc, Exc. 25 Com & 28 Hp) and 25/35 Hp (500/565cc) 5-32
  - 25-55 Hp (737cc) and 25-70 Hp (913cc).. 5-33
  - 40/50 Hp (815cc) EFI................ 5-35
  - 70 Hp (1298cc) EFI 5-35

5-2 LUBRICATION AND COOLING

LUBRICATION SYSTEMS (2-STROKE MOTORS)

Unlike 4-stroke engines, which contain a reservoir of oil that is re-circulated during engine operation, mixing oil with the fuel lubricates 2-stroke engines. Internal engine components of 2-stroke motors are lubricated as the fuel/oil mixture passes through the crankcase and the cylinder.

Generally speaking, there are 2 methods of adding oil to a 2-stroke outboard. The first is to pre-mix oil with the gasoline whenever the fuel tank is filled. The pre-mix method is generally used on smaller (lower horsepower) motors and on most commercial outboards. It is easiest to perform this on portable fuel tanks that can be agitated to ensure proper mixture, but it can be successfully accomplished on larger built-in tanks, as long as care is taken to properly measure the amounts of fuel/oil being added.

For ease of service and to ensure a constant supply of 2-stroke oil, many Johnson/Evinrude motors are equipped with an oil injection system. Most of the larger 2-stroke motors are rigged with some form of automatic oiling system from the factory or upon installation. Various forms of automatic oiling systems are used. The AccuMix system can be found on nearly any 2-stroke outboard (as it is has been available in versions that are both built-into a portable fuel tanks or independent, remote-mounted), but it is most often found on smaller hp motors (generally 521cc and smaller engines). The Variable Ratio Oiling (VRO2) system, is usually found on 25-55 Hp (737cc) and 25-70 Hp (913cc) motors, though there are some exceptions with certain special or commercial models. Lastly, the 25/35 Hp (500/565cc) 3-cylinder motors are equipped from the factory with an integral oil reservoir tank and oil mixing unit.

AccuMix Oil Injection System

◆ See Figure 1

The AccuMix oil injection system is designed to provide a fuel/oil ratio of 50/1 regardless of powerhead rpm.

The AccuMix system is located either entirely inside the portable fuel tank or inside a remote mount, self-contained housing found inline between the fuel tank and motor. On fuel tank mounted systems, a 1-1/2 quart reservoir canister contains enough oil for almost five tanks full of fuel. The remote mounted unit is of comparable size. An oil metering pump is located at the base of the canister. Pulses from the powerhead mounted fuel pump assembly activate the oil pump. The oil metering pump automatically blends fuel from the fuel pickup with oil in the reservoir canister.

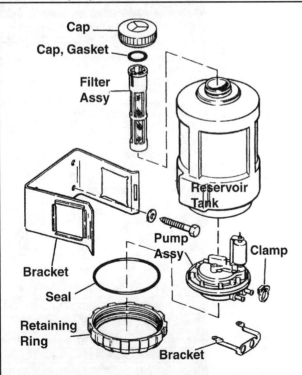

Fig. 1 Exploded view of a typical remote mount AccuMix assembly

A low-oil warning indicator activates a warning horn when the level of oil falls below one pint. If the operator sustains powerhead operation after the warning horn sounds, the fuel supply is automatically cut off to shutdown the powerhead. The powerhead cannot be restarted until oil has been added to the canister.

For fuel tank integral assemblies, a visual low oil level indicator is located on the reservoir canister cover. This indicator is of the "glass eye" type, similar to those found on newer style automotive batteries. Remote mount units may be equipped with a sight glass or might require the cap to be removed in order to check oil level.

■ Due to the inherent design of the fuel tank integrated version of the oil injection system, very few individual parts can be repaired or replaced, if found to be defective. As an example: if the low oil level float is found to be defective, unfortunately, the entire system must be replaced. Therefore, no "troubleshooting" procedures are given, because the manufacturer has provided no evaluating tests.

If any problem is encountered with the delivery of oil, follow the procedures outlined under AccuMix Service. The instructions deal mostly with cleaning out the system. If the problem persists, the only remedy on integral tank models is a new system. However components are usually available to service the remote mount system, see your local parts dealer for component availability.

ACCUMIX SERVICE

◆ See Figure 1

Maintenance of this type oil injection system is limited to draining and flushing the oil reservoir canister each season. On integral fuel tank units, to properly clean the canister and the integral oil filter, it should be removed from the fuel tank. On remote mounted units, the entire remote canister can be removed for flushing and cleaning, or the filter assembly can be removed once the cap and seal are removed from the top of the tank. In all cases, canister or filter element cleaning is accomplished using fresh gasoline or solvent.

Procedures for the canister service on the integral fuel tank units will be found in the following paragraphs.

When the reservoir canister has been removed from the fuel tank, the fuel line pickup screen can also be serviced.

Clean the vent screw and the area around the screw each time the fuel tank is filled. This vent screw must be fully open to allow powerhead operation and to permit air to enter the tank (taking the place of consumed fuel/oil). The vent screw and fuel tank cap cannot be serviced, other than cleaning. If defective for any reason, the assembly must be replaced as a unit.

Oil Reservoir Canister (Integral Fuel Tank Units)

◆ See Figures 2, 3 and 4

Removal (Integral Fuel Tank Units)

◆ See Figures 2, 3 and 4

1. Disconnect the electrical harness and the fuel line connector from the top of the canister. Remove the retaining screws (normally 8 slotted-head screws), securing the cover and canister to the fuel tank. Observe the three additional washers under the tank handle bracket, when removing the handle. Lift out the canister and cover together from the tank, taking care not to spill any oil remaining in the canister. Remove and discard the gasket between the canister flange and the tank.

2. Ease the cover from the canister. The fuel tube must be disengaged from the cavity in the cover. Remove and discard the O-ring between the cover and the canister.

■ Do not attempt to disassemble the oil pump at the canister base. No replacement parts are available. If defective the pump must be replaced as an assembly.

LUBRICATION AND COOLING 5-3

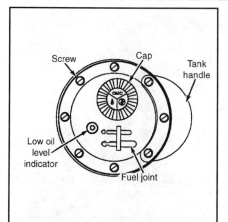

Fig. 2 Top view of a typical AccuMix oil injection system

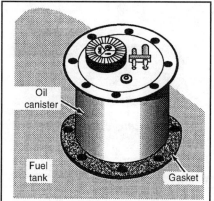

Fig. 3 A cork gasket seals the canister flange to the fuel tank

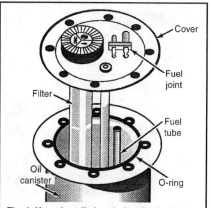

Fig. 4 Upon installation, index the fuel tube with the cavity directly under the fuel joint (and the canister/cover mounting holes)

Cleaning And Inspection (All Units)

Drain any oil from the canister. Obtain a container of solvent and "dunk" the oil filter a couple of times. Take care not to alter the low oil level indicator float. Use a shop towel moistened with solvent and wipe down the surfaces of the cover and the length of the low oil level indicator tube. Blow the cover assembly and filter dry with compressed air.

Pour some solvent into the canister to rinse any residue from the canister walls. Make certain no debris obstructs the oil pump pickup or the low oil cutoff float mounted on the pump.

Inspect and service the fuel line pickup screen, as necessary.

Assembly (Integral Fuel Tank Units)
◆ See Figures 2, 3 and 4

1. Secure the fuel pickup line into the clip at the base of the canister. Place a new gasket on the fuel tank surface and lower the canister into the tank. Align the canister flange holes with the gasket holes.
2. Apply a light coat of OMC Triple-Guard grease on both sides of the O-ring, and then position the ring onto the canister flange. Align the holes in the O-ring with the flange holes.
3. Lower the cover over the canister and make sure the fuel tube indexes with the cavity on the underside of the cover.
4. Position one washer at each of the three holes for the fuel tank handle bracket, and then hold the handle in place while these screws are started. Install the remaining screws finger-tight, then tighten all of the screws using a crisscross pattern to 10 inch lbs. (1.1 Nm).

Engage the electrical harness and the fuel connector to the fittings on top of the canister.

Variable Ratio Oil (VRO2) Oil Injection System

◆ See Figure 5

The VRO2 system consists of a remote oil reservoir (tank), a VRO2 oil/fuel pump and mixer assembly, an integral **No** or **Low Oil** warning circuit and the necessary hoses and fittings to connect the various items for efficient operation. All connections in the system must be airtight to prevent serious damage to the powerhead.

As the name implies, the VRO pump moves oil from the oil reservoir to the powerhead. However, it is a dual pump and also moves fuel. Pumping action stops automatically if fuel is not available at the pump for any reason. This automatic pump shutdown feature prevents the carburetors from filling with oil.

The warning circuits will vary slightly with rigging and whether or not the boat is equipped with the System Check Monitor. If the boat does NOT have the system check monitor gauge, the warning circuit is strictly audible (in the form of the warning horn usually located in the remote box). However, generally speaking, boats/motors rigged after 1995 are equipped with a System Check Monitor gauge which includes LED warning lights in addition to the horn. The audible signals will vary depending upon the system and the circumstances.

For boats/motors which DO NOT have the System Check Monitor, the audible warnings are as follows:
• First, as a low oil level warning; The horn will sound for 1/2 second every 20 seconds if the oil tank level drops below 1/4 of the tank's capacity.
• Secondly, the warning horn will cycle repeatedly on for 1/2 second, off for 1/2 second, on for 1/2 second, etc repeatedly if no oil is reaching the pump via the oil supply line.

For boats/motors which are equipped with the System Check Monitor, the audible and visual warnings are as follows:
• First, as a low oil level warning; The horn will sound for 10 seconds and the **Low Oil** LED in the System Check monitor gauge will illuminate if the oil tank level reaches 1/4 of the tank's capacity.
• Secondly, the warning horn will sound for 10 seconds and the **No Oil** LED in the System Check monitor gauge will illuminate if the no oil is reaching the pump via the oil supply line.

✳✳ WARNING

To continue powerhead operation after the NO OIL LED and warning horn sounds would almost certainly invite serious damage to internal moving parts and powerhead seizure!

VRO2 SYSTEM VERIFICATION & TROUBLESHOOTING

When equipped, the VRO2 system is vitally important to the life of an outboard. Upon initial installation or repair of the system, proper operation of the system and the warning circuits must be verified. If you have any reason to suspect the system, verify proper operation of the warning circuits. It is not a bad idea to go through the trouble of verifying proper warning circuit operation on an annual basis especially when the motor is being removed from storage.

■ If the NO or LOW OIL LEDs illuminate (on boats rigged with the System Check gauge) and/or the warning horn activates at anytime during engine operation, shut the engine down immediately and check the oil level. If adding oil to the tank does not rectify the situation, troubleshoot the problem before continuing to operate the powerhead. If necessary, pre-mix oil into the fuel supply to get the boat and motor safely to shore. An oil supply should be kept on board for this purpose, or if oil is present in the oil reservoir, it can be siphoned off for mixing in the tank (if sufficient oil is available to achieve a proper fuel/oil ratio.

■ If adding oil to the tank does not shut the warning Low Oil LED and/or audible signal off, remove the pickup assembly from the tank and check to see if the float is stuck. The float can be manually repositioned to see if the circuitry may be at fault as opposed to the physical fuel supply. But remember that the NO OIL warning on these models signals that insufficient oil pressure is reaching the pump.

Checking the LOW OIL Warning Circuit

◆ See Figures 6 and 7

Either remove the oil tank pickup and manually reposition the float so it is below the 1/4 level mark or siphon oil from the tank until there is less than 1/4 of tank capacity left in the reservoir. Turn the keyswitch on and wait for

5-4 LUBRICATION AND COOLING

Fig. 5 Most large 2-stroke (737cc and 913cc) motors are equipped with a VRO2 pump/oil injection system

Fig. 6 If necessary for testing or inspection, the oil tank pickup unit is easily removed from the tank...

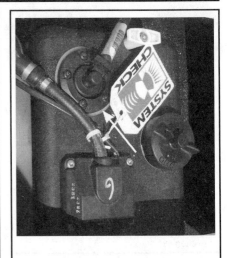

Fig. 7 Generally the pickup unit is secured by about 4 (usually Torx® head) screws

the System Check gauge to complete the self-test, then watch for the **Low Oil** LED to illuminate (System Check) or listen for the audible warning. Slowly add oil to the tank or manually reposition the float upward, the LED should go out and/or the audible warning should stop as soon as the float passes the 1/4 tank level.

■ If the warning circuit does not function properly, check the Warning System circuitry. Refer to the information for Testing Warning Circuits, found in the Warning System Troubleshooting section.

Checking the NO OIL Warning Circuit

◆ See Figure 8

Disconnect the oil line and fuel lines from the fuel tank or outboard, then connect a portable tank containing a 50:1 fuel/oil pre-mix to the engine's fuel inlet line. Connect a source of cooling water to the motor, then start and run the outboard at about 1500 rpm.

■ It may take a few minutes to consume the residual oil in the pump and oil line.

Once any residual oil is consumed, the **No Oil** LED should illuminate (System Check) and/or the warning horn should sound the no oil signal.

■ If the warning circuit does not function properly, check the Warning System circuitry. Refer to the information for Testing Warning Circuits, found in the Warning System Troubleshooting section.

Testing the VRO2 Pump for Self-Priming

◆ See Figure 8

To properly test if the VRO2 pump will self-prime you should start with a pump that is free of oil. If the pump was removed for service, this is no big deal, but if the pump remained installed it is best to empty the pump. The easiest way to accomplish this is to attach a pre-mixed source of fuel and run the motor without the oil line attached, refer to Checking the NO OIL Warning Circuit for more details.

1. If the oil line is empty, start with the hose disconnected from the VRO2 pump. Use the primer bulb to fill the oil line (by holding the end of the line upward while gently squeezing the bulb until oil appears in the line itself.

2. Connect the oil line to the VRO2 pump, but **do not** squeeze the primer bulb after the line has been connected (or you will manually prime the pump).

3. With the outboard connected to a source of cooling water and the fuel line attached to a portable tank of 50:1 pre-mix fuel/oil, start and run the engine at 1500 rpm until the **No Oil** warning LED (System Check) goes out and/or audible warning ceases. If the warning(s) remain, the pump is not self-priming properly.

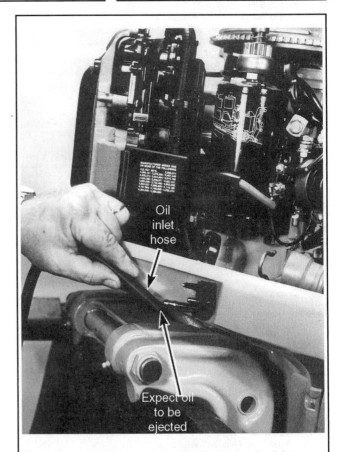

Fig. 8 One method of checking the NO OIL warning circuit is to disconnect the oil line from the motor and attach a tank of pre-mixed fuel/oil, then start and run the engine

Testing the VRO2 Fuel/Oil Pump Circuits

If trouble is suspected with the fuel or oil circuits of the VRO2 system functional tests can be conducted to check fuel pump or oil pump delivery.

LUBRICATION AND COOLING 5-5

Checking VRO2 Fuel Pump Delivery
◆ See Figures 9, 10 and 11

※※ WARNING

Use EXTREME care when working with fuel or flammable solvents. Refer to the cautions under General Information and under the Fuel System sections.

Because this test involves running the engine at speed and under load it must be conducted in a test tank with a suitable test wheel or on the boat (attached to a dock or using an assistant to navigate). You will also need a vacuum gauge, a T-fitting and about 8 in. (20cm) of clear vinyl hose to connect the gauge. Depending on test results, you might also need a fuel pressure gauge capable of reading 0-15 psi (0-103 kPa).

1. With the engine either mounted on a launched craft or in a test tank, start and run the engine at idle until it reaches normal operating temperature, then stop the engine.

※※ WARNING

When removing fuel hoses from fittings, always PUSH (never pull) on the hose itself to prevent the possibility of damaging the fitting. If pushing won't do it, use a small utility knife to carefully cut a slit in the end of the hose and peel it free of the fitting (the hose will then have to be trimmed or replaced upon reconnection).

2. Carefully install the vacuum gauge using the vinyl hose and T-fitting between the fuel inlet fitting on the lower engine cover and the fuel pump. The clear vinyl hose should run from the T-fitting to the fuel pump. Use wire-ties to secure the hoses on the fittings.

※※ WARNING

To prevent damage to the vacuum gauge the primer bulb and the manual or electric fuel primer should not be used once the gauge is installed.

3. Start the engine and run at Wide-Open Throttle (WOT) for at least 2 minutes. During this time, keep a close eye on the clear vinyl hose. There should be no signs of air or vapor bubbles. Also, monitor the reading on the vacuum gauge - it should not exceed 4 in. Hg. (13.5 kPa) at any point. Based on your observations of the hose and of the gauge readings proceed as follows:

4. If the gauge read more than 4 in. Hg. (13.5 kPa), check the fuel delivery system for restrictions. For more details, refer to information on fuel lines and fittings found under the Fuel System section. If you suspect the anti-siphon valve check it as follows:
  a. Remove the anti-siphon valve from the fuel tank.
  b. Connect a clear vinyl hose at least 25 in. (630mm) in length to the tank side of the valve using an adapter.
  c. Fill the clear hose with water until the height of water in the hose/valve (measured to the tank side flange of the anti-siphon valve) is 20 in. (500mm). No water should run through the valve.
  d. Continue to slowly add water until the measured height of the water in the hose/valve is 25 in. (630mm). Water must begin to run through the valve **before** the height reaches 25 in. (630mm) or the valve must be replaced.

5. If the gauge read less than 4 in. Hg (13.5 kPa) and air bubbles were present in the clear hose, check the entire fuel supply system for air leaks. For more details, refer to information on fuel lines and fittings found under the Fuel System section. Repair or replace any leaking hoses, fittings or connections.

■ **An air leak into a fuel supply line does not always create a visible fuel leak until the engine is turned off and the fuel system is pressurized (as it is naturally warmed say, in the heat of the afternoon sun).**

6. If the gauge read less than 4 in. Hg. (13.5 kPa) and no air bubbles were present in the clear hose, move the vinyl hose to the outlet side of the inline fuel filter. Start and run the engine again at WOT, watching for bubbles at this point in the hose as proceed as follows based on the results:
  • If bubbles appear, test or replace the inline filter.
  • If still no bubbles appear, remove the clear hose and install a 0-15 psi (0-103 kPa) fuel pressure gauge between the pump and the carburetors. Start and run the engine at 800 rpm in gear. Once fuel pressure stabilizes the gauge must not indicate less than 3 psi (21 kPa) when held level with the VRO fuel outlet fitting. If the pressure is less than 3 psi (21 kPa), service and check the Pulse Limiter as detailed in this section. If the pulse hose and pulse limiter are in good shape, refer to the accompanying VRO Pump Troubleshooting Chart.

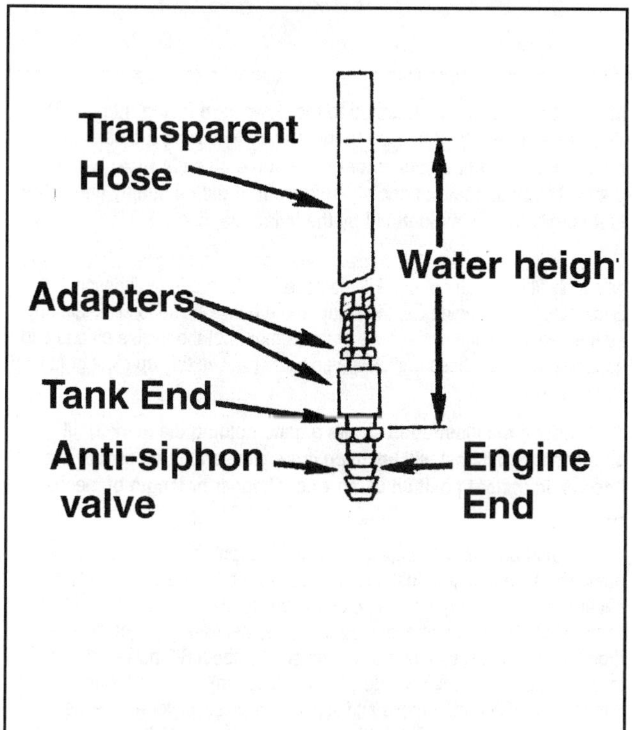

Fig. 9 Check the anti-siphon valve using a length of clear vinyl hose and water

Fig. 10 VRO2 fuel pump pressure is checked using a gauge between the pump and carburetors

5-6 LUBRICATION AND COOLING

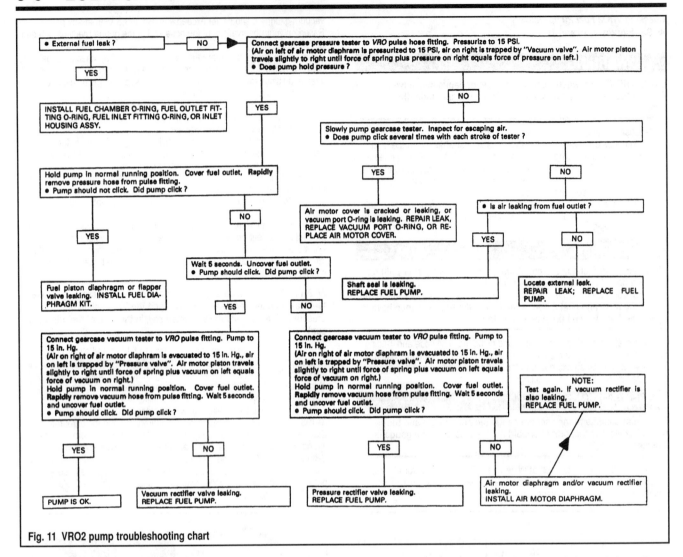

Fig. 11 VRO2 pump troubleshooting chart

Checking VRO2 Oil Pump Delivery

♦ See Figure 12

If a functional check of the pump is desired or, if warning circuits check ok, but trouble is still suspected a pump out check must be performed to prevent the possibility of SEVERE damage to the powerhead if the oil circuit of the pump is not functioning properly. In order to perform this test you will need a 10 in. (25cm) long piece of 1/4 in. (6.35mm) inner diameter (ID) clear vinyl tube. Starting several inches from the open end use a permanent marker to make gradations on the line every 1/2 in. (12.7mm) for a total length of 4 in. (101.6cm). These gradations will be used to monitor the amount of oil drawn into the pump during the functional test.

■ Since the engine is run, but NOT at Wide Open Throttle (WOT) for the VRO2 oil pump delivery check this test CAN be performed using a flush fitting or any other suitable source of cooling water.

1. Hook up a supply of cooling water to the engine, then start and run the engine until it reaches normal operating temperature.

※※ WARNING

When removing fuel or oil hoses from fittings, always PUSH (never pull) on the hose itself to prevent the possibility of damaging the fitting. If pushing won't do it, use a small utility knife to carefully cut a slit in the end of the hose and peel it free of the fitting (the hose will then have to be trimmed or replaced upon reconnection).

2. Shut the engine OFF, then carefully push the oil supply hose from the pump fitting.

■ The oil supply hose attaches to the downward facing fitting at the rear of the oil pump. It is easy to identify by following the oil supply hose from the engine connector (through the oil sight tube) to the pump. The hose that connects just in front of the oil supply line is the fuel supply line, it is identified by the inline fuel filter.

3. Install the 10 in. (25cm) clear vinyl hose (with the gradations) to the oil supply fitting on the pump. Fill the hose with Evinrude or Johnson 2-Stroke Outboard Engine Oil, then start and run the engine to eliminate any air trapped in the line or the base of the pump. Shut the engine off and add additional oil to the hose until it is even with or above the top-most gradation in the hose.

■ Although the illustration shows a hand holding the marked oil supply hose, this test will be much more pleasant if you secure the hose in an upright position using a coat hanger or length of mechanic's wire.

4. Start and run the engine at about 1500 rpm while monitoring the pump cycles. Although a fuel pressure gauge can be used to monitor fuel outlet pulses, light finger pressure on the outlet hose will accomplish the same result. Count the total number of pulses necessary for the oil level to drop 3 in. (76mm) within the hose. It must take about 6-8 pulses for the oil level to drop 3 in. (76mm) inside a 1/4 in. (6.35mm) inner diameter (ID) hose. If results are different, remove and service the pulse limiter and recheck before overhauling or replacing the pump.

5. Once tests or repairs are completed, reconnect the oil supply line to the pump making sure there is an air tight seal. Replace the oil supply line metal spring clamp if it has become weak, damaged or deformed.

LUBRICATION AND COOLING 5-7

✳✳ CAUTION

Keep in mind that an air leak in the oil supply line could allow the pump to loose prime and stop delivering oil to the motor. Although the warning circuit should alert the operator to this condition, a leak could allow the motor to be operated with borderline oil pressure for sometime before it drops below safety spec. Don't take the risk, make sure the oil supply line a sealed and check the fittings visually after each outing.

VRO2 SYSTEM COMPONENT SERVICING

VRO2 Pickup and Oil Supply Hose

◆ See Figures 8, 13 and 14

If the oil supply line is suspect of restrictions it should be removed and inspected/cleaned in 2 areas, the oil pickup and the oil supply line.

1. Remove the oil pickup assembly and checked for a clogged or contaminated filter as follows:
   a. Loosen the Torx® head (usually T25) retaining screws using a suitable driver, then carefully lift the oil pickup assembly from the tank.
   b. Carefully separate the pickup filter from the pickup assembly, then flush the filter using a mild solvent or some fresh fuel.

✳✳ WARNING

Use EXTREME care when working with fuel or flammable solvents. Refer to the cautions under General Information and under the Fuel System sections.

   c. Install the filter back onto the pickup assembly, taking care not to move the pickup assembly on its support rods (as the pickup assembly is specifically adjusted to tank height).

■ It is not necessary for the pickup assembly to be removed for the tank when checking the oil inlet line for restrictions. However, if both are being tested at the same time, it is best to leave the pickup out until it has been determined what repairs (if any) are necessary.

2. With the pickup assembly removed from the tank, perform a vacuum test on the system to check for air leaks as follows:

✳✳ WARNING

When removing fuel or oil hoses from fittings, always PUSH (never pull) on the hose itself to prevent the possibility of damaging the fitting. If pushing won't do it, use a small utility knife to carefully cut a slit in the end of the hose and peel it free of the fitting (the hose will then have to be trimmed or replaced upon reconnection).

   a. Carefully disconnect the oil hose from the pickup assembly and from the oil inlet fitting on the outboard.
   b. Use low-pressure air from a compressor (or your lungs) to purge the hose of oil.
   c. Carefully plug the pickup end of the hose (OMC part # 329661 is available for this purpose).
   d. Connect a hand-held vacuum pump tester to the outboard fitting end of the oil supply line, then apply 7 in. Hg (23.6 kPa) of vacuum to the line. Watch the gauge and make sure that the vacuum holds for at least 5 minutes. If it does not, make sure vacuum is not leaking at the plugged end of the hose otherwise, locate and repair the leak or replace the oil inlet line.

■ An easy way to locate vacuum leaks is by applying a small amount of clean engine oil to the hose at each of the fittings while watching the gauge. When oil is applied to a leaking fitting, the oil will temporarily seal the leak and stop the gauge from moving.

3. Once service has been completed, reconnect the oil lines and secure using clamps.
4. Check the total height of the pickup assembly to make sure it was not disturbed during service. The height from the bottom of the pickup flange to the end of the pickup tube should be 6.84-6.96 in. (174-177mm) for 1.8 gallon tanks or 8.74-8.86 in. (222-225mm) for 3.0 gallon tanks. If necessary, the pickup can be adjusted by sliding it on the support rods.

Pulse Limiter

■ If fuel pump pressure is less than 3 psi (21 kPa) when checked using a pressure gauge inline between the VRO2 pump assembly and the carburetors with the engine running at 800 rpm, service and check the pulse limiter to make sure there are no clogs or restrictions.

An inline pulse limiting check valve is installed between the motor and the VRO2 pump. It is designed to allow normal pressure/vacuum pulses from the engine to the pump, but to close in the event of a backfire, protecting the pump from damage. The check valve must be clean and free of clogs to work properly and ensure the engine receives a proper supply of fuel and oil.

1. Remove the pulse limiter from the crankcase and/or the pulse hose(s).
2. Visually check inside the limiter for signs of excessive carbon deposits.
3. If necessary use OMC Carburetor and Choke Cleaner to backflush the valve, removing carbon deposits. The valve must be replaced if it cannot be cleaned sufficiently.

■ If excessive carbon deposits are found, check the outboard for possible causes of excessive backfiring such as:

- Incorrect linkage adjustments
- Incorrect engine operating temperatures
- Crankcase air leaks

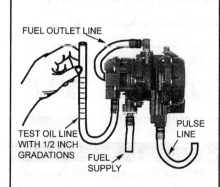

Fig. 12 VRO2 oil pump delivery is checked by counting the number of fuel pulses it takes to consume a specific amount of oil

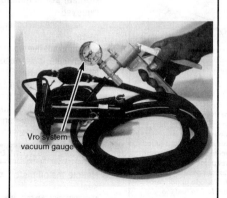

Fig. 13 An oil supply hose is checked by plugging one end of the hose and applying vacuum to the other

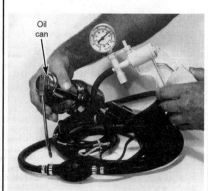

Fig. 14 If leaks are present, locate them by applying a small amount of engine oil. When oil is applied to a leaking fitting, the oil will temporarily seal the leak

5-8 LUBRICATION AND COOLING

- Carburetor problems
- Extended use of the flushing attachment
- Internal damage

4. On models equipped with a threaded pulse limiter, clean the threads and coat them using OMC Pipe Sealant with Teflon (or an equivalent threadsealant). Install the pulse limiter, then connect the pulse hose and secure using the clamp.

5. On models equipped with an inline pulse limiter, install the limiter with the metal end facing toward the powerhead hose. Secure both hoses using clamps.

VRO2 Pump Assembly Removal and Installation

◆ See Figure 15  MODERATE

1. Disconnect the negative battery cable for safety.

※ WARNING

When removing fuel or oil hoses from fittings, always PUSH (never pull) on the hose itself to prevent the possibility of damaging the fitting. If pushing won't do it, use a small utility knife to carefully cut a slit in the end of the hose and peel it free of the fitting (the hose will then have to be trimmed or replaced upon reconnection).

2. Tag and disconnect the lines from the VRO2 pump assembly as follows:
   a. Normally the fuel outlet line (attached to the top of the pump assembly) is secured using a plastic tie. The tie must be cut and then the line can be carefully pushed off the pump fitting.
   b. The pulse line from the powerhead is secured to the bottom of the pump assembly, at the front of the housing, it is normally secured by a metallic spring-type clamp. Use a pair of pliers to carefully compress the clamp tangs and slide it back over the hose, then push the hose from the pump fitting.
   c. The fuel inlet line is secured to the bottom, center of the pump assembly. It is usually secured by a spring-type metallic clamp and is removed in the same manner as the pulse hose.
   d. The oil inlet line is secured to the bottom, rear of the pump assembly. It is usually secured by a spring-type metallic clamp and is removed in the same manner as the pulse and fuel supply hoses.

3. Remove the pump wiring connector from the retainer and then disengage the connector from the wiring harness.
4. Loosen the pump-to-manifold bracket screws (normally 3). They are located on the back side of the bracket, facing the opposite direction from the Torx® head cover screws located around the perimeter of the pump cover.
5. Carefully remove the pump from the manifold bracket.
6. If the pump is to be overhauled, disassemble it as detailed in this section.

To install:

7. If the pump was overhauled, assemble it as follows:
8. Position the pump to the manifold bracket and install the retaining screws. Tighten the screws to 18-24 inch lbs. (2-3 Nm).
9. Engage the pump connector to the wiring harness, then position the connector into the wire loom retainer.
10. Connect the hoses to the pump assembly as tagged during removal (pump-to-carburetor line at the top, pulse line on bottom furthest from wiring and the fuel supply line bottom center). DO NOT connect the oil supply line at this time. Secure each of the hoses either using plastic ties or metallic spring-type clamps, whichever was used prior to removal. Check each of the metallic spring-type clamps to make sure they have not lost their spring tension and replace if damaged, worn or weak.

■ **The oil supply line should be left disconnected until the oil pump delivery has been confirmed by following the procedure in this section for Checking VRO2 Oil Pump Delivery.**

11. Connect the negative battery cable, then verify proper pump operation using the procedure in this section for Checking VRO2 Oil Pump Delivery.

VRO2 Pump Assembly Overhaul

◆ See Figures 16 thru 30  DIFFICULT

■ **For the various housing screws on most VRO2 pumps you will need Torx® T-10 and T-15 drivers or bits.**

1. Remove the VRO2 pump assembly from the powerhead as detailed in this section.
2. Place the pump assembly on a clean worksurface with the inlet housing assembly (the end of the pump with the wiring) facing downward and the air motor cover facing upward.

※ CAUTION

There is a compressed spring mounted under the air motor cover, be sure to hold the cover downward when removing the covers screws to prevent injury or damage.

3. While holding the air motor cover downward against spring pressure loosen and remove the Torx® head screws securing the cover to the intermediate housing. Once the screws are removed, slowly lift the cover from the housing, releasing spring pressure as it is removed.

■ **As with all overhaul procedures, lay out each component on the worksurface in the order and facing the same direction as it was removed.**

4. Place the cover, screws, vacuum passage O-ring and 2 springs aside.
5. Invert the pump assembly for access to the fuel inlet housing screws. Remove the Torx® head screws securing the fuel inlet housing to the intermediate housing.
6. Lift the fuel inlet housing straight up and off the intermediate housing. The oil piston should remain behind, so pulling the housing upward with withdraw it from the pump.
7. Carefully disconnect the oil piston and link from the fuel piston, then position the components aside.

※ WARNING

Handle the oil piston with great care to keep from denting the piston seal.

8. Hold the intermediate housing with the air motor piston **down**, then push the fuel piston **down**. Turn the air motor piston counterclockwise in order to loosen it, then spin the air motor piston off. The stem will unscrew from either the fuel or air motor piston.

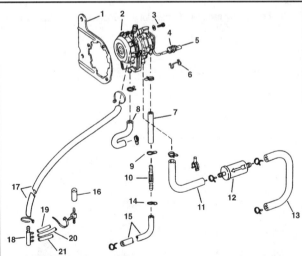

1 - Bracket, Fuel pump to manifold
2 - Fuel pump Assy
3 - Screw, fuel pump to mount
4 - Connector, 4 socket plug
5 - Terminal, socket
6 - Retainer, connector
7 - Hose assy, sight tube
8 - Pulse hose
9 - Clamp, connector
10 - Sight tube, oil
11 - Hose, filter to fuel pump
12 - Fuel filter assy
13 - Hose, fuel connector to filter
14 - Clamp, connector
15 - Hose, connector to sight tube
16 - Hose, fuel tess to middle carb.
17 - Hose, fuel pump to manifold
18 - Fuel, manifold assy
19 - Hose, manifold to upper carb.
20 - Hose, manifold to fuel tee
21 - Hose, manifold to lower carb.

Fig. 15 Exploded view of the VRO2 pump assembly mounting and hose connections

LUBRICATION AND COOLING 5-9

9. Remove the fuel piston and air motor piston from the intermediate housing. Remove the fuel inlet housing O-ring.

10. Remove the stem from either piston. Set the air motor piston aside, then discard the stem and steel washers. Set the nylon valve retainer, flapper valve, fuel piston and stem screw aside.

✱✱ WARNING

DO NOT use pliers to remove a stem unless it is to be replaced.

■ An oil check valve and an oil filter that should not require cleaning or replacement are located under the oil-inlet/electronics module. There are no service parts available for these components.

■ The fuel inlet assembly, oil pulse lever and oil inlet/electronics module are a matched set and must not be mixed with components from another pump.

11. If replaceable fittings are to be removed, proceed as follows:

Fig. 16 You'll need a T-10 or T-15 Torx® head driver to remove the housing cover screws

Fig. 17 Remove the air motor cover slowly, carefully releasing the spring pressure

Fig. 18 Unless there are signs of weepage, there is usually no reason to remove the electronics module from the top of the inlet housing

Fig. 19 Loosen the Torx® head screws securing the inlet housing to the intermediate housing...

Fig. 20 ... then lift the inlet housing, carefully pulling the pump off the oil piston

Fig. 21 Lay out each component on the work surface in the order and facing the same direction as it was removed

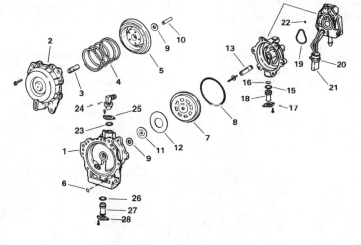

1- Intermediate Housing
2- Motor housing cover
3- Spring, poppet
4- Spring, air motor
5- Diaphragm, air motor
6- O-Ring, vacuum passage
7- Diaphragm, Fuel
8- Seal, fuel inlet housing
9- Washer, piston stem
10- Piston stem
11- Retainer, fuel valve
12- Valve, fuel
13- Piston & link, oil
14- inlet housing assy
15- Seal, fuel inlet
16- Valve, fuel inlet
17- Retainer, fuel inlet
18- Nipple, fuel inlet
19- Seal, Filter housing
20- Connector
21- Socket
22- Seal, actuator shaft
23- O-Ring, fuel outlet
24- Fitting, fuel outlet, V6
25- Retainer, fuel outlet
26- O-Ring, pulse fitting
27- Fitiing, pulse
28- Retainer, pulse fitting

Fig. 22 Exploded view of the VRO2 pump assembly

LUBRICATION AND COOLING

a. If removing a 90 degree fitting (such as the fuel outlet fitting on most models), matchmark or note the direction it faces before removal.

b. Remove the Torx® head screws securing the fitting retainer and fitting to the pump housing.

c. Carefully remove the retainer, fitting and O-ring from the housing. If removing the fuel inlet fitting, remove the valve located under the fitting as well.

To assemble:

12. If one or more of the replaceable fittings were removed, install them as follows:

a. If installing the fuel inlet fitting, position the fuel inlet valve with the hold over the alignment pin.

b. If installing the pulse or fuel inlet fittings, place a drop of clean 2-stroke engine oil on the **thin** O-rings to help ease installation.

c. Put the O-ring in position on the fitting. When installing the **thin** O-rings on the pulse or inlet fittings, they should be placed toward the end of the fitting.

d. Carefully push the fitting into the housing and twist slightly to seat it, then install the retainer.

■ **When thread-forming screws are reinstalled into a plastic component, first make sure the threads are clean and free of all oil. Next, position the screw against the threads and turn the screw backwards (counterclockwise) very slowly by hand until you can feel the screw fall down into the threads slightly, then turn the screw forward until it is finger-tight. Using this method the screw will find the old threads instead of cutting new ones, which might strip.**

e. If a 90 degree fitting must be aligned directly over a retainer screw, apply a drop of oil to the fitting (where it enters the retainer) and position it just past the screw, then tighten the one retainer screw securely. Place a Phillips screwdriver in the end of the 90 degree fitting and carefully rotate it into alignment over, then tighten the other retaining screw.

f. Secure the retainers by tighten the Torx® head screws only until the retainer is bottomed on the housing. DO NOT over-tighten the screws as they will strip easily.

■ **Always use new washers and stem when assembling the air motor and fuel pistons. Install the stem by hand (never use pliers or damage will likely occur).**

13. Working from the hex recess side, carefully insert the stem screw into the fuel piston. Place the flapper valve and valve retainer (with the recess facing outward) onto the screw. Place a new washer in the retainer recess, then thread the new stem onto the screw.

■ **Make sure the hex head of the stem screw is in the hex recess and that the flapper valve and washer are centered on the piston, before finger-tightening the stem.**

14. Apply a light coating of clean 2-stroke engine oil onto the stem, then insert the stem through the seal into the intermediate housing.

15. Place a new washer on the air motor piston screw. Hold the air motor piston under the intermediate housing with the poppet valve pushed fully upward.

■ **The screw hex head will be held in the hex recess ONLY when the poppet valve is pushed fully upward.**

16. Position the air motor piston screw upward against the stem (if necessary, the fuel piston can be pushed downward to extend the stem for better visibility). Hold the air motor screw up against the stem and turn the fuel piston a few revolutions to start the screw in the stem threads.

17. Push downward on the fuel piston while spinning the air motor piston onto the stem.

18. **Very lightly** tighten the 2 pistons against each other in order to tighten the stem on both screws.

■ **Do not risk damaging either piston or the stem by over-tightening the assembly. Besides, there is no need to over-tighten anything since, once the assembly is positioned between the housings, they cannot turn sufficiently to loosen again.**

19. Snap the oil piston and link carefully onto the stem of the fuel piston screw, then place a drop of clean 2-stroke engine oil on the seal at the tip of the piston.

Fig. 23 Remove the oil piston and link from the fuel piston

Fig. 24 Once the air motor and fuel pistons/diaphragms are compressed and unthreaded, carefully pull the air motor diaphragm (with the piston stem in this case) from the intermediate housing...

Fig. 25 ...then inspect the air motor (shown) and fuel diaphragms for damage and replace, as necessary

Fig. 26 During assembly, position O-rings carefully to keep them from getting pinched or cut

Fig. 27 Install the oil piston and link...

Fig. 28 ...then apply a drop of clean 2-stroke engine oil to the piston seal

LUBRICATION AND COOLING

20. Place an O-ring seal in position around the fuel inlet housing flange, then lower the housing over the oil pump piston while inserting the piston into the oil pump bore.
21. Push the fuel inlet housing and O-ring down against the intermediate housing, rotating back-and-forth to seat the O-ring in the groove and align the flange screw holes. Then install and tighten the Torx® head screws using care not to damage the screws or housings.

■ When thread-forming screws are reinstalled into a plastic component, first make sure the threads are clean and free of all oil. Next, position the screw against the threads and turn the screw backwards (counterclockwise) very slowly by hand until you can feel the screw fall down into the threads slightly, then turn the screw forward until it is finger-tight. Using this method the screw will find the old threads instead of cutting new ones, which might strip.

22. Put a light coating of a marine grade grease on the vacuum port O-ring, then place the O-ring in its bore. Place the air motor spring on the piston.
23. Position the poppet spring in the air motor cover (over the center boss), then twist while gently pushing downward to wedge it in place.
24. Position the cover over the air motor piston spring, then push the cover straight downward, compressing the spring and guiding the poppet spring into the center of the piston.
25. Hold the cover in position and carefully start a couple of the cover screws. Verify that the vacuum port O-ring has remained in position, then install the remaining cover screws and gently tighten in a crisscross pattern.
26. Install the VRO2 pump assembly to the powerhead, as detailed in this section.
27. Verify proper pump operation using the procedure in this section for Checking VRO2 Oil Pump Delivery

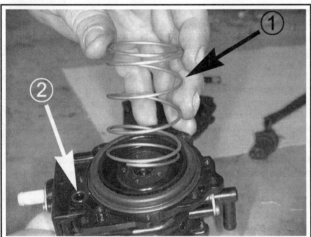

Fig. 29 Position the large spring (1) and the vacuum passage O-ring (2) to the intermediate housing...

Fig. 30 ...then install the cover while carefully compressing the spring and making sure the passage O-ring does not dislodge

Integral Oil Mixing Unit System (25/35 Hp 3-Cylinder Motors)

◆ See Figure 31

The 25/35 Hp (500/565cc) motors are equipped with a unique powerhead mounted oil tank and mixing unit. The system is combined with the motor's fuel delivery system to provide a source of blended oil and fuel to the 3 carburetors.

On these motors, 2 powerhead mounted fuel pumps work in conjunction with each other. One pump draws fuel from the boat's fuel tank and feeds it to the second pump, that pushes it through the oil mixing unit and to the fuel delivery manifold to the carburetors.

The system is protected by a level sensor sending unit that is mounted inside the oil tank. The sending unit circuit remains open when oil level is above the approximate 1/4 tank level. Once the sending unit float drops to or slightly below that level it will complete the **Low Oil** warning circuit for the System Check monitor. When the float drops to the lowest position on the pickup assembly, the sending unit will complete the **No Oil** circuit.

■ For more information on Warning System operation, please refer to the Ignition and Electrical Section.

The mixing unit system is generally reliable and maintenance free as long as the operator keeps an eye on the oil level to confirm proper sending unit operation.

TROUBLESHOOTING THE 25/35 HP MOTOR MIXING UNIT

No information is provided by the manufacturer for mixing unit testing. Although the unit is designed to prevent operation of the motor without proper oiling, care should be taken by the operator to monitor rates of fuel and oil usage to ensure that the system is continuing to operate properly. If performance problems dictate or if trouble is suspected with the mixing unit, the unit can be removed and disassembled for inspection or overhaul.

Sending Unit Resistance Tests

◆ See Figure 32

Sending unit operation is dependent upon the mechanical position of the float on the pickup assembly. If necessary, the unit itself can be checked using an ohmmeter to confirm the warning circuit signals will be closed at the appropriate points.

Although tests can be conduced by siphoning the oil from the tank, then adding oil gradually while watching resistance on the 2 sending unit circuits, it is advisable (and easy enough) to remove the sending unit from the tank for testing. Once the sending unit is removed from the oil reservoir tank you can perform both a visual check, as well as a physical inspection to ensure the float does not stick at any position.

1. Remove the sending unit from the oil reservoir tank, as detailed in this section.
2. Set a DVOM to read on the resistance or continuity scale.
3. First check the **Low Oil** circuit. Connect the meter probes to the terminals B and C (black and tan/black wires) of the sending unit connector, then slide the float by hand to each of the positions pictured in the accompanying illustration and watch for the following results:
   • Float at or above position A (about 1/4 of the distance from the bottom of float travel): meter should show no continuity (high or infinite resistance).
   • Float at position B (slightly below the 1/4 tank line, slightly closer to the bottom of the float travel): meter should show continuity (low or no resistance).
   • Float at position C (float at bottom of travel): meter should show continuity (low or no resistance).
4. Next check the **No Oil** circuit. Connect the meter probes to the terminals B and A (black and tan/yellow wires) of the sending unit connector, then slide the float by hand to each of the positions pictured in the accompanying illustration and watch for the following results:
   • Float at or above position A (about 1/4 of the distance from the bottom of float travel): meter should show no continuity (high or infinite resistance).
   • Float at position B (slightly below the 1/4 tank line, slightly closer to the bottom of the float travel): meter should show no continuity (high or infinite resistance).
   • Float at position C (float at bottom of travel): meter should show continuity (low or no resistance).

5-12 LUBRICATION AND COOLING

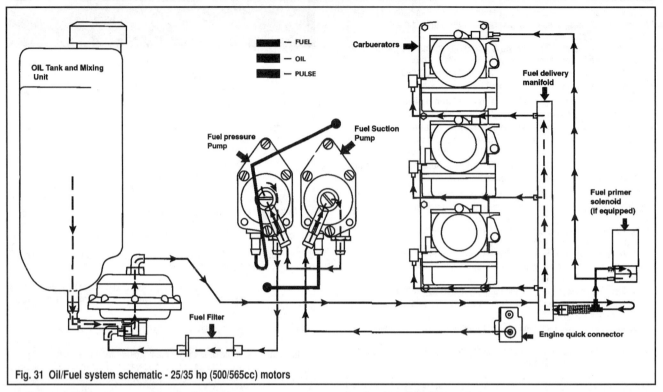

Fig. 31 Oil/Fuel system schematic - 25/35 hp (500/565cc) motors

5. If results differ, the sending unit should be replaced to ensure powerhead protection.

SERVICING THE 25/35 HP MOTOR MIXING UNIT

The sending unit can be removed with the oil tank still installed on the powerhead. Also, the oil tank and mixing unit can be removed from the powerhead as an assembly without draining the oil tank, but if mixing unit is to be disconnected, draining the tank is a good idea (to prevent a mess). There are alternates to draining the oil tank. For one, the oil tank feed line could be plugged with a plastic fitting or capped using a sturdy plastic bag and a wire tie. Also, a spare hose could be placed on the tank outlet fitting, the end of which could be suspended above the oil level in the tank to keep it from draining. But, with all of these alternates, be prepared to catch at least some amount of dripping oil. Keeping the oil tank firmly in place will make escaping oil flow a little slower, but it will flow nonetheless, so work fast if capping the outlet.

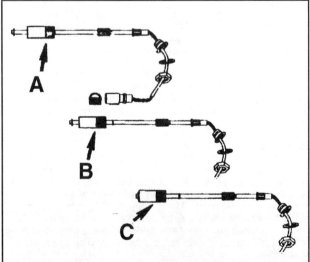

Fig. 32 Float positions for resistance testing, A (open circuits), B (Low Oil circuit closes), C (No Oil circuit closes)

Sending Unit Removal/Installation

◆ See Figure 33

1. Disconnect the negative battery cable for safety.
2. Separate the sending unit 3-pin wiring connector from the retainer, then disengage the connector from the engine harness.
3. Using a backup wrench to keep the sending unit from turning, loosen and remove the nut securing the sending unit to the top of the oil reservoir tank.

■ **If additional clearance is necessary for the wrenches or to withdraw the sending unit from the top of the oil tank, remove the screw fastening the tank to the upper mounting bracket and reposition the tank slightly for better access.**

4. Carefully lift the sending unit from the tank.
5. If necessary, note the positioning of the wires in the 3-pin connector, then use a pin removal tool to release the internal tangs and pull the wires from the connector shell. Pull the nut, washer and grommet off the sending unit lead.
6. Installation is the reverse of the removal. It is usually a good idea to install a new nut, washer and grommet to make certain the sending unit it properly mounted and sealed to the oil tank.
7. When installing the sending unit into the tank, let it rest on the bottom of the reservoir, then install the grommet to the tank, the washer on top of the grommet and thread the nut onto the sending unit (tightening it securely).
8. If the upper oil tank mounting screw was removed, apply OMC Locquic Primer, followed by OMC Ultra Lock (or equivalent threadlocking compounds) to the screw, then install and tighten to 216-240 inch lbs. (24.4-27.1 Nm)
9. Connect the wiring upon completion of assembly. If wires were removed from the connector, make sure they are installed to the proper pins as noted during removal.
10. Connect the negative battery cable.

Oil Tank And Mixing Unit Removal/Installation

◆ See Figures 33 and 34

As stated earlier the oil tank and mixing unit can be removed as an assembly without draining the oil tank. If the mixing unit is to be removed

LUBRICATION AND COOLING 5-13

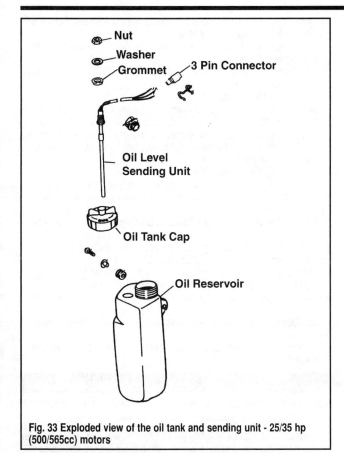

Fig. 33 Exploded view of the oil tank and sending unit - 25/35 hp (500/565cc) motors

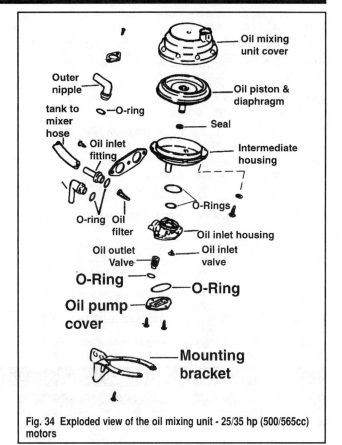

Fig. 34 Exploded view of the oil mixing unit - 25/35 hp (500/565cc) motors

individually or separated from the tank, it is probably best to drain the tank, but the tank outlet line could be capped or plugged as well.

1. Disconnect the negative battery cable for safety.
2. Remove the starboard side lower engine cover. For details, please refer to Engine Covers (Top and Lower Cases), in the Engine Maintenance section.
3. Separate the sending unit 3-pin wiring connector from the retainer, then disengage the connector from the engine harness.
4. Remove the screw fastening the top of the oil reservoir tank to the top bracket.
5. Loosen the fuel filter outlet hose clamp and slide it back on the fuel hose, past the fitting. Carefully push the fuel hose off the filter fitting using a small drain pan or rag to catch any fuel that might leak from the hose and filter.
6. Remove the screw securing the lower oil tank and mixing unit brackets to the powerhead, then pull the bracket away from the powerhead slightly for better access to the mixing unit.
7. Carefully cut the wire tie the secures the fuel hose to the mixing unit outlet nipple, then carefully push the hose off the nipple using a small drain pan or rag to catch any fuel that might leak from the hose and mixing unit.
8. If necessary, separate the mixing unit from the oil tank by carefully cutting the tie strap and carefully pushing the oil hose off the oil tank or mixing unit nipple. Plug the tank to prevent excessive oil dripping/loss or contamination. Overhaul, clean and inspect the mixing unit as detailed in this section or replace it, as necessary.

To install:

9. If removed, connect the mixing unit to the oil tank using the oil feed line, then secure using a new wire tie.
10. Connect the fuel outlet hose to the mixing unit assembly, then secure using a new wire tie.
11. Apply OMC Locquic Primer, followed by OMC Ultra Lock (or equivalent threadlocking compounds) to the lower mixing unit and tank bracket, as well as the upper oil tank bracket, mounting screws.
12. Position the oil reservoir tank and mixing unit assembly to the powerhead, then install tighten the lower bracket mounting screw to 180-204 inch lbs. (20.3-23 Nm)
13. Connect the fuel hose to the outlet side of the fuel filter and secure using the clamp.

14. Route the oil tank sending unit lead behind the upper oil tank bracket, then install and tighten the upper bracket screw to 216-240 inch lbs. (24.4-27.1 Nm)
15. Engage the oil tank sending unit wiring to the engine harness and secure in the retainer.
16. Install the starboard lower engine cover.
17. Connect the negative battery cable.

Mixing Unit Overhaul

♦ See Figure 34

1. Remove the mixing unit from the powerhead and oil tank, as described in this section.
2. Cut the wire tie, then carefully push the short fuel hose from the fuel inlet nipple.
3. To ensure proper alignment of all components during installation matchmark the various covers, fittings and housings before removal. This is especially important for the 90 degree fittings and for the oil pump cover, which must be installed in one direction only.
4. From underneath the mixing unit, remove the 2 screws securing the oil inlet housing to the Intermediate housing.
5. Separate the oil inlet housing from the intermediate housing, then remove and discard the large and small O-rings located between them.
6. Remove the 2 screws securing the oil pump cover to the oil inlet housing, then remove the cover. Remove and discard the elongated cover O-ring.
7. Remove the 3 screws securing the fitting retainer to the oil inlet housing, then separate the retainer from the housing. Remove the oil and fuel inlet fittings from the oil inlet housing and discard the O-rings.
8. Remove and discard the old oil filter.
9. Remove and discard the oil inlet valve.
10. **Push** the oil outlet valve and O-ring free of the housing. Do not **pull** the outlet valve from the housing, as this could damage the valve seat.
11. Remove the 2 screws securing the retainer for the fuel outlet fitting to the cover the oil mixing unit. Remove the retainer and fitting, then remove and discard the O-ring.
12. Pinch and roll the piston seal off the diaphragm and piston assembly.

5-14 LUBRICATION AND COOLING

13. Matchmark the mounting bracket to the mixing unit, then remove the 6 screws and the bracket from the unit.
14. Remove the 2 remaining screws and washers from the oil mixing unit cover, then remove the cover from the intermediate housing. Carefully remove the piston and diaphragm assembly, the small spring and large spring. Discard the old piston and diaphragm assembly.
15. Wash oil inlet housing components using clean solvent, then blow dry with low pressure compressed air (or shake the components free of most solvent and allow to air dry).

■ **Make sure all cleaning, inspection and assembly takes place in a clean, dust free environment.**

16. Clean the cover and oil inlet housing threads using OMC Cleaning Solvent, or an equivalent mild solvent/cleaner.

To assemble

■ **Apply a light coating of clean 2-stroke engine oil to all contact points and replacement components during assembly.**

17. Position the small spring on the end of the short stub on the piston and diaphragm assembly, then place the large spring inside the diaphragm.
18. Rest the diaphragm and piston assembly (with the springs facing upward) inside the intermediate housing.
19. Place the oil mixing unit cover over top of the piston and diaphragm assembly (along with the small and large springs). As you compress the large spring into the cover, make sure the small spring seats over the button on the underside of the cover.
20. Rotate the cover to align the cover and housing exactly as during disassembly, then press them together and secure using the 2 screws and washers. Then install the mixing unit bracket using the 6 screws. Tighten all 8 screws to 5-7 inch lbs. (0.6-0.8 Nm).
21. Using care not to damage the seal, install the oil piston seal into the groove on the oil piston.
22. Position the oil outlet valve into the oil inlet housing and secure using a NEW O-ring.
23. Install a NEW oil inlet valve.
24. Position a NEW elongated O-ring into the groove on the oil pump cover, then install the cover to the oil inlet housing aligning the matchmarks made during removal (this should ensure that the small round depression on the inside of the cover faces the oil inlet valve.) Install the cover screws and tighten to 5-7 inch lbs. (0.6-0.8 Nm).
25. Install a NEW oil inlet filter.
26. Install a NEW O-ring to the oil and fuel inlet ports, then install the fittings (aligning any matchmarks made for 90 degree fittings). Install the fitting retainer and tighten the screws to 5-7 inch lbs. (0.6-0.8 Nm).
27. Install a NEW large and NEW small O-ring into the oil inlet housing.
28. Connect the oil inlet housing to the intermediate housing, making sure the O-rings are not dislodged as they are seated, then install and tighten the retaining screws to 7-9 inch lbs. (0.8-1.0 Nm).
29. Place a NEW O-ring in the fuel outlet port (on top of the oil mixing unit cover), then install the outlet fitting (aligning any matchmarks made earlier). Install the retainer and tighten the screws to 5-7 inch lbs. (0.6-0.8 Nm).
30. Attach the short fuel inlet hose to the fuel inlet nipple, then secure using a new wire-tie.
31. Connect the mixing unit to the oil tank and install to the powerhead, as detailed in this section.

LUBRICATION SYSTEM (4-STROKE MOTORS)

Description and Operation

◆ See Figures 35 thru 39

The forced lubrication system used on Johnson/Evinrude 4-stroke outboards closely resembles the system used on other 4-stroke engines like the one in your car or truck. Its primary components consist of an oil sump located under the powerhead, an oil pickup tube submerged in the sump, an oil pump, an oil filter and various oil passage ways in the engine.

Oil is drawn from the sump by the pump and, on most models is then pumped through a filter, where dirt and metal particles are removed. We say most motors, because some of the smaller carbureted 4-strokes send oil from the pump to various parts of the motor, including the oil filter. It is true, however, that on all models this oil passes through the oil filter before it reaches the crankshaft.

After the pump (and filter in most cases) the oil is then delivered under pressure to various parts of the engine via oil galleys machined into the engine. Oil from the oil pump flows through drilled passages in the camshaft to the rockers and/or valves. The oil also travels through one or more galleries in the cylinder block. On all models, pressurized oil will travel through a passage to the powerhead mounted oil pressure switch, which holds the switch circuit open (preventing the warning system from activating) whenever sufficient oil pressure exists in the passage (hopefully, anytime the engine is running).

The oil pump itself is a Trochoid design and consists of an inner rotor, outer rotor and pump body. Normally, the inner rotor is driven by the camshaft while the outer rotor is free in the pump body and is driven by the inner rotor. As the rotors spin, the volume of the oil between them changes and provides the force to push oil out into the oil gallery. Most oiling systems are protected from over pressurization by a check valve that directs pressure back to the sump if it rises beyond a certain point.

Oil Pressure

TESTING

◆ See Figure 40

If there is any doubt as to whether or not a 4-stroke engine is receiving sufficient oil, the oil pressure should be checked using a suitable oil pressure gauge, 1/8 in. NPT adapter and a tachometer to monitor engine speeds. On carbureted motors, the oil pressure switch is removed and the gauge is installed in the powerhead switch bore, but on EFI motors a dedicated pressure test port is located near the spin-on oil filter.

■ **A gauge with a range of 0-100 psi (0-690 kPa) will be sufficient for all Johnson/Evinrude 4-strokes. A gauge with a more limited operating range may also be used on most carbureted models, but refer to the specifications and make sure there is at least 10% additional capacity in the gauge over the specified oil pressures for the engine being tested.**

To properly test the oil pressure the engine must be run both at idle and at or near Wide Open Throttle (WOT), therefore a flush fitting is not suitable for this test. Either mount the engine in a test tank (and use a suitable test wheel) or launch the craft using an assistant to navigate while you check oil pressures.

Locate the oil pressure switch (carbureted models) or pressure port (EFI models) as follows:
- For 5/6 hp (128cc) models, the switch is located under a rubber boot at the upper port side of the powerhead, immediately in front of the air intake for the carburetor.
- For 8/9.9 hp (211cc) models, the switch is located under a rubber boot, at the upper port side of the powerhead, immediately in front of the oil dipstick and behind the carburetor.
- For 9.9/15 hp (305cc) models, the switch is located under a rubber boot at the upper port side of the powerhead, immediately in front of the oil fill and right above the oil filter.
- For 40/50 hp (815cc) EFI models, the pressure port is located directly below the spin-on oil filter. The port lower engine cover must be removed for access.
- For 70 hp (1298cc) EFI models, the pressure port is located directly above the spin-on oil filter. The starboard lower engine cover must be removed for access.

■ **For carbureted models, verify switch location using the wire harness colors. For details, refer to the schematics provided in the Wiring Diagrams section.**

1. With the engine cold, locate and remove the oil pressure switch or port, as applicable. On carbureted motors, use tape to cover the pressure switch wiring terminal (to prevent it from grounding), then tie the wiring back out of the way of any moving components.
2. Following the tool manufacturer's instructions, connect a mechanical oil pressure gauge using a 1/8 in. NPT adapter in the switch or pressure port.

LUBRICATION AND COOLING 5-15

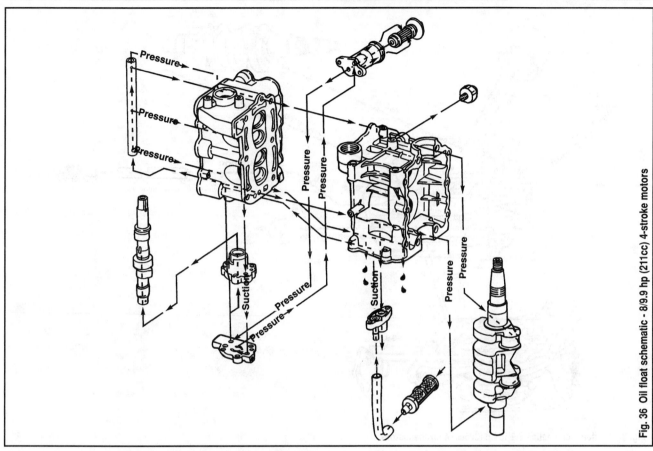

Fig. 36 Oil float schematic - 8/9.9 hp (211cc) 4-stroke motors

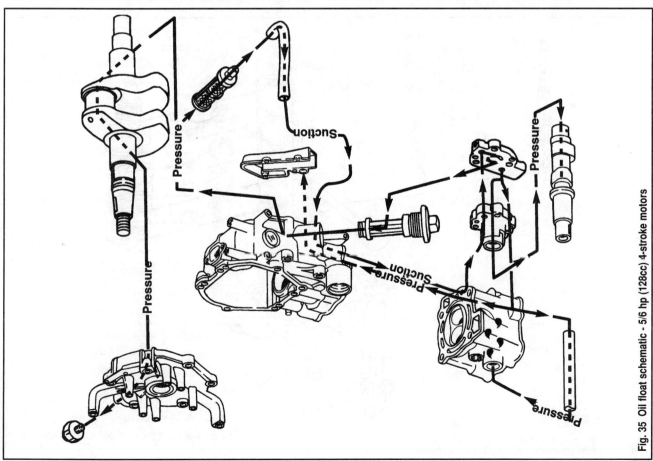

Fig. 35 Oil float schematic - 5/6 hp (128cc) 4-stroke motors

5-16 LUBRICATION AND COOLING

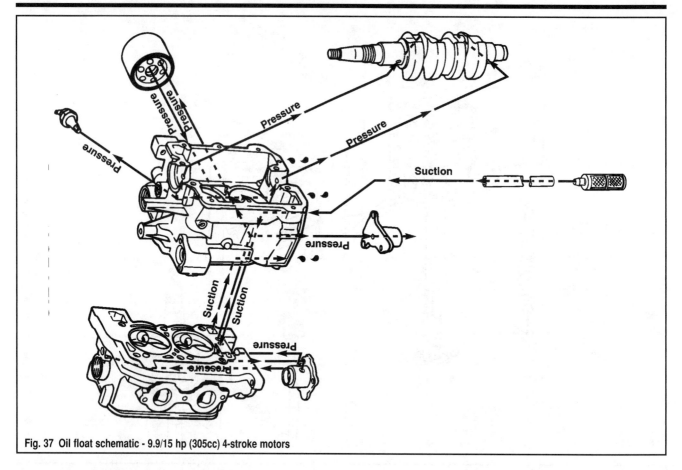
Fig. 37 Oil float schematic - 9.9/15 hp (305cc) 4-stroke motors

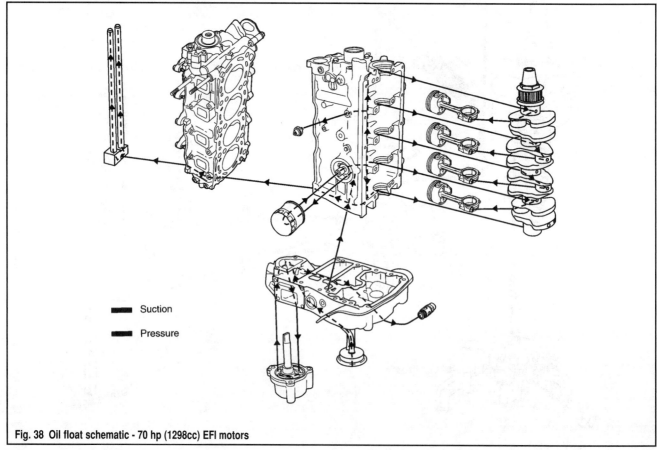
Fig. 38 Oil float schematic - 70 hp (1298cc) EFI motors

LUBRICATION AND COOLING 5-17

6. Allow the engine to idle until it reaches normal operating temperature while continuing to watch the oil gauge. It may take as much as 10 minutes to fully warm the motor.

7. Once the engine is fully warmed, record the oil pressure at the specified engine rpm as follows:
   • For 5/6 hp (128cc) motors pressure must be at least 15 psi (140 kPa) at 900 rpm and 35-40 psi (245-276 kPa) at 4200-5200 rpm.
   • For 8/9.9 hp (211cc) motors pressure must be at least 10 psi (70 kPa) at 900 rpm and 35-40 psi (245-276 kPa) at 4200-5200 rpm.
   • For 9.9/15 hp (305cc) motors pressure must be at least 20 psi (140 kPa) at 2000 rpm and 35-45 psi (245-314 kPa) at 5000-6000 rpm.
   • For 40/50 hp (815cc) motors pressure must be 42-54 psi (295-375 kPa) at 4000 rpm.
   • For 70 hp (1298cc) motors pressure must be 60-70 psi (420-490 kPa) at 3000 rpm.

■ Low oil pressure readings may be the result of low oil level, diluted oil, incorrect type or grade of oil, , a faulty oil pump or damage to the crankshaft and bearings (excessive clearance). High oil pressure readings typically indicate a damaged or clogged oil pressure relief valve, blockage at the powerhead mating surface or in cylinder oil passages. As usual with troubleshooting, inspect the simple things first before suspecting and inspecting the complicated items.

8. Once the oil pressure has been recorded at the appropriate throttle setting(s), slowly return the engine to idle, then shut the engine **OFF**.
9. Remove the oil pressure gauge and adapter, as well as the tachometer.
10. Wrap a layer of Teflon® tape around the oil pressure switch (carbureted motors) or pressure port plug (EFI motors) threads.
11. Carefully thread the switch or port into the opening by hand, then tighten until snug (but do not overtighten and damage the switch/plug or powerhead threads).
12. For carbureted motors, If a torque wrench and sensor socket is available on carbureted motors, tighten the switch to 120-168 inch lbs. (13.5-19 Nm). Then reconnect the switch wiring.

Oil Pump

The exact location and mounting of the oil pump varies slightly by model, and for the same reason, so do the service procedures. For the most part, the oil pump is mounted directly under the powerhead, either under the cylinder head or cylinder block, depending on the model. For all carbureted 4-strokes, as well as 70 hp EFI motors, the pump is driven off the end of the camshaft. The 40/50 hp EFI motors however utilize a pump that is driven off the crankshaft instead of the camshaft.

On all 9.9/15 hp (305cc) and larger 4-stroke motors, the positioning of the pump requires that the powerhead be removed (or conversely, the gearcase and, if equipped, the midsection be removed from the powerhead, usually whichever is preferred will work) for access to the pump assembly. However, on small carbureted 4-strokes, including the 5/6 hp (128cc) and 8/9.9 hp (211cc) models, the pump and mounting bolts can be accessed once the lower engine covers are removed.

REMOVAL, OVERHAUL & INSTALLATION

Oil pump component removal, inspection and installation instructions are provided in this section for the various Johnson/Evinrude 4-stroke motors. However, the manufacturer only has provided inspection and measurement specifications for the 40/50 hp EFI motors. On models where the manufacturer does not make inspection specifications available they advise technicians to disassemble the pump and check visually inspect for obvious signs of damage, contamination or excessive wear. For these models, repair or replace the pump if any of these signs are found, or if an oil pressure test reveals pressure below specification and all other causes (like low oil level, restricted pick-up or excessive bearing clearance) have been ruled out.

5/6 Hp (128cc) 4-Stroke Motors

◆ See Figure 41

The oil pump assembly for the 5/6 hp 4-stroke motors is bolted to the bottom side of the cylinder head so the oil pump shaft can be driven directly off the end of the camshaft. Access is tight, but possible with the powerhead installed.

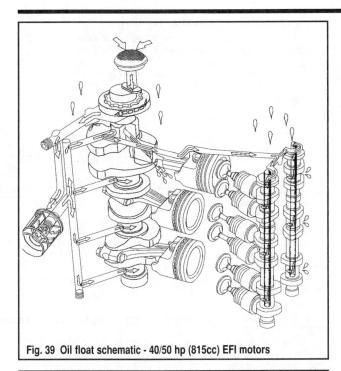

Fig. 39 Oil float schematic - 40/50 hp (815cc) EFI motors

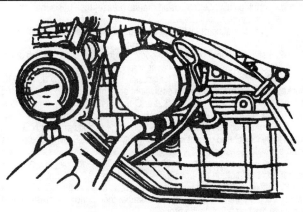

Fig. 40 Use a commonly available oil pressure gauge and a 1/8 in. NPT adapter to measure oil pressure at the oil pressure port

3. Following the tool manufacturer's instructions, connect a suitable shop tachometer.
4. Start the engine while watching the oil pressure gauge. Within 10 seconds of start-up, sufficient pressure must be established in order to warrant continuing to run and test the engine. Cold oil pressure is normally toward the top end of the specified range, and typically, pressure will drop slightly as the engine warms. Either way, sufficient pressure must be available or the engine cannot be safely run/tested.

✼✼ WARNING

Watch the gauge to make sure enough oil pressure is being generated to warrant continuing the test. If this test is a result of an activation of the warning circuit, DO NOT allow the engine to idle until it is warm if there is little or no oil pressure. If there is little or no pressure and the oil level is sufficient, oil pump and related components (pickup tube/inlet screen) must be removed for examination to determine the cause of the problem. Before performing extensive tear-down work, make sure there are no clogged passages (use low pressure compressed air to blow out the oil pressure test port and passage).

5. While the engine idles, check the gauge and fitting for leaks. If present, shut the engine OFF and correct the oil leak before proceeding.

5-18 LUBRICATION AND COOLING

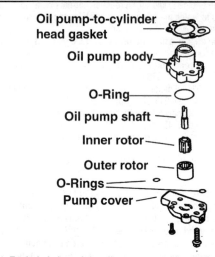

Fig. 41 Exploded view of the oil pump assembly - 5/6 hp (128cc) and 8/9.9 hp (211cc) 4-stroke motors

1. Disconnect the negative battery cable for safety.
2. Remove the lower engine covers for access. For details, please refer to Engine Covers (Top and Lower Cases), in the Engine Maintenance section.
3. Remove the bolts (usually 3) securing the oil pump body to the cylinder head. Remove and discard the old gasket from the mating surface.
4. To disassemble the pump for cleaning, inspection or repair:
   a. Remove the 2 Philips head screws and lockwashers from the top of the pump body.
   b. Lift the pump cover from the body, then remove and discard the cover O-ring along with the inlet and outlet O-rings.
   c. Matchmark the inner and outer rotors to ensure proper assembly, then remove the rotors from the pump body.
   d. Clean the rotors, housing, cover and shaft using a suitable solvent.
   e. Inspect the rotors, housing, cover and oil pump shaft for signs of obvious damage, contamination or excessive wear. Although the O-rings and gaskets are available, the rotors, driveshaft and body are a matched set and are normally replaced only as an assembly. Check with your local parts supplier to determine if a rebuild is kit is available for damaged pumps.

To install:

5. If the pump was disassembled for cleaning and inspection, assemble the pump components as follows:
   a. Apply a light coating of clean 4-stroke engine oil to the pump components, then assemble the inner and outer rotors, aligning the matchmarks made during removal.
   b. Coat the NEW O-rings using a marine grease, then position them on the cover, inlet and outlet fittings.
   c. Apply a coating of OMC Nut Lock, or an equivalent threadlocking compound to the threads of the cover screws.
   d. Install the pump cover to the body, making sure not to disturb or pinch the O-rings, then install the cover screws and tighten securely.
6. Apply a light coating of OMC Gasket Sealing Compound or an equivalent to the threads of the oil pump-to-cylinder head mounting bolts.
7. Install the oil pump to the cylinder head using a new gasket, then tighten the bolts to 84-106 inch lbs. (10-12 Nm).
8. Before installing the lower engine covers, connect the negative battery cable and install an oil pressure gauge, then perform a pressure test to ensure the pump is working properly.
9. After repairs and testing is completed, install the lower engine covers as detailed in the Engine Maintenance section.

8/9.9 Hp (211cc) 4-Stroke Motors

◆ See Figure 41

The oil pump and filter housing assembly for the 8/9.9 hp 4-stroke motors is bolted to the bottom side of the cylinder head so the oil pump shaft can be driven directly off the end of the camshaft. Access is tight, but possible with the powerhead installed.

1. Disconnect the negative battery cable for safety.
2. Remove the lower engine covers for access. For details, please refer to Engine Covers (Top and Lower Cases), in the Engine Maintenance section.
3. Remove the bolts (usually 3) securing the oil pump body and filter assembly to the cylinder head. Remove and discard the old gasket from the mating surface.
4. To disassemble the pump for cleaning, inspection or repair:
   a. Remove the bolts (usually 2) securing the oil filter housing to the pump body. Remove the filter housing from the pump, then remove and discard the O-ring.
   b. Remove the slotted screws and lockwashers (usually 2) securing the pump body cover.
   c. Lift the pump cover from the body, then remove and discard the cover O-ring along with the inlet and outlet O-rings.
   d. Matchmark the inner and outer rotors to ensure proper assembly, then remove the rotors from the pump body.
   e. Clean the rotors, housing, cover and shaft using a suitable solvent.
   f. Inspect the rotors, housing, cover and oil pump shaft for signs of obvious damage, contamination or excessive wear. Although the O-rings and gaskets are available, the rotors, driveshaft and body are a matched set and are normally replaced only as an assembly. Check with your local parts supplier to determine if a rebuild is kit is available for damaged pumps.

To install:

5. If the pump was disassembled for cleaning and inspection, assemble the pump components as follows:
   a. Apply a light coating of clean 4-stroke engine oil to the pump components, then assemble the inner and outer rotors, aligning the matchmarks made during removal.
   b. Coat the NEW O-rings using marine grease, then position them on the cover, inlet and outlet fittings.
   c. Apply a coating of OMC Nut Lock, or an equivalent threadlocking compound to the threads of the cover screws.
   d. Install the pump cover to the body, making sure not to disturb or pinch the O-rings, then install the cover screws and tighten securely.
   e. Coat the pump body-to-filter housing bolt threads lightly using OMC Gasket Sealing Compound or an equivalent sealant. Install the filter housing to the pump body using a new gasket or O-ring (as applicable), then tighten the bolts to 84-106 inch lbs. (10-12 Nm).
6. Coat the oil pump assembly-to-cylinder head bolts and the NEW pump-to-cylinder head gasket lightly using OMC Gasket Sealing Compound or an equivalent sealant.
7. Install the oil pump assembly to the cylinder head using the new gasket and then tighten the bolts to 84-106 inch lbs. (10-12 Nm).
8. Before installing the lower engine covers, connect the negative battery cable and install an oil pressure gauge, then perform a pressure test to ensure the pump is working properly.
9. After repairs and testing is completed, install the lower engine covers as detailed in the Engine Maintenance section.

9.9/15 Hp (305cc) 4-Stroke Motors

◆ See Figure 42

The oil pump assembly for the 9.9/15 hp 4-stroke motors is bolted to the bottom side of the cylinder head so the oil pump shaft can be driven directly off the end of the camshaft. A separate oil filter adapter is mounted to the side of the powerhead. Access is too tight when the powerhead is installed on the gearcase, so the powerhead must be removed (or alternated, the gearcase can be removed from the powerhead, as desired).

1. Disconnect the negative battery cable for safety.
2. For access, remove either the powerhead from the gearcase or the gearcase from the powerhead, as desired. For details refer to the appropriate procedures in the Powerhead and Gearcase sections.
3. Remove the bolts (usually 3) securing the pump cover and assembly to the bottom end of the cylinder head. This will free both the cover and the body.
4. Lift the pump cover from the body, then remove and discard the cover O-ring.
5. Matchmark the pump body, inner and outer rotors to ensure proper assembly, then remove body from the cylinder head along with the rotors and the pump shaft.
6. Clean the rotors, housing, cover and shaft using a suitable solvent.
7. Inspect the rotors, housing, cover and oil pump shaft for signs of obvious damage, contamination or excessive wear. Each of the pump

LUBRICATION AND COOLING 5-19

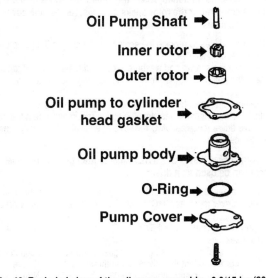

Fig. 42 Exploded view of the oil pump assembly - 9.9/15 hp (305cc) 4-stroke motors

components, including matched rotor sets should be available from your parts dealer. Replace worn or damaged components or the pump assembly, as necessary and desired.

To install:

8. Apply a light coating of clean 4-stroke engine oil to the inner and outer rotors and the oil pump shaft.

9. Install the pump shaft along with the inner and outer rotors, aligning the matchmarks made during removal.

10. Coat the new pump cover O-ring using a suitable marine grease, then position it and the cover over the pump body. Put the assembly down on a work surface either sitting on the cover or sitting on the housing with the cover straight up so it will not dislodge.

11. Apply a light coating of OMC Gasket Sealing Compound to the threads of the pump-to-cylinder head bolts and to the NEW pump body-to-cylinder head gasket.

12. Pick up the pump assembly, holding the cover firmly in place, then position the new gasket over the bottom of the pump body while aligning the bolt holes.

13. Position the pump body to the cylinder head by aligning the pin and slot in the camshaft and oil pump shaft.

14. Make sure the gasket and bolt holes of the cover, housing and cylinder head area all aligned, then install the cover bolts and tighten to 84-106 inch lbs. (10-12 Nm).

15. Before installing the lower engine covers, connect the negative battery cable and install an oil pressure gauge, then perform a pressure test to ensure the pump is working properly.

16. After repairs and testing is completed, install the lower engine covers as detailed in the Engine Maintenance section.

40/50 Hp (815cc) EFI Motors

◆ See Figures 43, 44 and 45

On 40/50 hp EFI motors the powerhead assembly is mounted to a midsection and engine holder that contains the oil pan (sump), oil pump pickup and filter screen. The engine mounts and gearcase are also secured to the midsection assembly.

The oil pump for these models is contained in an oil pump case that is bolted to the bottom of the powerhead, between the powerhead and the engine holder. Once the oil pump case is removed from the powerhead the oil pump cover can be removed for access to the rotors.

1. Disconnect the negative battery cable for safety.

2. For access, remove the powerhead from the gearcase. For details refer to the procedures in the Powerhead section.

3. Remove the bolts and carefully pry the oil pump cover off the bottom of the powerhead.

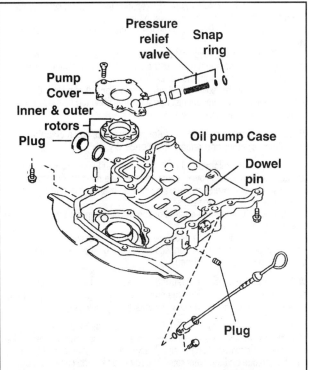

Fig. 43 Exploded view of the oil pump assembly - 40/50 hp (815cc) EFI motors

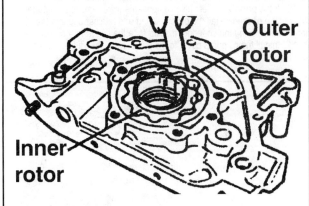

Fig. 44 Measure clearance between the outer rotor and the pump case

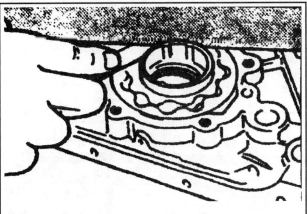

Fig. 45 Measure the oil pump side clearance between a straightedge and outer rotor

5-20 LUBRICATION AND COOLING

4. Remove the screws (usually 6) securing the oil pump cover to the pump case, then lift the oil pump cover off the case.
5. Matchmark the rotors, then lift them inner and outer rotor from the pump case.
6. If necessary disassemble the pressure relieve valve by removing the snapring from the bore in the side of the cover, then withdrawing the retainer cap, spring and piston from the bore.
7. Use a suitable solvent to clean the rotors and, if removed, the pressure relief valve components. Use a lint free shop rag soaked in solvent to wipe the cover and pump case clean.

■ Be careful not to remove the rotor matchmarks when cleaning or keep them together and remark them once dry.

8. Inspect the rotors and the oil pump housing for worn, discolored or damaged surfaces. If removed, check the pressure relief valve components for signs of wear (scoring on the piston or a weak spring) or damage. Replace all suspect components.
9. Place the inner and outer rotor into the oil pump housing, positioned to align the matchmarks.
10. Using a feeler gauge, measure the clearance between the outer rotor and the oil pump case. The rotors and case must be replaced if clearance is greater than 0.0122 in. (0.31mm).

■ Remember that when measuring clearance with a feeler gauge, the specified size (or smaller) gauge must pass through with a slight drag. A larger sized gauge must not fit or clearance is excessive.

11. Measure the rotor-to-cover side clearance using a precision straightedge. Position the straightedge across the case and both rotors, then, while pushing downward on the straightedge, measure clearance between the straightedge and the outer rotor. The rotors and pump case must be replaced if clearance is greater than 0.0059 in. (0.15mm).

To install:
12. Coat the inner and outer rotors with clean 4-stroke engine oil. Position them in the case aligning the matchmarks made earlier.
13. If the pressure relief valve components were removed, coat them with engine oil then install the piston, spring and retainer cap into the cover bore. Secure using the snapring, making sure the ring is fully seated.
14. Install the oil pump cover to the pump case, then tighten the screws securely.

■ Once the cover is tightened, rotate the inner rotor by hand and make sure if turns smoothly. If the rotor sticks or binds, remove the cover and check for causes.

15. It is a good idea to replace the seals before installing the pump case, if desired, proceed as follows:
  a. Remove the screws and washers securing the seal housing to the underside of the pump case, then remove the housing and discard the gasket.
  b. Remove the seals from the seal housing and pump case.
  c. Install a new housing seal with the spring facing away from the tool. Use a suitable driver (such as OMC #319875 or equivalent) to gently press the seal into position.

■ A smooth socket that is SLIGHTLY smaller than the outer-diameter of the seal can be used as a driver.

  d. Install a new case seal with the spring facing away from the tool. Use a suitable driver (such as OMC #326690 or equivalent) to gently press the seal into position.
  e. Apply a light coating of OMC Triple-Guard or an equivalent marine grease to the seal lips
  f. Install the seal housing using a new gasket. Apply a light coating of OMC Nut Lock or equivalent threadlock to the threads of the seal cover screws. Install the screws and tighten to 89-96 inch lbs. (9.5-11 Nm).
16. Apply a light coating of OMC Gasket Sealing Compound to the threads of the case screws and to the mating surface of the case and powerhead.
17. Apply a light coating of OMC Gel-Seal II or equivalent to the parting line of the cylinder block/head and to the mating flange of the oil pump case.
18. Install the oil pump case to the bottom of the powerhead while rotating the crankshaft slightly to align the drive with the oil pump.
19. Install and tighten the cover screws to 16-18 ft. lbs. (22-24 Nm) using a star-shaped crossing pattern starting from the inner fasteners and working outward.

70 Hp (1298cc) EFI Motors
◆ See Figure 46

On 70 hp EFI motors the powerhead assembly is mounted to a midsection and engine holder. The midsection is also the oil pan or sump,

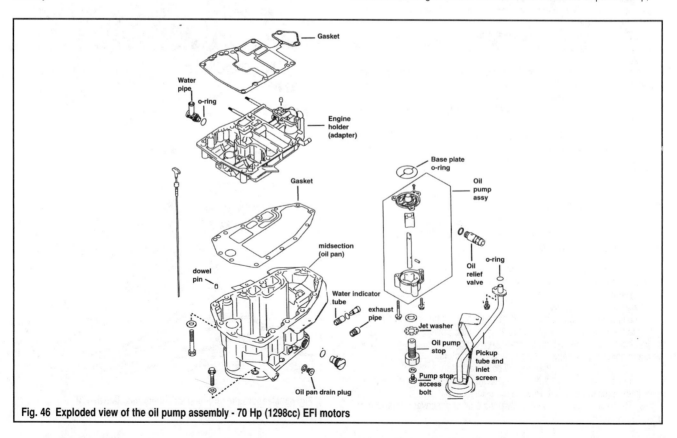

Fig. 46 Exploded view of the oil pump assembly - 70 Hp (1298cc) EFI motors

LUBRICATION AND COOLING 5-21

and houses the oil pump pickup with filter screen that is bolted to the engine holder. The oil pump assembly is also bolted to the bottom of the engine holder. An exhaust housing is bolted to the underside of the midsection (the gearcase then bolts to the exhaust housing).

Although the manufacturer provides procedures for removing the oil pump and components, it does not provide any information for inspection. Furthermore, parts books do not list individual components for replacement, so it is possible that only complete replacement pumps are available. Should a pressure check or a visual inspection of the pump components reveal problems, check with your local parts dealer to see what is currently available for pump repair or replacement.

※※ WARNING

To prevent damage to the camshaft or the oil pump driveshaft, the oil pump shaft must be properly aligned to the camshaft slot during powerhead installation. An oil pump stop is threaded into the midsection. A smaller stop bolt located in the large threaded stop is removed to allow rotation of the oil pump through the midsection as the powerhead is installed. For more details, refer to the Powerhead removal and installation procedure in the Powerhead section.

1. Disconnect the negative battery cable for safety.
2. For access, remove the powerhead from the gearcase. For details refer to the procedures in the Powerhead section.

■ **The oil pump is bolted to the bottom of the engine holder or engine adapter found on top of the midsection.**

3. Remove the bolts (usually 6) securing the engine holder to the top of the midsection (oil pan) assembly. The screws are threaded from underneath the midsection flange, upward into the engine holder.
4. Carefully break the gasket seal and lift the engine holder from the top of the midsection. The oil pump and the pickup tube are attached to the underside of the holder.
5. Invert the holder and place it carefully on a clean worksurface.
6. Inspect the oil pump pickup tube and inlet screen. If necessary the tube and screen can be removed for cleaning or replacement at this time.

■ **Before removing the pump rotors either identify the alignment marks on the outer and inner rotors or by matchmarking them to ensure proper installation. (Small dots are machined into the surface of one rotor "tooth" for each component. During assembly, these small dots must be aligned on the same side).**

7. Remove the oil pump cover-to-engine holder retaining screws (usually 3), then remove the cover, followed by the pump components (driveshaft, rotors and base plate). Remove and discard the old base plate O-ring.
8. Carefully remove all gasket material from the midsection (oil pan) and engine adapter.
9. Use a suitable solvent to clean the rotors, cover and base plate.

■ **Be careful not to remove the rotor matchmarks when cleaning or keep them together and remark them once dry.**

10. Inspect the rotors, cover and base plate for worn, discolored or damaged surfaces.

To install:
11. Place a NEW O-ring on the oil pump base plate.
12. Coat the inner and outer rotors with clean 4-stroke engine oil.
13. Install the oil pump components (aligning the matchmarks or the machined marks on the rotors) along with the oil pump cover.
14. Install the oil pump cover screws and tighten to 80-90 inch lbs. (9-10 Nm)
15. If removed, install the oil pump pickup tube and inlet screen using a NEW O-ring. Tighten the retaining bolt to 80-90 inch lbs. (9-10 Nm).
16. Install the engine adapter to the midsection using a new gasket. Install and tighten the retaining bolts (usually 6 located under the midsection flange) to 8-12 ft. lbs. (10.8-16.2 Nm).
17. Install the powerhead while carefully aligning the camshaft drive slot with the oil pump driveshaft. For details, please refer to the procedures in the Powerhead section.

Oil Pressure Switch and Warning System

For information on testing the oil pressure switch and warning LED and/or horn/buzzer circuitry, please refer to Warning System in the Ignition and Electrical System section.

COOLING SYSTEM

◆ See Figure 47

■ **For specific water flow diagrams showing the cooling system components and water passages for each Johnson/Evinrude outboard, please refer to Cooling System Schematics in this section.**

All Johnson/Evinrude outboard engines are equipped with a raw water cooling system, meaning that sea, lake or river water is drawn through a water intake in the gearcase lower unit and pumped through the powerhead by a water pump impeller. The exact mounting and location of the pump/impeller varies slightly on some models, but with only a few exceptions, it is mounted to the lower unit along the gearcase-to-intermediate section split line.

For many boaters, annual replacement of the water pump impeller is considered cheap insurance for a trouble-free boating season. This is probably a bit too conservative for most people, but after a number of trouble-free seasons, an impeller doesn't owe you anything and you should consider taking the time and a little bit of money necessary to replace it. Remember that should an impeller fail you'll be stranded. Worse, a worn impeller will simply supply less cooling water than required by specification, allowing the powerhead to run hot placing unnecessary stress on components and best or risking overheating the powerhead at worst.

All 5 hp and larger motors (except the single-cylinder 5 hp 2-stroke) are equipped with a thermostat that restricts the amount of cooling water allowed into the powerhead until the powerhead reaches normal operating temperature. The purpose of the thermostat is to increase engine performance and reduce emissions by making sure the engine warms as quickly as possible to operating temperature and remains there during use under all conditions. Running a motor without a thermostat may prevent it from fully warming, not only increasing emissions and reducing fuel economy, but it will likely lead to carbon fouling, stumbling and poor performance in general. It can even damage the motor, especially if the

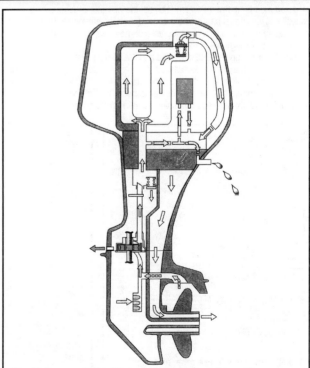

Fig. 47 Cut-away view of a typical outboard cooling system showing water flow - 40/50 hp EFI motors shown

5-22 LUBRICATION AND COOLING

motor is then run under load (such as full-throttle operation) without allowing it to thoroughly warm). A restricted thermostat can promote engine overheating. The good news is that should you be caught on the water with a restricted thermostat, you should be able to easily remove it and get back to shore, just make sure you replace it before the next outing.

The water intake grate and cooling passages throughout the powerhead and gearcase comprise the balance of the cooling system. Both components require the most simple, but most frequent maintenance to ensure proper cooling system operation. The water intake grate should be inspected before and after each outing to make sure it is not clogged or damaged. A damaged grate could allow debris into the motor that could clog passages or damage the water pump impeller (both conditions could lead to overheating the powerhead). Cooling passages have the tendency to become clogged gradually over time by debris and corrosion. The best way to prevent this is to flush the cooling system **after each use** regardless of where you boat (salt or freshwater). But obviously, this form of maintenance is even more important on vessels used in salt, brackish or polluted waters that will promote internal corrosion of the cooling passages.

Description and Operation

◆ See Figures 48 thru 52

The water pump uses an impeller driven by the driveshaft (on all but the smallest models, where it is driven by the propshaft) sealing between an offset housing and lower plate to create a flexing of the impeller blades. The rubber impeller inside the pump maintains an equal volume of water flow at most operating speeds.

At low speeds the pump acts like a full displacement pump with the longer impeller blades following the contour of the pump housing. As pump speed increases, and because of resistance to the flow of water, the impellers bend back away from the pump housing and the pump acts like a centrifugal pump. If the impeller blades are short, they remain in contact throughout the full RPM range, supplying full pressure.

✱✱ WARNING

The outboard should never be run without water, not even for a moment. As the dry impeller tips come in contact with the pump housing or insert, the impeller will be damaged. In most cases, damage will occur to the impeller in seconds.

On most powerheads, if the powerhead overheats, a warning circuit is triggered by a temperature switch to signal the operator of an overheat condition. This should happen before major damage can occur. Reasons for overheating can be as simple as a plastic bag over the water inlet, or as serious as a leaking head gasket.

Whenever the powerhead is started and the cooling system begins pumping water through the powerhead, a water indicator stream will appear from a cooling system indicator in the engine cover. The water stream fitting commonly becomes blocked with debris (especially when lazy operators fail to flush the system after each use, yes we said LAZY, does this mean

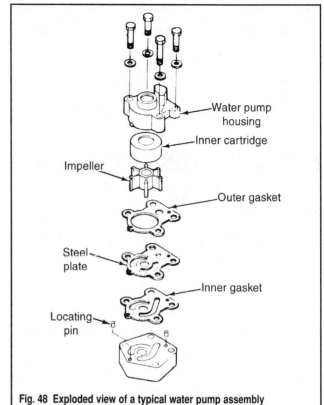

Fig. 48 Exploded view of a typical water pump assembly

YOU?) and ceases flowing. This leads one to suspect a cooling system malfunction. Clean the opening in the fitting using a stiff piece of wire before testing or inspecting other cooling system components.

EFI motors are equipped with a water pressure relief valve that allows additional water flow at higher engine speeds by providing an additional exit passage. Increased pump flow and pressure at higher engine speeds causes the valve to open.

Whenever water is pumped through the powerhead it absorbs and removes excessive heat. This means that anytime a motor begins to overheat, there is not enough (or no) water flowing to the powerhead. This can happen for various reasons, including a damaged or worn impeller, clogged intake or passages or a stuck closed/restricted thermostat. A sometimes overlooked cause of overheating is the inability of the linings of the cooling passage to conduct heat. Over time, large amounts of corrosion deposits will form, especially on engines that have not received sufficient maintenance. Corrosion deposits can insulate the powerhead passages from the raw water flowing through them.

Fig. 49 The water intake grate and cooling passages should be checked and cleaned with each use

Fig. 50 Most water pumps are mounted to the top of the gearcase lower unit

Fig. 51 The water pump impeller is the heart of the cooling system

LUBRICATION AND COOLING 5-23

Troubleshooting the Cooling System

◆ See Figures 48 thru 52

■ **When troubleshooting the cooling system, especially for overheat conditions, the motor should be run in a test tank or on a launched vessel (to simulate normal running conditions.). Running the motor on a flushing device may provide both a higher volume of water than the system would deliver (and often times water that is much colder to begin with).**

A water-cooled powerhead has a lot of problems to consider when talking about overheating. The most overlooked tends to be the simplest, clogged cooling passages or water intake grate. Although a visual inspection of the intake grate will go a long way, the cooling passage condition can really only be checked by operating the motor or disassembling it to observe the passages.

Damaged or worn cooling system components tend to cause most other problems. And since there are relatively few components, they are easy to discuss. The most obvious is a thermostat that is damaged or corroded will often cause the motor to run hot or cold (depending on the position in which the thermostat is stuck, closed or open). The water pump impeller is really the heart of the cooling system and it is easy to check, easy that is once it is accessed.

Periodic inspection and replacement of the water pump impeller is a mainstay for many mariners. There are those that wouldn't really consider launching their vessel at the beginning of a season without this. If the water pump is removed for inspection, check the impeller and housing for wear, grooves or scoring that might prevent proper sealing. Check for grooves in the driveshaft where the seal rides. Any damage in these areas may cause air or exhaust gases to be drawn into the pump, putting bubbles into the water. In this case, air does not aid in cooling. When inspecting the pump, consider the following:

• Is the pump inlet clear and clean of foreign material or marine growth? Check that the inlet screen is totally open. How about the impeller?

• Try and separate the impeller hub from the rubber. If it shows signs of loosening or cracking away from the hub, replace the impeller.

• Has the impeller taken a set, and are the blade tips worn down or do they look burned? Are the side sealing rings on the impeller worn away? If so, replace the impeller.

Remember that the life of the powerhead depends on this pump, so don't reuse any parts that look damaged. Are any parts of the impeller missing! If so, they must be found. Broken pieces will migrate up the water tube into the water jacket passages and cause a restriction that could block a water passage. It can be expensive or time consuming to locate the broken pieces in the water passages, but they must be found, or major damage could occur.

■ **Truth of the matter is that pump impeller materials have gotten better over the years, and some manufacturers even recommend going as long as 3 years on one impeller, provided the number of hours in each season are not excessive. We can't tell you how to make this decision, other than to know that you're taking a calculated risk which COULD cause you to tear down the powerhead or worse if you loose the bet.**

The best insurance against breaking the impeller is to replace it at the beginning of each boating season (are you sensing a pattern here?), and to NEVER run it out of the water. If installing a metal-bodied pump housing, coat all screws with non-hardening sealing compound to retard galvanic corrosion. The water tube carries the water from the pump to the powerhead. Grommets seal the water tube to the water pump and exhaust housing at each end of the tube, and can deteriorate. Also, the water tube(s) should be checked for holes through the side of the tube, for restrictions, dents, or kinks.

Overheating at high RPM, but not under light load, may indicate a leaking head gasket. If a head gasket is leaking, water can go into the cylinder, or hot exhaust gases may go into the water jacket, creating exhaust bubbles and excessive heat. Remember that aluminum heads have a tendency to warp, and usually need to be surfaced each time they are removed. If necessary, they can be resurfaces by using emery paper and a surface block moving in a figure-eight motion. Also, inspect the cylinders and pistons for damage. Other areas to consider are the exhaust cover gaskets and plate. Look for corrosion pin holes. This is rare, but if the outboard has been operated in salt water over the years, there may just be a problem.

After 60 seconds at 1500 rpm.

After 90 seconds at 1500 rpm.

After 30 seconds at 2000 rpm.

After 45 seconds at 2000 rpm.

After 60 seconds at 2000 rpm

Fig. 52 NEVER run the motor without a source of cooling water, an impeller can be destroyed in less than a minute

5-24 LUBRICATION AND COOLING

If the outboard is mounted too high on the transom, air may be drawn into the water inlet or sufficient water may not be available at the water inlet. When underway the outboard anti-ventilation plate should be running at or near the bottom of the boat and parallel to the surface of the water. This will allow undisturbed water to come to the lower unit, and the water pick-up should be able to draw sufficient water for proper cooling.

Whenever the outboard has been run in polluted, brackish or saltwater, the cooling system should be flushed. Follow the instructions provided under Flushing the Cooling System in the Engine Maintenance section for more details. But in most cases, the outboard flushed for at least five minutes. This will wash the salt from the castings and reduce internal corrosion. If the outboard is small and there is no flushing tool that will fit, run the outboard in a tank, drum, or bucket.

There is no need to run in gear during the flushing operation. After the flushing job is done, rinse the external parts of the outboard off to remove the salt spray.

When service work is done on the water pump or lower unit, all the bolts that attach the lower unit to the exhaust housing, and bolts that hold the water pump housing (unless otherwise specified), should be coated with nonhardening gasket sealing compound to guard against corrosion. If this is not done, the bolts may become seized by galvanic corrosion and may become extremely difficult to remove the next time service work is performed.

Last but not least, check to be sure the overheat warning system is working properly. By grounding the wire at the sending unit, the horn should sound, and/or a light should turn on. More details can be found under Warning System in the Ignition and Electrical System section.

■ On EFI motors, refer to the troubleshooting information found in the Electronic Fuel Injection (EFI) section for more hints on troubleshooting an overheating motor.

TESTING COOLING SYSTEM EFFICIENCY

◆ See Figure 53

If trouble is suspected, cooling system efficiency can be checked by running the motor in a test tank or on a launched boat (while an assistant navigates) and monitoring cylinder head temperatures. There are 2 common methods available to monitor cylinder head temperature, the use of a heat sensitive marker or an electronic pyrometer.

■ Because of the importance of accurate testing and the range of operating temperatures that can be checked, the manufacturer does not provide information for using a temperature sensitive marker on EFI motors. So for EFI motors go the extra mile and obtain an hand-held pyrometer.

The Stevens Instrument company markets a product known as the Markal Thermomelt Stik®. This is a physical marker that can be purchased to check different heat ranges. The marker is designed to leave a chalky mark behind on a part of the motor that will remain chalky until it is warmed to a specific temperature, at which point the mark will melt appearing liquid and glossy. When using a Thermomelt Stik or equivalent indicator, markers of 2 different heat specifications are necessary for this test. For all models you will want a 163°F (73°C) marker to check for overheating. You will also need either a 100°F (38°C) marker for 8 hp and smaller 2-strokes or a 125° (52°C) marker for all other motors to determine if the motor is failing to reach normal operating temperature.

Alternately, an electronic pyrometer may be used. Many DVOMs are available with thermo-sensor adapters that can be touched to the cylinder head in order to get a reading. Also, some instrument companies are now producing relatively inexpensive infra-red pyrometers (such as the Raytek® MiniTemp®) of a point-and-shoot design. These units are simply pointed toward the cylinder head while holding down the trigger and the electronic display will give cylinder head temperature. For ease of use and relative accuracy of information, it is hard to beat these infra-red pyrometers. Be sure to follow the tool manufacturer's instructions closely when using any pyrometer to ensure accurate readings.

To test the cooling system efficiency, obtain either a Thermomelt Stik (or equivalent temperature indicating marker) or a pyrometer and proceed as follows:

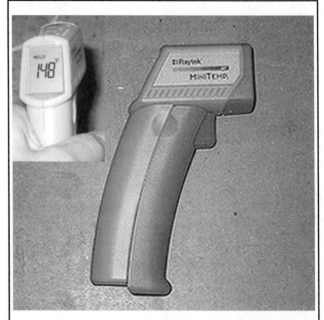

Fig. 53 By far the easiest way to check cylinder head temperature is with a hand-held pyrometer like the MiniTemp® from Raytek® pictured here

1. If available, install a shop tachometer to gauge engine speed during the test.
2. Make sure the proper propeller or test wheel is installed on the motor.
3. Place the motor in a test tank or on a launched craft.

■ In order to ensure proper readings, water temperature must be approximately 60-80°F (18-24°C).

4. Start and run the engine at 3000 rpm for **at least** five minutes.
5. Reduce engine speed to about 900 rpm as proceed as follows depending on the test equipment:
• If using Thermo-melt Stiks, make 2 marks on the cylinder head, one with the low-range marker and one with the high-range marker. Continue to operate the motor at 900 rpm. The low-range mark must turn liquid and glossy or the engine is being overcooled (if equipped, check the thermostat for a stuck open condition). The high-range mark must remain chalky, or the motor is overheating (if equipped, check the thermostat for a stuck closed condition and then check the cooling system passages and the water pump impeller).

■ For most models, temperature readings should be taken on the cylinder head. For EFI motors, the manufacturer specifically recommends taking temperature readings on the top of the thermostat housing. For Colt/Junior models the readings should be taken on the crankcase itself, right behind the point where the cylinder head attaches.

• If using a pyrometer, take temperature readings on the cylinder head. Temperature readings must be 125-155°F (53-67°C) for carbureted motors or 126-134°F (48-52°C) for EFI motors otherwise the engine is being over/under cooled. If equipped, check the thermostat first for either condition and then suspect the cooling water passages and/or the impeller.

6. Increase engine speed to 5000 rpm and continue to watch the markers or the reading on the pyrometer. The engine must not overheat at this speed either or the system components must be examined further.

■ When checking the engine at speed (5000 rpm) expect temperatures to vary slightly from the idle test. Some models will run slightly hotter and some slightly cooler due to the differences in volume of water delivered by the cooling system when compared with engine load. For 25-55 hp (737cc) motors engine temperatures should not exceed 120°F (50°C). For EFI motors, temperatures at speed must NOT exceed 134°F (52°C). On all other models, temperatures at speed must be exceed 160°F (71°C).

LUBRICATION AND COOLING 5-25

TESTING THE THERMOSTAT

All 5 hp and larger motors (except the single-cylinder 5 hp 2-stroke) are equipped with a thermostat that restricts the amount of cooling water allowed into the powerhead until the powerhead reaches normal operating temperature. The purpose of the thermostat is to prevent cooling water from reaching the powerhead until the powerhead has warmed to normal operating temperature. In doing this the thermostat will increase engine performance and reduce emissions.

However, this means that the thermostat is vitally important to proper cooling system operation. A thermostat can fail by seizing in either the open or closed positions, or it can, due to wear or deterioration, open or close at the wrong time. All failures would potentially affect engine operation.

A thermostat that is stuck open or will not fully close, may prevent a powerhead from ever fully warming, this could lead to carbon fouling, stumbling, hesitation and all around poor performance. Although these symptoms could occur at any speed, they are more likely to affect most motors at idle when high water flow through the open orifice will allow for more cooling than the lower production of heat in the powerhead requires.

A thermostat that is stuck closed will usually reveal itself right away as the engine will not only come up to temperature quickly, but the temperature warning circuit should be triggered shortly thereafter. However, thermostat that is stuck partially closed may be harder to notice. Cooling water may reach the powerhead and keep it within normal operating temperature range at various engine rpm, but allow heat to build up at other rpm. Generally speaking, engines suffering from this type of thermostat failure will show symptoms a part or full throttle, but problems can occur at idle as well. Symptoms, besides overheating, may include hesitation, stumbling, increased noise and smoke from the motor and, general, poor performance.

Testing a thermostat is a relatively easy proposition. Simple remove the thermostat from the powerhead and suspend it in a container of water, then heat the water watching for the thermostat element to move (open) and noting at what temperature it accomplishes this. Unfortunately, some of the thermostats used on carbureted Johnson/Evinrude engines are assembled in the thermostat housing on the powerhead. On some of these motors, changes in the thermostat element (or vernatherm) may not be obvious. If you suspect a faulty or inoperable thermostat and cannot seem to verify proper opening/closing temperatures, it may be a good idea (especially since you've already gone through the trouble of removing the thermostat) to simply replace it (it's a relatively low cost part, that performs an important function). Doing so should remove it from suspicion for at least a couple of seasons.

1. Locate and remove the thermostat from the powerhead, as detailed in this section.
2. Suspend the thermostat and a thermometer in a container of water. For most accurate test results, it is best to hang the thermostat and a thermometer using lengths of string so that they are not touching the bottoms or sides of the container (this ensures that both components remain at the same temperature as the water and not the container).
3. Slowly heat the water while observing the thermostat vernatherm for movement. The moment you observe movement, check the thermometer and note the temperature. If the water begins to boil (reaches about 212°F/100°C at normal atmospheric pressure) and NO movement has occurred, discontinue the test and throw the piece of junk thermostat away (if you are SURE there was no movement from the vernatherm).
4. Remove the source of heat and allow the water to cool (you can speed this up a little by adding some cool water to the container, but if you're using a glass container, don't add too much or you'll risk breaking the container). Observe the vernatherm again for movement as the water cools. When movement occurs, check the thermometer and record the temperature.
5. Compare the opening temperature with the following specifications:
 • For carbureted motors, the thermostat must open at or above 125°F (53°C). The thermostat must be fully open by 136-144°F (58-62°C).
 • For EFI motors, the thermostat must open at approximately 118-126°F (48-52°C).
6. Specifications for closing temperatures are not specifically provided by the manufacturer, but typically a thermostat must close a temperature close to, but below the temperature for the opening specification. Refer to the Cooling System Efficiency test to determine the operating temperatures for your motor. The thermostat MUST close below that temperature. A slight modulation (repeated opening and closing) of the thermostat can occur at borderline temperatures to make sure the powerhead remains in the proper operating range.
7. Replace the thermostat if it does not operate as described, or if you are unsure of the test results and would like to eliminate the thermostat as a possible problem. Refer to the removal and installation procedure for Thermostat in this section for more details.

Water Pump

On all Johnson/Evinrude motors, the water pump is attached to the inside of the gearcase. The pump itself usually consists of a composite material impeller attached to a portion of the driveshaft that runs through a pump housing and/or cover. On most models the housing is located on the gearcase lower unit-to-midsection or adapter section split line, but there are exceptions, like the 2.0-3.5 hp motors, on which the pump is located in the lower unit, directly behind the propeller. On models where the pump is located on the gearcase lower unit split line, the lower unit must be removed for access. For more details, refer to the removal and installation procedure for the model undergoing service.

■ For specific water flow diagrams showing the cooling system components and water passages for each Johnson/Evinrude outboard, please refer to Cooling System Schematics in this section.

REMOVAL & INSTALLATION

■ Replace the impeller, gaskets and any O-rings/seal whenever the water pump is removed for inspection or service. There is no reason to use questionable parts. Keep in mind that damage to the powerhead caused by an overheating condition (and the subsequent trouble that can occur from becoming stranded on the water) quickly overtakes the expense of a water pump service kit.

Colt/Junior Motors

◆ See Figures 54 thru 58

1. Remove the gearcase lower unit from the intermediate housing. For details, refer to the Gearcase section.
2. Remove the two screws, and then lift the water pump housing up and free of the driveshaft.
3. Remove and discard the water tube grommet.
4. Lift the impeller and slide it free of the driveshaft.

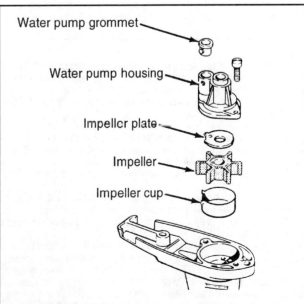

Fig. 54 Exploded view of the water pump assembly - Colt/Junior models

5-26 LUBRICATION AND COOLING

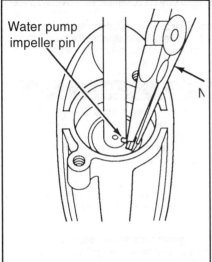

Fig. 55 Install the impeller drive pin into the driveshaft

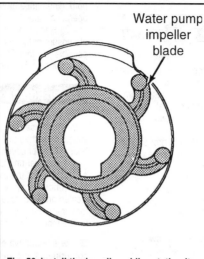

Fig. 56 Install the impeller while rotating it so the blades area facing the proper direction

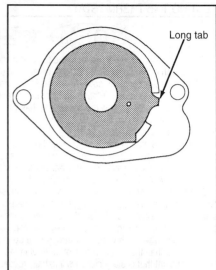

Fig. 57 Install the impeller plate so the long tabs are positioned as shown

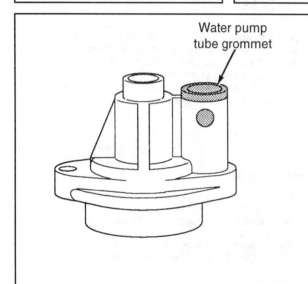

Fig. 58 Gently insert a new water tube grommet into the top of the pump

To install:

5. Using a pair of needle nose pliers, insert the water pump impeller pin into the driveshaft by aligning the "T" portion of the pin with the axis of the shaft. Press the pin inward until the pin bottoms against the shaft.

6. Install the plastic pump spacer into the lower unit pump cavity. The side of the spacer with the two small prongs toward the cavity. The prongs must enter the large main water intake hole in the lower unit casting.

7. Apply a very light coating of oil to the impeller blades. Rotate the impeller in a clockwise direction as the impeller is worked into the housing cup. Be sure the blades are seated in the direction shown.

8. Insert the impeller plate into the water pump body with the long tabs extending into the water outlet opening.

9. Apply a light coating of OMC Gasket Compound, or equivalent, to the exterior surface of the impeller cup. Install the cup, with the impeller in place, into the impeller pump body. Check to be sure the tabs on the cup index into the slot in the pump body.

10. Apply a light coating of oil to the outside and inside surfaces of the water pump tube grommet. Force the grommet into the water pump opening.

11. Apply a coating of OMC Gasket Sealing Compound, or equivalent, to the threads of the water pump retaining screws. Carefully slide the water pump down the driveshaft, and then align the mounting holes with the holes in the matching holes in the lower unit. Install and tighten the two screws to 25-35 inch lbs. (2.8-4.0 Nm).

12. Install the gearcase lower unit, as detailed in the Gearcase section.

2.0-3.5 Hp (78cc) Motors

◆ See Figures 59, 60 and 61

1. Remove the propeller. For details, refer to Propeller under Engine Maintenance.

2. Remove the 2 bolts securing the water pump cover to the lower unit.

3. Using 2 small screwdrivers - one working on each side-pry the water pump cover free of the lower unit. Take care not to mar the sealing surfaces of the pump cover or the mating lower unit. Take care not to mar the sealing surfaces of the pump cover or the mating lower unit.

■ Another way to remove the water pump cover is to rotate the cover about 90°, and then gently tap the ears of the cover using a soft-faced mallet to jar it free.

4. Slide the water pump impeller free of the propeller shaft. Pull the impeller drive pin out of the propeller shaft.

5. Thoroughly inspect the impeller, cover and bearing housing for signs of damage or wear and replace, as necessary. For more details, please refer to Inspection & Overhaul, in this section.

■ Regardless of condition, it is always a good idea to replace the impeller to ensure proper cooling system operation.

To install:

6. Push the water pump impeller drive pin into the hole in the propeller shaft.

7. Cover the impeller vanes with a light coating of engine oil. Install the water pump impeller onto the propeller shaft and slide it forward on the shaft until the groove in the impeller begins to index over the pin. Rotate the water pump impeller and shaft clockwise and at the same time (making sure the impeller blades seat properly), push the impeller into the bearing carrier cavity. The impeller is properly installed when the pin is indexed to the full depth of the groove.

8. Position the water pump cover over the impeller. Apply a coating of OMC Gasket Sealing Compound to the threads of the 2 water pump cover bolts. Secure the cover in place with the 2 bolts. Tighten the bolts to 60-84 inch lbs. (7-9 Nm).

9. Install the propeller.

3-4 Hp (87cc) Motors

◆ See Figures 62 thru 66

1. Remove the gearcase lower unit from the intermediate housing. For details, refer to the Gearcase section.

2. Remove the three bolts securing the water pump cover/housing to the lower unit, then slide the cover and water pump components up and off the driveshaft. If necessary, use a small prybar or screwdriver to carefully pry up evenly around the pump cover perimeter to release it from the plate.

LUBRICATION AND COOLING 5-27

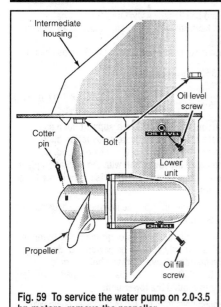

Fig. 59 To service the water pump on 2.0-3.5 hp motors, remove the propeller...

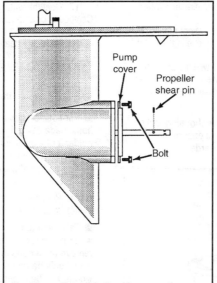

Fig. 60 ...then unbolt and remove the water pump cover...

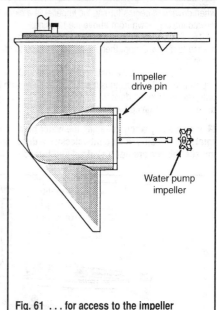

Fig. 61 ...for access to the impeller

3. Remove and discard the O-ring from the cover.
4. Remove and discard the old grommet from the top of the cover.
5. Using a pair of needle nose pliers, pull the pump impeller free of the cover. If the insert is damaged, pull the insert out of the cover with the needlenose pliers.
6. Remove the impeller drive pin from the flat on the driveshaft.
7. Using a thin flat blade screwdriver pry evenly around the perimeter of the pump impeller plate to release it from the lower unit surface. Lift off the plate, gasket, and the water intake screen. Discard the gasket and clean the intake screen.

■ **A used tooth brush is a very useful tool to clean the screen.**

8. Thoroughly inspect the impeller, cover, insert and impeller plate for signs of damage or wear and replace, as necessary. For more details, please refer to Inspection & Overhaul, in this section.

■ **Regardless of condition, it is always a good idea to replace the impeller (and often the insert as well) to ensure proper cooling system operation.**

To install:
9. Slide the water intake screen down over the driveshaft, with the 3 prongs on the screen facing up.
10. Apply a coating of OMC Gasket Sealing Compound, or equivalent sealant to both sides of a new gasket. Install the gasket and then slide the impeller plate down into position over the screen.
11. Apply a dab of OMC Needle Bearing Assembly Grease or another suitable lubricant onto the pump impeller drive pin, and then place it against the flat on the driveshaft.
12. Apply a light coating of oil to a new water pump grommet, then install it to the top of the cover.
13. If removed, lightly coat the outer diameter of a new insert cup using OMC Gasket Sealing Compound or an equivalent sealant, then push a new insert into the water pump cover.
14. Apply a light coating of oil to the new water pump impeller blades, then carefully position the impeller into the insert while turning the impeller counterclockwise. In this way each vane of the impeller is folded back slightly with the tip of the impeller blade facing clockwise when observed from underneath the pump housing. This is the same position the blades will be in

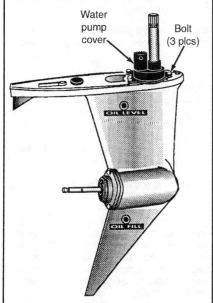

Fig. 62 To service the water pump on 3/4hp (87cc) motors, remove the 3 pump cover/housing bolts...

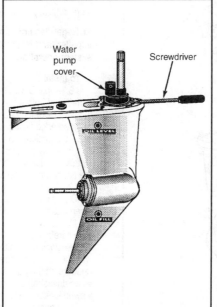

Fig. 63 ...then carefully break the gasket seal and remove the cover/housing from the lower unit

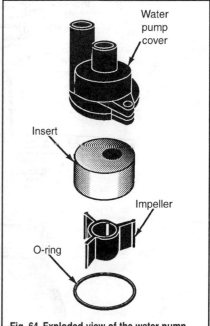

Fig. 64 Exploded view of the water pump cover/housing and components

5-28 LUBRICATION AND COOLING

when driveshaft begins turning the impeller in a clockwise direction (when viewed looking down from **above** the housing.)

15. Apply a light coating of OMC Triple-Guard or equivalent marine grease to a new O-ring, then install the O-ring into the groove in the water pump cover.

16. Position the driveshaft so the drive pin will align with the slot in the impeller when the housing is installed. Carefully slide the assembled water pump down onto the driveshaft and over the plate engaging the impeller drive pin and aligning the bolt holes for the housing.

■ If necessary when installing the water pump cover/housing, only rotate the housing in a counterclockwise direction to ensure the impeller blades are not dislodged and turned backwards.

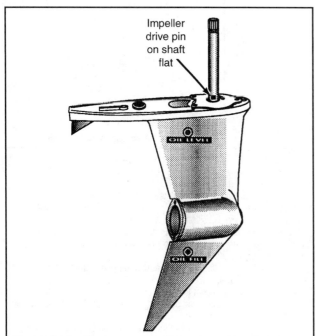

Fig. 65 The impeller is rotated by a drive pin attached to the flat on the driveshaft

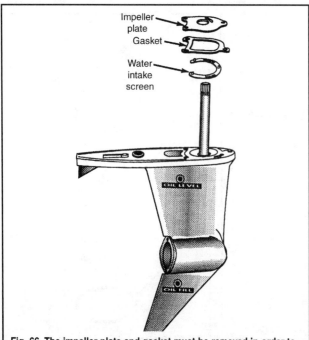

Fig. 66 The impeller plate and gasket must be removed in order to check the water intake screen

17. Coat the water pump bolt threads using OMC Gasket Sealing Compound or an equivalent sealant, then install and tighten to 25-35 inch lbs. (2.8-4 Nm).

18. Install the gearcase lower unit, as detailed in the Gearcase section.

4 Deluxe (87cc) and 5/6/8 Hp (164cc) Motors

◆ See Figures 67 thru 70

■ For ease of service mount the gearcase in a support such as a homemade cavitation plate holder. To fabricate a cavitation plate holder, cut a groove in a short piece of 2" x 6" piece of wood. Cut the groove so it can accommodate the lower unit with the cavitation plate resting on top of the wood. Clamp the wood in a vise to hold the lower unit securely during service.

1. Remove the gearcase lower unit from the intermediate housing. For details, refer to the Gearcase section.

■ Take time to notice the 4 bolts passing through the water pump. If a water pump replacement is the only work to be performed, do not remove the 3 bolts in the gearcase cover. If these bolts should mistakenly be removed, the shift rod will be dislodged - bad news. It would then be necessary to rebuild the lower unit.

2. Remove the 4 through-bolts from the top of the water pump. Work the water pump upward and free of the driveshaft.

■ The impeller may come off with the pump or may remain behind on the shaft.

3. If the impeller did not come off with the water pump work it free of the keyway and up/off the driveshaft. If the impeller remains in the housing, use a pair of needle nose pliers to pull the pump impeller free of the insert in the pump cover/housing.

4. Remove the impeller key.

5. If the insert is damaged, pull the insert out of the cover with the needle nose pliers.

■ Notice the plate and two gaskets under the water pump impeller, especially how the plate is installed with the beveled, or rounded side facing up. Both gaskets are identical and either one may be installed on top of the plate, but both gaskets should be replaced anytime the water pump is removed.

6. Remove the water pump plate along with the upper and lower gaskets.

7. Thoroughly inspect the impeller, pump cover/housing, insert and impeller plate for signs of damage or wear and replace, as necessary. For more details, please refer to Inspection & Overhaul, in this section.

■ Regardless of condition, it is always a good idea to replace the impeller (and often, the insert too) in order to ensure proper cooling system operation.

To install:

8. Apply OMC Gasket Sealing Compound or equivalent sealant to both surfaces of both water pump plate gaskets. Place one gasket in position on the surface of the gearcase cover. Place the water pump plate (with the rounded edge facing upward) into position on the gasket. Place the second gasket in position on the top of the water pump plate.

9. Apply a dab of OMC Needle Bearing Assembly Grease or another suitable lubricant onto the pump impeller drive key, and then place it against the flat on the driveshaft.

10. If removed, lightly coat the outer diameter of a new insert cup using OMC Gasket Sealing Compound or an equivalent sealant, then push a new insert into the water pump cover.

11. Apply a light coating of oil to the new water pump impeller blades, then carefully position the impeller into the insert while turning the impeller counterclockwise. In this way each vane of the impeller is folded back slightly with the tip of the impeller blade facing clockwise when observed from underneath the pump housing. This is the same position the blades will be in when driveshaft begins turning the impeller in a clockwise direction (when viewed looking down from **above** the housing.)

12. Apply a coating of OMC Gasket Sealing Compound or equivalent to the water pump bolt threads, then install the housing/cover and tighten the bolts evenly and alternately to 60-84 inch lbs. (7-9 Nm).

13. Install the gearcase lower unit, as detailed in the Gearcase section.

LUBRICATION AND COOLING 5-29

Fig. 67 A wooden fixture is easily fabricated to hold the lower unit during service

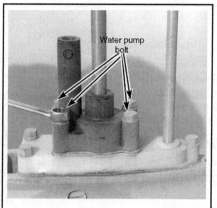

Fig. 68 If only servicing the water pump, ONLY remove the 4 pump bolts and NOT the 3 gearcase cover screws

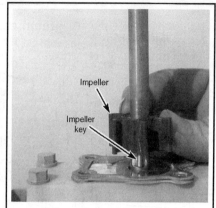

Fig. 69 The impeller is turned by the driveshaft using the impeller key

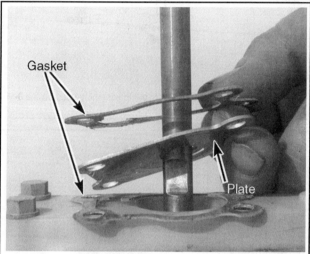

Fig. 70 Remove the impeller plate along with the upper and lower gaskets

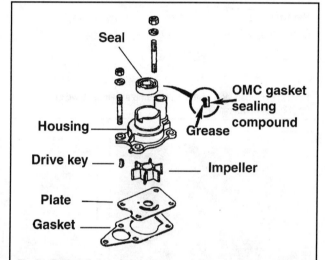

Fig. 71 Exploded view of the water pump assembly used on the 5 hp (109cc) motors

5 Hp (109cc) Motors

◆ See Figure 71

 MODERATE

1. Remove the gearcase lower unit from the intermediate housing. For details, refer to the Gearcase section.
2. Remove the nuts securing the water pump housing to the lower unit, then lift the housing from the gearcase.
3. Carefully work the impeller up and off the driveshaft.
4. Remove the impeller key.
5. Remove the impeller plate. Carefully clean all traces of gasket from the top of the gearcase housing and the bottom of the impeller plate.
6. If necessary, remove the driveshaft seal from the top of the water pump housing.
7. Thoroughly inspect the impeller, housing and impeller plate for signs of damage or wear and replace, as necessary. For more details, please refer to Inspection & Overhaul, in this section.

■ Regardless of condition, it is always a good idea to replace the impeller to ensure proper cooling system operation.

To install:

8. If removed, lightly coat the outer diameter of a new water pump driveshaft seal using OMC Gasket Sealing Compound or equivalent, then use a suitably sized socket or driver to carefully install the seal into the housing with the lips facing upward. Coat the seal lips using OMC Triple-Guard or equivalent marine grease.
9. Apply a coating of OMC Gasket Sealing Compound or equivalent sealant to both sides of a new impeller plate gasket, then position the gasket and plate onto the gearcase.

■ A small dab of marine grease can be used to help hold the key in position when installing the impeller.

10. Insert the impeller key to the driveshaft keyway, then slide the water pump impeller down the shaft and over top of the key.
11. Slide the water pump housing down the driveshaft until it is just above the impeller. Rotate the driveshaft **clockwise** while carefully pushing the pump housing down over the impeller and plate.
12. Install the water pump housing nuts and tighten to 36-60 inch lbs. (4-7 Nm).
13. Install the gearcase lower unit, as detailed in the Gearcase section.

5/6 Hp (128cc), 8/9.9 Hp (211cc), 9.9-15 Hp (216/255cc) and 9.9/15 Hp (305cc)

◆ See Figures 72 thru 76

MODERATE

■ There are some indications indication that the 10/14 hp (216/255cc) 2-stroke commercial models were usually equipped with the split-type gearcase used by 25 Commercial and 28 hp (521cc) motors covered later in this section. The quick and easy way to determine if you should instead follow THAT procedure would be to check the gearcase at the height of the prop shaft. If there is a seam (split line) that runs the horizontal length of the case, refer to the procedure for the 521cc Commercial motors later in this section.

1. Remove the gearcase lower unit from the intermediate housing. For details, refer to the Gearcase section.

5-30 LUBRICATION AND COOLING

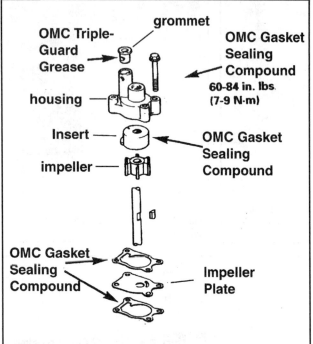

Fig. 72 Exploded view of the water pump assembly used on the 5/6 hp (128cc) 4-stroke motors

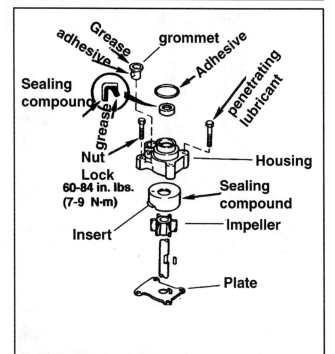

Fig. 73 Exploded view of the water pump assembly used on the 8-15 hp motors

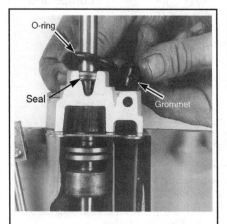

Fig. 74 The top of the water pump housing contains a water tube grommet, driveshaft O-ring and driveshaft seal

Fig. 75 Four bolts secure the water pump housing and impeller plate to the top of the lower unit

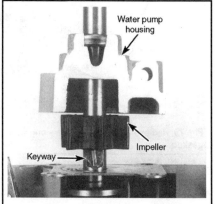

Fig. 76 This cutaway shows the difference in size between the impeller and housing. The impeller blades must bend to one side to fit within the housing

2. Remove the 4 bolts securing the water pump cover to the lower unit, then slide or work the cover up off the driveshaft.

■ Depending on whether or not the impeller pulls off the driveshaft with the housing, either remove it from the housing using a pair of needle-nose pliers or gently pry it upward from the shaft using a prybar.

3. If the impeller did not come off with the cover, work the impeller off the driveshaft.
4. Remove the impeller drive pin from the driveshaft.
5. Remove the impeller plate. Clean the plate and lower unit of any remaining gasket material and/or sealant.

■ The 5/6 hp models utilize a gasket on both sides of the impeller plate.

6. Remove and discard the grommet (for the water tube) from the top of the pump housing.
7. On 8-15 hp motors remove and discard the water pump housing O-ring and the driveshaft seal from the top of the housing.

8. If the insert is damaged, pull the insert out of the cover with the needle nose pliers.
9. Thoroughly inspect the impeller, pump cover/housing, insert and impeller plate for signs of damage or wear and replace, as necessary. For more details, please refer to Inspection & Overhaul, in this section.

■ Regardless of condition, it is always a good idea to replace the impeller (and often, the insert too) in order to ensure proper cooling system operation.

To install:
10. For 5/6 hp motors proceed as follows:
   a. Apply a light coating of OMC Gasket Sealing Compound, or equivalent sealant to both sides of the upper and lower impeller plate gaskets. Position one gasket on the gearcase and the other on the impeller plate, then slide the plate down over the lower gasket and onto the lower unit.
   b. Carefully install and seat the water tube grommet into the taller opening on top of the water pump housing. Lubricate the inside of the grommet using OMC Triple-Guard or equivalent marine grease.

LUBRICATION AND COOLING 5-31

c. If removed, coat the outer diameter of the impeller insert using OMC Gasket Sealing Compound, or equivalent sealant, then install the insert into the water pump housing.

d. Apply a light coat of engine or gear oil to the insert, then install the impeller to the insert while rotating the impeller counterclockwise. In this way the impeller blade tips will be facing the clockwise when looking at the underside of the pump housing (this is the same direction that they will face when the housing is installed and the driveshaft rotates clockwise, the normal direction of rotation).

11. For 8-15 hp motors proceed as follows:

a. Apply a light coating of OMC Gasket Sealing Compound, or equivalent sealant to the outer diameter (the metal casing) of a NEW driveshaft seal. Then install the seal into the top of the water pump housing with the seal lips facing downward (facing toward the impeller).

b. Apply a light coating of OMC Adhesive M or an equivalent sealant to the impeller housing O-ring and to the outer diameter of the water tube grommet. Install the O-ring onto the large round bore on top of the water pump housing and the grommet to the water tube (smaller round boss) on top of the housing.

c. Apply a light coating of OMC Triple-Guard or equivalent marine grease to the inner surface of the water tube grommet.

d. If removed, coat the outer diameter of the impeller insert using OMC Gasket Sealing Compound, or equivalent sealant, then install the insert into the water pump housing making sure the small bleed hole (if used) is positioned toward the water tube.

e. Apply a light coat of engine or gear oil to the insert, then install the impeller to the insert while rotating the impeller counterclockwise. In this way the impeller blade tips will be facing the clockwise when looking at the underside of the pump housing (this is the same direction that they will face when the housing is installed and the driveshaft rotates clockwise, the normal direction of rotation, when viewed from above).

f. Apply a thin bead of OMC Adhesive M or equivalent to machined mating surfaces of the gearcase and impeller plate. Carefully slide the impeller plate over the driveshaft and into position on the gearcase.

✱✱ WARNING
Use caution when applying sealant and installing the impeller plate. Excessive sealant may be pressed out from between the mating surface during installation. This sealant could clog (or partially clog a cooling passage) or worse, it could dry on an impeller contact surface, which would cause damage or accelerated wear to the impeller.

12. Apply a dab of OMC Needle Bearing Assembly Grease or equivalent to the impeller drive key to help hold it in place, then position the pin to the slot in the driveshaft.

13. Slide the impeller housing onto the driveshaft, turning the driveshaft clockwise slightly, as necessary to align the impeller with the drive key. Carefully seat the housing and impeller against the impeller plate.

✱✱ WARNING
Only rotate the driveshaft CLOCKWISE when viewed from the top of the water pump or the impeller could be damaged.

14. For 5/6 hp motors, apply a light coating of OMC Gasket Sealing Compound, or equivalent sealant to the threads of the water pump housing bolts.

15. For 8-15 hp motors, apply a light coating of OMC Nut Lock or equivalent threadlocking compound to the threads of the water pump housing bolts.

16. Install and finger-tighten the 4 water pump housing bolts, then tighten the bolts to 60-84 inch lbs. (7-9 Nm).

17. Install the gearcase lower unit, as detailed in the Gearcase section.

10/14 Hp (216/255cc), 25 Commercial And 2-Cylinder Motors

◆ See Figure 77

1. Remove the gearcase lower unit from the intermediate housing. For details, refer to the Gearcase section.

2. Remove the 4 bolts and washers from the top of the water pump. Work the water pump upward and free of the driveshaft.

■ **The impeller may come off with the pump or may remain behind on the shaft.**

3. If the impeller did not come off with the water pump work it free of the keyway and up/off the driveshaft. If the impeller remains in the housing, use a pair of needle nose pliers to pull the pump impeller free of the insert in the pump housing.

4. Remove the impeller key.

■ **If you've taken the trouble to go this far (to service a water pump) it is a good idea to replace the impeller, insert and seal to ensure proper pump operation.**

5. If the insert is damaged or if the driveshaft seal is to be replaced, pull the insert out of the cover with the needle nose pliers.

6. Remove the driveshaft seal from the underside of the water pump housing (located under the insert with the housing inverted).

7. Remove the water tube grommet from the round boss on the top of the pump housing.

■ **Take note of the plate and two gaskets under the water pump impeller. Both gaskets should be replaced anytime the water pump is removed.**

8. Carefully pry under the edge of the impeller plate to break the gasket seal, then remove the plate along with the upper and lower gaskets.

9. Thoroughly inspect the impeller, pump cover/housing, insert and impeller plate for signs of damage or wear and replace, as necessary. For more details, please refer to Inspection & Overhaul, in this section.

■ **Regardless of condition, it is always a good idea to replace the impeller (and often, the insert too) in order to ensure proper cooling system operation.**

To install:

10. Apply a light coating of OMC Gasket Sealing Compound, or equivalent sealant to the outer diameter (the metal casing) of a NEW driveshaft seal. Then install the seal into the water pump housing with the seal lips facing upward (away from the insert and impeller).

11. If removed, coat the outer diameter of the impeller insert using OMC Gasket Sealing Compound, or equivalent sealant, then install the insert into the water pump housing. There are 2 square tabs that will lock the insert into the housing.

12. Apply a light coat of engine or gear oil to the insert, then install the impeller to the insert while rotating the impeller counterclockwise. In this way the impeller blade tips will be facing the clockwise when looking at the underside of the pump housing (this is the same direction that they will face when the housing is installed and the driveshaft rotates clockwise, the normal direction of rotation, when viewed from above).

13. Apply a light coating of OMC Adhesive M or an equivalent sealant to a new water pump grommet, then install it into the water tube bore on top of the housing. Apply a light coating of OMC Triple-Guard or equivalent marine grease to the inner surface of the water tube grommet.

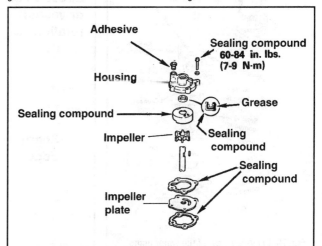

Fig. 77 Exploded view of the water pump assembly used on the 25 Commercial and 28 hp motors as well as some 10/14 Hp (216/255cc) motors (all with Split Gearcases)

5-32 LUBRICATION AND COOLING

14. Apply OMC Gasket Sealing Compound or equivalent sealant to both surfaces of both water pump plate gaskets. Place one gasket in position on the surface of the gearcase cover. Place the water pump plate (with the rounded edge facing upward) into position on the gasket. Place the second gasket in position on the top of the water pump plate.
15. Apply a dab of OMC Triple-Guard or equivalent marine grease to the impeller drive key to help hold it in place, then position the key to the slot in the driveshaft.
16. Slide the impeller housing onto the driveshaft, turning the driveshaft clockwise slightly, as necessary to align the impeller with the drive key. Carefully seat the housing and impeller against the impeller plate.

✱✱ WARNING

Only rotate the driveshaft CLOCKWISE when viewed from the top of the water pump or the impeller could be damaged.

17. Apply a light coating of OMC Gasket Sealing Compound, or equivalent sealant to the threads of the water pump housing bolts.
18. Install and finger-tighten the 4 water pump housing bolts, then tighten the bolts to 60-84 inch lbs. (7-9 Nm).
19. Install the gearcase lower unit, as detailed in the Gearcase section.

18 Jet-35 Hp (521cc, Exc. 25 Com & 28 Hp) and 25/35 Hp (500/565cc) Motors

◆ See Figures 78, 79 and 80

1. Remove the gearcase lower unit from the intermediate housing. For details, refer to the Gearcase section.
2. On 25/35 Hp (500/565cc) motors, remove and discard the long, thin seal from the gearcase located immediately behind the water pump housing.
3. Remove the 6 bolts and washers from the top of the water pump.
4. Pull upward on the driveshaft, removing the driveshaft and the water pump assembly from the lower unit.
5. Work the water pump and impeller free of the driveshaft.
6. Remove the impeller key.

■ If you've taken the trouble to go this far (to service a water pump) it is a good idea to replace the impeller, insert and seal to ensure proper pump operation.

7. If the insert is damaged or if the driveshaft O-ring is to be replaced, pull the insert out of the cover with the needle nose pliers.
8. Remove the driveshaft O-ring from the underside of the water pump housing (located under the insert with the housing inverted).
9. Remove the thick shift rod O-ring and the shift rod bushing from the underside of the pump housing.
10. Remove the water tube grommet from the round boss on the top of the pump housing.
11. Remove the split irregular shaped O-ring from the bottom of the water pump housing.

■ Take note of the plate and the gasket located under the plate. The gasket should be replaced anytime the water pump is removed.

12. Carefully pry under the edge of the impeller plate to break the gasket seal, then remove the plate along with the gasket. Remove all traces of gasket material from the plate and gearcase mating surfaces.
13. Thoroughly inspect the impeller, pump cover/housing, insert and impeller plate for signs of damage or wear and replace, as necessary. For more details, please refer to Inspection & Overhaul, in this section.

■ Regardless of condition, it is always a good idea to replace the impeller (along with the O-rings and insert too) in order to ensure proper cooling system operation.

To install:
14. Apply a light coating of oil to a NEW shift rod bushing, then install it in the small bore on the underside of the housing.
15. Apply a light coating of OMC Triple-Guard or equivalent marine grease to the 2 round housing O-rings (the small thick shift-rod O-ring and the large, round driveshaft O-ring). Install the O-rings to the housing.
16. If removed, coat the outer diameter of the impeller insert using OMC Gasket Sealing Compound, or equivalent sealant, then install the insert into the water pump housing. There are 2 tabs that will lock the insert into the housing.
17. Apply a light coat of engine or gear oil to the insert, then install the impeller to the insert while rotating the impeller counterclockwise. In this way the impeller blade tips will be facing the clockwise when looking at the underside of the pump housing (this is the same direction that they will face when the housing is installed and the driveshaft rotates clockwise, the normal direction of rotation, when viewed from above).
18. Install the water tube grommet to the top of the water pump housing as follows:
• For 2-cylinder motors, apply a light coating of either Scotch-Grip Rubber Adhesive or OMC Adhesive M (or an equivalent sealant) to a new water pump grommet, then install it into the water tube bore on top of the housing. Apply a light coating of OMC Triple-Guard or equivalent marine grease to the inner surface of the water tube grommet.
• For 3-cylinder motors, apply a light coating of OMC Adhesive M or equivalent sealant to the groove on the underside of the one-piece water and shift tube grommet, then position the grommet on the top of the water pump housing.
19. Apply a light coating of OMC Adhesive M or an equivalent sealant the grooves in the underside of the water pump housing, then install a NEW

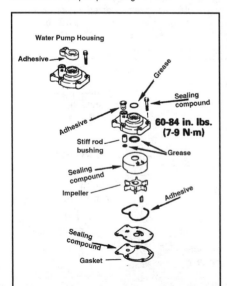

Fig. 78 Exploded view of the water pump assembly used on the 25/35 Hp (500/565cc) 3-cylinder and 18 Jet-35 Hp (521cc, except 25 commercial and 28 hp, 2-cylinder) motors

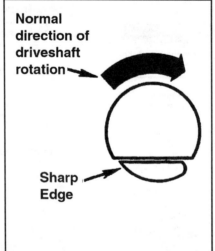

Fig. 79 Driveshaft and impeller drive key installation viewed from ABOVE the housing

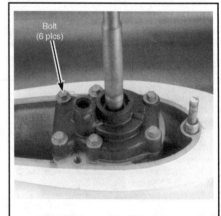

Fig. 80 The water pump housing is mounted to the lower unit by 6 bolts on all 18 Jet-35 hp (521cc) and 25/35 hp (500/565cc) motors, except 25 commercial and 28 hp 2-cylinder models

LUBRICATION AND COOLING 5-33

water pump housing-to-impeller plate O-ring (irregular shaped, split O-ring) into the groove.

■ In the next step, hold the impeller to keep it from being dislodged from the housing as the driveshaft is inserted.

20. Apply a light coating of gear oil to the driveshaft pinion splines, then slide the shaft through the water pump housing and impeller. Slide the shaft downward until the drive key flat is below the impeller.

21. Align flat on the driveshaft with the groove in the impeller, then insert the drive key with the sharp edge facing the normal direction of rotation for the driveshaft.

■ Remember the driveshaft normally rotates clockwise when viewed from ABOVE, but since you're working from below, the shaft would rotate counterclockwise.

22. Hold the key in position, then slide the key and driveshaft upward, into the impeller.

✱✱ WARNING

Make sure the impeller is securely engaged with the drive key or insufficient cooling and powerhead damage could result.

23. Apply OMC Gasket Sealing Compound or equivalent sealant to both surfaces of the NEW impeller plate gasket. Place the gasket in position on the surface of the gearcase cover, then install the impeller plate over the gasket.

24. Insert the driveshaft through the impeller plate, then slide the water pump and driveshaft assembly downward to engage the driveshaft splines with the pinion. Turn the driveshaft and housing together aligning the holes of the water pump housing with the plate, gasket and gearcase lower unit.

25. Apply a light coating of OMC Gasket Sealing Compound, or equivalent sealant to the threads of the water pump housing bolts.

26. Install and finger-tighten the 6 water pump housing bolts, then tighten the bolts to 60-84 inch lbs. (7-9 Nm).

27. On 25/35 Hp (500/565cc) motors, apply a light coating of OMC Adhesive M or equivalent sealant to a NEW seal for the gearcase (the thin seal that mounts immediately behind the water pump housing). Install the seal to the gearcase.

28. Install the gearcase lower unit, as detailed in the Gearcase section.

25-55 Hp (737cc) and 25-70 Hp (913cc) Motors

◆ See Figures 79 and 81 thru 91

1. Remove the gearcase lower unit from the intermediate housing. For details, refer to the Gearcase section.
2. Remove the 4 bolts securing the water pump housing to the lower unit, then slide or work the cover up off the driveshaft.

■ Depending on whether or not the impeller pulls off the driveshaft with the housing, either remove it from the housing using a pair of needle-nose pliers or gently pry it upward from the shaft using a prybar.

3. If the impeller did not come off with the cover, work the impeller off the driveshaft.
4. Remove the impeller drive pin, and the impeller O-ring (if equipped), from the driveshaft.
5. Remove the impeller plate. Clean the plate and lower unit of any remaining gasket material and/or sealant.
6. Disassemble the water pump housing by removing the impeller (if it did not come out when removing the pump), the insert, grommets and/or O-rings. Although people have been known to reuse some of these components, it just doesn't make sense. If you've come this far to fix a water pump problem (or for other service requirements) it is a good idea to replace the impeller, insert and all O-rings to ensure proper water pump operation.
7. Thoroughly inspect the impeller, cover, insert and impeller plate for signs of damage or wear and replace, as necessary. For more details, please refer to Inspection & Overhaul, in this section.

■ Regardless of condition, it is always a good idea to replace the impeller (as well as the insert and the O-rings) to ensure proper cooling system operation.

To install:

■ When a 6-blade impeller is installed on the 40RP, 40RW, 40WR, 45 and 55 hp 2-cylinder motors or any of the 3-cylinder motors, the water pump does not require installation of an impeller O-ring

8. For 2-cylinder motors, except the 40RP, 40RW, 40WR, 45 and 55 hp models:
   a. Apply a drop of OMC Adhesive M or an equivalent sealant to the 4 ribs driveshaft O-ring seal groove (located in the underside of the water pump housing, above the insert when it is installed).

✱✱ WARNING

Be careful not to get any sealant on the air bleed groove or the pump could loose its prime and fail to pump water in service.

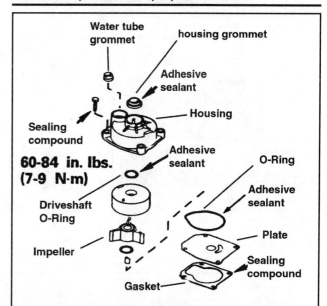

Fig. 81 Exploded view of the water pump assembly used on the most 2-cylinder 25-50 hp (737cc) motors, except 40WR, 40RP, 40RW and 45 hp motors (note some early-models utilize a 2-piece insert, when so equipped the driveshaft O-rings and the plate gasket may not be used)

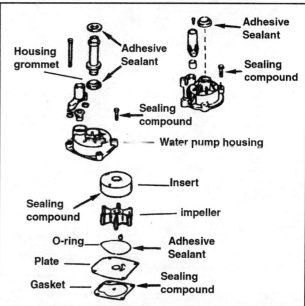

Fig. 82 Exploded view of the water pump assembly used on all 3-cylinder 25-70 hp (913cc) motors, and some 2-cylinder 40-55 hp motors, including the 40RP, 40RW, 40WR and 45 and 55 hp motors

5-34 LUBRICATION AND COOLING

b. Install the driveshaft O-ring to the groove in the underside of the water pump housing (above the insert).

9. If removed (and you had to in order to access the driveshaft O-ring on models so equipped), coat the outer diameter of the impeller insert using OMC Gasket Sealing Compound, or equivalent sealant, then install the insert into the water pump housing. Most models use a square index tab to lock the insert into place in the housing.

■ When installing the impeller to the water pump insert, be sure the drive pin slot is facing outward.

10. Apply a light coat of engine or gear oil to the insert, then install the impeller to the insert while rotating the impeller counterclockwise. In this way the impeller blade tips will be facing the clockwise when looking at the underside of the pump housing (this is the same direction that they will face when the housing is installed and the driveshaft rotates clockwise, the normal direction of rotation, when viewed from above).

11. Apply a thin bead of OMC Adhesive M or an equivalent sealant in the groove for the water pump housing-to-impeller plate O-ring (the irregular shaped O-ring that mounts in the bottom of the pump housing). Install the O-ring to the groove.

12. Install the water tube grommet into the impeller housing with the inside chamfered edge facing upward. If equipped with a water tube bracket, install and secure it to the housing (most models that use this bracket are equipped with a retaining screw).

13. Apply a thin bead of OMC Adhesive M or an equivalent sealant in the groove for the impeller housing grommet, then position the grommet in the top of the housing.

14. Install the impeller plate as follows:
• For all non-Jet models, apply a bead of OMC Gasket Sealing Compound or equivalent sealant to both sides of a new impeller plate gasket, then position the gasket and plate on the gearcase lower unit. If used, install the impeller O-ring over the driveshaft and slide it down into contact with the plate.

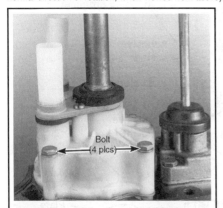

Fig. 83 Remove the 4 retaining bolts to separate the water pump from the lower unit

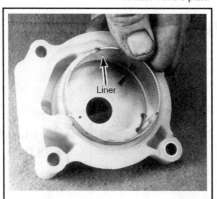

Fig. 84 Upon assembly, install a new insert (liner) to the pump housing . . .

Fig. 85 . . . then install the impeller while rotating counterclockwise (so the blades face clockwise, the normal direction of rotation)

Fig. 86 Apply a thin bead of sealant, then install the pump housing O-ring seal

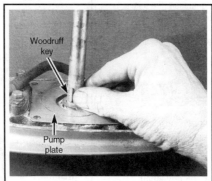

Fig. 87 Use a dab of grease to hold the impeller drive pin in position . . .

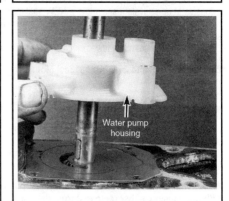

Fig. 88 . . . then slide the water pump housing and impeller down the shaft

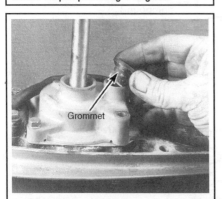

Fig. 89 If not done already, the water tube grommet should be replaced. . .

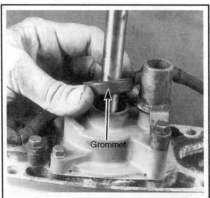

Fig. 90 . . .along with the housing grommet (secure both with adhesive sealant)

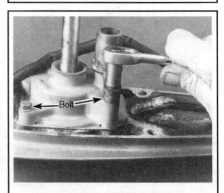

Fig. 91 With all parts installed and the housing aligned, install and tighten the pump bolts

LUBRICATION AND COOLING 5-35

■ Some early model gearcases may have sealed the plate directly to the gearcase with the sealing compound only and no gasket. If you pulled one off, be sure to use a replacement OR if the kit is upgraded to include a new gasket for your application. Otherwise, you should be good with just the sealing compound.

• For all Jet models, Apply a bead of OMC Gasket Sealing Compound or equivalent sealant directly to the impeller plate (along the gearcase mating surface), then install the impeller plate to the gearcase lower unit **without** using a gasket.

15. Apply a dab of OMC Needle Bearing Assembly Grease or another suitable lubricant onto the pump impeller drive pin, and then place it against the flat on the driveshaft. Insert the drive key with the sharp edge facing the normal direction of rotation for the driveshaft.

■ Remember the driveshaft normally rotates clockwise when viewed from ABOVE, but since you're working from below, the shaft would rotate counterclockwise.

16. Slide the water pump assembly down the driveshaft, aligning the impeller housing with the gearcase. Before the impeller comes to the drive pin, rotate the driveshaft to align the drive pin with the slot in the impeller and then slide the pump the rest of the way down against the impeller plate.

** WARNING

It is critical that the drive pin does not move out of position when sliding the pump/impeller down over the driveshaft.

17. Make sure the drive pin has properly engaged the impeller.
18. Apply a light coating of OMC Gasket Sealing Compound, or equivalent sealant to the threads of the water pump housing bolts.
19. Install and finger-tighten the 4 water pump housing bolts, then tighten the bolts to 60-84 inch lbs. (7-9 Nm).
20. Install the gearcase lower unit, as detailed in the Gearcase section.

40/50 Hp (815cc) EFI Motors

♦ See Figure 92

1. Remove the gearcase lower unit from the intermediate housing. For details, refer to the Gearcase section.
2. Remove the 4 nuts and lock-washers securing the water pump housing to the lower unit, then slide or work the cover up off the driveshaft.

■ Depending on whether or not the impeller pulls off the driveshaft with the housing, either remove it from the housing using a pair of needle-nose pliers or gently pry it upward from the shaft using a prybar.

3. If the impeller did not come off with the cover, work the impeller off the driveshaft.
4. Remove the impeller drive pin from the driveshaft.
5. If necessary, use a putty knife to help break the gasket seal, then remove the impeller plate. Clean the plate and lower unit of any remaining gasket material and/or sealant.
6. Disassemble the water pump housing by removing the impeller (if it did not come out when removing the pump), driveshaft seal and water tube grommet. It should be possible to carefully pry the seal and grommet from the top of the housing, but be sure not to damage the sealing surface of the housing. If necessary, use a suitable small-jawed puller to help remove the driveshaft seal.
7. Thoroughly inspect the impeller, housing and impeller plate for signs of damage or wear and replace, as necessary. For more details, please refer to Inspection & Overhaul, in this section.

■ Regardless of condition, it is always a good idea to replace the impeller and seals to ensure proper cooling system operation.

To install:
8. If removed, install a NEW driveshaft seal into the bore on top of the water pump housing. Use a suitably sized socket (make sure it is smooth to prevent seal damage and is the same size of the hard seal casing) to drive the seal into position with the seal lips facing the tool. Seat the seal in the bore, then remove the socket and apply a light coating of OMC Triple-Guard, or equivalent marine grease.

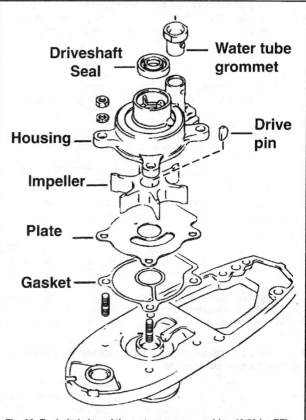

Fig. 92 Exploded view of the water pump assembly - 40/50 hp EFI motors

9. Install the NEW water tube grommet to the small bore in the top of the pump housing.
10. Apply OMC Gasket Sealing Compound or equivalent sealant to both surfaces of the NEW impeller plate gasket. Place the gasket into position on the surface of the gearcase cover and then install the impeller plate over the gasket.
11. Apply a dab of grease to the drive pin, then position it into the driveshaft. Slide the impeller down the driveshaft, aligning the groove in the impeller with the drive pin, then slide the impeller over the pin.

■ If impeller-to-pump housing installation is difficult, apply a light coating of marine grade grease to the impeller housing bore.

12. Slide the water pump housing down the driveshaft and over the impeller. Rotate the driveshaft by hand (clockwise when viewed from above) as the housing engages the impeller. Position the pump over the retaining studs.

** WARNING

Make sure the impeller drive pin does not become disengaged with the driveshaft or impeller while the housing is being installed or the pump will not operate properly. Problems with the cooling system can lead to serious powerhead damage due to overheating.

13. Apply a coating of OMC Gasket Sealing Compound or equivalent sealant to the threads of the water pump retaining studs. Install the lockwashers and nuts, then tighten nuts to 70-80 inch lbs. (8-9 Nm).
14. Install the gearcase lower unit, as detailed in the Gearcase section.

70 Hp (1298cc) EFI Motors

♦ See Figure 93

1. Remove the gearcase lower unit from the intermediate housing. For details, refer to the Gearcase section.
2. Remove the 4 bolts and washers securing the water pump housing to the lower unit, then slide or work the cover up off the driveshaft.

5-36 LUBRICATION AND COOLING

■ Depending on whether or not the impeller pulls off the driveshaft with the housing, either remove it from the housing using a pair of needle-nose pliers or gently pry it upward from the shaft using a prybar.

3. If the impeller did not come off with the cover, work the impeller off the driveshaft.
4. Remove the impeller drive pin from the driveshaft.
5. If necessary, use a putty knife to help break the gasket seal, then remove the impeller plate. Clean the plate and lower unit of any remaining gasket material and/or sealant.
6. Disassemble the water pump housing by removing the impeller (if it did not come out when removing the pump) and water tube grommet.
7. Thoroughly inspect the impeller, housing and impeller plate for signs of damage or wear and replace, as necessary. For more details, please refer to Inspection & Overhaul, in this section.

■ Regardless of condition, it is always a good idea to replace the impeller and seals to ensure proper cooling system operation.

To install:

8. Lubricate the NEW water tube grommet using a light coating of OMC Triple-Guard or equivalent marine grease, then install the grommet into the small bore in the top of the pump housing. While installing the grommet, be sure to align the raised bushing bosses with the holes in the pump housing.
9. Apply OMC Gasket Sealing Compound or equivalent sealant to both surfaces of the NEW impeller plate gasket. Place the gasket into position on the surface of the gearcase cover and then install the impeller plate over the gasket.

■ Be sure to align the gasket and impeller plate with the 2 dowels on the top of the gearcase lower unit.

10. Apply a light coating of oil or marine grade grease to the new water pump impeller blades, then carefully position the impeller into the insert while turning the impeller counterclockwise. In this way each vane of the impeller is folded back slightly with the tip of the impeller blade facing clockwise when observed from underneath the pump housing. This is the same position the blades will be in when driveshaft begins turning the impeller in a clockwise direction (when viewed looking down from **above** the housing.)
11. Apply a dab of grease to the drive pin, then position it into the driveshaft.
12. Install the water pump housing by carefully sliding it down the driveshaft assembly, aligning the bolt holes in the housing with the holes in the impeller plate and gearcase. As the pump and impeller reaches the impeller drive pin, turn the driveshaft slowly (in a clockwise direction when viewed from above) until the pin aligns with the groove in the impeller, then seat the pump, sliding the impeller over the pin.

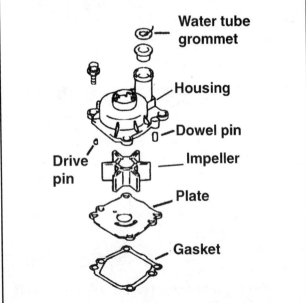

Fig. 93 Exploded view of the water pump assembly - 70 hp EFI motors

✴✴ WARNING

Make sure the impeller drive pin does not become disengaged with the driveshaft or impeller while the housing is being installed or the pump will not operate properly. Problems with the cooling system can lead to serious powerhead damage due to overheating.

13. Apply a coating of OMC Gasket Sealing Compound or equivalent threads of the water pump housing bolts, then install and tighten the bolts to 14 ft. lbs. (20 Nm).
14. Install the gearcase lower unit, as detailed in the Gearcase section.

INSPECTION & OVERHAUL

 MODERATE

Let's face it, you've gone through the trouble of removing the water pump for a reason. Either, you've already had cooling system problems and you're looking to fix it, or you are looking to perform some preventive maintenance. Although the truth is that you can just remove and inspect the impeller, replacing only the impeller (or even reusing the impeller if it looks to be in good shape). But WHY? The cost of a water pump rebuild kit is very little when compared with the even the time involved to get this far. If you've misjudged a component, or an O-ring (which by the way, never reseal quite the same way the second time), then you'll be taking this lower unit off again in the very near future to replace these parts. And, at best this will be because the warning system activated or you noticed a weak coolant indicator streams or, at worst, it will be cause you're dealing with the results of an overheated powerhead.

In short, if there is one way to protect you and your engine, it is to replace the impeller and insert (if used) along with all O-ring seals, grommets and gaskets, anytime the pump housing is removed. If not, take time to thoroughly clean and inspect the old impeller, housing and related components before assembly and installation. New components should be checked against the old. Seek explanations for differences with your parts supplier (but keep in mind that some rebuild kits may contain upgrades or modifications.

1. Remove and disassemble the water pump assembly as detailed in this section.
2. Carefully remove all traces of gasket or sealant material from components. If some material is stubborn, use a suitable solvent in the next steps to help clean material. Avoid scraping whenever possible, especially on plastic components whose gasket surfaces are easily scored and damaged.
3. Clean all metallic components using a mild solvent (such as Simple Green® cut with water or mineral spirits), then dry using compressed air (or allow them to air dry).
4. Clean plastic components using isopropyl alcohol or OMC Cleaning Solvent.

■ Skip the next step, we really mean it, don't INSPECT the impeller, REPLACE IT. Ok, we've been there before, if you absolutely don't want to replace the impeller. Let's say it's only been used one season or so and you're here for another reason, but you're just being thorough and checking the pump, then perform the next step.

5. Check the impeller for missing, brittle or burned blades. Inspect the impeller side surfaces and blade tips for cracks, tears, excessive wear or a glazed (or melted) appearance. Replace the impeller if these defects are found. Next, squeeze the vanes toward the hub and release them. The vanes should spring back to the extended position. Replace the impeller if the vanes are set in a curled position and do not spring back when released.
6. The water pump impeller should move smoothly up or downward on the driveshaft. If not, it could become wedged up against the housing or down against the impeller plate, causing undue wear to the top or bottom of the blades and hub. Check the impeller on the driveshaft and, if necessary, clean the driveshaft contact surface (inside the impeller hub) using emery cloth.
7. Inspect the water pump body or the lining inside the water pump body, as applicable, for burned, worn or damaged surfaces. Replace the impeller lining, if equipped, or the water pump body if any defects are noted.
8. Visually check the water pump housing (and insert, if equipped), along with the impeller wear plate for signs of overheating including warpage (especially on the plate) or melted plastic. Some wear is expected on the impeller plate and, if equipped, on the housing insert, but deep grooves (with edges) are signs of excessive wear requiring component replacement.

LUBRICATION AND COOLING 5-37

■ A groove is considered deep or edged if it catches a fingernail.

9. Check the water tube grommets and seals for a burned appearance or for cracked or brittle surfaces. Replace the grommets and seals if any of these defects are noted.

Thermostat

All 5 hp and larger motors (except the single-cylinder 5 hp 2-stroke) are equipped with a thermostat that restricts the amount of cooling water allowed into the powerhead until the powerhead reaches normal operating temperature. The purpose of the thermostat is to prevent cooling water from reaching the powerhead until the powerhead has warmed to normal operating temperature. In doing this the thermostat will increase engine performance and reduce emissions.

On all models so equipped, the thermostat components are mounted in a cooling passage, under an access cover that is sealed using a gasket or an O-ring. On most 2-stroke models the thermostat is mounted directly into the top of the cylinder head. On most 4-stroke models (except the 40/50 hp EFI motors) the thermostat is located somewhere other than the cylinder head. On some 4-strokes (specifically the 9.9/15 hp and 70 hp models) it is mounted in a bore on the top of the cylinder block and other 4-strokes (the 5/6 hp and 8/9.9 hp models) it is found on the top of the intake manifold.

REMOVAL & INSTALLATION

◆ See Figures 94 thru 105

On all models, the thermostat assembly is mounted under a cover on the powerhead. The size, shape and location of this cover, including the number of components and seals found underneath varies by model.

■ Some models are equipped with an assembled thermostat, meaning that the components of the thermostat (vernatherm, springs, diaphragm and cup) are removed individually. On these models it is imperative that you take note of the order and orientation of each component as you disassemble them to ensure proper installation and operation.

• For 5/6/8 hp (164cc), 9.9-15 hp (216/255cc), 25/35 hp (500/565cc), 25-55 hp (737cc) and 25-70 hp (913cc) motors, the thermostat is found under a small, irregular (on all except the 25/35 hp 3-cylinder motors) shaped cover facing aft, on top of the cylinder head. For these motors the thermostat assembly is normally an assembled unit that includes (in this order) the cover, seal/O-ring, washer, outer spring, diaphragm and cup assembly, vernatherm (thermostat), inner spring and housing. However, some early models (including some early 90's 5/6/8 hp, 25-55 hp [737cc] especially 45/55 Commercial models and 25-70 hp [913cc]) may be equipped with a one-piece, non-assembled thermostat. Just pay close attention to the order of components as they are removed for assembly reference.

• On 5/6 hp (128cc) 4-stroke motors, the thermostat is found under a round cover on the rear top of the intake manifold (on the port side of the cylinder head). The thermostat on these motors is an assembled unit consisting (in this order) of a cover, gasket, washer, outer spring, spacer/spring retainer, diaphragm and cup assembly, vernatherm (thermostat), inner spring and housing.

• On 8/9.9 hp (211cc) motors, the thermostat is found under a rounded irregular-shaped cover that is attached to the rear top of the intake manifold (port side). The thermostat on these motors is an assembled unit consisting (in this order) of a cover, gasket, washer, outer spring, diaphragm and cup assembly, vernatherm (thermostat), inner spring and housing.

• On 9.9/15 hp (305cc) motors, the thermostat is found under a rounded irregular-shaped cover that is attached to the top, rear of the cylinder block (starboard side). It is just in front of the ignition coil and slightly starboard of the engine-lifting bracket. The thermostat on these motors is a one-piece unit, mounted above a seal and directly below a cover that is sealed with a gasket.

• On 18 Jet-35 hp (521cc) motors, the thermostat is located on the top of the cylinder head, but the entire cylinder head cover must be removed for access. The thermostat on these motors is a one-piece unit mounted above a gasket and below a spring that contacts the cylinder head cover to hold the thermostat in position.

• On 40/50 hp EFI models, the thermostat is found under an irregular-shaped cover mounted to the top of the cylinder head, directly behind the engine-lifting bracket. The thermostat on these motors is a one-piece unit secured under the cover and sealed with a cover gasket.

• On 70 hp EFI models, the thermostat is found under a round cover at the aft, starboard top side of the cylinder block (directly behind the engine-lifting bracket). The thermostat on these motors is a one-piece unit secured under the cover by removable upper and lower stops. The cover is sealed with a gasket.

1. Disconnect the negative battery cable (if equipped) and/or remove the spark plug wire(s) from the plug(s) and ground them on the powerhead for safety.

2. Remove the engine top cover for access.

3. Locate the thermostat housing for the motor undergoing service. Refer to the accompanying paragraphs describing thermostat components and mounting location.

4. On some 4-stroke motors, a hose is attached to the thermostat housing cover. On these motors, the hose can either be disconnected or, in some cases if there is sufficient play in the hose, it can be left attached while the cover is removed and pushed aside for access to the thermostat. If necessary or desired, cut the wire tie or loosen the hose clamp, then carefully push the hose from the cover fitting.

■ On most 4-stroke motors, the cover can be installed facing different directions. For these models, matchmark the cover to the mating surface or otherwise make a note of cover orientation to ensure installation facing the proper direction. This is especially important for covers to which hoses attach (that have been removed).

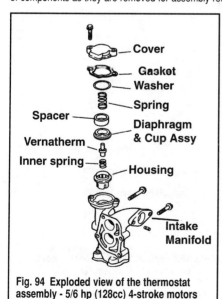

Fig. 94 Exploded view of the thermostat assembly - 5/6 hp (128cc) 4-stroke motors

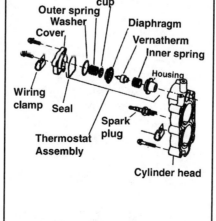

Fig. 95 Exploded view of the thermostat assembly - 5/6/8 hp (164cc) motors (note some early-models may use a 1-piece t-stat instead of the assembled type shown)

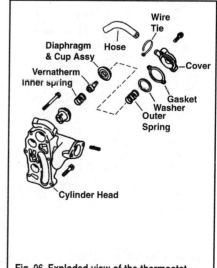

Fig. 96 Exploded view of the thermostat assembly - 8/9.9 hp (211cc) 4-stroke motors

5-38 LUBRICATION AND COOLING

5. Loosen and remove the thermostat cover as follows:
- For all models except the 25/35 hp motors, remove the cover bolts (on some models the bolts are Torx® head and require a suitably-sized Torx® driver). If necessary, tap around the outside of the cover using a rubber or plastic mallet to help loosen the seal, then remove the cover from the powerhead.
- For 25/35 hp motors, the cover is threaded in position and there is a large flat on the center of the cover. Use a suitably-sized wrench (or a large adjustable or large pair of slip-joint pliers) to loosen the cover and then unthread it from the cylinder head.

■ Remove the thermostat cover slowly, on most models (except for the 9.9/15 hp and larger 4-strokes) there is a spring and/or a washer located under the cover which may come loose when the cover is removed. Keep track of both the order and the orientation of all components as they are removed.

6. Check if the seal or gasket was removed with the cover. When a composite gasket and/or sealant was used, make sure all traces of gasket and sealant material are removed from the cover and the powerhead mounting surface.

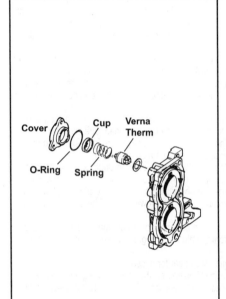

Fig. 97 Exploded view of the thermostat assembly - 9.9-15 hp (216cc) 2-stroke motors

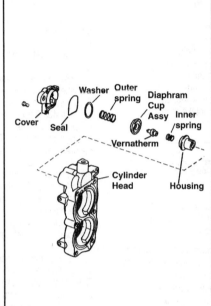

Fig. 98 Exploded view of the thermostat assembly - 9.9-15 hp (255cc) 2-stroke motors

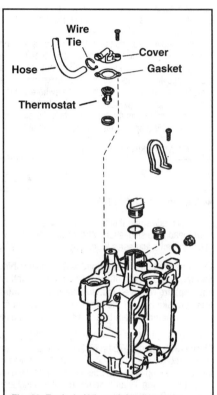

Fig. 99 Exploded view of the one-piece thermostat assembly - 9.9/15 hp (305cc) 4-stroke motors

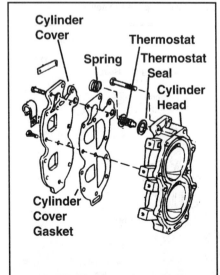

Fig. 100 Exploded view of the one-piece thermostat assembly - 18 Jet-35 hp (521cc) motors

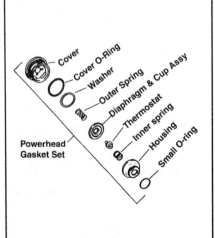

Fig. 101 Exploded view of the thermostat assembly - 25/35 hp (500/565cc) 2-stroke motors

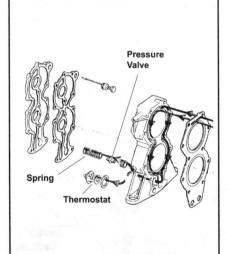

Fig. 102 Exploded view of the thermostat and pressure valve assemblies - 45/55 hp (737cc) 2-stroke Commercial motors through 1994

LUBRICATION AND COOLING 5-39

7. Remove the vernatherm and any mounting components. On most assembled thermostats that means removing the washer, outer spring, diaphragm and cup assembly, followed by the vernatherm, inner spring and housing. Refer to the accompanying descriptions and diagrams for more details. Pay close attention to each component's orientation. Lay each component out on the worksurface in order to ensure installation facing the same directions.

8. Visually inspect the thermostat for obvious damage including corrosion, cracks/breaks or severe discoloration from overheating. Make sure any springs have not lost tension. If necessary, refer to the Testing the Thermostat in this section for details concerning using heat to test thermostat function.

To install:

9. Install each of the thermostat components in the reverse of the removal procedure. Replace any gaskets, seals and/or O-rings. Pay close attention to the direction each component is installed. On most models with assembled thermostats, the pin end of the vernatherm should face outward.

■ On EFI models, the thermostat should be positioned with the spring and copper plug side of the assembly facing the engine NOT the cover.

10. For all models except the 9.9-15 hp 2-strokes and the 25/35 hp (500/565cc) motors, apply a thin coating of OMC Gasket Sealing Compound or an equivalent sealant to both side of a NEW composite gasket and position the gasket on the cover. Or, if an O-ring seal is used, apply a thin bead of the sealant to the cover groove, then install the seal into the groove and apply a thin coating on top of the seal.

11. For 9.9-15 hp 2-strokes and the 25/35 hp (500/565cc) motors position the new seal on the cover without using sealant.

12. Apply a light coating of OMC Gasket Sealing Compound or equivalent sealant to the threads of the cover bolts on carbureted models. For EFI models, apply a light coating of OMC Nut Lock to the cover bolt threads. For 25/35 hp (500/565cc) motors, do not treat the threads of cover housing.

13. Install the thermostat cover to the powerhead, making sure the gasket or seal and any springs, washers or other thermostat components are all properly seated. Tighten the cover bolts (or the cover itself on 25/35 hp 3-cylinder motors) securely.

■ For 18 Jet-35 hp (521cc) motors, the cylinder head cover (which also acts as the thermostat cover) bolts must be tightened using a criss-cross pattern, starting at the center and working outward. Be sure to tighten these bolts to 60-84 inch lbs. (7-9 Nm).

14. If removed on 4-stroke motors so equipped, connect the hose to the thermostat cover fitting and secure using the clamp or a new wire tie.

15. Connect the negative battery cable and/or spark plug lead(s), then verify proper cooling system operation.

Water Pressure Relief Valve

◆ See Figures 102 and 106

The job of the pressure relief valve is to open in response to system pressure providing relief and preventing potential damage from over-pressurization. On many thermostat equipped models (especially most assembled thermostats that are designed so one of the assembly springs is a pressure relief spring), the thermostat assembly acts as a water pressure relief valve. This method however is inefficient, as it can counteract the very job of the thermostat (to remain closed during engine warm-up/prevent overcooling).

EFI models are equipped with a separate water pressure relief valve that is capable of diverting water from the powerhead during periods of over-pressurization. This valve is located on the powerhead for 70 hp motors, or inside the gearcase on 40/50 hp motors.

A FEW of the carbureted models, like the 45/55 Commercial Twins through 1994 are also equipped with a pressure relief valve. In the specific case of the 45/55 Commercials, it is located toward the base of the cylinder head, under the cylinder head cover and next to the thermostat assembly. For removal and installation of this particular valve, follow the same procedures as necessary for Thermostat service, listed earlier in this section.

REMOVAL & INSTALLATION

◆ See Figure 106

 MODERATE

■ For 40/50 hp EFI motors, the water pressure relief valve is located inside the gearcase, underneath the oil pan/sump. The oil pan must be removed in order to access the valve. For this reason, please refer to the procedures located in the Gearcase section on 40/50 hp EFI motors.

■ For 45/55 hp Commercial twins the pressure relief valve is located toward the base of the cylinder head, under the cylinder head cover and next to the thermostat assembly. For removal and installation of this particular valve, follow the same procedures as necessary for Thermostat service, listed earlier in this section.

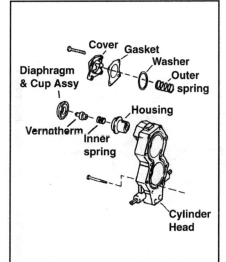

Fig. 103 Exploded view of the thermostat assembly - 25-55 hp (737cc) 2-stroke motors shown (except 45/55 Commercial models through 1994), 25-70 hp (913cc) 2-stroke motors same except cylinder head varies (and some early models may not use an assembled vernatherm, meaning there will be less components)

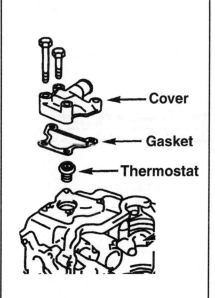

Fig. 104 Exploded view of the one-piece thermostat assembly - 40/50 hp EFI 4-stroke motors

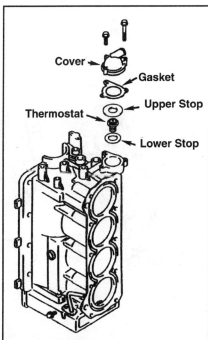

Fig. 105 Exploded view of the one-piece thermostat assembly - 70 hp EFI 4-stroke motors

5-40 LUBRICATION AND COOLING

Fig. 106 A water pressure relief valve is mounted under a thermostat-like housing on 70 hp EFI motors

For 70 hp motors, the water pressure relief valve is found under a cover on the upper, starboard side of the powerhead. It is located under a thermostat-like cover at the top of the exhaust manifold water cooling passage.

The valve is designed to open at 5.56-5.84 psi (38.22-40.18 kPa) of pressure, allowing about 5.3 gallons (20 liters) per minute to flow at 7 psi (49 kPa). Be especially careful when servicing it, as the manifold WILL overheat if it is installed backwards.

■ You've got a choice. On some installations, it may be possible to unbolt and reposition the cover with the hose attached. If not, you should matchmark the cover to the powerhead to ensure installation facing the proper direction.

1. If the hose is being removed, matchmark the cover, then squeeze the tabs of the spring-loaded clamp and slide the clamp back on the hose. Disconnect the hose from the water pressure relief valve cover fitting.
2. Remove the 2 cover retaining bolts, then carefully break the gasket seal and remove the cover from the powerhead. If necessary, use a rubber or plastic mallet to help break the seal.

■ Remove the cover slowly so the valve does not become dislodged. Remember that you need to note the direction in which it is currently installed in order to prevent manifold overheating if it is accidentally installed facing the wrong way.

3. Remove the valve from the coolant passage, noting the direction it is facing.
4. Check the valve for obviously damaged, worn or corroded surfaces. Check the valve spring for corrosion, damaged or debris. Replace the valve if any defects are noted or if it is suspect.
5. Carefully clean all traces of gasket or sealant from the valve cover mating surfaces.

To install:
6. Install the valve into the coolant passage, facing the same direction as noted during removal.
7. Apply a light coating of OMC Gasket Sealing Compound, or equivalent sealant to a NEW gasket cover, then install the cover and gasket. Be sure align the matchmarks made earlier, making sure the cover is facing in the same direction.
8. Install the retaining bolts and tighten securely.
9. If removed, connect the hose to the cover fitting, then position the spring clamp to secure the hose. As with all spring clamps, replace it before connecting the hose if it feels like it has lost its tension.

COOLING SYSTEM SCHEMATICS
◆ See Figures 107 thru 123

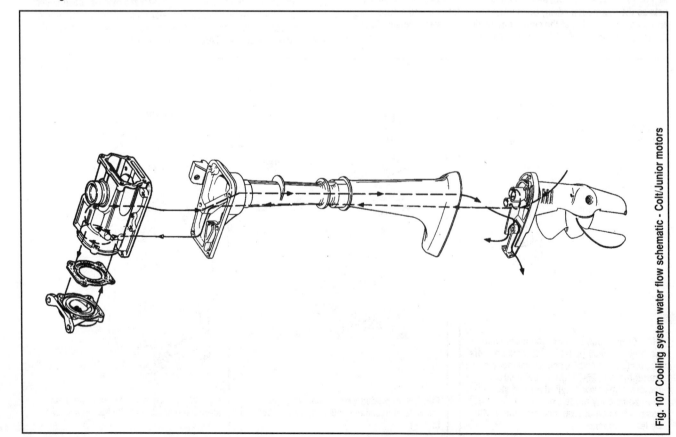

Fig. 107 Cooling system water flow schematic - Colt/Junior motors

LUBRICATION AND COOLING 5-41

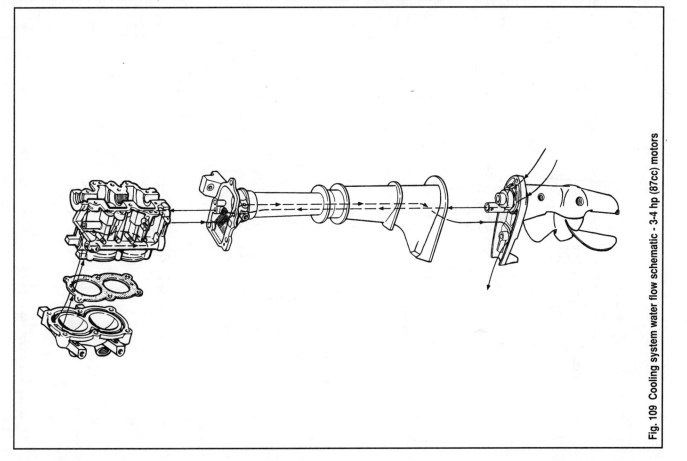

Fig. 109 Cooling system water flow schematic - 3-4 hp (87cc) motors

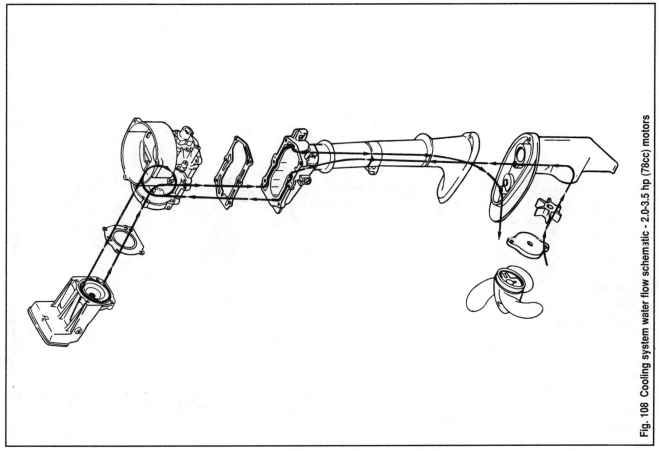

Fig. 108 Cooling system water flow schematic - 2.0-3.5 hp (78cc) motors

5-42 LUBRICATION AND COOLING

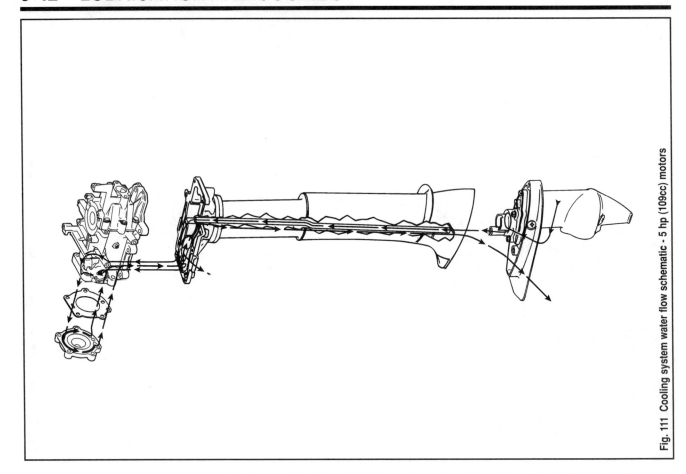

Fig. 111 Cooling system water flow schematic - 5 hp (109cc) motors

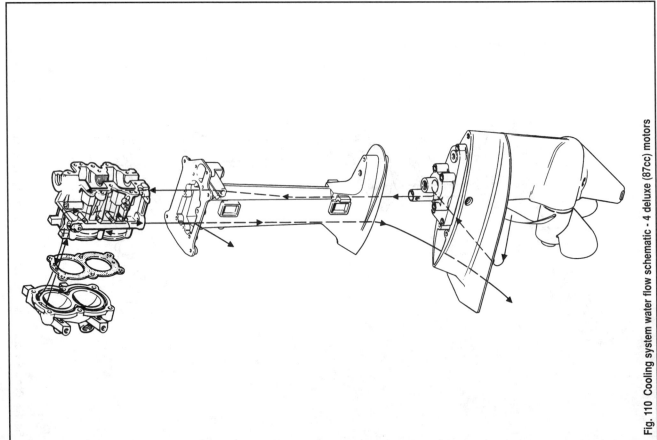

Fig. 110 Cooling system water flow schematic - 4 deluxe (87cc) motors

LUBRICATION AND COOLING 5-43

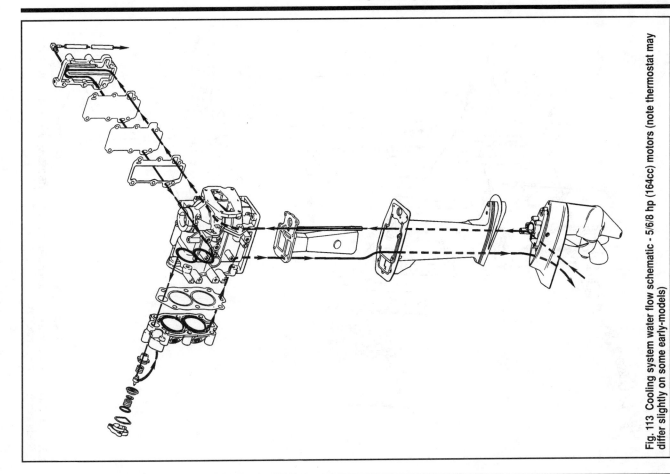

Fig. 113 Cooling system water flow schematic - 5/6/8 hp (164cc) motors (note thermostat may differ slightly on some early-models)

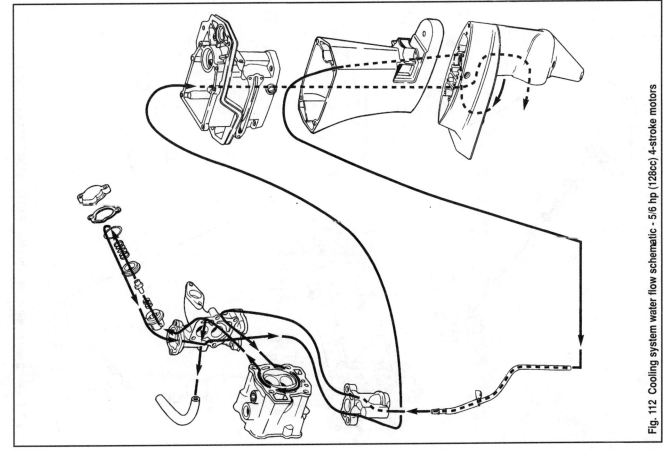

Fig. 112 Cooling system water flow schematic - 5/6 hp (128cc) 4-stroke motors

5-44 LUBRICATION AND COOLING

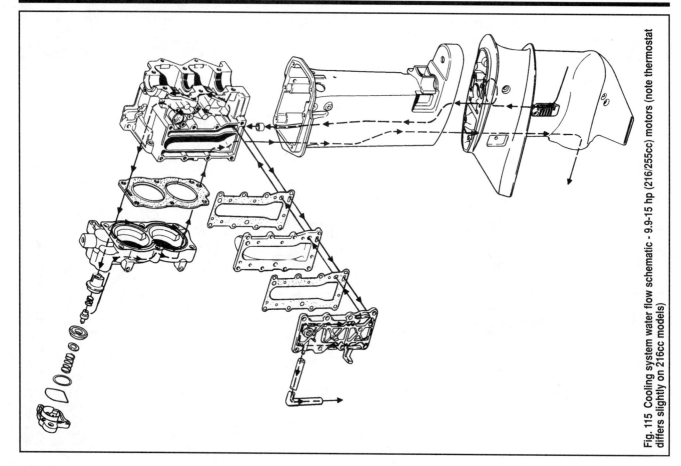

Fig. 115 Cooling system water flow schematic - 9.9-15 hp (216/255cc) motors (note thermostat differs slightly on 216cc models)

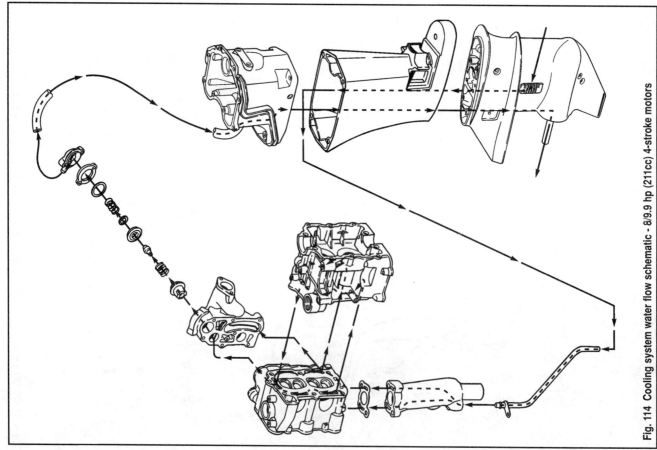

Fig. 114 Cooling system water flow schematic - 8/9.9 hp (211cc) 4-stroke motors

LUBRICATION AND COOLING 5-45

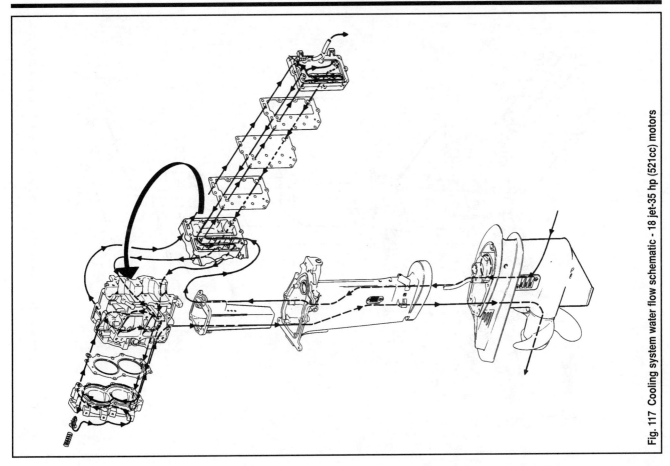

Fig. 117 Cooling system water flow schematic - 18 jet-35 hp (521cc) motors

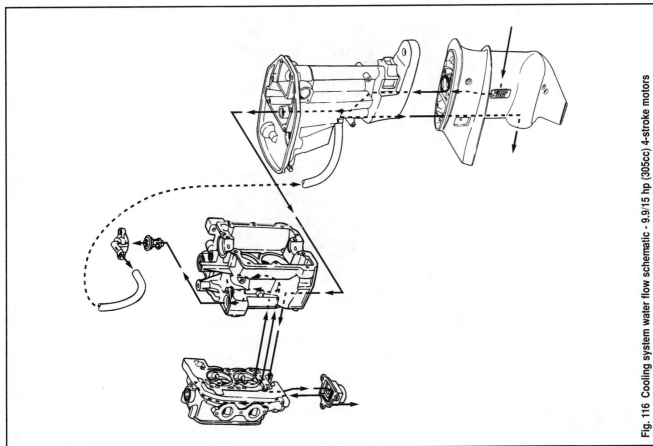

Fig. 116 Cooling system water flow schematic - 9.9/15 hp (305cc) 4-stroke motors

5-46 LUBRICATION AND COOLING

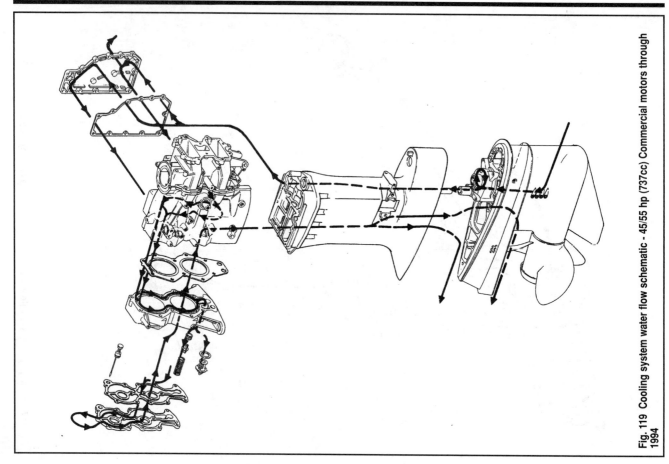

Fig. 119 Cooling system water flow schematic - 45/55 hp (737cc) Commercial motors through 1994

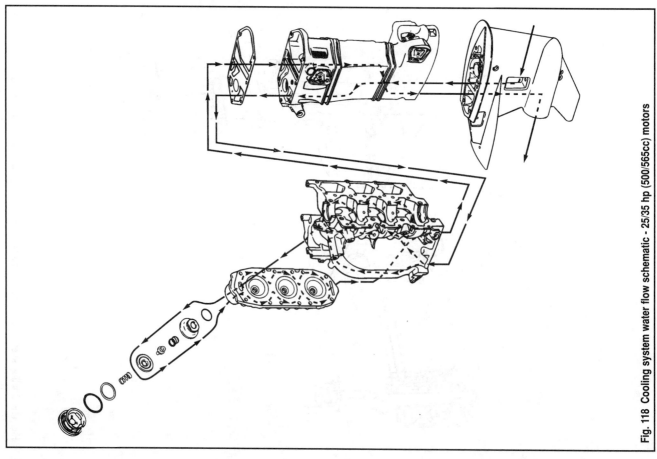

Fig. 118 Cooling system water flow schematic - 25/35 hp (500/565cc) motors

LUBRICATION AND COOLING 5-47

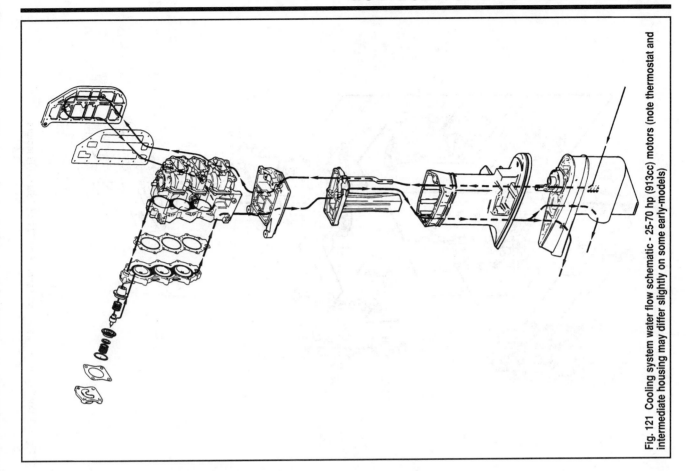

Fig. 121 Cooling system water flow schematic - 25-70 hp (913cc) motors (note thermostat and intermediate housing may differ slightly on some early-models)

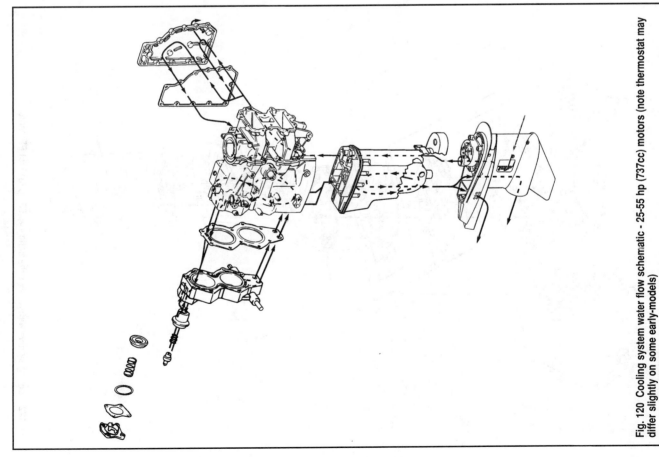

Fig. 120 Cooling system water flow schematic - 25-55 hp (737cc) motors (note thermostat may differ slightly on some early-models)

5-48 LUBRICATION AND COOLING

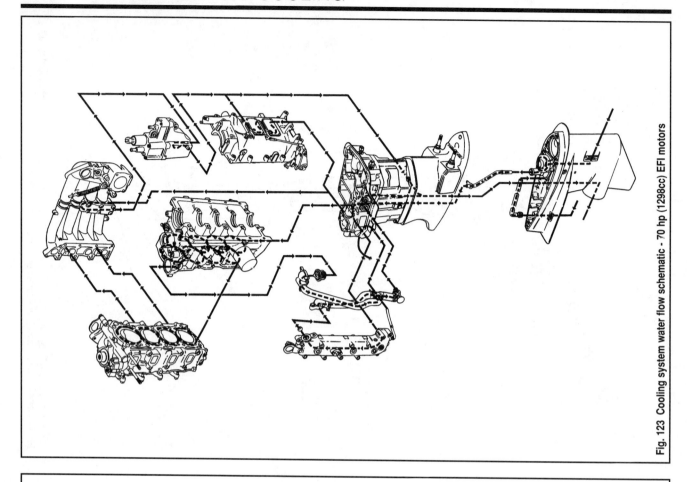

Fig. 123 Cooling system water flow schematic - 70 hp (1298cc) EFI motors

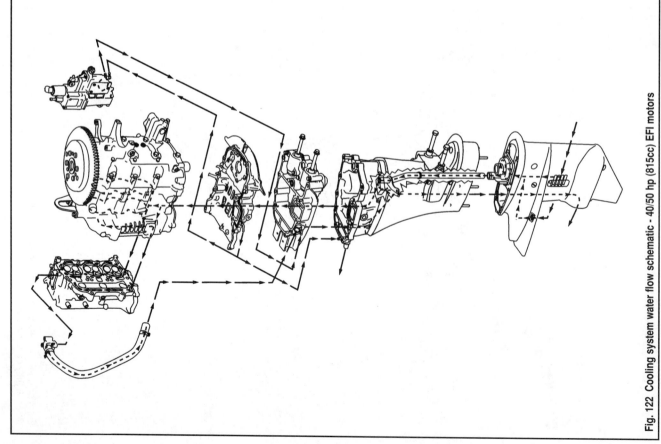

Fig. 122 Cooling system water flow schematic - 40/50 hp (815cc) EFI motors

6

POWERHEAD REPAIR & OVERHAUL

POWERHEAD	6-2
POWERHEAD BREAK-IN	6-87
SPECIFICATIONS	6-89

BREAKING IN A POWERHEAD......... 6-88
  2-STROKE MOTORS 6-88
    Colt/Junior and 2-8 Hp 6-88
    9.9-70 Hp Motors (Exc. 25/35 Hp 3-Cyl) .. 6-88
    25/35 Hp 3-Cyl................ 6-88
  CARBURETED 4-STROKE 6-88
  EFI 4-STROKE 6-89
POWERHEAD SERVICE................ 6-4
  COLT/JUNIOR...................... 6-5
    Disassembly & Assembly 6-5
    Removal & Installation 6-5
  2.0-3.5 HP (78cc) 6-7
    Disassembly & Assembly 6-7
    Removal & Installation 6-7
  5 HP (109cc)...................... 6-9
    Disassembly & Assembly 6-9
    Removal & Installation 6-9
  5/6 HP (128cc) 4-STROKE 6-11
    Cylinder Block Overhaul......... 6-14
    Cylinder Head Overhaul 6-13
    Powerhead Overhaul 6-12
    Removal & Installation 6-11
  3/4 HP & 4 DELUXE (87cc)......... 6-16
    Disassembly & Assembly 6-18
    Removal & Installation 6-16
  5/6/8 HP (164cc) 6-16
    Disassembly & Assembly 6-18
    Removal & Installation 6-16
  8/9.9 HP (211cc) 4-STROKE 6-23
    Cylinder Block Overhaul......... 6-26
    Cylinder Head Overhaul 6-25
    Powerhead Overhaul 6-24
    Removal & Installation 6-23
  9.9/10/14/15 HP (216cc) 6-28
    Disassembly & Assembly 6-29
    Removal & Installation 6-28
  9.9/10/15 HP (255cc)............. 6-28
    Disassembly & Assembly 6-29
    Removal & Installation 6-28
  9.9/15 HP (305cc) 4-STROKE 6-33
    Cylinder Block Overhaul......... 6-35
    Cylinder Head Overhaul 6-35
    Powerhead Disassembly & Assembly 6-34
    Removal & Installation 6-33
  18 JET-35 HP (521cc).............. 6-38
    Disassembly & Assembly 6-39
    Removal & Installation 6-38
  25/35 HP (500/565cc) 6-43
    Disassembly & Assembly 6-45
    Removal & Installation 6-43
  25-55 HP (737cc) 6-49
    Disassembly & Assembly 6-52
    Removal & Installation 6-49
  25-70 HP (913cc) 6-49
    Disassembly & Assembly 6-52

    Removal & Installation 6-51
  40/50 HP (815cc) EFI 6-58
    Cylinder Head Overhaul 6-61
    Cylinder Block Overhaul......... 6-61
    Powerhead Overhaul 6-60
    Removal & Installation 6-58
  70 HP (1298cc) EFI................ 6-65
    Removal & Installation 6-65
    Powerhead Disassembly & Assembly 6-67
    Cylinder Head Overhaul 6-68
    Cylinder Block Overhaul......... 6-69
  CLEANING & INSPECTION 6-72
    Cleaning 6-72
    Cylinder Head Component Inspection... 6-73
    Cylinder Block Component Inspection... 6-77
POWERHEAD........................ 6-2
  FLYWHEEL........................ 6-2
    Removal & Installation 6-2
  POWERHEAD...................... 6-4
    Service........................ 6-4
    Cleaning & Inspection 6-72
  TIMING BELT/CHAIN 6-84
    Removal & Installation 6-84
POWERHEAD BREAK-IN.............. 6-87
  BREAKING IN A POWERHEAD 6-88
    2-Stroke Motors 6-88
    Colt/Junior and 2-8 Hp Motors 6-88
    9.9-70 Hp Motors (Exc. 25/35 Hp 3-Cyl).. 6-88
    25/35 Hp 3-Cylinder Models 6-88
    Carbureted 4-Stroke Motors 6-88
    EFI 4-Stroke Motors 6-89
SPECIFICATIONS.................... 6-89
  COLT/JUNIOR (43cc) 1 CYL......... 6-89
  2.0-3.5 HP (78cc) 1 CYL........... 6-90
  5.0 (109cc) 1 CYL................ 6-90
  5/6 HP (128cc) 1 CYL.............. 6-91
  3/4/4 Deluxe HP (87cc) 2 CYL...... 6-91
  8/9.9 HP (211cc) 2 CYL............ 6-92
  5.0-8.0 HP (164cc) 2 CYL.......... 6-92
  9.9/10/14/15 HP (216cc) 2 CYL..... 6-93
  9.9/10/15 HP (255cc) 2 CYL....... 6-93
  9.9/15HP (305cc) 2 CYL........... 6-94
  18-35 HP (521cc) 2 CYL........... 6-94
  25-55 HP (737cc) 2 CYL........... 6-95
  25/35 HP (500/565cc) 3 CYL...... 6-95
  25-70 HP (913cc) 3 CYL........... 6-96
  40/50 HP (815cc) 3 CYL........... 6-96
  40/50 HP (815cc) 3 CYL........... 6-97
  70 (1298cc) 4 CYL................ 6-98
TIMING BELT/CHAIN 6-84
  REMOVAL & INSTALLATION 6-84
    5-15 Hp....................... 6-84
    40/50 Hp EFI.................. 6-85
    70 Hp EFI 6-86

POWERHEAD REPAIR AND OVERHAUL

POWERHEAD

You can compare the major components of an outboard with the engine and drivetrain of your car or truck. In doing so, the powerhead is the equivalent of the engine and the gearcase is the equivalent of your drivetrain (the transmission/transaxle). The powerhead is the assembly that produces the power necessary to move the vehicle, while the gearcase is the assembly that transmits that power via gears, shafts and a propeller (instead of tires).

Speaking in this manner, the powerhead is the "engine" or "motor" portion of your outboard. It is an assembly of long-life components that are protected through proper maintenance. Lubrication, the use of high-quality oils (2-stroke or 4-stroke) and proper fuel/oil ratios (2-stroke) or frequent oil inspection/changes (4-stroke) are the most important ways to preserve powerhead condition. Similarly, proper tune-ups that help maintain proper air/fuel mixture ratios and prevent pinging, knocking or other potentially damaging operating conditions are the next best way to preserve your motor. But, even given the best of conditions, components in a motor begin wearing the first time the motor is started and will continue to do so over the life of the powerhead.

Eventually, all powerheads will require some repair. The particular broken or worn component, plus the age and overall condition of the motor may help dictate whether a small repair or major overhaul is warranted. The complexity of the job will vary with 2 major factors. As much as you can generalize about mechanical work:
- The age of the motor (the older OR less well maintained the motor is) the more difficult the repair
- The larger and more complex the motor, the more difficult the repair.

Again, these are generalizations and, working carefully, a skilled do-it-yourself boater can disassemble and repair a 70 hp EFI powerhead, as well as a seasoned professional. But both DIYers and professionals must know their limits. These days, many professionals will leave portions of machine work (from cylinder block and piston disassembly, clean and inspection to honing and assembly up to a machinist). This is not because they are not capable of the task, but because that's what a machinist does day in and day out. A machinist is naturally going to be more experienced with the procedures.

If a complete powerhead overhaul is necessary on your outboard, we recommend that you find a local machine shop that has both an excellent reputation and that specializes in marine work. This is just as important and handy a resource to the professional as a DIYer. If possible, consult with the machine shop before disassembly to make sure you follow procedures or mark components, as they would desire. Some machine shops would prefer to perform the disassembly themselves. In these cases, you can usually remove the powerhead from the gearcase and deliver the entire unit to the shop for disassembly, inspection, machining and assembly.

If you decide to perform the entire overhaul yourself, proceed slowly, taking care to following instructions closely. Consider using a digital camera (if available) to help document assemblies during the removal and disassembly procedures. This can be especially helpful if the overhaul or rebuild is going to take place over an extended amount of time. If this is your first overhaul, don't even THINK about trying to get it done in one weekend, YOU WON'T. It is better to proceed slowly, asking help when necessary from your trusted parts counterman or a tech with experience on these motors.

Keep in mind that anytime pistons, rings and bearings have been replaced, the powerhead must be broken-in again, as if it were a brand-new motor. Once a major overhaul is completed, refer to the section on Powerhead Break-In for details on how to ensure the rings set properly without damage or scoring to the new cylinder wall or the piston surfaces. Careful break-in or a properly overhauled motor will ensure many years of service for the trusty powerhead.

Flywheel

On all models, the flywheel is secured to the top of the crankshaft. The flywheel is important to engine operation on multiple levels. First and foremost, it represents a means by which the crankshaft can be rotated (either by hand using a manual rewind starter or via an electric starter motor) for engine start-up. Mechanically, the flywheel is used as a means of continuing engine rotation and momentum between piston powerstrokes. Permanent magnets are mounted to all Johnson/Evinrude flywheels that are used with the charging and/or ignition systems to generate voltage in various coils mounted underneath the flywheel (for more details, refer to the Ignition and Electrical System section).

REMOVAL & INSTALLATION

◆ See Figures 1 thru 7

Flywheel removal and installation is a relatively straightforward procedure during which a tool of some sort it used to hold the flywheel (and therefore the crankshaft) from rotating while the retaining nut (carbureted motors) or bolt (EFI motors) is loosened. Then once the flywheel is unbolted, a universal puller is usually necessary to free it from the tapered portion of the crankshaft.

Various tools can be used to hold the flywheel on most models. On almost all models a suitably sized (large) strap wrench that fits around the entire flywheel can usually be used. But, the high torque value on the flywheel retainer, especially for EFI motors, may make this difficult. A flywheel holding tool designed to engage with multiple teeth along the curved circumference of the flywheel is a good alternative.

■ **Although mechanics have been known to do it, we don't recommend using a flywheel holding tool or prybar against just ONE flywheel tooth and a boss or bolt in the top of the powerhead. Placing that much stress on a single tooth can result in damage to the flywheel (such as breaking teeth from the ring gear) resulting in a need for flywheel replacement.**

Perhaps the best answer for holding the flywheel steady on most motors is to fabricate a holding tool from cold rolled steel, but the tool can be prohibitively large for some of the larger Johnson/Evinrude outboards. If a flywheel holder is fabricated, it must be strong enough to hold against the flywheel retainer torque value, even with holes cut into it for 3 flywheel holding bolts and a larger hole cut into the center for the socket that is used to loosen the flywheel retainer.

On motors through 35 hp (565cc and smaller) a flywheel holding tool can be easily fabricated using a length of 1/4 in. (6.3mm) cold roll steel that is about 15 in. (38cm) in length. Use measurements in the accompanying illustration to cut the appropriate sized holes (one for the socket and the other 3 for the bolts that will be threaded into the top of the flywheel in order to hold it steady when loosening the flywheel retainer). This tool is used along with 3 bolts that are threaded into the top of the flywheel to hold it securely.

※※ WARNING

Never strike the flywheel to loosen it otherwise the magnets may crack or dislodge resulting in poor charging and/or ignition system performance. Always use a puller to help dislodge a tight flywheel.

1. If equipped, remove the flywheel cover or hand-rewind starter/pulley from the top of the motor and the flywheel. For details on rope start models, refer to the Hand Rewind Starter section.
2. Secure the flywheel using a suitable holding tool. When using a fabricated holding tool, use a set of bolts (such as the puller bolts) to firmly seat the tool against the top of the flywheel.
3. Using a large breaker bar and suitably sized socket, turn the flywheel retaining nut (carbureted motors) or bolt (EFI motor) counterclockwise to loosen and remove it from the top of the flywheel.

■ **To protect the crankshaft from damage, apply a light coating of OMC Moly Lube or equivalent lubricant to the puller center (pressing) screw threads and to the center hole of the crankshaft**

4. Remove the flywheel holder and install a suitable puller to the top of the flywheel by aligning the puller slots with 3 bolt holes in the top of the flywheel. Install the puller bolts to secure the assembly, then installed center screw to the puller to drive the flywheel off the top of the crankshaft. Make sure the puller assembly is sitting level, parallel to the top of the flywheel (adjust the puller bolts as necessary).

■ **A universal threaded puller is best for this purpose, refrain from using a jawed puller, even if you have one that is big enough, because they do not spread the force out as well as a 3-bolt, threaded puller.**

5. Carefully tighten the puller center screw until the flywheel releases

POWERHEAD REPAIR AND OVERHAUL 6-3

from the crankshaft taper. Remove the flywheel and puller tool assembly and then remove the puller tool from the flywheel.

6. Thoroughly clean and inspect the flywheel assembly for signs of damaged/missing teeth or magnets and for any cracks, severe discoloration or other signs of damage. Replace the flywheel, if necessary.

7. Remove the flywheel key from the crankshaft taper. Check the end of the crankshaft and key for damage. If the key is damaged in any way, replace it to ensure proper engine operation.

To install:

8. Install the flywheel drive key (a NEW one, if necessary) into the key slot in the crankshaft. Be sure to face the rounded side of the key toward the crankshaft. On carbureted motors, pay close attention to key alignment. On all models that utilize a marked key, the upset mark should face downward. On all carbureted motors, except the 9.9-15 hp (216/255cc) 2-strokes, make sure the outside of the key (flat side) is positioned parallel to the centerline of the crankshaft (straight up and down). On 9.9-15 hp (216/255cc) 2-strokes,

Fig. 1 Use a large strap wrench . . .

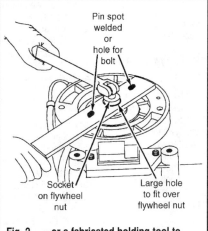

Fig. 2 . . . or a fabricated holding tool to hold the flywheel

Fig. 3 Once the retainer is removed, use a puller to free the flywheel

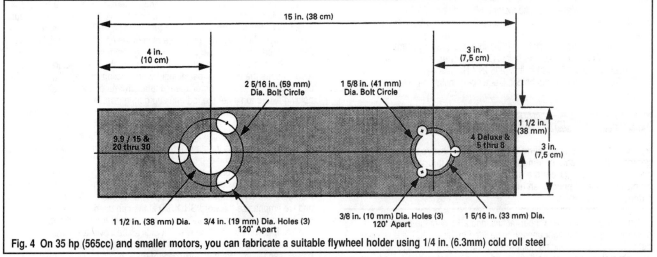

Fig. 4 On 35 hp (565cc) and smaller motors, you can fabricate a suitable flywheel holder using 1/4 in. (6.3mm) cold roll steel

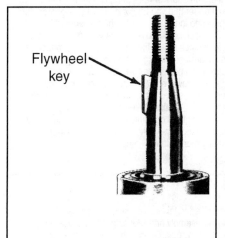

Fig. 5 Install the keyway to the crankshaft slot - 2-5 hp single-cylinder, 2-stroke shown

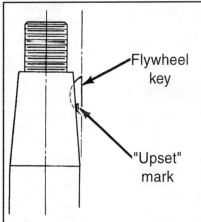

Fig. 6 On carbureted motors larger than the 2-5 hp single-cylinder, 2-strokes, face key the upset mark downward

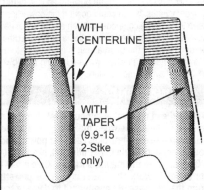

Fig. 7 Depending on the model the key should be installed either parallel to the shaft centerline (most models) or to the crankshaft taper (9.9-15 hp 216/255cc 2-strokes)

6-4 POWERHEAD REPAIR AND OVERHAUL

the flat side of the keyway should be positioned parallel to the crankshaft taper (top portion tilted slightly toward the center of the crankshaft).

■ On Colt/Junior or other breaker point models there is a cam drive pun mounted in the base of the exposed portion of the crankshaft (below and inline with the keyway). If the cam drive pin was removed, apply a light coating of OMC nut lock or equivalent threadlocker to the pin, then install the pin into the crankshaft. Also, on these models, the flywheel key itself may be marked with an end that says UP or TOP, and if so marked, the mark should face upward.

9. Clean the crankshaft and flywheel tapers using solvent and allow to air dry. The surfaces must be clean and dry to allow the proper taper locking once the flywheel is installed.

10. Lower the flywheel onto the crankshaft taper, turning the flywheel in order to align the keyway properly with the key. Once aligned, carefully push downward on the flywheel to seat it.

11. Verify that the keyway and key are properly engaged by rotating the flywheel slowly by hand in the normal direction of rotation (clockwise when viewed from above). The crankshaft must turn along with the flywheel or the key and keyway are not properly engaged (if so, remove the flywheel and realign it, inspect they key again to make sure there is no damage.)

12. On carbureted motors, apply a coating of OMC Gasket Sealing Compound or equivalent sealant to the threads of the flywheel nut.

13. Install the flywheel retaining nut (carbureted motors) or bolt (EFI motors) and washer (if used).

14. Using the flywheel holder, secure the flywheel from turning and tighten the flywheel retainer to:
  • Colt/Junior motors: 22-25 ft. lbs. (30-40 Nm).
  • 2-5 hp, single-cylinder 2-stroke motors: 29-33 ft. lbs. (40-45 Nm).
  • 3/4 hp and 4 Deluxe, two-cylinder 2-stroke motors: 30-40 ft. lbs. (40-54 Nm).
  • 5-8 hp, two-cylinder 2-stroke motors: 40-50 ft. lbs. (54-70 Nm).
  • 5-15hp 4-stroke motors and 9.9-15 hp 2-strokes: 45-50 ft. lbs. (60-70 Nm).
  • 18 Jet-70 hp 2-stroke motors: 100-105 ft. lbs. (135-140 Nm).
  • 40-50 EFI motors: 145-150 ft. lbs. (200-205 Nm).
  • 70 EFI motors: 150-160 ft. lbs. (205-217 Nm).

15. Remove the flywheel holder. If equipped, install the flywheel cover or hand rewind starter.

Powerhead

✱✱ **CAUTION** When removing the powerhead, always secure the gearcase in a suitable holding fixture to prevent injury damage if the powerhead releases from the gearcase suddenly (which often occurs on outboards that have been in service for some time or that are extensively corroded).

On most of the single cylinder motors, and a FEW of the twins, the powerhead is actually light enough to be lifted by hand. BUT, it is really a better idea to use an engine hoist when lifting the powerhead assembly. This not only makes sure that the powerhead is secured at all times, but helps prevent injuries, which could occur if the powerhead releases suddenly. Also, keep in mind the components such as the driveshaft could be damaged if the powerhead is removed or installed at an angle other than perfectly perpendicular to the gearcase. It is a lot easier to align the powerhead when it is supported, than when you are holding the powerhead and trying to raise it from or lower it into position.

✱✱ WARNING

If powerhead removal is difficult, first check to make sure there are no missed fasteners between the gearcase (midsection, exhaust housing, oil sump on some 4-strokes) and the powerhead itself. If none are found, apply suitable penetrating oil, like WD-40® or our new favorite, PB Blaster® to the mating surfaces. Give a few minutes or a few hours for the oil to work, then carefully pry the powerhead free while lifting it from the gearcase. Be careful not to damage the mating surfaces by using any sharp-tipped prybars or other by prying on thin/weak gearcase or powerhead bosses/surfaces.

When working on a powerhead, use either a very sturdy workbench or an engine stand to hold the assembly. Keep in mind that larger HP motors utilize powerheads that can weight several hundred pounds.

Although removal and installation is relatively straightforward on most models, the overhaul procedures can be quite involved. Whatever portion you decide to tackle, always, always, always, ALWAYS take good notes and tag as many parts/hoses/connections as you can during the removal process. As they are removed, arrange components along the worksurface in the same orientation to each other as they are when installed.

Most Johnson/Evinrude motors (especially the 2-stroke models) are equipped with needle/roller bearings for the crankshaft, connecting rod and/or wrist pins. We say most, but not all, because some models also use or instead use bearing liner inserts. Of the ones equipped with needle/roller bearings, some use bearings contained completely within cages, while others (many of the wrist pin bearings and some crankshaft bearings on smaller models) use loose needle assemblies. When working with loose needle bearings, care must be taken that none are lost during removal and that all are aligned during installation.

Most Johnson/Evinrude connecting rods are of the fractured cap design. This means that, during manufacture, the connecting rod and cap were once 1 piece, but they were broken apart. This style of manufacturer is common in many high-performance applications from marine engines to motorcycles and some automobiles. It affects service in 2 ways. For starters, on models so equipped, it becomes critical that connecting rod-to-cap alignment is maintained. Not only can you not install a connecting rod cap from a different rod, but also you can't install a given cap in the reverse direction on its correct rod. Although the manufacturer places alignment marks or dimples on the rod and cap of most fractured connecting rods, be sure to matchmark all connecting rod caps before removal. When installed, the fracture lines of the caps should all but disappear. To check alignment, run a fingernail or a pick across the fracture line. If the nail or pick catches, recheck the alignment marks or make sure the connecting rod cap is properly bolted in position.

When it comes to bolting the connecting rod cap to the rod end, the second issue arises. On all models, to ensure proper alignment, it is a good idea to loosen or tighten the bolts in stages, alternating from one bolt to the other, turning each bolt the same amount each time. Make sure that amount is less than one full turn (try 1/4 - 1/2 turn each time). This may seem tedious, but it helps ensure proper seating of the cap and, if used, bearing liner or needle cage. Proper rod and bearing alignment become critical on larger versions of these motors, so much so that the manufacturer recommends the use of a special tool for connecting rod cap alignment. For all 2-strokes 18 Jet and larger, the manufacturer recommends using a holding fixture to center the rod and cap during installation.

We are well aware than many shops use alternate methods to tighten the connecting rod caps. One popular method is to run the caps down (alternating the bolts until finger tight) then tightening the bolts in stages. Still only turning them less than one turn until the torque specification is reached. Once all connecting rod cap bolts are tightened using this method, a soft-faced (brass) mallet or hammer is held against one side of the rod and cap, then the other side is gently tapped with another hammer. The position of the hammers is then reversed and the tapping is repeated. This method **should** and we emphasize the word **SHOULD** fit the cap/bearing assembly to the rod. However, there is no way to be certain except to return the motor to service. If it worked, she'll be fine. If it didn't, she'll eventually wear the crankshaft journal, connecting rod and bearing, causing premature powerhead failure. We therefore, cannot really recommend this method with full confidence, and would prefer that you buy or borrow the alignment tool

Service procedures provided in this section for the powerhead include Removal & Installation, Disassembly & Assembly, as well as Cleaning & Inspection. Since multiple engines of varying hp are normally produced from the same basic mechanical design, sharing components and procedures across different hp models, most powerhead service procedures are divided by engines or engine families. However, Cleaning & Inspection procedures share much more features between motors of different families and are covered mostly under a common section. Differences in cleaning and inspection procedures will be pointed out, and will mostly occur in the lack or inclusion of different components (especially when it comes to 4-stroke models since they use many more internal components).

SERVICE

◆ See Figure 8

Service procedures provided in this section include Removal & Installation, Disassembly & Assembly and Cleaning & Inspection. Most powerhead service procedures are generally divided by engines or engine families (since many times multiple engines of varying hp are actually of the same basic mechanical design, sharing components and procedures across

POWERHEAD REPAIR AND OVERHAUL   6-5

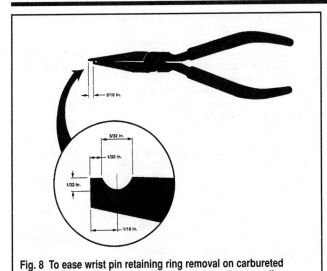

Fig. 8 To ease wrist pin retaining ring removal on carbureted Johnson/Evinrude motors, modify a pair of needle-nose pliers as shown

different models). However, Cleaning & Inspection procedures share much more features between motors and are therefore covered under their own section.

Colt/Junior Motors

Removal & Installation

▶ See Figure 9

1. Remove the manual rope starter assembly. For details, please refer to the Hand Rewind Starter section.
2. Remove the fuel tank and then remove the carburetor. For details, please refer to the Fuel System section.
3. Remove the flywheel, as detailed in this section.
4. Remove the ignition coil and related ignition components. For details, please refer to the Ignition and Electrical Systems Section.
5. Remove the 4 screws securing the starter mounting bracket. Start with the 2 rear hex/slotted screws which are on either side of the spark plug, threaded upward into the bracket from underneath. Then remove the 2 slotted screws from the front center of the bracket/cover assembly and lift the bracket from the powerhead.
6. Remove the air intake silencer.
7. Remove the carburetor and leaf plate assembly, as detailed in the fuel system section.
8. Remove the 6 bolts from underneath the powerhead-to-gearcase flange (there should be 3 on each side of the motor, one forward, one aft and one toward the center of each side).
9. Remove the powerhead by carefully lifting it straight upward and off the lower unit. If the powerhead is stuck, pry carefully along the edges and use a rubber mallet to help break the gasket seal and corrosive adhesion.
10. Remove and discard the gasket from the mating surfaces.

To install:
11. Apply a light coating of OMC Moly Lube or equivalent to the sides of the driveshaft splines (NOT to the end of the shaft as it could prevent the shaft from fully seating in the crankshaft).
12. Position a new gasket on the gearcase mating flange (install the gasket dry, without any sealer), then carefully lower the powerhead into place, making sure the crankshaft and driveshaft mate.
13. Apply a light coating of OMC Gasket Sealing Compound, or equivalent sealant to the threads of the powerhead retaining bolts, then install and tighten the bolts to 60-84 inch lbs. (7-9 Nm).
14. Install the ignition coil and ignition components that were removed from the powerhead.
15. Install the carburetor and leaf plate assembly.
16. Install the air intake silencer.
17. Install the starter bracket.
18. Install the fuel tank.
19. Install the hand rewind starter.
20. If a new or rebuilt powerhead was just installed, refer to Break-In in this section for details on how to ensure proper break-in of the new powerhead components.

Disassembly & Assembly

▶ See Figure 9

■ To simplify assembly, remember to layout all bolts, components and clamps in the order of removal. This is especially true for the wiring harness and related clamps.

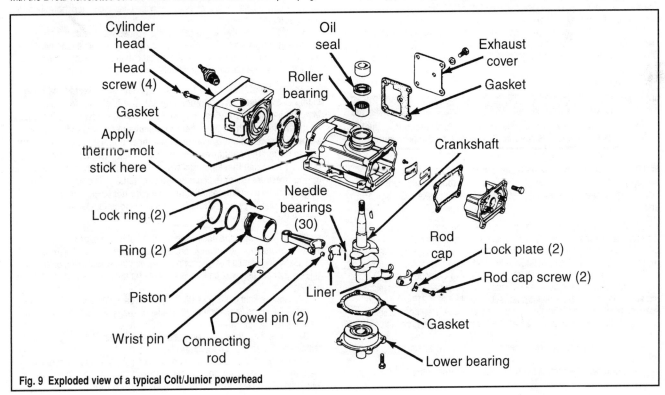

Fig. 9 Exploded view of a typical Colt/Junior powerhead

6-6 POWERHEAD REPAIR AND OVERHAUL

1. Remove the powerhead from the lower unit as detailed under Removal & Installation (including removal of the the fuel and ignition components as directed in the procedure).
2. Loosen the bolts, then remove the cylinder head and gasket.
3. Loosen the bolts, then remove the exhaust cover and gasket.
4. Loosen and remove the 2 bolts securing the connecting rod cap, then remove the cap and carefully collect the 30 loose needle bearings.
5. Carefully push the piston out through the top of the cylinder, making sure NOT to allow the connecting rod to contact the inside of the piston skirt or the cylinder wall.
6. Once the piston is removed, reinstall the rod cap finger tight.
7. Remove and discard the old piston rings.
8. Remove the screws retaining the lower bearing housing, then remove the crankshaft and housing as an assembly from the cylinder block.
9. Using a pair of snapring pliers (such as OMC #675032) reach into the bearing housing from the crankshaft side and compressing the retaining ring while sliding the crankcase head off the lower bearing.
10. Using a small, internal jawed puller (such as OMC #432131) remove the crankcase head seal from the top (crankcase side) of the lower bearing housing/crankcase head. Also remove and discard the O-ring.
11. Apply a light coating of OMC Gasket Sealing Compound, or equivalent sealant to the outer diamter of a new crankcase head seal. Drive the seal into position from the top (crankcase side) of the bearing housing using a suitable driver (such as OMC #326575 Pinion Installation Tool from OMC #391257 Pinion Bearing Kit). Be sure to install the seal with the lip facing the tool, then drive the seal carefully into position until seated.
12. Apply a light coating of OMC Triple-Guard, or an equivalent marine grade grease to the crankcase head seal lips and to the new crankcase head O-ring. Install the O-ring to the crankcase head.
13. If the lower main bearing or retaining ring requires replacement use a shop press and bearing separator to force the bearing off. However, ONLY remove the bearing if it requires replacement, as removal will destroy it.
14. Use the same small, internal jawed puller to remove the upper oil seal from the crankcase.
15. If the upper main bearing needs to be replaced, support the crankcase and use a suitable driver to carefully push the bearing out of position. Again, like the lower main bearing, removal will destroy it, so ONLY do it if it must be replaced.
16. Use a pair of needle-nose pliers to remove and discard the retaining rings from either side of the piston wrist pin.
17. Looking up the piston's skirt, check for one side of the wrist pin boss to be marked with an "L" which stands for the "Loose" side. Place the marked side facing upward in a piston cradle (such as OMC #326572), then use a suitable driver (such as OMC #326624, Wrist Pin Pressing Pin) to push the pin through the piston.
18. Refer to the information found under Cleaning & Inspection, in this section for details on inspecting the powerhead components to determine which should be replaced and to prepare the ones that are being reused for assembly.

To assemble:

✳✳ WARNING

During assembly, take your time and NEVER force a component unless a shop press or drive tool is specifically required by the procedure.

■ **Lightly coat all components (except those that are coated with sealant) with clean, fresh Johnson/Evinrude Outboard Lubricant or equivalent oil during assembly. Always replace all gaskets O-rings and seals.**

19. If removed, install a new upper main bearing through the bottom of the crankcase. Press it upward into position with the lettered side of the bearing case facing the driver. Continue to press until the bearing is recessed 0.062 in. (1.5mm) before the finished thrust face of the crankcase.
20. Apply a light coating of OMC Gasket Sealing Compound or equivalent sealant to the outer metal case of a new upper seal, then install the seal using a suitable seal installer (such as OMC #330219). Press the seal into position with the lip facing into the crankcase, then apply a light coating of OMC Triple-Guard or an equivalent marine grade grease to the seal lips.
21. If removed, install a new lower main bearing and retaining ring as follows:
    a. Make sure the bottom journal is thoroughly free of dirt, debris or corrosion.
    b. Place the retaining ring on the crankshaft with the sharp edge facing away from the crankshaft.
    c. Support the crank between the 2 counterweights.
    d. Coat the bearing journal of the crankshaft lightly with OMC HT400.
    e. Place the bearing on the crankshaft with the lettered side facing upward, away from the crankshaft.
    f. Press the bearing onto the shaft until seated.
22. Oil the lower main bearing, then slide the crankcase head over the bearing and secure using the retaining ring.
23. Oil the crankshaft, then position it into the crankcase, seating the crankcase head.
24. Apply a light coating of Permatex No. 2 to the threads of the crankcase head retaining bolts, then install and tighten the bolts to 60-84 inch lbs. (7-9 Nm).
25. Apply a light coating of clean engine oil or assembly lube to the wrist pin and the pin bore in the piston.
26. Support the piston in the shop press piston cradle, again with the marked loose (**L**) side facing upward. Use the same drift pin or wrist pin tool to slowly start pressing the wrist pin into the piston bore. Stop pressing when the pin appears at the first opening for the connecting rod.
27. Install 2 new wrist pin retaining rings.
28. If removed, place the needle bearing liners in the connecting rod and cap. Make sure the liner with the hole is placed in the cap and also be sure the dovetail ends of the liner match when the connecting rod and cap are assembled.

✳✳ CAUTION

The connecting rod cap can only be installed in one alignment with the rod. The cap and rod are embossed on one side each for alignment purposes to make sure the cap is installed properly.

29. Use the piston ring expander to carefully install the 2 piston rings (although inspection may reveal that the old rings can be reused, WHY do it? You've come this far, so install new rings). Spread the ring **just** enough to slide the each ring over the head of the piston and then down, into place on the piston.

■ **Pay attention to the dowel pin in each ring groove when installing the piston rings. The ring gaps must be positioned at the dowel pins or the cylinder and/or piston/rings could be damaged during installation into the crankcase.**

30. Apply a light coating of 2-stroke motor oil to the piston, bore, rings and a piston ring compressor tool.
31. Making sure the ring gaps remains aligned with the dowel pin, use a piston ring compressor to evenly compress the ring and insert the piston into the cylinder bore. Also make sure the piston exhaust deflector is facing the exhaust cavity. Remove the ring compressor as the piston is inserted.

■ **On small motors such as this one, a proper sized band-type hose clamp can be used as a ring compressor. The best type to use is a Mercruiser (yes, Mercruiser) exhaust bellows hose clamp, as it contains an inner ring on which the outer rings slides to compress it. Typical automotive hose clamps will rotate the inner ring, which can rotate the piston rings out of alignment (off their dowels). Also, it is often possible to install these pistons without a compressor, as the ring spring-pressure is not excessive and can usually be compressed by finger-pressure during installation. Use care not to cock and damage the rings, piston or bore though if this method is used. Take your time and oil the components well.**

32. To make sure the rings were not damaged or broken during installation, use a screwdriver to press on each ring through the exhaust port. Each ring should push lightly back against the screwdriver, otherwise it is likely that the ring was broken. If so, stop, remove the piston and check for damage.
33. Apply a light coating of OMC Needle Bearing Assembly Grease to the connecting rod liner and the 30 needle bearings. Place 14 bearings into the rod liner.
34. Move the connecting rod to engage the crankpin, then place the remaining 16 bearings on the crankpin.
35. Install the connecting rod cap. Apply a light coating of 2-stroke oil to the threads of the connecting rod cap screws, then install and tighten them alternating back and forth to 60-70 inch lbs. (7.0-7.5 Nm).

POWERHEAD REPAIR AND OVERHAUL 6-7

36. Insert a small rod or wire through the oil hole in the cap and check to make sure all the needles are in place. IF the needles are all properly seated it will NOT be possible to touch the actual crankpin. If you CAN touch the crankpin either the needles are out of position or the correct number of needles was not installed.

37. Apply a light coating of OMC Gasket Sealing Compound or equivalent to both sides of a new exhaust cover gasket. Install the exhaust cover and gasket, then tighten the screws to 60-84 inch lbs. (7-9 Nm).

38. Apply a LIGHT coating of OMC Gasket Sealing Compound or equivalent to a new cylinder head gasket. Install the cylinder head and gasket, then tighten the screws using multiple passes of a crossing pattern to 60-84 inch lbs. (7-9 Nm).

39. Install the powerhead and follow the break-in procedure to ensure proper operation.

2.0-3.5 Hp (78cc) Motors

Removal & Installation

MODERATE

1. Remove the engine covers. For details, please refer to Engine Covers (Top and Lower Cases) in the Maintenance and Tune-Up Section.
2. Remove the fuel tank and then remove the carburetor. For details, please refer to the Fuel System section.
3. Remove the manual rope starter assembly. For details, please refer to the Hand Rewind Starter section.
4. Remove the flywheel, as detailed in this section.
5. Remove the ignition coil and related ignition components. For details, please refer to the Ignition and Electrical Systems Section.
6. Remove the 6 bolts from underneath the powerhead-to-gearcase flange (there should be 3 on each side of the motor, one forward, one aft and one toward the center of each side).
7. Remove the powerhead by carefully lifting it straight upward and off the lower unit. If the powerhead is stuck, pry carefully along the edges and use a rubber mallet to help break the gasket seal and corrosive adhesion.
8. Remove and discard the gasket from the mating surfaces.

To install:

9. Apply a light coating of OMC Moly Lube or equivalent to the square end of the crankshaft.
10. Apply a light coating of OMC Gasket Sealing Compound, or equivalent sealant to **both** side of a NEW powerhead gasket.
11. Position the gasket on the gearcase mating flange, then carefully lower the powerhead into place, making sure the crankshaft and driveshaft mate.
12. Apply a light coating of OMC Gasket Sealing Compound, or equivalent sealant to the threads of the powerhead retaining bolts, then install and tighten the bolts to 60-84 inch lbs. (7-9 Nm).
13. Install the ignition coil and ignition components that were removed from the powerhead.
14. Install the hand rewind starter. Tighten the retaining screws to 60-84 inch lbs. (7-9 Nm).
15. Install the fuel tank and the carburetor.

16. Install the engine covers.
17. If a new or rebuilt powerhead was just installed, refer to Break-In in this section for details on how to ensure proper break-in of the new powerhead components.

Disassembly & Assembly

OEM ③ DIFFICULT

◆ See Figures 8 and 10 thru 13

■ To simplify assembly, remember to layout all bolts, components and clamps in the order of removal. This is especially true for the wiring harness and related clamps.

1. Remove the powerhead from the lower unit as detailed under Removal & Installation.
2. Loosen the bolts (4), then remove the cylinder head and gasket.
3. Remove bolts (2), then remove the lower crankcase head (the round boss located on the bottom of the cylinder).
4. Remove the main bearings bolts (6), securing the halves of the crankcase together around the crankshaft. Then use a rubber or plastic mallet to lightly tap the crankshaft upward until the halves just begin to separate. Lift upward and remove the crankcase half.
5. Remove the 2 screws securing the reed (leaf) valve and then remove the left stop and leaf valve from the crankcase.
6. Remove the upper and lower crankshaft seals.
7. Remove semi-circular (half-moon shaped) upper and lower bearing retainers from the cylinder block.
8. Use a suitable piston ring expander to remove the ring from the piston, being careful not to score or damage the piston.
9. Remove the wrist pin retaining rings from the bores in the side of the piston (at either end of the wrist pin). Use a pair of modified needle-nose pliers, as pictured. Discard the old retaining wrings.
10. Use a driver that is just smaller than the piston bore to carefully tap or press the wrist pin from the piston.
11. Check the wrist pin bearing for wear or damage and replace, if necessary. If you are in any way unsure of the condition of this bearing, replace it to ensure reliability.
12. If the crankshaft upper and/or lower ball bearings require replacement, use OMC #115316 or an equivalent bearing puller to pull the bearings free from the crankshaft. When using this puller, be sure to place the bearing-locating pin diagonally on the puller to prevent damage to the pin.

■ The OMC bearing puller is a split, shoulder design that contacts the bearing cage and two semi-circular points around either side of the cage. Then two bolts are threaded through the puller halves to push against the crankshaft counterweights. If another puller-type is used (such as a jawed puller) use great to make sure the crankshaft is not damaged. It may be necessary to place something over the end of the shaft to protect it from the puller screw.

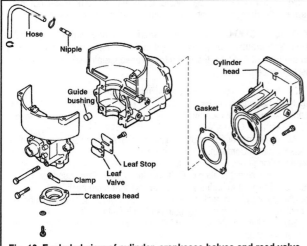

Fig. 10 Exploded view of cylinder, crankcase halves and reed valve assembly - 2.0-3.5 hp (78cc) motors

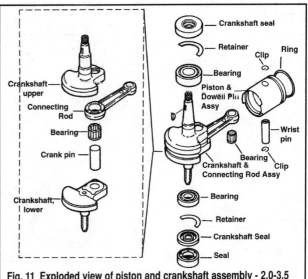

Fig. 11 Exploded view of piston and crankshaft assembly - 2.0-3.5 hp (78cc) motors

6-8 POWERHEAD REPAIR AND OVERHAUL

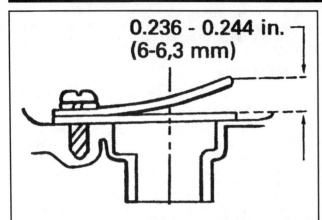

Fig. 12 Measure the reed valve opening at the end of the leaves - 2.0-3.5 hp (78cc) motors

13. Refer to the information found under Cleaning & Inspection, in this section for details on inspecting the powerhead components to determine which should be replaced and to prepare the ones that are being reused for assembly.

To assemble:

✶✶ WARNING

During assembly, take your time and NEVER force a component unless a shop press or drive tool is specifically required by the procedure.

■ Lightly coat all components (except those that are coated with sealant) with clean, fresh Johnson/Evinrude Outboard Lubricant or equivalent oil during assembly. Always replace all gaskets and seals.

14. If removed, use a suitable bearing driver (like OMC #115309 or equivalent) to install NEW upper and/or lower bearings to the crankshaft assembly. Face the bearings with the lettered sides towards the driver and support the crankshaft between the 2 counterweights during installation. If possible, use a shop press to install the bearings with smooth and constant pressure, but if necessary, you can tap the bearings gently into position.
15. Apply a light coat of oil to the piston wrist pin and the bore in the piston, then begin inserting it through one end (either will do) of the piston.
16. Apply a light coat of oil to the wrist pin bearing, then position it into the connecting rod.
17. Insert the connection rod in place inside the bottom of the piston skirt so the arrow mark on top of the piston is facing downward toward the bottom of the crankshaft. Insert the wrist pin the rest of the way through the connecting rod and piston (if necessary use a suitable driver to gently tap the wrist pin into position, but take great care not to damage the piston, bearing or rod).

Fig. 13 A Mercruiser exhaust bellows hose band clamp can be used as a ring compressor on some 2-stroke models

18. Install 2 new wristpin retaining rings.
19. Use a piston ring expander to carefully install the ring (although inspection may reveal that the old ring can be reused, WHY do it? You've come this far, so install a new ring). Spread the ring **just** enough to slide the wring over the head of the piston and then down, into place on the piston.

■ **Pay attention to the dowel pin when installing the piston ring. The ring gap must be positioned at the dowel pin or the cylinder and/or piston/ring could be damaged during installation into the crankcase.**

20. Coat the piston and cylinder bore with clean oil.
21. Making sure the ring gap remains aligned with the dowel pin, use a piston ring compressor to evenly compress the ring and insert the piston into the cylinder bore. Remove the ring compressor as the piston is inserted.

■ **On small motors such as this one, a proper sized band-type hose clamp can be used as a ring compressor. The best type to use is a Mercruiser (yes, Mercruiser) exhaust bellows hose clamp, as it contains an inner ring on which the outer rings slides to compress it. Typical automotive hose clamps will rotate the inner ring, which can rotate the piston rings out of alignment (off their dowels). Also, it is often possible to install these pistons without a compressor, as the ring spring-pressure is not excessive and can usually be compressed by finger-pressure during installation. Use care not to cock and damage the rings, piston or bore though if this method is used. Take your time and oil the components well.**

22. To make sure the ring was not damaged or broken during installation, use a screwdriver to press on the ring through the exhaust port. The ring should push lightly back against the screwdriver, otherwise it is likely that the ring was broken. Stop, remove the piston and check for damage.
23. Position the crankshaft assembly on the cylinder block, rotating the bearings so that the locating pins are positioned in the cylinder block notches.
24. Install the upper and lower (semi-circular) retainers in the cylinder block grooves.
25. Install NEW upper and lower crankshaft seals with the lip of each seal facing toward the crankcase.
26. Apply a light coat of OMC Screw Lock, or equivalent threadlock to the threads of the leaf valve screws, then install the leaf valve along with the stop. Tighten the valve retaining screws to 25-37 inch lbs. (3-4 Nm).
27. Inspect the leaf valve opening using a dial caliper or stack of feeler gauges. The opening at the end of the valve should be 0.236-0.244 in. (6-6.3mm). The valve should be removed and replaced if opening is out of specification.
28. Use OMC Cleaning Solvent or an equivalent solvent to thoroughly remove all traces of grease, oil or lubricant from the crankcase/cylinder block halve flanges. Allow the surfaces to air dry.
29. Apply a light coating of OMC Locquic Primer or equivalent to the crankcase half flange, and allow the primer to air dry.
30. Apply a light coating of OMC Gel-Seal II or equivalent to the cylinder block half flange. Be sure that the sealant coats the entire flange evenly, but be careful not to apply excessive amounts. The sealant must not be applied within 1/4 in. (6mm) of the bearings or the sealant could be forced out from between the flanges into contact with the bearings during assembly.

■ **Although OMC Gel-Seal II has a shelf life of one year when stored at room temperature, buy a new tube if in doubt. Keep in mind that using an old tube of Gel-Seal II could allow crankcase air leaks (leading to performance problems).**

31. Install the lower crankcase half into position over the crankshaft. Apply a light coating of OMC Gel-Seal II or equivalent to the threads of the crankcase/cylinder block halve flange bolts. Install the bolts and tighten to 90-120 inch lbs. (10-14 Nm)
32. Install the round, lower crankcase head. Apply a light coating of OMC Gasket Sealing Compound, or equivalent sealant to the threads of the 2 head retaining bolts and tighten to 60-84 inch lbs. (7-9 Nm)
33. Apply a light coating of OMC Gasket Sealing Compound, or equivalent sealant to both sides of a new cylinder head gasket, then install the cylinder head and gasket. Install the cylinder head bolts **dry** with no lubricant or sealant, then tighten the bolts to 60-84 inch lbs. (7-9 Nm).
34. Install the powerhead and follow the break-in procedure to ensure proper operation.

POWERHEAD REPAIR AND OVERHAUL

5 Hp (109cc) Motors

◆ See Figure 14

Removal & Installation

◆ See Figure 14

1. Remove the top engine cover.
2. Remove the manual rope starter assembly. For details, please refer to the Hand Rewind Starter section.
3. Remove the fuel tank and then remove the carburetor. For details, please refer to the Fuel System section.
4. Remove the flywheel, as detailed in this section.
5. Remove the ignition coil and related ignition components. For details, please refer to the Ignition and Electrical Systems Section.
6. Remove the 6 bolts from underneath the powerhead-to-gearcase flange (threaded upward from underneath the one-piece lower engine cover).
7. Remove the powerhead by carefully lifting it straight upward and off the lower unit. If the powerhead is stuck, use a rubber mallet to help break the gasket seal. Be very careful not to damage the lower engine cover while removing the powerhead.
8. Remove and discard the gasket from the mating surfaces.

To install:

9. Apply a light coating of OMC Moly Lube or equivalent to the driveshaft splines.
10. Apply a light coating of RTV sealant to **both** side of a NEW powerhead gasket.
11. Position the gasket on the lower engine cover (at the powerhead-mating surface), then carefully lower the powerhead into place, making sure the crankshaft and driveshaft splines are aligned.
12. Apply a light coating of RTV sealant to the threads of the powerhead retaining bolts, then install and tighten the bolts to 11-15 ft. lbs. (15-20 Nm).
13. Install the ignition coil and ignition components that were removed from the powerhead.
14. Install the fuel tank and the carburetor.
15. Install the hand rewind starter.
16. Install the top engine cover.
17. If a new or rebuilt powerhead was just installed, refer to Break-In in this section for details on how to ensure proper break-in of the new powerhead components.

Disassembly & Assembly

◆ See Figures 8 and 13 thru 17

■ To simplify assembly, remember to layout all bolts, components and clamps in the order of removal. This is especially true for the wiring harness and related clamps.

1. Remove the powerhead from the lower unit as detailed under Removal & Installation.
2. Loosen the clamp and remove the recirculation hose from the side of the powerhead.
3. Remove the 2 mounting bolts and then remove the starter brace from the top of the powerhead.
4. Remove the 2 screws, then remove the semi-circular (half-moon shaped) water jacket cover from the bottom of the powerhead. Remove and discard the gasket.
5. Remove the 5 cylinder head cover screws, then remove the cylinder head cover and discard the gasket.
6. Remove the bolts (6) securing the cylinder block and crankcase halves around the crankshaft. Then, use a rubber or plastic mallet to lightly tap the crankshaft upward until the halves just begin to separate. Lift upward and remove the crankcase half from the cylinder half.
7. If necessary, remove the 2 screws securing the reed (leaf) valve, then remove valve and stop from the crankcase.
8. Remove and discard the upper and lower crankshaft seals, along with the upper driveshaft seal.
9. Carefully lift the crankshaft, connecting rod and piston assembly from the cylinder block.
10. Remove the wrist pin retaining rings from the bores in the side of the piston (at either end of the wrist pin). Use a pair of modified needle-nose pliers, as pictured or, if not available, use a small pry tool to carefully pry the old rings free of the bore. Discard the old retaining wrings.
11. Use a driver that is just smaller than the piston bore to carefully tap or press the wrist pin from the piston.
12. Check the wrist pin bearing for wear or damage and replace, if necessary. If you are in any way unsure of the condition of this bearing, replace it to ensure reliability.
13. Use a suitable piston ring expander to remove the rings from the piston, being careful not to score or damage the piston.
14. If the crankshaft upper and/or lower ball bearings require replacement, use a suitable bearing puller to pull the bearings free from the crankshaft. When using a puller, be careful not to damage the crankshaft assembly. Do not remove the bearings unless they are going to be replaced.

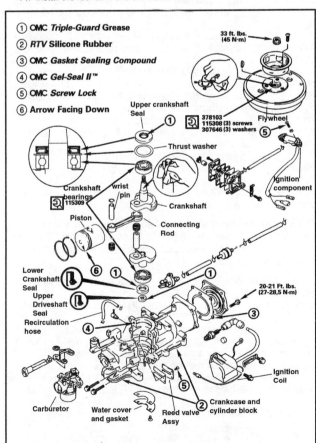

Fig. 14 Exploded view of the powerhead assembly with assembly sealant and lubricant instructions - 5 hp (109cc) motors

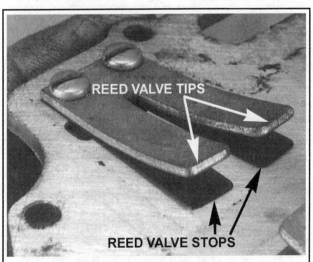

Fig. 15 Measure the reed valve tip-to-stop clearance to determine if the valve assembly should be replaced

6-10 POWERHEAD REPAIR AND OVERHAUL

15. Refer to the information found under Cleaning & Inspection, in this section for details on inspecting the powerhead components to determine which should be replaced and to prepare the ones that are being reused for assembly.

To assemble:

✱✱ WARNING

During assembly, take your time and NEVER force a component unless a shop press or drive tool is specifically required by the procedure.

■ Lightly coat all components (except those that are coated with sealant) with clean, fresh Johnson/Evinrude Outboard Lubricant or equivalent oil during assembly. Always replace all gaskets and seals.

16. If removed, install the reed valve to the crankcase half and tighten the retaining screws securely.
17. Check the reed valve tip-to-stop clearance using a set of feeler gauges. Clearance must be 0.008 in. (0.20mm) or less, otherwise the valve and stop assembly must be replaced.
18. If removed, use a suitable bearing driver (like OMC #115309 or equivalent) to install NEW upper and/or lower bearings to the crankshaft assembly. Face the bearings with the lettered sides towards the driver and support the crankshaft between the 2 counterweights during installation. If possible, use a shop press to install the bearings with smooth and constant pressure, but if necessary, you can tap the bearings gently into position.
19. Use a piston wring expander to carefully install the rings (although inspection may reveal that the old rings can be reused, WHY do it? You've come this far, so install new rings). Spread each ring **just** enough to slide the wring over the head of the piston and then down, into place in the ring groove.

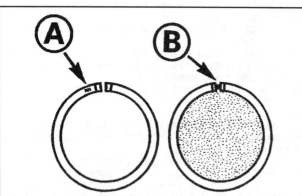

Fig. 16 When installing the rings make sure the marked side (A) faces upward and the gaps are located at the dowel pins (B)

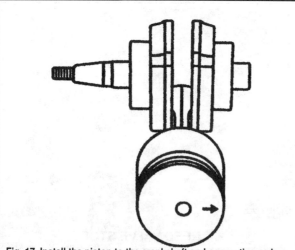

Fig. 17 Install the piston to the crankshaft and connecting rod so the marked arrow on the piston crown faces the bottom of the crankshaft

■ When installing rings, make sure that the marked side faces upward toward the head of the piston and the ring gap is centered on the dowel pin. Failing to place the gap around the dowel pin will result in cylinder and/or piston/ring damage during installation into the crankcase (or worse, during service if this is missed completely).

20. Apply a light coat of oil to the piston wrist pin and the bore in the piston, then begin inserting it through one end (either will do) of the piston.
21. If removed, apply a light coat of oil to the wrist pin bearing, then position it into the connecting rod.
22. Insert the connection rod in place inside the bottom of the piston skirt so the arrow mark on top of the piston is facing toward the bottom of the crankshaft. Insert the wrist pin the rest of the way through the connecting rod and piston (if necessary use a suitable driver to gently tap the wrist pin into position, but take great care not to damage the piston, bearing or rod).
23. Install 2 new wristpin retaining rings.
24. Coat the piston and cylinder bore with clean oil.
25. Making sure the ring gaps remain aligned with the dowel pin, use a piston ring compressor to evenly compress the ring and insert the piston into the cylinder bore. Remove the ring compressor as the piston is inserted.

■ On small motors such as this one, a proper sized band-type hose clamp can be used as a ring compressor. The best type to use is a Mercruiser (yes, Mercruiser) exhaust bellows hose clamp, as it contains an inner ring on which the outer rings slides to compress it. Typical automotive hose clamps will rotate the inner ring, which can rotate the piston rings out of alignment (off their dowels). Also, it is often possible to install these pistons without a compressor, as the ring spring-pressure is not excessive and can usually be compressed by finger-pressure during installation. Use care not to cock and damage the rings, piston or bore though if this method is used. Take your time and oil the components well.

26. To make sure the rings were not damaged or broken during installation, use a screwdriver to lightly press on each ring through the exhaust port. The ring should push back against the screwdriver, otherwise it is likely that the ring was broken. Stop, remove the piston and check for damage.
27. Position the crankshaft assembly on the cylinder block, rotating the bearings so that the locating pins are positioned in the cylinder block notches.
28. Apply a light coating of OMC Triple-Guard, or equivalent marine grease to the lips of NEW upper and lower crankshaft seals, as well as the NEW upper driveshaft seal. In all cases, install the seals with the lips facing downward, toward the gearcase. When installing the driveshaft seal, hold it in position and push it squarely into position.
29. Seat the crankshaft in the crankcase, making sure the thrust washer is properly located in the crankcase groove.
30. Apply a light coating of OMC Gel-Seal II or equivalent to the cylinder block half flange. Be sure that the sealant coats the entire flange evenly, but be careful not to apply excessive amounts. The sealant must not be applied within 1/4 in. (6mm) of the bearings or the sealant could be forced out from between the flanges into contact with the bearings during assembly.

■ Although OMC Gel-Seal II has a shelf life of one year when stored at room temperature, buy a new tube if in doubt. Keep in mind that using an old tube of Gel-Seal II could allow crankcase air leaks (leading to performance problems).

31. Install the lower crankcase half into position over the crankshaft. Apply a light coating of OMC Gel-Seal II or equivalent to the threads of the crankcase/cylinder block halve flange bolts. Install the bolts and tighten to 72-102 inch lbs. (8-12 Nm)
32. Apply a light coating of OMC Gasket Sealing Compound, or equivalent sealant to both sides of a new cylinder head gasket, then install the cylinder head and gasket. Install the cylinder head bolts **dry** with no lubricant or sealant, then tighten the bolts to 20-21 ft. lbs. (27-28.5 Nm).
33. Install the semi-circular (half-moon shaped) water jacket cover to the bottom of the powerhead using a new gasket. Tighten the retaining screws securely.
34. Install the starter brace to the top of the powerhead and tighten the 2 mounting bolts securely.
35. Connect the recirculation hose to the side of the powerhead and secure using the clamp.
36. Install the powerhead and follow the break-in procedure to ensure proper operation.

POWERHEAD REPAIR AND OVERHAUL 6-11

5/6 Hp (128cc) 4-Stroke Motors
Removal & Installation

MODERATE

◆ See Figures 18, 19 and 20

1. Remove the engine covers. For details, please refer to Engine Covers (Top and Lower Cases) in the Maintenance and Tune-Up Section.
2. Remove the manual rope starter assembly. For details, please refer to the Hand Rewind Starter section.
3. At the front of the motor, remove the spring clip from the slot in the shift lever, then lift the shift rod lever.
4. Also at the front of the motor, remove the cotter pin from the starter lockout lever (a black composite component just to the side of the shift rod/lever assembly from the previous step). Once the cotter pin is removed, separate the lever from the pin and disconnect from the lockout cable.

■ The lockout cable is released by pressing inward on the sides of the cable tangs, then pulling the cable free.

5. Remove the fuel hose from the cover connector and, remove the lower covers completely from the engine.
6. If the powerhead is being removed for overhaul, tag and disconnect all ignition and electrical system components from the powerhead. For details please refer to the procedures in the Ignition and Electrical System section including procedures for the coil, power pack, charge coil, sensor/trigger coil and oil pressure switch.
7. If the powerhead is being removed for overhaul, remove the fuel system components. For details, please refer to the Fuel System section for procedures on the carburetor and fuel pump.
8. Tag and disconnect the valve cover oil mist-to-carburetor recirculation hose, overboard coolant indicator hose and the exhaust tube-to-oil sump hose.
9. Remove the retaining bolts (4) securing the tiller control bracket to the upper mounts.
10. Remove the nuts and washers (4) securing the oil sump and powerhead to the gearcase midsection and then remove the powerhead and oil sump as an assembly.
11. Refer to Powerhead Disassembly & Assembly in this section for further service. Once separated, the cylinder head and/or cylinder block can also be disassembled for inspection or overhaul.

To install:
12. Thoroughly clean the mating surfaces of the oil sump and midsection.
13. Lower the powerhead carefully into position (placing the oil sump over the midsection studs), while aligning the driveshaft and mating surfaces.
14. Apply a coating of OMC Nut Lock or an equivalent threadlock to the stud threads, then install the nuts and lockwashers. Tighten the nuts to 15-17 ft. lbs. (20-23 Nm) using a crossing pattern as shown.
15. Apply a coating of OMC Nut Lock or an equivalent threadlock to the threads of the tiller handle bracket, then install the bolts and tighten to 108-132 inch lbs. (12-15 Nm).
16. Connect the hoses as tagged during removal. Be sure to secure the hoses using clamps or wire-ties as applicable.
17. Install the fuel system components that were removed. For details, please refer to the carburetor or fuel pump procedures in the Fuel System section.
18. Install the ignition and electrical components that were removed. For details, please refer to the Ignition and Electrical System section.

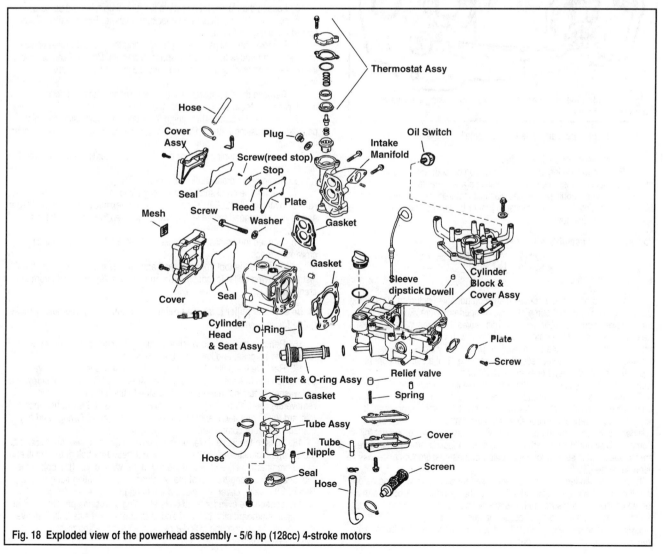

Fig. 18 Exploded view of the powerhead assembly - 5/6 hp (128cc) 4-stroke motors

6-12 POWERHEAD REPAIR AND OVERHAUL

Fig. 19 Remove the oil sump-to-gearcase nuts to free the powerhead and sump assembly

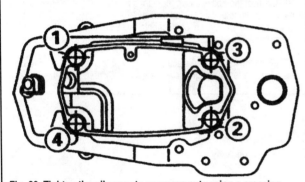

Fig. 20 Tighten the oil sump-to-gearcase nuts using a crossing pattern as shown

19. Install the rope starter assembly as detailed in the hand rewind starter section.
20. Install the lower engine covers.
21. Refer to the Timing and Synchronization procedures in the General Information section, then perform all applicable adjustments.
22. If a new or rebuilt powerhead was just installed, refer to Break-In in this section for details on how to ensure proper break-in of the new powerhead components.

Powerhead Disassembly & Assembly

◆ See Figures 18 and 21

Although the powerhead houses many components (including the starting, fuel and ignition systems) some of those components were likely removed during powerhead assembly removal. This leaves 2 major sub-assemblies, the cylinder head and the cylinder block, along with various components for other systems (such as the thermostat, oil pump and various tubes or housings). Although it is not necessary to remove all of these components to separate the cylinder head or cylinder block for further service, we recommend that if you've come this far you consider a complete disassembly and inspection of the powerhead, including ALL of these components. That is what we've provided in terms of a Disassembly & Assembly procedure. If access is necessary only to certain components (the cylinder head or the cylinder block), use care in determining which steps you will or will not follow. Complete Cylinder Head Overhaul and Cylinder Block Overhaul procedures are included in this section for disassembly of these assemblies once they've been separated.

Check the cylinder block and cylinder head for excessive warpage before installation. It is always a good idea to clean the cylinder block holes with a thread chaser, making sure the holes are clean and dry. Even small amounts of dirt or debris will adversely affect bolt torque.

Cylinder head gaskets are installed dry, do not apply sealant, but be sure that you always install a NEW gasket.

1. Remove the powerhead and oil sump assembly from the lower unit as detailed under Removal & Installation
2. Remove the bolts (2) securing the upper engine mounts to the oil sump and then remove the mount assemblies. Check the mounts for wear or damage and replace, as necessary.
3. Remove the oil sump-to-powerhead bolts (6) and the exhaust tube-to-cylinder head bolts (2). Remove the powerhead assembly from the oil sump.
4. Remove the Philips screws (2) securing the shift linkage assembly (consisting of an arm, base, pivot and linkage) to the top of the oil sump. Remove the assembly to inspect all components for wear or damage and replace, as necessary.
5. Remove the exhaust tube from aft portion of the oil sump. Remove the seal from the tube for inspection and replace, if necessary. Visually inspect the tube for corrosion or damage.
6. Remove the Philips screws (3) securing the exhaust cover to the side of the oil sump, then remove the cover and discard the O-ring (a new O-ring MUST be used during installation). Visually inspect the exhaust passages for damage or excessive corrosion.
7. Remove the bushing from the water tube, then remove the bolts (3) securing the exhaust/water tube cover. Remove the cover, then remove and discard the gasket.
8. Remove the flange screw retaining the water tube and then remove the tube and grommet. Inspect the tube, grommet and sump passages. Check the hole near the end of the tube for restriction. If necessary, clean it out using solvent or a small length of wire (taking care not to elongate the small hole).
9. Remove the oil pump and filter assembly for cleaning and inspection. For detail, please refer to the Lubrication and Cooling section.
10. Loosen the clamp, then remove the oil pickup tube and screen. Clean the assembly in solvent and check for damage. Replace the tube or screen if damage is found.
11. Remove the Philips screws (4) mounting the pressure relief valve housing, then remove the housing, spring and valve. Remove and discard the gasket, then inspect the valve housing, spring and valve damage or excessive wear.
12. Remove the thermostat assembly for cleaning and inspection. For detail, please refer to the Lubrication and Cooling section.
13. Remove the bolts (4) retaining the intake manifold using multiple passes of a crossing pattern. Remove the manifold, then remove and discard the gasket.
14. Remove the timing belt, for details please refer to the procedure in this section.
15. Remove the crankshaft sprocket as follows:
   a. Secure the crankshaft from turning using the OMC Crankshaft Pulley Nut removal tool (#342669). The tool is a large hex nut with an internal taper and keyway designed to use with a wrench to keep the crankshaft from turning.
   b. Loosen the crankshaft sprocket nut using a large adjustable or crowfoot wrench.
   c. Remove the tool from the shaft and the key from the keyway, then install a 2-jawed puller to loosen and remove the pulley/belt guide from the crankshaft.

■ Be sure the puller jaws are positioned under the crankshaft sprocket guide for removal.

16. Remove the bolt and washer securing the camshaft sprocket, then remove the sprocket and key from the camshaft.
17. Remove the bolts (4) securing the breather assembly to the valve cover, then remove the cover and discard the O-ring. Remove the screw and lockwasher securing the reed and reed stop to the breather plate, then remove the reed, stop and plate. Remove the O-ring and mesh breather (discard the O-ring and inspect the breather). Replace all O-rings and any damaged or worn components.
18. Loosen the valve cover bolts (4) using a crossing sequence. Start at the second highest cover bolt on the cylinder head (the bolt at the small end of the breather cover) and end at the top of the valve cover (the bolt in the rounded cutout along the top of the breather cover mounting flange). Remove the valve cover and discard the O-ring.
19. Loosen the cylinder head bolts (4) using a crossing pattern. Start at the upper, starboard side bolt and end at the upper port side bolt. Remove the cylinder head from the cylinder block, then remove and discard the cylinder head gasket.

POWERHEAD REPAIR AND OVERHAUL

20. As necessary, refer to Cylinder Head Overhaul for details on disassembly of the cylinder head and valve train components and/or to Cylinder Block Overhaul for details on disassembly for the piston, rings and crankshaft assembly.

To assemble:

21. Assemble the cylinder head and/or cylinder block, as detailed under the overhaul procedures in this section. Be sure to replace all gaskets and seals.
22. Apply a light coating of OMC Gasket Sealing Compound, or equivalent sealant to both sides of a new cylinder head gasket, then position the gasket on the cylinder block.
23. Apply a light coating of clean engine oil to the cylinder head bolt threads. Carefully position the cylinder head to the block, over the gasket, then install and finger tighten the bolts. Torque the bolts using multiple passes of a crossing pattern (that starts at the upper bolt on the port side and ends at the upper bolt on the starboard side) to 18-20 ft. lbs. (24-27 Nm).
24. Install the camshaft key with the outer key edge parallel to the camshaft centerline (outer edge straight up-and-down), then install the camshaft sprocket, but do not tighten the retaining bolt fully at this time.
25. Install the crankshaft key (outer edge straight up-and-down, parallel to the shaft centerline), then install the belt guide and sprocket. Use a socket as a driver to gently tap the sprocket over the crankshaft taper. Position the outer belt guide and lock plate. Apply a coating of OMC Nut Lock or equivalent threadlock to the nut, then install the crankshaft sprocket nut and tighten to 18-20 ft. lbs. (24-27 Nm) while using the nut tool to hold the shaft from turning (as during removal). Keep the tool in position while the camshaft pulley bolt is tightened after timing belt installation.
26. Install the timing belt, as detailed in this section.
27. Once the timing belt is properly positioned and the timing marks are properly aligned, remove the bolt from the camshaft sprocket and apply a coating of OMC Nut Lock or equivalent thread. Reinstall the bolt and tighten to 84-106 inch lbs. (10-12 Nm). Remove the crankshaft holder tool.
28. Adjust the valve lash as detailed in the Maintenance and Tune-Up section.
29. Apply a light coating of OMC Triple-Guard, or equivalent marine grease to a NEW valve cover O-ring, then position the O-ring in the cover groove. Apply a light coating of OMC Gasket Sealing Compound, or equivalent sealant to the threads of the valve cover bolts, then install the cover and finger-tighten the bolts.
30. Torque the valve cover bolts (4) to 84-106 inch lbs. (10-12 Nm) using multiple passes of a crossing sequence. Start at the top of the valve cover (the bolt in the rounded cutout along the top of the breather cover mounting flange) and end at the second highest bolt on the cylinder head (the bolt at the opposite end of the breather cover).
31. Apply a light coating of OMC Triple-Guard, or equivalent marine grease to NEW breather plate and cover O-rings, then install the O-rings (one in the valve cover and the other in the breather cover).
32. Apply a light coating of OMC Nut Lock, or equivalent threadlock to the threads of the reed stop screw and the breather cover bolts. Install the reed and stop, then tighten the screw to 15-22 inch lbs. (1.7-2.5 Nm). Install the breather cover and tighten the bolts to 36-60 inch lbs. (4-6 Nm).
33. Apply a light coating of OMC Gasket Sealing Compound, or equivalent sealant to the bolt threads and to both sides of a NEW intake manifold gasket. Install the manifold and gasket, then tighten the bolts to 84-106 inch lbs. (10-12 Nm) using multiple passes of a crossing sequence.
34. If the flush plug and washer was removed, install tighten to 60-84 inch lbs. (8-10 Nm).
35. Install the thermostat assembly, as detailed in the Lubrication and Cooling section.
36. Position the oil pressure relief valve spring into the valve, then place the valve orifice in the block.
37. Apply a light coating of OMC Gasket Sealing Compound, or equivalent sealant to a NEW relief valve cover gasket and a light coating of OMC Nut Lock, or equivalent threadlock to the threads of the cover screws. Install the cover and gasket, then tighten the screws to 36-60 inch lbs. (4-6 Nm).
38. Install the oil pickup tube and inlet screen, then secure using the spring-type clamp. If the screen was replaced, be sure to use a wire-tie to secure the screen to the hose.
39. Assemble and install the oil pump, as detailed in the Lubrication and Cooling section.
40. Install the water tube and grommet into the oil sump. Apply a coating of OMC Nut Lock or equivalent threadlock to the tube flange screw, then

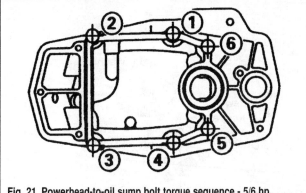

Fig. 21 Powerhead-to-oil sump bolt torque sequence - 5/6 hp (128cc) models

tighten to 36-60 inch lbs. (4-6 Nm). Finally, install the bushing to the water tube.

41. Apply a light coating of OMC Gasket Sealing Compound, or equivalent sealant to a NEW exhaust/water tube cover gasket. Apply a coating of OMC Nut Lock or equivalent threadlock to the threads of the cover bolts, then install the cover and gasket on the oil sump and tighten the bolts to 84-106 inch lbs. (10-12 Nm).
42. Apply a light coating of clean engine oil to a NEW exhaust passage O-ring. Apply a light coating of OMC Nut Lock or equivalent threadlock to the threads of the exhaust passage cover screws (3). Install the cover to the side of the oil sump and tighten the screws securely.
43. Using a new seal, install the exhaust tube in position on the exhaust/water tube cover.
44. Apply a light coating of OMC Nut Lock or equivalent threadlock to the threads of the shift linkage screws, then install the assembly and tighten the screws securely.
45. Apply a light coating of OMC Gasket Sealing Compound, or equivalent sealant to the new exhaust tube gasket and sump-to-powerhead gasket. Position the powerhead on the oil sump.
46. Apply OMC Ultra Lock or equivalent high-strength threadlock to the powerhead-to-sump bolts, then install and tighten to 36-60 inch lbs. (4-6 Nm) using the torque sequence shown.
47. Apply a light coating of OMC Gasket Sealing Compound, or equivalent sealant to the exhaust tube-to-cylinder head screws, then install and tighten the screws to 15-17 ft. lbs. (20-23 Nm).
48. Apply a light coating of OMC Nut Lock, or equivalent threadlock to the threads of the upper mount bolts (2), then install the upper engine mounts and tighten the bolts to 108-132 inch lbs. (12-15 Nm).
49. Install the powerhead, as detailed in this section.

Cylinder Head Overhaul

◆ See Figure 22

Before disassembly, refer to Cleaning and Inspection in this section for details on how to perform a Valve Leakage check. If leakage is present or there is an obvious defect, the valve train must be disassembled for inspection and component replacement.

■ **Disassembly and repair requires special tools and equipment. If not available, bring the cylinder head to a reputable machine shop with marine experience.**

Be sure to mark (and make a note, in case the marks come off during service) the mounting location and orientation of all cylinder head components prior to removal.

■ **Some Johnson/Evinrude outboards are equipped with progressive rate valve springs (that can only be installed in one direction). Just in case your model is too, mark the direction of installation (mark or note the side of the spring that faces the cylinder head when installed).**

1. Remove the cylinder head from the powerhead as detailed under Powerhead Disassembly & Assembly, in this section.
2. Mark the rocker arms to ensure installation in their original positions.

6-14 POWERHEAD REPAIR AND OVERHAUL

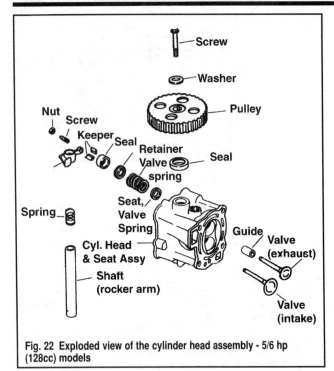

Fig. 22 Exploded view of the cylinder head assembly - 5/6 hp (128cc) models

3. If removed, thread the camshaft pulley bolt and washer into the end of the camshaft, then use the bolt to turn the camshaft until each rocker arm is in the unloaded position (resting on the base of the camshaft instead of the raised lobe). With rocker arm in the unloaded position loosen the valve adjustment nut and back off the tappet screw a few of turns to create additional valve lash.

■ OMC Tappet Adjustment Tool, #341444 is available for easy access to the tappet adjuster screws.

4. Slowly and carefully pull the rocker arm shaft from the cylinder head. As the shaft is withdrawn, remove each of the rocker arms, spacers and springs while noting the positioning of each component. Mark the components to ensure assembly in the correct positions.
5. Remove the camshaft from the cylinder head.
6. Use either a universal clamp/lever type valve spring compressor (like OMC #341446) or a screw type compressor, to remove each valve, spring and seal:
 a. If using a the OMC tool, place one clamp on the valve face and the other on the valve spring cap, then compress the spring **just** sufficiently to remove the keepers from the stem. Remove the keepers from the valve and slowly release the spring pressure.

■ Do not compress the spring fully.

 b. Remove the valve compressor tool and remove valve and spring from the cylinder head.

■ Keep all valve components together, springs, retainers, keepers and valves must not be mixed during installation.

 c. If replacement is necessary (since the manufacturer recommends the spring and seal are replaced whenever they are removed), use a pair of pliers to grasp and remove the valve seal from the cylinder head. Discard the old seal and spring.
 d. If replacement is necessary, remove the valve spring seat from the cylinder head.
 e. Repeat for the remaining valve(s).
7. If it is being replaced (and once again, we must recommend that you've come this far, you really should), remove the camshaft seal from the cylinder head using a slide hammer and large jaw puller.
8. Inspect and measure the cylinder head components as described in this chapter.

To assemble:
9. Install the valves to the cylinder head, as follows:
 a. If removed, install the NEW valve spring seat to the cylinder head.

 b. Apply a light coating of engine oil to the NEW valve seal, then press it carefully onto the valve guide, making sure it is properly seated.
 c. Apply a light coating of engine oil to the valve stem, then slide it into the valve guide. Place the valve fully against the seat, then install the valve spring and cap over the valve stem.
 d. Use the valve spring compressor to push downward on the spring, **just** sufficiently to install the valve keepers. Once the keepers are in position, slowly release the spring pressure, allowing the cap and keepers to take the pressure. Make sure the keepers are properly seated.
 e. Repeat to install the remaining valve.
10. Apply a light coating of engine oil to the camshaft (lobes and journals), then slide the camshaft carefully into the cylinder head.
11. Apply a light coating of engine oil to the rocker arm shaft, rocker arms (bores and faces), spacers, tappet adjusting screws and nuts.
12. Slide the rocker arm shaft slowly into the cylinder head, installing each arm, spacer and spring in the positions marked/noted during removal.

■ When installing the rocker arm shaft you must align the rocker shaft locating pin with the relief in the head.

■ Leave the valve lash adjusters loose until the cylinder head is installed. Be sure that the tappets remain free of contact with the valves thereby ensuring that no valve train components can be damaged during cylinder head installation and initial engine timing.

13. If removed, install a new camshaft oil seal. Apply a light coating of OMC Gasket Sealing Compound, or equivalent sealant to the seal casing. Apply a light coating of engine oil to the seal lips. Position the seal squarely over the camshaft and bore (the casing should face upward, away from the camshaft). Use a seal driver or a smooth socket that contacts only the seal casing to carefully drive the seal until it is flush with the head.

■ On some motors, the camshaft sprocket key, if left in place, can damage the seal during installation. If necessary, remove and retain the key to prevent damage.

14. Install the cylinder head as detailed under Powerhead Disassembly & Assembly, in this section.

Cylinder Block Overhaul

◆ See Figures 8, 18 and 23 thru 26

On 5/6 hp 4-stroke motors, the crankshaft is mounted through the crankcase cover (as opposed to underneath the cover on most larger motors and most automotive type inline motors). The piston and connecting rod must therefore be removed before the cover can be unbolted and separated from the cylinder block/case.

The crankshaft for this motor turns on main bearings found in the cover and the cylinder block. These bearings should not be removed unless they are going to be replaced. Unlike many 4-stroke motors, the connecting rods do not utilize bearing inserts, as the rod and cap itself contain machined bearing surfaces. Especially for this reason, it is critical that the connecting rod and end cap are kept as a matched set, so they are positioned in the exact same orientation (facing the same directions) as originally installed.

■ If you ONLY need to access the piston for inspection/service, follow just the first half of the procedure. Only follow the steps pertaining to piston removal.

1. Remove the powerhead from the gearcase as detailed under Removal & Installation, then remove the cylinder head as detailed under Powerhead Disassembly & Assembly.
2. If necessary, remove the dowels and the water deflector from the cylinder block-to-head mating surface.
3. Remove the Philips screws (2) securing the cylinder block access plate to the front of the block. Remove the plate, then remove and discard the gasket.
4. Rotate the crankshaft so the rod cap is accessible through the opening at the front of the block.
5. Bend back the retainer tabs, then remove the connecting rod bolts, washers, retainer and cap. Discard the retainers and cap bolts (they must be replaced anytime they are removed).

POWERHEAD REPAIR AND OVERHAUL 6-15

⚠ WARNING

The piston is removed by pushing on the connecting rod through the access hole in the front of the cylinder block. Work slowly and do NOT force the piston outward. Also, when pushing, take care not to allow the rod to contact the cylinder wall or the cylinder/rod could be damaged.

6. Using a 1/4 in. (6mm) diameter wooden dowel, gently tap on the rod in order to push the piston and rod upward, out of the block.

■ The rod and rod end cap are a matched set. Both must be replaced if either is damaged. Never substitute an end cap from a different connecting rod.

7. Matchmark the connecting rod to the piston to ensure installation with the same alignment.
8. Use a universal piston ring expander to remove the piston rings. Discard all the rings, they should not be reused.
9. Use a pair of needle nose pliers modified as illustrated to remove the wrist pin retaining rings from the piston.

■ The wrist pin-to-piston boss fit is loose on both sides, therefore little or no force should not be necessary to remove the wrist pin.

10. Use a driver that is slightly smaller than the wrist pin (like the OMC Wrist Pin Remover/Installer #342657) to carefully tap the wrist pin from the piston.

■ Stop here if ONLY the piston is being serviced. Refer to Cleaning & Inspection in this section for piston and cylinder bore inspection procedures.

11. Remove the six crankcase cover-to-cylinder block screws using a star-crossing pattern. If not done during flywheel removal, remove the key from the crankshaft.
12. Tap around the cover using a rubber or plastic mallet to help loosen the seal and then gently pry upward on the cover (simultaneously at bosses on each side using two small prybars). Gently break the gasket seal and remove the cover and crankshaft assembly from the block.

■ If necessary, you can tap gently on the bottom end of the crankshaft using a plastic or rubber mallet to help separate the crankshaft and cover assembly from the block.

13. Remove the crankshaft from the cover.

■ The cover and block bearings and/or seals should not be removed unless they are to be replaced. The seals can be removed independently of the bearings if the bearings are in good condition, but seal replacement is necessary or desired.

14. If necessary, remove the seal and bearing from the cover. Use a slide hammer and large jawed internal puller (a puller that attaches to the seal lips/casing from within the seal) to remove the seal. To remove the bearing, invert the cover and support it with the top (outside of the cover) facing down on a 1 3/8 in. (35mm) socket. Use a shop press and a 13/16 in. (21mm) socket as a driver to gently press the bearing free of the cover.
15. Remove the thrust washer from the inside the center back of the block.

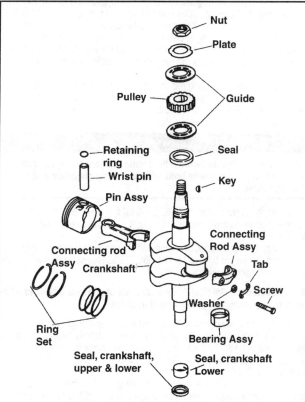

Fig. 23 Exploded view of the crankshaft, piston and main bearing assembly - 5/6 hp (128cc) models

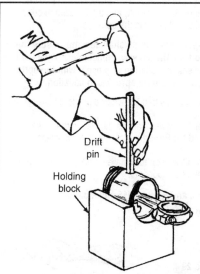

Fig. 24 The best method for removing wrist pins is using a holding block and a drift pin (driver) along with a shop press or hammer to push the wrist pin from the piston

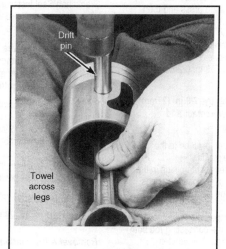

Fig. 25 If the wrist pin does not have an interference fit, you can support the piston on your legs with a shop cloth when carefully tapping the wrist pin free of the piston bore

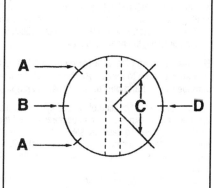

Fig. 26 Piston ring gap positioning - (A) segmented rings, (B) middle compression ring, (C) segmented ring and (D) top compression ring

POWERHEAD REPAIR AND OVERHAUL

16. If necessary, remove and replace any loose or damaged dowel pins.

17. If necessary, remove the seal and bearing from the block in the same manner as the cover seal and bearing (with the exception that the block does not need to be supported on a 1 3/8 in. (35mm) socket). When pressing the bearing out, invert the block and rest it on the cover mating surface.

18. Check the crankshaft sleeve. If damaged, replace it using a slide hammer and a small-jawed puller.

19. Refer to Cleaning & Inspection for details concerning cylinder block, piston and crankshaft inspection. Don't forget to visually inspect the block mating surface and dowel pins for damage (replace any loose or damaged pins). Also, remember to check all bearing surfaces for signs of excessive wear or damage. The oil pressure sender hole must be free of dirt or debris.

To assemble:

20. Apply a light coating of clean engine oil (or assembly lube) to the wrist pin and bore in the side of the piston.

21. Align the matchmarks made on the piston and connecting rod removal, then, since the wrist pin **should** be a loose fit in the piston, insert the wrist pin by hand and push it through the piston into the connecting rod.

22. Secure the wrist pin in the piston using the retaining ring. Place the ring in the end groove, then snap it into position using your finger. Turn the ring in the groove until the gap faces directly away (180° opposite) the semi-circular relief in the wrist pin bore.

■ **Apply a light coating of clean engine oil or assembly lube to each ring in order to help ease installation.**

23. Install the oil control ring assembly. Position the segmented ring into the bottom groove first, then insert the end of the first scraper ring under the segmented ring. Work carefully around the piston until the scraper ring is seated. Finally, insert the second scraper ring on top of the segmented ring and work around the piston to seat it.

24. Position the scraper ring end gaps each at 45° from each end of the wrist pin (the wrist pin is represented as a dotted line in the accompanying illustration). Make sure the segmented ring gap is at least 45 degrees from the wrist pin centerline, but on the opposite side of the piston and wrist pin centerline from the scraper ring gaps.

25. Install the middle ring into the middle piston ring groove with the number **2** stamp facing upwards and the white painted mark to the right side of the gap. Position the middle ring gap directly between the 2 scraper ring gaps (90° from the wrist pin centerline). Refer to the accompanying illustration for clarification.

26. Install the top ring in the top piston groove with the stamped **R** facing up and with the orange paint mark to the right side of the gap. Position the top ring gap directly opposite the middle compression ring (in the same general area of the segmented ring gap, but more precisely 90° from the wrist pin centerline). Again, refer to the illustration for details.

27. If removed, lubricate and install the bearing to the cylinder block and/or cover. Support the block or cover in a suitable shop press and then apply a light coating of clean engine oil or assembly lube to the bearing. Place the bearing in position over the bore, aligning the bearing tab with the slot, then use OMC #115311 or an equivalent sized driver to slowly press the bearing into the bore. When installing the block bearing, stop pressing when the bearing is inserted about halfway, then substitute a 7/8 in. (22mm) socket to finish pressing the bearing into position.

■ **When installing the cylinder block bearing using the 7/8 in. (22mm) socket, take care to make sure the socket does not contact and damage the locating pin for the thrust bushing.**

28. Apply a light coating of clean engine oil or assembly lube to the cylinder wall, an expandable ring compressor and the piston.

29. Verify the ring gaps are still in position, then secure the piston in the ring compressor. Tighten the compressor **just** sufficiently to push the rings into the grooves, but do not over-tighten it as the piston must still be free to slide into the bore.

30. Position the piston over the bore with the stamped arrow on the piston dome and the raised dot on the bottom end of the connecting rod both facing the flywheel side of the motor. Use a wooden dowel or wooden hammer handle to gently tap the piston down through the ring compressor and into the cylinder bore.

✳✳ WARNING

When tapping the piston into position MAKE SURE the connecting rod does not contact and damage the cylinder walls.

31. If removed, install a new crankshaft sleeve to the crankshaft OMC Crankshaft Sleeve Installer (#342657), or equivalent sized driver. Be sure to fully seat the sleeve and then apply a coating of OMC Moly Lube or equivalent assembly lube to the seal.

32. Install and seat the thrust washer in the bottom of the block, making sure the slot in the bottom of the washer aligns with the block locating pin.

■ **Position the piston with the crown just SLIGHTLY above the cylinder deck (making sure the rings remain in the bore) to provide clearance for crankshaft installation.**

33. Apply a light coating of clean engine oil or assembly lube to the thrust washer and bearing, along with the crankshaft main and rod bearing journals. Inset the crankshaft carefully into the block.

34. Apply a light coating of clean engine oil or assembly lube to the connecting rod and cap bearing surfaces. Install the cap to the connecting rod with the raised dot facing the flywheel (this should align any other marks made during removal).

35. Secure the connecting rod cap using NEW rod bolts and a NEW retainer. Tighten the bolts gradually (back and forth in stages) to 96-108 inch lbs. (11-12 Nm). Bend the retainer tabs up onto the bolt hex heads to lock them in position.

✳✳ WARNING

USE A TORQUE WRENCH when tightening the connecting rod cap bolts. Over-tightening these bolts will lead to cap distortion and bearing failure.

36. Apply a light coating of OMC Gel-Seal II or equivalent to the cylinder block and cover mating surfaces, then install the cover over the cylinder block and crankshaft.

37. Apply a light coating of engine oil to the cover bolt threads, then install and tighten the bolts to 15-17 ft. lbs. (20-23 Nm) using multiple passes of a star (crossing) sequence.

■ **Although OMC Gel-Seal II has a shelf life of one year when stored at room temperature, buy a new tube if in doubt.**

38. Apply a light coating of OMC Gasket Sealing Compound, or equivalent sealant to both sides of the NEW block access plate gasket and a light coating of OMC Nut Lock, or equivalent threadlock to the threads of the access plate cover screws. Install the access plate and gasket, then securely tighten the 2 Philips screws.

39. If removed, install a new upper or lower crankshaft seal to the cover or cylinder block, respectively. Apply a light coating of OMC Gasket Sealing Compound, or equivalent sealant to the outer diameter of the seal casing. Also, apply a light coating of clean engine oil or assembly lube to the seal lips. Position the seal over the shaft and into the bore, then use OMC Crankshaft Seal Installer #342215 (or equivalent driver with a large enough bore not to contact or damage the crankshaft) and a 1 in. (25mm) deep socket to gently tap the seal until it seats in the bore.

■ **When installing the upper (cover) crankshaft seal, make sure the crankshaft key is removed from the keyway. Immediately place the key back into the keyway after seal installation or place it with the sprocket so it will not be forgotten or lost during sprocket installation.**

40. If removed, install the dowels and the water deflector to the cylinder block mating surface.

41. Assemble and install the powerhead, as detailed in the Powerhead Disassembly & Assembly and the Powerhead Removal & Installation procedures in this section.

42. Refer to Break-In in this section for details on how to ensure proper break-in of the new powerhead components.

3/4 Hp and 4 Deluxe (87cc), and 5/6/8 Hp (164cc) Motors

◆ See Figures 27, 28 and 29

Removal & Installation

◆ See Figures 27, 28 and 29

1. Remove the engine top cover and, although not absolutely necessary on some smaller models, remove the front and rear lower covers for additional access. For details, please refer to Engine Covers (Top and Lower Cases) in the Maintenance and Tune-Up Section.

POWERHEAD REPAIR AND OVERHAUL

2. For 4-Deluxe models, remove the low speed knob and choke knobs.

3. Remove the manual rope starter assembly. For details, please refer to the Hand Rewind Starter section.

4. On models equipped with an integral fuel tank, remove the tank from the engine. For details, please refer to the Fuel System section.

5. If necessary for service (or desired to prevent possible damage) remove the carburetor. For details, please refer to the Fuel System section.

6. Remove the flywheel, as detailed in this section.

7. Remove the ignition coil and related ignition components. For details, please refer to the Ignition and Electrical Systems Section.

8. On 3/4 hp (87cc) motors, except on 4-Deluxe models, remove the 2 screws securing the shield to the midsection, then remove the shield.

9. Remove the 5 (some 1990 models) or 6 (most models) bolts from underneath the powerhead-to-gearcase flange (there should be 3 on each side of the motor, one forward, one aft and one toward the center of each side).

10. Remove the powerhead by carefully lifting it straight upward and off the exhaust housing/lower unit. If the powerhead is stuck, pry carefully along the edges and use a rubber mallet to help break the gasket seal and corrosive adhesion.

11. On 3/4 hp (87cc) motors, except on 4-Deluxe models, remove and discard the gasket from the mating surfaces.

12. On 4-Deluxe models, remove the bolts (4) attaching the inner exhaust housing to the powerhead. Remove the inner exhaust housing from the powerhead, then remove and discard the gasket. The water tube on these models is held in place by a retainer from the inside of the inner housing. If necessary, the retainer can be removed to free the water tube.

13. On 5/6/8 hp (164cc) models, remove the bolts (3) attaching the exhaust housing to the inner exhaust housing. Check the condition of the upper water tube grommet and remove the adapter plate. Finally, remove the bolts securing the inner exhaust housing to the powerhead, then remove the inner exhaust housing and remove/discard the gasket.

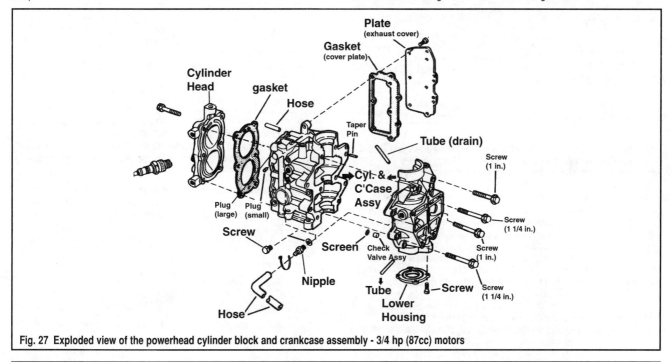

Fig. 27 Exploded view of the powerhead cylinder block and crankcase assembly - 3/4 hp (87cc) motors

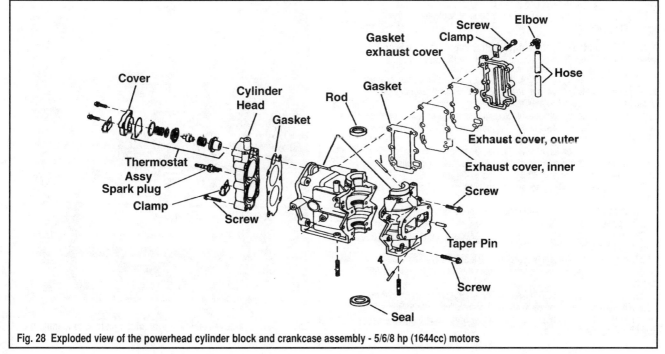

Fig. 28 Exploded view of the powerhead cylinder block and crankcase assembly - 5/6/8 hp (1644cc) motors

6-18 POWERHEAD REPAIR AND OVERHAUL

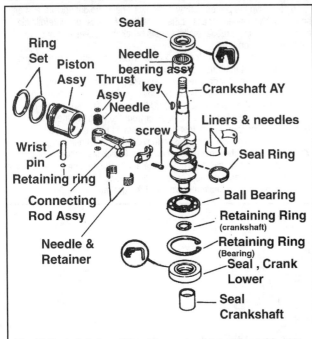

Fig. 29 Exploded view of the piston and crankshaft assembly - 3/4 hp and 4 Deluxe (87cc) motors and 5/6/8 hp (164cc) motors

To install:

14. On 4-Deluxe and 5/6/8 hp (164cc) models, proceed as follows to prepare the powerhead for installation:

   a. If removed, apply a light coating of OMC Triple-Guard, or equivalent marine grease to the water tube grommet, then install the water tube and grommet to the inner exhaust tube.

   b. Install the inner exhaust tube to the powerhead using a new gasket (do not apply sealant to the gasket). Then install and tighten the bolts to 60-84 inch lbs. (7-9 Nm) starting from the center and working outward.

15. On 4-Deluxe models only, proceed as follows:

   a. For ease of installation and to prevent, remove the gearcase from the exhaust housing. For details, refer to the Lower Unit section.

   b. Apply a light coating of OMC Triple-Guard, or equivalent marine grease to the end of the water tube, then install the powerhead to the exhaust housing. Install and tighten the bolts to 60-84 inch lbs. (7-9 Nm) starting at the center and working outward.

16. Apply a light coating of OMC Moly Lube or equivalent to the driveshaft splines.

■ DO NOT coat the end of the driveshaft with grease, as this could prevent the driveshaft from seating properly in the crankshaft.

17. Except for 4-Deluxe models, position a new gasket on the exhaust housing, then carefully lower the powerhead into position, aligning the crankshaft and driveshaft splines as the assembly is seated.

18. On 4-Deluxe models, install the gearcase to the exhaust housing, aligning the driveshaft/crankshaft splines and making sure that the water tube enters the water pump grommet.

19. Apply a light coating of OMC Gasket Sealing Compound, or equivalent sealant to the threads of the powerhead retaining bolts, then install and tighten the bolts to 60-84 inch lbs. (7-9 Nm).

20. For 5/6/8 hp (164cc) models, install the 3 inner exhaust housing-to-exhaust housing bolts.

21. Install the ignition coil and ignition components that were removed from the powerhead.

22. Install the carburetor, if removed for service or for safety during powerhead removal.

23. If equipped, install the integral fuel tank assembly.

24. Install the hand rewind starter. Tighten the retaining screws to 60-84 inch lbs. (7-9 Nm).

25. Install the top and, if removed, lower, engine covers.

26. If a new or rebuilt powerhead was just installed, refer to Break-In in this section for details on how to ensure proper break-in of the new powerhead components.

Disassembly & Assembly

◆ See Figures 8, 24, and 27 thru 41

■ To simplify assembly, remember to layout all bolts, components and clamps in the order of removal. This is especially true for the wiring harness and related clamps. Matchmark all component assemblies such as pistons, connecting rods to ensure installation of the correct pairs and installation in the correct orientation. Most of all, take your time.

1. Remove the powerhead from the lower unit as detailed under Removal & Installation.

2. If not removed to prevent the possibility of damage to it during powerhead removal, remove the carburetor. For details, refer to the procedure under the Fuel System section.

3. On 5/6/8 hp (164cc) motors, if only the lower crankshaft seal requires replacement, use a punch to carefully puncture and pry the seal from the powerhead. Be very careful not to contact and damage the crankshaft or the powerhead sealing surfaces.

4. Remove the bolts (2) and remove the intake manifold from the crankcase.

5. There are usually 2 bolts securing the leaf plate (reed valve plate) to the crankcase, remove the bolts and remove the plate.

■ Do not bend and leaf valves by hand as this could damage them so they either do not seal properly at best, or at worst, will break off during service. Do not disassemble the leaf plate unless some portion of the plate and valve assemblies is corroded or damaged and requires replacement.

6. Inspect the leaf plate and, if necessary, disassemble it as follows:

   a. Visually check for obvious signs of distortion on the plate. Inspect the valve tips for signs of cracks, chips or other obvious damage. Make sure the valve tips are not distorted or loose. Make sure the leaf stops are not distorted or loose. If there are any obvious defects, that component must be replaced.

Fig. 30 Before removal, label the tops of the pistons to identify the Top and Bottom (or No. 1 and No. 2 pistons)

POWERHEAD REPAIR AND OVERHAUL 6-19

b. Use a machinist's straightedge and a feeler gauge to check the leaf plate for distortion. Lay the straightedge across the plate only (not the leaf valves) at various points, then see if a 0.003 in. (0.08mm) or large feeler gauge can be inserted underneath the straightedge at any point. If it does, the plate is warped and must be replaced.

c. Visually inspect the check valve on the plate for damage or corrosion.

d. Visually check all gasket surfaces to make sure they are smooth and free of nicks. If necessary, a gasket surface may be dressed to smooth shallow nicks using a fine emery cloth.

e. Use the straightedge and feeler gauge to check for warpage across each of the gasket surfaces. Check at the center of the straightedge to make sure a 0.003 in. (0.08mm) or large feeler gauge cannot be inserted underneath the straightedge.

f. If disassembly is necessary, matchmark the leaf valves to their stops, then remove the screws and then separate the leaf valves from the plate. If they are being reused, do not allow the leaves and stops to become mismatched.

7. For 5/6/8 hp (164cc) motors, remove the thermostat cap and components from the powerhead. For details, please refer to the thermostat procedures in the Lubrication and Cooling section.

8. Loosen the cylinder bolts (6), then remove the cylinder head and discard the old gasket.

9. Use a permanent marker to label the No. 1 and No. 2 pistons (or T for Top and B for bottom, as desired).

10. On the side of the cylinder block, loosen the bolts (usually 9 on 3/4 hp motors or 6 on 5/6/8 hp motors, but check in case your motor varies), then remove the exhaust cover. Remove and discard the gasket from the mating surface.

■ If the exhaust cover is stuck on the side of the powerhead, gently tap around the edges using a rubber or plastic mallet to help break the gasket seal.

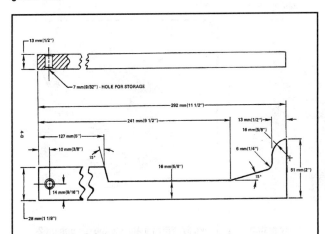

Fig. 31 You can fabricate a crankcase taper pin tool using a piece of 1 1/8 in. (28mm) wide x 1/2 in. (13mm) thick cold roll steel

11. On 5/6/8 hp motors, remove the inner exhaust cover. Remove and discard the gasket from the mating surface.

12. On 3/4 hp and 4 Deluxe models, remove the bolts (usually 4), then remove the round lower crankcase head from the bottom of the powerhead. Remove the seal from the crankcase bore (under the lower head mounting point)

13. Use a pair of snapring pliers to remove the lower crankshaft retaining ring.

14. Either use a fabricated taper pin tool (as shown in the accompanying illustration) to carefully push the crankcase taper pin toward the intake manifold surface or use a punch with a diameter larger than the pin. The taper pin can be accessed at the lower portion of the flywheel end of the crankcase (on 3/4 hp and 4 Deluxe models, it is found just below the flywheel end of the exhaust cover mounting area if the cylinder bores are facing upward. On 5/6/8 hp models the pin is visible in a boss on the exhaust cover side of the crankcase.

✷✷ WARNING

NEVER use a tool smaller than the taper pin bore to remove the pin or damage could occur.

15. Remove the main crankcase-to-cylinder block bolts (6) along with the flange bolts (4). For installation purposes (since bolts are of different lengths), arrange the bolts in holes punched into a cardboard box resembling the bolt pattern from which they were removed.

16. Loosely install the cylinder head back on the cylinder bore to keep the pistons from falling out as the crankshaft is removed.

17. Rotate the powerhead so the intake manifold surface is facing upward, then use a plastic or rubber mallet to tap upward on the exposed portion of the crankshaft (at a 90° angle to the shaft). Tap lightly until the crankcase half **just** starts to separate from the cylinder block half. Lift and remove the crankcase from the bottom of the cylinder block (exposing the crankshaft).

18. Remove the center main bearing liner, then collect and retain the 30 (3/4 hp and 4 Deluxe) or 23 (5/6/8 hp) needle bearings.

■ Keep in mind that all wear parts (such as pistons, connecting rods and needle bearings) MUST be installed in their original positions (and with the same orientations) if they are to be reused. Mark each component with a permanent marker to ensure proper assembly.

19. Matchmark and remove the connecting rod caps along with the connecting rod cap bearings.

20. Lift the crankshaft assembly carefully out of the cylinder block.

21. Once the crankshaft has been removed, position the connecting rod caps and cap bearings back onto the rods. Be sure to position them **precisely** as marked and noted before removal. This will help ensure no parts or lost or inadvertently mixed. Install the cap bolts finger-tight to secure the caps.

■ If the powerhead is on a workbench instead of an engine stand, position the powerhead securely on one end for the next step.

22. Rotate the powerhead until the cylinder head is facing upward, then remove the cylinder head again. Remove each of the pistons from the cylinder bore by pushing upward on the connecting rod from underneath.

Fig. 32 On models equipped with a taper pin, drive the pin from the bore in the crankcase

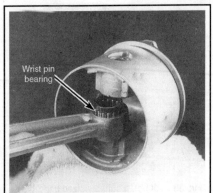

Fig. 33 When removing the wrist pin, take care not to loose the needle bearings (they are loose on some models)

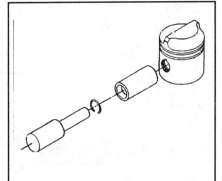

Fig. 34 If directed, use a shouldered driver to install the wrist pin retaining ring inside one wrist pin bore

6-20 POWERHEAD REPAIR AND OVERHAUL

✱✱ WARNING

DO NOT allow the connecting rods to contact and damage the cylinder walls or the insides of the piston skirts during removal.

23. Disassemble the crankshaft further for inspection and component replacement, as follows:
    a. Remove the seal ring from the center of the crankshaft.
    b. Remove the upper crankshaft seal and main bearing from the crankshaft.
    c. Remove the lower crankshaft seal (if not done already on 5/6/8 hp motors).
    d. ONLY remove the lower crankshaft bearing if it requires replacement.
    e. Visually inspect the crankshaft seal/sleeve assembly for signs of wear or damage. If replacement is necessary use a grind or cutting tool to carefully cut the sleeve free of the shaft.

✱✱ WARNING

Take EXTREME care when grinding or cutting the old crankshaft seal/sleeve assembly. DO NOT nick, mark, scratch or otherwise bugger-up your crankshaft.

24. Disassemble each piston for further inspection and component replacement, as follows:
    a. Use a universal ring expander to remove the rings from the pistons. If the rings are to be reused, mark them and keep them in sets to ensure installation in the original locations and positions.

■ We've said it before. We know there will be some circumstances that will prompt people to reuse rings, but for the most part, if you've come THIS far, it is probably a good idea to bite the bullet and replace the rings to ensure long-lasting performance from the powerhead.

   b. Using a pair of needle-nosed pliers (regular pliers may work on some 3/4 hp or 4 Deluxe models, otherwise use a pair that's been modified as shown in the accompanying illustration), remove the wrist pin retaining rings. Discard the old retaining rings.

■ A piston cradle can be fabricated from a large block of wood. The important design features of the cradle are that it is semi-circular on top to hold (or cradle, get it?) the piston and that there is a bore at the bottom center into which the wrist pin can be pressed. The final important feature of the cradle is that it is STURDY to withstand the force of the shop press and the wrist pin fit.

   c. On 5/6/8 hp motors the pistons must be heated in an oven or an oil bath to 200-400°F (93-204°C) in order to remove the wrist pins. Use great care and suitable protection (in the form of multiple oven mitts and possibly a soft-jawed tool such as slip-joint pliers) when handling the heated pistons.

■ DO NOT use a torch to heat the pistons for wrist pin removal as uneven heating can cause unequal expansion and contraction rates, damaging the pistons.

   d. Position the piston in a suitable piston cradle. On 3/4 hp and 4 Deluxe motors, be sure to position it with the side of the piston marked **L** for loose facing upward.

■ On 3/4 hp and 4 Deluxe models, look under the piston skirt at the 2 wrist pin bosses to identify the side marked L for loose. The mark should be stamped on one of the 2 bosses.

   e. Using a suitable drift pin (like OMC #326624 for 3/4 hp and 4 Deluxe models or #333141 for 5/6/8 hp motors) and shop press to slowly and smoothly drive the wrist pin free of the piston bore. If the OMC tool is not available, be sure to use a drift pin that is just slightly smaller than the diameter of the wrist pin itself.

Fig. 35 Use a universal ring expander to spread the rings JUST enough to slip over the piston

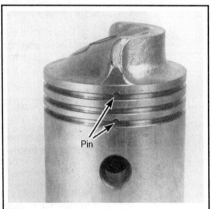

Fig. 36 Johnson/Evinrude 2-stroke piston ring grooves are equipped with dowels on which to align the ring gaps

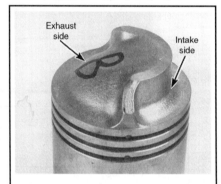

Fig. 37 When installing the pistons on 2-stroke motors, be sure the curved intake side faces the intake port

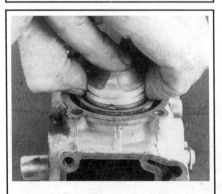

Fig. 38 On smaller motors, finger pressure can usually be used to compress the rings during assembly...

Fig. 39 ...if not, a suitably sized band-type hose clamp (such as a Mercruiser exhaust bellows clamp) can substitute for a ring compressor on 2-stroke motors

Fig. 40 Gently tap the piston into position, through the ring compressor (or hose clamp, shown)

POWERHEAD REPAIR AND OVERHAUL   6-21

■ On 3/4 hp and 4 Deluxe motors, as the wrist pin is removed from the piston you must retain the 21 loose needle bearings and the 2 thrust washers from the connecting rod bore. If you can, stop pressing once the wrist pin JUST clears the bottom of the connecting rod bore. Then carefully withdraw the rod along with the needle bearings (place a thin bladed tool like a putty knife or your finger under the rod hole as it is pulled free of the piston bore/wrist pin).

25. Refer to the information found under Cleaning & Inspection, in this section for details on inspecting the powerhead components to determine which should be replaced and to prepare the ones that are being reused for assembly.

To assemble:

※※ WARNING

During assembly, take your time and NEVER force a component unless a shop press or drive tool is specifically required by the procedure.

■ Lightly coat all components (except those that are coated with sealant) with clean, fresh Johnson/Evinrude Outboard Lubricant or equivalent oil during assembly. Always replace all gaskets, O-rings and seals.

26. On 3/4 hp and 4 Deluxe motors, make sure the water deflector is in position around the tops of the cylinder bores on the cylinder block-to-head mating surface.

27. If disassembled for inspection and/or component replacement, assemble each piston as follows:

■ When using a driver to install the wrist pin bearings it MUST contact the full diameter of the outer bearing cage.

   a. On 5/6/8 hp motors if removed, use a suitable driver such as OMC Bearing Remover/Installer #327645 or an equivalent driver to install the wrist pin bearings into the wrist pin end of the connecting rod. Make sure the oil hole in the bearing aligns with the oil hole on the end of the wrist pin. Apply a light coating of engine oil or assembly lube to the bearing and wrist pin bore before installation.

   b. On 5/6/8 hp motors the pistons must be heated in an oven or an oil bath to 200-400°F (93-204°C) in order to install the wrist pins. Use great care and suitable protection (in the form of multiple oven mitts and possibly a soft-jawed tool such as slip-joint pliers) when handling the heated pistons. Before heating the pistons, install a new snapring in the far end of the wrist pin bore from where the pin will be inserted. Make sure the snapring fully seats in the groove (the recommended method of installation is to use a shouldered driver such as OMC cone 333142 and 333141).

   c. Apply a light coating of clean engine oil or assembly lube to the wrist pin and the pin bore in the piston.

   d. Support the piston in the shop press piston cradle, again with the marked loose (**L**) side facing upward for 3/4 hp and 4 Deluxe models. Use the same drift pin or wrist pin tool to slowly start pressing the wrist pin into the piston bore. Stop pressing when the pin appears at the first opening for the connecting rod.

   e. For 3/4 hp and 4 Deluxe models, assemble the 21 needle bearings into position in the connecting rod end, using OMC Needle Bearing Assembly grease (or equivalent) to hold them in position.

   f. Place one washer over the slightly exposed head of the wrist pin and position the connecting rod so the opening aligns with the pin. Continue pressing the pin slowly through the connecting rod, until the wrist pin appears at the other end of the piston, but this time, don't let it come out far enough to be exposed yet. Use a pair of needle-nose pliers to hold the second thrust washer in the small clearance between the second wrist pin bore and the connecting rod, then slowly finish pressing the wrist pin through the connecting rod, thrust washer and second pin bore on the piston.

   g. Install the wrist pin retaining rings (or the remaining ring on 5/6/8 hp motors).

   h. On 5/6/8 hp motors, check the piston using a micrometer to ensure it has not become distorted during assembly. Measure the piston at 2 points, 90° apart from each other, approximately 1/8 - 1/4 in. (3-6mm) above the bottom edge of the piston skirt. The piston is considered non-serviceable (out-of-round) if the difference between the 2 measurements is 0.002 in. (0.05mm).

   i. Use a ring expander to install the pistons rings on each piston. Each piston ring groove should contain a dowel pin, around which the ring gap must be situated. If not the rings will likely break during installation (or if by they don't by some miracle, during the first start-up).

■ When using the ring expander, spread each ring JUST enough to slip over the piston.

28. Install each piston as follows:
   a. Apply a light coating of clean engine oil or assembly lube to the piston and piston bore.
   b. Use a suitable ring compressor or a band-type clamp to evenly compress the rings (oiling the inside of the compressor or clamp will also help ease installation). Make sure the rings are still properly positioned with the gaps over the dowels.

■ On small motors such as this one, a proper sized band-type hose clamp can be used as a ring compressor. The best type to use is a Mercruiser (yes, Mercruiser) exhaust bellows hose clamp, as it contains an inner ring on which the outer rings slides to compress it. Typical automotive hose clamps will rotate the inner ring, which can rotate the piston rings out of alignment (off their dowels). Also, it is often possible to install these pistons without a compressor, as the ring spring-pressure is not excessive and can usually be compressed by finger-pressure during installation. Use care not to cock and damage the rings, piston or bore though if this method is used. Take your time and oil the components well.

   c. Start the piston into the bore with the deflector side of the piston (circular sharp edge) facing the intake port side of the cylinder block.
   d. Slowly insert the piston, making sure the connecting rod does not strike and damage the piston skirt, or the strike/scrape and damage the cylinder wall. If the piston hangs up on a ring, don't force it, push the piston back upward and remove the compressor to check for potential problems, such as a ring that has shifted the gap away from the dowel.

29. Once the pistons are installed in the bores, temporarily install the cylinder head again to keep them from falling out.

30. If disassembled, reassemble the crankshaft as follows:
   a. If removed, lubricate a new crankshaft seal/sleeve using isopropyl alcohol, then support the crankshaft in a shop press between the 2 lower counterweights and carefully press the seal/sleeve into position.
   b. If removed, oil a new lower main bearing, then support the crankshaft in a shop press between the 2 lower counterweights and carefully press the bearing into position. Install the retaining ring with the flat side facing outward.
   c. Apply a light coating of clean engine oil or assembly lube to the upper main bearing, then install the bearing to the shaft with the lettered side facing downward (toward the shaft).
   d. Apply a light coating of OMC Gasket Sealing Compound, or equivalent sealant to the casing (outer diameter) of a new upper main bearing seal, then install the seal on top of the crankshaft with the lip facing downward.
   e. Install a new seal ring to the center of the crankshaft assembly.

■ If the connecting rod end-caps were reinstalled to prevent loss, make sure they are matchmarked, then remove them so the crankshaft assembly can be installed. Position them aside ensuring that they will not be mixed-up during assembly.

31. For 3/4 hp and 4 Deluxe models, install the crankshaft and center main needle bearing assembly, as follows:
   a. Place the main bearing liner half that contains the hole into the cylinder block journal.
   b. Apply a light coating of OMC Needle Bearing Assembly Grease, or equivalent to the 30 needle bearings for the main bearing liner. Position 14 of those needles in the main bearing liner half which you installed in the cylinder block during the previous step.
   c. Apply a light coating of clean engine oil or assembly lube to the crankshaft, then install the crankshaft into the block, taking care not to displace the 14 needle bearings in the main bearing liner half.

■ When placing the crankshaft into the cylinder block, make sure the upper main bearing dowel aligns with the crankcase recess.

32. For 5/6/8 hp models, install the crankshaft and center mail needle bearing assembly, as follows:
   a. Apply a light coating of OMC Needle Bearing Assembly Grease, or

6-22 POWERHEAD REPAIR AND OVERHAUL

equivalent to the 23 needle bearings for the main bearing liner. Position the needles around the center main journal on the crankshaft, then position sleeves around the bearings. The retaining ring end of the sleeves should face away from the flywheel. Use the retaining ring to secure the sleeves.

b. Apply a light coating of clean engine oil or assembly lube to the crankshaft, then install the crankshaft into the block with the seal ring gap facing upward and while aligning the upper main bearing dowel with the crankshaft recess. Also, align the center main bearing dowel pin hole with the pin in the cylinder block.

33. Move the pistons toward the crankshaft. Secure each connecting rod to the crankshaft by placing one connecting rod bearing half in the connecting rod journal and the other in rod cap journal. Seat the rod journals against the crankpins, then align the matchmarks and install the caps. Oil and install the connecting rod cap bolts. Tighten the bolts in multiple passes to 60-70 inch lbs. (7-7.5 Nm).

34. For 3/4 hp and 4 Deluxe motors, place the balance of the 30 needles for the crankshaft main bearing (there should be 16 left) onto the center main bearing journal. Cover the needles using the other bearing liner half, aligning the dovetails.

35. Clean and degrease the mating flange of the crankcase and cylinder block halves using OMC Cleaning Solvent, or an equivalent solvent. Allow the flanges to air dry.

36. Apply a light coating of OMC Locquic Primer or equivalent to the mating flange of the crankcase half and allow it to air dry.

37. Apply a light coating of OMC Gel-Seal II or equivalent to the cylinder block half flange. Be sure that the sealant coats the entire flange evenly, but be careful not to apply excessive amounts. The sealant must not be applied within 1/4 in. (6mm) of the bearings or the sealant could be forced out from between the flanges into contact with the bearings during assembly.

■ Although OMC Gel-Seal II has a shelf life of one year when stored at room temperature, buy a new tube if in doubt. Keep in mind that using an old tube of Gel-Seal II could allow crankcase air leaks (leading to performance problems).

38. Carefully lower the crankcase half into place over the cylinder block half.

39. Apply a light coating of OMC Gel-Seal II or equivalent to the threads of the 6 crankcase-to-cylinder block (main bearing) bolts, then install and finger-tighten the bolts.

40. Once the crankcase if fully seated on the cylinder block, install the crankcase taper pin and make sure it seats fully. Use a plastic or rubber mallet to gently tap the bottom of the crankshaft to fully seat the lower main bearing.

41. Tighten the 6 crankcase-to-cylinder block (main bearing) bolts using multiple passes starting with the center bolts and working outward in a spiral pattern. Torque them to 60-84 inch lbs. (7-9 Nm) for 3/4 hp and 4 Deluxe motors, or to 144-168 inch lbs. (16-19 Nm) for 5/6/8 hp motors.

42. Install and tighten the 4 flange bolts to 60-84 inch lbs. (7-9 Nm).

43. Install the lower bearing retaining ring to the powerhead with the **beveled edge facing outward**. Also, make sure the retaining ring gap is centered around the oil hole in the front of the crankcase.

44. Apply a light coating of OMC Gasket Sealing Compound, or equivalent sealant to the casing (outer diameter) of a NEW lower crankshaft seal. Install the seal either with the lip facing downward to the gearcase (on 3/4 hp and 4 Deluxe models) or with the seal lips facing upward toward the flywheel (on 5/6/8 hp motors), as applicable, until the seal bottoms on the retaining ring.

45. Before proceeding, place the flywheel in position over the keyway and check for binding between the crankshaft, connecting rods and bearings by rotating the crankshaft (using the flywheel). Remove the flywheel again until it is installed later in this procedure.

46. If temporarily installed to retain the pistons while the connecting rod caps were unbolted, remove the cylinder head.

47. Apply a light coating of OMC Gasket Sealing Compound, or equivalent sealant to both sides of a NEW cylinder head gasket. Position the gasket on the head with the tab on one end facing upward (toward the top of the powerhead) and the numbered side of the gasket facing the cylinder block.

■ **The cylinder head bolt threads should NOT be coated with sealant.**

48. Install the cylinder head and gasket, then install and finger-tighten the bolts. Tighten the bolts using multiple passes of the torque sequence to 60-84 inch lbs. (7-9 Nm) for 3/4 hp and 4 Deluxe motors, or to 144-168 inch lbs. (16-19 Nm) for 5/6/8 hp motors.

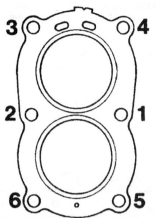

Fig. 41 Cylinder head torque sequence as positioned with the powerhead in a vertical position - 3/4 hp and 4 Deluxe (87cc) motors and 5/6/8 hp (164cc) motors (gasket for 3/4 hp and 4 Deluxe shown)

49. For 5/6/8 hp motors, install the thermostat assembly and cover. For details, please refer to the Thermostat procedures in the Lubrication and Cooling section.

50. Apply a light coating of OMC Gasket Sealing Compound, or equivalent sealant to both sides of a NEW exhaust cover gasket, then install the cover and secure using the bolts. Tighten the bolts to 25-35 inch lbs. (3-4 Nm) for 3/4 hp and 4 Deluxe motors, or to 60-84 inch lbs. (7-9 Nm) for 5/6/8 hp motors.

51. If disassembled for component replacement, assemble the leaf plate assembly as follows:

a. Place the leaf valves on the plate. If new valves do not seat on the plate, try turning them over.

✶✶ WARNING

NEVER turn over used valves on the plate, as they may break when they are returned to service. If used valves do not seat in their original direction of installation, they must be replaced.

b. Check the valves, if any leaves are standing open, apply light pressure using the eraser end of a pencil (the valve should close with light pressure). If not, check the plate for high spots or burrs.

✶✶ WARNING

NEVER lap the leaf plate. An plate that is too smooth may cause the leaf valves to stick closed when returning a motor to service after winterization.

c. Apply a light coating of OMC Locquic Primer or equivalent to the valve screws and allow it to air dry.

d. Apply a light coating of OMC Screw Lock, or equivalent threadlock to the mounting screw threads, wiping off any excessive adhesive compound.

e. Assemble the leaf valve shim (if used) and the leaf stop, then install and finger-tighten the retaining screws.

f. Center the leaf valves on the plate using the matchmarks made during removal, then tighten the screws to 25-35 inch lbs. (2.8-4 Nm).

52. Install the intake manifold and new leaf plate assembly using new gaskets. Tighten the retaining bolts to 96-120 inch lbs. (10.8-13.6 Nm) for 3/4 hp motors or to 60-84 inch lbs. (7-9 Nm) for 4 Deluxe and 5/6/8 hp motors.

■ **On 3/4 hp motors, refer to the Ignition and Electrical System section for details on Ignition Module Removal & Installation. The ignition module-to-flywheel air gap must be checked whenever the intake manifold is installed.**

53. For 3/4 hp and 4 Deluxe motors, install the lower crankcase head to the bottom of the powerhead. Tighten the bolts to 25-35 inch lbs. (2.8-4 Nm) using a crossing (X) pattern.

54. Install the powerhead assembly, as detailed in this section, then refer to Break-In in this section for details on how to ensure proper break-in of the new powerhead components.

POWERHEAD REPAIR AND OVERHAUL 6-23

8/9.9 Hp (211cc) 4-Stroke Motors

◆ See Figure 42

Removal & Installation

◆ See Figures 42 and 43

1. Either disconnect the negative battery cable (electric start models) and/or tag and disconnect the spark plug wires (rope start models), for safety.
2. Remove the engine covers. For details, please refer to Engine Covers (Top and Lower Cases) in the Maintenance and Tune-Up Section.
3. Remove the manual rope starter assembly. For details, please refer to the Hand Rewind Starter section. In order to remove the covers completely, don't forget the following:
   • Disconnect the manual choke knob from the choke link.
   • Once the screws are removed (including the screw hidden behind the water indicator) and the covers are separated, remove the cotter pin from the shift handle rod-to-link pin assembly, then remove the pin. The shift handle rod and link pin assembly is found at the lower front of the powerhead. If you have trouble identifying it, rock the shifter back and forth and watch what moves, then trace the rod to the cotter pin and link.
   • Remove the fuel hose from the cover connector.
4. If the powerhead is being removed for overhaul, tag and disconnect all ignition and electrical system components from the powerhead. For details please refer to the procedures in the Ignition and Electrical System section including procedures for the coil, power pack, sensor/trigger coil, charge coil, battery charge coil (if applicable) and oil pressure switch.
5. If the powerhead is being removed for overhaul, remove the fuel system components. For details, please refer to the Fuel System section for procedures on the carburetor, fuel pump and, if equipped, fuel primer solenoid.
6. Tag and disconnect the valve cover oil mist-to-carburetor recirculation hose, overboard coolant indicator hose and the thermostat to oil sump hose.
7. Remove the retaining bolts (4) securing the steering bracket to the upper mounts.
8. Remove the nuts and washers (4) securing the oil sump and powerhead to the gearcase midsection (they are located on either side of the motor, just above the oil sump-to-midsection split line), then remove the powerhead and oil sump as an assembly.

■ Lift the powerhead straight up and off the midsection. If necessary, use a plastic or rubber mallet to help break the seal. If necessary, you can assist removal with a small prybar, but take great care not to score and damage the mating surfaces.

9. Refer to Powerhead Disassembly & Assembly in this section for further service. Once separated, the cylinder head and/or cylinder block can also be disassembled for inspection or overhaul.

To install:

10. Thoroughly clean the mating surfaces of the oil sump and midsection.
11. Lower the powerhead carefully into position (placing the oil sump over the midsection studs), while aligning the driveshaft and mating surfaces.
12. Apply a coating of OMC Nut Lock or an equivalent threadlock to the stud threads, then install the nuts and lockwashers. Tighten the nuts to 15-17 ft. lbs. (20-23 Nm) using multiple passes of the proper crossing pattern, as shown.
13. Apply a coating of OMC Nut Lock or an equivalent threadlock to the threads of the steering bracket, then install the bolts and tighten to 108-132 inch lbs. (12-15 Nm).
14. Connect the hoses as tagged during removal. Be sure to secure the hoses using clamps or wire-ties as applicable.
15. Install the fuel system components that were removed. For details, please refer to the carburetor, fuel pump and, if applicable, fuel primer solenoid procedures in the Fuel System section.
16. Install the ignition and electrical components that were removed. For details, please refer to the Ignition and Electrical System section.
17. Install the rope starter assembly as detailed in the hand rewind starter section.

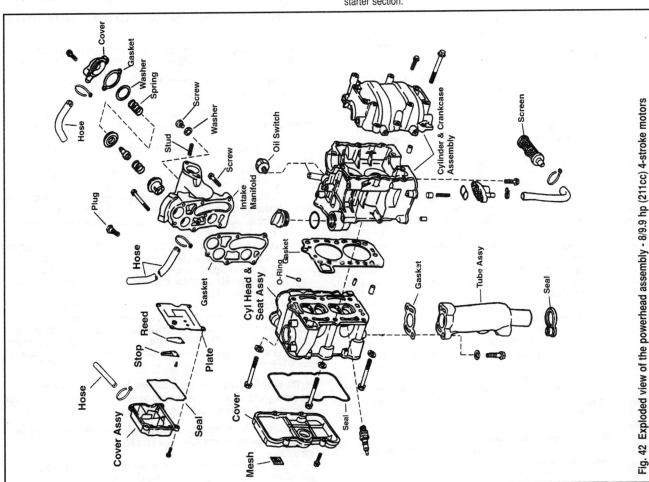

Fig. 42 Exploded view of the powerhead assembly - 8/9.9 hp (211cc) 4-stroke motors

6-24 POWERHEAD REPAIR AND OVERHAUL

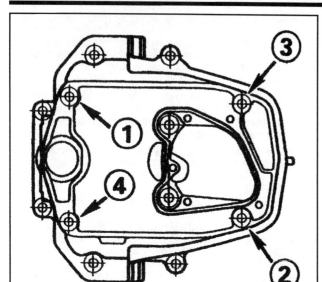

Fig. 43 Tighten the powerhead and oil-sump-to-midsection nuts in the sequence shown - 8/9.9 hp (211cc) 4-stroke motors

18. Install the lower engine covers.
19. Connect the negative battery cable and/or the spark plug wires, as applicable.
20. Refer to the Timing and Synchronization procedures in the General Information section, then perform all applicable adjustments.
21. If a new or rebuilt powerhead was just installed, refer to Break-In in this section for details on how to ensure proper break-in of the new powerhead components.

Powerhead Disassembly & Assembly

◆ See Figures 42 and 44

Although the powerhead houses many components (including the starting, fuel and ignition systems) some of those components were likely removed during powerhead assembly removal. This leaves 2 major sub-assemblies, the cylinder head and the cylinder block, along with various components for other systems (such as the thermostat, oil pump, breather assembly components and various tubes or housings). Although it is not necessary to remove all of these components to separate the cylinder head or cylinder block for further service, we recommend that if you've come this far you consider a complete disassembly and inspection of the powerhead, including ALL of these components. That is what we've provided in terms of a Disassembly & Assembly procedure. If access is necessary only to certain components (the cylinder head or the cylinder block), use care in determining which steps you will or will not follow. Complete Cylinder Head Overhaul and Cylinder Block Overhaul procedures are included in this section for disassembly of these assemblies once they've been separated.

Check the cylinder block and cylinder head for excessive warpage before installation. It is always a good idea to clean the cylinder block holes with a thread chaser, making sure the holes are clean and dry. Even small amounts of dirt or debris will adversely affect bolt torque.

Cylinder head gaskets are installed dry, do not apply sealant, but be sure that you always install a NEW gasket.

1. Remove the powerhead and oil sump assembly from the lower unit as detailed under Removal & Installation.
2. Remove the bolts (2) securing the upper engine mounts to the oil sump and remove the mount assemblies. Check the mounts for wear or damage and replace, as necessary.
3. Remove the bolts (5) and lock-washers retaining the exhaust deflector and water tube to the powerhead. Remove the exhaust deflector and water tube, then remove and discard the O-ring seal and the water tube/exhaust tube seal.
4. If necessary, remove the oil pump and the oil filter housing. For details, refer to the procedures in the Lubrication and Cooling section. It is a good idea to disassemble the oil pump housing to inspect the rotor assembly and determine if it should be replaced.

5. Remove the bolts (2) securing the exhaust tube to the bottom of the cylinder head. Lower the tube from the cylinder head, then remove and discard the old gasket.
6. Remove the nuts (8) and lockwashers securing oil sump to the powerhead. Remove the sump and the old gasket from the powerhead. Discard the gasket and visually check the sump for signs of contamination. Clean the sump and inspect for damage.
7. Loosen the clamp and remove the oil pickup tube along with the inlet screen. Clean the tube and screen using a mild solvent and check for damage.
8. Remove the bolts (2) mounting the pressure relief valve housing (to which the pickup tube was attached). Remove the housing, along with the spring and valve. Remove and discard the seal. Check the valve housing, spring and valve for excessive wear or damage.
9. Remove the oil pan and gasket from the powerhead.
10. If necessary, remove the thermostat cover and remove the thermostat components for inspection. For details, please refer to the Thermostat procedures in the Lubrication and Cooling Section.
11. Remove the bolts (7) securing the intake manifold to the powerhead. To help prevent possible warpage of the manifold, loosen the bolts in multiple passes of a crossing pattern that starts at the center and works outward. Remove the manifold, then remove and discard the gasket.
12. Remove the timing belt. For details, please refer to the procedure in this section.
13. If necessary, remove the crankshaft sprocket as follows:
   a. Secure the crankshaft from turning using the OMC Crankshaft Pulley Nut removal tool (#342212). The tool is a large hex nut with an internal taper and keyway designed to use with a wrench to keep the crankshaft from turning.
   b. Loosen the crankshaft sprocket nut using a large adjustable or crow-foot wrench.
   c. Remove the tool from the shaft and the key from the keyway.
   d. Remove the lock plate and the belt guide from the sprocket.
   e. Remove the bolts (3) and the starter bracket for access to the sprocket.

■ Inspect the dowel pins for damage, then remove and replace them if found.

   f. Install a 2-jawed puller to loosen and remove the pulley/belt guide from the crankshaft. Remove and retain the sprocket key from the keyway.

■ Be sure the puller jaws are positioned under the crankshaft sprocket guide for removal.

14. If necessary for cylinder head service, remove the bolt and washer retaining the camshaft sprocket. Remove the sprocket and key from the shaft.
15. If necessary, disassemble and inspect the breather assembly as follows:
   a. Remove the breather cover-to-valve cover retaining bolts (4) and then remove the breather cover. Remove and discard the O-ring.
   b. Remove the screws (2) and lock washers holding the reed, stop and plate. Remove the components, then remove and discard the plate O-ring.
   c. Remove, clean and inspect the mesh breather. Replace the mesh breather if it is damaged.
16. Loosen the valve cover bolts (4) using a crossing sequence. Start at the lower bolt on the port side of the head and end at the top bolt on the port side of the head (the slightly lower of the two bolts in the rounded cutout along the sides of the breather cover mounting flange). Remove the valve cover and discard the O-ring.
17. Loosen the cylinder head bolts (6) using multiple passes of a counterclockwise, spiraling pattern that works from the outside inward. Start at the lowest bolt on the port side, then to the lowest bolt on the starboard side. Next to the highest bolt on the starboard side, then the highest bolt on the port side. Finally to the middle bolt on the port side followed by the middle bolt on the starboard side.

■ The cylinder head bolts are of different lengths. As they are removed, place each of them through round cutouts in a cardboard box to help keep track of the bolt pattern and ensure ease of installation.

18. Remove the cylinder head from the cylinder block. If necessary, use a rubber or plastic mallet to help break the gasket seal. Remove and discard the old head gasket.
19. Remove and discard the O-rings from the dowel pins in the cylinder block. Check for loose or damaged dowel pins and replace, as necessary.

POWERHEAD REPAIR AND OVERHAUL   6-25

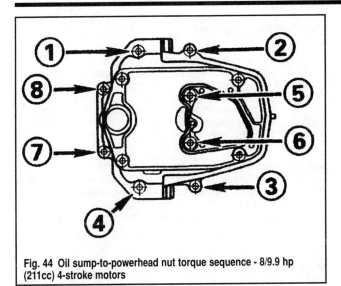

Fig. 44 Oil sump-to-powerhead nut torque sequence - 8/9.9 hp (211cc) 4-stroke motors

20. As necessary, refer to Cylinder Head Overhaul for details on disassembly of the cylinder head and valve train components and/or to Cylinder Block Overhaul for details on disassembly for the piston, rings and crankshaft assembly.

To assemble:

21. Assemble the cylinder head and/or cylinder block, as detailed under the overhaul procedures in this section. Be sure to replace all gaskets and seals.
22. Apply a light coating of OMC Gasket Sealing Compound, or equivalent sealant to both sides of a new cylinder head gasket, then position the gasket on the cylinder block.
23. Install new O-rings on the cylinder block dowel pins, but do not coat them with grease.
24. Apply a light coating of clean engine oil to the cylinder head bolt threads. Carefully position the cylinder head to the block, over the gasket, then install and finger tighten the bolts. Torque the bolts using multiple passes of a clockwise spiraling pattern working from the center outwards. Starts at the middle bolt on the starboard, then move to the middle bolt on the port side. Next tighten the upper bolt on the port side, followed by the upper bolt on the starboard side. Finally, tighten the lower bolt on the starboard side then the lower bolt on the port side. Use multiple passes of this sequence to tighten the bolts to 18-20 ft. lbs. (24-27 Nm).
25. Install the camshaft key with the outer key edge parallel to the camshaft centerline (outer edge straight up-and down), then install the camshaft sprocket, but do not tighten the retaining bolt fully at this time.
26. Install the crankshaft sprocket key (outer edge straight up-and-down, parallel to the shaft centerline), then install the belt guide and sprocket. Ideally, the sprocket can be gently pushed into place using the retaining nut and a crow-foot wrench, but you'll have to keep the crankshaft from turning to accomplish this. You can use the OMC Crankshaft Pulley Nut removal tool (#342212) or an equivalent tool to grab onto flywheel key. However, if necessary, you can use a socket and a rubber mallet to gently tap the sprocket over the crankshaft taper. Just take care not to score or otherwise damage the shaft during installation.
27. Apply a light coating of OMC Gasket Sealing Compound, or equivalent sealant, to the threads of the starter bracket bolts. Install the starter bracket and tighten the bolts to 84-106 inch lbs. (10-12 Nm).
28. Install the outer belt guide and lock plate (if you used the crankshaft sprocket nut to push the sprocket into position, you'll have to loosen and remove the nut for this). Apply a coating of OMC Nut Lock or an equivalent threadlock to the nut, then install the crankshaft sprocket nut and tighten to 18-20 ft. lbs. (24-26 Nm) while using the nut tool to hold the shaft from turning (as during removal). Keep the tool in position while the camshaft pulley bolt is tightened after timing belt installation.
29. Install the timing belt, as detailed in this section.
30. Once the timing belt is properly positioned and the timing marks are properly aligned, remove the bolt from the camshaft sprocket and apply a coating of OMC Nut Lock or equivalent thread. Reinstall the bolt and tighten to 84-106 inch lbs. (10-12 Nm). Remove the crankshaft holder tool.
31. Adjust the valve lash as detailed in the Maintenance and Tune-Up section.

32. Apply a light coating of OMC Triple-Guard, or equivalent marine grease to a NEW valve cover O-ring, then position the O-ring in the cover groove. Apply a light coating of OMC Gasket Sealing Compound, or equivalent sealant to the threads of the valve cover bolts, then install the cover and finger-tighten the bolts.
33. Torque the valve cover bolts (4) to 84-106 inch lbs. (10-12 Nm) using multiple passes of a crossing sequence. Start at the top bolt on the port side of the cover and end on the bottom bolt on the same side.
34. Apply a light coating of OMC Triple-Guard, or equivalent marine grease to NEW breather plate and cover O-rings, then install the O-rings (one in the valve cover and the other in the breather cover).
35. Apply a light coating of OMC Ultra Lock, or equivalent high-strength threadlock to the threads of the reed stop screws. Install the reed, stop and plate, then tighten the screw to 15-22 inch lbs. (7-2.5 Nm).
36. Apply a light coating of OMC Net Lock, or equivalent threadlock to the threads of the breather cover bolts. Install the breather cover and tighten the bolts to 36-60 inch lbs. (4-6 Nm).
37. Apply a light coating of OMC Gasket Sealing Compound, or equivalent sealant to the bolt threads and to both sides of a NEW intake manifold gasket. Install the manifold and gasket, then tighten the bolts to 84-106 inch lbs. (10-12 Nm) using multiple passes of a clockwise spiraling sequence that works inward from the outer bolts. Start at the upper bolt toward front of the motor, then the upper bolt toward the rear of the motor, next the lower bolt at the rear and the lower bolt at the front, then the middle bolt at the front, followed by the middle bolt at the rear. Finish with the most central bolt on the manifold.
38. If the flush plug and washer was removed, install tighten to 60-84 inch lbs. (8-10 Nm).
39. Install the thermostat assembly, as detailed in the Lubrication and Cooling section.
40. Apply a light coating of OMC Gasket Sealing Compound, or equivalent sealant to a NEW oil pan gasket, then position the pan and gasket to the powerhead.
41. Apply a light coating of OMC Gasket Sealing Compound, or equivalent sealant, to the threads of the oil pressure relief housing screws. Install a NEW O-ring to the housing, then install the housing along with the spring and relief valve. Tighten the screws to 84-106 inch lbs. (10-12 Nm).
42. Install the oil pickup tube and inlet screen and then secure using the spring-type clamp. If the screen was replaced, be sure to use a wire-tie to secure the screen to the hose.
43. Apply a light coating of OMC Gasket Sealing Compound, or equivalent sealant to both sides of a NEW oil sump-to-powerhead gasket, as well as to the stud threads. Install the gasket and sump, followed by the nuts and lock-washers. Tighten the nuts to 84-106 inch lbs. (10-12 Nm) using multiple passes of the torque sequence shown (a clockwise spiraling pattern working first the outside nuts and then the inner nuts).
44. Apply a light coating of OMC Gasket Sealing Compound, or equivalent sealant to both sides of a NEW exhaust tube gasket, and to the bolt threads as well. Install the exhaust tube and gasket, then tighten the bolts to 15-17 ft. lbs. (20-23 Nm).
45. If removed, assemble and install the Oil Pump, as detailed in the Lubrication and Cooling section.
46. Apply a light coating of OMC Adhesive M, or equivalent, to a NEW exhaust deflector O-ring, then position the O-ring.
47. Install a new water tube/exhaust tube seal in position on the exhaust tube.
48. Install the exhaust deflector and water tube and tighten the bolts to 36-60 inch lbs. (6-8 Nm).
49. Apply a light coating of OMC Nut Lock, or equivalent threadlock to the threads of the upper mount bolts (2), then install the upper engine mounts and tighten the bolts to 108-132 inch lbs. (12-15 Nm).
50. Install the powerhead, as detailed in this section.

Cylinder Head Overhaul

◆ See Figure 45

Cylinder head disassembly procedures are identical on the 5/6 hp and 8/9.9 hp 4-stroke Johnson/Evinrude outboards. For details regarding Cylinder Head Overhaul, please refer to the procedure under 5/6 hp motors in this section.

6-26 POWERHEAD REPAIR AND OVERHAUL

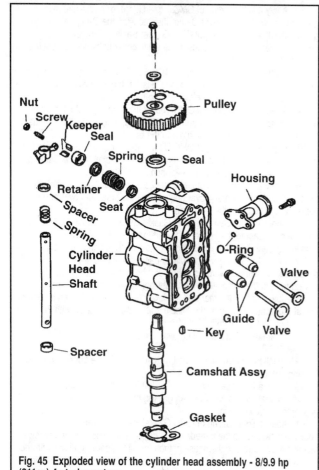

Fig. 45 Exploded view of the cylinder head assembly - 8/9.9 hp (211cc) 4-stroke motors

Cylinder Block Overhaul

◆ See Figures 8, 24, 25, 26, 46 and 47

The crankshaft for this motor turns on replaceable main bearing insert halves found in the cylinder block and crankcase. However, unlike many 4-stroke motors, the connecting rods do not utilize bearing inserts, as the rod and cap itself contain machined bearing surfaces. Especially for this reason, it is critical that the connecting rod and end cap are kept as a matched set, so they are positioned in the exact same orientation (facing the same directions) as originally installed.

During service, the crankshaft assembly can be removed from the block without piston removal or vice-versa. If either is attempted, take EXTREME care to keep the journal surfaces from being damaged. When free of the crankshaft journals, the connecting rods should be restrained from contacting the piston bores as well. One method to accomplish this is to stuff a clean shop towel under the piston skirt. Another is to use rubber bands to hold the rods in position.

Cylinder block work is more easily performed on an engine stand, especially since this allows rotation of the motor to access pistons from above and below the cylinder block. But, work can also be performed on a clean workbench, if care is taken not to damage the mating surfaces or, in the case of the cylinder head mating surface, the dowel pins. When working on a bench, take care to prevent the block from falling over during service. Also, if the case is stood on end at any point, take great care to keep the crankshaft from falling out of the block during service. A crankshaft can suffer a fatal blow from such a fall, even from a short height.

1. Remove the powerhead from the gearcase as detailed under Removal & Installation, then remove the cylinder head as detailed under Powerhead Disassembly & Assembly.
2. Invert the cylinder block so the cylinder head mating surface is facing downward.

■ **If you're not working on an engine stand, take care not to damage the cylinder head dowel pins.**

3. Loosen the 10 crankcase-to-cylinder block (and main bearing) bolts using multiple passes in the reverse of the torque sequence.
4. Use a plastic or rubber mallet to tap gently upward on the crankcase, breaking the gasket seal, then, remove the crankcase from the block.
5. Note the location and orientation of each bearing half in the journals at either end of the crankcase. Remove and retain the bearings. Visually check the crankcase mating surfaces, the bearing surfaces and the seal surfaces for damage or wear (no scratches big enough to catch a fingernail should be present on any of the surfaces).

■ **Each connecting rod cap is unique and is matched for installation to only ONE connecting rod and only in ONE direction. Always mark the alignment of each rod cap before removal and, keep the caps off the rods for as short a time period as possible (to prevent mix-ups). Loosely install the caps as soon as possible during procedures.**

6. Matchmark each rod cap to its connecting rod in order to ensure installation with the same alignment.

■ **If a permanent marker is used, keep the rod cap installed on the connecting rod during cleaning, in case the mark should wear off. That way, a new mark could be made after the components dry.**

7. Bend the tabs on the connecting rod cap bolt retainers back so the bolts can be removed.
8. Loosen and remove the connecting rod cap bolts. To prevent possible damage to the cap and bearing surface, loosen each cap by alternating between bolts, giving them about 1/4 - 1/2 turn at a time.

■ **The connecting rod bolts and retainers are NOT reusable on these motors. You can thread the bolts back in place to prevent the caps from becoming lost or mixed, but they must be replaced during assembly.**

9. Once the bolts are removed, the cap should easily come free, but if necessary, tap it gently using a plastic or rubber mallet to free it from the rod. Remove the rod cap and then remove the bearing half from the shaft or cap. Keep the bearing with the rod cap.
10. Push the pistons and connecting rods downward in their bores away from the crankshaft. Remove the crankshaft from the cylinder block.
11. Remove the main bearing halves from the cylinder block. Note their locations and orientations for measurement and assembly purposes.
12. Check the dowel pins in the cylinder block-to-crankcase mating surface. Remove and replace any loose or damaged pins from the cylinder block.
13. Remove and discard the crankshaft seals from either end of the shaft.
14. Remove and discard the O-ring from the bore in the end of the crankshaft.
15. Check the crankshaft sleeve. If damaged, replace it using a slide hammer and a small-jawed puller.
16. Mark the top of each piston for location and orientation.
17. If the pistons are being removed, place the cylinder block with the head mating surface and the tops of the pistons either facing upward or sideways. (Actually, we find that upward at a slight angle is the best of both worlds, but you'll need and engine stand and you have to take extra care to make sure the connecting rods do not hit the cylinder walls). To remove the pistons, use a large wooden dowel (or a large wooden hammer handle) to carefully push on the connecting rod until the piston comes out of the top of the cylinder block.

■ **When removing the piston, take extra care to make sure the connecting rod does not contact and damage the cylinder walls. If the crankshaft was not removed, take extra care to make sure the shaft journals are not damaged either.**

18. If temporarily reinstalled to prevent loss, remove the connecting rod caps from the rods.
19. Use a universal piston ring expander to remove the piston rings. Retain all rings in order (top, second, bottom, etc) for identification or diagnostic purposes, but they should not be reused.
20. Matchmark the connecting rod to the piston to ensure installation with the same alignment.
21. Use a pair of needlenose pliers modified as illustrated to remove the wrist pin retaining rings from the piston.

POWERHEAD REPAIR AND OVERHAUL 6-27

■ The wrist pin-to-piston boss fit is loose on both sides, therefore little or no force should not be necessary to remove the wrist pin.

22. Use a driver that is slightly smaller than the wrist pin (like the OMC Wrist Pin Remover/Installer #342657) to carefully tap the wrist pin from the piston.

23. Refer to Cleaning & Inspection for details concerning cylinder block, piston and crankshaft inspection. Don't forget to visually inspect the block mating surface and dowel pins for damage (replace any loose or damaged pins). Also, remember to check all bearing surfaces for signs of excessive wear or damage. The oil pressure sender hole must be free of dirt or debris.

To assemble:

24. Apply a light coating of clean engine oil (or assembly lube) to the wrist pin and bore in the side of the piston.

25. Align the matchmarks made on the piston and connecting rod removal, then, since the wrist pin **should** be a loose fit in the piston, insert the wrist pin by hand and push it through the piston into the connecting rod.

26. Secure the wrist pin in the piston using the retaining ring. Place the ring in the end groove and snap it into position using your finger. Turn the ring in the groove until the gap faces directly away (180° opposite) the semi-circular relief in the wrist pin bore.

■ Apply a light coating of clean engine oil or assembly lube to each ring in order to help ease installation.

27. Install the oil control ring assembly. Position the segmented ring into the bottom groove first, then insert the end of the first scraper ring under the segmented ring. Work carefully around the piston until the scraper ring is seated. Finally, insert the second scraper ring on top of the segmented ring and work around the piston to seat it.

28. Position the scraper ring end gaps each at 45° from each end of the wrist pin (the wrist pin is represented as a dotted line in the accompanying illustration). Make sure the segmented ring gap is at least 45 degrees from the wrist pin centerline, but on the opposite side of the piston and wrist pin centerline from the scraper ring gaps.

29. Install the middle ring into the middle piston ring groove with the number **2** stamp facing upwards and the white painted mark to the right side of the gap. Position the middle ring gap directly between the 2 scraper ring gaps (90° from the wrist pin centerline). Refer to the accompanying illustration for clarification.

30. Install the top ring in the top piston groove with the stamped **R** facing up and with the orange paint mark to the right side of the gap. Position the top ring gap directly opposite the middle compression ring (in the same general area of the segmented ring gap, but more precisely 90° from the wrist pin centerline). Again, refer to the illustration for details.

31. Apply a light coating of clean engine oil or assembly lube to the cylinder wall, an expandable ring compressor and the piston.

32. Verify the ring gaps are still in position, then secure the piston in the ring compressor. Tighten the compressor **just** sufficiently to push the rings into the grooves, but do not over-tighten it as the piston must still be free to slide into the bore.

33. Position the piston over the bore with the stamped arrow on the piston dome and the **F** on the connecting rod both facing the flywheel side of the motor (and any other alignment marks made during removal oriented as you intended). Use a wooden dowel or wooden hammer handle to gently tap the piston down through the ring compressor and into the cylinder bore.

✳✳ WARNING

When tapping the piston into position MAKE SURE the connecting rod does not contact and damage the cylinder walls.

34. If removed, install a new crankshaft sleeve with the O-ring on the end of the crankshaft using the OMC Crankshaft Sleeve Installer (#342662), or equivalent sized driver. Be sure to fully seat the sleeve and then apply a coating of OMC Moly Lube or equivalent assembly lube to the O-ring.

35. Apply a light coating of clean engine oil or assembly lube to the NEW main bearing halves, then position them in the cylinder block and crankcase. Position each bearing edge flush with the block.

36. Apply a light coating of clean engine oil or assembly lube to the crankshaft main bearing and connecting rod journals.

37. Apply a light coating of clean engine oil or assembly lube to the connecting rod and cap bearing surfaces.

38. Carefully position the crankshaft into the cylinder block on the main bearings.

39. Push each piston slowly upward in the bore, toward the crankshaft. Make sure the connecting rods do not contact and damage the cylinder walls or the crankshaft journals.

40. Install each connecting rod cap using NEW bolts and retainers. Tighten the bolts back and forth in stages (about 1/4-1/2 turn at a time) to 96-108 inch lbs. (11-12 Nm). Bend the retainer tabs up onto the bolt hex heads to lock them in position.

✳✳ WARNING

Over-tightening the connecting rod cap screws will likely render the connecting rods and caps unusable due to cap distortion.

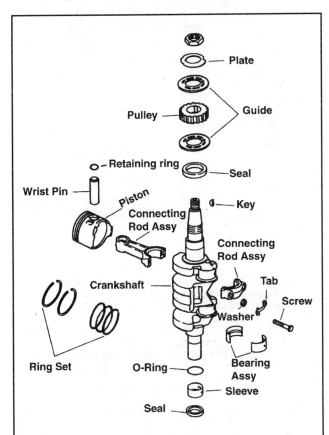

Fig. 46 Exploded view of the crankshaft, piston and main bearing assembly - 8/9.9 hp (211cc) 4-stroke motors

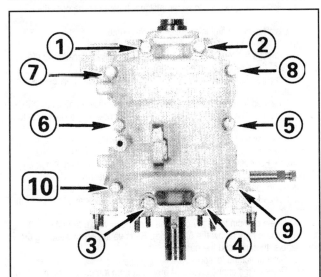

Fig. 47 Crankcase-to-cylinder block and main bearing torque sequence - 8/9.9 hp (211cc) 4-stroke motors

6-28 POWERHEAD REPAIR AND OVERHAUL

41. Apply a light coating of OMC Gel-Seal II or equivalent to the mating flanges of the cylinder block and crankcase. Be sure that the sealant coats the entire flange evenly, but be careful not to apply excessive amounts that could be forced out from between the flanges during assembly.

■ Although OMC Gel-Seal II has a shelf life of one year when stored at room temperature, buy a new tube if in doubt. Using an old tube of Gel-Seal II could prevent the crankcase from sealing properly.

42. Apply a light coating of clean engine oil to the threads of the 10 cylinder block-to-crankcase (main bearing) bolts. Install the bolts and finger-tighten, then use multiple passes of the torque sequence to tighten the bolts to specification. Tighten the small bolts to 84-106 inch lbs. (10-12 Nm) and the large bolts to 18-21 ft. lbs. (24-28 Nm).

43. Apply a light coating of OMC Gasket Sealing Compound, or equivalent sealant to the casings of a NEW upper and a NEW lower crankshaft seal. Apply a light coating of engine oil or assembly lube to the seal lips.

■ Don't loose the crankshaft sprocket and flywheel keys. When installing the upper crankshaft seal, place the keys with the sprocket and flywheel for use during installation.

44. Remove the keys from the top of the crankshaft (to prevent possible seal damage or interference with the seal installer), then position the upper seal over the crankshaft with the casing facing outward. Make sure the seal lips fit properly and start the casing squarely into the bore, then use a suitable driver (large enough to fit over the crankshaft, but still contact only the seal casing, such as OMC #342661) to press the seal slowly into position.

■ The OMC seal installer tool is designed for use with the crankshaft sprocket nut to slowly push the seal into position. Turn the nut slowly until the seal is fully seated, then remove the nut and installer tool.

45. Position the NEW lower crankshaft seal squarely over the crankshaft with the casing facing outward, then use a seal installer (such as OMC #342215 or an equivalent driver) to gently tap the seal into the bore until it contacts the powerhead.

46. Assemble and install the powerhead, as detailed in the Powerhead Disassembly & Assembly and the Powerhead Removal & Installation procedures in this section.

47. Refer to Break-In in this section for details on how to ensure proper break-in of the new powerhead components.

9.9/10/14/15 Hp (216cc) and 9.9/10/15 Hp (255cc) Motors

◆ See Figure 48

Removal & Installation

◆ See Figure 48

1. On electric start models, disconnect the negative battery cable for safety.
2. Remove the engine cover(s) for access. For details, please refer to Engine Covers (Top and Lower Cases) in the Maintenance and Tune-Up Section.
3. For 9.9/15 hp 216cc models, proceed as follows:
   a. Remove the shift lock lever screw from the side of the powerhead, just a little in front of the fuel pump. Separate the lever from the side of the powerhead and lay it down inside the lower engine cover (which cannot be removed yet on this model).
   b. Disconnect the fuel inlet hose from the pump.
   c. Disconnect the stop switch and ignition module ground leads from the ignition coil mounting screw. Then disconnect the single pin Amphenol connector for the stop switch.
   d. Remove the low speed adjustment knob, air silencer cover and finally the air silencer.
   e. Disconnect the choke knob from the knob detent, then pull the knob from the lower engine cover. Lift the knob detent from the carb.
   f. Tie a slip knot in the starter rope at the pulley, then remove the starter handle and bumper from the rope. Remove the bolt from the top center of the manual starter assembly. Lift the starter from the powerhead and thread a 3/8-16 nut on the mounting screw to secure the assembly.
   g. Remove the 3 nuts and six washers from the lower pan support studs.
   h. Disconnect the cooling water indicator hose from the powerhead nipple.

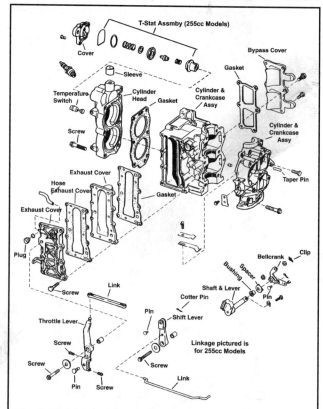

Fig. 48 Exploded view of the powerhead assembly (cylinder head, block and crankcase) - 9.9-15 hp (255cc) motors shown, 9.9-15 hp (216cc) motors very similar, but peripherals such as T-stat or shift/throttle linkage may vary

4. For 10/14 hp 216cc models, proceed as follows:
   a. Disconnect the stop switch and ignition module ground leads from the ignition coil mounting screw. Then disconnect the single pin Amphenol connector for the stop switch.
   b. For 10RS and 14 hp models, disconnect the choke knob from the knob detent.
   c. For 10KC or KS models, tag and disconnect the fuel hoses from the fuel primer.
   d. Tie a slip knot in the starter rope at the pulley just in front of the starter assembly, then allow the knot to engage the starter pulley. Remove the starter handle and bumper from the rope.
   e. Disconnect the cooling water indicator hose from the powerhead nipple.
   f. Disconnect the fuel inlet hose from the pump.
   g. Disconnect the throttle cable from the powerhead.
   h. Remove the four nuts from the powerhead studs (located on the underside of the cowling, 2 on each side).
5. If desired for overhaul purposes, remove the flywheel, as detailed in this section.
6. If desired for overhaul purposes, remove all ignition and electrical components from the powerhead, as necessary. For details, please refer to the Ignition and Electrical Systems Section.
7. If necessary for service (or desired to prevent possible damage) remove the carburetor and any other fuel system components from the powerhead. For details, please refer to the Fuel System section. If not, on 255cc models, you'll have to at least make sure the fuel hose is disconnected (on 216cc models, the hose was disconnected earlier).
8. For 255cc models, either disconnect the shift lever and linkage from the powerhead, or remove it completely, as follows:
   a. Use a drift pin to gently push the clip and drive pin from the shift handle (located right below the hand grip). Remove the screw securing the shift handle, then remove the handle.
   b. Disconnect the lockout rod.
   c. Remove the choke lever and grommet from the cutout at the front of the motor.
   d. Remove the pin, shift rod and grommet.
   e. Remove the screw and detent springs.

POWERHEAD REPAIR AND O...

f. Bend back the locktab, then loosen the bolt and remove the lock tab from the shift arm.

g. Remove the shift arm and spacer.

h. Remove the cotter pin, pin and shift lever. If necessary, remove the two shift arm bushing.

9. Remove the 6 bolts (only 5 three port and two starboard on 10/14 hp 216cc models) from underneath the powerhead-to-gearcase flange (there should be 3 on each side of the motor, one forward, one aft and one toward the center of each side).

10. Remove the powerhead by carefully lifting it straight upward and off the exhaust housing/lower unit. If the powerhead is stuck, pry carefully along the edges and use a rubber mallet to help break the gasket seal and corrosive adhesion.

11. Remove the exhaust and water tube assembly from the bottom of the powerhead, then remove and discard the gasket.

To install:

12. For all except the 10/14 hp 216cc models the manufacturer recommends removing the gearcase in order to ease the installation of the water tube into the water pump grommet. For details, refer to the Lower Unit section.

13. Position a new powerhead gasket on the exhaust housing (do NOT use any sealer on any models EXCEPT the 10/14 hp 216cc motors, on which the gasket should be coated on both sides lightly with OMC Gasket Sealing Compound).

14. If removed on all but 10/14 hp 216cc motors, install both grommets and water tubes in the inner water tube. Be sure to center the water tube in the grommets. Do NOT allow water tubes to touch the inner exhaust tube.

15. On all except 10/14 hp 216cc motors, install the inner exhaust tube on the powerhead, then install and tighten the screws to 60-84 inch lbs. (7-9 Nm).

16. Apply a generous amount of Permatex No. 2® or equivalent sealant to the machined diameter of the lower crankcase head.

17. Slowly lower the powerhead into the exhaust housing while carefully aligning the crankshaft-to-driveshaft splines. On all but 10/14 hp 216cc motors, also make sure:

• To install the long water tube is installed against the exhaust tube. A rubber band can be put in place to hold them together.

• To guide the long water tube into the exhaust housing's water tube opening.

• To guide the short water tube into the wider water tube opening in the rear of the exhaust housing.

18. On 9.9/15 hp (216cc) models only, before the powerhead is completely down:

a. Place a small washer over a large washer on the 3 lower pan support studs.

b. Guide the vertical control shaft into the vertical control shaft gear.

c. Place the remaining small and large washers onto the lower pan support studs, then start the 3 nuts BUT DO NOT TIGHTEN at this time. Be sure to place the ground lead on the rear stud.

d. Finish lowering the powerhead, tighten the 3 nuts securely.

19. Apply a light coating of OMC Gasket Sealing Compound, or equivalent sealant to the threads of the 6 (or 5 on 10/14 hp 216cc motors) powerhead retaining screws. Install the screws and tighten to 60-84 inch lbs. (7-9 Nm).

20. If removed, install the gearcase to the exhaust housing/midsection. For details, please refer to the procedures found in the Lower Unit section.

21. For 255cc models, either connect the shift lever and linkage to the powerhead, or if removed completely, install as follows:

a. If removed, install the 2 shift arm bushings.

b. Install the pin and shift lever, then secure using the cotter pin.

c. Install the shift arm and spacer.

d. Place the lock tab on the shift arm, then install the bolt and tighten securely. Bend the locktab down onto the bolt hex to prevent it from loosening.

e. Position the detent springs and secure using the screw.

f. Install the shift rod and secure using the pin, then install the grommet.

g. Attach the choke lever to each linkage, then position the grommet in the cutout.

h. Install the lockout rod.

i. Install the shift handle and secure using the screw.

j. Install the drive pin and clip into the shift handle.

22. Install the carburetor and any fuel system components that were removed.

23. Install all ignition and electrical c...

24. Install the flywheel, as detailed...

25. On 216cc models, reinstall the... order of removal, as noted earlier.

26. On electric start models, con... check operation of the neutral safety... properly, check and adjust it as foll...

a. Disconnect the battery cab... equipment.

b. Place the shifter in **neutral**, then disconnect the ... wire at the motor.

c. Connect an ohmmeter set to read continuity to the neutral start switch wire at the electrical box and the red starter wire, then press the start button.

d. The meter should show continuity only when the starter button is pressed and the shifter is in **neutral**.

e. If adjustment is necessary, loosen the switch adjustment screw (the small, flat-head screw at the base of the shifter handle, slightly above and slightly closer to the motor than the shift handle bolt). With the shifter and gearcase still in **neutral** the switch (located in the handle, just closer to the motor still than the adjustment screw) should be centered in the switch bore. Adjust the position as necessary and tighten the flat-head screw, then verify proper switch operation.

27. Install the engine cover(s).

28. If a new or rebuilt powerhead was just installed, refer to Break-In in this section for details on how to ensure proper break-in of the new powerhead components.

Disassembly & Assembly

◆ See Figures 8, 24, 25, 30 thru 40 and 48 thru 52

If the pistons are to be disassembled you will need a drift pin or driver that is slightly smaller than the wrist pin bore to carefully remove or install the wrist pins. The manufacturer makes a tool for this purpose, OMC #392511, but due to a slight decrease in the wrist pin inner diameter for some motors, a small amount of material might have to be removed from this tool before use. Check the tool fit, and if necessary, remove 0.015 in. (0.38mm) from the guide end of the tool.

■ To simplify assembly, remember to layout all bolts, components and clamps in the order of removal. This is especially true for the wiring harness and related clamps. Matchmark all component assemblies such as pistons, connecting rods to ensure installation of the correct pairs and installation in the correct orientation. Most of all, take your time.

1. Remove the powerhead from the lower unit as detailed under Removal & Installation.

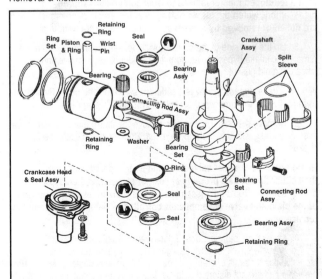

Fig. 49 Exploded view of the crankshaft, piston and main bearing assembly - 9.9/15 hp (255cc) motors

POWERHEAD REPAIR AND OVERHAUL

Fig. 50 Stuff rags inside the piston skirt to protect it from damage by the connecting rods anytime the rod is removed from the crankshaft journal

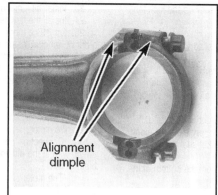

Fig. 51 Some Johnson/Evinrude connecting rods comes with alignment marks...

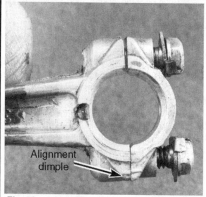

Fig. 52 ... but if not, be sure to matchmark the caps before removal and align them during assembly

2. If not removed to prevent the possibility of damage to it during powerhead removal, remove the carburetor. For details, refer to the procedure under the Fuel System section.

3. Remove the bolts (3) securing the lower crankcase head to the bottom of the powerhead, then remove the lower head. Use a punch to drive the seals from the lower crankcase head. Remove and discard the O-ring from head flange.

4. On the starboard side of the cylinder block, loosen the bolts (8), then remove the inner and outer exhaust covers. Remove and discard the gaskets from the mating surfaces. Check the inner exhaust cover for warpage or for pitting. If either condition exists, the inner exhaust cover must be replaced.

■ If the exhaust cover is stuck on the side of the powerhead, gently tap around the edges using a rubber or plastic mallet to help break the gasket seal.

5. Remove the thermostat cap and components from the top of the cylinder head. For details, please refer to the thermostat procedures in the Lubrication and Cooling section.

6. Loosen the cylinder bolts (6), using multiple passes of a counterclockwise spiraling sequence that starts at the outside bolts and works inwards. Start at the lower bolt on the port side and then move to the lower bolt on the starboard side. Next loosen the upper bolt on the starboard side, followed by the upper bolt on the port side. Lastly, loosen the middle bolt on the port side, followed by the middle bolt on the starboard side. Remove the cylinder head and discard the old gasket.

7. Using a felt-tipped marker or numbered/lettered punches, label the piston domes.

■ If a felt-tipped marker is used to label components during disassembly, watch the marks during cleaning. Replace any marks that are dulled or removed by solvent as soon as the component is dry.

8. On the port side of the cylinder block, loosen the bolts (6), then remove the intake bypass cover from the cylinder block. Remove and discard the old gasket.

■ One intake bypass screw is a different size than the other five. Note the location of the odd sized screw for installation purposes.

9. Remove the bolts (6) and then remove the intake manifold, leaf plate assembly and gaskets from the powerhead. Remove and discard the 2 gaskets.

■ Do not bend and leaf valves by hand as this could damage them so they either do not seal properly at best, or at worst, will break off during service. Do not disassemble the leaf plate unless some portion of the plate and valve assemblies is corroded or damaged and requires replacement.

10. Inspect the leaf plate and, if necessary, disassemble it as follows:
   a. Visually check for obvious signs of distortion on the plate. Inspect the valve tips for signs of cracks, chips or other obvious damage. Make sure the valve tips are not distorted or loose. Make sure the leaf stops are not distorted or loose. If there are any obvious defects, that component must be replaced.
   b. Check the valve disk (a small round valve and screen mounted one end of the plate, near the valve tips), it must be intact and able to move freely in the valve. Check the screen for debris or damage. Replace the valve, as necessary.
   c. Use a machinist's straightedge and a feeler gauge to check the leaf plate for distortion. Lay the straightedge across the plate only (not the leaf valves) at various points, then see if a 0.003 in. (0.08mm) or large feeler gauge can be inserted underneath the straightedge at any point. If it does, the plate or the manifold is warped and must be replaced.
   d. Check the leaf plate screws to make sure they are tight. If not, remove the screws one at a time, clean the threads. Apply a light coating of OMC Screw Lock, or equivalent threadlock to the threads of each screw before it is reinstalled and tightened securely.
   e. Visually check all gasket surfaces to make sure they are smooth and free of nicks. If necessary, a gasket surface may be dressed to smooth shallow nicks using a fine emery cloth.
   f. Use the straightedge and feeler gauge to check for warpage across the intake manifold gasket surfaces. Check at the center of the straightedge to make sure a 0.003 in. (0.08mm) or large feeler gauge cannot be inserted underneath the straightedge.
   g. If disassembly is necessary, matchmark the leaf valves to their stops, then remove the screws and separate the leaf valves, shims and stops from the plate. If they are being reused, do not allow the leaves, shims and stops to become mismatched.

✱✱ WARNING

Do not lift or bend the leaf valves during disassembly or they could become damaged. If a damaged leaf valve is installed it could break in use.

11. Either use a fabricated taper pin tool (as shown in the accompanying illustration) to carefully push the crankcase taper pin toward the intake manifold surface or use a punch with a diameter larger than the pin. The taper pin can be accessed at the upper portion of the crankcase (flywheel end).

✱✱ WARNING

NEVER use a tool smaller than the taper pin bore to remove the pin or damage could occur.

12. Remove the main crankcase-to-cylinder block bolts (6) along with the flange bolts (4) using multiple passes of a spiraling sequence that starts with the outer bolts and works towards the inner bolts. For installation purposes (since bolts are of different lengths), arrange the bolts in holes punched into a cardboard box resembling the bolt pattern from which they were removed.

13. Rotate the powerhead so the intake manifold surface is facing upward, then use a plastic or rubber mallet to tap upward on the exposed portion of the crankshaft (at a 90° angle to the shaft). Tap lightly until the crankcase half **just** starts to separate from the cylinder block half. Lift and remove the crankcase from the bottom of the cylinder block (exposing the crankshaft).

14. Loosely install the cylinder head back on the cylinder bore to keep the pistons from falling out as the crankshaft is removed.

POWERHEAD REPAIR AND OVERHAUL 6-31

■ Keep in mind that all wear parts (such as pistons, connecting rods, needle bearings and bearing liners) MUST be installed in their original positions (and with the same orientations) if they are to be reused. Mark each component with a permanent marker to ensure proper assembly.

15. Matchmark the connecting rod caps, then use a 5/32 in. hex wrench to loosen the connecting rod cap bolts. Loosen the bolts on a rod cap alternately, no more than one turn at a time. Remove the caps along with the connecting rod cap bearings.

16. Lift the crankshaft assembly carefully out of the cylinder block.

17. Once the crankshaft has been removed, position the connecting rod caps and cap bearings back onto the rods. Be sure to position them **precisely** as marked and noted before removal. This will help ensure no parts or lost or inadvertently mixed. Install the cap bolts finger-tight to secure the caps.

18. Disassemble the crankshaft further for inspection and component replacement, as follows:

    a. Use a pair of diagonal-cutting pliers to grasp and pry the key from the crankshaft. The pliers should be positioned with the blades across the key with the handle pointing up at an angle, then use the handle as a lever, pivoting on the bottoms of the blades.

    b. Slide the upper crankshaft seal and main bearing from the crankshaft.

    c. Remove the retaining ring from the groove in the center main bearing, then separate the halves and remove the 23 loose needle bearings.

    d. ONLY remove the lower crankshaft bearing if it requires replacement. First remove the retaining ring using a pair of snapring pliers, then support the bearing in a shop press and press the bearing off the shaft.

■ If the powerhead is on a workbench instead of an engine stand, position the powerhead securely on one end for the next step.

19. Rotate the powerhead until the cylinder head is facing upward (or upward at a slight angle to ease access to the connecting rods, then remove the cylinder head again. Remove each of the pistons from the cylinder bore by pushing upward on the connecting rod from underneath.

✱✱ WARNING

DO NOT allow the connecting rods to contact and damage the cylinder walls or the insides of the piston skirts during removal. One of the best ways to protect the piston is to pack the skirt with shop rags.

20. Disassemble each piston for further inspection and component replacement, as follows:

    a. Use a universal ring expander to remove the rings from the pistons. The manufacturer does not recommend reusing the rings. Retain them in sets for inspection purposes (marking their original locations), but discard them before installation.

■ We've said it before. We know there will be some circumstances that will prompt people to reuse rings, but for the most part, if you've come THIS far, it is probably a good idea to bite the bullet and replace the rings to ensure long-lasting performance from the powerhead.

    b. Using a pair of modified needle-nosed pliers as shown in the accompanying illustration, remove the wrist pin retaining rings. Discard the old retaining rings.

■ If necessary, a piston cradle can be fabricated from a large block of wood. The important design features of the cradle are that it is semi-circular on top to hold (or cradle, get it?) the piston and that there is a bore at the bottom center into which the wrist pin can tapped or pressed. The final important feature of the cradle is that it is STURDY enough to withstand the force of the shop press, BUT, keep in mind that the fit is loose on these pistons and significant force should not be necessary.

    c. The piston-to-wrist pin fit should be loose on both sides. Support the piston and use a suitable drift pin (like OMC #392511 to carefully and smoothly tap or push the wrist pin free of the piston bore). If the OMC tool is not available, be sure to use a drift pin that is just slightly smaller than the diameter of the wrist pin itself.

■ As the wrist pin is removed from the piston you must retain the 22 loose needle bearings and the 2 thrust washers from the connecting rod bore. If you can, stop tapping or pressing once the wrist pin JUST clears the bottom of the connecting rod bore. Then carefully withdraw the rod along with the needle bearings (place a thin bladed tool like a putty knife or your finger under the rod hole as it is pulled free of the piston bore/wrist pin). Inspect the bearings, if any of the needles are lost or worn, replace the entire bearing assembly.

21. Refer to the information found under Cleaning & Inspection, in this section for details on inspecting the powerhead components to determine which should be replaced and to prepare the ones that are being reused for assembly.

To assemble:

✱✱ WARNING

During assembly, take your time and NEVER force a component unless a shop press or drive tool is specifically required by the procedure.

■ Lightly coat all components (except those that are coated with sealant) with clean, fresh Johnson/Evinrude Outboard Lubricant or equivalent oil during assembly. Always replace all gaskets, O-rings and seals.

22. If disassembled for inspection and/or component replacement, assemble each piston as follows:

    a. Use a suitable shouldered driver (such as OMC Bearing Remover/Installer #392511) to install a new snapring in the far end of the wrist pin bore from where the pin will be inserted. Make sure the snapring fully seats in the groove. Turn the ring in the groove until the gap faces directly away (180° opposite) the semi-circular relief in the wrist pin bore.

    b. Apply a light coating of clean engine oil or assembly lube to the wrist pin and the pin bore in the piston.

    c. Support the piston (ideally in a piston cradle for ease of assembly) with the side opposite the installed wrist pin facing upward, then use the same drift pin or wrist pin tool to slowly start pressing the wrist pin into the piston bore. Stop pressing when the pin appears at the upper opening for the connecting rod.

    d. Working on the outside of the piston, use the wrist pin to help assemble the connecting rod bearing assembly. Slide one of the thrust washers down over the wrist pin, then slide the connecting rod down over the wrist pin. Apply a light coating of OMC Needle Bearing Assembly grease (or equivalent) to the wrist pin bearing needles, then insert the 22 needles between the wrist pin and the connecting rod.

    e. Place a small prytool under the washer and lift upward while sliding the connecting rod and bearings off the wrist pin (the washer and assembly grease should hold the bearings in position).

    f. Place the other thrust washer over top of the connecting rod and then insert the connecting rod, washer and needles into the piston (continuing to support the lower washer). Carefully press the wrist pin through the assembly. Work slowly, making sure none of the bearings are dislodged.

    g. Install the retaining wrist pin retaining ring and make sure it fully seats in the groove.

    h. Use a ring expander to install the pistons rings on each piston. Each piston ring groove should contain a dowel pin, around which the ring gap must be situated. If not the rings will likely break during installation (or if by they don't by some miracle, during the first start-up). Be sure to install the tapered ring in the top groove.

■ When using the ring expander, spread each ring JUST enough to slip over the piston.

23. Install each piston as follows:

    a. Apply a light coating of clean engine oil or assembly lube to the piston and piston bore.

    b. Use a suitable ring compressor or a band-type clamp to evenly compress the rings (oiling the inside of the compressor or clamp will also help ease installation). Make sure the rings are still properly positioned with the gaps over the dowels.

6-32 POWERHEAD REPAIR AND OVERHAUL

■ On small motors such as this one, a proper sized band-type hose clamp can be used as a ring compressor. The best type to use is a Mercruiser (yes, Mercruiser) exhaust bellows hose clamp, as it contains an inner ring on which the outer rings slides to compress it. Typical automotive hose clamps will rotate the inner ring, which can rotate the piston rings out of alignment (off their dowels). Also, it is often possible to install these pistons without a compressor, as the ring spring-pressure is not excessive and can usually be compressed by finger-pressure during installation. Use care not to cock and damage the rings, piston or bore though if this method is used. Take your time and oil the components well.

    c. Start the piston into the bore with the deflector side of the piston (circular sharp edge) facing the intake port side of the cylinder block.
    d. Slowly insert the piston, making sure the connecting rod does not strike and damage the piston skirt, or the strike/scrape and damage the cylinder wall. If the piston hangs up on a ring, don't force it, push the piston back upward and remove the compressor to check for potential problems, such as a ring that has shifted the gap away from the dowel.
  24. Once the pistons are installed in the bores, temporarily install the cylinder head again to keep them from falling out.
  25. If disassembled, reassemble the crankshaft as follows:
    a. If removed, apply a light coating of clean engine oil or assembly lube to a NEW lower main bearing, then support the crankshaft in a shop press between the 2 lower counterweights. Carefully press the bearing into position with the lettered side of the bearing facing upward toward the press. Install the retaining ring with the sharp edge of the ring facing away from the bearing.
    b. Apply a light coating of clean engine oil or assembly lube to the upper main bearing, then install the bearing to the shaft with the lettered side facing upward toward the flywheel.
    c. Apply a light coating of OMC Needle Bearing Assembly Grease, or equivalent to the 23 needles of the center main crankshaft bearing. Position the needles around the center crankshaft journal and secure using the sleeves. The retaining ring side of the sleeves should face the flywheel end of the crankshaft. Fasten the sleeves using the retaining ring.

■ If the connecting rod end-caps were reinstalled to prevent loss, make sure they are matchmarked, then remove them so the crankshaft assembly can be installed. Position them aside ensuring that they will not be mixed-up during assembly.

  26. Remove the rod caps and install the connecting rod bearings. Place one liner half in each connecting rod and the other in the cap. Apply a light coating of clean engine oil or assembly lube to the bearing surfaces.
  27. Apply a light coating of clean engine oil or assembly lube to the crankshaft journals, then carefully lower the shaft into the cylinder block. Take care not to allow the connecting rods to contact or damage the shaft journals. Align the crankshaft center main bearing dowel pin hole with the dowel pin in the cylinder block. Roll the upper main bearing pin into the cylinder block groove.
  28. Move the pistons toward the crankshaft. Seat the rod journals against the crankpins, then align the matchmarks and install the caps. Oil and install the connecting rod cap bolts, then finger-tighten the bolts alternating from side-to-side on the rod cap.
  29. Use a pencil or your fingernail to determine if the rod caps are fully aligned and seated. Scrape the pencil across the mating surface to see if it catches. If necessary, tap the cap lightly using a plastic or rubber mallet to fully seat it and check again (if it will not seat, double check the alignment marks). Once you are sure the cap is fully seated, tighten the bolts back and forth to 60-70 inch lbs. (7-7.5 Nm).
  30. Clean and degrease the mating flange of the crankcase and cylinder block halves using OMC Cleaning Solvent, or an equivalent solvent. Allow the flanges to air dry.
  31. Apply a light coating of OMC Locquic Primer or equivalent to the mating flange of the crankcase half and allow it to air dry.
  32. Apply a light coating of OMC Gel-Seal II or equivalent to the cylinder block half flange. Be sure that the sealant coats the entire flange evenly, but be careful not to apply excessive amounts. The sealant must not be applied within 1/4 in. (6mm) of the bearings or the sealant could be forced out from between the flanges into contact with the bearings during assembly.

■ Although OMC Gel-Seal II has a shelf life of one year when stored at room temperature, buy a new tube if in doubt. Keep in mind that using an old tube of Gel-Seal II could allow crankcase air leaks (leading to performance problems).

  33. Carefully lower the crankcase half into place over the cylinder block half.
  34. Apply a light coating of OMC Gel-Seal II or equivalent to the threads of the 6 crankcase-to-cylinder block (main bearing) bolts, then install and finger-tighten the bolts.
  35. Once the crankcase if fully seated on the cylinder block, install the crankcase taper pin and make sure it seats fully. Use a plastic or rubber mallet to gently tap the bottom of the crankshaft to fully seat the lower main bearing.
  36. Apply a light coating of OMC Gasket Sealing Compound, or equivalent sealant to the metal case of a NEW upper main bearing seal and light coating of oil to the seal lips. Position the seal with the lips facing inward and then use a suitable driver (such as OMC #391060) to gently tap the seal into position.
  37. Apply a light coating of OMC Gasket Sealing Compound, or equivalent sealant to the metal cases of the 2 NEW lower head seals. Apply a light coating of engine oil to the seal lips. Install the smaller diameter seal with the lips facing outward (facing the installation tool) first, then install the larger diameter seal with the lips facing inward (facing the smaller seal).

■ When installing the lower head seals either use OMC Seal Installer (#330251) or an equivalent driver. If using the OMC seal installer, one side is designed for the smaller diameter seal and the other side for the larger seal. If the OMC installer is not available, use care, especially when installing the smaller diameter seal to push on the casing and NOT the seal lips.

  38. Apply about 3cc of OMC Moly Lube between the lips of the 2 NEW lower head seals.
  39. Apply a light coating of engine oil or assembly lube on a NEW O-ring, then position it in the groove around the lower crankcase head.
  40. Apply a light coating of OMC Gasket Sealing Compound, or equivalent sealant to the threads of the lower crankcase head bolts, then install the head and finger-tighten the bolts.
  41. Tighten the 6 main bearing bolts in multiple passes of a spiraling pattern that works from the center screws and moves outward. Tighten the 6 bolts to 144-168 inch lbs. (16-19 Nm). Then tighten the smaller screws in a similar fashion to 60-84 inch lbs. (7-9 Nm).
  42. Tighten the lower crankcase head bolts to 60-84 inch lbs. (7-9 Nm).
  43. Apply a light coating of Permatex No. 2® or equivalent sealant to the 2 ribs in the cylinder head (they are at the top of the cylinder head, immediately adjacent to the thermostat housing).
  44. Apply a light coating of OMC Gasket Sealing Compound, or equivalent sealant to both sides of a NEW cylinder head gasket. Install the cylinder head and gasket, then install and finger-tighten the bolts.

■ Install the cylinder head bolts with the threads dry, they should NOT be coated with sealant.

  45. Tighten the cylinder head bolts using multiple passes of a clockwise spiraling pattern that starts at the center and works outward. Start at the center bolt on the starboard side and then tighten the center bolt on the port side. Move to the upper bolt on the port side, followed by the upper bolt on the starboard side. Finally tighten the lower bolt on the starboard side and the port bolt on the starboard side. Continue the spiral sequence until all bolts are tightened to 216-240 inch lbs. (24-27 Nm).
  46. Install the thermostat assembly and cover. For details, please refer to the Thermostat procedures in the Lubrication and Cooling section.
  47. Apply a light coating of OMC Gasket Sealing Compound, or equivalent sealant to both sides of the 2 NEW exhaust cover gaskets, then install the inner and outer covers. Install and tighten the retaining bolts to 60-84 inch lbs. (7-9 Nm) starting with the center screws and working outward.
  48. If disassembled for component replacement, assemble the leaf plate assembly as follows:
    a. Place the leaf valves on the plate. If new valves do not seat on the plate, try turning them over.

✱✱ WARNING

NEVER turn over used valves on the plate, as they may break when they are returned to service. If used valves do not seat in their original direction of installation, they must be replaced.

    b. Check the valves, if any leaves are standing open, apply light pressure using the eraser end of a pencil (the valve should close with light pressure). If not, check the plate for high spots or burrs.

POWERHEAD REPAIR AND OVERHAUL

※※ WARNING

NEVER lap the leaf plate. An plate that is too smooth may cause the leaf valves to stick closed when returning a motor to service after winterization.

    c. Apply a light coating of OMC Locquic Primer or equivalent to the valve screws and allow it to air dry.
    d. Apply a light coating of OMC Screw Lock, or equivalent threadlock to the mounting screw threads, wiping off any excessive adhesive compound.
    e. Assemble the leaf valve shim (if used) and the leaf stop, then install and finger-tighten the retaining screws.
    f. Center the leaf valves on the plate using the matchmarks made during removal (and the indexing marks that are placed on the leaf plate between the tips of the stops for each pair, then tighten the screws securely.

49. Install the intake manifold and leaf valve assembly using new gaskets. Install the gaskets dry, without sealer. Tighten the retaining screws to 60-84 inch lbs. (7-9 Nm).

50. Install the powerhead assembly, as detailed in this section, then refer to Break-In in this section for details on how to ensure proper break-in of the new powerhead components.

9.9/15 Hp (305cc) 4-Stroke Motors
◆ See Figure 53

Removal & Installation
◆ See Figure 53

1. Either disconnect the negative battery cable (electric start models) and/or tag and disconnect the spark plug wires (rope start models), for safety.
2. Remove the engine covers. For details, please refer to Engine Covers (Top and Lower Cases) in the Maintenance and Tune-Up Section.
3. Remove the manual rope starter assembly. For details, please refer to the Hand Rewind Starter section. In order to remove the covers completely, don't forget the following:

    • Remove the choke cable clamp and choke knob.
    • Remove all of the cover screws (2 at the top, 5 on the sides, and the one hidden behind the cooling stream indicator).
    • At the front base of the powerhead, remove the clip retaining the shift rod clevis to the shift rod link, then slide the clevis from the link.
    • Also at the front of the powerhead, at the base of the shifter arm, remove the shoulder screw and wave washer securing the shift lever to the tiller arm.
    • Disconnect the fuel hose from the fuel pump inlet.

4. If the powerhead is being removed for overhaul, tag and disconnect all ignition and electrical system components from the powerhead. For details please refer to the procedures in the Ignition and Electrical System section including procedures for the coil, power pack, stator, sensor/trigger coil and oil pressure switch.

5. If the powerhead is being removed for overhaul, remove the fuel system components. For details, please refer to the Fuel System section for procedures on the carburetor, fuel pump and, if equipped, fuel primer solenoid.

6. If equipped with a tiller control, remove the retaining bolts (usually 4), then disconnect any necessary components and remove the steering arm assembly.

7. Disconnect the shift rod from the arm.

8. Remove the bolts (6) securing the powerhead to the gearcase midsection/exhaust housing (there are 3 located on either side of the motor, just below the exhaust housing flange, threaded upward into the powerhead). Carefully lift the powerhead up and off the exhaust housing.

■ **Lift the powerhead straight up and off the midsection. If necessary, use a plastic or rubber mallet to help break the seal. If necessary, you can assist removal with a small prybar, but take great care not to score and damage the mating surfaces.**

9. Refer to Powerhead Disassembly & Assembly in this section for further service. Once separated, the cylinder head and/or cylinder block can also be disassembled for inspection or overhaul.

To install:

10. Thoroughly clean the mating surfaces of the oil sump and the midsection/exhaust housing. Remove all traces of the old gasket.

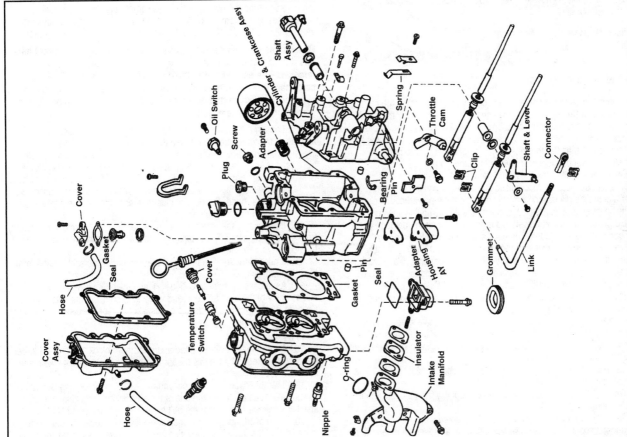

Fig. 53 Exploded view of the powerhead assembly (cylinder block, crankcase and cylinder head mounted components) - 9.9/15 hp (305cc) 4-stroke motors

6-34 POWERHEAD REPAIR AND OVERHAUL

11. Install a NEW powerhead base gasket with the bead facing the exhaust housing.
12. Check to make sure the exhaust seal and the water seal O-ring are both in proper position. Apply a light coating of OMC Gasket Sealing Compound, or an equivalent sealant.
13. Slowly lower the powerhead into position, while aligning the driveshaft and mating surfaces.
14. Apply a coating of OMC Nut Lock or an equivalent threadlock to the threads of the powerhead retaining bolts. Install the bolts and tighten to 15-17 ft. lbs. (20-23 Nm).
15. Connect the shift rod to the arm.
16. If equipped install and secure the tiller arm using the retainers. Tighten the retainers securely.
17. Install the fuel system components that were removed. For details, please refer to the carburetor, fuel pump and, if applicable, fuel primer solenoid procedures in the Fuel System section.
18. Install the ignition and electrical components that were removed. For details, please refer to the Ignition and Electrical System section.
19. Install the rope starter assembly as detailed in the hand rewind starter section.
20. Install the lower engine covers, paying particular attention to the following:
    • Connect and secure the fuel inlet hose to the fuel pump using a new wire tie.
    • Apply a light coating of OMC Nut Lock, or equivalent threadlock to the threads of the shift lever-to-steering arm shoulder screw. Install the shift lever and secure using the screw and wave washer, then tighten to 84-106 inch lbs. (9-12 Nm).
    • Install the shift rod clevis to the link and secure it using the clip.
    • Install the cover screws, including the one located behind the cooling stream indicator. Secure the cooling stream indicator and hose.
    • Connect the choke cable and knob.
21. Connect the negative battery cable and/or the spark plug wires, as applicable.
22. Refer to the Timing and Synchronization procedures in the General Information section, then perform all applicable adjustments.
23. If a new or rebuilt powerhead was just installed, refer to Break-In in this section for details on how to ensure proper break-in of the new powerhead components.

Powerhead Disassembly & Assembly

◆ See Figures 53, 54 and 55

Although the powerhead houses many components (including the starting, fuel and ignition systems) some of those components were likely removed during powerhead assembly removal. This leaves 2 major sub-assemblies, the cylinder head and the cylinder block, along with various components for other systems (such as the thermostat, oil pump, breather assembly components and various tubes or housings). Although it is not necessary to remove all of these components to separate the cylinder head or cylinder block for further service, we recommend that if you've come this far you consider a complete disassembly and inspection of the powerhead, including ALL of these components. After all, checking the oil pickup line and screen now could prevent a second powerhead removal to repair the damage done if these components should fail. For this reason, a complete teardown procedure is what we've provided in terms of a Disassembly & Assembly procedure. If access is necessary only to certain components (the cylinder head or the cylinder block), use care in determining which steps you will or will not follow. Complete Cylinder Head Overhaul and Cylinder Block Overhaul procedures are included in this section for disassembly of these assemblies once they've been separated.

Check the cylinder block and cylinder head for excessive warpage before installation. It is always a good idea to clean the cylinder block holes with a thread chaser, making sure the holes are clean and dry. Even small amounts of dirt or debris will adversely affect bolt torque.

Cylinder head gaskets are installed dry, do not apply sealant, but be sure that you always install a NEW gasket.

1. Remove the powerhead and oil sump assembly from the lower unit as detailed under Removal & Installation
2. Tag and remove the cooling system (overboard indicator), breather and thermostat hoses from the powerhead. Inspect all hoses for deterioration or cracking and replace upon assembly, if necessary.
3. If equipped, remove the emissions testing tube from the powerhead (when equipped it can be found immediately underneath the intake manifold).
4. Remove the oil filter, pressure switch, fill plug and level dipstick (all located on the port side of the powerhead).
5. Loosen the clamp and remove the oil pickup tube along with the inlet screen. Clean the tube and screen using a mild solvent and check for damage.
6. Remove the timing belt. For details, please refer to the procedure in this section.
7. If not done when removing the timing belt (and if necessary for further repairs) remove the crankshaft sprocket, as follows:
    a. Secure the crankshaft from turning using the OMC Crankshaft Pulley Nut removal tool (#342212). The tool is a large hex nut with an internal taper and keyway designed to use with a wrench to keep the crankshaft from turning.
    b. Loosen the crankshaft sprocket nut using a large adjustable or crow-foot wrench.
    c. Remove the tool from the shaft and the key from the keyway.
    d. Remove the nut and then remove the upper belt guide from the sprocket.
    e. Install a universal puller (such as an automotive steering wheel puller) to the crankshaft sprocket using 2 1/4 - 20 X 4 in. screws threaded through the puller and into the sprocket. Use the puller's pressing screw to slowly draw the sprocket from the crankshaft.
8. If necessary for cylinder head service, remove the bolt and washer retaining the camshaft sprocket. Remove the sprocket and key from the shaft.
9. Remove the intake manifold bolts and remove the manifold. Remove and discard the O-ring seals. Check the manifold for signs of excessive corrosion, leakage or damage.
10. Remove the bolts holding the exhaust adapter to the underside of the powerhead. Remove the adapter, then remove and discard the O-ring seal. Check the adapter for damage or excessive corrosion.
11. Remove the Torx® head bolts (3) securing the pressure relief valve to the underside of the cylinder block. Remove the valve and discard the gasket, then inspect the valve for damage or excessive corrosion.
12. Loosen the valve cover retaining bolts (8) using multiple passes in the reverse of the torque sequence (basically that results in a crossing pattern that works from the outside bolts to the inside bolts). Remove the valve cover from the cylinder head, then remove and discard the gasket.

■ Because the cylinder head bolts are different lengths, prevent mix-ups during installation by keeping track of them now as they are removed. The easiest method to do this is to punch holes in a cardboard box. Each hole should be placed to represent the hole in the cylinder head which that bole was threaded. Arrange each bolt in the cardboard, as it is removed.

13. Loosen the cylinder head bolts (6) using multiple passes in the reverse of the torque sequence (like the valve cover, this results in a crossing pattern that works from the outside bolts to the inside bolts). Gently tap on the cylinder head using a plastic or rubber mallet to help break the gasket seal, then carefully remove the cylinder head from the cylinder block.
14. If necessary for inspection or repair, remove the oil pump assembly from the cylinder head. For details, please refer to the Oil Pump procedure in the Lubrication and Cooling section.
15. If necessary for inspection or repair, remove the thermostat assembly from the cylinder block. For details, refer to the Thermostat procedure in the Lubrication and Cooling section.
16. If necessary, remove the lifting bracket and/or the shift linkage from the cylinder block.
17. As necessary, refer to Cylinder Head Overhaul for details on disassembly of the cylinder head and valve train components and/or to Cylinder Block Overhaul for details on disassembly for the piston, rings and crankshaft assembly.

To assemble:

18. Assemble the cylinder head and/or cylinder block, as detailed under the overhaul procedures in this section. Be sure to replace all gaskets and seals.
19. If removed, apply a light coating of OMC Gasket Sealing Compound, or equivalent sealant to the threads of the shift linkage retaining bolt, then install the linkage and tighten the bolt to 84-106 inch lbs. (9-12 Nm).
20. If removed, apply a light coating of OMC Nut Lock, or equivalent threadlock to the threads of the lifting bracket screws, then install the bracket and tighten the screws 84-106 inch lbs. (9-12 Nm).

POWERHEAD REPAIR AND OVERHAUL   6-35

21. If removed, install the Oil Pump assembly to the cylinder head as detailed in the Lubrication and Cooling section.
22. If removed, install the thermostat. For details, please refer to the Lubrication and Cooling section.
23. Apply a light coating of OMC Gasket Sealing Compound, or equivalent sealant to both sides of a new cylinder head gasket, then position the gasket on the cylinder block.

■ Remember that the cylinder head bolts are different lengths. Install them as arranged or noted during removal. What, you forgot to or worse, someone removed them from the arrangement, no fear. The longer bolts are installed on the starboard side of the head.

24. Apply a light coating of clean engine oil to the cylinder head bolt threads. Carefully position the cylinder head to the block, over the gasket, then install and finger tighten the bolts. Torque the bolts to 18-20 ft. lbs. (24-27 Nm) using multiple passes of the torque sequence, as shown in the accompanying illustration.
25. Apply a light coating of OMC Gasket Sealing Compound, or equivalent sealant to the NEW gasket for the pressure relief valve. Apply a light coating of clean engine oil to the threads of the relief valve bolts, then install the bolts and tighten to 84-106 inch lbs. (9-12 Nm).
26. Install a NEW seal ring in the exhaust adapter, then apply a light coating of OMC Gasket Sealing Compound, or equivalent sealant to the bolt threads and the mating surface. Install the adapter and tighten the bolts to 84-106 inch lbs. (9-12 Nm).
27. Apply a light coating of clean engine oil to NEW intake manifold O-rings, then install them to the grooves on the intake manifold. Apply a light coating of OMC Gasket Sealing Compound, or equivalent sealant to the threads of the intake manifold bolts, then install the manifold and tighten the bolts to 84-106 inch lbs. (9-12 Nm).
28. If removed, install the camshaft key with the outer edge parallel to the camshaft's centerline (outer edge straight up and down like the shaft). Position the camshaft sprocket and secure using the bolt and washer. Tighten the bolt finger-tight only at this time.
29. If removed, position the crankshaft sprocket and belt guide onto the crankshaft, then thread the nut onto the shaft, on top of the assembly. Install the crankshaft key with the outer edge parallel to the centerline of the crankshaft (again, outer edge of the key should be straight up and down).
30. Install the OMC Crankshaft Pulley Nut Removal/Installation tool #342212 or equivalent to hold the crankshaft from turning while you push the sprocket down into position using a crow-foot wrench on the sprocket nut. Once the sprocket is in position, loosen the nut again, removing the nut and belt guide so the timing belt can be installed.
31. Install the timing belt. For details, please refer to the procedure in this section.
32. If the camshaft sprocket bolt was only temporarily tightened earlier, back the bolt out of the pulley sufficiently to apply a light coating of OMC Nut Lock, or an equivalent threadlock, then install and tighten the bolt to 84-106 inch lbs. (9-12 Nm).
33. Adjust the valve lash as detailed in the Maintenance and Tune-Up section.
34. Install a new valve cover seal to the cover, and apply a light coating of OMC Gasket Sealing Compound, or equivalent sealant to the threads of valve cover retaining bolts. Install the cover and tighten the bolt using multiple passes of the illustrated torque sequence to 84-106 inch lbs. (10-12 Nm).
35. If the oil pressure switch was removed from the cylinder block, make sure the port is clean and free of debris. Apply a light coating of OMC Pipe Sealant or equivalent sealant with Teflon to the threads of the pressure switch. Install the switch and tighten to 120-168 inch lbs. (13.5-19 Nm).
36. If removed, install the oil fill plug, dipstick and a NEW oil filter.
37. If equipped, install the exhaust probe tube to the cylinder block, then tighten the fitting to 12-15 ft. lbs. (16-20 Nm).
38. Install and secure the cooling system indicator, breather and thermostat hoses to the powerhead as tagged or noted during removal.
39. If removed, install the oil pump pickup and inlet screen. Secure using a clamp or wire tie, as applicable.
40. Install the powerhead, as detailed in this section.

Cylinder Head Overhaul

◆ See Figure 56

Cylinder head disassembly procedures are identical on the 5/6 hp and 9.9/15 hp 4-stroke Johnson/Evinrude outboards. For details regarding Cylinder Head Overhaul, please refer to the procedure under 5/6 hp motors in this section.

Cylinder Block Overhaul

◆ See Figures 8, 24, 25, 26, 57 and 58

The crankshaft for this motor turns on replaceable main bearing insert halves found in the cylinder block and crankcase. However, unlike many 4 stroke motors, the connecting rods do not utilize bearing inserts, as the rod and cap itself contain machined bearing surfaces. Especially for this reason, it is critical that the connecting rod and end cap are kept as a matched set and are returned to the same crankshaft journal and in the exact same orientation (facing the same directions) as originally installed.

During service, the crankshaft assembly can be removed from the block without piston removal or vice-versa. If either is attempted, take EXTREME care to keep the journal surfaces from being damaged. When free of the crankshaft journals, the connecting rods should be restrained from contacting the piston bores as well. One method to accomplish this is to stuff a clean shop towel under the piston skirt. Another is to use rubber bands to hold the rods in position.

Cylinder block work is more easily performed on an engine stand, especially since this allows rotation of the motor to access pistons from above and below the cylinder block. But, work can also be performed on a clean workbench, if care is taken not to damage the mating surfaces or, in

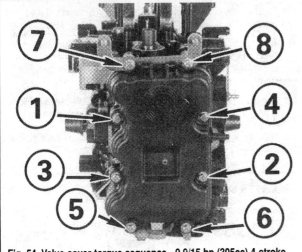

Fig. 54 Valve cover torque sequence - 9.9/15 hp (305cc) 4-stroke motors

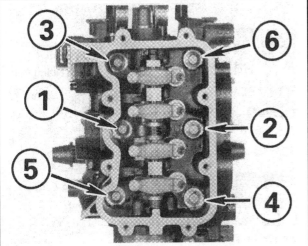

Fig. 55 Cylinder head torque sequence - 9.9/15 hp (305cc) 4-stroke motors

6-36 POWERHEAD REPAIR AND OVERHAUL

the case of the cylinder head mating surface, the dowel pins. When working on a bench, take care to prevent the block from falling over during service. Also, if the case is stood on end at any point, take great care to keep the crankshaft from falling out of the block during service. A crankshaft can suffer a fatal blow from such a fall, even from a short height.

1. Remove the powerhead from the gearcase as detailed under Removal & Installation, then remove the cylinder head as detailed under Powerhead Disassembly & Assembly.
2. Invert the cylinder block so the cylinder head mating surface is facing downward.

■ If you're not working on an engine stand, take care not to damage the cylinder head dowel pins.

3. Loosen the 10 crankcase-to-cylinder block (and main bearing) bolts using multiple passes in the reverse of the torque sequence.
4. Remove the lower crankcase rubber mounts, then use a plastic or rubber mallet to tap gently upward on the crankcase bosses, breaking the gasket seal. Carefully lift and remove the crankcase from the block.
5. Note the location and orientation of each bearing half in the journals at either end of the crankcase. Remove and retain the bearings. Visually check the crankcase mating surfaces, the bearing surfaces and the seal surfaces for damage or wear (no scratches big enough to catch a fingernail should be present on any of the surfaces).

■ Each connecting rod cap is unique and is matched for installation to only ONE connecting rod and only in ONE direction. Always mark the alignment of each rod cap before removal and, keep the caps off the rods for as short a time period as possible (to prevent mix-ups). Loosely install the caps as soon as possible during procedures.

6. Matchmark each rod cap to its connecting rod in order to ensure installation with the same alignment.

■ If a permanent marker is used, keep the rod cap installed on the connecting rod during cleaning, in case the mark should wear off. That way, a new mark could be made after the components dry.

7. Bend the tabs on the connecting rod cap bolt retainers back so the bolts can be removed.
8. Loosen and remove the connecting rod cap bolts. To prevent possible damage to the cap and bearing surface, loosen each cap by alternating between bolts, giving them about 1/4 - 1/2 turn at a time.

■ The connecting rod bolts and retainers are NOT reusable on these motors. You can thread the bolts back in place to prevent the caps from becoming lost or mixed, but they must be replaced during assembly.

9. Once the bolts are removed, the cap should easily come free, but if necessary, tap it gently using a plastic or rubber mallet to free it from the rod. Remove the rod cap and then remove the bearing half from the shaft or cap. Keep the bearing with the rod cap.
10. Push the pistons and connecting rods downward in their bores away from the crankshaft. Remove the crankshaft and thrust bearing from the cylinder block.
11. Remove the main bearing halves from the cylinder block. Note their locations and orientations for measurement and assembly purposes.
12. Check the dowel pins in the cylinder block-to-crankcase mating surface. Remove and replace any loose or damaged pins from the cylinder block.
13. Remove and discard the crankshaft seals from either end of the shaft.
14. Remove and discard the O-ring from the bore in the end of the crankshaft.
15. Check the crankshaft sleeve. If damaged, replace it using a slide hammer and a small-jawed puller.
16. Mark the top of each piston for location and orientation.
17. If the pistons are being removed, place the cylinder block with the head mating surface and the tops of the pistons either facing upward or sideways. (Actually, we find that upward at a slight angle is the best of both worlds, but you'll need and engine stand and you have to take extra care to make sure the connecting rods do not hit the cylinder walls). To remove the pistons, use a large wooden dowel (or a large wooden hammer handle) to carefully push on the connecting rod until the piston comes out of the top of the cylinder block.

■ When removing the piston, take extra care to make sure the connecting rod does not contact and damage the cylinder walls. If the crankshaft was not removed, take extra care to make sure the shaft journals are not damaged either.

18. Use a universal piston ring expander to remove the piston rings. Retain all rings in order (top, second, bottom, etc) for identification or diagnostic purposes, but they should not be reused.
19. Matchmark the connecting rod to the piston to ensure installation with the same alignment.

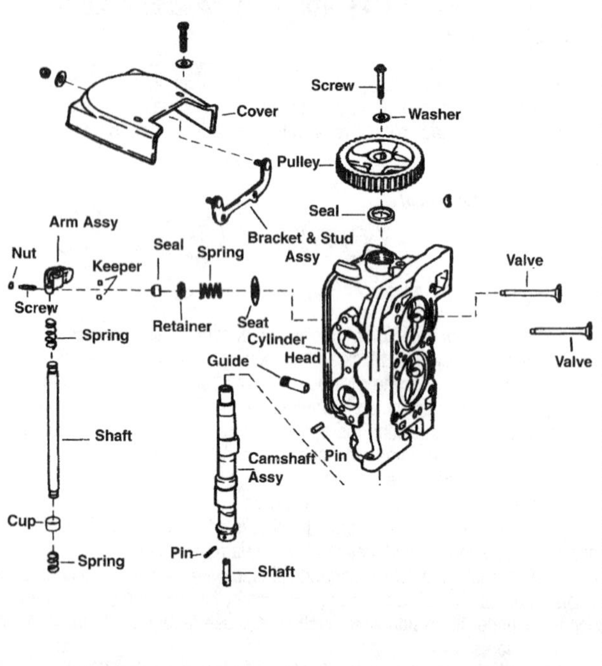

Fig. 56 Exploded view of the cylinder head assembly - 9.9/15 hp (305cc) 4-stroke motors

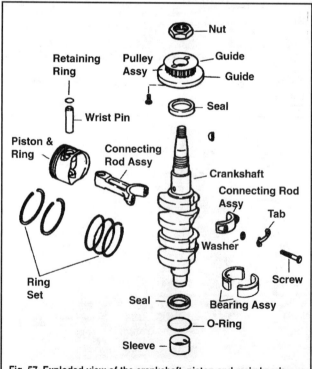

Fig. 57 Exploded view of the crankshaft, piston and main bearing assembly - 9.9/15 hp (305cc) 4-stroke motors

POWERHEAD REPAIR AND OVERHAUL 6-37

20. Use a pair of needle nose pliers modified as illustrated to remove the wrist pin retaining rings from the piston.

■ **The wrist pin-to-piston boss fit is loose on both sides, therefore little or no force should not be necessary to remove the wrist pin.**

21. Use a driver that is slightly smaller than the wrist pin (like the OMC Wrist Pin Remover/Installer #342657) to carefully tap the wrist pin from the piston.
22. Refer to Cleaning & Inspection for details concerning cylinder block, piston and crankshaft inspection. Don't forget to visually inspect the block mating surface and dowel pins for damage (replace any loose or damaged pins). Also, remember to check all bearing surfaces for signs of excessive wear or damage. The oil pressure sender hole must be free of dirt or debris.

To assemble:

23. Apply a light coating of clean engine oil (or assembly lube) to the wrist pin and bore in the side of the piston.
24. Align the matchmarks made on the piston and connecting rod removal, then, since the wrist pin **should** be a loose fit in the piston, insert the wrist pin by hand and push it through the piston into the connecting rod. Turn the ring in the groove until the gap faces directly away (180° opposite) the semi-circular relief in the wrist pin bore.
25. Secure the wrist pin in the piston using the retaining rings. Place the ring in the end groove, then carefully drive it into position using OMC Wrist Pin Retaining Ring installer (#341441) or an equivalent driver. If substituting a driver other than #341441, make sure it contacts the ring around the ring's entire circumference, but make sure it is **slightly** smaller than the outer diameter of the ring so it won't contact and damage the wrist pin bore in the piston.

■ **Apply a light coating of clean engine oil or assembly lube to each ring in order to help ease installation.**

26. Install the oil control ring assembly. Position the segmented ring into the bottom groove first, then insert the end of the first scraper ring under the segmented ring. Work carefully around the piston until the scraper ring is seated. Finally, insert the second scraper ring on top of the segmented ring and work around the piston to seat it.
27. Position the scraper ring end gaps each at 45° from each end of the wrist pin (the wrist pin is represented as a dotted line in the accompanying illustration). Make sure the segmented ring gap is at least 45 degrees from the wrist pin centerline, but on the opposite side of the piston and wrist pin centerline from the scraper ring gaps.
28. Install the middle compression ring (thicker compression ring) into the middle piston ring groove with the number stamping facing upwards and the white painted mark to the right side of the gap. Position the middle ring gap directly between the 2 scraper ring gaps (90° from the wrist pin centerline). Refer to the accompanying illustration for clarification.
29. Install the top ring in the top piston groove with the stamping facing up and with the orange paint mark to the right side of the gap. Position the top ring gap directly opposite the middle compression ring (in the same general area of the segmented ring gap, but more precisely 90° from the wrist pin centerline). Again, refer to the illustration for details.
30. Apply a light coating of clean engine oil or assembly lube to the cylinder wall, an expandable ring compressor and the piston.
31. Verify the ring gaps are still in position, then secure the piston in the ring compressor. Tighten the compressor **just** sufficiently to push the rings into the grooves, but do not over-tighten it as the piston must still be free to slide into the bore.
32. Position the piston over the bore with the stamped arrow on the piston dome facing the flywheel side of the motor (and any other alignment marks made during removal oriented as you intended). Use a wooden dowel or wooden hammer handle to gently tap the piston down through the ring compressor and into the cylinder bore.

※※ WARNING

When tapping the piston into position MAKE SURE the connecting rod does not contact and damage the cylinder walls.

33. If removed, install a new crankshaft sleeve with the O-ring on the end of the crankshaft using the OMC Crankshaft Sleeve Installer (#342210), or equivalent sized driver. Be sure to fully seat the sleeve, then apply a coating of OMC Moly Lube or equivalent assembly lube to the O-ring.
34. Apply a light coating of clean engine oil or assembly lube to the NEW main bearing halves, then position them in the cylinder block and crankcase.

Position each bearing edge flush with the block.
35. Apply a light coating of clean engine oil or assembly lube to the crankshaft main bearing and connecting rod journals.
36. Apply a light coating of clean engine oil or assembly lube to the connecting rod and cap bearing surfaces.
37. Carefully position the crankshaft and thrust bearing (with the reliefs facing the crank) into the cylinder block.
38. Push each piston slowly upward in the bore, toward the crankshaft. Make sure the connecting rods do not contact and damage the cylinder walls or the crankshaft journals.
39. Install each connecting rod cap using NEW bolts and retainers. Tighten the bolts back and forth in stages (about 1/4 - 1/2 turn at a time) to 96-108 inch lbs. (11-12 Nm). Bend the retainer tabs up onto the bolt hex heads to lock them in position.

※※ WARNING

Over-tightening the connecting rod cap screws will likely render the connecting rods and caps unusable due to cap distortion.

40. Apply a light coating of OMC Gel-Seal II or equivalent to the mating flanges of the cylinder block and crankcase. Be sure that the sealant coats the entire flange evenly, but be careful not to apply excessive amounts that could be forced out from between the flanges during assembly.

■ **Although OMC Gel-Seal II has a shelf life of one year when stored at room temperature, buy a new tube if in doubt. Using an old tube of Gel-Seal II could prevent the crankcase from sealing properly.**

41. Apply a light coating of clean engine oil to the threads of the 10 cylinder block-to-crankcase (main bearing) bolts. Install the bolts and finger-tighten, then use multiple passes of the torque sequence to tighten the bolts to specification. Tighten the small bolts to 84-106 inch lbs. (10-12 Nm) and the large bolts to 18-21 ft. lbs. (24-28 Nm).
42. Apply a light coating of OMC Gasket Sealing Compound, or equivalent sealant to the casings of a NEW upper and a NEW lower crankshaft seal. Apply a light coating of engine oil or assembly lube to the seal lips.

■ **Don't loose the crankshaft sprocket and flywheel keys. When installing the upper crankshaft seal, place the keys with the sprocket and flywheel for use during component installation.**

43. Remove the keys from the top of the crankshaft (to prevent possible seal damage or interference with the seal installer), then position the upper seal over the crankshaft with the casing facing outward. Make sure the seal lips fit properly and start the casing squarely into the bore, then use a suitable driver (large enough to fit over the crankshaft, but still contact only the seal casing, such as OMC #342213) to press the seal slowly into position.

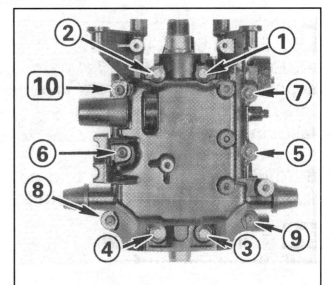

Fig. 58 Crankcase-to-cylinder block (main bearing) bolt torque sequence - 9.9/15 hp (305cc) 4-stroke motors

6-38 POWERHEAD REPAIR AND OVERHAUL

■ The OMC seal installer tool is designed for use with the crankshaft sprocket nut to slowly push the seal into position. Turn the nut slowly until the seal is fully seated, then remove the nut and installer tool.

44. Position the NEW lower crankshaft seal squarely over the crankshaft with the casing facing outward, then use a seal installer (such as OMC #342215 or an equivalent driver) to gently tap the seal into the bore until it contacts the powerhead.

45. Assemble and install the powerhead, as detailed in the Powerhead Disassembly & Assembly and the Powerhead Removal & Installation procedures in this section.

46. Refer to Break-In in this section for details on how to ensure proper break-in of the new powerhead components.

18 Jet-35 Hp (521cc) Motors

◆ See Figure 59

Removal & Installation

◆ See Figure 59

1. On electric start models, disconnect the negative battery cable for safety.
2. Remove the top engine cover for access.
3. Remove the rope starter assembly, as detailed in the Hand Rewind Starter section.
4. At the linkage on the port side of the motor, disconnect the spring from the pin on the throttle control lever, then remove the washer.
5. Remove the screws (4) from the throttle control lever pivot bearing clamps (the semi-circular clamps on both sides of the throttle control lever, just in front of the spring disconnected in the previous step).
6. Separate the pin from the top of the throttle control rod, then lift slotted armature plate link from the top of the throttle control lever. Move the throttle control lever aside.

7. Either remove the center bolt and separate the fuel pump cover from the pump body (so the inlet screen can be checked for damage) or simply cut the wire tie and disconnect the fuel hose from the pump.

8. Disconnect the cooling system (overboard) indicator hose from the fitting on the exhaust cover. Free the hose from the retaining clamp.

9. Tag and disconnect the primer hoses from the carburetor and from the intake manifold. There should be 2 hoses run to the carburetor and one to the intake.

10. For UFI Ignition Models, disengage the 1-pin connector for the stop switch, then remove the stop switch and ignition module ground leads from the ignition coil screw.

11. For tiller control models with CDI, disengage the 5-pin connector, then using a pin removal tool, remove the stop-switch wire (terminal **E**) from the connector. For more details on wire color or terminal **E** identification, refer to the schematics or the information on stop switch testing in the Ignition and Electrical System section.

12. Disconnect the ground lead from the powerhead.

13. On the starboard side of the motor, disconnect the shift lever. Insert a paper clip into the plunger and then remove the pin and washer from the actuator cam link. Disconnect the link from the shift actuator cam.

14. Using a suitably sized small box wrench (such as OMC #322700) remove the powerhead retaining nuts facing upward at the base of the powerhead (in the lower engine cover) on the port and starboard sides. Access to these nuts is very tight. Be patient.

15. From underneath the engine cover, remove the bolts (4) securing the exhaust housing to the powerhead.

16. With the bolts and nuts removed, the powerhead can be lifted from the gearcase assembly. Although it can be lifted by hand, we'd recommend an engine hoist to prevent damage to the assembly. If necessary, rock the powerhead back and forth slightly to break the seal.

17. Remove the inner exhaust tube from the bottom of the powerhead. Remove and discard the gasket.

18. Mount the powerhead on an engine stand. If necessary, strip it for overhaul by removing the carburetor (refer to Fuel System) and any electrical or ignition components (refer to Ignition and Electrical Systems).

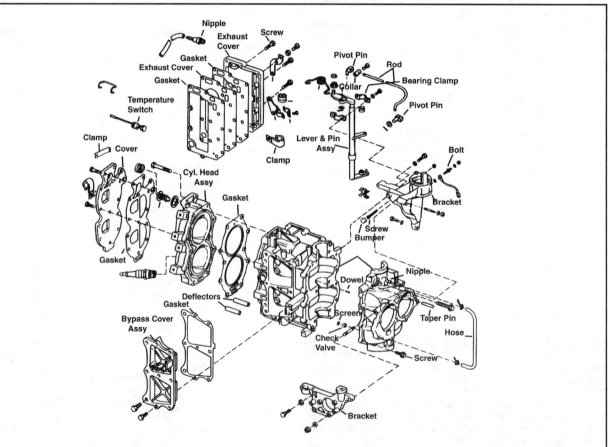

Fig. 59 Exploded view of the powerhead assembly - 18 jet-35 hp (521cc) motors (Note, port bracket and bracket mounted components shown are for electric start models, rope start models differ)

POWERHEAD REPAIR AND OVERHAUL 6-39

To install:

19. Apply a light coating of OMC Triple-Guard, or equivalent marine grease to a NEW O-ring and the outside bore of the lower crankcase head. Install the O-ring to the groove in the lower head.

20. Apply a light coating of OMC Moly Lube, or equivalent lubricant to the crankshaft splines. But, **do not** allow the lubricant to pack in the end of the splines, as the hydraulic pressure could prevent the driveshaft from seating in the crankshaft.

21. Check that the O-ring is present in the crankshaft sleeve.

22. Apply a light coating of OMC Triple-Guard, or equivalent marine grease to the upper outside diameter of the water tube, then make sure the water tube washer is still positioned on the shoulder near the end of the tube.

23. Position a new powerhead gasket (dry, without using any sealer), the install the inner exhaust housing and tighten the retaining bolts to 96-120 inch lbs. (11-14 Nm).

24. Carefully lower the powerhead onto the exhaust housing. Proceed slowly aligning the driveshaft and crankshaft splines, along with the water tube and/or inner exhaust housing. It may be necessary to rotate the powerhead back and forth slightly to align the crankshaft and driveshaft splines. Once everything is aligned, seat the powerhead on the exhaust housing.

25. Install the powerhead bolts (4) from underneath the exhaust housing flange (under the lower cover), then tighten to bolts to 191-216 inch lbs. (22-24 Nm).

26. Apply a light coating of OMC Nut Lock, or equivalent threadlock to the port and starboard exhaust housing studs. Then install and tighten the retaining nuts securely.

27. Connect the ground lead to the powerhead, then apply a coating of OMC Black Neoprene Dip or equivalent sealant.

28. Install the carburetor and any ignition or electrical system components stripped during overhaul.

✱✱ CAUTION

Pay close attention when routing wires and hoses. Make sure all components are routed in their original clamps and locations. They should be secured so that no contact will occur with moving components. Failure to pay close attention to this detail could cause significant injury or damage from a resultant fire or explosion.

29. For all rope or tiller electric start models:
   a. Install the neutral start plunger and spring through the component bracket, then use the pin to secure the plunger.
   b. Double-check the ignition plate routing, then install the component bracket, neutral start plunger, lockout lever using the pivot screw and flat washer. Tighten the pivot screw to 60-84 inch lbs. (7-9 Nm), then remove the pin securing the plunger.

30. Have an assistant slowly rotate the propeller shaft by hand while you shift the engine into **reverse**. Then place the actuator cam link into the shift actuator cam, securing the link with the washer and pin.

31. Seat the shift lever socket onto the lever, then snap the pocket onto the stud. Secure the socket using the washer and a cotter pin.

32. Route the cooling system indicator hose through the clamp and secure it to the exhaust cover fitting.

33. Either install the fuel pump cover (with inlet screen) or reconnect the fuel hose to the pump nipple and secure using a new wire tie, whichever is applicable.

34. Install the throttle control lever. Slide the throttle control rod through the nylon block on the control lever (the offset in the nylon block must face forward). Then, insert the pin in the end of the rod

35. Place the nylon bushing on the stud of the control lever and then engage the armature plate link on the stud. Place the washer over the stud and secure using the spring.

36. Install the starter lockout cable bracket.

37. Install the rope starter assembly.

38. Adjust the shift lever, as follows:
   a. Have an assistant slowly rotate the propeller shaft by hand while you shift the engine into **neutral**. Loosen both shift lever bolts (there is one above and one below the pivot).
   b. Move the shift actuator cam back and forth until the neutral shift lockout lever centers in the notch of the shift actuator cam.
   c. Hold everything in this position and tighten the bolts to 60-84 inch lbs. (7-9 Nm).

39. On electric start motors, connect the negative battery cable, then check neutral safety switch operation. It must only allow the starter to kick over when the shifter is in neutral. If necessary, adjust the neutral start switch, as follows:
   a. Disconnect the negative battery cable again for safety.
   b. Disconnect the yellow/red lead from the neutral start switch.
   c. With the shift handle set in **neutral** adjust the distance between the bottom of the switch and the top of the plunger until it is 1/16 in. (1.6mm). The easiest way to accomplish this is to loosen the 2 switch screws and insert a 1/16 in. (1.6mm) drill bit between the bottom of the switch and the top of the plunger. Move the switch until the switch and plunger just contact the drill bit, then tighten the screws securely.
   d. After moving the switch, verify adjustment using a DVOM set to read continuity. The switch must show continuity ONLY with the shifter in **neutral**.

40. If not done already (and if applicable) reconnect the stop switch wiring.

41. Install the top engine cover.

42. If a new or rebuilt powerhead was just installed, refer to Break-In in this section for details on how to ensure proper break-in of the new powerhead components.

Disassembly & Assembly

◆ See Figures 8, 24, 25, 30 thru 40, 50, 51, 52 and 59 thru 63

Although the majority of the powerhead overhaul procedure for 18 Jet-35 hp motors is fairly easy and straightforward, the connecting rod cap bolt torque procedure requires the use of a holding fixture in order to ensure that the connecting rod/cap and bearing are properly centered. Attempting to center and tighten the connecting rod caps without this fixture may cause unexpected, total powerhead failure sometime after the motor is returned to service. That is not to say people have not succeeded in centering and tightening the caps without the fixture, but to say that we cannot recommend it, as there are no guarantees.

■ **To simplify assembly, remember to layout all bolts, components and clamps in the order of removal. This is especially true for the wiring harness and related clamps. Matchmark all component assemblies such as pistons, connecting rods to ensure installation of the correct pairs and installation in the correct orientation. Most of all, take your time.**

1. Remove the powerhead from the lower unit as detailed under Removal & Installation.

2. Remove the water tube grommet by pressing the grommet tab, then pushing it out of the inner exhaust housing. To prevent an oversight during assembly, install the replacement grommet at this time. Apply a light coating of OMC Gasket Sealing Compound, or equivalent sealant to the outer diameter of the grommet, then position the grommet, aligning it with the housing bore. Use OMC #304148 or a suitably sized driver to gently position and seat the grommet.

3. Use a pair of diagonal-cutting pliers to grasp and pry the key from the crankshaft. The pliers should be positioned with the blades across the key with the handle pointing up at an angle, then use the handle as a lever, pivoting on the bottoms of the blades.

4. Remove the bolts (3) securing the crankcase lower head to the bottom of the crankcase. Remove the lower head by pulling outward evenly at all sides (use 2 small prybars and reposition them often as the head is gently worked out of the crankcase. Remove and discard the O-ring and the 3 bolts (which must be replaced after they are removed).

5. Using a punch, carefully drive the seal from the crankcase lower head, then clean the seal bore in the head, removing all traces of sealant.

6. Loosen and remove the cylinder head water cover screws (usually 14) using multiple passes of a pattern that starts with the outside screws and works toward the center. Once the screws are removed, use a rubber or plastic mallet to tap around the edge, breaking the gasket seal. Remove the cover and discard the gasket.

7. Remove the spring, thermostat and seal from the cylinder head.

8. Loosen and remove the cylinder head retaining bolts (10) using multiple passes of a counterclockwise spiraling pattern that starts at the outside, lower bolts and work towards the 2 center bolts. Once the bolts are removed, break the gasket seal and remove the cylinder head from the cylinder block.

6-40 POWERHEAD REPAIR AND OVERHAUL

9. Using a felt-tipped marker or numbered/lettered punches, label the piston domes.

■ If a felt-tipped marker is used to label components during disassembly, watch the marks during cleaning. Replace any marks that are dulled or removed by solvent as soon as the component is dry.

10. On the port side of the cylinder block, loosen the bolts, securing the outer and inner exhaust covers. Remove the cover and separate, remove and discard the gaskets. The inner exhaust cover must be checked for pitting, warpage or excessive corrosion and replaced, if any of these conditions are found.

11. On the starboard side of the cylinder block, loosen the bolts (9) and remove the intake bypass cover. Remove and discard the gasket from the mating surfaces.

12. Remove the bolts (7 hex-head on most models, but 2 hex-head bolts and 5 nuts on some early-models) and screw (1 flat-head) from the intake manifold, then remove the intake manifold. Remove and discard the gasket.

13. Remove the Phillip's head screw and remove the leaf plate. Remove and discard the gasket.

■ Do not bend and leaf valves by hand as this could damage them so they either do not seal properly at best, or at worst, will break off during service. Do not disassemble the leaf plate unless some portion of the plate and valve assemblies is corroded or damaged and requires replacement.

14. Inspect the leaf plate and, if necessary, disassemble it as follows:
    a. Visually check for obvious signs of distortion on the plate. Inspect the valve tips for signs of cracks, chips or other obvious damage. Make sure the leaf stops are not distorted or loose. If there are any obvious defects, that component must be replaced.
    b. Use a machinist's straightedge and a feeler gauge to check the leaf plate for distortion. Lay the straightedge across the plate only (not the leaf valves) at various points, then see if a 0.003 in. (0.08mm) or large feeler gauge can be inserted underneath the straightedge at any point. If it does, the plate or the manifold is warped and must be replaced.
    c. Visually check all gasket surfaces to make sure they are smooth and free of nicks. If necessary, a gasket surface may be dressed to smooth shallow nicks using a fine emery cloth.

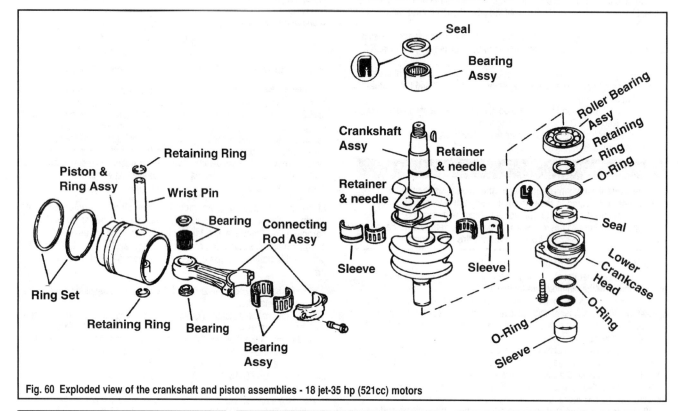

Fig. 60 Exploded view of the crankshaft and piston assemblies - 18 jet-35 hp (521cc) motors

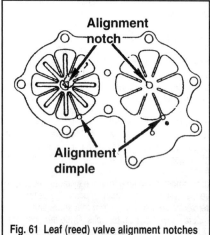

Fig. 61 Leaf (reed) valve alignment notches should be positioned as shown

Fig. 62 DO NOT face the notches toward the alignment dimples

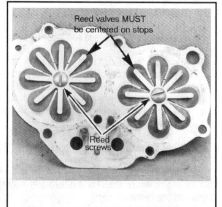

Fig. 63 Center the valves over the stops before tightening the retaining screw

d. Use the straightedge and feeler gauge to check for warpage across the intake manifold gasket surfaces. Check at the center of the straightedge to make sure a 0.003 in. (0.08mm) or large feeler gauge cannot be inserted underneath the straightedge.

e. If disassembly is necessary, matchmark the leaf valve assembly, then remove the screw and nut securing each assembly to the plate. If they are being reused, do not allow the leaves and stops to become mismatched.

✷✷ WARNING

Do not lift or bend the leaf valves during disassembly or they could become damaged. If a damaged leaf valve is installed it could break in use.

15. Either use a fabricated taper pin tool (as shown in the accompanying illustration) to carefully push the crankcase taper pin toward the intake manifold surface or use a punch with a diameter larger than the pin. The taper pin can be accessed at the upper portion of the crankcase (flywheel end).

✷✷ WARNING

NEVER use a tool smaller than the taper pin bore to remove the pin or damage could occur.

16. Remove the main crankcase-to-cylinder block bolts (6) along with the flange bolts (8) using multiple passes of a spiraling sequence that starts with the outer bolts and works towards the inner bolts. Loosen the flange bolts first and then loosen the main crankcase bolts. For installation purposes (since bolts are of different lengths), arrange the bolts in holes punched into a cardboard box resembling the bolt pattern from which they were removed.

17. Rotate the powerhead so the intake manifold surface is facing upward, then use a plastic or rubber mallet to tap upward on the exposed portion of the crankshaft (at a 90° angle to the shaft). Tap lightly until the crankcase half **just** starts to separate from the cylinder block half. Lift and remove the crankcase from the bottom of the cylinder block (exposing the crankshaft).

18. Loosely install the cylinder head back on the cylinder bore to keep the pistons from falling out as the crankshaft is removed.

■ Keep in mind that all wear parts (such as pistons, connecting rods, needle bearings and bearing liners) MUST be installed in their original positions (and with the same orientations) if they are to be reused. Mark each component with a permanent marker to ensure proper assembly.

19. Matchmark the connecting rod caps and then use a suitable socket (such as OMC Torque Socket #331638) to loosen the connecting rod cap bolts. Loosen the bolts on a rod cap alternately, no more than one turn at a time. Remove each cap along with the connecting rod cap needle bearing and cage.

20. Lift the crankshaft assembly carefully out of the cylinder block.

21. Once the crankshaft has been removed, position the connecting rod caps and cap bearings back onto the rods. Be sure to position them **precisely** as marked and noted before removal. This will help ensure no parts or lost or inadvertently mixed. Install the cap bolts finger-tight to secure the caps.

22. Disassemble the crankshaft further for inspection and component replacement, as follows:

a. Slide the upper crankshaft seal and main bearing from the crankshaft.

b. Remove and discard the O-ring from the end of the crankshaft sleeve. Install a new O-ring during assembly.

23. Check the crankshaft sleeve. If damaged, remove it using a slide hammer and a small-jawed puller.

a. ONLY remove the lower crankshaft bearing if it requires replacement. First remove the retaining ring using a pair of snapring pliers, then support the bearing in a shop press and press the bearing off the shaft.

■ If the powerhead is on a workbench instead of an engine stand, position the powerhead securely on one end for the next step.

24. Rotate the powerhead until the cylinder head is facing upward (or upward at a slight angle to ease access to the connecting rods, then remove the cylinder head again. Remove each of the pistons from the cylinder bore by pushing upward on the connecting rod from underneath.

✷✷ WARNING

DO NOT allow the connecting rods to contact and damage the cylinder walls or the insides of the piston skirts during removal. One of the best ways to protect the piston is to pack the skirt with shop rags.

25. Disassemble each piston for further inspection and component replacement, as follows:

a. Use a universal ring expander to remove the rings from the pistons. The manufacturer warns NOT to reuse the rings. Retain them in sets for inspection purposes (marking their original locations), but discard them before installation.

■ We've said it before. We know there will be some circumstances that will prompt people to reuse rings (for example, if for some reason the motor was disassembled after only a few hours of usage). But for the most part, if you've come THIS far, it is probably a good idea to bite the bullet and replace the rings to ensure long-lasting performance from the powerhead.

b. Using a pair of modified needle-nosed pliers as shown in the accompanying illustration, or a suitable pair of snapring pliers, remove the wrist pin retaining rings. Discard the old retaining rings.

■ If necessary, a piston cradle can be fabricated from a large block of wood. The important design features of the cradle are that it is semi-circular on top to hold (or cradle, get it?) the piston and that there is a bore at the bottom center into which the wrist pin can tapped or pressed. The final important feature of the cradle is that it is STURDY enough to withstand the force of the shop press, BUT, keep in mind that the fit is loose on these pistons and significant force should not be necessary.

c. The piston-to-wrist pin fit should be loose on both sides. Support the piston and use a suitable drift pin (like OMC #326356 to carefully and smoothly tap or push the wrist pin free of the piston bore). If the OMC tool is not available, be sure to use a drift pin that is just slightly smaller than the diameter of the wrist pin itself.

■ As the wrist pin is removed from the piston you must retain the 28 loose needle bearings and the 2 thrust washers from the connecting rod bore. If you can, stop tapping or pressing once the wrist pin JUST clears the bottom of the connecting rod bore. Then carefully withdraw the rod along with the needle bearings (place a thin bladed tool like a putty knife or your finger under the rod hole as it is pulled free of the piston bore/wrist pin). Inspect the bearings, if any of the needles are lost or worn, replace the entire bearing assembly.

26. Refer to the information found under Cleaning & Inspection, in this section for details on inspecting the powerhead components to determine which should be replaced and to prepare the ones that are being reused for assembly.

To assemble:

✷✷ WARNING

During assembly, take your time and NEVER force a component unless a shop press or drive tool is specifically required by the procedure.

■ Lightly coat all components (except those that are coated with sealant) with clean, fresh Johnson/Evinrude Outboard Lubricant or equivalent oil during assembly. Always replace all gaskets, O-rings and seals.

27. There are 3 water deflectors mounted in the cylinder block, in the water passages around the cylinder bores. They are mounted, more or less flush with the cylinder head mating surface. If removed, install new deflectors, lubricate the deflectors using STP Oil Treatment® (which should be available at most automotive parts supply stores).

■ There are 3 deflectors (2 short and 1 long). Pay attention to their mounting locations when removed. If it is too late, the short one normally is installed on the lower port side, opposite one of the long deflectors. When installing new deflectors, the tops of the deflectors may protrude slightly above the gasket surface, that's ok, DON'T SHORTEN them.

POWERHEAD REPAIR AND OVERHAUL

28. If disassembled for inspection and/or component replacement, assemble each piston as follows:

    a. Apply a light coating of OMC Needle Bearing Assembly grease (or equivalent) to the wrist pin bearing needles, then insert the 28 needles around the inner circumference of the connecting rod wrist pin bore. Install the OMC Wrist Pin Bearing tool #336660 into the bore to align the needles and hold them in place during assembly.

■ The wrist pin bearing tool is, essentially, a short round pin, roughly the same outer diameter of the wrist pin, but very short, just long enough to insert it through the connecting rod wrist pin bore and install the thrust washers on either side. A similarly sized dowel pin can be substituted, or a old connecting rod wrist pin from the same size piston could be used. If substituting another rod or drift pin, make sure the surface is smooth and clean/free of all corrosion. Also, make sure the outer diameter of the tool is smaller than the wrist pin and bore to prevent the possibility of binding or damage.

    b. With the bearings and the wrist pin tool in position, install the thrust washers. Again, a dab of grease should hold the thrust washers in position.

■ The manufacturer states that the wrist pin thrust washers should be positioned with the flat sides FACING the piston.

    c. Apply a light coating of clean engine oil or assembly lube to the wrist pin and the pin bore in the piston.
    d. Support the piston (ideally in a piston cradle for ease of assembly but the wrist pin fit should be loose on both sides so this is not absolutely necessary). Insert the wrist pin into the bore and gently press it through the bore until it just starts to appear inside the piston skirt. Stop pressing, then position the connecting rod with the needle bearings, thrust washers and wrist pin bearing installation tool inside the piston skirt. Continue to press the wrist pin into the position, through the thrust washers and bearings, pushing the wrist pin bearing installer through the other side and out of the piston.
    e. Once the wrist pin is properly positioned, use a small shouldered driver to install a new snapring in each end of the wrist pin bore. Make sure the snapring fully seats in the groove. Turn the ring in the groove until the gap faces downward, directly away (180° opposite) from the semi-circular relief in the top of the wrist pin bore.
    f. Use a ring expander to install the piston rings on each piston, making sure to install the tapered ring in the top groove. Each piston ring groove should contain a dowel pin, around which the ring gap must be situated. If not the rings will likely break during installation (or if by they don't by some miracle, during the first start-up). Be sure to install the tapered ring in the top groove.

■ When using the ring expander, spread each ring JUST enough to slip over the piston.

29. If disassembled, reassemble the crankshaft as follows:

    a. If removed, apply a light coating of clean engine oil or assembly lube to a NEW lower main bearing, then support the crankshaft in a shop press between the 2 lower counterweights. Carefully press the bearing into position with the lettered side of the bearing facing upward toward the press.

■ OMC Crankshaft Bearing/Sleeve installer #339749 can be used to safely seat the lower bearing and the sleeve in position. If not available, be sure to use a suitable driver that contacts the entire diameter of the bearing cage or the sleeve and does NOT contact or damage the crankshaft.

    b. If removed, apply a light coating of clean engine oil or assembly lube to the NEW crankshaft sleeve, then install the sleeve and gently drive or press the sleeve onto the shaft until the installer contacts the lower main bearing.

✷✷ WARNING

Make sure the sleeve surface is not nicked or damaged during installation or it must be removed and discarded (and another NEW sleeve installed).

    c. Apply a light coating of clean engine oil or assembly lube to a NEW O-ring, then install the O-ring into the crankshaft sleeve.
    d. Apply a light coat of OMC Moly Lube to the crankshaft splines.
    e. Using a pair of snapring pliers, carefully install the lower bearing retaining ring with the sharp edge of the ring facing outward, away from the bearing.
    f. Apply a light coating of clean engine oil or assembly lube to the upper main bearing, then install the bearing to the shaft with the lettered side facing upward toward the flywheel.
    g. Oil the 2 roller bearing assemblies and position them around the crankshaft center journal, then install the center main bearing sleeve with its ring end facing the lower end of the crankshaft. Secure the sleeves using the retaining ring.

■ If the connecting rod end-caps were reinstalled to prevent loss, make sure they are matchmarked, then remove them so the crankshaft assembly can be installed. Position them aside ensuring that they will not be mixed-up during assembly.

30. If removed, install each piston as follows:

    a. Apply a light coating of clean engine oil or assembly lube to the piston and piston bore.
    b. Use a suitable ring compressor or a band-type clamp to evenly compress the rings (oiling the inside of the compressor or clamp will also help ease installation). Make sure the rings are still properly positioned with the gaps over the dowels.
    c. Start the piston into the bore with the deflector side of the piston (circular sharp edge) facing the intake port side of the cylinder block. Also, make sure the connecting rods are centered in the piston to prevent damage.
    d. Slowly insert the piston, making sure the connecting rod does not strike and damage the piston skirt, or the strike/scrape and damage the cylinder wall. If the piston hangs up on a ring, don't force it, push the piston back upward and remove the compressor to check for potential problems, such as a ring that has shifted the gap away from the dowel.

31. Once the pistons are installed in the bores, temporarily install the cylinder head again to keep them from falling out.
32. Remove the rod caps and install the connecting rod bearings. Place one liner half in each connecting rod and the other in the cap. Apply a light coating of clean engine oil or assembly lube to the bearing surfaces.
33. Apply a light coating of clean engine oil or assembly lube to the crankshaft journals, then carefully lower the shaft into the cylinder block. Take care not to allow the connecting rods to contact or damage the shaft journals. Align the top and center main bearings with the pins in the cylinder block.
34. Move the pistons toward the crankshaft. Seat the rod journals against the crankpins, move the bearings from the caps to the connecting rod journals. Finally, align the matchmarks and position the caps over the bearings.

■ Besides any matchmarks made during removal, the connecting rod caps on these models are normally equipped with dots on one end of one face (directly below one connecting rod cap screw). These dots should also be aligned.

35. Oil and install the connecting rod cap bolts, then finger-tighten the bolts alternating from side-to-side on the rod cap.
36. Use a pencil or your fingernail to determine if the rod caps are fully aligned and seated. Scrape the pencil across the mating surface to see if it catches.
37. Center and align the connecting rod caps using the OMC Rod Cap Alignment Fixture #396749 as follows:

    a. Rotate the knob on the top of the tool until the flat marked **SET** aligns with the arrow on the fixture frame. Rotate the adjustment knob 180° to the lock position.
    b. Make sure the retaining jaw (with the flat head screw in the frame bore) and the forcing jaw (with the hex key at the base and the threaded flat protruding from the frame) are positioned on the fixture tool.
    c. Tighten the connecting rod cap screws to 25-30 inch lbs. (2-3 Nm).
    d. Apply a light coating of engine oil to corners of the connecting rod and cap where the fixture will be installed.
    e. Position the frame to the so the contact area of the forcing jaw is centered on the connecting rod and cap. Tighten the forcing screw until the jaws contact the connecting rod, then slide the frame downward until the adjustment stop (at the center of the frame) just contacts the top of the connecting rod cap. Verify that the groove lines in the jaws are centered on the rod/cap diameter.

✷✷ WARNING

The frame MUST be squarely in position as described in order to perform its function properly. Failure to center the frame, and therefore the connecting rod/cap, will lead to premature crankshaft, rod and bearing failure.

POWERHEAD REPAIR AND OVERHAUL   6-43

f. With the frame centered, tighten the forcing screw to 23 inch lbs. (2.5 Nm).

g. Loosen both connecting rod cap bolts 1/4 turn, then tighten each bolt to 40-60 inch lbs. (5-7 Nm).

h. Next, tighten the connecting rod cap bolts to 15-17 ft. lbs. (20-23 Nm), and finally tighten them to 30-32 ft. lbs. (40-43 Nm).

i. Loosen the forcing screw, remove the frame from the connecting rod/cap assembly, then repeat for the remaining rod/cap.

38. Clean and degrease the mating flange of the crankcase and cylinder block halves using OMC Cleaning Solvent, or an equivalent solvent. Allow the flanges to air dry.

39. Apply a light coating of OMC Locquic Primer or equivalent to the mating flange of the crankcase half and allow it to air dry.

40. Apply a light coating of OMC Gel-Seal II or equivalent to the cylinder block half flange. Be sure that the sealant coats the entire flange evenly, but be careful not to apply excessive amounts. The sealant must not be applied within 1/4 in. (6mm) of the labyrinth seal or bearings otherwise the sealant could be forced out from between the flanges into contact with the bearings during assembly.

■ Although OMC Gel-Seal II has a shelf life of one year when stored at room temperature, buy a new tube if in doubt. Keep in mind that using an old tube of Gel-Seal II could allow crankcase air leaks (leading to performance problems).

41. Carefully lower the crankcase half into place over the cylinder block half.

42. Apply a light coating of OMC Gel-Seal II or equivalent to the threads of the 6 crankcase-to-cylinder block (main bearing) bolts, then install and finger-tighten the bolts.

43. Once the crankcase if fully seated on the cylinder block, install the crankcase taper pin and make sure it seats fully.

44. Install and finger-tighten the crankcase flange screws.

45. Use a plastic or rubber mallet to gently tap the bottom of the crankshaft to fully seat the lower main bearing. Once seated, temporarily install the flywheel and rotate slowly to check for binding between the crankshaft and connecting rod bearings.

46. Apply a light coating of OMC Gasket Sealing Compound to the machined mating surface of the crankcase lower head and to the threads of the 3 lower head retaining bolts. Apply a light coating of OMC Triple-Guard, or equivalent marine grease to a new lower head O-ring, then position the O-ring in the groove. Install the lower head to the crankcase and finger-tighten the bolts.

47. Tighten the 6 main bearing bolts in multiple passes of a spiraling pattern that works from the center screws and moves outward. Tighten the 6 main bolts to 144-168 inch lbs. (16-19 Nm). Then tighten the 8 smaller screws in a similar fashion starting with the center and working outward to 60-84 inch lbs. (7-9 Nm).

48. Tighten the 3 crankcase lower head retaining bolts to 60-84 inch lbs. (7-9 Nm).

49. Apply a light coating of OMC Gasket Sealing Compound, or equivalent sealant to the metal case of a NEW upper main bearing seal and light coating of oil to the seal lips. Position the seal with the lips facing downward and then use a suitable driver (such as OMC #321539) to gently tap the seal into position.

50. Apply a light coating of OMC Gasket Sealing Compound, or equivalent sealant to both sides of a NEW cylinder head gasket. Install the cylinder head and gasket, then install and finger-tighten the bolts.

■ Install the cylinder head bolts with the threads dry, they should NOT be coated with sealant.

51. Tighten the cylinder head bolts using multiple passes of a clockwise spiraling pattern that starts at the center and works outward. Continue the spiral sequence until all bolts are tightened to 216-240 inch lbs. (24-27 Nm).

52. Install the seal, thermostat assembly and spring into the bore at the top of the cylinder head.

53. Apply a light coating of OMC Gasket Sealing Compound, or equivalent sealant to both sides of a NEW cylinder head water cover gasket. Install the cylinder head water cover and gasket, then install and finger-tighten the bolts. Tighten the water cover bolts using multiple passes of a clockwise spiraling pattern that starts at the center and works outward. Continue the spiral sequence until all bolts are tightened to 60-84 inch lbs. (7-9 Nm).

54. Apply a light coating of OMC Adhesive M, or equivalent to both sides of a new bypass cover gasket, then install the cover and tighten the bolts to 60-84 inch lbs. (7-9 Nm).

55. If disassembled for component replacement, assemble the leaf plate assembly as follows:

a. Place the leaf valves on the plate. If new valves do not seat on the plate, try turning them over.

✱✱ WARNING

NEVER turn over used valves on the plate, as they may break when they are returned to service. If used valves do not seat in their original direction of installation, they must be replaced.

b. Check the valves, if any leaves are standing open, apply light pressure using the eraser end of a pencil (the valve should close with light pressure). If not, check the plate for high spots or burrs.

✱✱ WARNING

NEVER lap the leaf plate. An plate that is too smooth may cause the leaf valves to stick closed when returning a motor to service after winterization.

c. Visually check leaf valve-to-plate alignment. When viewed from the leaf valve side of the plate, the alignment notches (small cutouts in the valve screw bore) must face in the proper direction. The notch on the port side leaf valve should face about 1 o'clock (about 90° away from the alignment mark). The notch on the starboard side valve must face about 11 o'clock (a little less than 180° measured counterclockwise from the alignment mark).

✱✱ WARNING

Premature valve failure will likely result if the leaves are not aligned properly, as directed in the previous step.

d. With each of the leave valves aligned as noted and centered over the valve stops, install and tighten the retaining screw to 25-35 inch lbs. (3-4 Nm).

56. Apply a light coating of OMC Gel-Seal II or equivalent to the threads of the leaf plate Philips screw, then install the plate to the crankcase using a new gasket. Once the crankcase, gasket and plate are properly aligned, install and securely tighten the Philips screw.

57. Install the intake manifold using a new gasket (dry, without any sealant), then install and tighten the retaining bolts/nuts to 60-84 inch lbs. (7-9 Nm).

58. Position a new powerhead gasket (dry, without any sealant), then install and tighten the inner exhaust housing bolts to 96-120 inch lbs. (11-14 Nm).

59. Install the powerhead assembly, as detailed in this section, then refer to Break-In in this section for details on how to ensure proper break-in of the new powerhead components.

25/35 Hp (500/565cc) Motors

◆ See Figure 64

Removal & Installation

◆ See Figure 64

1. On electric start models, disconnect the negative battery cable for safety.

2. Remove the top engine cover for access.

■ The engine covers and related components on these models use a number of various size and type fasteners. For assembly purposes, lay out each of the fasteners in a logical order as they are removed. Then, once the covers are removed, insert the fasteners back into their holes in order to ease identification later.

3. Remove the air intake silencer by loosening the retainer knob found at the center of the housing, then lifting the silencer for access to the drain hose. Disconnect the hose from the nipple at the bottom of the silencer and remove it completely from the engine.

4. For tiller models disconnect the choke cable pin from the retainer, then remove the clip securing the choke cable to the bracket on the powerhead (just in front of the carburetors). Separate the cable from the bracket.

5. For remote models, locate and disconnect the bullet connectors (3) for the trim/tilt wiring harness. The bullets are found just at the top of the

6-44　POWERHEAD REPAIR AND OVERHAUL

lower cover lip, near the spark plugs. If necessary, refer to the diagrams under Wiring Diagrams in the Ignition and Electrical System section for more details.

  6. Remove the lower cover retaining screws (6) from the starboard side of the motor. There are 3 screws at the front of the motor and 3 at the rear. At either end, one bolt is at the top of the cover, one at the center and one at the bottom.

  7. Pull the lower port engine cover away from the engine sufficiently to access and cut the wire tie strap securing the fuel hose to the connector. Position a shop rag over the fitting (as there could still pressurized fuel in the system), then disengage the hose, freeing the cover completely.

  8. Remove the bolt and clamp securing the wiring harness to the lower starboard engine cover.

  9. For electric start models, then remove the 2 screws holding the electrical connector cover to the engine, then remove the cover.

  10. For remote models, proceed as follows:
    a. Tag and disengage the 6 wiring connectors found under the electrical connector cover.
    b. Free the engine harness and grommet from the lower starboard motor cover, then remove the cover completely from the engine.

  11. For tiller models, proceed as follows:
    a. Disengage the wiring connector for the stop switch, then remove the stop switch wire from the ground stud.
    b. For tiller electric models, disengage the 6-pin connector then locate and remove the 2 yellow/red neutral safety switch wires from the connector.

Note the pin locations (2 and 6) to ensure installation of the terminals to the proper positions.
    c. Free the throttle cable ball socket from the ball on the throttle lever (carefully pry the socket free, or try using the OMC Ball Socket Remover #342226).
    d. Remove the clip securing the shift rod to the lever and separate the rod and lever.
    e. Remove the screws (2) holding the trunnion anchor block to the side of the powerhead.
    f. Remove the bolts (2) holding the steering arm to the tiller arm. Remove the steering arm, shift and throttle cables, along with the electrical cables and grommet from the engine as a complete assembly.
    g. Completely remove the starboard lower motor cover.

  12. Remove the screw holding the shift lever to the engine.

  13. For tiller models, unplug to 2-pin connector for the low oil light (found on the ignition module at the center of the port side of the motor, behind the carburetors, slightly forward of the cylinder head).

  14. Remove the bolts (3) securing the manual starter assembly or the flywheel cover (as applicable) to the powerhead. Remove the manual starter or flywheel cover.

  15. Secure a lifting eye bracket to the flywheel using 3 bolts to evenly spread and support the weight. Either use OMC #396748 or an equivalent eye attached to a bracket (such as a very sturdy adjustable puller) that can bolt to 3 flywheel bolt holes. Tighten the bolts just sufficiently to ensure proper thread engagement. DO NOT thread the bolts all the way through the flywheel and damage components mounted underneath.

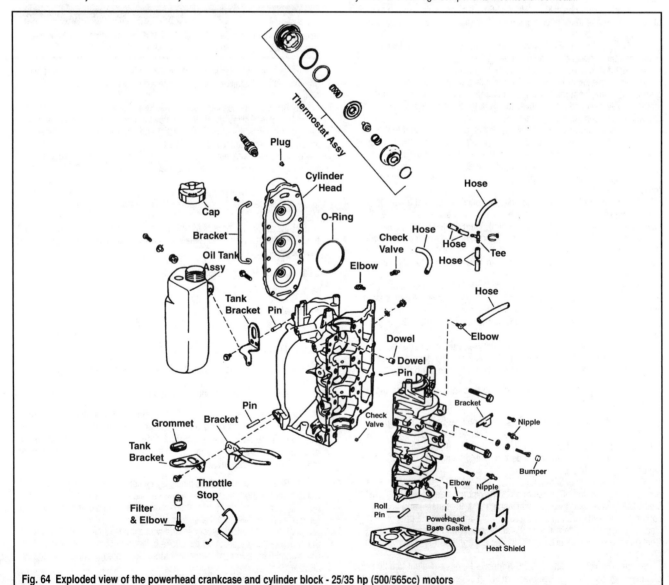

Fig. 64 Exploded view of the powerhead crankcase and cylinder block - 25/35 hp (500/565cc) motors

POWERHEAD REPAIR AND OVERHAUL 6-45

■ If possible, measure the depth of the threads using a small bore gauge or small straightedge, then thread the fixture bolts in to a depth that would make their ends equal to or just slightly above the bottom of the flywheel.

16. Remove the bolts (7) from around the lower perimeter of the powerhead (just under the exhaust housing upper flange, threaded upward into the powerhead) securing the powerhead to the exhaust housing.

17. Use an engine hoist to slowly and carefully break the seal and lift the powerhead assembly from the gearcase. If necessary, use prybars along the flange to carefully help separate the exhaust housing. Discard the old gasket and carefully clean the mating surfaces of debris, corrosion or sealant.

18. Mount the powerhead on an engine stand. If necessary, strip it for overhaul by removing the carburetors (refer to Fuel System) and any electrical or ignition components (refer to Ignition and Electrical Systems).

To install:

19. Apply a light coating of OMC Gasket Sealing Compound, or equivalent sealant to a new powerhead-to-exhaust housing gasket, then position the gasket on the exhaust housing.

20. Apply a light coating of OMC Moly Lube, or equivalent lubricant to the crankshaft splines. But, **do not** allow the lubricant to pack in the end of the splines, as the hydraulic pressure could prevent the driveshaft from seating in the crankshaft.

21. Use the lifting eye threaded into the flywheel to carefully lift the powerhead from the workbench or engine stand and lower it slowly onto the exhaust housing.

22. Apply a light coating of OMC Gasket Sealing Compound, or equivalent sealant to the threads of the 7 exhaust housing-to-powerhead bolts. Install the bolts from underneath exhaust housing flange, then tighten to bolts to 180-204 inch lbs. (20-23 Nm).

23. Reconnect the shift arm to the lever shaft and secure using the washer/screw. Tighten the screw to 84-106 inch lbs. (9-12 Nm).

24. Position the manual starter assembly or the flywheel cover (as applicable) to the top of the powerhead. Apply a light coating of OMC Locquic Primer or equivalent to the threads of the 3 retaining bolts and allow it to air dry. Then, apply a light coating of OMC Nut Lock, or equivalent threadlock to the bolt threads, install the bolts and tighten to 60-84 inch lbs. (7-9 Nm).

25. For tiller models, engage the 2-pin connector for the low oil light, then secure the connector to the tabs on the ignition module.

26. Position the starboard lower motor cover.

27. For tiller control models, proceed as follows:

   a. Connect the steering arm to the tiller arm, then install the 2 bolts and washers. Tighten the bolts to 216-240 inch lbs. (24-27 Nm). Thread 2 locknuts onto the threaded ends of the 2 tiller arm screws, then tighten the nuts to 216-240 inch lbs. (24-27 Nm)

   b. Insert the shift and electrical cables in the lower motor cover grommet, then secure the lower cover in position.

   c. Apply a light coating of OMC Locquic Primer or equivalent to the threads of 2 trunnion anchor screws and allow it to air dry. Then, apply a light coating of OMC Nut Lock, or equivalent threadlock to the bolt threads. Position the trunnion anchor block to the crankcase, then install the bolts and tighten to 84-106 inch lbs. (9-12 Nm).

   d. Connect the throttle cable ball socket to the throttle lever, connect the shift cable to the shift lever and secure using the clip.

   e. For tiller electric models, reinstall the 2 neutral safety switch wires to the 6-pin connector, then reconnect both the 6-pin and 1-pin harness connector halves and position them in the electrical connector block.

28. For remote models, proceed as follows:

   a. Insert the main engine cables into the motor cover grommet and then position the starboard lower motor cover. Route the cables along the side of the cover and back to the electrical bracket.

   b. Engage each of the connectors as tagged during removal, then secure them to the tabs in the electrical bracket.

29. Connect all ground wires to the ground stud and tighten the nut to 60-84 inch lbs. (7-9 Nm).

30. Coat all electrical connections with OMC Black Neoprene, or equivalent sealant to protect them from loosening and help fight corrosion.

31. Install the electrical connector cover and secure using the 2 screws. Tighten the screws to 24-36 inch lbs. (3-4 Nm).

32. For remote models, secure the main engine cable to the starboard cover using the J-clamp and screw. Tighten the screw to 84-106 inch lbs. (9-12 Nm).

33. For tiller electric models, if removed, connect the positive battery cable to the starter solenoid. Tighten the nut securely and position the plastic cover over the terminal/nut.

34. For all tiller models, secure the tiller arm electrical cables (and battery cables, if applicable) to the starboard lower motor cover using the J-clamp and screw. Tighten the screw to 84-106 inch lbs. (9-12 Nm).

35. Connect the fuel line to the fitting on the port side lower engine cover, then secure using a new wire tie. Position the cover and tighten the 6 screws to 84-106 inch lbs. (9-12 Nm).

36. For tiller models, position the choke cable back into the bracket and secure using the clip. Then, attach the choke cable retainer in the linkage.

37. For remote models, reconnect the 3 trim/tilt bullet connectors, making sure they are properly routed away from interference with any moving components.

38. Connect the drain hose to the nipple on the bottom of the air intake silencer, then install the silencer to the powerhead and tighten the silencer lock.

39. Install the top engine cover.

40. If equipped, connect the negative battery cable.

41. If a new or rebuilt powerhead was just installed, refer to Break-In in this section for details on how to ensure proper break-in of the new powerhead components.

Disassembly & Assembly

◆ See Figures 8, 25, 30 thru 40, 50 and 64 thru 67

Although the majority of the powerhead overhaul procedure for 25/35 hp 3-cylinder motors is fairly easy and straightforward, the connecting rod cap bolt torque procedure requires the use of a holding fixture in order to ensure that the connecting rod/cap and bearing are properly centered. Attempting to center and tighten the connecting rod caps without this fixture may cause unexpected, total powerhead failure sometime after the motor is returned to service. That is not to say people have not succeeded in centering and tightening the caps without the fixture, but to say that we cannot recommend it, as there are no guarantees.

■ To simplify assembly, remember to layout all bolts, components and clamps in the order of removal. This is especially true for the wiring harness and related clamps. Matchmark all component assemblies such as pistons, connecting rods to ensure installation of the correct pairs and installation in the correct orientation. Most of all, take your time.

1. Remove the powerhead from the lower unit as detailed under Removal & Installation.

2. Strip the powerhead for overhaul by removing all fuel system, ignition system and electrical components from the powerhead. For details, refer to the procedures in the Fuel System and the Ignition and Electrical System sections.

3. Tag and disconnect the single crankcase oiling hose and the 3 re-circulating hoses.

4. On remote models, remove the screws (2) fastening the trunnion block to the crankcase.

5. Remove the locknut and flat washer, then remove the throttle lever, spacer, shaft and both bushings from the crankcase.

6. Locate the throttle linkage at the center of the powerhead on the starboard side, then remove the shoulder bolt, 2 washers and throttle linkage from the crankcase.

7. Remove the Intake manifold bolts (0), then remove the intake manifold and discard the old gasket.

■ Leaf plate assemblies on these models cannot be disassembled to replace individual valve components. Inspect the assemblies and, if defective or worn, replace them.

8. If necessary, remove the 3 leaf plate assemblies for inspection or replacement, as follows:

   a. Remove the leaf plate assembly screws (there are 2 screws per assembly), then remove the assembly and gasket. Discard the old gasket.

✱✱ WARNING

Do NOT lift or bend leaf valves that have been in service or they may be damaged. Even if not visible, they could be damaged, causing them to break sometime after they are returned to service.

6-46 POWERHEAD REPAIR AND OVERHAUL

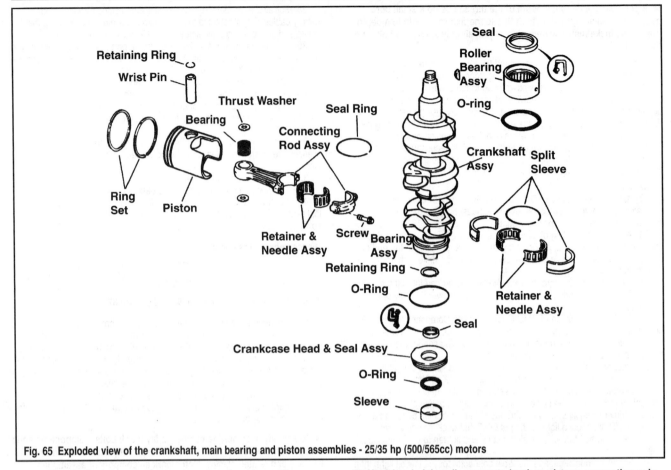

Fig. 65 Exploded view of the crankshaft, main bearing and piston assemblies - 25/35 hp (500/565cc) motors

b. Visually check the valve tips for signs of cracks, chips or other obvious damage. Make sure the leaf stops are not distorted or loose. If there are any obvious defects, the entire valve assembly must be replaced.

c. Check the leaf plate screws to make sure they are tight. If they have loosened at all, retighten each one individually as follows. Loosen the screw a turn or two, just enough to apply a light coating of OMC Screw Lock or an equivalent threadlock, then tighten the screw to 25-35 inch lbs. (3-4 Nm) and proceed on to the next screw.

d. Visually check all gasket surfaces to make sure they are smooth and free of nicks. If necessary, a gasket surface may be dressed to smooth shallow nicks using a fine emery cloth.

e. Use a machinist's straightedge and a feeler gauge to check for warpage across the intake manifold gasket surfaces. Lay the straightedge in all directions across the manifold mating surface, checking at the center of the straightedge each time. Make sure a 0.004 in. (0.10mm) or large feeler gauge cannot be inserted underneath the straightedge in any direction.

9. Loosen and remove the crankcase flange screws (12), then loosen and remove the 8 main bearing screws using a spiraling pattern that works from the outer toward the inner bolts.

10. With the crankcase facing upward carefully lift and separate the crankcase from the cylinder block. If the gasket seal does not come loose easily, use a plastic or rubber mallet to tap upward on the exposed portion of the crankshaft (at a 90° angle to the shaft). Tap lightly until the crankcase half **just** starts to separate from the cylinder block half. Lift and remove the crankcase from the bottom of the cylinder block (exposing the crankshaft).

11. Loosen and remove the cylinder head retaining bolts using multiple passes of a spiraling pattern that starts at the outside and work towards the center bolts. Once the bolts are removed, remove the cylinder head. Remove and discard the cylinder head O-rings.

12. Using a felt-tipped marker or numbered/lettered punches, label the piston domes. Remember that the No. 1 cylinder is the cylinder closest to the flywheel.

■ If a felt-tipped marker is used to label components during disassembly, watch the marks during cleaning. Replace any marks that are dulled or removed by solvent as soon as the component is dry.

■ Keep in mind that all wear parts (such as pistons, connecting rods, needle bearings and bearing liners) MUST be installed in their original positions (and with the same orientations) if they are to be reused. Mark each component with a permanent marker to ensure proper assembly.

13. Matchmark the connecting rod caps and then use a suitable socket (such as OMC Torque Socket #342664) to loosen the connecting rod cap bolts. Work on one piston at a time, supporting the piston at the dome with one hand while loosening the bolts on that piston's rod cap alternately, no more than one turn at a time. Remove each cap along with the connecting rod cap needle bearing and cage. Carefully push the piston from the bore and out the top of the cylinder block.

✱✱ WARNING

When removing each piston, take great care to prevent the connecting rod from contacting and damaging the crankshaft journal, cylinder wall and, finally, once removed, the cylinder piston skirt. One method to help prevent this is to stuff rags in the skirt before or after the piston is removed to keep the connecting rod centered.

14. Repeat the previous step for each of the other pistons. To prevent any possible mix-up reinstall each bearing assembly and connecting rod cap to the piston and connecting rod as soon as it is removed from the cylinder block.

15. Lift the crankshaft assembly carefully out of the cylinder block.

16. If the lower crankshaft seal and/or the bearing must be replaced:
 a. Remove the lower bearing seal housing assembly.
 b. Remove and discard the O-ring.
 c. Use a drift punch or small prytool as a lever to remove the housing seal, then discard the seal.
 d. Inspect the housing for wear or damage and replace, if necessary.

17. Remove the upper crankshaft bearing and seal assembly, then remove and discard the O-ring. Support the bearing assembly in a shop press or on top of a vise (adjusted so the housing only sits on the jaws, then use a punch to remove and discard the old seal.

POWERHEAD REPAIR AND OVERHAUL 6-47

18. Remove O-ring from the end of the crankshaft sleeve. Inspect the O-ring and, if worn or damaged, install a new O-ring during assembly.

19. Check the crankshaft sleeve. If damaged, remove it using a slide hammer and a small-jawed puller.

20. ONLY remove the lower crankshaft bearing if it requires replacement. First remove the retaining ring using a pair of snapring pliers, then support the bearing in a shop press and press the bearing off the shaft.

21. Remove both sets of main bearings and split sleeves. Remove the crankcase seal rings for inspection.

22. Disassemble each piston for further inspection and component replacement, as follows:

   a. Use a universal ring expander to remove the rings from the pistons. The manufacturer warns NOT to reuse the rings. Retain them in sets for inspection purposes (marking their original locations), but discard them before installation.

■ We've said it before. We know there will be some circumstances that will prompt people to reuse rings (for example, if for some reason the motor was disassembled after only a few hours of usage). But for the most part, if you've come THIS far, it is probably a good idea to bite the bullet and replace the rings to ensure long-lasting performance from the powerhead.

   b. Using a pair of modified needle-nosed pliers as shown in the accompanying illustration, or a suitable pair of snapring pliers, remove the wrist pin retaining rings. Discard the old retaining rings.

■ If necessary, a piston cradle can be fabricated from a large block of wood. The important design features of the cradle are that it is semi-circular on top to hold (or cradle, get it?) the piston and that there is a bore at the bottom center into which the wrist pin can tapped or pressed. The final important feature of the cradle is that it is STURDY enough to withstand the force of the shop press, BUT, keep in mind that the fit is loose on these pistons and significant force should not be necessary.

   c. The piston-to-wrist pin fit should be loose on both sides. Support the piston and use a suitable drift pin (like OMC #342657 to carefully and smoothly tap or push the wrist pin free of the piston bore). If the OMC tool is not available, be sure to use a drift pin that is just slightly smaller than the diameter of the wrist pin itself.

■ As the wrist pin is removed from the piston you must retain the 24 loose needle bearings and the 2 thrust washers from the connecting rod bore. If you can, stop tapping or pressing once the wrist pin JUST clears the bottom of the connecting rod bore. Then carefully withdraw the rod along with the needle bearings (place a thin bladed tool like a putty knife or your finger under the rod hole as it is pulled free of the piston bore/wrist pin). Inspect the bearings, if any of the needles are lost or worn, replace the entire bearing assembly.

23. If necessary, remove the neutral detent retaining ring from the bore in the side of the crankcase using a small prytool. Remove the detent ball and spring.

24. Refer to the information found under Cleaning & Inspection, in this section for details on inspecting the powerhead components to determine which should be replaced and to prepare the ones that are being reused for assembly.

To assemble:

✳✳ WARNING

During assembly, take your time and NEVER force a component unless a shop press or drive tool is specifically required by the procedure.

■ Lightly coat all components (except those that are coated with sealant) with clean, fresh Johnson/Evinrude Outboard Lubricant or equivalent oil during assembly. Always replace all gaskets, O-rings and seals.

25. If removed, install the neutral detent ball and spring into the crankcase bore, then secure using a retaining ring. Seat the ring using a suitable driver (that is just smaller than the bore) such as OMC #342658.

26. If removed, apply a light coating of clean engine oil or assembly lube to a NEW lower main bearing and to the end of the crankshaft, then support the crankshaft in a shop press between the 2 lower counterweights. Carefully press the bearing into position with the lettered side of the bearing facing upward toward the press.

■ OMC Crankshaft Bearing/Sleeve installer #342222 can be used to safely seat the lower bearing and the sleeve in position. If not available, be sure to use a suitable driver that contacts the entire diameter of the bearing cage or the sleeve and does NOT contact or damage the crankshaft.

27. If removed, apply a light coating of clean engine oil or assembly lube to the NEW crankshaft sleeve, then install the sleeve and gently drive or press the sleeve onto the shaft until the installer contacts the lower main bearing.

✳✳ WARNING

Make sure the sleeve surface is not nicked or damaged during installation or it must be removed and discarded (and another NEW sleeve installed).

■ If the installer sticks on the sleeve after installation, use a slide hammer to pull it free.

28. Apply a light coating of clean engine oil or assembly lube to a crankshaft O-ring, then install the O-ring into the crankshaft sleeve.

29. Apply a light coat of OMC Moly Lube to the crankshaft splines.

30. Using a pair of external snapring pliers, carefully install the lower bearing retaining ring with the sharp edge of the ring facing outward, away from the bearing.

31. Oil the 2 center main roller bearing and split sleeve assemblies, then install them around the crankshaft center journals in their original positions. Be sure to face the split sleeve ring grooves away from flywheel.

32. Install the crankshaft seal rings, making sure the ends of the rings overlap.

33. Apply a light coating of OMC Gasket Sealing Compound, or equivalent sealant to the metal casing of a NEW lower housing seal. Place the seal, with the exposed lip facing into the housing, then press against the outer metal seal casing until the edge of the case is flush with the inside surface of the lower housing.

34. Apply a light coating of OMC Gasket Sealing Compound, or equivalent sealant to the lower housing O-ring, O-ring groove and flange, then install the O-ring into the lower housing groove.

35. Apply a light coating of engine oil to the seal lip, then install the lower bearing seal housing onto the end of the crankshaft.

36. Apply a light coating of OMC Gasket Sealing Compound, or equivalent sealant to the outside casing of a NEW upper crankshaft bearing seal, then place the seal into the upper bearing housing. Press the seal into place until it seats, then apply a light coating of clean engine oil or assembly lube to the seal lips.

37. Apply a light coating of clean engine oil or assembly lube to the upper main bearing and to a NEW O-ring. Install the O-ring onto the upper bearing and seal assembly and then install the assembly onto the crankshaft.

■ The pistons and connecting rods for these motors should be installed, not only in their original positions, but in their original orientations (with the same sides facing the flywheel as before disassembly). Luckily, these components are marked. Pistons are marked with the word UP on their domes, this word should be positioned toward the flywheel during installation. Similarly, connecting rods/caps are equipped with alignment marks (on either side of the rod/cap, near the cap split line) that must be faced toward the flywheel. Finally, the part numbers on the upper end of the connecting rod itself should also be faced toward the flywheel.

38. If disassembled for inspection and/or component replacement, assemble each piston as follows:

   a. Apply a light coating of OMC Needle Bearing Assembly grease (or equivalent) to the wrist pin bearing needles, then insert the 24 needles around the inner circumference of the connecting rod wrist pin bore. Install the OMC Wrist Pin Bearing tool #342217 into the bore to align the needles and hold them in place during assembly.

■ The wrist pin bearing tool is, essentially, a short round pin, roughly the same outer diameter of the wrist pin, but very short, just long enough to insert it through the connecting rod wrist pin bore and install the thrust washers on either side. A similarly sized dowel pin can be substituted, or a old connecting rod wrist pin from the same size

6-48 POWERHEAD REPAIR AND OVERHAUL

piston could be used. If substituting another rod or drift pin, make sure the surface is smooth and clean/free of all corrosion. Also, make sure the outer diameter of the tool is smaller than the wrist pin and bore to prevent the possibility of binding or damage.

    b. With the bearings and the wrist pin tool in position, install the thrust washers. Again, a dab of grease should hold the thrust washers in position.

    c. Apply a light coating of clean engine oil or assembly lube to the wrist pin and the pin bore in the piston.

    d. Support the piston (in a piston cradle if available, but a clean workbench surface or your lap is fine for these motors, as the wrist pins must be installable without force on these motors). Insert the wrist pin into the bore and gently press it by hand through the bore until it just starts to appear inside the piston skirt. Stop, then position the connecting rod with the needle bearings, thrust washers and wrist pin bearing installation tool inside the piston skirt. Continue to push the wrist pin into the position, through the thrust washers and bearings, pushing the wrist pin bearing installer through the other side and out of the piston.

■ **No force should be required to install the wrist pins on these motors. DO NOT install a wrist pin that requires force, instead, inspect the wrist pin for burrs or carbon deposits and remove them using a fine grit sand paper or steel wool. Remove all traces of loose metal or particles from the wrist pin using solvent and re-oil the pin before attempting to install it again.**

    e. Once the wrist pin is properly positioned, use a small driver to install a new snapring in each end of the wrist pin bore. Make sure the snapring fully seats in the groove. Turn the ring in the groove until the gap faces directly away (180° opposite) from the semi-circular relief in the side of the wrist pin bore.

    f. Use a ring expander to install the pistons rings on each piston. Each piston ring groove should contain a dowel pin, around which the ring gap must be squarely situated. If not the rings will likely break during installation (or if by they don't by some miracle, during the first start-up). Be sure to install the tapered ring in the top groove.

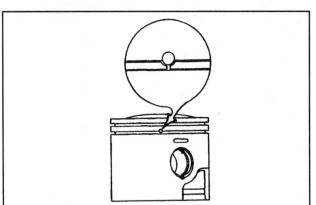

Fig. 66 Each piston ring must be centered squarely around the dowel pin

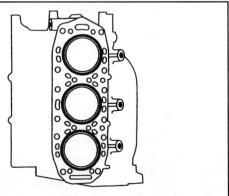

Fig. 67 Apply a bead of GE RTV sealant around each of the 3 water passages on the cylinder block

■ **When using the ring expander, spread each ring JUST enough to slip over the piston.**

39. Inspect the cylinder block-to-crankcase mating surface to ensure there are NO traces of sealant left. If necessary, remove any remaining traces of hardened sealant to prevent potential problems with bearing alignment.

40. If working on an engine stand, rotate the cylinder block so the head surface is facing upward, horizontal to the floor. If working on a bench, position the block securely on one end so both the cylinder head and crankcase mating surfaces are accessible.

■ **The manufacturer makes different ring compressor tools for both standard and oversize versions of the pistons used in 500cc and the pistons used in 565cc motors. Be sure to obtain the correct one for the pistons being installed, or use a universal compressor (or special hose clamp as described in this section).**

41. If removed, install each piston as follows:
    a. Apply a light coating of clean engine oil or assembly lube to the piston and piston bore.

    b. Use a suitable ring compressor or a band-type clamp with a solid inner ring to evenly compress the rings (oiling the inside of the compressor or clamp will also help ease installation). Make sure the rings are still properly positioned with the gaps over the dowels. Also, make sure the connecting rod remains centered in the piston to prevent possible damage to the cylinder walls or piston skirt.

    c. Start the piston into the bore with the **UP** marking on the piston dome facing the flywheel.

    d. Slowly insert the piston, making sure the connecting rod does not strike and damage the piston skirt, or the strike/scrape and damage the cylinder wall. If the piston hangs up on a ring, don't force it, push the piston back upward and remove the compressor to check for potential problems, such as a ring that has shifted the gap away from the dowel.

42. Once the pistons are installed in the bores, apply a light coating of OMC Triple-Guard, or equivalent marine grease to 3 NEW cylinder head O-rings, then install them in the cylinder block grooves.

43. Apply a 1/16 (2mm) bead of GE RTV or equivalent sealant around each cylinder water passage, keeping the sealant off the O-rings. Then position the cylinder head on the block, over the sealant and O-rings.

44. Install the cylinder head bolts (dry, without any sealant) and tighten them to 180-204 inch lbs. (20-23 Nm) using multiple passes of a spiraling pattern that starts at the center and works outward.

45. Rotate the cylinder block the head faces downward and the crankcase mating surface is facing upward.

46. Carefully push all pistons downward (up in their bores) to Top Dead Center (TDC).

47. Remove the connecting rod end caps and cap bearing halves (positioning them aside to keep track of their location and orientations). If matchmarks are not readily visible, matchmark the connecting rods to the caps and bearing halves before removal to ensure proper installation.

48. If not already done, apply a light coating of clean engine oil or assembly lube to the crankshaft journals, as well as to the connecting rod bearings.

49. Lower the shaft into the position. Be sure the crankcase seal ring gaps are facing upward (toward the crankcase). Also, make sure the tab on the lower seal bearing housing is in the crankcase recess and that each center main bearing split sleeve is positioned on the dowel pin.

50. Starting with the No. 1 piston, install each of the pistons to the crankshaft in turn, as follows:

    a. Slowly and carefully pull the connecting rod up until the bearing in the rod contacts the crankshaft journal (crankpin).

    b. Seat the rod journal against the crankpin, move the bearings from the caps to the connecting rod journals. Finally, align the matchmarks and position the caps over the bearings.

■ **Besides any matchmarks made during removal, the connecting rod caps on these models are normally equipped with dots on the connecting rod/cap. These dots should also be aligned. Remember, these dots must face the flywheel.**

    c. Oil and install the connecting rod cap bolts, then finger-tighten the bolts alternating from side-to-side on the rod cap.

    d. Use a pencil or your fingernail to determine if the rod caps are fully aligned and seated. Scrape the pencil across the mating surface to see if it catches.

POWERHEAD REPAIR AND OVERHAUL 6-49

■ The connecting rod cap must be aligned/centered using OMC Rod Cap Alignment Fixture #396749 and OMC Connecting Rod Alignment Jaw #437273. Follow the next steps closely to ensure powerhead durability.

e. Rotate the knob on the top of the tool until the flat marked **SET** aligns with the arrow on the fixture frame. Rotate the adjustment knob 180° to the lock position.

■ When using the OMC Rod Screw Torque Socket #342664 and the OMC alignment fixture, the outer diameter of the socket must be ground down to allow proper engagement with the rod screw.

f. Make sure the retaining jaw (with the flat head screw in the frame bore) and the forcing jaw (with the hex key at the base and the threaded flat protruding from the frame) are positioned on the fixture tool.

g. Tighten the connecting rod cap screws to 25-30 inch lbs. (2-3 Nm).

h. Apply a light coating of engine oil to corners of the connecting rod and cap where the fixture will be installed.

i. Position the frame to the so the contact area of the forcing jaw is centered on the side of the connecting rod and cap. Tighten the forcing screw until the jaws contact the connecting rod, then slide the frame downward until the adjustment stop (at the center of the frame) just contacts the top of the connecting rod cap. Verify that the groove lines in the jaws are centered on the rod/cap diameter.

✳✳ WARNING

The frame MUST be squarely in position as described in order to perform its function properly. Failure to center the frame, and therefore the connecting rod/cap, will lead to premature crankshaft, rod and bearing failure.

j. With the frame centered, tighten the forcing screw to 23 inch lbs. (2.5 Nm).

k. Loosen both connecting rod cap bolts 1/4 turn, then tighten each bolt to 40-60 inch lbs. (5-7 Nm).

l. Next, tighten the connecting rod cap bolts to 84-106 inch lbs. (9-12 Nm), and finally tighten them to 170-190 inch lbs. (19-21 Nm).

m. Loosen the forcing screw, remove the frame from the connecting rod/cap assembly, then repeat for the remaining rods/caps.

51. Clean and degrease the mating flange of the crankcase and cylinder block halves using OMC Cleaning Solvent, or an equivalent solvent. Allow the flanges to air dry.

52. Apply a light coating of OMC Locquic Primer or equivalent to the mating flange of the crankcase half and allow it to air dry.

53. Apply a light coating of OMC Gel-Seal II or equivalent to the cylinder block flange. Be sure that the sealant coats the entire flange evenly, but be careful not to apply excessive amounts. The sealant must not be applied within 1/4 in. (6mm) of the labyrinth seal or bearings otherwise the sealant could be forced out from between the flanges into contact with the bearings during assembly.

■ Although OMC Gel-Seal II has a shelf life of one year when stored at room temperature, buy a new tube if in doubt. Keep in mind that using an old tube of Gel-Seal II could allow crankcase air leaks (leading to performance problems).

54. Carefully lower the crankcase half into place over the cylinder block half.

55. Apply a light coating of OMC Locquic Primer or equivalent to the threads of the 8 main bearing bolts and allow it to air dry.

56. Apply a light coating of OMC Ultra Lock, or equivalent high-strength threadlock to the threads of the 8 main bearing bolts and a coating OMC Gel-Seal II or equivalent to the underside of each washer face (the side that contacts the crankcase). Install the bolts and washers finger tight.

■ The 2 longer main bearing bolts are use at the top (flywheel) end of the crankcase.

57. Tighten the main bearing screws to 216-240 inch lbs. (24-27 Nm) using multiple passes of a spiraling sequence that starts with the center screws and works outward.

58. If removed, install the lower drain nipple.

59. Apply a light coating of OMC Locquic Primer or equivalent to the threads of the 12 crankcase flange bolts and allow it to air dry.

60. Apply a light coating of OMC Ultra Lock, or equivalent high-strength threadlock to the threads of the 12 crankcase flange bolts, then install and tighten the bolts to 60-84 inch lbs. (7-9 Nm).

61. Peel the backing off a NEW intake manifold gasket, then position the gasket to the leaf plate side of the manifold.

62. If removed, install the 3 leaf plate assemblies to the intake manifold as follows:

a. Apply a light coating of OMC Locquic Primer or equivalent to the threads of the leaf plate retaining screws, and allow the primer to air dry.

b. Apply a light coating of OMC Ultra Lock, or equivalent high-strength threadlock to the threads of the leaf plate screws, then install each leaf plate assembly and tighten the screws to 25-35 inch lbs. (3-4 Nm).

63. Apply a light coating of OMC Locquic Primer or equivalent to the threads of the 8 intake manifold bolts and allow it to air dry.

64. Apply a light coating of OMC Ultra Lock, or equivalent high-strength threadlock to the threads of the 8 intake manifold bolts. Install the intake manifold and tighten the bolts to 60-84 inch lbs. (7-9 Nm).

65. Install the throttle/spark lever to the side of the crankcase using the 2 flat washers and shouldered bolt, making sure to position the larger of the 2 washers between the lever and the crankcase. Tighten the bolt to 60-84 inch lbs. (7-9 Nm).

66. Apply a light coating of OMC Triple-Guard, or equivalent marine grease to the inside and outside of the 2 throttle lever bushings and to the throttle lever shaft.

67. Install the 2 bushing into the bosses on the crankcase, then install the lever shaft in the bushings, placing the spacer on the end of the shaft. Position the throttle lever on the shaft end along with the flat washer, install and tighten the nut to 60-84 inch lbs. (7-9 Nm).

68. For remote models, apply a light coating of OMC Locquic Primer or equivalent to the threads of the 2 trunnion anchor screws and allow it to air dry. Next, apply a light coating of OMC Nut Lock, or equivalent threadlock to the 2 screws and secure the trunnion block to the crankcase using the screws. Tighten to 84-106 inch lbs. (9-12 Nm).

69. Reconnect the one oiling hose and the three re-circulating hoses to the powerhead as tagged during removal.

70. Install all fuel, ignition and electrical system components as detailed in the Fuel System and Ignition and Electrical System sections.

71. Install the powerhead assembly, as detailed in this section, then refer to Break-In in this section for details on how to ensure proper break-in of the new powerhead components.

25-55 Hp (737cc) and 25-70 Hp (913cc) Motors

◆ See Figures 68 and 69

Removal & Installation - 2-Cylinder Motors

◆ See Figures 68 and 70 thru 74

1. On electric start models, disconnect the negative battery cable for safety.

2. Remove the top engine cover for access and, for models equipped with 2-piece lower covers, remove the lower covers as detailed under Engine Covers (Top and Lower Cases) in the Maintenance and Tune-Up Section.

3. Remove the shift rod screw (from the shift rod lever at the lower front of the powerhead).

4. From a cutout in the rear exhaust housing, remove the nut from the rear lower powerhead stud (below the cylinder head).

5. From underneath the exhaust housing flange, loosen and remove the exhaust housing-to-powerhead bolts (8).

6. Attach an engine hoist the lifting hook. If necessary for better balance, install a second lifting hook to the top of the powerhead (on many Johnson/Evinrude motors a hook can be secured to the top of the flywheel via the 3 puller bolt holes).

7. Slowly and carefully lift the powerhead assembly straight up and off the exhaust housing.

8. Remove and discard the old powerhead gasket.

9. Mount the powerhead on an engine stand. If necessary, strip it for overhaul by removing the carburetors (refer to Fuel System) and any electrical or ignition components (refer to Ignition and Electrical Systems).

To install:

10. Position a new powerhead gasket in place without any sealant.

6-50 POWERHEAD REPAIR AND OVERHAUL

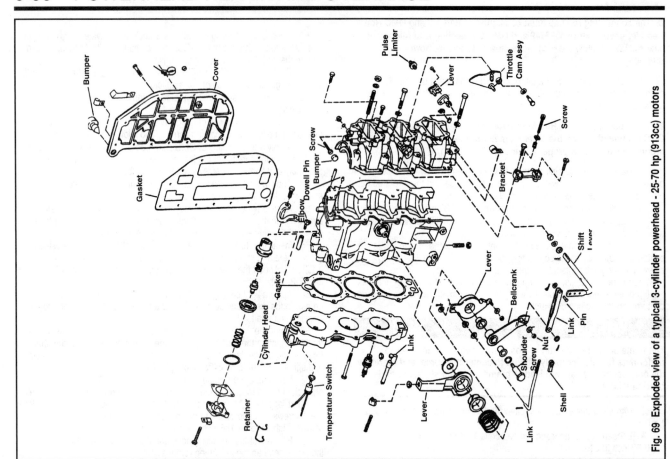

Fig. 69 Exploded view of a typical 3-cylinder powerhead - 25-70 hp (913cc) motors

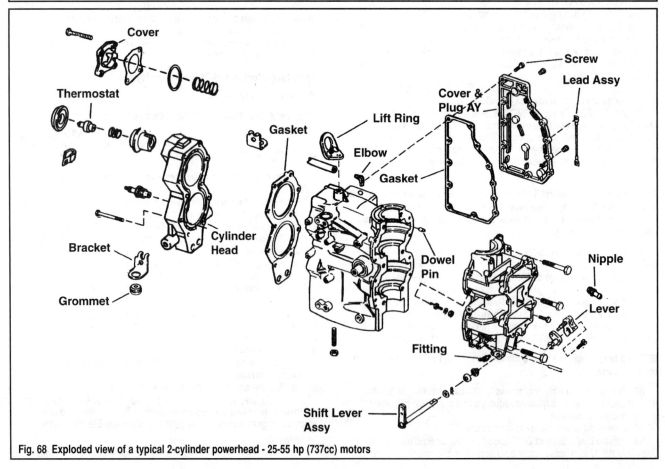

Fig. 68 Exploded view of a typical 2-cylinder powerhead - 25-55 hp (737cc) motors

POWERHEAD REPAIR AND OVERHAUL 6-51

Fig. 70 On models equipped with 2-piece lower covers remove the retaining bolts...

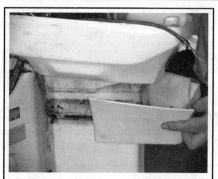
Fig. 71 ...the separate the lower covers for access

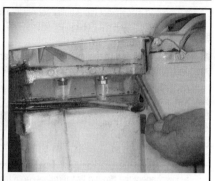
Fig. 72 Loosen and remove the powerhead flange bolts...

Fig. 73 ...and remove the nut from the powerhead stud in the cutout

Fig. 74 Attach a suitable engine lifting device in order to remove the powerhead

11. Apply a light coating of OMC Moly Lube, or equivalent lubricant to the driveshaft splines. But, **do not** coat the top surface of the shaft, as the resultant hydraulic pressure could prevent it from seating in the crankshaft.

12. Use the lifting eye(s) to carefully lift the powerhead from the workbench or engine stand and lower it slowly onto the exhaust housing.

13. Apply a light coating of OMC Gasket Sealing Compound, or equivalent sealant to the threads of the 8 exhaust housing-to-powerhead bolts. Install the bolts from underneath exhaust housing flange, then tighten to bolts to 18-20 ft. lbs. (25-27 Nm).

14. Install the exhaust housing-to-powerhead stud nut, then tighten the nut to 18-20 ft. lbs. (25-27 Nm).

15. Connect the shift rod to the shift rod link, then install and tighten the retaining screw securely.

16. If equipped with 2-piece lower covers, install the lower engine covers.

17. If equipped, connect the negative battery cable.

18. Perform the Timing and Synchronization adjustments found in the Maintenance and Tune-Up section.

19. Install the top engine cover.

20. If a new or rebuilt powerhead was just installed, refer to Break-In in this section for details on how to ensure proper break-in of the new powerhead components.

Removal & Installation - 3-Cylinder Motors

◆ See Figures 68 and 70 thru 74    DIFFICULT

1. On electric start models, disconnect the negative battery cable for safety.

2. Remove the top engine cover for access.

3. For rope start models, remove the manual starter assembly. For details, refer to the procedures in the Hand Rewind Starter section.

4. If applicable, and if necessary, remove the 3 bolts securing the starter ratchet to the top of the flywheel, then remove the ratchet.

5. If equipped, disconnect the starter lockout cable clamp, then separate the cable from the shift lever.

6. For tiller control models, disconnect the throttle cable pivot and the cable bracket.

7. Disconnect the shift lever.

8. Disconnect the stop switch ground lead, then remove the lead terminal from position **E** of the 5-pin connector. A pin removal tool should be used, along with isopropyl alcohol for lubrication. To help with switch, connector and wire identification, please refer to the Wiring Diagrams in the Ignition and Electrical System section.

9. On manual primer models, cut the tie straps, then tag and disconnect the hoses from the manual primer assembly.

10. Remove the screw from the shift rod at the front of the powerhead assembly.

11. Tag and disconnect the fuel inlet hose from the fuel pump.

12. If equipped with the VRO2 system, remove the retaining clip, then push the shift linkage toward the powerhead in order to release the shift rod.

13. On the inside of the lower cover, at the back of the powerhead, just below the cylinder head), remove the 2 lower cover screws (there should be one on either side of the top cover latch). Then, just to the port and slightly above the port side lower cover screw, loosen and remove the screw retaining the powerhead ground lead.

14. On the rear, outside of the lower cover, remove the 4 horizontal lower cover retaining screws, then remove the rear lower cover.

15. Back inside the lower cover, locate and remove the 2 forward lower cover retaining screws (mounted vertically through the lower cover into the exhaust housing), then remove the forward lower cover.

16. Remove the power and starboard lower pan support bolts.

17. From underneath the exhaust housing flange, remove the exhaust housing-to-powerhead flange bolts (they are 6 screws threaded upward into the powerhead).

18. From the cutout, underneath the lower cover, remove the nut from the exhaust housing-to-powerhead stud.

19. For most models through 1998, proceed as follows:
   a. Disconnect the power trim/tilt harness connector.
   b. Remove the wire loom, then disconnect the 2-pin Amphenol connector located just below the fuel filter.

6-52 POWERHEAD REPAIR AND OVERHAUL

   c. Tag and disconnect the 2 bullet connectors for the tilt limit switch.
   d. Tag and disconnect the 3 bullet connectors for the power trim/tilt switch.

■ Refer to the Wiring Diagrams in the Ignition and Electrical System section for details, as necessary, on wire and connector identification.

   e. Tag and disconnect the hoses at the fuel filter, VRO2 sight tube and overboard indicator nipple (all located at or near the fuel filter).

※※ WARNING

Unless the VRO2 pump is being replaced, DO NOT disconnect any hoses from the pump.

   f. Disconnect the ground flag terminal, from the powerhead, just behind the fuel filter.

■ **Although most models are equipped with a lifting bracket, the manufacturer's instructions include installing a lifting bracket to the top of the flywheel to lift the motor. Although the manufacturer makes a special lifting hook adapter that attaches to their universal puller, a lifting eye that is directly bolted to the 3 flywheel bolt holes can be substituted.**

  20. Install OMC Universal Puller an Lift Eye #321537 along with OMC Lifting Fixture 396748, or equivalent to the top of the flywheel assembly.
  21. Slowly and carefully lift the powerhead assembly straight up and off the exhaust housing.
  22. Remove and discard the old powerhead gasket.
  23. Mount the powerhead on an engine stand. If necessary, strip it for overhaul by removing the carburetors (refer to Fuel System) and any electrical or ignition components (refer to Ignition and Electrical Systems).

To install:
  24. Position a new powerhead gasket in place without any sealant.
  25. Apply a light coating of OMC Moly Lube, or equivalent lubricant to the driveshaft splines. But, **do not** coat the top surface of the shaft, as the resultant hydraulic pressure could prevent it from seating in the crankshaft.
  26. Use the lifting eye(s) to carefully lift the powerhead from the workbench or engine stand and lower it slowly onto the exhaust housing. If necessary, turn the powerhead slowly clockwise (as viewed from above) to help align the crankshaft-to-driveshaft splines.
  27. Apply a light coating of OMC Gel-Seal II, or equivalent to the threads of the 6 exhaust housing-to-powerhead bolts. Install the bolts from underneath exhaust housing flange, then tighten the bolts in stages to 18-20 ft. lbs. (25-27 Nm).
  28. Install the exhaust housing-to-powerhead stud nut, then tighten the nut to 60-84 inch lbs. (7-9 Nm).
  29. For most models through 1998, proceed as follows:
   a. Connect the ground flag terminal, to the powerhead, just behind the fuel filter.
   b. Connect the hoses at the fuel filter, VRO2 sight tube and overboard indicator nipple as tagged during removal. Make sure all hoses are routed to avoid interference with moving components.
   c. Connect the 3 bullet connectors for the power trim/tilt switch as tagged during removal.
   d. Connect the 2 bullet connectors for the tilt limit switch as tagged during removal.
   e. Connect the 2-pin Amphenol connector located just below the fuel filter, then secure using the wire loom.
   f. Connect the power trim/tilt harness connector.
  30. Install the lower pan support bolts and washers to the starboard and port brackets. Tighten the bolts to 60-84 inch lbs. (7-9 Nm).
  31. Install the forward lower cover and tighten the screws securely.
  32. Position the rear lower cover and loosely install the 4 horizontal retaining screws from outside the housing. Install the 2 vertical rear cover screws (from inside the lower cover, near the base of the cylinder head). Once the rear lower cover is in position and all screws are started, tighten all 6 screws securely.
  33. Secure the powerhead ground lead using the bolt located near the port side lower cover screw (near the base of the cylinder head).
  34. If removed, install the screw to the shift rod at the front of the powerhead assembly. Tighten the shoulder screw to 60-84 inch lbs. (7-9 Nm). If equipped with the VRO2 system, with the shift rod and lever aligned, pull the lever away from the powerhead and install the retainer clip. On these models, the clip must compress the flat and wave washers.

  35. Reconnect the fuel inlet hose to the fuel pump.
  36. On manual primer models, reconnect the hoses to the manual primer assembly as tagged, then secure them using new wire ties.
  37. Reconnect the stop switch ground lead, and install the terminal back into position **E** of the 5-pin connector.
  38. Reconnect the shift lever.
  39. For tiller control models, connect the throttle cable pivot and the cable bracket.
  40. If equipped, connect the starter lockout cable to the shift lever and secure using the clamp.
  41. If removed, install the starter ratchet to the top of the flywheel and tighten the 3 bolts to 120-144 inch lbs. (14-16 Nm).
  42. For rope start models, install the manual starter assembly. For details, refer to the procedures in the Hand Rewind Starter section.
  43. If equipped, connect the negative battery cable.
  44. Perform the Timing and Synchronization adjustments found in the Maintenance and Tune-Up section.
  45. Install the top engine cover.
  46. If a new or rebuilt powerhead was just installed, refer to Break-In in this section for details on how to ensure proper break-in of the new powerhead components.

Disassembly & Assembly

◆ See Figures 8, 24, 25, 30 thru 37, 39, 40, 50, 51, 52, 68, 69 and 75 thru 91

  Although the majority of the powerhead overhaul procedure 25-55 hp (737cc) and 25-70 hp (913cc) motors is fairly easy and straightforward, the connecting rod cap bolt torque procedure requires a special holding fixture. The procedure and fixture are used in order to ensure that the connecting rod/cap and bearing are properly centered. Attempting to center and tighten the connecting rod caps without this fixture may cause unexpected, total powerhead failure sometime after the motor is returned to service. That is not to say people have not succeeded in centering and tightening the caps without the fixture, but to say that we cannot recommend it, as there are no guarantees.

■ **To simplify assembly, remember to layout all bolts, components and clamps in the order of removal. This is especially true for the wiring harness and related clamps. Matchmark all component assemblies such as pistons, connecting rods to ensure installation of the correct pairs and installation in the correct orientation. Most of all, take your time.**

  1. Remove the powerhead from the lower unit as detailed under Removal & Installation.
  2. If not done already, strip the powerhead for overhaul by removing all fuel system, ignition system and electrical components from the powerhead. For details, refer to the procedures in the Fuel System and the Ignition and Electrical System sections.
  3. Remove the bolts from the intake manifold and remove the intake manifold. Remove and discard the gasket.

■ **Do not bend and leaf valves by hand as this could damage them so they either do not seal properly at best, or at worst, will break off during service. Do not disassemble the leaf plate unless some portion of the plate and valve assemblies is corroded or damaged and requires replacement. ALSO, check part availability before starting, since the valves ARE replaced and not serviced on some models.**

  4. Inspect the leaf plate assemblies and the intake manifold, as follows:
   a. Two screws secure each leaf plate to the manifold. If necessary, loosen the screws and remove the assembly. Remove and discard the gaskets.
   b. Visually check for obvious signs of distortion on the valves or plates. Inspect the valve tips for signs of cracks, chips or other obvious damage. Make sure the leaf stops are not distorted or loose. If there are any obvious defects, that component must be replaced.
   c. Use a machinist's straightedge and a feeler gauge to check the leaf plate for distortion. Lay the straightedge across the plate only (not the leaf valves) at various points, then see if a 0.003 in. (0.08mm) or large feeler gauge can be inserted underneath the straightedge at any point. If it does, the plate is warped and must be replaced.

POWERHEAD REPAIR AND OVERHAUL 6-53

d. Check the leaf plate screws to make sure they are tight. If they have loosened at all, retighten each one individually as follows. Loosen the screw a turn or two, just enough to apply a light coating of OMC Screw Lock or an equivalent threadlock, then tighten the screw to 25-35 inch lbs. (3-4 Nm) and proceed on to the next screw.

e. Visually check the intake manifold gasket surfaces to make sure they are smooth and free of nicks. If necessary, a gasket surface may be dressed to smooth shallow nicks using a fine emery cloth.

f. Use the straightedge and feeler gauge to check for warpage across the intake manifold gasket surfaces. Check at the center of the straightedge to make sure a 0.004 in. (0.10mm) or large feeler gauge cannot be inserted underneath the straightedge.

g. Check the manifold balance passage to make sure it is free of restrictions.

h. If disassembly is necessary, matchmark the components and remove the leaf stop screws, then remove the leaf valves and stops. If they are being reused, do not allow the leaves and stops to become mismatched.

✽✽ WARNING

Do not lift or bend the leaf valves during disassembly or they could become damaged. If a damaged leaf valve is installed it could break in use.

5. On 2-cylinder motors, remove the starboard lower engine cover mount bracket.

6. On 45/55 Commercial 2-cylinder models through 1994, remove the 14 screws retaining the cylinder head water jacket cover (using multiple passes of a sequence that starts at the edges and works its way inwards), then remove the cover and discard the gasket.

7. Using multiple passes in the reverse of the illustrated torque sequence, loosen and remove the cylinder head retaining bolts. Remove the cylinder head from the block. As necessary, remove the cylinder head cover and disassemble the thermostat. For details, refer to the Thermostat procedures in the Lubrication and Cooling System section.

8. Loosen and remove the exhaust cover retaining screws and then remove the exhaust cover from the side of the cylinder block. Remove and discard the gasket.

9. Loosen and remove the crankcase lower head retaining screws and then remove the lower head. Use a small puller and puller bridge to remove the seal(s).

■ Lower head and seal design vary slightly with model and year. Most late model motors are equipped with a single crankshaft seal in the lower head, while most early model motors contain both a driveshaft and a crankshaft seal in the lower head. Refer to the accompanying illustrations for more details.

Fig. 75 Remove the bolts and loosen the exhaust cover gasket seal using a rubber mallet...

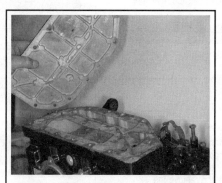
Fig. 76 ...then lift the exhaust cover from the powerhead

Fig. 77 Loosen the main bearing and then the crankcase flange (shown) bolts

Fig. 78 To remove the lower head, first loosen the retaining bolts...

Fig. 79 ...then, if necessary, carefully pry at the seam...

Fig. 80 ...and remove the crankcase lower head

Fig. 81 Label all components before removal

Fig. 82 Loosen the connecting rod cap bolts while alternating from side-to-side...

Fig. 83 Then push upward on the cap...

6-54 POWERHEAD REPAIR AND OVERHAUL

Fig. 84 ... to remove the cap and bearing from the crankpin

Fig. 85 Carefully lift the crankshaft from the cylinder block

Fig. 86 Use care when removing the pistons

10. Either use a fabricated taper pin tool (as shown in the accompanying illustration) to carefully push the crankcase taper pin toward the intake manifold surface or use a punch with a diameter larger than the pin. The taper pin can be accessed at the lower portion of the crankcase (gearcase end) on 2-cylinder motors or at the upper portion of the crankcase (flywheel end) for 3-cylinder motors.

✱✱ WARNING

NEVER use a tool smaller than the taper pin bore to remove the pin or damage could occur.

11. Using multiple passes of a spiraling pattern that starts from the outward retainers and works toward the center, loosen the main bearing bolts and, on some models, nuts.
12. Loosen and remove the crankcase flange bolts.
13. Rotate the powerhead so the crankcase is facing upward, then use a plastic or rubber mallet to tap upward on the exposed portion of the crankshaft (at a 90° angle to the shaft). Tap lightly until the crankcase half **just** starts to separate from the cylinder block half. Lift and remove the crankcase from the bottom of the cylinder block (exposing the crankshaft).

■ Keep in mind that all wear parts (such as pistons, connecting rods, needle bearings and bearing liners) MUST be installed in their original positions (and with the same orientations) if they are to be reused. Mark each component with a permanent marker to ensure proper assembly.

14. Using a felt-tipped marker or numbered/lettered punches, label the piston domes. Remember that the No. 1 cylinder is the cylinder closest to the flywheel.

■ If a felt-tipped marker is used to label components during disassembly, watch the marks during cleaning. Replace any marks that are dulled or removed by solvent as soon as the component is dry.

15. Matchmark the connecting rod caps and then use a suitable socket (such as OMC Torque Socket #331638) to loosen the connecting rod cap bolts. Work on one piston at a time, supporting the piston at the dome with one hand while loosening the bolts on that piston's rod cap alternately, no more than one turn at a time. Remove each cap along with the connecting rod cap needle bearing and cage. Carefully push the piston from the bore and out the top of the cylinder block.

✱✱ WARNING

When removing each piston, take great care to prevent the connecting rod from contacting and damaging the crankshaft journal, cylinder wall and, finally, once removed, the cylinder piston skirt. One method to help prevent this is to stuff rags in the skirt before or after the piston is removed to keep the connecting rod centered.

16. Repeat the previous step for each of the other pistons. To prevent any possible mix-up reinstall each bearing assembly and connecting rod cap to the piston and connecting rod as soon as it is removed from the cylinder block.
17. Lift the crankshaft assembly carefully out of the cylinder block.

Fig. 87 Before crankshaft installation, push the pistons to the top of their bores (placing the connecting rods so they won't interfere with the shaft

18. Remove the upper main crankshaft bearing by sliding it off the shaft, then remove and discard the O-ring. If necessary (and we advise it), remove and discard the seal.
19. Remove the retaining rings, then remove the center main bearings and split sleeves. Be sure to matchmark/identify all components and keep the assemblies separate if they are going to be reused.
20. For 1999 and later models, remove O-ring from the end of the crankshaft sleeve. Inspect the O-ring and, if worn or damaged, install a new O-ring during assembly. On these models, check the crankshaft sleeve. If damaged, remove it using a large-jawed puller and either a puller bridge or a slide hammer.
21. ONLY remove the lower crankshaft bearing if it requires replacement. First remove the retaining ring using a pair of external snapring pliers, support the bearing in a shop press using a bearing separator tool, and then press the bearing off the shaft.
22. Disassemble each piston for further inspection and component replacement, as follows:
   a. Use a universal ring expander to remove the rings from the pistons. The manufacturer warns NOT to reuse the rings. Retain them in sets for inspection purposes (marking their original locations), but discard them before installation.

■ We've said it before. We know there will be some circumstances that will prompt people to reuse rings (for example, if for some reason the motor was disassembled after only a few hours of usage). But for the most part, if you've come THIS far, it is probably a good idea to bite the bullet and replace the rings to ensure long-lasting performance from the powerhead.

   b. Using a pair of modified needle-nosed pliers as shown in the accompanying illustration, or a suitable pair of snapring pliers, remove the wrist pin retaining rings. Discard the old retaining rings.

POWERHEAD REPAIR AND OVERHAUL 6-55

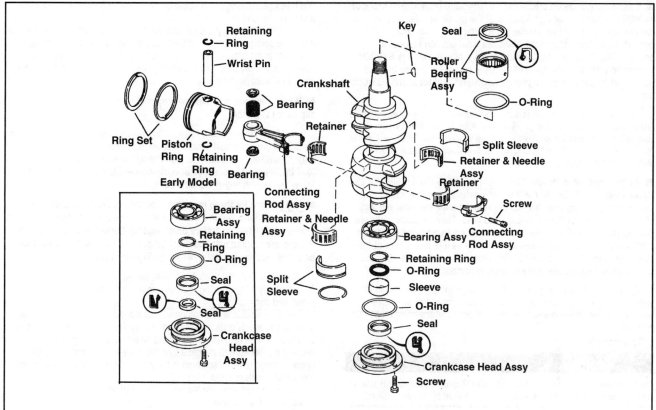

Fig. 88 Exploded view of a the crankshaft, main bearing and piston assembly - 25-55 hp (737cc) motors (late-model shown, early model differs in lower head and seal assembly)

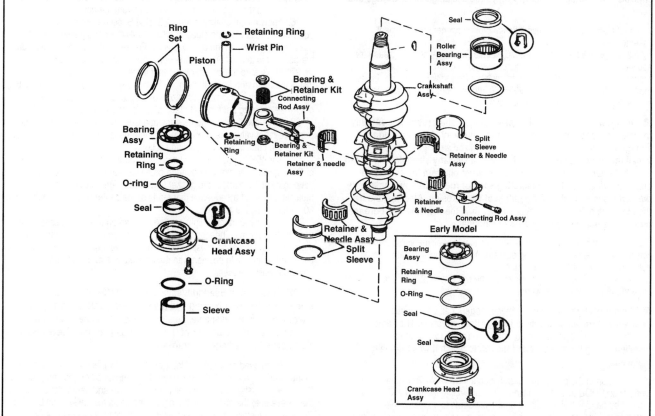

Fig. 89 Exploded view of a the crankshaft, main bearing and piston assembly - 25-70 hp (913cc) motors (late-model shown, early model differs in lower head and seal assembly)

POWERHEAD REPAIR AND OVERHAUL

■ If necessary, a piston cradle can be fabricated from a large block of wood. The important design features of the cradle are that it is semi-circular on top to hold (or cradle, get it?) the piston and that there is a bore at the bottom center into which the wrist pin can tapped or pressed. The final important feature of the cradle is that it is STURDY enough to withstand the force of the shop press, BUT, keep in mind that the fit is loose on these pistons and significant force should not be necessary.

    c. The piston-to-wrist pin fit should be loose on both sides. Support the piston and use a suitable drift pin (like OMC #326356 to carefully and smoothly tap or push the wrist pin free of the piston bore). If the OMC tool is not available, be sure to use a drift pin that is just slightly smaller than the diameter of the wrist pin itself.

■ As the wrist pin is removed from the piston you must retain the 28 loose needle bearings and the 2 thrust washers from the connecting rod bore. If you can, stop tapping or pressing once the wrist pin JUST clears the bottom of the connecting rod bore. Then carefully withdraw the rod along with the needle bearings (place a thin bladed tool like a putty knife or your finger under the rod hole as it is pulled free of the piston bore/wrist pin). Inspect the bearings, if any of the needles are lost or worn, replace the entire bearing assembly.

    23. Refer to the information found under Cleaning & Inspection, in this section for details on inspecting the powerhead components to determine which should be replaced and to prepare the ones that are being reused for assembly.

To assemble:

✱✱ WARNING

During assembly, take your time and NEVER force a component unless a shop press or drive tool is specifically required by the procedure.

■ Lightly coat all components (except those that are coated with sealant) with clean, fresh Johnson/Evinrude Outboard Lubricant or equivalent oil during assembly. Always replace all gaskets, O-rings and seals.

    24. If disassembled for inspection and/or component replacement, assemble each piston as follows:
    a. Apply a light coating of OMC Needle Bearing Assembly grease (or equivalent) to the wrist pin bearing needles, then insert the 28 needles around the inner circumference of the connecting rod wrist pin bore. Install the OMC Wrist Pin Bearing tool #336660 into the bore to align the needles and hold them in place during assembly.

■ The wrist pin bearing tool is, essentially, a short round pin, roughly the same outer diameter of the wrist pin, but very short, just long enough to insert it through the connecting rod wrist pin bore and install the thrust washers on either side. A similarly sized dowel pin can be substituted, or a old connecting rod wrist pin from the same size piston could be used. If substituting another rod or drift pin, make sure the surface is smooth and clean/free of all corrosion. Also, make sure the outer diameter of the tool is smaller than the wrist pin and bore to prevent the possibility of binding or damage.

    b. With the bearings and the wrist pin tool in position, install the thrust washers with the flat sides facing outward. Again, a dab of grease should hold the thrust washers in position.
    c. Apply a light coating of clean engine oil or assembly lube to the wrist pin and the pin bore in the piston.

■ The connecting rod itself on these models has no specific orientation and can be installed facing either direction. The same is NOT true for the caps, which are matched to the rods and must be installed facing the same side of the rod to which it was originally installed.

    d. Support the piston (in a piston cradle if available, though a clean workbench SHOULD suffice). Insert the wrist pin into the bore and gently press it by hand through the bore until it just starts to appear inside the piston skirt. Stop, then position the connecting rod with the needle bearings, thrust washers and wrist pin bearing installation tool inside the piston skirt. Continue to push the wrist pin into the position, through the thrust washers and bearings, pushing the wrist pin bearing installer through the other side and out of the piston.
    e. Once the wrist pin is properly positioned, use a small driver to install a NEW snapring in each end of the wrist pin bore. Make sure the snapring fully seats in the groove. Turn the ring in the groove until the gap faces downward, directly away (180° opposite) from the semi-circular relief in the side of the wrist pin bore.

■ When using the ring expander, spread each ring JUST enough to slip over the piston.

    f. Use a ring expander to install the pistons rings on each piston. Each piston ring groove should contain a dowel pin, around which the ring gap must be squarely situated. If not the rings will likely break during installation (or if by they don't by some miracle, during the first start-up). Be sure to install the tapered ring in the top groove.

■ The pistons for these motors should be installed, not only in their original positions, but in their original orientations (with the same sides facing the flywheel as before disassembly). Luckily, they are marked. Pistons are marked with the word UP on their domes, this word should be positioned toward the flywheel during installation.

    25. If removed, install each piston as follows:
    a. Apply a light coating of clean engine oil or assembly lube to the piston and piston bore.
    b. Use a suitable ring compressor or a band-type clamp with a solid inner ring to evenly compress the rings (oiling the inside of the compressor or clamp will also help ease installation). Make sure the rings are still properly positioned with the gaps over the dowels. Also, make sure the connecting rod remains centered in the piston to prevent possible damage to the cylinder walls or piston skirt.
    c. Start the piston into the bore with the **UP** marking on the piston dome facing the flywheel.
    d. Slowly insert the piston, making sure the connecting rod does not strike and damage the piston skirt, or the strike/scrape and damage the cylinder wall. If the piston hangs up on a ring, don't force it, push the piston back upward and remove the compressor to check for potential problems, such as a ring that has shifted the gap away from the dowel.
    26. Apply a light coating of OMC Gasket Sealing Compound, or equivalent sealant to both sides of a NEW cylinder head gasket, then position the gasket on the cylinder block. For 2-cylinder motors, make sure the gasket tab is positioned on the port side of the cylinder block.
    27. Make sure the cylinder head bolt threads are clean and dry, then install the head and finger-tighten the bolts. Torque the bolts to 18-20 ft. lbs. (25-27 Nm) using multiple passes of the illustrated torque sequence (a clockwise spiraling pattern that starts at the center and works outward).
    28. If removed, install the thermostat assembly and, if applicable, the pressure relief valve assembly to the cylinder head/cover. For details, please refer to the Lubrication and Cooling section.
    29. For 45/55 hp Commercial models through 1994, apply a light coating of OMC gasket sealing compound to both sides of a new cylinder head water cover gasket, then install the gasket and cover. Tighten the bolts using multiple passes of a sequence that starts at the center and works outward to 60-84 inch lbs. (7-9 Nm).
    30. If removed, apply a light coating of clean engine oil or assembly lube to a NEW lower main bearing and to the end of the crankshaft, then support the crankshaft in a shop press between the 2 lower counterweights. Carefully press the bearing into position with the lettered side of the bearing facing upward toward the press.

■ OMC Crankshaft Bearing/Sleeve installer #342686 can be used to safely seat the lower bearing and the sleeve in position on 1999 and later models. If not available, be sure to use a suitable driver that contacts the entire diameter of the bearing cage or the sleeve and does NOT contact or damage the crankshaft.

    31. If removed on 1999 and later models (for replacement or for access to the bearing), apply a light coating of clean engine oil or assembly lube to the NEW crankshaft sleeve. Install the new sleeve and gently drive or press the sleeve onto the shaft until the installer contacts the lower main bearing. Remove the installer, then apply a light coating of OMC Moly Lube or equivalent assembly lube to the NEW O-ring (which should be used if a new sleeve is installed) and position it in the end of the sleeve.

POWERHEAD REPAIR AND OVERHAUL 6-57

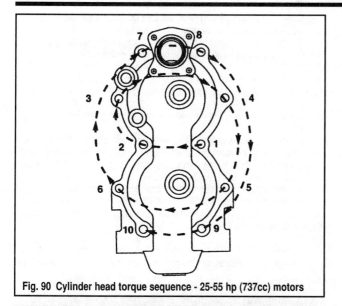

Fig. 90 Cylinder head torque sequence - 25-55 hp (737cc) motors

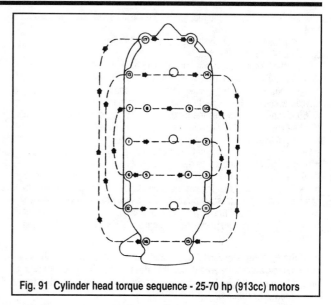

Fig. 91 Cylinder head torque sequence - 25-70 hp (913cc) motors

✱✱ WARNING

Make sure the sleeve surface is not nicked or damaged during installation or it must be removed and discarded (and another NEW sleeve installed).

■ If the installer sticks on the sleeve after installation, use a slide hammer to pull it free.

32. Install the lower bearing retaining ring in the crankshaft groove with the sharp (square) edge facing away from the bearing.
33. Oil the center main roller bearing and split sleeve assemblies, then install them around the crankshaft center journals in their original positions. Be sure to face the split sleeve ring grooves away from flywheel. Secure the sleeves using the retaining rings.
34. Apply a light coating of OMC Gasket Sealing Compound, or equivalent sealant to the outside diameter of a NEW upper crankshaft seal and the lower crankshaft seal (or for 1998 and earlier models, the 2 NEW crankshaft/driveshaft seal).
35. Use suitably sized drivers to install the seals to the upper bearing and to the crankcase lower head with the seal lips properly positioned. Upper seal casings should face upward (toward the flywheel), while lower seal casings (including both the crankshaft and driveshaft seals on models so equipped) should face downward toward the gearcase. Use the following drivers or equivalents:
• Upper seal: use #334500 for 2-cylinder models through 1998 or #326567 for all other models.
• Lower seal: use #339752 on all except the 1990-93 2-cylinder models or 1990-98 45 and 55 hp 2-cylinder models (or #334998 on all 1990-93 2-cylinder models or 1990-98 45 and 55 hp 2-cylinder models).
36. Once installed, lubricate the upper crankshaft seal lips using a light coating of OMC Triple-Guard, or equivalent marine grease. Lubricate the lower seal lips liberally using OMC Moly Lube or equivalent assembly lube.
37. Apply a light coating of OMC Triple-Guard, or equivalent marine grease to a NEW upper seal O-ring and a NEW lower crankcase head O-ring. Position the O-rings in the grooves in the upper bearing and the crankcase lower head, respectively.
38. Apply a light coating of clean engine oil or assembly lube to the upper main bearing and to the top end of the crankshaft, then slide the seal/bearing assembly onto the top of the shaft with the seal end facing the flywheel.
39. Inspect the cylinder block-to-crankcase mating surface to ensure there are NO traces of sealant left. If necessary, remove any remaining traces of hardened sealant to prevent potential problems with bearing alignment.
40. Rotate the cylinder block the head faces downward and the crankcase mating surface is facing upward.
41. Carefully push all pistons downward (up in their bores) to Top Dead Center (TDC) in order to ease crankshaft installation.

42. Remove the connecting rod end caps and cap bearing halves (positioning them aside to keep track of their location and orientations). If matchmarks are not readily visible, matchmark the connecting rods to the caps and bearing halves before removal to ensure proper installation.

■ Leave the connecting rod bearing halves (as opposed to the rod cap halves) in position in the rods.

43. If not already done, apply a light coating of clean engine oil or assembly lube to the crankshaft journals, as well as to the connecting rod bearings.
44. Lower the shaft into the position. Be sure the upper O-ring aligns with the groove and the main bearings align with their locating pin(s).
45. Slowly and carefully pull each connecting rod up until the bearing in the rod contacts the crankshaft journal (crankpin).
46. Seat each rod journal against the crankpin and then move the bearings from the caps to the connecting rod journals. Finally, align the matchmarks and position the caps over the bearings.

■ Besides any matchmarks made during removal, the connecting rod caps on these models are normally equipped with dots on the connecting rod/cap. These dots should also be aligned.

47. Oil and install the connecting rod cap bolts, then finger-tighten the bolts alternating from side-to-side on the rod cap.
48. Use a pencil or your fingernail to determine if the rod caps are fully aligned and seated. Scrape the pencil across the mating surface to see if it catches.
49. Center and align each of the connecting rod caps in turn using the OMC Rod Cap Alignment Fixture #396749 as follows:
  a. Rotate the knob on the top of the tool until the flat marked **SET** aligns with the arrow on the fixture frame. Rotate the adjustment knob 180° to the lock position.
  b. Make sure the retaining jaw (with the flat head screw in the frame bore) and the forcing jaw (with the hex key at the base and the threaded flat protruding from the frame) are positioned on the fixture tool.
  c. Tighten the connecting rod cap screws to 25-30 inch lbs. (2-3 Nm).
  d. Apply a light coating of engine oil to corners of the connecting rod and cap where the fixture will be installed.
  e. Position the frame to the so the contact area of the forcing jaw is centered on the connecting rod and cap. Tighten the forcing screw until the jaws contact the connecting rod, then slide the frame downward until the adjustment stop (at the center of the frame) just contacts the top of the connecting rod cap. Verify that the groove lines in the jaws are centered on the rod/cap diameter.

✱✱ WARNING

The frame MUST be squarely in position as described in order to perform its function properly. Failure to center the frame, and therefore the connecting rod/cap, will lead to premature crankshaft, rod and bearing failure.

6-58 POWERHEAD REPAIR AND OVERHAUL

f. With the frame centered, tighten the forcing screw to 23 inch lbs. (2.5 Nm).

g. Loosen both connecting rod cap bolts 1/4 turn, then tighten each bolt to 40-60 inch lbs. (5-7 Nm).

h. Next, tighten the connecting rod cap bolts to 15-17 ft. lbs. (20-23 Nm), and finally tighten them to 30-32 ft. lbs. (40-43 Nm).

i. Loosen the forcing screw, remove the frame from the connecting rod/cap assembly, then repeat for the remaining rod(s)/cap(s).

50. Clean and degrease the mating flange of the crankcase and cylinder block halves using OMC Cleaning Solvent, or an equivalent solvent. Allow the flanges to air dry.

51. Apply a light coating of OMC Locquic Primer or equivalent to the mating flange of the crankcase half and allow it to air dry.

52. Apply a light coating of OMC Gel-Seal II or equivalent to the cylinder block flange. Be sure that the sealant coats the entire flange evenly, but be careful not to apply excessive amounts. The sealant must not be applied within 1/4 in. (6mm) of the labyrinth seal or bearings otherwise the sealant could be forced out from between the flanges into contact with the bearings during assembly.

■ Although OMC Gel-Seal II has a shelf life of one year when stored at room temperature, buy a new tube if in doubt. Keep in mind that using an old tube of Gel-Seal II could allow crankcase air leaks (leading to performance problems).

53. Carefully lower the crankcase half into place over the cylinder block half.

54. Install and finger-tighten the main bearing bolts and, if applicable, nuts.

55. Make sure the crankcase is stead, then install and seat the crankcase taper pin.

56. Using a plastic or rubber faced mallet, gently tap on the bottom of the crankshaft to seat the lower main bearing.

57. Apply a light coating of OMC Gasket Sealing Compound, or equivalent sealant to the threads of the crankcase lower head retaining bolts. For 2-cylinder motors through 1998, also apply the sealant to the mating surface of the lower head flange.

58. Install the crankcase lower head and finger-tighten the bolts

59. Tighten the main bearing screws to 18-20 ft. lbs. (24-27 Nm) using multiple passes of a spiraling sequence that starts with the center screws and works outward.

60. Temporarily position the flywheel and key on the end of the crankshaft, then rotate the crankshaft clockwise (when viewed from above the flywheel) to check for binding. The crankshaft and main bearings must move smoothly or the cause must be found and remedied before returning the motor to service.

61. Install the crankcase flange screws and, if applicable, the pan support brackets, then tighten to 60-84 inch lbs. (7-9 Nm).

62. Tighten the crankcase lower head screws to 60-84 inch lbs. (7-9 Nm) for models through 1998 or to 96-120 inch lbs. (11-14 Nm) for 1999 and later models.

63. Apply a light coating of OMC Gasket Sealing Compound, or equivalent sealant to both sides of a new exhaust cover gasket, then install the cover and gasket. Tighten the retaining bolts to 60-84 inch lbs. (7-9 Nm) using a spiraling pattern that starts at the center and works outward.

64. If disassembled for component replacement, prepare and assemble the leaf plates as follows:

a. Place the leaf valves on the plate. If new valves do not seat on the plate, try turning them over.

✱✱ WARNING

NEVER turn over used valves on the plate, as they may break when they are returned to service. If used valves do not seat in their original direction of installation, they must be replaced.

b. Check the valves, if any leaves are standing open, apply light pressure using the eraser end of a pencil (the valve should close with light pressure). If not, check the plate for high spots or burrs.

✱✱ WARNING

NEVER lap the leaf plate. An plate that is too smooth may cause the leaf valves to stick closed when returning a motor to service after winterization.

c. Apply a light coating of OMC Locquic Primer or equivalent to the valve screws and allow it to air dry.

d. Apply a light coating of OMC Screw Lock, or equivalent threadlock to the mounting screw threads, wiping off any excessive adhesive compound.

e. Assemble the leaf valve shim (if used) and the leaf stop, then loosely install (but do not tighten) the retaining screws.

f. Hold the leaf valve assembly in a horizontal position (this will usually align the leaf valves over the ports in the leaf plate). Tighten the 2 screws evenly.

g. Mark the edges of the leaf valve using a sharp pencil, then use the pencil eraser to gently open the leaf valves to check alignment over the port. Once again, the valves must be spaced evenly over the port. If not, loosen the mounting screws and reposition the valve.

h. Once centered, tighten all the screws to 25-35 inch lbs. (2.8-4 Nm).

i. Apply a light coating of OMC Locquic Primer or equivalent to the valve assembly-to-manifold screw threads and allow it to air dry. Then, apply a light coating of OMC Nut Lock, or equivalent threadlock to the screw threads.

j. Install the leaf valve assemblies to the manifold using NEW gaskets (dry, without sealer) and tighten the retaining screws to 25-35 inch lbs. (2.8-4 Nm).

65. Install the intake manifold using a new gasket (dry, without any sealant), then install and tighten the retaining bolts to 60-84 inch lbs. (7-9 Nm).

66. For 2-cylinder motors, install the starboard lower engine cover mount bracket and tighten the retainers to 60-84 inch lbs. (7-9 Nm).

67. Install all fuel, ignition and electrical system components that were removed for access/overhaul, as detailed in the Fuel System and Ignition and Electrical System sections.

68. Install the powerhead assembly, as detailed in this section, then refer to Break-In in this section for details on how to ensure proper break-in of the new powerhead components.

40/50 Hp (815cc) EFI Motors

◆ See Figure 92

Removal & Installation

◆ See Figures 92 and 93

1. For safety, properly relieve the fuel system pressure. Please refer to Fuel System Pressurization under Fuel Injection for details.

2. Disconnect the negative battery cable for safety.

3. Remove the fuel inlet hose at the connector.

4. Remove the flywheel cover.

5. Remove the air intake silencer. For details, please refer to the Fuel System section.

6. Remove the lower engine covers. For details, please refer to the Engine Covers (Top and Lower Cases) in the Maintenance and Tune-Up Section.

7. Remove the Electronic Control Unit (ECU) cover and then disconnect the blue and green power trim/tilt relay wires. Disengage the trim/tilt sender connector.

8. At the front of the motor, remove the screws retaining the front panel, then move the panel for access, sliding it back from the linkage. Remove the screws securing the shift linkage and the screws retaining the shift linkage. Disconnect the shift arm from the shift rod.

9. Squeeze the tabs of the spring-loaded clamp for the cooling system indicator hose on the powerhead nipple, then slide it up the hose, off the fitting. Tag and disconnect the hose from the nipple.

10. Remove the retaining pins and the lower cover seal.

11. Tag and disconnect the thermostat-to-adapter hose.

12. Loosen and remove the small bolts (3), large bolts (5) and the nut retaining the powerhead and oil pump case.

■ The powerhead retaining bolts are threaded up from underneath the engine holder flange.

13. Attach a suitable engine hoist to the hook provided on top of the cylinder block, then carefully lift the powerhead straight up and off the engine holder. Keep the powerhead as level as possible while lifting it to avoid damage to the driveshaft or gearcase components.

14. Mount the powerhead in a suitable engine stand.

POWERHEAD REPAIR AND OVERHAUL 6-59

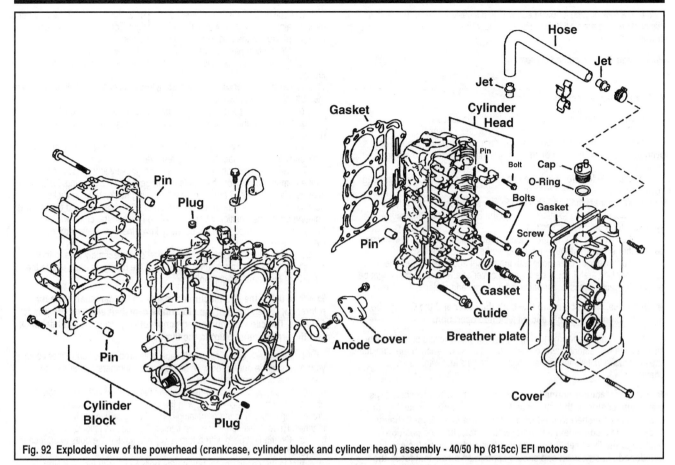

Fig. 92 Exploded view of the powerhead (crankcase, cylinder block and cylinder head) assembly - 40/50 hp (815cc) EFI motors

15. If the oil pump is to be serviced, loosen the clamp and disconnect the water hose from the nipple, then loosen the bolts (7) securing the oil pump case to the bottom of the powerhead. Tap the case with a rubber mallet to break the gasket seal and then carefully pry the case from the bottom of the powerhead. Refer to the Oil Pump procedures in the Lubrication and Cooling section for more details.

16. Remove all traces of gasket or sealant from the powerhead, oil pump case and/or the engine holder. Clean and inspect the mating surfaces for signs of damage or extensive pitting/corrosion. Replace the case or holder if damage or corrosion is extensive.

To install:

■ Before installing the oil pump case, make sure the 2 dowel pins are installed in the lower end of the cylinder block. For details, refer to the accompanying illustration.

17. If removed, install the oil pump case to the powerhead. Follow the steps of the Oil Pump procedure regarding application of sealant to the bolts and mating surfaces, as well as for the bolt torque pattern (a star-shaped crossing pattern working from the inner bolts outward). For details, refer to the Lubrication and Cooling section.

18. Connect the water hose and secure using a clamp.
19. Apply a light coating of OMC Triple-Guard, or equivalent marine grease to a NEW O-ring and install to the tube on the engine holder.
20. Check the engine holder dowel pins for looseness or damage and remove/replace, as necessary. Position a new powerhead gasket over the engine holder dowel pins.
21. Apply a light coating of OMC Moly Lube or equivalent to the crankshaft splines.
22. Use the engine hoist to carefully lower the powerhead onto the engine holder. Work slowly, taking care to keep the powerhead as level as possible to prevent damaging the driveshaft or other gearcase components.
23. Apply a light coating of OMC Gel-Seal II to the threads of the powerhead bolts and mounting stud.
24. Install and finger-tighten the retainers, then torque the retainers to specification. Tighten the small bolts to 16-17 ft. lbs. (22-23 Nm) and the large bolts (as well as the stud nut) to 35-37 ft. lbs. (47-50 Nm).

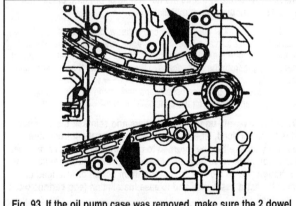

Fig. 93 If the oil pump case was removed, make sure the 2 dowel pins are installed in the lower cylinder block mating surface, as shown

25. Connect the thermostat-to-sump hose, then secure using the clamp.
26. Position the lower engine cover seal, then route the cooling system indicator hose through the seal and connect it to the fitting. Secure the hose using a clamp.
27. Install the shift arm on the rod. Slide the link through the linkage, then install the screws and tighten securely. Install the front panel and tighten the screws securely.
28. Connect the trim/tilt relay wires (blue on the top and green on the bottom) using the lock-washers and nuts. Tighten the nuts securely.
29. Install the lower engine covers and attach the fuel inlet hose to the connector.
30. Install the air intake silencer assembly.
31. Install the flywheel cover.
32. Connect the negative battery cable and properly pressurize the fuel system. Thoroughly inspect the system for fuel leaks.
33. Install the top engine cover.

6-60 POWERHEAD REPAIR AND OVERHAUL

34. If a new or rebuilt powerhead was just installed, refer to Break-In in this section for details on how to ensure proper break-in of the new powerhead components.

Powerhead Disassembly & Assembly

◆ See Figures 92 and 94

1. Remove the powerhead, as detailed in this section.
2. Remove the timing chain, as detailed in this section.
3. If not done already (while striping the powerhead for removal or overhaul), remove any components that will interfere with valve cover removal:
   a. Remove the breather hose and ignition coils from the cylinder head cover.
   b. Disconnect the harness from the camshaft position sensor. For details, please refer to the information on the Electronic Fuel Injection system in the Fuel System section.
   c. Loosen and remove the cylinder head cover bolts using a spiraling pattern that starts at the outer bolts and works towards the inside.
   d. Remove the cover along with the gasket and O-rings from the powerhead. Be careful not to damage the cover or any O-rings that might be reused. Inspect the gasket for damaged surfaces and replace, if necessary.

■ Of course, we recommend that both the cover and the O-rings be replaced to ensure reliable and oil leak free operation.

4. Position the powerhead with the cylinder head facing upward, then either locate the factory marks or make matchmarks on each of the camshaft caps to ensure installation in their original locations and with the same orientation.

■ The manufacturer normally placed marks on these camshaft caps that would ensure both location and orientation during assembly. Each cap should be etched with either an I for Intake or an E for Exhaust. They should also contain a number which indication the position relative to the flywheel (higher the number, further away the cap is positioned). Lastly, each cap should contain an arrow that faces the flywheel. If any of these marks are not visible or present, place your own marks to ensure proper installation.

5. Loosen the all of the camshaft cap bolts (working on both shafts simultaneously) 1/4 turn at a time starting at the inner bolt for the center caps, then moving to the outer bolts for the center caps. Repeat for each of the caps, moving from the center caps to the outer caps, then repeat, until the bolts are all finger-loose.
6. Remove each of the camshaft caps and arrange them on a clean worksurface, then carefully lift the camshafts from the cylinder head.

■ Don't loose or mix-up the valve tappets and shims. If the cylinder head is to be serviced, label each one and remove them from their bores. They must either be returned to their original bores, or replaced during a complete valve lash adjustment procedure during installation. Even though you've labeled them, keep them sorted or arranged after removal for extra assurance and to ease installation (egg cartons work well for this).

7. Remove the intake manifold, as detailed in the Fuel System section.
8. Using a deep well socket (normally 10mm on these motors), loosen and remove the cylinder head bolts using multiple passes in the reverse of the torque sequence.
9. Carefully lift the cylinder head from the block. If the gasket seal is difficult to break, start by using a rubber or plastic mallet to tap around the edges and then carefully pry to help free the head.
10. Refer to Cylinder Head Overhaul, in this section for details on disassembling the head for component inspection and replacement.

To assemble:

11. Make sure the cylinder block is at No. 1 TDC, meaning that the No. 1 cylinder is at or near the top of its travel. Check the crankshaft sprocket to ensure that the dimple is facing the block timing mark that is opposite the cylinder block.
12. Position a NEW head gasket over the 2 dowel pins in the cylinder block.
13. Carefully lower the cylinder head onto the cylinder block over the gasket and dowels. Make sure you do not disturb the gasket positioning when seating the head.

14. Apply a light coating of clean engine oil to the threads of the cylinder head retaining bolts, then install them finger-tight. Torque the bolts using multiple passes of the torque sequence as follows:
   a. On the first pass, tighten all bolts to 21 ft. lbs. (30 Nm).
   b. On the next pass, use the reverse of the torque sequence too loosen each of the bolts again.
   c. On the next pass, use the normal torque sequence again and tighten the bolts to 21 ft. lbs. (30 Nm).
   d. On the final pass, tighten all bolts to 43-44 ft. lbs. (59-60 Nm).
15. Make sure the camshaft cap dowel pins properly installed, as noted during removal.
16. If the valve tappets and shims were removed, install them back into their original bores as labeled during removal.
17. With the camshaft sprockets installed (loosely in the case of the intake camshaft), position the shafts for proper No. 1 TDC timing. Hold the shaft just over the head and turn it so the timing mark on the sprocket is pointing down (toward the mark on the head) when installed and the arrow is pointing up (to the position where the colored links in the chain will be installed). In this timed position, carefully lower the camshafts into the cylinder head.
18. Apply a light coating of clean engine oil or assembly lube to the camshaft journals, to the bearing surfaces of the camshaft caps and to the threads of the camshaft cap bolts.

■ When using the manufacturer's marks on the camshaft caps for reference, remember that the I for Intake camshaft side and the E is for Exhaust. The caps are numbered consecutively from the flywheel back and the arrows must point to the flywheel.

19. Install each of the camshaft caps in its proper position (either using your marks made during removal or the letters, numbers and arrows that were on the components before assembly).
20. Install the bolts and finger-tighten them starting with the middle caps on both camshafts (and starting with the inner bolt on each cap) then proceeding outward to the remaining caps of both camshafts in the same manner. Repeat this sequence and tighten the bolts as follows:
   a. On the first pass with a torque wrench tighten the bolts to 30 inch lbs. (3.4 Nm).
   b. On the second pass, tighten the bolts to 60 inch lbs. (6.8 Nm).
   c. On the final pass, tighten the bolts to 90 inch lbs. (10.2 Nm).
21. Check that the camshafts and crankshaft are still in the No. 1 TDC position. Timing marks must align, as detailed under the Timing Chain procedures in this section. Also, a quick check of the camshaft lobes for the No. 1 cylinder will give you your answer on the cylinder head. At TDC of the No. 1 cylinder, all valves (intake and exhaust) must be closed for engine compression). Therefore, the camshaft lobes must be facing out, away from the tappets on that cylinder.

■ This is an interference motor without screw adjustable valves, the only way to safely time the motor is with the cylinder head off or ALL of the valve tappets removed.

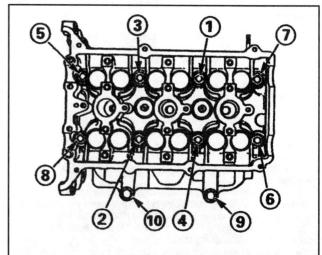

Fig. 94 Cylinder head torque sequence - 40/50 hp (815cc) EFI motors

POWERHEAD REPAIR AND OVERHAUL   6-61

22. Install the timing chain, as detailed in this section.
23. If valve train components or the camshafts were replaced, check and adjust the valve lash as detailed in the Maintenance and Tune-Up section.
24. Apply GM Silicone Rubber Sealer to the O-rings and gasket (we recommend that you use NEW O-rings and a NEW gasket, regardless of the old condition).
25. Install the valve cover using the O-rings and gasket, then install and tighten the bolts to 84-90 inch lbs. (9.5-10.2 Nm) using multiple passes of a spiraling pattern that starts at the center bolts and works outwards.
26. Install the intake manifold, as detailed in the Fuel System section.
27. Install the Powerhead, as detailed in this section.
28. Be sure to reconnect the camshaft position sensor wiring, the ignition coils and the valve cover breather hose.
29. If a new or rebuilt powerhead was just installed, refer to Break-In in this section for details on how to ensure proper break-in of the new powerhead components.

Cylinder Head Overhaul

◆ See Figures 92 and 95 thru 97

Before disassembly, refer to Cleaning and Inspection in this section for details on how to perform a Valve Leakage check. If leakage is present or there is an obvious defect, the valve train must be disassembled for inspection and component replacement.

■ **Disassembly and repair requires special tools and equipment. If not available, bring the cylinder head to a reputable machine shop with marine experience.**

Be sure to mark (and make a note, in case the marks come off during service) the mounting location and orientation of all cylinder head components prior to removal.

■ **These motors ARE equipped with progressive rate valve springs (that can only be installed in one direction). Be sure to mark the direction of installation during removal (mark or note the side of the spring that faces the cylinder head when installed).**

1. Remove the cylinder head from the powerhead as detailed under Powerhead Disassembly & Assembly, in this section.
2. If not done during powerhead disassembly, use a permanent marker to label the location and orientation of each tappet and shim, then remove them from the cylinder head.
3. Use the marker to label the location and orientation of each valve before it is removed from the cylinder head.
4. Use a valve spring compressor tool (most universal valve spring tools should work, but threaded clamp or lever type compressor is recommended) to remove each valve, spring and seal as follows:
   a. Place one clamp on the valve face and the other on the valve spring cap

■ **Do not compress the spring fully.**

b. Compress the spring **just** sufficiently to free the keepers from the stem. Remove the keepers from the valve, then slowly release the spring pressure.
c. Remove the valve compressor tool and remove valve and spring from the cylinder head. Remember to mark the spring orientation (which side faces the cylinder head).

■ **Keep all valve components together as they should not be mixed or transferred to other valves during assembly.**

d. Carefully pry the seal from the valve guide, then discard the seal.
e. Remove the spring seat.
5. If not already done, remove the thermostat from the cylinder head for inspection or replacement. For details, refer to the Thermostat procedures in the Lubrication and Cooling section.
6. Inspect and measure the cylinder head components as described in this section.

To assemble:

7. If removed, install the thermostat assembly. For details, refer to the procedures in the Lubrication and Cooling section.
8. Install the valves to the cylinder head, as follows:
   a. Position the valve spring seat to the cylinder head.
   b. Apply a light coating of engine oil to the NEW valve stem seal, then press it carefully onto the valve guide using finger pressure, making sure it is properly seated.
   c. Apply a light coating of engine oil to the valve stem, then slide it into the valve guide. Place the valve fully against the seat, then install the valve spring and cap over the valve stem. Make sure the narrow spring coil faces downward against the valve seat (as marked during removal).
   d. Use the valve spring compressor to push downward on the spring, **just** sufficiently to install the valve keepers. Once the keepers are in position, slowly release the spring pressure, allowing the cap and keepers to take the pressure. Make sure the keepers are properly seated.
   e. Repeat to install the remaining valves.
9. Apply a light coating of clean engine oil or assembly lube to each tappet and shim, then position them into the cylinder head as marked during removal.
10. Install the cylinder head as detailed in Powerhead Disassembly & Assembly in this section.

Cylinder Block Overhaul

◆ See Figures 8, 24, 25, 26, 50, 92 and 98 thru 104

The crankshaft for this motor turns on replaceable main bearing insert halves found in the cylinder block and crankcase. Like many 4-stroke motors, the connecting rods also utilize bearing inserts. All machined and wear components, including the pistons, connecting rods, rings and bearings MUST be returned to their original positions when they are reused. Matchmark all components to installation in the proper locations and in the same orientations (facing the same direction).

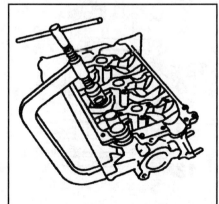

Fig. 95 Use a valve compressor tool to remove the keepers and gently release spring pressure

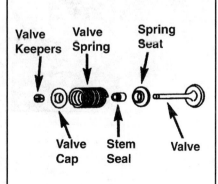

Fig. 96 Exploded view of a valve assembly - EFI motors

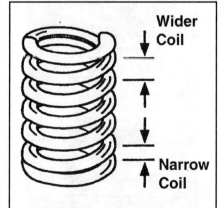

Fig. 97 Install the narrow coil end of the spring toward the cylinder head and valve seat

6-62 POWERHEAD REPAIR AND OVERHAUL

During service, the crankshaft assembly can be removed from the block without piston removal or vice-versa. If either is attempted, take EXTREME care to keep the journal surfaces from being damaged. When free of the crankshaft journals, the connecting rods should be restrained from contacting the piston bores as well. One method to accomplish this is to stuff a clean shop towel under the piston skirt. Another is to use rubber bands to hold the rods in position.

Cylinder block work is more easily performed on an engine stand, especially since this allows rotation of the motor to access pistons from above and below the cylinder block. But, work can also be performed on a clean workbench, if care is taken not to damage the mating surfaces or, in the case of the cylinder head mating surface, the dowel pins. When working on a bench, take care to prevent the block from falling over during service. Also, if the case is stood on end at any point, take great care to keep the crankshaft from falling out of the block during service. A crankshaft can suffer a fatal blow from such a fall, even from a short height.

1. Remove the powerhead from the gearcase as detailed under Removal & Installation, then remove the cylinder head as detailed under Powerhead Disassembly & Assembly.

2. Invert the cylinder block so the cylinder head mating surface is facing downward.

■ If you're not working on an engine stand, take care not to damage the cylinder head mating surface.

3. Remove the bolts (5), then carefully break the gasket seal and remove the upper seal housing from the top of the crankcase. Remove and discard the gasket.

4. Loosen the crankcase-to-cylinder block flange bolts (8) and the main bearing bolts (8) using multiple passes in the reverse of the torque sequence. This will essentially result in a crossing pattern that starts with the outer flange bolts moving toward the center flange bolts, then moves to the outer main bearing bolts and finally moves to the inner main bearing bolts.

5. Separate the crankcase from the cylinder block. If the seal is tough to break, locate the notches cast into the mating surfaces, then use 2 prybars (blunt or coated with electrical tape to help protect the surfaces) to apply light pressure helping to separate the components.

6. Note the location and orientation of each bearing half in the crankcase journals. Remove and retain the bearings. Visually check the crankcase mating surfaces, the bearing surfaces and the seal surfaces for damage or wear (no scratches big enough to catch a fingernail should be present on any of the surfaces).

7. Take note of the direction the crankshaft seal lips are facing, then carefully pry the seals free. Remove and discard the old seals.

■ To help prevent mix-ups during disassembly or inspection, once the crankshaft is removed from the block (or if the shaft is remaining in place, once each piston is removed), reinstall the connecting rod cap and bearing inserts loosely to the rods.

Fig. 98 Place lengths of rubber hose over the connecting rod studs to protect the crankshaft and cylinder walls...

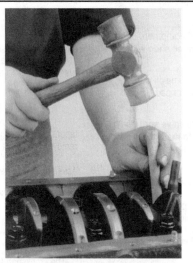

Fig. 99 ...then use a hammer and dowel to gently tap the piston to the top of the bore

Fig. 100 During installation, use a piston wring compressor to safely insert the piston and rings

Fig. 101 Check main and crankpin bearing clearances by applying a strip of gauging material, bolting the bearings in place...

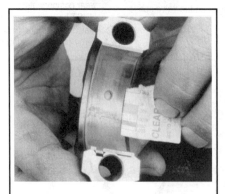

Fig. 102 ...then removing the bearings again and comparing the thickness of the material to the scale

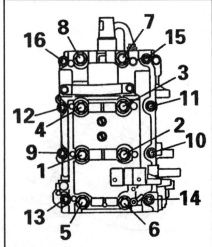

Fig. 103 Main bearing and crankcase flange bolt torque sequence - 40/50 hp (815cc) EFI motors

POWERHEAD REPAIR AND OVERHAUL 6-63

8. Using a permanent marker, label each piston, connecting rod and rod end-cap to ensure installation. Although it is unclear whether or not the connecting rods and caps are of the same fractured design used on other Johnson/Evinrude motors (especially since the EFI motors are built by Suzuki), matchmark the connecting rods to the caps and maintain the same orientation, just to be safe.

■ **If working on an engine stand, bolt a piece of wood or metal over the cylinder head surface (use a few of the cylinder head bolts) to keep the pistons from dislodging and falling free once the connecting rod caps are unbolted.**

9. Loosen and remove the connecting rod cap nuts. To prevent possible damage to the cap and bearing surface, loosen each cap by alternating between nuts, giving them about 1/4 - 1/2 turn at a time.
10. Once the nuts are removed, the cap should easily come free, but if necessary, tap it gently using a plastic or rubber mallet to free it from the rod. Remove the rod cap and then remove the bearing half from the shaft or cap. Keep the bearing with the rod cap.

■ **Wrap the connecting rod studs using tape to help protect the cylinder walls from damage.**

11. Push the pistons and connecting rods downward in their bores away from the crankshaft. Remove the crankshaft and thrust bearings from the cylinder block.
12. Remove the main bearing halves from the cylinder block. Note their locations and orientations for measurement and assembly purposes.
13. If the pistons are being removed, place the cylinder block with the head mating surface and the tops of the pistons either facing upward or sideways. (Actually, we find that upward at a slight angle is the best of both worlds, but you'll need and engine stand and you have to take extra care to make sure the connecting rods do not hit the cylinder walls). To remove the pistons, use a large wooden dowel (or a large wooden hammer handle) to carefully push on the connecting rod until the piston comes out of the top of the cylinder block.

■ **When removing the piston, take extra care to make sure the connecting rod and, on these models, especially a rod stud, does not contact and damage the cylinder walls. If the crankshaft was not removed, take extra care to make sure the shaft journals are not damaged either.**

14. Use a universal piston ring expander to remove the piston rings. Retain all rings in order (top, second, bottom, etc) for identification or diagnostic purposes, but they should not be reused.
15. Matchmark the connecting rod to the piston to ensure installation with the same alignment.
16. Use a pair of needle-nose pliers modified as illustrated to remove the wrist pin retaining rings from the piston.

■ **The wrist pin-to-piston boss fit is loose on both sides, therefore little or no force should not be necessary to remove the wrist pin.**

17. Use a driver that is slightly smaller than the wrist pin (a 12mm deep socket is usually the right size on these motors) to carefully tap the wrist pin from the piston.
18. Remove the screws (2) securing the anode cover to the block and then remove the cover to expose the anode. Remove and discard the gasket. Loosen the screw and remove the anode.
19. Refer to Cleaning & Inspection for details concerning cylinder block, piston and crankshaft inspection. Don't forget to visually inspect the block mating surface and dowel pins for damage (replace any loose or damaged pins). Also, remember to check all bearing surfaces for signs of excessive wear or damage. All oil passages must be clean and free of dirt or debris.

To assemble:
20. If removed, install the anode (if the old anode is worn in any way, it is a good idea to replace it at this time) and tighten the retaining bolt securely. Then install the anode cover using a new gasket and tighten the bolts securely.
21. Assemble and prepare the pistons for installation, as follows:
   a. Apply a light coating of clean engine oil (or assembly lube) to the wrist pin and bore in the side of the piston.

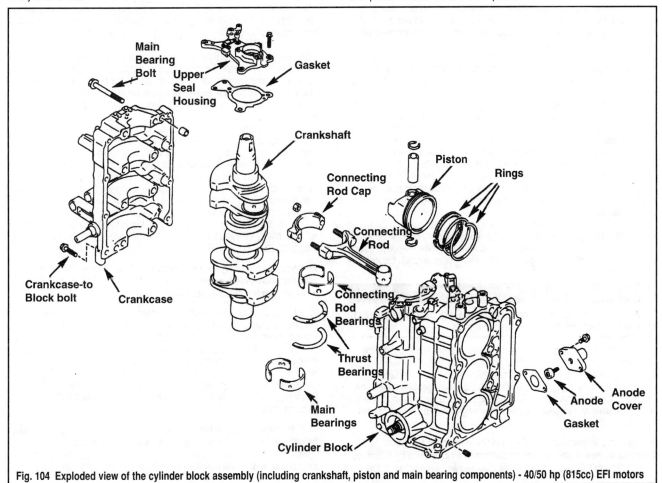

Fig. 104 Exploded view of the cylinder block assembly (including crankshaft, piston and main bearing components) - 40/50 hp (815cc) EFI motors

6-64 POWERHEAD REPAIR AND OVERHAUL

  b. Align the matchmarks made on the piston and connecting rod removal. Also, note that a mark on the top of the piston dome is provided and it should align with the oil hole at the base of the connecting rod.

  c. Since the wrist pin **should** be a loose fit in the piston, insert the wrist pin by hand and push it through the piston into the connecting rod. If necessary, a small socket can be used to help tap the wrist pin into position, but significant force should not be necessary.

  d. Install NEW wrist pin retaining rings, then turn the rings in the groove so the ring gaps face away, at a right angle (90°) from the semi-circular relief in the wrist pin bore.

✱✱ WARNING

Spread the rings only enough to allow the rings to slide over the piston, as they will likely crack or break if they are spread too much during installation. Though some rings (especially oil control rings) may be installed by hand, a universal ring expander will likely make this much easier. Take your time and be careful, don't force the rings during installation.

  e. Install a new set of rings using a universal ring expander. Either refer to the instructions that came with the ring set (if other than OE) or use the instructions provided here in the following steps for Original Equipment (OE) Johnson/Evinrude ring sets.

■ **On Johnson/Evinrude parts and MOST aftermarket rings, the mark sides of the rings should be positioned facing upward.**

  f. Apply a light coating of clean engine oil or assembly lube to each ring in order to help ease installation.

  g. Install the oil control ring assembly. Position the segmented ring into the bottom groove first, then insert the end of the first scraper ring under the segmented ring. Work carefully around the piston until the scraper ring is seated. Finally, insert the second scraper ring on top of the segmented ring and work around the piston to seat it.

  h. Position the scraper ring end gaps each at 45° from each end of the wrist pin (the wrist pin is represented as a dotted line in the accompanying illustration). Make sure the segmented ring gap is at least 45 degrees from the wrist pin centerline, but on the opposite side of the piston and wrist pin centerline from the scraper ring gaps.

  i. Install the middle compression ring into the middle piston ring groove with the stamping **2N** facing upwards. Position the middle ring gap directly between the 2 scraper ring gaps (90° from the wrist pin centerline). Refer to the accompanying illustration for clarification.

  j. Install the top ring in the top piston groove with the stamping **N** facing up. Position the top ring gap directly opposite the middle compression ring (in the same general area of the segmented ring gap, but more precisely 90° from the wrist pin centerline). Again, refer to the illustration for details.

  22. Apply a light coating of clean engine oil or assembly lube to the cylinder wall, an expandable ring compressor and the piston.

  23. Verify the ring gaps are in proper position, then secure the piston in the ring compressor. Tighten the compressor **just** sufficiently to push the rings into the grooves, but do not overtighten it as the piston must still be free to slide into the bore.

  24. Position the piston over the bore with the stamped mark on the piston dome facing the flywheel side of the motor (and any other alignment marks made during removal oriented as you intended). Use a wooden dowel or wooden hammer handle to gently tap the piston down through the ring compressor and into the cylinder bore.

✱✱ WARNING

When tapping the piston into position MAKE SURE the connecting rod does not contact and damage the cylinder walls.

  25. Install the connecting rod bearing halves into the rod ends and the caps. Align bearing inserts with the oil holes with the oil holes in the connecting rods during installation. If the bearings are to be checked for proper oil clearance using Plastigage® or an equivalent gauging compound, DO NOT oil them at this time. If the clearances have already been checked (during cleaning and inspection), then apply a light coating of clean engine oil or assembly lube.

■ **The same instructions for the previous step go for the main bearings as well. If you've already confirmed oil clearance by following that portion of this procedure during cleaning and inspection of the crankshaft and connecting rod components, then apply a light coating of clean engine oil or assembly lube as each bearing insert is installed. If not, leave the bearings dry by AVOID rotating the components on the dry surface until measurements are completed and the surfaces are properly pre-lubricated.**

  26. Block off the cylinder head area to keep the pistons from falling out when the cylinder block is inverted, then turn the block so the crankcase mating flange is facing upward.

  27. Install the main bearing halves into the crankcase and the cylinder block, making sure to align the oil holes in the bearing inserts with the oil holes in the block. Also, make sure the inserts are flush with the crankcase-to-block mating surfaces. Again, if the bearings are to be checked for proper oil clearance using Plastigage® or an equivalent gauging compound, DO NOT oil them at this time. If the clearances have already been checked (during cleaning and inspection), then apply a light coating of clean engine oil or assembly lube.

  28. Apply a light coating of clean engine oil or assembly lube to NEW thrust bearings and position them in the cylinder block with the grooves facing outward.

  29. Apply a light coating of clean engine oil or assembly lube to 2 NEW crankshaft seals, then install them to the lower crankshaft seal housing. Press until seated with the top facing outward, then lubricate and install a NEW large seal until seated. Finally, lubricate and install a NEW O-ring.

  30. Position all the piston's and connecting rods at or near TDC to ease crankshaft installation. Again, unless you are about to check clearances using a gauging material, apply a light coating of clean engine oil or assembly lube to the crankshaft main journals and connecting rod crankpins.

  31. Carefully lower the crankshaft into position, making sure the shaft does not contact (and become damaged by) any connecting rods or studs.

  32. Pull each connecting rod slowly and carefully into position and secure it to the crankshaft either to check bearing oil clearance or for installation purposes, as follows:

  a. Carefully push/pull the piston into position on the crankshaft. It is usually easiest to rotate the crankshaft slightly so the crankpin is centered when accomplishing this. BUT, if bearings are not oiled (because you are checking clearances), lift up slightly on the shaft itself to unload it, don't allow the journals to scrape or rub along the dry bearing inserts.

  b. If the clearance is about to be checked, make sure the crankpin is clean and free of any material or oil, then apply a short length section of Plastigage® or equivalent compression gauging material (according to the material manufacturer's instructions) across the top of the crankpin journal. Make sure the material is applied directly inline with the crankshaft (from flywheel-to-driveshaft at one level, NOT across the curved bearing surface from side-to-side).

  c. Ensure the cap is correctly positioned by aligning the matchmarks. The rod cap should contain an arrow mark from the factory that faces the flywheel. Once aligned, install the cap and bearing onto the rod. Finger-tighten the NEW rod nuts, alternating back and forth between the nuts.

  d. Check the connecting rod cap alignment using a pick or the tip of a sharp pencil. Run the tip across the rod-to-cap mating surface to make sure it does not catch on the lip. If it does, double-check the factory rod/cap alignment marks, as well as any matchmarks made during removal.

  e. Tighten the nuts by alternating back and forth between the 2 nuts, no more than 1 turn at a time. Torque the nuts first to 21 ft. lbs. (30 Nm) and then to 25-26 ft. lbs. (34-35 Nm).

■ **If you are checking the oil clearance, DO NOT rotate the crankshaft with the Plastigage® installed or the application will be ruined, you'll have remove the cap and start over.**

  f. If you are checking clearance, loosen the nuts (alternating again) and remove the cap. Compare the flattened size of the gauging material with the scale provided on or in the packaging. Each connecting rod must have an oil clearance less than 0.0026 in. (0.065mm) or the bearing must be replaced. If the clearance was acceptable, remove all traces of the gauging material and move on to the next connecting rod. Once all connecting rod bearing and crankshaft main bearing clearances have been verified properly oil the bearings, journals and crankpins, then follow the proper installation and torquing steps again.

■ **If an out of specification reading occurs, double-check your results before condemning the bearing. If the reading is still out of specification, try another bearing before turning to the crankshaft and connecting rod as the possible culprits of an excessive clearance.**

POWERHEAD REPAIR AND OVERHAUL 6-65

33. If not performed during cleaning and inspection, check the connecting rod side clearance (rod-to-crankshaft clearance) by determining what size feeler gauge will pass between the rod and crankshaft with a slight drag. As a check, the next larger size gauge must NOT pass at all and the next smaller size should pass without interference/contact). Clearance should be 0.004-0.014 in. (0.10-0.35mm) or the connecting rod and/or crankshaft must be replaced As usual, double-check all out of specification measurements before replacing a component.

34. If the main bearing oil clearances must still be checked, proceed in roughly the same fashion as the check occurred on the connecting rods. Make sure the crankshaft main bearing journals are clean and free of any material or oil, then apply a short length section of Plastigage® or equivalent compression gauging material (according to the material manufacturer's instructions) across the top of each crankshaft main bearing journal. Make sure the material is applied directly inline with the crankshaft (from flywheel-to-driveshaft at one level, NOT across the curved journal bearing surface from side-to-side).

■ **The crankcase-to-cylinder block mating surfaces must be COMPLETELY free of old sealant, dirt, corrosion or debris in order to ensure proper seal and main bearing fitting. Once installed, be absolutely certain that the surfaces are in direct contact.**

35. If you have ALREADY checked the main bearing oil clearances and are performing the final installation of the crankshaft (meaning you've oiled the bearing surfaces and the crankshaft journals/crankpins), apply a light coating of OMC Gel-Seal II or equivalent to the crankcase and cylinder block mating surfaces. Do NOT over-apply the sealant and keep it away from the bearing surfaces.

■ **Although OMC Gel-Seal II has a shelf life of one year when stored at room temperature, buy a new tube if in doubt. Keep in mind that using an old tube of Gel-Seal II could prevent proper crankcase sealing.**

36. If not done already, make sure any dowel pins that are used are in position in the cylinder block or crankcase mating surfaces.
37. Carefully lower the crankcase (with the bearing inserts in position) over the crankshaft and cylinder block.
38. Apply a light coating of clean engine oil to the threads of the main bearing and crankcase-to-cylinder block flange bolts. Install each of the bolts and finger-tighten them.
39. Using at least 3 passes of the torque sequence (a sequence that includes both sets of bolts) tighten the main bearing and crankcase-to-cylinder block flange bolts to specification as follows:
   a. On the first pass, tighten the main bearing bolts (which are longer) to 7 ft. lbs. (10 Nm) and the crankcase flange bolts (which are shorter) to 45 inch lbs. (5 Nm).
   b. On the second pass tighten the main bearing bolts to 29 ft. lbs. (40 Nm) and the crankcase flange bolts to 14 ft. lbs. (19 Nm).
   c. On the final pass tighten the main bearing bolts to 35-37 ft. lbs. (47-50 Nm) and the crankcase flange bolts to 17-19 ft. lbs. (23-26 Nm).
40. If you're checking the main bearing clearances, DO NOT rotate the crankshaft (or the gauging material will be smeared and the applications will be ruined). Loosen and remove the main bearing and crankshaft flange bolts using at LEAST as many passes in the reverse of the sequence. Then remove the crankcase and measure the gauging material on each of the main bearing journals. Each journal must have a bearing clearance less than 0.0026 in. (0.065mm) or the bearing must be replaced. If the clearance was acceptable, remove all traces of the gauging material and properly oil all of the bearing, journal and crankpin surfaces, then apply sealant and install the crankcase for real. Basically you should go back to the connecting rod cap installation steps (unless they were oiled and tightened after earlier after you checked their clearances, in which case you only need to go back to the Gel-Seal II step) and properly finish installation.

■ **If an out of specification reading occurs, double-check your results before condemning the bearing. If the reading is still out of specification, try another bearing before turning to the crankshaft as the possible culprit of an excessive clearance.**

41. Once the all the journals and bearings are properly oiled and the crankcase/bearings are installed/tightened, slowly rotate the crankshaft to make sure there is no binding, roughness or tight spot. If any binding is noted, the source must be located, which will likely require the removal of the crankcase again.

42. Install a new seal into the upper crankcase seal housing. Position the seal to the housing with the spring side facing a suitable driver (like OMC #339752 or a smooth suitably sized socket or piece of pipe). Carefully press or drive the seal into the housing, then apply a light coating of OMC Triple-Guard, or an equivalent marine grade grease to the seal lips.
43. Install the upper crankcase seal housing to the powerhead using a new gasket. Tighten the retaining bolts to 48-60 inch lbs. (5.4-6.8 Nm).
44. Check that the camshafts and crankshaft are both in the No. 1 TDC position. Timing marks must align, as detailed under the Timing Chain procedures in this section. Also, a quick check of the camshaft lobes for the No. 1 cylinder will give you your answer on the cylinder head. At TDC of the No. 1 cylinder, all valves (intake and exhaust) must be closed for engine compression). Therefore, the camshaft lobes must be facing out, away from the tappets on that cylinder.

■ **This is an interference motor without screw adjustable valves, the only way to safely time the motor is with the cylinder head off or ALL of the valve tappets removed.**

45. With the cylinder head and cylinder block properly timed, assemble the powerhead as detailed in this section.
46. Remember to follow all break-in procedures whenever wear components of the cylinder block have been replaced.

70 Hp (1298cc) EFI Motors

◆ See Figures 105 and 106

On fuel injected engines, always relieve system pressure prior to disconnecting any high-pressure fuel circuit component, fitting or fuel line. For details, please refer to Fuel System Pressurization under Fuel Injection.

Removal & Installation

◆ See Figures 105 and 106

1. For safety, properly relieve the fuel system pressure. Please refer to Fuel System Pressurization under Fuel Injection for details.
2. Disconnect the negative battery cable for safety.
3. Remove the lower engine covers. For details, please refer to the Engine Covers (Top and Lower Cases) in the Maintenance and Tune-Up Section.
4. Loosen the shift lever shaft nut in order to disengage the rod from the lever.
5. Remove the screws and nuts retaining the exhaust manifold. Next, tag and disconnect the hoses to the thermostat, adapter and the oil sump, then remove the hoses along with the manifold assembly. Remove and discard the old gasket.
6. Tag and disconnect the adapter-to-intake manifold and the vapor separator/crankcase-to-port sump hoses.
7. Disconnect the trim motor wires from the relays, then pull the wires from the front panel. If necessary, refer to the schematics in the Wiring Diagram section to help identify the wires.
8. Loosen and remove the powerhead-to-engine holder bolts (10 bolts that are threaded up into the bottom of the powerhead from underneath the engine holder).
9. At the rear of the motor, remove the cylinder head-to-holder nuts (2) and washers.
10. Place two nuts on the top exhaust manifold stud and then tighten the 2 nuts against each other to loosen and remove the stud. Remove the lifting eye from the exhaust manifold and attach it to the vacant stud hole using the same bolt and 2 powerhead-to-engine holder bolt washers. Be sure to install the lifting eye with the bend facing outward to avoid interference between the lifting chain and the powerhead.
11. Connect an engine hoist to the engine lift brackets (one at the front, port side of the motor and the other at the rear starboard side) using a spreader bar to keep the chains separated.
12. Carefully lift the powerhead straight up and off the engine holder. Keep the powerhead as level as possible while lifting it to avoid damage to the driveshaft or gearcase components. Remove and discard the old powerhead-to-holder gasket.
13. Mount the powerhead in a suitable engine stand.
14. If necessary, access the oil pump and related components by removing the bolts (6) threaded upward from the oil pan into the engine holder and by removing the steering arm bolts, then removing the adapter from the oil pan and steering arm. Remove and discard the adapter-to-oil pan gasket, then inspect/clean/service the oil pick-up, pump and pressure

6-66 POWERHEAD REPAIR AND OVERHAUL

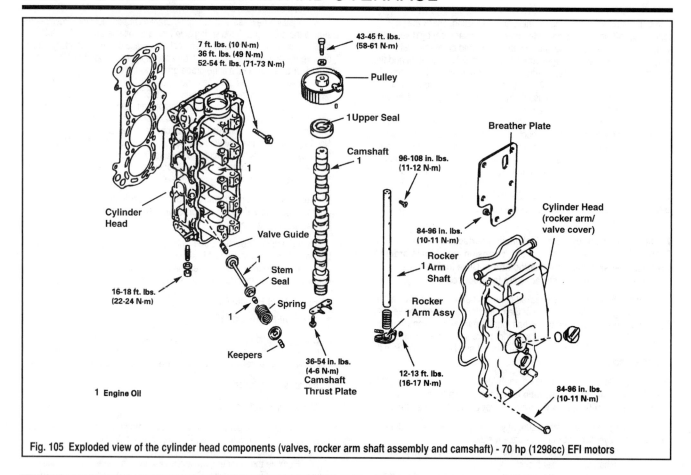

Fig. 105 Exploded view of the cylinder head components (valves, rocker arm shaft assembly and camshaft) - 70 hp (1298cc) EFI motors

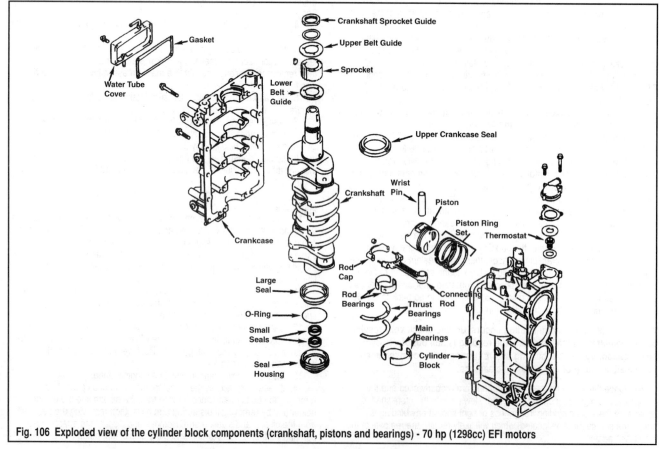

Fig. 106 Exploded view of the cylinder block components (crankshaft, pistons and bearings) - 70 hp (1298cc) EFI motors

POWERHEAD REPAIR AND OVERHAUL 6-67

relief valve. For more details, please refer to the Oil Pump procedures in the Lubrication and Cooling section. Inspect and clean the oil sump.

15. Remove all traces of gasket or sealant from the powerhead, oil sump and/or the engine holder. Clean and inspect the mating surfaces for signs of damage or extensive pitting/corrosion. Replace the oil sump or engine holder if damage or corrosion is extensive.

To install:

✱✱ WARNING

The 70 hp EFI motor utilizes an oil pump that is mounted to the bottom side of the engine holder. During powerhead installation, the oil pump shaft must be aligned to the slot. The oil pump and/or camshaft will be damaged if the powerhead is installed with these components misaligned. For this reason, an oil pump access screw and retainer bolt is provided on the rear of the engine holder.

16. If removed, install the engine holder to the oil sump using a NEW gasket. Install the oil sump-to-engine holder bolts (6) and tighten to 16-18 ft. lbs. (22-24 Nm). Apply a light coating of OMC Nut Lock or equivalent threadlock to the threads of the steering arm bolts, then install and tighten the bolts to 40-47 ft. lbs. (56-64 Nm).

17. Locate the oil pump access and retainer bolts at the rear of the motor, under the engine holder flange (the access screw is a smaller bolt that is threaded into the larger oil pump retainer bolt). Remove the access bolt and bend back the tangs that are securing the retainer bolt. Partially unthread the retainer bolt from the engine holder.

18. Position a NEW powerhead-to-engine holder gasket, then raise the powerhead assembly (using the hoist) and lower it ALMOST completely into position of the engine holder. Insert a screwdriver through the opening in the oil pump retainer bolt and align the oil pump shaft to the camshaft by turning slighting or wiggling the shaft. Once they are aligned, seat the powerhead on the engine holder.

19. Use the screwdriver to hold the oil pump shaft in position, then seat the retainer bolt. Tighten the retainer bolt to 34-38 ft. lbs. (46-51 Nm), then bend one of the retainer tangs upward and another downward to lock the bolt in position.

20. Install the access bolt to the head of the retainer bolt and tighten to 108-120 inch lbs. (12-14 Nm).

21. Remove the engine lifting eye and reinstall the exhaust manifold stud.

22. Install the cylinder head-to-engine holder nuts and tighten to 16-18 ft. lbs. (22-24 Nm).

23. Install the engine holder-to-powerhead bolts and washers, then tighten to 35-37 ft. lbs. (47-50 Nm).

24. Install the exhaust manifold using a NEW gasket, then install and tighten the bolts and nuts to 16-18 ft. lbs. (22-24 Nm).

25. Connect all hoses and tagged during removal.

26. Connect the shift rod to the lever, then tighten the nut to 35-37 ft. lbs. (47-50 Nm).

27. Reconnect the trim/tilt wiring.

28. Install the lower engine covers.

29. Connect the negative battery cable and properly pressurize the fuel system. Thoroughly inspect the system for fuel leaks.

30. Install the top engine cover.

31. If a new or rebuilt powerhead was just installed, refer to Break-In in this section for details on how to ensure proper break-in of the new powerhead components.

Powerhead Disassembly & Assembly

◆ See Figures 105, 106 and 107

1. Remove the powerhead as detailed in this section and mount it on a suitable engine stand.
2. Remove the timing belt, as detailed in this section.
3. Remove the intake manifold, as detailed in the Fuel System section.
4. If not already done, remove the following components from the cylinder head or head (rocker arm/valve) cover:
   • Remove the low pressure fuel pump from the cylinder head cover as described in the Fuel System section.
   • Tag and disconnect the breather hose from the cylinder head cover.
   • Remove the ignition coils from the cylinder head. For details, refer to the Ignition System section.

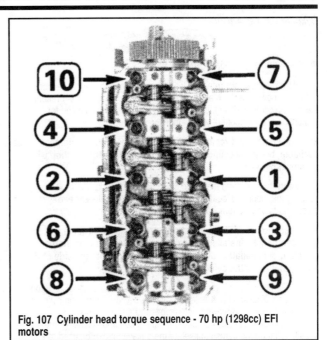

Fig. 107 Cylinder head torque sequence - 70 hp (1298cc) EFI motors

5. Support the cylinder head (rocker arm/valve cover) and remove the six cover bolts using a crossing pattern.
6. Pull the cover from the cylinder head, then remove and discard the cover gasket.
7. If necessary, loosen the bolts and remove the breather plate from the cover.
8. Loosen the cylinder head bolts gradually, using multiple passes in the reverse of the torque sequence.
9. Carefully lift the cylinder head from the block. If the gasket seal is difficult to break, start by using a rubber or plastic mallet to tap around the edges, then carefully pry to help free the head.
10. Refer to Cylinder Head Overhaul, in this section for details on disassembling the head for component inspection and replacement.

To assemble:

11. For ease of assembly and to prevent possible damage to components, make sure the cylinder block and cylinder head are each timed to No. 1 TDC. This means that the No. 1 cylinder is at or near the top of its travel and the valves for the No. 1 cylinder in the head are fully closed (the rockers are resting on the bases not the raised portion of the camshaft lobes.

12. Position a NEW cylinder head gasket onto the block mating surface, then carefully lower the head onto the cylinder block over the gasket. Make sure you do not disturb the gasket positioning when seating the head.

13. Install the cylinder head retaining bolts and tighten using at least 3 passes of the torque sequence as follows:
   a. Tighten the bolts to 7 ft. lbs. (10 Nm).
   b. Next, tighten the bolts to 36 ft. lbs. (49 Nm).
   c. On the final pass, tighten the bolts to 52-54 ft. lbs. (71-73 Nm).

14. If removed, install the breather plate to the cylinder head cover and tighten the retaining bolts to 84-96 inch lbs. (9.5-11 Nm).

15. If the cylinder head was overhauled or the valve lash was disturbed in any way, properly adjust the valves, as detailed under Valve Clearance in the Maintenance and Tune-Up section.

16. Install the intake manifold, as detailed under Fuel System. Be sure to use a new gasket and tighten the bolts using the proper torque sequence to 16-18 ft. lbs. (22-24 Nm).

17. Install the timing belt, as detailed in this section.

18. Install the cylinder head (rocker arm/valve) cover to the cylinder head using a NEW seal, then tighten the cover bolts to 84-96 inch lbs. (9.5-11 Nm) using a crossing pattern.

19. Install the ignition coils to the cylinder head, as detailed in the Ignition and Electrical System section.

20. Connect the breather hose to the cylinder head cover fitting, then install the fuel pump.

21. Install the powerhead, as detailed in this section.

22. If a new or rebuilt powerhead was just installed, refer to Break-In in this section for details on how to ensure proper break-in of the new powerhead components.

6-68 POWERHEAD REPAIR AND OVERHAUL

Cylinder Head Overhaul

◆ See Figures 105, 95, 96, 97 and 108

Before disassembly, refer to Cleaning and Inspection in this section for details on how to perform a Valve Leakage check. If leakage is present or there is an obvious defect, the valve train must be disassembled for inspection and component replacement.

■ **Disassembly and repair requires special tools and equipment. If not available, bring the cylinder head to a reputable machine shop with marine experience.**

Be sure to mark (and make a note, in case the marks come off during service) the mounting location and orientation of all cylinder head components prior to removal.

■ **These motors ARE equipped with progressive rate valve springs (that can only be installed in one direction). Be sure to mark the direction of installation during removal (mark or note the side of the spring that faces the cylinder head when installed).**

1. Remove the cylinder head from the powerhead as detailed under Powerhead Disassembly & Assembly, in this section. Clamp the cylinder head securely in a soft-jawed vise attached to a sturdy workbench.
2. Remove the camshaft pulley retaining bolt and washer (there is usually a flat on the camshaft itself or the pulley that can be used to keep the shaft from turning while loosening the retaining bolt). Once the bolt and washer are withdrawn, remove the camshaft pulley and drive pin.

■ **To ensure installation in the original locations and to prevent mix-up, use a permanent marker to identify the rocker arms shafts, springs, and rocker arm assemblies. The shaft ends themselves differ slightly, and the differences can be used for identification, but matchmarks and labels will make it easier. If the rocker arm assemblies are to be reused, they MUST be returned to their original positions.**

3. Remove the screws securing the rocker arm shafts, then slowly withdraw the shafts, removing the springs and rocker arm assemblies. If the shafts are disassembled, remember to arrange the rocker arm assemblies and springs to prevent mix-up of components and to ensure ease of installation in their original locations.
4. Puncture the camshaft seal case using a punch or awl, then carefully pry it from the cylinder head bore.

※※ WARNING

When piercing and prying the camshaft seal from the bore, DO NOT contact (and thereby damage) either the bore or camshaft sealing surfaces.

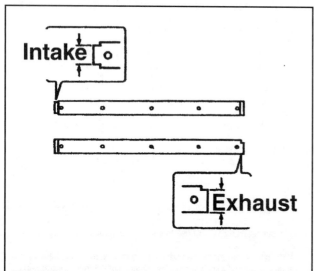

Fig. 108 The intake and exhaust rocker arm shafts are identified by the shoulders on that shaft ends

5. Remove the 2 bolts securing the camshaft thrust plate to the bottom end (opposite end from the pulley) of the cylinder head. Remove the plate.
6. Slowly withdraw the camshaft through the front (pulley end) opening in the cylinder head.
7. Use the marker to label the location and orientation of each valve before it is removed from the cylinder head.
8. Use a valve spring compressor tool (most universal valve spring tools should work, but threaded clamp or lever type compressor is recommended) to remove each valve, spring and seal as follows:
   a. Place one clamp on the valve face and the other on the valve spring cap

■ **Do not compress the spring fully.**

   b. Compress the spring **just** sufficiently to free the keepers from the stem. Remove the keepers from the valve, then slowly release the spring pressure.
   c. Remove the valve compressor tool and remove valve and spring from the cylinder head. Remember to mark the spring orientation (which side faces the cylinder head).

■ **Keep all valve components together as they should not be mixed or transferred to other valves during assembly.**

   d. Carefully pry the seal from the valve guide, then discard the seal.
   e. Remove the spring seat.
9. Inspect and measure the cylinder head components as described in this chapter.

To assemble:
10. Install the valves to the cylinder head, as follows:
   a. Position the valve spring seat to the cylinder head.
   b. Apply a light coating of engine oil to the NEW valve stem seal, then press it carefully onto the valve guide using finger pressure, making sure it is properly seated.
   c. Apply a light coating of engine oil to the valve stem, then slide it into the valve guide. Place the valve fully against the seat, then install the valve spring and cap over the valve stem. Make sure the narrow spring coil faces downward against the valve seat (as marked during removal).
   d. Use the valve spring compressor to push downward on the spring, **just** sufficiently to install the valve keepers. Once the keepers are in position, slowly release the spring pressure, allowing the cap and keepers to take the pressure. Make sure the keepers are properly seated.
   e. Repeat to install the remaining valves.
11. Apply a light coating of clean engine oil or assembly lube to the journals on the camshaft and the journal bearing surfaces in the cylinder head bores. Carefully insert the camshaft through the pulley end of the head.
12. Install the camshaft thrust plate to the rear of the head and tighten the retaining bolts to 36-54 inch lbs. (4-6 Nm).
13. Apply a light coating of clean engine oil to the lips of a NEW camshaft seal, then align the seal with the bore in the end of the crankshaft. Carefully press or tap the seal into position (make sure the installation tool does NOT contact the camshaft or cylinder head). Continue to press the seal into position until it is 0.02 in. (0.5mm) **below** the surface of the cylinder head.
14. Apply a light coating of clean engine oil or assembly lube to the rocker arm, spring and shaft assemblies. Install the shafts, making sure the shafts are installed on the correct sides of the cylinder head. Refer to the accompanying illustration to help identify the shafts (using the shoulders at the shaft ends). The intake shaft has a shoulder that occurs on both sides of the shaft end, while the exhaust shaft has a shoulder on only one side of the shaft end.

■ **If the rocker arms, springs and shafts are disassembled, be sure that all components are installed in their original positions during assembly.**

15. Install the rocker arm shaft retaining screws and tighten to 96-108 inch lbs. (11-12 Nm).
16. Install the drive pin to the camshaft, then install the pulley, locating it over the pin. Install the camshaft pulley bolt and washer, then tighten to 43-45 ft. lbs. (58-61 Nm).
17. Install the cylinder head as detailed in Powerhead Disassembly & Assembly in this section.

POWERHEAD REPAIR AND OVERHAUL 6-69

Cylinder Block Overhaul

◆ See Figures 24, 26, 50, 98 thru102, 106, 109 and 110

The crankshaft for this motor turns on replaceable main bearing insert halves found in the cylinder block and crankcase. Like many 4-stroke motors, the connecting rods also utilize bearing inserts. All machined and wear components, including the pistons, connecting rods, rings and bearings MUST be returned to their original positions when they are reused. Matchmark all components to installation in the proper locations and in the same orientations (facing the same direction).

During service, the crankshaft assembly can be removed from the block without piston removal or vice-versa. If either is attempted, take EXTREME care to keep the journal surfaces from being damaged. When free of the crankshaft journals, the connecting rods should be restrained from contacting the piston bores as well. One method to accomplish this is to stuff a clean shop towel under the piston skirt. Another is to use rubber bands to hold the rods in position.

Cylinder block work is more easily performed on an engine stand, especially since this allows rotation of the motor to access pistons from above and below the cylinder block. But, work can also be performed on a clean workbench, if care is taken not to damage the mating surfaces or, in the case of the cylinder head mating surface, the dowel pins. When working on a bench, take care to prevent the block from falling over during service. Also, if the case is stood on end at any point, take great care to keep the crankshaft from falling out of the block during service. A crankshaft can suffer a fatal blow from such a fall, even from a short height.

1. Remove the powerhead from the gearcase as detailed under Removal & Installation, then remove the cylinder head as detailed under Powerhead Disassembly & Assembly.
2. If not done already, remove the crankshaft pulley as follows:
   a. Use OMC Crankshaft Holder #345827 or equivalent tool over the crankshaft taper and aligned with the flywheel keyway to keep the crankshaft from turning.
   b. With the crankshaft held steady by the tool, loosen the pulley retaining nut.
   c. Remove the tool, the pulley nut, the upper belt guide, pulley, pulley key and lower belt guide.
3. If necessary, remove the thermostat assembly, for details refer to the procedures in the Lubrication and Cooling section.
4. Invert the cylinder block so the cylinder head mating surface is facing downward.

■ If you're not working on an engine stand, take care not to damage the cylinder head mating surface.

5. Loosen the crankcase-to-cylinder block flange bolts (10) and the main bearing bolts (10) using multiple passes in the reverse of the torque sequence. This will essentially result in a crossing pattern that starts with the outer flange bolts moving toward the center flange bolts, then moves to the outer main bearing bolts and finally moves to the inner main bearing bolts.
6. Separate the crankcase from the cylinder block. If the seal is tough to break, locate the notches cast into the mating surfaces, then use 2 prybars (blunt or coated with electrical tape to help protect the surfaces) to apply light pressure helping to separate the components.
7. Note the location and orientation of each bearing half in the crankcase journals. Remove and retain the bearings. Visually check the crankcase mating surfaces, the bearing surfaces and the seal surfaces for damage or wear (no scratches big enough to catch a fingernail should be present on any of the surfaces).
8. Remove the bolts securing the water jacket covers to the crankcase, then remove the covers and discard the gaskets.

■ To help prevent mix-ups during disassembly or inspection, once the crankshaft is removed from the block (or if the shaft is remaining in place, once each piston is removed, reinstall the connecting rod cap and bearing inserts loosely to the rods.

9. Using a permanent marker, label each piston, connecting rod and rod end-cap to ensure installation. Although it is unclear whether or not the connecting rods and caps are of the same fractured design used on other Johnson/Evinrude motors (especially since the EFI motors are built by Suzuki), matchmark the connecting rods to the caps and maintain the same orientation, just to be safe.

■ If working on an engine stand, bolt a piece of wood or metal over the cylinder head surface (use a few of the cylinder head bolts) to keep the pistons from dislodging and falling free once the connecting rod caps are unbolted.

10. Loosen and remove the connecting rod cap nuts. To prevent possible damage to the cap and bearing surface, loosen each cap by alternating between nuts, giving them about 1/4 - 1/2 turn at a time.
11. Once the nuts are removed, the cap should easily come free, but if necessary, tap it gently using a plastic or rubber mallet to free it from the rod. Remove the rod cap, then remove the bearing half from the shaft or cap. Keep the bearing with the rod cap.

■ Wrap the connecting rod studs using tape to help protect the cylinder walls from damage.

12. Push the pistons and connecting rods downward in their bores away from the crankshaft, then remove the thrust bearings.
13. Remove the crankshaft, upper seal and lower seal housing assembly from the cylinder block.
14. Remove the large seal, O-ring and 2 smaller seals from the crankshaft lower seal housing assembly.
15. Remove the main bearing halves from the cylinder block. Note their locations and orientations for measurement and assembly purposes.
16. If the pistons are being removed, place the cylinder block with the head mating surface and the tops of the pistons either facing upward or sideways. (Actually, we find that upward at a slight angle is the best of both worlds, but you'll need and engine stand and you have to take extra care to make sure the connecting rods do not hit the cylinder walls). To remove the pistons, use a large wooden dowel (or a large wooden hammer handle) to carefully push on the connecting rod until the piston comes out of the top of the cylinder block.

■ When removing the piston, take extra care to make sure the connecting rod and, on these models, especially a rod stud, does not contact and damage the cylinder walls. If the crankshaft was not removed, take extra care to make sure the shaft journals are not damaged either.

17. Use a universal piston ring expander to remove the piston rings. Retain all rings in order (top, second, bottom, etc) for identification or diagnostic purposes, but they should not be reused.
18. Matchmark the connecting rod to the piston to ensure installation with the same alignment. Although there should be alignment marks from the factory, additional marks are never a bad idea). If one component (piston or connecting rod) is replaced, the marks should be transferred onto the replacement part.

■ Unlike most Johnson/Evinrude motors, the wrist pins on this model are not retained using snap-rings. On these motors, the wrist pins are press fit and must be removed or installed only using suitable equipment. DO NOT remove the wrist pin unless the piston or connecting rod is damaged and is to be replaced.

19. Use a OMC Piston Pin Remover/Installer #345826 (or an equivalent driver that is slightly smaller than the wrist pin) and a shop press with a suitable piston cradle to slowly and carefully force the wrist pin from the piston.
20. Refer to Cleaning & Inspection for details concerning cylinder block, piston and crankshaft inspection. Don't forget to visually inspect the block mating surface and dowel pins for damage (replace any loose or damaged pins). Also, remember to check all bearing surfaces for signs of excessive wear or damage. All oil passages must be clean and free of dirt or debris.

To assemble:

21. Assemble and prepare the pistons for installation, as follows:
    a. Apply a light coating of clean engine oil (or assembly lube) to the wrist pin and bore in the side of the piston.
    b. Align the matchmarks made on the piston and connecting rod removal. Also, note that a mark on the top of the piston dome is provided and it should align with the oil hole at the base of the connecting rod (for details, refer to the accompanying illustration).
    c. Using a suitable shop press, driver and piston cradle, carefully press the wrist pin through the piston and connecting rod. Orient the piston with the dome mark facing downward in the piston cradle (away from the press/driver).

6-70 POWERHEAD REPAIR AND OVERHAUL

※※ **WARNING**

Spread the rings only enough to allow the rings to slide over the piston, as they will likely crack or break if they are spread too much during installation. Though some rings (especially oil control rings) may be installed by hand, a universal ring expander will likely make this much easier. Take your time and be careful, don't force the rings during installation.

    d. Install a new set of rings using a universal ring expander. Either refer to the instructions that came with the ring set (if other than OE) or use the instructions provided here in the following steps for Original Equipment (OE) Johnson/Evinrude ring sets.

■ On Johnson/Evinrude parts and MOST aftermarket rings, the mark sides of the rings should be positioned facing upward.

    e. Apply a light coating of clean engine oil or assembly lube to each ring in order to help ease installation.
    f. Install the oil control ring assembly. Position the segmented ring into the bottom groove and insert the end of the first scraper ring under the segmented ring. Work carefully around the piston until the scraper ring is seated. Finally, insert the second scraper ring on top of the segmented ring and work around the piston to seat it.
    g. Position the scraper ring end gaps each at 45° from each end of the wrist pin (the wrist pin is represented as a dotted line in the accompanying illustration). Make sure the segmented ring gap is at least 45 degrees from the wrist pin centerline, but on the opposite side of the piston and wrist pin centerline from the scraper ring gaps.
    h. Install the middle compression ring (the thick ring) into the middle piston ring groove with the stamping **R** facing upwards. Position the middle ring gap directly between the 2 scraper ring gaps (90° from the wrist pin centerline). Refer to the accompanying illustration for clarification.
    i. Install the top compression ring (the thin compression ring) in the top piston groove. Position the top ring gap directly opposite the middle compression ring (in the same general area of the segmented ring gap, but more precisely 90° from the wrist pin centerline). Again, refer to the illustration for details.

22. Apply a light coating of clean engine oil or assembly lube to the cylinder wall, an expandable ring compressor and the piston.
23. Verify the ring gaps are in proper position, then secure the piston in the ring compressor. Tighten the compressor **just** sufficiently to push the rings into the grooves, but do not overtighten it as the piston must still be free to slide into the bore.
24. Position the piston over the bore with the stamped mark on the piston dome facing the flywheel side of the motor (and any other alignment marks made during removal oriented as you intended). Use a wooden dowel or wooden hammer handle to gently tap the piston down through the ring compressor and into the cylinder bore.

※※ **WARNING**

When tapping the piston into position MAKE SURE the connecting rod does not contact and damage the cylinder walls.

25. Install the connecting rod bearing halves into the rod ends and the caps. Align bearing inserts with the oil holes in the oil holes in the connecting rods during installation. If the bearings are to be checked for proper oil clearance using Plastigage® or an equivalent gauging compound, DO NOT oil them at this time. If the clearances have already been checked (during cleaning and inspection), then apply a light coating of clean engine oil or assembly lube.

■ The same instructions for the previous step go for the main bearings as well. If you've already confirmed oil clearance by following that portion of this procedure during cleaning and inspection of the crankshaft and connecting rod components, then apply a light coating of clean engine oil or assembly lube as each bearing insert is installed. If not, leave the bearings dry by AVOID rotating the components on the dry surface until measurements are completed and the surfaces are properly pre-lubricated.

26. Block off the cylinder head area to keep the pistons from falling out when the cylinder block is inverted, then turn the block so the crankcase mating flange is facing upward.

27. Install the main bearing halves into the crankcase and the cylinder block, making sure to align the oil holes in the bearing inserts with the oil holes in the block. Also, make sure the inserts are flush with the crankcase-to-block mating surfaces. Again, if the bearings are to be checked for proper oil clearance using Plastigage® or an equivalent gauging compound, DO NOT oil them at this time. If the clearances have already been checked (during cleaning and inspection), then apply a light coating of clean engine oil or assembly lube.
28. Apply a light coating of clean engine oil or assembly lube to NEW thrust bearings and position them in the cylinder block with the grooves facing outward.
29. If the crankshaft lower seal housing was removed to install new seals, proceed as follows:
    a. Apply a light coating of clean engine oil or assembly lube to 2 NEW small seals, then install them to the lower crankshaft seal housing. Press until seated with the top facing outward.
    b. Lubricate and install a NEW large seal until seated.
    c. Lubricate and install a NEW O-ring.
    d. Lubricate and install the lower seal housing assembly onto the crankshaft.
30. Apply a light coating of clean engine oil or assembly lube to a new upper crankshaft seal, then slide the seal onto the shaft with the spring side facing inward.

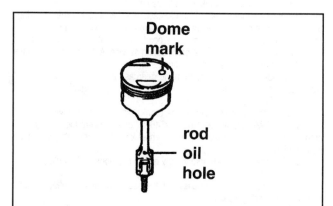

Fig. 109 The piston dome mark must align with the connecting rod oil hole as shown

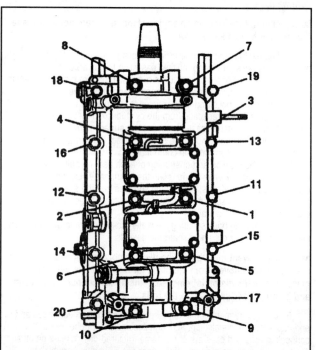

Fig. 110 Main bearing and crankcase flange bolt torque sequence - 70 hp (1298cc) EFI motors

POWERHEAD REPAIR AND OVERHAUL 6-71

31. Position all the piston's and connecting rods at or near TDC to ease crankshaft installation. Again, unless you are about to check clearances using a gauging material, apply a light coating of clean engine oil or assembly lube to the crankshaft main journals and connecting rod crankpins.

32. Carefully lower the crankshaft into position, making sure the shaft does not contact (and become damaged by) any connecting rods or studs. When lowering the crankshaft into position, make sure the tabs of the upper seal and lower housing each fit into the cylinder block grooves.

33. Pull each connecting rod slowly and carefully into position and secure it to the crankshaft either to check bearing oil clearance or for installation purposes, as follows:

   a. Carefully push/pull the piston into position on the crankshaft. It is usually easiest to rotate the crankshaft slightly so the crankpin is centered when accomplishing this. BUT, if bearings are not oiled (because you are checking clearances), lift up slightly on the shaft itself to unload it, don't allow the journals to scrape or rub along the dry bearing inserts.

   b. If the clearance is about to be checked, make sure the crankpin is clean and free of any material or oil, then apply a short length section of Plastigage® or equivalent compression gauging material (according to the material manufacturer's instructions) across the top of the crankpin journal. Make sure the material is applied directly inline with the crankshaft (from flywheel-to-driveshaft at one level, NOT across the curved bearing surface from side-to-side).

   c. Ensure the cap is correctly positioned by aligning the matchmarks. The rod cap should contain an arrow mark from the factory that faces the flywheel. Once aligned, install the cap and bearing onto the rod. Finger-tighten the NEW rod nuts, alternating back and forth between the nuts.

   d. Check the connecting rod cap alignment using a pick or the tip of a sharp pencil. Run the tip across the rod-to-cap mating surface to make sure it does not catch on the lip. If it does, double-check the factory rod/cap alignment marks, as well as any matchmarks made during removal.

   e. Tighten the nuts by alternating back and forth between the 2 nuts, no more than 1 turn at a time. Torque the nuts first to 13 ft. lbs. (18 Nm) and then to 25-26 ft. lbs. (34-35 Nm).

■ **If you are checking the oil clearance, DO NOT rotate the crankshaft with the Plastigage® installed or the application will be ruined, you'll have remove the cap and start over.**

   f. If you are checking clearance, loosen the nuts (alternating again) and remove the cap. Compare the flattened size of the gauging material with the scale provided on or in the packaging. Each connecting rod must have an oil clearance less than 0.0031 in. (0.079mm) or the bearing must be replaced. If the clearance was acceptable, remove all traces of the gauging material and move on to the next connecting rod. Once all connecting rod bearing and crankshaft main bearing clearances have been verified properly oil the bearings, journals and crankpins, then follow the proper installation and torquing steps again.

■ **If an out of specification reading occurs, double-check your results before condemning the bearing. If the reading is still out of specification, try another bearing before turning to the crankshaft and connecting rod as the possible culprits of an excessive clearance.**

34. If not performed during cleaning and inspection, check the connecting rod side clearance by determining what size feeler gauge will pass between the rod and crankshaft with a slight drag. As a check, the next larger size gauge must NOT pass at all and the next smaller size should pass without interference/contact). Clearance should be 0.004-0.014 in. (0.10-0.35mm) or the connecting rod and/or crankshaft must be replaced As usual, double-check all out of specification measurements before replacing a component.

35. If removed, install the water jacket covers to the crankcase using NEW gaskets. Install and tighten the retaining bolts to 36-54 inch lbs. (4-6 Nm).

36. If the main bearing oil clearances must still be checked, proceed in roughly the same fashion as the check occurred on the connecting rods. Make sure the crankshaft main bearing journals are clean and free of any material or oil, then apply a short length section of Plastigage® or equivalent compression gauging material (according to the material manufacturer's instructions) across the top of each crankshaft main bearing journal. Make sure the material is applied directly inline with the crankshaft (from flywheel-to-driveshaft at one level, NOT across the curved journal bearing surface from side-to-side).

■ **The crankcase-to-cylinder block mating surfaces must be COMPLETELY free of old sealant, dirt, corrosion or debris in order to ensure proper seal and main bearing fitting. Once installed, be absolutely certain that the surfaces are in direct contact.**

37. If you have ALREADY checked the main bearing oil clearances and are performing the final installation of the crankshaft (meaning you've oiled the bearing surfaces and the crankshaft journals/crankpins), apply a light coating of OMC Gel-Seal II or equivalent to the crankcase and cylinder block mating surfaces. Do NOT over-apply the sealant and keep it away from the bearing surfaces.

■ **Although OMC Gel-Seal II has a shelf life of one year when stored at room temperature, buy a new tube if in doubt. Keep in mind that using an old tube of Gel-Seal II could prevent proper crankcase sealing.**

38. If not done already, make sure any dowel pins that are used are in position in the cylinder block or crankcase mating surfaces.

39. Carefully lower the crankcase (with the bearing inserts in position) over the crankshaft and cylinder block.

40. Apply a light coating of clean engine oil to the threads of the main bearing and crankcase-to-cylinder block flange bolts. Install each of the bolts and finger-tighten them.

41. Using at least 3 passes of the torque sequence (a sequence that includes both sets of bolts) tighten the main bearing and crankcase-to-cylinder block flange bolts to specification as follows:

   a. On the first pass, tighten the main bearing bolts (which are longer) to 7 ft. lbs. (10 Nm) and the crankcase flange bolts (which are shorter) to 45 inch lbs. (5 Nm).

   b. On the second pass tighten the main bearing bolts to 29 ft. lbs. (40 Nm) and the crankcase flange bolts to 14 ft. lbs. (19 Nm).

   c. On the final pass tighten the main bearing bolts to 35-37 ft. lbs. (48-50 Nm) and the crankcase flange bolts to 17-19 ft. lbs. (23-26 Nm).

42. If removed, install the thermostat assembly, as detailed under Thermostat in the Lubrication and Cooling section. Be sure to use sealant on the gasket and threadlocking compound on the threads of the cover screws.

43. Install the crankshaft timing belt pulley assembly. Position the lower belt guide, followed by the crankshaft key and the pulley, next position the upper belt guide, washer and nut. Keep the crankshaft from turning and tighten the nut to 50-52 ft. lbs. (68-70 Nm).

44. If you're checking the main bearing clearances, DO NOT rotate the crankshaft (or the gauging material will be smeared and the applications will be ruined). Loosen and remove the main bearing and crankshaft flange bolts using at LEAST as many passes in the reverse of the sequence. Then remove the crankcase and measure the gauging material on each of the main bearing journals. Each journal must have a bearing clearance less than 0.0024 in. (0.060mm) or the bearing must be replaced. If the clearance was acceptable, remove all traces of the gauging material and properly oil all of the bearing, journal and crankpin surfaces, then apply sealant and install the crankcase for real. Basically you should go back to the connecting rod cap installation steps (unless they were oiled and tightened after earlier after you checked their clearances, in which case you only need to go back to the Gel-Seal II step) and properly finish installation.

■ **If an out of specification reading occurs, double-check your results before condemning the bearing. If the reading is still out of specification, try another bearing before turning to the crankshaft as the possible culprit of an excessive clearance.**

45. Once the all the journals and bearings are properly oiled and the crankcase/bearings are installed/tightened, slowly rotate the crankshaft to make sure there is no binding, roughness or tight spot. If any binding is noted, the source must be located, which will likely require the removal of the crankcase again.

46. Check that the camshaft and crankshaft are both in the No. 1 TDC position. Timing marks must align, as detailed under the Timing Belt procedures in this section. Also, a quick check of the camshaft lobes for the No. 1 cylinder will give you your answer on the cylinder head. At TDC of the No. 1 cylinder, all valves (intake and exhaust) must be closed for engine compression). Therefore, the camshaft lobes must be facing away from the rocker arms on that cylinder.

47. With the cylinder head and cylinder block properly timed, assemble the powerhead as detailed in this section.

48. Remember to follow all break-in procedures whenever wear components of the cylinder block have been replaced.

POWERHEAD REPAIR AND OVERHAUL

CLEANING & INSPECTION

All powerhead components must be clean and free of gasket material, oil and carbon deposits before they are inspected. Take your time when cleaning components. Before using solvent to clean something, make sure the chemical is compatible with the material of which the component is constructed. Also, check each component for matchmarks or ID marks before cleaning (as some solvents may remove even permanent marker). If necessary, re-ID the component as soon it has been cleaned and dried (before moving onto the next component). When 2 removed components are matchmarked to ensure exact alignment during installation, if possible, fasten them together during cleaning (in case the marks come off). In this way, they can be re-matchmarked after the cleaning process is through.

When it comes to inspection, this section will provide information on how to check various components. But, keep in mind that not all components will be found on all motors. If in doubt whether a component is used (or should be checked) refer to the Disassembly & Assembly or Overhaul procedures AND to the Specifications Chart for your motor. If a component is not listed in the specification chart, it is either not used, or, does not have a specific tolerance for which it must be inspected. Whether or not a tolerance is provided, all components must be clean and free of obvious defects (deep cracks, scoring, excessive carbon deposits that cannot be removed, warpage, etc). If in doubt whether or not a component is serviceable, seek someone with more Johnson/Evinrude experience than yourself.

■ The 4-stroke motor, by design, tends to be much more complicated than a 2-stroke motor, containing a greater number of precision machined parts that require inspection. This is especially true in the cylinder head and related components (camshaft and valve train) most of which are not found on 2-strokes.

Generally speaking, the larger the hp model, the more involved will be the cleaning and inspection process. The EFI 4-stroke Johnson/Evinrude motors are the most involved, basically being the marine equivalents of small, modern, fuel injected, automotive motors.

When inspection involves precision measurements, not only must the components be clean, but they must be measured roughly at room temperature, using the appropriate measurement equipment to ensure accurate results.

It is very important to keep in mind that all wear components (pistons, rings, shafts, springs, valves, bearings, etc) **must** be reinstalled in their original locations whenever they are being reused. Wear patterns form on all contact surfaces during use. Mismatching wear patterns will accelerate wear, while matching wear patterns helps ensure a durable and reliable repair.

Cleaning

MODERATE

◆ See Figures 111 thru 116

✱✱ WARNING

Avoid removing excessive amounts of metal when removing carbon deposits.

1. Use a blunt-tip scraper or dulled chisel to loosen carbon deposits from various components of the combustion chamber and ports/valves. Work slowly and carefully to prevent damaging or excessively scoring the surfaces. Then use a Scotchbrite pad and mild solvent to remove most/all of the remaining deposits. Remove deposits from the following components, as applicable:
   • Remove carbon from the combustion chambers in the cylinder head. Use great care on 4-stroke motors, since the valve seating surface will be exposed after disassembly. DO NOT score or damage the valve seating surfaces. Another option is to protect the valve seating surface by leaving the valves in place while you are removing the carbon deposits, but the same care must be taken to prevent deep scores on the bottoms of the valves.
   • On 2-stroke motors, remove all carbon deposits from the areas around exhaust ports.

Fig. 111 Although a blunt chisel is preferred, a wire brush can be used WITH CARE to clean most carbon deposits

Fig. 112 Pistons CAN be cleaned while still installed, but again use care, and leave NO metallic deposits behind

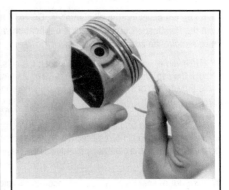

Fig. 113 The preferred method for cleaning piston grooves is to use the filed end of a broken ring. . .

Fig. 114 . . . but a ring groove cleaning tool can be used, with care to prevent damaging the piston

Fig. 115 The cylinder walls must show an obvious cross-hatching (tiny grooves, criss-crossed in a pattern around the bore). . .

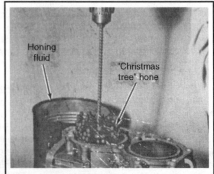

Fig. 116 . . . otherwise a cylinder hone must be used to break the smooth glazed surface into a cross-hatch pattern

POWERHEAD REPAIR AND OVERHAUL 6-73

• If equipped, remove carbon deposits or corrosion from the exhaust cover.

■ Pistons can be removed when still installed in the bores. This is handy when the cylinder head is removed for service without completely disassembling the crankcase. If this is to be accomplished, position the piston to be cleaned at TDC and cover the other piston bore(s) using rags and plastic. Thoroughly clean all debris using solvent and compressed air (WHILE WEARING SAFETY GLASSES) before moving to the next piston.

• Remove carbon deposits from the top of the piston(s). When working on the piston domes, use a light touch to prevent scratching, or worse, gouging the piston.

✱✱ WARNING

Wire brushes are not recommended for cleaning piston domes since particles of steel could become lodged into the piston surface. If this occurs, they could glow hot when the piston is returned to service, causing pre-ignition or detonation that could damage the piston and combustion chamber.

• Remove carbon from the ring grooves either using a ring groove cleaner, or, better yet, using a broken piece of the piston ring with an angle ground on the end. When using a ring to clean the piston grooves, use the ring actually removed from the groove (if it is being replaced) or one from the same groove on another piston.

✱✱ WARNING

When cleaning piston ring grooves, use the same caution as with the pistons. Do NOT remove excessive amount of material or the piston will be damaged beyond use. Some ring groove cleaning tools are heavy duty and will easily remove too much material, so use them with care. Believe it or not, the manufacturer recommends the filed broken ring method and we concur.

2. Inspect all water passages for corrosion deposits, debris or blockage. Remove debris and clean corrosion as needed and accessible. Use low pressure compressed air to blow out all water passages.

3. Clean and degrease all regularly oiled surfaces (including the crankshaft, pistons and connecting rods) using solvent or degreaser. Use low pressure compressed air to remove all build-up from shaft and rod oil holes.

4. Remove all traces of gasket or sealant using OMC Gel Seal and Gasket Remover or equivalent gasket removing solvent. Whenever possible, avoid the use of gasket scrapers to help avoid the possibility of scoring and damaging the gasket mating surfaces.

■ Honing the cylinder walls can wait until after measurements are taken to determine if boring for oversized pistons/rings will be necessary. If honing is pushed off until after inspection, be sure to follow the remaining steps in order to finish cleaning the block, then repeat the appropriate steps again, after honing.

5. Check the cylinder walls for glazing (a smooth, glassy appearance) and, if found, hone the cylinder walls using a medium grit cylinder hone. Use a slow rpm while raising and lowering the hone through the cylinder in order to cross-hatch the cylinder walls for maximum oil retention.

✱✱ WARNING

Use the cylinder hone slowly, carefully and as little as possible to avoid the possibility of removing too much material from the cylinder walls. If the bores are over-honed, it could cause the pistons to be below specification for the new cylinder measurement, or worse, cause the cylinder bores to be overspec for what pistons are available.

6. Wash the entire cylinder block, crankcase and head using warm, soapy water to remove all traces of contaminants. Use low pressure compressed air to blow dry all passageways.

7. Apply a light coating of clean engine oil to all machined surfaces that are not about to be measured right away. When you return to the task of measuring components that have been oiled, use a solvent covered rag to wipe away the oil before measurements are taken.

8. Cover all components using a plastic sheet to keep dust, dirt or debris from contaminating the cleaned and especially the oiled surfaces.

Cylinder Head Component Inspection

The VAST majority of cylinder head inspection procedures apply only to 4-stroke motors. This is because a cylinder head on a 2-stroke motor is usually nothing more than a head, a cover that is bolted in place on top of the pistons. It is really used more as an access to the pistons themselves than anything else. For 2-stroke motors, the only applicable procedure is to check cylinder head flatness to make sure it is not warped. On the other hand, the cylinder heads on 4-stroke motors contain the camshaft, rockers or valve shims/tappets and valve train, all of which require inspection.

■ Before going any further, if you haven't already, check the Engine Specifications chart for the motor on which you are working. Keep it handy when reviewing the inspection process, both to determine which checks are applicable to your motor and to provide the required specification.

Checking for Valve Leakage

The cylinder heads on 4-stroke motors should be checked for valve leakage before disassembly for overhaul.

Check the cylinder head for valve leakage as described under Cleaning & Inspection in this section. If leakage or an obvious defect is present, disassemble the cylinder head (as detailed in this section under the various Cylinder Head Overhaul procedures) and inspect the valve seating surfaces.

1. Place the cylinder head on a suitable work surface. Make sure the intake manifold ports are located above the valves.

2. Pour solvent or kerosene into the intake and exhaust ports, and watch the valves. There should be little or no leakage from a closed valve.

3. If significant leakage is present, check the valve seats, valve faces and springs for signs of defects, wear or damage.

4. Pour the solvent or kerosene out of the cylinder head before proceeding.

Checking the Cylinder Head

◆ See Figures 117 and 118

For all motors (2- and 4-stroke alike), check the cylinder head flatness by measuring the amount of warp across the head at various directions across the head mating surfaces. As with all inspection procedures, make sure the head is completely free of dirt, debris, oil or gasket materials. A machinist's straightedge and a feeler gauge set are necessary for this check. Be sure to hold the straightedge firmly on the head and use the feeler gauge to measure the gap at the midpoint. The warpage measurement is the largest feeler gap that can be inserted between the straightedge and the head with a slight drag. The next larger gauge should not pass and the next smaller should pass without drag. Check the Engine Specifications chart for maximum allowed warpage for the motor on which you are working.

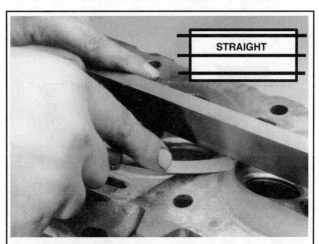

Fig. 117 Use a machinist's straightedge and a feeler gauge set to check the cylinder head for warpage across the mating surfaces, both straight across. . .

6-74 POWERHEAD REPAIR AND OVERHAUL

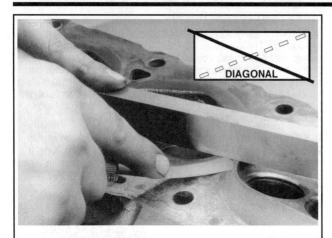

Fig. 118 ... and at diagonals across the head to be sure the surface is within specification for reuse

If the surface is just slightly out of spec, you can usually true it using a surfacing plate and 600-grit sandpaper as follows:

1. Cover a large, completely flat metal plate (a surfacing plate) with abrasive sandpaper (600-grit) with the abrasive surface facing upward.
2. Place the cylinder head on the plate, over the paper with the gasket mating surface downward.
3. Move the cylinder head in a figure-eight motion over the paper while pushing downward with gentile pressure.

✱✱ WARNING

DO NOT remove too much material from the cylinder head mating surface, as you will affect (increase) engine compression. Significant changes in engine compression from cylinder head or block resurfacing and the increase of carbon deposits (which will occur naturally in use) can lead to performance problems (such as pinging/pre-ignition).

4. Clean and recheck the surface constantly, until the cylinder head is back within specification.
5. When finished, clean the surface thoroughly with hot soapy water, then dry using low pressure compressed air.

Checking the Camshaft

◆ See Figures 119 thru 122

All 4-stroke motors utilize a camshaft that is timed to the crankshaft via a timing chain (40/50 EFI motors only) or belt. The camshaft is used to open and close the intake/exhaust valves at the appropriate time allowing the engine to draw air/fuel mixtures into the combustion chamber, seal the chamber when producing power, and open the chamber afterwards to release the unburned remains of the air/fuel charge. The height of in the intake and exhaust lobes are probably the most important specification on a camshaft, as they determine the maximum possible amount the shaft can open the valves. A shaft with worn lobes will not fully open the affected valves, leading to performance problems (by preventing the combustion chamber from breathing properly, preventing it from drawing in sufficient air/fuel charges and/or from completely venting the exhaust gasses).

In addition to camshaft lobe specs, many more camshaft specifications are available for the EFI motors than are available for the smaller, carbureted Johnson/Evinrudes.

On all engines, the general condition of the shaft itself must be checked. Look for scored or worn areas, signs of discoloration from heat or excessive wear possibly caused by a lack of oiling. This is especially important on EFI motors, as experience in the field has shown them to be particularly susceptible to premature camshaft wear, especially if the oil is not changed with sufficient frequency or if the wrong grade/type is used. Replace the camshaft is obvious defects are found or if any surfaces measure out of specification.

As with all inspection procedures, make sure the surfaces are clean and dry.

1. Using a micrometer, measure the height of each camshaft lobes. When the lobe (raised portion of the camshaft) is positioned upward, the height is the vertical measurement from base to lobe tip. Record each measurement and compare with the Engine Specifications chart.
2. For 5/6 and 8/9.9 hp motors, check the compression relief pin. The pin is found on the side, towards the end of the shaft. Use a small, flat-blade screwdriver or pry tool to check if the pin slides easily and smoothly (without sticking). If the pin sticks or requires significant effort to move, the camshaft should be replaced.
3. For EFI motors, check the camshaft run-out, as follows:
   a. Support the camshaft on V-blocks under the top and bottom (end) journals.
   b. Position a dial indicator with the tip on one of the middle journals and zero the dial.
   c. Rotate the camshaft slowly and record the maximum movement on the dial.
   d. Compare the reading to the Engine Specifications chart. Replace the camshaft if readings are greater than specified.
4. For EFI motors, check the bearing surfaces and oil clearances, as follows:
   a. Using a micrometer, measure the thickness of the camshaft bearing journals (at the point where they contact the cylinder head surface). Compare the measurements with the Engine Specifications chart. If the journal thickness is within specification, the camshaft is good, check the bearing surfaces in the cylinder head.
   b. For 40/50 hp EFI models, install and tighten the camshaft caps as detailed in the Powerhead Disassembly & Assembly procedure in this section.
   c. Using a micrometer or bore gauge, measure the cylinder head camshaft bore (journal holder) at each journal surface and record the results. Compare the measurements with the Engine Specifications chart. If the measurement is out of specification, the cylinder head must be replaced.
   d. Oil clearance is the amount of gap that exists between the journal and bore (for oil). In this case, the measurement is equal to the size of the bore, minus the size of the journal. Subtract each journal measurement from

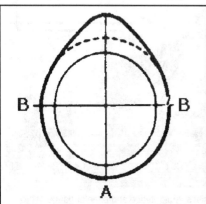

Fig. 119 Cut-away view of a camshaft lobe showing height (A) versus base width (B)

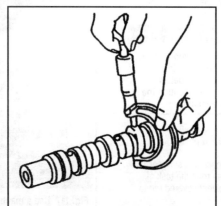

Fig. 120 Measure camshaft lobe height using a micrometer as shown

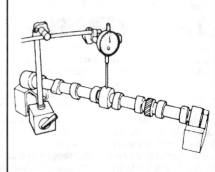

Fig. 121 Use a dial indicator and pair of V-blocks to check camshaft run-out on EFI motors

POWERHEAD REPAIR AND OVERHAUL

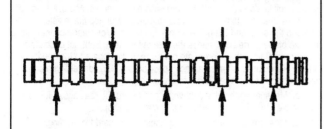

Fig. 122 Also on EFI motors, measure the bearing journals (shown) and cylinder head journal holders (bores) to see if they are in spec and to determine oil clearance

the bore size and compare to the camshaft oil clearance specifications. If the camshaft is within specification, replace the cylinder head to correct excessive oil clearance (or vice-versa).

■ As with all precision measurements, double-check your work before condemning a part (especially a big one like the cylinder head or the camshaft.

Checking the Tappets (40/50 EFI only) or Rocker Arms and Shafts

◆ See Figures 123 and 124

Various valve train configurations are used on Johnson/Evinrude 4-stroke motors. Most of them use a single camshaft that actuates the valves through shaft mounted, adjustable rocker arms. However, the 40/50 hp motors are equipped with dual-camshafts mounted directly over the valves and actuate them using shim/tappet assemblies. As with all inspection procedures, make sure the surfaces are clean and dry. Refer to the Engine Specifications chart for the motor on which you are working and the appropriate inspection steps in the following procedure:

1. For all 4-stroke motors, visually check for signs of obviously damaged (cracked, chipped), discolored or excessively worn surfaces. On models that utilize rocker arm/shaft assemblies with spacers and springs, check the springs for weakness or distortion, check the shafts for obvious warpage. Replace damaged or excessively worn parts.

■ During inspection, expect to find some wear on the rocker arm face. But, there should be no pitting. If wear is bad enough to show pitting on the face, replace the rocker arm to ensure component durability.

2. For 40/50 hp EFI motors, refer to the Engine Specifications chart, then measure the tappet and tappet bore as follows:

   a. Use a micrometer to measure the outer diameter of each valve tappet. Record the measurements and compare to specification. Replace any tappets that measure less than spec.

   b. Using a bore gauge, measure the cylinder head tappet bore diameters. Again, record the measurements and compare to specification. The cylinder head must be replaced if tappet bores are worn beyond spec.

   c. Oil clearance is the amount of gap that exists between the tappet and bore (for oil). In this case, the measurement is equal to the size of the bore, minus the size of the tappet. Subtract each tappet measurement from the appropriate bore size and compare to the tappet oil clearance specifications. If the bore is within specification, replace the tappet to correct excessive oil clearance. But, if the tappets are within spec and no slightly thicker tappet is available to correct oil clearance due to bore wear the cylinder head must be replaced.

3. For 70 hp EFI motors, check the rocker arm and shaft dimensions as follows:

   a. Using a micrometer, measure the rocker arm shaft at each rocker arm contact surface. Compare the reading to the Engine Specifications chart and replace the shaft if out of spec.

   b. Using a bore gauge, measure each the rocker arm bore diameter. Compare the reading to spec and replace the rocker arm, if the measurement is greater than allowed.

   c. Oil clearance is the amount of gap that exists between the shaft and arm bore (for oil). In this case, the measurement is equal to the size of the bore, minus the size of the shaft. Subtract each shaft measurement from the appropriate bore size and compare to the service limit for the shaft-to-rocker arm clearance specifications. If the bores are within specification, replace the shaft to correct excessive oil clearance (or vice-versa).

   d. As a final check of the shaft check it for excessive warpage/run-out using V-blocks and a dial indicator in the same fashion that the camshaft was checked (for more details, please refer to Checking the Camshaft). Support the ends of the shaft on V-block, then position and zero a dial indicator to the center of the shaft. Slowly rotate the rocker arm shaft and record the highest indication on the dial. Replace the shaft if the run-out exceeds the spec for maximum run-out.

Checking the Valve Assemblies

◆ See Figures 125 thru 130

Probably the most important key to cylinder head overhaul is the measurement and resurfacing or replacement (if necessary) of the valves, seats and guides. The tools and experience necessary to properly perform this phase of powerhead overhaul are USUALLY limited to machine shops. A talented mechanic can successfully perform these measurements, and with access to the right tools, repair a damaged cylinder head. But, most wont bother, instead referring the work to a reputable marine machine shop, since the time and effort is not worth the risk balance.

If a machine shop is not available or desired, take your time during this procedure. Make ALL listed checks and, when specs are available, measurements.

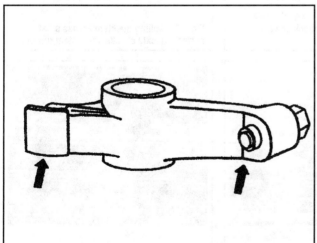

Fig. 123 For all models except 40/50 hp EFI motors, check the rocker arm faces for pitting and replace, as necessary

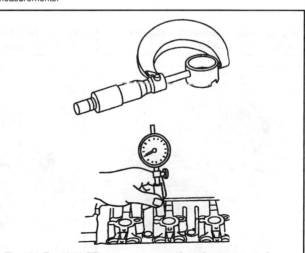

Fig. 124 For 40/50 EFI motors, measure the valve tappets and cylinder head tappet bores

1. On all motors, visually check the valve faces and seats for signs of pitting, cracked, corroded or damaged surfaces. Then check the valve stems for signs of obvious wear or damage.

2. For carbureted 4-stroke motors, check the following:

   a. Check the valve stems for signs of warpage. There are 2 acceptable methods for this, either use a flat piece of glass (such as a table top without and edge) to roll the valve stems (with the faces hanging over the edge to allow the full stem to contact the glass surface). You should be able to feel warpage as the stem is rolled. The other method is to use a pair of V-blocks and a dial-gauge (in a fashion similar to checking the camshaft or rocker arm shaft for run-out). However, keep in mind that Johnson/Evinrude does not publish specifications for maximum allowable run-out, so if a stem is questionable, discuss/show it to a machine shop.

   b. If not done already, CAREFULLY remove carbon deposits from the valve guides. Visually check them for signs of damage or excessive wear. If guide replacement is necessary, refer it to a machine shop.

   c. Valve seat concentricity can only be checked using special tools. If these tools are not available, refer it to a machine shop.

   d. The manufacturer does NOT provide spring specifications for these motors and, instead, advises that they be replaced whenever they are removed.

3. For EFI motors, check the following:

   a. Using a small bore gauge, measure the inside diameter of each guide and record the measurement. Compare to the Engine Specifications chart. If guide replacement is necessary, refer it to a reputable machine shop.

   b. Using a micrometer, measure the valve stem outer diameter at various points along the shaft. Record each measurement and use them to determine shaft-to-guide oil clearance (the amount of gap that exists between the stem and the guide for oil). In this case, oil clearance is equal to the size of the guide bore, minus the size of the stem. Subtract each stem measurement from the appropriate bore size and compare to the service limit for the guide-to-stem clearance specifications. If guide replacement is necessary, refer it to a reputable machine shop.

   c. Measure each valve seat width and record the readings. If it is out of spec, have the valve seat surfaced at a machine shop.

   d. Measure each valve head thickness and record the readings. Replace any valve with a thickness below spec.

   e. For each valve stem, measure the distance between the keeper groove and the very end of the valve stem. Replace any valve whose distance is below spec.

   f. Use a dial-indicator and a set of V-blocks to check the run-out of each valve stem and valve head. Position and zero the dial indicator at the center of the stem, then slowly rotate the valve and record the stem run-out. Move the indicator to the edge of the valve head and zero it again, then slowly rotate the valve and record the head Run-out. Compare the readings and replace any valve whose head or stem exceeds the radial run-out specs.

   g. Use a sliding caliper to measure free length (completely uncompressed length) of each valve spring. Record the results and compare to the Engine Specifications chart. Replace any spring that is less than the minimum spec.

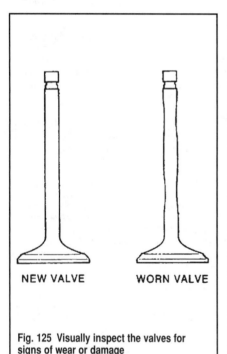

Fig. 125 Visually inspect the valves for signs of wear or damage

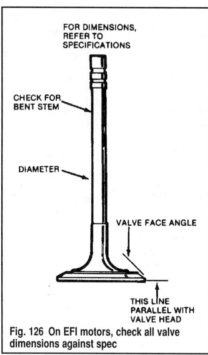

Fig. 126 On EFI motors, check all valve dimensions against spec

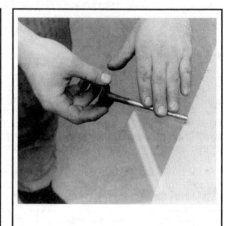

Fig. 127 Rolling the stem across a flat surface quickly checks valve stem run-out

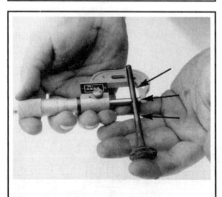

Fig. 128 On EFI motors, measure the stem thickness at various points

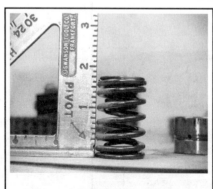

Fig. 129 Check the valve spring for distortion using a square

Fig. 130 On EFI motors, make sure the free length is not below the minimum spec

POWERHEAD REPAIR AND OVERHAUL 6-77

h. Using a spring tension measurement device, check the tension and the specified height. Replace the valve spring is tension is less than specified for the given height.

i. Position each valve spring on a flat surface, then slide a square up next to each to measure the deviation from squareness (the maximum distance the top of the spring is from the square when the bottom is butt up against the square). Replace any spring that is out of spec.

Cylinder Block Component Inspection

Make sure the work area is suitable to cylinder block reconditioning. This means it will have to be relatively clean and free of dust, dirt, debris or moisture. The presence of any one of the aforementioned contaminants will force you to take steps to protect the components. For instance, mop a dirty floor to help prevent kicking up dust and debris when you are working. Or, if necessary, place some large, flattened cardboard boxes down over the otherwise dusty floor (this will help when oil or assembly lube is invariably spilled or dripped during the rebuilding process. Make sure there is sufficient light, especially when you are checking for cracks or damage on the component surfaces.

Make sure that you are comfortable with your ability to read the various precision measuring equipment including micrometers and bore gauges. If necessary, you might want to consider handing the block over to a reputable marine machine shop for inspection and overhaul and assembly.

■ **All components must be clean and dry before taking precision measurements. Remember too that all specifications, unless otherwise noted, are for components at about room temperature. Temperature variations will also cause differences in measurements.**

As with cylinder head procedures, not all components are used by all models. For instance most of the 2-stroke motors use needle-roller bearings, while 4-stroke models may instead be equipped with bearing liner inserts. There are some important differences in how you should inspect these 2 types of bearings. Similarly, the manufacturer does not supply the same specifications for all motors, especially the smaller, lower hp models. In some cases this is because the specification flat out would not apply, such as a main bearing oil clearance on a needle-roller bearing (that dog just won't hunt). But when using bearing liner inserts, that oil clearance is a critically important measurement that must be taken to ensure the durability of a rebuild. In some cases, the manufacturer recommends that components are not to be reused (such as rings) even though many traditional rebuilds would include measuring and possibly reusing these components. Just keep in mind that in the end, following the manufacturer's recommendations will not normally cause a problem, but the same can't be said for ignoring them.

Checking the Cylinder Block

◆ See Figure 131

Once clean, visually inspect the cylinder bores for cracks, glazing (a smooth surface), scoring or deep gouges. A cylinder hone is used to clean up a glazed or lightly scratched/scored surface. However, a hone cannot usually help with deep gouges. If deep cuts are found in the cylinder walls you'll either have to replace the block or, if an oversize piston is available (see a friendly parts supplier), have a machine shop bore the cylinder oversize. Another option that **might** be available on some models is to have the cylinder sleeved (a process in which a new cylinder wall is pressed into place in the block, giving a fresh wear surface that can be bored or honed to match the piston size).

After inspecting the cylinders, turn your attention to the crankcase, cylinder block, head, exhaust cover (if applicable) and, intake or exhaust manifold (again, as applicable) mating surfaces. These surfaces must be clean and free of all dirt, debris or sealant. Visually inspect these surfaces for signs of deep scratches, cracks or other damage. If dowel pins are used, make sure they are not loose or damaged, and replace, as necessary.

■ **White almost powder-like deposits are sometimes formed when water enters the combustion chamber. If such contamination is noted before cleaning the cylinder head, check the cylinder walls and cylinder head for cracks. Also, think back to the condition of the cylinder head gasket or seal, as that too can be the culprit.**

Visually check all Inspect all bolts, studs, nuts and bolt holes for cracks, corrosion or damaged threads. Using the proper size tap or die is usually a good idea to make sure they are completely free of dirt, debris or corrosion. The main bearing and the cylinder head bolts/holes are probably the most critical, so pay special attention to them.

■ **Although threaded inserts can sometimes be used to repair bolt holes, a block that is heavily corroded or experiences problems in multiple holes should be replaced. If corrosion has caused threads to fail in multiple holes, the rest are probably not far behind.**

The cylinder block and the crankcase are normally a matched set and must be replaced as an assembly. For details, refer to your parts supplier.

Once the block is cleaned and otherwise ready for use, check the cylinder head gasket surface for flatness by measuring the amount of warp across the block at various directions across the head mating surfaces. As with all inspection procedures, make sure the block is completely free of dirt, debris, oil or gasket materials. A machinist's straightedge and a feeler gauge set are necessary for this check. Be sure to hold the straightedge firmly on the block and use the feeler gauge to measure the gap at the midpoint. The warpage measurement is the largest feeler gap that can be inserted between the straightedge and the head with a slight drag. The next larger gauge should not pass and the next smaller should pass without drag. Check the Engine Specifications chart for maximum allowed warpage for the motor on which you are working (note that the EFI motors have a **very** slightly different spec for the block than the head, all other motors use the same spec for both block and head).

If the surface is just slightly out of spec, you can usually true it using a surfacing plate and 600-grit sandpaper as follows:

1. Cover a large, completely flat metal plate (a surfacing plate) with abrasive sandpaper (600-grit) with the abrasive surface facing upward.

2. Place the cylinder block on the plate, over the paper, with the mating surface facing downward.

3. Move the cylinder block in a figure-eight motion over the paper while pushing downward with gentile pressure.

✱✱ WARNING

DO NOT remove too much material from the cylinder block-to-head mating surface, as you will affect (increase) engine compression. Significant changes in engine compression from cylinder head or block resurfacing and the increase of carbon deposits (which will occur naturally in use) can lead to performance problems (such as pinging/pre-ignition).

4. Clean and recheck the surface constantly, until the surface is back within specification.

5. When finished, clean the surface thoroughly with hot soapy water, then dry using low pressure compressed air.

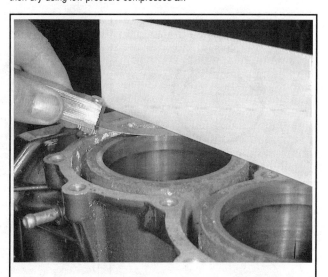

Fig. 131 Check the cylinder block along the head gasket mating surfaces for flatness

6-78 POWERHEAD REPAIR AND OVERHAUL

Checking the Cylinder Bores
◆ See Figures 132 thru 136

A cylinder bore gauge or a telescoping gauge must be used to take various measurements of the cylinder bore size. The major reason for measuring the cylinder bore is simply to determine how much wear has occurred in the motor. The most basic check of bore size is to make sure that the bore is the proper size for the pistons (that it has not worn beyond use). A piston that has too loose a fit in the cylinder block will allow combustion gasses to escape past the rings, losing both compression and power. On 4-stroke engines, this also prevents the oil rings from properly performing their jobs, as they will not fully contact the cylinder walls. Increased oil consumption is a symptom of an overly worn cylinder bore in 4-stroke motors. A lack of engine compression is another symptom of a worn bore.

Additionally, you must check to make sure that the cylinder has not worn unevenly. So, it is not enough to measure the bore at one point (vertically or horizontally), because the cylinder could be worn in a slightly oval shape (out-of-round) giving a larger or smaller measurement across the diameter in when the measurement is taken in different directions. Similarly, because the cylinder could be wider at the top or bottom (tapered) it must be measured at different depths.

It is critical to the success of the rebuild that these measurements are taken accurately. Only use precision gauges, with all gauges and the cylinder block at room temperature. Practice using the gauges and double or triple-check all measurements. Work slowly and carefully. Most bore gauges will have a thumb-wheel that is designed to free-wheel over a certain torque. If equipped, be sure to use it, as it will prevent pre-loading the gauge to the point where it would give inaccurate readings.

If either the proper gauges or the confidence to use them is not available, then refer this task to a reputable marine machine shop.

1. If not done during the initial cleaning, hone the cylinder bore lightly to remove any glazing prior to measurement. Remember that a cross-hatching will be necessary for proper oil retention and honing will change the measurements slightly, so it would be pointless to measure, then hone, since re-measurement will be necessary after honing.

■ When using a cylinder hone follow any instructions from the tool manufacturer. Also, keep in mind the following points. Always use an appropriate honing oil, keep the hone perfectly parallel to the depth of the bore and be sure to move it slowly in and out of the cylinder.

2. Using the cylinder bore gauge or telescoping gauge, measure the cylinder bores at 2 depths, the first about 1/4 in. (6mm) below the deck of the cylinder bore. The second depth should be in the lower area of piston travel (at least halfway down the bore). On 2-stroke motors, make sure it is slightly above the ports. At each depth, measure the bore 2 times. The first time, measure across the bore (side-to-side of the block, at a right angle to the centerline of the engine), and the second time, straight through the bore (from end-to-end of the block, parallel to the centerline of the engine).

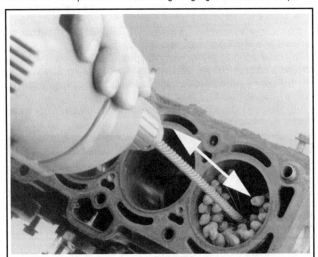

Fig. 132 If the cylinder's are glazed, use a hone to achieve a cross-hatching

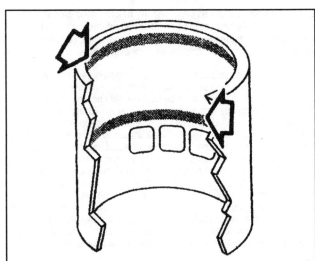

Fig. 133 Measure the cylinder bores at 2 heights (2-stroke shown, 4-strokes do not have the cylinder wall ports)

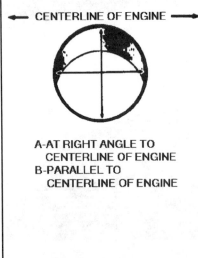
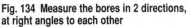

Fig. 134 Measure the bores in 2 directions, at right angles to each other

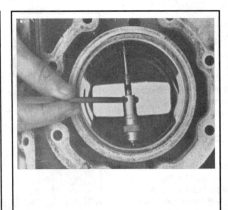

Fig. 135 Using a telescoping gauge to measure across the bore

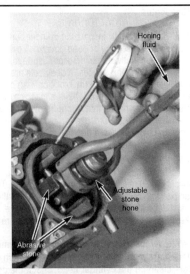

Fig. 136 A machinist can use a hone to cut the bore oversize

Compare the measurements to each other and the Engine Specifications chart, as follows:

    a. All of the measurements must be either within bore specification, or must not exceed oversize limit.

    b. In order to calculate cylinder taper, subtract the smallest diameter measured at the lower point in the bore from the largest diameter measured near the top of the bore.

    c. In order to calculate cylinder out-of-roundness, subtract the smaller of the measurements taken at each depth from the larger of the measurements taken at the **same** depth.

    d. Compare all measurements to the Engine Specifications chart to determine if the cylinder is usable or if it must be honed, sleeved (if possible) or discarded. Speak with your parts supplier and machine shop regarding oversize piston availability and how to proceed if boring is necessary.

■ **Remember that there are always variances in production. Be certain to obtain the oversize pistons and provide them, along with the block, to the machine shop for boring. Do NOT bore the cylinder without having the pistons on hand for matching.**

Checking the Crankshaft

◆ See Figures 137, 138 and 139

Once the crankshaft has been cleaned, it should be thoroughly inspected to make sure it is not damaged or excessively worn. Obvious defects, such as visible warpage or cracks are signs of un-serviceability. Similarly, significant etching are discolored areas may be signs that the crankshaft should not be reused.

Specifications will vary greatly from engine-to-engine. On most motors, specs are provided to help gauge the amount of wear on crankshaft or crankpin journals. A crankshaft that contains one or more journals, which are now out of spec, must be replaced. Some motors include specifications for crankshaft run-out. Journal taper and out-of-round specifications are available for the EFI 4-stroke motors. Refer to the Engine Specifications chart for the motor on which you are working and proceed as follows:

1. Visually inspect the crankshaft surfaces for signs of damage. Look closely at the bearing surfaces (journals) for signs of heat discoloration, etching or cracks. If you are uncertain of the condition, consult a reputable machine shop for advice before condemning the shaft.

2. If run-out specifications are available proceed as follows:

    a. Support the crankshaft on the outer journals (the bearing surfaces at either end of the shaft) using a set of V-blocks.

    b. Mount a dial indicator so it contacts the center journal surface on the shaft and then zero the dial.

■ **On single-cylinder motors, there is no center main journal. If a run-out specification is supplied for these motors, check 2 times, once with the dial mounted on the top journal and once with the dial mounted on the bottom journal.**

    c. Rotate the crankshaft slowly, while watching the gauge, then record the greatest amount of movement shown on the dial. Compare that reading to the specification.

    d. If there is more than one central journal, repeat at the other journal closet to center.

    e. If run-out exceeds specification, the crankshaft should be replaced.

3. If main journal and crankpin diameter specifications are available, use an outside micrometer to measure the diameter the journals and crankpin(s). On EFI motors, take multiple measurements across a single surface from front-to-back of the shaft, and at 90° angles around the shaft. Subtract the smaller readings from the larger readings (of the front-to-back measurements and of the measurements at the same point but rotated 90° around the shaft) in order to determine taper and out-of-round. The crankshaft must be replaced if any readings are out of spec (lower than specified diameter and, for EFI motors, greater amounts of taper or out-of-round than allowed, if it cannot be turned to correct it).

4. If top/bottom/center or main bearing journal specifications are available, use an outside micrometer to measure the diameter of the crankshaft journals. For EFI motors, again take measurements back-and-forth along the same journal (to determine taper) and at 90° angles around the journal at the same point forward-to-back (to determine out-of-round). Replace any crankshaft whose journal dimensions are below specification and, for EFI motors, whose journals are more tapered or out-of-round than allowed (unless it can be turned to correct it).

5. Double-check all out of spec readings and seek the advice of a reputable marine machine shop if the crankshaft fails these checks.

Checking the Pistons

◆ See Figures 140 thru 147

Visually check the pistons for signs of erosion at the edges of the dome or for cracks or physical damage to the dome. Check the ring grooves signs of erosion as well. Check the piston skirt for scoring or obvious damage. The piston must be replaced if any of these signs of damage are found.

The types of specifications provided by the manufacturer vary by engine. Refer to the Engine Specifications chart for the motor on which you are working and follow the applicable steps.

1. Visually inspect the pistons as noted for signs of obvious damage.

2. On all except the 25/35 hp (500/565cc) 3-cylinder and all EFI motors, check the sides of the piston skirt, as the factory pistons were equipped with a ground surface. If the surface has worn smooth in a triangular patter (refer to the accompanying illustration), the piston must be replaced.

Fig. 137 For all motors, visually inspect the crankshaft for damage or excessive wear

Fig. 138 Although V-blocks are best for checking run-out, in a pinch, you can usually use the cylinder block

Fig. 139 If specs are available, measure the main bearing and crankpin surfaces using a micrometer

6-80 POWERHEAD REPAIR AND OVERHAUL

3. If piston diameter specifications are available, measure the diameter of the piston skirt using a micrometer. For 2-stroke motors, the measurement should be taken at about 1/8 - 1/4 (3 - 6mm) from the bottom of the skirt. On EFI motors, the measurement should be taken 0.6 in. (15mm) from the bottom of the skirt. Record the measurement(s).

■ When measuring the pistons on 18 jet-35 hp (521cc) motors, be aware that they may be equipped with pistons from multiple manufacturers. The can usually be identified visually by a notch along the centerline of the dome on Art and Rightway pistons. The Zollner piston domes do not have the aforementioned notch. Identification is important, as specifications will vary between the manufacturers. For more details, refer to the accompanying illustrations.

4. If piston out-of-round specifications are available, repeat the diameter measurement around the base of the piston at a 90° angle from the previous measurement. Subtract the smaller measurement from the larger to determine out-of-roundness.

■ On 18 jet-35 hp (521cc) motors, the major diameter (measured at a 90° angle from the wrist pin hole) MUST be larger than the minor diameter (measured along the wrist pin centerline) by the cam dimension. Essentially, the piston, MUST be out-of-round by the cam spec for these motors. Again, refer to the accompanying illustrations.

5. For EFI motors, make the following measurements and calculations:

Fig. 140 Visually inspect the pistons for signs of damage or scoring

Fig. 141 On pistons with a ground surface, replace them if a smooth wear pattern has formed as shown

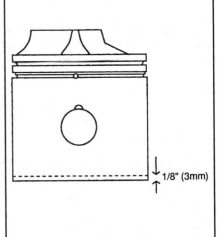
Fig. 142 On carbureted models with piston diameter specs, measure the diameter here

Fig. 143 Measure the piston outer diameter using a micrometer

Fig. 144 On EFI models, measure the wrist pin bores. . .

Fig. 145 . . . and wrist pin diameters to determine clearance

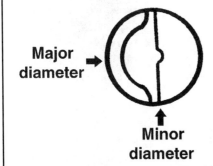
Fig. 146 18 jet-35 hp (521cc) motors may be equipped with Art, Rightway. . .

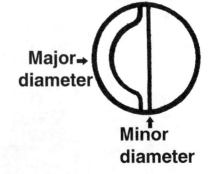
Fig. 147 . . . or Zollner pistons (Zollner pistons do not have the notch in the dome center)

POWERHEAD REPAIR AND OVERHAUL 6-81

a. Determine piston clearance by subtracting the piston diameter from the relevant cylinder bore diameter. Clearance must not exceed the maximum specification.

b. Use a bore gauge or sliding caliper to measure the wrist pin bore inner diameter (take measurements on either side of the piston). Measurements must fall within spec.

c. Use a micrometer to measure the outside diameter of each piston wrist pin. Take multiple measurements along the length of the pin, recording the largest and smallest diameters. Replace the wrist pin if the measurements do not fall within spec.

d. Calculate the piston pin-to-piton bore clearance by subtracting the largest of the pin diameter measurements from the smallest of the bore measurements. The piston and/or pin must be replaced if clearance is out of spec.

e. Use a set of feeler gauges to measure the ring groove widths. As with all feeler gauges, remember the reading is the gauge that passes with a slight drag (the next larger should not pass and the next smaller should pass without a drag). If the readings are below spec, carefully clean the grooves of carbon deposits. If readings are above spec, the piston must be replaced.

Checking the Piston Rings

◆ See Figures 148 thru 151

The manufacturer recommends replacing ALL piston rings anytime the powerhead is disassembled for service and we concur. Unless the powerhead is disassembled for some bizarre reason during very low hours of usage, there is no reason to skip this critical part of the rebuild. Replacing the rings will help with power, performance and durability.

Each of the rings must be measured to check gap and, if specifications are provided, groove clearance. Once a new set of rings has been measured and confirmed within spec either install them on the piston or keep the sorted with the piston to ensure installation only in the bore and on the corresponding piston for which they were measured.

1. On EFI motors, specifications are provided for the thickness of the top 2 (compression) rings. Measure the rings to make sure they are within spec. Double-check part numbers or mounting locations if readings are out of spec. For 70 hp motors, note that the thinner of the 2 rings is the top compression ring.

■ **Ring gap cannot be measured until after all honing is completed. Remember that honing will remove additional amounts of metal, which would change the ring gap.**

2. Measure the installed ring gap for the top two rings (compression rings) for each piston/cylinder bore as follows:

a. Apply a light coating of engine oil or assembly lube to the walls of the cylinder bore (this will help prevent scuffing or damage to the bore surface).

b. Carefully squeeze the ring and install it to the top of the bore, then use the piston dome (inverted and facing downward toward the ring) to slowly and carefully push the ring into the bore until square. For most motors, the ring gap can be checked with the rings towards the top of piston travel, but on EFI motors the manufacturer recommends positioning the ring about 1 in. (25.4mm) from the bottom of the cylinder bore.

c. Use a feeler gauge set to measure the gap between the ring ends. Remember the thickness of the feeler gauge that passes through with a slight drag is the measurement (the next larger should not pass and the next smaller should pass freely).

■ **If the gap is too large/small, try a different ring. DO NOT use rings with an incorrect gap.**

3. If a ring-to-groove clearance spec is provided for the compression rings (the top 2 rings on motors with more than 2, such as 4-strokes), check it as follows:

a. Either install the over the piston into the groove, or hold the ring into the groove by hand. In either case, be CERTAIN that you are checking the ring in the correct groove and for the correct piston/cylinder bore that you've just measured its end gap (and determined that it was in spec).

b. Use a feeler gauge to check the clearance between the ring and the piston groove. Again, the measurement is equal to the gauge that passes with a slight drag.

c. If clearance is insufficient, recheck the piston groove for carbon deposits and clean, CAREFULLY if found. If the clearance is too great, the piston must be replaced.

4. For all carbureted motors, install the top 2 rings (the compression rings) to the piston, then lay a machinists straight edge across the side of the piston (from dome-to-skirt). The rings must NOT hold the straight edge away from the piston. If they do, check for additional carbon deposits in the grooves or recheck for the part number to ensure they are the correct rings.

Fig. 148 Position each ring squarely in its intended bore...

Fig. 149 ...then use a feeler gauge to check end-gap

Fig. 150 If a spec is available, check the ring-to-groove clearance

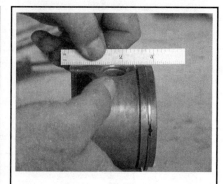

Fig. 151 On all except EFI motors, use a straight-edge to double-check the installed rings

POWERHEAD REPAIR AND OVERHAUL

Checking the Connecting Rod(s)

◆ See Figures 152 and 153

Once a connecting rod has been cleaned, it should be thoroughly inspected to make sure it is not damaged, bent or excessively worn. Obvious defects, such as visible warpage or cracks are signs of un-serviceability. Specifications are available on the smallest and the largest of the motors covered by this manual. On 2-5 hp, single cylinder 2-stroke motors, connecting rod deflection should be measured using a dial gauge. On the EFI motors, the connecting rod big-end (crankshaft end) must be checked for proper width, oil clearance and rod-to-crankshaft clearance. Although the width is measured using a micrometer, the oil clearance is normally measured using Plastigage® or an equivalent gauging material while the rod-to-crankshaft clearance is measured using a feeler gauge.

1. Visually inspect a connecting rod for twisted, bend or otherwise damaged surfaces for signs of damage. Look closely for cracks, chips, pitting or other signs of rough/damaged surfaces. Replace any damaged rod. If you are uncertain of the condition, consult a reputable machine shop for advice before condemning the rod.

2. For 2-5 hp, single cylinder 2-stroke motors, assemble the connecting rod to the crankshaft and use a dial gauge to measure deflection (the amount of movement) at the piston end of the rod. Push the rod gently toward one end of the crankshaft and hold, then place the gauge against the rod and zero it. Gently rock the rod back and forth, recording the maximum movement shown on the gauge and compare it to the Engine Specifications chart.

3. For EFI motors, proceed as follows:

   a. Measure the connecting rod big-end (crankshaft end) width using a micrometer.

   b. Either temporarily assembly the connecting rods to the crankshaft (using the intended bearings) and a gauging material (in order to determine connecting rod big end oil clearance or hold off and perform this during cylinder block assembly. In both cases, refer to Cylinder Block Overhaul for details.

   c. Again, with the connecting rod temporarily installed on the crankshaft, use a feeler gauge to check the rod-to-crankshaft side clearance. As with all feeler gauges, the measurement is the gauge that passes through with a slight drag. The next larger gauge should not pass at all and the next smaller gauge should pass without interference.

   d. Compare each of the measurements to the Engine Specifications chart in order to help determine in the connecting rod must be replaced. Repeat for each rod.

Checking the Rod and Main Bearings

◆ See Figures 154 and 155

Johnson/Evinrude outboards are equipped with various types of connecting rod and main bearings. Most motors are equipped with needle roller bearings. Some of those bearing assemblies are mounted in bearing cages, while others, although enclosed by halves or by a journal surface (such as the wrist-pin end of many connecting rods) are loose. A few motors, specifically the EFI engines, use automotive style bearing insert liners (smooth bearing halves, which are pressure lubricated by the 4-stroke oil pump.

In all cases, bearings and bearing surfaces should be checked for obvious signs of wear or damage. Bearing surfaces (journals and inserts) should be smooth, free of scratches, cracks, pitting, scoring or other damage. Needle bearings and cages should also be free of damage. Check needles for signs of uneven wear. Needle bearings, especially bearings that use loose needles, must be checked closely for missing needles.

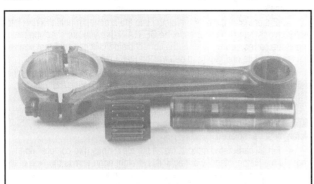

Fig. 152 Visually inspect the connecting rod for signs of damage

Fig. 154 Most carbureted motors use caged (shown) or loose needle roller bearings

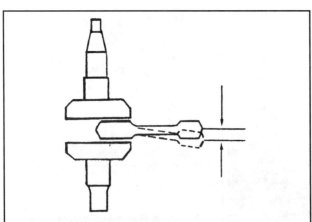

Fig. 153 On 2-5 hp 1-cylinder motors, use a dial gauge to measure rod deflection

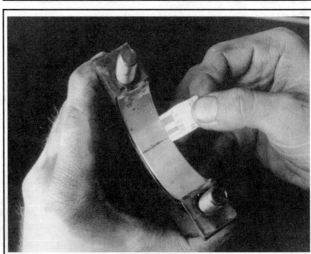

Fig. 155 Some motors (such as EFI models) use bearing inserts, which must be measured with a gauging material

POWERHEAD REPAIR AND OVERHAUL

All bearings are subjected to heat in use. Keep a close eye out for heat discoloration.

To ensure durability and dependability, replace bearings unless they appear like new.

Unless the crankshaft and/or cylinder block is replaced, always purchase replacement bearings of the same size and type. This is especially important on EFI motors, whose bearings are marked (with color codes) from the factory. For more details, please refer to Selecting Replacement Bearing Inserts (EFI Motors) in this section.

Check thrust bearings for signs of damage, including highly polished or heat discolored surfaces and, replace if found. On EFI motors, the thrust bearings should be measured using a micrometer and replaced, if below spec. Refer to the Engine Specifications chart.

Checking the Cylinder Block Exhaust Covers

Although most Johnson/Evinrude outboards use some form of exhaust cover mounted to the side of the cylinder block, no specifications for warpage are available. However, each time the cover is removed, take some time to inspect it for signs of excessive corrosion, damage or warpage. A cover that is damaged or warped cannot ensure an affective seal and should be replaced. Be especially careful when working on a motor that has suffered an overheat condition as exhaust covers are prone to warpage under these conditions.

■ Improper removal or installation procedures can sometimes warp an exhaust cover. Use of a crossing or spiraling pattern can help prevent damage.

Selecting Replacement Bearing Inserts (EFI Motors)

◆ See Figures 156 thru 165

The EFI motors covered by this manual are manufactured using bearing liner inserts for the crankshaft main and connecting rod bearings. Whenever bearing inserts such as these are used (as opposed to needle bearings) there is little room to move from specified oil clearance. Just like the bearings used in most automotive engines today, these precision machined surfaces are designed to operate within a specific and tight clearance in order to maintain proper oiling of the bearing surface by the 4-stroke oil pump.

However, also like most automotive engines, small deviations in machining will occur, which necessitates properly matching the bearing insert to the crankshaft journal surface. In order to help select the proper replacement bearings, the cylinder block and crankshaft are stamped with identification codes that can be used to help select the proper main bearings. Although the crankpin bearing oil clearance must also be checked, no stamped codes are provided to assist in the process.

During a rebuild, unless the crankshaft is machined or replaced, you should almost always use the same color code bearing when installing it back into the original cylinder block. If the colors are no longer visible, use the stampings on the flyweight and cylinder block to determine the correct color codes, and therefore the correct bearing size.

If OE bearings are not available, use the bearing thickness charts at a starting point to help select appropriate aftermarket bearings for your motor.

1. Determine the crankshaft journal outer diameters from the numeric codes stamped on the flyweight closet to the flywheel taper. Each code's position on the flyweight indicates for which main journal the code is intended. The outermost position (see the illustration) is for the No. 1 bearing (the bearing journal closest the flywheel), the next inboard codes is for the No. 2 bearing (the next closest), and so on. On 40/50 hp motors, there are only 4 main bearing journals, so there will only be 4 codes stamped on the crankshaft. The code itself is a number from 1 to 3. Compare it to the accompanying table for your motor to determine the original crankshaft journal outside diameter (double-check this measurement using a micrometer such as detailed under Checking the Crankshaft, in this section). Record the journal number and code for each of the main bearing journals.

2. Next, determine the cylinder block/crankcase holder inside diameter by locating the letter codes (A, B or C) stamped into the cylinder block on the starboard side. The first position (closet to the flywheel) represents the code for the No. 1 bearing holder, the second position is for the No. 2 bearing holder, and so on. There should be one code, per main bearing used by the block. Record the letter code for each corresponding main bearing journal bore.

3. Using the letter and number codes and the accompanying cross-reference chart, select the bearing inserts with the proper color code for each of the individual locations.

■ If testing shows the crankshaft journal outer diameters to be different from the original codes, assign them new numeric codes based on your measurements. Like with all critical measurements, measure twice and be certain before proceeding. Also, verify all bearing clearances during installation.

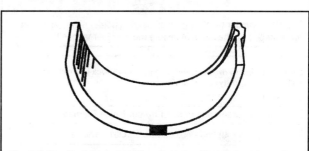

Fig. 156 Bearing inserts on EFI motors should contain color codes

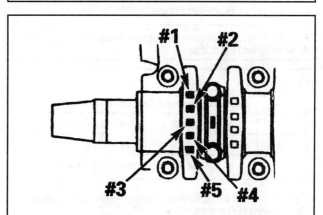

Fig. 157 Crankshaft journal diameter codes are stamped on the flywheel (numbers indicate journal position)

1 Numbers in drawing indicate crankshaft journal location (#1 is closest to flywheel end).

Standard:

Code	Crankshaft Journal Outside Diameter
1	44.994 - 45.000 mm (1.7714 - 1.7717 in.)
2	44.988 - 44.994 mm (1.7712 - 1.7714 in.)
3	44.982 - 44.988 mm (1.7709 - 1.7712 in.)

Fig. 158 Crankshaft journal diameter codes - 40/50 hp (815cc) EFI motors

1 Standard:

Code	Crankshaft Journal Outside Diameter
1	51.994 - 52.000 mm (2.0470 - 2.0472 in.)
2	51.988 - 51.994 mm (2.0467 - 2.0470 in.)
3	51.982 - 51.988 mm (2.0465 - 2.0467 in.)

Fig. 159 Crankshaft journal diameter codes - 70 hp (1298cc) EFI motors

6-84 POWERHEAD REPAIR AND OVERHAUL

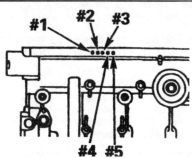

Fig. 160 Cylinder block/crankcase bearing holder diameter codes are stamped on the block (numbers indicate journal position)

2 Numbers in drawing indicate cylinder block journal location (#1 is closest to flywheel end).

Standard:

Code	Crankcase Bearing Holder Inside Diameter
A	49.000 - 49.006 mm (1.9291 - 1.9294 in.)
B	49.006 - 49.012 mm (1.9294 - 1.9296 in.)
C	49.012 - 49.018 mm (1.9296 - 1.9298 in.)

Fig. 161 Cylinder block/crankcase bearing holder diameter codes - 40/50 hp (815cc) EFI motors

2 Standard:

Code	Crankcase Bearing Holder Inside Diameter
A	56.000 - 56.006 mm (2.2047 - 2.2050 in.)
B	56.006 - 56.012 mm (2.2050 - 2.2052 in.)
C	56.012 - 56.018 mm (2.2052 - 2.2054 in.)

Fig. 162 Cylinder block/crankcase bearing holder diameter codes - 70 hp (1298cc) EFI motors

Bearing Selection Table

		Crankshaft Journal Outside Diameter		
	Code	1	2	3
Crankcase bearing holder inside diameter	A	Green	Black	No Paint
	B	Black	No Paint	Yellow
	C	No Paint	Yellow	Blue

Fig. 163 Bearing code-to-bearing color cross-reference chart - EFI motors

3 Standard:

Code	Crankshaft Bearing Thickness
Green	1.999 - 2.003 mm (0.0787 - 0.0789 in.)
Black	2.002 - 2.006 mm (0.0788 - 0.0789 in.)
No Paint	2.005 - 2.009 mm (0.0789 - 0.0791 in.)
Yellow	2.008 - 2.012 mm (0.0790 - 0.0792 in.)
Blue	2.011 - 2.015 mm (0.0792 - 0.0793 in.)

Fig. 164 Standard bearing thickness by color - 40/50 hp (815cc) EFI motors

3 Standard:

Code	Crankshaft Bearing Thickness
Green	1.998 - 2.0026 mm (0.0787 - 0.0788 in.)
Black	2.001 - 2.005 mm (0.0788 - 0.0789 in.)
No Paint	2.004 - 2.008 mm (0.0789 - 0.0791 in.)
Yellow	2.007 - 2.011 mm (0.0790 - 0.0792 in.)
Blue	2.010 - 2.014 mm (0.0791 - 0.0793 in.)

Fig. 165 Standard bearing thickness by color - 70 hp (1298cc) EFI motors

Timing Belt/Chain (4-Stroke Motors Only)

All 4-stroke motors are equipped with a timing belt or chain that is used to ensure that the camshaft(s) and crankshaft rotate together, in proper synchronization or timing. Proper camshaft-to-crankshaft timing is necessary to ensure that the valves open and close at the proper instants in order to sustain for 4-stroke engine operation. It is these changes in the combustion chamber that occur due to the crankshaft and valve train which allow a 4-stroke motor to operate. Remember that for every complete cycle of a 4-stroke motor, the crankshaft makes 2 complete revolutions (meaning that every piston goes up and down 2 times for each time the camshaft turns once). Whether or not a piston moving downward is on the intake stroke or the power stroke is simply a result of the camshaft position and whether or not the intake valves are open at the time of the piston's motion.

All motors, except the 40/50 hp EFI 4-strokes are equipped with a timing belt. The belt is mounted on sprockets at the top of the motor, just under the flywheel. For these motors, the belt should be inspected periodically to ensure it is in good condition. Remember that a snapped belt will instantly stop the motor and strand the boat. For motors equipped with a timing belt, the powerhead does not need to be removed in order to access or service the belt.

On 40/50 hp EFI motors, the crankshaft-to-camshaft timing (and there are dual-camshafts on these motors) is accomplished by a chain instead of a belt. The chain itself is a long-life component that does not require periodic inspection or replacement, which is a good thing really, since it is mounted underneath the powerhead, just above the gearcase. Obviously, on 40/50 hp EFI motors, the powerhead must be removed for access to the timing chain.

✱✱ WARNING

Some 4-stroke motors are of the interference design, meaning that if the crankshaft or camshaft is rotated independently a piston and a valve could strike, damaging or destroying one or both. For this reason, don't risk it. If the cylinder head is installed, never rotate one without the other.

REMOVAL & INSTALLATION

5-15 Hp 4-Stroke Motors

◆ See Figures 166 and 167

The carbureted 4-stroke motors covered by this manual are all equipped with a timing belt that should be inspected periodically and replaced if worn or damaged. Belt service is a relatively simple matter of accessing the belt (on the top of the powerhead), and making sure the engine is properly timed during removal or installation.

On some motors it is possible to remove the timing belt (for replacement) without removing the flywheel. In order to accomplish this, remove the manual starter assembly/flywheel cover and camshaft cover for access, then slide the belt carefully off the camshaft pulley. Once the belt is free of the camshaft pulley, use the slack created to free the belt from the crankshaft pulley and slide it over the flywheel. If the belt is difficult to remove, loosen the bolt and remove the camshaft pulley in order to free the belt. To install the replacement belt, slide it carefully over the flywheel and onto the crankshaft pulley teeth, finally, pull it over the camshaft sprocket.

POWERHEAD REPAIR AND OVERHAUL 6-85

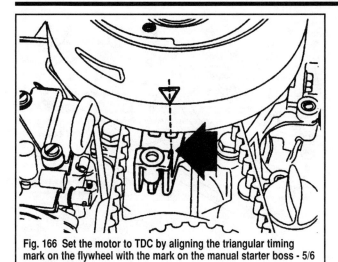

Fig. 166 Set the motor to TDC by aligning the triangular timing mark on the flywheel with the mark on the manual starter boss - 5/6 hp motors

Fig. 167 Set the motor to TDC by aligning the triangular timing mark on the flywheel with the pointer on the manual starter boss - 8/9.9 hp motors

1. Remove the rope starter assembly, for details, please refer the procedures in the Hand Rewind Starter section.
2. If equipped, remove the camshaft pulley cover.
3. On 5/6 and 8/9.9 hp motors, rotate the flywheel clockwise (slowly and by hand) to align the flywheel mark with the protrusion on the manual starter boss (between the flywheel and camshaft pulley).
4. Remove the flywheel, as detailed in this section.
5. Make sure the motor is set to Top Dead Center (TDC) of the No. 1 cylinder as follows:
   • On 9.9/15 hp motors, once the flywheel is removed, rotate the camshaft pulley until the pointer aligns with the raised boss on the cylinder head (the boss is directly between the 2 legs of the small, semi-circular bracket bolted to the top of the cylinder block. A white mark on the crankshaft lower belt guide should then align with the port side cylinder head-to-cylinder block split line. This will position the engine at TDC.
   • On 5/6 and 8/9.9 hp motors, verify that the camshaft pulley mark is aligned with the cylinder head protrusion (found at the head mating surface split line, between the cam and crank pulleys).

✸✸ WARNING

To prevent the possibility of damage (on interference motors) and to make installation easier, once the timing belt is removed from the engine, DO NOT rotate the crankshaft or camshaft for any reason (unless specifically required by a repair or testing procedure, and even then, follow the instructions closely.)

6. If the timing belt is being removed for access and NOT to be replaced, mark the belt to indicate the direction of rotation.
7. Carefully pull upward on the timing belt to free it from the camshaft sprocket. Once the belt off the pulley sprocket, use the slack to release it from crankshaft pulley.

■ In some cases, the belt can be difficult to remove. If so, remove the camshaft sprocket bolt in order to free the sprocket and create the necessary slack.

8. Use low pressure compressed air (if available) or a soft-bristled brush (if not) to loosen and remove dirt, debris or worn rubber from the pulleys.
9. Visually check the belt for worn or damaged surfaces as detailed under Timing Belt in the Maintenance and Tune-Up section.

To install:
10. Double-check that the timing marks are still aligned (the engine is still at TDC). If not, proceed as follows to align the timing marks:
   • For 5/6 and 8/9.9 hp motors, if removed, lower the flywheel back in position over the crankshaft taper (but do not install the retaining nut). Slowly turn the crankshaft clockwise, using the flywheel, until the mark on the flywheel aligns with the protrusion on the manual starter boss. Next, slowly turn the camshaft pulley clockwise until the camshaft pulley mark is aligned with the cylinder head protrusion (found at the head mating surface split line, between the cam and crank pulleys).
   • For 9.9/15 hp motors, slowly rotate the camshaft pulley clockwise until the pointer aligns with the raised boss on the cylinder head. Next, slowly rotate the crankshaft clockwise, by hand, until the white mark on the lower belt guide aligns with the port side cylinder head-to-cylinder block split line.
11. Install the timing belt around the crankshaft pulley, aligning the teeth of the belt with the pulley.

■ If you are reusing the previous belt, be sure to install it in the same direction of rotation, as marked during removal.

12. If the flywheel was removed, install the upper belt guide and retaining nut to make sure the belt does not move while it is positioned over the camshaft sprocket.
13. Make sure the timing marks are all still in alignment, then slip the belt over the camshaft pulley. Be sure that you do NOT rotate the camshaft or crankshaft pulleys while the belt is being installed or timing will be off.
14. If the camshaft pulley bolt was removed, apply a light coating of OMC Nut Lock, or equivalent threadlock to the bolt threads, then install and tighten it to 84-106 inch lbs. (10-12 Nm). If the bolt was loosened, but installed again with the pulley before or during belt installation, back it out sufficiently to apply threadlock to the threads, then install and tighten the bolt to specifications.
15. Turn the flywheel slowly, by hand, through 2 complete rotations and check to make sure the timing marks are still perfectly aligned. If not, remove the belt from the camshaft pulley, realign it and repeat to ensure proper installation.
16. If equipped, install the camshaft pulley cover.
17. If removed, install the flywheel as detailed in this section.
18. Install the rope starter assembly, as detailed in the Hand Rewind Starter section.

40/50 Hp EFI Motors

♦ See Figure 168

Unlike any other Johnson/Evinrude outboards, the 40/50 hp EFI 4 stroke motors are equipped with a long-life timing chain, mounted on sprockets underneath the powerhead. Therefore, also unlike the other 4-stroke motors, access to the timing chain requires removal and partial disassembly of the powerhead.

Although the camshaft sprockets can be removed, the crankshaft timing sprocket is integrated into the shaft.

✸✸ WARNING

This is an INTERFERENCE motor, meaning that if the crankshaft or a camshaft is rotated OUT OF timing with the other, valves and pistons could strike, damaging or destroying each other. The safest way to rotate this engine is with the timing chain properly installed. The only other way to rotate these shafts if to create sufficient valve lash by removing all off the valve tappets so the camshaft lobes DO NOT open the valves when the shafts are turned.

6-86 POWERHEAD REPAIR AND OVERHAUL

1. Remove the powerhead as described in this section.
2. Remove the oil pump case, as described under Powerhead in this section and under Oil Pump in the Lubrication and Cooling section.

■ Always turn the crankshaft in the normal direction of engine rotation. This is clockwise when looking down at the flywheel from above the powerhead, or counterclockwise when looking at the timing chain from below the powerhead.

3. Rotate the crankshaft until the mark on the end of the crankshaft is pointing directly away from the cylinder head/camshafts and aligns with the raised boss on the cylinder block. At this point, the colored links of the timing chain must align with the arrows on the camshaft sprockets **and** the marks on the camshaft sprockets (which are on the opposite side of the sprockets from the arrows) must align with the raised bosses on the cylinder head.

✷✷ WARNING

Unless the valve tappets are removed, to prevent them valves from opening or closing and/or the cylinder head is removed from the cylinder block, NEVER rotate the camshafts from this position once the timing chain is removed. This is an interference motor, meaning that if the crankshaft or camshaft is turned independent of one another, it is likely that a piston and a valve will come into contact and one or both will be damaged.

4. Remove the bolts (2) from the timing chain tensioner link (the short, rectangular bracket with the tension adjuster).
5. Remove the bolt, washer and spacer from the curved timing chain tensioner guide. Remove the tensioner guide from the powerhead.
6. Remove both bolts from the straight chain guide, on the opposite side of the chain from the tensioner guide. Remove the straight guide.
7. Remove the 2 bolts, then remove the small chain guide located between the camshaft sprockets.
8. Hold the intake camshaft from turning, then loosen the bolt removing the sprocket, dowel pin and timing chain.
9. Clean all components, then visually check for worn, corroded or otherwise damaged parts. Replace any part of which you are unsure. Remember it took some effort to get here.

To install:
10. Make sure the exhaust camshaft sprocket mark is still facing the raised boss (extrusion) on the cylinder head. Also, confirm that the No. 1 cylinder is at TDC of the compression stroke, with the crankshaft sprocket dimple facing the mark on the crankcase that is on the far side of the sprocket from the cylinder head.

✷✷ WARNING

If the engine is not properly timed, either the camshafts or all of the valve lash tappets must be removed in order to rotate the crankshaft safely back into timing.

11. Install the dowel pins and the camshaft sprocket to the intake camshaft.
12. Hold the intake camshaft and sprocket in position, then route the timing chain around the remaining camshaft and crankshaft sprockets. The yellow link in the timing chain must align with the crankshaft timing mark (the same mark that is facing away from the cylinder head and towards the mark on the cylinder block). Also, make sure the blue links align with the arrows (opposite the timing marks) on the camshaft sprockets.
13. Apply a light coating of clean engine oil to the threads of the camshaft sprocket bolts. Continue to hold the camshaft, sprocket and timing chain from turning while installing the bolts (3) into the sprocket and camshaft. Tighten the camshaft sprocket bolts to 84-90 inch lbs. (9.5-10.2 Nm).
14. Install the small timing chain guide between the camshaft sprockets, then tighten the bolts to 84-90 inch lbs. (9.5-10.2 Nm).
15. Apply a light coating of clean engine oil or assembly lube to the straight timing chain guide and the curved tensioner guide. Install both guides using the bolts, spacer(s) and washers, then tighten the bolts securely.
16. Press on the piston of the tensioner link and move the ratchet latch until the piston slides up completely, then place the tensioner link in position on the powerhead, allowing the piston to ratchet out until it contacts the link. Install the retaining bolts and tighten to 84-90 inch lbs. (9.5-10.2 Nm).

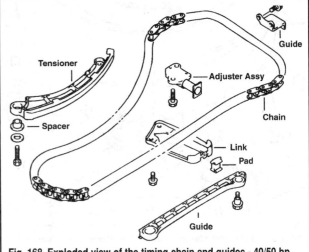

Fig. 168 Exploded view of the timing chain and guides - 40/50 hp (815cc) EFI motors

■ Some tensioner link and adjuster assemblies have a hole in the latch/body assembly in which a small pick or straightened paper clip can be inserted to hold the tension adjuster in position during installation. The pick or clip is then removed once the components are installed and bolts are threaded.

17. Slowly and carefully rotate the crankshaft 2 complete revolutions to verify correct camshaft timing and chain tension. After 2 complete revolutions, all timing marks and colored links must align properly as before removal and during installation.
18. If the cylinder head or any valve train components were replaced, check and adjust the valve lash as detailed under Valve Lash in the Maintenance and Tune-Up section.
19. Install the oil pump case to the powerhead as detailed under Oil Pump in the Lubrication and Cooling section. Pay attention to sealant and torque instructions.
20. Install the powerhead, as detailed in this section.

70 Hp EFI Motors

◆ See Figures 169 and 170

✷✷ WARNING

To prevent potential engine damage, do not allow the crankshaft to rotate once the timing belt is removed.

1. Remove the flywheel, as detailed in this section.
2. Remove the bolts retaining the stator and the starter motor bracket, then remove the stator, disconnect the MAP sensor and relocate the wires for access.
3. Make sure the motor is set to Top Dead Center (TDC) of the No. 1 cylinder by slowly turning the crankshaft sprocket (clockwise by hand) until the timing mark (dot) on the face of the camshaft pulley aligns with the protrusion on the cylinder head. At this point the timing mark (small holes in the crankshaft pulley belt guides) will align with the protrusion on the cylinder block.

■ When the camshaft and crankshaft pulley's are aligned with their timing mark protrusions, they will also both be facing inward towards each other.

4. Loosen the bolts (2) securing the belt tensioner to the cylinder block, but do not remove them. Allow the tensioner to move and retighten the bolts only after the belt has sufficient slack.

■ If the tensioner does not move enough to create slack sufficient for belt removal, force it back by hand before tightening the fasteners.

5. If the belt is to be reused, either mark the direction of rotation (clockwise) or note any factory marks, which may already be present on the belt for installation purposes.

POWERHEAD REPAIR AND OVERHAUL   6-87

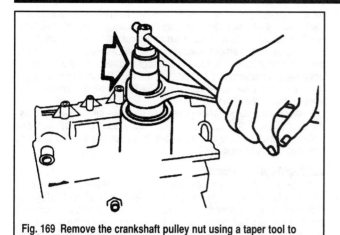

Fig. 169 Remove the crankshaft pulley nut using a taper tool to hold the shaft from turning

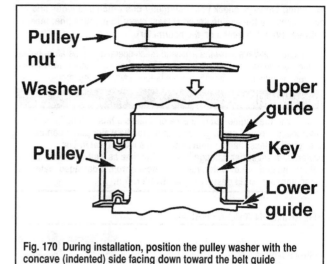

Fig. 170 During installation, position the pulley washer with the concave (indented) side facing down toward the belt guide

6. Remove the belt from the camshaft pulley by pulling upward, then free the belt from the crankshaft pulley and belt guides.

7. If the crankshaft pulley must be removed for seal or crankshaft service, proceed as follows:

■ An OMC crankshaft holder (#345827) or equivalent tool designed to fit over the crankshaft taper and keyway to hold the crankshaft from turning is required in order to remove the crankshaft pulley.

   a. Place the flywheel key into the slot in the crankshaft. Align the slot in the crankshaft adapter with the key, and seat the adapter to the crankshaft taper.
   b. Use a wrench to loosen the crankshaft pulley sprocket while holding the crankshaft adapter stationary using a breaker bar and socket, as shown.
   c. Remove the tool, nut, washer, upper belt guide, pulley, key and lower belt guide from the crankshaft, in that order.
   d. Carefully clean all dirt or debris from the crankshaft taper using solvent and a rag. Check the pulley and guides for damage or signs of excessive wear and replace, as necessary.

8. Use low pressure compressed air (if available) or a soft-bristled brush (if not) to loosen and remove dirt, debris or worn rubber from the pulleys.

9. Visually check the belt for worn or damaged surfaces as detailed under Timing Belt in the Maintenance and Tune-Up section.

To install:

10. If the crankshaft pulley was removed, install it as follows:
   a. Install the lower belt guide onto the shaft, so the curved edge is facing the cylinder block and the timing mark (hole) facing the cylinder block protrusion (toward the camshaft pulley).
   b. Insert the pulley key into the crankshaft slot, then install the pulley while aligning the pulley slot with the key.
   c. Install the upper belt guide, making sure the curved edge faces away from the cylinder block and the timing mark (hole) faces the toward the cylinder block protrusion (toward the camshaft pulley).
   d. Install the pulley washer with the concave (indented) side facing down toward the belt guide.
   e. Install the pulley nut, then tighten the nut to 50-52 ft. lbs. (68-70 Nm) using the Crankshaft Taper tool (and the flywheel key) to keep the crankshaft from turning.

11. Double-check that the timing marks are still aligned (the engine is still at TDC). If not, proceed as follows to align the timing marks as follows:
   a. If the camshaft or crankshaft pulley moved only slightly (just a couple of teeth) realign them now.
   b. If the engine (due to service or disassembly) is off TDC by anything more than a couple of teeth, loosen all of the valve lash adjusters (so all valves remain closed regardless of camshaft position). That way, you can safely rotate the camshaft and the crankshaft independently without fear of pistons/valves colliding and becoming damaged. For more details, please refer to the Valve Lash adjustment procedures under the Maintenance and Tune-Up section.

12. Install the timing belt over the crankshaft pulley first and then over the camshaft pulley. Make sure to align any directional arrows made or noted during removal or any factory made directional arrows. (Remember the belt/motor rotates clockwise when viewed from above).

■ When installed properly, there should be no slack on the non-tensioner side of the belt.

13. Loosen the tensioner bolts, allowing the spring tension to push the tensioner against the belt, tightening it on the pulleys. Tighten the tensioner bolts to 16-18 ft. lbs. (22-24 Nm).

14. Check the timing mark alignment, then slowly rotate the crankshaft by hand through 2 complete turns and recheck that the marks are still in alignment. If not, loosen the tensioner, adjust the belt position and repeat until the belt and motor remain in timing after the crankshaft is turned 2 complete revolutions.

15. Install the starter motor bracket and tighten the retainers to 16-18 ft. lbs. (22-24 Nm).

16. Apply a coating of OMC Locquic Primer and OMC Nut Lock or equivalent threadlock to the threads of the stator retaining bolts, then install the stator and tighten the bolts to 24-36 inch lbs. (3-4 Nm).

17. Connect the MAP sensor wiring, making sure it is secured where it will not be damaged by moving components.

18. Install the flywheel, as detailed in this section.

19. If the valve train was touched (during service or to re-time the motor) follow the Valve Lash adjustment procedure in the Maintenance and Tune-Up section.

POWERHEAD BREAK-IN

Anytime a new or rebuilt powerhead is installed (this includes a powerhead whose wear components such as pistons and rings, main bearings or, in the case of 4-stroke motors, cylinder head components have been replaced), the motor must undergo proper break-in.

By following break-in procedures largely consisting of specific engine operating limitations during the first 10 hours of operation, you will help you will help ensure a long and trouble-free life. Failure to follow these recommendations may allow components to seat improperly, causing accelerated wear and premature powerhead failure.

On all motors, special attention is required to the engine oil during initial break-in. For 2-stroke motors, pay close attention to the special fuel/oil mixture requirements. On 4-stroke motors, expect increased oil consumption during break-in. Don't be alarmed, but be sure to check the 4-stroke engine oil level frequently to prevent the possibility of powerhead damage from operation with a low oil level. Oil consumption should lessen as the piston rings seat to the cylinder walls.

Especially during break-in, pay close attention to all pre and post operation checks. This goes double when checking for fuel, oil or water leaks. At each start-up and frequently during operation, check for presence of the cooling indicator stream.

At the completion of break-in, double-check the tightness of all exposed engine fasteners.

While each engine requires slightly different steps for powerhead break-in, one procedure is common. During the entire first 20 hours of engine operation, **vary** the engine speed. This allows parts to wear in under conditions throughout the powerband, not just at idle or mid-throttle.

POWERHEAD REPAIR AND OVERHAUL

■ During break-in, check your hourmeter or a watch frequently and be sure to change the engine speed at least every 15 minutes (that means between every 2-3 tenths on the hourmeter).

Be sure to **always** allow the engine to reach operating temperature before setting the throttle anywhere above idle. This means you should always start and run the motor for at least 5 minutes before advancing the throttle.

** WARNING

NEVER run the engine out of the water, unless a flush fitting is used to provide a source of cooling. Remember that the water pump can be destroyed in less than a minute just from a lack of water. The powerhead will suffer damage in very little time as well, but even if it is not overheated out of water, reduced cooling from a damaged water pump impeller could destroy it later. Don't risk it.

BREAKING IN A POWERHEAD

2-Stroke Motors

Colt/Junior and 2-8 Hp Motors

For the first 5 hours of engine operation, use a 25:1 fuel/oil mixture. If equipped with the AccuMix R® oiling system, add a 50:1 fuel/oil mixture to the primary fuel tank which, when combined with the oiling of the AccuMix system will result in a 25:1 fuel/oil mixture. For more details, please refer to the information for Engine Oil (2-Stroke) in the Maintenance and Tune-Up section.

During break-in, observe the following time-table and limitations on engine operation.

■ Check the cooling indicator stream repeatedly to ensure proper engine cooling.

• During the first 10 minutes, operate the engine in gear at **only** fast idle.
• During the next 50 minutes, operate the engine in gear **below** 3500 rpm. If the boat planes easily, use **full** throttle to quickly bring the boat on plane, then immediately reduce throttle to 1/2 or less, but making sure the boat remains on plane. Vary the engine speed at least every 15 minutes.
• During the second hour of break-in, use full throttle to quickly plane the boat, then immediately reduce throttle to 3/4 or less, but make sure the boat remains on plane. Continue to vary the engine speed at least every 15 minutes. At various intervals, operate the engine at **full throttle** for 1-2 minutes, then reduce the throttle to 3/4 for an additional minute or two in order to allow the pistons to cool off slowly. Don't just drop from Wide Open Throttle (WOT) to idle.
• For the next 3 hours, continue to vary the engine speed and avoid continuous full-throttle operation.
• After the first 20 hours, the powerhead should be fully broken-in, follow the steps of the 20-hour service.

9.9-70 Hp Motors (Except the 25/35 Hp 3-Cylinder Models)

For the first 12 gallons of fuel, use a 25:1 fuel/oil mixture. If equipped with either the AccuMix R® or VRO2<ref.> oiling system, add a 50:1 fuel/oil mixture to the first 12 gallons of fuel run through the fuel tank. When combined with the oiling output of the oiling system this pre-mix will result in a 25:1 fuel/oil mixture. For more details, please refer to the information for Engine Oil (2-Stroke) in the Maintenance and Tune-Up section.

During break-in, observe the following time-table and limitations on engine operation.

■ Check the cooling indicator stream repeatedly to ensure proper engine cooling.

• During the first 20 minutes, operate the engine in gear at **only** fast idle. **do not** exceed 1500 rpm during the first 20 minutes.
• During the next 40 minutes, operate the engine in gear **below** 3500 rpm. If the boat planes easily, use **full** throttle to quickly bring the boat on plane, then immediately reduce throttle to 1/2 or less, but making sure the boat remains on plane. Vary the engine speed at least every 15 minutes.
• During the next 9 hours of break-in, use the throttle to quickly plane the boat, then immediately reduce throttle to 3/4 or less, making sure the boat remains on plane and making sure engine speed does **not** exceed 4500 rpm. Continue to vary the engine speed at least every 15 minutes. About every 30 minutes, operate the engine at **full throttle** for about 1 minute, then reduce the throttle to 3/4 for about an additional minute in order to allow the pistons to cool off slowly. Don't just drop from Wide Open Throttle (WOT) to idle. Throughout this time period, avoid continuous full-throttle operation.
• If equipped with either the AccuMix R® or VRO2® oiling system, check the oil tank after the first 12 hours of operation and verify that the oil level has dropped (indicating that the system is working). Do **not** stop running premix in the primary fuel tank unless you are certain the oiling system is working. Top off the oil tank.
• After the first 20 hours, the powerhead should be fully broken-in, follow the steps of the 20-hour service. Be sure properly retorque the cylinder head bolts (this should be done after the engine is run, but only after the cylinder head has cooled to the touch).

25/35 Hp 3-Cylinder Models

For the first 10 hours of engine operation, you **must** use a 50:1 pre-mix fuel/oil mixture in the primary fuel tank **in addition** to the oil supplied by this motors oiling system. After 10 hours, verify that the oil level has dropped in the mixing system tank before ceasing the use of pre-mix in the primary fuel tank. For more details, please refer to the information for Engine Oil (2-Stroke) in the Maintenance and Tune-Up section.

■ If, for any reason during the first 10 hours of engine operation, the mixing system is disconnected, you MUST run a 25:1 fuel/oil ratio premix in the primary fuel tank.

During break-in, observe the following time-table and limitations on engine operation.

■ Check the cooling indicator stream repeatedly to ensure proper engine cooling.

• During the first 10 minutes, operate the engine in gear at **only** fast idle.
• During the next 50 minutes, operate the engine in gear **below** 2700 rpm. If the boat planes easily, use **full** throttle to quickly bring the boat on plane, then immediately reduce throttle to 1/2 or less, but making sure the boat remains on plane. Vary the engine speed at least every 15 minutes.
• During the second hour of break-in, use full throttle to quickly plane the boat, then immediately reduce throttle to 3/4 or less, but make sure the boat remains on plane. Continue to vary the engine speed at least every 15 minutes. At various intervals, operate the engine at **full throttle** for 1-2 minutes, then reduce the throttle to 3/4 for an additional minute or two in order to allow the pistons to cool off slowly. Don't just drop from Wide Open Throttle (WOT) to idle.
• For the next 8 hours, continue to vary the engine speed. Be sure to avoid continuous full-throttle operation for long periods.
• After the first 20 hours, the powerhead should be fully broken-in, follow the steps of the 20-hour service. Be sure properly retorque the cylinder head bolts (this should be done after the engine is run, but only after the cylinder head has cooled to the touch).

Carbureted 4-Stroke Motors

For the first 20 hours of engine operation, use OMC Four-Stroke Outboard Break-In Lubricant in order to ensure proper break-in leading to long and trouble-free engine life.

During break-in, observe the following time-table and limitations on engine operation.

■ Check the cooling indicator stream repeatedly to ensure proper engine cooling.

• During the first 5 minutes, operate the engine in **Neutral** at slow idle to allow proper warm-up.
• During the next 15 minutes, operate the engine in gear at the slowest speed possible.
• During the balance of the first 2 hours, vary the engine speed at least every 15 minutes, but **do not** exceed 1/2 throttle at any point. If the first 2 hours do not occur in one outing, repeat the idle and warm-up procedure each time.
• During the third hour of engine operation, vary the engine speed at least every 15 minutes, but **do not** exceed 3/4 throttle at any point. If the first 3 hours do not occur in one outing, repeat the idle and warm-up procedure each time.
• During the balance of the first 10 hours, vary the engine speed at least every 15 minutes, but **do not** run at full-throttle continuously for more than 5 minutes. Allow at least 5 minutes of engine warm-up each time the boat is launched.

- After the first 20 hours, the powerhead should be fully broken-in, follow the steps of the 20-hour service.

EFI 4-Stroke Motors

For the first 20 hours of engine operation pay close attention to engine oil levels. Check it frequently and add, as necessary.

During break-in, observe the following time-table and limitations on engine operation.

■ **Check the cooling indicator stream repeatedly to ensure proper engine cooling.**

• During the first 20 minutes, operate the engine in gear **only** at fast idle. **do not** exceed 1500 rpm.

• During the next 40 minutes, vary the engine speed at least every 15 minutes, but **do not** exceed either 1/2 throttle for 40/50 hp motors, or 3500 rpm for 70 hp motors.

■ **If the boat planes easily, use full throttle to quickly accelerate the boat onto plane, then immediately slow throttle to 1/2 or less, while making sure the boat remains on plane at this throttle setting.**

• During the next 9 hours of engine operation, vary the engine speed at least every 15 minutes. Continue to bring the boat quickly on plane using full throttle, then reduce the throttle setting below 3/4 throttle and **do not** exceed 4500 rpm. About every 30 minutes, run the engine at full throttle for about one minute, then slow the motor to 2<over throttle or less, allowing the pistons to cool.

• After the first 20 hours, the powerhead should be fully broken-in, follow the steps of the 20-hour service.

SPECIFICATIONS

ENGINE SPECIFICATIONS - Colt/Junior (43cc) SINGLE CYLINDER ENGINES

Component	U.S. (in.) ①	Metric (mm) ①
Cylinder Bore		
Standard Bore Diameter	1.5643-1.5650	39.74-39.75
Oversize Service Limit	0.002 Max	0.05 Max
Out-of-round		
Service Limit	0.003 Max	0.08 Max
Taper		
Service Limit	0.002 Max	0.05 Max
Cylinder Head		
Gasket Surface Warpage	0.004 Max	0.10 Max
Crankshaft		
Top Journal		
Diameter	0.7497-0.7502	19.01-19.06
Bottom Journal		
Diameter	0.6691-0.6695	17.00-17.01
Crankpin		
Diameter	0.6695-0.6700	17.01-17.02
Piston		
Diameter	1.5620-1.5625	39.67-39.69
Out-of-round	0.002 Max	0.05 Max
Rings		
Groove clearance		
Service Limit	0.004 Max	0.010 Max
Gap		
Production	0.015-0.025	0.38-0.64

① Unless otherwise noted

6-90 POWERHEAD REPAIR AND OVERHAUL

ENGINE SPECIFICATIONS - 5.0 HP (109cc) SINGLE CYLINDER ENGINES

Component	U.S. (in.) ①	Metric (mm) ①
Cylinder Bore		
Standard Bore Diameter	2.1654-2.1659	55.000-55.015
Oversize Service Limit	0.002 Max	0.05 Max
Out-of-round		
Service Limit	0.003 Max	0.08 Max
Taper		
Service Limit	0.002 Max	0.05 Max
Cylinder Head		
Gasket Surface Warpage	0.003 per Inch	0.08 per 25.4mm
Crankshaft		
Connecting rod		
Rod Deflection	0.16 Max	4.0 Max
Crankshaft Run-Out	0.000-0.002	0.00-0.05
Piston		
Diameter	2.1630-2.1636	54.940-54.955
Out-of-round	0.002 Max	0.05 Max
Ring Gap		
Production Top	0.006-0.012	0.15-0.30
2nd	0.006-0.012	0.15-0.30
Service Limit	0.031 Max	0.80 Max

① Unless otherwise noted

ENGINE SPECIFICATIONS - 2.0-3.5 HP (78cc) SINGLE CYLINDER ENGINES

Component	U.S. (in.) ①	Metric (mm) ①
Cylinder Bore		
Standard Bore Diameter	1.8890-1.8906	48.00-48.02
Oversize Service Limit	0.002 Max	0.05 Max
Out-of-round		
Service Limit	0.003 Max	0.08 Max
Taper		
Service Limit	0.002 Max	0.05 Max
Cylinder Head		
Gasket Surface Warpage		
thru 1998	0.004 Max	0.10 Max
1999-01	0.003 per Inch	0.08 per 25.4mm
Crankshaft		
Top Journal		
Diameter	0.7875-0.7878	20.002-20.010
Bottom Journal		
Diameter	0.5906-0.5910	15.001-15.011
Crankpin		
Diameter	0.6299-0.6301	16.000-16.005
Connecting rod		
Rod Deflection	0.022-0.056	0.6-1.5
Crankshaft Run-Out	0.000-0.002	0.00-0.05
Piston		
Diameter	1.8868-1.8873	47.9247-47.9374
Out-of-round	0.002 Max	0.05 Max
Rings		
Groove clearance		
Service Limit	0.0026 Max	0.066 Max
Gap Production		
Top	0.0059-0.0138	0.15-0.35
2nd	0.0059-0.0138	0.15-0.35
Valve (Leaf/Reed) Opening	0.236-0.244	6.0-6.3

① Unless otherwise noted

POWERHEAD REPAIR AND OVERHAUL

ENGINE SPECIFICATIONS - 5/6 HP (128cc) SINGLE CYLINDER ENGINES

Component		U.S. (in.)①	Metric (mm)①
Camshaft			
Lobe Height	Intake	0.952-0.956	24.172-24.272
	Exhaust	0.953-0.957	24.212-24.312
Cylinder Bore			
Diameter		2.2244-2.2248	56.50-56.51
Oversize Service Limit		0.002 Max	0.05 Max
Out-of-round			
Service Limit		0.003 Max	0.08 Max
Taper			
Service Limit		0.002 Max	0.05 Max
Cylinder Head			
Gasket Surface Warpage			
1995-98		0.004 Max	0.10 Max
1999-01		0.003 per Inch	0.08 per 25.4mm
Crankshaft			
Top and Bottom Journals			
Diameter 1997-98		1.2200-1.2204	30.99-31.00
1999-01		0.9837-0.9842	24.98-25.00
Crankpin			
Diameter 1997-98		1.1398-1.1401	28.95-28.96
1999-01		1.1791-1.1795	29.95-29.96
Piston			
Compression Rings			
Groove clearance			
Service Limit		0.004 Max	0.10 Max
Gap			
Production (Both Rings)		0.006-0.014	0.15-0.35
Valves			
Seat Concentricity			
Service Limit		0.002 Max	0.05 Max

① Unless otherwise noted

ENGINE SPECIFICATIONS - 3/4/4 Deluxe HP (87cc) TWO CYLINDER ENGINES

Component		U.S. (in.)①	Metric (mm)①
Cylinder Bore			
Standard Bore Diameter		1.5643-1.5650	39.74-39.75
Oversize Service Limit		0.002 Max	0.05 Max
Out-of-round			
Service Limit		0.003 Max	0.08 Max
Taper			
Service Limit		0.002 Max	0.05 Max
Cylinder Head			
Gasket Surface Warpage			
thru 1998		0.004 Max	0.10 Max
1999-01		0.003 per Inch	0.08 per 25.4mm
Crankshaft			
Top Journal			
Diameter		0.7515-0.7520	19.08-19.10
Center Journal			
Diameter		0.6685-0.6690	16.98-16.99
Bottom Journal			
Diameter		0.6691-0.6695	17.00-17.01
Crankpin			
Diameter		0.6695-0.6700	17.01-17.02
Piston			
Diameter		1.5625-1.5631	39.69-39.70
Out-of-round			
Service Limit		0.002 Max	0.05 Max
Rings			
Groove clearance			
Service Limit		0.004 Max	0.10 Max
Gap	Top	0.005-0.015	0.13-0.38
Production	2nd	0.005-0.015	0.13-0.38
Piston pin			
Diameter		1.5625-1.5631	39.69-39.70

① Unless otherwise noted

6-92 POWERHEAD REPAIR AND OVERHAUL

ENGINE SPECIFICATIONS - 8/9.9 HP (211cc) TWO CYLINDER ENGINES

Component	U.S. (in.) ①	Metric (mm) ①
Camshaft		
Lobe Height		
Intake	0.952-0.956	24.172-24.272
Exhaust	0.953-0.957	24.212-24.312
Cylinder Bore		
Diameter	2.2244-2.2248	56.50-56.51
Oversize Service Limit	0.002 Max	0.05 Max
Out-of-round		
Service Limit	0.003 Max	0.08 Max
Taper Service Limit	0.002 Max	0.05 Max
Cylinder Head		
Gasket Surface Warpage		
1995-98	0.004 Max	0.10 Max
1999-01	0.003 per Inch	0.08 per 25.4mm
Crankshaft		
Top and Bottom Journals		
Diameter	1.2200-1.2204	30.99-31.00
Crankpin Diameter	1.1398-1.1401	28.95-28.96
Piston		
Compression Rings		
Groove clearance		
Service Limit	0.004 Max	0.10 Max
Gap		
Production (Both Rings)	0.006-0.014	0.15-0.35
Valves		
Seat Concentricity		
Service Limit	0.002 Max	0.05 Max

① Unless otherwise noted

ENGINE SPECIFICATIONS - 5.0-8.0 HP (164cc) TWO CYLINDER ENGINES

Component		U.S. (in.) ①	Metric (mm) ①
Cylinder Bore			
Standard Bore Diameter		1.9373-1.9380	49.21-49.23
Oversize Service Limit		0.002 Max	0.05 Max
Out-of-round			
Service Limit		0.003 Max	0.08 Max
Taper Service Limit		0.002 Max	0.05 Max
Cylinder Head			
Gasket Surface Warpage			
thru 1998		0.004 Max	0.10 Max
1999-01		0.003 per Inch	0.08 per 25.4mm
Crankshaft			
Top Journal Diameter		0.8762-0.8767	22.26-22.27
Center Journal Diameter			
1990-95		0.8127-0.8132	20.64-20.65
1996-01		0.8120-0.8125	20.62-20.64
Bottom Journal Diameter		0.6691-0.6695	17.00-17.01
Crankpin Diameter		0.6695-0.6700	17.01-17.02
Piston			
Diameter		1.9345-1.9355	49.14-49.16
Out-of-round		0.002 Max	0.05 Max
Rings			
Groove clearance			
Service Limit		0.004 Max	0.10 Max
Gap Production	Top	0.005-0.015	0.13-0.38
	2nd	0.005-0.015	0.13-0.38

① Unless otherwise noted

POWERHEAD REPAIR AND OVERHAUL

ENGINE SPECIFICATIONS - 9.9/10/14/15 HP (216cc) 2-CYLINDER ENGINES

Component	U.S. (in.) ①	Metric (mm) ①
Cylinder Bore		
Standard Bore Diameter	2.1875-2.1883	55.56-55.58
Oversize Service Limit	0.002 Max	0.05 Max
Out-of-round		
Service Limit	0.003 Max	0.08 Max
Taper		
Service Limit	0.002 Max	0.05 Max
Cylinder Head		
Gasket Surface Warpage	0.004 Max	0.10 Max
Crankshaft		
Top Journal		
Diameter	0.8757-0.8762	22.24-22.26
Center Journal		
Diameter	0.8120-0.8125	20.63-20.64
Bottom Journal		
Diameter	0.7870-0.7874	19.98-19.99
Crankpin		
Diameter	0.8120-0.8125	20.63-20.64
Piston		
Diameter	2.1845-2.1850	55.49-55.50
Out-of-round	0.002 Max	0.05 Max
Rings		
Groove clearance		
Service Limit	0.004 Max	0.10 Max
Gap		
Production Top	0.005-0.015	0.13-0.38
2nd	0.005-0.015	0.13-0.38

① Unless otherwise noted

ENGINE SPECIFICATIONS - 9.9/10/15 HP (255cc) TWO CYLINDER ENGINES

Component	U.S. (in.) ①	Metric (mm) ①
Cylinder Bore		
Standard Bore Diameter	2.3745-2.3750	60.31-60.33
Oversize Service Limit	0.02 Max	0.508 Max
Out-of-round		
Service Limit	0.003 Max	0.08 Max
Taper		
Service Limit	0.002 Max	0.05 Max
Cylinder Head		
Gasket Surface Warpage		
thru 1998	0.004 Max	0.10 Max
1999-01	0.003 per Inch	0.08 per 25.4mm
Crankshaft		
Top Journal		
Diameter	0.8757-0.8762	22.24-22.26
Center Journal		
Diameter	0.8120-0.8125	20.63-20.64
Bottom Journal		
Diameter	0.7870-0.7874	19.98-19.99
Crankpin Diameter	0.8120-0.8125	20.63-20.64
Piston		
Out-of-round	0.002 Max	0.05 Max
Rings		
Groove clearance		
Service Limit	0.004 Max	0.10 Max
Gap		
Production Top	0.005-0.015	0.13-0.38
2nd	0.005-0.015	0.13-0.38

① Unless otherwise noted

POWERHEAD REPAIR AND OVERHAUL

ENGINE SPECIFICATIONS - 9.9/15 HP (305cc) TWO CYLINDER ENGINES

Component	U.S. (in.) [1]	Metric (mm) [1]
Camshaft		
Lobe Height Intake and Exhaust	0.959-0.967	24.360-24.560
Cylinder Bore		
Diameter	2.5590-2.5600	65.00-65.03
Oversize Service Limit	0.002 Max	0.05 Max
Out-of-round Service Limit	0.003 Max	0.08 Max
Taper Service Limit	0.002 Max	0.05 Max
Cylinder Head		
Gasket Surface Warpage 1995-98	0.004 Max	0.10 Max
1999-01	0.003 per Inch	0.08 per 25.4mm
Crankshaft		
Top and Bottom Journals Diameter	1.2990-1.2994	32.99-33.01
Crankpin Diameter	1.1801-1.1815	29.98-30.01
Piston		
Compression Rings		
Groove clearance Service Limit	0.004 Max	0.10 Max
Gap Production (Both Rings)	0.006-0.020	0.15-0.51
Valves		
Seat Concentricity Service Limit	0.002 Max	0.05 Max

[1] Unless otherwise noted

ENGINE SPECIFICATIONS - 18-35 HP (521cc) TWO CYLINDER ENGINES

Component	U.S. (in.) [1]	Metric (mm) [1]
Cylinder Bore		
Standard Bore Diameter	2.9995-3.0005	76.19-76.21
Oversize Service Limit	0.003 Max	0.08 Max
Out-of-round Service Limit	0.003 Max	0.08 Max
Taper Service Limit	0.002 Max	0.05 Max
Cylinder Head		
Gasket Surface Warpage thru 1998	0.004 Max	0.10 Max
1999-01	0.003 per Inch	0.08 per 25.4mm
Crankshaft		
Top Journal Diameter	1.2510-1.2515	31.78-31.79
Center Journal Diameter	1.1833-1.1838	30.06-30.07
Bottom Journal Diameter	0.9842-0.9846	25.00-25.01
Crankpin Diameter	1.1823-1.1828	30.03-30.04
Piston		
Major Diameter [2]		
Zollner Pistons	2.9956-3.0266	76.11-76.88
Art, Rightway Pistons	2.9969-3.0269	76.12-79.96
Cam Dimension [3]		
Zollner Pistons	0.005-0.007	0.12-0.18
Art, Rightway Pistons	0.0015-0.0025	0.04-0.06
Rings		
Groove clearance (lower ring only) Service Limit	0.004 Max	0.10 Max
Gap Production Top	0.007-0.017	0.18-0.43
2nd	0.007-0.017	0.18-0.43

[1] Unless otherwise noted
[2] Major diameter is measured at a 90 degree angle to the wrist pin centerline, while minor diameter is inline with wrist pin
[3] Cam dimension (similar to out-of-round) is the measured difference between major and minor diameters

POWERHEAD REPAIR AND OVERHAUL

ENGINE SPECIFICATIONS - 25/35 HP (500/565cc) THREE CYLINDER ENGINES

Component		U.S. (in.)①	Metric (mm)①
Cylinder Bore			
Standard Bore Diameter			
25 Hp Motors		2.3495-2.3505	59.68-59.70
35 Hp Motors		2.4995-2.5005	63.49-63.51
Oversize Service Limit			
1995-98		0.004 Max	0.10 Max
1999-01		0.003 Max	0.08 Max
Out-of-round	Service Limit	0.004 Max	0.10 Max
Taper	Service Limit	0.002 Max	0.05 Max
Cylinder Head			
Gasket Surface Warpage			
1995-98		0.004 Max	0.10 Max
1999-01		0.003 per Inch	0.08 per 25.4mm
Crankshaft			
Top Journal	Diameter	1.4979-1.4984	38.05-38.06
Center Journal	Diameter	1.3748-1.3752	34.92-34.93
Bottom Journal	Diameter	1.1810-1.1815	30.00-30.01
Crankcase Seal Ring	Thickness (Service Limit)	0.1 Min	2.5 Min
Crankpin	Diameter	1.1823-1.1828	30.03-30.04
Piston			
Diameter			
25 Hp Motors		2.3440-2.3450	59.54-59.56
35 Hp Motors		2.4940-2.4950	63.35-63.37
Out-of-round		0.003 Max	0.08 Max
Ring Gap	Production		
Top		0.005-0.020	0.13-0.51
2nd		0.005-0.020	0.13-0.51

① Unless otherwise noted

ENGINE SPECIFICATIONS - 25-55 HP (737cc) TWO CYLINDER ENGINES

Component		U.S. (in.)①	Metric (mm)①
Cylinder Bore			
Standard Bore Diameter		3.1870-3.1880	80.95-80.98
Oversize Service Limit		0.003 Max	0.08 Max
Out-of-round		0.002 Max	0.05 Max
Taper	Service Limit	0.004 Max	0.10 Max
		0.003 per Inch	0.08 per 25.4mm
Cylinder Head			
Gasket Surface Warpage			
thru 1998			
1999-01			
Crankshaft			
Top Journal	Diameter		
1990 (and 1991 25D/40/48/50 hp models)		1.4974-1.4979	38.03-38.04
1992-01 (1991 40/45/55 commercial models)		1.4986-1.4991	38.06-30.08
Center Journal	Diameter		
1990 (and 1991 25D/40/48/50 hp models)		1.3748-1.3752	34.92-34.93
1992-01 (1991 40/45/55 commercial models)		1.3745-1.3749	34.91-34.92
Bottom Journal	Diameter	1.1810-1.1815	30.00-30.01
Crankpin	Diameter	1.1823-1.1828	30.03-30.04
Piston			
Major Diameter ②		3.1831-3.2131	80.85-81.61
Out-of-round ③		0.004 Max	0.10 Max
Ring Gap	Production (Both Rings)		
1990-94 ④		0.007-0.017	0.18-0.43
1995 and later			
Standard 25D, 40, 48 and 50/50SPL Hp models		0.019-0.031	0.48-0.79
Other Models ⑤		0.010-0.022	0.25-0.56

① Unless otherwise noted
② Major diameter is measured in line with the upper piston ring dowel pin
③ To determine out-of-round measure piston diameter at several other locations around the piston, each measurement must be smaller than Major diameter, but no more than out-of-round dimension
④ These models years also have a ring side clearance specification of 0.004 in. (0.10mm)
⑤ Includes all Commercial models as well as 40RS, 40RP, 40RW, 40WR, 45 and 55

POWERHEAD REPAIR AND OVERHAUL

ENGINE SPECIFICATIONS - 40/50 HP (815cc) THREE CYLINDER ENGINES

Component		U.S. (in.) [1]	Metric (mm) [1]
Camshaft			
Journal			
Outer Diameter [2]	Standard	0.9029-0.9037	22.934-22.955
	Service Limit	0.897 Min	22.784 Min
Inner Diameter [3]	Standard	0.9055-0.9063	23.000-23.021
	Service Limit	0.9122 Max	23.171 Max
Oil Clearance	Standard	0.0018-0.0034	0.045-0.087
	Service Limit	0.0047 Max	0.120 Max
Lobe Height - 40 Hp Motors			
Intake	Standard	1.4776-1.4839	37.53-37.69
	Service Limit	1.4736 Min	37.43 Min
Exhaust	Standard	1.4858-1.4921	37.74-37.90
	Service Limit	1.4819 Min	37.64 Min
Lobe Height - 50 Hp Motors			
Intake	Standard	1.5051-1.5114	38.23-38.39
	Service Limit	1.5012 Min	38.13 Min
Exhaust	Standard	1.4858-1.4921	37.74-37.90
	Service Limit	1.4819 Min	37.64 Min
Shaft Runout Service Limit		0.004 Max	0.10 Max
Cylinder Bore			
Diameter		2.7953-2.7961	71.00-71.02
Oversize Service Limit		0.004 Max	0.10 Max
Out-of-round			
Taper	Service Limit	0.004 Max	0.10 Max
Cylinder Block			
Gasket Surface Warpage		0.0024 Max	0.06 Max
Cylinder Head			
Gasket Surface Warpage		0.002 Max	0.05 Max
Crankshaft			
End-Play	Standard	0.0043-0.0122	0.11-0.31
	Service Limit	0.0138 Max	0.35 Max
Main Bearings			
Case Bore Size			
Journal		1.9291-1.9298	49.000-49.018

ENGINE SPECIFICATIONS - 25-70 HP (913cc) THREE CYLINDER ENGINES

Component		U.S. (in.) [1]	Metric (mm) [1]
Cylinder Bore			
Standard Bore Diameter		3.1870-3.1880	80.95-80.97
Oversize Service Limit		0.003 Max	0.08 Max
Out-of-round	Service Limit	0.003 Max	0.08 Max
Taper	Service Limit	0.002 Max	0.05 Max
Cylinder Head			
Gasket Surface Warpage			
thru 1998		0.004 Max	0.10 Max
1999-01		0.003 per Inch	0.08 per 25.4mm
Crankshaft			
Top Journal	Diameter	1.4974-1.4979	38.03-38.05
Center Journal	Diameter	1.3748-1.3752	34.92-34.93
Bottom Journal	Diameter	1.1810-1.1815	30.00-30.01
Crankpin	Diameter	1.1823-1.1828	30.03-30.04
Piston			
Diameter		3.1806-3.1841	80.79-80.88
Out-of-round		0.003 Max	0.08 Max
Ring Gap			
Production (Both Rings)			
1990-94 [2]		0.007-0.017	0.18-0.43
1995 and later			
Except 65 Hp Models		0.019-0.031	0.48-0.79
65 Hp Models		0.010-0.022	0.25-0.56

[1] Unless otherwise noted
[2] These models years also have a ring side clearance specification of 0.004 in. (0.10mm)

POWERHEAD REPAIR AND OVERHAUL

ENGINE SPECIFICATIONS - 40/50 HP (815cc) THREE CYLINDER ENGINES

Component	U.S. (in.) ①	Metric (mm) ①
Connecting Rod		
Big-End Width	1.7709-1.7717	44.982-45.000
Taper and Out-of-Round	0.0026 Max	0.065 Max
Oil Clearance Standard	0.0008-0.0016	0.020-0.040
Service Limit	0.0026 Max	0.065 Max
Thrust Bearing Thickness	0.0016 Max	0.04 Max
Run-Out (at Center Journal)	0.097-0.099	2.47-2.52
Crankpin		
Journal Diameter	1.4594-1.4961	37.982-38.000
Taper and Out-of-Round	0.0026 Max	0.065 Max
Big-End Width	0.870-0.874	22.10-22.20
Rod-to-Crankshaft Side Clearance	0.004-0.014	0.10-0.35
Intake Manifold		
Gasket Surface Warpage	0.004 Max	0.10 Max
Pistons		
Clearance Production	0.0008-0.0016	0.020-0.040
Service Limit	0.0039 Max	0.100 Max
Ring Groove Width		
Compression Ring Grooves	0.0398-0.0406	1.01-1.03
Oil Ring Groove	0.0594-0.0602	1.51-1.53
Out-of-round	0.003 Max	0.08 Max
Skirt Outer Diameter ④	2.7941-2.7949	70.970-70.990
Compression Rings		
Groove Clearance (Production) Top	0.0008-0.0020	0.02-0.05
2nd	0.0008-0.0023	0.02-0.06
Groove Clearance (Service Limit)	0.0039 Max	0.10 Max
Gap (Production) Top	0.004-0.010	0.10-0.25
2nd	0.010-0.018	0.25-0.40
Ring Thickness	0.0382-0.0390	0.97-0.99
Piston pin		
Bore Inner Diameter	0.7089-0.7092	18.006-18.014
Pin Outer Diameter	0.7085-0.7087	17.996-18.000
Clearance Production	0.0002-0.0007	0.006-0.018
Service Limit	0.0016 Max	0.040 Max
Valve system		
Guides		
Inner Diameter	0.2185-0.2170	5.500-5.512
Intake Guide-to-Stem Clearance		
Production	0.0008-0.0019	0.020-0.047
Service Limit	0.0028 Max	0.070 Max
Exhaust Guide-to-Stem Clearance		

ENGINE SPECIFICATIONS - 40/50 HP (815cc) THREE CYLINDER ENGINES

Component	U.S. (in.) ①	Metric (mm) ①
Production	0.0018-0.0028	0.045-0.072
Service Limit	0.0035 Max	0.090 Max
Heads/Seats		
Head Thickness Intake	0.028	0.70
Exhaust	0.020	0.50
Head-to-Seat Contact Width		
Intake	0.071-0.087	1.80-2.20
Exhaust	0.065-0.098	1.65-2.05
Seat Concentricity	0.002 Max	0.050 Max
Radial Head Runout	0.003 Max	0.080 Max
Springs		
Free Length Production	1.30	33.1
Service Limit	1.25 Min	31.8 Min
Squareness	0.080	2.0
Pressure Production	21.4-24.9 lb. @ 1.12 in.	9.7-11.3 kg @ 28.5mm
Service Limit	19.6 lb. @ 1.12 in. Min	8.9 kg @ 28.5mm Min
Stems		
End Length	0.1260 Min	3.20 Min
Height Above Deck	0.430	11.0
Stem Outer Diameter Intake	0.2152-0.2157	5.465-5.480
Exhaust	0.2142-0.2148	5.440-5.455
Stem Run-Out	0.002 Max	0.05 Max
Tappets		
Bore Inner Diameter	1.0630-1.0638	27.000-27.021
Tappet Outer Diameter	1.0614-1.0620	26.959-26.975
Tappet-to-Bore Clearance		
Production	0.0010-0.0024	0.025-0.062
Service Limit	0.0059 Max	0.150 Max

① Unless otherwise noted
② Specification is for outer diameter of camshaft journal
③ Specification is for the inner diameter of the cylinder head camshaft journal holder
④ Skirt outer diameter is measured 0.75 in. (19mm) from bottom

6-98 POWERHEAD REPAIR AND OVERHAUL

ENGINE SPECIFICATIONS - 40/50 HP (815cc) THREE CYLINDER ENGINES

Component		U.S. (In.) [1]	Metric (mm) [1]
Heads/Seats			
Head Thickness	Production	0.0018-0.0028	0.045-0.072
	Service Limit	0.0035 Max	0.090 Max
Head Thickness	Intake	0.028	0.70
	Exhaust	0.020	0.50
Head-to-Seat Contact Width	Intake	0.071-0.087	1.80-2.20
	Exhaust	0.065-0.098	1.65-2.05
Seat Concentricity		0.002 Max	0.050 Max
Radial Head Runout		0.003 Max	0.080 Max
Springs			
Free Length	Production	1.30	33.1
	Service Limit	1.25 Min	31.8 Min
Squareness		0.080	2.0
Pressure	Production	21.4-24.9 lb. @ 1.12 in.	9.7-11.3 kg @ 28.5mm
	Service Limit	19.6 lb. @ 1.12 in. Min	8.9 kg @ 28.5mm Min
Stems			
End Length		0.1260 Min	3.20 Min
Height Above Deck		0.430	11.0
Stem Outer Diameter	Intake	0.2152-0.2157	5.465-5.480
	Exhaust	0.2142-0.2148	5.440-5.455
Stem Run-Out		0.002 Max	0.05 Max
Tappets			
Bore Inner Diameter		1.0630-1.0638	27.000-27.021
Tappet Outer Diameter		1.0614-1.0620	26.959-26.975
Tappet-to-Bore Clearance	Production	0.0010-0.0024	0.025-0.062
	Service Limit	0.0059 Max	0.150 Max

[1] Unless otherwise noted
[2] Specification is for outer diameter of camshaft journal
[3] Specification is for the inner diameter of the cylinder head camshaft journal holder
[4] Skirt outer diameter is measured 0.75 in. (19mm) from bottom

ENGINE SPECIFICATIONS - 70 HP (1298cc) FOUR CYLINDER ENGINES

Component		U.S. (In.) [1]	Metric (mm) [1]
Camshaft			
Journal			
Outer Diameter [2]	1	1.7687-1.7697	44.925-44.950
	2	1.7608-1.7618	44.724-44.750
	3	1.7530-1.7539	44.526-44.550
	4	1.7451-1.7461	44.325-44.351
	5	1.7372-1.7382	44.125-44.150
Inner Diameter [3]	1	1.7717-1.7723	45.001-45.016
	2	1.7638-1.7644	44.800-44.816
	3	1.7559-1.7565	44.600-44.615
	4	1.7480-1.7487	44.400-44.417
	5	1.7402-1.7408	44.201-44.216
Oil Clearance	Standard	0.0020-0.0036	0.050-0.091
	Service Limit	0.0059 Max	0.150 Max
Lobe Height	Intake Standard	1.4815-1.4878	37.63-37.79
	Service Limit	1.4776 Min	37.531 Min
	Exhaust Standard	1.4814-1.4877	37.63-37.79
	Service Limit	1.4775 Min	37.529 Min
Shaft Runout Service Limit		0.004 Max	0.10 Max
Cylinder Bore			
Diameter		2.9134-2.9142	74.00-74.02
Oversize Service Limit		0.004 Max	0.10 Max
Out-of-round			
Taper	Service Limit	0.004 Max	0.10 Max
Cylinder Block			
Gasket Surface Warpage		0.0024 Max	0.06 Max
Cylinder Head			
Gasket Surface Warpage		0.002 Max	0.05 Max
Crankshaft			
End-Play	Standard	0.0043-0.0122	0.11-0.31
	Service Limit	0.0150 Max	0.38 Max
Main Bearings			
Case Bore Size		2.2047-2.2054	56.000-56.018
Journal	Diameter	2.0465-2.0472	51.982-52.000

ENGINE SPECIFICATIONS - 70 HP (1298cc) FOUR CYLINDER ENGINES

Component	U.S. (in.) ①	Metric (mm) ①
Taper and Out-of-Round	0.0016 Max	0.040 Max
Oil Clearance		
Standard	0.0006-0.0014	0.016-0.036
Service Limit	0.0024 Max	0.060 Max
Run-Out (at Center Journal)	0.002 Max	0.06 Max
Thrust Bearing Thickness	0.097-0.099	2.47-2.52
Crankpin		
Journal		
Diameter	1.6528-1.6535	41.982-42.000
Taper and Out-of-Round	0.0016 Max	0.040 Max
Big-End Width	0.870-0.874	22.10-22.20
Connecting Rod		
Big-End Width	0.864-0.866	21.95-22.00
Big-End Oil Clearance		
Standard	0.0008-0.0020	0.020-0.050
Service Limit	0.0031 Max	0.079 Max
Rod-to-Crankshaft Side Clearance	0.004-0.014	0.10-0.35
Pistons		
Clearance		
Production	0.0008-0.0016	0.020-0.040
Service Limit	0.0039 Max	0.100 Max
Ring Groove Width		
Top Compression Ring Groove	0.048-0.049	1.22-1.24
Bottom Compression Ring Groove	0.059-0.060	1.51-1.53
Oil Ring Groove	0.111-0.112	2.81-2.83
Out-of-round	0.003 Max	0.08 Max
Skirt Outer Diameter ④	2.9122-2.9130	73.970-73.990
Compression Rings		
Groove Clearance (Production)		
Top	0.0012-0.0027	0.03-0.07
2nd	0.0008-0.0024	0.02-0.06
Groove Clearance (Service Limit)		
Top	0.0047 Max	0.12 Max
2nd	0.0039 Max	0.10 Max
Gap (Production)		
Top	0.006-0.028	0.15-0.70
2nd	0.008-0.028	0.20-0.70
Ring Thickness		
Top	0.046-0.047	1.17-1.19
2nd	0.058-0.059	1.47-1.49
Piston pin		
Bore Inner Diameter	0.6694-0.6697	17.003-17.011
Pin Outer Diameter	0.6691-0.6693	16.995-17.000
Clearance		
Production	0.0001-0.0006	0.003-0.015
Service Limit	0.0016 Max	0.040 Max

ENGINE SPECIFICATIONS - 70 HP (1298cc) FOUR CYLINDER ENGINES

Component	U.S. (in.) ①	Metric (mm) ①
Valve system		
Guides		
Inner Diameter	0.2756-0.2762	7.000-7.015
Intake Guide-to-Stem Clearance		
Production	0.0008-0.0020	0.020-0.050
Service Limit	0.0028 Max	0.070 Max
Exhaust Guide-to-Stem Clearance		
Production	0.0018-0.0030	0.045-0.075
Service Limit	0.0035 Max	0.090 Max
Heads/Seats		
Head Thickness		
Intake	0.028	0.70
Exhaust	0.020	0.50
Head-to-Seat Contact Width		
Intake	0.071-0.087	1.80-2.20
Exhaust	0.065-0.098	1.65-2.05
Seat Concentricity	0.002 Max	0.050 Max
Radial Head Runout	0.003 Max	0.080 Max
Springs		
Free Length		
Production	1.94	49.3
Service Limit	1.89 Min	48.1 Min
Squareness	0.080	2.0
Pressure		
Production	54.7-64.3 lb. @ 1.63 in.	24.8-29.2 kg @ 41.5mm
Service Limit	50.2 lb. @ 1.63 in. Min	22.8 kg @ 41.5mm Min
Stems		
End Length	0.2380 Min	6.05 Min
Height Above Deck	0.550	14.0
Stem Outer Diameter		
Intake	0.2742-0.2748	6.965-6.980
Exhaust	0.2732-0.2738	6.940-6.955
Stem Run-Out	0.002 Max	0.05 Max
Rocker Arms and Shafts		
Rocker Arm Inner Diameter	0.6299-0.6306	16.000-16.017
Rocker Arm Shaft Outer Diameter	0.6289-0.6294	15.974-15.987
Shaft-to-Rocker Arm Clearance		
Production	0.0005-0.0018	0.012-0.045
Service Limit	0.0035 Max	0.090 Max
Shaft Runout	0.005 Max	0.120 Max

① Unless otherwise noted
② Specification is for outer diameter of camshaft journal
③ Specification is for the inner diameter of the cylinder head camshaft journal holder
④ Skirt outer diameter is measured 0.75 in. (19mm) from bottom

GEARCASE	7-2
DESCRIPTION AND OPERATION	7-2
GEARCASE ASSEMBLY	7-2
REMOVAL & INSTALLATION	7-2
Colt/Junior (43cc)	7-2
2.0-3.5 Hp (78cc)	7-3
5 Hp (109cc)	7-4
3/4 Hp (87cc)	7-4
4 Deluxe (87cc)	7-7
5/6 Hp (128cc)	7-7
5/6/8 Hp (164cc)	7-7
8/9.9 Hp (211cc)	7-8
9.9-15 Hp (216/255cc)	7-8
9.9/15 Hp (305cc)	7-8
10/14 (216/255cc)	7-10
20-35 Hp (521cc)	7-10
25/35 Hp (500/565cc)	7-10
25-55 Hp (737cc)	7-13
25-70 Hp (913cc) (Exc. Jet)	7-13
40/50 Hp (815cc) EFI	7-15
70 Hp (1298cc) EFI	7-17
PROPELLER SHAFT SEAL	7-18
REPLACEMENT	7-18
Colt/Junior (43cc)	7-18
2.0-3.5 Hp (78cc)	7-18
5 Hp (109cc)	7-18
3/4 Hp (87cc)	7-19
4 Deluxe (87cc)	7-20
5/6 Hp (128cc)	7-20
5/6/8 Hp (164cc)	7-20
8/9.9 Hp (211cc)	7-21
9.9-15 Hp (216/255cc)	7-21
9.9/15 Hp (305cc)	7-21
10/14 Hp (211cc/255cc)	7-23
20-35 Hp (521cc)	7-23
25/35 Hp (500/565cc)	7-23
25-55 Hp (737cc)	7-25
25-70 Hp (913cc)	7-25
40/50 Hp (815cc) EFI	7-26
70 Hp (1298cc) EFI	7-27
JET DRIVE	7-29
DESCRIPTION & OPERATION	7-29
IMPELLER	7-29
Removal, Shimming & Installation	7-29
JET DRIVE ASSEMBLY	7-31
Removal & Installation	7-31
REVERSE GATE	7-32
Gate Adjustment	7-32
Remote Cable Adjustment	7-32

7

LOWER UNIT

GEARCASE	7-2
JET DRIVE	7-29

7-2 LOWER UNIT

GEARCASE

Description and Operation

◆ See Figure 1

The gearcase is considered that part of the outboard below the midsection/exhaust housing. The gearcase contains the propeller shaft, the driven and pinion gears, the driveshaft from the powerhead and the water pump. On models equipped with shifting capabilities, the forward and reverse gears, together with the clutch, shift assembly, and related linkage, are all housed within the case.

The single most important task for proper gearcase maintenance is inspecting it for signs of leakage after each use. If oil can get out, then water can get in. And, water, mixing with or replacing the oil in the gearcase will wreak havoc with the shafts and gears contained within the housing.

The second most important task for proper gearcase maintenance is checking and maintaining the oil inside the case. Not only is it important to make sure the oil is at the proper level (not above or below), but it is important to check the oil for signs of contamination from moisture. Water entering the gearcase will usually cause the oil to turn a slightly milky-white color. Also, significant amounts of water mixed with the oil will give the appearance of an overfilled condition.

If you suspect water in the gearcase, start by draining and closely inspecting the fluid (refer to the procedures found in the Maintenance and Tune-Up section). Then, refill the unit with fresh oil and test the outboard (by using it!). Watch the fluid level closely after the test, and for the first few outings. If any oil leaks out or water enters, either the propeller shaft seal must be replaced or the gearcase must be disassembled, inspected and completely overhauled. To be honest, a complete overhaul is recommended, because corrosion and damage may have occurred if moisture was in the gearcase long enough. But, in some cases, if the leak was caught in time, and there is no significant wear, damage or corrosion in the gearcase, the propeller shaft seal can usually be replaced with the gearcase still installed to the outboard.

The last, most important task you can perform to help keep your gearcase in top shape, is to flush the inside and outside of the gearcase after each use. Rinse the outside of the unit with a hose to remove any sea life, salt, chemicals or other corrosion inducing substances that you may have picked up in the water. Cleaning the gearcase will also help you spot potential trouble, such as gearcase oil leaks, cracks or damage that may have occurred during use. Remove any sand, silt or dirt that could potentially damage seals or clog passages. Once you've rinsed the outside, hook up a flushing device and do the same for the inside. Again, details are found in the Maintenance and Tune-Up section, look under Flushing the Cooling System.

Gearcase Assembly

REMOVAL & INSTALLATION

The most common reason for removing and installing the gearcase is to perform service (inspect or replace) the water pump impeller. On all motors except the 2.0-3.5 hp (78cc) motors, the water pump is found on the gearcase-to-midsection (sometimes known as the intermediate or exhaust housing) split line. On 2.0-3.5 hp (78cc) motors the pump is mounted just in front of the propeller, so the gearcase does not have to be removed on these small motors in order to service the pump.

Removal and installation procedures are provided here for each of the gearcases used on these Johnson/Evinrude motors. Exploded views are also provided, in case disassembly and overhaul are required.

Colt/Junior (43cc) Motors
◆ See Figures 2 and 3

1. For safety, disconnect the spark plug lead, then ground it to the cylinder head.
2. If necessary for service or access, remove the propeller, for details refer to the procedure in the Maintenance and Tune-Up section.
3. Remove the two screws securing the lower unit to the exhaust housing.
4. Taking Care not to damage the driveshaft and the water tube, separate the lower unit from the exhaust housing by pulling straight downward.
5. If necessary for service or overhaul, drain the gear oil from the gearcase.
6. Thoroughly inspect the gearcase and exhaust housing for signs of damage. Make sure all mating surfaces are clean and free of debris, corrosion or damage.

To install:

7. Apply a light coating of OMC Moly Lube, or equivalent assembly lubricant to the driveshaft splines. Be sure to coat only the SIDES of the splines and not the top of the shaft, as that could hydraulically prevent the driveshaft from fully seating in the crankshaft spline.
8. Apply a light coat of clean liquid soap to the water tube grommet.
9. Apply a light coating of OMC Nut Lock, or equivalent threadlock to the threads of the 2 gearcase mounting screws.
10. Install the gearcase, while carefully aligning the water tube in the grommet and the driveshaft splines to the crankshaft shaft. If necessary, turn the propeller shaft slowly clockwise (when viewed from the shaft end) to align the splines.

Fig. 1 A neglected lower unit cannot be expected to perform to maximum efficiency, compared with a unit receiving TLC (tender loving care)

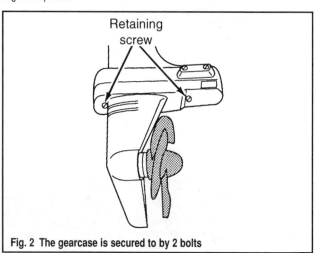

Fig. 2 The gearcase is secured to by 2 bolts

LOWER UNIT

■ ALWAYS spin the propeller or propeller shaft clockwise, when viewed from the shaft end, in order to prevent possible damage to the water pump impeller.

11. Install the 2 gearcase mounting screws and tighten to 60-84 inch lbs. (7-9 Nm).
12. Apply a light coating of OMC Triple-Guard, or equivalent marine grease to the propeller shaft and to a new propeller drive pin.
13. Install the propeller and secure using a new cotter pin. For more details, please refer to Propeller under the Maintenance and Tune-Up section.
14. Properly refill the gearcase with lubricant. For details, please refer to the procedure in the Maintenance and Tune-Up section.
15. Reconnect the spark plug lead.

2.0-3.5 Hp (78cc) Motors

◆ See Figures 4, 5 and 6

■ Remember, the gearcase does not have to be removed from the outboard in order to service the water pump on these models.

1. For safety, disconnect the spark plug lead, then ground it to the cylinder head.
2. Remove the propeller, for details refer to the procedure in the Maintenance and Tune-Up section.
3. Remove the 2 bolts securing the gearcase to the exhaust housing. One bolt is threaded upward from under the cavitation plate (just above the propeller, and behind the water intake). The other bolt is threaded downward, from the very front of the outboard on a boss, above the cavitation plate.
4. Separate the gearcase assembly from the outboard by pulling carefully straight downward, withdrawing the driveshaft and water tube from the exhaust housing.
5. If necessary, carefully pull the tube and upper driveshaft (transmission shaft) from the gearcase.
6. If necessary for service or overhaul, drain the gear oil from the gearcase.
7. Thoroughly inspect the gearcase and exhaust housing for signs of damage. Make sure all mating surfaces are clean and free of debris, corrosion or damage.

To install:

8. If removed, insert the transmission tube into the exhaust housing, pushing the tube into the lower crankshaft seal.
9. Apply a light coating of OMC Moly Lube, or equivalent assembly lubricant to both ends of the transmission shaft and to the driveshaft splines. Insert the shaft into the tube and push the shaft onto the end of the crankshaft.
10. Apply a light coat of clean engine oil to the water tube grommet.
11. Apply a light coating of OMC Nut Lock, or equivalent threadlock to the threads of the 2 gearcase mounting screws.
12. Install the gearcase, while carefully aligning the water tube in the grommet and the driveshaft splines to the transmission shaft. If necessary, turn the propeller shaft slowly clockwise (when viewed from the shaft end) to align the splines.

■ ALWAYS spin the propeller or propeller shaft clockwise, when viewed from the shaft end, in order to prevent possible damage to the water pump impeller.

13. Install the 2 gearcase mounting screws and tighten to 89-115 inch lbs. (10-13 Nm).

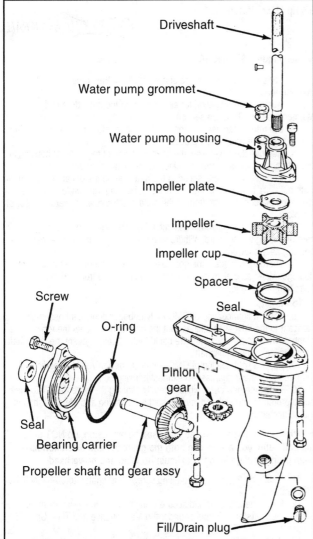

Fig. 3 Exploded view of the gearcase assembly - Colt/Junior motors

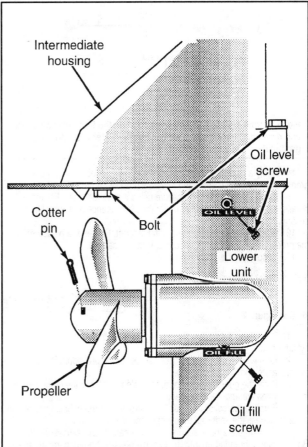

Fig. 4 The gearcase is secured by 2 bolts, one above and one below the cavitation plate

7-4 LOWER UNIT

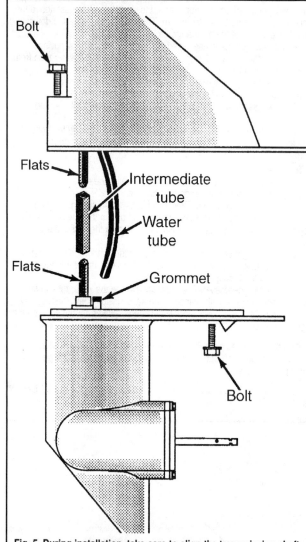

Fig. 5 During installation, take care to align the transmission shaft and driveshaft flats, while also guiding the water tube into the grommet

14. Apply a light coating of OMC Triple-Guard, or equivalent marine grease to the propeller shaft and to a new propeller drive pin.
15. Install the propeller and secure using a new cotter pin. For more details, please refer to Propeller under the Maintenance and Tune-Up section.
16. Properly refill the gearcase with lubricant. For details, please refer to the procedure in the Maintenance and Tune-Up section.
17. Reconnect the spark plug lead.

5 Hp (109cc) Motors

◆ See Figure 7

1. For safety, disconnect the spark plug lead, then ground it to the cylinder head.
2. Remove the propeller, for details refer to the procedure in the Maintenance and Tune-Up section.
3. Drain the gearcase oil. For details, please refer to the procedure in the Maintenance and Tune-Up section.
4. Remove the cover from the starboard side of the exhaust housing in order to access the shift rod, then loosen the shift rod screw.
5. Remove the 2 bolts securing the gearcase to the exhaust housing. Both bolts are threaded up into the exhaust housing from underneath the cavitation plate, one at the front of the plate and the other, toward the rear of the plate.

6. Separate the gearcase assembly from the outboard by pulling carefully straight downward, withdrawing the driveshaft from the exhaust housing as it is lowered.

■ Check the water pump to see if the water tube bushing stayed behind on the tube or if it is contained in the pump housing.

7. Thoroughly inspect the gearcase and exhaust housing for signs of damage. Make sure all mating surfaces are clean and free of debris, corrosion or damage.

To install:
8. Apply a light coating of OMC Moly Lube, or equivalent assembly lubricant to the driveshaft splines.
9. Apply a light coating of RTV sealant to the mating surfaces of the gearcase and the exhaust housing.
10. Apply a light coating of OMC Gasket Sealing Compound, or equivalent sealant to the threads of the 2 gearcase-to-exhaust housing bolts.
11. Install the gearcase, while carefully aligning the dowel pins, water tube, water pump seal and driveshaft.
12. Install the 2 gearcase mounting screws and tighten to 11-15 ft. lbs. (15-20 Nm).
13. Install the propeller and secure using a new cotter pin. For more details, please refer to Propeller under the Maintenance and Tune-Up section.
14. Properly refill the gearcase with lubricant. For details, please refer to the procedure in the Maintenance and Tune-Up section.
15. Reconnect the spark plug lead.

3/4 Hp (87cc) Motors

◆ See Figures 8, 9 and 10

1. For safety, disconnect the spark plug lead, then ground it to the cylinder head.
2. Remove the propeller, for details refer to the procedure in the Maintenance and Tune-Up section.
3. Drain the gearcase oil. For details, please refer to the procedure in the Maintenance and Tune-Up section.
4. Remove the cover from the starboard side of the exhaust housing in order to access the shift rod, then loosen the shift rod screw.
5. Remove the 2 bolts securing the gearcase to the exhaust housing. Both bolts are threaded up into the exhaust housing from underneath the cavitation plate, one at the front of the plate and the other, toward the rear of the plate.
6. Separate the gearcase assembly from the outboard by pulling carefully straight downward, withdrawing the driveshaft from the exhaust housing as it is lowered.
7. Thoroughly inspect the gearcase and exhaust housing for signs of damage. Make sure all mating surfaces are clean and free of debris, corrosion or damage.

To install:
8. Apply a light coating of OMC Moly Lube, or equivalent assembly lubricant to the driveshaft splines. But, be careful not to coat the top surface of the driveshaft, as lubricant there could hydraulically prevent the driveshaft from seating in the crankshaft.
9. Apply a light coating of OMC Nut Lock, or equivalent threadlock to the threads of the 2 gearcase-to-exhaust housing bolts.
10. Install the gearcase, while carefully aligning the water tube into the pump grommet, the shift rod into the shift rod bushing and the driveshaft to the crankshaft.

■ When installing a gearcase that uses an extension, do NOT tighten the bolts until you are certain both the gearcase and extension have been properly aligned to the exhaust housing and powerhead.

11. Install the 2 gearcase mounting screws and tighten to 60-84 inch lbs. (7-9 Nm).
12. Install the propeller and secure using a new cotter pin. For more details, please refer to Propeller under the Maintenance and Tune-Up section.
13. Properly refill the gearcase with lubricant. For details, please refer to the procedure in the Maintenance and Tune-Up section.
14. Reconnect the spark plug lead.

LOWER UNIT 7-5

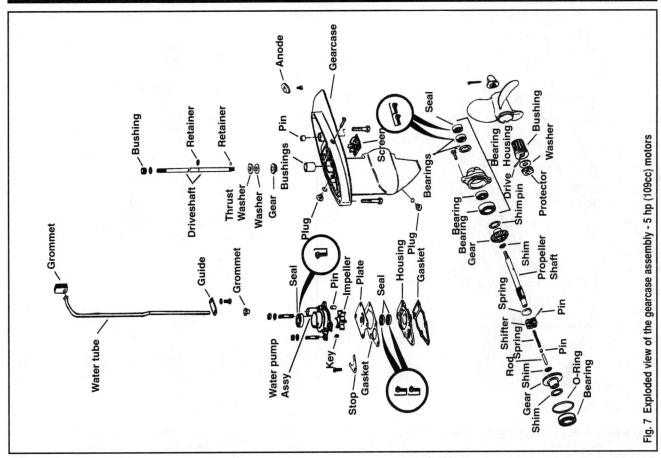

Fig. 7 Exploded view of the gearcase assembly - 5 hp (109cc) motors

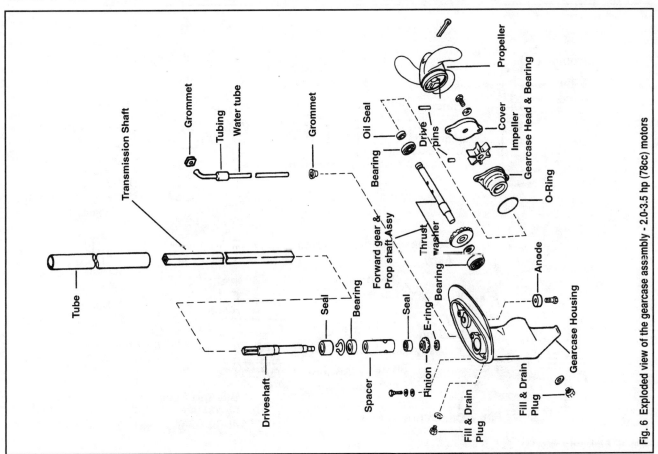

Fig. 6 Exploded view of the gearcase assembly - 2.0-3.5 hp (78cc) motors

7-6 LOWER UNIT

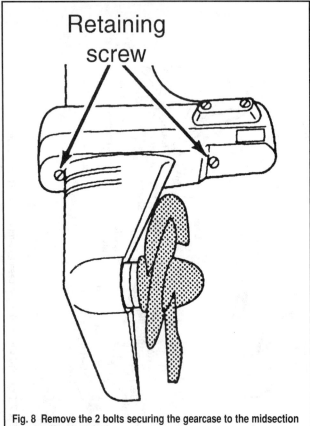

Fig. 8 Remove the 2 bolts securing the gearcase to the midsection (intermediate housing)...

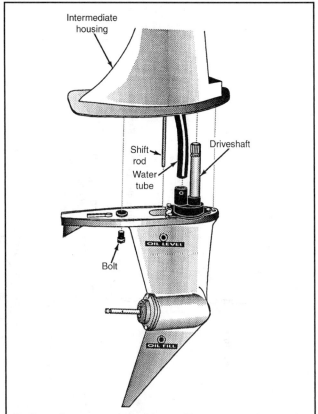

Fig. 9 ... then remove the gearcase, taking care not to damage the water tube, shift rod and driveshaft

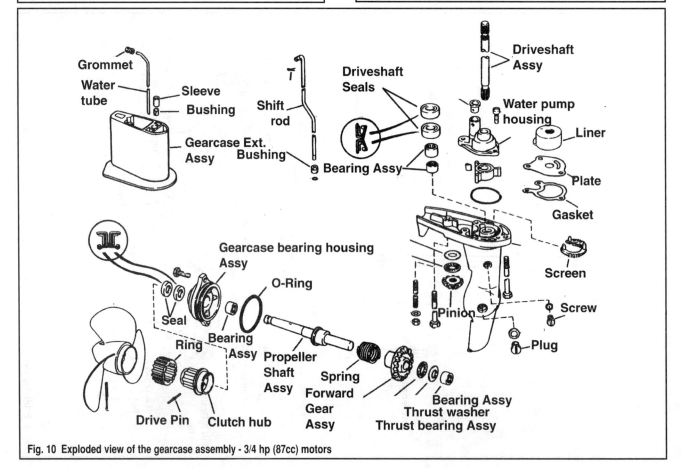

Fig. 10 Exploded view of the gearcase assembly - 3/4 hp (87cc) motors

LOWER UNIT 7-7

4 Deluxe (87cc), 5/6 Hp (128cc) and 5/6/8 Hp (164cc) Motors

◆ See Figures 11, 12 and 13

1. For safety, disconnect the spark plug lead(s), then ground it(them) to the cylinder head.
2. Remove the propeller, for details refer to the procedure in the Maintenance and Tune-Up section.
3. Drain the gearcase oil. For details, please refer to the procedure in the Maintenance and Tune-Up section.
4. Rotate the propeller shaft slowly by hand and shift the gearcase into **forward**.
5. Remove the 3 bolts securing the gearcase to the exhaust housing. One bolt is threaded downward into the gearcase on top of a boss at the front of the exhaust housing. The other 2 bolts are threaded upward into the exhaust housing from underneath the cavitation plate (immediately adjacent to the anode).

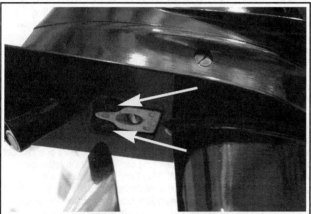

Fig. 11 On these models, 2 of the gearcase bolts are found adjacent to the anode

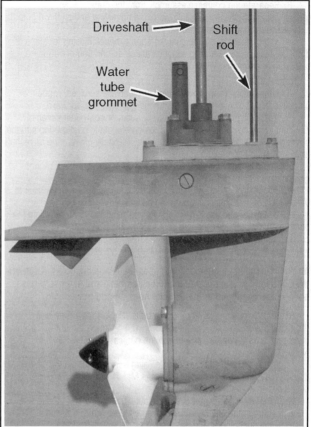

Fig. 12 When removing the gearcase, take care not to damage the water tube, shift rod or driveshaft

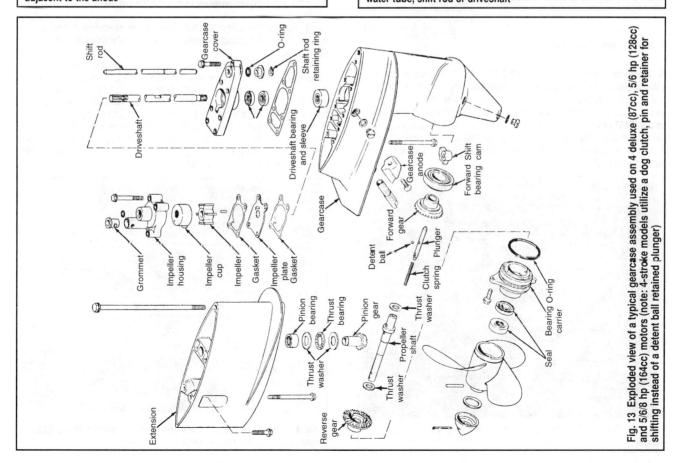

Fig. 13 Exploded view of a typical gearcase assembly used on 4 deluxe (87cc), 5/6 hp (128cc) and 5/6/8 hp (164cc) motors (note: 4-stroke models utilize a dog clutch, pin and retainer for shifting instead of a detent ball retained plunger)

7-8 LOWER UNIT

6. Separate the gearcase assembly from the outboard by pulling carefully straight downward, withdrawing the driveshaft and shift rod from the exhaust housing as it is lowered.
7. Thoroughly inspect the gearcase and exhaust housing for signs of damage. Make sure all mating surfaces are clean and free of debris, corrosion or damage.

To install:
8. If the gearcase was overhauled or otherwise serviced, rotate the propeller shaft slowly by hand and shift the gearcase into **forward**.
9. Apply a light coating of OMC Moly Lube, or equivalent assembly lubricant to the driveshaft splines. But, be careful not to coat the top surface of the driveshaft, as lubricant there could hydraulically prevent the driveshaft from seating in the crankshaft.
10. Apply a light coating of OMC Triple-Guard, or equivalent marine grease to the lower outside diameter of the water tube.
11. Apply a light coating of OMC Nut Lock, or equivalent threadlock to the threads of the 3 gearcase retaining bolts.
12. Install the gearcase, while carefully aligning the water tube into the pump grommet, the shift rod into position and the driveshaft to the crankshaft. If necessary, slowly rotate the flywheel clockwise (when viewed from above) in order to align the crankshaft and driveshaft splines.

■ When installing a gearcase that uses an extension, do NOT tighten the bolts until you are certain both the gearcase and extension have been properly aligned to the exhaust housing and powerhead.

13. Install the 3 gearcase mounting bolts. Tighten the 1 front bolt (threaded downward into the gearcase) to 120-144 inch lbs. (14-16 Nm) and the 2 rear bolts (threaded upward into the gearcase) to 60-84 inch lbs. (7-9 Nm).
14. Install the propeller and secure using a new cotter pin. For more details, please refer to Propeller under the Maintenance and Tune-Up section.
15. Properly refill the gearcase with lubricant. For details, please refer to the procedure in the Maintenance and Tune-Up section.
16. Reconnect the spark plug lead(s).

8/9.9 Hp (211cc), 9.9-15 Hp (216/255cc) and 9.9/15 Hp (305cc) Motors

◆ See Figures 14 thru 19

■ There are some indications that the 10/14 hp (216/255cc) 2-stroke commercial models were usually equipped with the split-type gearcase used by 25 Commercial and 28 hp (521cc) motors covered later in this section. The quick and easy way to determine if you should instead follow THAT procedure would be to check the gearcase at the height of the prop shaft. If there is a seam (split line) that runs the horizontal length of the case, refer to the procedure for the 521cc Commercial motors later in this section.

1. For safety, disconnect the spark plug leads and ground them to the cylinder head. For electric start models, disconnect the negative battery cable for safety as well.
2. Remove the propeller, for details refer to the procedure in the Maintenance and Tune-Up section.
3. Drain the gearcase oil. For details, please refer to the procedure in the Maintenance and Tune-Up section.
4. For 8/9.9 hp (211cc) 4-stroke motors, remove the clip from shift shaft, then tap the shaft in order to free the clevis from the shaft. Finally, remove the snap pin from the clevis retaining the pin to the shift rod (remove the pin and clevis).
5. For 9.9/15 hp (305cc) 4-stroke motors, remove the clip from the shift lever connector (at the front of the motor, right below the ignition module). Then remove the shift connector from the lever (unscrew the connector from the shift rod).
6. Remove the 6 bolts securing the gearcase to the exhaust housing (or extension on 20 and 25 in. models). There are 3 bolts threaded upward from below the gearcase flange on either side of the outboard.
7. Separate the gearcase assembly from the outboard by pulling carefully straight downward, withdrawing the driveshaft and shift rod from the exhaust housing as it is lowered. On 9.9-15 hp (216/255cc) motors, once the gearcase is **just** separated from the exhaust housing, loosen and remove the shift rod connector screw, then pull the gearcase completely free.
8. Thoroughly inspect the gearcase and exhaust housing for signs of damage. Make sure all mating surfaces are clean and free of debris, corrosion or damage.

To install:
9. On models with a gearcase extension, make sure the driveshaft tube extension, water tube extension and/or gearcase exhaust seals are all in place, as applicable. Apply a light coating of OMC Triple-Guard or equivalent marine grease to the end of the water tube coming out of the extension.
10. Apply a light coating of OMC Moly Lube, or equivalent assembly lubricant to the driveshaft splines. But, be careful not to coat the top surface of the driveshaft, as lubricant there could hydraulically prevent the driveshaft from seating in the crankshaft.
11. For 9.9-15 hp (216/255cc) motors, pull upward on the shift rod to place the gearcase into **reverse**, then move the shift lever on the outboard to **reverse** also.
12. Apply a light coating of OMC Triple-Guard, or equivalent marine grease to the lower outside diameter of the water tube or water pump seal (as applicable).
13. Apply a light coating of OMC Nut Lock, or equivalent threadlock to the threads of the 6 gearcase retaining bolts. For 9.9-15 hp (216/255cc) motors, apply a light coating of OMC Gasket Sealing Compound, or equivalent sealant to the threads of the shift rod connector screw. Place all bolts/screws aside on a clean surface for use in the next steps.
14. Install the gearcase, while carefully aligning the water tube into the pump grommet, the shift rod (either into the lower motor cover grommet for 4-stroke motors or into position at the shift rod connector on 2-stroke motors) and the driveshaft to the crankshaft splines.

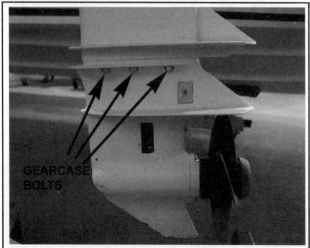

Fig. 14 To remove the gearcase, remove the 3 bolts from either side

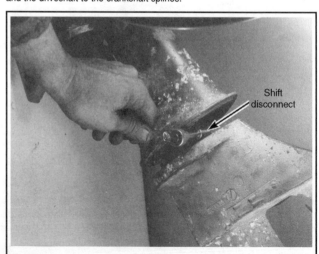

Fig. 15 On 2-stroke models, pull the gearcase down sufficiently to disconnect the shift rod linkage

LOWER UNIT 7-9

■ On 2-stroke motors, don't fully seat the gearcase to the exhaust housing or extension housing until the shift rod is fastened in the next step.

15. For 9.9-15 hp (216/255cc) motors, align the groove in the lower shift rod with the screw hole in the connector, then install and tighten the screw to 60-84 inch lbs. (7-9 Nm).

■ When installing a gearcase that uses an extension, do NOT tighten the bolts until you are certain both the gearcase and extension have been properly aligned to the exhaust housing and powerhead.

16. Install the 6 gearcase mounting bolts and tighten to 96-120 inch lbs. (11-14 Nm).

17. For 9.9/15 hp (305cc) 4-stroke motors, thread the shift rod connector onto the shift rod until it bottoms, then turn the connector back just until the hole in it aligns with the shift arm. Secure the shift rod connector onto the shift arm pin using a clip.

18. For 8/9.9 hp (211cc) 4-stroke motors, position the clevis and pin on the shift rod, then secure using the snap pin. Next, align the clevis with the shaft and push the shaft through both sides of the clevis. Install the clip on the shaft.

19. Grease the shaft, then properly install and secure the propeller. For more details, please refer to Propeller under the Maintenance and Tune-Up section.

20. Properly refill the gearcase with lubricant. For details, please refer to the procedure in the Maintenance and Tune-Up section.

21. Reconnect the spark plug leads.

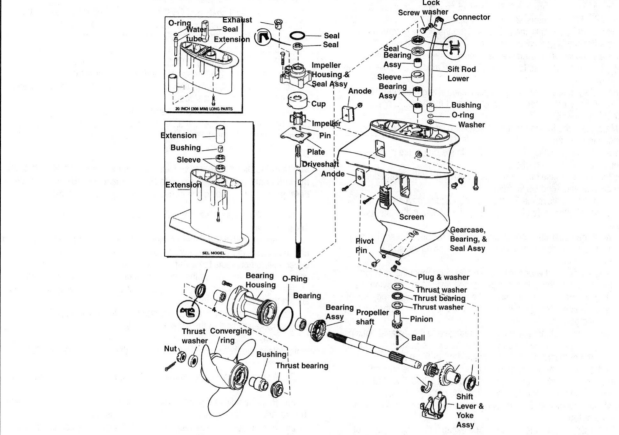

Fig. 16 Exploded view of a typical gearcase assembly used on 8/9.9 hp (211cc), 9.9-15 hp (216/255cc) and 9.9/15 hp (305cc) motors (note the shift rod on 9.9/15 hp 4-stroke motors is straight, and not cocked forward as on other models, also most 1990-95 models use dual back-to-back propshaft seals)

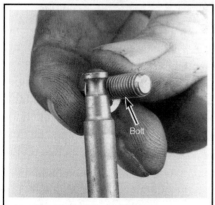

Fig. 17 On 2-strokes, the shift rod connector bolt must fit in the groove. . .

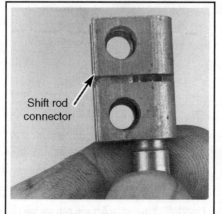

Fig. 18 . . . place the connector in position. . .

Fig. 19 . . . then thread and tighten the bolt locking it to the groove

7-10 LOWER UNIT

10/14 (216/255cc), 20-35 Hp (521cc) and 25/35 Hp (500/565cc) Motors
Except Jet, 10/14 Hp, 28 Hp, 30 SPL and Some 25 Hp Comm Models

◆ See Figures 20 thru 25  MODERATE

The gearcase used on most 20-35 hp (521cc) motors (not including Jet, 28 hp, 30 SPL and some commercial 25 hp models) is the identical unit used by the 25/35 hp (500/565cc) 3-cylinder motors. The water intake screens on this gearcase provide access to the shift rods for disconnection before the gearcase is removed. Otherwise, the unit is similar to most of the other Johnson/Evinrude gearcases used by both larger and smaller inline motors. The gearcase itself is a 1-piece housing that can be disassembled by removing shaft components from the top and from the end of the housing.

However, some 10/14 hp (216/255cc) and 25 hp Commercial and all 28 hp and 30 SPL (521cc) motors are equipped with a split lower gearcase, that is easily identified by the split line that runs along the housing around the propeller shaft. On this gearcase, the access to the propshaft components is available once the lower housing/skeg is unbolted from the rest of the gearcase. Service procedures for the split gearcase models are covered later in this section.

1. For safety, disconnect the spark plug leads and ground them to the cylinder head. For electric start models, disconnect the negative battery cable for safety as well.
2. Matchmark the trim tab to the gearcase for installation purposes, then remove the trim tab.
3. Remove the propeller, for details refer to the procedure in the Maintenance and Tune-Up section.
4. Drain the gearcase oil. For details, please refer to the procedure in the Maintenance and Tune-Up section.
5. Loosen and remove the 2 Phillips screws and remove the water intake screens (at either side of the gearcase) for access to the shit rod assembly.
6. Check the upper shift rod for signs of paint residue. If there is any paint on the upper shit rod, remove it before disengaging the shift rod connectors.
7. Using 2 open-end wrenches, loosen and disconnect the shift rod through the opening in the side of the gearcase provided by removing the water intake screens. Remove the plastic keeper from the upper rod, then if possible, slip the upper fitting nut off the rod. If not, leave the upper fitting nut in place until the gearcase is removed.
8. Remove the fasteners securing the gearcase to the exhaust housing. There are either 3 bolts threaded upward from below the gearcase flange on either side of the outboard (for most 521cc models) or there are 4 bolts (two on either side) and a nut (on top of the exhaust housing). The second configuration (4 bolts and a nut) is normally found on 500/565cc models).
9. Separate the gearcase assembly from the outboard by pulling carefully straight downward. Be careful not to damage the driveshaft, shift rod or the water tube.
10. Thoroughly inspect the gearcase and exhaust housing for signs of damage. Make sure all mating surfaces are clean and free of debris, corrosion or damage.

To install:

11. Check the water tube for damage. Visually inspect the outer diameter for signs of dents or burrs and remove if the water tube is being reused. Make sure the water tube washer is positioned on the tube.
12. Apply a light coating of OMC Triple-Guard, or equivalent marine grease to the upper and lower outside diameter of the water tube then install the straight end of the tube through the guide into the inner exhaust tube.
13. Apply a light coating of OMC Triple-Guard, or equivalent marine grease to 2 new driveshaft spacer O-rings, then install the O-rings on the driveshaft spacer and position the spacer in the exhaust housing with the tabs facing the rear of the housing.
14. Apply a light coating of OMC Triple-Guard, or equivalent marine grease to the upper shift rod, then temporarily install the upper shift rod connector (fitting nut) on the gearcase lower shift rod (since it usually cannot be positioned once the gearcase is installed).
15. Apply a light coating of OMC Moly Lube, or equivalent assembly lubricant to the driveshaft splines. But, be careful not to coat the top surface of the driveshaft, as lubricant there could hydraulically prevent the driveshaft from seating in the crankshaft.
16. Apply a light coating of OMC Gasket Sealing Compound, or equivalent sealant to the threads of the 5 or 6 gearcase fasteners (6 bolts or 4 bolts and a stud/nut, as applicable). Place all fasteners aside on a clean surface for use in the next steps.
17. Install the gearcase, while carefully aligning the water tube into the pump grommet and the driveshaft to the crankshaft splines. If necessary, slowly rotate the flywheel clockwise (when viewed from above) to align the crankshaft splines with the driveshaft.

■ **When installing a gearcase that uses an extension, do NOT tighten the bolts until you are certain both the gearcase and extension have been properly aligned to the exhaust housing and powerhead.**

18. Install the gearcase mounting fasteners and tighten to 192-216 inch lbs. (22-25 Nm).
19. Slide the shift rod upper connector (fitting) half up onto the upper rod and then install the plastic keeper below it. Reposition the shift lever until the upper shift rod contacts the lower shift rod connector. Hold the lower connector while you thread the upper connector by hand, use an open-end wrench to hold the lower connector while using another wrench to securely tighten the upper connector.
20. Install the water intake screens to both the port and starboard side of the gearcase and tighten the screws securely.
21. Grease the shaft, then properly install and secure the propeller. For more details, please refer to Propeller under the Maintenance and Tune-Up section.
22. Align the matchmarks made earlier, then securely reinstall the trim tab.
23. Properly refill the gearcase with lubricant. For details, please refer to the procedure in the Maintenance and Tune-Up section.
24. Reconnect the spark plug leads and, if applicable, connect the negative battery cable.
25. If necessary, adjust the shift lever and/or neutral start switch.

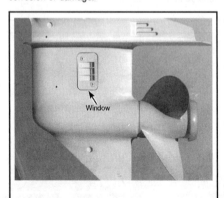
Fig. 20 The water intake grates act as a window...

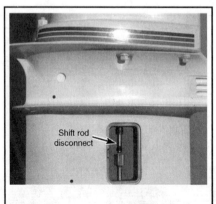
Fig. 21 ...through which you can access the shift rod

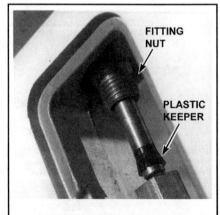
Fig. 22 Unthread the upper fitting nut and remove the keeper

LOWER UNIT 7-11

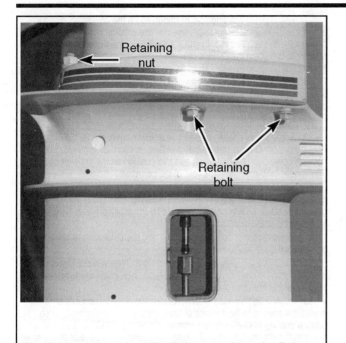

Fig. 23 Loosen the retainers (3-cylinder motor shown, 2-cylinder motor normally use 6 bolts)

Fig. 24 During removal or installation, take care to prevent damaging the shift rod, water tube and driveshaft

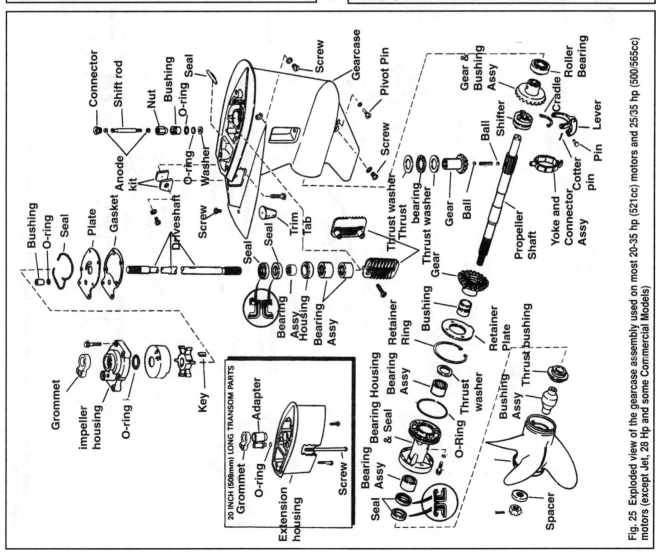

Fig. 25 Exploded view of the gearcase assembly used on most 20-35 hp (521cc) motors and 25/35 hp (500/565cc) motors (except Jet, 28 Hp and some Commercial Models)

7-12 LOWER UNIT

28 Hp, 30 SPL and Some 10/14 Hp, 25 Hp Comm Models (With Split Lower Gearcases)

◆ See Figures 17 thru 19 and 26

MODERATE

Unlike the more conventional 1-piece gearcase housing used on most Johnson/Evinrude motors (including most 521cc models), some 25 hp Commercial and all 28 hp and 30 SPL (521cc) motors are equipped with a split lower gearcase. Models with the split gearcase are easily identified by the split line that runs along the housing in alignment with the propeller shaft or the 6 bolt heads that are visible at the bottom of the gearcase threaded upward on either side of the skeg. On this gearcase, the access to the propshaft components is available once the lower housing/skeg is unbolted from the rest of the gearcase.

1. For safety, disconnect the spark plug leads and ground them to the cylinder head. For electric start models, disconnect the negative battery cable for safety as well.
2. Remove the propeller, for details refer to the procedure in the Maintenance and Tune-Up section.
3. Loosen the retainers, then remove the cover plate and gasket from the port side of the exhaust housing for access to the shift rod connector.
4. Shift the outboard into **forward** gear, then loosen and remove the lower bolt from the shift rod connector.
5. Remove the 4 bolts (two on either side) and the nut (on top of the exhaust housing) securing the gearcase assembly.
6. Separate the gearcase assembly from the outboard by pulling carefully straight downward. Be careful not to damage the driveshaft, shift rod or the water tube.
7. The impeller housing spacer (mounted on the driveshaft, directly above the impeller housing) may remain in the exhaust housing or may come out with the gearcase. In either situation, locate the spacer, then remove and discard the O-rings (as new O-rings must be used during assembly).
8. Thoroughly inspect the gearcase and exhaust housing for signs of damage. Make sure all mating surfaces are clean and free of debris, corrosion or damage.

To install:

9. Apply a light coating of clean gearcase oil over the 2 NEW impeller housing spacer O-rings, then install the O-rings into their grooves. Slide the housing down over the driveshaft and into position on the impeller housing. The spacer tabs must face the rear of the gearcase.
10. Apply a light coating of OMC Moly Lube, or equivalent assembly lubricant to the driveshaft splines. But, be careful not to coat the top surface of the driveshaft, as lubricant there could hydraulically prevent the driveshaft from seating in the crankshaft.
11. Make sure that both the gearcase and the outboard are shifted into the **forward** gear position.
12. If removed, apply a light coating of OMC Triple-Guard, or equivalent marine grease to both ends of the water tube, then install the tube with the straight end in the inner exhaust housing guide.
13. Apply a light coating of OMC Gasket Sealing Compound, or equivalent sealant to the threads of the 4 gearcase retaining bolts. Also, apply a coating of OMC Locquic Primer followed by OMC Nut Lock, or equivalent high-strength threadlocking compound to the threads of the stud. Place all fasteners aside on a clean surface for use in the next steps.

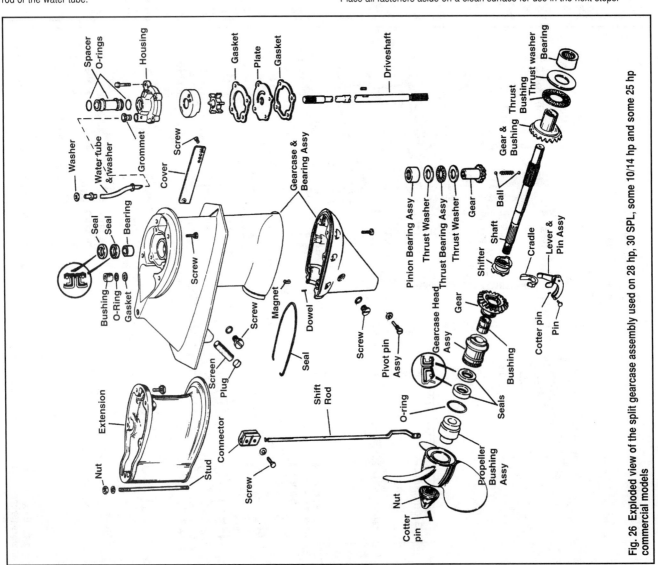

Fig. 26 Exploded view of the split gearcase assembly used on 28 hp, 30 SPL, some 10/14 hp and some 25 hp commercial models

LOWER UNIT 7-13

14. Install the gearcase, while carefully aligning the water tube into the grommet, the driveshaft to the crankshaft splines and the lower shift rod into the shift rod connector. If necessary, slowly rotate the flywheel clockwise (when viewed from above) to align the crankshaft splines with the driveshaft.

■ When installing a gearcase that uses an extension, do NOT tighten the bolts until you are certain both the gearcase and extension have been properly aligned to the exhaust housing and powerhead.

15. Install the gearcase mounting fasteners, then tighten the bolts to 192-216 inch lbs. (22-25 Nm) and the stud nut to 45-50 ft. lbs. (61-68 Nm).
16. Align the lower shift rod groove with the lower hole in the shift rod connector, then install the bolt and tighten to 120-144 inch lbs. (14-16 Nm).
17. Apply a light coating of OMC Adhesive M, or equivalent, to a NEW exhaust housing cover plate gasket, then install the plate and gasket. Tighten the retainers securely.
18. Grease the shaft, then properly install and secure the propeller. For more details, please refer to Propeller under the Maintenance and Tune-Up section.
19. If the gearcase was drained for overhaul, properly refill the gearcase with lubricant. It's not a bad idea to perform this service regardless, to make sure fresh oil is in the gearcase anyway. For details, please refer to the procedure in the Maintenance and Tune-Up section.
20. Reconnect the spark plug leads and, if applicable, connect the negative battery cable.

25-55 Hp (737cc) and 25-70 Hp (913cc) Motors (Except Jet Models)

Some 2-Cylinder Motors Including 40 (Except 40RW, 40RP and 40WR), 48 and 50 Hp Models

◆ See Figure 27

Although most 2- and 3-cylinder motors use the same gearcase assembly, a few of the lighter duty 2-cylinder motors, including the 40 (except 40RW, 40RP and 40 WR), 48 and 50 hp models use a unique assembly. On these models, a bearing housing anode is mounted between the propeller and the bearing carrier (on all other 2- and 3-cylinder motors, the bearing housing anode is bolted to the bottom of the bearing carrier).

The propeller shaft seal can be found in the bearing carrier, directly behind the anode. Although some earlier production models utilize a dual-seal assembly, most of these motors utilize a single propeller shaft seal.

1. For safety, disconnect the spark plug leads and ground them to the cylinder head. For electric start models, disconnect the negative battery cable for safety as well.
2. Remove the propeller, for details refer to the procedure in the Maintenance and Tune-Up section.
3. Remove the shift rod screw (protruding under the top engine cover, above the exhaust housing).

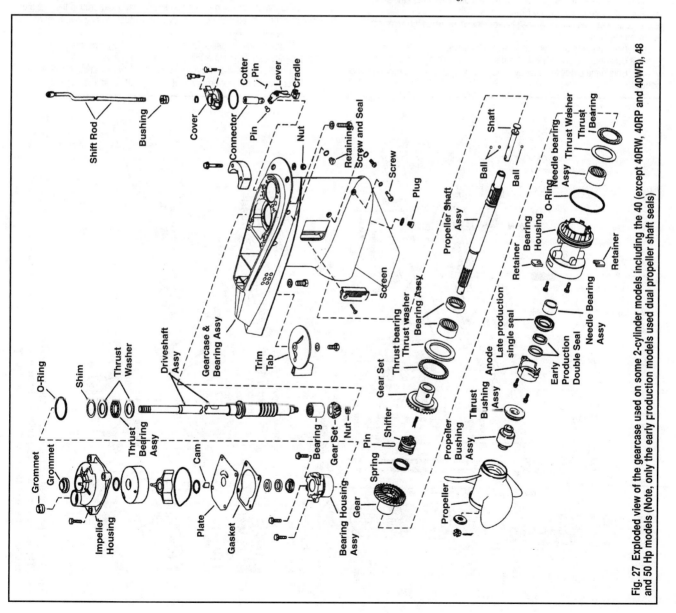

Fig. 27 Exploded view of the gearcase used on some 2-cylinder models including the 40 (except 40RW, 40RP and 40WR), 48 and 50 Hp models (Note, only the early production models used dual propeller shaft seals)

7-14 LOWER UNIT

4. Using a thin-walled 5/8 in. socket, remove the gearcase bolt threaded straight upward underneath the cavitation plate (just in front of the trim-tab).

5. Remove the 4 remaining bolts securing the gearcase to the exhaust housing (there are 2 on either side of the outboard, threaded upward from the gearcase into the exhaust housing).

6. Separate the gearcase assembly from the outboard by pulling carefully straight downward. Be careful not to damage the driveshaft, shift rod or the water tube.

7. Thoroughly inspect the gearcase and exhaust housing for signs of damage. Make sure all mating surfaces are clean and free of debris, corrosion or damage.

To install:

8. Apply a light coating of OMC Moly Lube, or equivalent assembly lubricant to the driveshaft splines. But, be careful not to coat the top surface of the driveshaft, as lubricant there could hydraulically prevent the driveshaft from seating in the crankshaft.

9. Apply a light coating of OMC Nut Lock, or equivalent threadlock, to the threads of the 5 gearcase fasteners then place them aside on a clean surface for use in the next steps.

10. Install the gearcase, while carefully aligning the exhaust housing inner water tube into the pump and the driveshaft to the crankshaft splines. If necessary, slowly rotate the flywheel clockwise (when viewed from above) to align the crankshaft splines with the driveshaft.

11. Install the 5 gearcase mounting fasteners. Tighten the four (3/8 in.) flange bolts to 18-20 ft. lbs. (24-27 Nm) and the one (7/16 in.) cavitation plate bolt to 28-30 ft. lbs. (38-40 Nm).

12. Connect the shift rod to the shift rod link, then install the shift rod screw and tighten securely.

13. Verify the neutral detent adjustment as follows:
   a. Loosen the 2 powerhead detent screws.
   b. Rotate the propeller shaft by hand and verify that the gearcase is in **neutral**.
   c. Verify that the powerhead shift linkage is in **neutral**.
   d. Tighten the 2 powerhead detent screws to 60-84 inch lbs. (7-9 Nm).

14. Grease the shaft, then properly install and secure the propeller. For more details, please refer to Propeller under the Maintenance and Tune-Up section.

15. If the gearcase was drained for overhaul, properly refill the gearcase with lubricant. It's not a bad idea to perform this service regardless, to make sure fresh oil is in the gearcase anyway. For details, please refer to the procedure in the Maintenance and Tune-Up section.

16. Reconnect the spark plug leads and, if applicable, connect the negative battery cable.

3-Cylinder Motors and Most 2-Cylinder Motors Except 40, 40RS, 48 and 50 Hp Models

◆ See Figures 28 thru 33

Most of the larger Johnson/Evinrude 2- and 3-cylinder motors (except the 2-cylinder 40, 40RS, 48 and 50 hp models) use the same heavy-duty gearcase assembly. These assemblies can be identified by the fact that the bearing housing anode is bolted to the bottom of the bearing carrier (instead of between the propeller and bearing carrier as it is on the lighter duty models).

These models utilize a dual, back-to-back propeller shaft seal assembly mounted to the bearing carrier.

1. For safety, disconnect the spark plug leads and ground them to the cylinder head. For electric start models, disconnect the negative battery cable for safety as well.

2. Remove the propeller, for details refer to the procedure in the Maintenance and Tune-Up section.

3. For 2-cylinder motors, remove the shift lever retaining pin. Push the shift lever toward the powerhead to disengage the shift lever from the rod.

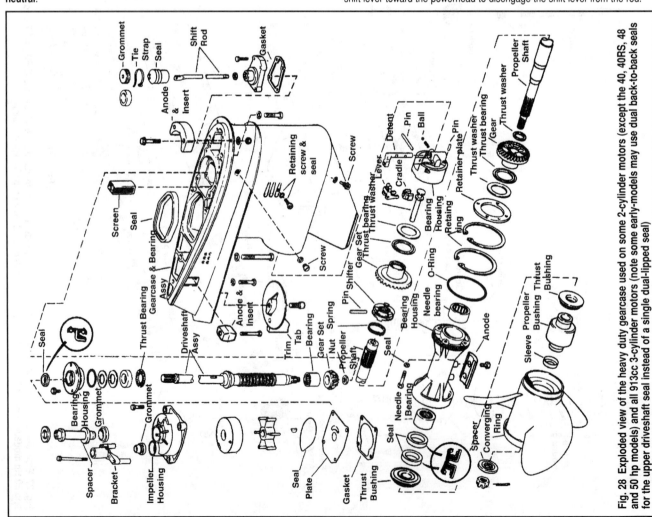

Fig. 28 Exploded view of the heavy duty gearcase used on some 2-cylinder motors (except the 40, 40RS, 48 and 50 hp models) and all 913cc 3-cylinder motors (note some early-models may use dual back-to-back seals for the upper driveshaft seal instead of a single dual-lipped seal)

LOWER UNIT 7-15

4. For 3-cylinder motors, if equipped, remove the shift rod screw. If not equipped with a shift rod screw, an assembly similar to the 2-cylinder motor is used, remove the lever retaining pin, and then disengage the lever from the rod by pushing it back toward the powerhead.
5. Matchmark the trim tab to the gearcase for installation purposes, then remove the trim tab.
6. Using a thin-walled 5/8 in. socket, remove the gearcase bolt threaded straight upward underneath the cavitation plate (just in front of the trim-tab).
7. Remove the bolt found inside the trim tab cavity.
8. Remove 4 remaining gearcase bolts (two on either side of the motor) from either side of the gearcase-to-exhaust housing flange.
9. Separate the gearcase assembly from the outboard by pulling carefully straight downward. Be careful not to damage the driveshaft, shift rod or the water tube.
10. Thoroughly inspect the gearcase and exhaust housing for signs of damage. Make sure all mating surfaces are clean and free of debris, corrosion or damage.

To install:

11. Apply a light coating of OMC Moly Lube, or equivalent assembly lubricant to the driveshaft splines. But, be careful not to coat the top surface of the driveshaft, as lubricant there could hydraulically prevent the driveshaft from seating in the crankshaft.
12. For 3-cylinder motors, check the condition of the inner exhaust housing seal and replace, if necessary. Apply a light coating of OMC Adhesive M or equivalent to the inside surface of the seal on the exhaust housing. Then, apply a light coating of OMC Triple-Guard, or equivalent marine grease, to the seal's outer surface.
13. Apply a light coating of OMC Gasket Sealing Compound, or equivalent sealant to the threads of the gearcase fasteners, then place aside on a clean surface for use in the next steps.
14. Install the gearcase, while carefully aligning the water tube into the water pump and the driveshaft to the crankshaft splines. If necessary, slowly rotate the flywheel clockwise (when viewed from above) to align the crankshaft splines with the driveshaft.
15. Install the gearcase mounting fasteners. Tighten the 3/8 in. bolts to 18-20 ft. lbs. (24-27 Nm) and the 7/16 in. bolts to 28-30 ft. lbs. (38-40 Nm).
16. Engage the shift rod as follows:
  • For 2-cylinder motors, engage the rod to the link, then install the pin.
  • For 3-cylinder motors that utilize a pin, place the shift rod in **neutral**, then align the rod with the lever. Pull the lever away from the powerhead to engage the rod and install the retaining pin. Make sure the pin compresses the flat and spring washers against the casing.
  • For 3-cylinder motors that utilize a screw, engage the shift rod into the rod link, then tighten the screw securely.
17. Grease the shaft, then properly install and secure the propeller. For more details, please refer to Propeller under the Maintenance and Tune-Up section.
18. Align the matchmarks made earlier, then reinstall the trim tab. Tighten the retaining bolt to 18-20 ft. lbs. (24-27 Nm) for 2-cylinder motors or to 28-30 ft. lbs. (38-40 Nm) for 3-cylinder motors.
19. If the gearcase was drained for overhaul, properly refill the gearcase with lubricant. It's not a bad idea to perform this service regardless, to make sure fresh oil is in the gearcase anyway. For details, please refer to the procedure in the Maintenance and Tune-Up section.
20. Reconnect the spark plug leads and, if applicable, connect the negative battery cable.

40/50 Hp (815cc) EFI Motors

◆ See Figure 34

The 40/50 hp EFI Johnson/Evinrude motors utilize a gearcase that is not unlike the larger 2-stroke Johnson/Evinrude motors. Removal and installation of the gearcase assembly is a relatively straightforward procedure. However the gearcase itself contains a precision shimmed assembly of gears, bearings and shaft components. Disassembly/assembly is not recommended without significant experience and the use of precision measuring tools to ensure proper gear/bearing shimming.

1. For safety, disconnect the negative battery cable.
2. Remove the propeller, for details refer to the procedure in the Maintenance and Tune-Up section.
3. Make sure the gearcase shifter is in the **neutral** position.
4. Access the shift rod locknut and turn buckle at the front of the outboard (it may be necessary to remove the lower engine covers, as detailed under Maintenance and Tune-Up). Loosen the shift rod locknut, then loosen the turn buckle, counting the number of turns of the turn buckle necessary to disconnect the shift rod. Write this number down, as you will thread the turn buckle the same number of turns during assembly to ensure proper adjustment.
5. Remove 6 bolts that are threaded upward from the gearcase flange into the exhaust housing/midsection. There are 3 bolts on either side of the outboard.

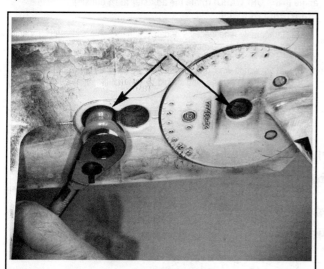

Fig. 29 A thin-walled socket is needed to access the recessed gearcase bolt, and usually, the trim-tab bolt

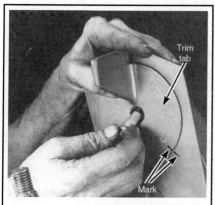

Fig. 30 Matchmark the trim tab and remove it for access...

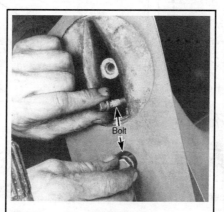

Fig. 31 ...then remove the 2 gearcase bolts found under the cavitation plate

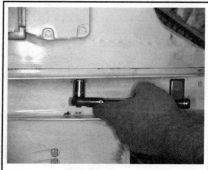

Fig. 32 Next, remove the 4 gearcase flange bolts...

7-16 LOWER UNIT

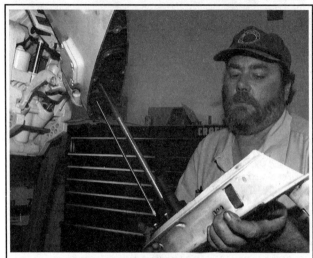

Fig. 33 ... and carefully lower the gearcase from the powerhead taking care not to damage the driveshaft, shift rod or water tube

6. Separate the gearcase assembly from the outboard by pulling carefully straight downward. Be careful not to damage the driveshaft, shift rod or the water tube.

7. Thoroughly inspect the gearcase and exhaust housing for signs of damage. Make sure all mating surfaces are clean and free of debris, corrosion or damage.

To install:

8. Apply a light coating of OMC Moly Lube, or equivalent assembly lubricant to the driveshaft splines. But, be careful not to coat the top surface of the driveshaft, as lubricant there could hydraulically prevent the driveshaft from seating in the crankshaft.

9. Apply a light coating of OMC Adhesive M, or equivalent, and install the dowel pins to the front and rear of the gearcase flange.

10. Apply a light coating of RTV sealant to the exhaust housing mating surface.

11. Make sure the gearcase is in **neutral**.

12. Apply a light coating of OMC Gasket Sealing Compound, or equivalent sealant to the threads of the gearcase fasteners, then place aside on a clean surface for use in the next steps.

13. Install the gearcase, while carefully aligning the water tube into the water pump outlet and the driveshaft to the crankshaft splines. If necessary, slowly rotate the flywheel clockwise (when viewed from above) to align the crankshaft splines with the driveshaft.

14. Install the gearcase mounting fasteners and tighten them to 16 ft. lbs. (23 Nm).

15. Reconnect the shift linkage by threading the turnbuckle onto the rod, the exact same number of turns as noted during removal, then secure using the locknut.

16. If the gearcase was drained for overhaul, properly refill the gearcase with lubricant. It's not a bad idea to perform this service regardless, to make sure fresh oil is in the gearcase anyway. For details, please refer to the procedure in the Maintenance and Tune-Up section.

17. Double-check the shift rod adjustment using an assistant to turn the propeller shaft by hand while you shift the lever from **neutral** to **forward** and then from **neutral** to **reverse**. The gears must engage at approximately the same angle from **neutral** in each direction. If not, loosen the shift rod locknut and adjust the turnbuckle until neutral is centered.

■ If the FORWARD gear engages too early (at a smaller angle than REVERSE) thread the adjuster (turn buckle) CLOCKWISE until the angles are equal. If the REVERSE gear engages too early (at a smaller angle than FOREWARD) rotate the adjuster (turn buckle) COUNTERCLOCKWISE.

18. Temporarily reconnect the negative battery cable and verify proper shifter adjustment with the motor running on a source of cooling water. The propeller shaft must not rotate in **neutral**, but should begin rotating in the appropriate direction when the gearcase is shifted into **forward** or **reverse**. Once this has been verified, you should disconnect the cable again for safety until all service has been finished.

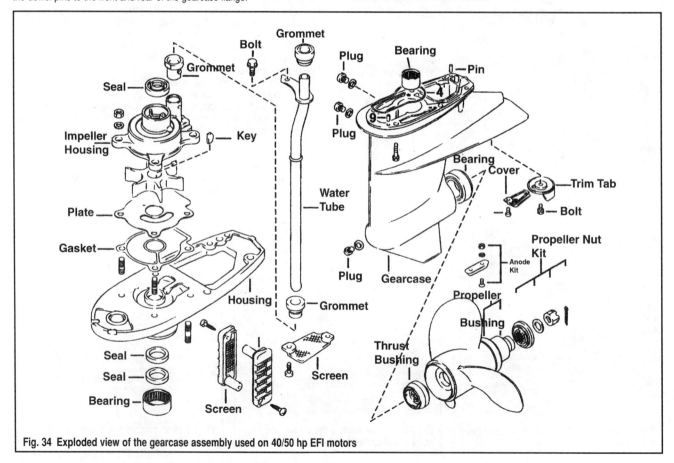

Fig. 34 Exploded view of the gearcase assembly used on 40/50 hp EFI motors

LOWER UNIT 7-17

※※ WARNING

When the shaft is turning in gear, shift into NEUTRAL and allow it to naturally stop rotating before attempting to shift it again. Remember that when you're on the water the propeller produces a significant amount of drag to slow the propeller shaft down as soon as power is removed from shaft (as soon as the gearcase is shifted into NEUTRAL).

19. Grease the shaft, then properly install and secure the propeller. For more details, please refer to Propeller under the Maintenance and Tune-Up section.
20. If removed, install the lower engine covers, as detailed in the Maintenance and Tune-Up section.
21. Reconnect the negative battery cable.

70 Hp (1298cc) EFI Motors

◆ See Figure 35

The 70 hp EFI Johnson/Evinrude motor utilizes a gearcase similar to the heavy-duty units found on the larger Johnson/Evinrude 2-stroke motors. Removal and installation of the gearcase assembly is a relatively straightforward procedure. However the gearcase itself contains a precision shimmed assembly of gears, bearings and shaft components. Disassembly/assembly is not recommended without significant experience and the use of precision measuring tools to ensure proper gear/bearing shimming.

1. For safety, disconnect the negative battery cable.
2. Remove the propeller, for details refer to the procedure in the Maintenance and Tune-Up section.
3. Make sure the gearcase shifter is in the **neutral** position.
4. Access the shift rod connector at the front of the motor, just below the swivel bracket. Remove the cotter pin and disengage the connector. Discard the old cotter pin, as a new one must be used during installation to ensure safe and trouble-free operation.
5. Matchmark the trim tab to the gearcase to preserve adjustment, then remove the retaining bolt and the trim tab.

■ The trim tab must be removed for access to the gearcase bolt hidden within the trim tab cavity.

6. Remove 7 bolts that are threaded upward from the gearcase into the exhaust housing/midsection. There are 3 bolts on each side of the outboard and one in the trim tab cavity.
7. Separate the gearcase assembly from the outboard by pulling carefully straight downward. Be careful not to damage the driveshaft, shift rod or the water tube.
8. Thoroughly inspect the gearcase and exhaust housing for signs of damage. Make sure all mating surfaces are clean and free of debris, corrosion or damage.

To install:

9. Apply a light coating of OMC Moly Lube, or equivalent assembly lubricant to the driveshaft splines. But, be careful not to coat the top surface of the driveshaft, as lubricant there could hydraulically prevent the driveshaft from seating in the crankshaft.
10. Apply a light coating of OMC Gasket Sealing Compound, or equivalent sealant to the threads of the 7 gearcase fasteners and to the dowel pins. Place the fasteners aside on a clean surface for use later, then install the dowel pins to the front and rear of the gearcase flange.
11. Apply a light coating of RTV sealant to the exhaust housing mating surface.
12. Make sure the gearcase is in **neutral**.
13. Install the gearcase, while carefully aligning the water tube into the water pump outlet and the driveshaft to the crankshaft splines. If necessary, slowly rotate the flywheel clockwise (when viewed from above) to align the crankshaft splines with the driveshaft.
14. Install the gearcase mounting fasteners and tighten them to 40 ft. lbs. (55 Nm).
15. Reconnect the upper and lower shift rods using the hex-head connector and secure using a new cotter pin.
16. Align the matchmarks made earlier and install the trim tab. Tighten the retaining bolt to 14 ft. lbs. (20 Nm).
17. If the gearcase was drained for overhaul, properly refill the gearcase with lubricant. It's not a bad idea to perform this service regardless, to make sure fresh oil is in the gearcase anyway. For details, please refer to the procedure in the Maintenance and Tune-Up section.

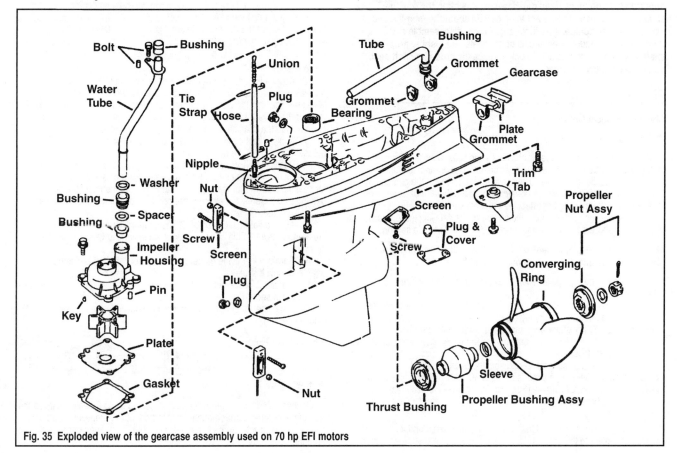

Fig. 35 Exploded view of the gearcase assembly used on 70 hp EFI motors

7-18 LOWER UNIT

18. Temporarily reconnect the negative battery cable and verify proper shifter adjustment with the motor running on a source of cooling water. The propeller shaft must not rotate in **neutral**, but should begin rotating in the appropriate direction when the gearcase is shifted into **forward** or **reverse**. Once this has been verified, you should disconnect the cable again for safety until all service has been finished.

✳✳ WARNING

When the shaft is turning in gear, shift into NEUTRAL and allow it to naturally stop rotating before attempting to shift it again. Remember that when you're on the water the propeller produces a significant amount of drag to slow the propeller shaft down as soon as power is removed from shaft (as soon as the gearcase is shifted into NEUTRAL).

19. Grease the shaft, then properly install and secure the propeller. For more details, please refer to Propeller under the Maintenance and Tune-Up section.
20. If removed, install the lower engine covers, as detailed in the Maintenance and Tune-Up section.
21. Reconnect the negative battery cable.

Propeller Shaft Seal

If the gearcase oil is contaminated by water or there are signs of leakage at the propshaft seal, the gearcase REALLY should be disassembled, thoroughly inspected and assembled again, using new seals. However, if the propeller shaft seal is the culprit AND you are certain that there is no damage (due to lack of proper oiling or from corrosion) inside the gearcase, you CAN replace just the prop shaft seal. Furthermore, this task can usually be accomplished with the gearcase still attached to the outboard. This is especially handy if you've been diligent about inspecting the gearcase and notice a damaged seal (perhaps from tangled fishing line or the like) immediately upon removing the boat/motor from the water after an excursion.

■ **Read and believe, replacing ONLY the propeller shaft seal is rarely the right way to handle the situation (unless you've discovered the problem before most of the oil is lost and before moisture has had the opportunity to do much damage). However, it is often the way the situation is handled. If you want to be certain about the long-life and condition of the components in your gearcase, disassemble and thoroughly overhaul it. Replace all of the seals, not just the propeller shaft. If, however, you decide to only replace the propeller shaft seal, you have been fairly warned.**

REPLACEMENT

Colt/Junior (43cc) Motors

◆ See Figure 3

On these models, the propeller shaft is sealed using a single seal, installed in the rear of the gearcase head/bearing carrier mounted just in front of the propeller (with the seal lips facing inward toward the gearcase). The seal can be replaced with the gearcase installed, but the bearing carrier removed from the housing to protect the propeller shaft.

1. Remove the propeller from the outboard. For details, please refer to the procedure located in the Maintenance and Engine Tune-Up section.

■ **To prevent loss of gearcase oil during this procedure either drain the gearcase or position the outboard at full tilt.**

2. Remove the 2 propeller shaft bearing housing (bearing carrier) retaining bolts.
3. Carefully slide the gearcase head (bearing carrier) off the propeller shaft. It may be necessary to gently tap on the carrier ears using a rubber mallet (or to gently pry under both ears using 2 small prybars). Be careful when removing the carrier to prevent damage to the gearcase and carrier mating/sealing surfaces.
4. Remove and discard the O-ring from the bearing carrier.
5. Using a puller bridge plate (such as OMC #432127) and an internal jawed puller (such as OMC #432130), carefully remove the seal from the backside of the bearing carrier.

To install:

6. Apply a light coating of OMC Gasket Sealing Compound, or equivalent sealant to the outer surface of the NEW propeller shaft seal, then install the seal into the bearing carrier with the lips facing inward toward the bearing carrier and gearcase. Use a suitably sized socket or bearing driver and a mallet to gently tap the seal into the back of the carrier housing.

7. Apply a light coating of OMC Triple-Guard, or equivalent marine grease to the lips of the new seal.
8. Apply a light coating of OMC Triple-Guard, or equivalent marine grease to a new bearing carrier O-ring, then install the O-ring to the carrier groove.
9. Carefully install the carrier over the propeller shaft and into the gearcase, until it seats.
10. Apply a light coating of OMC Gasket Sealing Compound, or equivalent sealant to the threads of the gearcase head retaining bolts, then install the bolts and tighten to 60-84 inch lbs. (7-9 Nm).
11. If drained, properly refill the gearcase, as detailed in the Maintenance and Tune-Up section.
12. Install the propeller, as detailed in the Maintenance and Tune-Up section.

2.0-3.5 Hp (78cc) Motors

◆ See Figures 6 and 36 thru 40

On these models, the propeller shaft seal is located in the bearing carrier mounted just below the water pump impeller.

1. Remove the water pump from the gearcase. For details, please refer to the procedure in the Lubrication and Cooling section.

■ **To prevent loss of gearcase oil during this procedure either drain the gearcase or position the outboard at full tilt.**

2. If not done already, remove the impeller drive pin from the propeller shaft.
3. Carefully slide the gearcase head (bearing carrier) off the propeller shaft. It may be necessary to gently tap on the carrier ears using a rubber mallet (or to gently pry under both ears using 2 small prybars). Be careful when removing the carrier to prevent damage to the gearcase and carrier mating/sealing surfaces.
4. Remove and discard the O-ring from the bearing carrier.
5. Using a puller bridge plate (such as OMC #432127), backing plate (such as OMC #115312) and an internal jawed puller (such as OMC #432130), carefully remove the bearing, then using the same tools, remove the seal from the backside of the bearing carrier.

To install:

6. Apply a light coating of OMC Gasket Sealing Compound, or equivalent sealant to the outer surface of the NEW propeller shaft seal, then install the seal into the bearing carrier with the lettered side facing outward (toward the driver). Use a suitably sized socket or bearing driver and a mallet to gently tap the seal into the back of the carrier housing.
7. Apply a light coating of OMC Triple-Guard, or equivalent marine grease to the lips of the new seal.
8. Position a NEW bearing into the carrier with the lettered side facing outward, away from the seal. Use the same socket or driver that was used for seal installation (but making sure that the socket contacts the outer bearing race). Gently tap or press the bearing down into the bearing carrier until it is seated.
9. Apply a light coating of OMC Triple-Guard, or equivalent marine grease to a new bearing carrier O-ring, then install the O-ring to the carrier groove.
10. Carefully install the carrier over the propeller shaft and into the gearcase, until it seats.
11. If drained, properly refill the gearcase, as detailed in the Maintenance and Tune-Up section.
12. Install the water pump assembly, as detailed in the Lubrication and Cooling section.

5 Hp (109cc) Motors

◆ See Figure 7

On these models, the propeller shaft is sealed using 2 seals, installed the rear of the bearing carrier mounted just in front of the propeller. The seals can be replaced with the gearcase installed, but the bearing carrier removed from the housing to protect the propeller shaft.

LOWER UNIT 7-19

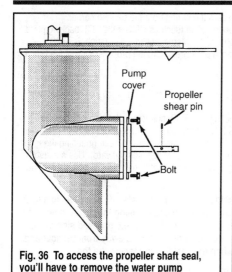

Fig. 36 To access the propeller shaft seal, you'll have to remove the water pump cover...

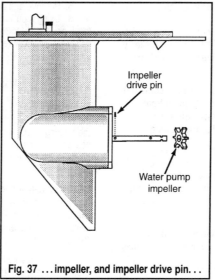

Fig. 37 ...impeller, and impeller drive pin...

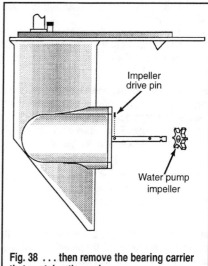

Fig. 38 ...then remove the bearing carrier that contains the seal

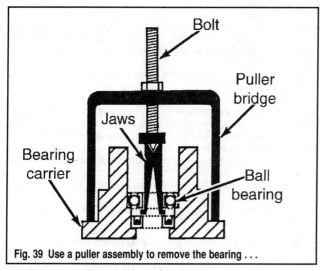

Fig. 39 Use a puller assembly to remove the bearing...

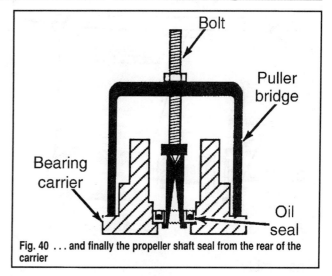

Fig. 40 ...and finally the propeller shaft seal from the rear of the carrier

1. Remove the propeller from the outboard. For details, please refer to the procedure located in the Maintenance and Engine Tune-Up section.

■ To prevent loss of gearcase oil during this procedure either drain the gearcase or position the outboard at full tilt.

2. Remove the 2 propeller shaft bearing housing (bearing carrier) retaining bolts.

3. Carefully insert 2 small prybars under the bearing carrier lip, and gently pry outward to remove the housing and propeller shaft as an assembly.

4. Remove the bearing carrier from the propeller shaft, leaving the shim, reverse gear and thrust washer behind. Take care not to loose any of these components.

5. Remove and discard the O-ring from the bearing carrier.

6. Using a puller bridge plate (such as OMC #432127), backing plate (such as OMC #115312) and an internal jawed puller (such as OMC #432131), carefully remove both seals from the propeller side of the bearing carrier.

To install:

7. Apply a light coating of OMC Gasket Sealing Compound, or equivalent sealant to the outer surface of the 2 NEW propeller shaft seals, then install both seals into the bearing carrier with the seal lips facing outward (toward the driver and the propeller). Use a suitably sized socket or bearing driver and a mallet to gently tap the seals into the bearing carrier.

8. Apply a light coating of OMC Triple-Guard, or equivalent marine grease to the lips of the new seals, to the NEW bearing carrier O-ring and to the push-rod on the tip of the propeller shaft.

9. Install the NEW bearing carrier O-ring to the carrier groove, then insert the propeller shaft (with the thrust washer, reverse gear and shim) through the carrier.

10. Carefully install the carrier and propeller shaft assembly into the gearcase.

11. Install the bearing carrier retaining bolts and tighten to 58-84 inch lbs. (6-10 Nm).

12. If drained, properly refill the gearcase, as detailed in the Maintenance and Tune-Up section.

13. Install the propeller, as detailed in the Maintenance and Tune-Up section.

3/4 Hp (87cc) Motors

◆ See Figures 10 and 41 thru 44

On these models, the propeller shaft is sealed using 2 back-to-back seals, installed the rear of the bearing carrier mounted just in front of the propeller. The seals can only be replaced with the gearcase installed if the bearing carrier is carefully removed from the propeller shaft. Otherwise, the gearcase should be removed so the driveshaft can be withdrawn, allowing the propeller shaft to come out with the bearing carrier.

1. Remove the propeller from the outboard. For details, please refer to the procedure located in the Maintenance and Engine Tune-Up section.

■ To prevent loss of gearcase oil during this procedure either drain the gearcase or position the outboard at full tilt.

LOWER UNIT

2. Remove the 2 propeller shaft bearing housing (bearing carrier) retaining bolts.

3. Use a rubber or plastic mallet to gently tap one of the bearing carrier ears, until it is cocked, exposing the back of the ears. Then, use the mallet to tap alternately on the back of each carrier ear, driving the carrier slowly back and out of the gearcase.

■ If the driveshaft has been removed, the propeller shaft will come out with the carrier. If not, protect the pinion and driveshaft gears from damage by keeping pressure inward on the propeller shaft while tapping the carrier. This will keep the force applied to the bearing carrier from transferring to the gears.

4. If the gearcase is removed from the outboard (meaning you've removed the driveshaft and the propeller shaft came out with the bearing carrier), remove the propeller shaft and components from the bearing carrier. Take note of all propeller shaft component positions and take care not to misplace any parts.

■ If no further service is being conducted on the gearcase and the prop shaft was removed, the propeller shaft and shaft components may be installed back into the case at this point.

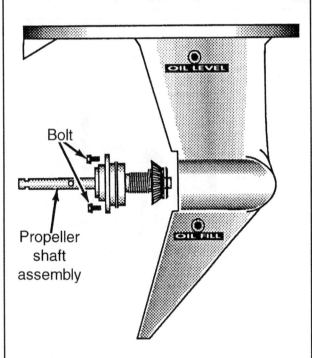

Fig. 41 You'll have to remove the bearing carrier to access the 2 propeller shaft seals

5. Remove and discard the O-ring from the bearing carrier.
6. Using a puller bridge plate (such as OMC #432127), backing plate (such as OMC #115312) and an internal jawed puller (such as OMC #432131), carefully remove both seals from the propeller side of the bearing carrier.

To install:

7. Carefully clean the bearing carrier O-ring groove of all traces of sealant.
8. Apply a light coating of DPL Penetrating Lubricant, or equivalent to the outer surface of the 2 NEW propeller shaft seals. Install both seals into the bearing carrier with the casings facing back-to-back (meaning the inner seal lip faces inward and the outer seal lip faces outward). Use a suitably sized socket or bearing driver and a mallet to gently tap the seals into the bearing carrier.

■ It is usually a good idea to install the seals one at a time, to avoid applying excessive force to the outer seal using the driver. If you use the OMC seal installer (#327572), make sure the grooved side of the installer is used when the seal lips are facing away from the tool and the flat side is used when the seal lips are facing the tool.

9. Apply a very light coating of OMC Triple-Guard, or equivalent marine grease to the lips of the new seals, to the NEW bearing carrier O-ring and to the propeller shaft.
10. If the driveshaft and propeller shaft were both removed, install propeller shaft, shaft components and driveshaft back into the gearcase.
11. Place a piece of clear, cellophane tape (such as Scotch Tape®) over the propeller shaft groove to keep it from damaging the seals when you install the bearing carrier.
12. Install the NEW bearing carrier O-ring to the carrier groove. Wipe the carrier surface clean on either side of the groove.
13. Apply a light coating of OMC Gasket Sealing Compound, or equivalent sealant to the threads of the bearing carrier screws AND to the bearing carrier housing, on either side of the O-ring.
14. Carefully install the carrier over the propeller shaft and into the gearcase, seating the carrier to the case while aligning the bolt ears.
15. Install the bearing carrier retaining bolts and tighten to 60-84 inch lbs. (7-9 Nm).
16. If drained, properly refill the gearcase, as detailed in the Maintenance and Tune-Up section.
17. Install the propeller, as detailed in the Maintenance and Tune-Up section.

4 Deluxe (87cc), 5/6 Hp (128cc) and 5/6/8 Hp (164cc) Motors

◆ See Figures 13 and 45 thru 47

On these models, the propeller shaft is sealed using a single seal, installed into the rear of the bearing carrier mounted just in front of the propeller and seal saver. The seal can only be replaced with the gearcase installed if the bearing carrier is carefully removed from the propeller shaft. Otherwise, the gearcase should be removed so the driveshaft can be withdrawn, allowing the propeller shaft to come out with the bearing carrier.

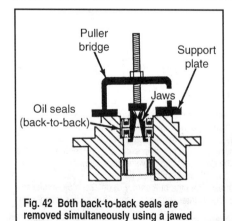

Fig. 42 Both back-to-back seals are removed simultaneously using a jawed puller

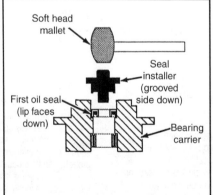

Fig. 43 When installing the dual seals, the inner seal lips face inward . . .

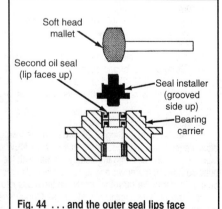

Fig. 44 . . . and the outer seal lips face outward

LOWER UNIT 7-21

Fig. 45 To access the seal, remove the bearing carrier bolts...

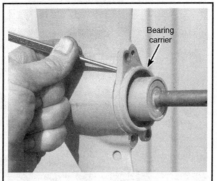

Fig. 46 ...tap on the ears gently to rotate them out of alignment with the case...

Fig. 47 ...then gently pull the bearing retainer off the prop shaft and out of the gearcase

1. Remove the propeller from the outboard. For details, please refer to the procedure located in the Maintenance and Engine Tune-Up section.

■ **To prevent loss of gearcase oil during this procedure either drain the gearcase or position the outboard at full tilt.**

2. Remove the 2 propeller shaft bearing housing (bearing carrier) retaining bolts.
3. Use a small driver to gently tap one of the bearing carrier ears, until it is cocked about 10-15° from alignment with the gearcase, exposing the back of the ears. Then, use a soft mallet to tap alternately on the back of each carrier ear, driving the carrier slowly back and out of the gearcase.

■ **If the driveshaft has been removed, the propeller shaft will come out with the carrier. If not, protect the pinion and driveshaft gears from damage by keeping pressure inward on the propeller shaft while tapping the carrier. This will keep the force applied to the bearing carrier from transferring to the gears.**

4. If the gearcase is removed from the outboard (meaning you've removed the driveshaft and the propeller shaft came out with the bearing carrier), remove the propeller shaft and components from the bearing carrier. Take note of all propeller shaft component positions and take care not to misplace any parts.

■ **If no further service is being conducted on the gearcase, the propeller shaft and shaft components may be installed back into the case at this point.**

5. Remove and discard the O-ring from the bearing carrier.
6. Remove the seal saver from the bearing carrier.
7. Temporarily reinstall the bearing carrier back into the gearcase, then use a small jawed internal puller (such as OMC #432131) and a slide-hammer to remove the seal from the bearing carrier. Remove the carrier from the housing again. If the gearcase (and therefore the propeller shaft) was not removed, you'll have to mount the bearing carrier in a fixture in order to use the slide hammer. We'd recommend you bolt it to a metal plate or a sturdy block of wood and then secure the plate or wood into a vise.

※※ **WARNING**

Make sure the puller jaws contact the seal and NOT the bushing inside the bearing housing.

To install:

8. Carefully clean the bearing carrier O-ring groove of all traces of sealant.
9. Apply a light coating of DPL Penetrating Lubricant, or equivalent to the outer surface of the NEW propeller shaft seal. Install the seal into the bearing carrier with the letters on the casing facing outward (back towards the driver). Use a suitably sized socket or bearing driver and a mallet to gently tap the seal into the bearing carrier.
10. Apply a light coating of OMC Triple-Guard, or equivalent marine grease to the lips of the new seal and to the NEW bearing carrier O-ring.
11. If the gearcase and driveshaft were removed, install the propeller shaft, shaft components and driveshaft back into the gearcase.
12. Install the NEW bearing carrier O-ring to the carrier groove. Wipe the carrier surface clean on either side of the groove.

13. Apply a light coating of OMC Gasket Sealing Compound, or equivalent sealant to the threads of the bearing carrier screws AND to the flange around the O-ring on the bearing carrier housing.
14. Carefully install the carrier over the propeller shaft and into the gearcase, seating the carrier to the case while aligning the bolt ears.
15. Install the bearing carrier retaining bolts and tighten to 60-84 inch lbs. (7-9 Nm).
16. Install the seal saver onto the propeller shaft with the lip facing outward and press it securely into position.
17. If drained, properly refill the gearcase, as detailed in the Maintenance and Tune-Up section.
18. Install the propeller, as detailed in the Maintenance and Tune-Up section.

8/9.9 Hp (211cc), 9.9-15 Hp (216/255cc) and 9.9/15 Hp (305cc) Motors

◆ See Figures 16 and 48 thru 54

■ **There are some indications that the 10/14 hp (216/255cc) 2-stroke commercial models were usually equipped with the split-type gearcase used by 25 Commercial and 28 hp (521cc) motors covered later in this section. The quick and easy way to determine if you should instead follow THAT procedure would be to check the gearcase at the height of the prop shaft. If there is a seam (split line) that runs the horizontal length of the case, refer to the procedure for the 521cc Commercial motors later in this section.**

The propeller shaft for these motors is usually sealed using a single seal pressed into a large bearing carrier (bearing and seal housing assembly) that is mounted in the rear of the gearcase. Well, usually that is, AFTER 1995, as from 1990 to 1995 these gearcases usually used dual back-to-back mounted seals. Either way, the propeller shaft is positioned through the carrier. It is usually possible to replace the seal with the gearcase installed, but **only** if the propshaft is held securely in place during carrier removal and installation. If the propshaft is allowed to shift, the shift detent spring and balls could dislodge, requiring the gearcase to be removed so the driveshaft can be withdrawn (freeing the propeller shaft) in order to reassemble the components.

If you decide to try and replace the seal without disturbing the propeller shaft, a second set of hands is recommended to help steady the propeller shaft during service. Once the bearing carrier is removed, pack the opening in the gearcase using clean, lint-free shop rags to help keep the shaft from shifting.

1. Remove the propeller from the outboard. For details, please refer to the procedure located in the Maintenance and Engine Tune-Up section.

■ **To prevent loss of gearcase oil during this procedure either drain the gearcase or position the outboard at full tilt.**

2. Remove the 2 propeller shaft bearing housing (bearing carrier) retaining bolts.
3. Install a suitable flywheel or steering wheel puller to the bearing carrier (or use OMC #386631), but do not thread the bolts into the gearcase. The puller must be designed so that you can push against the end of the propeller shaft, while the puller legs or jaws pull against the bearing carrier, drawing it from the gearcase.

7-22 LOWER UNIT

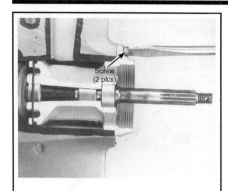
Fig. 48 Cut-away view of the bearing carrier - remove the bolts...

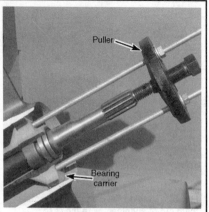
Fig. 49 ...then use a puller to loosen it...

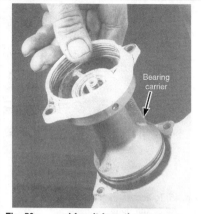
Fig. 50 ...and free it from the gearcase

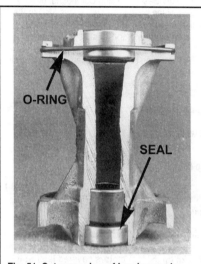
Fig. 51 Cut-away view of bearing carrier showing seal and O-ring

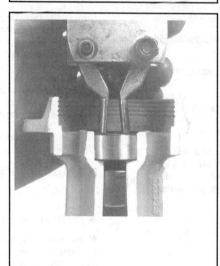
Fig. 52 Preferred seal removal is using an internal jawed puller...

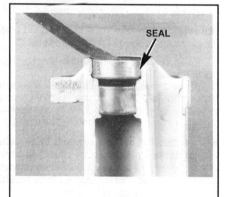
Fig. 53 ...but a pry tool can usually be used with caution

4. Remove and discard the O-ring from the bearing carrier.
5. Remove the seal or seals (as applicable) from the rear (propeller side) of the bearing carrier using an internal jawed puller (such as OMC #432131, along with puller bridge #432127).

To install:
6. Carefully clean the bearing carrier flange and O-ring groove of all traces of sealant.

■ Unless otherwise noted, all bearings, shafts and gears must be lubricated with OMC Ultra-HPF, or equivalent gearcase lubricant, during assembly.

7. For models (usually through 1995) which utilize dual back-to-back seals, apply a light coating OMC Gasket Sealing Compound to the outer surface of the NEW propeller shaft seals. Install the seals, one at a time into the bearing carrier. For the inner seal, face the lips inward toward the gearcase (away from the tool). For the outer seal, face the lips outward (facing the back toward the driver).
8. For models (usually 1996 and later) which utilize a single dual-lipped seal, apply a light coating of DPL Penetrating Lubricant, or equivalent to the outer surface of the NEW propeller shaft seal. Install the seal into the bearing carrier with the exposed lips facing the back toward the driver (the larger outer-diameter set of lips faces inward toward the gearcase). Use a driver that is shouldered so it contacts only the seal casing around the exposed set of lips, as shown.
9. Apply a light coating of OMC Triple-Guard, or equivalent marine grease to the lips of the new seal(s) and to the NEW bearing carrier O-ring.
10. Install the NEW bearing carrier O-ring to the carrier groove. Wipe the carrier surface clean on either side of the groove.
11. Apply a light coating of OMC Gasket Sealing Compound, or equivalent sealant to the threads of the bearing carrier screws AND to the aft support flange of the bearing carrier housing.

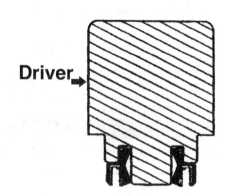
Fig. 54 When installing the seal on extended lip models, make sure the driver contacts the case around the exposed lip side of the seal (the protected seal lips should face downward into the carrier).

12. Carefully install the carrier over the propeller shaft and into the gearcase, seating the carrier to the case while aligning the bolt ears.
13. Install the bearing carrier retaining bolts and tighten to 60-80 inch lbs. (7-9 Nm).
14. If drained, properly refill the gearcase, as detailed in the Maintenance and Tune-Up section.
15. Install the propeller, as detailed in the Maintenance and Tune-Up section.

LOWER UNIT 7-23

10/14 Hp (211cc/255cc), 20-35 Hp (521cc) and 25/35 Hp (500/565cc) Motors
Except Jet, 10/14 Hp, 28 Hp, 30 SPL and Some 25 Hp Comm Models

◆ See Figures 25, 43, 44 and 55 thru 61

The propeller shaft for these motors is sealed using dual back-to-back seals that are mounted in a large bearing carrier (bearing and seal housing assembly) that is mounted in the rear of the gearcase. The propeller shaft is positioned through the carrier. It is usually possible to replace the seal with the gearcase installed, but **only** if the propshaft is held securely in place during carrier removal and installation. If the propshaft is allowed to shift, the shift detent spring and balls could potentially dislodge, requiring the gearcase to be removed so the driveshaft can be withdrawn (freeing the propeller shaft) in order to reassemble the components. Luckily, the reverse gear and bushing is secured by a plate and retaining ring, that also have the affect of helping to steady the propshaft.

1. For safety, disconnect the spark plug leads and ground them to the cylinder head. For electric start models, disconnect the negative battery cable for safety as well.
2. Matchmark the trim tab to the gearcase for installation purposes, then remove the trim tab.
3. Remove the propeller, for details refer to the procedure in the Maintenance and Tune-Up section.

■ **To prevent loss of gearcase oil during this procedure either drain the gearcase or position the outboard at full tilt.**

4. Remove the 2 propeller shaft bearing housing (bearing carrier) retaining bolts.
5. Install a suitable flywheel or steering wheel puller to the bearing carrier (or use OMC #378103). Do not thread the bolts into the retainer-to-gearcase bolt holes, instead use the alternate pair of holes that are present in the face of the retainer at about 5 and 11 o'clock. The puller must be designed so that you can push against the end of the propeller shaft, while the puller legs or jaws pull against the bearing carrier, drawing it from the gearcase.

■ **You should use two 1/4 -20 x 6 in. screws to thread a universal puller to the bearing carrier. The thrust washer may come out with the carrier, if so, keep track of it.**

6. Remove and discard the O-ring from the bearing carrier.
7. Remove the 2 back-to-back seals from the rear (propeller side) of the bearing carrier using an internal jawed puller (such as OMC #432131, along with puller bridge #432127).

✶✶ WARNING

BE CAREFUL to remove ONLY the seals and not the bearings unless they are to be replaced. Once a bearing is removed it must be replaced.

To install:
8. Carefully clean the bearing carrier aft support flange of all traces of sealant.

■ **Unless otherwise noted, all bearings, shafts and gears must be lubricated with OMC Ultra-HPF, or equivalent gearcase lubricant, during assembly.**

9. Apply a light coating of OMC Gasket Sealing Compound, or equivalent sealant to the outer surface of the 2 NEW propeller shaft seals. Install both seals into the bearing carrier with the casings facing back-to-back (meaning the inner seal lip faces inward and the outer seal lip faces outward). Use a suitably sized socket or bearing driver and a mallet to gently tap the seals into the bearing carrier.

■ **It is usually a good idea to install the seals one at a time, to avoid applying excessive force to the outer seal using the driver. If you use the OMC seal installer (#335821), make sure the grooved side of the installer is used when the seal lips are facing away from the tool and the flat side is used when the seal lips are facing the tool.**

10. Apply a light coating of OMC Triple-Guard, or equivalent marine grease to the lips of the new seals and to the NEW bearing carrier O-ring.
11. Install the NEW bearing carrier O-ring to the carrier groove. Wipe the carrier surface clean on either side of the groove.
12. If the thrust washer came out with the bearing carrier, apply a light coating of OMC Needle Bearing Assembly Grease, or equivalent to the back of the bearing carrier housing, then position the thrust washer.
13. Apply a light coating of OMC Gasket Sealing Compound, or equivalent sealant to the aft support flange of the bearing carrier housing, to the threads of the bearing carrier retaining bolts and to the bolt O-rings.
14. To ensure proper alignment during installation, thread the OMC Guide Pin (#383175) into the retainer plate (through one of the carrier retaining screw holes). If the guide pin is not available, obtain a long screw of the same or slightly smaller thread than the bearing carrier retainer bolts. Make a guide pin by grinding the threads off the end of the bolt at the point that will stick out from the back of the bearing carrier to the very end of the bolt itself. Make sure this portion of the bolt is ground and polished completely smooth, with no ridges or defects that could catch on the gearcase threads.
15. Carefully position the carrier to the propeller shaft and into the gearcase with the word **UP** toward the top of the gearcase. Use a brass punch and mallet to gently seat the carrier to the case. As the carrier is tapped in place, make sure the guide pin is inserting through the appropriate bearing carrier-to-gearcase screw hole. If you are using a fabricated pin, make sure the pin does not seat before the carrier, keeping the carrier from properly seating or damaging the gearcase as the pin is forced inward.
16. Install one of the bearing carrier retaining bolts and finger-tighten, then remove the guide pin and install the second carrier bolt. Finally, tighten both carrier retaining bolts to 60-84 inch lbs. (7-9 Nm).
17. If drained, properly refill the gearcase, as detailed in the Maintenance and Tune-Up section.

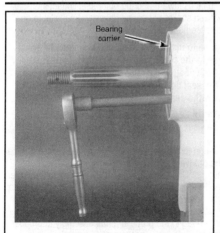

Fig. 55 To replace the propeller shaft seals, unbolt the bearing carrier...

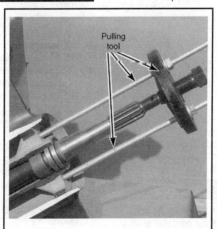

Fig. 56 ... then use a puller to remove it from the gearcase

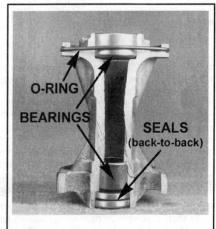

Fig. 57 Cut-away view of the bearing carrier on these models

7-24 LOWER UNIT

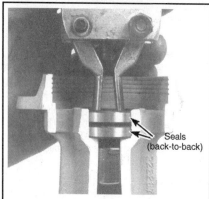

Fig. 58 Remove the seals using a suitable puller . . .

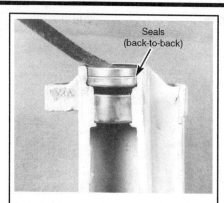

Fig. 59 . . . or, if necessary, carefully using a prybar

Fig. 60 Install each of the seals one at a time, using a driver

Fig. 61 During installation make sure the marking UP is facing toward the top of the gearcase

1. For safety, disconnect the spark plug leads and ground them to the cylinder head. For electric start models, disconnect the negative battery cable for safety as well.
2. Remove the propeller, for details refer to the procedure in the Maintenance and Tune-Up section.
3. Drain the gearcase oil, as detailed in the Maintenance and Tune-Up section.
4. Remove the shift lever pivot pin threaded into the lower starboard side of the lower gearcase housing.
5. Remove the 6 screws securing the lower gearcase housing to the upper gearcase housing, then using a rubber or plastic mallet, tap gently on the side of the skeg to loosen the seal. Carefully remove the lower gearcase housing, supporting the propeller shaft assembly, as necessary as it is removed. Remove and discard the lower gearcase housing seal from the mating surface.
6. Remove the cradle from underneath the propeller shaft, as necessary to lower the shaft, either completely from the housing, or just sufficiently to withdraw the gearcase head/seal carrier from the propeller shaft.
7. Secure the gearcase head in a suitable bearing press or on a retainer plate, then use a puller bridge (such as #432127) and an internal small jawed puller (such as #432131) to remove the 2 seals from the rear of the head/carrier.

To install:
8. Carefully clean the bearing carrier (outer and inner mating surfaces) of all traces of sealant.

■ **Unless otherwise noted, all bearings, shafts and gears must be lubricated with OMC Ultra-HPF, or equivalent gearcase lubricant, during assembly.**

9. Apply a light coating of OMC Gasket Sealing Compound, or equivalent sealant to the outer surface of the 2 NEW propeller shaft seals. Install both seals into the bearing carrier with the casings facing back-to-back (meaning the inner seal lip faces inward and the outer seal lip faces outward). Use a suitably sized socket or bearing driver and a mallet to gently tap the seals into the gearcase head/seal carrier.

■ **It is usually a good idea to install the seals one at a time, to avoid applying excessive force to the outer seal using the driver. If you use the OMC seal installer (#330655), make sure the grooved side of the installer is used when the seal lips are facing away from the tool and the flat side is used when the seal lips are facing the tool.**

10. Apply a light coating of OMC Triple-Guard, or equivalent marine grease to the lips of the new seals and to the NEW gearcase head/seal carrier O-ring.
11. Install the NEW O-ring to the carrier groove. Wipe the carrier surface clean on either side of the groove.
12. Apply a light coating of OMC Adhesive M, or equivalent, to the upper gearcase assembly, along the gearcase head mating surface, right in front of (inboard) of the O-ring contact area (along a line across the upper gearcase that includes the upper gearcase-to-gearcase head retaining pin).
13. Install the gearcase head to the propeller shaft assembly and upper gearcase. Be sure to align the hole in the gearcase head with the pin in the upper gearcase. If the entire propeller shaft assembly was removed, make sure the cradle is properly positioned on the end of the shifter lever and over the shaft.

18. Grease the shaft, then properly install and secure the propeller. For more details, please refer to Propeller under the Maintenance and Tune-Up section.
19. Align the matchmarks made earlier, then securely reinstall the trim tab.
20. Reconnect the spark plug leads and, if applicable, connect the negative battery cable.

28 Hp, 30 SPL and Some 10/14 Hp, 25 Hp Comm Models

◆ See Figures 26, 43 and 44

The propeller shaft for these motors is sealed using dual back-to-back seals that are mounted in a small gearcase head or seal carrier mounted in the rear of the gearcase. Because of the split design of the gearcase on these models, access to the gearcase head is only possible once the skeg and lower gearcase housing has been unbolted from the upper gearcase housing. It is easiest to service this assembly with the gearcase completely removed from the outboard, drained and inverted, but it should be possible to perform a propeller shaft seal replacement with the motor at least at a full tilt, but if possible, laying completely on its side or partially downward. If the gearcase is remaining installed, you'll still want to drain it first to prevent a big mess when the lower gearcase housing is removed.

✱✱ WARNING

When servicing the gearcase while still installed on the motor, at any angle short of powerhead downward, take care that the propeller shaft does not dislodge suddenly and fall once the lower gearcase housing is removed. Should this occur, both damage to the components and injury (from falling components) could occur.

LOWER UNIT 7-25

■ There should be about 1/32 in. (0.8mm) gap between the forward bearing and the thrust washer before the next step.

14. Using the OMC Gearcase Alignment Tool (#390880) to press against the end of the propeller shaft, firmly seat the forward gear and thrust bearing against the gearcase. If using any other tool to perform this, make SURE the shaft assembly is properly centered in the gearcase. Whatever tool is used must continue to apply pressure to lock the shaft in position as the lower gearcase half is installed.

15. Obtain a length of gearcase seal and cut it to 12 7/8 in. (32cm) in length.

16. Apply a light coating of OMC Adhesive M, or equivalent, to the machines mating surfaces of the upper and lower gearcase halves, as well as to the exposed area of the gearcase head O-ring.

17. Position the gearcase seal in the groove on the lower gearcase, then trim the ends so the seal fits EXACTLY in the groove.

18. Apply a light coating of RTV sealant to about 1/2 in. (12mm) of each end of the seal.

19. Apply a light coating of OMC Gasket Sealing Compound, or equivalent sealant to the threads of the lower gearcase retaining screws.

■ In the next step, push rearward on the lower gearcase using steady pressure and the lower-to-upper gearcase screws and tightened, drawing the lower gearcase into position.

20. Install the lower gearcase on the upper gearcase, making sure the seals and sealant all remain in position, then install the front and rear screws. Alternately tighten those 4 screws until the lower gearcase has been drawn down into position on the upper gearcase. Next, install the remaining 2 screws and alternately tighten them as well. Tighten all screws alternating from side-to-side and while working from the front of the gearcase toward the rear of the gearcase to 60-80 inch lbs. (7-9 Nm). Remove the alignment tool.

21. Apply a light coating of OMC Triple-Guard, or equivalent marine grease to a NEW pivot pin O-ring.

22. Apply a light coating of OMC Nut Lock, or equivalent threadlock to the threads of the pivot pin, then install the pin (with the O-ring) and tighten to 48-84 inch lbs. (5-9 Nm). If it is difficult to threads the pin, move the shift rod in or out slightly to center the hole in the shifter lever with the pin.

23. Properly refill the gearcase, as detailed in the Maintenance and Tune-Up section.

24. Grease the shaft, then properly install and secure the propeller. For more details, please refer to Propeller under the Maintenance and Tune-Up section.

25. Reconnect the spark plug leads and, if applicable, connect the negative battery cable.

25-55 Hp (737cc) and 25-70 Hp (913cc) Motors

Some 2-Cylinder Motors Including 40 (Except 40RW, 40RP and 40WR), 48 and 50 Hp Models

◆ See Figures 27, 43 and 44

Although most 2- and 3-cylinder motors use the same gearcase assembly, a few of the lighter duty 2-cylinder motors, including the 40 (except 40RW, 40RP and 40 WR), 48 and 50 hp models use a unique assembly. On these models, a bearing housing anode is mounted between the propeller and the bearing carrier (on all other 2- and 3-cylinder motors, the bearing housing anode is bolted to the bottom of the bearing carrier).

The propeller shaft seal can be found in the bearing carrier, directly behind the anode. Although some earlier production models utilize a dual-seal assembly, most of these motors utilize a single propeller shaft seal. In both cases, seal removal can be easily accomplished with the gearcase installed, as long as a universal puller is used to carefully free the bearing carrier from the gearcase housing.

1. Remove the propeller from the outboard. For details, please refer to the procedure located in the Maintenance and Engine Tune-Up section.

■ To prevent loss of gearcase oil during this procedure either drain the gearcase or position the outboard at full tilt.

2. Remove the 2 bearing housing anode bolts and then remove the anode.

3. Remove the 2 bearing carrier bolts and retainers.

■ You should use two 1/4 -20 x 8 in. screws to thread a universal puller to the bearing carrier.

4. Install a suitable flywheel or steering wheel puller to the bearing carrier (or use OMC #378103). The puller must be designed so that you can push against the end of the propeller shaft, while the puller legs or jaws pull against the bearing carrier, drawing it from the gearcase.

5. Remove the bearing carrier, then remove and discard the O-ring from the bearing carrier groove.

■ If the thrust washer comes out with the bearing carrier, keep track of it for installation purposes.

6. Remove the seal or seals (as applicable) from the rear (propeller side) of the bearing carrier using an internal jawed puller (such as OMC #432130, along with puller bridge #432127 or a slide hammer such as #432128).

✱✱ WARNING

DO NOT remove the bearings unless you plan on replacing them. The action of removing the bearings will normally stress or damage them enough to make them unserviceable (and why take the risk).

To install:

■ Unless otherwise noted, all bearings, shafts and gears must be lubricated with OMC Ultra-HPF, or equivalent gearcase lubricant, during assembly.

7. Using solvent, carefully clean the bearing carrier of all traces of sealant. After the solvent has dried, lubricate the bearings using gearcase oil.

8. On early production models equipped with dual, back-to-back seals, install the 2 seals as follows. Apply a light coating of OMC Gasket Sealing Compound, or equivalent sealant to the outer surface of the 2 NEW propeller shaft seals. Install both seals into the bearing carrier with the casings facing back-to-back (meaning the inner seal lip faces inward and the outer seal lip faces outward). Use a suitably sized socket or bearing driver and a mallet to gently tap the seals into the gearcase head/seal carrier.

■ It is usually a good idea to install the seals one at a time, to avoid applying excessive force to the outer seal using the driver. If you use the OMC seal installer (#326556), make sure the grooved side of the installer is used when the seal lips are facing away from the tool and the flat side is used when the seal lips are facing the tool.

9. Most models are equipped with a single, extended lip seal, to install a single seal proceed as follows. Apply a light coating of DPL Penetrating Lubricant, or equivalent to the outer surface of the NEW propeller shaft seal. Install the seal into the bearing carrier with the exposed lips facing the back toward the driver (the larger outer-diameter set of lips faces inward toward the gearcase). Use a driver that is shouldered so it contacts only the seal casing around the exposed set of lips, as shown.

10. Apply a light coating of OMC Triple-Guard, or equivalent marine grease to the lips of the new seal and to the NEW bearing carrier O-ring.

11. Install the NEW bearing carrier O-ring to the carrier groove. Wipe the carrier surface clean on either side of the groove.

12. Apply a light coating of OMC Gasket Sealing Compound, or equivalent sealant to the bearing carrier O-ring flange and aft support flange.

13. If the thrust washer came out with the bearing carrier during disassembly, apply a light coating of OMC Needle Bearing Assembly Grease, or equivalent to the thrust washer, then place it in the back of the bearing carrier.

14. Carefully install the carrier over the propeller shaft and into the gearcase with the carrier bolt holes positioned vertically and the drain slot facing downward.

15. Apply a light coating of OMC Nut Lock, or equivalent threadlock to the threads of the bearing carrier retaining bolts, then install and tighten to 120-144 inch lbs. (14-16 Nm).

16. Install the bearing housing anode and tighten the bolts securely.

17. If drained, properly refill the gearcase, as detailed in the Maintenance and Tune-Up section.

18. Install the propeller, as detailed in the Maintenance and Tune-Up section.

7-26 LOWER UNIT

3-Cylinder Motors and Most 2-Cylinder Motors Except 40, 40RS, 48 and 50 Hp Models

OEM — **2** — **MODERATE**

◆ See Figures 28 and 55 thru 61

Most of the larger Johnson/Evinrude 2- and 3-cylinder motors (except the 2-cylinder 40, 40RS, 48 and 50 hp models) use the same heavy-duty gearcase assembly. These assemblies can be identified by the fact that the bearing housing anode is bolted to the bottom of the bearing carrier (instead of between the propeller and bearing carrier as it is on the lighter duty models).

These models utilize a dual, back-to-back propeller shaft seal assembly mounted to the bearing carrier. Seal replacement can be easily accomplished with the gearcase installed, as long as a universal puller is used to carefully free the bearing carrier from the gearcase housing.

1. Remove the propeller from the outboard. For details, please refer to the procedure located in the Maintenance and Engine Tune-Up section.

■ **To prevent loss of gearcase oil during this procedure either drain the gearcase or position the outboard at full tilt.**

2. Using a thin-walled 5/16 in. socket, remove the 4 bearing carrier bolts and O-rings. Discard the O-rings and clean all sealing compound from the bolt threads.

■ **You should use two 5/16 -18 x 8 in. screws to thread a universal puller to the bearing carrier.**

3. Install a suitable flywheel or steering wheel puller to the bearing carrier (or use OMC #378103). The puller must be designed so that you can push against the end of the propeller shaft, while the puller legs or jaws pull against the bearing carrier, drawing it from the gearcase.

4. Remove the bearing carrier, then remove and discard the O-ring from the bearing carrier groove.

5. Remove the seals from the rear (propeller side) of the bearing carrier using an internal large-jawed puller (such as OMC #432129, along with puller bridge #432127).

** WARNING

DO NOT remove the bearings unless you plan on replacing them. The action of removing the bearings will normally stress or damage them enough to make them unserviceable (and why take the risk).

6. Inspect the bearing housing anode and replace if it is worn to 2/3 or less of the original size.

To install:

■ **Unless otherwise noted, all bearings, shafts and gears must be lubricated with OMC Ultra-HPF, or equivalent gearcase lubricant, during assembly.**

7. Using solvent, carefully clean the bearing carrier of all traces of sealant. After the solvent has dried inspect the bearings to make sure they are still serviceable. Rotate the needles by hand, feeling for freedom of movement. Also, check the bearing housing O-ring groove for sharp edges that could cut the O-ring and prevent sealing. After inspection is complete, lubricate the bearings using gearcase oil.

8. Apply a light coating of OMC Gasket Sealing Compound, or equivalent sealant to the outer surface of the 2 NEW propeller shaft seals. Install both seals into the bearing carrier with the casings facing back-to-back (meaning the inner seal lip faces inward and the outer seal lip faces outward). Use a suitably sized socket or bearing driver and a mallet to gently tap the seals into the gearcase head/seal carrier.

■ **It is usually a good idea to install the seals one at a time, to avoid applying excessive force to the outer seal using the driver. If you use the OMC seal installer (#326551), make sure the grooved side of the installer is used when the seal lips are facing away from the tool and the flat side is used when the seal lips are facing the tool.**

9. Apply a light coating of OMC Triple-Guard, or equivalent marine grease to the lips of the new seal and to the NEW bearing carrier O-ring.

10. Install the NEW bearing carrier O-ring to the carrier groove. Wipe the carrier surface clean on either side of the groove.

■ **The OMC guide pins #383175 are used to align the bearing carrier with the gearcase during installation. They are essentially long, smooth shafts with threads on the very ends. A pair of alignment pins can be easily fabricated using a pair of long bolts (with the heads cut off) or studs with thread matching the bearing carrier retaining bolts. Simply polish 9/10 of the threads down to a smooth surface, leaving threads only on the last inch or so of the shaft to thread it into the gearcase.**

11. Install 2 OMC guide pins (#383175) into the gearcase housing. Do NOT thread them further than 2 turns into the housing retainer plate.

12. Apply a light coating of OMC Gasket Sealing Compound, or equivalent sealant to the bearing carrier O-ring flange and aft support flange. Make sure that no sealant contacts the forward thrust surface or the bearings found inside the housing.

13. Carefully align the carrier over the guide pins and the propeller shaft with the word **UP** facing toward the top of the gearcase, then gently tap the carrier into the gearcase using a soft-faced mallet.

14. Install 2 of the 4 bearing carrier retaining bolts and finger-tighten, then remove the guide pins, install and finger-tighten the remaining 2 bolts. Finally, tighten the 4 bolts to 120-140 inch lbs. (14-16 Nm).

15. If drained, properly refill the gearcase, as detailed in the Maintenance and Tune-Up section.

16. Install the propeller, as detailed in the Maintenance and Tune-Up section.

40/50 Hp (815cc) EFI Motors

◆ See Figure 62

The 40/50 hp EFI Johnson/Evinrude motors utilize a gearcase that is not unlike the larger 2-stroke Johnson/Evinrude motors. Removal and installation of the gearcase assembly is a relatively straightforward procedure. However the gearcase itself contains a precision shimmed assembly of gears, bearings and shaft components. Disassembly/assembly is not recommended without significant experience and the use of precision measuring tools to ensure proper gear/bearing shimming.

These models utilize a bearing carrier that houses two seals. The seals are both installed with their lips facing outward (back toward the propeller). Although gearcase disassembly calls for removal of the propshaft and bearing carrier as an assembly, it appears that this is just for complete overhaul purposes. It should be possible to draw only the bearing carrier from the end of the propshaft, if a puller is used to hold the shaft in place while drawing the carrier outward. Should you encounter extreme resistance, remove the carrier and propshaft as an assembly using a slide-hammer, then separate the components and reinstall the propshaft.

1. Remove the propeller from the outboard. For details, please refer to the procedure located in the Maintenance and Engine Tune-Up section.

■ **To prevent loss of gearcase oil during this procedure either drain the gearcase or position the outboard at full tilt.**

2. Remove the 2 bearing carrier retaining bolts.

3. Install a suitable flywheel or steering wheel puller to the bearing carrier. The puller must be designed so that you can push against the end of the propeller shaft, while the puller legs or jaws pull against the bearing carrier, drawing it from the gearcase.

■ **If removal of the entire shaft and carrier assembly is desired, use the OMC Propeller Shaft Retainer Removal Tool (#345831) and OMC Slide Hammer (#391008) to remove the assembly from the gearcase.**

4. Remove the bearing carrier, then remove and discard the O-ring from the bearing carrier groove.

5. Remove the seals from the rear (propeller side) of the bearing carrier using an internal large-jawed puller (such as OMC #432129, along with puller bridge #432127).

** WARNING

DO NOT remove the bearings unless you plan on replacing them. The action of removing the bearings will normally stress or damage them enough to make them unserviceable (and why take the risk).

LOWER UNIT 7-27

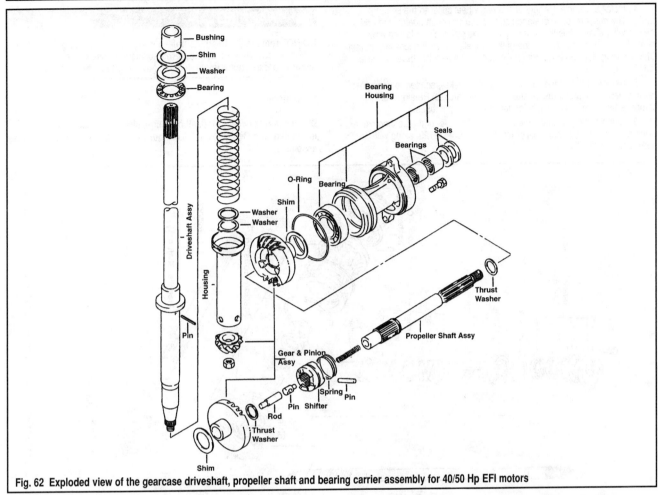

Fig. 62 Exploded view of the gearcase driveshaft, propeller shaft and bearing carrier assembly for 40/50 Hp EFI motors

To install:

■ Unless otherwise noted, all bearings, shafts and gears must be lubricated with OMC Ultra-HPF, or equivalent gearcase lubricant, during assembly.

6. Using solvent, carefully clean the bearing carrier of all traces of sealant.

7. Install 2 NEW propeller shaft seals to the bearing carrier, **dry** with no sealant or lubricant on the seal casings. Use a suitably sized socket or driver (such as OMC Seal Installer #345830) to carefully install each of the seals with their lips facing upward (outward toward the propeller).

■ It is usually a good idea to install the seals one at a time, to avoid applying excessive force to the outer seal using the driver.

8. Apply a light coating of OMC Triple-Guard, or equivalent marine grease to the lips of the new seal and to the NEW bearing carrier O-ring.

9. Install the NEW bearing carrier O-ring to the carrier groove. Wipe the carrier surface clean on either side of the groove.

10. Apply a light coating of OMC Gasket Sealing Compound, or equivalent sealant to the bearing carrier O-ring flange, the carrier support flange and the threads of the bearing carrier retaining bolts.

11. Carefully align over the propeller shaft and gently tap it into position using a soft-faced mallet. In order to prevent possibly cocking the carrier in the gearcase and damaging the case, carrier, shaft or bearings, be sure to gently tap around the entire perimeter of the carrier during installation. Also, make sure the retaining bolt ears remain aligned with the gearcase bolt holes.

12. Install the carrier retaining bolts and tighten to 12 ft. lbs. (17 Nm).

13. If drained, properly refill the gearcase, as detailed in the Maintenance and Tune-Up section.

14. Install the propeller, as detailed in the Maintenance and Tune-Up section.

70 Hp (1298cc) EFI Motors

◆ See Figures 63, 34 and 65

The 70 hp EFI Johnson/Evinrude motor utilizes a gearcase similar to the heavy-duty units found on the larger Johnson/Evinrude 2-stroke motors. Removal and installation of the gearcase assembly is a relatively straightforward procedure. However the gearcase itself contains a precision shimmed assembly of gears, bearings and shaft components. Disassembly/assembly is not recommended without significant experience and the use of precision measuring tools to ensure proper gear/bearing shimming.

These models utilize a bearing carrier that houses two seals. The seals are both installed with their lips facing outward (back toward the propeller). The one unique feature of this gearcase, comes in the tabbed bearing carrier retainer which must be removed using some form of a heavy duty spanner wrench (capable of generating torque of 115 ft. lbs. (160 Nm). To be honest, that would be one heck of a spanner, and most adjustable spanners will not do the trick, so you'll need to luck out with one that fits exactly or buy manufacturer's tool. Once the retainer is out of the way, the carrier assembly is removed with a puller, in the same basic fashion as on most Johnson/Evinrude outboards.

1. Remove the propeller from the outboard. For details, please refer to the procedure located in the Maintenance and Engine Tune-Up section.

■ To prevent loss of gearcase oil during this procedure either drain the gearcase or position the outboard at full tilt.

2. Bend the tabs of the lockwasher back away from the carrier retainer.

3. Remove the carrier retainer by rotating it counterclockwise. You'll need a spanner wrench capable of grabbing the teeth of the retainer or, more likely, we'd recommend you use the OMC Propeller Shaft Retainer Remover/Installer Tool (#342688).

LOWER UNIT

4. Install a suitable flywheel or steering wheel puller to the bearing carrier. The puller must be designed so that you can push against the end of the propeller shaft, while the puller legs or jaws pull hook under the tabs located in the carrier on either side of the propeller shaft. If necessary, obtain OMC Bearing Housing Puller Kit (#5000004) and the OMC Universal Puller (#378103).

5. Remove the bearing carrier, along with the key, spacer, reverse gear, reverse gear shim and thrust washer from the gearcase. Retain all of components in the same orientation for installation purposes.

6. Remove the seals from the rear (propeller side) of the bearing carrier using an internal large-jawed puller (such as OMC #432129, along with puller bridge #432127).

✳✳ WARNING

DO NOT remove the bearings unless you plan on replacing them. The action of removing the bearings will normally stress or damage them enough to make them unserviceable (and why take the risk).

To install:

■ Unless otherwise noted, all bearings, shafts and gears must be lubricated with OMC Ultra-HPF, or equivalent gearcase lubricant, during assembly.

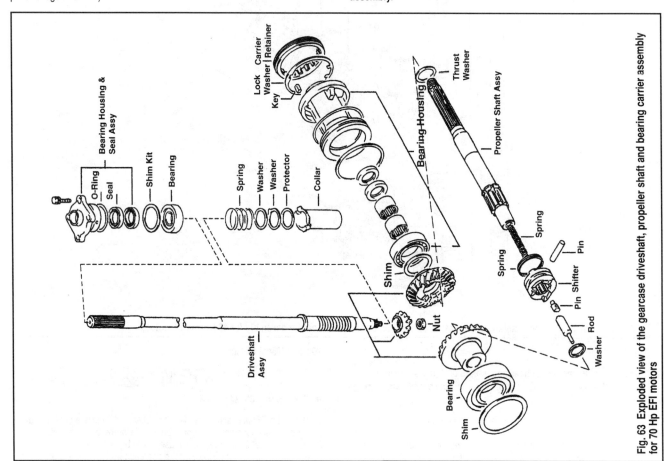

Fig. 63 Exploded view of the gearcase driveshaft, propeller shaft and bearing carrier assembly for 70 Hp EFI motors

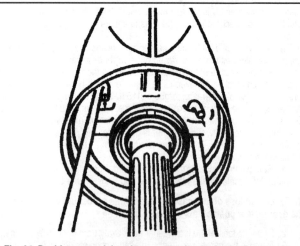

Fig. 64 Besides a special tool to remove the carrier retainer, you'll need a puller with hooked jaws or legs to attach to the bearing carrier as illustrated

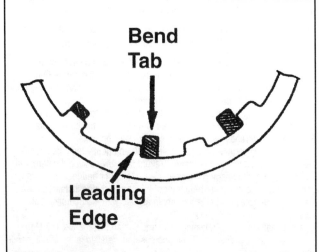

Fig. 65 Bend a tab of the lock washer that is just in contact with the leading edge of a retainer tooth

LOWER UNIT 7-29

7. Using solvent, carefully clean the bearing carrier of all traces of sealant.

8. Install 2 NEW propeller shaft seals to the bearing carrier, **dry** with no sealant or lubricant on the seal casings. Use a suitably sized socket or driver (such as OMC Seal Installer #5000001) to carefully install each of the seals with their lips facing upward (outward toward the propeller).

■ It is usually a good idea to install the seals one at a time, to avoid applying excessive force to the outer seal using the driver.

9. Apply a light coating of OMC Triple-Guard, or equivalent marine grease to the lips of the new seal and to the NEW bearing carrier O-ring.

10. Install the NEW bearing carrier O-ring to the carrier groove. Wipe the carrier surface clean on either side of the groove.

11. Install the reverse gear thrust washer, reverse gear, the gear shim and the spacer to the propeller shaft.

12. Apply a light coating of OMC Gasket Sealing Compound, or equivalent sealant to the bearing carrier O-ring flange, the carrier support flange and the threads of the bearing carrier retainer and gearcase.

13. Carefully align over the propeller shaft and gently tap it into position using a soft-faced mallet. In order to prevent possibly cocking the carrier in the gearcase and damaging the case, carrier, shaft or bearings, be sure to gently tap around the entire perimeter of the carrier during installation.

14. Install the tabbed lock washer by aligning the boss on the propeller shaft with the notch in the washer.

15. Thread the bearing retainer into the gearcase with the **OFF** mark facing outward. Tighten the retainer to 115 ft. lbs. (160 Nm) using the Retainer Remover/Installer Tool.

16. Bend one tab of the lock washer where there is no gap between it and the leading edge of the retainer (to ensure the retainer cannot loosen, even slightly, in service).

17. If drained, properly refill the gearcase, as detailed in the Maintenance and Tune-Up section.

18. Install the propeller, as detailed in the Maintenance and Tune-Up section.

JET DRIVE

Description & Operation

◆ See Figures 66, 67 and 68

The Outboard Jet Drive provides reliable propulsion with a minimum of moving parts. The units operate under the same laws and principles employed for the other jet drive units. Simply stated, water is drawn into the unit through an intake grille by an impeller driven by a driveshaft off the crankshaft of the powerhead. The water is immediately expelled under pressure through an outlet nozzle directed away from the stern of the boat.

As the speed of the boat increases and reaches planing speed, the jet drive discharges water freely into the air, and only the intake grille makes contact with the water.

No gears are used in the jet drive assembly. Because of this, anytime the motor is operating the crankshaft and anything attached to it rotates. This means the driveshaft and the impeller are always turning when the motor is operating. **forward**, **neutral** or **reverse** is determined by the positioned of a linkage controlled gate. In the normal position the gate leaves the nozzle completely uncovered, allowing thrust to move the boat forward. When the shifter is placed in the neutral position, the gate moves partially over the nozzle so it deflects the thrust downward (which usually results in very slow creeping of the craft forward or rearward, depending on trim position). With the shifter in the reverse position the gate moves to completely cover the nozzle, deflecting thrust in the exact opposite direction of normal operation.

Conventional controls are used for powerhead speed, movement of the boat, shifting and power trim and tilt.

Jet Drive Servicing includes impeller removal, shimming and installation, jet drive assembly removal and installation, and reverse gate adjustment.

Impeller

REMOVAL, SHIMMING & INSTALLATION

◆ See Figures 69 thru 75

The jet drive impeller is designed to draw water through an opening and force it outward through a smaller nozzle, thus pressurizing the water flow and producing thrust to move the boat. In order for the jet drive to operate properly (create maximum thrust), there must be the proper amount of clearance between the outer edge of the jet drive impeller and the water intake housing cone wall. The jet drive impeller should be periodically checked for nicks, burrs, damage or wear that could adversely affect clearance. For more details, please refer to Jet Drive Impeller in the Maintenance and Tune-Up section.

Clearance must be adjusted anytime a new impeller is installed or anytime inspection reveals that wear has caused the gap to increase beyond spec. Adjustment is a relatively simple matter of removing the impeller and removing shims mounted below the impeller to a spot above the impeller. Moving the shims will place the impeller slightly lower in the tapered housing, decreasing clearance. If, upon installation of a new impeller you find insufficient clearance, move the impeller slightly upward, into the tapered housing, by removing shims from above the impeller and placing them below it.

■ Always check impeller clearance before removal. If clearance is within specification, keep track of all shims as they are removed and return them to their original positions (above and/or below the impeller).

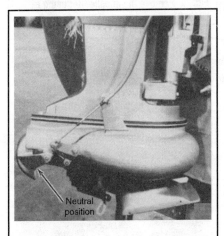

Fig. 66 Forward, Neutral or Reverse is determined by gate position on a jet drive

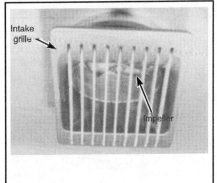

Fig. 67 The impeller is protected by an intake grate

Fig. 68 The impeller is the heart of the jet drive

7-30 LOWER UNIT

1. For safety, disconnect the spark plug leads and ground them to the cylinder head. For electric start models, disconnect the negative battery cable for safety as well.
2. Remove the bolts (usually 6) securing the intake housing (grille) to the jet housing assembly.
3. Using a screwdriver or small prytool, carefully bend the locktab washer ears back off the impeller nut, then loosen and remove the nut and washer. Discard the locktab washer and replace with a new one.

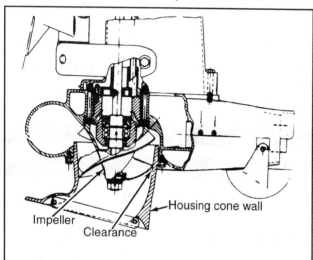

Fig. 69 Jet drive impeller clearance is adjusted by shimming the impeller higher or lower in the housing. Usually, moving the impeller downward (moving shims from below to above the impeller) decreases clearance

■ The locktab washer prevents the possibility of the impeller nut loosening during service (which could lead to a catastrophic failure of the jet drive). Don't risk damage to the unit or your safety on the water by reusing a locktab washer, as the bendable tabs may become brittle and break off after multiple uses. Replace the locktab washer anytime it is removed.

4. Remove the impeller, along with any shims that are installed below it. Keep track of all shims for installation in their original positions (even if the impeller is being replaced, use these shims as a starting point for adjustment).
5. Remove the impeller key and sleeve, then remove any shims that were installed above the impeller. Again, keep track of these shims for installation purposes.

To install:
6. Install the upper shims, key and sleeve (bushing) followed by the impeller, lower shims, a NEW tabbed washer and nut. Finger-tighten the nut and check impeller clearance. It must be 0.020-0.030 in. (0.5-0.8mm) or the nut must be loosened and shims must be moved (either taken from above the impeller and moved below, or vice versa).
7. Once proper clearance has been verified or obtained, tighten the impeller nut to 16-18 ft. lbs. (22-24 Nm).
8. Bend the tabs of the new tabbed lock-washer down against the impeller nut to lock it in position.
9. Apply a light coating of OMC Gasket Sealing Compound, or equivalent sealant to the threads of the intake housing (grille) retaining bolts. Install the intake housing and finger-tighten the bolts, then check to make sure there is no binding or rubbing between the casing liner and impeller.

■ If there is greater clearance on one side of the intake housing than the other, loosen the retaining bolts and make sure the housing is centered before tightening the bolts.

10. Once you've verified that the housing is centered, tighten the intake housing (grille) retaining bolts to 60-84 inch lbs. (7-9 Nm).
11. Recheck housing and impeller clearance to make sure no components have shifted during installation.

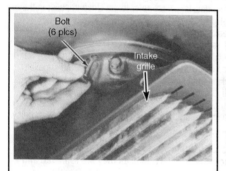

Fig. 70 To remove the impeller, first remove the intake grille...

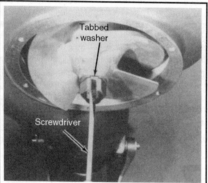

Fig. 71 ...then bend the tab washer ears off the nut

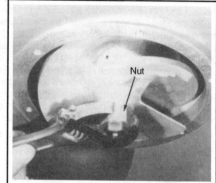

Fig. 72 Remove the impeller nut using a wrench...

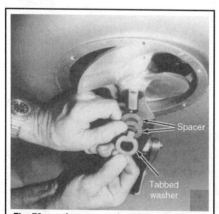

Fig. 73 ...the remove the tab washer and any spacers

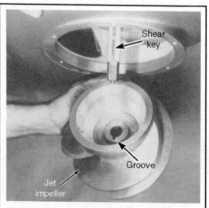

Fig. 74 Lower the impeller from the driveshaft...

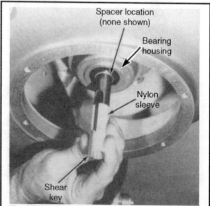

Fig. 75 ...then remove the key, sleeve and any other shims

LOWER UNIT 7-31

12. Reconnect the spark plug leads and, if applicable, connect the negative battery cable.

Jet Drive Assembly

REMOVAL & INSTALLATION

◆ See Figures 76, 77 and 78

Most of the jet drive assembly retaining bolts are found in the impeller cavity (above the impeller). For this reason, the impeller must be removed in order to remove the drive assembly.

1. For safety, disconnect the spark plug leads and ground them to the cylinder head. For electric start models, disconnect the negative battery cable for safety as well.
2. Remove the Impeller as detailed in this section.
3. Disconnect the shift linkage from the reverse gate. On some models this involves removing the retaining clip, washer and spring from the lower shift rod, then pulling the rod out of the gate cam. On other models, you'll have to remove the nut and flat washer in order to free the cable end from the reverse gate cam. On these models the cable is then removed by rotating the cable to the vertical position and removing the adjusting nut and cable from the anchor.
4. Remove the adapter-to-jet housing bolt (external bolt threaded downward into the top of the jet housing from the adapter flange).

■ In the next step, keep close track of the bolts, as they are different lengths on some models and should be returned to their original bores.

5. Remove the 4 internal bolts and lockwashers that are threaded upward into the adapter from underneath the jet drive assembly (inside the impeller cavity).

■ Do NOT mistake the 4 bearing housing bolts for the jet drive assembly retaining bolts. The jet drive bolts are found outboard from the bearing housing.

6. Once the jet drive assembly bolts are removed, carefully lower the assembly (including the jet drive housing, bearing housing, driveshaft and water pump) from the adapter.
7. If the adapter or exhaust housing requires service, remove the bolts (usually 5) threaded upward from below the adapter, then lower the adapter from the housing.

To install:

8. If the adapter was removed, apply a light coating of OMC Gasket Sealing Compound or equivalent sealant to the threads of the adapter bolts, then install and tighten the bolts to 192-216 inch lbs. (22-25 Nm).
9. If equipped (and if removed), install the water tube extension on the tube.
10. Apply a light coating of OMC Moly Lube, or equivalent assembly lubricant to the driveshaft splines. But, be careful not to coat the top surface of the driveshaft, as lubricant there could hydraulically prevent the driveshaft from seating in the crankshaft.
11. Apply a light coating of OMC Gasket Sealing Compound, or equivalent sealant to the threads of the 5 jet drive housing fasteners (if they're of different lengths don't mix them up now). Place the fasteners aside on a clean surface for use in the next steps.
12. Install the jet drive housing assembly, while carefully aligning the water tube into the water pump outlet and the driveshaft to the crankshaft splines. If necessary, slowly rotate the flywheel clockwise (when viewed from above) to align the crankshaft splines with the driveshaft.
13. Install the 4 jet drive housing bolts from underneath the housing (in the impeller cavity) and finger-tighten. If the bolts are of different lengths, be sure to return them to their original bores.

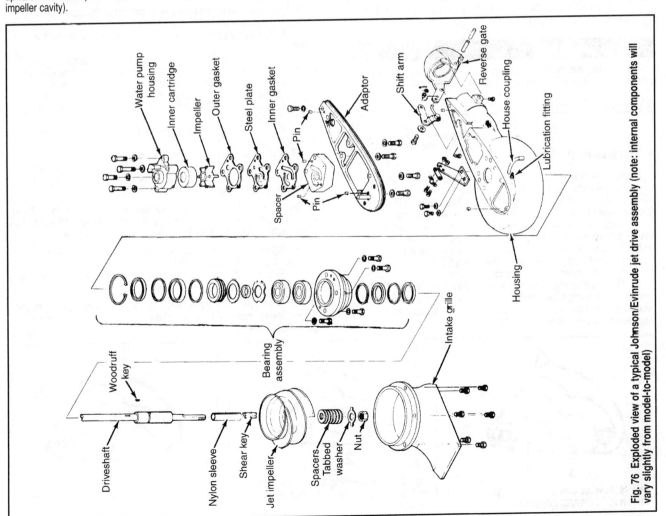

Fig. 76 Exploded view of a typical Johnson/Evinrude jet drive assembly (note: internal components will vary slightly from model-to-model)

7-32 LOWER UNIT

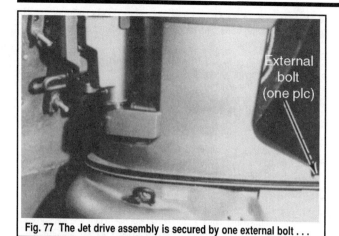

Fig. 77 The Jet drive assembly is secured by one external bolt . . .

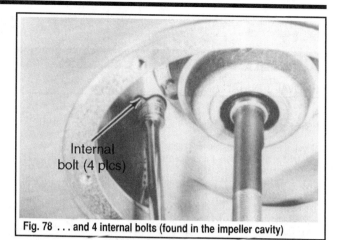

Fig. 78 . . . and 4 internal bolts (found in the impeller cavity)

■ **WHAT?** You didn't read that note earlier telling you to sort them (or your buddy kicked the tray over). That's ok. Normally, when the bolts are of different lengths, the two 5/16 - 16 x 2 1/2 in. bolts are threaded into the forward holes, while the two 5/16 - 16 x 3 in. bolts are threaded into the rear holes. If you're unsure, threads the bolts slowly by hand and finger-tighten to ensure they are in their proper bores.

14. Install the upper retaining bolt (threaded downward through the adapter to the top of the housing assembly). Tighten all 5 of the jet drive housing assembly retaining bolts to 10-12 ft. lbs. (14-16 Nm).
15. Install the Impeller, as detailed in this section.
16. Reconnect the spark plug leads and, if applicable, the negative battery cable.
17. Follow the Jet Drive Bearing lubrication procedure, as detailed in the Maintenance and Tune-Up section.

Reverse Gate

After repairs, parts replacement or rigging, it is possible that the reverse gate or remote cable will require adjustment to ensure proper operation.

GATE ADJUSTMENT

◆ See Figure 79

To adjust the reverse gate on the jet drive housing, proceed as follows:
1. Set the shifter handle to the **neutral** detent.
2. Hold the reverse gate up and check to make sure there is a 9/16 in. (23.9mm) gap between the top of the gate and the water flow passage.
3. If adjustment is necessary, have an assistant continue to hold the reverse gate up and check the clearance, while you loosen the cam screw (1) and rotate the eccentric nut (2) until the clearance is correct.
4. Tighten the cam screw, then recheck the adjustment.
5. Move the shifter handle to the **forward** detent.
6. With the gate now in the **forward** detent try to lift up on the reverse gate by hand (moving it back toward **neutral**). The gate should NOT move. If necessary, adjust the eccentric nut again, so the **neutral** gap is correct, but the gate will not move upward in the **forward** position.

REMOTE CABLE ADJUSTMENT

◆ See Figures 80 and 81

Proper cable adjustment is necessary in order to keep water pressure (from boat movement) from moving the reverse gate upward blocking normal thrust.

※※ **CAUTION**

Should be reverse gate shift suddenly and unexpectedly over the thrust nozzle, the boat will stop violently, shifting and possibly ejecting passengers and cargo. Proper adjustment is critical.

This procedure starts with the cable disconnected, for replacement, re-rigging or installation of a jet drive assembly.
1. Pull upward on the shift cam (1) until the roller is in the far end of the **forward** range (2).
2. Shift the remote control to the **forward** position. Temporarily, push the insert the cable guide onto the shift cam stud, then pull firmly on the cable casing to remove all free-play. Next, adjust the cable trunnion until it aligns with the anchor bracket, then remove the cable from the cam stud.
3. Connect the cable trunnion to the anchor bracket, then turn 90° to lock the trunnion in position.
4. Push the cable guide back onto the cam stud, then install the washer and finger-tighten the locknut.
5. Shift the remote control to the **neutral** position (3). The cam roller must snap into the **neutral** detent when you pull upward on the reverse gate with moderate pressure. If the gate does not shift properly into **neutral**, lengthen the cable slightly and recheck the adjustment.

■ Remember, it is CRITICAL that the reverse gate remains locked into the FORWARD position when the remote is shifted into FORWARD. You should NOT be able to move the gate out of this position by hand.

6. Tighten the cam locknut securely, then loosen it 1/8 - 1/4 of a turn in order to allow free movement of the cam.
7. Secure the cable loosely to the steering cable using a wire tie, but be sure NOT to restrict shift cable movement.

■ With the engine running in NEUTRAL and the warm-up lever raised, the engine will run rough due to exhaust restriction from the reverse gate. This is normal.

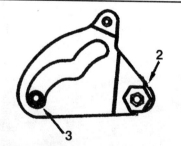

Fig. 79 Reverse gate adjustment (gate shown in FORWARD position)

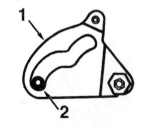

Fig. 80 When adjusting the remote cable, first move the cam (1) to the FORWARD position (2) . . .

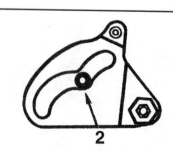

Fig. 81 . . . next, check that the cam snaps into the NEUTRAL position (3)

TRIM/TILT SYSTEMS 8-2
   DESCRIPTION AND OPERATION 8-3
      Conventional Large Motor System 8-5
      Conventional Small Motor System 8-6
      Fastrak System 8-3
      Manually Tilting A FasTrak Engine 8-5
   INTRODUCTION 8-2
      General Information 8-2
   SYSTEM IDENTIFICATION 8-2
      Small-Motor Conventional System 8-2
      FasTrak System 8-3
      Large-Motor Conventional System 8-3
   TROUBLESHOOTING 8-7
      Checking Current Draw with an Ammeter 8-8
      Checking the Motor No Load Operation 8-9
      Checking the Relays - Except EFI 8-11
      Checking the Trim Gauge 8-12
      Checking the Trim Sender 8-12
      Conventional Large Motor System 8-7
      Conventional Small Motor System 8-8
      FasTrak System 8-7
      Power Supply 8-10
   TILT LIMIT SWITCH (FASTRAK ONLY) 8-12
      Adjustment 8-12
   TRIM SENDING UNIT ADJUSTMENT 8-12
      FasTrak Systems 8-13
      Large-Motor Conventional Systems 8-13
      Small-Motor Conventional Systems 8-14
   TRIM/TILT ASSEMBLY - FASTRAK
   SYSTEMS 8-14
      Assembly 8-17
      Cleaning and Inspection 8-17
      Disassembly 8-16
      Overhaul 8-16
      Removal & Installation 8-14
   TRIM/TILT ASSEMBLY - LARGE-MOTOR CONVENTIONAL SYSTEMS 8-17
      Assembly 8-19
      Disassembly 8-19
      Removal & Installation 8-17
   TRIM/TILT ASSEMBLY - SMALL-MOTOR CONVENTIONAL SYSTEMS 8-20
      Removal & Installation 8-20
TRIM/TILT WIRING 8-22
   Conventional Systems 8-22
   Fastrak Systems 8-22

8

TRIM & TILT

TRIM/TILT SYSTEMS 8-2
TRIM/TILT WIRING 8-22

8-2 TRIM AND TILT

TRIM/TILT SYSTEMS

This section covers basic system troubleshooting and major component removal, replacement and installation for the 3 major power trim/tilt systems found on inline Johnson/Evinrude motors.

Introduction

GENERAL INFORMATION

Trim

Motor trim is generally defined as the angle of the gearcase when compared with the vertical line of the transom. Adjusting motor trim will affect a boat's handling. Loosely speaking, the bow tends to moves in the direction of the trim.

When the motor is trimmed up (outward, away from the transom) the bow will tend to move upward. This can be used to raise the bow slightly to increase fuel economy or performance, or it can be used to help stabilize the boat when running **with** chop. However, in bow up positions, the boat may tend to pull to the port side. Be careful as excessive bow-up trim may cause propeller ventilation (resulting in slippage) and, could cause the bow to rise suddenly or excessively when encountering a wake or rough surf.

When the motor is trimmed down (toward the transom) the bow will tend to move downward. This position is best for quick acceleration onto plane and for maximum towing power (such as for water skiing). This position is generally used for climbing onto plane from a standing start or idle operation or for stabilizing the boat when running **against** chop. With bow down trim, the motor will tend to push boat starboard slightly and the boat will have more of a tendency to plow (as more of the bow will be pushed into the water).

✱✱ CAUTION

Trim must be carefully set for safe boat operation. Running the boat with the motor trimmed excessively up or down will cause dangerous conditions. Excessive upward trim could cause the boat to react violently to surf or wakes. Excessive downward trim, especially at speed, may cause plowing that could turn or spin the boat suddenly. Either condition could eject passengers.

Trim should be adjusted to suit operational, boat/passenger load and weather conditions. Generally speaking, small adjustments in or out from vertical will usually achieve the desired results of optimum handling and fuel economy for the boat and conditions.

Tilt

One advantage to an outboard is the ability to fully raise the motor out of the water for beaching, trailering or even storage or docking in salt water (to keep the motor out of the corrosive water over long periods of time). Generally speaking, smaller motors are tilted manually while larger motors are tilted using a power trim/tilt system. However, this is a generalization, as many larger Johnson/Evinrude outboards are equipped with a gas-charge assisted manual tilt system. And, at the same time, some smaller motors (usually remote models) are equipped with power trim/tilt systems.

When equipped with a manual system, a lever on the stern/swivel bracket must be released and the engine case is then grasped in order to manually raise or lower the motor. A tilt pin may be used to manually adjust downward trim position on these motors.

With power systems, a tilt rod can be used to lift the motor upward, past operating trim levels or it can be used to lower the motor back to operating levels.

In ALL cases, some manual locking bracket must be used to secure the motor when trailering. The power trim/tilt systems available for Johnson/Evinrude motors include an integral trailering bracket.

Some Johnson/Evinrude outboards are equipped with a power trim tilt system in order to assist in trimming the motor or lifting the motor for trailering, shallow water operation or beaching. Using the power trim function, the operator can adjust gearcase angle (tilt from vertical at the transom) to improve handling. While the motor is trimmed up (outward, away from the transom), force from the motor will tend to push the bow up. This

System Identification

◆ See Figures 1, 2 and 3

Many of the Johnson/Evinrude motors (mostly remote models) covered here are equipped with a power trim/tilt system from the factory. Also, power trim/tilt systems may be added at the time of motor installation/boat rigging. There are essentially 3 basic versions of the trim-tilt system that might be found on a Johnson/Evinrude outboard and, since systems may be changed or added when rigged, it is nearly impossible to identify the type of trim tilt system used based on solely on engine type. However, the 3 systems used can be identified visually.

■ **Although system identification is necessary for parts replacement and for specifications during some test procedures, the actual operation (especially the wiring) is virtually the same for all 3 systems. Differences generally occur in the manifold/valve designs.**

TYPES OF TRIM/TILT SYSTEMS

◆ See Figures 1, 2 and 3

Small-Motor Conventional System

◆ See Figure 1

Generally speaking, this system may be found on 25-50 hp motors. The small-motor conventional system is made up of a single-ram and is visually identified by the fact that the motor is mounted onto a manifold next to a single cylinder. Both other systems used by Johnson/Evinrude contained dual-rams, which makes identification of this system relatively easy in the scheme of things. A version of this system is normally installed on 40-50 hp EFI motors.

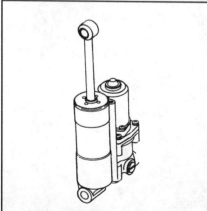

Fig. 1 The small motor conventional system is easy to spot, as it utilizes a single piston for tilt/trim

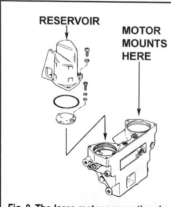

Fig. 2 The large motor conventional system is identified by the motor and reservoir mounted vertically onto the cylinder/manifold mating surface

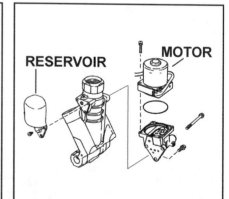

Fig. 3 The FasTrak® system is similar to the large motor conventional, but the motor/valve assembly and reservoir bolt horizontally to the cylinder body

TRIM AND TILT 8-3

Large-Motor Conventional System

◆ See Figures 2, 4 and 5

Generally speaking, this system may be found on some 50-70 hp motors (NOT including EFI models). The large-motor conventional system utilizes a dual-ram trim system, along with a single tilt cylinder/shock absorber, all of which are mounted to a single manifold assembly. These characteristics make it similar to the FasTrak® system used on most large motors. The major difference is that, on the conventional system, the motor and reservoir are bolted vertically down onto the manifold (on the FasTrak® system, those components are bolted horizontally to the manifold).

When testing or repairing the electric motor on large-motor conventional systems, it is important to identify the type of motor installed on the system (Prestolite, Bosch or one of 3 Showa models) in order to ensure you have the proper specifications. Refer to the visual and statistical information in the accompanying illustration to help make the correct motor identification for testing and replacement purposes.

FasTrak® System

◆ See Figures 3 and 6

The FasTrak® system is found on all of the larger Johnson/Evinrude outboards, including the V motors not covered here. It may be found on the largest of the inline motors, generally the 913cc 2-stroke motors, as well as the 70 hp EFI 4-stroke motor. Like the large-motor conventional system the FasTrak® system utilizes a dual-ram trim system, along with a single tilt cylinder/shock absorber mounted to a single cylinder body. The major difference between the FasTrak® and the conventional system is that on the FasTrak® assembly, the pump/valve manifold assembly and the reservoir, although positioned vertically like on the conventional system, are actually fastened to the cylinder body horizontally (bolted from the side). A less obvious difference at first glance is the fact that the cylinder body does not contain the manifold valves on the FasTrak® system, as a separate valve body is bolted between the cylinder body and the motor.

■ The trim/tilt system used by the 70 hp EFI motor is not an exact duplicate of the system used for other Johnson/Evinrude motors. It differs mostly in the relay and control wiring, and these differences are reflected in certain separate troubleshooting procedures.

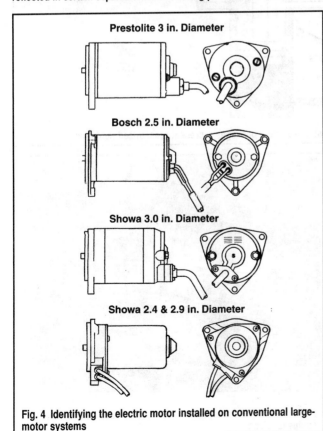

Fig. 4 Identifying the electric motor installed on conventional large-motor systems

Description and Operation

FASTRAK® SYSTEM

◆ See Figures 7, 8 and 9

The FasTrak® system is completely self-contained in an assembly that is mounted to the engine stern bracket. The main system components include an electric motor, a fluid reservoir and pump/valve manifold assembly and a cylinder body.

The FasTrak® system utilizes a cylinder body with 2 trim cylinder rams and one combination tilt cylinder/shock absorber. Whenever the trim button is pressed upward the electric motor rotates clockwise, (as viewed from the pump end) pumping hydraulic fluid to force the cylinders upward. Because of a mechanical advantage the trim cylinders perform **most**, but not all of the work. Once the motor is raised 21° (to the end of the trim cylinder travel), the tilt cylinder moves the engine through the remaining 54° of the tilt range.

■ While in the tilt range (above the trim range) the engine will only operate at idle/low throttle.

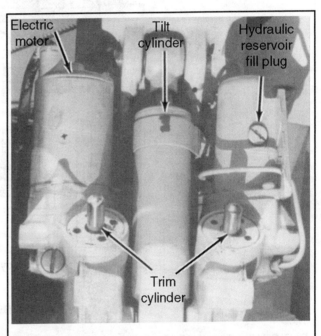

Fig. 5 Major components of conventional large-motor systems

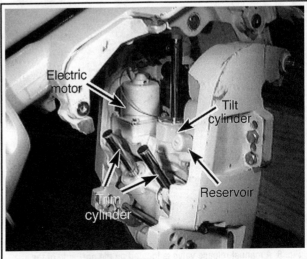

Fig. 6 Major components of FasTrak® systems

8-4 TRIM AND TILT

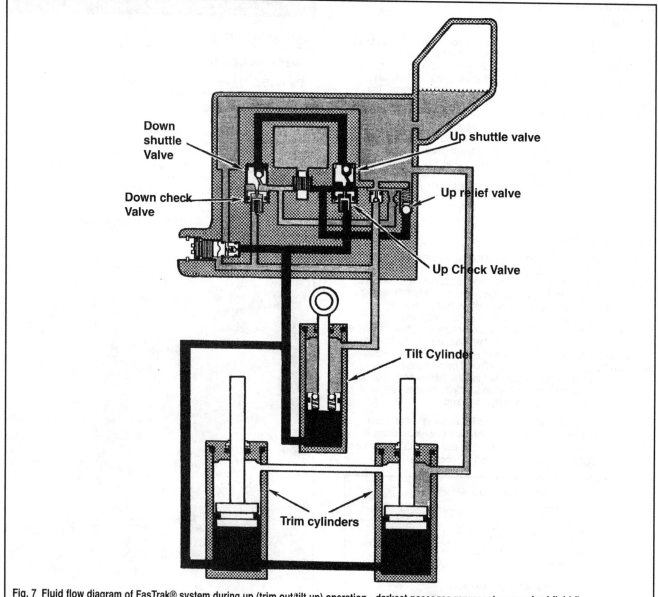

Fig. 7 Fluid flow diagram of FasTrak® system during up (trim out/tilt up) operation - darkest passages represent pressurized fluid flow

Fig. 8 A manual release valve is located on the port side of the FasTrak® system

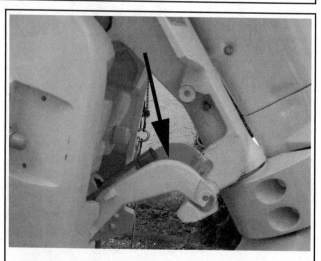

Fig. 9 An integral trailering bracket provides support when the engine is raised

TRIM AND TILT 8-5

When the down switch is activated, fluid is pumped only to the top of the tilt cylinder whose piston moves the engine downward (pushing fluid from underneath the tilt piston as it travels downward). The force of the engine moving down pushes against the trim cylinder rams, forcing fluid out from underneath the trim cylinder pistons as they return to the bottom of their travel.

A manual release valve is provided in order to provide tilt/trim function in the event of certain electrical or mechanical system failures. The valve is normally located on the port side of the trim/tilt bracket assembly and is accessed with a screwdriver through a hole on the outside of the stern bracket.

An integral trailering bracket is provided to protect the system from stress and shock. The bracket should be used whenever the boat is towed or in storage.

A tilt limit switch can be set to prevent the engine from contacting the boat's motor well during trim/tilt operation. However, setting the trim limit switch will not prevent the engine from striking the motor well if moved manually or due to a severe impact.

Manually Tilting A FasTrak® Engine

To manually tilt a FasTrak® engine, open the manual release valve by rotating it about 3 1/2 turns **counterclockwise** using a screwdriver inserted through the access hole on the outside of the stern bracket. Turn lightly until the valve **just** contacts the retaining ring.

With the valve open you can manually raise the engine (which will allow fluid to flow from the top of the tilt cylinder to the reservoir and to the underside of the tilt cylinder).

✱✱ WARNING

Once the release valve is opened, the engine must be supported when it is lifted, either using a hoist or the trailering bracket. DO NOT attempt to hold the engine up, above the trim range using the tilt cylinder and manual release valve. Similarly, do not allow the engine to drop suddenly once the support is removed.

To trim the engine out, start by releasing the valve and raising the motor further than the final desired trim position. Close the release valve and gently allow the tilt cylinder to take the motor's weight; then slowly open the release valve, bleeding off pressure using the weight of the engine. Close the valve again once the desired trim level has been reached.

■ Remember that the manual release valve must be closed (turned gently inward) in order for the FasTrak® system to operate properly.

CONVENTIONAL LARGE MOTOR SYSTEM

◆ See Figure 10

The large motor conventional trim/tilt system is completely self-contained in an assembly that is mounted to the engine stern bracket. The entire assembly is secured to the engine stern brackets by 3 bolts on either side. Mounted on or in the main manifold assembly are all system components including an electric motor, a fluid reservoir and 2 trim cylinder/ram assemblies and a single tilt cylinder/shock absorber.

Whenever the trim button is pressed upward the electric motor rotates clockwise, (as viewed from the pump end) pumping hydraulic fluid to force the cylinders upward. Because of a mechanical advantage the trim cylinders perform **most**, but not all of the work. Once the motor is raised 15° (to the end of the trim cylinder travel), the tilt cylinder moves the engine through the remaining 50° of the tilt range.

■ Although the engine can be operated above the 15° trim tilt level (for shallow water driving), at speeds above 1500 rpm, the trim-up relief valve will open, automatically lowering the motor back to the fully trimmed out (15° position). Also, when the motor is running above 1500 rpm, it will NOT tilt upward past the 15° position.

When the down switch is activated, fluid is pumped to the top of the trim and tilt cylinders whose pistons move the engine downward (pushing fluid from underneath the pistons as they travel downward).

A manual release valve is provided in order to provide tilt/trim function in the event of certain electrical or mechanical system failures. The valve is normally located on the starboard side of the trim/tilt bracket assembly and is accessed with a slotted screwdriver through a hole on the outside of the stern bracket.

✱✱ WARNING

NEVER open the manual release valve more than 1 1/2 turns. And, the valve must be closed again in order for the shock absorption system and reverse thrust system to work.

A trim gauge is provided with the motor and a sending unit is attached to the port stern bracket. The switch can be accessed with the motor fully tilted.

An integral trailering bracket is provided to protect the system from stress and shock. The bracket should be used whenever the boat is towed or in storage.

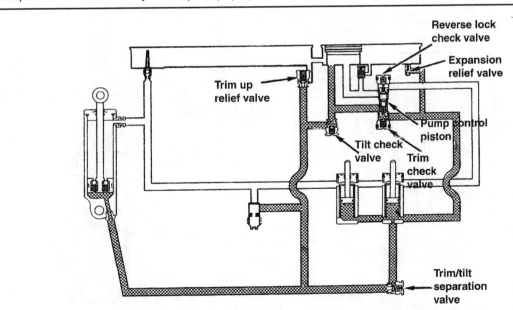

Fig. 10 Fluid flow diagram of the conventional large motor system during up (trim out/tilt up) operation - shaded passages represent pressurized fluid flow

8-6 TRIM AND TILT

CONVENTIONAL SMALL MOTOR SYSTEM

◆ See Figures 11, 12 and 13

The small motor conventional trim/tilt system is completely self-contained in an assembly that is mounted to the engine stern bracket. The hydraulic unit consists of a manifold that contains all valving, the fluid reservoir, pump, motor and single combination trim/tilt cylinder. During operation, the hydraulic unit pivots on the lower thrust rod and the piston rod that attaches to the underside of the engine swivel bracket.

Whenever the trim button is pressed upward the electric motor rotates clockwise, (as viewed from the pump end) pumping hydraulic fluid to force the cylinders upward. As the piston in the trim/tilt cylinder begins to extend, the first 15° of engine movement is considered trim (standard engine operational drive), as the engine is still supported by both the port and starboard stern brackets. But, as the piston continues to push upward, the final 50° of engine movement are considered the tilt range.

When the down switch is activated, fluid is pumped to the top of the tilt cylinder, whose piston moves the engine downward (pushing fluid from underneath the piston as it travels downward).

A manual release valve is provided on the assembly in order to provide tilt/trim function in the event of certain electrical or mechanical system failures. In order to move the engine by hand, loosen the manual release valve a minimum of 3 turns. Once the engine is positioned as desired, the

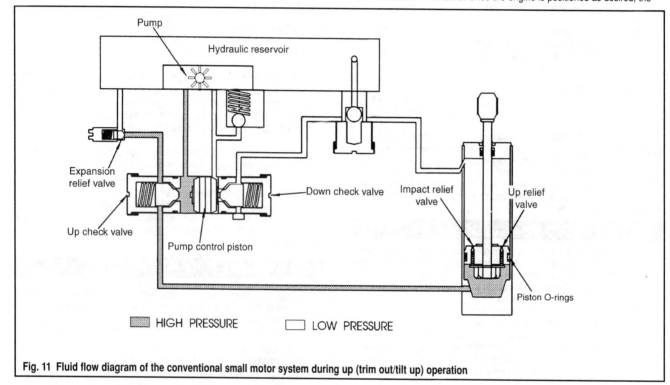

Fig. 11 Fluid flow diagram of the conventional small motor system during up (trim out/tilt up) operation

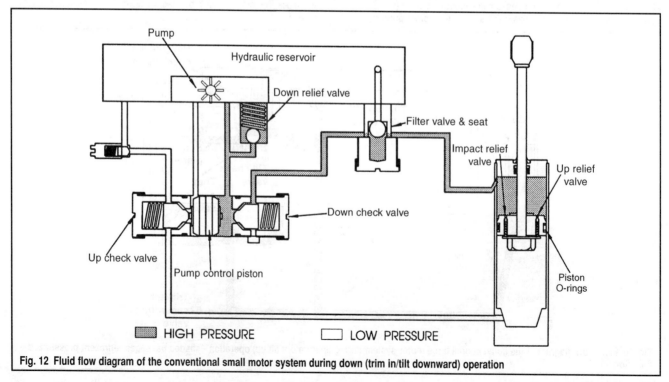

Fig. 12 Fluid flow diagram of the conventional small motor system during down (trim in/tilt downward) operation

TRIM AND TILT 8-7

valve must closed and tightened to 45-55 inch lbs. (5-6 Nm). The electro-hydraulic system will not work again until the valve is closed and tightened.

A trim gauge can be added to this system. When equipped, the sending unit is mounted on the port side of the swivel bracket.

Engines equipped with this system are normally equipped with integral trailering locks to protect the system from stress and shock. The locks MUST be used whenever the boat is towed or in storage. When setting the locks, be sure to retract the cylinder until the locks are fully engaged.

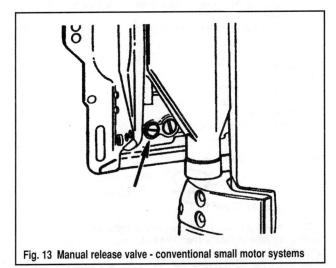

Fig. 13 Manual release valve - conventional small motor systems

Troubleshooting

TRIM/TILT SYSTEM DIAGNOSIS

FasTrak® System

◆ See Figures 14 and 15

Should the FasTrak® system malfunction, use the accompanying symptom and service chart to help determine the cause. Unfortunately, some of the service steps require "substituting" a known good part, which is often not available unless purchased. Occasionally, substituting a known good part will fix the problem. But, more often than not, it only proves that the original part was fine. For this reason, substitution should be avoided unless absolutely necessary (unless the parts counter person is a real buddy and will take back the good part if it doesn't fix the problem). If one of the service steps involves substitution, skip it and eliminate the other possibilities first.

Also, keep in mind these basic points for system troubleshooting:
• If the system does not work **and** the motor does not run or make any noise, then refer to the Power Supply procedure in this section.
• If the engine tilts part of the way upward, but does not move smoothly or with a constant sound, there is probably air in the system (usually caused by low fluid level). Check and refill the reservoir (as detailed in the Maintenance and Tune-Up section). Bleed air from the system by running the engine fully up and down with the trim/tilt motor for at least 5 complete cycles, pausing between each cycle to recheck fluid level and top-off, as necessary.
• If the motor seems to be binding mechanically, open the manual release valve and tilt the motor manually up and downward to check for smooth operation or binding.
• If the engine does not tilt as high as it should or the motor stops operating at the maximum position (does not sound like it is stalled, but shuts off), check the Tilt Limit Switch Adjustment, as detailed in this section.
• If, after a repair or manual release valve replacement, the motor will not lower from the shallow water drive position at any throttle setting, make sure the proper manual release valve is installed. The manual release valve used on inline motors should **not** have a shallow groove in the valve face. If a groove is present, the valve is designed for V6 engines.

Conventional Large Motor System

If problems are encountered with a conventional large motor system, use the following list of symptoms and possible causes to help determine the problem. Before starting the troubleshooting procedure, make sure all basic system checks are completed as follows:
• If the system does not work **and** the motor does not run or make any noise, then refer to the Power Supply procedure in this section.
• If the engine tilts part of the way upward, but does not move smoothly or with a constant sound, there is probably air in the system (usually caused by low fluid level). Check and refill the reservoir (as detailed in the Maintenance and Tune-Up section). Bleed air from the system by running the engine in ten second spurts fully up and down with the trim/tilt motor for at least 5 complete cycles, pausing between each cycle to recheck fluid level and top-off, as necessary.
• If the motor seems to be binding mechanically, open the manual release valve and tilt the motor manually up and downward to check for smooth operation or binding.

1. If the only the tilt cylinder leaks down, check the following:
• Oil lines
• Manual relief valve
• Trim-up relief valve
• Tilt pistons and/or seals
• Tilt check valve

2. If both the tilt and trim cylinders leak down, check the following:
• Trim pistons and seals
• Trim cylinder sleeves
• Trim check valve
• Expansion relief valve

3. If there is no reverse lock, check the following:
• Manual release valve
• Filter valve
• Tilt pistons and/or seals
• Reverse lock check valve
• Oil lines

4. If the unit works, but slower than normal, check the following:
• Make sure there is no mechanical binding
• Hydraulic pump
• Electric motor
• Trim-down relief valve
• Expansion relief valve

5. If the motor runs but the system does not move the engine, check the following:
• Fluid level
• Pump coupler
• Hydraulic pump

6. If there is no trim when under load, check the following:
• Manual release valve
• Trim relief valve
• Trim pistons and/or seals
• Trim cylinder sleeves
• Expansion relief valve

7. If the tilt down function does not work, check the following:
• Manual release valve
• Filter valve
• Trim-down relief valve
• Pump control piston

8. If the unit becomes locked in the tilt up position, check the expansion relief valve.

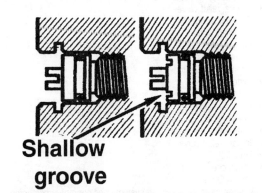

Fig. 14 Make sure the proper manual release valve is installed - the face of the manual release valve for inline engines should not have a groove on FasTrak® equipped inline motors

8-8 TRIM AND TILT

Symptoms	Service Steps
Unit will not move in either direction	① ② ③ ④ ⑥
Unit runs slowly in one direction, normal speed in other direction	① ② ⑥
Unit runs slowly in both directions (if low hours)	① ② ⑦
(if high hours)	① ② ⑤ ⑥ ⑦
Unit leaks DOWN and/or will not hold trim position against thrust in forward.	① ② ⑤ ⑦
Unit leaks UP, will not hold a trim position against propeller thrust in reverse. Does not leak down or lose trim position in forward.	① ② ⑥
Unit leaks both UP and DOWN – Leaks down in tilt range and/or will not hold trim position against thrust in forward or reverse.	① ② ⑦
Unit will not trim/tilt one way, but works okay the other way. Unit will not run DOWN – but runs UP okay; or it runs DOWN okay – but will not run UP.	② ③ ⑥

Service Steps

A. Be sure the manual release valve is closed. If NOT, tighten firmly and retest.
B. Temporarily, install a known good manual release valve and retest. If symptoms remain, original valve is not the problem. Reinstall original valve.
C. If symptoms disappear, the original valve was faulty. Remove temporary valve and replace with correct valve.
D. Go to **Power Supply** in this section to determine if problem is power supply.
E. Remove trim motor and check condition of drive coupling. If coupling is damaged, replace it.
F. Temporarily, install a known good pump manifold assembly and retest. If symptoms remain, original assembly is not the problem. Reinstall original assembly.
G. If symptoms disappear, the original pump manifold assembly was faulty. Remove temporary assembly and replace it with correct one.
H. Install replacement pump manifold and retest. If all symptoms are not corrected, reconsider the problem using the new symptoms.
I. Install O-ring kit. Look for any cylinder damage. Look for chips in fluid or impact valves. Look for other abnormal conditions. If all symptoms are not corrected, reconsider the problem using the new symptoms.

Fig. 15 FasTrak® system diagnosis - perform the appropriate service steps in order for the applicable symptom

Conventional Small Motor System

If problems are encountered with a conventional small motor system, use the following list of symptoms and possible causes to help determine the problem. Before starting the troubleshooting procedure, make sure all basic system checks are completed as follows:

• If the system does not work **and** the motor does not run or make any noise, then refer to the Power Supply procedure in this section.
• If the engine tilts part of the way upward, but does not move smoothly or with a constant sound, there is probably air in the system (usually caused by low fluid level). Check and refill the reservoir (as detailed in the Maintenance and Tune-Up section). Bleed air from the system by running the engine fully up and down with the trim/tilt motor for at least 5 complete cycles, pausing between each cycle to recheck fluid level and top-off, as necessary. When running the motor upward or downward to bleed the system, hold the switch (with the motor running) for an additional 5-10 seconds after the unit reaches the top or bottom of it's travel, and then activate the switch in the opposite direction.
• If the motor seems to be binding mechanically, open the manual release valve and tilt the motor manually up and downward to check for smooth operation or binding.

1. If the cylinder leaks down, check the following:
• Manual relief valve
• External fluid leaks
• Up check valve
• Impact relief valve
• Up relief valve
• Expansion relief valve
• Piston O-rings

2. If there is no reverse lock, check the following:
• External fluid leaks
• Manual release valve
• Filter valve and seat
• Impact relief valve
• Up relief valve
• Down check valve
• Piston O-rings

3. If the motor runs but the system does not move the engine, check the following:
• Fluid level
• Manual release valve
• Pump coupler
• Hydraulic pump

4. If the motor does not tilt downward, check the following:
• Manual release valve
• Down relief valve
• Filter valve and seat
• Pump control piston
• Down check valve

5. If the unit works, but slower than normal, check the following:
• Fluid level
• Make sure there is no mechanical binding
• Hydraulic pump
• Electric motor
• Manual release valve
• Down relief valve
• Expansion relief valve

6. If the unit becomes locked in the tilt up position, check the following:
• Expansion relief valve
• Make sure there is no mechanical binding
• Pump control piston

Checking Current Draw with an Ammeter

◆ See Figures 16, 17 and 18

One of the quickest ways to narrow down system problems is through the use of an ammeter to check current draw by the motor during system operation. In order accomplish this test, you'll need a fully charged battery with at least a 360 CCA (50 amp-hour). Also, you'll need a stopwatch and an ammeter. The capacity of the ammeter will vary with the system you are testing. You'll need one capable of reading 0-100 amps (for FasTrak® systems), 0-60 amps (for large-motor conventional systems) or 0-25 amps (for small-motor conventional systems). The test should occur with the boat either on a trailer or sitting dockside/moored with little or no current.

■ **If testing a large-motor conventional system, refer to System Identification in this section for information on identifying the type of electric motor installed. Specifications for the types of electric motors installed on large-motor conventional systems will vary by motor type.**

1. Connect a suitable ammeter in series, between the battery side of the starter solenoid and red 14 gauge lead to the trim/tilt motor/junction box, as shown.

■ **In the next step, time only the trim up for FasTrak® systems, both the time for trim up and, separately, the time for tilt up on large-motor conventional systems or the combined time for trim and tilt up for small-motor conventional systems.**

2. Start with the motor trimmed fully inward (toward the transom), then operate the trim/tilt switch to move the motor fully out through the trim range (FasTrak® motors) or the trim and tilt range (conventional motors). Watch the ammeter while the motor runs through range and use the stopwatch to determine the amount of time it takes. On FasTrak® systems, the motor should move through trim range (NOT the tilt range) in 9 seconds and draw approximately 22 amps. On small-motor conventional systems, the motor should take 13-19 seconds to run to the upward stall position, but no specification is available for current draw during this process. For large-motor conventional systems, refer to the accompanying chart to determine trim up acceptable trim up and tilt up current draws and times.

■ **For dual-ram systems (both FasTrak® and large-motor conventional systems) the motor has reached the end of the trim-out (up) range once it begins to JUST come off the 2 trim rams.**

3. For FasTrak® systems, run the engine the rest of the way up through tilt range, the amount of time does not matter.

4. Once the motor reaches stall height, check the ammeter reading with the motor operating against the stall. At the stall-up position, current draw should be approximately 60-75 amps for FasTrak® systems or 7-12 amps for small-motor conventional systems. For large-motor conventional systems, refer to the accompanying chart, as specifications vary with the type of electric motor installed.

5. Run the motor back down in the same manner as up. That means for small-motor conventional systems, you'll time the entire tilt and trim range

TRIM AND TILT

	Prestolite 3 in. Dia.		Bosch 2.5 in. Dia.		Showa 3 in. Dia.		Showa 2.4 in. Dia.	
	Normal Current Draw	Time in Seconds	Normal Current Draw	Time in Seconds	Normal Current Draw	Time in Seconds	Normal Current Draw	Time in Seconds
ⓑ Trim Up	11-15	7-9	7-10	7-9	5-8	8-10	7-9	8-10
ⓒ Tilt Up	11-15	7-9	7-10	6-9	8-10	7-9	9-12	6-7
Stall Up	30-35	•	30-35	•	19-23	•	25-29	•
Stall Down	21-25	•	18-32	•	14-18	•	12-17	•
Full Range Up	•	15-20	•	14-16	•	16-18	•	14-16
Full Range Down	•	15-20	•	15-20	•	16-18	•	15-17

Fig. 16 On large-motor conventional systems, motor test specifications will vary with the type of electric motor installed

downward (it should take 10-16 seconds and there are no current draw specifications for the trip down). For large-motor systems you'll time the trim and tilt ranges separately, while watching the ammeter both times. For FasTrak® systems, run the motor back down until it **just** contacts the trim rams. This is the beginning of the trim-down range. Then, run the motor down the trim range until it is fully trimmed inward, while watching the ammeter and using the stopwatch to time it. The electric motor on FasTrak® systems should draw approximately 16 amps and take 9 seconds to run inward through the trim range.

6. Once the motor is fully trimmed inward (at the trimmed-in stall position), continue to operate the system while observing the ammeter. At the stall-down position, current draw should be approximately 35-45 amps for FasTrak® systems or 15-20 amps for small-motor conventional systems. For large-motor conventional systems, refer to the accompanying chart, as specifications vary with the type of electric motor installed.

7. If there is a normal current draw, but there is a slow operating speed on FasTrak® systems, check the following:
- Damaged impact valves
- Malfunctioning check or shuttle valves
- Damaged manual release valve

8. If there is a normal current draw, but there is a slow operating speed on small- or large-motor conventional systems, check the following:
- Damaged pump control piston
- Malfunctioning check valve

9. If there is a low current draw, check the following:
- Leaking valves or O-rings
- Weak relief valve springs
- Damaged pump
- Fouled or damaged check valves (FasTrak® only)
- Damaged manual release valve (FasTrak® only)

10. If there is a high current draw, check the following:
- Binding pump
- Binding motor
- Sticking valves
- Damaged relief valve springs (on FasTrak® systems, this should only cause stall up malfunctions)

Checking the Motor No Load Operation

◆ See Figure 19

The electric motor used on these systems can be tested as a stand-alone component, but you will need an ammeter and a vibration or mechanical tachometer. The ammeter must be capable of reading 0-50 amps for FasTrak® and large-motor conventional systems, or 0-25 amps for small-motor conventional systems. Also, you will need a fully charged battery with at least a 360 CCA (50 amp-hour) rating, and a vise or mounting to hold the motor steady during the test.

The motor must first be removed from the assembly.

■ **If testing a large-motor conventional system, refer to System Identification in this section for information on identifying the type of electric motor installed. Specifications for the types of electric motors installed on large-motor conventional systems will vary by motor type.**

1. Remove the Trim/Tilt motor from the valve body/manifold assembly, as detailed in this section.
2. Mount the trim/tilt motor in a suitable holding fixture or a soft-jawed vise.

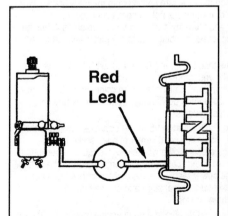

Fig. 17 Wire an ammeter between the solenoid and the trim/tilt motor power feed to check current draw

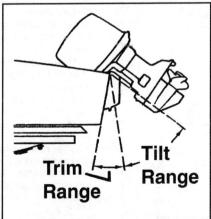

Fig. 18 Current draw and performance should be timed in the trim range (while the motor is in contact with the trim rams)

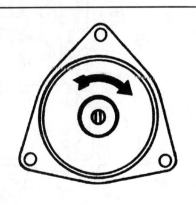

Fig. 19 The motor must rotate clockwise (viewed from pump) with power applied to the BLUE lead

8-10 TRIM AND TILT

3. Connect an ammeter in series between a freshly charged battery (of at least 360 CCA/50 amp hour rating) and the motor during the next steps of the test. The ammeter red lead should be connected to the battery.

4. While holding the vibration or mechanical tachometer against the motor to determine rotational speeds, connect the battery to the motor's **blue** lead. The motor must rotate clockwise when viewed from the pump end. Note the readings on both the ammeter and the tachometer.

5. While holding the vibration or mechanical tachometer against the motor to determine rotational speeds, connect the battery to the motor's **green** lead. The motor must rotate counterclockwise when viewed from the pump end. Note the readings on both the ammeter and the tachometer.

6. The motor must turn the correct direction for each test, as noted in the previous steps. The motors must draw the appropriate maximum amperage (or less) while turning the appropriate minimum speed as follows:

- For FasTrak® systems, the motor must draw no more than 10 amps while rotating at least 7000 rpm.
- For small-motor conventional systems, the motor must draw no more than 4.5 amps while rotating at least 5000 rpm.
- For large-motor conventional systems specifications vary with the type of electric motor installed.

On large motor systems, refer to the following motor specifications:

- Prestolite motors should draw no more than 7 amps while turning at least 4700 rpm.
- Bosch motors should draw no more than 4.5 amps while turning at least 5450 rpm.
- 3.0 and 2.4 diameter Showa motors should draw no more than 4.5 amps while turning at least 5000 rpm.
- 2.9 in diameter Showa motors should draw no more than 10 amps while turning at least 7000 rpm.

Power Supply

◆ See Figure 20

In order to properly check the power supply to the trim/tilt motor, make sure the battery is in good condition and fully charged. Also make sure the keyswitch is turned to the **OFF** position, with the key removed and the safety lanyard is removed from the switch.

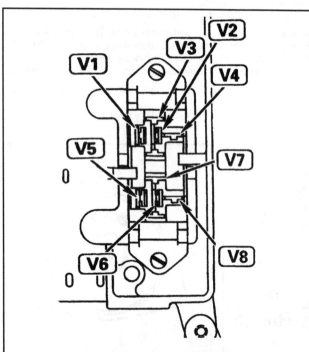

Fig. 20 Although the bracket (and relay positioning within the bracket) may vary, the relative positions of the terminals does not change. V1 and V2 (or V5 and V6) are always the pair of parallel terminals, when compared to the other 3 in the socket

Checking Power Supply - Except EFI Motors

The following tests are made on the relay side of the trim/tilt relay sockets. Terminals may be identified using the wire colors. Remember that the up relay/up circuit contains blue motor wiring while the down relay/down circuit contains green circuit wiring. The referenced terminals can also be identified using the Trim/Tilt Wiring Diagrams in this section.

1. Connect the DVOM black lead to a good engine ground, then check for voltage at the battery side of the starter solenoid by probing with the red DVOM lead. If there is no voltage, check the battery and wiring, otherwise proceed with the next step.

2. Mark the UP and DOWN relays, then remove them from the relay sockets.

3. Check the UP circuit, start by connecting the red DVOM lead to V1 (up circuit/red wire) and the black lead to V2 (up circuit/black wire at center of relay socket). There should be battery voltage present, otherwise check the red wire between the relay socket and starter solenoid (FasTrak® and large motor) or the terminal strip (small motor). If the power feed is not the problem check the ground wire between relay socket (or terminal strip) and engine ground. If power is present, proceed with the next step.

4. Connect the black DVOM lead to a good engine ground and the red lead to V3 (blue/white wire). Push the UP trim button and watch for battery voltage on the DVOM. If no voltage is present, check the remote trim switch and wiring (if they check out, proceed with the next step). If voltage was present at V3, proceed with the next step.

5. Push trim/tilt UP button on the lower motor cover and check for voltage at V3. If voltage is present, skip the remainder of this and the entire next step. If there is no voltage, trace the blue/white wire out of the trim switch on the lower motor cover to the bullet connector and check for voltage there. If there is still no voltage present, check for voltage at the red bullet connector (or red/purple connector for small motor systems) that supplies the motor cover trim switch. If switch and wire is good, proceed with the next step.

6. Some models (such as most/all FasTrak® systems and some conventional systems) contain a tilt/limit switch. On models so equipped, if there was no voltage at V3 in one of the previous steps, try isolating the tilt limit switch. Disconnect the blue/white tilt limit switch bullet connectors, then reconnect them, bypassing the tilt limit switch and repeat the previous step to test for voltage at V3 when pressing the trim/tilt up button on the lower motor cover. If voltage is now present adjust or replace the tilt limit switch.

7. Install the up relay to the bracket, **but** only slide the blades halfway into the relay socket so that you can still access them with the DVOM probes.

8. Connect the red DVOM lead to relay blade M for terminal V4 (the bottom blade that slides into the terminal socket for the blue motor wire) and the black lead to a good engine ground. Press the UP trim button and watch for voltage on the blade. If there is no voltage, perform the Checking the Relay procedure in this section. If voltage was present, proceed with the next step.

9. Push the relay the rest of the way into the bracket, now disconnect the blue and green wire connector from the trim motor. Using the red DVOM lead, check for voltage on the blue wire coming from the relay bracket while the UP button is pressed. If voltage is not present, check the blue wire between the connector and the relay bracket. If voltage is present, but the motor was not operating, check the blue lead to the motor and the motor itself.

10. Repeat the same test, but for the DOWN circuit (which never uses a tilt limit switch). We'll take you through it again though, so as not to get confused with the terminal identifications and wire colors that differ slightly. Start by removing both the up and down relays from the relay bracket (if installed after the previous steps).

11. Connect the red DVOM lead to V5 (down circuit/red wire) and the black lead to V6 (up circuit/black wire at center of relay socket). There should be battery voltage present, otherwise check the red wire between the relay socket and starter solenoid (FasTrak® and large motor) or the terminal strip (small motor). If the power feed is not the problem check the ground wire between relay socket (or terminal strip) and engine ground. If power is present, proceed with the next step.

12. Connect the black DVOM lead to a good engine ground and the red lead to V7 (green/white wire). Push the DOWN trim button and watch for battery voltage on the DVOM. If no voltage is present, check the remote trim switch and wiring (if they check out, proceed with the next step). If voltage was present at V7, proceed with the next step.

13. Push trim/tilt DOWN button on the lower motor cover and check for voltage at V7. If voltage is present, proceed with the next step. If there is no voltage, trace the green/white wire out of the trim switch on the lower motor cover to the bullet connector and check for voltage there. If there is still no voltage present, check for voltage at the red bullet connector (or red/purple connector on small motor systems) that supplies the motor cover trim switch.

14. Install the down relay to the bracket, **but** only slide the blades halfway into the relay socket so that you can still access them with the DVOM probes.

15. Connect the red DVOM lead to relay blade M for terminal V8 (the bottom blade that slides into the terminal socket for the green motor wire) and the black lead to a good engine ground. Press the DOWN trim button and watch for voltage on the blade. If there is no voltage, perform the Checking the Relay procedure in this section. If voltage was present, proceed with the next step.

16. Push the relay the rest of the way into the bracket, now disconnect the blue and green wire connector from the trim motor. Using the red DVOM lead, check for voltage on the green wire coming from the relay bracket while the DOWN button is pressed. If voltage is not present, check the green wire between the connector and the relay bracket. If voltage is present, but the motor was not operating, check the green lead to the motor and the motor itself.

Checking Power Supply - EFI Motors

1. Remove the electrical component cover for access to the relays and wiring.

2. Connect the red DVOM lead to the positive (forward) terminal for the top (which happens to be the up relay, as identified by the blue motor wire) relay and connect the black DVOM lead to a good engine ground. Repeat for the lower relay (which happens to be the down relay as identified by the green motor wire). If either test fails to show battery voltage, check the wiring between the battery, starter solenoid and the motor relays.

3. Check continuity between each relay ground post (the lower post on each relay) and a good engine ground. If either test shows no continuity, check and repair the ground circuit.

4. Connect the red DVOM to the top terminal of the upper relay and the black lead to the top terminal of the lower relay. Activate the engine mounted trim switch UP, watching the meter for approximate battery voltage, then activate the switch DOWN, again, watching for approximate battery voltage. If battery voltage is present in both tests, perform the Motor No Load Operation Test, as detailed in this section. If voltage was **not** present for either or both tests, proceed with the next step to test the switch and harness.

5. Connect the DVOM red lead to the light blue lead and the black DVOM lead to a good engine ground. Push the UP trim switch and watch the meter for battery voltage. Repeat the test on the pink wire, but push the switch in the DOWN position. If battery voltage was present at both leads, the switch and harness is operating correctly. If voltage was not present at one or both tests, check the fuse, then proceed with the next step.

6. Disconnect the pink wire bullet connector, then connect the DVOM red lead to the switch side of the pink wire and the black lead to a good engine ground. Push the trim down button, watching the meter for battery voltage. Reconnect the pink wire, then repeat the test at the light blue bullet connector, but pushing the switch up button. If voltage was not present at either or both tests, replace the trim/tilt switch. If voltage was present, proceed with the next step to test the relay.

7. Connect the red DVOM lead to the down relay motor (green wire) stud and the black DVOM lead to a good engine ground. Press the trim down button, watching for battery voltage. Repeat the test for the up relay motor (blue wire) stud and a good engine ground, but this time press the trim button up while watching for voltage. If voltage is not present at either test, replace the appropriate relay.

8. If necessary, check the trim relay resistance as follows:

   a. To test the down relay, disconnect the pink wire bullet connector. Connect one DVOM lead to a good engine ground and the other to the relay side of the pink wire. Resistance should be 4-5 ohms at an ambient temperature of about 68°F (20°C).

   b. To test the up relay, disconnect the blue wire bullet connector. Connect one DVOM lead to a good engine ground and the other to the relay side of the blue wire. Resistance should be 4-5 ohms at an ambient temperature of about 68°F (20°C).

Checking the Relays - Except EFI Motors

◆ See Figure 21

A relay is essentially a remote controlled switch. Typically speaking, a relay works when a switch circuit is energized, pulling the contacts of a normally open electro-magnetic switch closed, completing the circuit that the relay controls. Therefore, 2 wires from this type of relay will connect to the control circuit (the **S** terminals). One of those wires will be ground for the control circuit (black in this case) and one of them will be colored (either blue/white for the UP circuit or green/white for the DOWN circuit). Two other terminals of these relays are connected to the battery (one to battery positive or **B+**, the red wire and the other to battery negative or **B-**, the black wire). The battery circuit blades are the center blade and the other blade directly above it or to the side (depending on how the relay is oriented) that is parallel to the center blade. The final blade terminal, known as the **M** terminal, is the wire to the controlled motor (blue for the UP circuit or green for the DOWN circuit).

Because relays consist of various internal wiring connections (of the electro-magnetic switch), it can be tested using a DVOM to check resistance or voltage during various test conditions. In addition to a DVOM, you'll need a fully-charged 12 volt battery and jumper leads that are used to apply battery voltage to the signal circuit.

Use the accompanying illustrations to for quick and easy terminal identification.

1. Remove the relay to be tested from the relay bracket.

2. Connect the DVOM meter leads (set to read resistance) across the **B-** and **M** terminals. The meter must show no continuity at this time.

3. Connect the DVOM meter leads (set to read resistance) across the **B+** and **M** terminals. The meter must show no continuity at this time.

4. Connect the DVOM meter leads (set to read resistance) across the 2 **S** terminals. The meter must show 70-100 ohms resistance.

5. Using the set of jumper wires, apply 12 volts to the **S** terminals, while checking for continuity between the **B+** and **M** terminals. There must now be continuity.

■ **Remember, when power is applied to the signal circuit (S terminals) the internal relay switch contacts should close, providing power to the relay controlled circuit (battery power to the motor).**

6. With power still applied to the **S** terminals, check for continuity between the **B-** and **M** terminals, there should still be NO continuity.

7. Replace the relay if any test results vary.

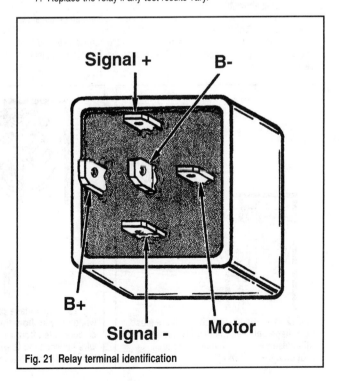

Fig. 21 Relay terminal identification

8-12 TRIM AND TILT

Checking the Trim Gauge

◆ See Figures 22 and 23

The trim gauge can be quickly checked using a voltmeter and a jumper wire. To determine if a problem is with the gauge or the circuit, proceed as follows:

1. Locate the trim switch wiring. Refer to the accompanying diagrams, the wiring in this section or the wiring under Wiring Diagrams in the Ignition and Electrical Systems section, as necessary.
2. Turn the keyswitch to the **ON** position, then use the DVOM to check for voltage between the trim gauge **I** (purple wire) and **G** (black wire) terminals.
3. If there was no voltage in the previous step, suspect the instrument harness, keyswitch or 20-amp fuse.
4. If voltage is present, disconnect the white/tan lead from the gauge **S** terminal. With the keyswitch still **ON**, the gauge should indicate full-downward trim position. Now, us a jumper wire to connect the **S** (white/tan wire) and **G** (black wire) terminals. If power is applied to the circuit and the gauge is operating properly, it will now indicate full-upward trim position.
5. Replace the gauge if it does not operate properly with power applied. If the gauge checks out in the previous step, refer to Checking the Trim Sender, in this section, as the sender or wiring is the likely culprit.

Checking the Trim Sender

◆ See Figure 24

The trim sender can be checked using an ohmmeter. To ensure accurate test readings, a digital meter is recommended. Also, remember that resistance specifications are for readings taken at ambient temperatures of about 68°F (20°C) and readings taken with components at other temperatures will vary.

1. Locate the 3-pin connector for the trim control harness between the instrument and engine trim harness. The connector normally contains the white/tan lead from the trim sender, as well as the blue/white and green/white motor control circuit wires.
2. With the keyswitch **OFF** to prevent possible damage to the meter, connect the DVOM (set to read resistance) between the white/tan wire (terminal C of the 3-pin connector) and a good engine ground.
3. With the engine in the full-downward trim position, the meter must show a reading above 80 ohms.
4. With the engine in the full-upward trim position, the meter must show a reading below 10 ohms.
5. If readings differ, replace the trim sender. If the sender tests good, suspect the trim gauge or circuit, refer to Checking the Trim Gauge in this section.

Tilt Limit Switch (FasTrak® Only)

◆ See Figures 25 and 26

Some models (normally including all motors equipped with the FasTrak® system) are equipped with a tilt limit switch mounted to the assembly at the tilt/swivel bracket. The switch can be adjusted to electronically limit the maximum amount of tilt, therefore protecting the operator from striking the motor well with the engine using the tilt feature. However, keep in mind that this switch only stops the motor from operating past a certain point, and does nothing to mechanically stop the motor from raising upward, past that point due to manual intervention. This means that, in the case of a severe impact, the motor could be thrust upward, striking the motor well. If possible, the motor should be repositioned on the transom bracket to prevent such possible collisions. In the event that this is not possible, the tilt limit switch provides some small measure of protection during normal raising of the motor via the tilt system.

ADJUSTMENT

◆ See Figure 27

If switch adjustment is necessary, proceed as follows

1. Using the trim/tilt switch, trim the motor all the way inward (toward the boat).
2. With the motor trimmed in, locate the switch and adjustment tab on the tilt bracket. It is accessible through a housing where part of the tilt pivot passes into the bracket. Although the adjustment may vary on some models, typically you:
   - Push the lower adjustment tab upward to reduce the maximum amount of tilt electronically allowed by the system.
   - Push the upper adjustment tab upward to increase the maximum amount of tilt electronically allowed by the system.

■ **If your limit switch varies, move the tab(s) in the opposite direction, as necessary, to properly adjust the system.**

3. Verify the adjustment using the trim/tilt switch and repeat, as necessary for optimum adjustment (to stop the motor at a point before it strikes the motor well).

Trim Sending Unit

Component repair, overhaul or replacement may require the trim sending unit to be adjusted. Similarly, if the motor is rigged to a different boat, the unit will have to be adjusted.

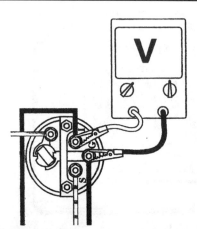

Fig. 22 With the keyswitch ON, there must be voltage between the I and G terminals otherwise the gauge cannot operate properly

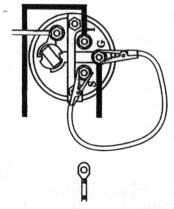

Fig. 23 With voltage present and the white/tan wire from the harness disconnected from the gauge, use a jumper wire between the G and S terminals, and see if the gauge reads trim fully-upward

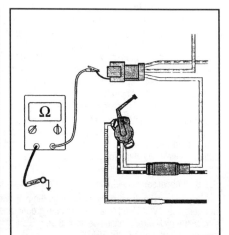

Fig. 24 The trim sender is checked using an ohmmeter to take resistance readings between the white/tan lead and a good engine ground

TRIM AND TILT 8-13

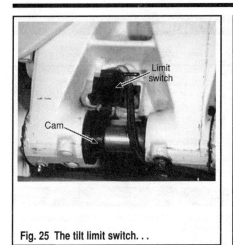

Fig. 25 The tilt limit switch...

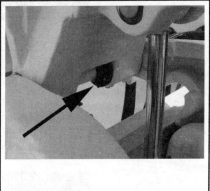

Fig. 26 ...and cam are mounted near main tilt tube

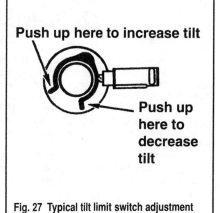

Fig. 27 Typical tilt limit switch adjustment

ADJUSTMENT

FasTrak® Systems

◆ See Figures 28 and 29

1. Tilt the engine and engage the trailering bracket.
2. Temporarily install a thrust rod (heavy drift pin or OMC #436541) into the No. 3 hole on the stern bracket.

■ If a thrust rod is not available, note the current location of the trim stop pin, then remove the spiral spring clip securing it and temporarily move the trim stop pin itself to the No. 3 hole in the bracket.

3. Locate the sending unit by tracing the wiring (normally there's a white/tan, a black/tan and a green wire). Refer to the Wiring Diagrams, either in this section or the Ignition and Electrical Systems section for more details.

■ The sending unit is normally located immediately adjacent to the lower tilt pin assembly.

4. Disengage the trailering bracket, then lower the motor against the thrust rod. Check the trim gauge needle, it should show a centered position.
5. If the gauge is not reading a centered position, tilt the engine up slightly, loosen the sending unit screws and adjust the sending unit (by pivoting it up or down). Then lower the engine back against the thrust rod to check the adjustment. Repeat until the gauge shows the motor is centered when it is sitting against the thrust rod.
6. After adjustment is correct, tighten the screws, making sure the sending unit does not move.
7. Remove the temporary thrust rod/drift pin.

Large-Motor Conventional Systems

◆ See Figure 30

On these models, the sending unit is initially adjusted to the center position, however on some motors, additional adjustment may be necessary for the gauge to read full trim up when the motor reaches the upward limit of trim adjustment (upward movement on the trim rams).

To adjust the sending unit, proceed as follows:

1. Turn the keyswitch **ON**.
2. Using the trim/tilt assembly, raise the motor to the maximum trim-up position (the point at which the tilt cylinder would just start to lift the motor off the trim rams, but make sure the motor is still sitting on the rams).
3. For 3-cylinder motors, move the horizontally mounted angle adjusting rod to the center hole.
4. Loosen the sending unit screws, leaving them snug, but not completely tightened (in this way the unit can be gently pivoted, but should not shift on its own accord).
5. On 3-cylinder motors, lower the outboard against the angle adjusting rod, then check the trim gauge needle. It should show the center position. If necessary, adjust the sending unit using a screwdriver to gently pivot it up or down slightly to center the needle.
6. Raise the engine back up to the top of the trim range again and tighten the sending unit screws securely.
7. Remove the angle adjusting rod from the center hole and return it to its former position (normally the hole closest to the transom).

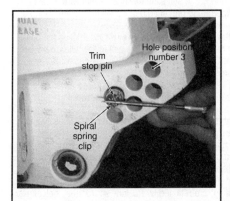

Fig. 28 Temporarily install a thrust rod (or the trim pin) in the bracket No. 3 hole..

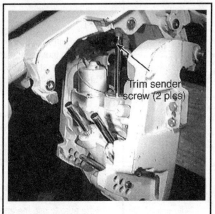

Fig. 29 ...then lower the motor onto the pin and carefully adjust the switch

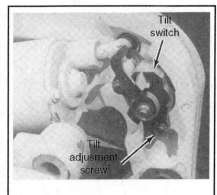

Fig. 30 Trim sending unit on large-motor conventional systems

8-14 TRIM AND TILT

Small-Motor Conventional Systems

◆ See Figures 31 and 32

On these motors, the gauge is considered properly adjusted when the eccentric cam on the sending unit is positioned so the gauge needle aligns with the lowest trim-down mark when the motor is at the bottom (downward or inward) most point of its travel.

1. Using the trim/tilt system, move the motor to the lowest point of its travel and verify that the gauge needle aligns with that point. If the needle is above or below that point, proceed with the next step in order to adjust the sending unit.
2. Locate the sending unit and eccentric cam. When the motor is trimmed all the way down the sending unit lever must touch the cam just forward of the top. Refer to the accompanying illustration.
3. Raise the engine and engage the trailering lock for access and safety, then loosen cam screw in order to make the necessary adjustment.
4. If the needle was above the lowest mark on the gauge, move the thick part of the cam **A** in the accompanying illustration **toward** the contact point **B**.
5. If the needle was below the lowest mark on the gauge, move the thick part of the cam **A** in the accompanying illustration **away** from the contact point **B**.
6. Tighten the cam screw, then recheck the reading with the motor trimmed all the way in. Repeat as necessary to achieve proper adjustment.

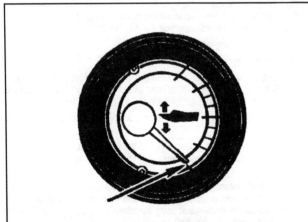

Fig. 31 On small-motor systems, the sending unit must be adjusted so the gauge needle aligns precisely with the lowest line when the motor is trimmed fully downward

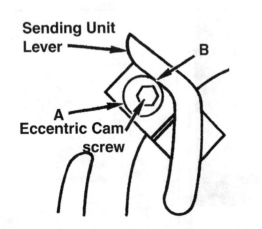

Fig. 32 With the motor trimmed downward the sending unit lever should touch the eccentric cam (A) just forward of the top of the cam, at point (B)

Trim/Tilt Assembly - FasTrak® Systems

REMOVAL & INSTALLATION

FasTrak® Systems

Carbureted Motors

◆ See Figure 33

※※ WARNING

Anytime a steel tool is struck with a hammer, there is the possibility of chips flying which could cause serious eye injury. Therefore, wear safety glasses while removing the tilt cylinder pin.

1. Disconnect the negative battery cable for safety.
2. Remove the harness clamp (wire loom), then disengage the system 2-pin connector.
3. Disengage the 2 bullet connectors for the tilt limit switch.
4. Disconnect the remaining power trim/tilt wiring connector, then remove the rubber grommet. Using a small-bladed terminal removal tool, carefully depress the locktabs and remove each of the wire terminals from the connector. Note each of the wiring locations for installation purposes.
5. Pull the wires through the hole in the port side of the lower engine cover.
6. Using a screwdriver inserted through the hole in the port side bracket, unscrew the manual release valve, then raise the engine and lock in place using the tilt support. For safety, install a holding strap or sturdy rope around the motor to make sure it will not accidentally drop if disturbed during service.
7. Disconnect the ground wire from the trim/tilt unit.
8. Separate the trim/tilt unit wires that are located in the braided tube, so they can be removed through the hole in the stern bracket.
9. Using a pair of snapring pliers, carefully remove the external snaprings from the tilt upper pin, then, using a using a punch, carefully push the upper pin out of the bore.
10. Manually push the tilt piston downward into the cylinder for clearance.
11. Using the snapring pliers, carefully remove the snaprings from the lower pin, then use the punch to gently drive the lower pin from the bracket.
12. Remove the trim/tilt unit from the stern brackets.
13. Refer to the overhaul procedures and accompanying exploded views, if disassembly or component replacement is necessary.
14. Keep the following points in mind when servicing the hydraulic trim/tilt assembly:

• Before disassembly, thoroughly clean and degrease the unit. All outer surfaces should be cleaned with a stiff synthetic needle (not wire) brush and hot soapy water. It is important to prevent dirt or debris from entering the unit during service.
• Before removing the manual release valve or disassembling the unit, temporarily connect the motor wiring and run the unit to the full upward trim position then jog the unit DOWN and loosen the reservoir cap 1 full turn to equalize system pressures. But, remember that there could be significant residual hydraulic pressure left behind some components, so always wear safety goggles and loosen fittings gradually, allowing pressure to bleed off before removal.
• NEVER apply heat to the cylinder body or cylinders as excessive heat can lead to high pressure leaks or component failure during operation.
• Never paint individual components while the unit is disassembled. The fear is that some portion of the paint might flake off and enter a hydraulic passage during assembly.
• Some of the components in the assembly (such as the reservoir or valve/pump body mounting) are retained by Pozidriv screws that look similar to Philips screws, but would damaged by use of an improper (Philips) driver. Be sure to use the appropriate size/type of driver on all screws.
• Always use clean, lint-free shop cloth when handling assembly components.
• During assembly, replace all seals and O-rings to ensure proper, trouble-free operation.

TRIM AND TILT 8-15

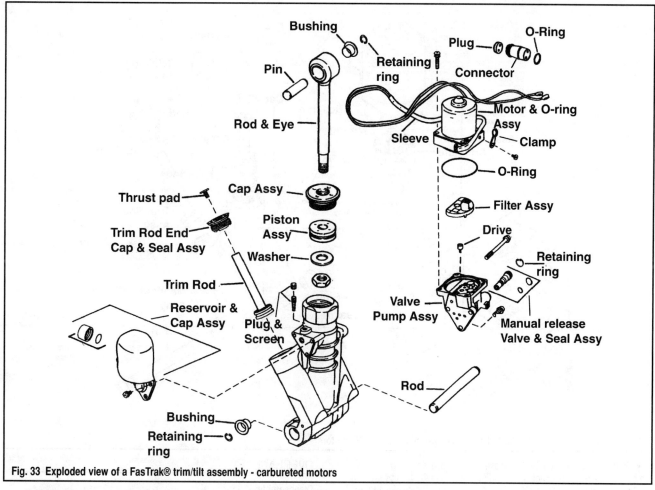

Fig. 33 Exploded view of a FasTrak® trim/tilt assembly - carbureted motors

• During valve/pump body installation (once the body is attached to the hydraulic unit), fill the pump cavity up to the top of the 2 bosses using Evinrude/Johnson Power Trim/Tilt and Steering Fluid (or equivalent). Then, using a slotted-screwdriver, rotate the pump-to-motor coupler back and forth to bleed air from the pump. Continue to rotate it a few turns in each direction until bubbles stop coming out of the pump.

To install:

15. Apply a light coating of Evinrude/Johnson Triple-Guard, or equivalent marine grease to the lower and upper pins. Place the pins on a clean plastic bag until they are installed.

16. Position the trim/tilt assembly to the stern bracket, then carefully insert the lower pin to secure it.

17. If the tilt tube nut was loosened, tighten it to 50-54 ft. lbs. (68-73 Nm), then back off one nut 1/8 - 1/4 turn.

18. Install the external snap-rings to the lower pin with the sharp edge facing outward.

19. Manually extend the tilt cylinder rod until it aligns with the holes in the swivel bracket, then install the upper pin. Secure using the external snap-rings (also with the sharp edge facing outward).

20. Reconnect the wiring:
    a. Place the trim/tilt wires back into the braided tube, then install them through the hole in the stern bracket.
    b. Reconnect the ground wire to the trim/tilt unit.
    c. Release the tilt support and lower the engine for ease of access, then firmly retighten the manual release valve.
    d. Install the 2 wires through the hole in the port side of the lower engine cover.
    e. Connect the 2 trim limit switch bullet connectors, the 2-pin system connector and the trim/tilt connector (after the terminals are properly reinstalled). Install the rubber grommet and the wire loom, to the appropriate connectors.

21. Reconnect the negative battery cable, then run the engine up and down through several cycles, checking for proper operation and proper fluid level, top-off, as necessary.

70 Hp EFI Motors

◆ See Figure 34

1. Tilt the outboard to the fully tilted position, the engage the tilt lock to secure the motor. For safety, install a holding strap or sturdy rope around the motor to make sure it will not accidentally drop if disturbed during service.

2. Disconnect the negative battery cable for safety.

3. Using a pair of snapring pliers, carefully remove the external snap-rings from the tilt upper pin, then, using a using a punch, carefully push the upper pin out of the bore.

4. Manually push the tilt piston downward into the cylinder for clearance.

5. Remove the electrical component cover for access to the motor wiring. Loosen and remove the nuts retaining the blue and green motor wires to the relays.

6. Open the clamp on the lower engine cover, then carefully pull the wires and sleeve from the cover grommet. Pull the wires and sleeve through the hole in the transom bracket.

7. Remove the nut from the thrust rod (mounted across the bottom of the transom bracket). Mark the hole in which the rod is installed to ensure installation in the same location.

8. Using the snapring pliers, carefully remove the snaprings from the lower pin, then use the punch to gently drive the lower pin from the bracket.

9. Remove the trim/tilt unit from the stern brackets.

10. Refer to the overhaul procedures and accompanying exploded views, if disassembly or component replacement is necessary.

11. Keep the following points in mind when servicing the hydraulic trim/tilt assembly:

• Before disassembly, thoroughly clean and degrease the unit. All outer surfaces should be cleaned with a stiff synthetic needle (not wire) brush and hot soapy water. It is important to prevent dirt or debris from entering the unit during service.

8-16 TRIM AND TILT

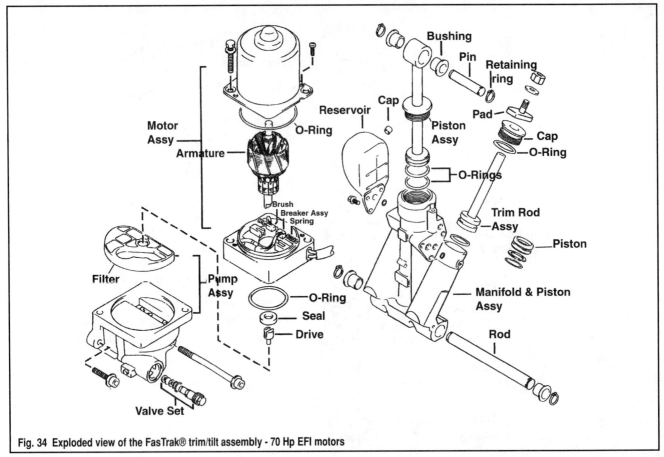

Fig. 34 Exploded view of the FasTrak® trim/tilt assembly - 70 Hp EFI motors

- Before removing the manual release valve or disassembling the unit, temporarily connect the motor wiring and run the unit to the full upward trim position then jog the unit DOWN and loosen the reservoir cap 1 full turn to equalize system pressures. But, remember that there could be significant residual hydraulic pressure left behind some components, so always wear safety goggles and loosen fittings gradually, allowing pressure to bleed off before removal.
- NEVER apply heat to the cylinder body or cylinders as excessive heat can lead to high pressure leaks or component failure during operation.
- Never paint individual components while the unit is disassembled. The fear is that some portion of the paint might flake off and enter a hydraulic passage during assembly.
- Always use clean, lint-free shop cloth when handling assembly components.
- During assembly, replace all seals and O-rings to ensure proper, trouble-free operation.
- During valve/pump body installation (once the body is attached to the hydraulic unit), fill the pump cavity up to the top of the 2 bosses using OMC Power Trim/Tilt and Steering Fluid (or equivalent). Then, using a slotted-screwdriver, rotate the pump-to-motor coupler back and forth to bleed air from the pump. Continue to rotate it a few turns in each direction until bubbles stop coming out of the pump.

To install:

12. Apply a light coating of OMC Triple-Guard, or equivalent marine grease to the lower and upper pins. Place the pins on a clean plastic bag until they are installed.
13. Position the trim/tilt assembly to the stern bracket, then carefully insert the lower pin and secure using the external snap-rings.
14. Install the thrust rod to the same hole from which it was removed (to ensure proper adjustment). Tighten the thrust rod retaining nut securely.
15. Route and reconnect the wiring.
16. Manually extend the tilt cylinder rod until it aligns with the holes in the swivel bracket, then install the upper pin. Secure using the external snaprings.
17. Reconnect the negative battery cable, then run the engine up and down through several cycles, checking for proper operation and proper fluid level.

OVERHAUL

Disassembly

◆ See Figures 33 and 34

Thoroughly clean the external surfaces of all dirt and scale build-up before disassembling. Clean the unit with a wire brush and plenty of soap and water, to prevent any contamination of internal components.

1. Remove the reservoir fill cap; invert the unit over a drain pan; and drain all fluid from the unit.
2. Slide out the two nylon bushings from the end of the tilt cylinder.
3. Remove the three Phillips head screws securing the reservoir to the cylinder body. Lift off the reservoir and discard the O-ring in the cylinder body. Using a 6mm hex key wrench, remove the hex plug from the cylinder body and lift out the filter.
4. Tighten the manual release valve until it is snug. Using a pair of snap ring pliers, remove the internal snap ring at the end of the cavity. Back-out the manual release valve and withdraw it from manifold.
5. Remove the screw securing the motor wire harness clamp to the side of the pump body. Remove the three remaining bolts securing the pump motor to the manifold. Lift off the motor assembly and remove the O-ring from the end of the motor housing.
6. Using a hex wrench, remove the three Allen head screws securing the pump manifold assembly to the cylinder body. Lift off the pump manifold and remove the five O-rings from the cylinder body.
7. Secure the lower body of the trim/tilt unit in a vise with soft jaws to prevent damaging the finish on the unit. Using a universal spanner wrench or a Tilt Cylinder End Cap Tool (#326485), loosen and back the end cap out of the tilt cylinder.
8. Remove the trim/tilt unit from the vise and drain any fluid in the cylinder into a suitable container. Pull up on the tilt cylinder piston and withdraw it from the cylinder.
9. Remove and discard the large O-rings from the end cap and piston.

TRIM AND TILT   8-17

10. Place the end of the tilt cylinder in a vise with the piston end up. Remove the nut and washer from the end of the rod and slide the piston off the rod.

11. Lift out the spring, plunger, and ball from each bore in the piston. Make a note of the position from which each spring - size and number of coils - is removed from the piston.

12. Using a screwdriver, pry the scraper seal out of the end cap. Lift out the O-ring inside the cap.

13. Secure the lower body of the trim/tilt unit in a vise with soft jaws to prevent damaging the finish on the unit. Using a universal spanner wrench or a Trim Cylinder End Cap Tool (#436710), loosen and back out the trim rod end cap from the trim cylinder. Repeat this step for the other trim cylinder end cap.

14. Remove the unit from the vise and drain any fluid in the cylinders into an appropriate container. Pull the trim rod out from the cylinder and remove the O-rings from the end cap. Remove the two split rings and O-ring from the piston end of the trim rod.

15. Place the trim cylinder into a Rod Holder Tool (#983213) or similar device and clamp the unit in a vise. Remove the wear plate from the end of the rod and slide the end cap off the rod.

16. Using a screwdriver, pry the scraper seal out of the end cap. Lift out the quad O-ring inside the trim end cap.

■ The trim/tilt pump, valve body, and motor do not contain any serviceable components. Therefore, do not attempt to disassemble these items. If one of these components are suspected of malfunctioning, a new replacement item must be obtained and installed.

Cleaning and Inspection

◆ See Figures 33 and 34

Wash all disassembled components and parts in solvent and blow them dry with low-pressure compressed air. Always use a lint free shop cloth when handling trim/tilt components.

Never use O-rings a second time. Always replace the O-rings with new ones. During assembling always lubricate new O-rings and seals with Evinrude/Johnson Power Trim/Tilt and Power Steering Fluid or GM Dexron II automatic transmission fluid.

Inspect the two nylon rings on the manual release valve for cuts or a split. If the nylon rings are damaged, a new manual release valve will have to be purchased. Check and replace the O-ring on the end of the valve.

Check the machined surfaces on the reservoir and cylinder body for nicks and scratches. Minor scratches may be removed with crocus cloth.

Inspect the bores in the trim and tilt cylinders for excessive scoring. If the bores are heavily scored, such a condition will result in excessive wear on new piston O-rings and wiper seals, causing internal leakage.

Clean the threaded end of the tilt rod with a wire brush to remove all traces of the nut locking agent. Keep the threads clean, dry and free of power trim fluid.

Check the nylon bushings in the end of the tilt cylinder for excessive wear. If the bushings are worn, replace them.

If the unit requires painting, wait until the unit is completely assembled. Plug all exposed fittings, ports, and electrical connector pins. Tape any exposed portions of the tilt and trim rods. Painting individual components is not recommended, because such action may allow paint chips to contaminate the fluid and possibly block the small hydraulic ports and valves.

Assembly

◆ See Figures 33 and 34

1. Lubricate a new tilt cylinder scraper seal and O-ring with power trim/tilt fluid. Insert the O-ring, and the scraper seal with the lip of the seal facing out. Place Scotch Tape over the threads on the piston end of the rod to protect the scraper seal from being damaged. Slide the end cap down onto the tilt rod with the threaded end facing up.

2. Place the tilt rod into a vise with the piston end facing up. Set the piston onto the end of the rod and a new O-ring. Insert the check ball, plunger and springs back into the piston bores, as noted during disassembly.

3. Apply a coating of Evinrude/Johnson Locquic Primer on the threads of the piston nut and the tilt rod threads and allow it to dry. Apply Evinrude/Johnson Nut Lock to the nut and rod, and then slide the washer over the piston and install the nut. Tighten the nut to 58-87 ft. lbs. (79-118 Nm).

4. Lubricate a new trim cylinder scraper seal and quad ring for the trim rod end cap with Power Trim/Tilt Fluid. Insert the quad ring and the scraper seal with the lip of the seal facing out.

5. Place the trim rod into a Rod Holder Tool (#983213) or similar device and clamp the unit in a vise. Slide the end cap onto the trim rod and thread the wear plate into the end of the rod. Tighten the wear plate to 84-108 inch lbs. (9.9-12 Nm). Remove the trim rod from the vise and holding fixture.

6. Lubricate and slide a new O-ring onto the outside groove of the trim rod end cap.

7. Lubricate a new O-ring and two split rings, and then slide the O-ring onto the piston. Insert one split ring on each side of O-ring. Position the ends of the split rings 180° apart. Repeat these steps for the other trim rod.

8. Place a new or clean filter into the pump manifold cavity.

9. Lubricate five new O-rings and install them into the cylinder body. Position the pump manifold against the cylinder body and align the three bolt holes. Install the three Allen head bolts and tighten them alternately until snug. Tighten the three bolts to 60-84 inch lbs.(7-9 Nm).

10. Lubricate and insert the manual release valve into the pump manifold. Thread the valve in until it is seated, and then install the snap ring into the end of the opening, using a pair of snap ring pliers.

11. Mount the cylinder body upright into a vise with soft jaws. Lubricate the O-rings and back-up rings on the trim rods and insert the trim rods into the trim cylinders. Fill the cylinder with Evinrude/Johnson Power Trim/Tilt Steering Fluid. Thread the end cap into the trim cylinder, using a universal spanner wrench or a Trim Cylinder End Cap tool (#436710). Tighten the trim cylinder end cap to a torque value of 60-70 ft. lbs. (81-95 Nm). Repeat this step for the other trim rod and cylinder.

12. Place a new O-ring on the piston and end cap of the tilt rod. Lower the tilt rod into the tilt cylinder bore. Fill the cylinder with Evinrude/Johnson Power Trim/Tilt Steering Fluid. Thread the end cap into the tilt cylinder, using a universal spanner wrench or a Tilt Cylinder End Cap tool (#326485). Tighten the end cap to 58-87 ft. lbs. (79-118 Nm).

13. Lubricate and slip a new O-ring over the lip on the pump motor. Fill the pump cavity with Evinrude/Johnson Power Trim/Tilt Steering Fluid and lower the motor onto the pump. Install the three bolts and tighten them to 35-52 inch lbs. (4-6 Nm). Attach the motor wire leads to the side of the motor with the clamp and screw.

14. Insert a new or clean filter into the cylinder body. Install the 6mm hex plug and tighten it securely. Lubricate a new O-ring and place it into the bore of the cylinder body. Align the reservoir with the cylinder body and secure it with three Allen head screws. Tighten the Allen head screws to 35-52 inch lbs.(4-6 Nm).

15. Install two nylon bushings in the end of the tilt cylinder and the end of cylinder body.

Trim/Tilt Assembly - Large-Motor Conventional Systems

REMOVAL & INSTALLATION

◆ See Figure 35

※※ WARNING

Anytime a steel tool is struck with a hammer, there is the possibility of chips flying which could cause serious eye injury. Therefore, wear safety glasses while removing the tilt cylinder pin.

1. Disconnect the negative battery cable for safety.

2. Matchmark the angle of the adjusting rod location, then remove the rod.

3. Manually lift the engine and engage the tilt support. For safety, install a holding strap or sturdy rope around the motor to make sure it will not accidentally drop if disturbed during service.

4. Disconnect the blue and green wires from the pump motor connector housing.

5. Using a punch and mallet, carefully remove the spring clip from the tilt cylinder pin.

8-18 TRIM AND TILT

6. Manually push the tilt piston downward into the cylinder for clearance.

7. Remove the bolts (usually 3 on each side) securing the assembly to the port and stern brackets.

8. Remove the trim/tilt unit from the stern brackets while pulling the cable through the bracket.

9. Refer to the Disassembly and Assembly procedures, along with the accompanying exploded views if overhaul or component replacement is necessary.

10. Keep the following points in mind when servicing the hydraulic trim/tilt assembly:

• Before disassembly, thoroughly clean and degrease the unit. All outer surfaces should be cleaned with a stiff synthetic needle (not wire) brush and hot soapy water. It is important to prevent dirt or debris from entering the unit during service.

• Before disassembling the unit, temporarily connect the motor wiring and run the unit until the trim rams and tilt piston are in the complete down position. Then, momentarily operate it in the reverse direction (upward) in order to equalize internal pressures. But, remember that there could be significant residual hydraulic pressure left behind some components, so always wear safety goggles and loosen fittings gradually, allowing pressure to bleed off before removal.

• NEVER apply heat to the cylinder body or cylinders as excessive heat can lead to high pressure leaks or component failure during operation.

• Never paint individual components while the unit is disassembled. The fear is that some portion of the paint might flake off and enter a hydraulic passage during assembly.

• Always use clean, lint-free shop cloth when handling assembly components.

• During assembly, replace all seals and O-rings to ensure proper, trouble-free operation.

• During pump assembly installation, position the pump and install the 3 bolts (2 Allen® head and 1 hex-head). Tighten each of the 3 bolts gradually (one turn at a time) alternating between the bolts to draw the pump evenly into position.

To install:

11. Apply a light coating of Evinrude/Johnson Triple-Guard, or equivalent marine grease to the tilt rod.

12. Position the trim/tilt assembly and insert the tilt cylinder pin.

13. Pull the wiring through the stern brackets.

14. Insert the spring clip to the tilt cylinder pin.

15. Apply a light coating of Evinrude/Johnson Nut Lock, or equivalent threadlock to the threads of the stern bracket-to-trim/tilt assembly manifold bolts. Install and tighten the bolts (usually 3 on each side) to 18-20 ft. lbs. (24-27 Nm).

16. Align the marks made during removal and install the angle adjusting rod in the same position from which it was removed.

17. If the tilt tube nuts were loosened, tighten them to 50-54 ft. lbs. (68-73 Nm), then back off one nut 1/4 turn.

18. Reconnect the negative battery cable, then run the engine up and down through several cycles, checking for proper operation and proper fluid level.

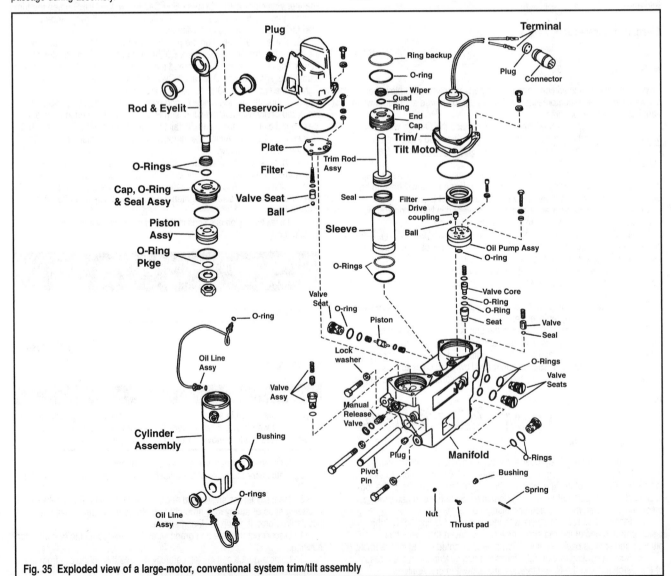

Fig. 35 Exploded view of a large-motor, conventional system trim/tilt assembly

TRIM AND TILT

DISASSEMBLY

◆ See Figure 35

This section contains complete detailed procedures to disassemble the complete unit. However, open the system and remove only the necessary parts to inspect, replace, and restore the system to satisfactory service.

■ TAG EVERYTHING as it is removed, in order to ensure proper assembly.

Drain the system by removing the reservoir plug and draining the hydraulic fluid into a suitable container. Observe any local restrictions on the disposal of this type material.

■ A holding bracket (fixture) can be quickly and easily made from scrap plywood to hold the trim/tilt unit while service work is being performed. The drawing at the top of the previous column gives dimensions and a rough plan. The illustration at the bottom of the same column shows the unit mounted in such a fixture.

Valves and Pistons

1. Remove the manual release valve by first using a pair of snapring pliers to remove the retaining ring. Next, use the trim switch to run the motor fully down and loosen the manual release valve one full turn. Tap the trim switch up and down a few times, then slowly remove the valve.
2. Tag and remove each of the external (outside) valves using a proper size drag link socket to fit the slot properly.
3. The letdown control piston can be reached by first removing the impact letdown valve, and then carefully removing the piston with a pair of needle nose pliers.
4. The pump control piston and springs are removed by first removing the reverse lock check valve and then lifting the piston and springs free with a pair of needle nose pliers. The pump control piston can only be removed from the aft end of the hole.

Reservoir and Valves

1. The reservoir is removed by first removing the upper and lower hydraulic lines and fittings. Next, remove the screws from the reservoir flange and lift the reservoir free.
2. Remove and discard the O-ring. If further disassembly of the reservoir is desired, hold down on the reservoir manifold plate and at the same time remove the three screws securing the plate. Lift the plate free.
3. Remove the relief valve and impact sensor valve assemblies by lifting them free of the body with a pair of needle nose pliers. The filter valve may be lifted out with a small stiff hook.
4. Because damage usually occurs during removal, the filter valve seat and O-ring must be replaced, if they are removed.

Motor and Pump

1. Remove the screws on the motor flange and lift the motor free. Remove and discard the O-ring.
2. Remove the hydraulic pump filter. The hydraulic pump may be removed by simply removing the attaching screws and lifting the pump free. Lift the trim down pump relief valve free of the body.
3. The expansion relief valve core can now be lifted out with a small hook shaped piece of wire.

Trim Cylinders

1. Obtain special tool Trim Cylinder End Cap Remover (#324958) or an adjustable spanner wrench. Use the special tool or the spanner wrench to remove the end cap.
2. Lift the trim piston assembly free of the cylinder. The sleeve fits snugly. Therefore, Trim Sleeve Remover (#325065) or a screwdriver with the tip bent 90° must be used to remove the sleeve. Slip the tool in under the sleeve, and then remove the sleeve.

Tilt Cylinder

Disconnect the upper and lower hydraulic lines at the tilt cylinder and at the manifold. Push the cylinder pivot pin to one side and remove the cylinder

■ Two types of tilt cylinders are normally found on these motors, one manufactured by Showa and the other by Prestolite. Identification is important due to minor differences in service. The Showa tilt cylinder is identified by an "S" after the part number stamped on the side of the cylinder. The Prestolite cylinder is identified with a "P" stamped on the bottom side by the part number. When ordering replacement parts, be sure to identify the specific cylinder being serviced.

1. Clamp the cylinder in a vise at the flats of the cylinder end. Obtain and use special End Cap Remover (#326485) with a 1/2 in. breaker bar to loosen the end cap assembly. Removal of the end cap is not an easy task, even by using the special tools mentioned.
2. After the end cap is removed, the piston and rod assembly may be withdrawn free of the cylinder.
3. The piston contains four valves. These valves cannot be serviced separately. If the valves are worn or are not functioning properly, the piston must be replaced as an assembly.
4. For Presolite pistons, remove and discard the O-ring from around the piston. Heat with a torch, a vise and bar through the rod end will be necessary to break the Locktite bond on the thread of the piston and piston rod. Clamp the piston securely in a vise, apply the heat, and then use the bar through the rod end to unscrew the rod from the piston. Clean all traces of old sealant from the piston rod threads.
5. For Showa pistons, clamp the piston rod end in a vise and remove the nut securing the piston to the rod. Clean all traces of sealant from the piston rod threads.

Cleaning and Inspection

◆ See Figure 35

Discard all used O-rings and seals. Clean all parts in solvent and blow them dry with compressed air. Inspect the cylinders and sleeves for any sign of excessive wear or scoring. Inspect all parts for dirt, chips, and damage. Replace any damaged valve seats or other questionable parts.

ASSEMBLY

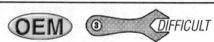

◆ See Figure 35

Use the accompanying exploded views, along with the following procedures to reassemble the trim/tilt assembly.
Lubricate all internal parts with Evinrude/Johnson Power Trim/Tilt Fluid prior to assembling.

Tilt Cylinder

1. Lubricate a new O-ring and seal with Evinrude/Johnson Power Trim/Tilt Fluid, and then install both into the cylinder end cap. Install a Seal Protector (#326005) onto the threads of the piston rod and then install the end cap onto the rod. If the seal protector is not available, wrap the threads with tape as a protection against damaging the seal when the end cap is installed to the rod. Remove the tape after the end cap is installed.
2. If assembling a Prestolite cylinder (which was disassembled), clean the piston rod threads with Evinrude/Johnson Locquic Primer (#384884), or equivalent. Clamp the rod in a vise holding the pin end. Coat the rod threads with Evinrude/Johnson Ultra Lock (#388517), or equivalent. Thread the piston assembly onto the rod. Take care not to damage the surface of the piston. Use a flywheel strap wrench to bring the piston up snug.
3. If assembling a Showa cylinder (which was disassembled), clean the piston rod threads with Evinrude/Johnson Locquic Primer (#384884) and install the small O-ring, washer, and piston onto the piston rod. The small holes on the piston must face upward. Apply Evinrude/Johnson Nut Lock to the piston rod threads. Secure the rod in a vise and then install and tighten the nut to a torque value of 58-87 ft. lbs. (79-118 Nm).
4. Lubricate the piston assembly. Install a new O-ring to the outside diameter of the piston. Carefully insert the piston assembly into the cylinder. Tighten the end cap assembly using an End Cap Remover (#326485), or an adjustable spanner wrench.
5. If the band has been removed, slide it into the cylinder. Use a new O-ring on the fitting and screw the fitting into the band, with the pilot on the fitting indexing into the hole in the cylinder.
Lubricate the cylinder pivot pin and pivot pin bushings with Evinrude/Johnson Triple-Guard Grease. Install the cylinder to the manifold assembly. Attach the hydraulic line fitting on the starboard side.

8-20 TRIM AND TILT

Trim Cylinder

1. Insert the piston and rod assembly into the end cap. Carefully slide the piston into the trim cylinder until the piston and cylinder sleeve are butted against the end cap.
2. Slide the assembled unit into the cylinder cavity. Obtain an End Cap Remover (#324958) or an adjustable spanner wrench and tighten the end cap to 30-40 ft. lbs. (40-54 Nm).

Reservoir and Valves

1. To replace a valve, first install the valve seat. Next install the ball, core, and spring in that order. Install the filter valve ball, and then insert the filter valve seat with the O-ring end facing up.
2. Slide the filter into the manifold plate. Shift the manifold plate until the filter is on top of the filter valve and the attaching screw holes are aligned. Thread the screws into place, and then tighten them alternately and evenly to keep the valve spring positioned properly.
3. Place a new O-ring in position. Secure the reservoir in place with the attaching screws. Tighten the screws securely.

Pistons and Valves

1. Slide the letdown control piston into the starboard cavity with the rounded end going in first. Install the impact letdown valve. Slide the pump control piston into the port cavity from the rear with the small end going in first.
2. Install the reverse lock check valve and the trim check valve. Install the tilt check valve and the trim/tilt separation valve (long body valve). Tighten the valve securely.
3. Install the manual release valve with a new O-ring. Tighten the valve to provide operator with shock absorber protection. Install the snap ring with the flat side facing out.

Pump Assembly and Motor

1. Install the pump relief valve and spring. Install a new O-ring. Check to be sure the pump drive tang indexes with the hole directly opposite the round locating boss.
2. Install the pump with the locating boss indexed into the pump cavity recess. Secure the pump in place with the three attaching screws. Tighten the screws securely. Install the pump filter and fill the filter cavity and the area over the pump with Evinrude/Johnson Power Trim/Tilt hydraulic fluid.

■ **The pump cavity must be filled with hydraulic fluid during assembling, or the unit will not operate.**

3. Install a new O-ring onto the trim motor. Install the motor and at the same time rotate the motor shaft until the shaft engages with the pump shaft. Position the motor with the cable on the port side of the assembly. Install the three attaching screws. Tighten the screws securely.
4. Fill the reservoir with Evinrude/Johnson Power Trim/Tilt Fluid and purge the system of air.

Trim/Tilt Assembly - Small-Motor Conventional Systems

REMOVAL & INSTALLATION

◆ See Figures 36 and 37

1. Tilt the outboard to the fully tilted position, the engage the tilt lock to secure the motor. For safety, install a holding strap or sturdy rope around the motor to make sure it will not accidentally drop if disturbed during service.
2. Disconnect the negative battery cable for safety.
3. Disconnect the blue and green wires from the pump motor connector housing.
4. Remove the spring clip from the end of the cylinder pin.
5. Using a slide-hammer and an adapter (such as OMC #340624) remove the pin from the stern bracket and assembly.
6. Remove one of the 3/4 in. locknuts from the angle adjustment rod, then remove the rod from the stern brackets.
7. Pull the trim/tilt assembly away from the stern brackets to provide access to disconnect the ground lead. Remove ground from the motor mounting screw and then remove the assembly completely from the outboard.
8. Refer to the accompanying exploded views if disassembly or component replacement is necessary.
9. Keep the following points in mind when servicing the hydraulic trim/tilt assembly:

• Before disassembly, thoroughly clean and degrease the unit. All outer surfaces should be cleaned with a stiff synthetic needle (not wire) brush and hot soapy water. It is important to prevent dirt or debris from entering the unit during service.

• Before disassembling the unit, temporarily connect the motor wiring and run the unit until the trim rams and tilt piston are in the complete down position. Then, momentarily operate it in the reverse direction (upward) in order to equalize internal pressures. But, remember that there could be significant residual hydraulic pressure left behind some components, so always wear safety goggles and loosen fittings gradually, allowing pressure to bleed off before removal.

• NEVER apply heat to the cylinder body or cylinders as excessive heat can lead to high pressure leaks or component failure during operation.

• Never paint individual components while the unit is disassembled. The fear is that some portion of the paint might flake off and enter a hydraulic passage during assembly.

• Always use clean, lint-free shop cloth when handling assembly components.

• During assembly, replace all seals and O-rings to ensure proper, trouble-free operation.

• During pump assembly installation, position the pump and thread the bolts, then use the bolts to draw the pump evenly in position. Tighten each bolt, one turn at a time, alternating between the bolts to press the pump evenly into the housing.

To install:

10. Apply a light coating of OMC Triple-Guard, or equivalent marine grease to the trim/tilt rod, the angle adjustment rod and to the thrust rod bushings.
11. If removed, install the bushings.
12. Connect the ground lead, then position the trim/tilt assembly between the stern brackets.
13. Install the angle adjustment rod, then tighten the locknut(s) to 20-25 ft. lbs. (27-34 Nm).
14. Connect the pump motor wires.
15. Align the cylinder with the swivel bracket and secure using the cylinder pin. Secure the pin using the spring clip.
16. Reconnect the negative battery cable, run the engine up and down through several cycles and then check for proper operation and proper fluid level.

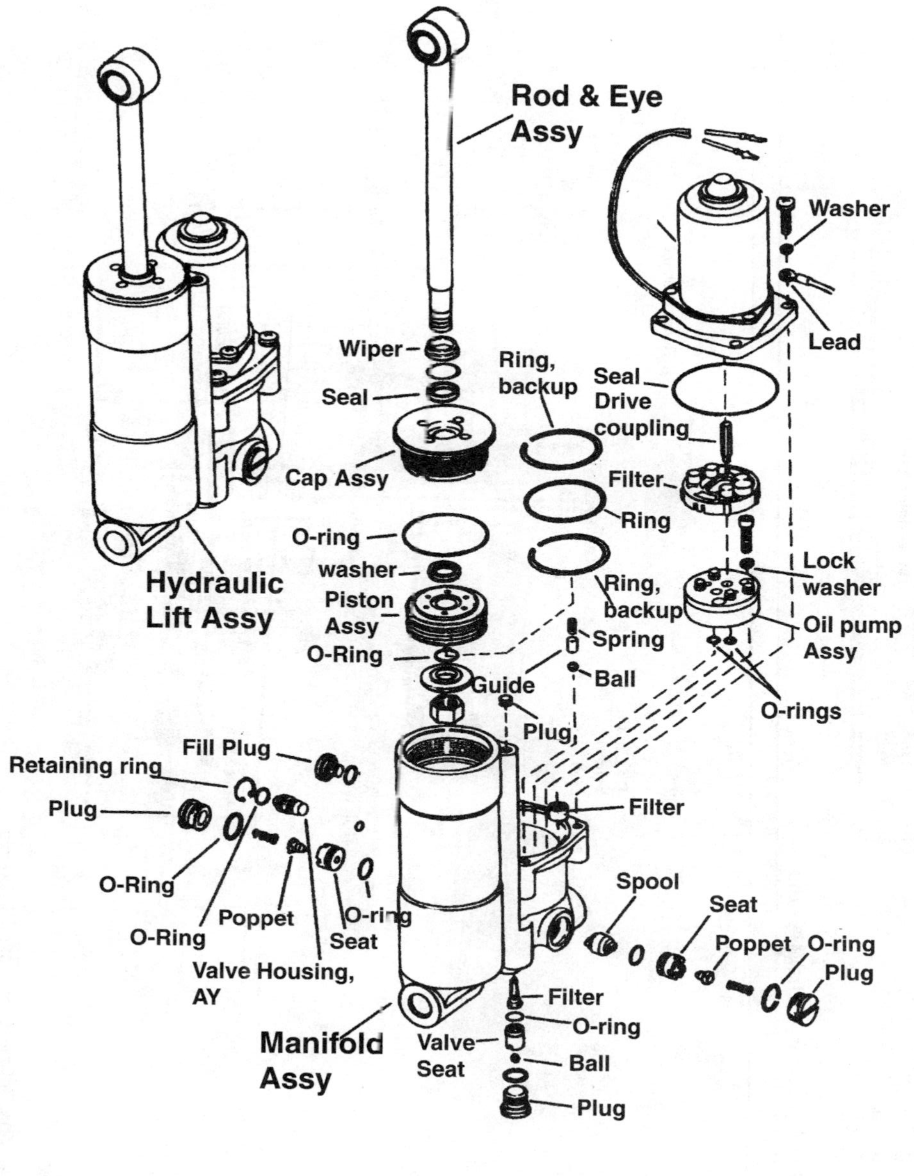

TRIM AND TILT 8-21

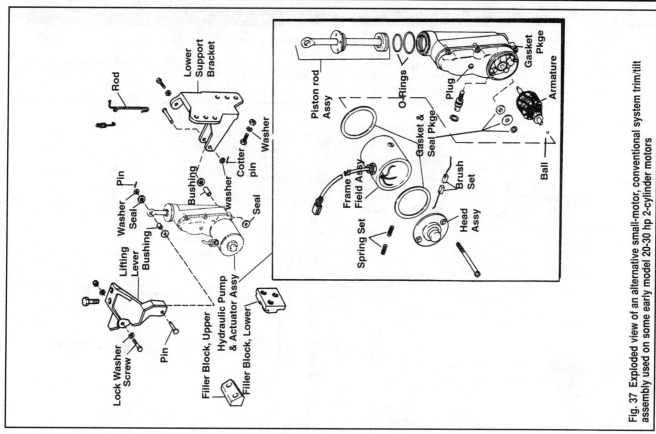

Fig. 37 Exploded view of an alternative small-motor, conventional system trim/tilt assembly used on some early model 20-30 hp 2-cylinder motors

Fig. 36 Exploded view of a typical small-motor, conventional system trim/tilt assembly (see next illustration for alternative system used on some 20-30 hp twins)

8-22 TRIM AND TILT

TRIM/TILT WIRING

For power trim/tilt systems installed at the factory as part of original motor equipment, details on the exact system used for that specific outboard are included in the individual engine schematics found under Wiring Diagrams in the Ignition and Electrical Systems section. However, most wiring colors and connectors should be standardized on Johnson/Evinrude outboards and the accompanying system diagrams should also be applicable to all, but the EFI motors.

For all motors where the OE trim/tilt system was added during rigging, use the accompanying diagrams.

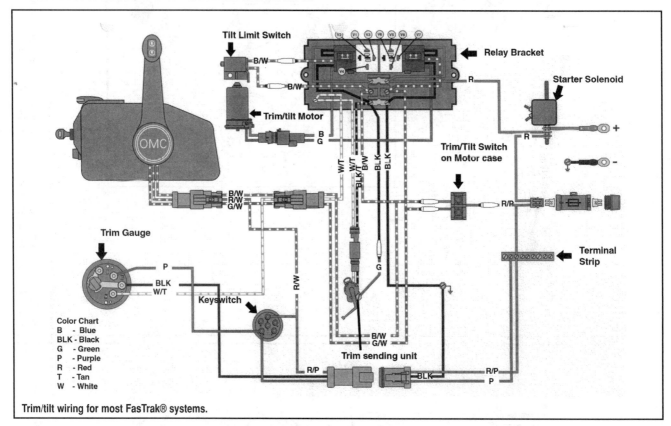

Trim/tilt wiring for most FasTrak® systems.

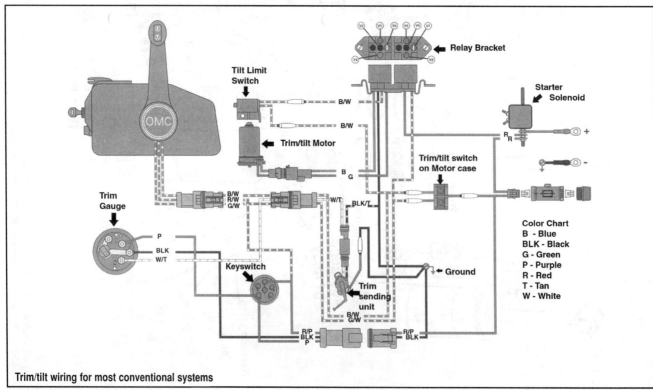

Trim/tilt wiring for most conventional systems

CONTROL CABLES	9-14
RIGGING	9-14
REMOTE CONTROLS	**9-2**
CONCEALED SIDE MOUNT REMOTE	9-9
Disassembly & Assembly	9-10
Replacing The Trim/Tilt Switch	9-9
DESCRIPTION & OPERATION	9-2
Accessory Connector	9-3
Control Handle	9-2
Emergency Stop Switch	9-3
Fast Idle Lever	9-3
Key Switch And Primer	9-2
Neutral Lockout	9-2
Neutral Start Switch	9-3
Throttle Friction Adjustment Screw/Knob	9-3
Trim/Tilt Switch	9-3
Warning Horn	9-3
DUAL HANDLE SURFACE MOUNT REMOTE FOR SINGLE MOTORS	9-7
Overhaul	9-7
INTRODUCTION	9-2
PRE-WIRED BINNACLE MOUNT REMOTE	9-11
Overhaul	9-11
SMALL MOTOR STANDARD SURFACE REMOTE	9-6
Overhaul	9-6
STANDARD SURFACE MOUNT REMOTE	9-3
Overhaul	9-3

9

REMOTE CONTROLS

REMOTE CONTROL	9-2
CONTROL CABLES	9-14

9-2 REMOTE CONTROLS

REMOTE CONTROLS

Introduction

◆ See Figures 1 and 2

A remote control unit is seldom sold with just an outboard unit. In most cases, the control box is sold separately as an option or it is included with a "package" deal - boat, outboard, control box, and trailer.

If the control box was included in the "package," the unit will most likely be one of the latest production models from OMC at the time - with Johnson or Evinrude colors and decals. But, the final decision on what control unit is mounted rests in the hands of the boat manufacturer and the dealer that performs the initial rigging. For this reason, caution must be used when following the procedures contained here to ensure that they apply to unit installed on the boat. Newer motors rigged to older boats (and vice versa) could be equipped with aftermarket or older/newer units that are not detailed here for obvious reasons. We've included procedures for the various Johnson/Evinrude remotes that were produced and sold at the same time these motors were manufactured.

This section covers overhaul (disassembly and assembly procedures) which includes removal of the unit from the boat, separating the two halves and replacement of switches/warning horns contained within. We also provide a few additional words on lubrication.

Like with many marine components, non-use is absolutely the greatest enemy of the control unit. The large number of eccentrics, cams, levers, linkages, etc. should be operated at regular intervals - as often as once a month and the interior parts lubricated whenever the components begin to show signs of stiffness and binding.

■ WOULD YOU BELIEVE, as much as 90% of steering and shifting problems are directly caused by the system simply not being operated. Without movement, steering and shifting cables and linkages have a tendency to "freeze." Would you also believe, service shops report that well over 50% of boat cables replaced every year is due to lack of movement.

Perhaps the most important thing you can do to preserve the function of your remote control unit is to use it. Therefore, during off-season, when the boat is laid up in a yard, or on a trailer alongside the house, take time to go aboard and operate the steering from hard-over to hard-over. Also, shift the remote control unit through the full range several times to ensure corrosion does not develop causing a fitting or joint to "freeze" preventing proper movement.

Description & Operation

The following components and features may be incorporated in the OMC remote control box (such as the one pictured in the photos for this section) and are usually associated with the larger OMC outboard units covered in this manual.

Locations of the various parts in or on the control box are identified within the accompanying illustrations.

The function of the item is usually described in the name.

CONTROL HANDLE

As the name implies, this handle controls the gear position of the lower unit - Neutral, Forward, and Reverse. It also controls the powerhead rpm. From the vertical position (straight up), movement about 32° forward shifts the lower unit into forward gear and movement of the handle about 32° aft of vertical shifts the unit into reverse gear. Further movement past the 32° position in either forward or reverse will increase powerhead rpm.

KEY SWITCH AND PRIMER

◆ See Figure 3

Rotating the key switch clockwise to the first detent energizes all powerhead accessories. Further movement clockwise to the second detent will energize the solenoid to activate the cranking motor. Pushing the key inward at the first or second detent position will energize the fuel primer solenoid to choke the carburetor.

Rotating the key back to the full counterclockwise position will cut power to the ignition system and all powerhead-controlled accessories.

NEUTRAL LOCKOUT

This safety feature prevents the control handle from moving to the forward or reverse positions when the control handle is in the neutral position.

With the outboard operating, the helmsperson simply depresses the knob upward against the handle to release the handle for movement - either forward or aft. The neutral lockout will automatically engage when the control handle is returned to the neutral position.

THROTTLE FRICTION ADJUSTMENT SCREW/KNOB

The throttle friction adjustment screw/knob does exactly what the name implies - places friction on the control handle to prevent unwanted "creep" of

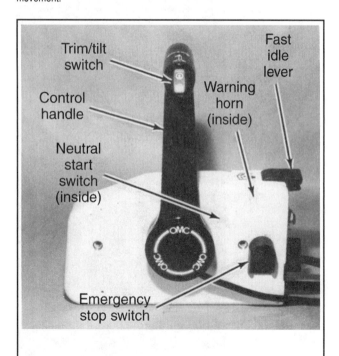

Fig. 1 Typical Johnson/Evinrude side-mount remote control unit

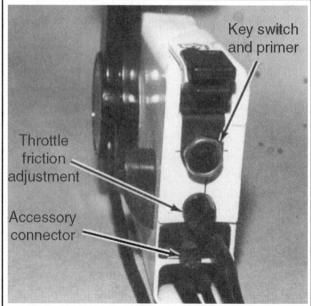

Fig. 2 Some Johnson/Evinrude remote control units incorporate the keyswitch and primer on the housing instead of utilizing a separate switch on the dash

REMOTE CONTROLS 9-3

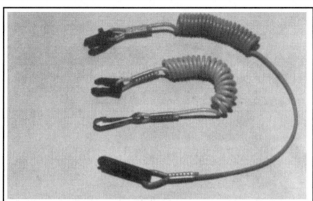

Fig. 3 Two different emergency switch lanyards used with the Johnson/Evinrude remote control units. Remember - on some models, it is not possible to start the powerhead if the lanyard is not in place

the control handle while the powerhead is operating and/or the boat is underway. Rotating the knob clockwise will increase friction on the control handle. Logically, rotating the knob counterclockwise will decrease friction and the handle movement will be more "free."

NEUTRAL START SWITCH

Again, the function of this switch is obvious - to prevent the cranking motor circuit from being energized except when the control handle is in the neutral position. On most Johnson/Evinrude remote controls, this is a mechanical switch. Stated another way, the start switch will only allow the key switch to be rotated to the start position when the control handle is in the neutral position. On other remotes, this switch is electrical, breaking the circuit regardless of keyswitch position unless the control lever is in Neutral.

EMERGENCY STOP SWITCH

This switch prevents operation of the powerhead unless a safety lanyard is in place holding the switch button in an outward position. The other end of the lanyard should be attached to an item of clothing or the life jacket worn by the helmsperson. Should the individual be thrown overboard or away from the control station the clip will be pulled from the switch and the powerhead will immediately shut down.

WARNING HORN

This audible warning device is intended to alert the operator that a critical operating condition has developed; the powerhead should be shut down; and the cause determined before extensive and expensive damage is done.

Depending on the powerhead being serviced, the warning device may emit:
- Continuous Tone indicating a powerhead overheat condition.
- Continuous Short Pulse Tone indicating a no oil condition - bad news - very bad news.
- Short Pulse Tone - every 20-seconds indicating a low oil condition.
- Continuous Tone - at or near Wide Open Throttle (WOT) indicating a restriction in the fuel supply.

FAST IDLE LEVER

When equipped, this lever controls powerhead rpm when the control handle is in the neutral position. The lever should be raised to assist during powerhead startup and favorable idle speed. The lever must be returned to the run position before moving the control handle out of the neutral position.

ACCESSORY CONNECTOR

The accessory connector permits easy access to the powerhead accessory and tachometer circuits. Several OMC tachometers and wiring kits provide a mating connector simplifying accessory installation. Maximum draw on the accessory circuit must not exceed 5 amps.

TRIM/TILT SWITCH

This switch is included on control boxes with power trim/tilt systems installed. The switch permits the helms-person to raise or lower the outboard through about 0-21° while the unit is operating in Forward gear. When the powerhead is shut down or is operating below 1500 rpm, this same switch allows and controls the outboard to tilt about 22-75°.

Standard Surface Mount Remote Units

OVERHAUL

◆ See Figures 4 thru 12  MODERATE

Refer to the accompanying illustration for parts identification.

1. Disconnect the battery cables for safety. This will help prevent the possibility of potential burns or shorts that could occur while working on the control unit.

✱✱ CAUTION

Disconnecting the battery cables will also make sure that the engine is not accidentally cranked or started (which could lead to injury or damage).

2. Loosen the retainers and separate the remote from the boat with the cable still attached.

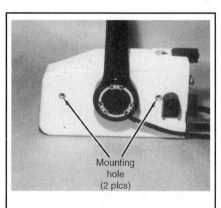

Fig. 4 Standard remote housings are normally mounted using 2 screws

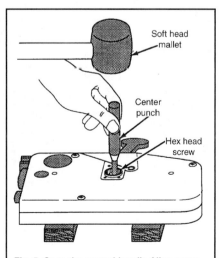

Fig. 5 Once the control handle Allen screw is loosened, dislodge the handle splines as shown

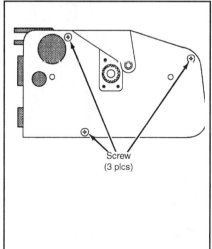

Fig. 6 Remove the housing-to-cover screws...

9-4 REMOTE CONTROLS

3. Place the control housing so the control handle side (the face of the housing) is positioned downward on small wooden blocks. Position the blocks on either side of the handle, contacting the housing face.

4. From the back of the housing (now facing upward), loosen the Allen head screw (for the remote handle) 3 complete turns.

■ The remote control Allen head screw can be found at the base of the handle. It is accessed through the small round bore on the back of the remote housing.

5. Place a punch inside the head of the Allen screw, then gently tap on the punch using a soft-faced mallet to dislodge the control handle splines from the hub splines.

6. Finish unthreading the Allen screw, then remove the remote handle.

7. Remove the housing-to-housing cover screws (3) from the back of the housing. Carefully separate the housing and cover.

8. Remove the shift cable pin from the shift control clevis (at the bottom of the shift lever). Disconnect the shift cable.

9. Lift the throttle cable trunnion from the housing pocket and then pull back on the cable to expose the pin. Remove the cable and pin from the throttle lever.

10. Remove the screw and retainer plate hooked into the top of the housing assembly.

11. Remove the friction adjustment lever along with the knob/screw assembly.

12. Remove the screws (2) securing the neutral start switch and then remove the switch from the housing.

✱✱ CAUTION

The detent spring is compressed and may fly free when removing the shift and throttle plate from the housing. In order to prevent the possibility of injury, be sure to wear safety glasses.

13. Slowly lift the shift and throttle plate from the housing.
14. Remove the detent roller, shoe and spring.
15. Lift the shift lever assembly from the control housing.
16. Remove and discard the flat-head screw at the center of the throttle lever assembly. Remove the countersunk washer from the lever as well, then lift the throttle lever assembly and spacer from the housing.
17. Remove the screw and locknut fastening the shift lockout cam to the fast idle lever. Discard the locknut, as it must be replaced once it is removed.
18. Remove the keeper plate screw at the top of the housing (directly above the shift lockout lever). Lift the shift lockout cam and lever from the housing and then remove the spring washer and fast idle lever from the housing.
19. If necessary, remove the ignition keyswitch assembly for replacement.
20. If control handle disassembly is desired (such as for access to the trim/tilt switch), proceed as follows:

   a. Remove the screws (2) fastening the slide control lever plate to the base of the handle.
   b. Lift the neutral lock slide out of the control handle, then remove the screws (3) fastening the handle cover and the one screw fastening the handle knob. Remove the cover and knob.
   c. Remove the screw securing the trim/tilt switch to the cover and then remove the switch assembly.

21. Clean and inspect the remote control components as follows:

   a. Wipe all metallic, non-electrical, parts with a clean rag soaked in a mild solvent. Electrical and plastic components should be cleaned using a dry rag and/or low pressure compressed air.
   b. Check the flange bosses inside the shift lockout cam for signs of rounding at the edges. If found, the cam should be replaced.
   c. Check the slots inside the fast idle lever for wear and replace the lever, if necessary.
   d. Visually check the balance of the mechanical components for signs of cracks, damage or excessive wear. Replace any worn or damaged components to ensure proper remote operation.

22. Assembly is essentially the reverse of disassembly. Refer to the accompanying illustration for critical torque values and lubrication points. Also, be sure to pay particular attention to the following:

   a. Sliding surfaces and pivot points must be properly lubricated using OMC Moly Lube, or an equivalent assembly grease to ensure long-life and trouble/binding-free operation.
   b. Apply a light coating of OMC Screw Lock or equivalent threadlock to the threads of the screws noted in the accompanying illustration.
   c. All wiring must be connected and secured at the points noted during removal. Wiring must be routed in a manner that will prevent contact with moving components and must not be pinched between the housing and cover during assembly.

Fig. 7 . . . then separate the housing from the cover

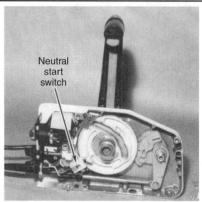

Fig. 8 Once separated, components such as the neutral start switch...

Fig. 9 . . . ignition keyswitch and warning horn are accessible

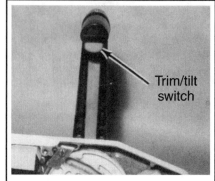

Fig. 10 The trim/tilt switch is inside the handle, above the neutral lockout slider

REMOTE CONTROLS

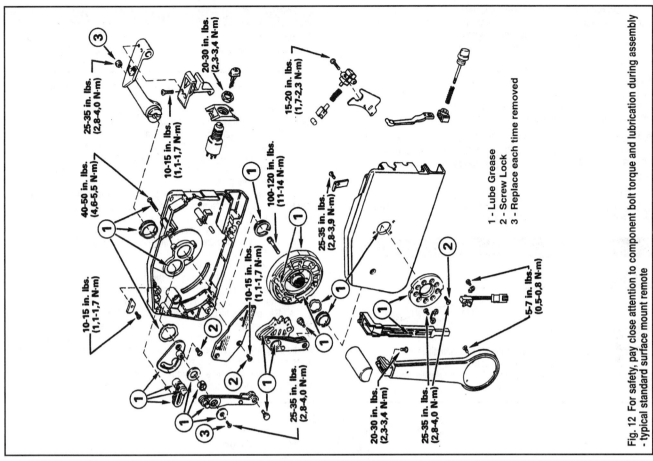

Fig. 12 For safety, pay close attention to component bolt torque and lubrication during assembly - typical standard surface mount remote

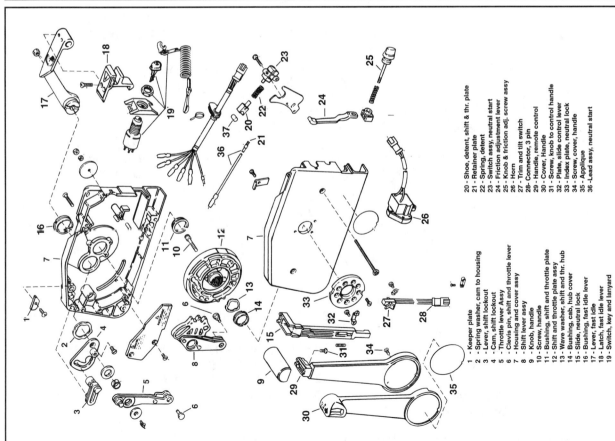

Fig. 11 Exploded view of a typical Johnson/Evinrude standard surface mount remote that has been pre-wired for a keyswitch and the System Check® monitor wiring.

9-6 REMOTE CONTROLS

d. **Do not** over or under tighten the fasteners for the remote housing components. Loose fasteners could allow components to shift come loose or worse, wedge stuck in a given position, this could lead to loss or throttle or shift control. Similarly over-tightened fasteners may lead to binding or breaking components.

e. Once assembly is complete, slowly move the remote handle through the range of operation. Feel for smooth and free movement. Adjust the friction screw/knob assembly so the throttle does not change in response to vibration from engine or boat operation. Do not set the friction adjuster so tight as to bind the control handle causing jerky shift/throttle adjustments.

■ When adjusting the friction screw/knob, turn the knob CLOCKWISE to increase friction or COUNTERCLOCKWISE to decrease friction. NEVER over-tighten the screw, which would lock the throttle in position.

f. Refer to the Timing and Synchronization adjustments in the Maintenance and Tune-Up section for information on additional information on control cable adjustment and lubrication.

Small Motor Standard Surface Remote Units

OVERHAUL

◆ See Figures 13 and 14

Refer to the accompanying illustration for parts identification.
1. Disconnect the battery cables for safety. This will help prevent the possibility of potential burns or shorts that could occur while working on the control unit.

✷✷ CAUTION
Disconnecting the battery cables will also make sure that the engine is not accidentally cranked or started (which could lead to injury or damage).

2. Loosen the retainers and separate the remote from the boat with the cable still attached.
3. Loosen the flat-head screws (usually 4), then remove the control housing cover (back plate) from the back of the housing.
4. Place the control handle in **neutral**.
5. Remove the shift and throttle cable trunnion plates (the plates that fasten the cable housings to the end of the remote housing).
6. Disconnect the shift and throttle cables from the remote assembly by removing the cotter pins.
7. If equipped, remove the cotter pin from throttle friction adjustment screw. Remove the adjustment assembly retaining screws and then remove the assembly.
8. If equipped, and if necessary for service, remove the neutral switch.
9. Carefully pry the cover out from the base of the remote handle, then remove the handle bolt.
10. Remove the neutral lock plate.
11. Loosen the bolt fastening the throttle lever, then remove the lever with the screw, washer, spring and arm. Remove the driveshaft.
12. If necessary, disassemble the internal drive components as follows:
   a. Remove the friction adjustment plug and then carefully separate the decal from the handle side of the remote housing assembly.
   b. Hold the drive components in position (by pushing on the shift and throttle support) and remove the retaining screws (3) on the outside of the cover, on the decal mounting surface.
   c. Remove the shift and throttle support, the shift lever gear and pin, drive gear, neutral lock ball, detent roller and leaf springs.

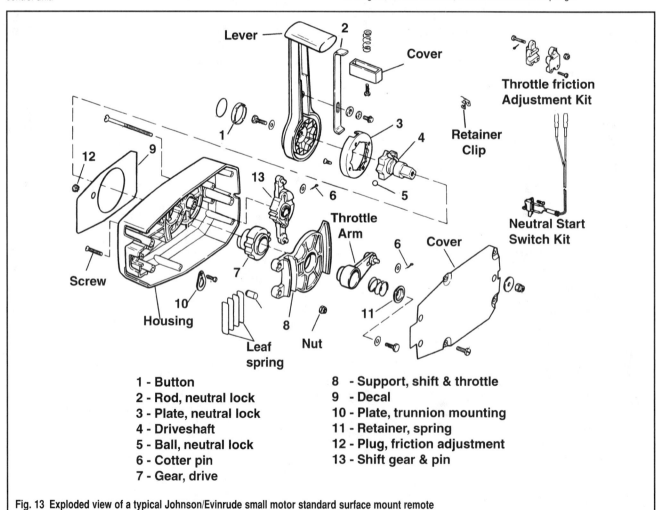

1 - Button
2 - Rod, neutral lock
3 - Plate, neutral lock
4 - Driveshaft
5 - Ball, neutral lock
6 - Cotter pin
7 - Gear, drive
8 - Support, shift & throttle
9 - Decal
10 - Plate, trunnion mounting
11 - Retainer, spring
12 - Plug, friction adjustment
13 - Shift gear & pin

Fig. 13 Exploded view of a typical Johnson/Evinrude small motor standard surface mount remote

REMOTE CONTROLS 9-7

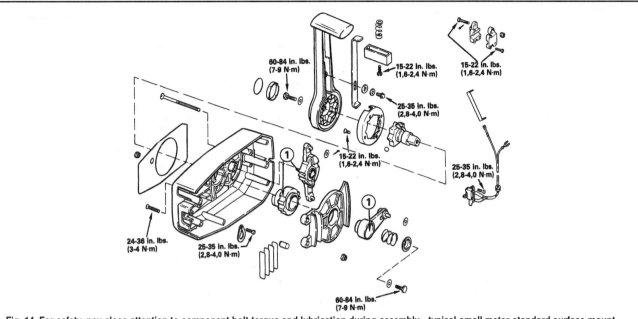

Fig. 14 For safety, pay close attention to component bolt torque and lubrication during assembly - typical small motor standard surface mount remote

13. Clean and inspect the remote control components as follows:
    a. Wipe all metallic, non-electrical, parts with a clean rag soaked in a mild solvent. Electrical and plastic components should be cleaned using a dry rag and/or low pressure compressed air.
    b. Visually check all mechanical components for signs of cracks, damage or excessive wear. Replace any worn or damaged components to ensure proper remote operation.
14. Assembly is essentially the reverse of disassembly. Refer to the accompanying illustration for critical torque values and lubrication points. Also, be sure to pay particular attention to the following:
    a. Sliding surfaces and pivot points must be properly lubricated using OMC Moly Lube, or an equivalent assembly grease to ensure long-life and trouble/binding-free operation.
    b. **Do not** over or under tighten the fasteners for the remote housing components. Loose fasteners could allow components to shift come loose or worse, wedge stuck in a given position, this could lead to loss or throttle or shift control. Similarly over-tightened fasteners may lead to binding or breaking components.
    c. The shift and throttle support is retained by 3 locking nuts, replace any of the nuts that have lost their locking action.
    d. When installing the throttle lever (using the spring, washer and screw), be sure to check the patch lock screw and replace it, if it has lost its locking action. Also, be sure that the throttle lever points away from the trunnion pockets.
    e Apply a very light coating of OMC Adhesive M or equivalent to hold the decal to the handle side of the control housing.
    f. The neutral lock plate was installed with thread-forming screws. The safe way to install them is to turn each of the thread-forming screws backwards (counterclockwise) until they just drop into the original threads, and then tighten them by turning clockwise.
    g All wiring must be connected and secured at the points noted during removal. Wiring must be routed in a manner that will prevent contact with moving components and must not be pinched between the housing and cover during assembly.
    h. Once assembly is complete, slowly move the remote handle through the range of operation. Feel for smooth and free movement. Adjust the friction screw so the throttle does not change in response to vibration from engine or boat operation. Do not set the friction adjuster so tight as to bind the control handle causing jerky shift/throttle adjustments.

■ The friction screw is located under the small cover on the housing, just in front of the remote handle on starboard installations, or just behind the remote handle on port installations. Turn the screw CLOCKWISE to increase friction or COUNTERCLOCKWISE to decrease friction. NEVER over-tighten the screw, which would lock the throttle in position.

    i. Refer to the Timing and Synchronization adjustments in the Maintenance and Tune-Up section for information on additional information on control cable adjustment and lubrication.

Dual Handle Surface Mount Remote Units for Single Motors

OVERHAUL

◆ See Figures 15 and 16

MODERATE

Refer to the accompanying illustration for parts identification.
1. Disconnect the battery cables for safety. This will help prevent the possibility of potential burns or shorts that could occur while working on the control unit.

✶✶ CAUTION

Disconnecting the battery cables will also make sure that the engine is not accidentally cranked or started (which could lead to injury or damage).

2. Loosen the retainers and separate the remote from the boat with the cable still attached.
3. Remove the 2 screws from the lower corners of the housing, then carefully separate the cover and housing, while removing the control housing spacer.
4. Remove the screw and washer securing the shift lever
5. Lift the lever from the cover, remove the bushing from the lever and remove the plastic spring and washer from the lever hub.
6. Remove the screw and locknut fastening the shift cable to the casing guide and then remove the cable. To separate the casing guide from the shift lever, gently push the insert out.
7. Loosen the throttle friction adjustment knob.
8. Remove the screw and washer that secure the throttle lever assembly.
9. Gently lift the throttle lever from the housing, then remove the bushing from the lever and the two washers from the hub.
10. Remove the screw and locknut fastening the throttle cable in the casing guide and then remove the cable. To separate the casing guide from the throttle lever, gently push the insert out.
11. Fully unscrew the throttle friction adjustment knob, then pull the throttle friction detent arm (with spring and bushing) from the housing.

9-8 REMOTE CONTROLS

12. Remove the idle stop lever adjustment screw, plastic washer, lever and bushing from the housing.
13. Fully unscrew the idle stop adjustment knob, then remove the retainer and spring from the housing.
14. Clean and inspect the remote control components as follows:
    a. Wipe all metallic, non-electrical, parts with a clean rag soaked in a mild solvent. Electrical and plastic components should be cleaned using a dry rag and/or low pressure compressed air.
    b. Visually check all mechanical components for signs of cracks, damage or excessive wear. Replace any worn or damaged components to ensure proper remote operation.
15. Assembly is essentially the reverse of disassembly. Refer to the accompanying illustration for critical torque values and lubrication points. Also, be sure to pay particular attention to the following:

    a. Sliding surfaces and pivot points must be properly lubricated using OMC Moly Lube, or an equivalent assembly grease to ensure long-life and trouble/binding-free operation.

    b. **Do not** over or under tighten the fasteners for the remote housing components. Loose fasteners could allow components to shift come loose or worse, wedge stuck in a given position, this could lead to loss or throttle or shift control. Similarly over-tightened fasteners may lead to binding or breaking components.

    c. Once assembly is complete, slowly move the remote handle through the range of operation. Feel for smooth and free movement. Adjust the friction knob so the throttle does not change in response to vibration from engine or boat operation. Do not set the friction adjuster so tight as to bind the control handle causing jerky shift/throttle adjustments.

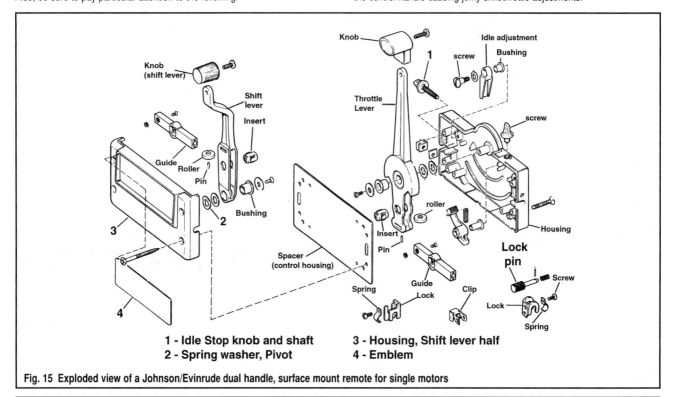

Fig. 15 Exploded view of a Johnson/Evinrude dual handle, surface mount remote for single motors

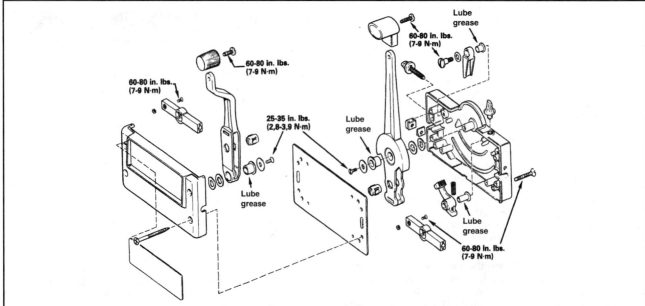

Fig. 16 For safety, pay close attention to component bolt torque and lubrication during assembly - dual handle, surface mount remote for single motors

REMOTE CONTROLS 9-9

■ Turn the friction adjustment knob CLOCKWISE to increase friction or COUNTERCLOCKWISE to decrease friction. NEVER overtighten the knob, which would lock the throttle in position.

 d. Refer to the Timing and Synchronization adjustments in the Maintenance and Tune-Up section for information on additional information on control cable adjustment and lubrication. To set Idle rpm, turn the idle adjustment knob **clockwise** to increase engine idle or **counterclockwise** to decrease idle.

Concealed Side Mount Remote Units

REPLACING THE TRIM/TILT SWITCH

◆ See Figures 17 and 18

Refer to the accompanying illustration for parts identification.
 1. Disconnect the battery cables for safety. This will help prevent the possibility of potential burns or shorts that could occur while working on the control unit.

✳✳ CAUTION

Disconnecting the battery cables will also make sure that the engine is not accidentally cranked or started (which could lead to injury or damage).

■ Take note of the wire routing for installation purposes.

 2. Cut the wire tie securing the trim switch leads to the lower corner of the control body. Disconnect the trim switch leads from the cable connector.
 3. Remove the rubber boot from the warm-up knob at the base of the remote handle. Remove the retaining screw and knob.
 4. Move the control handle fully into the **reverse** position and insert a 1/8 in. hex-key into the opening at the bottom of the control handle. Loosen the set-screw (using the hex key), then carefully pull the control handle from the shaft while freeing the trim switch wires from the mounting plate.
 5. Remove the screws (2) fastening the cover to the back of the remote handle and then separate the cover from the remote handle.

✳✳ CAUTION

The neutral lock rod spring is under tension and could fly free when removing the lockrod. Wear safety glasses to protect your eyes.

 6. Carefully remove the neutral lock slide and spring, while keeping track of the spring.
 7. Remove and discard the retaining clip from the back of the trim/tilt switch, but use care not to damage or gouge the housing in the head of the remote handle.
 8. With the retaining clip free, carefully pull the switch and wire leads from the handle. Take note of the wire routing to ensure trouble-free installation.
 9. Installation is essentially the reverse of removal, but pay particular attention to the following:
 a. The wires molded into the trim switch are off-center, so be sure to position them toward the top of the handle when inserting the switch into the cover.

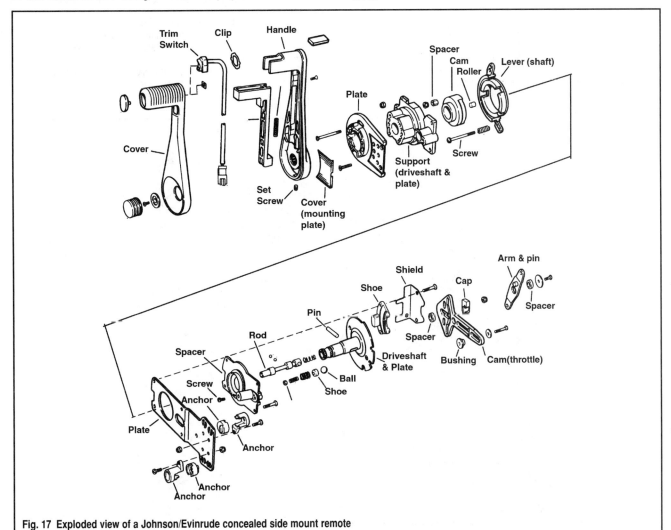

Fig. 17 Exploded view of a Johnson/Evinrude concealed side mount remote

9-10 REMOTE CONTROLS

b. The switch retaining clip must be replaced whenever it is removed, so be sure to use a new retaining clip when securing the switch to the housing in the remote handle.

c. Route the switch wires through the hole in the lockout lever as noted during removal.

d. Once the remote handle is fully assembled, make sure the housing is still in the full **reverse** setting, then carefully push the handle onto the shaft. Verify that the neutral lock slide drops into the slot when shifting the handle up into **neutral**, then tighten the set-screw.

e. Route the switch wires as noted during removal and secure with a new wire tie.

DISASSEMBLING AND ASSEMBLING THE HOUSING

◆ See Figures 17 and 18

Refer to the accompanying illustration for parts identification.

1. Disconnect the battery cables for safety. This will help prevent the possibility of potential burns or shorts that could occur while working on the control unit.

※※ CAUTION

Disconnecting the battery cables will also make sure that the engine is not accidentally cranked or started (which could lead to injury or damage).

2. Remove the remote handle from the control assembly, as detailed in Replacing the Trim Switch, in this section. Unless the switch or components in the remote handle are faulty, there is no need to disassemble the handle.

3. Remove the Philips screws (3) securing the remote assembly to the mounting plate and then remove the control from the boat (with the cables attached).

4. Disconnect the throttle and shift cables from the control assembly.

5. Position the remote handle back on the control assembly in order to shift the assembly into **neutral**, then remove the handle once again.

6. Remove the screw and washer securing the throttle arm and then remove the throttle arm.

7. Remove the throttle arm spacer, then remove the screw and washer securing the throttle cam. Remove the friction control cap and locknut from the far end of the throttle cam and then remove the throttle cam from the assembly.

8. Remove the 2 spacers from the driveshaft and plate.

9. Remove the 2 screws and locknuts from the neutral switch shield. Remove the shield, shoe and the neutral switch, taking care not to loose the detent ball and related components.

10. Remove the final 2 screws and nuts (on the other end of the assembly from the shield and shield screws), then lift out the driveshaft and plate along with the detent spacer as an assembly.

11. Lift the cable mounting plate off the support.

12. If the driveshaft neutral detent components must be inspected, lubricated or replaced, disassemble them as follows:

a. Press inward on the neutral warm-up rod (at the end of the shaft) and hold it down while withdrawing the shift cam drive pin from the base of the shaft.

b. Separate the shaft from the spacer and remove the detent ball. Be ready to catch the detent ball as it will likely fall free when the shaft is removed from the spacer.

c. Remove the neutral warm-up rod, driveshaft ball and spring from the shaft.

d. Remove the detent shoe, inner spring, outer spring and the left-hand thread screw and nut (which is removed by turning clockwise). Don't mix this left-hand thread screw up with other fasteners, as only this screw or another left-hand thread screw, must be installed during assembly.

13. If the driveshaft shift cam and roller assembly must be inspected, lubricated or replaced, disassemble the balance of the control assembly, as follows:

a. Remove the shift lever from the center of the driveshaft support.

b. Remove the throttle friction screw and spring from the end of the driveshaft support.

c. Remove the 3 shift cam rollers, followed by shifter cam.

d. Remove the ball from the bore in the side of the shifter cam.

e. Remove the 3 hex spacers and locknuts from the holes in the driveshaft support.

14. Clean and inspect the remote control components as follows:

a. Wipe all metallic, non-electrical, parts with a clean rag soaked in a mild solvent. Electrical and plastic components should be cleaned using a dry rag and/or low pressure compressed air.

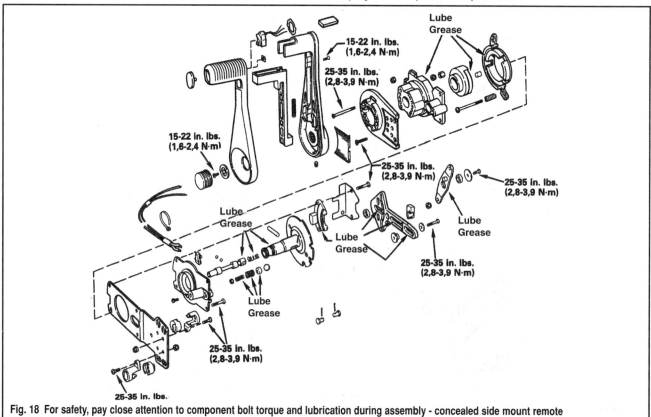

Fig. 18 For safety, pay close attention to component bolt torque and lubrication during assembly - concealed side mount remote

b. Visually check all mechanical components for signs of cracks, damage or excessive wear. Replace any worn or damaged components to ensure proper remote operation.

15. Assembly is essentially the reverse of disassembly. Refer to the accompanying illustration for critical torque values and lubrication points. Also, be sure to pay particular attention to the following:

   a. Sliding surfaces and pivot points must be properly lubricated using OMC Moly Lube, or an equivalent assembly grease to ensure long-life and trouble/binding-free operation.

   b. **Do not** over or under tighten the fasteners for the remote housing components. Loose fasteners could allow components to shift come loose or worse, wedge stuck in a given position, this could lead to loss or throttle or shift control. Similarly over-tightened fasteners may lead to binding or breaking components.

   c. Don't forget the left-hand threaded screw and nut used in the neutral detent assembly. DO NOT substitute a normal, right-hand thread screw if they are lost or damaged, obtain a suitable left-hand thread replacement.

   d. Once assembly is complete, slowly move the remote handle through the range of operation. Feel for smooth and free movement. Turn the throttle friction adjustment screw (located in the bore on the mounting plate, nearest to the remote handle) so the throttle does not change in response to vibration from engine or boat operation. Do not set the friction adjuster so tight as to bind the control handle causing jerky shift/throttle adjustments.

■ **Turn the friction adjustment screw CLOCKWISE to increase friction or COUNTERCLOCKWISE to decrease friction. NEVER overtighten the screw, which would lock the throttle in position.**

   e. Use the shifter detent friction screw (mounted through the bore in the mounting plate, furthest from the remote handle), to adjust the shifter feel. Turn the screw **clockwise** to increase friction or **counterclockwise** to decrease it. As with the throttle friction screw adjustment, do not over-tighten the screw to lock the shifter in position. The best way to set the detent friction screw is to adjust it while slowly moving the shifter slowly back and forth from **forward** to **neutral** and **reverse**, until the desired amount of friction is obtained at the handle.

   f. Refer to the Timing and Synchronization adjustments in the Maintenance and Tune-Up section for information on additional information on control cable adjustment and lubrication.

Pre-wired Binnacle Mount Remote Units

OVERHAUL

◆ **See Figures 19 thru 22**

Refer to the accompanying illustration for parts identification.

1. Disconnect the battery cables for safety. This will help prevent the possibility of potential burns or shorts that could occur while working on the control unit.

✸✸ CAUTION

Disconnecting the battery cables will also make sure that the engine is not accidentally cranked or started (which could lead to injury or damage).

2. Loosen the retainers and separate the remote from the boat with the cable still attached.
3. If equipped, remove the nut securing the ignition keyswitch, then remove the cover.
4. Disconnect the throttle and shift cables from the control assembly.
5. Position the shifter handle in **neutral**, then carefully pry the screw cover from the base of the handle. Remove the screw and washer, and then gently pry the remote handle from the driveshaft splines.
6. If necessary, disassemble and inspect the handle.

■ **The trim/tilt switch assembly is mounted in the top of the handle.**

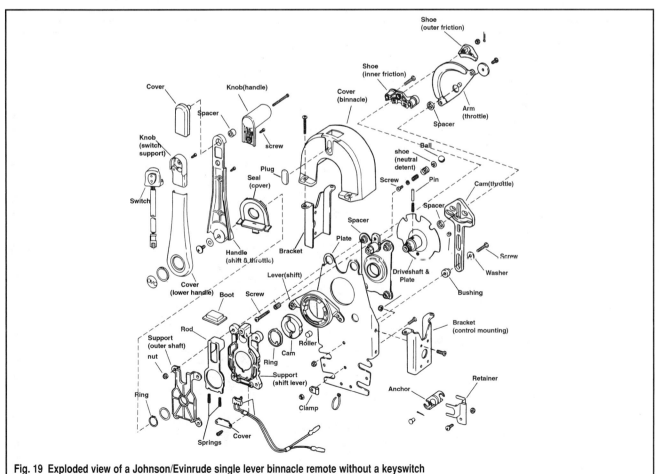

Fig. 19 Exploded view of a Johnson/Evinrude single lever binnacle remote without a keyswitch

9-12 REMOTE CONTROLS

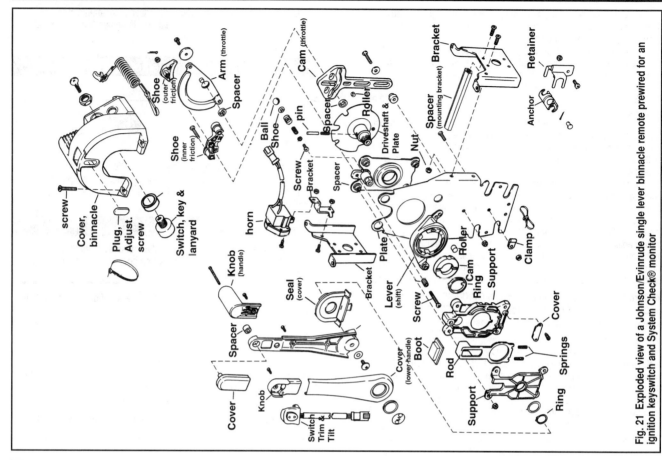

Fig. 21 Exploded view of a Johnson/Evinrude single lever binnacle remote prewired for an ignition keyswitch and System Check® monitor

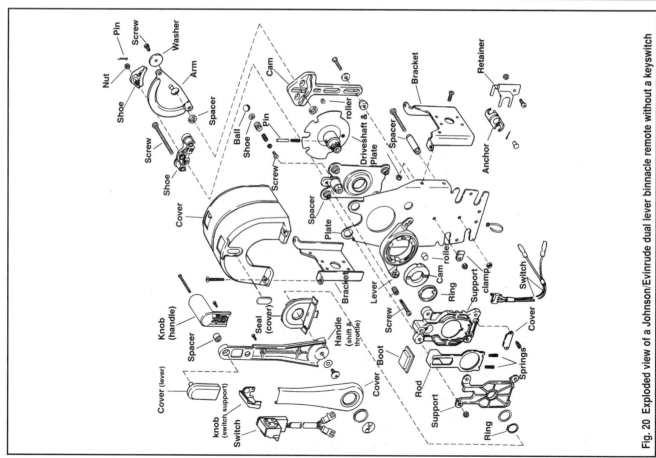

Fig. 20 Exploded view of a Johnson/Evinrude dual lever binnacle remote without a keyswitch

REMOTE CONTROLS

7. If equipped, remove the screws (2), then lift off the clamp and the neutral switch from the shift lever support. Cut the wire tie and remove the screw securing the wire clamp (but be sure to take note of the wire routing for installation purposes).

8. If equipped, remove the warning horn mounting screw, then remove the horn and wiring harness assembly. Again, note the wiring routing for installation purposes.

9. Remove the cover mounting brackets from either side of the housing.

10. Remove the cotter pin from the friction adjustment screw (at the top of the throttle lever plate), then loosen the screw and remove it along with the nut and outer friction shoe.

11. Remove the screw, washer, throttle lever plate and bushing.

12. Remove the screw, washer and nut from the throttle cam and bushing. Remove the throttle cam, bushing, spacer and throttle pin roller.

13. Using a pair of snapring pliers, remove the ring from the splined end of the driveshaft.

14. Remove the screws and nuts fastening the inner friction shoe and then remove the shoe.

15. Remove the screws and nuts (usually 4) holding the rest of the assembly together and then pry the outer shaft support carefully off the driveshaft.

16. Wearing safety glasses (to protect from injury) slowly remove the neutral warm-up rod, keeping pressure on the springs at the bottom of the rod until spring tension is released. This will help prevent the springs from flying free and either hitting someone in the eyes or disappearing in the process. Fully remove the shift lever support.

17. Remove the friction adjustment screw (with spring) from the shift lever support or the neutral detent plate spacer.

18. Remove the cable mounting plate and shift lever from the driveshaft, detent plate and shifter cam.

19. Remove the shift lever from the cable mounting plate and then remove the rollers from the lever.

20. If necessary for inspection, lubrication or repair, disassemble the driveshaft and neutral detent plate spacer components as follows:

 a. Push the drive pin into the shaft using the neutral warm-up ring and then carefully lift the ring from the cam.

■ Use finger pressure over the drive pin to hold it in place as the cam is lifted free, otherwise the pin will become a projectile.

 b. Slowly release the spring pressure on the pin once the cam is removed, then remove the pin and spring.

 c. Carefully pry the shaft from the detent plate, making sure to retain the detent ball located between them.

 d. Remove the ball (if it did not fall out earlier), shoe, springs, nut and screw from the detent plate.

21. Clean and inspect the remote control components as follows:

 a. Wipe all metallic, non-electrical, parts with a clean rag soaked in a mild solvent. Electrical and plastic components should be cleaned using a dry rag and/or low pressure compressed air.

 b. Visually check all mechanical components for signs of cracks, damage or excessive wear. Replace any worn or damaged components to ensure proper remote operation.

22. Assembly is essentially the reverse of disassembly. Refer to the accompanying illustration for critical torque values and lubrication points. Also, be sure to pay particular attention to the following:

 a. Sliding surfaces and pivot points must be properly lubricated using OMC Moly Lube, or an equivalent assembly grease to ensure long-life and trouble/binding-free operation.

 b. **Do not** over or under tighten the fasteners for the remote housing components. Loose fasteners could allow components to shift come loose or worse, wedge stuck in a given position, this could lead to loss or throttle or shift control. Similarly over-tightened fasteners may lead to binding or breaking components.

 c. During installation, be sure to attach both cables to the **handle side** of the throttle lever plate. First, insert one cable pin through the throttle lever, then through the throttle cable. Next insert the other pin through the shift cable, then through the shift lever.

 d. All wiring must be connected and secured at the points noted during removal. Wiring must be routed in a manner that will prevent contact with moving components and must not be pinched between components during assembly.

 e. Once assembly is complete, slowly move the remote handle through the range of operation. Feel for smooth and free movement. Adjust the throttle friction adjustment screw so the throttle does not change in response

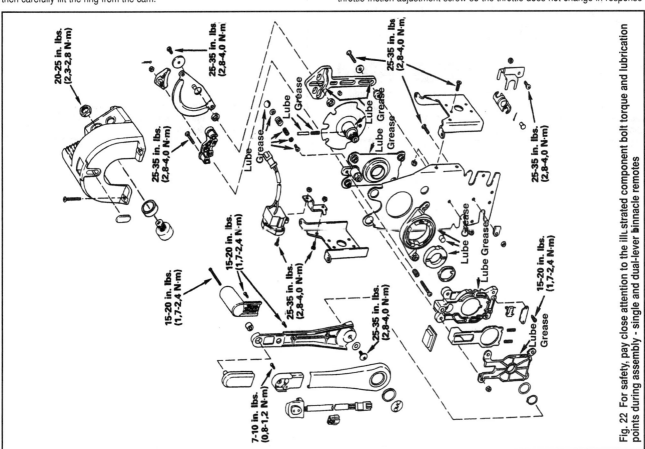

Fig. 22 For safety, pay close attention to the illustrated component bolt torque and lubrication points during assembly - single and dual-lever binnacle remotes

9-14 REMOTE CONTROLS

to vibration from engine or boat operation. Do not set the friction adjuster so tight as to bind the control handle causing jerky throttle adjustments.

■ **The throttle friction adjustment screw is located under a small cover on the side, top of the housing (just below the peak of the curve). There are actually 2 screws which can be accessed through an oval cutout in the side of the housing, the top of these is the throttle friction screw (the lower of these is the shifter detent screw). Turn the either friction adjustment screw CLOCKWISE to increase friction or COUNTERCLOCKWISE to decrease friction. NEVER overtighten the screw, which would lock the throttle or shifter in position.**

f. Once the throttle friction screw is set, adjust the shifter detent friction screw (at the bottom of the same oval cutout) to adjust the shifter feel. The best way to set the detent friction screw is to adjust it while slowly moving the shifter slowly back and forth from forward to neutral and reverse, until the desired amount of friction is obtained at the handle. As with the throttle friction screw adjustment, do not over-tighten the screw to lock the shifter in position.

g. Refer to the Timing and Synchronization adjustments in the Maintenance and Tune-Up section for information on additional information on control cable adjustment and lubrication.

CONTROL CABLES

Rigging

◆ See Figure 23

The control cables should be replaced if inspection reveals any signs of damage, wear or even fraying at the exposed ends. Remember that loss of one or more cables while underway could cause loss of control at worst or, at best, strand the boat. Check cable operation frequently and inspect the cables at each service. Replace any that are hard to move or if excessive play is noted. Never replace just one cable (unless a freak accident caused damage to the cable), if one cable is worn or has failed, assume the other is in like condition and will soon follow.

Before removal, mark the cable mounting points on the remote control using a permanent marker. This will help ensure easy and proper installation of the replacement. Unless there is a reason to doubt the competence of the person who originally rigged the craft, match the cable lengths as closely as possible. Otherwise, re-measure and determine proper cable lengths as if this was a new rigging.

When rigging an engine to a new boat, determine cable length by measuring from the center point of the motor along the intended cable route to either the side-mount or center-console remote location (refer to the accompanying illustration). Add 3 ft. 4 in. (1.02 M) to the measurement and purchase a cable of the same length (or one that is **slightly** longer than that measurement. Johnson/Evinrude replacement cables are normally available in one-foot increments from 5-20 ft. and in two-foot increments to 50 ft.

※※ **CAUTION**

To prevent the danger of cable binding or other conditions that could cause a loss of steering control when underway, take care to route all cables with the fewest number and most gentle bends possible. A bend should NEVER have a radius of less than 6 in. (15cm).

Follow the procedures for remote service in this section and the Timing and Synchronization adjustments in the Maintenance and Tune-Up section whenever removing/installing or replacing the throttle and shift cables.

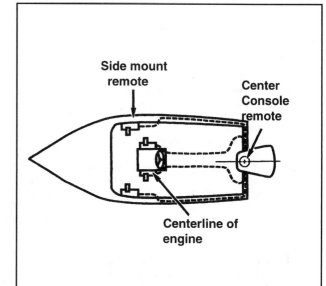

Fig. 23 For cable replacement or rigging, measure the distance from the centerline of the motor to the remote (and add 3 ft. 4 in./1.02 meters to the measurement)

HAND REWIND STARTERS..10-2
  INTRODUCTION 10-2
  GENERAL INFORMATION 10-2
  2.0-3.5 HP (78cc) MOTORS 10-2
    Assembly & Installation 10-4
    Cleaning & Inspection 10-4
    Description 10-2
    Removal & Disassembly 10-3
  5 HP (109cc) MOTORS 10-5
    Description 10-5
    Disassembly & Assembly 10-5
    Removal & Installation 10-5
    Rope Replacement 10-5
  9.9-15 HP (216cc) MOTORS...10-6
    Adjustment 10-9
    Description 10-6
    Disassembly & Assembly 10-8
    Removal & Installation 10-6
    Rope Replacement 10-7
  COLT/JUNIOR AND 3-35 HP (UNDER 600cc) MOTORS (EXC. 5 HP/109cc, 4
  DELUXE, 9.9-15 HP/216cc AND 18 JET-35 HP/521cc MODELS) 10-9
    Disassembly & Assembly..10-12
    Removal & Installation 10-9
    Rope Replacement 10-12
  4 DELUXE (87cc) MOTORS 10-13
    Description 10-13
    Disassembly & Assembly 10-15
    Rope Replacement 10-13
    Removal & Installation 10-14
  18 JET-35 HP (521cc) MOTORS..10-16
    Disassembly & Assembly 10-16
    Removal & Installation 10-16
  25-55 HP (737cc) AND 25-70 HP (913cc) MOTORS 10-19
    Description 10-19
    Disassembly & Assembly 10-19
    Removal & Installation 10-19
  EFI MOTORS...10-21
EMERGENCY STARTING **10-22**
  General Information 10-22

10

HAND REWIND STARTER

HAND REWIND STARTERS 10-2
EMERGENCY STARTING 10-22

HAND REWIND STARTER

HAND REWIND STARTERS

Introduction

◆ See Figures 1, 2 and 3

Rope start Johnson/Evinrude outboards utilize a hand rewind starter assembly to turn the flywheel for engine startup and to automatically recoil the rope afterwards. Smaller hp models that are equipped with an electric start will normally also be equipped with the hand rewind starter assembly for use as a back up in case of electric motor or battery failure. Furthermore, almost all Johnson/Evinrude outboards through 70 hp motors **can** be started using a pull rope in emergency situations. Larger models are equipped with an emergency rope mounted in the engine case that can be temporarily connected to the flywheel for emergency purposes. We emphasize the word contain the word **can** because only someone the size of a NFL linebacker will typically succeed pull-starting a larger outboard by hand.

Most of the hand rewind starters installed on the Johnson/Evinrude outboards consist of a flat disc type assembly that is mounted atop the flywheel. These starters engage a ratchet plate that turns the flywheel when the rope is pulled. However, there are a few other types of hand rewind starters which are found on various models, especially the earlier models covered here such as:

- The 4 Deluxe model, utilizes a swing arm/drive gear type hand starter assembly that instead mounts to the side of the powerhead and swings upward to engage gear teeth with the flywheel when the rope is pulled.
- The 9.9-15 hp (216cc) 2-strokes (produced through 1992) utilize a disc that engages the teeth of the flywheel mounted horizontally next to the flywheel.

General Information

◆ See Figure 4

Essentially, the hand rewind starter is a strictly a mechanical device used to crank the powerhead for starting. Because it is such a basic mechanical device, one normally encounters very few problems with the assembly. The spring will last an incredibly long time, if used properly. The greatest enemy of the spring is the operator.

Three causes can contribute to starter failure. Two may be prevented, the third cannot.

The most common problem is the result of the operator pulling the starter rope too far outward. If the operator places one hand on the powerhead and pulls the rope with the other hand, it is physically impossible, in this position, to pull the rope too far. Problems develop when the operator uses both hands to pull on the rope, with no control on how far the rope can be extended. The rope may be broken or the knot released from the starter disc. In either case, the spring rewinds with tremendous speed and in almost all cases travels past its normal rewind position bending the end of the spring in reverse. Therefore, more maintenance work is involved than merely replacing the rope.

Another bad habit, while using the hand rewind starter, is the operator releasing his/her grip on the rope when it is in the extended position, allowing the rope to freely rewind. The operator should never release his grip, but hold onto the rope and slowly feed it back into the assembly, thus controlling the rewind. The owner should remain alert to any wear on the rope and replace it long before the possibility of breaking might occur. When a rope breaks, the spring usually rewinds with incredible speed, the same as if the rope is suddenly released, and usually causes damage to the spring or other starter assembly components.

The third cause of spring failure cannot be prevented - age. As the outboard continues to perform year after year, the age of the spring steel will finally take its toll.

■ **Depending on the model and the powerhead, anywhere from 6 to 12 feet of spring steel length is wound into about a 4 in. diameter. This places the spring under unbelievable tension - a real "tiger in a cage" - making it a potentially dangerous force. Therefore, any time the hand starter is serviced, especially during work on the spring, safety glasses should be worn and the work performed with the utmost care. The procedures must be followed exactly as presented for each starter, to prevent possible injury to the worker or others in the area.**

Any time the rope is broken, the starter spring will rewind with incredible speed. Such action will cause the spring to rewind past its normal travel and the end of the spring will be bent back out of shape. Therefore, if the rope has been broken, the starter should be completely disassembled and the spring repaired or replaced.

2.0-3.5 Hp (78cc) Motors

DESCRIPTION

◆ See Figure 5

The hand rewind starter for these small powerheads is a simple coiled spring unit mounted on top of the powerhead just in front of the fuel tank.

Fig. 1 The most common Johnson/Evinrude hand rewind starters are mounted atop the flywheel

Fig. 2 Swing arm type hand rewind starter mounted on the port side of a Model 4D powerhead

Fig. 3 Hand rewind starter mounted flat (horizontally), next to the flywheel on 9.9-15 hp (216cc) powerheads.

HAND REWIND STARTER 10-3

Fig. 4 The pull rope broke on this starter assembly causing the spring to rewind with such incredible speed, it actually bent back in a reverse direction

REMOVAL & DISASSEMBLY

◆ See Figures 5 thru 8

 MODERATE

1. Remove the engine covers. For details, please refer to Engine Cover (Top and Lower Cases) in the Maintenance and Tune-Up Section.
2. For safety, disconnect the spark plug lead, then ground it to the cylinder head for safety.
3. Remove the 2 screws securing the fuel tank to the powerhead. The forward screw also secures the aft leg of the rewind starter. Remove the other screws securing the other 2 legs of the starter.
4. Lift the fuel tank slightly and remove the rewind starter assembly from the powerhead.
5. Pull on the rope handle to gain some slack in the rope; then either hold the rope/pulley to keep them from rewinding or unbolt the cover panel and tie a slip knot to lock the rope from rewinding; and then remove the handle.
After the handle is removed allow the pulley to slowly rewind pulling the rope into the housing.
6. Gently push or pry the circlip (E-ring) out of the groove in the center shaft.
7. Lift the thrust washer off the friction plate and free of the center shaft.
8. Lift the friction plate slightly and snap the friction spring out of the hole in the center shaft. Remove the friction plate. The return spring will come with the plate. Remove the spring cover, and then the friction spring and ratchet.

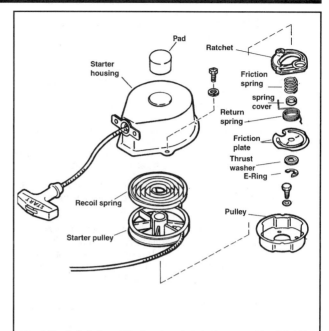

Fig. 5 Exploded view of the hand rewind starter assembly - 2.0-3.5 hp (78cc) motors

✱✱ WARNING

The rewind spring is a potential hazard. The spring is under tremendous tension when it is wound - a real "tiger" in a cage! If the spring is suddenly released, severe personal injury could result from being struck by the spring with force. Therefore, the following 2 steps must be performed with care to prevent personal injury to you and others in the area.

9. Carefully "rock" the pulley back and forth until the spring disengages from the bottom of the pulley, then carefully life the pulley straight off the housing, leaving the spring behind.
10. Now, slowly turn the housing over and gently place it on the floor (or workbench) with the spring facing the floor. Tap the top of the housing with a mallet and the spring will fall free of the housing and unwind almost instantly and with force, but be contained within the housing. Tilt the container with the opening away from you. The spring will be released from the housing and unwind rapidly. Turn the housing over and unhook the end of the spring from the peg in the housing.
11. Untie the knot in the end of the old starter rope and pull the rope free.

Fig. 6 The hand rewind starter on 2.0-3.5 hp (78cc) motors

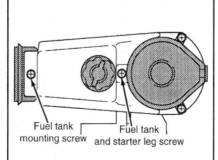

Fig. 7 The fuel tank must be repositioned slightly for access to the starter

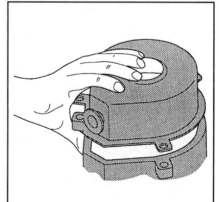

Fig. 8 Removing or installing the starter assembly

HAND REWIND STARTER

CLEANING & INSPECTION

◆ See Figures 5 and 9

 MODERATE

Wash all metallic parts in solvent, and then blow them dry with low-pressure compressed air.

Remove any trace of corrosion and wipe all metal parts with an oil-dampened cloth.

Inspect the rope. Replace the rope if it appears to be weak or frayed. If the rope is frayed, check the pulley and housing (especially the holes through which the rope passes) for rough edges or burrs. Remove the rough edges or burrs with a file and polish the surface until it is smooth.

■ Replacement ropes on these models should be trimmed to a length of about 53 in. (135cm). Be sure to fuse both ends to help prevent fraying.

Inspect the starter spring end hooks. Replace the spring if it is weak, corroded or cracked. Inspect the tab on the spring retainer plate. This tab is inserted into the inner loop of the spring. Therefore, be sure it is straight and solid. Inspect the inside surface of the pulley rewind recess for grooves or roughness. Grooves may cause erratic rewinding of the starter rope.

ASSEMBLY & INSTALLATION

◆ See Figures 5 and 6 thru 10

 MODERATE

⁂ CAUTION

Wear a good pair of gloves while winding and installing the spring. The spring will develop tension and the edges of the spring steel usually pretty sharp. The gloves will help prevent cuts to the hands and fingers.

⁂ CAUTION

We strongly recommend wearing a pair of safety goggles or a face shield while the spring is being installed. As the work progresses a "tiger" is being forced into a cage, over 14 ft. (4.3m) of spring steel wound into about 4 in. (10.2cm) circumference. If the spring is accidentally released, it will lash out with tremendous ferocity and very likely could cause personal injury to the installer or other persons nearby.

1. Apply a light coating of OMC Triple-Guard, or equivalent marine grease (or even Lubriplate 777, if available) to the spring, spring cavity and spindle.

2. Hold the spring in a coil in one gloved hand, with the other hand, hook the looped end of the spring over the peg in the housing. Coil the spring counterclockwise into the housing (when viewed looking down into an inverted housing). The easiest way to accomplish this task is to hold the spring in one gloved hand and with the other gloved hand, rotate the housing. Continue working the spring until it is all confined within the housing.

3. Feed one end of a new rope through the hole in the pulley, and then tie a figure "8" knot in the end of the rope. Pull on the rope until the knot is confined inside the pulley recess. Wind the entire pull rope counterclockwise around the pulley.

4. Carefully lower the pulley down over the center shaft of the housing. As the pulley goes into the housing the inner spring hook must catch the pulley slot. Once the hook is indexed, the seat the pulley in the housing.

5. Position the ratchet onto the pulley, with the flat side of the ratchet against the pulley and the round hole indexed over the post.

6. Slide the friction spring down the center shaft. This spring serves as a spacer and exerts an upward pressure on the friction plate. Install the spring cover on top of the spring.

7. Hook the end of the return spring into the friction plate slot, then insert the other end of the return spring into the hold in the pulley. Install the return spring and friction plate assembly over the spindle.

8. Position the thrust washer, then push down on the friction plate and snap the Circlip into the groove on the center shaft to secure the plate and associated parts in place.

■ Be sure to install the circlip with the sharp side facing outward.

9. Place the pull rope into the notch and turn the pulley 3 turns counterclockwise from the relaxed position.

10. Feed the rope through the eyelet on the starter housing and, when there is about 12 in. (30cm) of rope outside the start housing tie a slipknot to hold the rope in place.

11. If removed, install the motor cover panel onto the starter housing.

12. Feed the free end of the rope through the handle and tie a figure "8" knot in the end. Tuck the knot into the handle (excess rope beyond the knot can be cut off, if necessary).

Release the slipknot, then relax the tension on the pulley and allow it to slowly rewind until the handle is against the starter housing.

13. Lift the fuel tank slightly and install the hand rewind starter assembly onto the powerhead.

14. Install and finger-tighten the retaining screws for the starter housing and the fuel tank, then tighten the screws to 60-84 inch lbs. (7-9 Nm).

15. Reconnect the spark plug lead that was disconnected for safety.

16. Install the engine covers. For details, please refer to Engine Cover (Top and Lower Cases) in the Maintenance and Tune-Up Section.

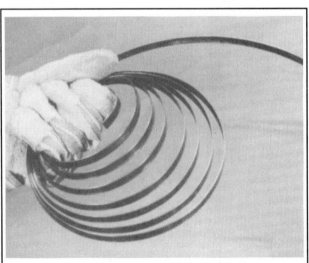

Fig. 9 Inspect the spring for wear, damage or corrosion - handle with care

Fig. 10 Use this knot (a "figure 8" knot) to retain the starter rope ends

HAND REWIND STARTER 10-5

5 Hp (109cc) Motors

DESCRIPTION

◆ See Figure 11

The hand rewind starter for the single-cylinder 5 hp 2-stroke powerhead is a simple coiled spring unit mounted atop the flywheel.

REMOVAL & INSTALLATION

◆ See Figure 11

1. For safety, disconnect the spark plug lead, then ground it to the cylinder head for safety.
2. Remove the screws (3) retaining the hand rewind starter assembly to the powerhead.
3. Loosen the adjustment nut, then remove the starter lockout cable from the housing and lockout arm.

To install:
4. Position the hand rewind starter assembly into position on the powerhead.
5. Install the 3 retaining screws and tighten to 72-102 inch lbs. (8-12 Nm).
6. Install the cable into the lockout arm and tighten the adjusting nuts to secure it to the starter housing.
7. Verify proper lockout operating and adjust, as necessary, to ensure the starter assembly can only be operated with the engine in **neutral**.

ROPE REPLACEMENT

◆ See Figure 10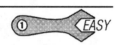

■ Replacement ropes for this model should be trimmed to a length of about 52 in. (132cm). Be sure to fuse both ends to help prevent fraying.

1. Remove the starter assembly from the top of the powerhead.
2. Pull the rope completely out, then allow the pulley to retract just until the hole aligns with the lockout slot.

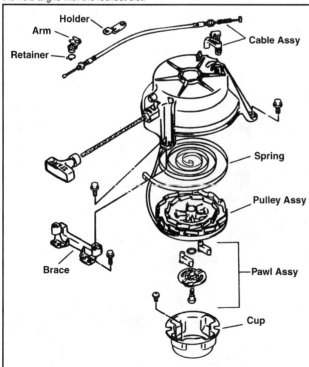

Fig. 11 Exploded view of the hand rewind starter assembly - 5 hp (109cc) motors

3. Place a small drift pin or length of wire through the lockout slot and hole to retain the pulley. Secure the wire or drift pin to prevent them from becoming dislodged during service.
4. Invert the housing for access, then free the rope knot from the pulley and untie or cut the old rope free.
5. Free the rope knot from the handle and untie or cut the old rope free. Remove the rope completely from the handle and housing.

✱✱ CAUTION

If the pulley must be turned in the next step, the wire or drift pin used to secure it earlier must first be removed. Use extreme care to prevent the spring tension from releasing suddenly.

6. To install the new rope, make sure the cavity in the pulley aligns with the housing rope holes. If necessary, carefully turn the pulley counterclockwise until the cavity aligns as noted, then place a small drift pin or a length of wire through the hole in the underside of the pulley (furthest away from the handle end of the housing) to secure the pulley.
7. Carefully thread the rope through the holes, then tie a figure "8" knot (as shown) to secure the rope. Tuck the knot into the pulley recess.
8. Insert the rope through the starter handle and anchor, then tie a second figure "8" knot to secure.
9. While holding tension on the rope, remove the wire or pin and slowly release the spring tension allowing it to rewind the new rope into the housing.

DISASSEMBLY & ASSEMBLY

◆ See Figure 11

1. Remove the starter assembly from the top of the powerhead, as detailed in this section.
2. If necessary, remove the rope from the assembly, as detailed in this section.
3. With the housing inverted on a workbench, remove the bolt and pawl retainer from the bottom center of the pulley.

■ Be sure to note the spring over the pulley pin.

4. Remove the pawls from the pulley, and then inspect the pawl assembly components for wear or damage.
5. Slowly and carefully remove the pulley from the housing, while making sure to keep your fingers away from the uncoiling spring.
6. Face the spring and underside of the housing downward on the workbench, and then tap gently using a mallet to free the spring from the housing. Once the spring has finished uncoiling, carefully lift the housing by tilting it so the opening faces away from people or breakables (there will still be some tension to the spring).
7. If the lockout is damaged or worn, remove the retainer, then remove and replace the lockout.

To assemble:
8. Wash all metallic parts in solvent, and then blow them dry with low-pressure compressed air. Wipe all other components using a clean shop rag.
9. Remove any trace of corrosion and wipe all metal parts with an oil-dampened cloth.
10. Inspect the rope. Replace the rope if it appears to be weak or frayed. If the rope is frayed, check the pulley and housing (especially the holes through which the rope passes) for rough edges or burrs. Remove the rough edges or burrs with a file and polish the surface until it is smooth.

■ Replacement ropes on these models should be trimmed to a length of about 52 in. (132cm).

11. Inspect the starter spring end hooks. Replace the spring if it is weak, corroded or cracked. Inspect the pawl retainer, retainer ring and screw for wear or damage, and replace, as necessary.

✱✱ CAUTION

Wear a good pair of gloves while winding and installing the spring. The spring will develop tension and the edges of the spring steel usually pretty sharp. The gloves will help prevent cuts to the hands and fingers.

10-6 HAND REWIND STARTER

✴✴ CAUTION

We strongly recommend wearing a pair of safety goggles or a face shield while the spring is being installed. As the work progresses a "tiger" is being forced into a cage, a long steel sprig is wound into about 4 in. (10.2cm) circumference. If the spring is accidentally released, it will lash out with tremendous ferocity and very likely could cause personal injury to the installer or other persons nearby.

12. Apply a light coating of OMC Triple-Guard, or equivalent marine grease (or even Lubriplate 777, if available) to the spring, spring cavity and shoulder screw.
13. If the spring was removed, hold the spring in a coil in one gloved hand, with the other, carefully insert the spring into the pulley so the hook engages the pulley slot, then carefully wind the spring into the pulley.
14. Install the pulley and spring into the housing, while engage the spring hook onto the hub in the housing.
15. Install the springs and pawls into the pulley, then install the retainer and tighten the bolt to 72-102 inch lbs. (8-12 Nm).
16. Wind the pulley about 4 turns counterclockwise and align the rope guide hole on the housing with the pulley cavity, then place a small drift pin or length of wire in the rearmost pulley hole (furthest from the handle portion of the housing) to secure the pulley.
17. Install the rope as detailed under Rope Replacement, in this section.
18. Install the hand rewind starter assembly, as detailed in this section.

9.9-15 hp (216cc) Motors

DESCRIPTION

This pinion gear starter is a design employing the principles of an automotive-type starter motor. A nylon pinion gear slides upward and engages the flywheel ring gear as the starter rope is pulled. The pinion gear automatically disengages when the engine starts. The ratio between the pinion gear and the ring gear was selected to provide maximum cranking speed with minimum pulling effort to ensure fast and easy powerhead start.

A lockout pawl linked to the cam follower prevents manual start engagement if the throttle is advanced beyond the start position.

REMOVAL & INSTALLATION

◆ See Figures 12 thru 16

1. Disconnect the high-tension leads from the spark plugs. Ground the high-tension leads. Pull the rope out far enough, and then tie a knot in the rope. Allow the rope to rewind until the knot is against the front of the cowling.
2. Remove the handle rope anchor and then the rope from the handle. Remove the rubber bumper.
3. Untie the knot in the rope and allow the rope to slowly wind onto the pulley and the spring to unwind. When the rope is about to pass through the opening in the cowling, hold the pulley, and then allow your grip to slip to permit the rope to continue winding onto the pulley, and the spring to unwind slowly. Do not release the grip on the pulley completely. If the spring rewinds too rapidly the spring may be damaged along with other parts.

✴✴ WARNING

The rewind spring is under tremendous tension and is a potential hazard. Therefore, safety glasses should be worn and extreme care exercised to follow the procedures carefully during removal, disassembling, and assembling work with the starter.

4. Hold the pulley and cup together, and at the same time, loosen the center mounting bolt and back it out of the intake manifold. Do not remove the center bolt from the starter. Continue to hold the starter firmly together and carefully remove it from the powerhead. Take care not to damage the starter pawl.
5. If the starter is only removed in order to accomplish other work, install a 3/8 x 16 nut onto the far side of the thru-bolt to hold the starter together and prevent the spring from escaping.

To install:

6. Hold the starter pulley and cup together and position the assembly in place on the powerhead. Thread the mounting screw into the manifold and tighten the screw securely.

7. Thread the rope through the front cowling and pull it all the way out. With the rope fully extended, the starter spring end must be free to extend a minimum of 1/2 in. from the cup. If the spring is not free to extend 1/2 in. from the cup, allow the starter rope to fully rewind onto the pulley.
8. After the rope has rewound onto the pulley, release one full turn of the rope from the pulley and make the test again. Repeat the procedure until the spring is free to extend a minimum of 1/2 in. from the cup when the rope is fully extended. With the rope still through the front of the cowling tie a knot in the rope. Allow the rope to rewind until the knot is tight against the cowling.
9. Install the rubber bumper, handle and anchor onto the rope.
10. Secure the rope by pressing the anchor into the handle.
11. Remove the knot from the rope and allow the rope to rewind onto the pulley. When the rope is fully rewound, the handle should be up tight against the cowling.

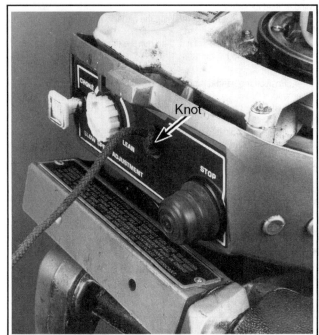

Fig. 12 Tie a knot in the rope, then allow the rope to rewind against the knot

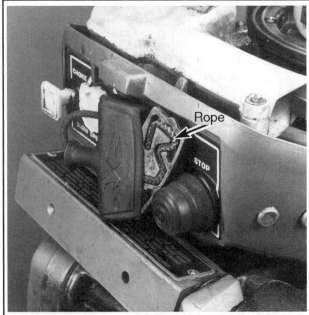

Fig. 13 Remove the rope anchor, then separate the rope and remove the bumper

HAND REWIND STARTER   10-7

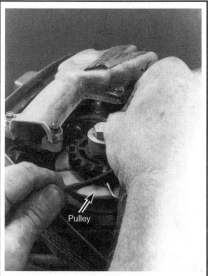
Fig. 14 Carefully allow the spring to slowly unwind

Fig. 15 Hold the pulley and cup together as you loosen the mounting bolt

Fig. 16 Install a 3/8 in. x 16 nut to hold the starter assembly together

ROPE REPLACEMENT

◆ See Figures 17 thru 25

■ Replacement ropes for must be trimmed to the proper length, then about 1/2 in. (13mm) of each end should be fused to help prevent fraying.

Pay close attention to replacement rope length; too long a rope can damage the starter assembly, while too short a rope will lead to difficult starting. Replacement ropes for most of these models should be cut to 65 in. (165cm) in length. However, be sure to measure the old rope (or pieces of the old rope to confirm), as some models may be equipped with slightly longer ropes.

1. Remove the starter from the powerhead as detailed earlier in this section.
2. Clamp the starter in a vise, as shown. Tighten the vise just slightly onto the cup bushing. Remove the center bolt.
3. Remove the pinion spring and gear from the pulley.
4. Exercise care during this next procedure. Slide a putty knife or other similar flat tool in between the bottom side of the pulley and top side of the spring. Using the tool, work the pulley off the spring without allowing the spring to escape from the cup.
5. Remove the four screws from the back side of the pulley, and then separate the pulley. Remove the rope.

To install:

■ The length and diameter of the starter rope required will vary depending on the horsepower size of the model being serviced. Therefore, check the Hand Starter Rope Specifications in the Appendix, and then purchase a quality nylon piece of the proper length and diameter size. Only with the proper rope, will you be assured of efficient operation following installation. Each end of the nylon rope should be "fused" by burning them slightly with a very small flame (a match flame will do) to melt the fibers together. After the end fibers have been "fused" and while they are still hot, use a piece of cloth as protection and pull the end out flat to prevent a "glob" from forming.

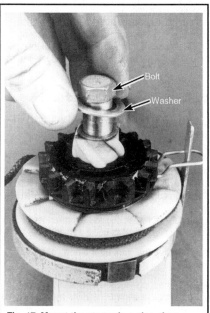
Fig. 17 Mount the starter in a vise, then remove the center bolt

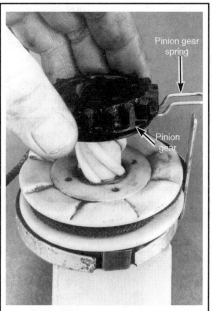
Fig. 18 Remove the pinion spring and gear from the pulley

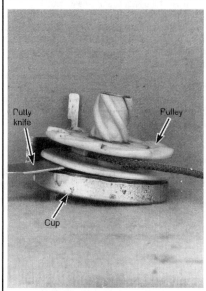
Fig. 19 CAREFULLY work the pulley off the spring without allowing the spring to escape from the cup

10-8 HAND REWIND STARTER

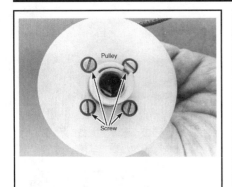

Fig. 20 Remove the 4 screws from the back of the pulley to separate it and remove the rope

Fig. 21 Install the replacement rope in the pulley

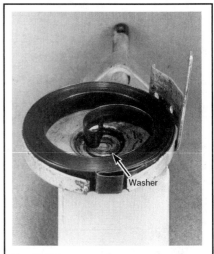

Fig. 22 If the cup washer was removed, place the washer into the cup under

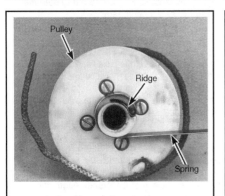

Fig. 23 View showing the spring indexed into the pulley. Installation is not performed in this manner.

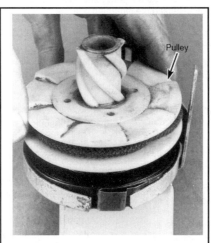

Fig. 24 Installing the pulley over the spring

Fig. 25 Wind the rope onto the pulley COUNTERCLOCKWISE

6. Tie a figure 8 knot as close to the end of the rope as possible, as shown. After the knot has been tied, feed the knot into the recess of the pulley.

7. Position the other half of the pulley over the rope. Rotate the cap slightly to align the four holes in the cap with the holes in the other half of the pulley. Secure the pulley together with the four retaining screws. Tighten the screws securely.

8. If the cup washer was removed, place the washer into the cup under the inner loop of the spring.

9. Install the pulley down over the top of the spring. Check to be sure the end of the spring is engaged in the pulley slot where the rope is installed. Do not lubricate the pulley or the pinion gear.

10. Install the pinion gear and the pinion spring onto the pulley.

11. Lubricate the mounting screw and washer with OMC outboard oil, and then install the screw through the pulley and cup.

12. Thread a 3/8 x 16 nut onto the mounting bolt to hold the pulley and spring in the cup until the starter is installed on the powerhead.

13. Wind the rope onto the pulley counterclockwise.

14. Install the starter assembly as detailed earlier in this section.

DISASSEMBLY & ASSEMBLY

◆ See Figures 26 thru 39

1. After the starter has been removed, as detailed earlier in this section, thread a 3/8 x 16 nut onto the center thru-bolt. Clamp the starter in a vise with the mounting screw secured between the vise jaws.

2. Use a screwdriver to engage the spring loop, and then pull the spring out of the cup until it is completely unwound.

3. Release the starter from the vise. Remove the nut, mounting bolt, and washer. Remove the pinion gear, pinion spring, pulley, rewind spring, and cup washer. As the pulley is removed from the cup, the end of the spring may remain attached to the pulley. If it is still attached, snap it loose with a screwdriver.

To assemble:
Wash all parts except the rope in solvent and then blow them dry with compressed air.

Remove any trace of corrosion and wipe all metal parts with an oil-dampened cloth.

Inspect the starter spring end loops. Replace the spring if it is weak, corroded or cracked.

Inspect the rope. Replace the rope if it appears to be weak or frayed. If the rope is frayed, check the hole through which the rope passes for rough edges or burrs. Remove the rough edges or burrs with a file, and polish the surface until it is smooth.

Inspect the pinion gear and pulley for wear, chipped or broken teeth. Inspect the cup for corrosion or damage, such as being warped out of shape.

■ The following procedures pickup the work after a new rope has been installed.

4. Coat the inside surface of the cup with OMC Type A Lubricant. Position the rewind spring into the cup, as shown. Place the cup washer into the cup.

5. Install the pulley with the spring loop engaging the pulley, as shown. Check to be sure the spring feeds out of the cup slot.

HAND REWIND STARTER 10-9

6. Install the pinion gear onto the pulley.
7. Coat the threads of the mounting screws with OMC Outboard Oil. Insert the center thru-bolt through the pulley and cup.
8. Thread a 3/8 x 16 nut onto the bolt to hold the parts together.
9. Hold the cup and at the same time wind the spring into the cup by rotating the pulley counterclockwise as viewed from the top of the pulley. As soon as excessive resistance is felt during the winding, feed the spring into the cup through the slot to relieve spring tension. Continue to wind and feed the spring into the cup until the loop on the end of the spring is drawn up tight against the side of the cup.
10. Wind the rope counterclockwise around the pulley. Install the pinion gear spring. Hold the spring in place with a rubber band or piece of string.
11. Install and adjust the starter as detailed in the other procedures in this section.

ADJUSTMENT

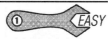

1. Place the shift lever in the neutral position, and the throttle in the start position. Check to see if the lockout pawl clears the highest point on the starter pulley by 0.050-0.110 in. (1.27-2.79mm). This clearance is required to allow powerhead start with the shift mechanism in gear and the throttle in the start position.
2. To adjust, loosen the nut and rotate the adjusting screw inward or outward until the proper clearance is obtained. Tighten the nut securely to hold the adjustment.
3. To check, the hand starter should crank the powerhead with the shift mechanism in either forward or reverse gear and with the throttle in the start position.
4. Move the throttle to the fast position. The starter lockout pawl should prevent the starter from cranking the powerhead.

■ To restart a powerhead from the throttle fast position, the throttle must be moved to the slow position and then advanced to the start position in order to remove backlash from the lockout linkage.

Colt/Junior and 3-35 hp (under 600cc) Motors (Except 5 hp/109cc, 4 Deluxe, 9.9-15 hp/216cc and 18 Jet-35 hp/521cc Models)

REMOVAL & INSTALLATION

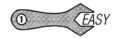

◆ See Figures 40 and 41

1. For safety, disconnect the spark plug lead(s), then ground it (or them, as applicable) to the cylinder head.
2. If equipped with starter lockout linkage or a cable, proceed as appropriate based on the model:

Fig. 26 Remove the starter, thread a nut onto the thru-bolt, secure the assembly in a vise with the jaws on the bolt/nut

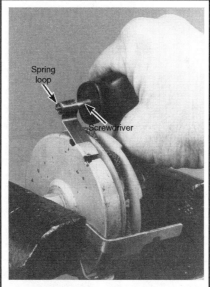

Fig. 27 Use a screwdriver to grab and pull the spring out

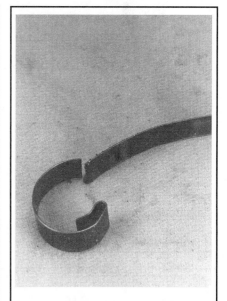

Fig. 28 Starter spring with broken end.

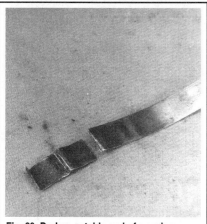

Fig. 29 Broken outside end of a spring.

Fig. 30 A new spring.

Fig. 31 A distorted starter cup no longer fit for service.

10-10 HAND REWIND STARTER

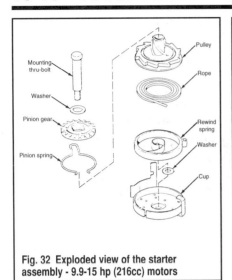

Fig. 32 Exploded view of the starter assembly - 9.9-15 hp (216cc) motors

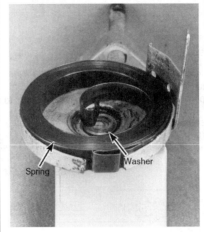

Fig. 33 Install the spring and washer into the cup as shown

Fig. 34 Install the pulley so the spring loop engages, making sure the spring feeds out the cup slot

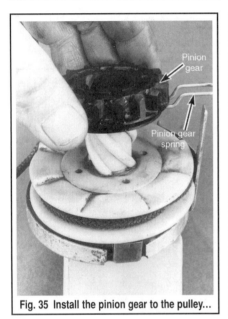

Fig. 35 Install the pinion gear to the pulley...

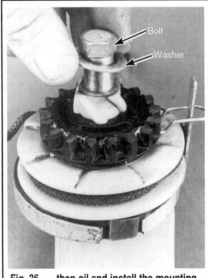

Fig. 36 ...then oil and install the mounting screw

Fig. 37 Use a nut to temporarily hold the assembly together

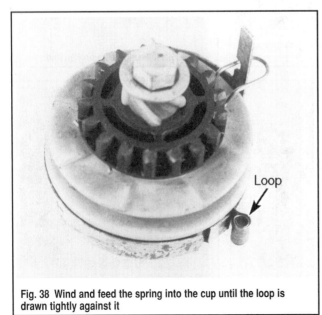

Fig. 38 Wind and feed the spring into the cup until the loop is drawn tightly against it

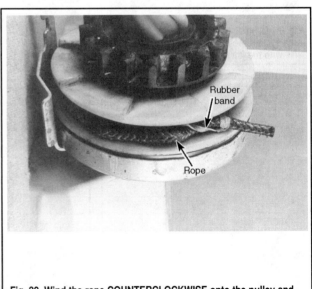

Fig. 39 Wind the rope COUNTERCLOCKWISE onto the pulley and use a rubber band to hold it

HAND REWIND STARTER 10-11

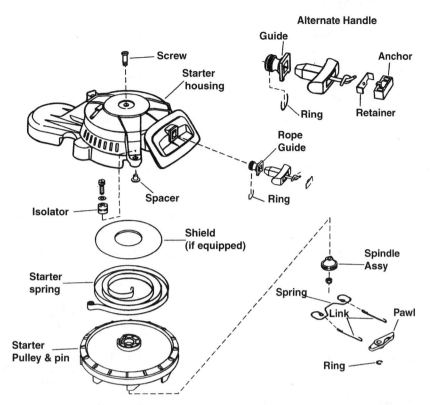

Fig. 40 Exploded view of a typical hand rewind starter assembly used on most Colt/Junior and 3-35 hp motors (housing and handle designs may vary slightly, but basic mechanism should be the same)

• For 5/6 hp (128cc) and 8/9.9 hp (211cc) 4-stroke motors, disconnect the starter lockout cable from the shift detent and/or the manual starter housing by squeezing the tabs while pulling the cable up and out of the housing.

■ **When a cable is used for the starter lockout, take note of the cable routing and of the location of any tie straps, if used, for installation purposes.**

• For 5/6/8 hp (164cc) motors, disconnect the neutral lockout cable by pushing on the side of the locking tab while pulling the cable out of the starter housing.
• For 9.9-15 hp (255cc) motors, loosen the starter lockout cam screw (found on the port side of the handle housing) several turns. Separate the link from the lockout cam (though it is usually easier to wait until the bolts are removed from the housing and the housing is lifted slightly).
• For 9.9/15 hp (305cc) 4-stroke motors, follow the lockout linkage down to the end opposite the starter assembly and remove the clevis end of the link from the bellcrank. Also, at this time, unsnap the fuse connection from the starter housing.
• For 25/35 hp (500/565cc) motors disengage the 2-pin connector for the low oil lead from the engine harness. The lockout cam link must be removed from the housing, but it is usually easier to remove this once the housing is unbolted and lifted slightly, in the next step.

3. For most models, remove the 3 bolts securing the hand rewind starter assembly to the powerhead, then carefully lift it from the powerhead. On motors equipped with lockout linkage, it is usually easiest to disengage the linkage at this time. For the following models, additional steps or fasteners may be used as noted:
• For Colt/Junior models, remove the screw from the rear of the cover and remove the lower cover, then remove the filler cap from the fuel tank. Then access the 4 mounting screws for the starter assembly.

✱✱ WARNING

DO NOT remove the large screw from the top of the engine cover at this time or the starter spring will disengage and unwind forcibly when removing the engine cover.

Fig. 41 The starter assembly is usually secured to the powerhead using 3 retaining bolts - cover designs and lockout controls vary

• For early 1990 3/4 hp (87cc) models, remove the screw from lower engine cover for access. If equipped with an integral tank, remove the filler cap from the fuel tank. Then access the 3 or 4 mounting screws for the starter assembly. On some 4 hp models (Ultra/Excel) you must unbolt the throttle cable trunnion and, after unbolting the starter, unthread the cable from the throttle knob.

4. To service the lockout assembly components, proceed as follows:
• On lockout cable equipped motors, remove the springs and plungers for cleaning, inspection and replacement.
• On lockout linkage equipped motors, remove the retaining pin, then remove the plunger, spring, tappet and cam from the housing (remove the screw from the end of the cam).

To install:

5. If equipped with starter lockout linkage (except 9.9/15 hp 4-stroke motors), hold the assembly in position above the powerhead and engage the linkage.

10-12 HAND REWIND STARTER

6. Lower the starter housing onto the powerhead, while aligning the bolt holes.
7. For 9.9/15 hp (255cc) motors, turn the starter lockout cam screw inward carefully just until the screw head contacts the cam. Do not overtighten, which could damage the threads or the cam.
8. Install and tighten the mounting bolts (usually 3, but sometimes 4) to 60-84 inch lbs. (7-9 Nm).
9. Using the hand starter assembly, slowly pull the rope outward to ensure smooth starter engagement and proper operation. Slowly allow the rope to retract back into the housing, then complete installation.
10. For 9.9/15 hp (305cc) 4-stroke motors, connect the clevis end of the starter lockout link to the bellcrank. Also for these motors, snap the fuse connection back into the starter cover.
11. For models equipped with a lockout cable, route it as noted during removal. Connect the cable to the starter assembly and/or to the shift detent (as removed) by pushing inward at the cable ends until the locking tabs engage. If removed, install a wire tie to the original location to ensure there is no interference with moving components.
12. Now, double-check starter lockout operation. Make sure the starter can only be operated in **neutral**.
13. For 25/35 hp (500/565cc) motors, engage the 2-pin connector for the low-oil wiring, then secure the holder to the ignition module/power pack.
14. Reconnect the spark plug lead(s).

ROPE REPLACEMENT

◆ See Figures 40 and 42

■ **Replacement ropes for must be trimmed to the proper length, then about 1/2 in. (13mm) of each end should be fused to help prevent fraying.**

Pay close attention to replacement rope length; too long a rope can damage the starter assembly, while too short a rope will lead to difficult starting. Replacement ropes for most of these models should be cut to 59-59 1/2 in. (150-151cm) in length. However, be sure to measure the old rope (or pieces of the old rope to confirm), as some models may be equipped with slightly longer ropes.

In order to replace the starter rope, the assembly must be removed from the powerhead and locked in position with the rope fully extended. The manufacturer's Starter Spring Installer Kit (No. 342682) contains a pin that can be used for this purpose, but a suitably sized/sturdy drift pin can be substituted.

■ **On 9.9/15 hp (305cc) 4-stroke motors the starter lockout linkage can also be used to retain the pulley without the need for a drift pin or the referenced kit.**

1. Remove the starter assembly from the top of the powerhead, as detailed in this section.

Fig. 42 Some rope handles are equipped with an anchor to which the rope is secured

2. Slowly pull the starter rope out of the housing until it is fully extended, then insert a drift pin or the locking pin from OMC (#342682) to lock the pulley into position. On 3/4 hp (87cc) motors, use a wire tie around the locking pin and the starter housing leg to keep the pulley from turning.

■ **When the starter housing is inverted (resting with the top of the housing facing downward onto the workbench) and the handle housing towards the 6 o'clock position, a cutout at the 3 o'clock position can be used to secure the pulley using the drift or lock pin.**

3. Free the rope knot from the channel in the pulley, then untie or cut the old rope free. Remove the rope and handle completely from the housing.
4. Remove the cap from the starter handle, then push the knot (and anchor housing, if equipped) free of the handle. Untie or cut the old rope free.
5. To install the new rope, carefully thread the rope through the housing and into the pulley, then tie a figure "8" knot (as shown) to secure the rope. Tuck the knot into the pulley channel.
6. Insert the rope through the starter handle. Install the rope to the anchor (if used) or tie a second figure "8" knot, as applicable to secure.
7. While holding tension on the rope, remove the pin (or release the starter lockout on 9.9/15 hp 4-strokes, as applicable) and slowly release the spring tension allowing it to rewind the new rope into the housing.

DISASSEMBLY & ASSEMBLY

◆ See Figures 40 and 43 thru 45

In order to service the starter assembly, the starter rope must first be removed using a drift pin or the locking pin (or the starter lockout on 9.9/15 hp 4-strokes) from the manufacturer's Starter Spring Installer Kit (#342682) to lock the pulley in position. Also, although the kit is not absolutely necessary for disassembly, it does make assembly a lot easier. Read the Assembly procedure in this section before deciding to proceed without the kit.

✶✶ CAUTION

The spring is under a tremendous amount of tension, wear safety glasses and heavy gloves to protect yourself.

1. Remove the starter assembly from the top of the powerhead, as detailed in this section.
2. Remove the rope from the starter housing assembly, as detailed in this section.

■ **It is not necessary to remove the rope from the handle, unless the rope must be replaced.**

3. Remove the stop tool and allow the pulley to slowly unwind (by keeping some drag or tension on the pulley itself), until spring tension is released.
4. With the housing inverted on a workbench, remove the retaining ring, pawl and linkage from the housing.
5. For all models, except the 9.9/15 hp (305cc) 4-stroke motors, loosen the spindle screw (on the top center of the housing), while holding the spindle locknut (on the bottom center of the housing). Hold the pulley in position while withdrawing the spindle screw.
6. For 9.9/15 hp (305cc) 4-stroke motors, remove the retaining ring from the top of the spindle (the retaining ring is used in place of a spindle screw and locknut on these motors). Hold the pulley in position while withdrawing the spindle.

✶✶ CAUTION

The spring is still under tremendous tension. DO NOT attempt to remove the spring or release that tension with the housing facing upward.

7. Carefully lift the pulley from the housing, leaving the spring in position.

✶✶ CAUTION

When using the housing to protect against the uncoiling spring, keep your hands and fingers away from the edge of the assembly.

HAND REWIND STARTER 10-13

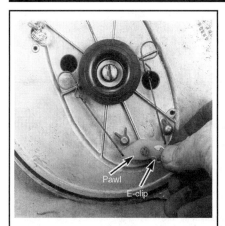

Fig. 43 The pawl is secured by a small retaining ring which . . .

Fig. 44 . . . once removed, frees the pawl and linkage

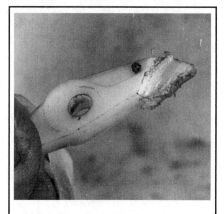

Fig. 45 Inspect the pawl for damage or wear (this one must be replaced)

8. Position the housing so the spring is facing downward toward the workbench, then tap on the housing using a rubber mallet to free the spring. It should uncoil under the housing.

9. Once the spring has finished uncoiling, carefully lift the housing by tilting it so the opening faces away from people or breakables (there will still be some tension to the spring).

To assemble:
Wash all metallic parts in solvent, and then blow them dry with low-pressure compressed air.

Remove any trace of corrosion and wipe all metal parts with an oil dampened cloth.

Inspect the rope. Replace the rope if it appears to be weak or frayed. If the rope is frayed, check the pulley and housing (especially the holes through which the rope passes) for rough edges or burrs. Remove the rough edges or burrs with a file and polish the surface until it is smooth.

■ **Replacement ropes on these models should be trimmed to a length of about 59 in. (150cm).**

Inspect the starter spring end hooks. Replace the spring if it is weak, corroded or cracked. Inspect the inside surface of the pulley rewind recess for grooves or roughness. Grooves may cause erratic rewinding of the starter rope.

Inspect the pawl and links for wear.

If applicable, check the lockout assembly components for wear or damage and replace, as necessary.

In order to service the starter assembly, the pulley must be locked in position using a drift pin or the locking pin from the manufacturer's Starter Spring Installer Kit (#342682).

The Starter Spring Installer Kit (#342682) makes hand rewind starter assembly much easier, providing a steady mount on which the housing can be secured in an inverted position. It also provides a temporary spindle on which the pulley can be secured for safety, while still allowing the pulley to be rotated in order to wind the spring. The kit is more essential for 9.9/15 hp (305cc) 4-stroke motors, as they do not have bolt on which the pulley can be loosely installed. Of course, a bolt of the proper size and strength could be obtained and substituted for these models.

10. If the kit is available, secure the mandrel from the kit in a vice and, for all except 3/4 hp (87cc) motors, position the spring around it. Align the spring in a direction so that if you followed the spring coils from the center outward, you'd be traveling in a clockwise spiraling direction.

11. Position the housing over the mandrel (if using the kit) or over the spindle screw on the workbench (a longer screw of the same thread as the spindle screw could be used like the mandrel, to secure in a vise to help steady the assembly).

12. For all except 3/4 hp (87cc) motors, insert the inner hooked end of the rewind spring through the slot in the starter assembly housing.

13. Place the pulley over the spring, engaging the hooked end of the spring into the slot in the pulley. If the kit is not available, loosely install the spindle and locknut, just enough to keep the pulley from going anywhere.

■ **If using the kit, the manufacturer recommends (for all except 3/4 hp (87cc) motors) actually securing the pulley to the mandrel using the pulley retainer (with the taper facing the pulley) and the left-hand thread nut from the kit. Once the pulley is secured pull lightly on the spring to take up any slack, then turn pulley counterclockwise slowly (using the handle from the kit) until the hooked end of the spring engages the slot in the pulley.**

14. Slowly rotate the pulley counterclockwise until the spring is pulled fully into the housing, then allow the pulley to slowly unwind until all tension is relieved.

15. Turn the pulley counterclockwise from the relaxed position about 3 1/2 turns on most motors or 4 1/2 for 3/4 hp (87cc) motors. On the last half-turn, rotate it just to the point where the rope channel in the pulley aligns with the opening in the housing, then secure in place. Use a drift pin or the locking pin from the kit (or the starter lockout on 9.9/15 hp 4-stroke motors).

16. Install the rope, as detailed in this section.

17. If using the kit, remove the nut and pulley retainer from the mandrel, then, while keeping downward pressure on the pulley, remove the starter housing assembly from the mandrel. Install the spindle, screw and locknut (or retaining ring on 9.9/15 hp (305cc) 4-stroke motors).

18. For models equipped with a spindle screw and locknut, hold the screw from turning while tightening the locknut to specification. For all motors except 3/4 hp (87cc), tighten the locknut to 120-144 inch lbs. (14-16 Nm). For 3/4 hp (87cc) motors tighten the locknut to 60-84 inch lbs. (7-9 Nm).

19. Install the pawl and linkage, then secure by snapping the retaining ring into position with the sharp edge facing outward.

20. If the lockout parts were removed, install them at this time.

4 Deluxe (87cc) Motors

DESCRIPTION

◆ See Figure 46

The 4 Deluxe utilizes a swing arm type starter mounted on the port side of the powerhead. The drive gear works on an axis. As the rope is pulled, a swing arm moves the drive gear upward to engage with the teeth of the flywheel ring gear. A coil spring winds and tightens as the rope unwinds. The spring then coils the rope around a pulley as the rope handle is returned to the control panel.

ROPE REPLACEMENT

◆ See Figure 10

■ **Replacement ropes for must be trimmed to the proper length of approximately 59 1/2 in. (151cm) for these models, then about 1/2 in. (13mm) of each end should be fused to help prevent fraying.**

Pay close attention to replacement rope length; too long a rope can damage the starter assembly, while too short a rope will lead to difficult starting.

1. For safety, remove the spark plugs and ground the high tension leads.

10-14 HAND REWIND STARTER

2. Pull the starter rope out, and then tie a knot in the rope behind the handle. Allow the rope to rewind until the knot is against the cowling. Untie the knot in the end of the rope, and then remove the handle and the rubber bumper.

3. Remove the knot tied in the previous steps, then allow the rope to slowly wind into the starter. Before the rope end passes the cowling, firmly grasp the starter pulley, and then allow the starter to unwind using pressure on the pulley to control it.

4. Turn the pulley counterclockwise until the knot is visible through the access window in the side of the pulley housing. If necessary, turn the pulley back clockwise until the knot is visible. Use a short length of rope placed between the pulley and idler gears to hold the spring from unwinding, then carefully pull the old rope out of the pulley through the access window.

5. Matchmark the sides of the pulley to the rope hole as a reference.

6. Wind the pulley by hand 2 full turns clockwise, then place the end of the rope in the gear mesh point.

7. Tie a figure "8" knot in one end of the replacement rope as shown in the accompanying illustration. Use a small pick to push the other end of the rope through the access window in the side of the pulley. Pull the rope all the way through the pulley until the knot seats firmly in the cavity.

8. Use a small wire hook to thread the rope past the rope guard on the bottom and idler on the top. Wind the rope around the pulley clockwise, as many turns as possible until only a short length is extending out the front of the pulley above the rope guard.

9. Pull the rope out of the pulley as you thread it through the lower motor cover eyelet, then tie a slipknot in the rope to hold it in position.

10. Pull the rope through the handle and secure using another figure "8" knot, then release the slipknot, allowing the slack to slowly rewind into the starter.

11. Spring pre-tension is checked by pulling the rope until it is fully outward, then pulling out the spring end (from the opposite side of the pulley) until tight. The spring length should be 8-18 in. (20-46cm) when pulled as noted. If the spring is out more than 18 in. (46cm) the rope must be wound at least one additional turn around the pulley. If the spring is out less than 8 in. (20cm), the rope must be unwound at least one additional turn from the pulley.

REMOVAL & INSTALLATION

◆ See Figures 47 thru 51

1. For safety, remove the spark plugs and ground the high tension leads.

2. Pull the starter rope out, and then tie a knot in the rope behind the handle. Allow the rope to rewind until the knot is against the cowling. Untie the knot in the end of the rope, and then remove the handle and the rubber bumper.

3. Remove the knot tied in the previous steps, then allow the rope to slowly wind into the starter. Before the rope end passes the cowling, firmly grasp the starter pulley, and then allow the starter to unwind using pressure on the pulley to control it.

4. At this point, check the back side of the starter. Notice the hook of the starter spring protruding out of a hole in the starter. Grasp the spring hook with a pair of needle-nose pliers, and then pull the spring out as far as possible, to relieve tension on the spring.

5. If necessary for access, remove or loosen and reposition the port ignition coil from the powerhead. For details, refer to the procedures in the Ignition and Electrical System section.

6. Remove the bolt (usually 3/8 in.) from the starter bracket (which was located just behind the ignition coil).

7. Remove the starter assembly through-bolt from the center of the pulley housing.

Fig. 46 A unique swing arm type hand rewind starter is found on the port side of 4 Deluxe models

Fig. 47 The rope handle is secured using a knot

Fig. 48 The pulley is used to draw the rope into the housing

Fig. 49 Pull the spring out of the housing to relieve tension

Fig. 50 A bracket bolt is found just behind the port ignition coil

Fig. 51 The starter assembly through-bolt is threaded through the center of the assembly

HAND REWIND STARTER   10-15

8. Hold the starter assembly together while removing it from the powerhead.

To install:

9. Apply a light coating of OMC Screw Lock, or equivalent threadlock to the shoulder screw threads.

10. Position the starter assembly to the powerhead, then finger-tighten the bolts.

11. Install the rope, as detailed under Rope Replacement, in this section.

■ **The idler gear arm travel is controlled by 2 stops. The upper stop determines the depth of engagement between the flywheel and idler gear teeth. A slot in the cut and stop assembly bracket provides for adjustment.**

12. Hold the idler gear arm stop against the cup stop, then fully engage the idler gear teeth in the flywheel. While holding the assembly in this position, securely tighten the bolts.

13. If removed or repositioned, install the port ignition coil to the powerhead.

14. Install the sparks plugs and connect the leads.

DISASSEMBLY & ASSEMBLY

◆ See Figures 52, 53 and 54

1. Carefully lift the idler gear arm, idler gear and the idler gear arm spring from the housing.

✱✱ CAUTION

The next step could be dangerous. Removing the pulley from the cup must be done with care to prevent personal injury. Use safety glasses to protect your eyes and sturdy gloves to protect your hands.

2. Lift the pulley slightly and then use a screwdriver and work the spring free of the pulley. Do not allow the spring to be released from the cup. After the pulley has been removed, notice the position of the spring loop. Remove the rope from the pulley. Notice how the rope unwinds from the pulley counterclockwise.

3. Remove the bushings from the idler gear arm and the bushings installed one on each side of the pulley.

4. Two different methods are suggested to remove the spring from the starter cup. One method involves pulling continuously on the end of the spring that contains the loop. The second method is to simply toss the cup a safe distance onto carpeting or a lawn, allowing the spring to be released instantly from the cup.

If this second method is used, be sure the spring will not cause a threat to any individual in the area when it is released.

To assemble:

Wash all parts except the rope in solvent and then blow them dry with compressed air.

Remove any trace of corrosion and wipe all metal parts with an oil-dampened cloth.

Inspect the starter spring end loops. Replace the spring if it is weak, corroded or cracked.

Inspect the rope. Replace the rope if it appears to be weak or frayed. If the rope is frayed, check the hole through which the rope passes for rough edges or burrs. Remove the rough edges or burrs with a file, and polish the surface until it is smooth.

Inspect the dog ears of the pulley gears to be sure they are not worn and are free of burrs. Check the idler gear for cracks and missing teeth.

Inspect the pulley and bushing in cup and stop assembly. Replace if excessively worn, cracked or chipped.

5. Apply a light coating of OMC Triple-Guard, or equivalent marine grease (or even Lubriplate 777, if available) to the inside of the cup and stop assembly and to the spring.

6. If removed, insert the bushing in the idler gear arm.

7. Position the small washer into the cup and stop assembly.

8. If removed, position the spring retainer onto the pulley, then hook the spring end loop to the roll pin on the pulley.

9. Install the cup and stop assembly with washer over the pulley, making sure the spring is positioned through the tabs.

10. Temporarily install the through-bolt into the center of the cup and stop assembly, loosely installing a nut to secure the assembly.

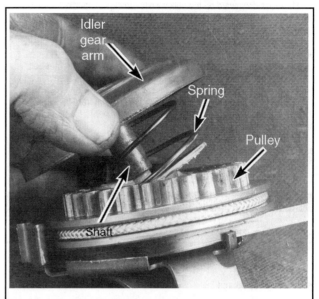

Fig. 52 Separating the idler gear assembly from the pulley

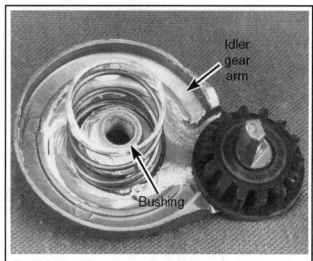

Fig. 53 Inspect all components for signs of wear or damage and replace, as necessary

Fig. 54 During assembly, be sure to position the spring hook over the pulley roll pin

10-16 HAND REWIND STARTER

11. Press downward on the pulley while turning counterclockwise to wind the spring into the cup and stop assembly. Continue winding until the spring's outside loop engages the assembly face.
12. Slowly release the pulley, allowing the spring tension to gradually ease.
13. Hold the starter assembly together and remove the through-bolt.
14. Position the idler gear arm spring onto the gear arm, then place the gear onto the arm.
15. Remove all traces of adhesive, dirt or corrosion from the screw threads, then apply a light coat of OMC Locquic primer to the threads of the shoulder screw.
16. Insert the shoulder screw through the idler gear arm, then install the arm assembly over the pulley making sure the gear arm stop is positioned between the 2 tabs on the cup and stop assembly.
17. Hold the assembly together and install to the powerhead.

18 Jet-35 Hp (521cc) Motors

REMOVAL & INSTALLATION

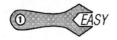

◆ See Figures 55 and 56

1. For safety, disconnect the spark plug leads, then ground them to the cylinder head.
2. Remove the 3 bolts securing the hand rewind starter assembly to the powerhead, then carefully lift it from the powerhead.
3. To disconnect the lockout cable, pull back on the cable housing while using a screwdriver to push downward on 1 side of the lockout cable retaining tab. Pull gently back until the tab is clear of the housing, then separate the cable from the assembly.
4. If necessary, remove the lockout spring and plunger for inspection or replacement.

To install:
5. Visually check the starter mount for damage or wear. If noted, remove the mount using a crescent wrench, then apply OMC Locquic Primer and OMC Screw Lock (or equivalent threadlocking compound) to the threads of the new mount, install and tighten securely. Position the caps on the starter mounts.
6. Connect the starter lockout cable to the housing, then lower the housing onto the powerhead, while aligning the bolt holes.

■ **The short starter housing bolt is normally mounted on the port side of the motor.**

7. Install and tighten the 3 mounting bolts to 48-72 inch lbs. (5.4-8 Nm).
8. Using the hand starter assembly, slowly pull the rope outward to ensure smooth starter engagement and proper operation. Slowly allow the rope to retract back into the housing. Now, double-check starter lockout operation. Make sure the starter can only be operated in **neutral**.
9. Reconnect the spark plug leads.

DISASSEMBLY & ASSEMBLY

◆ See Figures 55 and 57 thru 62

No special tools are required in order to disassemble the hand rewind starter assembly for these motors. However, the OMC Starter Spring Winder and Installer (#392093) is recommended for installation. Please review the assembly procedure before proceeding.

※※ **CAUTION**

The spring is under a tremendous amount of tension, wear safety glasses and heavy gloves to protect yourself.

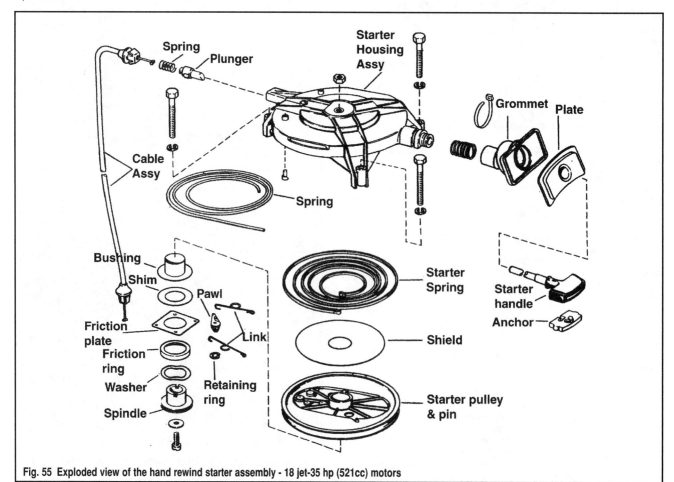

Fig. 55 Exploded view of the hand rewind starter assembly - 18 jet-35 hp (521cc) motors

HAND REWIND STARTER 10-17

1. Remove the starter assembly from the top of the powerhead, as detailed in this section.

2. Pull the starter rope partially out of the pulley, then tie a slipknot in order to hold it in this position and keep tension off the starter handle.

3. Carefully pry the rope anchor out of the handle, then separate the rope from the anchor and the handle.

4. While holding the rope, release the slipknot, then allow the rope to rewind into the pulley until spring tension is fully released.

5. Remove the spindle nut from the center top of the housing, then invert the housing so the underside is facing upward and remove the spindle screw, washer, spindle, spring washer and friction ring.

6. Carefully pry the retaining ring free of each pawl, then remove both pawls along with the links and friction plate from the underside of the pulley as an assembly.

7. Remove the spindle shim and bushing.

✱✱ CAUTION

Wear safety glasses and sturdy gloves when dealing with the spring. It is under tremendous tension and severe personal injury could result from mishandling.

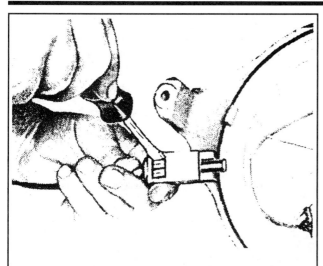

Fig. 56 Use a screwdriver to release the retaining tab on the lockout cable

Fig. 57 On these models the starter rope is secured to an anchor seated within the starter handle

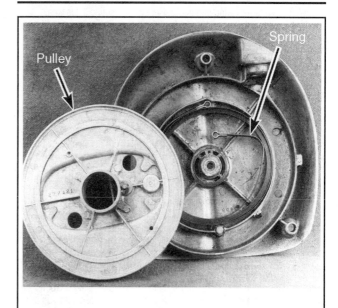

Fig. 58 Typical pulley and spring assembly, as positioned to the housing assembly

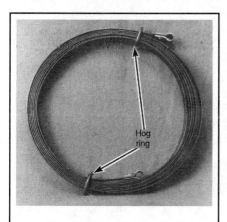

Fig. 59 When installing a new pre-wound spring, remove only the hog ring near the outer loop...

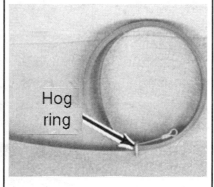

Fig. 60 ...then pull on the outer loop until the spring is small enough to fit in the housing

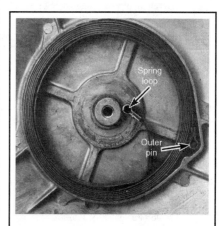

Fig. 61 When installed, make sure the spring is coiled clockwise when followed from inner toward outer loop

10-18 HAND REWIND STARTER

8. Use a gloved hand to hold the pulley into the housing while inverting it again, placing it pulley side down on the workbench. Keep all hands, fingers, toes, eyes, and other precious body parts away from the housing, then gently tap on the housing to dislodge the pulley and allow the spring to completely unwind.

9. Once the spring has finished uncoiling, carefully lift the housing by tilting it so the opening faces away from people or breakables (there may still be some tension to the spring).

10. Carefully remove the pulley and the spring from the inside of the housing, but only after all tension is released.

To assemble:

Wash all metallic parts in solvent, and then blow them dry with low-pressure compressed air.

Remove any trace of corrosion and wipe all metal parts with an oil-dampened cloth.

Inspect the rope. Replace the rope if it appears to be weak or frayed. If the rope is frayed, check the pulley and housing (especially the holes through which the rope passes) for rough edges or burrs. Remove the rough edges or burrs with a file and polish the surface until it is smooth.

■ **Replacement ropes on these models should be trimmed to a length of about 73 1/2 in. (186.6cm). Use a heat source or small open flame to fuse 1/2 in. (12mm) of each end to help prevent fraying.**

Inspect the starter spring end hooks. Replace the spring if it is weak, corroded or cracked. Inspect the inside surface of the pulley rewind recess for grooves or roughness. Grooves may cause erratic rewinding of the starter rope.

Inspect the pawl and links for wear.

Check the friction plate, spindles and links for signs of wear or damage.

Check the starter assembly spring shield for signs of excessive wear and replace, if necessary.

Check the lockout assembly components for wear or damage and replace, as necessary.

The outer spring hook (loop) is installed on the pin located in the starter housing found about halfway between the center of the assembly and one of the assembly legs. The inner spring hook is attached to the pulley. The spring itself should be positioned so that if you travel along the coil from the inner hook to the outer hook (when the spring is positioned in the starter housing) you will travel in a clockwise spiral.

During assembly there are multiple ways to achieve this spring installation. The safest and easiest method is to use the OMC Starter Spring Winder and Installer (#392093). The next best method is to install a pre-wound replacement spring with the outer hook over the pin in the housing, then release the hog ring or metal retaining pin using a metal plate or wooden block to keep the spring in position. Lastly, the spring can be wound by gloved hands (an extra set of them will come in handy for this) using a pin to retain the inner loop of the spring while the rest of the spring is wound around it. If the last method is used, take your time and make sure you are winding the spring in the same direction it was previously wound.

■ **Replacement ropes on these models should be trimmed to a length of about 73 1/2 in. (186.6cm). Use a heat source or small open flame to fuse 1/2 in. (12mm) of each end to help prevent fraying.**

11. Prepare the spring for installation by winding it either by hand or by using the OMC Winder and Installer (#392093), as follows:

   a. Clamp the spring winder base in a vise and then insert the adapter release plate into the winder base.

   b. Apply a light coating of OMC Triple-Guard, or equivalent marine grease (or even Lubriplate 777, if available) to the spring.

   c. Position the inner spring loop in the winder base, then insert the pin of the crank and pin assembly (a rectangular plate with a handle on one end and a pin toward the center, on the opposite face of the plate from the handle) into the inner spring loop.

   d. Using the crank retainer screw, bolt the crank and pin assembly to the winder base.

   e. Turn the crank assembly slowly in the direction indicated on the tool (should be clockwise) in order to draw the spring into the base and wind it. Continue until the outer loop contacts the winder base.

   f. Unbolt and remove the crank retainer screw, then remove the crank and pin assembly from the winder base and spring.

   g. Lift the adapter release plate, along with the spring, from the winder base.

12. If installing a pre-wound replacement spring, they are usually retained by 2 hog rings. Cut the one ring adjacent to the outer loop end of the spring, but leave the other hog ring in place. See if the spring, as it is wound, will fit in the housing, if not, gently pull on the outer loop end while holding the inner loop end to make the would circle tighter. Once the circle is small enough, wrap the outer end back around the spring and the spring is ready to install.

13. Place the starter spring shield onto the pulley.

14. Position the wound spring into the starter assembly housing so the outer spring loop is located over the pin in the housing. If using the adapter tool, press through the holes in the tool to push the spring out of the tool and into the housing.

15. If installing a pre-wound spring, place a wooden block or metal plate against the spring, to keep it from escaping when the second hog ring is cut free, then carefully cut and remove the ring. Remove the block or plate slowly, to make sure the spring won't jump free.

16. Bend the inner spring loop toward the center of the starter assembly in order to align the loop with the pin on the pulley, then slowly lower the pulley into the position, making sure the pin properly engages the loop. Rotate the pulley very slightly against the spring pressure to ensure proper engagement and then slowly rotate the pulley back into the pressure-relieved position.

17. Apply a light coating of OMC Triple-Guard, or equivalent marine grease (or even Lubriplate 777, if available) to the spindle, spindle bushing and the pulley pawl pins.

18. Install the bushing and shim into the starter pulley.

19. Position the friction plate, links and pawls, then secure the assembly using the pawl retaining rings.

■ **When installing the friction ring, keep in mind that the flat section of the ring must align with the flat section of the spindle.**

20. Install the spindle, friction ring and spring washer.

21. Clean all traces of threadlock, corrosion or debris from the spindle screw and nut threads, then install the screw and washer into the starter housing and tighten the screw to 120-145 inch lbs. (14-16 Nm). Next, apply a light coating of OMC Locquic Primer to the portion of the screw threads that protrude through the starter housing. Apply a light coating of OMC Nut Lock, or equivalent threadlock to the nut threads, then install the nut and tighten to 120-145 inch lbs. (14-16 Nm), while holding the screw.

22. If removed, tie a figure "8" knot as illustrated into one end of the replacement starter rope.

23. Position the starter housing upside down on the workbench with the pulley facing upward, then wind the pulley counterclockwise by hand until the spring is tight. Back off the spring about 1/2 to 1 turn and thread the rope through the hole in the pulley **behind** the roll pin and back out through the starter housing.

24. While still holding the pulley from turning (and extra set of hands is handy here too [pun intended]), apply a light coating of OMC Triple-Guard or an equivalent marine grease to the very end of the starter rope. Feed the end of the rope through the starter handle and then press the rope into the rope anchor channel so the rope end firmly contacts the end of the channel. Finally, press the anchor into the handle and give a tug on the rope to seat the anchor into the pulley.

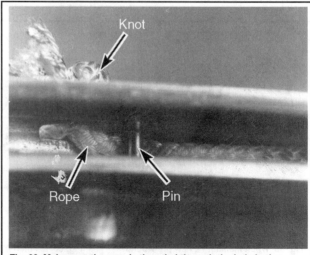

Fig. 62 Make sure the rope is threaded through the hole in the pulley BEHIND the roll pin and that the knot is fully seated

HAND REWIND STARTER 10-19

25. Holding spring tension via the starter rope now, allow the pulley to slowly rewind the rope into the starter housing.

26. Pull the rope back outward, while observing the pawls and feeling for smooth operation. Make sure the pawls extend when the starter rope is pulled and retract when the rope is rewound.

27. Install the assembly, as detailed in this section.

25-55 Hp (737cc) and 25-70 Hp (913cc) Motors

DESCRIPTION

◆ See Figure 63

This hand rewind starter is found on the largest of the carbureted Johnson/Evinrude inline outboards, so it is therefore the largest, most heavy-duty of hand rewind starter assemblies. Similar to most Johnson/Evinrude outboards, the starter is mounted atop of the flywheel on three mounting legs. The unit has a large rectangular pawl plate visible on the underside surface.

Because this starter is used on fairly large horsepower powerheads, the actual number in the field is rather small. As can be imagined, the operator of such a motor must be somewhat stout.

REMOVAL & INSTALLATION

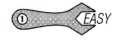

◆ See Figure 63

1. For safety, disconnect the spark plug leads, then ground them to the cylinder head.
2. Remove the screw securing the starter lockout cable clamp to the rear of the housing, then separate the lockout slide from the housing.
3. Remove the 3 bolts (along with lockwashers and round washers) securing the hand rewind starter assembly to the powerhead.
4. Remove the 2 bolts securing the starter handle bracket to the powerhead, then carefully lift the starter housing assembly from the powerhead

To install:

5. Position the starter over the flywheel with the three legs aligned over the holes in the powerhead for the retaining bolts. Make sure the lifting ring is place on the rear screw mounting boss.
6. Install the 5 retaining bolts (3 for the housing and 2 for the handle bracket). Make sure that the washers are placed between the starter housing rubber mounts and the motor.
7. Tighten the 3 starter housing bolts to 120-144 inch lbs. (14-16 Nm), then tighten the 2 handle bracket screws to 60-84 inch lbs. (7-9 Nm).
8. Apply a light coating of OMC Triple-Guard, or equivalent marine grease (or even Lubriplate 777, if available) to the starter lockout slide channel on the starter housing. Next, install the lockout slide into the channel and loosely secure the cable to the housing using the clamp and screw.
9. With the gearcase in **neutral** adjust the lockout cable so the lockout slide is centered on the lever, then tighten the cable clamp screw securely to hold this adjustment.
10. Shift the lower unit into forward and then reverse gear and make a check of the starter lockout system. The starter must be locked, and unable to rotate when the lower unit is in any gear other than neutral.
11. If the starter fails this test, loosen the cable clamp and adjust the position of the slide until no motion of the starter is possible when the lower unit is in forward or reverse gear.
12. Reconnect the spark plug leads.

DISASSEMBLY & ASSEMBLY

◆ See Figures 9 and 63 thru 66

1. Pull the rope out enough to tie a knot in the rope. Tie a slipknot, and then allow the rope to rewind to the knot, holding the rope out of the housing so you can remove the handle.
2. Work the rope anchor out of the rubber covered handle, and then remove the rope from the anchor. Remove the handle from the rope. Untie the knot in the rope, but hold and prevent the disc pulley from rotating. Now, slack the hold on the pulley, and permit it to turn thus winding the rope back onto the pulley slowly. Continue to allow the spring in the pulley to unwind slowly until all tension has been released.

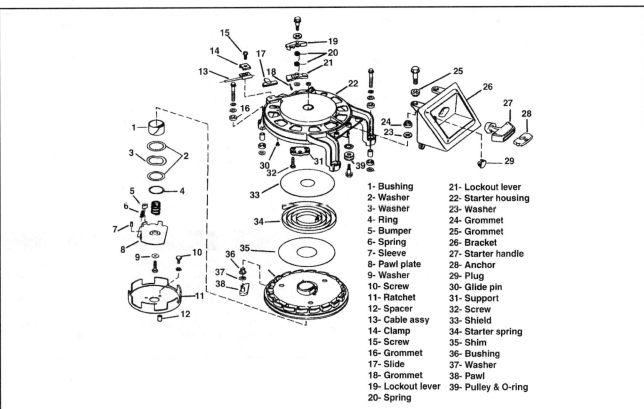

Fig. 63 Exploded view of the hand rewind starter assembly - 3-cylinder motor shown (2-cylinder motors very similar)

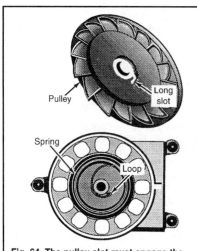

Fig. 64 The pulley slot must engage the inner hook of the spring to ensure correct starter function

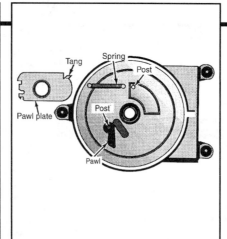

Fig. 65 The pawl plate return spring is positioned between the plate tang and pulley post

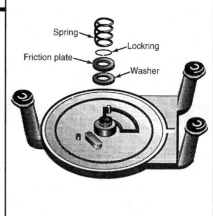

Fig. 66 The friction plates (lower not pictured) and spring washer are secured using the pulley lock-ring

3. If necessary, unbolt the starter handle bracket from the housing assembly.

4. Remove the shoulder screw retaining the rope guide pulley to the edge of the starter housing, then remove the pulley and O-ring.

5. Remove the bolt retaining the lockout lever, the lockout lever springs, housing washer and, on the other side of the housing, the pulley support with screw. Remove the components.

6. Place one hand under the housing to prevent all the large pawl plate and other small parts from flying loose. Keep one finger over the center bolt.

7. Remove the large center nut and washer from atop the starter housing. Carefully turn the starter housing over without disturbing any parts.

✳✳ CAUTION

The rewind spring is a potential hazard. The spring is under tremendous tension when it is wound - a real "tiger" in a cage! If the spring should accidentally be released, severe personal injury could result from being struck by the spring with force. Therefore, the following steps must be performed with care to prevent personal injury to self and others in the area.

8. Lift off the center bolt and washer, and then lift off the large pawl plate and the plate return spring from the side of the starter housing.

9. Remove the starter pawl and spring washer from the pawl pivot post.

10. Remove the center spring and then use a screwdriver to pry out the pulley lock-ring from under the spring. Lift off the friction plate and the spring washer beneath the plate.

11. Hold the pulley and the housing together tightly and turn the complete assembly over with the legs extending downward in the normal manner. Now, lower the complete assembly to the floor. When the legs make contact with the floor, release your grip. The pulley will fall and the spring will be released from the housing almost instantly and with considerable force. However, the three legs should contain the spring and prevent it from lashing out causing possible injury to self or others in the area. If the spring was not released from the housing as just described, the only safe method is to again jar the three legs on the floor and dislodge the spring.

12. If necessary, unwind the rope out of the pulley groove.

To assemble:

Wash all metallic parts in solvent, and then blow them dry with low-pressure compressed air.

Remove any trace of corrosion and wipe all metal parts with an oil-dampened cloth.

Inspect the rope. Replace the rope if it appears to be weak or frayed. If the rope is frayed, check the pulley and housing (especially the holes through which the rope passes and the rope guide pulley) for rough edges or burrs. Remove the rough edges or burrs with a file and polish the surface until it is smooth.

■ **Replacement ropes on these models should be trimmed to a length of about 96 1/2 in. (245cm). Use a heat source or small open flame to fuse 1/2 in. (12mm) of each end to help prevent fraying.**

Inspect the starter spring end hooks. Replace the spring if it is weak, corroded or cracked. Inspect the inside surface of the pulley rewind recess for grooves or roughness. Grooves may cause erratic rewinding of the starter rope.

Check all starter components for signs of wear or damage. Be sure to check the pawl for wear and replace, as necessary.

Check the lockout assembly components for wear or damage and replace, as necessary.

The outer spring hook (loop) is installed on the pin located in the starter housing found about halfway between the center of the assembly and one of the assembly legs. The inner spring hook is attached to the pulley. The spring itself should be positioned so that if you travel along the coil from the inner hook to the outer hook (when the spring is positioned in the starter housing) you will travel in a clockwise spiral.

During assembly there are multiple ways to achieve this spring installation. The safest and easiest method is to use the OMC Starter Spring Winder and Installer (#392093). The next best method is to hand wind the spring (using gloved hands, and keeping in mind than an extra set of them will come in handy for this). When using this method, you actually wind the spring starting at the outer loop instead of the inner and work in the opposite direction (turning it in the opposite direction too) towards center). If this method is used, take your time and make sure you are winding the spring in the same direction it was previously wound. Lastly, in some cases, it may be possible to install a pre-wound replacement spring with the outer hook over the pin in the housing, then release the hog ring or metal retaining pin using a metal plate or wooden block to keep the spring in position.

13. Place the starter housing upside-down (spring side facing upward) on the workbench and then install the starter spring shield into the starter housing.

✳✳ CAUTION

Wear a good pair of gloves while installing the spring. The spring will develop tension and the edges of the spring steel are sharp. The gloves will prevent cuts on hands and fingers. It is also strongly recommended that you wear a pair of safety goggles or a face shield while the spring is being installed. As the work progresses a "tiger" is being forced into a cage. If the spring is accidentally released, it will lash out with tremendous ferocity and very likely could cause personal injury to the installer or other persons nearby.

HAND REWIND STARTER   10-21

14. If available, prepare the spring for installation by winding using the OMC Winder and Installer (#392093), as follows:

   a. Clamp the spring winder base in a vise and then insert the release plate into the winder base.

   b. Apply a light coating of OMC Triple-Guard, or equivalent marine grease (or even Lubriplate 777, if available) to the spring.

   c. Position the inner spring loop in the winder base (with the loop facing inward). Next, insert the pin of the crank and pin assembly (a rectangular plate with a handle on one end and a pin toward the center, on the opposite face of the plate from the handle) into the inner spring loop.

   d. Using the crank retainer screw, bolt the crank and pin assembly to the winder base.

   e. Turn the crank assembly slowly in the direction indicated on the tool (should be clockwise) in order to draw the spring into the base and wind it. Continue until the outer loop contacts the winder base.

   f. Unbolt and remove the crank retainer screw, then remove the crank and pin assembly from the winder base and spring.

   g. Lift the adapter release plate, along with the spring, from the winder base.

   h. Position the wound spring into the starter assembly housing so the outer spring loop is located over the pin in the housing. Press through the holes in the tool to push the spring out of the tool and into the housing.

15. If the spring winder tool is not available, wind it by hand, as follows:

   a. Slide the spring onto the outer pin and then start the spring from the outside edge of the housing and insert it into the housing **counterclockwise**. In the end, this will have the same final result in spring winding (i.e. that you would follow the spring in a clockwise patter from the inner loop toward the outer loop).

   b. Work the first turn into the housing, and then hold the spring down with one hand and continue to wind the spring into the housing.

■ Patience and time are required to work the spring completely into the housing.

16. If installing a pre-wound replacement spring, proceed as follows:

   a. Replacement springs are usually retained by 2 hog rings. Cut the one ring adjacent to the outer loop end of the spring, but leave the other hog ring in place.

   b. See if the spring, as it is wound, will fit in the housing, if not, gently pull on the outer loop end while holding the inner loop end to make the would circle tighter. Once the circle is small enough, wrap the outer end back around the spring and the spring is ready to install.

   c. Position the wound spring into the starter assembly housing so the outer spring loop is located over the pin in the housing. If using the adapter tool, press through the holes in the tool to push the spring out of the tool and into the housing.

17. If installing a pre-wound spring, place a wooden block or metal plate against the spring, to keep it from escaping when the second hog ring is cut free, then carefully cut and remove the ring. Remove the block or plate slowly, to make sure the spring won't jump free.

18. Apply a light coating of OMC Triple-Guard, or equivalent marine grease (or even Lubriplate 777, if available) to the pulley bushing and to the boss of the starter pawl. Install the bushing onto the pulley, then put the shim in place.

■ The starter pawl boss should be installed in the pulley, but, if it was removed or has become dislodged, reserve it for installation later.

19. Bend the inner spring loop toward the center of the starter assembly in order to align the loop with the slot in the pulley, then slowly lower the pulley into the position, making sure the slot properly engages the loop. Rotate the pulley very slightly against the spring pressure to ensure proper engagement, then slowly rotate the pulley back into the pressure-relieved position.

20. Install the friction plate spring washer between the 2 friction plates and position the assembly over the pulley hub. Secure the assembly by positioning the lock-ring over the plate and gently tapping around the circumference of the ring until it is well seated into the groove of the hub. Place the starter housing spring over the lock-ring.

21. Make sure the lubricated starter pawl boss is in place, or install it at this time, then position the spring washer and install the pawl.

22. Make sure the starter housing spring is positioned in the center of the pulley.

■ Read ahead, it is easier to clean the threads of the pawl plate retaining screw and nut at this point.

23. Hook one end of the pawl plate return spring on the pawl plate tang, then pressure the other end of over the pulley post. Slide the pawl plate into position to align with the pulley hub.

24. Clean all traces of threadlock, corrosion or debris from the pawl plate retaining screw and nut threads, then install the screw and washer though the pawl plate and thread it into the starter housing. Tighten the pawl plate screw to 120-144 inch lbs. (14-16 Nm). Next, apply a light coating of OMC Locquic Primer to the portion of the screw threads that protrude through the starter housing. Apply a light coating of OMC Nut Lock, or equivalent threadlock to the nut threads, then install the nut and tighten securely.

25. Install the rope guide pulley and O-ring, then secure using the shoulder screw. Tighten the screw securely.

26. If removed, install the starter handle bracket to the housing, making sure to place the washers between the rubber mounts of the bracket and the housing. Tighten the shoulder screws securely.

27. If removed, tie a figure "8" knot as illustrated into one end of the replacement starter rope.

28. Position the starter housing upside down on the workbench with the pulley facing upward, then wind the pulley counterclockwise by hand until the spring is tight. Back off the spring just until the rope cavity of the pulley is aligned with the rope guide, then thread the rope through the pulley, guide and starter handle bracket. Make sure the knot in the rope fully seats in the pulley.

29. Tie a slipknot in the rope to hold the pulley in position (or use the help of an assistant), then apply a light coating of OMC Triple-Guard or an equivalent marine grease to the very end of the starter rope. Thread the rope through the starter handle (a Starter Rope Threading Tool {#378774} is available to make this task easier). Then press the rope into the rope anchor channel so the rope end firmly contacts the end of the channel. Next, press the anchor into the handle and give a tug on the rope to seat the anchor into the pulley.

30. Untie the slipknot and hold spring tension via the starter rope and then allow the pulley to slowly rewind the rope into the starter housing.

31. Pull the rope back outward, while observing the pawl and feeling for smooth operation. Make sure the pawl extends when the starter rope is pulled and retracts when the rope is rewound.

32. Install the assembly, as detailed in this section.

33. Install the starter lockout lever, springs (with the lower tang of the upper spring positioned on the casting), and thick washer onto the housing. Install the bolt with the thin washer under the bolt head and tighten securely. Apply a thin coating of OMC Nut Lock or equivalent threadlock to the threads of the support screw, then install the pulley support and tighten the screw to 60-84 inch lbs. (7-9 Nm).

34. Install the hand rewind starter assembly as detailed in this section.

EFI Motors

The EFI Johnson/Evinrude motors are not equipped with a hand rewind starter assembly. However, a cord is normally supplied with the tool kit for emergency starting. For more details, please refer to Emergency Starting, in this section.

10-22 HAND REWIND STARTER

EMERGENCY STARTING

General Information

◆ See Figure 10, 67 and 68

✱✱ CAUTION

If an emergency rope is used on any powerhead never wrap the rope end around your hand for a better grip. If the powerhead should happen to backfire, the sudden jerk on the rope would severely injure an individual's hand.

✱✱ CAUTION

The manufacturer does not recommend the use of jumper cables or booster batteries on Johnson/Evinrude motors, especially because of the danger of sparks igniting the explosive vapors normally found in and around a discharged battery.

Failures to the starting system for your outboard can leave you stranded on the water. In these circumstances, these Johnson/Evinrude outboards can be pull-started using an emergency procedure (provided that all other systems are functioning properly). This is even true on the larger electric start models (including the EFI motors).

Most of the smaller electric start motors will also retain the manual hand rewind starter assembly, which can be used in the event of a failure of the electric start system. But, if the rope breaks, or for electric start models that do not have hand rewind starter housing installed, an emergency rope can be used to crank the motor.

For smaller motors, you'll have to use a piece of the broken starter rope or a substitute piece of 9/16 in. (3.66mm) cord which is about 4 ft. (1.2mm) in length. Larger motors should have an emergency pull rope (complete with handle) in the tool kit or mounted in a compartment on the engine top cover (refer to your owner's manual for more details).

Before attempting to pull start a motor, first check the following:
- Perform all pre-use checks. For details, please refer to the information under Engine Maintenance and Tune-Up.
- Make sure all other systems are functioning properly and properly connected. There must be fuel and, for remote oil 2-stroke, oil, in the tank(s).
- Make sure the battery switch (for electric start models) is **ON** and functioning properly. Make sure the battery is properly connected.
- Loosen the fuel tank filler cap to vent any pressure or vacuum.
- Make sure the engine is in the normal operating trim.
- The fuel pump primer bulb must be firm.
- The gearshift must be in **neutral**
- The fast idle lever (if applicable) is raised to the best start position.

To start the engine using an emergency pull rope, proceed as follows:
1. Remove the engine top cover.
2. If equipped, remove the hand rewind starter assembly. For details, please refer to the procedures in this section.
3. For electric models so equipped, remove the flywheel cover (this includes all EFI motors).
4. Locate a suitable length or rope or an emergency pull start rope (supplied with most electric only start motors).
5. If the end of the rope is not already tied, make a figure "8" knot in the end of the rope, as shown in the accompanying illustration.
6. Follow all normal starting procedures and make sure the keyswitch is in the **ON** position.

■ For EFI motors, listen when the keyswitch is turned on to hear if the electric fuel pump runs to prime the system. If necessary, cycle the ignition OFF, wait a few seconds, then back on once or twice to ensure the system is primed.

7. Place the rope knot in the notch on the flywheel pulley and then wrap the rope around the flywheel in a clockwise direction (when viewed from above).

■ On electric start models, make sure the rope and knot will clear the starter motor pinion when pulled.

8. On manual prime motors, move the primer lever upward to the manual start (prime) position, squeeze the primer bulb once, then move the primer lever back downward to the normal run position.
9. Make sure you (and anyone else on board) are seated (for safety). Also, if there is no handle on the rope end, **make sure you do not wrap the rope around your hand**.
10. Pull the starter cord quickly to rotate the engine. On EFI models, the first pull is needed to energize the electrical system. For these models you will most definitely have to rewind the cord and pull again.
11. Continue to rewind and pull the cord until the motor starts (or someone sees you, has pity, and offers a tow).

■ If the electrical system on electric start models is functioning properly, the battery should now begin to charge. BUT, if the battery was low or dead because of a problem with the system no charging will occur. Furthermore, on EFI engines, the motor will only run until voltage drops to a level that cannot sustain ignition and then the motor will stall.

Fig. 67 For emergency starts, tuck the rope knot into the flywheel notch and wind it clockwise around the flywheel

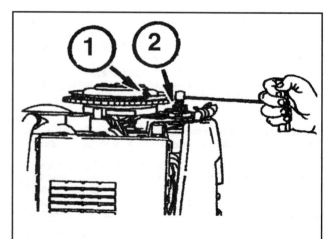

Fig. 68 On electric start models, make sure the rope and knot (1), will clear the starter motor pinion (2)

MASTER INDEX   10-23

A

AC LIGHTING COIL.. 4-66
  Removal & Installation... 4-67
  Testing... 4-66
ACCUMIX OIL INJECTION SYSTEM..................................... 5-2
  Assembly.. 5-3
  Cleaning And Inspection... 5-3
  Removal... 5-2
AIR INTAKE SILENCER AND FLAME ARRESTER........................... 3-51
  Removal & Installation.. 3-51
ANCHORS... 1-10
ANODES (ZINCS).. 2-35
  Inspection.. 2-35
  Servicing... 2-35

B

BATTERY... 4-67
BAILING DEVICES... 1-10
BOAT MAINTENANCE... **2-37**
  Batteries... 2-37
  Fiberglass Hull... 2-39
BOATING EQUIPMENT (NOT REQUIRED BUT RECOMMENDED).............. **1-10**
  Anchors... 1-10
  Bailing Devices... 1-10
  Compass... 1-10
  First Aid Kit... 1-10
  Tools And Spare Parts... 1-12
  VHF-FM Radio.. 1-10
BOATING SAFETY... **1-4**
  Courtesy Marine Examinations.................................... 1-10
  Regulations For Your Boat....................................... 1-4
  Required Safety Equipment....................................... 1-5
BOLTS, NUTS AND OTHER THREADED RETAINERS......................... 1-26
BREAKER POINT IGNITION SYSTEMS................................... 2-47
  General Information... 2-47
  Inspection & Testing.. 2-47
  Replacement... 2-48
BREAKING IN A POWERHEAD.. 6-88
  2-Stroke Motors... 6-88
  Carbureted 4-Stroke... 6-88
  EFI 4-Stroke.. 6-89

C

CAMSHAFT POSITION (CMP) SENSOR................................... 3-60
  Removal & Installation.. 3-61
  Testing... 3-60
CARBURETED FUEL SYSTEM.. **3-11**
  Description And Operation....................................... 3-11
  Troubleshooting... 3-12
  Fuel Pump... 3-34
  Manual Fuel Primer.. 3-38
  Electric Fuel Primer.. 3-39
CARBURETOR
  2.0-3.5 Hp Motors... 3-14
  5 Hp (109cc) Motors... 3-16
  Colt/Junior, 3-4 Hp & 4 Deluxe Motors........................... 3-18
  5/6 Hp 4-Stroke Motors.. 3-20
  8/9.9 & 9.9/15 Hp 4-Stroke Motors............................... 3-22
  9.9/10/14/15 Hp (216/255cc) Motors Thru 1993 & All 18 Jet-35 Hp
  (521cc) Motors.. 3-24
  25/35 Hp (500/565cc) Motors..................................... 3-28
  All Except Above, Including 5/6/6.5/8 (164cc), 9.9/10/15 (255cc, 1994
  And Later), 25-55 (737cc) & 25-70 (913cc) Motors................ 3-30
  Cleaning & Inspection... 3-33
CHARGE COIL... 4-24
  Removal & Installation.. 4-29
  Testing... 4-24
CHARGING CIRCUIT.. **4-57**
  AC Lighting Coil.. 4-66
  Battery... 4-67
  Charging System Identification.................................. 4-58
  General Information... 4-57
  Rectifier... 4-62
  Regulator/Rectifier... 4-64
  Service Precautions... 4-57
  Stator/Battery Charge Coil...................................... 4-61
  Troubleshooting... 4-58
CHARGING SYSTEM IDENTIFICATION................................... 4-58
CHEMICALS... 1-17
  Cleaners.. 1-18
  Lubricants & Penetrants... 1-17
  Sealants.. 1-18
CLEARING A SUBMERGED MOTOR.................................... **2-83**
CLOSED THROTTLE POSITION (CTP) SWITCH............................ 3-67
  Removal & Installation.. 3-67
  Testing... 3-67
COOLING SYSTEM... **5-21**
  Cooling System Schematics....................................... 5-40
  Description & Operation... 5-22
  Flushing.. 2-16
  Thermostat.. 5-37
  Troubleshooting... 5-23
  Water Pressure Relief Valve..................................... 5-39
  Water Pump.. 5-25
COOLING SYSTEM SCHEMATICS.. 5-40
  Colt/Junior Motors.. 5-40
  2.0-3.5 hp (78cc)... 5-41
  3-4 hp (87cc)... 5-41
  4 deluxe (87cc)... 5-42
  5 hp (109cc).. 5-42
  5/6 hp (128cc) 4-stroke... 5-43
  5/6/8 hp (164cc).. 5-43
  8/9.9 hp (211cc) 4-stroke....................................... 5-44
  9.9-15 hp (216/255cc)... 5-44
  9.9/15 hp (305cc) 4-stroke...................................... 5-45
  18 jet-35 hp (521cc).. 5-45
  25/35 hp (500/565cc).. 5-46
  45/55 hp (737cc) Commercial..................................... 5-46
  25-55 hp (737cc).. 5-47
  25-70 hp (913cc).. 5-47
  40/50 hp (815cc) EFI.. 5-48
  70 hp (1298cc) EFI.. 4-48
COMPASS... 1-10
COMPRESSION TESTS... 2-41
  Leakage Check... 2-42
  Tune-Up Check... 2-41
CONTROL CABLES... **9-14**
  Rigging... 9-14
COURTESY MARINE EXAMINATIONS..................................... 1-10
CRANKSHAFT POSITION (CKP) SENSOR................................. 3-61
  Removal & Installation.. 3-62
  Testing... 3-62

D

DE-CARBONING THE PISTONS... 2-41
DUAL HANDLE SURFACE MOUNT REMOTE FOR
SINGLE MOTORS.. 9-7
  Overhaul.. 9-7

E

EFI SYSTEM RELAY... 3-67
  Testing & Service... 3-67
ELECTRIC FUEL PRIMER... 3-39
  Function Test... 3-39
  Removal And Installation.. 3-40
  Solenoid Check.. 3-40
ELECTRONIC (CDI/UFI) IGNITION.................................... 2-51
  Inspection.. 2-51
ELECTRONIC FUEL INJECTION (EFI)............................... **3-41**
  Air Intake Silencer And Flame Arrester.......................... 3-51
  Camshaft Position (CMP) Sensor.................................. 3-60
  Closed Throttle Position (CTP) Switch........................... 3-67
  Crankshaft Position (CKP) Sensor................................ 3-61
  Description & Operation... 3-41
  EFI System Relay.. 3-67
  Engine Control Unit (ECU)....................................... 3-62
  Engine Symptom Diagnostic Charts................................ 3-47
  Fuel Rail And Injectors... 3-58
  Idle Air Control (IAC) Valve.................................... 3-68
  Intake Manifold... 3-52

MASTER INDEX

Low Pressure Fuel Pump ... 3-53
Manifold Absolute Pressure (MAP) Sensor ... 3-66
Neutral (Safety) Switch ... 3-69
Self Diagnostic System ... 3-43
Sensor And Circuit Resistance/Output Tests ... 3-45
Temperature Sensors ... 3-63
Throttle Body ... 3-51
Troubleshooting ... 3-41
Vapor Separator Tank & Fuel Pump ... 3-55
ELECTRONIC TOOLS ... **1-23**
  Battery Chargers ... 1-24
  Battery Testers ... 1-23
  Gauges ... 1-24
  Multi-Meters (DVOMS) ... 1-24
EMERGENCY STARTING ... **10-22**
  General Information ... 10-22
ENGINE CONTROL UNIT (ECU) ... **3-62**
  Removal & Installation ... 3-63
ENGINE COVERS ... **2-15**
  Removal & Installation ... 2-15
ENGINE IDENTIFICATION ... **2-2**
  Engine Serial Numbers ... 2-3
ENGINE MAINTENANCE ... **2-15**
  Anodes (Zincs) ... 2-35
  Cooling System ... 2-16
  Engine Covers ... 2-15
  Engine Oil (2-Stroke) ... 2-18
  Engine Oil/Filter (4-Stroke) ... 2-20
  Fuel Filter ... 2-25
  Gearcase (Lower Unit) Oil ... 2-23
  Jet Drive Impeller ... 2-32
  Propeller ... 2-29
  RescuePro Rotor ... 2-33
  Timing Belt ... 2-36
ENGINE OIL (2-STROKE) ... **2-18**
  Filling ... 2-19
  Recommendations ... 2-18
ENGINE OIL/FILTER (4-STROKE) ... **2-20**
  Checking Oil Level ... 2-20
  Oil/Filter Change ... 2-21
  Recommendations ... 2-20
ENGINE SYMPTOM DIAGNOSTIC CHARTS ... **3-47**
  Engine Cranks But Won't Run ... 3-47
  Engine Idles Improperly ... 3-48
  Engine Overheats ... 3-50
  Engine Runs Rough/Lacks Power ... 3-49
  Engine Won't Crank ... 3-47
ENGINE TEMPERATURE SENSORS (EFI) ... **4-101**
ENGINE TEMPERATURE SWITCHES (CARB) ... **4-101**
  Removal & Installation ... 4-102
  Testing ... 4-101

F
FASTENERS, MEASUREMENTS AND CONVERSIONS ... **1-26**
  Bolts, Nuts And Other Threaded Retainers ... 1-26
  Standard And Metric Measurements ... 1-27
  Torque ... 1-27
FLYWHEEL ... **6-2**
  Removal & Installation ... 6-2
FLYWHEEL & BREAKER POINT ... **4-11**
  Installation ... 4-15
  Removal ... 4-11
FUEL ... **3-2**
  Alcohol-Blended Fuels ... 3-3
  Checking ... 3-3
  High Altitude Operation ... 3-3
  Octane Rating ... 3-3
  Recommendations ... 3-3
  Vapor Pressure ... 3-3
FUEL FILTER ... **2-25**
  Carbureted Motors ... 2-26
  EFI Motors ... 2-28
FUEL LINES AND FITTINGS ... **3-8**
  Service ... 3-9
  Testing ... 3-8

FUEL PRIMER
  Manual ... 3-38
  Electric ... 3-39
FUEL PUMP - CARB ... 3-34
  Overhaul ... 3-37
  Removal & Installation ... 3-26
  Testing ... 3-34
FUEL RAIL AND INJECTORS ... 3-58
  Injector Operational Test ... 3-59
  Injector Resistance Test ... 3-59
  Injector Signal Test ... 3-59
  Removal & Installation ... 3-59
FUEL SYSTEM BASICS ... **3-2**
  Fuel ... 3-2
  Fuel System Pressurization ... 3-4
  Fuel System Service Cautions ... 3-2
FUEL SYSTEM PRESSURIZATION ... **3-4**
  Pressurizing (Checking For Leaks) ... 3-5
  Relieving Pressure (EFI Only) ... 3-5
FUEL SYSTEM SERVICE CAUTIONS ... **3-2**
FUEL TANK ... **3-6**
  Service ... 3-6
FUEL TANK AND LINES ... **3-6**
  Fuel Lines And Fittings ... 3-8
  Fuel Tank ... 3-6

G
GEARCASE (LOWER UNIT) OIL ... **2-23**
  Checking ... 2-24
  Draining & Filling ... 2-25
  Recommendations ... 2-24
GEARCASE ... **7-2**
  Checking ... 2-24
  Description And Operation ... 7-2
  Draining & Filling ... 2-25
  Gearcase Assembly ... 7-2
  Propeller Shaft Seal ... 7-18
  Recommendations ... 2-24
GEARCASE REMOVAL & INSTALLATION ... **7-2**
  Colt/Junior (43cc) ... 7-2
  2.0-3.5 Hp (78cc) ... 7-3
  5 Hp (109cc) ... 7-4
  3/4 Hp (87cc) ... 7-4
  4 Deluxe (87cc) ... 7-7
  5/6 Hp (128cc) ... 7-7
  5/6/8 Hp (164cc) ... 7-7
  8/9.9 Hp (211cc) ... 7-8
  9.9-15 Hp (216/255cc) ... 7-8
  9.9/15 Hp (305cc) ... 7-8
  10/14 (216/255cc) ... 7-10
  20-35 Hp (521cc) ... 7-10
  25/35 Hp (500/565cc) ... 7-10
  25-55 Hp (737cc) ... 7-13
  25-70 Hp (913cc) (Exc. Jet) ... 7-13
  40/50 Hp (815cc) EFI ... 7-15
  70 Hp (1298cc) EFI ... 7-17

H
HAND REWIND STARTERS ... **10-2**
  Introduction ... 10-2
  General Information ... 10-2
  2.0-3.5 Hp (78cc) Motors ... 10-2
  5 Hp (109cc) Motors ... 10-5
  9.9-15 Hp (216cc) Motors ... 10-6
  Colt/Junior And 3-35 Hp (Under 600cc) Motors (Exc. 5 Hp/109cc, 4 Deluxe, 9.9-15 Hp/216cc And 18 Jet-35 Hp/521cc Models) ... 10-9
  4 Deluxe (87cc) Motors ... 10-13
  18 Jet-35 Hp (521cc) Motors ... 10-16
  25-55 Hp (737cc) And 25-70 Hp (913cc) Motors ... 10-19
  EFI Motors ... 10-21
HAND TOOLS ... **1-19**
  Hammers ... 1-22
  Pliers ... 1-21
  Screwdrivers ... 1-22
  Socket Sets ... 1-19

MASTER INDEX

Wrenches	1-21

HOW TO USE THIS MANUAL ... **1-2**
- Avoiding The Most Common Mistakes ... 1-3
- Avoiding Trouble ... 1-2
- Can You Do It? ... 1-2
- Directions And Locations ... 1-2
- Maintenance Or Repair? ... 1-2
- Professional Help ... 1-3
- Purchasing Parts ... 1-3
- Where To Begin ... 1-2

I

IDLE AIR CONTROL (IAC) VALVE ... 3-68
- Removal & Installation ... 3-68
- Testing ... 3-68

IGNITION COIL & ARMATURE - BREAKER POINT ... 4-11
- Installation ... 4-14
- Removal ... 4-11

IGNITION COILS - ELECTRONIC ... 4-51
- Description & Operation ... 4-51
- Removal & Installation ... 4-54
- Testing ... 4-52

IGNITION SYSTEMS (BREAKER POINT MAGNETO) ... **4-8**
- Cleaning & Inspection ... 4-13
- Flywheel & Breaker Point ... 4-11
- General Information ... 4-8
- Ignition Coil & Armature ... 4-11
- Ignition Service ... 4-10
- Top Seal ... 4-13
- Troubleshooting ... 4-8

IGNITION SYSTEMS (ELECTRONIC) ... **4-15**
- Charge Coil ... 4-24
- Description And Operation ... 4-15
- Ignition Coils ... 4-51
- Power Coil ... 4-39
- Power Pack (Ignition Module) - Carb Only ... 4-45
- Sensor/Trigger Coil ... 4-41
- Troubleshooting ... 4-16

INTAKE MANIFOLD ... 3-52
- Removal & Installation ... 3-52

INTEGRAL OIL MIXING UNIT SYSTEM (25/35 HP 3-CYL) ... 5-11
- Overhaul ... 5-13
- Removal/Installation ... 5-12

J

JET DRIVE ... **7-29**
- Description & Operation ... 7-29
- Impeller ... 7-29
- Jet Drive Assembly ... 7-31
- Reverse Gate ... 7-32

JET DRIVE ASSEMBLY ... 7-31
- Removal & Installation ... 7-31

JET DRIVE BEARING ... 2-7
- Recommended Lubricant ... 2-7
- Daily Bearing Lubrication ... 2-7
- Grease Replacement ... 2-7

JET DRIVE IMPELLER ... 2-32
- Checking ... 2-32
- Inspection ... 2-32

L

LOW PRESSURE FUEL PUMP - EFI ... 3-53
- Overhaul ... 3-55
- Pump Pressure Test ... 3-54
- Removal & Installation ... 3-54

LUBRICATION SERVICE ... **2-6**
- Electric Starter Motor Pinion ... 2-6
- Engine Cover Latches ... 2-7
- Engine Mount Clamp Screws ... 2-7
- Jet Drive Bearing ... 2-7
- Power Trim/Tilt Reservoir ... 2-8
- Linkage, Cables And Shafts ... 2-8
- Steering ... 2-13
- Swivel Bracket ... 2-14
- Tilt Assembly ... 2-14

LUBRICATION SYSTEMS (2-STROKE) ... **5-2**
- Accumix Oil Injection System ... 5-2
- Integral Oil Mixing Unit System (25/35 Hp 3-Cyl) ... 5-11
- Oil Tank And Mixing Unit ... 5-12
- Pulse Limiter ... 5-7
- Sending Unit ... 5-12
- System Verification & Troubleshooting ... 5-3
- Troubleshooting ... 5-11
- Variable Ratio Oil (VRO2) Oil Injection System ... 5-3
- VRO2 Pickup And Oil Supply Hose ... 5-7
- VRO2 Pump ... 5-8

LUBRICATION SYSTEM (4-STROKE) ... **5-14**
- Description And Operation ... 5-14
- Oil Pressure ... 5-14
- Oil Pressure Switch And Warning System ... 5-21
- Oil Pump ... 5-17

M

MANIFOLD ABSOLUTE PRESSURE (MAP) SENSOR ... 3-66
- Removal & Installation ... 3-66
- Testing ... 3-66

MANUAL FUEL PRIMER ... 3-38
- Assembly/Installation ... 3-39
- Cleaning and Inspection ... 3-39
- Function Test ... 3-38
- Primer Check ... 3-38
- Removal/Disassembly ... 3-38

MEASURING TOOLS ... 1-25
- Depth Gauges ... 1-26
- Dial Indicators ... 1-25
- Micrometers & Calipers ... 1-25
- Telescoping Gauges ... 1-26

N

NEUTRAL (SAFETY) SWITCH ... 3-69
- Removal & Installation ... 3-70
- Testing ... 3-69

O

OIL LEVEL SWITCH (OIL INJECTED 2-STROKES) ... 4-103

OIL PRESSURE SWITCH (4-STROKE ONLY) ... 4-102
- Removal & Installation ... 4-103
- Testing ... 4-103

OIL PUMP - 4-STROKE ... 5-17
- Removal, Overhaul & Installation ... 5-17

P

POWER COIL ... 4-39
- Removal & Installation ... 4-41
- Testing ... 4-39

POWERHEAD ... **6-2**
- Flywheel ... 6-2
- Powerhead ... 6-4
- Timing Belt/Chain ... 6-84

POWERHEAD ... 6-4
- Service ... 6-4
- Cleaning & Inspection ... 6-72

POWERHEAD BREAK-IN ... **6-87**
- Breaking In A Powerhead ... 6-88

POWERHEAD SERVICE ... 6-4
- Colt/Junior ... 6-5
- 2.0-3.5 Hp (78cc) ... 6-7
- 5 Hp (109cc) ... 6-9
- 5/6 Hp (128cc) 4-Stroke ... 6-11
- 3/4 Hp & 4 Deluxe (87cc) ... 6-16
- 5/6/8 Hp (164cc) ... 6-16
- 8/9.9 Hp (211cc) 4-Stroke ... 6-23
- 9.9/10/14/15 Hp (216cc) ... 6-28
- 9.9/10/15 Hp (255cc) ... 6-28
- 9.9/15 Hp (305cc) 4-Stroke ... 6-33
- 18 Jet-35 Hp (521cc) ... 6-38
- 25/35 Hp (500/565cc) ... 6-43
- 25-55 Hp (737cc) ... 6-49
- 25-70 Hp (913cc) ... 6-49
- 40/50 Hp (815cc) Efi ... 6-58
- 70 Hp (1298cc) Efi ... 6-65
- Cleaning & Inspection ... 6-72

10-25

10-26 MASTER INDEX

POWER PACK (IGNITION MODULE) - CARB ONLY . . . 4-45
Removal & Installation . . . 4-51
Testing . . . 4-46
POWER TRIM/TILT RESERVOIR . . . 2-8
Fluid Level/Condition . . . 2-8
Recommended Lubricant . . . 2-8
PRE-WIRED BINNACLE MOUNT REMOTE . . . 9-11
Overhaul . . . 9-11
PROPELLER SHAFT SEAL REPLACEMENT . . . 7-18
Colt/Junior (43cc) . . . 7-18
2.0-3.5 Hp (78cc) . . . 7-18
5 Hp (109cc) . . . 7-18
3/4 Hp (87cc) . . . 7-19
4 Deluxe (87cc) . . . 7-20
5/6 Hp (128cc) . . . 7-20
5/6/8 Hp (164cc) . . . 7-20
8/9.9 Hp (211cc) . . . 7-21
9.9-15 Hp (216/255cc) . . . 7-21
9.9/15 Hp (305cc) . . . 7-21
10/14 Hp (211cc/255cc) . . . 7-23
20-35 Hp (521cc) . . . 7-23
25/35 Hp (500/565cc) . . . 7-23
25-55 Hp (737cc) . . . 7-25
25-70 Hp (913cc) . . . 7-25
40/50 Hp (815cc) EFI . . . 7-26
70 Hp (1298cc) EFI . . . 7-27
PROPELLER . . . 2-29
Inspection . . . 2-29
Removal & Installation . . . 2-30
PULSE LIMITER . . . 5-7

R
RECTIFIER . . . 4-62
Removal & Installation . . . 4-63
Testing . . . 4-63
REGULATIONS FOR YOUR BOAT . . . 1-4
Capacity Information . . . 1-4
Certificate Of Compliance . . . 1-4
Documenting Of Vessels . . . 1-4
Hull Identification Number . . . 1-4
Length Of Boats . . . 1-4
Numbering Of Vessels . . . 1-4
Registration Of Boats . . . 1-4
Sales And Transfers . . . 1-4
Ventilation . . . 1-5
Ventilation Systems . . . 1-5
REGULATOR/RECTIFIER . . . 4-64
Removal & Installation . . . 4-66
Testing . . . 4-64
REMOTE CONTROLS . . . 9-2
Concealed Side Mount Remote . . . 9-9
Description & Operation . . . 9-2
Dual Handle Surface Mount Remote For Single Motors . . . 9-7
Introduction . . . 9-2
Pre-Wired Binnacle Mount Remote . . . 9-11
Small Motor Standard Surface Remote . . . 9-6
Standard Surface Mount Remote . . . 9-3
REQUIRED SAFETY EQUIPMENT . . . 1-5
Fire Extinguishers . . . 1-6
Personal Flotation Devices . . . 1-7
Sound Producing Devices . . . 1-8
Visual Distress Signals . . . 1-8
Types Of Fires . . . 1-5
Visual Distress Signals . . . 1-8
Warning System . . . 1-7
RESCUEPRO ROTOR . . . 2-33
RE-COMMISSIONING . . . 2-83
REVERSE GATE . . . 7-32
Gate Adjustment . . . 7-32
Remote Cable Adjustment . . . 7-32

S
SAFETY IN SERVICE . . . 1-12
Do's . . . 1-12
Don'ts . . . 1-13
SAFETY TOOLS . . . 1-17
Eye And Ear Protection . . . 1-17
Work Clothes . . . 1-17
Work Gloves . . . 1-17
SELF DIAGNOSTIC SYSTEM . . . 3-43
Reading & Clearing Codes . . . 3-43
SENSOR AND CIRCUIT RESISTANCE/OUTPUT TESTS . . . 3-45
Testing EFI Components . . . 3-45
SENSOR/TRIGGER COIL . . . 4-41
Removal & Installation . . . 4-45
Testing . . . 4-41
SHOP EQUIPMENT . . . 1-17
Chemicals . . . 1-17
Safety Tools . . . 1-17
SMALL MOTOR STANDARD SURFACE REMOTE . . . 9-6
Overhaul . . . 9-6
SPARK PLUGS . . . 2-43
Heat Range . . . 2-43
Inspection & Gapping . . . 2-46
Reading . . . 2-45
Removal & Installation . . . 2-44
SPARK PLUG WIRES . . . 2-47
Removal & Installation . . . 2-47
Testing . . . 2-47
SPECIFICATIONS
Capacities - Two-Stroke Engines . . . 2-90
Capacities - Four-Stroke Engines . . . 2-91
Carburetor Set-Up . . . 3-71
Conversion Factors . . . 1-28
Engine - Colt/Junior (43cc) 1 Cyl . . . 6-89
Engine - 2.0-3.5 Hp (78cc) 1 Cyl. . . . 6-90
Engine - 5.0 (109cc) 1 Cyl . . . 6-90
Engine - 5/6 Hp (128cc) 1 Cyl. . . . 6-91
Engine - 3/4/4 Deluxe Hp (87cc) 2 Cyl . . . 6-91
Engine - 8/9.9 Hp (211cc) 2 Cyl . . . 6-92
Engine - 5.0-8.0 Hp (164cc) 2 Cyl. . . . 6-92
Engine - 9.9/10/14/15 Hp (216cc) 2 Cyl . . . 6-93
Engine - 9.9/10/15 Hp (255cc) 2 Cyl. . . . 6-93
Engine - 9.9/15hp (305cc) 2 Cyl . . . 6-94
Engine - 18-35 Hp (521cc) 2 Cyl. . . . 6-94
Engine - 25-55 Hp (737cc) 2 Cyl. . . . 6-95
Engine - 25/35 Hp (500/565cc) 3 Cyl . . . 6-95
Engine - 25-70 Hp (913cc) 3 Cyl. . . . 6-96
Engine - 40/50 Hp (815cc) 3 Cyl. . . . 6-96
Engine - 40/50 Hp (815cc) 3 Cyl. . . . 6-97
Engine - 70 (1298cc) 4 Cyl . . . 6-98
General Engine . . . 2-85
General Engine System . . . 2-87
Ignition Testing Specifications - Carb Motors . . . 4-141
Lubrication Services . . . 2-89
Magneto Breaker Point Gap . . . 2-84
Maintenance Intervals . . . 2-89
Torque Values . . . 1-27
Tune-Up Specifications . . . 2-91
Two-Stroke Motor Fuel:Oil Ratio . . . 2-90
Valve Clearance . . . 2-94
STANDARD AND METRIC MEASUREMENTS . . . 1-27
STANDARD SURFACE MOUNT REMOTE . . . 9-3
Overhaul . . . 9-3
STARTER MOTOR . . . 4-86
Cleaning And Inspection . . . 4-95
Disassembly/Assembly . . . 4-89
Removal & Installation . . . 4-87
Testing . . . 4-87
STARTER MOTOR SOLENOID/RELAY SWITCH . . . 4-97
Removal & Installation . . . 4-98
Testing The Solenoid . . . 4-97
STARTING CIRCUIT . . . 4-69
Description And Operation . . . 4-69
Starter Motor . . . 4-86
Starter Motor Solenoid/Relay Switch . . . 4-97
Troubleshooting . . . 4-69
STATOR/BATTERY CHARGE COIL . . . 4-61
Removal & Installation . . . 4-62
Testing . . . 4-61

STORAGE	2-81	**TROUBLESHOOTING**	1-13
Re-Commissioning	2-83	Basic Operating Principles	1-13
Winterization	2-81	Breaker Point ignition	4-8
		Carburetor	3-12
T		Charging system	4-58
TEMPERATURE SENSORS	3-63	Cooling System	5-24
Removal & Installation	3-65	EFI	3-41
Testing	3-64	Electrical systems	4-6
THERMOSTAT	5-37	Electronic Ignition	4-16
Removal & Installation	5-37	Lubrication System - 2-stroke	5-11
THROTTLE BODY	3-51	Starting circuit	4-69
Removal & Installation	2-51	Thermostat	5-25
TILT LIMIT SWITCH (FASTRAK ONLY)	8-12	Trim/Tilt	8-7
Adjustment	8-12	VRO2	5-3
TIMING AND SYNCHRONIZATION	**2-51**	Warning system	4-99
HOMEMADE SYNCHRONIZATION TOOL	2-52	**TUNE-UP**	**2-40**
2.0-3.5 HP MODELS	2-53	Breaker Points Ignition Systems	2-47
5 HP (109cc) MODELS	2-53	Compression Tests	2-41
COLT/JUNIOR, 3/4 HP & 4 DELUXE MODELS	2-53	De-Carboning The Pistons	2-41
5-8 HP (164cc)	2-	Electronic (CDI/UFI) Ignition	2-51
STROKE MODELS	2-54	Introduction	2-40
5/6 HP (128cc) & 8/9.9 HP (211cc) 4-STROKE MODELS	2-56	Spark Plugs	2-43
9.9/10/14/15 HP (216cc) & 9.9/10/15 HP (255cc) 2-STROKE		Spark Plug Wires	2-47
MODELS	2-57	Tune-Up Sequence	2-40
9.9/15 HP (305cc) 4-STROKE MODELS	2-59	TUNE-UP SEQUENCE	2-40
20 HP (521cc) MODELS	2-61		
18 JET & 25-35 HP (521cc) MODELS	2-63	**U**	
25-55 HP (737cc)	2-CYL	**UNDERSTANDING AND TROUBLESHOOTING**	
MODELS	2-65	**ELECTRICAL SYSTEMS**	**4-2**
25/35 HP (500/565cc) 3-CYL	2-69	Basic Electrical Theory	4-2
25-70 HP (913cc) 3-CYL	2-72	Electrical Components	4-2
40/50 HP 4-STROKE MODELS	2-76	Electrical Testing	4-6
70 HP 4-STROKE MODELS	2-77	Precautions	4-7
TIMING BELT/CHAIN		Test Equipment	4-4
Inspection	2-36	Troubleshooting	4-6
Removal & Installation	6-84	Wire And Connector Repair	4-7
TOOLS	**1-19**		
Electronic Tools	1-23	**V**	
Hand Tools	1-19	**VALVE CLEARANCE (4-STROKE)**	**2-78**
Measuring Tools	1-25	Valve Lash Adjustment	2-78
Other Common Tools	1-22	VAPOR SEPARATOR TANK & FUEL PUMP	3-55
Special Tools	1-23	Removal & Installation	3-57
TOP SEAL	4-13	Testing	3-56
Replacement	4-13	VARIABLE RATIO OIL (VRO2) OIL INJECTION SYSTEM	5-3
TRIM SENDING UNIT ADJUSTMENT	8-12	VRO2 PICKUP AND OIL SUPPLY HOSE	5-7
FasTrak Systems	8-13	VRO2 PUMP	5-8
Large-Motor Conventional Systems	8-13	Overhaul	5-8
Small-Motor Conventional Systems	8-14	Removal and Installation	5-8
TRIM/TILT ASSEMBLY - FASTRAK SYSTEMS	8-14		
Assembly	8-17	**W**	
Cleaning and Inspection	8-17	WARNING HORN OR BUZZER	4-103
Disassembly	8-16	Testing	4-103
Overhaul	8-16	**WARNING SYSTEM**	**4-98**
Removal & Installation	8-14	Description And Operation	4-98
TRIM/TILT ASSEMBLY - LARGE-MOTOR CONVENTIONAL		Engine Temperature Sensors (EFI)	4-101
SYSTEMS	8-17	Engine Temperature Switches (Carb)	4-101
Assembly	8-19	Oil Level Switch (Oil Injected 2-Strokes)	4-103
Disassembly	8-19	Oil Pressure Switch (4-Stroke Only)	4-102
Removal & Installation	8-17	Troubleshooting	4-99
TRIM/TILT ASSEMBLY - SMALL-MOTOR CONVENTIONAL		Warning Horn Or Buzzer	4-103
SYSTEMS	8-20	WATER PRESSURE RELIEF VALVE	5-39
Removal & Installation	8-20	Removal & Installation	5-39
TRIM/TILT SYSTEMS	**8-2**	Water Pump	5-25
Description And Operation	8-3	Removal & Installation	5-25
Introduction	8-2	Inspection & Overhaul	5-36
System Identification	8-2	WINTERIZATION	2-81
Troubleshooting	8-7	**WIRING DIAGRAMS**	**4-104**
Tilt Limit Switch (Fastrak Only)	8-12	INDEX	4-104
Trim Sending Unit Adjustment	8-12	1990-95 Breaker Point Models	4-105
Trim/Tilt Assembly - Fastrak Systems	8-14	1996-01 Rope Start/CDI 2.0-3.5 Hp (78cc) 1-Cyl	4-105
Trim/Tilt Assembly - Large-Motor Conventional Systems	8-17	1999-01 Rope Start 5 Hp (109cc) 1-Cyl/2-Stroke	4-106
Trim/Tilt Assembly - Small-Motor Conventional Systems	8-20	1997-01 Rope Start 5/6 hp (128cc) 1-Cyl/4-Stroke	4-106
TRIM/TILT WIRING	**8-22**	1990 4 Hp (87cc) Excel 4/Ultra 4 2-Cyl	4-107
Conventional Systems	8-22	1990-01 Rope Start 3/4 Hp (87cc) 2-Cyl	4-107
Fastrak Systems	8-22	1990-01 Rope Start 6/8 Hp (164cc) Sail 2-Cyl	4-108

10-28 MASTER INDEX

1990-93 Rope Start 4Deluxe (87cc), 5/6/8 Hp (164cc) & 9.9-15 Hp (216/255cc) 2-Cyl/2- Stroke w/UFI & AC power.................. 4-109
1991-01 Rope Start 4Deluxe (87cc), 5/6/6.5/8 Hp (164cc) and 9.9-15 Hp (255cc) 2-Cyl/2-Stroke w/CDI............................... 4-109
1996-01 Rope Start 8/9.9 Hp (211cc) 2-Cyl/4-Stroke............. 4-110
1997-98 Tiller Electric 9.9 Hp (211cc) 2-Cyl/4-Stroke 4-110
1997-98 Remote Electric 9.9 Hp (211cc) 2-Cyl/4-Stroke 4-111
1990-92 Tiller Electric 9.9-15 Hp (211cc) and
Electric Sail 2-Cyl/2-Stroke w/UFI............................. 4-111
1990-92 Tiller Electric 9.9-15 Hp (211cc) 2-Cyl/2-Stroke w/UFI 4-112
1993-01 Tiller Electric 9.9-15 Hp (255cc) and
Electric Sail 2-Cyl/2-Stroke................................. 4-112
1993-01 Remote Electric 9.9-15 Hp (255cc) 2-Cyl/2-Stroke......... 4-113
1995-01 Rope Start 9.9/15 Hp (305cc) 2-Cyl/4-Stroke............. 4-114
1995-01 Tiller Electric 9.9 Hp (305cc) High Thrust 2-Cyl/4-Stroke ... 4-114
1995 Remote Electric 9.9 Hp (305cc) High Thrust 2-Cyl/4-Stroke .. 4-115
1996-01 Remote Electric 9.9 Hp (305cc) High Thrust 2-Cyl/4-Stroke. 4-115
1995-01 Remote Electric 15 Hp (305cc) 2-Cyl/4-Stroke 4-116
1990-92 Rope Start 20-30 Hp (521cc) 2-Cyl w/UFI & AC power 4-117
1990-92 Tiller Electric 20-30 Hp (521cc) 2-Cyl w/UFI 4-118
1990-92 Remote Electric 20-30 Hp (521cc) 2-Cyl w/UFI 4-118
1993-01 Rope Start 18 Jet and 20-35 Hp (521cc) 2-Cyl w/CDI & AC. 4-119
1993-01 Tiller Electric 18 Jet and 20-35 Hp (521cc) 2-Cyl w/CDI ... 4-119
1993-01 Remote Electric 18 Jet and 20-35 Hp (521cc) 2-Cyl 4-120
1990-93 Rope Start 25D-55 Hp (737cc) 2-Cyl w/UFI & AC 4-121
1990-93 Tiller Electric 25D-55 Hp (737cc) 2-Cyl w/UFI 4-121
1990-93 Remote Electric 25D-55 Hp (737cc) 2-Cyl............... 4-122
1992-01 Rope Start 25D-55 Hp (737cc) 2-Cyl w/CDI & AC 4-123
1992-95 Remote Electric 25DE and 48E (737cc) 2-Cyl
w/CDI & VRO, Manual Trim/Tilt................................. 4-124
1992-95 Remote Electric 25DTL (737cc) 2-Cyl w/CDI, VRO & PTT .. 4-124
1996-98 Remote Electric 48 Hp (737cc) E Model and
55 Hp (737cc) Com 2-Cyl................................... 4-125
1992-98 Tiller Electric 40-55 Hp (737cc) 2-Cyl 4-125
1992-95 Electric Start 40/50 Hp (737cc) TTL 2-Cyl w/CDI,
VRO & PTT 4-126
1996-01 Electric Start 40 Hp (737cc) TTL 2-Cyl w/PTT............ 4-127
1992-01 Remote Electric 40-55 Hp (737cc) 2-Cyl................. 4-128
1996-01 Rope Start 25/35 Hp (500/565cc) 3-Cyl.................. 4-130
1996-01 Tiller Electric 25/35 Hp (500/565cc) 3-Cyl............... 4-130
1996-01 Remote Electric 25/35 Hp (500/565cc) 3-Cyl............. 4-131
1999-01 Remote Electric 40/50 Hp (815cc) 3-Cyl/4-Stroke w/PTT... 4-132
1990-01 Rope 65 Hp (913cc) Commercial 3-Cyl w/AC 4-132
1990-92 Electric 60-70 Hp (913cc) TTL 3-Cyl w/PTT.............. 4-133
1993-98 Electric 50-60 Hp (913cc) TTL 3-Cyl/2-Stroke w/PTT...... 4-133
1990-92 Remote Electric 60-70 Hp (913cc) EL 3-Cyl
w/pre-mix & manual trim/tilt 4-134
1993-98 Remote Electric 50-70 Hp (913cc) EL and
65 Hp (913cc) WMLE Com 3-Cyl/2-Stroke 4-135
1993-01 Remote Electric 65 Hp (913cc) WML,
WMLW & WE Com 3-Cyl 4-136
1990-92 Remote Electric 60-70 Hp (913cc) TL & TX 3-Cyl w/PTT... 4-137
1993-95 Remote Electric 50-70 Hp (913cc) D, DT, PL, TL,
TX & TY 3-Cyl w/PTT.. 4-137
1996-01 Remote Electric 40-70 Hp (913cc) D, DT, PL, TL,
TX & TY Model 3-Cyl w/PTT.................................. 4-138
1999-01 Remote Electric 70 Hp (1298cc) 4-Cyl w/Electric Trim/Tilt .. 4-139
1996-01 Remote Control/Keyswitch for MWS Wiring Harness...... 4-140
1996-01 MWS Instrument Wiring Harness for Remotes 4-140
1996-01 Remote Control/Keyswitch for MWS Wiring Harness...... 4-140
Conventional Trim/Tilt Systems 8-22
Fastrak Systems 8-22

SELOC PUBLISHING'S FULL-LINE MASTER LIST

ISBN	PART NO.		TITLE/DESCRIPTION	YEARS

OUTBOARDS

ISBN		PART NO.	TITLE/DESCRIPTION	YEARS
089330	018-7	1000	Chrysler Outboards, All Engines	1962-84
089330	055-1	1100	Force Outboards, All Engines	1984-99
089330	048-9	1200	Honda Outboards, All Engines	1978-01
089330	007-1	1300	Johnson/Evinrude Outboards, 1-2 Cyl	1956-70
089330	008-X	1302	Johnson/Evinrude Outboards, 1-2 Cyl	1971-89
089330	009-8	1306	Johnson/Evinrude Outboards, 3-4 Cyl	1958-72
089330	010-1	1308	Johnson/Evinrude Outboards, 3, 4 & 6 Cyl	1973-91
089330	063-2	1311	Johnson/Evinrude Outboards - All V Engines	1992-01
089330	052-7	1312	Johnson/Evinrude Outboards, All In-line engines/2 & 4 Stroke	1996-01
089330	015-2	1400	Mariner Outboards, 1-2 Cyl	1977-89
089330	016-0	1402	Mariner Outboards, 3, 4 & 6 Cyl	1977-89
089330	012-8	1404	Mercury Outboards, 1-2 Cyl	1965-91
089330	013-6	1406	Mercury Outboards, 3-4 Cyl	1965-89
089330	014-4	1408	Mercury Outboards, 6 Cyl	1965-89
089330	051-9	1416	Mercury/Mariner Outboards, All Engines	1990-00
089330	050-0	1600	Suzuki Outboards, All Engines	1988-99
089330	064-0	1701	Yamaha Outboards, All Engines	1984-96
089330	065-9	1703	Yamaha Outboards, All Engines	1997-03
089330	066-7	1705	Yamaha/Mercury/Mariner 4-Stroke Engines	1995-04

STERN DRIVES

ISBN		PART NO.	TITLE/DESCRIPTION	YEARS
089330	029-2	3000	Marine Jet Drive	1961-96
089330	005-5	3200	Mercruiser Stern Drives	1964-91
089330	053-5	3206	Mercruiser Stern Drives	1992-01
089330	004-7	3400	OMC Stern Drives	1964-86
089330	056-X	3404	OMC Stern Drives	1986-98
089330	011-X	3600	Volvo/Penta Stern Drives	1968-91
089330	038-1	3602	Volvo/Penta Stern Drives	1992-93
089330	057-8	3606	Volvo/Penta Stern Drives	1992-03

INBOARDS

ISBN		PART NO.	TITLE/DESCRIPTION	YEARS
089330	049-7	7400	Yanmar Inboards	1975-98

PERSONAL WATERCRAFT

ISBN		PART NO.	TITLE/DESCRIPTION	YEARS
089330	032-2	9200	Kawasaki	1973-91
089330	042-X	9202	Kawasaki	1992-97
089330	045-4	9400	Polaris	1992-97
089330	033-0	9000	Sea-Doo/Bombardier	1988-91
089330	043-8	9002	Sea-Doo/Bombardier	1992-97
089330	034-9	9600	Yamaha	1987-91
08933	044-6	9602	Yamaha	1992-97

Seloc-On-Line (Internet Access)

ISBN		PART NO.	TITLE/DESCRIPTION	YEARS
089330	075-6	5000	One Mfg/Model - Subscription per user 3 years	1990-01

PROFESSIONAL TECHNICIANS MANUALS

ISBN		PART NO.	TITLE/DESCRIPTION	YEARS
089330	060-8	4500	Labor Guide - Johnson and Evinrude	1980-00
089330	061-6	4550	Labor Guide - Yamaha	1980-00
089330	062-4	4600	Labor Guide - Mercury	1980-00

BRIGGS AND STRATTON

ISBN		PART NO.	TITLE/DESCRIPTION	YEARS
096376	610-4	610	Briggs & Stratton 4 Stroke Horizontal Crankshaft	1950+
096376	611-2	611	Briggs & Stratton 4 Stroke Vertical Crankshaft	1950+
096376	612-0	612	Briggs & Stratton 4 Stroke Overhead Crankshaft	1950+

SelocOnLine

SelocOnLine is a maintenance and repair database accessed via the Internet.
Always up-to-date, skill level and special tool icons, quick access buttons to wiring diagrams, specification charts, maintenance charts, and a parts database.
Contact your local marine dealer or see our demo at www.seloconline.com

SelocOnLine

Step 1
Select your manufacturer
Year / Model
Engine

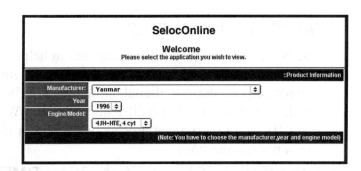

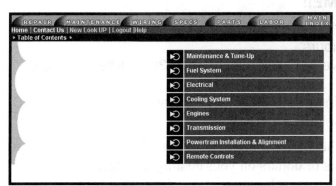

Step 2
Select an Engine System

Step 3
Select the repair

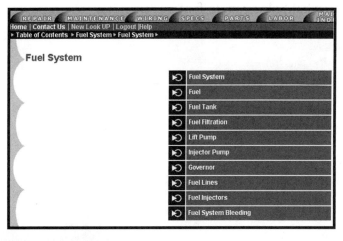

Step 4
Print the repair procedure

Seloc OnLine Solutions (SOS)

FEATURES AND BENEFITS

Seloc Publishing has been an innovator in helping boaters and technicians maintain and repair outboard and inboard engines and/or drive systems for over 30 years. From performing actual teardown procedures to digital photography, Seloc has been there all the way. Now Seloc wants to assist you in solving those difficult problems. We are as close as your fingertips or your phone. Not 100% sure about that procedure? Email or phone us, we can help you with your engine and boat related questions.

REAL WORLD SOLUTIONS IN REAL TIME!!

- Technical questions answered by factory trained technicians.
- Response to questions in less that 24 hours.
- Priority Response System (ability to have technician call you in 30 minutes or less).
- Technical documentation and diagrams
- Monthly newsletter containing helpful hints, and "How to" features.
- Future access to on-line repository of SelocSOS inquiries

One visit to SelocSOS could potentially save you hundreds of dollars. Above all, our priority is to get you back in the water. You have invested thousands of dollars on your boating investment. Now it is time to get a return on your investment. Visit us at www.selocsos.com or selocmarine.com and see what Seloc can do for you.

Seloc PRO
Real World Solutions In Real Time

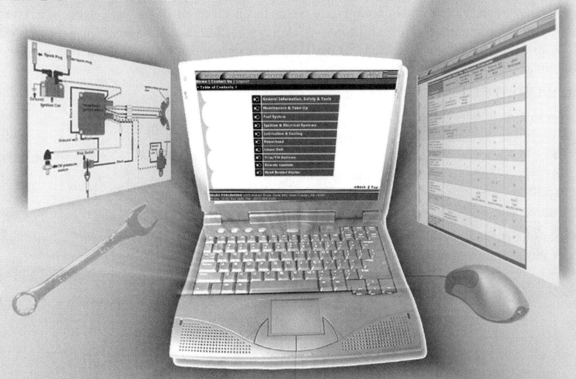

Seloc Pro is a mechanical repair database that is accessed via the internet. There is no other tool in your shop that will save you as much time or provide you with as much productivity as Seloc Pro. Unlike manufacturer information you may presently use, Seloc Pro allows you to navigate the same way thru our database regardless of the manufacturer. Our database is written so that each procedure is 1-3 pages in length. New content or changes to content can be added in minutes.

Seloc Pro is always up-to-date.

Features Include:
Quick access buttons to databses for Wiring Diagrams, Specifications, Parts and Labor Times

Seloc Labor Times - Real world freshwater and saltwater times for hundreds of operations for Johnson, Evinrude, Yamaha, and Mercury from 1980 thru 2000

Hyper-linked index for quick access to unit repair sections

Mfgs covered include Force, Honda, Johnson, Evinrude, Mercruiser, Mercury, Suzuki, Yamaha and Yanmar.
Coming during the 4th quarter are Volvo Penta and OMC Stern Drive

Contact your local distributor for a demo
or Seloc at 866-735-6255

Seloc Publishing®

Seloc OnLine Solutions (SOS)

FEATURES AND BENEFITS

Seloc Publishing has been an innovator in helping boaters and technicians maintain and repair outboard and inboard engines and/or drive systems for over 30 years. From performing actual teardown procedures to digital photography, Seloc has been there all the way. Now Seloc wants to assist you in solving those difficult problems. We are as close as your fingertips or your phone. Not 100% sure about that procedure? Email or phone us, we can help you with your engine and boat related questions.

REAL WORLD SOLUTIONS IN REAL TIME!!

- Technical questions answered by factory trained technicians.
- Response to questions in less that 24 hours.
- Priority Response System (ability to have technician call you in 30 minutes or less).
- Technical documentation and diagrams
- Monthly newsletter containing helpful hints, and "How to" features.
- Future access to on-line repository of SelocSOS inquiries

One visit to SelocSOS could potentially save you hundreds of dollars. Above all, our priority is to get you back in the water. You have invested thousands of dollars on your boating investment. Now it is time to get a return on your investment. Visit us at www.selocsos.com or selocmarine.com and see what Seloc can do for you.

SELOC PUBLISHING PRESENTS

SELOC LABOR GUIDES

Johnson & Evinrude 1980-2000
Yamaha 1984-2000
Mercury 1980-2000

Seloc labor manuals cover all engines and thousands of operations. Our real world times include both freshwater and saltwater times.

Contact your local distributor or call us at 1-866-735-6255